Testing of Digital Systems

Device testing represents the single largest manufacturing expense in the semiconductor industry, costing over \$40 billion a year. The most comprehensive and wide ranging book of its kind, *Testing of Digital Systems* covers everything you need to know about this vitally important subject. Starting right from the basics, the authors take the reader through automatic test pattern generation, design for testability and built-in self-test of digital circuits before moving on to more advanced topics such as I_{DDQ} testing, functional testing, delay fault testing, CMOS testing, memory testing, and fault diagnosis. The book includes detailed treatment of the latest techniques including test generation for various fault models, discussion of testing techniques at different levels of the integrated circuit hierarchy and a chapter on system-on-a-chip test synthesis. Written for students and engineers, it is both an excellent senior/graduate level textbook and a valuable reference.

Niraj K. Jha is Professor of Electrical Engineering at Princeton University and head of the Center of Embedded System-on-a-Chip Design, where his current research is focussed on the synthesis and testing of these devices. He is a fellow of IEEE, associate editor of *IEEE Transactions on Computer-Aided Design* and the *Journal of Electronic Testing: Theory and Applications* (*JETTA*) and a recipient of the AT&T Foundation award and the NEC preceptorship award for research excellence.

Sandeep Gupta is an Associate Professor in the Department of Electrical Engineering at the University of Southern California. He is Co-Director of the M.S. Program in VLSI Design, with research interests in the area of VLSI testing and design. He is a member of IEEE.

Testing of Digital Systems

N. K. Jha and S. Gupta

CAMBRIDGE
UNIVERSITY PRESS

PUBLISHED BY THE PRESS SYNDICATE OF THE UNIVERSITY OF CAMBRIDGE
The Pitt Building, Trumpington Street, Cambridge, United Kingdom

CAMBRIDGE UNIVERSITY PRESS
The Edinburgh Building, Cambridge CB2 2RU, UK
40 West 20th Street, New York, NY 10011-4211, USA
477 Williamstown Road, Port Melbourne, VIC 3207, Australia
Ruiz de Alarcón 13, 28014, Madrid, Spain
Dock House, The Waterfront, Cape Town 8001, South Africa

http://www.cambridge.org

First published 2003

Printed in the United Kingdom at the University Press, Cambridge

Typeface Times 10.5/14pt.   *System* LATEX 2_ε [DBD]

A catalogue record of this book is available from the British Library

ISBN 0 521 77356 3 hardback

To family and friends for their encouragement and understanding, to teachers and colleagues for sharing their wisdom, to students for sharing their curiosity.

Contents

16   System-on-a-chip test synthesis

Preface

The fraction of the industrial semiconductor budget that manufacturing-time testing consumes continues to rise steadily. It has been known for quite some time that tackling the problems associated with testing semiconductor circuits at earlier design levels significantly reduces testing costs. Thus, it is important for hardware designers to be exposed to the concepts in testing which can help them design a better product. In this era of system-on-a-chip, it is not only important to address the testing issues at the gate level, as was traditionally done, but also at all other levels of the integrated circuit design hierarchy.

This textbook is intended for senior undergraduate or beginning graduate levels. Because of its comprehensive treatment of digital circuit testing techniques, it can also be gainfully used by practicing engineers in the semiconductor industry. Its comprehensive nature stems from its coverage of the transistor, gate, register-transfer, behavior and system levels of the design hierarchy. In addition to test generation techniques, it also covers design for testability, synthesis for testability and built-in self-test techniques in detail. The emphasis of the text is on providing a thorough understanding of the basic concepts; access to more advanced concepts is provided through a list of additional reading material at the end of the chapter.

The contents of the book are such that it contains all the material required for a first, one-semester, course in Testing (approximately 40 hours of teaching). The chapters are organized such that seven of the chapters contain mandatory material, while a selection from the remaining chapters may optionally be included.

Each chapter contains a set of exercises with different difficulty levels which can be used for in-class, as well as take-home exercises or tests.

In addition, the chapters contain many examples and a summary.

Chapter 1 introduces the readers to basic concepts in testing, such as faults, errors, tests, failure rate, fault coverage, and test economics.

Chapter 2 deals with fault models at various levels of the integrated circuit design hierarchy, e.g., behavioral, functional, structural, switch-level and geometric fault models. It also discusses different types of delay models, and inductive fault analysis.

Chapter 3 describes how fault-free and faulty circuit elements can be represented. It discusses a logic simulation algorithm and ways to accelerate logic simulation. It then proceeds to fault simulation, starting with well-known fault collapsing and fault dropping concepts. It discusses the following fault simulation paradigms in detail: parallel fault simulation, parallel-pattern single-fault propagation simulation, deductive fault simulation, concurrent fault simulation, and critical path tracing. It also provides a brief background into approximate, low-complexity fault simulation approaches.

Chapter 4 covers test generation for combinational circuits. It starts with a discussion of composite circuit representation and value systems. Then it proceeds to basic concepts in test generation and implication procedures. Structural test generation algorithms and testability analysis techniques are targeted next. These include the D-algorithm, PODEM and their enhancements. Next, non-structural algorithms such as those based on satisfiability and binary decision diagrams are covered. Static and dynamic test compaction techniques and test generation algorithms for reduced heat and noise complete the chapter.

Chapter 5 deals with test generation for sequential circuits. It first classifies sequential test generation methods and faults. This is followed by discussion of fault collapsing and fault simulation. Test generation methods covered are those that start from a state table or gate-level implementation. Testing of asynchronous sequential circuits is also included. The chapter ends with a discussion of sequential test compaction methods.

Chapter 6 discusses I_{DDQ} testing. For combinational circuits, it targets testing of leakage faults and unrestricted bridging faults as well as test compaction. For sequential circuits, it targets test generation, fault simulation and test compaction. Under fault diagnosis, it covers analysis, diagnostic fault simulation and diagnostic test generation. It then introduces built-in current sensors. Under advanced I_{DDQ} testing concepts, it discusses i_{DD} pulse response testing, dynamic current testing, depowering, current signatures, and applicability of I_{DDQ} testing to deep submicron designs. It ends with a discussion on economics of I_{DDQ} testing.

Chapter 7 covers universal test sets, various types of pseudoexhaustive testing, and iterative logic array testing under the umbrella of functional testing.

Chapter 8 describes delay fault testing methods in detail. It first gives a classification of various types of delay faults. Then it provides test generation and fault simulation methods for combinational and sequential circuits containing the different types of delay faults. It also discusses some pitfalls of delay fault testing and how to overcome them. It closes with some advanced delay fault testing techniques.

Chapter 9 deals with test generation methods for static and dynamic CMOS circuits for stuck-open and stuck-on faults. It discusses the test invalidation problem for static CMOS circuits and design for testability methods to avoid them.

Chapter 10 covers fault diagnosis techniques. It includes both cause–effect and effect–cause diagnosis methods. It also discusses diagnostic test generation.

Chapter 11 describes design for testability methods. It discusses scan design, both full and partial, in considerable detail. It goes on to describe organization and use of scan chains. It then discusses boundary scan. It finally describes design for testability techniques for delay faults and low heat dissipation.

Chapter 12 discusses built-in self-test (BIST) techniques. It starts with the basic concepts such as pattern generation, computation of test length, response compression, and aliasing analysis. It then proceeds to BIST methodologies, both in-situ and scan-based.

Chapter 13 concentrates on synthesis for testability techniques at the gate level. It covers many techniques under stuck-at and delay fault testability of combinational and sequential circuits.

Chapter 14 deals with memory testing. The topics covered are: reduced functional faults, traditional memory tests, March tests, pseudorandom memory tests, and BIST for embedded memories.

Chapter 15 discusses high-level test synthesis. It deals with the register-transfer and behavior levels of the design hierarchy. It describes hierarchical and high-level test generation techniques first. Then it discusses design for testability, synthesis for testability, and BIST techniques at the register-transfer and behavior levels.

Chapter 16 covers the modern topic of system-on-a-chip test synthesis. It discusses core-level test, core test access and core test wrappers.

The core chapters are Chapters 1, 2, 3, 4, 5, 11 and 12. These describe fault models, fault simulation and test generation for combinational and sequential circuits, design for testability and BIST. In a one semester course, if test generation needs to be emphasized, then this material can be augmented with a subset of Chapters 6, 7, 8, 9, 10 and 14. These chapters deal with I_{DDQ} testing, functional testing, delay fault testing, CMOS testing, fault diagnosis and memory testing, respectively. Alternatively, if synthesis for testability and testing at register-transfer, behavior and system levels need to be emphasized, the instructor can choose from Chapters 13, 15 and 16.

This book would not have been possible without help from many people. We first want to acknowledge Prof. Ad van de Goor who wrote the Introduction and Memory Testing chapters. We would like to thank our colleagues throughout the world who have used preprints of this book and given valuable feedback. These include Profs. S. Blanton, K.-T. Cheng, S. Dey, D. Ha, J. P. Hayes, J. Jacob, P. Mazumder, I. Pomeranz, D. Pradhan, S. M. Reddy, and S. Seth. We are indebted to our copy editor, F. Nex, who discovered some latent bugs. Finally, our thanks go to the students at Princeton University and University of Southern California who helped make the book better through their helpful suggestions.

Niraj K. Jha
Sandeep Gupta

Gate symbols

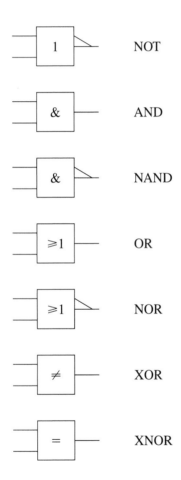

1   NOT

&   AND

&   NAND

$\geqslant 1$ OR

$\geqslant 1$ NOR

$\neq$ XOR

$=$ XNOR

1 Introduction

by Ad van de Goor

We introduce some basic concepts in testing in this chapter. We first discuss the terms fault, error and failure and classify faults according to the way they behave over time into permanent and non-permanent faults.

We give a statistical analysis of faults, introducing the terms failure rate and mean time to failure. We show how the failure rate varies over the lifetime of a product and how the failure rates of series and parallel systems can be computed. We also describe the physical and electrical causes for faults, called failure mechanisms.

We classify tests according to the technology they are designed for, the parameters they measure, the purpose for which the test results are used, and the test application method.

We next describe the relationship between the yield of the chip manufacturing process, the fault coverage of a test (which is the fraction of the total number of faults detected by a given test) and the defect level (the fraction of bad parts that pass the test). It can be used to compute the amount of testing required for a certain product quality level.

Finally, we cover the economics of testing in terms of time-to-market, revenue, costs of test development and maintenance cost.

1.1 Faults and their manifestation

This section starts by defining the terms failure, error and fault; followed by an overview of how faults can manifest themselves in time.

1.1.1 Failures, errors and faults

A system **failure** occurs or is present when the *service* of the system differs from the specified service, or the service that should have been offered. In other words: the system *fails* to do what it has to do. A failure is caused by an *error*.

There is an **error** in the system (the system is in an erroneous state) when its *state* differs from the state in which it should be in order to deliver the specified service. An *error* is caused by a *fault*.

A **fault** is present in the system when there is a *physical difference* between the 'good' or 'correct' system and the current system.

Example 1.1   A car cannot be used as a result of a flat tire. The fact that the car cannot be driven safely with a flat tire can be seen as the *failure*. The failure is caused by an *error*, which is the erroneous state of the air pressure of the tire. The *fault* that caused the erroneous state was a puncture in the tire, which is the physical difference between a good tire and an erroneous one.

Notice the possibility that a fault does not (immediately) result in a failure; e.g., in the case of a very slowly leaking tire. □

1.1.2   Fault manifestation

According to the way faults manifest themselves in time, two types of faults can be distinguished: *permanent* and *non-permanent* faults.

1.1.2.1 Permanent faults

The term **permanent fault** refers to the presence of a fault that affects the functional behavior of a system (chip, array or board) *permanently*. Examples of permanent, also called *solid* or *hard*, faults are:

- Incorrect connections between integrated circuits (ICs), boards, tracks, etc. (e.g., missing connections or shorts due to solder splashes or design faults).
- Broken components or parts of components.
- Incorrect IC masks, internal silicon-to-metal or metal-to-package connections (a manufacturing problem).
- Functional design errors (the implementation of the logic function is incorrect).

Because the permanent faults affect the logic values in the system permanently, they are easier to detect than the non-permanent faults which are described below.

1.1.2.2 Non-permanent faults

Non-permanent faults are present only part of the time; they occur at random moments and affect the system's functional behavior for finite, but unknown, periods of time. As a consequence of this random appearance, detection and localization of non-permanent faults is difficult. If such a fault does not affect the system during test, then the system appears to be performing correctly.

The non-permanent faults can be divided into two groups with different origins: *transient* and *intermittent* faults.

Transient faults are caused by environmental conditions such as cosmic rays, α-particles, pollution, humidity, temperature, pressure, vibration, power supply fluctuations, electromagnetic interference, static electrical discharges, and ground loops.

Transient faults are hard to detect due to their obscure influence on the logic values in a system. Errors in random-access memories (RAMs) introduced by transient faults are often called **soft errors**. They are considered non-recurring, and it is assumed that no permanent damage has been done to the memory cell. Radiation with α-particles is considered a major cause of soft errors (Ma and Dressendorfer, 1989).

Intermittent faults are caused by non-environmental conditions such as loose connections, deteriorating or ageing components (the general assumption is that during the transition from normal functioning to worn-out, intermittent faults may occur), critical timing (hazards and race conditions, which can be caused by design faults), resistance and capacitance variations (resistor and capacitor values may deviate from their specified value initially or over time, which may lead to timing faults), physical irregularities, and noise (noise disturbs the signals in the system).

A characteristic of intermittent faults is that they behave like permanent faults for the duration of the failure caused by the intermittent fault. Unfortunately, the time that an intermittent fault affects the system is usually very short in comparison with the application time of a test developed for permanent faults, which is typically a few seconds. This problem can be alleviated by continuously repeating the test or by causing the non-permanent fault to become permanent. The natural transition of non-permanent faults into permanent faults can take hours, days or months, and so must be accelerated. This can be accomplished by providing specific environmental *stress conditions* (temperature, pressure, humidity, etc.). One problem with the application of stress conditions is that new faults may develop, causing additional failures.

1.2   An analysis of faults

This section gives an analysis of faults; it starts with an overview of the frequency of occurrence of faults as a function of time; Section 1.2.2 describes the behavior of the failure rate of a system over its lifetime and Section 1.2.3 shows how the failure rate of series and parallel systems can be computed. Section 1.2.4 explains the physical and electrical causes of faults, called *failure mechanisms*.

1.2.1   Frequency of occurrence of faults

The frequency of occurrence of faults can be described by a theory called *reliability theory*. In-depth coverage can be found in O'Connor (1985); below a short summary is given.

The point in time t at which a fault occurs can be considered a random variable u. The probability of a failure *before* time t, $F(t)$, is the *unreliability* of a system; it can be expressed as:

$$F(t) = P(u \leq t). \tag{1.1}$$

The **reliability** of a system, $R(t)$, is the probability of a correct functioning system at time t; it can be expressed as:

$$R(t) = 1 - F(t), \tag{1.2}$$

or alternatively as:

$$R(t) = \frac{\text{number of components surviving at time } t}{\text{number of components at time } 0}. \tag{1.3}$$

It is assumed that a system initially will be operable, i.e., $F(0) = 0$, and ultimately will fail, i.e., $F(\infty) = 1$. Furthermore, $F(t) + R(t) = 1$ because at any instance in time either the system has failed or is operational.

The derivative of $F(t)$, called the **failure probability density function** $f(t)$, can be expressed as:

$$f(t) = \frac{dF(t)}{dt} = -\frac{dR(t)}{dt}. \tag{1.4}$$

Therefore, $F(t) = \int_0^t f(t)dt$ and $R(t) = \int_t^\infty f(t)dt$.

The **failure rate**, $z(t)$, is defined as the conditional probability that the system fails during the time-period $(t, t + \Delta t)$, given that the system was operational at time t.

$$z(t) = \lim_{\Delta t \to 0} \frac{F(t + \Delta t) - F(t)}{\Delta t} \cdot \frac{1}{R(t)} = \frac{dF(t)}{dt} \cdot \frac{1}{R(t)} = \frac{f(t)}{R(t)}. \tag{1.5}$$

Alternatively, $z(t)$ can be defined as:

$$z(t) = \frac{\text{number of failing components per unit time at time } t}{\text{number of surviving components at time } t}. \tag{1.6}$$

$R(t)$ can be expressed in terms of $z(t)$ as follows:

$$\int_0^t z(t)dt = \int_0^t \frac{f(t)}{R(t)}dt = -\int_{R(0)}^{R(t)} \frac{dR(t)}{R(t)} = -\ln \frac{R(t)}{R(0)},$$

or, $R(t) = R(0)e^{-\int_0^t z(t)dt}$. $\tag{1.7}$

The **average lifetime** of a system, θ, can be expressed as the mathematical expectation of t to be:

$$\theta = \int_0^\infty t \cdot f(t)dt. \tag{1.8}$$

For a non-maintained system, θ is called the **mean time to failure (*MTTF*)**:

$$MTTF = \theta = -\int_0^\infty t \cdot \frac{dR(t)}{dt} dt = -\int_{R(0)}^{R(\infty)} t \cdot dR(t).$$

Using partial integration and assuming that $\lim_{T \to \infty} T \cdot R(T) = 0$:

$$MTTF = \lim_{T \to \infty} \left\{ -t \cdot R(t) \mid_0^T + \int_0^T R(t) dt \right\} = \int_0^\infty R(t) dt. \tag{1.9}$$

Given a system with the following reliability:

$$R(t) = e^{-\lambda t}, \tag{1.10}$$

the failure rate, $z(t)$, of that system is computed below and has the constant value λ:

$$z(t) = \frac{f(t)}{R(t)} = \frac{dF(t)}{dt} / R(t) = \frac{d(1 - e^{-\lambda t})}{dt} / e^{-\lambda t} = \lambda e^{-\lambda t} / e^{-\lambda t} = \lambda. \tag{1.11}$$

Assuming failures occur randomly with a constant rate λ, the *MTTF* can be expressed as:

$$MTTF = \theta = \int_0^\infty e^{-\lambda t} dt = \frac{1}{\lambda}. \tag{1.12}$$

For illustrative purposes, Figure 1.1 shows the values of $R(t)$, $F(t)$, $f(t)$ and $z(t)$ for the life expectancy of the Dutch male population averaged over the years 1976–1980 (Gouda, 1994). Figure 1.1(a) shows the functions $R(t)$ and $F(t)$; the maximum age was 108 years, the graph only shows the age interval 0 through 100 years because the number of live people in the age interval 101 through 108 was too small to derive useful statistics from. Figures 1.1(b) and 1.1(c) show $z(t)$ and Figure 1.1(d) shows $f(t)$ which is the derivative of $F(t)$. Notice the increase in $f(t)$ and $z(t)$ between the ages 18–20 due to accidents of inexperienced drivers, and the rapid decrease of $z(t)$ in the period 0–1 year because of decreasing infant mortality.

1.2.2 Failure rate over product lifetime

A well-known graphical representation of the failure rate, $z(t)$, as a function of time is shown in Figure 1.2, which is known as the **bathtub curve**. It has been developed to model the failure rate of mechanical equipment, and has been adapted to the semiconductor industry (Moltoft, 1983). It can be compared with Figure 1.1(d). The bathtub curve can be considered to consist of three regions:

- Region 1, with decreasing failure rate (infant mortality).
  Failures in this region are termed *infant mortalities*; they are attributed to poor quality as a result of variations in the production process.
- Region 2, with constant failure rate; $z(t) = \lambda$ (working life).
  This region represents the 'working life' of a component or system. Failures in this region are considered to occur randomly.

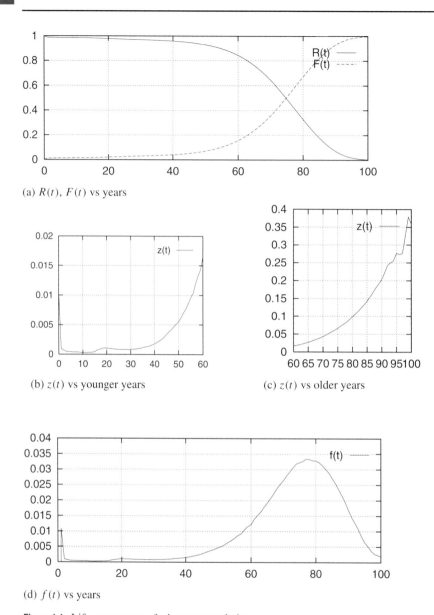

(a) $R(t)$, $F(t)$ vs years

(b) $z(t)$ vs younger years

(c) $z(t)$ vs older years

(d) $f(t)$ vs years

Figure 1.1   Life expectancy of a human population

- Region 3, with increasing failure rate (wearout).

  This region, called 'wearout', represents the end-of-life period of a product. For electronic products it is assumed that this period is less important because they will not enter this region due to a shorter economic lifetime.

From Figure 1.2 it may be clear that products should be shipped to the user only after they have passed the infant mortality period, in order to reduce the high field repair cost. Rather than ageing the to-be-shipped product for the complete infant mortality

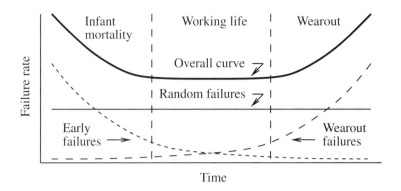

Figure 1.2 Bathtub curve

period, which may be several months, a shortcut is taken by increasing the failure rate. The failure rate increases when a component is used in an 'unfriendly' environment, caused by a **stress condition**. An important stress condition is an increase in temperature which accelerates many physical–chemical processes, thereby accelerating the ageing process. The accelerating effect of the temperature on the failure rate can be expressed by the experimentally determined **equation of Arrhenius**:

$$\lambda_{T_2} = \lambda_{T_1} \cdot e^{(E_a(1/T_1 - 1/T_2)/k)}, \tag{1.13}$$

where:

 T_1 and T_2 are absolute temperatures (in Kelvin, K),
 λ_{T_1} and λ_{T_2} are the failure rates at T_1 and T_2, respectively,
 E_a is a constant expressed in electron-volts (eV), known as the **activation energy**,
 and k is Boltzmann's constant ($k = 8.617 \times 10^{-5}$ eV/K).

From Arrhenius' equation it can be concluded that the failure rate is exponentially dependent on the temperature. This is why temperature is a very important stress condition (see example below). Subjecting a component or system to a higher temperature in order to accelerate the ageing process is called **burn-in** (Jensen and Petersen, 1982). Practical results have shown that a burn-in period of 50–150 hours at 125 °C is effective in exposing 80–90% of the component and production-induced defects (e.g., solder joints, component drift, weak components) and reducing the initial failure rate (infant mortality) by a factor of 2–10.

Example 1.2 Suppose burn-in takes place at 150 °C; given that $E_a = 0.6$ eV and the normal operating temperature is 30 °C. Then the acceleration factor is:

$$\lambda_{T_2}/\lambda_{T_1} = e^{0.6(1/303 - 1/423)/8.617 \times 10^{-5}} = 678,$$

which means that the infant mortality period can be reduced by a factor of 678. □

1.2.3   Failure rate of series and parallel systems

If all components of a system have to be operational in order for the system to be operational, it is considered to be a **series system**. Consider a series system consisting of n components, and assume that the probability of a given component to be defective is independent of the probabilities of the other components. Then the reliability of the system can be expressed (assuming $R_i(t)$ is the reliability of the ith component) as:

$$R_s(t) = \prod_{i=1}^{n} R_i(t). \tag{1.14}$$

Using Equation (1.7), it can be shown that:

$$z_s(t) = \sum_{i=1}^{n} z_i(t). \tag{1.15}$$

A **parallel system** is a system which is operational as long as at least one of its n components is operational; i.e., it only fails when *all* of its components have failed. The unreliability of such a system can be expressed as follows:

$$F_p(t) = \prod_{i=1}^{n} F_i(t). \tag{1.16}$$

Therefore, the reliability of a parallel system can be expressed as:

$$R_p(t) = 1 - \prod_{i=1}^{n} F_i(t). \tag{1.17}$$

1.2.4   Failure mechanisms

This section describes the physical and electrical causes for faults, called **failure mechanisms**. A very comprehensive overview of failure mechanisms for semiconductor devices is given in Amerasekera and Campbell (1987), who identify three classes (see Figure 1.3):

1 Electrical stress (in-circuit) failures:

These failures are due to poor design, leading to electric overstress, or due to careless handling, causing static damage.

2 Intrinsic failure mechanisms:

These are inherent to the semiconductor die itself; they include crystal defects, dislocations and processing defects. They are usually caused during wafer fabrication and are due to flaws in the oxide or the epitaxial layer.

3 Extrinsic failure mechanisms:

These originate in the packaging and the interconnection processes; they can be attributed to the metal deposition, bonding and encapsulation steps.

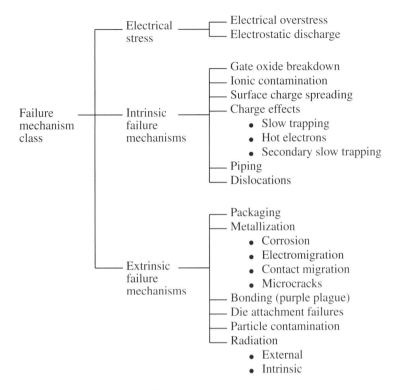

Figure 1.3 Classification of failure mechanisms

Over time, the die fabrication process matures, thereby reducing the intrinsic failure rate, causing the extrinsic failure rate to become more dominant. However, it is very difficult to give a precise ordering of the failure mechanisms; some are dominant in certain operational and environmental conditions, others are always present but with a lower impact.

An important parameter of a failure mechanism is E_a, the activation energy, describing the temperature dependence of the failure mechanism. E_a typically varies between 0.3 and 1.5 eV. Temperatures between $125\,°C$ and $250\,°C$ have been found to be effective for burn-in, without causing permanent damage (Blanks, 1980). The exact influence of the temperature on the failure rate (i.e., the exact value of E_a) is very hard to determine and varies between manufacturers, batches, etc. Table 1.1 lists experimentally determined values for the activation energies of the most important failure mechanisms, which are described next.

Corrosion is an electromechanical failure mechanism which occurs under the condition that moisture and DC potentials are present; Cl^- and Na^+ ions act as a catalyst. Packaging methods (good sealing) and environmental conditions determine the corrosion process to a large extent; CMOS devices are more susceptible due to their low power dissipation.

Table 1.1. *Activation energies of some major failure mechanisms*

Failure mechanism	Activation energy E_a
Corrosion of metallization	0.3–0.6 eV
Electrolytic corrosion	0.8–1.0 eV
Electromigration	0.4–0.8 eV
Bonding (purple plague)	1.0–2.2 eV
Ionic contamination	0.5–1.0 eV
Alloying (contact migration)	1.7–1.8 eV

Electromigration occurs in the Al (aluminum) metallization tracks (lines) of the chip. The electron current flowing through the Al tracks causes the electrons to collide with the Al grains. Because of these collisions, the grains are dislocated and moved in the direction of the electron current. Narrow line widths, high current densities, and a high temperature are major causes of electromigration, which results in open lines in places where the current density is highest.

Bonding is the failure mechanism which consists of the deterioration of the contacts between the Au (gold) wires and the Al pads of the chip. It is caused by interdiffusion of Au–Al which causes open connections.

Ionic contamination is caused by mobile ions in the semiconductor material and is a major failure mechanisms for MOS devices. Na^+ ions are the most mobile due to their small radius; they are commonly available in the atmosphere, sweat and breath. The ions are attracted to the gate oxide of a FET transistor, causing a change in the threshold voltage of the device.

Alloying is also a form of Al migration of Al into Si (silicon) or Si into Al. Depending on the junction depth and contact size, the failure manifests itself as a shorted junction or an open contact. As device geometries get smaller, alloying becomes more important, because of the smaller diffusion depths.

Radiation (Ma and Dressendorfer, 1989) is another failure mechanism which is especially important for dynamic random-access memories (DRAMs). Trace impurities of radioactive elements present in the packaging material of the chip emit α-particles with energies up to 8 MeV. The interaction of these α-particles with the semiconductor material results in the generation of electron–hole pairs. The generated electrons move through the device and are capable of wiping out the charge stored in a DRAM cell, causing its information to be lost. This is the major cause of soft errors in DRAMs. Current research has shown that high-density static random-access memories (SRAMs) also suffer from soft errors caused by α-particles (Carter and Wilkins, 1987).

1.3 Classification of tests

A test is a procedure which allows one to distinguish between good and bad parts. Tests can be classified according to the technology they are designed for, the parameters they measure, the purpose for which the test results are used, and the test application method.

1.3.1 Technology aspect

The type of tests to be performed depends heavily on the technology of the circuit to be tested: analog, digital, or mixed-signal.

Analog circuits have the property that the domain of values of the input and output signals is analog; i.e., the signals can take on *any* value in a given range (this range is delimited by a lower and an upper bound (e.g., in the case of voltage levels those bounds may be determined by the supply voltage, resulting in a range of 0 to +5 V)). Analog tests aim at determining the values of the analog parameters such as voltage and current levels, frequency response, bandwidth, distortion, etc. The generation of the test input stimuli, the processing of these stimuli by the circuit, as well as the determination of the values of the test response signals are inherently imprecise due to the analog nature of the signals (infinite accuracy does not exist). Therefore, the determination whether a circuit satisfies its requirements is not based on a single value, but on a range of values (i.e., an interval), for each of the test response signals.

Digital circuits have the property that the domain of values of the input and output signals is binary (usually referred to as *digital*); i.e., the signals can only take on the value 'logic 0' or 'logic 1'. Tests for digital circuits determine the values of the binary test response signals, given binary test stimuli which are processed digitally by the to-be-tested circuit; this can be done precisely due to the binary nature of the signal values. These tests are called *logical tests* or *digital tests*.

Mixed-signal circuits have the property that the domain of values of the input signals is digital (analog) while the domain of the values of the output signals is analog (digital), e.g., digital-to-analog converter (DAC) and analog-to-digital converter (ADC) circuits. Testing mixed-signal circuits is based on a combination of analog and digital test techniques.

This book mainly focuses on testing digital circuits.

1.3.2   Measured parameter aspect

When testing digital circuits, a classification of tests can be made based on the nature of the type of measurement which is performed on the value of the binary signal. When the measurement aims at verifying the logical correctness of the signal value one speaks about logical tests; when it concerns the behavior of the signal value in time, or its voltage level and/or drive capability, one speaks about electrical tests.

Logical tests: **Logical tests** aim at finding faults which cause a change in the logical behavior of the circuit: deviations from the good device are only considered faults when a response signal level is a 'logic 0' instead of the expected 'logic 1', or vice versa. These faults may be anywhere in the circuit and are not considered to be time-dependent (i.e., they are permanent faults).

Electrical tests: **Electrical tests** verify the correctness of a circuit by measuring the values of electrical parameters, such as voltage and current levels, as well as their behavior over time. They can be divided into *parametric* tests and *dynamic* tests.

Parametric tests are concerned with the *external behavior* of the circuit; i.e., voltage/current levels and delays on the input and output pins of the chip. The specifications of the signal values on the input and output pins of a chip have a time-independent part (voltage and current levels) and a time-dependent part (rise and fall times). The qualification of a chip via the verification of the time-independent properties of the signal values on the input and output pins is called **DC parametric testing**, whereas the qualification via the verification of the time-dependent properties is called **AC parametric testing**. For a comprehensive treatment of DC and AC parametric tests the reader is referred to Stevens (1986) and van de Goor (1991).

A special case of DC parametric testing is the I_{DDQ} **test** method (Hawkins and Soden, 1986); this test method has been shown to be capable of detecting logical faults in CMOS circuits and can be used to measure the reliability of a chip. I_{DDQ} is the quiescent power supply current which is drawn when the chip is not switching. This current is caused by sub-threshold transistor leakage and reverse diode currents and is very small; on the order of tens of nA. I_{DDQ} tests are based on the fact that defects like shorts and abnormal leakage can increase I_{DDQ} by orders of magnitude. It has been shown that the I_{DDQ} test method is effective in detecting faults of many fault models (see Chapter 2 for a description of fault models).

Dynamic tests are aimed at detecting faults which are time-dependent and internal to the chip; they relate to the speed with which the operations are being performed. The resulting failures manifest themselves as logical faults. Delay tests, which verify whether output signals make transitions within the specified time, and refresh tests for DRAMs, belong to the class of dynamic tests.

The main emphasis of this book is on logical tests; however, a chapter has been included on I_{DDQ} fault testing because of its elegance and fault detection capabilities, and a chapter on delay fault testing (which is a form of dynamic test) has been included because of its importance owing to the increasing speed of modern circuits.

1.3.3   Use of test results

The most obvious use of the result of a test is to distinguish between good and bad parts. This can be done with a test which **detects faults**. When the purpose of testing is repair, we are interested in a test which **locates faults**, which is more difficult to accomplish.

Testing can be done during normal use of the part or system; such tests are called **concurrent tests**. For example, byte parity, whereby an extra check bit is added to every eight information bits, is a well known concurrent error detection technique capable of detecting an odd number of errors. Error correcting codes, where, for example, seven check bits are added to a 32-bit data word, are capable of correcting single-bit errors and detecting double-bit errors (Rao and Fujiwara, 1989). Concurrent tests, which are used extensively in fault-tolerant computers, are not the subject of this book. **Non-concurrent tests** are tests which cannot be performed during normal use of the part or system because they do not preserve the normal data. They have the advantage of being able to detect and/or locate more complex faults; these tests are the subject of this book.

Design for testability (DFT) is a technique which allows non-concurrent tests to be performed faster and/or with a higher fault coverage by including extra circuitry on the to-be-tested chip. DFT is used extensively when testing sequential circuits. When a test has to be designed for a given fault in a combinational circuit with n inputs, the search space for finding a suitable test stimulus for that fault consists of 2^n points. In case of a sequential circuit containing f flip-flops, the search space increases to 2^{n+f} points, because of the 2^f states the f flip-flops can take on. In addition, since we cannot directly control the present state lines of a sequential circuit nor directly observe the next state lines, a justification sequence is required to get the circuit into the desired state and a propagation sequence is required to propagate errors, if any, from the next state lines to primary outputs. This makes test generation a very difficult task for any realistic circuit. To alleviate this problem, a DFT technique can be used which, in the test mode, allows all flip-flops to form a large shift register. As a shift register, the flip-flops can be tested easily (by shifting in and out certain 0–1 sequences); while at the same time test stimuli for the combinational logic part of the circuit can be shifted in (scanned in), and test responses can be shifted out (scanned out), thus reducing the sequential circuit testing problem to combinational circuit testing. This particular form of DFT is called *scan design*.

When the amount of extra on-chip circuitry for test purposes is increased to the extent that test stimuli can be generated and test responses observed on-chip, we speak of **built-in self-test (BIST)**, which obviates the requirement for an automatic test equipment (ATE) and allows for at-speed (at the normal clock rate) testing.

1.3.4 Test application method

Tests can also be classified, depending on the way the test stimuli are applied to the part and the test responses are taken from the part.

An ATE can be used to supply the test stimuli and observe the test responses; this is referred to as an **external test**. Given a board with many parts, the external test can be applied in the following ways:

1 Via the normal board connectors:

This allows for a simple interface with the ATE and enables the board to be tested at normal speed; however, it may be difficult (if not impossible) to design tests which can detect all faults because not all circuits are easy to reach from the normal connectors. At the board level, this type of testing is called **functional testing**; this is the normal way of testing at this level.

2 Via a special fixture (a board-specific set of connectors):

A special board-specific connector is used to make all signal lines on the board accessible. High currents are used to drive the test stimuli signals in order to temporarily overdrive existing signal levels; this reduces the rate at which tests can be performed to about 1 MHz. This type of testing is called **in-circuit testing**. It enables the location of faulty components.

1.4 Fault coverage requirements

Given a chip with potential defects, an important question to be answered is how extensive the tests have to be. This section presents an equation relating the *defect level* of chips to the *yield* and fault coverage (Williams and Brown, 1981; Agrawal *et al.*, 1982; McCluskey and Buelow, 1988; Maxwell and Aitken, 1993). This equation can be used to derive the required test quality, given the process yield and the desired defect level.

The **defect level**, *DL*, is the fraction of bad parts that pass all tests. Values for *DL* are usually given in terms of defects per million, DPM; desired values are less than 200 DPM, which is equal to 0.02% (Intel, 1987).

The **process yield**, *Y*, is defined as the fraction of the manufactured parts that is defect-free. The exact value of *Y* is rarely known, because it is not possible to detect all faulty parts by testing all parts for all possible faults. Therefore, the value of *Y* is usually approximated by the ratio: number-of-not-defective-parts/total-number-of-parts; whereby the number-of-not-defective-parts is determined by counting the parts which pass the used test procedure.

The **fault coverage**, *FC*, is a measure to grade the quality of a test; it is defined as the ratio of: actual-number-of-detected-faults/total-number-of-faults (where the faults are assumed to belong to a particular fault model, see Chapter 2). In practice, it will be impossible to obtain a complete test (i.e., a test with $FC = 1$) for a VLSI part because of: (a) imperfect fault modeling: an actual fault (such as an open connection in an IC) may not correspond to a modeled fault or vice versa, (b) data dependency of faults: it may not be sufficient to exercise all functions (such as ADD and SUB, etc. in a microprocessor) because the correct execution of those functions may be data-dependent (e.g., when the carry function of the ALU is faulty), and (c) testability limitations may exist (e.g., due to pin limitations not all parts of a circuit may be accessible). This implies that if the circuit passes the test, one cannot guarantee the absence of faults.

Assume that a given chip has exactly n stuck-at faults (SAFs). An SAF is a fault where a signal line has a permanent value of logic 0 or 1. Let m be the number of faults detected ($m \leq n$) by a test for SAFs. Furthermore, assume that the probability of a fault occurring is independent of the occurrence of any other fault (i.e., there is no clustering), and that all faults are equally likely with probability p. If A represents the event that a part is free of defects, and B the event that a part has been tested for m defects while none were found, then the following equations can be derived (McCluskey and Buelow, 1988).

The fault coverage of the test is defined as: $FC = m/n$. $\qquad$ (1.18)

The process yield is defined as: $Y = (1 - p)^n = P(A)$. $\qquad$ (1.19)

$$P(B) = (1 - p)^m. \qquad (1.20)$$

$$P(A \cap B) = P(A) = (1 - p)^n. \qquad (1.21)$$

$$P(A|B) = P(A \cap B)/P(B) = (1 - p)^n/(1 - p)^m$$
$$= (1 - p)^{n(1-m/n)} = Y^{(1-FC)}. \qquad (1.22)$$

The defect level can be expressed as:

$$DL = 1 - P(A|B) = 1 - Y^{(1-FC)}. \qquad (1.23)$$

Figure 1.4 (Williams and Brown, 1981) shows curves for *DL* as a function of *Y* and *FC*; it is a nonlinear function of *Y*. However, for large values of *Y*, representing manufacturing processes with high yield, the curve approaches a straight line.

Example 1.3 Assume a manufacturing process with an yield of $Y = 0.5$ and a test with $FC = 0.8$; then $DL = 1 - 0.5^{(1-0.8)} = 0.1295$ which means that 12.95% of all shipped parts are defective. If a *DL* of 200 DPM (i.e., $DL = 0.0002 = 0.02\%$) is required, given $Y = 0.5$, the fault coverage has to be: $FC = 1 - (\log(1 - DL)/\log Y) = 0.999\,71$, which is 99.971%. $\qquad \square$

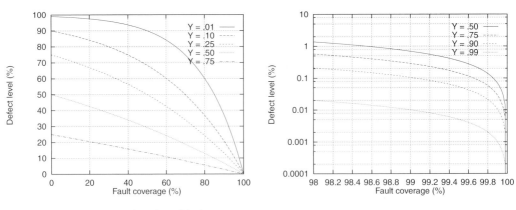

Figure 1.4 Defect level as a function of fault coverage

1.5   Test economics

This section discusses the economics of testing; it starts with a subsection highlighting the repair cost during product phases; thereafter the economics and liability of testing are considered. The third subsection gives an analysis of the cost/benefit of test development, and the last subsection discusses the field maintenance cost.

1.5.1   Repair cost during product phases

During the lifetime of a product, the following phases can be recognized: component manufacture (chips and bare boards), board manufacture, system manufacture, and working life of a product (when it is in the field). During each phase, the product quality has to be ensured. The relationship of the test and repair costs during each of these phases can be approximated with the *rule-of-ten* (Davis, 1982) (see Figure 1.5): if the test and repair cost in the component manufacturing phase is R, then in the board manufacture phase it is $10R$, in the system manufacturing phase it is $100R$, and during the working life phase it is $1000R$. This is due to the increase in the difficulty level of locating the faulty part, the increase in repair effort (travel time and repair time) and the larger volume of units involved.

1.5.2   Economics and liability of testing

Section 1.5.1 explained the rule-of-ten for the test and repair cost. From this it is obvious that by eliminating faults in the early phases of a product, great savings can be obtained. In addition, a good test approach can reduce the development time, and therefore the time-to-market, as well as the field maintenance cost. However, test development also takes time and therefore increases the development time. The

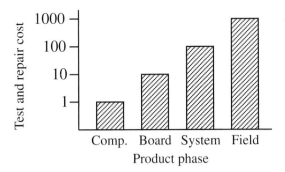

Figure 1.5 Rule-of-ten

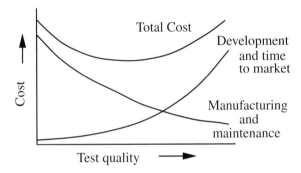

Figure 1.6 Influence of test quality on cost

right amount of test development effort therefore is a tradeoff (see Figure 1.6). With increasing test quality: (a) the cost of test development, as well as the elapsed test development time (which can be translated into a later time-to-market), increases, and (b) the manufacturing test and repair cost decreases. Therefore, the total cost of the product (during all its product phases) is minimized for some test quality which is not the highest test quality.

Another important reason for testing is the liability aspect when defects during the working life period cause breakdowns which may lead to personal injury, property damages and/or economic loss. Well-known examples from the car industry (accidents) and the medical practice (overdose of radiation) show civil and punitive lawsuits where the equipment manufacturer has been liable for large amounts (millions of dollars) of money.

Looking at the lifetime of a product, one can recognize several economic phases (see Figure 1.7(a)): the development phase, the market growth phase, and the market decline phase.

During the development phase, an associated cost is incurred, resulting in an initial loss. Initially, the development team is small and grows when the product is better defined such that more disciplines can participate. Later on, the development effort

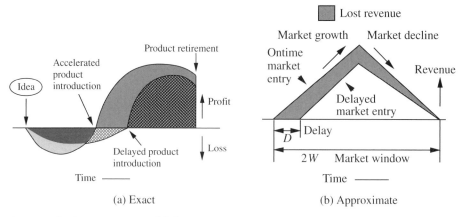

Figure 1.7 Product revenue over lifetime

decreases until the development has been completed. The area of the curve under the zero-line of Figure 1.7(a) represents the total product development cost.

After development, the product is sold, resulting in a profit. Initially, the market share is low and increases with increasing product acceptance (market growth); thereafter, due to competition/obsolescence, the market share decreases (market decline) until the product is taken out of production (product retirement). The total product revenue is the difference in the two areas marked 'profit' and 'loss', respectively.

For typical digital systems, such as mainframes, workstations, and personal computers, the development period may be on the order of one to two years, while the *market window* (consisting of the market growth and market decline parts) is on the order of four years (for personal computers it is closer to two years!). Usually, the development cost is a very small part of the total lifetime cost of a system (on the order of 10%) and is rather constant for a given system. Ignoring this cost and approximating the market growth and market decline areas by straight lines, the life-cycle model of Figure 1.7(a) may be approximated with that of Figure 1.7(b).

Digital technology evolves at a rate of about $1.6X$/year, which means that the speed (in terms of clock frequency) increases, and the cost (in terms of chip area) decreases, at this rate. The result is that when a product is being designed with a particular technology in mind, the obsolescence point is more or less fixed in time. Therefore, when a delay (D) is incurred in the development process, this will have a severe impact on the revenue of the product (see Figure 1.7).

From Figure 1.7(b), and assuming that the maximum value of the market growth is M and reached after time W, the revenue lost due to a delay, D, (hatched area) can be computed as follows. The expected revenue (ER) in case of on-time market entry is: $ER = \frac{1}{2} \cdot 2W \cdot M = W \cdot M$. The revenue of the delayed product (RDP) is: $RDP = \frac{1}{2} \cdot (2W - D)\left(\frac{(W-D)}{W} \cdot M\right)$.

The lost revenue (*LR*) is: $LR = ER - RDP$

$$= W \cdot M - \frac{2W^2 - 3D \cdot W + D^2}{2W} \cdot M$$

$$= ER \cdot \frac{D(3W - D)}{2W^2}. \tag{1.24}$$

It should be noted that a delay '*D*' may be incurred due to excessive test development, but more likely, due to insufficient test development because of more repairs during later product phases and a longer development phase due to the longer time to get the product to the required quality level.

1.5.3   Cost/benefit of test development

Many factors contribute to the cost of developing a test (Ambler *et al.*, 1992). These factors can be divided into per unit cost and cost incurred due to changes in the product development process (i.e., schedule and performance consequences).

1   Engineering cost, C_E:

This is the time spent in developing the test, and possibly also modifying the design, in order to obtain the required fault coverage. For an application-specific integrated circuit (ASIC), C_E can be calculated as:

$$C_E = (\text{number of Eng. days}) \cdot (\text{cost per day})/(\text{number of ASICs}). \tag{1.25}$$

Example 1.4   If it takes 20 days to develop the test, while the cost/day is $600 and 2000 ASICs are produced; then

$$C_E = 20 \cdot \$600/2000 = \$6/\text{ASIC}.$$

2   Increased ASIC cost, C_A:

In order to obtain the required fault coverage, DFT and/or BIST techniques may have to be used. This increases the chip area and reduces the yield such that the cost of the ASIC increases. When the additional DFT/BIST circuitry causes the die area to increase from A_O to A_E, then the cost of the die will increase by the following amount (C_O is the original die cost):

$$C_A = C_O \cdot \{A_E/A_O \cdot Y^{(1 - \sqrt{A_E/A_O})} - 1\}. \tag{1.26}$$

Example 1.5   Assume an ASIC cost of $50 without on-chip test circuits. Furthermore, assume that 50% of this cost is due to the die; the remainder is due to packaging, testing and handling. If the additional circuits increase the die area by 15%, then with an yield of 60% the cost of the die will increase by $C_A = \$4.83$. The cost of the ASIC will increase from $50 to $54.83.

3 Manufacturing re-work cost, C_M:

If the ASIC test program has a fault coverage of less than 100% then some ASICs may be defective at the board level. These faulty ASICs have to be located and replaced; this involves de-soldering the defective part and soldering its replacement. When surface mount technology is used, this may be (nearly) impossible such that the board has to be scrapped. C_M can be expressed as follows: $C_M =$ (board repair cost per ASIC) · (defect level 'DL', representing the fraction of faulty components) · (number of ASICs on the board) · (number of boards).

Example 1.6 Consider a single-board system with 25 ASICs which have been produced with a yield of 0.6 and tested with a fault coverage of 0.95, while the board repair cost per ASIC is $500. Then using Equation (1.23), C_M can be computed as follows:

$$C_M = (1 - 0.6^{(1-0.95)}) \cdot 25 \cdot \$500 = \$315/\text{system}.$$

4 Time-to-market cost, C_T:

This is a complex issue since testing increases the time-to-market because of test development and (possible) redesign time. On the other hand, testing decreases the time-to-market due to the fact that the manufacturing re-work time and cost will decrease. Given the normal expected revenue, the product delay 'D' and the market window 'W'; the time-to-market cost, C_T, can be approximated using Equation (1.24).

Example 1.7 Suppose we are given a $10 000 system of which 2000 are projected to be sold over a time period of three years (i.e., $2W = 36$) which is delayed by two months due to insufficient testing (i.e., $D = 2$). Using Equation (1.24), C_T will be:

$$C_T = \$20\,000\,000 \cdot \{2(3 \cdot 18 - 2)/2 \cdot 18^2\} = \$3\,209\,877.$$

Thus, $C_T/2000 = \$1605/\text{system}$.

5 Cost of speed degradation, C_S:

On-chip test circuitry may cause a degradation in speed due to increased fanin and fanout of circuits and/or due to the addition of one or more logic levels in performance-critical paths. This will make the product less competitive such that its price (and profit margin) may decrease. In order to maintain a constant price/performance ratio, the price of the degraded system has to drop from the price of the original system 'P_O' proportionately with the speed degradation such that ($Perf_D$ = performance of degraded system):

$$C_S = P_O \cdot (1 - Perf_D/Perf_O). \tag{1.27}$$

Example 1.8 Assume a system with a price of $10 000 (i.e., $P_O = \$10\,000$) which

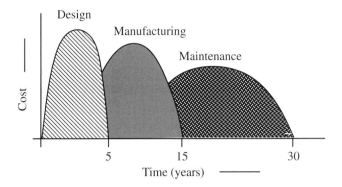

Figure 1.8 Life-cycle cost

incurs a speed degradation of 5% due to on-chip test circuitry, then the decrease in system price will be: $C_S = \$10\,000 \cdot (1 - 0.95) = \$500$.

6 ATE cost, C_T:

The cost of testing a chip, using ATE, is small compared with the costs of points (1)–(5) described above. It can be approximated to a fixed cost per second (e.g., \$0.20/s). Except for special cases where very long test times are required, its contribution to the total cost can be ignored.

1.5.4 Field maintenance cost

Industry analysts estimate that 26% of the revenue and 50% of the profit are generated by the field service centers of equipment vendors (Ambler *et al.*, 1992). In addition, repeated sales depend to a large extent on the quality of service (good/fast maintenance) which reduces the down-time. The importance of field service can furthermore be deduced from the rule-of-ten shown in Figure 1.5.

The life-cycle cost of an industrial electronic system, such as a telephone system or a control or a defense system, is depicted in Figure 1.8 (Ambler *et al.*, 1992). It consists of three parts: design cost (typically on the order of 10%), manufacturing cost (30%), and maintenance cost (60%). The user's 60% maintenance cost, which does not include the cost of down-time, relates to the vendor's 50% profit due to field service.

The maintenance cost can be reduced as follows:

1 Reduction in the number of failures:

This can be done by including fault tolerance techniques (such as error correcting codes for memories and triple modular redundancy for central processing units) in the design.

2 Reduction of the down-time:

This can be accomplished by reducing the travel time and/or the repair time.

The travel time can be reduced by allowing for remote diagnosis (i.e., the use of telecommunication facilities to diagnose remotely) such that the correct replacement part can be taken along on the repair visit.

The repair time can be reduced by reducing the diagnostic time (through DFT/BIST and/or better test software) and the part replacement time (by having spare parts stocked locally). On-line documentation, product modularity, etc., are other aspects which can reduce the down-time.

Remote repair, where the system is diagnosed remotely and the defective (software) part replaced remotely, is another way of reducing down-time and repair cost. In addition, remote diagnostics/repair allows for centers of competence such that rarely occurring faults may be diagnosed more quickly.

Summary

- A failure means that a system does not perform; this is caused by an error, which means that the system is in a wrong state; the error is caused by a fault which is the physical difference between the good and the faulty system.
- Permanent faults affect the correct operation of the system all the time; a non-permanent fault only does this part of the time. Non-permanent faults can be classified as transient faults, caused by environmental conditions, or as intermittent faults, caused by non-environmental conditions such as ageing components.
- A system with a reliability $R(t) = e^{-\lambda t}$ has a constant failure rate, i.e., $z(t) = \lambda$, and $MTTF = 1/\lambda$.
- The bathtub curve shows the failure rate over the lifetime of a product. The infant mortality period can be reduced by using burn-in techniques. The activation energy is a measure of the sensitivity of a failure mechanism to increased temperature.
- Failure mechanisms are the physical and/or electrical causes of a fault. The most important failure mechanisms are: corrosion, electromigration, bonding, ionic contamination, alloying and radiation.
- Tests can be classified according to: technology aspects (analog, digital and mixed-signal circuits), measured parameter aspect (logical and electrical tests; where electrical tests can be subdivided into parametric tests (DC parametric and AC parametric), and dynamic tests), use of test results (fault detection versus fault location, concurrent versus non-concurrent tests), and the test application method (functional versus in-circuit).
- DFT techniques use extra chip circuitry to allow for accelerating of, and/or obtaining a higher fault coverage of, a non-concurrent test. When the amount of on-chip circuitry is such that the non-concurrent test can be performed without the use of an ATE, one speaks of BIST.
- A relationship exists between the defect level, the process yield and the fault coverage.

- The cost of repair increases by an order of magnitude for every product phase.
- The optimal test quality is a tradeoff between the time-to-market and the manufacturing and maintenance cost.
- The lifetime cost of an industrial product for the user is dominated by the maintenance cost. This can be reduced by reducing the number of failures and/or reducing the downtime.

Exercises

1.1 Suppose we are given a component operating at a normal temperature of $30\,°C$. The goal is to age this component such that, when shipped, it already has an effective life of one year. The values for E_a can be taken from Table 1.1 where the midpoint of the interval has to be taken.

   (a) Assume a burn-in temperature of $125\,°C$; what is the burn-in time assuming corrosion of metallization is the dominant failure mode?

   (b) Same question as (a); now with a burn-in temperature of $150\,°C$.

   (c) Assume a burn-in temperature of $125\,°C$; what is the burn-in time assuming alloying is the dominant failure mode?

   (d) Same question as (c); now with a burn-in temperature of $150\,°C$.

1.2 Suppose we are given a system consisting of two subsystems connected in series. The first subsystem is a parallel system with three components; the second subsystem is a parallel system with two components. What is the reliability, $R(t)$, of this system given that the failure rate, $z(t)$, of each of the five components is λ?

1.3 Given a manufacturing process with a certain yield and a certain required defect level for the shipped parts, compute the required fault coverage:

   (a) Given $DL = 20$ DPM and $Y = 0.6$;

   (b) Given $DL = 20$ DPM and $Y = 0.8$;

   (c) Given $DL = 4$ DPM and $Y = 0.6$;

   (d) Given $DL = 4$ DPM and $Y = 0.8$.

1.4 Suppose a team is designing a product consisting of three boards, each containing 10 ASICs. The process producing the ASICs has an yield of 0.7; while tests have a fault coverage of 95%. The ASIC die price is $30, the cost of a packaged and tested ASIC (without test circuits) is $60. The engineering cost is $600/day and the board-level manufacturing re-work cost is $500 per ASIC. The targeted product life cycle is 48 months; the system price is $10 000 including a profit margin of 20% and the expected number of units sold is 5000.

   In order to improve the fault coverage to 99%, on-chip circuitry has to be added with the following consequences: the die area increases by 10%, the speed of the chip decreases by 5%, and the extra test development cost due to the on-chip circuits takes one month (assume a month has 22 engineering days).

(a) Given the projected system:
  i. What is the manufacturing re-work cost (C_M) and what is the expected revenue (ER)?
  ii. What is the manufacturing re-work cost when the yield is increased to 0.8?
  iii. What is the manufacturing re-work cost when the yield is increased to 0.8 and the fault coverage is increased to 99%?
(b) Given the system, enhanced with test circuitry.
  i. What are the new values for C_E, C_A, C_M, and C_S?
  ii. What is the value of the lost revenue (LR)?
  iii. What is the difference in total profit owing to the inclusion of the test circuitry?
(c) Assume that the test circuitry reduces the maintenance cost by 20%. Furthermore, assume that the original system price ($10 000) was only 40% of the life-cycle cost for the customer due to the maintenance cost (see Figure 1.8). How much is the original life-cycle cost and what is the new life-cycle cost given the system enhanced with test circuitry?

References

Agrawal, V.D., Seth, S.C., and Agrawal, P. (1982). Fault coverage requirements in production testing of LSI circuits. *IEEE J. of Solid-State Circuits*, **SC-17**(1), pp. 57–61.

Ambler, A.P., Abadir, M., and Sastry, S. (1992). *Economics of Design and Test for Electronic Circuits and Systems*. Ellis Horwood: Chichester, UK.

Amerasekera, E.A. and Campbell, D.S. (1987). *Failure Mechanisms in Semiconductor Devices*. John Wiley & Sons: Chichester, UK.

Blanks, H.S. (1980). Temperature dependence of component failure rate. *Microelec. and Rel.*, **20**, pp. 219–246.

Carter, P.M. and Wilkins, B.R. (1987). Influences of soft error rates in static RAMs. *IEEE J. of Solid-State Circuits*, **SC-22** (3), pp. 430–436.

Davis, B. (1982). *The Economics of Automated Testing*. McGraw-Hill: London, UK.

van de Goor, A.J. (1991). *Testing Semiconductor Memories, Theory and Practice*. John Wiley & Sons: Chichester, UK.

Gouda (1994). *Life Table for Dutch Male Population 1976–1980*. Gouda Life Insurance Company (R.J.C. Hersmis): Gouda, The Netherlands.

Jensen, F. and Petersen, N.E. (1982). *Burn-in*. John Wiley & Sons: Chichester, UK.

Hawkins, C.F. and Soden, J.M. (1986). Reliability and electrical properties of gate oxide shorts in CMOS ICs. In *Proc. Int. Test Conference*, Washington D.C., pp. 443–451.

Intel (1987). *Components Quality Reliability Handbook*. Intel Corporation: Santa Clara, CA.

Ma, T.P. and Dressendorfer, P.V. (1989). *Ionizing Radiation Effects in MOS Devices and Circuits*. John Wiley & Sons: New York, NY.

McCluskey, E.J. and Buelow, F. (1988). IC quality and test transparency. In *Proc. Int. Test Conference*, pp. 295–301.

Maxwell, P.C. and Aitken, R.C. (1993). Test sets and reject rates: all fault coverages are not created equal. *IEEE Design & Test of Computers* **10**(1), pp. 42–51.

Moltoft, J. (1983). Behind the 'bathtub' curve – a new model and its consequences. *Microelec. and Rel.*, **23**, pp. 489–500.

O'Connor, P.D.T.O. (1985). *Practical Reliability Engineering.* John Wiley & Sons: Chichester, UK.

Rao, T.R.N. and Fujiwara, E. (1989). *Error-Control Coding for Computer Systems.* Prentice-Hall: Englewood Cliffs, NJ.

Stevens, A.K. (1986). *Introduction to Component Testing.* Addison-Wesley: Reading, MA.

Williams, T.W. and Brown, N.C. (1981). Defect level as a function of fault coverage. *IEEE Trans. on Computers*, **C-30**, pp. 987–988.

2 Fault models

In order to alleviate the test generation complexity, one needs to model the actual defects that may occur in a chip with fault models at higher levels of abstraction. This process of fault modeling considerably reduces the burden of testing because it obviates the need for deriving tests for each possible defect. This is made possible by the fact that many physical defects map to a single fault at the higher level. This, in general, also makes the fault model more independent of the technology.

We begin this chapter with a description of the various levels of abstraction at which fault modeling is traditionally done. These levels are: behavioral, functional, structural, switch-level and geometric.

We present various fault models at the different levels of the design hierarchy and discuss their advantages and disadvantages. We illustrate the working of these fault models with many examples.

There is currently a lot of interest in verifying not only that the logical behavior of the circuit is correct, but that its temporal behavior is also correct. Problems in the temporal behavior of a circuit are modeled through delay faults. We discuss the main delay fault models.

We discuss a popular fault modeling method called inductive fault analysis next. It uses statistical data from the fabrication process to generate physical defects and extract circuit-level faults from them. It then classifies the circuit-level faults based on how likely they are to occur.

Finally, we describe some interesting relationships among some of the fault models.

2.1 Levels of abstraction in circuits

Circuits can be described at various levels of abstraction in the design hierarchy. These levels are behavioral, functional, structural, switch-level, and geometric (see Figure 2.1).

A **behavioral description** of a digital system is given using a hardware description language, such as VHDL or Verilog. It depicts the data and control flow. Increasingly, the trend among designers is to start the synthesis process from a behavioral description.

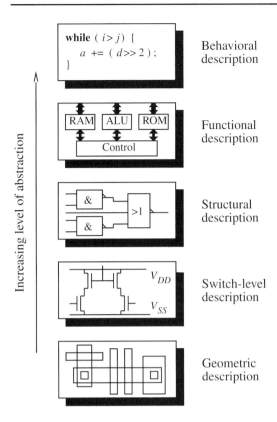

Figure 2.1 Levels of abstraction

A **functional description** is given at the register-transfer level (RTL). This description may contain registers, modules such as adders and multipliers, and interconnect structures such as multiplexers and busses. This description is sometimes the product of **behavioral synthesis** which transforms a behavioral description into an RTL circuit.

A **structural description** is given at the logic level. It consists of logic gates, such as AND, OR, NOT, NAND, NOR, XOR, and interconnections among them. This is the most commonly used description level.

A **switch-level description** establishes the transistor-level details of the circuit. In CMOS technology, each logic gate is described using an interconnection of a pMOS and an nMOS network. These networks themselves consist of an interconnection of several transistors. These transistors are usually connected in series-parallel fashion, although non-series-parallel structures are also possible.

A **geometric description** is given at the layout level. From this description, one can determine line widths, inter-line and inter-component distances, and device geometries.

2.2   Fault models at different abstraction levels

Fault modeling is the process of modeling defects at higher levels of abstraction in the design hierarchy (Timoc *et al.*, 1983; Hayes, 1985; Shen *et al.*, 1985; Abraham and Fuchs, 1986). The advantage of using a fault model at the lowest level of abstraction is that it closely corresponds to the actual physical defects, and is thus more accurate. However, the sheer number of defects that one may have to deal with under a fault model at this level may be overwhelming. For example, a chip made of 50 million transistors could have more than 500 million possible defects. Therefore, to reduce the number of faults and, hence, the testing burden, one can go up in the design hierarchy, and develop fault models which are perhaps less accurate, but more practical. In fact, a good strategy may be to first derive tests for fault models at higher levels, and then determine what percentage of faults at the lower levels are covered by these tests. Since typically the number of faults becomes smaller as we go up in the level of abstraction, the time taken for test generation becomes smaller. Also, since a fault at a higher level models many faults at a lower level, such a test set should cover a high percentage of lower-level faults. However, because of imperfect modeling, many lower-level faults may remain undetected by this higher-level test set. Such faults can be specifically targeted at the lower level of the design hierarchy. This approach can result in significant savings in the time required to derive the tests. The above scenario assumes that the design description of the circuit is available at various levels of abstraction, which is unfortunately not always the case.

Fault models have been developed at each level of abstraction, i.e., behavioral, functional, structural, switch-level, and geometric.

Behavioral fault models are defined at the highest level of abstraction. They are based on the behavioral specification of the system. For example, if a digital system is described using a hardware description language, such as VHDL or Verilog, one could inject various types of faults in this description. The collection of these faults will constitute a behavioral fault model. Precisely what types of faults are included in a behavioral fault model depends on the ease with which they allow detection of realistic faults at the lower levels of abstraction.

Functional fault models are defined at the functional block level. They are geared towards making sure that the functions of the functional block are executed correctly. In addition, they should also make sure that unintended functions are not executed. For example, for a block consisting of random-access memory (RAM), one type of functional fault we may want to consider is when one or more cells are written into, other cells also get written into. This type of fault is called *multiple writes*. Various other types of functional faults can also be defined for RAMs. For a microprocessor, a functional fault model can be defined at the instruction level or RTL.

Structural fault models assume that the structure of the circuit is known. Faults under these fault models affect the *interconnections* in this structure. The most well-known fault model under this category is the *single stuck-at* fault model. This is the most widely used fault model in the industry. Indeed, because of its longevity, it is sometimes termed the classical fault model. Its popularity depends on the fact that it can be applied to various semiconductor technologies, and that detection of all single stuck-at faults results in the detection of a majority of realistic physical defects (in many cases, up to 80–85% of the defects are detected). However, the current trend is towards augmenting this fault model with other fault models which allow detection of defects that the stuck-at fault model is unable to cover.

Switch-level fault models are defined at the transistor level. The most prominent fault models in this category are the *stuck-open* and *stuck-on* fault models. If a transistor is permanently non-conducting due to a fault, it is considered to be **stuck-open**. Similarly, if a transistor is permanently conducting, it is considered to be **stuck-on**. These fault models are specially suited for the CMOS technology.

Geometric fault models assume that the layout of the chip is known. For example, knowledge of line widths, inter-line and inter-component distances, and device geometries are used to develop these fault models. At this level, problems with the manufacturing process can be detected. The layout information, for example, can be used to identify which lines or components are most likely to be shorted due to a process problem. The *bridging* fault model thus developed leads to accurate detection of realistic defects. With shrinking geometries of very large scale integrated (VLSI) chips, this fault model will become increasingly important.

2.2.1 Behavioral fault models

Behavioral fault models are associated with high-level hardware descriptions of digital designs (Levendel and Menon, 1982; Ghosh and Chakraborty, 1991). They are related to failure modes of the constructs in the hardware description language, such as VHDL (Navabi, 1993) or a subset of C (Davidson and Lewandowski, 1986). The importance of these models comes from the increasing desire among designers to start the synthesis process from a behavioral specification.

Various behavioral fault models can be derived as follows (Ghosh and Chakraborty, 1991):

- A variable R in the description of the digital system may be assumed to be at either value V_L or V_H permanently rather than the correct intermediate value V_I, where V_L and V_H, respectively, represent the lower and higher extremes of the logical value system. Even though R could be held permanently at some intermediate value due to a fault, such faults are not considered because their extremely large number would make handling them impractical.
- A *call* to a function may be assumed to fail in such a way that it always returns V_L or V_H.

- The "**for** (CC) $\{B\}$" clause of the language could fail such that the body $\{B\}$ is always executed or never executed, irrespective of whether condition CC evaluates to *true*.
- The "**switch** (Id)" clause could fail in one of the following ways:
  - all of the specified cases are selected,
  - cases corresponding to the lower and higher extreme values of the switch identifier's value system are selected,
  - none of the specified cases are selected.
- An "**if** (Y) **then** $\{B_1\}$ **else** $\{B_2\}$" construct could fail in one of the following ways:
  - the set of statements $\{B_1\}$ is never executed and set $\{B_2\}$ is always executed,
  - set $\{B_1\}$ is always executed and set $\{B_2\}$ is never executed,
  - set $\{B_1\}$ is executed when the Boolean expression Y evaluates to false and set $\{B_2\}$ is executed when Y evaluates to *true*.
- The assignment statement "$X := Y$" could fail in one of the following ways:
  - the value of X remains unchanged,
  - X assumes V_L or V_H rather than V_I,
  - X assumes the value V_L or V_H based on some probability function.
- The "**waitfor** S" synchronizing construct could fail such that the clause may always be executed or never be executed irrespective of the state of signal S.

Each of the above behavioral fault models can be related to an actual defect in the corresponding hardware realization. Test sets derived from these fault models have been found to detect a high percentage (around 85%) of faults belonging to fault models defined at the lower levels of abstraction, such as stuck-at faults.

2.2.2   Functional fault models

A functional fault model ensures that the given functional block executes the function it was designed for, and does not execute any unintended function. Such fault models are usually ad hoc, geared towards specific functional blocks.

Consider a multiplexer. One can derive the following functional fault model for it.

- A 0 and a 1 cannot be selected on each input line.
- When an input is being selected, another input gets selected instead of or in addition to the correct input (in the latter case, the two inputs could be thought to produce the AND or OR of the two logic values).

A functional fault model can be considered good if it is not too complex for test generation purposes, and the resulting functional test set provides a high coverage of lower-level fault models. Based on this criterion, the above fault model is indeed a good one, because it leads to a high coverage of stuck-at and bridging faults.

Another type of functional fault model assumes that the truth table of the functional block can change in an arbitrary way. This usually leads to exhaustive testing, where, for example, for an n-input combinational circuit, all the 2^n vectors need to be applied.

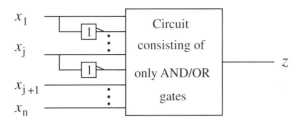

Figure 2.2 A restricted implementation

Clearly, this becomes impractical for large values of n. However, in many cases, the number of vectors can be reduced. Consider a multi-output circuit with n inputs, where output z_i depends on n_i primary inputs. Let $n_i < n$ for all i. Then a pseudoexhaustive test set can be derived which feeds all 2^{n_i} vectors to the corresponding subcircuit which realizes z_i.

If a minor restriction is imposed on the kinds of circuits that are synthesized for the given Boolean function, then the above truth table based fault model can also be used in conjunction with a concept called **universal test set** (Betancourt, 1971), (Akers, 1973), (Reddy, 1973), (Gupta and Jha, 1988). Such a test set detects any fault which changes the truth table in any implementation meeting this restriction as long as the circuit remains combinational in the presence of the fault, hence the name *universal*. The restriction requires that the circuit be implemented as shown in Figure 2.2. No assumption is made regarding the structure or the number of levels of gates in the internal part of the circuit, other than that it consists only of AND and OR gates. This kind of circuit can be easily obtained by logic synthesis tools, such as SIS (Sentovich *et al.*, 1992). In Figure 2.2, we have assumed that the circuit is **binate** in inputs $x_1, \ldots, x_j$ (i.e., it depends on both the input and its complement), and **unate** in inputs $x_{j+1}, \ldots, x_n$ (i.e., only the input or its complement is required). If there are u unate inputs and $n - u$ binate inputs, then the size of the universal test set can be at most $U(u) \cdot 2^{n-u}$ (Akers, 1973), where

$$U(u) = \left(\begin{array}{c} u + 1 \\ \lfloor (u + 1)/2 \rfloor \end{array} \right).$$

Although the universal test set is very powerful, it does suffer from the problem that the size of the test set becomes exhaustive when there are no unate inputs. For example, for the function $z = x_1 x_2 x_3 + \bar{x}_1 \bar{x}_2 \bar{x}_3$, the universal test set consists of all eight three-bit vectors. Even when not all inputs are binate, the size of the test set can be quite large.

Let us next consider iterative logic arrays (ILAs). An ILA consists of a repeated interconnection of identical logical cells. One of the simplest examples of an ILA is a ripple-carry adder, as shown in Figure 2.3 for the four-bit case, where each full adder (FA) is assumed to be a cell. A functional fault model, called the **single cell fault model** (SCFM), is frequently used for ILAs. SCFM assumes that the interconnection

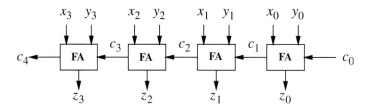

Figure 2.3　A four-bit ripple-carry adder

structure of the cells in the ILA is known, although the logic implementation of the cell is unknown. Under this fault model, functions of a single cell are assumed to fail in an arbitrary way, as long as the faulty cell remains combinational. For the ripple-carry adder, for example, the fault is assumed to change the truth tables of the sum and carry functions of a cell in an arbitrary fashion. Therefore, under SCFM, each cell is fed its exhaustive test set and errors propagated to observable outputs. For our example, this means that a test set needs to be derived which will feed all eight vectors to each of the full adders. It is known that, irrespective of the number of cells in the ripple-carry adder, a test set of size eight can be derived to accomplish the above job. ILAs with this property are said to be **C-testable** (Friedman, 1973).

Efficient functional fault models have also been developed for sequential circuit testing, memory testing, and microprocessor testing. These fault models are discussed in the respective chapters.

2.2.3　Structural fault models

In structural testing, we need to make sure that the interconnections in the given structure are fault-free and are able to carry both logic 0 and 1 signals. The stuck-at fault (SAF) model is directly derived from these requirements. A line is said to be stuck-at 0 (SA0) or stuck-at 1 (SA1) if the line remains fixed at a low or high voltage level, respectively (assuming positive logic). An SAF does not necessarily imply that the line is shorted to ground or power line. It could be a model for many other cuts and shorts internal or external to a gate. For example, a cut on the stem of a fanout may result in an SA0 fault on all its fanout branches. However, a cut on just one fanout branch results in an SA0 fault on just that fanout branch. Therefore, SAFs on stems and fanout branches have to be considered separately.

We next show how the SAF model can detect many realistic defects. Consider the TTL NAND gate shown in Figure 2.4. Fault modeling for opens and shorts in such a gate was done in Beh *et al.* (1982). The results of this modeling are shown in Table 2.1. In the left column, various possible defects are given. In the right column, the SAF, which models the defects, is shown. Of these defects, T_1, T_2 and T_3 base–emitter shorts do not directly map to the corresponding SAF, yet are detected when the SAF is. All

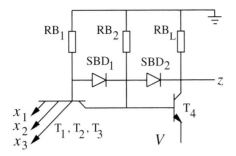

Figure 2.4   A three-input TTL NAND gate (Beh *et al.*, 1982) (© 1982 IEEE)

Table 2.1. *Defect detection in the TTL NAND gate (Beh* et al.*, 1982) (© 1982 IEEE)*

Physical defect	Defect detection
RB_1 open	z SA0
RB_2, R_L, SBD_1, SBD_2 open, T_2, T_3 base open	undetectable
T_1 emitter open, base open, base–emitter short	x_1 SA1
T_2 emitter open, base–emitter short	x_2 SA1
T_3 emitter open, base–emitter short	x_3 SA1
T_4 emitter open, base open, collector open, base–emitter short	z SA1
T_1, T_2, T_3 collector open	z SA0
T_1, T_2, T_3 collector–emitter short	undetectable
T_4 collector–emitter short, collector-base short	z SA0
T_1, T_2, T_3 collector-base short	z SA0
T_2, T_3 base open	undetectable

other detectable defects directly map to the corresponding SAF. The defects which are labeled undetectable cannot be detected just by monitoring the logic value at output z. However, they may be detectable through other means. For example, the defect RB_2 open, which significantly increases the delay of the gate, can be detected if we measure the delay of the circuit. Such faults are called delay faults, and are considered later in this chapter. From Table 2.1, it is clear that the SAF model can detect a large number of defects, yet the table also points to the need for augmenting this fault model with other fault models if we want a still higher coverage of defects. Similar conclusions are also valid for other technologies.

If the SAF is assumed to occur on only one line in the circuit, it is said to belong to the **single SAF model**. Otherwise, if SAFs are simultaneously present on more than one line in the circuit, the faults are said to belong to the **multiple SAF model**. If the circuit has k lines, it can have $2k$ single SAFs, two for each line. However, the number of multiple SAFs is $3^k - 1$ because there are three possibilities for each line (SA0, SA1, fault-free), and the resultant 3^k cases include the case where all lines are fault-free. Clearly, even for relatively small values of k, testing for all multiple SAFs is

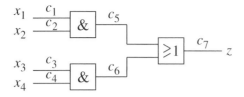

Figure 2.5 A logic circuit with SAFs

impossible. However, as we shall see in Chapter 13 (Section 13.1.4), synthesis methods exist which generate a multi-level circuit from a two-level circuit for which the single SAF test set of the two-level circuit also detects all multiple SAFs in the multi-level circuit.

A subset of the multiple SAF model, called the **unidirectional SAF model**, is also sometimes used. In this model, all the affected lines are assumed to be stuck at the same logic value, either 0 or 1.

Consider the circuit shown in Figure 2.5. Assume first that only line c_5 has an SA0 fault. To test for this single SAF, we can apply $(x_1, x_2, x_3, x_4; z) = (1, 1, 0, 1; 1)$ to the circuit. In the presence of the fault, $z = 0$, and the fault is detected. If c_5 SA0 and c_6 SA0 are simultaneously present, then we have a unidirectional SA0 fault. If, in addition, line c_3 SA1 is present, then we have a multiple SAF. Both the unidirectional and multiple SAFs are coincidentally also detected by vector $(1, 1, 0, 1; 1)$. In fact, one can check that any vector which makes $z = 1$ in the fault-free case will detect these two faults.

2.2.4   Switch-level fault models

Switch-level fault modeling deals with faults in transistors in a switch-level description of a circuit. This fault model has mostly been used with MOS technologies, specifically CMOS technology. The most prominent members in this category are the *stuck-open* and *stuck-on* fault models (Wadsack, 1978).

2.2.4.1 The stuck-open fault model

A **stuck-open fault (SOpF)** refers to a transistor which becomes permanently non-conducting due to some defect. Consider the two-input static CMOS NOR gate shown in Figure 2.6(a). This gate consists of an nMOS network containing transistors Q_1 and Q_2, and a pMOS network containing transistors Q_3 and Q_4. An nMOS (pMOS) transistor conducts when logic 1 (0) is fed to its input, otherwise it remains non-conducting. Suppose that defect d_1 causes an open connection in the gate, as shown. This prevents Q_1 from conducting and is thus said to result in an SOpF in Q_1. Let us see what happens when we apply the exhaustive set of input vectors to the faulty

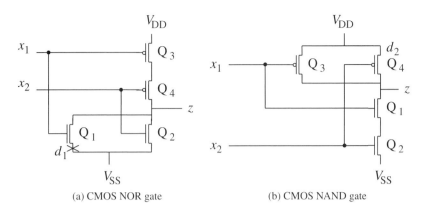

(a) CMOS NOR gate   (b) CMOS NAND gate

Figure 2.6 Two-input static CMOS gates

gate in the following sequence: $(x_1, x_2; z) = \{(0, 0; 1), (0, 1; 0), (1, 0; 0), (1, 1; 0)\}$ (in this sequence, z denotes the fault-free value). When $(0, 0; 1)$ is applied, Q_3 and Q_4 conduct, and output $z = 1$. Next, with the application of $(0, 1; 0)$, z gets pulled down to logic 0 through Q_2. When $(1, 0)$ is applied, there is no conduction path from z to V_{SS} because of the SOpF in Q_1. Therefore, z retains its previous logic value, which is 0. Finally, with the application of vector $(1, 1)$, $z = 0$ because of the conduction path through Q_2. We, therefore, see that we obtain the correct outputs at z in the presence of the SOpF even after the application of the exhaustive test set containing all two-bit input vectors. This is due to the fact that the SOpF has forced the gate to behave in a sequential fashion.

In order to test the circuit for an SOpF, we need a sequence of vectors. Usually, **two-pattern tests**, consisting of an **initialization vector** and a **test vector**, are used. Because the CMOS gate can retain its previous logic value at its output in the presence of an SOpF, the initialization vector is used to initialize the output to the logic value which is the complement of the logic value expected when the SOpF is tested. For example, to test the SOpF caused by defect d_1 in the NOR gate mentioned above, one needs to try to activate a conduction path through the faulty transistor, without activating any parallel path. There is only one such test vector: $(1, 0; 0)$. Since in the fault-free case, for this input vector we expect $z = 0$, the initialization vector should make $z = 1$. There is only one such vector: $(0, 0; 1)$. Therefore, $\{(0, 0; 1), (1, 0; 0)\}$ is a unique two-pattern test for this SOpF. When the fault is present, we get logic 1 at the output when $(1, 0; 0)$ is applied. Thus, the fault is detected. Similarly, the two-pattern test $\{(0, 0; 1), (0, 1; 0)\}$ can detect the SOpF in transistor Q_2. For detecting SOpFs in transistors Q_3 or Q_4, $\{(0, 1; 0), (0, 0; 1)\}$ or $\{(1, 0; 0), (0, 0; 1)\}$ can be used. Therefore, one possible test sequence which detects all four SOpFs in the NOR gate is $\{(0, 0), (0, 1), (0, 0), (1, 0)\}$.

In order to test the circuit for SOpFs, the test sequence needs to be applied at a rate sufficiently faster than those associated with the leakage current time constants

(Wadsack, 1978), which are of the order of milliseconds. Otherwise, leakage currents can invalidate the test. For example, consider the two-pattern test $\{(0, 0; 1), (1, 0; 0)\}$ for the SOpF in Q_1. If the application of $(1, 0; 0)$ is sufficiently delayed, the charge at output z may be lost due to leakage currents, and the correct output ($z = 0$) may be observed for this input vector even in the presence of the SOpF.

2.2.4.2 The stuck-on fault model

If a transistor has become permanently conducting due to some defect, it is said to have a **stuck-on fault (SOnF)**. Consider the two-input NAND gate shown in Figure 2.6(b). Suppose that defect d_2, such as missing polysilicon that forms the gate of transistor Q_4, results in an SOnF in this transistor. In order to try to test for this fault, the only vector we could possibly apply to the NAND gate is $(1, 1; 0)$. In the presence of the fault, transistors Q_1, Q_2 and Q_4 will conduct. This will result in some intermediate voltage at the output. The exact value of this voltage will depend on the on-resistances of the nMOS and pMOS transistors. If this voltage maps to logic 1 at the output, then the SOnF is detected, otherwise not. Now suppose that the only fault present in the gate is an SOnF in transistor Q_2. In order to try to test for this fault, the only vector we could possibly apply is $(1, 0; 1)$. In the presence of the fault, again the same set of transistors, Q_1, Q_2 and Q_4, will conduct. However, this time we would like the intermediate voltage to map to logic 0 in order to detect the fault. Since the same set of transistors is activated in both the cases, the resultant voltage at the output will be the same. Therefore, because of the contradictory requirements for the detection of the SOnFs in Q_4 and Q_2, only one of these two faults can be detected, not both.

The above example illustrates that just monitoring the logic value at the output of the gate, called **logic monitoring**, is not enough if we are interested in detecting all single SOnFs in it. Fortunately, a method based on **current monitoring** is available, which measures the current drawn by the circuit and can ensure the detection of all SOnFs (Malaiya and Su, 1982; Reddy *et al.*, 1984a; Malaiya, 1984). This method is based on the fact that whenever there is a conduction path from V_{DD} to V_{SS} due to an SOnF, the current drawn by the circuit increases by several orders of magnitude compared to the fault-free case. Thus, with the help of a current monitor, such faults can be detected. The disadvantage of current monitoring is that it is slow since it may be possible to feed vectors only at a few KHz rate, whereas in logic monitoring, it may be possible to apply vectors at tens or hundreds of MHz rate. However, in the recent past, much research effort has been directed towards making current monitoring based methods much faster.

2.2.5 Geometric fault models

Geometric fault models are derived directly from the layout of the circuit. Such models exploit the knowledge of line widths, inter-line and inter-component distances, and de-

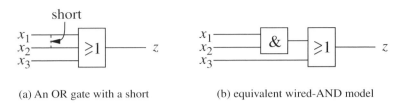

(a) An OR gate with a short   (b) equivalent wired-AND model

Figure 2.7 Wired-AND modeling of a short

vice geometries to determine what defects are most likely. At this level, one deals with opens and shorts. With shrinking geometries, the percentage of defects causing shorts, also called **bridging faults (BFs)**, can be expected to increase even more. BF models can be derived at various levels of abstraction. At the geometric level, such a fault model is the most accurate. However, BF models can also be defined at the structural or switch levels. For non-CMOS technologies, a BF between two lines is assumed to result in an AND or OR function being realized on the two lines. These kinds of faults are referred to as **wired-AND** and **wired-OR**, respectively (Kodandapani and Pradhan, 1980). Consider the three-input OR gate shown in Figure 2.7(a). If the short between inputs x_1 and x_2 is modeled as a wired-AND then the equivalent circuit would be as shown in Figure 2.7(b).

To see why not all shorts will map to a wired-AND or wired-OR model in a CMOS circuit, consider the short between lines c_1 and c_2 in the circuit shown in Figure 2.8. For some input vectors, this short will create a conducting path from V_{DD} to V_{SS}. For example, for $(x_1, x_2, x_3) = (1, 1, 0)$ there is a path from V_{DD} to V_{SS} through the pMOS network of the inverter, the short, and the nMOS network of the NAND gate. This will result in an intermediate voltage at the shorted nodes. Whether this creates a logic 0 or 1 at these nodes depends on the relative impedances of the two networks. The resultant logic value may also differ from one vector to another. For example, $(0, 1, 1)$ also creates a conduction path from V_{DD} to V_{SS}. However, it is possible that the shorted nodes have logic 1 for vector $(1, 1, 0)$ and logic 0 for vector $(0, 1, 1)$. Furthermore, different gates fed by the shorted nodes may interpret the intermediate voltage on these nodes as different logic values.

Even though it is clear that shorts in CMOS circuits cannot be just mapped to either wired-AND or wired-OR model, they can be detected by current monitoring since they activate a path from V_{DD} to V_{SS} (Levi, 1981; Malaiya and Su, 1982; Acken, 1983). If current monitoring is not possible for some reason, such faults could be considered twice for test generation, once assuming the wired-AND model and the other time the wired-OR model. Although this would not guarantee detection of the fault, it would increase the probability of detection.

BFs are also sometimes categorized as *feedback* or *non-feedback* faults. If one or more feedback paths are created in the circuit due to the fault, then it is called a **feedback fault**, otherwise a **non-feedback fault**. Another categorization is in terms

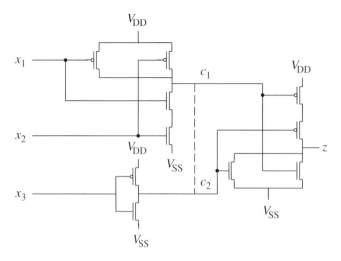

Figure 2.8   BF in a static CMOS circuit

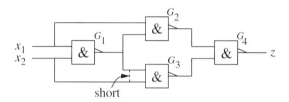

Figure 2.9   An input BF

of *input* versus *non-input BFs*. An **input BF** is a short among two or more inputs of the same gate in the circuit. Shorts not in this category are **non-input BFs**. An input BF is not always of the non-feedback type. Consider the four-NAND XOR function implementation shown in Figure 2.9. The short at the two inputs of gate G_3 actually creates a feedback BF.

2.2.6   Delay fault models

Instead of affecting the logical behavior of the circuit, a fault may affect its temporal behavior only; such faults are called **delay faults (DFs)**. DFs adversely affect the propagation delays of signals in the circuit so that an incorrect logic value may be latched at the output. With the increasing emphasis on designing circuits for very high performance, DFs are gaining wide acceptance.

Two types of DF models are usually used:

- The gate delay fault model: a circuit is said to have a **gate delay fault (GDF)** in some gate if an input or output of the gate has a lumped DF manifested as a slow $0 \rightarrow 1$ or $1 \rightarrow 0$ transition (Hsieh *et al*, 1977; Storey and Barry, 1977).

- The path delay fault model: a circuit is said to have a **path delay fault (PDF)** if there exists a path from a primary input to a primary output in it which is slow to propagate a $0 \rightarrow 1$ or $1 \rightarrow 0$ transition from its input to its output (Lesser and Shedletsky, 1980), (Smith, 1985).

A GDF can be further subdivided into two categories: **gross gate delay fault (G-GDF)** and **small gate delay fault (S-GDF)**. A GDF is a G-GDF if its **delay defect size** (i.e., the amount of time by which it delays the transition) is greater than the system clock period. Otherwise, it is an S-GDF. Thus, G-GDFs result in DFs in all the paths going through the faulty gate, and are, hence, catastrophic in nature. They are also sometimes called **transition faults (TFs)**. S-GDFs are not catastrophic, and have a delay defect size, smaller than the system clock period, associated with them. If this delay defect size is such that it results in a temporal failure in at least one path going through the faulty gate, then the S-GDF can be detected, otherwise not. For each path or gate, two tests are needed, one for the rising transition and another one for the falling transition.

Clearly, the PDF model is the more general of the two models as it models the cumulative effect of the delay variations of the gates and wires along the path. However, because the number of paths in a circuit can be very large, the PDF model may require much more time for test generation and test application than the GDF model. As a compromise between the two extremes, one can use the **segment delay fault (SDF)** model (Heragu *et al.*, 1996). This model considers slow $0 \rightarrow 1$ and $1 \rightarrow 0$ transitions on path segments of length L or smaller. Such segments consist of L or fewer lines from the different paths in the logic circuit. L can be chosen based on the available statistics about the types of manufacturing defects. L can be as small as one (GDFs) or as large as the maximum length of any path through the circuit (PDFs). Unlike the PDF model, the SDF model prevents an explosion in the number of faults being considered. At the same time, a delay defect over a path segment may be large enough that it creates a DF in every path passing through it. This assumption is more realistic than the assumption behind the G-GDF (or TF) model that requires that a delay defect on just one line be large enough to create a DF in every path passing through it.

Because of the need to propagate a transition, the DFs also require *two-pattern tests*, just like SOpFs. As before, the first vector of the two-pattern test is called the *initialization vector*, and the second vector is called the *test vector*. A test for a DF is said to be **robust** if it is independent of the delays in the rest of the circuit, otherwise it is said to be **non-robust** (Smith, 1985; Lin and Reddy, 1987). The word *robust* was first used with this meaning for SOpF detection in Reddy *et al.* (1984b). A robust test (RT) for a DF is said to be a **hazard-free robust test (HFRT)** if no hazards can occur on the tested path, regardless of the gate delay values (Savir and McAnney, 1986; Pramanick and Reddy, 1990; Devadas and Keutzer, 1992). A robust test, which is not hazard-free, called **non-hazard-free robust test**, allows hazards on the side inputs of

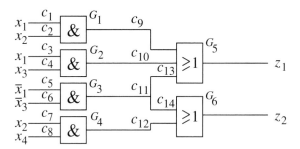

Figure 2.10 A circuit for illustrating DFs

a gate along the tested path when the path under test has a *controlling value* for that gate during initialization (Lin and Reddy, 1987). A **controlling value** refers to the input logic value of a gate which determines its output logic value, irrespective of the other input values (e.g., 0 for an AND or NAND gate, and 1 for an OR or NOR gate). Non-hazard-free robust tests are also sometimes called **general robust tests**. The logic values at the primary inputs are always assumed to be hazard-free.

If a DF test propagates the transition through more than one path to the output, it is called multiple-path propagating (MPP) (Pramanick and Reddy, 1990). On the other hand, if the transition propagation is done through a single path, the test is called single-path propagating (SPP). If from the initialization vector to the test vector, only one input changes (i.e., the one corresponding to the path under test), the test is called a single-input change (SIC) test, else it is called a multiple-input change (MIC) test.

Even if a PDF is not robustly testable, it may still result in an erroneous output value at the output sampling time for some input transitions. We can use the concept of *validatable non-robust test* to detect some of these faults (Reddy *et al.*, 1987). A set of two-pattern tests, T, is said to be a **validatable non-robust test** for a path P if and only if no member of T is a robust test for P, and if the circuit under test passes all tests in T then it can be concluded that the desired transition propagates along P in the allowed time. There may be other testable PDFs which are not testable, however, in either robust or validatable non-robust fashion. Such faults are simply said to be non-robustly testable. The remaining faults are said to be non-testable.

Let us use the two-level circuit shown in Figure 2.10 to develop tests using some of the above concepts, as follows.

- An SIC-SPP-HFRT for PDF $x_1 c_3 c_{10} z_1$ for the rising transition at input x_1 is $\{(0, 0, 1, 0), (1, 0, 1, 0)\}$. The reason is that the side inputs of the gates along the path, i.e., input x_3 of gate G_2 and inputs c_9 and c_{13} of gate G_5, maintain a hazard-free non-controlling value. A two-pattern test for a PDF also tests for G-GDFs along the path. For example, the above test also robustly tests for G-GDFs in gates G_2 and G_5 for the rising transition.

- An MIC-SPP-HFRT for PDF $x_2 c_7 c_{12} z_2$ for the rising input transition at input x_2 is

$\{(1, 0, 0, 1), (1, 1, 1, 1)\}$ because again the side inputs of the path maintain a hazard-free non-controlling value.

- An MIC-MPP-HFRT for PDFs $x_2c_7c_{12}z_2$ and $x_4c_8c_{12}z_2$ for the rising transitions at their inputs is $\{(1, 0, 1, 0), (1, 1, 1, 1)\}$. The reason is that side input c_{14} has a hazard-free non-controlling value, and excessive delay along either of the two paths will result in an incorrect 0 at output z_2 at sample time.

- An MIC-SPP-RT for PDF $x_1c_1c_9z_1$ for the falling transition at input x_1 is $\{(1, 1, 0, 0), (0, 1, 1, 0)\}$. The reason this test is not HFRT is that side inputs c_{10} and c_{13} may have a hazard for this test. However, if the above PDF is present, then output z_1 will be an incorrect 1 at sampling time. Hence, in this case, the hazards at the side inputs do not matter. However, PDF $x_1c_1c_9z_1$ for the rising transition does not have an RT or an HFRT. For example, if we were to use $\{(0, 1, 1, 0), (1, 1, 0, 0)\}$, then the hazards on side inputs c_{10} and c_{13} will now become relevant, because one may obtain the expected 1 at output z_1 owing to these hazards even when a DF in the above path is present. Thus, the above test is only a non-robust test.

- We saw above that PDF $x_1c_1c_9z_1$ for the rising transition at input x_1 is not robustly testable. However, a validatable non-robust test for it exists. Consider the set of two-pattern tests $[\{(0, 0, 0, 0), (1, 0, 0, 0)\}, \{(0, 1, 0, 0), (1, 1, 0, 0)\}]$. We can check that the first two-pattern test is an SIC-MPP-HFRT for PDFs $\bar{x}_1c_5c_{11}c_{13}z_1$ and $\bar{x}_1c_5c_{11}c_{14}z_2$ for the falling transition at their inputs. Suppose the circuit passes this test. Then when the second two-pattern test is applied and the circuit again passes the test, we know that PDF $x_1c_1c_9z_1$ for the rising transition cannot be present. Therefore, the above set of two-pattern tests is a validatable non-robust test for this fault.

2.3   Inductive fault analysis

A systematic way to approach the problem of fault modeling is through *inductive fault analysis* (IFA) (Shen *et al.*, 1985; Ferguson and Shen, 1988). IFA actually takes into account the fabrication technology and defect statistics in addition to the layout. The salient features of IFA are as follows:

- physical defects are generated using statistical data from the fabrication process,
- circuit-level faults caused by these defects are extracted,
- faults are then classified and ranked based on the likelihood of their occurrence.

The rank ordering of faults is done based on the percentage of physical defects that cause a particular fault. This ranked fault list can be used to determine the effectiveness of test sets derived for other fault models, or generate more effective test sets.

IFA assumes that *spot or point defects* are present in the chip. The density of such defects per unit area and the probability density function of the defect sizes are based on data obtained from actual experiments. Only single point defects are considered and these defects are generated randomly based on the above defect characterization.

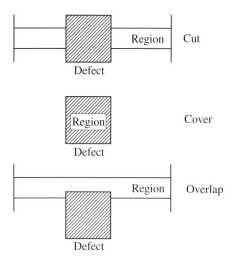

Figure 2.11 Geometric operators

The shape of the defects is assumed to be round or square. The point defect implies that the corresponding region either has extra material or missing material.

In Ferguson and Shen (1988), an automated fault extraction process for a single metal-layer, P-well CMOS technology is described. The defect can be a single transparent or opaque spot on any of the seven photolithographic masks: p-well, field oxide, polysilicon, n-implant, p-implant, contact cut, and metal. This results in fourteen possible defect types. Three geometric operators, which act on different regions, are defined as follows: *cut*, *cover*, and *overlap*, as shown in Figure 2.11. The application of technology analysis with the help of these operators is shown to result in four types of primitive faults in this technology: (a) breaks in conducting regions, (b) bridges between conducting regions, (c) missing transistors, and (d) new transistors. Examples of primitive faults caused by extra polysilicon defects are shown in Figure 2.12. Primitive faults can similarly be extracted for the remaining thirteen defect types.

As an example, for IFA performed on the CMOS layout of a 4 × 4 multiplier, Table 2.2 shows how different defects map to various faults. If we model a BF between a node and V_{DD} or V_{SS} as an SAF, then the mapping from extracted BFs to SAFs for the multiplier is as shown in Table 2.3. For example, 26% of the extra metal defects that caused faults mapped to SA1 faults, 19% to SA0 faults, and 55% to neither. Of all such metal defects, 96% were bridges between two nodes and 4% between more than two nodes. A similar analysis of non-bridge faults showed that about 50% of them can be modeled as SAFs. Hence, overall slightly less than 50% of all extracted faults mapped to SAFs. However, since a test set derived for one fault model frequently detects many non-modeled faults, the key question is what percentage of all extracted faults get detected using a test set derived for the classical single SAF fault model.

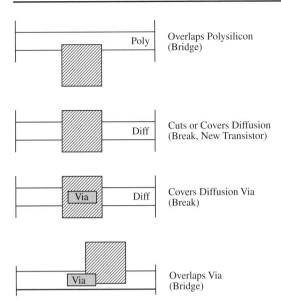

Figure 2.12 Primitive faults caused by extra polysilicon defects

Table 2.2. *Number of defects resulting in faults (in thousands) for the multiplier (Ferguson and Shen, 1988) (© 1988 IEEE)*

	miss. cont.	extra cont.	miss. metal	extra metal	miss. poly.	extra poly.	miss. diff.	extra diff.	miss. p-well	extra p-well	% of faults
Bridges	–	111	–	257	–	125	–	12	62	78	48%
Breaks	29	–	281	–	230	4	21	–	–	–	42%
SOnF	–	–	–	–	134	–	–	–	–	–	10%

Table 2.3. *Mapping of extracted BFs (Ferguson and Shen, 1988) (© 1988 IEEE)*

	miss. cont.	extra cont.	miss. metal	extra metal	miss. poly.	extra poly.	miss. diff.	extra diff.	miss. p-well	extra p-well	all types
SA1	–	35%	–	26%	–	8%	–	31%	–	62%	26%
SA0	–	24%	–	19%	–	5%	–	17%	53%	–	18%
Non-SAF	–	41%	–	55%	–	87%	–	52%	47%	38%	56%
two nodes	–	100%	–	96%	–	96%	–	99%	98%	100%	97%
> two nodes	–	–	–	4%	–	4%	–	1%	2%	< 1%	3%

Experiments from Ferguson and Shen (1988) seem to indicate that this is around 80%.

The above numbers should be interpreted cautiously since they were derived for a small number of small examples, and for a certain type of circuit and design style. However, clearly, IFA is a powerful and relatively accurate method for obtaining realistic fault models.

2.4   Relationships among fault models

Many interesting relationships have been established among the different fault models introduced in earlier sections. Some of these relationships are mentioned next.

- Suppose SPP-HFRTs have been found for all paths in a given combinational circuit. Then the same test set also robustly detects all SOpFs and detectable GDFs in a hazard-free fashion. Note that for many circuits, an SPP-HFRT does not exist for all the paths. For such circuits, the above results are not applicable. However, many methods are now known which can synthesize combinational (and even sequential) circuits to be completely testable by SPP-HFRTs. For such circuits, the above results are relevant.

- An SPP-HFRT for a path in a circuit is also an SPP-RT for that path. The reason is that the conditions for obtaining an SPP-HFRT are more restrictive than the conditions for obtaining an SPP-RT.

- A test set consisting of SPP-RTs for all the paths also robustly detects all the detectable GDFs in the combinational circuit. Such a test set also robustly detects all the SOpFs in the circuit, and all the single SAFs as well.

- A test set which detects all single SAFs in a static CMOS combinational circuit consisting of any interconnection of primitive (i.e., NOT, NOR and NAND) or complex gates also detects all multiple SOnFs in it if current monitoring is done (Jha and Tong, 1990). The same is true for a test set which detects all SOpFs, whether robustly or non-robustly. If the static CMOS circuit is made up of an interconnection of only primitive gates, then the test set which detects all single SAFs also detects all multiple SOnFs, all multiple input BFs and combinations of such multiple faults using current monitoring (Jha and Tong, 1990). For this result it is assumed that the input BFs do not result in any feedback connection (as shown in Figure 2.9).

The above relationships are summarized in Figure 2.13. The acronyms used in this figure can be expanded as follows:

- SPP-HFRPDFT: single-path propagating hazard-free robust PDF testability
- HFRGDFT: hazard-free robust GDF testability
- GRPDFT: general robust PDF testability
- GRGDFT: general robust GDF testability
- RSOPFT: robust SOpF testability
- SSAFT: single SAF testability
- MSONFT: multiple SOnF testability
- MIBFT: multiple input BF testability
- c.m.: current monitoring
- pr.: circuit with primitive gates only

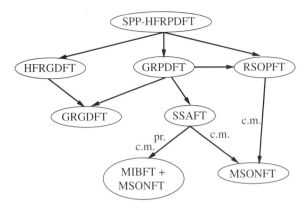

Figure 2.13 Relationships among different fault models

Summary

- Circuits, and correspondingly fault models, can be defined at various levels of abstraction: behavioral, functional, structural, switch-level, and geometric. Temporal fault models are also gaining popularity.
- A behavioral fault model includes variables assuming the wrong value, failure in the *for, switch, if-then-else-if* or *waitfor* constructs, or a failure in an assignment statement.
- A functional fault model makes sure that the functional block under test executes only the desired function and no other function. Functional fault models are frequently used for ILAs and microprocessors.
- The structural fault model consisting of SAFs is the most widely used fault model in the industry. This model is, therefore, sometimes also called the classical fault model.
- SOpFs in static CMOS circuits require two-pattern tests. These tests are called robust if they cannot be invalidated by timing skews at the primary inputs and/or arbitrary delays through the circuit.
- To guarantee detection of all SOnFs in a static CMOS circuit, one needs to monitor the current being drawn by it. BFs can also be detected through current monitoring.
- To verify the temporal behavior of a circuit, one can use the GDF or PDF model, the former leading to less complex test generation, whereas the latter giving a more comprehensive coverage of actual temporal defects.
- To generate realistic fault models from layouts, one can use inductive fault analysis which generates physical defects using statistical data from the manufacturing process.
- Complete coverage of all PDFs in a circuit also guarantees complete coverage under many other well-known fault models.

Exercises

2.1 What are the advantages of behavioral fault modeling? Relate the following behavioral fault models to actual faults that may occur in a circuit derived from the behavioral description.

(a) A "for (CC) $\{B\}$" clause fails.

(b) A "switch (Id)" clause fails.

(c) An "if (Y) then $\{B_1\}$ else $\{B_2\}$" construct fails.

(d) An assignment statement "$X := Y$" fails.

(e) A "waitfor S" synchronizing construct fails.

2.2 Derive a test set for a two-to-one multiplexer for the functional fault model defined in Section 2.2.2. Show that this functional test set also detects all single SAFs in a two-level AND–OR logic implementation of the multiplexer.

2.3 What are the advantages and disadvantages of universal test sets?

2.4 If the single cell fault model is used for testing an iterative logic array, is the corresponding test set guaranteed to detect all single SAFs in the circuit?

2.5 For the four-bit ripple-carry adder shown in Figure 2.3, derive a minimal test set under the single cell fault model. How would you extend this test set to any n-bit ripple-carry adder?

2.6 Derive all test vectors which will detect the multiple SAF consisting of c_3 SA1, c_5 SA0 and c_6 SA0 in the circuit shown in Figure 2.5.

2.7 Derive a minimal test set to detect all single SOpFs in the two-input NAND gate shown in Figure 2.6(b).

2.8 Assuming current monitoring is done, derive a minimal test set for all single SOnFs in the two-input NOR gate shown in Figure 2.6(a).

2.9 Assuming current monitoring is done, derive all test vectors which will detect the BF shown in Figure 2.8.

2.10 For the circuit shown in Figure 2.10, derive the following tests.

(a) SIC-SPP-HFRT for the PDF $\bar{x}_1 c_5 c_{11} c_{13} z_1$ for the falling transition at input $\bar{x}_1$.

(b) HFRT for the G-GDF in G_1 for the rising transition at its output.

References

Abraham, J.A. and Fuchs, W.K. (1986). Fault and error models for VLSI. *Proc. of IEEE*, **74** (5), pp. 639–654.

Acken, J.M. (1983). Testing for bridging faults (shorts) in CMOS circuits. In *Proc. Design Automation Conference*, pp. 717–718.

Akers, S.B. (1973). Universal test sets for logic networks. *IEEE Trans. on Computers*, **C-22** (9), pp. 835–839.

Beh, C.C., Arya, K.H., Radke, C.E., and Torku, K.E. (1982). Do stuck fault models reflect manufacturing defects. In *Proc. Int. Test Conference*, pp. 35–42.

Betancourt, R. (1971). Derivation of minimum test sets for unate logical circuits. *IEEE Trans. on Computers*, **C-20** (11), pp. 1264–1269.

Devadas, S. and Keutzer, K. (1992). Synthesis of robust delay fault testable circuits: theory. *IEEE Trans. on Computer-Aided Design*, **11** (1), pp. 87–101.

Davidson, S. and Lewandowski, J. (1986). ESIM/AFS – a concurrent architectural level fault simulator. In *Proc. Int. Test Conference*, pp. 375–383.

Ferguson, F.J. and Shen, J.P. (1988). A CMOS fault extractor for inductive fault analysis. *IEEE Trans. on Computer-Aided Design*, **7** (11), pp. 1181–1194.

Friedman, A.D. (1973). Easily testable iterative systems. *IEEE Trans. on Computers*, **C-22** (12), pp. 1061–1064.

Ghosh, S. and Chakraborty, T.J. (1991). On behavior fault modeling for digital systems. *J. of Electronic Testing: Theory and Applications*, **2**, pp. 135–151.

Gupta, G. and Jha, N.K. (1988). A universal test set for CMOS circuits. *IEEE Trans. on Computer-Aided Design*, **7** (5), pp. 590–597.

Hayes, J.P. (1985). Fault modeling. *IEEE Design & Test of Computers*, **2** (2), pp. 88–95.

Heragu, K., Patel, J.H., and Agrawal, V.D. (1996). Segment delay faults: a new fault model. In *Proc. VLSI Test Symposium*, pp. 32–39.

Hsieh, E.P., Rasmussen, R.A., Vidunas, L.J., and Davis, W.T. (1977). Delay test generation. In *Proc. Design Automation Conference*, pp. 486–491.

Jha, N.K. and Tong, Q. (1990). Detection of multiple input bridging and stuck-on faults in CMOS logic circuits using current monitoring. *Int. Journal of Computers & Electrical Engg.*, **16** (3), pp. 115–124.

Kodandapani, K.L. and Pradhan, D.K. (1980). Undetectability of bridging faults and validity of stuck-at fault test sets. *IEEE Trans. on Computers*, **C-29** (1), pp. 55–59.

Lesser, J.P. and Shedletsky, J.J. (1980). An experimental delay test generator for LSI logic. *IEEE Trans. on Computers*. **C-29** (3), pp. 235–248.

Levendel, Y. and Menon, P.R. (1982). Test generation algorithms for computer hardware description languages. *IEEE Trans. on Computers*, **C-31** (7), pp. 577–588.

Levi, M.W. (1981). CMOS is most testable. In *Proc. Int. Test Conference*, pp. 217–220.

Lin, C.J. and Reddy, S.M. (1987). On delay fault testing in logic circuits. *IEEE Trans. on Computer-Aided Design*, **6** (9), pp. 694–703.

Malaiya, Y.K. (1984). Testing stuck-on faults in CMOS integrated circuits. In *Proc. Int. Conference on Computer-Aided Design*, pp. 248–250.

Malaiya, Y.K. and Su, S.Y.H. (1982). A new fault model and testing technique for CMOS devices. In *Proc. Int. Test Conference*, pp. 25–34.

Navabi, Z. (1993). *VHDL: Analysis and Modeling of Digital Systems*, McGraw-Hill, New York.

Pramanick, A.K. and Reddy, S.M. (1990). On the design of path delay fault testable combinational circuits. In *Proc. Int. Symposium on Fault-Tolerant Computing*, pp. 374–381.

Reddy, S.M. (1973). Complete test sets for logic functions. *IEEE Trans. on Computers*, **C-22** (11), pp. 1016–1020.

Reddy, S.M., Agrawal, V.D., and Jain, S.K. (1984a). A gate-level model for CMOS combinational logic circuits with application to fault detection. In *Proc. Design Automation Conference*, pp. 504–509.

Reddy, S.M., Reddy, M.K., and Agrawal, V.D. (1984b). Robust tests for stuck-open faults in CMOS combinational circuits. In *Proc. Int. Symposium Fault-Tolerant Computing*, pp. 44–49.

Reddy, S.M., Lin, C.J., and Patil, S. (1987). An automatic test pattern generator for the detection of path delay faults. In *Proc. Int. Conference on Computer-Aided Design*, pp. 284–287.

Savir, J. and McAnney, W.H. (1986). Random pattern testability of delay faults. In *Proc. Int. Test Conference*, pp. 263–273.

Sentovich, E.M., Singh, K.J., Moon, C., Savoj, H., Brayton, R.K., and Sangiovanni-Vincentelli, A. (1992). Sequential circuit design using synthesis and optimization. In *Proc. Int. Conference on Computer Design*, pp. 328–333.

Shen, J.P., Maly, W., and Ferguson, F.J. (1985). Inductive fault analysis of MOS integrated circuits. *IEEE Design & Test of Computers*, **2** (6), pp. 13–26.

Smith, G.L. (1985). A model for delay faults based on paths. In *Proc. Int. Test Conference*, pp. 342–349.

Storey, T.M. and Barry, J.W. (1977). Delay test simulation. In *Proc. Design Automation Conference*, pp. 492–494.

Timoc, C. *et al.* (1983). Logical models of physical failures. In *Proc. Int. Test Conference*, pp. 546–553.

Wadsack, R.L. (1978). Fault modeling and logic simulation of CMOS and MOS integrated circuits. *Bell System Technical Journal*, **57** (5), pp. 1449–1474.

3 Combinational logic and fault simulation

In this chapter, we discuss logic and fault simulation methods for combinational circuits.

We begin by defining what constitutes a test for a fault and defining the main objectives of fault simulation algorithms. We then define some basic concepts and describe the notation used to represent the behavior of fault-free as well as faulty versions of a circuit.

We then describe logic simulation algorithms, including event-driven and parallel algorithms.

Next, we present a simple fault simulation algorithm and some basic procedures used by most fault simulation algorithms to decrease their average run-time complexity. This is followed by a description of the five fault simulation paradigms: parallel fault, parallel-pattern single-fault, deductive, concurrent, and critical path tracing.

Finally, we present some low complexity approaches for obtaining an approximate value of fault coverage for a given set of vectors.

3.1 Introduction

The objectives of fault simulation include (i) determination of the quality of given tests, and (ii) generation of information required for *fault diagnosis* (i.e., location of faults in a chip). In this chapter, we describe fault simulation techniques for combinational circuits. While most practical circuits are sequential, they often incorporate the full-scan design-for-testability (DFT) feature (see Chapter 11). The use of full-scan enables test development and evaluation using only the combinational parts of a sequential circuit, obtained by removing all flip-flops and considering all inputs and outputs of each combinational logic block as primary inputs and outputs, respectively. If test vectors are applied using the full-scan DFT features and the test application scheme described in Chapter 11, the reported test quality is achieved.

Let C be a combinational logic circuit with n primary inputs, $x_1, x_2, \ldots, x_n$, and m primary outputs, $z_1, z_2, \ldots, z_m$. We assume that the circuit is an interconnect of single-output gates where each gate implements an arbitrary logic function. The output of a gate is assumed to (a) directly drive the input of another gate, (b) drive the inputs of several gates via one or more *fanout systems*, or (c) be a primary output of the circuit. (See Section 3.2 for a more detailed description of the assumed circuit model.) Let $c_1, c_2, \ldots, c_k$ be the internal lines of the circuit, where c_j is either the output of a single-output gate, G, or a fanout of another line (in either case, the line must not be a primary output). The primary inputs, internal lines, and primary outputs of the circuit will be collectively referred to as **circuit lines**. Since most lines in any large circuit are internal lines, when it is clear from the context, we will use symbols such as c_i to also refer to all circuit lines. When it is necessary to distinguish between all circuit lines and internal lines, we will use symbols such as b and b_i to denote all circuit lines and symbols such as c and c_i to denote internal lines. Let L denote the total number of lines in the circuit including the primary inputs, internal lines, and primary outputs, i.e., $L = n + k + m$.

For a combinational circuit, a forward traversal – via gates and fanout systems – from the output of a gate cannot reach any of that gate's inputs.

Let $F = \{f_1, f_2, \ldots, f_{N_f}\}$ be the **fault list**, i.e., the set of faults of interest. In this and many subsequent chapters, we will primarily consider fault lists that contain all possible single stuck-at faults (SAFs) in the circuit or some subset thereof. (The SAF model is described in Section 2.2.3.) Let C^{f_i} denote a copy of circuit C that has fault f_i.

Let P be an n-tuple $(p_1, p_2, \ldots, p_n)$, where $p_j \in \{0, 1\}$, $1 \leq j \leq n$. Each p_j is called a **component** of the n-tuple. An n-tuple P whose component values, $p_1, p_2, \ldots, p_n$, are to be applied, respectively, to the primary inputs $x_1, x_2, \ldots, x_n$ of a circuit is called an **input vector** or, simply, a **vector**. A vector is sometimes also called a **pattern**. A vector P is said to **detect** or **test** fault f_i in circuit C if the response obtained at at least one of the outputs of C, say z_j, due to the application of P at its inputs, is the complement of that obtained at the corresponding output of C^{f_i}. Any vector that detects a fault f_i in a circuit C is said to be a **test vector** or a **test pattern** or, simply, a **test**, for f_i in C. If no test vector exists for a fault, then the fault is said to be **untestable**.

Consider the circuit shown in Figure 3.1(a) where, for each line b, the value $v(b)$ obtained due to the application of input vector $P = (0, 1)$ is shown next to the line. In other words, vector P **implies** value $v(b)$ at line b. Next, consider a faulty version of the above circuit that contains fault f_{22}, a stuck-at 1 (SA1) fault at line c_9, as shown in Figure 3.1(b). Also shown in this figure are the values $v^{f_{22}}(b)$ implied by the same input vector, $P = (0, 1)$, at each line of the faulty circuit, $C^{f_{22}}$. It can be seen that vector P is a test for this fault, since the output response to P for the faulty circuit is the complement of that for the fault-free circuit.

Figure 3.2 illustrates the conditions under which a vector P is a test. This figure depicts the fault-free circuit C as well as the faulty circuit C^{f_i}. The component values of vector P are applied to the corresponding inputs of C as well as C^{f_i}. (Since the two versions of the circuit are shown together, to avoid confusion, we use primed versions of the names assigned to primary inputs and outputs of C for the corresponding lines

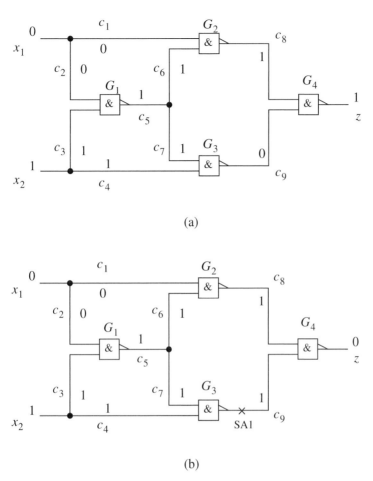

Figure 3.1   An example illustrating a test vector: (a) the fault-free circuit, and (b) a faulty version of the circuit

of C^{f_i}.) An **auxiliary circuit**, C_{aux}, that uses m two-input XOR gates and an m-input OR gate, combines the values obtained at the outputs of C with those obtained at the corresponding outputs of C^{f_i} into a single logic signal, Ω. Since the output of a two-input XOR gate is logic 1 if and only if its inputs have complementary logic values, the output of the jth XOR gate is logic 1 if and only if vector P implies complementary values at the jth outputs of the good and faulty circuits, z_j and z'_j, respectively. Hence, a logic 1 appears at the output of the auxiliary circuit, Ω, if and only if vector P is a test for fault f_i in circuit C.

One objective of fault simulation is to determine the faults in the fault list that are detected by the vectors in a given set. In some application scenarios, one of the outputs of a fault simulator is the set of faults detected by the given set of vectors. In other scenarios, only the number of faults detected may be reported. **Fault coverage** of a set of vectors is the number of faults in the fault list detected by its constituent vectors.

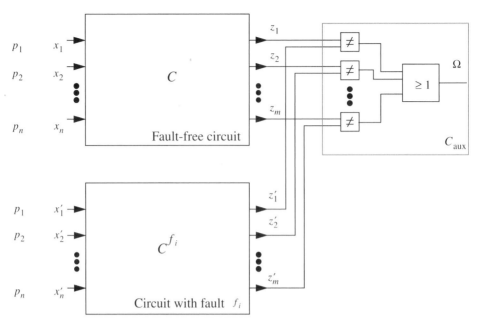

Figure 3.2 Representation of the conditions under which a vector $P = (p_1, p_2, \ldots, p_n)$ is a test for a fault f_i in a circuit C

Fault coverage is often reported as a percentage of the total number of faults in the fault list.

A second objective of fault simulation is to determine, for each vector in the given set of vectors, the values implied at the outputs for each faulty version of the circuit.

3.2 Preliminaries

We begin by precisely defining the class of combinational logic circuits that will be considered in this and some of the subsequent chapters.

3.2.1 Fault-free circuit elements

Any gate-level description of a combinational logic circuit can be viewed as an interconnection of two types of **circuit elements** shown in Figure 3.3, namely a **single-output gate** and a **fanout system**.

We use a circuit model in which a gate may have an arbitrary number of inputs and may implement any arbitrary logic function of these inputs. Most algorithms presented ahead can handle circuits containing such gates. Most examples will, however, be restricted to **primitive gates**, namely, AND, NAND, OR, NOR, and NOT. Sometimes we will also discuss some **complex gates**, such as XOR.

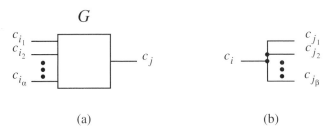

(a)   (b)

Figure 3.3 The basic circuit elements: (a) a single-output gate, (b) a fanout system

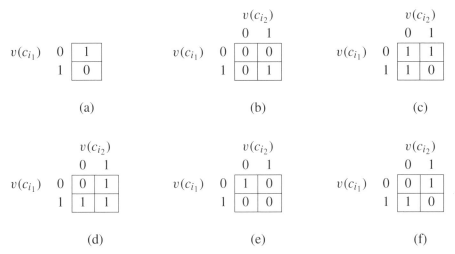

(a)   (b)   (c)

(d)   (e)   (f)

Figure 3.4 Two-valued truth tables describing the behavior of commonly used gates: (a) NOT, (b) two-input AND, (c) two-input NAND, (d) two-input OR, (e) two-input NOR, and (f) two-input XOR

3.2.1.1 Two-valued truth tables

The behavior of a fault-free gate can be described using a two-valued truth table. Figure 3.4 shows the two-valued truth tables for two-input primitive gates and the XOR gate. The logic values associated with inputs c_{i_1} and c_{i_2} and output c_j will be stored in variables $v(c_{i_1})$, $v(c_{i_2})$, and $v(c_j)$, respectively. Depending on the context, variables such as $v(c_{i_l})$ will represent variables associated with, or the values assigned to, lines c_{i_l}.

An α-input AND gate can be modeled by using a circuit comprised of $\alpha - 1$ two-input AND gates as shown in Figure 3.5(a). An α-input OR and an α-input XOR gate can be modeled in a similar fashion using a two-input version of the corresponding gate. A similar model can also be obtained for an α-input NAND gate, as shown in Figure 3.5(b). An α-input NOR gate can be modeled via a similar modification of the model of the α-input OR gate. Such models enable iterative use of the truth tables of two-input gates to describe the behavior of multi-input gates.

In the following, the two-valued truth table of a line c_j, which is the output of a gate G with inputs $c_{i_1}, c_{i_2}, \ldots, c_{i_\alpha}$, is denoted by function $\mathcal{G}_j^2(v(c_{i_1}), v(c_{i_2}), \ldots, v(c_{i_\alpha}))$.

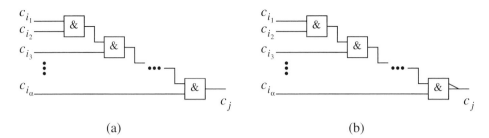

Figure 3.5 Models for α-input gates in terms of two-input gates: (a) AND, and (b) NAND

$v(c_i)$	$v(c_{j_1})$	$\cdots$	$v(c_{j_l})$	$\cdots$
0	0	$\cdots$	0	$\cdots$
1	1	$\cdots$	1	$\cdots$

Figure 3.6 Two-valued truth tables describing the behavior of a fanout system

For example, for a two-input NAND gate with output c_j and inputs c_{i_1} and c_{i_2}, $\mathcal{G}_j^2(v(c_{i_1}), v(c_{i_2}))$ denotes the behavior described by the truth table shown in Figure 3.4(c).

Now consider the other circuit element, namely a fanout system with input c_i and outputs $c_{j_1}, c_{j_2}, \ldots, c_{j_\beta}$. Input c_i and each output c_{j_l} are called the **stem** and a **branch** of the fanout system, respectively. Figure 3.6 shows the two-valued truth-table for each fanout branch. The two-valued truth table for each output c_{j_l} is also denoted by function $\mathcal{G}_{j_l}^2(v(c_i))$.

Given the values implied at the inputs of a circuit element, the two-valued truth table of one of its outputs enables the computation of the value implied at the output via table look-up. The size of the table can, however, grow very rapidly with the number of inputs of the circuit element. The behavior of a gate with a large number of inputs is often described using models of the type shown in Figure 3.5 to keep the size of truth tables manageable. In such cases, multiple look-ups are required to compute the value implied at the output of a gate with many inputs.

3.2.1.2 Logic expressions to simulate two-valued behavior of a circuit element

Logic expressions can also be used to describe the logic behavior of each output of each type of circuit element. A logic expression requires lower storage space than the corresponding truth table, especially for a gate with a large number of inputs. However, the computation of the value implied at a circuit line requires the evaluation of a logic expression describing its behavior. Any processor used to execute *logic or fault simulation* contains an arithmetic and logic unit (ALU) that efficiently performs several logic operations, including not, and, and or. Hence, computation of the logic value implied at a line can typically be performed efficiently.

Table 3.1. *Logic expressions describing the behavior of commonly used gates*

Gate driving line c_j	Logic expression $\mathcal{G}_j^2()$
NOT	$v(c_j) = \text{not } v(c_{i_1})$
AND	$v(c_j) = v(c_{i_1}) \text{ and } v(c_{i_2})$
NAND	$v(c_j) = \text{not } [v(c_{i_1}) \text{ and } v(c_{i_2})]$
OR	$v(c_j) = v(c_{i_1}) \text{ or } v(c_{i_2})$
NOR	$v(c_j) = \text{not } [v(c_{i_1}) \text{ or } v(c_{i_2})]$
XOR	$v(c_j) = \{v(c_{i_1}) \text{ and } [\text{ not } v(c_{i_2})]\} \text{ or } \{[\text{ not } v(c_{i_1})] \text{ and } v(c_{i_2})\}$

Logic expressions used to compute $v(c_j)$, the value implied at line c_j that is the output of a logic gate with inputs $c_{i_1}, c_{i_2}, \ldots, c_{i_\alpha}$, are shown in Table 3.1. Similar expressions can be derived for a gate that implements an arbitrary logic function by using any of the well-known techniques for the design of a two-level AND-OR circuit that implements a given logic function. Whenever logic expressions are used to describe the behavior of an output c_j of a circuit element, function $\mathcal{G}_j^2()$ refers to the corresponding logic expression.

In fact, most ALUs can perform each bit-wise logic operation on multi-bit inputs in as much time as they perform the operation on single-bit inputs. As will be seen later, this parallelism provided by ALUs is exploited by simulation algorithms.

3.2.1.3 Three-valued truth tables

In many scenarios, the value of one or more components of an n-tuple may be left *unspecified* – a fact that is often denoted by symbol x. In some such cases, the component's value is yet unspecified but may be specified as a logic 0 or 1 by a later step of an algorithm. In some other cases, such a component value represents a **don't care** and may be specified arbitrarily as logic 0 or 1. In other cases, an n-tuple with unspecified component values is used to denote a **cube** which represents a set of vectors. The set of vectors denoted by the cube can be obtained by enumerating all combinations of logic 0 and 1 values for each of the components of the n-tuple whose value is unspecified. For example, the three-tuple cube $(x, 0, x)$ represents the set of vectors $\{(0, 0, 0), (0, 0, 1), (1, 0, 0), (1, 0, 1)\}$ obtained by enumerating all combinations of logic 0 and 1 at the first and third components, whose values are unspecified in the three-tuple. Due to the extensive use of the unspecified value x, we need to extend the description of a gate's behavior to three-valued truth tables.

Logic 0 and 1 can be viewed as the two **basic values** for the *two-valued* as well as the *three-valued system*. Each **symbol** in the **two-valued system** can be interpreted as a set of basic values as $0 = \{0\}$ and $1 = \{1\}$. (Note that symbol '0' is used to denote a basic value as well as a symbol of the two-valued system. The meaning should be

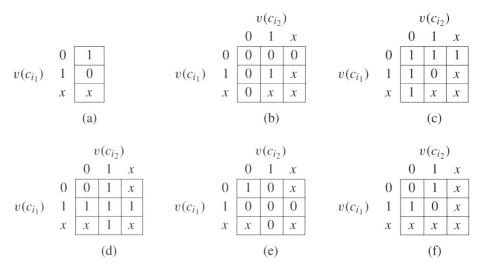

Figure 3.7 Three-valued truth tables of commonly used gates: (a) NOT, (b) two-input AND, (c) two-input NAND, (d) two-input OR, (e) two-input NOR, and (f) two-input XOR

clear from the context.) Each symbol in the **three-valued system**, namely, 0, 1, and x, can also be represented as a set of basic values. In this case, $0 = \{0\}$, $1 = \{1\}$, and $x = \{0, 1\}$.

Value x is said to be **unspecified** or, in general, **incompletely-specified**, since a line whose value is denoted by symbol x may be assigned 0 or 1, i.e., the symbols denoting $\{0\}$ and $\{1\}$ which are the sets of basic values that are subsets of the set denoted by x. In contrast, symbols $0 = \{0\}$ and $1 = \{1\}$ are said to represent **completely specified** values.

The above set notation is useful since it enables the derivation of truth tables for any value system using the truth tables for the basic values, provided each symbol of the new value system is represented as a set of given basic values.

Figure 3.7 shows the three-valued truth tables for all the gates whose two-valued truth tables are shown in Figure 3.4. For each type of gate, the function $\mathcal{G}_j^3(v(c_{i_1}), v(c_{i_2}), \ldots, v(c_{i_\alpha}))$ denotes the behavior described by the corresponding three-valued truth table shown in the figure, assuming that the gate output is c_j and its inputs are $c_{i_1}, c_{i_2}, \ldots, c_{i_\alpha}$. Each of these truth tables can be derived from the corresponding two-valued truth table in Figure 3.4 using the following procedure.

Procedure 3.1 [ComputeHigherValuedTruthTable( )]

1 Represent each symbol in the given logic value system by the corresponding set of basic values.

For example, in the three-valued system, represent symbol 0 as $\{0\}$, 1 as $\{1\}$, and x as $\{0, 1\}$.

2 Consider any given n-tuple of above values as a cube $\mathcal{C}$. Expand the cube into

the corresponding set of vectors $\mathcal{P}$ by enumerating all combinations of completely specified values that each component of the cube can take.

3 Use the truth table of the gate for the basic value system to compute the value at the gate output for each vector in set $\mathcal{P}$. Let $\mathcal{O}(\mathcal{C})$ be the set containing all distinct output values obtained in this manner. This set denotes the value of the gate output for the input value denoted by cube $\mathcal{C}$. The value at the output can be represented by the symbol that represents this set. (If this set is not identical to the set representation of any of the symbols in the value system, then it is denoted by an incompletely-specified value whose set representation is the smallest set that is a superset of this set.)

Example 3.1   Consider the case where the two-tuple cube $\mathcal{C} = (0, x)$ is applied to the inputs of a two-input AND gate. The set of vectors corresponding to this cube can be computed as $\mathcal{P} = \{(0, 0), (0, 1)\}$. The two-valued truth table for the AND gate can be used to obtain the output value for each vector in set $\mathcal{P}$. The output set for the AND gate can then be computed as the union of the output values obtained above. Hence, in this case, $\mathcal{O}_{\text{AND}}(\mathcal{C}) = \{0\} = 0$. Similarly, for a two-input OR gate, $\mathcal{O}_{\text{OR}}(\mathcal{C}) = \{0, 1\} = x$.

In other words, for input cube $\mathcal{C} = (0, x)$, the values at the outputs of two-input AND and OR gates are given by symbols 0 and x, respectively. $\square$

The reader is invited to use the above procedure to verify each entry of each three-valued truth table shown in Figure 3.7. The behavior represented by each of these truth tables is denoted by the function $\mathcal{G}_j^3(v(c_{i_1}), v(c_{i_2}))$, for output c_j of the corresponding gate.

In addition to enabling simulation of gates with one or more inputs that are assigned the unspecified value, the three-valued system also provides compact representations of the truth tables for some types of gates, especially primitive gates. The **singular cover** of a gate G (Roth, 1966) is one such compact representation of the function implemented by gate G. (The singular cover as described here is only a part of the singular cover as defined in Roth (1966).) The **on-set** (**off-set**) of the gate, i.e., the set containing all vectors which when applied to the inputs of the gate give 1 (0) at the output of the gate, are explicitly described by its singular cover.

The singular cover of a gate is obtained by combining multiple input vectors that belong to its on-set (similarly, those that belong to its off-set) into cubes. Any procedure that identifies prime implicants of a function can be used to obtain these cubes. Consider the singular cover of a two-input OR gate as an example. The three input vectors that comprise its on-set, namely $\{(0, 1), (1, 0), (1, 1)\}$, are combined into two cubes, namely $(x, 1)$ and $(1, x)$. Recall that cube $(x, 1)$ can be expanded into two vectors, $\{(0, 1), (1, 1)\}$, and cube $(1, x)$ can be expanded into vectors $\{(1, 0), (1, 1)\}$.

Figure 3.8 shows the singular covers for a few commonly used gates. In each of these singular covers, the cubes representing the on-set of the gate are shown first,

$v(c_{i_1})$	$v(c_j)$
0	1
1	0

(a)

$v(c_{i_1})$	$v(c_{i_2})$	$v(c_j)$
1	1	1
0	x	0
x	0	0

(b)

$v(c_{i_1})$	$v(c_{i_2})$	$v(c_j)$
0	x	1
x	0	1
1	1	0

(c)

$v(c_{i_1})$	$v(c_{i_2})$	$v(c_j)$
x	1	1
1	x	1
0	0	0

(d)

$v(c_{i_1})$	$v(c_{i_2})$	$v(c_j)$
0	0	1
x	1	0
1	x	0

(e)

$v(c_{i_1})$	$v(c_{i_2})$	$v(c_j)$
0	1	1
1	0	1
0	0	0
1	1	0

(f)

Figure 3.8 Singular covers for commonly used gates: (a) NOT, (b) two-input AND, (c) two-input NAND, (d) two-input OR, (e) two-input NOR, and (f) two-input XOR

$v(c_i)$	$v(c_{j_1})$	$\cdots$	$v(c_{j_l})$	$\cdots$
0	0	$\cdots$	0	$\cdots$
1	1	$\cdots$	1	$\cdots$
x	x	$\cdots$	x	$\cdots$

Figure 3.9 Three-valued truth tables describing the behavior of a fanout system

followed by the cubes representing the gate's off-set. Since a gate implements a completely specified logic function, the description of either of these sets is sufficient to completely describe the behavior of a gate. Despite this fact, often the on-set and the off-set are both described to decrease the run-time complexity of some algorithms.

The compactness of the singular cover representation is more evident for primitive gates with a large number of inputs. For example, the two-valued truth table of an eight-input OR gate contains 256 terms while its singular cover contains only nine cubes, eight describing the gate's on-set and one describing its off-set.

Now consider the other circuit element, namely a fanout system with input c_i and outputs $c_{j_1}, c_{j_2}, \ldots, c_{j_\beta}$. Figure 3.9 shows the three-valued truth-table for each fanout branch. The truth table for branch c_{j_l} is also denoted by the function $\mathcal{G}^3_{j_l}(v(c_i))$.

Table 3.2. *One commonly used encoding for the three-valued system*

Symbol	v	u
0	0	0
1	1	1
x	1	0
−	0	1

3.2.1.4 Controlling value of primitive gates

The singular cover of each primitive gate shows that there exists a value which, when applied to one input of the gate, fully specifies the value at its output. Such a value is called the gate's **controlling value**. The controlling value is 0 for AND and NAND gates and 1 for OR and NOR gates. The fully specified value that appears at the output of a gate G when its controlling value is applied at one or more of its inputs will be called its **controlled response** and denoted by $R(G)$. The controlled response is 0 for AND and NOR gates and 1 for NAND and OR gates.

Note that there exists no controlling value for some gates, such as XOR. For such gates, both the controlling value and controlled response are said to be undefined.

3.2.1.5 Logic expressions to simulate the three-valued behavior of circuit elements

As in the case with two-valued simulation, logic expressions may be used to describe and simulate three-valued behavior of circuit elements. Like their two-valued counterparts, three-valued expressions may reduce the space required to store truth tables and enable parallel simulation using bit-wise operations on multi-bit values.

Encoding of values: A minimum of two bits are required to encode the symbols 0, 1, and x. One common encoding uses two variables. In this representation, the value at line c_i is represented by two variables, $v(c_i)$ and $u(c_i)$, which collectively represent the symbols 0, 1, and x using the encoding shown in Table 3.2.

Using two variables to represent the value assigned to each line and the above encoding, the logic expressions describing the three-valued behavior of various gates are shown in Table 3.3.

3.2.2 Model of a combinational circuit

Any combinational logic circuit can be implemented as an interconnection of the two types of circuit elements discussed before: gates and fanout systems. Figure 3.10 shows an example circuit, C_{16}, which will be used in the following discussion to illustrate the definitions of some terms.

Table 3.3. *Logic expressions describing the three-valued behavior of commonly used gates*

Gate driving line c_j	Logic expressions $\mathcal{G}_j^3()$
NOT	$v(c_j) = \text{not } u(c_{i_1})$ $u(c_j) = \text{not } v(c_{i_1})$
AND	$v(c_j) = v(c_{i_1}) \text{ and } v(c_{i_2})$ $u(c_j) = u(c_{i_1}) \text{ and } u(c_{i_2})$
NAND	$v(c_j) = \text{not } [u(c_{i_1}) \text{ and } u(c_{i_2})]$ $u(c_j) = \text{not } [v(c_{i_1}) \text{ and } v(c_{i_2})]$
OR	$v(c_j) = v(c_{i_1}) \text{ or } v(c_{i_2})$ $u(c_j) = u(c_{i_1}) \text{ or } u(c_{i_2})$
NOR	$v(c_j) = \text{not } [u(c_{i_1}) \text{ or } u(c_{i_2})]$ $u(c_j) = \text{not } [v(c_{i_1}) \text{ or } v(c_{i_2})]$
XOR	$v(c_j) = \{v(c_{i_1}) \text{ and } [\text{ not } u(c_{i_2})]\} \text{ or } \{[\text{ not } u(c_{i_1})] \text{ and } v(c_{i_2})\}$ $u(c_j) = \{u(c_{i_1}) \text{ and } [\text{ not } v(c_{i_2})]\} \text{ or } \{[\text{ not } v(c_{i_1})] \text{ and } u(c_{i_2})\}$

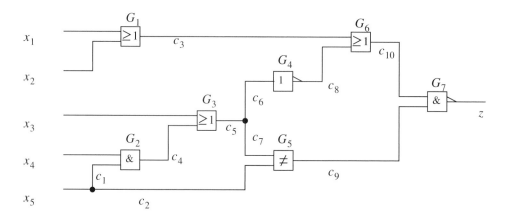

Figure 3.10 An example circuit C_{16} (Roth, 1966) (© 1966 IBM)

Our circuit model allows only single-output gates. If the output of a gate drives inputs of multiple gates, then one or more fanout systems are used to explicitly model the connections. For example, since output c_5 of gate G_3 drives inputs of multiple gates, a fanout system is used to connect c_5 to the inputs of gates it drives.

A line c_j is said to be a **fanout** of line c_i if (a) c_i is the stem of a fanout system and c_j is one of its branches, or (b) c_i is an input of a gate with output c_j. For example, in C_{16}, c_1 and c_2 are each a fanout of primary input x_5; also, c_6 and c_7 are fanouts of c_5. Also, c_5 is considered a fanout of primary input x_3 as well as line c_4. A line c_j is said to be in the **transitive fanout** of line c_i if starting from c_i one can reach c_j via a sequence of lines $(c_{\gamma_1}, c_{\gamma_2}, \ldots, c_{\gamma_l})$, where $c_{\gamma_1} = c_i$, c_{γ_δ} is a fanout of $c_{\gamma_{\delta-1}}$, for all

$\delta \in \{2, 3, \ldots, l\}$, and $c_{\gamma_l} = c_j$. Also, for convenience, line c_i will be said to be in its own transitive fanout. In C_{16}, lines $x_3, c_5, c_6, c_7, c_8, c_9, c_{10}$, and z are each in the transitive fanout of primary input x_3. Note that gates G_3, G_4, G_5, G_6, and G_7 are also said to be in the transitive fanout of x_3.

A line c_i is said to be a **fanin** of line c_j, if (a) c_i is the stem of a fanout system and c_j is one of its branches, or (b) c_i is an input of a gate whose output is c_j. In the latter case, c_i is also said to be a fanin of gate G whose output is c_j. **Transitive fanin** can be defined in a manner similar to the definition of transitive fanout – using the concept of fanin instead of fanout. All the gates, lines, and primary inputs in the transitive fanin of a line are collectively referred to as the line's **input cone** or, simply, its **cone**. In C_{16}, c_2 and c_7 are each a fanin of c_9 as well as gate G_5. Primary input x_5 is a fanin of lines c_1 and c_2. Furthermore, x_5 is in the transitive fanin of c_4 since one can write the sequence of lines (c_4, c_1, x_5) where each line in the sequence is a fanin of the previous line in the sequence. Finally, the circuit comprised of primary inputs x_3, x_4, and x_5, gates G_2, G_3, and G_4, and lines c_1, c_4, c_5, c_6, and c_8 are said to constitute the cone of line c_8. All the primary inputs, gates, and lines in C_{16} are said to belong to the cone of z.

A **path from a line c_i to line c_j in its transitive fanout** is the sequence of lines starting with c_i and ending with c_j where each line in the sequence is a fanout of the previous line in the sequence. The gates and fanout systems flanked by the above sequence of lines are also said to belong to the path. In C_{16}, the sequence of lines (c_4, c_5, c_7, c_9, z) is one of the paths from c_4 to primary output z; gates G_3, G_5, and G_7 are also said to belong to the above path. A **path from a line c_j to line c_i in its transitive fanin** can be defined as the reverse of the sequence of lines in the corresponding path from c_i to c_j. For example, the sequence of lines (z, c_9, c_7, c_5, c_4) is one of the paths from z to c_4. In some cases, where the order of lines in a sequence is not of consequence, a path from c_i to c_j can be viewed as being identical to the corresponding path from c_j to c_i.

3.2.3 Description of faulty circuit elements

The single SAF model has been used in practice for decades with reasonable success. Thus, it provides a good target for fault simulation and test generation. A single SAF in a faulty circuit is associated with a specific circuit line, called the **fault site**. The fault site may be a gate input, a gate output, a fanout stem, or a fanout branch.

Note that a circuit line may fall into more than one of the above categories. For example, in the circuit shown in Figure 3.1, c_5 is the output of gate G_1 as well as the stem of a fanout system; c_9 is the output of gate G_3 as well as an input of gate G_4. If any such line is the site of a fault, the fault can be associated with either of the circuit elements.

Two facts define the behavior of an SAF. First, logic value w is associated with the line that is the site of an SAw fault, independent of the value implied at the line by

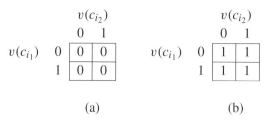

Figure 3.11 The two-valued truth tables of a two-input gate with an SAF at its output: (a) SA0, and (b) SA1

the input vector. Second, the SAF is assumed to not affect the lines that are not in the transitive fanout of the fault site. This implies, for example, that an SA0 fault at line c_6 in Figure 3.1 has no affect on the value at line c_5. Hence, an input vector may imply a logic 1 at line c_5 which in turn may imply a logic 1 at line c_7. Similarly, an SA1 fault at c_9 has no affect on the inputs of gate G_3, namely, lines c_4 and c_7. Clearly, the values at lines in the transitive fanout of the fault site may be affected by the fault via *fault effect propagation*.

In the case of a single SAF, the two-valued truth table of a faulty gate can be derived by using the truth table of the fault-free version of the gate and the behavior of the fault as described above. During simulation of, or test generation for, a circuit with a faulty element, the truth tables for the element with the particular fault are used to compute the value at each of its outputs.

3.2.3.1 Behavior of a gate with an SAF at its output

The two-valued truth table of a two-input NAND gate whose output is SA0 is shown in Figure 3.11(a). The truth table of the gate with output SA1 fault is shown in Figure 3.11(b). Since the presence of an SAF at the output of a gate causes it to have a constant value, each truth table shown in Figure 3.11 represents the behavior of any two-input gate with the corresponding SAF at its output, independent of the logic function implemented by the fault-free version of the gate.

3.2.3.2 Behavior of a gate with an SAF at one of its inputs

Consider an SA1 fault at one of the inputs, say c_{i_1}, of a two-input NAND gate with inputs c_{i_1} and c_{i_2} and output c_j. Figure 3.12 shows the gate with the SA1 fault under different input vectors. Due to the presence of the fault at c_{i_1}, the value at the corresponding input is always interpreted as a logic 1 by the NAND gate, independent of the value applied to line c_{i_1}. This is indicated by symbol 1 as shown in Figure 3.12(a) on the portion of line c_{i_1} between the mark indicating the fault and the NAND gate. This fact about the behavior of the fault can be used, in conjunction with the two-valued truth table of the fault-free version of a two-input NAND gate, to obtain

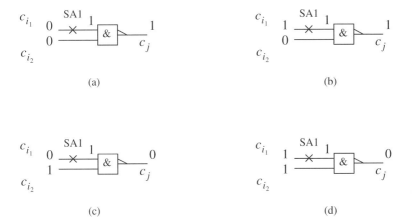

Figure 3.12   A two-input NAND gate with SA1 fault at one of its inputs

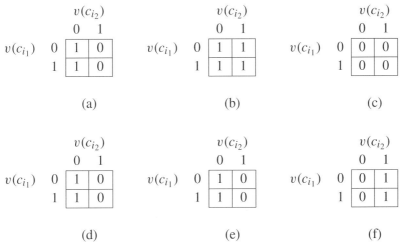

Figure 3.13   Two-valued truth tables of two-input gates with SAFs at input c_{i_1}: (a) NAND with SA1, (b) NAND with SA0, (c) NOR with SA1, (d) NOR with SA0, (e) XOR with SA1, and (f) XOR with SA0

the two-valued truth table of the faulty gate, shown in Figure 3.13(a). Two-valued truth tables of some other faulty gates are also shown in Figure 3.13.

3.2.3.3 Behavior of a gate with a general fault

Now consider a more general fault that may occur at a gate. If the presence of the fault alters the logic behavior of the gate, and if the output of the faulty gate can only take logic values (0 and 1) when any completely specified vector is applied at its inputs, then a two-valued truth table of the faulty gate can be written and used with the techniques described ahead.

$$v(c_{i_2})$$

		0	1
$v(c_{i_1})$	0	1	0
	1	0	0

Figure 3.14 Two-valued truth table of a two-input NAND gate with a fault that causes it to behave like a NOR gate

Figure 3.14 shows the truth table of a faulty two-input NAND gate where the presence of the fault has altered its logic behavior to be identical to that of a fault-free two-input NOR gate.

3.2.3.4 Behavior of a faulty fanout system

The above approach can be used to derive the truth tables of a fanout system shown in Figure 3.3(b) with SAFs at its inputs and outputs. Figure 3.15 shows some of the cases.

3.3   Logic simulation

In this section, we will first describe a basic algorithm for *logic simulation*. This will be followed by a description of the key concepts used to accelerate logic simulation.

Accurate simulation of logic values implied at circuit lines due to the application of a vector, as well as timing of each logic value transition and hazard caused, requires consideration of logic behavior as well as delays of each constituent logic element. In the case of synchronous sequential circuits, however, logic simulation is often performed primarily to determine the steady-state logic values implied at circuit lines by each vector being simulated. The temporal aspects of the circuit behavior, including times of occurrence of transitions and hazards and widths of hazards, are the purview of timing analysis. During test generation and fault simulation, it is even more common to focus exclusively on the steady-state logic behavior of the circuit under consideration.

A logic or fault simulation algorithm can be used to simulate each combinational logic block of a synchronous sequential circuit. The value at each output of a combinational logic block computed above is what will be latched at the next appropriate clock event into the latch/flip-flop driven by the output in the sequential circuit, *provided the delay of the circuit is within the requirements posed by the clocking strategy employed.* Such simulation algorithms are sometimes called *cycle-based* simulators. They are also called *zero-delay* simulators, since they do not consider delay values.

For the following discussion, we define **logic simulation** for a combinational logic circuit as the determination of steady-state logic values implied at each circuit line by

$v(c_i)$	$v(c_{j_1})$	$v(c_{j_2})$	$\cdots$	$v(c_{j_l})$	$\cdots$
0	1	1	$\cdots$	1	$\cdots$
1	1	1	$\cdots$	1	$\cdots$

(a)

$v(c_i)$	$v(c_{j_1})$	$v(c_{j_2})$	$\cdots$	$v(c_{j_l})$	$\cdots$
0	0	0	$\cdots$	0	$\cdots$
1	0	0	$\cdots$	0	$\cdots$

(b)

$v(c_i)$	$v(c_{j_1})$	$v(c_{j_2})$	$\cdots$	$v(c_{j_l})$	$\cdots$
0	0	0	$\cdots$	1	$\cdots$
1	1	1	$\cdots$	1	$\cdots$

(c)

$v(c_i)$	$v(c_{j_1})$	$v(c_{j_2})$	$\cdots$	$v(c_{j_l})$	$\cdots$
0	0	0	$\cdots$	0	$\cdots$
1	1	1	$\cdots$	0	$\cdots$

(d)

Figure 3.15 Two-valued truth tables of a fanout system with various single SAFs: (a) c_i SA1, (b) c_i SA0, (c) c_{j_l} SA1, and (d) c_{j_l} SA0

the vector applied to its primary inputs. In many simulation scenarios, each component of the input vector, $P = (p_1, p_2, \ldots, p_n)$, is fully specified, i.e., p_i can only take two values, 0 or 1. Such a simulation scenario will be referred to as **two-valued** logic simulation. It is, however, also quite common to perform simulation where one or more components of the input vector are incompletely specified, i.e., p_i may take one of three values, 0, 1, or x. Such a scenario will be referred to as **three-valued** logic simulation.

3.3.1 A simple logic simulation algorithm

In its simplest form, a zero-delay logic simulation algorithm for a combinational logic block reads an input vector $P = (p_1, p_1, \ldots, p_n)$, assigns each component p_i of the vector to the corresponding primary input x_i, and computes the new logic value implied at each line. These three steps are repeated for each vector for which simulation is desired.

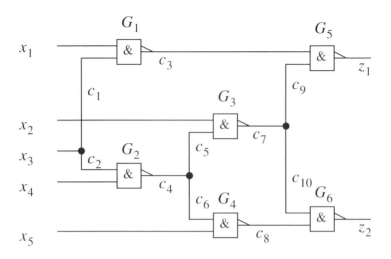

Figure 3.16   An example circuit C_{17}

In the above simple algorithm, the value at an output of a circuit element should not be computed until the value implied by the current vector at each input of the circuit element has been computed. For example, in the circuit shown in Figure 3.16, the value at line c_7, which is the output of gate G_3, should be computed only after the values implied by the current vector at inputs of G_3, namely lines x_2 and c_5, have been computed.

3.3.1.1   Levelization of circuit lines to determine the order of evaluation

For a combinational circuit, circuit lines can be ordered a priori such that if the values at the lines are computed in this order, then the above requirement is satisfied. This order can be determined by first computing a value called *input level* for each line c in the circuit.

The **input level** of line c_i in a combinational circuit, $\eta_{\text{inp}}(c_i)$, is the maximum number of circuit elements that are traversed along any path from any of the circuit's primary inputs to the line. The value of input level for each circuit line can be computed by using Procedure **InputLevelize( )** outlined next.

Procedure 3.2   [InputLevelize( )]
1 *Initialization:* For each circuit line c, $\eta_{\text{inp}}(c) = $ undefined.
2 For each primary input x_i, $\eta_{\text{inp}}(x_i) = 0$.
3 While there exist one or more logic elements such that (i) η_{inp} is defined for each of the element's inputs, and (ii) η_{inp} is undefined for any of its outputs, select one such element.
   (a) If the selected element is a gate with inputs $c_{i_1}, c_{i_2}, \ldots, c_{i_\alpha}$ and output c_j, then assign

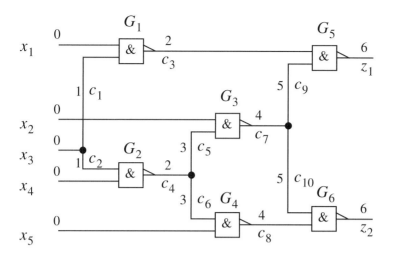

Figure 3.17 Computation of input levels of circuit lines

$$\eta_{\text{inp}}(c_j) = \max[\eta_{\text{inp}}(c_{i_1}), \eta_{\text{inp}}(c_{i_2}), \ldots, \eta_{\text{inp}}(c_{i_\alpha})] + 1.$$

(b) If the selected element is a fanout system with stem c_i and branches $c_{j_1}, c_{j_2}, \ldots, c_{j_\beta}$, then for each output c_{j_l}, where $l = 1, 2, \ldots, \beta$, assign

$$\eta_{\text{inp}}(c_{j_l}) = \eta_{\text{inp}}(c_i) + 1.$$

Step 3 of the above procedure can be implemented in an efficient manner such that the overall complexity of the procedure is $O(L)$, where L is the total number of lines in the circuit.

Once η_{inp} is determined for each circuit line, the lines can be sorted in ascending order of their η_{inp} values in an **ordered list** $Q_{\eta_{\text{inp}}}$ such that if $\eta_{\text{inp}}(c_i) < \eta_{\text{inp}}(c_j)$, then c_i appears earlier in list $Q_{\eta_{\text{inp}}}$ than c_j. According to the above definition, two lines that have the same η_{inp} value may appear in any order relative to each other.

Example 3.2 The input levels obtained for the circuit in Figure 3.17 are shown next to each line. For this circuit, one possible order in which the circuit lines may appear in the list is $Q_{\eta_{\text{inp}}} = (x_1, x_2, x_3, x_4, x_5, c_1, c_2, c_3, c_4, c_5, c_6, c_7, c_8, c_9, c_{10}, z_1, z_2)$. □

Since each input of a circuit element is always assigned a lower input level than each of its outputs, if values implied at circuit lines are computed in the order in which the lines appear in the ordered list $Q_{\eta_{\text{inp}}}$, then the value implied by the vector at each input of a circuit element is guaranteed to be computed before the value implied at any of its outputs is computed.

3.3.1.2 Computation of logic values

Let the logic value implied at line c by the current vector under simulation be denoted by $v(c)$. Simulation of vector $P = (p_1, p_2, \ldots, p_n)$ begins by assigning its components to the corresponding circuit inputs, i.e., $v(x_1) = p_1, v(x_2) = p_2, \ldots, v(x_n) = p_n$. Subsequently, the values at the other circuit lines, including primary outputs, are evaluated in the order in which the lines appear in list $Q_{\eta_{\text{inp}}}$. Consider the evaluation of $v(c_j)$, the value at a line c_j in the circuit. First, consider the case where c_j is the output of a gate with inputs $c_{i_1}, c_{i_2}, \ldots, c_{i_\alpha}$. In such a case, $v(c_j)$ is computed as

$$v(c_j) = \mathcal{G}_j^2(v(c_{i_1}), v(c_{i_2}), \ldots, v(c_{i_\alpha})), \tag{3.1}$$

where $\mathcal{G}_j^2()$ refers to the two-valued truth table or logic expression that describes the behavior of the gate driving line c_j. Next, consider the case where c_j is one of the branches of a fanout system with stem c_i. In this case,

$$v(c_j) = \mathcal{G}_j^2(v(c_i)), \tag{3.2}$$

where $\mathcal{G}_j^2()$ refers to the truth table or logic expression that describes the behavior of the fanout branch c_j.

If three-valued simulation is desired, then for each line, the corresponding function $\mathcal{G}_j^3()$ is used in Equations (3.1) and (3.2) instead of function $\mathcal{G}_j^2()$.

3.3.2   The algorithm SimpleLogicSimulation( )

The above concepts are combined in the following procedure, SimpleLogicSimulation( ).

Procedure 3.3 [SimpleLogicSimulation( )]
1 *Preprocessing:* Use Procedure InputLevelize( ) (Procedure 3.2) to compute the input level $\eta_{\text{inp}}(c_i)$ of each line c_i in the circuit as well as to obtain $Q_{\eta_{\text{inp}}}$, the ordered list of circuit lines.
2 While there exists a vector to be simulated, read a vector $P = (p_1, p_2, \ldots, p_n)$.
   (a) For each primary input x_i, assign $v(x_i) = p_i$.
   (b) In the order in which lines appear in $Q_{\eta_{\text{inp}}}$, for each line c_i in $Q_{\eta_{\text{inp}}}$, compute the value $v(c_i)$.
       Use Equation (3.1) if c_i is the output of a gate and Equation (3.2) if it is a branch of a fanout system.

The complexity of the above algorithm for simulation of one vector is O(L), where L is the the number of lines in the circuit.

Example 3.3 The application of the above procedure to the circuit shown in Figure 3.16 with vector $P = (1, 1, 1, 0, 1)$ gives the values shown in Figure 3.18. □

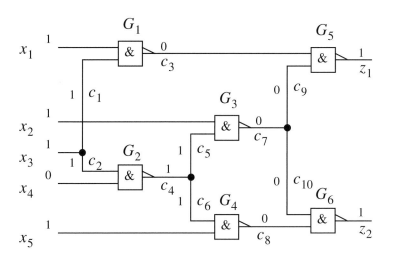

Figure 3.18   An example of simple logic simulation

3.3.3   Techniques to accelerate logic simulation

Two techniques are commonly employed to accelerate logic simulation. The first utilizes an important relationship between values implied at circuit lines by different vectors. The second one exploits the availability, in the processor on which simulation is performed, of multi-bit integers and hardware that efficiently performs bit-wise logic operations on bits of multi-bit integers. These techniques are discussed next.

3.3.3.1   Event-driven simulation

Figure 3.18 shows an example circuit and values implied at various lines by vector $(1, 1, 1, 0, 1)$. Now consider a scenario where the next vector to be simulated is $(1, 1, 0, 1, 0)$. Procedure SimpleLogicSimulation( ) can be used again to determine the values implied at circuit lines by this vector. These values are shown in Figure 3.19.

For the same scenario, Figure 3.20 shows the value implied at each line by the previous vector, $(1, 1, 1, 0, 1)$, as well as the current vector, $(1, 1, 0, 1, 0)$. At some lines, the value implied by the previous vector and that implied by the current vector are shown separated by an arrow, e.g., $0 \rightarrow 1$ at line c_3 and $1 \rightarrow 1$ at c_4. At the remaining lines, both vectors imply identical values and a single logic value is used to show the values implied at the line by the previous and current vectors, e.g., 1 at c_5.

Prior to the computation of the value implied by the current vector at any line c_i, variable $v(c_i)$ contains the value implied at c_i by the previous vector. For example, consider NAND gate G_1. In the above scenario, the value implied at this gate's input x_1, $v(x_1)$, by the current vector is identical to that implied by the previous vector. The value at the gate's other input c_1, $v(c_1)$, changes from logic 1 to logic 0. In contrast, the value implied at every input of gate G_3 by the current vector is identical to that implied by the previous vector.

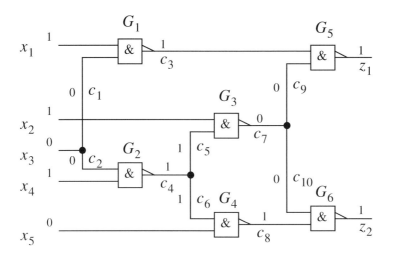

Figure 3.19 Simple logic simulation for a different vector

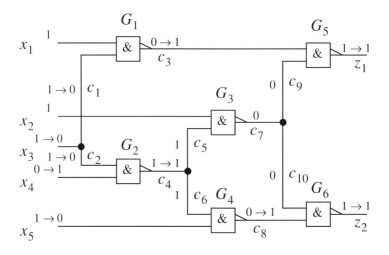

Figure 3.20 Values obtained by consecutive simulation of two vectors

Clearly, if the value implied by the current vector at every input of a circuit element is identical to that implied by the previous vector, then the value at each output of the circuit element will be identical to that computed for the previous vector. It is hence unnecessary to compute the values at the outputs of such a circuit element; the value assigned by previous vector can be retained and used to simulate gates in the fanout of each of its outputs.

In the scenario illustrated in Figure 3.20, it can be seen that values $v(c_1)$, $v(c_2)$, $v(c_3)$, $v(c_4)$, $v(c_8)$, $v(z_1)$, and $v(z_2)$ need to be computed, while $v(c_5)$, $v(c_6)$, $v(c_7)$, $v(c_9)$, and $v(c_{10})$ need not be recomputed. Furthermore, note that many of the values that are computed turn out to be identical to old values, e.g., the value at c_4, which is shown as $1 \rightarrow 1$ in the figure. (A single value is shown at lines whose value was not computed during the simulation of the current vector; two values separated by an arrow indicate that the value at the line was computed for the current vector.)

Events and selective simulation: An **event** is said to occur at a circuit line during the simulation of the current vector if the value implied at the line by the current vector is different from that implied by the previous vector simulated.

As described above, values at outputs of a circuit element need to be computed only if an event occurs at one or more inputs of the element during the simulation of the current vector. Such a simulation is called **event-driven simulation** and can reduce the average time complexity of simulation. This reduction occurs due to the fact that simulation of a vector typically causes events only at a fraction of circuit lines. For the scenario described above, the values at only seven out of 12 lines (not counting the primary inputs) in the circuit need to be computed in an event-driven simulation. However, the worst-case complexity of simulating one vector using the event-driven approach remains $O(L)$, where L is the number of lines in the circuit.

Event-driven simulation algorithm: The key feature of event-driven simulation that enables reduction in average simulation complexity is selective simulation. Selective simulation necessitates (i) identification of events, and (ii) maintaining lists of lines at which simulation must be performed.

An event at a line under consideration can be identified by comparing the value implied at the line by the current vector with that implied by the previous vector. First, when the value of each component of the current vector, $P = (p_1, p_2, \ldots, p_n)$, is read, *prior* to the assignment of value p_i to $v(x_i)$, p_i is compared with $v(x_i)$. $p_i \neq v(x_i)$ indicates the occurrence of an event at input x_i. Second, during the computation of the value implied at line c_i, the newly computed value is compared with the old value $v(c_i)$ *prior* to its assignment to $v(c_i)$. Again, if the newly computed value is different from the value contained in $v(c_i)$, an event is identified at c_i.

Occurrence of an event at any line indicates the need to compute the value at each output of the circuit element that it drives. This fact can be recorded by adding a suitable entry to a **task list**, which is an ordered list of tasks to be performed.

A **levelized task list** (Lee and Ha, 1991) Q is a set of ordered lists, Q_1, $Q_2, \ldots, Q_{\eta_{\max}}$, where $\eta_{\max}$ is the maximum value of input level assigned to any line in the circuit under simulation. List Q_i is used to store lines at input level i at which the values implied by the current vector need to be computed.

When an event is identified at an input x_i, then each fanout c_j of x_i is added to list $Q_{\eta_{\mathrm{inp}}(c_j)}$, provided that line c_j is *not* already in that list.

After the component values of the current vector are read, one line, say c_i is drawn from list Q_1, provided Q_1 is not empty. The new value at line c_i is then computed, whether an event occurs at c_i is identified, and then the computed value is assigned to

$v(c_i)$. If an event is identified at c_i, then each line c_j in the fanout of c_i is added to list $Q_{\eta_{inp}(c_j)}$, if c_j does not already exist in that list. The above steps are repeated until Q_1 is empty. The entire process is then repeated for Q_2, Q_3, and so on. The simulation of the current vector is complete when all the lists are empty.

Procedure 3.4 [EventDrivenSimulation( )]

1 *Preprocessing:* Use Procedure InputLevelize( ) to compute $\eta_{inp}(c_i)$ for all lines c_i in the circuit. Let η_{max} be the maximum value of input level assigned to any line.
2 *Initialization:*
    (a) For each circuit line c_i, assign $v(c_i) = x$.
    (b) Create task lists $Q_1, Q_2, \ldots, Q_{\eta_{max}}$. Initialize each list to be empty.
3 For each vector $P = (p_1, p_2, \ldots, p_n)$ that needs to be simulated
    (a) For $i = 1, 2, \ldots, n$
        i.   Read p_i.
        ii. If $p_i \neq v(x_i)$, then add each fanout c_j of x_i to list $Q_{\eta_{inp}(c_j)}$ if $c_j \notin Q_{\eta_{inp}(c_j)}$.
        iii. $v(x_i) = p_i$.
    (b) For $i = 1, 2, \ldots, \eta_{max}$
        While Q_i is not empty
        i.   Remove a line c_j from Q_i.
        ii. Compute the new value to be assigned to line c_j.
        iii. If the value computed above is not equal to $v(c_j)$, then add each line c_l that is a fanout of c_j to list $Q_{\eta_{inp}(c_l)}$, if $c_l \notin Q_{\eta_{inp}(c_l)}$.
        iv. Assign the newly computed value for line c_j to $v(c_j)$.

Example 3.4 The execution of Procedure EventDrivenSimulation( ) on the circuit shown in Figure 3.16 for vector $(1, 1, 1, 0, 1)$ followed by $(1, 1, 0, 1, 0)$ is illustrated in Figure 3.21. When the first vector is simulated, an event occurs at each line, since the value at each line is initialized to x. Subsequently, when the second vector is simulated, only the values at lines $c_1, c_2, c_3, c_4, c_8, z_1$, and z_2 are evaluated. The reader can verify that during the simulation of the second vector, before the value is computed for any line in the task list Q_i, the task list contains the following lines.

List	Lines
Q_1	c_1, c_2
Q_2	c_3, c_4
Q_3	
Q_4	c_8
Q_5	
Q_6	z_1, z_2

□

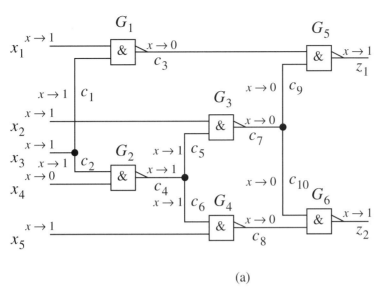

(a)

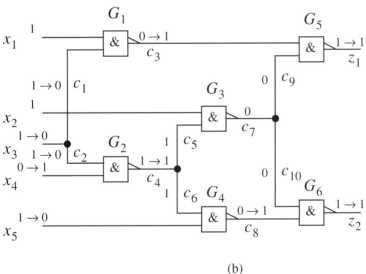

(b)

Figure 3.21 Illustration of event-driven simulation of two vectors: simulation of (a) the first vector, and (b) the second vector

3.3.3.2 Parallel simulation of multiple vectors

Logic values assigned to circuit lines are often stored in variables that are defined as integers. In all programming languages, an integer represents a W-bit storage, where typically, $W = 32$ or 64. For this reason, it is possible to represent values implied at a circuit line by W distinct vectors using a single integer. For example, the jth bit of the W bits of $V(c_i)$, the integer that represents the logic value at line c_i, can be used to represent value $v_j(c_i)$, where $v_j(c_i)$ is the value implied at line c_i by the jth vector, P_j. Hence, the single integer $V(c_i)$ can be used to represent the values as illustrated next.

Bit	Value
1	$v_1(c_i)$
2	$v_2(c_i)$
$\vdots$	$\vdots$
W	$v_W(c_i)$

Bit-wise logic operations, such as logic AND, OR, and NOT, are also supported by most high-level programming languages and performed efficiently by the ALUs of most processors. Typically, a bit-wise logic operation on W bits takes the same amount of time as the corresponding operation on single bits. Hence, if the function implemented by an output c_j of a logic element with inputs $c_{i_1}, c_{i_2}, \ldots, c_{i_\alpha}$, $\mathcal{G}_j^2()$, is computed using logic expressions, then the use of the above bit-wise logic operations allows parallel computation of the values $v_1(c_j), v_2(c_j), \ldots, v_W(c_j)$, if the values at the inputs and output of the circuit element are represented using the bit-encoding described above. Therefore, the use of multi-bit integers to represent values in parallel and bit-wise logic operations to evaluate values in parallel can accelerate Procedure SimpleLogicSimulation( ) by a factor of W, since W vectors can be read and simulated by a single execution of the instructions in Step 2.

3.3.4 Comments

Two important facts should be noted about the parallel simulation approach. First, parallel simulation is efficient only when the logic behavior of circuit elements is computed using logic expressions that use operations supported by the processor on which the simulation is performed. The reader can verify that if truth tables or singular covers are used, then either the computational complexity or the storage complexity associated with parallel simulation becomes prohibitively high.

Second, the concepts of parallel and event-driven simulations can be combined. However, some of the advantages accrued when each concept is applied individually may be diminished. This is due to the fact that if the two approaches are combined, then the value at each output of a circuit element must be computed as long as the value at any of the inputs of the logic element changes for *any one or more of the vectors* being simulated in parallel. Consequently, on an average, events may occur at a much larger fraction of circuit lines during parallel, event-driven simulation.

3.4 Fault simulation essentials

A very straightforward fault simulation algorithm can be obtained by repeated use of any logic simulation algorithm. For each vector, simulation of the fault-free version of the circuit, C, can be followed by simulation of each of the faulty versions of the circuit, C^{f_i}, where f_i is a fault in the fault list. Simulation of C^{f_i} is identical to that of C with the exception that the value at each output of the circuit element that is the site of fault f_i is computed using functions that capture the behavior of the circuit element in the presence of fault f_i.

The above approach is embodied in the following simple fault simulator.

Procedure 3.5 [SimpleFaultSimulation( )]
For each vector P
(a) Simulate the fault-free version of circuit C and compute $v(c_j)$ for all lines c_j.
(b) For each fault $f_i \in F$
    i. Simulate C^{f_i}, the faulty version of the circuit with fault f_i, to compute value $v^{f_i}(c_j)$ for each line c_j.
    ii. If $v^{f_i}(z_l)$ is the complement of $v(z_l)$ for any output z_l, then mark fault f_i as detected.

Since the number of single SAFs is proportional to L, the number of lines in a circuit, the complexity of the above fault simulation approach is $O(L^2)$. It has been shown in Harel and Krishnamurthy (1987) that the worst-case complexity of all known fault simulation algorithms is $O(L^2)$. It is further demonstrated that the fault simulation problem is equivalent to some intensively researched problems, such as matrix multiplication. Given the worst-case complexity of the best known algorithm for that problem, Harel and Krishnamurthy (1987) conclude that there is little hope of finding a fault simulation algorithm with linear complexity. All the approaches presented ahead attempt to reduce the average complexity as well as the value of the constants associated with the worst-case complexity expression.

Fault collapsing and fault dropping are two approaches that are commonly used to reduce the complexity of fault simulation.

3.4.1 Fault collapsing

The number of faulty versions of a circuit that need to be simulated can be decreased by exploiting two relations between two faults f_i and f_j: *fault equivalence* and *fault dominance*.

Two faults f_i and f_j in a circuit C are said to be **equivalent** (McCluskey and Clegg, 1971) if the corresponding faulty versions of the circuit, C^{f_i} and C^{f_j}, respectively, have identical input–output logic behavior. Clearly, if C^{f_i} is simulated then C^{f_j} need not be simulated, and vice versa.

In general, if faults $f_{i_1}, f_{i_2}, \ldots, f_{i_l}$ are equivalent, assuming that each of the above faults occurs by itself, then only one of these faults, say f_{i_1}, needs to be retained in the fault list used for fault simulation. If a vector detects f_{i_1}, then it also detects each of the single faults $f_{i_2}, f_{i_3}, \ldots, f_{i_l}$. Furthermore, for any vector, the response at the corresponding outputs of circuits $C^{f_{i_1}}, C^{f_{i_2}}, \ldots, C^{f_{i_l}}$ will be identical.

Note that fault equivalence is a **symmetric relation**, i.e., if fault f_i is equivalent to f_j, then f_j is equivalent to f_i. Furthermore, the relation is *transitive*, i.e., if f_i is equivalent to f_j and f_j equivalent to f_l, then f_i is equivalent to f_l. These properties are exploited by procedures that perform *fault collapsing*.

Next, consider a pair of faults f_i and f_j in a circuit C. Let $\mathcal{V}^{f_i}$ be the set of all vectors that detect fault f_i. Similarly, let $\mathcal{V}^{f_j}$ be the set of all vectors that detect fault f_j. Fault f_i is said to **dominate** (Poage and McCluskey, 1964) fault f_j if (a) $\mathcal{V}^{f_i} \supseteq \mathcal{V}^{f_j}$, and (b) each vector that detects f_j implies identical values at the corresponding outputs of C^{f_i} and C^{f_j}. Hence, if fault f_i dominates fault f_j, then any vector that detects f_j, not only detects f_i, but also generates the same response at the outputs of the faulty circuit C^{f_i} as it does at the corresponding outputs of C^{f_j}. It should, however, be noted that a vector that does not detect f_j may detect f_i. This is the case for any vector in the set $\mathcal{V}^{f_i} - \mathcal{V}^{f_j}$.

To reduce the run-time complexity of fault simulation, a fault f_i that dominates another fault f_j is very often deleted from the fault list. Note, however, that in this case, it is possible that for a set of vectors, fault simulation may report that f_j is not detected. In that case, no conclusion may be drawn about the detection of f_i.

The fault dominance relation is transitive but, in general, not symmetric. That is, if fault f_i dominates f_j and f_j dominates f_l, then f_i dominates f_l. However, in general, if f_i dominates f_j, then f_j may not dominate f_i. In fact, if fault f_i dominates f_j *and* f_j dominates f_i, then f_i and f_j are equivalent.

3.4.1.1 Some examples of fault equivalence and dominance

Consider the faulty versions of a two-input NAND gate shown in Figure 3.22(a) and (b). For each of these faulty versions, any possible combination of values at the gate inputs implies logic 1 at its output. Now consider a scenario in which a two-input NAND gate with inputs c_{i_1} and c_{i_2} and output c_j is used in an arbitrary combinational circuit C. Since the behavior of the faulty version of the gate with fault f_1, SA1 at its output c_j, is identical to that of the version with fault f_2, SA0 at input c_{i_1}, it can be demonstrated that the input–output behaviors of circuits C^{f_1} and C^{f_2} are identical. Hence, faults f_1 and f_2 are equivalent. A similar relationship can also be shown between either of these faults and the SA0 fault shown in Figure 3.22(c).

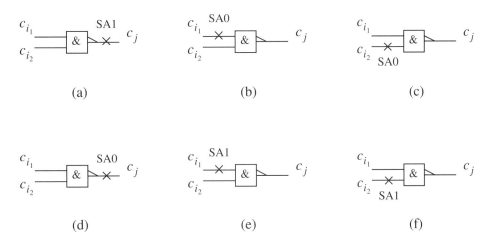

Figure 3.22 Various faulty versions of a two-input NAND gate

In general, for any NAND gate in any circuit, an SA1 fault at its output is equivalent to an SA0 fault at any of its inputs. Also, the SA0 fault at any one of its inputs is equivalent to the SA0 at any other input. In short, if the NAND gate inputs are $c_{i_1}, c_{i_2}, \ldots, c_\alpha$ and its output is c_j, then any two faults in the set $\{c_{i_1} \text{ SA0}, c_{i_2} \text{ SA0}, \ldots, c_{i_\alpha} \text{ SA0}, c_j \text{ SA1}\}$ are equivalent. A similar equivalence relationships can be derived for faults at inputs and outputs of other primitive gates and are shown in Table 3.4.

Consider next the SA0 fault at the output and the SA1 fault at one of the inputs of the two-input NAND gate, shown respectively in Figures 3.22(d) and (e). An erroneous value appears at gate output c_j with the SA0 fault if any of the following combination of values is applied to the gate inputs: $(0, 0)$, $(0, 1)$ or $(1, 0)$. In each of these cases, logic 0 appears at the gate output c_j instead of 1. For the faulty version with an SA1 fault at input c_{i_1} of the gate, an identical erroneous value appears at the gate output for only one of these combinations of input values, namely $(0, 1)$.

Let the above NAND gate be embedded within an arbitrary combinational circuit C. It can then be shown that if a vector detects fault f_3, SA1 at input c_{i_1}, it also detects fault f_4, SA0 at the gate output c_j. Furthermore, it can be shown that for any vector that detects f_3, the output responses of the faulty versions of the circuit, C^{f_3} and C^{f_4}, are identical. Hence, even when a NAND gate appears in an arbitrary circuit, an SA0 fault at its output dominates an SA1 fault at any one of its inputs. The SA0 fault at the output also dominates the SA1 at the other gate input shown in Figure 3.22(f). Table 3.4 shows dominance relationships between faults in other primitive gates.

Finally, consider any two faulty versions of an XOR gate with inputs c_{i_1} and c_{i_2} and output c_j, each with a distinct single SAF. As shown in Table 3.5, for each pair of faulty versions of the gate, there exists one combination of values at the gate inputs for which an erroneous value appears at the output for the first faulty version but not for the

Table 3.4. *Equivalence and dominance relations between faults located at the inputs* $(c_{i_1}, c_{i_2}, \ldots, c_\alpha)$ *and the output* (c_j) *of a gate*

Gate	Sets of equivalent faults	Dominance relations
NOT	$\{c_{i_1}\ \text{SA0}, c_j\ \text{SA1}\}$ $\{c_{i_1}\ \text{SA1}, c_j\ \text{SA0}\}$	
AND	$\{c_{i_1}\ \text{SA0}, c_{i_2}\ \text{SA0}, \ldots, c_{i_\alpha}\ \text{SA0}, c_j\ \text{SA0}\}$	c_j SA1 dominates c_{i_1} SA1 c_j SA1 dominates c_{i_2} SA1 $\vdots$ c_j SA1 dominates c_{i_α} SA1
NAND	$\{c_{i_1}\ \text{SA0}, c_{i_2}\ \text{SA0}, \ldots, c_{i_\alpha}\ \text{SA0}, c_j\ \text{SA1}\}$	c_j SA0 dominates c_{i_1} SA1 c_j SA0 dominates c_{i_2} SA1 $\vdots$ c_j SA0 dominates c_{i_α} SA1
OR	$\{c_{i_1}\ \text{SA1}, c_{i_2}\ \text{SA1}, \ldots, c_{i_\alpha}\ \text{SA1}, c_j\ \text{SAl}\}$	c_j SA0 dominates c_{i_1} SA0 c_j SA0 dominates c_{i_2} SA0 $\vdots$ c_j SA0 dominates c_{i_α} SA0
NOR	$\{c_{i_1}\ \text{SA1}, c_{i_2}\ \text{SA1}, \ldots, c_{i_\alpha}\ \text{SA1}, c_j\ \text{SA0}\}$	c_j SA1 dominates c_{i_1} SA0 c_j SA1 dominates c_{i_2} SA0 $\vdots$ c_j SA1 dominates c_{i_α} SA0
XOR	–	–

Table 3.5. *Fault detection in an XOR gate*

Fault	Input values that generate erroneous value at the gate output
SA0 at c_{i_1}	$(1, 0), (1, 1)$
SA1 at c_{i_1}	$(0, 0), (0, 1)$
SA0 at c_{i_2}	$(0, 1), (1, 1)$
SA1 at c_{i_2}	$(0, 0), (1, 0)$
SA0 at c_j	$(0, 1), (1, 0)$
SA1 at c_j	$(0, 0), (1, 1)$

second faulty version. Hence, there are no fault equivalence or dominance relationships between SAFs at the inputs and output of an XOR gate that hold in general.

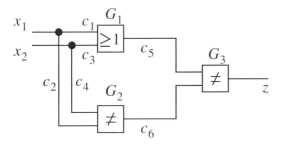

Figure 3.23 An example illustrating incompleteness of the equivalence and dominance relations

3.4.1.2 A fault collapsing procedure

The equivalence and dominance relations between SAFs at inputs and outputs of primitive gates may be used to formulate a simple procedure that can typically eliminate many faults from the complete fault list that contains each possible single SAF in a circuit. Since such a procedure decreases the size of a fault list, it is often called **fault collapsing**.

It should be noted that the equivalence and dominance relations identified above are incomplete in the sense that it is possible for two faults f_i and f_j, each located at an input or the output of a particular gate in a particular circuit, to be equivalent or dominant even though no such relation between these faults has been identified above. For example, no universally true relation can be derived between any two single SAFs associated with the inputs and the output of an XOR gate. Yet, the reader can verify that in the circuit shown in Figure 3.23, an SA1 fault at input c_6 of the XOR gate G_3 dominates the SA0 fault at the output z of the gate.

In addition, SAFs that are not associated with the inputs and output of a single gate can also be equivalent. The relations mentioned so far do not address such scenarios and the following procedure is not guaranteed to find all such relations.

In general, the complexity of identifying *all* fault equivalence and dominance relations is high. Hence, in practice, the equivalence and dominance relations identified between single SAFs associated with inputs and outputs of gates are used in an iterative fashion to achieve significant fault collapsing.

Procedure 3.6 [FaultCollapsing( )]
1 Starting at each primary output PO, compute the value of *output level* $\eta_{\text{out}}(c_i)$ for each line c_i.

The **output level** of a line c_i in a combinational circuit is the maximum number of circuit elements that are traversed along any path from the line to any primary output. The values $\eta_{\text{out}}(c_i)$ can be computed for each line c_i by using Procedure OutputLevelize( ) which can be obtained by modifying Procedure InputLevelize( ).

2 Sort the circuit lines in the order of non-decreasing values of η_{out} in an ordered list $Q_{\eta_{\text{out}}}$.

3 For each circuit line c_i, create a *local fault set* $F(c_i)$ containing all single SAFs that may be located at the line, i.e., $F(c_i) = \{c_i \text{ SA0}, c_i \text{ SA1}\}$.

4 While there exists at least one line in $Q_{\eta_{\text{out}}}$ that is the output of a gate G, remove the first line c_j from the ordered list $Q_{\eta_{\text{out}}}$.

 (a) Let $c_{i_1}, c_{i_2}, \ldots, c_{i_\alpha}$ be the inputs of gate G with output c_j.

 (b) Let $F_{\text{gate}} = F(c_j) \cup F(c_{i_1}) \cup F(c_{i_2}) \cup \cdots \cup F(c_{i_\alpha})$ be the set of single SAFs located at the inputs and the output of gate G.

 (c) Use Table 3.4 to identify the relations between the faults in the set F_{gate}. This will result in the identification of equivalence and dominance relations between pairs of faults.

 (d) Create *equivalent fault sets* $E_1, E_2, \ldots, E_\gamma$.

  Due to the symmetric and transitive nature of the fault equivalence relation, the faults in F_{gate} can be *partitioned* into **equivalent fault sets**, each containing one or more faults such that each pair of faults in an equivalent fault set is equivalent, and all faults that are equivalent to a fault f also belong to the equivalent fault set to which f belongs.

  Note that the equivalent fault sets form a **partition** of F_{gate}, i.e., $F_{\text{gate}} = E_1 \cup E_2 \cup \cdots \cup E_\gamma$, and no fault belongs to more than one equivalent fault set.

 (e) If any fault f_1 in F_{gate} dominates any other fault f_2 in F_{gate}, then delete f_1 from the local fault set to which it belongs.

 (f) For each equivalent fault set E_l that contains more than one fault, retain one fault and delete the others from the local fault set to which they belong.

  If any one of the faults in E_l is located at a line that is one of the inputs of gate G and is also the output of another primitive gate, then that fault may be retained. Otherwise, the fault on an arbitrarily selected input of the gate may be retained.

5 The collapsed fault list is obtained by taking the union of the local fault sets at each circuit line.

Example 3.5 Consider the circuit shown in Figure 3.24. First, $\eta_{\text{out}}(c_i)$ is computed for each circuit line, c_i. The values obtained are shown in Figure 3.24(a). In the increasing order of output level, the lines are added to an ordered list to obtain $Q_{\eta_{\text{out}}} = (z_2, z_1, c_{10}, c_9, c_8, c_3, c_7, c_6, c_1, x_5, x_1, c_5, x_2, c_4, c_2, x_4, x_3)$.

The local fault sets are then created for each line, where the local fault set is comprised of SA0 and SA1 faults at that line. Each of these sets is pictorially depicted in Figure 3.24(b), where an unshaded and a shaded box, respectively, depict an SA0 and an SA1 fault at the corresponding line.

The first line in the ordered list $Q_{\eta_{\text{out}}}$, z_2, is removed from the list. Since z_2 is the output of gate G_6, in this step all faults associated with the inputs and output of gate

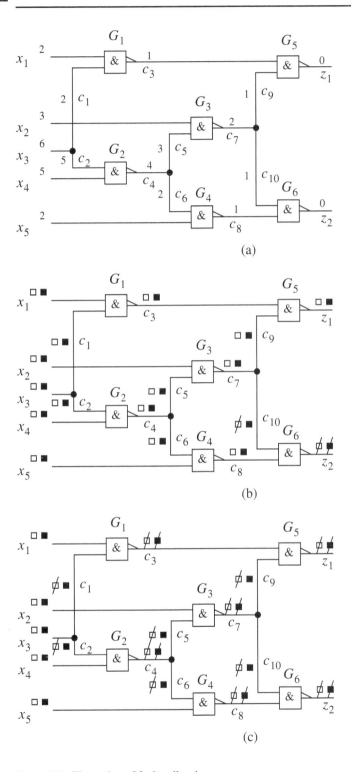

Figure 3.24 Illustration of fault collapsing

G_6 are collected to form the set of faults $F_{gate} = \{z_2\ \text{SA0}, z_2\ \text{SA1}, c_{10}\ \text{SA0}, c_{10}\ \text{SA1},$ $c_8\ \text{SA0}, c_8\ \text{SA1}\}$. Then Table 3.4 is used to find that fault z_2 SA0 dominates fault c_{10} SA1 as well as c_8 SA1. Also, the remaining three faults in F_{gate} are identified as being equivalent to each other.

F_{gate} is partitioned into equivalent fault sets $E_1 = \{z_2\ \text{SA1}, c_{10}\ \text{SA0}, c_8\ \text{SA0}\}$, $E_2 = \{z_2\ \text{SA0}\}$, $E_3 = \{c_{10}\ \text{SA1}\}$, and $E_4 = \{c_8\ \text{SA1}\}$. Only the first set contains more than one fault.

Next, fault z_2 SA0 is removed from the local fault set at line z_2 since it dominates other faults. Finally, among the faults in the equivalent fault set E_1, fault c_8 SA0 is retained and the other two faults are deleted from their corresponding local fault sets. The above deletions are depicted in Figure 3.24(b).

The main loop of the above procedure is then repeated for each gate and the local fault sets obtained at the completion of the algorithm are depicted in Figure 3.24(c). □

3.4.2 Fault dropping

In general, fault simulation may be performed for a circuit and a given sequence of patterns to (1) compute the fault coverage, and/or (2) determine the response for each faulty version of the circuit for each vector. In a majority of cases, however, fault simulation is performed to only compute the fault coverage. In such cases, fault simulation can be further accelerated via *fault dropping*.

Fault dropping is the practice in which faults detected by a vector are deleted from the fault list prior to the simulation of any subsequent vector. The decrease in complexity of fault simulation is due to the decrease in the average number of faults that are simulated for each vector.

The reduction in complexity can be quantified using the fault coverage vs. test length curve shown in Figure 3.25. The complexity of fault simulation without fault dropping can be approximated as $N_v \times N_f$ which is the area of the entire graph shown in Figure 3.25. In contrast, the complexity of fault simulation with fault dropping is approximately proportional to the shaded area shown in the graph. The reduction in the complexity of fault simulation can be especially large due to the fact that a very large fraction of all target faults are detected by a small fraction of vectors. The reduction can be significant if (a) a majority of faults in the circuit are testable, and (b) fault simulation is performed for a large number of vectors.

3.5 Fault simulation paradigms

In this section, we will discuss five main fault simulation paradigms for combinational circuits. Most fault simulation algorithms for combinational circuits are based on one or more of these paradigms.

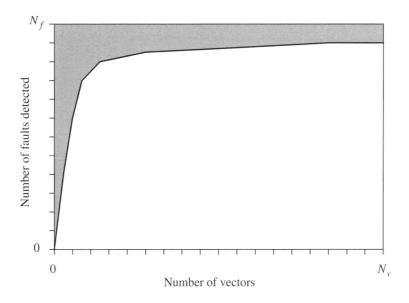

Figure 3.25 A typical fault coverage vs. number of test vector curve

3.5.1   Parallel fault simulation

Parallel fault simulation (Seshu, 1965) takes advantage of multi-bit representation of data and the availability of bit-wise operations. In this case, in each *pass* of simulation, the fault-free circuit as well as $W - 1$ faulty versions are simulated in parallel for a given vector, where W is the number of bits in the data representation on which bit-wise operations can be performed in parallel.

3.5.1.1 Encoding of values

If q faults are to be simulated for a vector, $P = (p_1, p_2, \ldots, p_n)$, then this vector is simulated for $\lceil q/(W - 1) \rceil$ passes. In the first pass, the fault-free circuit and circuit versions with faults $f_1, f_2, f_3, \ldots, f_{W-1}$ are simulated. In the next pass, the fault-free circuit and versions with faults $f_W, f_{W+1}, \ldots, f_{2(W-1)}$ are simulated, and so on. In general, in the lth pass, the fault-free circuit is simulated along with faulty versions with faults $f_{(l-1)(W-1)+1}, f_{(l-1)(W-1)+2}, \ldots, f_{l(W-1)}$.

In the lth pass, the values $v(c_i), v^{f_{(l-1)(W-1)+1}}(c_i), v^{f_{(l-1)(W-1)+2}}(c_i), \ldots, v^{f_{l(W-1)}}(c_i)$ are each represented using a distinct bit of $V(c_i)$ as

$$
V(c_i) = \begin{pmatrix} v(c_i) \\ v^{f_{(l-1)(W-1)+1}}(c_i) \\ v^{f_{(l-1)(W-1)+2}}(c_i) \\ \vdots \\ v^{f_{l(W-1)}}(c_i) \end{pmatrix}.
$$

Table 3.6. *Encoding of the ith bit of a fault mask for parallel fault simulation*

i^{th} bit simulates	i^{th} bit of mask	
	$M_Z(c_j)$	$M_O(c_j)$
Fault-free version of the circuit	1	0
Faulty circuit with an SA0 fault at c_j	0	0
Faulty circuit with an SA1 fault at c_j	1	1
Faulty circuit with a fault **not** located at c_j	1	0

At each primary input x_i, component p_i of the given vector P is assigned in the following manner. If p_i is 0, then 0 is assigned to each bit of $V(x_i)$, the integer associated with x_i, as

$$V(x_i) = \begin{pmatrix} 0 \\ 0 \\ 0 \\ \vdots \\ 0 \end{pmatrix}.$$

In the lth pass, this indicates $v(x_i) = 0$, $v^{f_{(l-1)(W-1)+1}}(x_i) = 0$, $v^{f_{(l-1)(W-1)+2}}(x_i) = 0, \ldots, v^{f_{l(W-1)}}(x_i) = 0$. If value $p_i = 1$, then 1 is assigned to each bit of $V(x_i)$, indicating $v(x_i) = 1$, $v^{f_{(l-1)(W-1)+1}}(x_i) = 1$, $v^{f_{(l-1)(W-1)+2}}(x_i) = 1, \ldots, v^{f_{l(W-1)}}(x_i) = 1$.

During a simulation pass, the corresponding values are computed for each line c_i, i.e., $v(c_i)$, $v^{f_{(l-1)(W-1)+1}}(c_i)$, $v^{f_{(l-1)(W-1)+2}}(c_i), \ldots, v^{f_{l(W-1)}}(c_i)$. Once all the faults in the fault list are simulated, the entire process is repeated for the next vector for which simulation is desired.

3.5.1.2 Fault insertion

In the lth pass, the first bit of value V at each line corresponds to the value implied at the line in the fault-free circuit, the second bit corresponds to that implied in the version of the circuit with fault $f_{(l-1)(W-1)+1}$, the third bit to that with fault $f_{(l-1)(W-1)+2}$, and so on.

Since each bit represents a different version of the circuit, it is **not** possible to simply replace the functions of one or more circuit elements by those describing the behaviors of faulty circuit elements. Instead, for each fault, an appropriate *fault mask* is used to insert the effect of the fault at its site.

A **fault mask** associated with line c_j, $M(c_j)$, is comprised of two W-bit integers, $M_Z(c_j)$, and $M_O(c_j)$. The encoding shown in Table 3.6 is used for bit i of each of these integers.

Consider an example scenario where the circuit shown in Figure 3.16 is being

Table 3.7. *Fault masks for parallel simulation of the fault-free circuit and three faulty versions with faults f_1: x_1 SA0, f_2: x_1 SA1, and f_3: x_2 SA0, respectively*

Fault mask	Line							
	x_1		x_2		x_3		$\cdots$	
	$M_Z(x_1)$	$M_O(x_1)$	$M_Z(x_2)$	$M_O(x_2)$	$M_Z(x_3)$	$M_O(x_3)$	$M_Z(\cdots)$	$M_O(\cdots)$
Bit-1	1	0	1	0	1	0	1	0
Bit-2	0	0	1	0	1	0	1	0
Bit-3	1	1	1	0	1	0	1	0
Bit-4	1	0	0	0	1	0	1	0

simulated for vector $P = (1, 1, 0, 1, 0)$. Assume that $W = 4$ and, in the pass under consideration, the fault-free version of the circuit is being simulated along with three faulty versions with faults f_1: x_1 SA0, f_2: x_1 SA1, and f_3: x_2 SA0, respectively. In this case, the fault mask at each line has $W = 4$ bits. The values of the four bits of each mask are shown in rows labeled Bit-1 through Bit-4 in Table 3.7.

3.5.1.3 Simulation

Once the values at the primary inputs and fault masks are set up as described above, the parallel logic simulation algorithm can be used with one modification, which is required for the use of fault masks at each line during the computation of the value implied at that line.

A pass of parallel fault simulation begins with assignment of values at each primary input x_i, where $V(x_i)$ is computed as described in Section 3.5.1.1. The values at circuit lines are then evaluated in the order in which lines appear in list $Q_{\eta_{inp}}$. The following two facts should be noted.

1. Multi-bit bit-wise operations are used to evaluate the values implied at output c_j of a circuit element using the multi-bit values $V(c_i)$ at each input of the element and logic equations describing the behavior of the output.
2. If any of the faulty circuit versions being simulated in the pass correspond to a fault site at c_j, then $V(c_j)$ is updated using the mask as

$$V(c_j) = \left[V(c_j) \text{ and } M_Z(c_j) \right] \text{ or } M_O(c_j),$$

where **and** and **or** are the bit-wise AND and OR operations, respectively. Note that the mask values are defined in such a manner that the above update of value $V(c_j)$ can be performed at all lines, including the fault-free lines.

The overall algorithm is comprised of subsequent passes for the same vector for other faulty versions of the circuit followed by repetition of the entire process for each of the other vectors. The following example illustrates parallel fault simulation.

Table 3.8. *Bit-values obtained due to parallel simulation of the fault-free version of the circuit shown in Figure 3.16 along with three faulty versions with faults f_1, f_2, and f_3 for vector $(1, 1, 0, 1, 0)$*

								Line								
Value	x_1	x_2	x_3	x_4	x_5	c_1	c_2	c_3	c_4	c_5	c_6	c_7	c_8	c_9	c_{10}	z_1	z_2
$v()$	1	1	0	1	0	0	0	1	1	1	1	0	1	0	0	1	1
$v^{f_1}()$	0	1	0	1	0	0	0	1	1	1	1	0	1	0	0	1	1
$v^{f_2}()$	1	1	0	1	0	0	0	1	1	1	1	0	1	0	0	1	1
$v^{f_3}()$	1	0	0	1	0	0	0	1	1	1	1	1	1	1	1	0	0

Example 3.6 Consider a scenario where parallel fault simulation is to be performed (a) on the circuit shown in Figure 3.16, (b) for vectors $P_1 = (1, 1, 0, 1, 0)$, $P_2 = (0, 0, 1, 1, 1)$, $P_3 = (0, 0, 0, 0, 0)$, and $P_4 = (1, 1, 1, 1, 1)$, and (c) for a fault list comprised of the faults f_1: x_1 SA0, f_2: x_1 SA1, f_3: x_2 SA0, f_{20}: c_5 SA1, and f_{28}: c_9 SA1.

Assuming $W = 4$, for the first vector, two passes are required for the simulation of five faults. In the first pass, the fault-free circuit along with three faulty versions with faults f_1, f_2, and f_3, respectively, may be simulated. The fault masks for this simulation pass are as shown in Table 3.7. This simulation pass gives the values shown in Table 3.8 for these versions of the circuit at various circuit lines. Clearly, fault f_3 is detected by this vector, since the values at both the outputs, $v^{f_3}(z_1)$ and $v^{f_3}(z_2)$, are complements of the values implied by the vector at the outputs in the fault-free circuit. The fault masks are then computed for the second pass to simulate the fault-free circuit along with two faulty versions of the circuit with faults f_{20} and f_{28}, respectively. The reader can verify that this pass determines that fault f_{28} is detected by the vector.

Prior to the simulation of the next vector, P_2, the faults detected by the first vector, namely f_3 and f_{28}, may be dropped from the fault list. Since only three faults remain in the fault list, only one pass is required for the second vector. In this pass, first the fault masks are determined to enable parallel simulation of the fault-free circuit along with the three faulty versions with faults f_1, f_2, and f_{20}, respectively. The reader can verify that parallel simulation of vector P_2 determines that only fault f_2 is detected by this vector. Subsequent simulation of P_3 determines that none of the faults that remain in the fault list is detected by P_3. Finally, simulation of P_4 identifies that both the remaining faults, f_1 and f_{20}, are detected by the vector.

The order in which simulation is performed for the above example scenario, along with the values implied at circuit outputs, are shown in Table 3.9.   □

3.5.2   Parallel-pattern single-fault propagation (PPSFP) simulation

An alternative strategy (Waicukauski *et al.*, 1985) in which the parallelism provided by the programming language and ALU may be exploited is via parallel simulation of

Table 3.9. *Results of execution of parallel fault simulation for the scenario in Example 3.6*

Pass no.	Circuits simulated	Vector simulated	Values computed
1	C	P_1	$v_1(z_1) = 1; v_1(z_2) = 1$
	C^{f_1}		$v_1^{f_1}(z_1) = 1; v_1^{f_1}(z_2) = 1$
	C^{f_2}		$v_1^{f_2}(z_1) = 1; v_1^{f_2}(z_2) = 1$
	C^{f_3}		$v_1^{f_3}(z_1) = 0; v_1^{f_3}(z_2) = 0$
2	C	P_1	$v_1(z_1) = 1; v_1(z_2) = 1$
	$C^{f_{20}}$		$v_1^{f_{20}}(z_1) = 1; v_1^{f_{20}}(z_2) = 1$
	$C^{f_{28}}$		$v_1^{f_{28}}(z_1) = 0; v_1^{f_{28}}(z_2) = 1$
3	C	P_2	$v_2(z_1) = 0; v_2(z_2) = 0$
	C^{f_1}		$v_2^{f_1}(z_1) = 0; v_2^{f_1}(z_2) = 0$
	C^{f_2}		$v_2^{f_2}(z_1) = 1; v_2^{f_2}(z_2) = 0$
	$C^{f_{20}}$		$v_2^{f_{20}}(z_1) = 0; v_2^{f_{20}}(z_2) = 0$
4	C	P_3	$v_3(z_1) = 0; v_3(z_2) = 0$
	C^{f_1}		$v_3^{f_1}(z_1) = 0; v_3^{f_1}(z_2) = 0$
	$C^{f_{20}}$		$v_3^{f_{20}}(z_1) = 0; v_3^{f_{20}}(z_2) = 0$
5	C	P_4	$v_4(z_1) = 1; v_4(z_2) = 0$
	C^{f_1}		$v_4^{f_1}(z_1) = 0; v_4^{f_1}(z_2) = 0$
	$C^{f_{20}}$		$v_4^{f_{20}}(z_1) = 1; v_4^{f_{20}}(z_2) = 1$

multiple vectors. In this paradigm, a **batch of vectors**, $P_1 = (p_{1,1}, p_{2,1}, \ldots, p_{n,1})$, $P_2 = (p_{1,2}, p_{2,2}, \ldots, p_{n,2}), \ldots, P_W = (p_{1,W}, p_{2,W}, \ldots, p_{n,W})$ are simulated in parallel.

If the fault list contains q faults during the simulation of a batch of W vectors, then their simulation is carried out in a total of $q + 1$ passes. In the first pass, parallel simulation is performed for the fault-free version of the circuit for the W vectors in the batch. In each subsequent pass for the batch of vectors, parallel simulation is performed for a faulty version of the circuit containing a unique fault that belongs to the current fault list.

The faults detected by the vectors in the batch are identified and dropped from the fault list, and the entire process repeated for the next batch of vectors. The complete process is repeated until either all faults in the fault list are detected, or simulation is completed for all vectors for which simulation is desired.

3.5.2.1 Simulation of a fault-free circuit for a batch of vectors

Fault-free simulation is identical to parallel logic simulation. Since W vectors are simulated in parallel, in the first pass, the bits of the variable associated with line c_i denote the values

$$
V(c_i) = \begin{pmatrix} v_1(c_i) \\ v_2(c_i) \\ \vdots \\ v_W(c_i) \end{pmatrix}.
$$

The simulation begins with the following assignment at each primary input x_i,

$$
V(x_i) = \begin{pmatrix} p_{i,1} \\ p_{i,2} \\ \vdots \\ p_{i,W} \end{pmatrix}.
$$

Subsequently, parallel logic simulation is performed. Recall that, to facilitate parallel simulation of W vectors, the values implied at a line are computed using logic expressions that capture the behavior of the corresponding output of the fault-free version of the logic element that drives the line. Also recall that multi-bit bit-wise logic operations are used to compute the values implied at the line by each vector in the batch.

At the conclusion of this phase, the values implied at each output z_i of the circuit by each vector in the batch, i.e., $v_1(z_1)$, $v_2(z_1)$, $\ldots$, $v_W(z_1)$, $v_1(z_2)$, $v_2(z_2)$, $\ldots$, $v_W(z_2)$, $\ldots$, $v_1(z_m)$, $v_2(z_m)$, $\ldots$, $v_W(z_m)$, are computed. These values are stored for comparison with the corresponding values computed for each faulty version of the circuit simulated in the subsequent passes for the same batch of vectors.

3.5.2.2 Simulation of a faulty version of the circuit for a batch of vectors

Simulation of each faulty circuit version is identical to that for the fault-free circuit with one important difference. At the circuit element that is the site of the fault, logic expressions describing the behavior of the corresponding faulty version of the element are used to compute the value implied at the outputs of the element.

For example, during the simulation of the fault-free version of the circuit shown in Figure 3.16, the behavior of each NAND gate with inputs c_{i_1} and c_{i_2} and output c_j is described using the expression $V(c_j) = \text{not } [V(c_{i_1}) \text{ and } V(c_{i_2})]$. More specifically, the behavior of line c_3 is described using the expression $V(c_3) = \text{not } [V(x_1) \text{ and } V(c_1)]$. Now consider a scenario where the faulty version of the circuit with fault f_2: x_1 SA1 is being simulated. During this simulation, the above expression may be modified to describe the behavior of the SA1 fault at input x_1 as $V(c_3) = \text{not } V(c_1)$. Note that when circuit C^{f_2} with this single SAF is simulated, only expressions describing the behavior of c_3 are changed; for each of the other logic elements in the circuit, expressions describing its fault-free behavior are used.

At the end of the simulation of the faulty version of the circuit C^{f_j}, the values implied at each output for each vector in the batch, i.e., $v_1^{f_j}(z_1), v_2^{f_j}(z_1), \ldots, v_W^{f_j}(z_1)$, $v_1^{f_j}(z_2), v_2^{f_j}(z_2), \ldots, v_W^{f_j}(z_2), \ldots, v_1^{f_j}(z_m), v_2^{f_j}(z_m), \ldots, v_W^{f_j}(z_m)$, are computed. Each of these values is compared with the corresponding value implied by the vector for the fault-free circuit. If there exists one or more vector P_i in the batch and output z_l, such that value $v_i(z_l)$ determined during fault-free simulation is the complement of the corresponding value $v_i^{f_j}(z_l)$ during the simulation of the faulty version of the circuit with fault f_j, then fault f_j is marked as detected and dropped from the fault list.

Example 3.7 Consider the simulation scenario used in Example 3.6 to illustrate parallel fault simulation, where fault simulation is to be performed (a) on the circuit shown in Figure 3.16, (b) for vectors $P_1 = (1, 1, 0, 1, 0)$, $P_2 = (0, 0, 1, 1, 1)$, $P_3 = (0, 0, 0, 0, 0)$, and $P_4 = (1, 1, 1, 1, 1)$, and (c) for a fault list comprised of the faults f_1: x_1 SA0, f_2: x_1 SA1, f_3: x_2 SA0, f_{20}: c_5 SA1, and f_{28}: c_9 SA1. Assuming $W = 4$, all the four vectors can be considered as one batch. In the first pass, the fault-free circuit is simulated and the values implied by the vectors at outputs z_1 and z_2 of the circuit by each vector are computed. These values, shown in Table 3.10, are stored for comparison with corresponding values computed in subsequent passes for each faulty version of the circuit.

In the second pass, simulation for the batch of vectors is repeated for the faulty version of the circuit C^{f_1} to obtain the corresponding values. Table 3.10 shows the order in which simulations are performed and values implied at outputs. Clearly, all faults in the given fault list are detected by this set of vectors.

A comparison of Tables 3.9 and 3.10 clearly reveals the difference between the order in which vectors and circuit versions are simulated in parallel fault simulation and PPSFP simulation.   □

3.5.3   Deductive fault simulation

The deductive fault simulation paradigm (Armstrong, 1972) utilizes a dynamic data structure to represent the response of each faulty version of the circuit for which simulation is desired. It comprises two main steps, which are typically performed in an interleaved manner. In one step, fault-free circuit simulation is performed for the given vector to compute the value implied at every line in the fault-free version of the circuit. In the other step, the value implied by the vector at every line in each faulty circuit version is deduced. Here, we present this approach assuming that each vector is completely specified.

Table 3.10. *Results of execution of PPSFP simulation for the scenario in Example 3.7*

Pass no.	Circuit simulated	Vectors simulated	Values computed
1	C	P_1	$v_1(z_1) = 1; v_1(z_2) = 1$
		P_2	$v_2(z_1) = 0; v_2(z_2) = 0$
		P_3	$v_3(z_1) = 0; v_3(z_2) = 0$
		P_4	$v_4(z_1) = 1; v_4(z_2) = 0$
2	C^{f_1}	P_1	$v_1^{f_1}(z_1) = 1; v_1^{f_1}(z_2) = 1$
		P_2	$v_2^{f_1}(z_1) = 0; v_2^{f_1}(z_2) = 0$
		P_3	$v_3^{f_1}(z_1) = 0; v_3^{f_1}(z_2) = 0$
		P_4	$v_4^{f_1}(z_1) = 0; v_4^{f_1}(z_2) = 0$
3	C^{f_2}	P_1	$v_1^{f_2}(z_1) = 1; v_1^{f_2}(z_2) = 1$
		P_2	$v_2^{f_2}(z_1) = 1; v_2^{f_2}(z_2) = 0$
		P_3	$v_3^{f_2}(z_1) = 0; v_3^{f_2}(z_2) = 0$
		P_4	$v_4^{f_2}(z_1) = 1; v_4^{f_2}(z_2) = 0$
4	C^{f_3}	P_1	$v_1^{f_3}(z_1) = 0; v_1^{f_3}(z_2) = 0$
		P_2	$v_2^{f_3}(z_1) = 0; v_2^{f_3}(z_2) = 0$
		P_3	$v_3^{f_3}(z_1) = 0; v_3^{f_3}(z_2) = 0$
		P_4	$v_4^{f_3}(z_1) = 1; v_4^{f_3}(z_2) = 0$
5	$C^{f_{20}}$	P_1	$v_1^{f_{20}}(z_1) = 1; v_1^{f_{20}}(z_2) = 1$
		P_2	$v_2^{f_{20}}(z_1) = 0; v_2^{f_{20}}(z_2) = 0$
		P_3	$v_3^{f_{20}}(z_1) = 0; v_3^{f_{20}}(z_2) = 0$
		P_4	$v_4^{f_{20}}(z_1) = 1; v_4^{f_{20}}(z_2) = 1$
6	$C^{f_{28}}$	P_1	$v_1^{f_{28}}(z_1) = 0; v_1^{f_{28}}(z_2) = 1$
		P_2	$v_2^{f_{28}}(z_1) = 0; v_2^{f_{28}}(z_2) = 0$
		P_3	$v_3^{f_{28}}(z_1) = 0; v_3^{f_{28}}(z_2) = 0$
		P_4	$v_4^{f_{28}}(z_1) = 1; v_4^{f_{28}}(z_2) = 0$

3.5.3.1 Representation of values

In this paradigm, the values implied by the vector being simulated at a line c_i are captured using a combination of the value implied at the line in the fault-free circuit, $v(c_i)$, and a *deductive fault list* (DFL). A **deductive fault list** for line c_i, denoted as $\mathcal{F}(c_i)$, is a set of faults, where a fault f_l appears in $\mathcal{F}(c_i)$, if and only if

1 f_l belongs to the current fault list, and
2 the value implied by the current vector at line c_i in the faulty version of circuit C^{f_l}, $v^{f_l}(c_i)$, is the complement of that implied in the fault-free circuit C, $v(c_i)$.

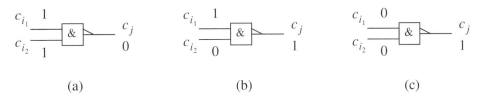

Figure 3.26 A two-input NAND with different logic values

3.5.3.2 Circuit traversal and computation of values

Zero-delay deductive fault simulation for a vector, $P = (p_1, p_2, \ldots, p_n)$, begins with the assignment of the components of the vector to the corresponding primary inputs, i.e., the assignment $v(x_i) = p_i$, for each primary input x_i.

Second, for each primary input x_i, all the faults associated with x_i that still exist in the fault list are examined to determine the *active local fault*, if any, associated with the line. If x_i SA0 belongs to the fault list and $v(x_i) = 1$, then this fault is called the **active local fault** and is the only fault in the **local deductive fault list** for the line, i.e., $\mathcal{F}_{\text{loc}}(x_i) = \{x_i \text{ SA0}\}$. If x_i SA1 belongs to the fault list and $v(x_i) = 0$, then that fault is the only active local fault and $\mathcal{F}_{\text{loc}}(x_i) = \{x_i \text{ SA1}\}$. If neither of the above cases holds, then $\mathcal{F}_{\text{loc}}(x_i) = \{\}$.

Third, at each primary input x_i, the deductive fault list is assigned the local deductive fault list, i.e., $\mathcal{F}(x_i) = \mathcal{F}_{\text{loc}}(x_i)$.

The circuit lines are then traversed in the order in which they are listed in $Q_{\eta_{\text{inp}}}$. This ensures that when a line c_j is visited, then for each of its fanins c_i, the fault-free circuit value, $v(c_i)$, as well as the DFL, $\mathcal{F}(c_i)$, have already been computed. Whenever a line c_j is visited, three operations are performed.

1 The value implied at c_j in the fault-free version of the circuit, $v(c_j)$, by vector P is computed. This computation may be performed in a manner similar to simple logic simulation.

2 Local faults at line c_j are analyzed and if an active local fault exists, then it is added to the local DFL, $\mathcal{F}_{\text{loc}}(c_j)$. Otherwise, $\mathcal{F}_{\text{loc}}(c_j) = \{\}$. This process is similar to that described above for a primary input.

3 The non-local faults that may propagate from the DFLs of one or more fanins of c_j to the DFL of c_j, $\mathcal{F}(c_j)$, are then identified. These faults and the faults in $\mathcal{F}_{\text{loc}}(c_j)$ are then combined to obtain $\mathcal{F}(c_j)$. This process is carried out using appropriate set operations described next.

3.5.3.3 Propagation of faults in DFLs via a circuit element

Consider the two-input NAND gate with fault-free circuit values shown in Figure 3.26(a). Also associated with inputs c_{i_1} and c_{i_2} are DFLs $\mathcal{F}(c_{i_1})$ and $\mathcal{F}(c_{i_2})$, respectively.

Note that, in general, a fault f_l may belong (a) only to $\mathcal{F}(c_{i_1})$, (b) only to $\mathcal{F}(c_{i_2})$, (c) to $\mathcal{F}(c_{i_1})$ as well as $\mathcal{F}(c_{i_2})$, or (d) to neither $\mathcal{F}(c_{i_1})$ nor $\mathcal{F}(c_{i_2})$. For the particular circuit values shown in Figure 3.26(a), namely, $v(c_{i_1}) = 1$ and $v(c_{i_2}) = 1$, it can be verified that in any of the first three scenarios listed above, fault f_l will also appear in the DFL at c_j, $\mathcal{F}(c_j)$. In other words, for this combination of fault-free circuit values for the NAND gate, $\mathcal{F}(c_j) = \mathcal{F}(c_{i_1}) \cup \mathcal{F}(c_{i_2}) \cup \mathcal{F}_{\text{loc}}(c_j)$, where $\mathcal{F}_{\text{loc}}(c_j)$ is the set containing the active local fault at c_j, if any.

Now consider the NAND gate shown in Figure 3.26(b). In this case, a non-local fault f_l can change the value at c_j only if it alters the value at c_{i_2} from 0 to 1. This is due to the fact that if the value at c_{i_2} remains 0 in the presence of a non-local fault f_l, the value at c_j will remain 1, i.e., at its fault-free value. Also note that the non-local fault can alter $\mathcal{F}(c_j)$ only if it does *not* alter the value at c_{i_1}. If that change occurs, then the value at c_{i_1} changes to 0, and causes the value at c_j to be 1, the fault-free value. It can be proven that for this combination of fault-free values, $\mathcal{F}(c_j)$ can be computed as $\mathcal{F}(c_j) = \left[\mathcal{F}(c_{i_2}) - \mathcal{F}(c_{i_1})\right] \cup \mathcal{F}_{\text{loc}}(c_j)$. Similar arguments can be used to show that for the NAND gate in Figure 3.26(c), $\mathcal{F}(c_j) = \left[\mathcal{F}(c_{i_1}) \cap \mathcal{F}(c_{i_2})\right] \cup \mathcal{F}_{\text{loc}}(c_j)$.

Consider the general case where c_j is the output of a primitive gate. Let $NCS(c_j)$ denote the inputs of the gate with output c_j at which the current vector implies the gate's non-controlling value in the fault-free version of the circuit. Let $CS(c_j)$ denote the inputs at which the vector implies the gate's controlling value. If there exists at least one input with controlling value, i.e., if $CS(c_j) \neq \{\}$, then the DFL for the output is given by

$$\mathcal{F}(c_j) = \left[\left\{\bigcap_{\forall c_{i_l} \in CS(c_j)} \mathcal{F}(c_{i_l})\right\} - \left\{\bigcup_{\forall c_{i_l} \in NCS(c_j)} \mathcal{F}(c_{i_l})\right\}\right] \bigcup \mathcal{F}_{\text{loc}}(c_j).$$

Otherwise, i.e., if $CS(c_j) = \{\}$, then

$$\mathcal{F}(c_j) = \left\{\bigcup_{\forall c_{i_l} \in NCS(c_j)} \mathcal{F}(c_{i_l})\right\} \bigcup \mathcal{F}_{\text{loc}}(c_j).$$

For an XOR gate with inputs $c_{i_1}, c_{i_2}, \ldots, c_{i_\alpha}$ and output c_j,

$$\mathcal{F}(c_j) = \left\{f_l \mid f_l \text{ appears in an odd number of sets} \right.$$
$$\left. \mathcal{F}(c_{i_1}), \mathcal{F}(c_{i_2}), \ldots, \mathcal{F}(c_{i_\alpha})\right\} \bigcup \mathcal{F}_{\text{loc}}(c_j).$$

For any other gate, arguments similar to those described above can be used to derive an expression for $\mathcal{F}(c_j)$ in terms of the fault-free circuit values and the DFL of each of its fanins c_{i_l}, i.e., in terms of $v(c_{i_l})$ and $\mathcal{F}(c_{i_l})$.

Finally, if c_j is a branch of a fanout system with stem c_i, then

$$\mathcal{F}(c_j) = \mathcal{F}(c_i) \cup \mathcal{F}_{\text{loc}}(c_j).$$

3.5.3.4 Detected faults and faulty circuit response

The existence of fault f_l in DFL $\mathcal{F}(z_i)$, where z_i is a primary output, indicates that in the circuit with fault f_l, C^{f_l}, the current vector implies at z_i a value that is the complement of $v(z_i)$, the value implied at the output in the fault-free version of the circuit. Hence, a fault f_l is detected by vector P if and only if there exists at least one output z_i such that $f_l \in \mathcal{F}(z_i)$, where $\mathcal{F}(z_i)$ is the DFL computed by performing deductive fault simulation for vector P. The values implied by the vector at line c_i in C^{f_l} can hence be computed as

$$
v^{f_l}(c_i) = \begin{cases} v(c_i), & \text{if } f_l \notin \mathcal{F}(c_i), \\ \overline{v(c_i)}, & \text{if } f_l \in \mathcal{F}(c_i). \end{cases}
$$

This relation can be used to compute the response at each primary output for each faulty version of the circuit.

Example 3.8 Consider the circuit shown in Figure 3.27(a). Procedure FaultCollapsing( ) provides the following fault list: $F = \{f_1\colon x_1 \text{ SA0}, f_2\colon x_1 \text{ SA1}, f_3\colon x_2 \text{ SA0}, f_4\colon x_2 \text{ SA1}, f_5\colon x_3 \text{ SA0}, f_6\colon x_3 \text{ SA1}, f_7\colon x_4 \text{ SA0}, f_8\colon x_4 \text{ SA1}, f_9\colon x_5 \text{ SA0}, f_{10}\colon x_5 \text{ SA1}, f_{12}\colon c_1 \text{ SA1}, f_{14}\colon c_2 \text{ SA1}, f_{20}\colon c_5 \text{ SA1}, f_{22}\colon c_6 \text{ SA1}, f_{28}\colon c_9 \text{ SA1}, f_{30}\colon c_{10} \text{ SA1}\}$.

Figure 3.27(a) shows the values implied at each circuit line by the current vector $(1, 1, 0, 1, 0)$ in the fault-free version of the circuit. Figure 3.27(b) shows the DFL computed for each line in the circuit using the above method.

At each primary input, the DFL only contains the active local fault. All other DFLs can be computed using the relations described above. Consider, as an example, the computation of $\mathcal{F}(z_1)$. All the faults that appear in $\mathcal{F}(c_9)$ *except* f_6 are included in $\mathcal{F}(z_1)$. f_6 is not included since it also appears in $\mathcal{F}(c_3)$, indicating that the presence of f_6 will change the value at c_3 from 1 to 0 and change the value at c_9 from 0 to 1, leaving z_1 at the fault-free value.

The DFLs at primary outputs indicate that the vector detects faults f_3, f_6, f_{14}, f_{28}, and f_{30}.

The outputs of the circuit with fault f_3 are $v^{f_3}(z_1) = 0$ and $v^{f_3}(z_2) = 0$. Also, for the circuit with fault f_6, $v^{f_6}(z_1) = 1$ and $v^{f_6}(z_2) = 0$.   □

3.5.3.5 Comparison and extensions

Comparison: Deductive fault simulation can be viewed as a variant of parallel fault simulation that uses dynamic lists, $\mathcal{F}$, to store the faults being simulated instead of a fixed array of values, v^{f_l}. A fault being simulated does *not* appear in list $\mathcal{F}(c_i)$ if the value implied by the current vector at line c_i in the corresponding faulty version of the circuit is identical to that for the fault-free version of the circuit. (Hence, loosely speaking, the dynamic list is analogous to a sparse matrix representation while the

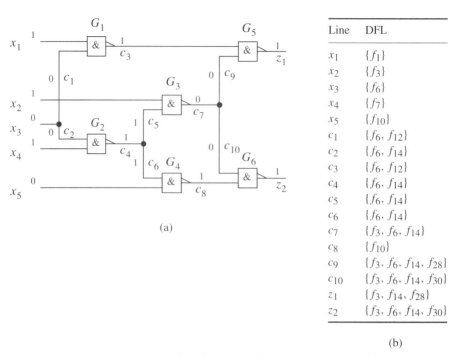

Line	DFL
x_1	$\{f_1\}$
x_2	$\{f_3\}$
x_3	$\{f_6\}$
x_4	$\{f_7\}$
x_5	$\{f_{10}\}$
c_1	$\{f_6, f_{12}\}$
c_2	$\{f_6, f_{14}\}$
c_3	$\{f_6, f_{12}\}$
c_4	$\{f_6, f_{14}\}$
c_5	$\{f_6, f_{14}\}$
c_6	$\{f_6, f_{14}\}$
c_7	$\{f_3, f_6, f_{14}\}$
c_8	$\{f_{10}\}$
c_9	$\{f_3, f_6, f_{14}, f_{28}\}$
c_{10}	$\{f_3, f_6, f_{14}, f_{30}\}$
z_1	$\{f_3, f_{14}, f_{28}\}$
z_2	$\{f_3, f_6, f_{14}, f_{30}\}$

(a)

(b)

Figure 3.27 Illustration of deductive fault simulation: (a) logic values implied by the vector at lines of the fault-free circuit, and (b) deductive fault lists for each line

fixed array of values used in parallel fault simulation corresponds to the complete description of a matrix.)

In parallel fault simulation, one bit (or two bits, if three-valued simulation is performed) is required at each line for each fault. This memory requirement is predictable and hence storage may be allocated statically, leading to faster access. In contrast, in deductive fault simulation, a fault may or may not appear in $\mathcal{F}$ at a line. For each fault that does appear in the DFL, an identification tag for the fault must be stored, which typically requires multiple bits. Hence, the storage required for deductive fault simulation may be smaller or greater than that for parallel fault simulation, depending on the average value of the fraction of all possible faults that appear in a list. Also, the number of faults that would appear in a list is unpredictable. Hence, storage must be allocated dynamically, leading to higher data access times due to the need to perform additional pointer operations.

The main run-time complexity advantage of deductive fault simulation is due to the fact that in this algorithm a fault effect is processed only at the circuit elements to whose inputs the corresponding fault effect propagates. In contrast, in parallel fault simulation, each fault effect is processed at each gate.

Extensions: Deductive fault simulation was originally implemented for *three-valued simulation* (Armstrong, 1972). One key modification required is the extension of DFLs to accommodate scenarios where the unspecified value may be assigned to a line either in the fault-free or a faulty circuit, while a fully-specified value is assigned in the other case. Also, the set operations used to compute the DFLs at each output of a circuit element must be extended to consider scenarios in which combinations of fully specified and unspecified values are implied at the gate's inputs.

Deductive fault simulation may also be extended to perform operations in an event-driven fashion. In that case, at a circuit line c_i, value $v(c_i)$ and DFL $\mathcal{F}(c_i)$ are computed only if (a) there is a change in the logic value at one or more fanins of c_i, or (b) there is a change in the DFL of one or more fanins of c_i, or (c) one or more of the local faults associated with line c_i have been dropped since the previous simulation.

Finally, deductive fault simulation may be extended to consider delays.

3.5.4   Concurrent fault simulation

In the concurrent fault simulation paradigm (Ulrich and Baker, 1973), the fault-free version of the circuit and each of its faulty versions are *concurrently* simulated for a given vector. **Concurrent** simulation of different circuit versions is characterized by the following.

1 The values implied by the given vector at a line c_i in each circuit version are computed at the same time.

2 In a particular faulty version of the circuit, if (a) the value at each fanin of line c_i is identical to the corresponding value in the fault-free circuit, **and** (b) the corresponding fault is not located at c_i or one of its fanins, then the value at line c_i in the particular faulty version is not computed explicitly. Instead the corresponding value computed in the fault-free version of the circuit is implicitly assumed to have been implied at the line in the particular faulty version of the circuit. (This is similar to how such faults are handled by a deductive fault simulation algorithm.)

3 For each faulty version of the circuit where the above conditions are not satisfied for line c_i, the value implied at the line is computed. However, the value implied at the line in each such faulty version is computed independently. (This is unlike parallel fault simulation, where bit-wise parallel operations are used to compute the values implied at a line in each of its versions in parallel. This is also unlike deductive fault simulation, where set operations on DFLs of the fanin of c_i are used to collectively compute the DFL at c_i.)

Like deductive fault simulation, the concurrent fault simulation paradigm also exploits the fact that the values implied by a vector at a large fraction of lines in a faulty version of a circuit are typically identical to those implied at the corresponding lines in the fault-free version of the circuit. However, the manner in which concurrent simulation takes advantage of this fact necessitates event-driven simulation.

3.5.4.1 Representation of circuit versions and values

In this simulation paradigm, one *complete* copy of the circuit is used to represent the fault-free version of the circuit, C. At each input and output of each gate, values implied by the current vector are stored. In addition, a version of circuit C^{f_l} is used for each fault f_l that still remains in the fault list. However, a faulty circuit version C^{f_l} may be *incomplete*, i.e., not all gates may exist in this copy. In C^{f_l}, a copy of gate G_j, denoted as $G_j^{f_l}$, exists if *one or both* of the following conditions are satisfied.

1 Fault f_l is local to the gate, i.e., it is associated with an input or output of the gate.
2 The value implied at at least one input or output of the gate is different from that implied at the corresponding line in the fault-free version of the circuit.

Example 3.9 Figure 3.28 illustrates the complete copy of the fault-free version of the circuit along with incomplete copies of the circuit corresponding to four faulty versions, each with a single SAF f_2: x_1 SA1, f_3: x_2 SA0, f_8: x_4 SA1, and f_{14}: c_3 SA1, respectively. (Faults f_2, f_3, f_8, and f_{14} are associated with the gates G_2, G_1, G_3, and G_3, respectively.) Since values at all lines are initialized to the unspecified value, x, only the copy of the gate with which each fault is associated exists in the corresponding faulty version of the circuit. For example, in circuit C^{f_3}, the partial copy of the circuit only contains a copy of gate G_1 in the form $G_1^{f_3}$. Since explicit copies of gates G_2, G_3, and G_4 do not exist for this faulty version, the corresponding copies in the fault-free circuit implicitly denote the copies of gates $G_2^{f_3}$, $G_3^{f_3}$, and $G_4^{f_3}$.   □

The interconnection between circuit elements is only stored in the fault-free copy of the circuit, C. The same interconnection is assumed to exist in each faulty circuit version.

Example 3.10 In Figure 3.28, the output of G_1 drives the inputs of gates G_2 and G_3 via the fanout system with stem c_1. In faulty circuit C^{f_3}, the output of gate $G_1^{f_3}$ is also assumed to drive the copies of gates $G_2^{f_3}$ and $G_3^{f_3}$ via a fanout system. Note that since copies of gates $G_2^{f_3}$ and $G_3^{f_3}$ are not explicitly included in C^{f_3} for this vector, the corresponding gates in the fault-free circuit, G_2 and G_3, respectively, denote these gates in C^{f_3}.   □

The values implied by the current vector at each input and output of each copy of a gate are stored in a manner that is identical to that used to store the corresponding values in the copy of the gate in C.

Another way to visualize the circuit copies and values stored in concurrent fault simulation is as follows. Start with a complete copy of the fault-free circuit as well as each faulty version of the circuit. Associate with each gate input and output a variable to store logic values implied at the line by the current vector. Next, reduce the complexity of the circuit representation by eliminating the interconnects between

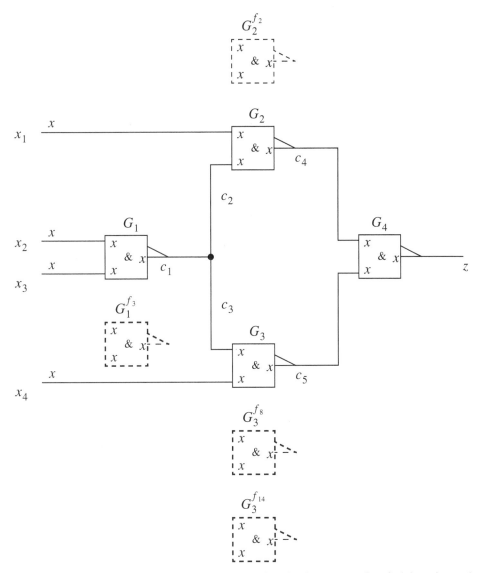

Figure 3.28 Initial status of an example circuit illustrating circuit representation, fault insertion, and values used by concurrent fault simulation

gates in each faulty version of the circuit, since these interconnects are identical to those between the corresponding gates in the fault-free version of the circuit. Finally, from each faulty circuit copy, remove each gate (a) that corresponds to a fault not associated with the gate, **and** (b) at which the current vector implies values at each of its inputs and output that are identical to those implied at the corresponding lines in the fault-free version of the circuit. This can be done since the behavior of as well as the values associated with each such copy of the gate are identical to those of the copy of the gate in the fault-free version of the circuit.

3.5.4.2 Fault insertion

In concurrent fault simulation, each single SAF fault is associated with a gate input or output. This can be achieved easily for all faults, except for a fault at a primary input that is a fanout stem (e.g., x_3 in the circuit shown in Figure 3.16) or a fault at a fanout branch that is a primary output. In such cases, a non-inverting buffer can be added at the fault site and the fault associated with that non-inverting buffer.

As mentioned earlier, once a fault f_l is associated with a gate G_j, then a copy $G_j^{f_l}$ of gate G_j is created in the corresponding faulty version of the circuit. Fault f_l is said to be a **local fault** of gate G_j. For each fault f_l that is local to gate G_j, the copy of gate $G_j^{f_l}$ is retained in circuit C^{f_l} until f_l is detected and dropped from the fault list.

During the computation of values implied by the vector at lines in C^{f_l}, whenever the value implied at the output of the gate with which the fault is associated is computed, the logic expression or the truth table that describes the behavior of the gate in the presence of fault f_l is used. For each of the other gates in C^{f_l}, if any, the expression or the truth table that describes the behavior of the fault-free version of the gate is used.

Example 3.11 In the circuit shown in Figure 3.28, in C^{f_3}, the expression or truth table that captures the behavior of a two-input NAND gate with an SA0 fault at its input x_2 is used to evaluate the value at output c_1 of gate G_1. For each of the other gates, expressions or truth tables describing the behavior of a fault-free two-input NAND gate are used. □

3.5.4.3 Events in concurrent fault simulation

Benefits of concurrent fault simulation can only be accrued when the algorithm is implemented in an event-driven manner. Next, we describe the types of events and their semantics.

Types of events: In concurrent fault simulation, conceptually, each version of the circuit is simulated concurrently. However, whenever possible, the values computed at lines in the fault-free version of the circuit are implicitly used to denote the values implied at the corresponding lines of each faulty circuit version.

In general, a set of events may occur at a circuit line c_i after the values implied at the line in each version of the circuit are computed. The events in this set represent the changes that occur in the value or the status of the line in each copy of the circuit. Three types of events may appear in a set.

Fault-free circuit event: This event occurs whenever the value implied at line c_i by the current vector is different from that implied at the line by the previous vector in the fault-free circuit. If the current vector implies value $v(c_i)$ at the line, then the event can be written as $(c_i, v(c_i))$. When line c_i at which the event occurs is clear from

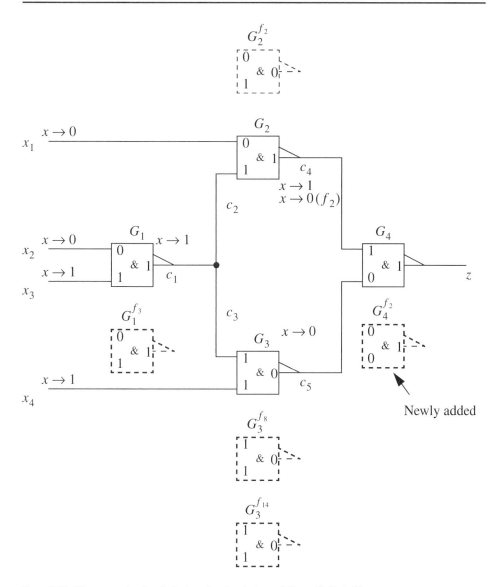

Figure 3.29 The example circuit during the simulation of $P_1 = (0, 0, 1, 1)$

the context, this event can be simply denoted by the values implied at the line by the previous and current vectors separated by an arrow ($\rightarrow$). An example of such an event at line c_4 of the circuit is shown in Figure 3.29, where the value at c_4 changes from the unspecified value to logic 1 in the fault-free circuit and is denoted by the notation $x \rightarrow 1$.

Faulty circuit event: If the value implied at line c_i in the faulty version of circuit C^{f_i} by the current vector is different from that implied by the previous vector, then such an event occurs. It can be written as $(c_i, v^{f_l}(c_i), f_l)$, where $v^{f_l}(c_i)$ is the value implied by the current vector at line c_i of the faulty version of the circuit with fault f_l. Again, when line c_i at which the event occurs is clear from the context, such an event will be simply denoted by the values implied at the line by the previous and current vectors separated by an arrow, along with the identification tag of the fault, f_l.

An example of such an event at line c_4 of the circuit is shown in Figure 3.29, where the value at the line changes from the unspecified value to logic 0 in the faulty version of the circuit with fault f_2. This event is denoted by notation $x \rightarrow 0\ (f_2)$.

If a vector implies identical values at a line c_i in the fault-free circuit and a faulty circuit with fault f_l, then c_i is said to be **invisible** in the circuit with fault f_l. Otherwise, the line is said to be **visible**. If a line c_i in the faulty circuit with fault f_l was invisible for the previous vector but becomes visible for the current vector, then the line is said to be **newly visible**.

Sometimes, a newly visible line also has an event. An example is line c_4 in Figure 3.29, where the line in the circuit with fault f_2 is made newly visible by the current vector. In this case, an $x \rightarrow 0$ event occurs at the line in C^{f_2}. All such cases of a line becoming newly visible are covered by the events in this category.

Divergence of the value at a line in a faulty circuit: Sometimes, even though the current vector implies the same value at a line c_i in circuit C^{f_l} as the previous vector, a change in the value implied at the line in the fault-free circuit can cause line c_i in C^{f_l} to become newly visible. When this occurs, the fact that the line has become newly visible is signaled to the gates in the fanout of line c_i via an event that is written as $(c_i, v^{f_l}(c_i), f_l)$. Again, when line c_i at which the event occurs is clear from the context, such an event will be simply denoted by the values implied at the line by the previous and current vectors separated by an arrow, along with the identification tag of the fault, f_l.

An example of such an event at line c_5 of the circuit is shown in Figure 3.30. The value at the line remains constant at logic 0 in the circuit with fault f_3, but the line is newly visible due to the fact that the current vector implies at line c_5 in the fault-free circuit a value that is different from that implied by the previous vector. This event is simply denoted by notation $0 \rightarrow 0\ (f_3)$ next to line c_5.

Event semantics: Each event at a line associated with a faulty circuit version is assumed to occur only in the corresponding faulty circuit version. In contrast, the event at the line associated with the fault-free circuit is assumed to occur not only at the line in the fault-free circuit but also in each faulty circuit in which (a) the line was invisible in the previous vector, **and** (b) the line continues to be invisible for the current vector.

Consider as an example event $x \rightarrow 0$ shown in Figure 3.29 at line c_5 that occurs due to the application of $P_1 = (0, 0, 1, 1)$ following the initialization of the values at circuit lines shown in Figure 3.28. In this case, the event is assumed to occur in the

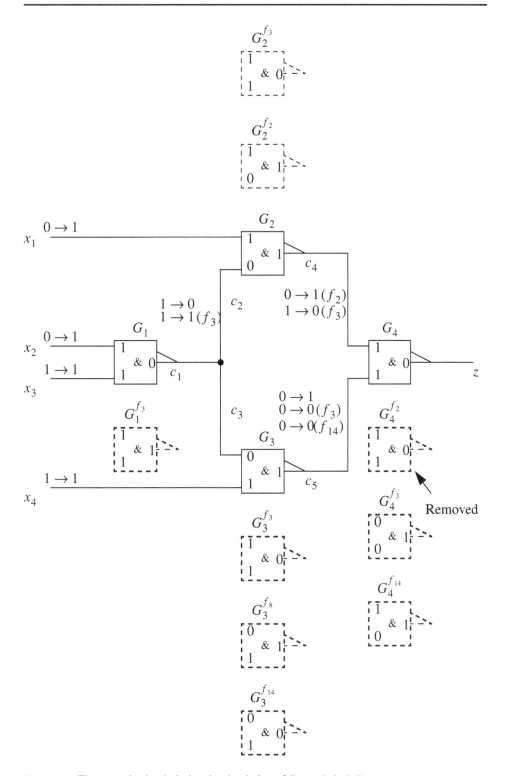

Figure 3.30 The example circuit during the simulation of $P_2 = (1, 1, 1, 1)$

fault-free circuit C as well as the faulty circuit versions C^{f_2}, C^{f_3}, C^{f_8}, and $C^{f_{14}}$, since, for each of these faulty circuit versions, line c_5 was invisible for the previous vector (initialization) and remains so for the current vector.

In contrast, event $x \rightarrow 1$ at line c_4 in the fault-free circuit following the application of P_1 is assumed to occur only in the fault-free circuit and versions C^{f_3}, C^{f_8}, and $C^{f_{14}}$, since the line becomes visible in faulty circuit C^{f_2} following the application of P_1.

3.5.4.4 Simulation algorithm

The main step of the concurrent simulation algorithm is comprised of concurrent computation of values implied by the vector at a line c_j in the fault-free as well each faulty circuit version.

Analysis of events at the inputs of the gate with output c_j**:** First, the events at each input of gate G with output c_j are analyzed. The fault-free circuit event at each input is applied to the corresponding input in the fault-free circuit. Also, an event at an input line in the circuit with fault f_l is applied to that input in the faulty circuit version. The above syntax and semantics are also used to identify the faulty circuit versions in which a fault-free circuit event is to be implicitly applied. Given the above conditions, this requires the determination of the visibility of the line for the previous and current vectors.

Determining visibility of an input for the previous vector: An input c_i of gate G in the circuit with fault f_l was invisible in the previous vector if either (a) a copy of gate G did not exist in the faulty circuit C^{f_l} for the previous vector, or (b) such a copy of the gate, G^{f_l}, existed for the previous vector, but the previous vector had implied identical values at line c_i in the fault-free circuit C and faulty circuit C^{f_l}.

Determining visibility of an input for the current vector: If an input c_i of gate G with output c_j is determined as being invisible for the previous vector in the circuit with fault f_l, its visibility for the current vector needs to be determined. This can be accomplished by simply checking whether the set of events at line c_i includes one designated for the faulty circuit C^{f_l}. If no such event exists, then input c_i in C^{f_l} remains invisible for the current vector.

An event at input c_i of gate G in the fault-free circuit is then slated for application to the line in each copy of gate G^{f_l} that exists in the circuit representation and at whose input c_i is determined as being invisible for the previous as well as current vectors. The event is also slated for application to the line in the fault-free version of the circuit. Finally, each event at a line c_i in C^{f_l} is slated for application to that line in the corresponding copy of the gate. Note that in some instances, this slates events for application to faulty copies of the gate that currently do not exist in the representation.

Example 3.12 Consider the circuit in Figure 3.29. The event at each input of gate G_1 is determined as being applicable to the fault-free circuit as well as to each faulty version. However, since an explicit copy of the gate exists only for the circuit with local fault f_3, the event at each input is slated for application to the corresponding inputs of each copy of the gate, namely G_1 and $G_1^{f_3}$. The events at the inputs of gates G_2 and G_3 are analyzed and are similarly slated. Next, consider the events at input c_4 of gate G_4. In this case, the fault-free circuit event is slated for application to line c_4 in the fault-free as well as each faulty circuit version, except the version with fault f_2, which is newly visible.

Since only the fault-free copy of gate G_4 existed for the previous vector (see Figure 3.28 and note that no faulty copies of G_4 existed at the end of initialization), the fault-free event $x \to 1$ is slated for application to input c_4 of G_4 in the fault-free circuit. Also, event $x \to 0$ (f_2) is slated to be applied to input c_4 of $G_4^{f_2}$, the copy of the gate in the circuit with fault f_2. $\square$

Note that, in the above example, the copy of gate $G_4^{f_2}$ does not exist in the circuit and will be added to the circuit when this event is executed, as described ahead.

Execution of events: Once the events have been analyzed, they are executed via their application to the circuit lines.

Determination of values implied by the current vector: For each copy of the gate that exists in the circuit, the value implied at the input by the current vector is determined as (a) the new value from the event slated at the input, or (b) if no event is slated at the input, then the value implied at the line by the previous vector. These values are then used to compute the value implied at output c_j of the gate. Any event that occurs in this process is then identified and recorded for transmission to the gates in its fanout.

Example 3.13 Consider gate G_2 in the circuit in Figure 3.29. As described above, events $x \to 0$ and $x \to 1$ are slated for inputs x_1 and c_2, respectively, in each copy of gate G_2. For each copy of the gate, the values implied by the current vector at these inputs are determined as 0 and 1, respectively. In the fault-free copy of gate G_2, the value implied at its output is computed as logic 1 by using its truth table. Since the faulty copy of the gate, $G_2^{f_2}$, corresponds to a local fault f_2, namely input x_1 SA1, the truth table for that particular faulty version of the gate is used to compute the value implied by the current vector at the output as logic 0.

Next, consider gate G_4 in the circuit in Figure 3.29. In the copy of the gate in the fault-free circuit, the values implied by the current vector at its inputs c_4 and c_5 are determined as 1 and 0, respectively. The value implied at its output is computed using the behavior of an appropriate version of the gate and determined to be 1. At this gate, events are also slated for application to the copy of the gate in the circuit with fault f_2, even though the gate does not exist in the circuit representation.   □

Divergence (creation of a copy of the gate): If an event is slated for a faulty version of a gate that currently does not exist in the circuit representation, then a **divergence** is said to occur at that gate. When divergence occurs, a copy of the gate is created and added to the corresponding faulty version of the circuit. The events slated at the input are then processed as in any other copy of the gate.

Example 3.14 Let us return to the above example of the events slated for application to the copy of gate G_4 with fault f_2 in Figure 3.29. Since the corresponding copy of the gate does not exist, this is an example of divergence. A copy of gate $G_4^{f_2}$ is hence added to the representation. The values at both the inputs of this copy of the gate are then determined to be logic 0. Since the corresponding fault f_2 is not local to G_4, the truth table of a fault-free NAND gate is used to compute the value at its output as logic 1.   □

Convergence (removal of a copy of a gate): Recall that a copy of a gate G in the faulty circuit C^{f_l} exists only if either (a) f_l belongs to the fault list and is local to gate G, or (b) if the value implied by the current vector at at least one input or output of G is different from that implied at the corresponding line in the fault-free circuit.

When values implied at the output of a copy of a gate G in a circuit with fault f_l are computed, the values at each of its inputs and output may turn out to be identical to those implied at the corresponding lines in the fault-free circuit. If fault f_l is not local to gate G, then the faulty copy of the gate, G^{f_l}, is said to **converge** and is dealt via removal of gate G^{f_l} from the circuit. On the other hand, if fault f_l is local to gate G, then the copy of gate G^{f_l} is retained in the circuit until fault f_l is dropped. Note that when gate G^{f_l} converges, fault f_l becomes invisible at the output of gate G.

Consider the scenario in Figure 3.30. First, consider the computation of values at line c_4, the output of gate G_2. The $0 \to 1$ event at its input x_1 is slated at that input for the fault-free as well as each faulty version of the gate, since the effect of none of the faults is visible at x_1 in the previous or the current vector. The $1 \to 0$ event at input c_2 is slated for the line in the fault-free as well as each faulty version, except the version with fault f_3 whose effect is visible at the line for the current vector. (Note that the effect of f_3 became visible at c_2 even though the value implied by the current vector at c_2 in the circuit with fault f_3 did not change.)

The slated events and the values implied at input lines are analyzed to determine the values implied at each input of each copy of the gate. In the fault-free copy, G_2, as well as in the copy in the circuit with fault f_2, $G_2^{f_2}$, the values at inputs x_1 and c_2 are determined as logic 1 and 0, respectively. In the first case, the truth table of a fault-free NAND gate is used to compute the value implied at the output as 1. The same value is computed for the output of gate $G_2^{f_2}$, using the truth table of a copy of the NAND gate with an SA1 fault at its input x_1.

Since the effect of f_3 has become visible at input c_2 of gate G_2 and no copy of G_2 existed in the circuit with fault f_3 for the previous vector, a divergence has occurred. A new copy of the gate is created and added to the circuit description as $G_2^{f_3}$. The events slated at inputs x_1 and c_2 of $G_2^{f_3}$, namely $0 \rightarrow 1$ at x_1 (which was slated for the input in each copy of gate G_2) and $1 \rightarrow 1$ at c_2 (which was slated only for the input in copy $G_2^{f_3}$), are analyzed to determine that logic 1 is implied by the current vector at both the inputs of $G_2^{f_3}$. Since fault f_3 is not local to G_2, the truth table of a fault-free two-input NAND gate is used to determine that logic 0 is implied at output c_4 of $G_2^{f_3}$.

Finally, an example of convergence can be seen at G_4 in Figure 3.30. When the values implied at output z of gate G_4 are computed, it is discovered that the values implied at the inputs and output of $G_4^{f_2}$ are identical to those implied at the corresponding lines in the copy of the gate in the fault-free circuit. Since f_2 is not local to gate G_4, this signals a convergence of values and the copy of the gate in the circuit with fault f_2, $G_4^{f_2}$, is removed.

Computation of events at the output of a gate: The events at the output of each copy of the gate are computed during the execution of events. If the value implied by the current vector at the output of a copy of a gate is different from that implied by the previous vector, then an event is generated at that output. In addition, if the value implied at the copy of the gate in the fault-free circuit changes, then the value implied at the output of each copy of the gate is compared with that implied at the corresponding output in the fault-free circuit. This process is used to identify whether the effect of a fault has become newly visible in a faulty copy of the gate. When such outputs are found, then appropriate events are added to the set of events at the output.

An example of such an event can be seen at output c_1 of gate G_1 in Figure 3.30. Even though the application of vector P_2 does not change the value implied at c_1 in the circuit with fault f_3, a change in the value implied at c_1 in the fault-free circuit causes the effect of fault f_3 to become newly visible at line c_1. Event $1 \rightarrow 1$ (f_3) is hence added to the set of events at c_1 to signal the newly visible fault f_3.

For each vector, events caused at the primary inputs are added to the event list. The above steps are then used to process events and schedule resulting events. In this process, new copies of gates may be created and old ones may be removed. Simulation is complete when the event list becomes empty.

3.5.4.5 Comparison

The main advantage of the concurrent fault simulation approach is that it can be easily extended to cases where the results of fault simulation depend on timing of events, especially when timing may vary from one copy of the circuit to another. Note that such a simulation cannot be handled by most versions of the other simulation algorithms described above.

3.5.5   Critical path tracing

The critical path tracing paradigm differs from the paradigms described earlier in two main ways. First, the paradigm *implicitly* targets all faults within certain parts of a circuit. Hence, the complexity of fault simulation under this paradigm is *potentially* independent of the number of faults. This is in contrast to all the paradigms discussed earlier, where the fault simulation complexity is roughly proportional to the number of faults. Second, in its strictest form, the paradigm can only be applied to **fanout-free** circuits, i.e., circuits that contain no fanouts.

Since almost all useful circuits contain fanouts, critical path tracing must be used in conjunction with some other simulation approach. The use of critical path tracing is then confined mainly to the *fanout-free regions* of the circuit. We begin with a description of the critical path tracing approach for fanout-free circuits and an explanation of why its application is limited largely to fanout-free regions of a circuit. This will be followed by descriptions of complete fault simulation approaches obtained by combining critical path tracing with other simulation methods.

3.5.5.1 Key concepts

Consider a fanout-free circuit comprised solely of gates, i.e., one that does not contain any fanout system, and where the output of a gate either drives an input of one other gate or is a primary output. Each of these gates may have an arbitrary number of inputs and each gate's output may implement an arbitrary logic function of its inputs.

One key characteristic of a fanout-free circuit is that the effect of a single SAF at any line can propagate to no more than one input of any gate. For example, the effect of the single SA0 fault at input x_1 in the circuit of Figure 3.31 can only propagate to input c_1 of gate G_6 and c_6 of gate G_8; it cannot propagate to any of the inputs of any other gate in the circuit.

The above observation forms the basis of the critical path tracing method that considers implicitly all single SAFs in a fanout-free circuit (Hong, 1978; Abramovici *et al.*, 1984). Critical path tracing is performed in two passes. The first pass is logic simulation, in which the given vector, $P = (p_1, p_2, \ldots, p_n)$, is simulated and the value, $v(c_i)$, implied at each line c_i in the fault-free circuit is determined. Any logic simulation algorithm may be used in this phase. The second pass entails a backward, breadth-first traversal of circuit as described next.

A line c_i in a circuit C is said to be **critical** (Wang, 1975) for a vector P if and only if the vector detects the SA$\overline{w}$ fault at the line, where $w \in \{0, 1\}$ is the value implied at the line by the vector, i.e., $w = v(c_i)$. The criticality of a line c_i, denoted as $Cr(c_i)$, is

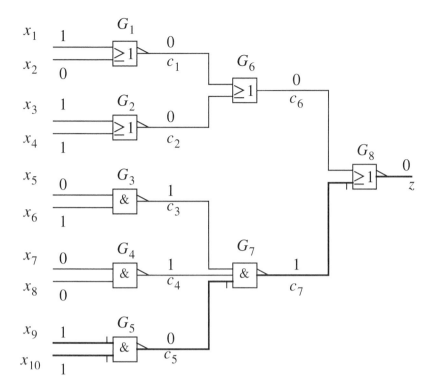

Figure 3.31 An example of critical path tracing on an example fanout-free circuit

assigned a value as follows:

$$Cr(c_i) = \begin{cases} 1, & \text{if vector } P \text{ detects the SA}\overline{w} \text{ fault at } c_i, \\ 0, & \text{otherwise.} \end{cases} \tag{3.3}$$

In other words, a line c_i is critical if and only if $Cr(c_i) = 1$.

The objective of backward traversal is to identify the criticality of each line in the circuit for the given vector. This is accomplished by first determining the *sensitivity* of each input of each gate. The criticality of the gate output and the sensitivity of each of its inputs are then used to determine the criticality of each of the gate's inputs.

An input c_{i_l} of a gate G with inputs $c_{i_1}, c_{i_2}, \ldots, c_{i_l}, \ldots, c_{i_\alpha}$ and output c_j is said to be **sensitive** (Abramovici *et al.*, 1984) for a vector P if the vector implies values at the gate inputs and output such that if the value at input c_{i_l}, $v(c_{i_l})$, is complemented to $\overline{v(c_{i_l})}$, while the values at other gate inputs remain unchanged, then the value at c_j is complemented to $\overline{v(c_j)}$. The sensitivity of a gate input c_{i_l} for the current vector is denoted as $sens(c_{i_l})$, which is assigned a value as follows.

$$sens(c_{i_l}) = \begin{cases} 1, & \text{if } c_{i_l} \text{ is sensitive for the vector,} \\ 0, & \text{otherwise.} \end{cases} \tag{3.4}$$

The value at input c_{i_l} of a gate G may change in the manner described above due

Table 3.11. *Conditions under which a vector P sensitizes input c_{i_l} of a gate with inputs $c_{i_1}, c_{i_2}, \ldots, c_{i_l}, \ldots, c_{i_\alpha}$*

Gate	Values that make input c_{i_l} sensitive					
	$v(c_{i_1})$	$v(c_{i_2})$	$\cdots$	$v(c_{i_{l-1}})$	$v(c_{i_{l+1}})$	$\cdots$	$v(c_{i_\alpha})$
AND	1	1	$\cdots$	1	1	$\cdots$	1
NAND	1	1	$\cdots$	1	1	$\cdots$	1
OR	0	0	$\cdots$	0	0	$\cdots$	0
NOR	0	0	$\cdots$	0	0	$\cdots$	0
XOR	x	x	$\cdots$	x	x	$\cdots$	x

to the presence of a single SAF at (a) line c_{i_l}, or (b) at a line in the transitive fanin of c_{i_l}, whose fault effect has successfully propagated to line c_{i_l} for the current vector. If c_{i_l} is sensitive, then the effect of either of the above two faults would propagate to the output of gate G, i.e., to line c_j.

In a fanout-free circuit, since a single SAF can affect no more than one gate input, only the sensitivity of each gate input needs to be computed. This assumes that the values at other gate inputs remain constant at those implied by vector P in the fault-free circuit. These sensitivity values can be computed easily for any arbitrary gate and are especially easy to identify for primitive gates where a gate input is sensitive for a vector if and only if the vector implies the gate's non-controlling value at each of its other inputs. Furthermore, each input of an XOR gate is sensitive for any vector. Table 3.11 shows the conditions under which a vector P makes input c_{i_l} of a gate sensitive.

If the sensitivity of a gate input c_{i_l}, $sens(c_{i_l})$, and the criticality of the gate's output c_j, $Cr(c_j)$ are known, then the criticality of c_{i_l} is given by

$$Cr(c_{i_l}) = sens(c_{i_l})Cr(c_j). \tag{3.5}$$

3.5.5.2 The critical path tracing procedure

The procedure for critical path tracing for a fanout-free circuit can now be written (Abramovici *et al.*, 1984).

Procedure 3.7 [CriticalPathTracingFanoutFree( )]

1 For the given vector, $P = (p_1, p_2, \ldots, p_n)$, perform logic simulation and determine, for each line c_i, the value, $v(c_i)$, implied at the line in the fault-free version of the circuit.

2 *Initialization:* For each line c_i, assign $Cr(c_i) = 0$.

3 Let z be the primary output of the fanout-free circuit.
   (a) Set $Cr(z) = 1$.
   (b) Set $Q = (z)$. /* Q is an ordered *task list* that contains lines to be processed */.

4 While there exists a line c_j in list Q

(a) Remove c_j from Q.

(b) For each input c_{i_l} of gate G with output c_j

    i.   Determine $sens(c_{i_l})$.

    ii. $Cr(c_{i_l}) = sens(c_{i_l})Cr(c_j)$.

    iii. If $Cr(c_{i_l}) = 1$ and c_{i_l} is not a primary input, then add c_{i_l} to the task list Q.

The above procedure determines the criticality, $Cr(c_j)$, of each line c_j for a given vector, P. If P implies a logic value $w \in \{0, 1\}$ at line c_j, i.e., $v(c_j) = w$, then the SA$\overline{w}$ fault at line c_j is detected by P if and only if $Cr(c_j) = 1$. Otherwise, neither SAF at line c_j is detected by the vector.

Example 3.15 The above procedure can be illustrated by using the fanout-free circuit shown in Figure 3.31. The procedure begins by performing logic simulation for the given vector $P = (1, 0, 1, 1, 0, 1, 0, 0, 1, 1)$ to determine the value implied at each circuit line in the fault-free circuit. The values obtained are shown in the figure. Next, each line is initialized as not being critical, i.e., for each line c_j, $Cr(c_j) = 0$. The output is then marked as being critical, via the assignment $Cr(z) = 1$. Line z is then added to the ordered task list Q.

Line z is then removed from Q and input c_6 of the gate with z as the output, namely G_8, is considered. An analysis reveals that c_6 is not sensitive, and hence $sens(c_6) = 0$ and $Cr(c_6) = 0$. The other input of G_8, c_7, is then examined. In this case, $sens(c_7) = 1$ and hence $Cr(c_7) = sens(c_7)Cr(z) = 1$. Since c_7 is found to be critical, it is added to Q. Next, line c_7 is removed from Q and its inputs c_3, c_4, and c_5 are analyzed successively. Only c_5 is found to be sensitive and critical. c_5 is hence added to Q only to be retrieved from the list in the next step. Inputs of G_5, x_9 and x_{10}, are then analyzed successively. Both are found to be sensitive and hence critical. The procedure ends by providing $Cr(x_9) = 1$, $Cr(x_{10}) = 1$, $Cr(c_5) = 1$, $Cr(c_7) = 1$, and $Cr(z) = 1$. In Figure 3.31, a notch is placed at each gate input that is identified as sensitive, and each line that is identified as being critical is highlighted.

The procedure hence reports that vector $(1, 0, 1, 1, 0, 1, 0, 0, 1, 1)$ detects the faults x_9 SA0, x_{10} SA0, c_5 SA1, c_7 SA0, and z SA1.   □

3.5.5.3 Difficulty of applying critical path tracing to circuits with fanouts

A circuit with reconvergent fanouts can be viewed as a number of *fanout-free regions* interconnected using fanout systems. A circuit can be decomposed into a set of **maximal fanout-free regions** (FFR) (Abramovici *et al.*, 1984) by disconnecting each branch of each fanout system from the corresponding stem to divide the circuit into one or more disconnected subcircuits. Each such subcircuit is called a maximal FFR. The maximal FFRs of the circuit shown in Figure 3.32 obtained in this manner are shown in Figure 3.33. Clearly, each FFR has a single output which is either the stem of a fanout system or a primary output. Hence, each FFR can be uniquely identified by

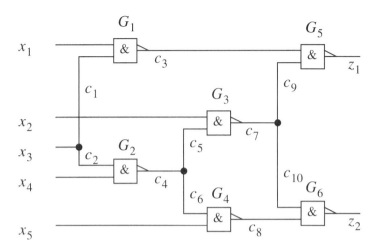

Figure 3.32 An example circuit to illustrate critical path tracing

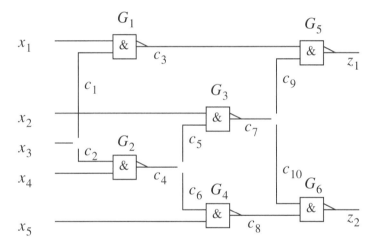

Figure 3.33 An example of identification of maximal FFRs of a circuit with reconvergent fanouts

its output. For example, the FFR with output z_2 will be referred to as FFR(z_2). Once a circuit with fanout systems is divided into FFRs, critical path tracing can be applied to each FFR.

The main difficulty in the application of critical path tracing to a circuit with fanouts is that the criticality of the stem of a fanout system does not bear any simple universal relationship to the criticality of its branches. The following example illustrates this fact.

Example 3.16 Consider the circuit and vector shown in Figure 3.34(a) (Abramovici *et al.*, 1984). This circuit can be divided into two FFRs, one comprised only of primary input x_2 and the other of the remaining circuit in which branches c_1 and c_2 of the fanout system in the original circuit are viewed as independent lines detached from their stem, x_2. Procedure CriticalPathTracingFanoutFree( ) can be applied to the larger of the above two FFRs and the criticality of all lines in the circuit, with the exception of stem x_2 of the fanout system, can be identified as shown in the figure. In this case, one of the branches, c_1, of the fanout system is critical. The reader can verify that fault f_3: x_2 SA0 is not detected by this vector. This is due to the fact that the effect of the SA0 fault at x_2 propagates to c_1 as well as c_2. The fault effect at c_2 propagates to c_3 and then to c_5, while the fault effect at c_1 propagates to c_4. Hence, in the version of the circuit with the single SAF f_3, the values at inputs c_4 and c_5 of gate G_4 are logic 0 and 1, respectively. Since the values at the two inputs of gate G_4 have been complemented by the presence of the fault, the results of critical path tracing at the gate cannot be used to deduce whether or not the fault is detected by the vector. This is due to the fact that the analysis performed during critical path tracing is valid only for cases where the fault effect alters the value at no more than one input of a gate. In this case, fault f_3 alters the value at both inputs of G_4 canceling each other's effect. Hence, the fault effect does not propagate to z.

In contrast, for the same circuit the same branch, c_1, of the fanout is critical for vector $(1, 1, 0)$, as shown in Figure 3.34(b). In this case, however, fault f_3 is detected.

Finally, consider the circuit shown in Figure 3.35. For the vector shown in the figure, neither of the branches of the fanout is critical. The reader can verify that the SA1 fault at x_2 is detected by this vector, however.   □

The above example illustrates that a fanout stem may sometimes be not critical even when one or more of its branches are critical, while at other times it may be critical even when none of its branches is critical. Clearly, no simple universal relationship can be assumed for the criticality of a stem in terms of the criticality of its branches.

3.5.5.4 Complete fault simulation based on critical path tracing

The inability of critical path tracing to determine criticality of fanout stems must be addressed to obtain a *complete* critical path tracing based fault simulation algorithm that is applicable to circuits with fanouts. The process of identifying the criticality of a fanout stem is called **stem analysis** (Abramovici *et al.*, 1984) and is the key problem that needs to be solved for the development of such an algorithm.

One simple way to perform stem analysis at a stem c_i, at which current vector P implies a value $w = v(c_i)$, is to use any of the other fault simulation algorithms to perform *explicit fault simulation* to identify whether fault $SA\overline{w}$ at stem c_i is detected by the current vector. Next, we describe techniques that attempt to exploit characteristics of the structure of the circuit to reduce the complexity of such explicit fault simulation.

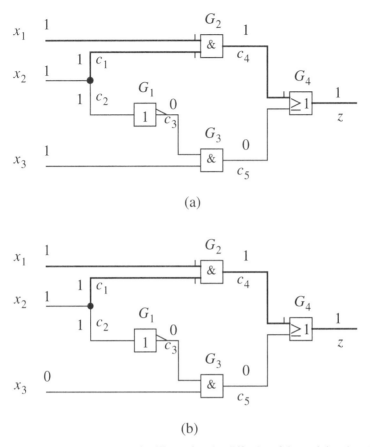

Figure 3.34 Example scenarios illustrating the difficulty of determining the criticality of a stem in terms of criticality of its fanout branches (Abramovici *et al.*, 1984) (© 1984 IEEE)

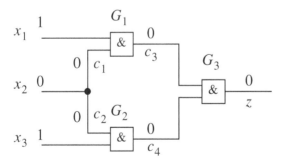

Figure 3.35 Another example scenario illustrating the difficulty of determining the criticality of a stem in terms of criticality of its fanout branches

Independent branches of a fanout system: One of the scenarios under which the criticality of a branch of a fanout system may imply the criticality of the stem is when the fanout branch is *independent* (Brglez, 1985; Antreich and Schulz, 1987).

Let $Trfan(c)$ be the set of primary outputs of a circuit that are in the transitive fanout of line c in the circuit. A branch c_{j_l} of a fanout system with stem c_i and branches c_{j_1}, $c_{j_2}, \ldots, c_{j_l}, \ldots, c_{j_\beta}$ is said to be **independent** if

$$Trfan(c_{j_l}) \bigcap \left[\bigcup_{\gamma=1,2,\ldots,\beta;\gamma \neq l} Trfan(c_{j_\gamma}) \right] = \{\}.$$

In other words, outputs in the transitive fanout of branch c_{j_l} are distinct from those in the transitive fanout of any of the other branches, $c_{j_1}, c_{j_2}, \ldots, c_{j_{l-1}}, c_{j_{l+1}}, \ldots, c_{j_\beta}$, of the fanout system. This implies that any line in the transitive fanout of c_{j_l} is not in the transitive fanout of any of the other branches. In such a case, if branch c_{j_l} is critical, then stem c_i of the fanout system is also critical. This can be demonstrated through the following observations. First, any fault effect that propagates to c_i does propagate to c_{j_l}. Second, the fact that branch c_{j_l} is critical implies that the appropriate SAF at c_{j_l} is detected at one or more of the circuit outputs. Third, the only difference between the propagation of the fault effect at stem c_i and that of an appropriate fault at branch c_{j_l} is that the effects of the former are also propagated to the other branches, $c_{j_1}, c_{j_2}, \ldots, c_{j_{l-1}}, c_{j_{l+1}}, \ldots, c_{j_\beta}$, of the fanout system. However, the fact that c_{j_l} is independent implies that the propagation of fault effects at $c_{j_1}, c_{j_2}, \ldots, c_{j_{l-1}}$, $c_{j_{l+1}}, \ldots, c_\beta$, does not affect in any way the propagation of the effect at c_{j_l}. Hence, stem c_i of a fanout system is critical, i.e., $Cr(c_i) = 1$, if any of its independent fanout branches c_{j_l} is critical, i.e., if $Cr(c_{j_l}) = 1$.

The above result shows that there is no need for explicit fault simulation for the appropriate fault at the stem. Since explicit fault simulation can be potentially computationally expensive, this can help reduce the overall complexity of fault simulation.

Consider the circuit in Figure 3.32. Each branch c_9 and c_{10} of the fanout system with stem c_7 is independent. Hence, in this circuit, stem c_7 is critical if and only if one or more of its branches is critical. Therefore, explicit fault simulation is **not** required to determine the criticality of stem c_7. In contrast, none of the branches of the other two fanout systems in this circuit is independent.

Dominators in a single-output cone: Some critical path tracing based algorithms simulate each output cone of the circuit separately. In such a case, the concept of *dominator* (Kirkland and Mercer, 1987; Antreich and Schulz, 1987; Tarjan, 1974) of a stem can help reduce the complexity of explicit fault simulation during stem analysis.

A line c_j in the cone of output z_l of a circuit is said to be a **dominator** (Tarjan, 1974) of another line c_i in the same cone if all paths from line c_i to output z_l pass via line c_j. Let $Dom(c_i, z_l)$ denote the set of all lines that are dominators of c_i in the cone of output z_l. The fact that c_j is a dominator of c_i in cone z_l is denoted as

$c_j \in Dom(c_i, z_l).$

If paths from c_i to one of its dominators $c_j \in Dom(c_i, z_l)$ do not contain any other dominator of line c_i in the cone of output z_l, then c_j is said to be the **immediate dominator** of c_i in that cone. Whenever this is the case, the relation between c_i and c_j is represented as

$$c_j = Dom_{imm}(c_i, z_l).$$

A line c_i is said to be **critical at another line** c_j for a given vector P if the effect of $SA\overline{w}$ fault at c_i, where w is the value implied at line c_i by the vector in the fault-free circuit, is propagated to line c_j. This can be denoted by using the notation $cr(c_i, c_j)$ which is assigned logic 1 if this is the case and logic 0 otherwise.

Note that

$$Cr(c_i) = cr(c_i, z_1) + cr(c_i, z_2) + \cdots cr(c_i, z_m).$$

Using this notation, explicit fault simulation for the analysis of stem c_i can be described as evaluation of $cr(c_i, c_j)$, for one line c_j after another. Consider a scenario where (a) $c_j \in Dom(c_i, z_q)$, (b) critical path tracing based simulator has been applied to the cone of output z_q and has computed the criticality of c_j at z_q, $cr(c_j, z_q)$, and (c) explicit fault simulation is being performed to determine the criticality of stem c_i. Explicit fault simulation determines, line by line for each line c_l in the transitive fanout of c_i, the value of $cr(c_i, c_l)$, which denotes whether the effect of the fault at c_i propagates to that line. In general, explicit simulation of a fault at stem c_i may need to be carried out until the effect of the fault propagates to a primary output. However, if c_j is an immediate dominator of c_i in the cone of output z_q, then

$$cr(c_i, z_q) = cr(c_i, c_j)cr(c_j, z_q).$$

When this is the case, the effect of the fault at c_i reduces to a single fault effect at c_j, its dominator in the cone of z_q. Hence, when critical path tracing is performed on each output cone separately, i.e., when value $cr(c_j, z_q)$ is computed for each line c_j and output z_q, explicit simulation of a stem fault must only be performed until the fault effect reaches (or fails to reach) its immediate dominator (Harel *et al.*, 1987) in the output cone being simulated. This helps reduce the number of circuit lines where explicit fault simulation must be performed. Since the complexity of explicit fault simulation is proportional to the number of lines in the circuit region simulated, this helps reduce the complexity of explicit fault simulation.

Stem regions: As described above, the immediate dominator of a fanout stem may be used to reduce the complexity of explicit fault simulation only if the criticality of each line at each output is computed. This is equivalent to performing fault simulation of a circuit one output cone at a time. This can be computationally wasteful for circuits with many output cones that share a large number of gates. The notion of a *stem region* (Maamari and Rajski, 1988, 1990a) is defined to similarly reduce the complexity of

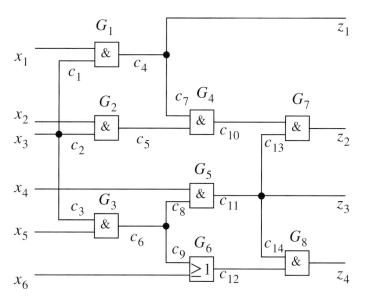

Figure 3.36 A second example circuit to illustrate the stem region of a stem (Maamari and Rajski, 1990a) (© 1990 IEEE)

explicit fault simulation performed for stem analysis in scenarios where all circuit cones are simulated simultaneously.

The stem region of a fanout stem c_i is defined using the following series of definitions adapted from Maamari and Rajski (1990a).

If each branch of a fanout system is independent, then its stem is said to be a **non-reconvergent** fanout stem, otherwise it is said to be **reconvergent**.

A gate G is said to be a **primary reconvergence gate** of a fanout stem c_i if there exist at least two disjoint paths from c_i to output line c_j of gate G. Two paths from c_i to c_j are said to be **disjoint** if they do not share any lines other than c_i and c_j.

Example 3.17 In the circuit in Figure 3.32, x_3 and c_4 are each reconvergent fanout stems while c_7 is a non-reconvergent fanout stem. For x_3, gate G_5 is the only primary reconvergence gate, since two disjoint paths, (x_3, c_1, c_3, z_1) and $(x_3, c_2, c_4, c_5, c_7, c_9, z_1)$, exist from stem x_3 to output z_1 of gate G_5. Gate G_6 is the only primary reconvergence gate for fanout stem c_4.

In the circuit in Figure 3.36 (Maamari and Rajski, 1990a), x_3 is a reconvergent stem. Gates G_4 and G_7 are each a primary reconvergence gate for stem x_3. Stem c_6 is the other reconvergent stem in this circuit with a single primary reconvergence gate G_8. The other two fanout stems, c_4 and c_{11}, are non-reconvergent stems.

Finally, in the circuit in Figure 3.37 (Lee and Ha, 1991), both fanout stems, x_3 and c_5, are reconvergent. Each has a single primary reconvergence gate; gate G_3 for stem x_3 and gate G_6 for stem c_5.   □

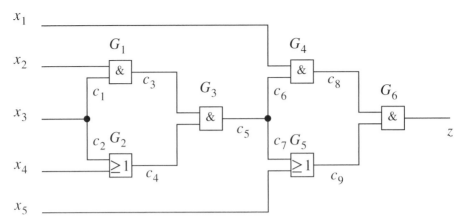

Figure 3.37　A third example circuit to illustrate the stem region of a stem (Lee and Ha, 1991) (© 1991 IEEE)

The notion of *secondary reconvergence* gates of a fanout stem will be defined next. However, first reconvergent stems need to be classified into *narrow* and *wide*.

If no reconvergent fanout stems exist between a reconvergent stem and any of its primary reconvergence gates, then the stem is called a **narrow reconvergent stem**. Otherwise, the stem is called a **wide reconvergent stem**.

Example 3.18　In the circuit in Figure 3.32, reconvergent stem x_3 is a wide reconvergent stem, since reconvergent stem c_4 lies on the path from x_3 to its primary reconvergence gate G_5. On the other hand, reconvergent stem c_4 is narrow, since the only fanout stem that lies between c_4 and its primary reconvergence gate, G_6, is c_7, which is a non-reconvergent stem. Similarly, in the circuit in Figure 3.36, x_3 is a wide reconvergent stem and c_6 is a narrow reconvergent stem. Finally, in Figure 3.37, both reconvergent stems, x_3 and c_5, are narrow. □

If a narrow reconvergent stem c_j is located between a stem c_i and any of its primary reconvergence gates, then all the primary reconvergence gates of c_j that are not primary reconvergence gates of c_i are called **secondary reconvergence gates** of c_i.

If a wide reconvergent stem c_j is located between a stem c_i and any of its primary reconvergence gates, then all the primary and secondary reconvergence gates of c_j that are not primary reconvergence gates of c_i are **secondary reconvergence gates** of c_i.

The above sequence of definitions can be used to identify all primary and secondary reconvergence gates of each reconvergent stem in the following order. First, all primary reconvergence gates can be identified for each reconvergent stem. Second, wide reconvergent stems whose reconvergence paths only contain narrow reconvergent stems are analyzed and the above definitions used to identify all secondary reconvergent gates of each such stem. Third, wide reconvergent stems whose reconvergence paths contain narrow stems and/or only those wide reconvergent stems that are analyzed in the above steps are processed and all their secondary reconvergence gates identified. The above iterative process can be repeated until all primary and secondary reconvergence gates are identified for each reconvergent stem.

Once all reconvergence gates of a fanout stem are identified, its *stem region* is identified using the notion of *closing reconvergence gates* of the stem. A primary or secondary reconvergence gate of a stem c_i, whose transitive fanout does not contain another primary or secondary reconvergence gate of c_i, is said to be a **closing reconvergence gate** of stem c_i.

The **stem region** of a reconvergent fanout stem c_i is the set of circuit lines and circuit elements that are (a) in the transitive fanout of c_i, **and** (b) in the transitive fanin of one or more of its closing reconvergence gates. In addition, all outputs of any circuit element in the stem region are also said to belong to the stem region.

Any output line of a stem region that is not an input of any other circuit element in the stem region is said to be an **exit line** of the stem region.

Example 3.19 In the circuit in Figure 3.32, stem x_3 has a single primary reconvergence gate G_5. Stem c_4 is a narrow reconvergent stem with a single primary reconvergence gate G_6. Since reconvergence paths from x_3 contain c_4, the primary reconvergence gate G_6 of c_4 is a secondary reconvergence gate for x_3. The closing reconvergence gates of stem x_3 are G_5 and G_6. The stem region of x_3 is comprised of lines $x_3, c_1, c_2, c_3, c_4, c_5, c_6, c_7, c_8, c_9, c_{10}, z_1$, and z_2; gates G_1, G_2, G_3, G_4, G_5, and G_6; and fanout systems with stems x_3, c_4, and c_7. The exit lines in this case are z_1 and z_2. Reconvergent stem c_4 has a single reconvergence gate, G_6. Its stem region is comprised of lines $c_4, c_5, c_6, c_7, c_8, c_9, c_{10}$, and z_2; gates G_3, G_4, and G_6; and fanout systems with stems c_4 and c_7. Note that line c_9 is the output of the fanout system with stem c_7. Since this stem belongs to the stem region of stem c_4, c_9 is said to belong to the stem region, despite the fact that it is not in the transitive fanin of the closing reconvergence gate G_6 of stem c_4. The exit lines of c_4 are lines c_9 and z_2.

Similarly, the stem region of stem x_3 in Figure 3.36 are the lines $x_3, c_1, c_2, c_3, c_4, c_5, c_6, z_1, c_7, c_8, c_9, c_{10}, c_{11}, c_{12}, c_{13}, z_3, c_{14}, z_2$, and z_4; gates G_1 to G_8; and fanout systems with stems x_3, c_4, c_6, and c_{11}. The exit lines of the stem region of x_3 are z_1, z_2, z_3, and z_4. The stem region of c_6 is comprised of lines $c_6, c_8, c_9, c_{11}, c_{12}, c_{13}, z_3, c_{14}$, and z_4; gates G_5, G_6 and G_8; and fanout systems with stems c_6 and c_{11}. The exit lines of the stem region of c_6 are c_{13}, z_3 and z_4.

Finally, in Figure 3.37, the stem region of stem x_3 is comprised of lines x_3, c_1, c_2, c_3, c_4, and c_5; gates G_1, G_2 and G_3; and the fanout system with stem x_3. The exit line of stem x_3 is c_5. The stem region of c_5 is comprised of lines c_5, c_6, c_7, c_8, c_9 and z; gates G_4, G_5 and G_6; and the fanout system with stem c_5. The exit line of this stem region is z. $\qquad\square$

Let $c_{e_1}, c_{e_2}, \ldots, c_{e_q}$ be the exit lines of the stem region of a reconvergent fanout stem c_i in a circuit. It has been shown in Maamari and Rajski (1990a) that no line in the circuit is in the transitive fanout of more than one of these exit lines. Hence, if the effect of the fault at a stem propagates to one of its exit lines c_{e_l}, then further propagation of the fault effect from c_{e_l} is not affected by the presence or absence of the fault effect at any of the other exit lines of the stem region. Therefore, the criticality of a stem c_i can be obtained in terms of its criticality at each of its exit lines and the criticality of the exit lines themselves as

$$Cr(c_i) = cr(c_i, c_{e_1})Cr(c_{e_1}) + cr(c_i, c_{e_2})Cr(c_{e_2}) + \cdots$$
$$+ cr(c_i, c_{e_l})Cr(c_{e_l}) + \cdots + cr(c_i, c_{e_q})Cr(c_{e_q}),$$

where '+' in the above expression denotes a logical OR operation.

Recall that explicit fault simulation can be used to determine criticality $cr(c_i, c_j)$ of a stem c_i at another circuit line c_j. The above relation based on the stem region and exit lines of stem c_i provides an opportunity to reduce the explicit fault simulation complexity in two main ways. First, explicit fault simulation needs to be performed only until its criticality is determined at each of its exit lines, $c_{e_1}, c_{e_2}, \ldots, c_{e_l}, \ldots, c_{e_q}$. Furthermore, if the criticality of c_i is determined at an exit line c_{e_l} and it is found that $cr(c_i, c_{e_l}) = 1$, then if $Cr(c_{e_l}) = 1$, stem c_i can be declared critical. Whenever this happens, explicit fault simulation for stem c_i can be immediately terminated.

Each of the above techniques requires the analysis of the circuit structure for identification of lines with certain characteristics. Since none of these properties depend upon the values implied by the current vector or specific target faults, the circuit structure needs to be analyzed only once in a *pre-processing* phase before the simulation begins to identify independent branches of each fanout system, the immediate dominator of each stem for each output, and/or the stem region and exit lines of each reconvergent stem. The properties identified during the pre-processing step can then be used during the simulation of each vector. We will loosely refer to such techniques to reduce explicit fault simulation complexity as **static**, since they exploit properties identified during pre-processing.

It should be noted that further reduction in explicit fault simulation may be realized by taking advantage of the values computed during the simulation of a particular vector. An example of such a reduction in explicit fault simulation complexity is when stem c_i is found to be critical at one of its exit lines (i.e., $cr(c_i, c_{e_l}) = 1$) and that particular exit line is critical (i.e., $Cr(c_{e_l}) = 1$). Such reductions in complexity of explicit fault simulation will be loosely referred to as **dynamic**, since they exploit properties of combination of values implied by a specific vector.

A dynamic technique to reduce explicit fault simulation complexity: When explicit fault simulation is used to analyze a stem, typically values are computed at lines in increasing order of input levels. During this process, if at any circuit level, it is determined that the fault effect only exists at a single line, say c_j, in the level, then the criticality of stem c_i can be immediately determined, since under these conditions (Abramovici *et al.*, 1984),

$$Cr(c_i) = cr(c_i, c_j)Cr(c_j).$$

This is due to the fact that the fault effect can propagate to any line, whose input level is higher than that of c_j, only via c_j. Hence, there exists no other path along which the fault effect may propagate and alter the propagation of the effect at c_j in any manner.

Note that in the above scenario, explicit fault simulation has already determined $cr(c_i, c_j) = 1$. Hence, if $Cr(c_i)$ value is known, the critically of stem c_i can be determined immediately.

The notion of a *dynamic dominator* has been defined and used to accelerate fault simulation in Becker *et al.* (1991).

3.5.5.5 Overall critical path tracing based simulation procedures

A complete fault simulation procedure can be obtained by combining critical path tracing with some other fault simulation approach for stem analysis. Some or all of the above techniques to reduce the complexity of explicit fault simulation may be utilized at this stage.

PPSFP simulation paradigm has emerged as the predominant choice for stem analysis (Antreich and Schulz, 1987; Maamari and Rajski, 1990a; Lee and Ha, 1991). Typically, it is efficient to analyze one reconvergent stem at a time. This is even more true in light of the above techniques which provide mechanisms for early termination of explicit fault simulation for a stem fault. PPSFP enables the analysis of one stem fault at a time, by inserting the appropriate fault at the stem.

PPSFP can also exploit the availability of multi-bit words and constant-time multi-bit bit-wise operations by simulating W patterns in parallel. This can ideally be combined with a parallel logic simulation and a parallel-pattern version of critical path tracing, which can be obtained easily by making small modifications to the procedure described earlier.

Different simulators have proposed different orders in which (a) critical path tracing of various FFRs, and (b) analysis of various stems, are carried out. In Abramovici *et al.* (1984), one output cone is simulated at a time. First, critical path tracing is performed on the FFR whose output is the primary output whose cone is being simulated. Next, the stems of fanout systems whose branches drive this FFR are analyzed. The FFRs with these stems as outputs are analyzed next. The process continues in this fashion. In a macroscopic sense, this order of critical path tracing and stem analysis is backward. The FFRs and stems are also analyzed in a backward order in Maamari and Rajski (1990a), where all cones in the circuit are considered together.

In Antreich and Schulz (1987), critical path tracing is first performed on each FFR in the circuit. This is followed by an analysis of stems. The stems that are closest to the outputs, i.e., those with the lowest values of output level (η_{out}) are analyzed first, followed by the stems that have the next lowest η_{out} value, and so on. The key advantage of this method is that the criticality of each input of each FFR at the output of the FFR is known. These criticality values are used to dynamically reduce the complexity of explicit fault simulation during stem analysis.

In Lee and Ha (1991), stem analysis and critical path tracing are performed in an interleaved fashion starting at the primary inputs and proceeding in increasing order of input levels (η_{inp}). This sometimes avoids unnecessary tracing at some FFRs at higher levels. However, in some circumstances, many dynamic techniques for the reduction of explicit fault simulation cannot be used.

3.5.5.6 Extensions of critical path tracing

As mentioned above, critical path tracing can be extended to efficiently consider W patterns in parallel (Antreich and Schulz, 1987), especially for commonly-used logic gates, namely primitive, XOR and XNOR gates. It can also be extended easily for three-valued simulation. Such an extension requires a modification of the algorithms to consider the three-valued system and appropriate encoding of values. Also, many of the definitions presented above need to be modified.

For circuits that only use primitive gates, several techniques to accelerate critical path tracing have been reported in Abramovici *et al.* (1984).

Finally, faults detected by vectors already simulated may be dropped to reduce the complexity of simulation of subsequent vectors. However, since criticality of some nodes is used to derive the criticality of other nodes, the interdependence of criticality values of different faults should be analyzed. Only those faults may be dropped whose criticality values are not required for the computation of the criticality of any of the faults that are not yet detected (Maamari and Rajski, 1990b). Note that the interdependence of the criticality values depends on the order in which FFRs and stems are processed. In the case where the stem and FFRs are processed in an interleaved fashion starting at the primary inputs, a fault may be dropped as soon as it is detected. In contrast, when the circuit is traversed backward starting at the primary outputs, e.g., Maamari and Rajski (1990b), interdependences between criticality values of faults must be considered.

3.6   Approximate, low-complexity fault simulation

The worst-case complexity of simulation of all single SAFs in a circuit with L lines is $O(L^2)$. This complexity may be very high when the circuit under consideration has millions of lines. Several low-complexity **approximate fault simulation** techniques to compute an approximate value of the fault coverage for any given set of test vectors have been proposed. In addition, techniques to *estimate* fault coverage for a set containing a given number of randomly generated vectors have also been proposed. The techniques in this category are discussed in Chapter 12. Here, we discuss the approximate fault simulation techniques and their applications. The techniques discussed are sometimes called **fault graders**.

We also discuss another approximate simulation technique that identifies faults that are *not* detected by a given vector. The intended purpose of this technique is to reduce the number of faults for which explicit fault simulation must be performed for a given vector.

3.6.1   Fault grading approaches

As was shown earlier, critical path tracing can implicitly target faults in FFRs. However, the criticality of stems cannot be accurately deduced in an implicit manner. This necessitates explicit analysis of each stem to determine its criticality, thereby increasing the complexity of fault simulation.

The fault grader proposed in Brglez *et al.* (1984) and Brglez (1985) uses an approximation to apply critical path tracing to a combinational circuit with reconvergent fanouts. In this method, critical path tracing is performed via backward traversal of circuit lines starting at primary outputs. It stops at fanout branches, until the criticality of each branch $c_{j_1}, c_{j_2}, \ldots, c_{j_\beta}$, of the fanout system is determined. The criticality of stem c_i, $Cr(c_i)$, is then assigned the value

$$Cr(c_i) \approx Cr(c_{j_1}) + Cr(c_{j_2}) + \cdots + Cr(c_{j_\beta}),$$

where '+' denotes the logic OR operation. The backward critical path tracing then continues in the FFR with output c_i. This continues until the primary inputs are reached.

As was illustrated earlier, the above relation used to estimate the criticality of a stem is not always correct. In some cases, e.g., the scenario depicted in Figure 3.34(a), this will declare a stem to be critical when it is not. In some other cases, such as the scenario shown in Figure 3.35, it will declare a stem to be not critical when it is indeed critical. Hence, the fault coverage reported by this technique does not have a systematic error, in the sense that the reported fault coverage may be an underestimate for some cases and an overestimate for others. However, since this method relies exclusively on critical path tracing, its complexity is O(L).

The method in Abramovici *et al.* (1984) performs critical path tracing as well as stem analysis. However, stem analysis is performed only at stems of fanout systems at least one of whose branches is identified as critical. In this approach, the fault coverage reported in cases such as the one illustrated in Figure 3.35 could be lower than the actual fault coverage. However, unlike the previous fault grading approach, the fault coverage can only be underestimated. Clearly, this is a safe scenario. In cases where this method reports acceptable fault coverage, the real coverage can only be higher; in cases where the coverage reported by the method is insufficient, explicit fault simulation can be performed for faults not reported by the method as detected. In this technique, the reduction in complexity is achieved because of the fact that the approximation used allows backward critical path tracing and stem analysis to stop wherever lines are declared as not being critical.

3.6.2 A method to identify faults not detected by a vector

A method to identify some of the faults not detected by a vector is presented in Akers *et al.* (1990). Following logic simulation, in $O(L)$ complexity, this method can identify a large proportion of faults that are not detected by the vector. The information provided can then be used to temporarily remove these faults from the list of faults for which explicit fault simulation is performed for the current vector, providing overall reduction in fault simulation complexity.

The key concept behind the method is the notion of a *partial assignment*. For a given fully specified vector, a **partial assignment** is a set of values at a subset of all circuit lines that are sufficient to imply the fully specified value implied by the vector at each primary output of the circuit. Note that the set of values obtained at all circuit lines by simulation of a fully specified vector can also be viewed as a special case of partial assignment.

Consider the circuit shown in Figure 3.38 and vector $(1, 1, 1, 0, 1)$ (Akers *et al.*, 1990). Figure 3.38(a) shows the values obtained at various lines via logic simulation.

Next, a partial assignment can be obtained using the following procedure. First, the value at each output is marked as *required* (using an $*$) and the value at each of the remaining lines in the circuit is marked as *not required* (no $*$ is used). Second, the gate outputs are processed in the order of increasing output level values. If the value at the output of a gate is marked as required, and if the values assigned to any subset of its inputs is sufficient to imply that value at its output, then the value at each input in the subset is marked as required while the value at each of the remaining gate inputs is left with its initial mark (not required). On the other hand, if the value at the output is marked as not required, then the value at each input is left with its initial mark (not required). For example, if the values at inputs c_{i_1} and c_{i_2} and output c_j of a two-input NAND gates are $v(c_{i_1}) = 0$, $v(c_{i_2}) = 0$, and $v(c_j) = 1$ and the value at the output is marked as required, then the value at either of the inputs is sufficient to imply the value at the output. Hence, either of the input values may be marked as required and the other input left with its initial mark (not required). If the values at the inputs and outputs of the same gate were $v(c_{i_1}) = 1$, $v(c_{i_2}) = 0$, and $v(c_j) = 1$, then the value at input c_{i_2} is marked as required and that at input c_{i_1} is left at its initial mark (not required). Finally, if $v(c_{i_1}) = 1$, $v(c_{i_2}) = 1$, and $v(c_j) = 0$, then the values at both inputs must be marked as required.

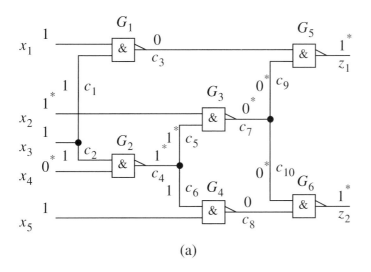

(a)

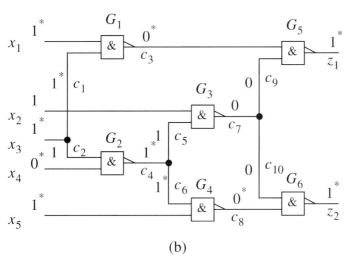

(b)

Figure 3.38 Two example partial assignments for a vector for an example circuit (Akers *et al.*, 1990) (© 1990 IEEE)

Example 3.20 Figure 3.38(a) shows the result of one execution of the above procedure. Note that the above procedure is non-deterministic, since whenever there exist multiple ways of marking values at gate inputs as required and not required, it allows an arbitrary choice to be made. For the same circuit and the same vector, another set of choices gives the values shown in Figure 3.38(b). □

Clearly, the single SA1 fault at a line is not detected by the vector if logic 1 is assigned to the line by the logic simulation step. Similarly, the single SA0 fault at a line that is assigned logic 0 during logic simulation is not detected by the vector. Each partial assignment helps identify additional single SAFs that cannot be detected by the vector applied at the inputs.

The partial assignment in Figure 3.38(a) identifies additional single SAFs as not detected by the vector. For example, the value at line x_1 is marked as not required in this figure. This implies that, for this vector, the value implied at each output can be obtained even if the value at x_1 was not specified. In particular, this indicates that a single SA0 fault at x_1, which would change the logic 1 value implied by the current vector at the line to logic 0, would fail to change the value at any primary output, and hence is not detected by the vector. In general, if for a given vector, the value at a line is marked as not required in any of the partial assignments, then neither of the single SAFs associated with the line is detected by the vector. Collectively, the two partial assignments shown in Figure 3.38 show that (a) the single SA0 faults at each output, z_1 and z_2, **are detected** by the vector, and (b) the single SA1 fault at line x_4 and the single SA0 fault at c_4 **may be detected** by the vector. All other single SAFs are shown to be not detected by the vector. Hence, explicit fault simulation needs to be performed only for the two single SAFs, x_4 SA1 and c_4 SA0, to determine whether or not either of these faults is really detected by the current vector.

The procedure for identifying a partial assignment, given the result of logic simulation, is non-deterministic and can potentially provide a large number of distinct partial assignments. While each partial assignment can help identify additional faults as being not detected by the vector, the complexity of the procedure grows with the number of partial assignments enumerated. An $O(L)$ procedure that typically helps identify a large number of single SAFs as not being detected by a given vector is presented in Akers *et al.* (1990).

Summary

- A description of the interconnections between elements of a given circuit in conjunction with two-valued truth tables of the elements (represented in the form of tables, logic expressions, or singular covers) is sufficient to perform two-valued as well as three-valued logic simulation.

- The values implied by a given vector at circuit lines should be computed in such an order that the value implied at each input of an element is computed before the value implied at its outputs. This order can be determined using circuit levelization.

- Bit-wise, multi-bit operations can be used to simulate multiple vectors in parallel, provided the logic behavior of each circuit element is captured using logic expressions.

- Event-driven simulation computes the value implied by a vector at the outputs of a circuit element only if the current vector implies, at one or more inputs of the element, a value that is different from that implied by the previous vector. Such a strategy can reduce the average complexity of simulation.

- Fault collapsing and fault dropping reduce the average number of faults that must be simulated per vector.
- The worst-case complexity of simulation of each possible single SAF in a circuit with L lines is $O(L^2)$.
- The parallel fault simulation algorithm uses multi-bit logic operations to simulate in parallel the fault-free version of the given circuit along with a number of distinct faulty versions for a given vector.
- The parallel-pattern single-fault propagation simulation algorithm simulates a batch of multiple vectors in parallel. Subsequently, simulation is performed for the batch of vectors for each faulty version of the circuit.
- The deductive fault simulation algorithm computes for each line in the circuit the set of single SAFs that are detected at the line by the given vector. At each line, the set of faults is computed by performing appropriate set operations on the sets of faults detected and logic values implied at each of its fanins. Deductive fault simulation considers the effect of a fault only at lines at which it is detected.
- A concurrent fault simulator dynamically creates multiple, partial copies of circuit versions (fault-free as well as faulty) and computes the values implied by the current vector at each line in each version. The values at each line in each circuit are computed concurrently, but independently.
- Critical path tracing implicitly simulates every single SAF located at lines within a fanout-free region via circuit traversal. The worst-case complexity of critical path tracing based fault simulation for a circuit with L lines that has no reconvergent fanout is $O(L)$.
- Critical path tracing must be combined with some other fault simulation approach to perform fault simulation for an arbitrary combinational circuit. Several techniques exist to reduce the overall complexity of such a composite approach. The worst-case complexity for a circuit with L lines remains $O(L^2)$, however.
- Low-complexity approaches compute the approximate value of fault coverage provided by a given set of vectors.
- A low-complexity approach that identifies faults not detected by a vector can be used as a pre-processing step to fault simulation.

Additional reading

Compiled simulators perform logic and fault simulation for a given circuit by recompiling the main simulator program along with a description of the circuit being simulated (Barzilai *et al.*, 1987). Under this simulator, the code must be recompiled for every circuit for which simulation is desired, but potentially such simulation can be much faster. **Demand-driven** simulation (Smith *et al.*, 1987, 1988) is based on the observation that the value implied at a circuit line needs to be computed only if the value is required to compute some other value. A binary-decision diagram (BDD) based method for simulation of multiple faults is presented in Takahashi *et al.* (1991).

Graph-based methods have been proposed in Antreich and Schulz (1987), Harel and Krishnamurthy (1988) and Becker *et al.* (1991) to reduce the complexity of fault simulation. A method that restricts fault-free circuit simulation and fault-effect propagation to smaller portions of the circuit as faults are dropped is presented in Maamari and Rajski (1990b).

A comparison of fault-simulation paradigms is presented in Goel and Moorby (1984). Analyses of complexity of fault simulation and test generation are presented in Goel (1980) and Harel and Krishnamurthy (1987).

Simulation techniques for synchronous sequential circuits are discussed in Chapter 5.

Exercises

3.1 Model an α-input XNOR gate in terms of two-input gates (similar to the models shown in Figure 3.5). How many table lookups are required before the value at the output is computed for a given combination of input values?

3.2 Write logic expressions of the type shown in Table 3.1 describing two-valued logic behavior of a gate that implements each of the following functions.
  (a) A three-input XOR.
  (b) The carry output of a full adder.

3.3 Derive a three-valued truth table for a gate whose two-valued logic behavior is described by the following truth table.

$$v(c_{i_2})$$

		0	1
$v(c_{i_1})$	0	1	0
	1	1	1

3.4 Use Procedure ComputeHigherValuedTruthTable( ) to derive three-valued truth tables of gates implementing the following functions.
  (a) $v(c_j) = v(c_{i_1}) \, v(c_{i_2}) + \overline{v(c_{i_1})} \, \overline{v(c_{i_2})}$.
  (b) $v(c_j) = v(c_{i_1}) \, v(c_{i_2}) + v(c_{i_2}) \, v(c_{i_3}) + v(c_{i_1}) \, v(c_{i_3})$.
  (c) $v(c_j) = v(c_{i_1}) + \overline{v(c_{i_2})}$.

3.5 Derive the three-valued singular cover for each of the functions in Problem 3.2.

3.6 Derive three-valued singular covers of eight-input versions of a (a) NAND, (b) NOR, and (c) XOR gate. For each gate, compare the complexity of the two-valued truth table with that of its three-valued singular cover.

3.7 Consider a two-input logic gate with inputs c_{i_1} and c_{i_2} and output c_j that implements the following function:

$$v(c_j) = v(c_{i_1}) \, \overline{v(c_{i_2})}.$$

(a) Write the two-valued truth table of the gate.

(b) Does this gate have a controlling value? Can the notion of controlling value be generalized to be applicable to this gate?

(c) Does this gate have a controlled response? Can the notion of controlled response be generalized using the above generalized notion of controlling value?

3.8 Repeat the above problem for gates implementing the following functions.

(a) $v(c_j) = v(c_{i_1}) \, v(c_{i_2}) + \overline{v(c_{i_1})} \, \overline{v(c_{i_2})}$.

(b) $v(c_j) = v(c_{i_1}) \, v(c_{i_2}) + v(c_{i_2}) \, v(c_{i_3}) + v(c_{i_1}) \, v(c_{i_3})$.

(c) $v(c_j) = v(c_{i_1}) + v(c_{i_2}) + \overline{v(c_{i_3})}$.

(d) $v(c_j) = \overline{v(c_{i_1}) \, v(c_{i_2}) + v(c_{i_3}) \, v(c_{i_4})}$.

3.9 Consider a scenario in which three-valued simulation is desired. Assume that the value at each line c_j is represented by two variables, $v(c_j)$ and $u(c_j)$, which collectively represent the symbols 0, 1, and x using the encoding shown in Table 3.2. Furthermore, assume that the programming language used to write your simulator and the CPU used to perform logic simulation support logic operations AND, OR, and NOT.

Derive logic expressions (similar to those shown for primitive gates in the text) for the values $v(\ )$ and $u(\ )$ associated with the output of each of the following gates.

(a) A three-input XOR gate.

(b) A two-input XNOR gate.

(c) The carry function of a full adder.

3.10 Assume that in addition to logic operations AND, OR, and NOT, your programming language and CPU also support the operation XOR. Can you use this operation to more efficiently evaluate the value at the output of a two-input XOR gate under the following two scenarios?

(a) Two-valued simulation is desired.

(b) Three-valued simulation is desired and the encoding shown in Table 3.2 is used.

Comment on how much reduction in the complexity of gate evaluation may be achieved due to the use of this logic operation.

3.11 Derive the two-valued truth table of the faulty versions of the following gates.

(a) A two-input NOR gate with inputs c_{i_1} and c_{i_2} and output c_j with an SA0 fault at c_{i_2}.

(b) A two-input AND gate with inputs c_{i_1} and c_{i_2} and output c_j with an SA1 fault at c_{i_2}.

(c) A four-input gate with inputs $c_{i_1}, c_{i_2}, c_{i_3}$, and c_{i_4} that implements the function $\overline{v(c_{i_1}) \, v(c_{i_2}) + v(c_{i_3}) \, v(c_{i_4})}$, with an SA0 fault at input c_{i_3}.

(d) A gate that implements the function $v(c_j) = v(c_{i_1}) + \overline{v(c_{i_2})}$ with an SA1 fault at input c_{i_1}.

(e) A gate that implements the function $v(c_j) = v(c_{i_1}) + \overline{v(c_{i_2})}$ with an SA1 fault at input c_{i_2}.

3.12 Derive a three-valued truth table for each faulty gate in the above problem.

3.13 Perform simple logic simulation for vector $(1, 0, 0, 0, 1)$ for the circuit shown in Figure 3.10. Clearly describe how to determine the order in which to consider lines for computation of the value implied by the given vector. Show the order in which lines are considered and the value obtained at each line by simulation.

3.14 Consider the circuit C_{17} shown in Figure 3.16. Assume that vector $P_1 = (0, 0, 0, 0, 0)$ is simulated followed by simulation of vector $P_2 = (1, 1, 1, 1, 1)$ using Procedure SimpleLogicSimulation( ). Identify each circuit line at which an event occurs during the simulation of P_2.

Identify the lines at which the value must be recomputed during event-driven simulation of P_2. Compare the number of lines at which values must be computed for P_2 for the two types of logic simulation.

3.15 Rewrite Procedure InputLevelize( ) so as to minimize its complexity. (Hint: the complexity of Procedure 3.2 is high since at each step it becomes necessary to examine various circuit elements until one element is identified such that the level at its output is *undefined* and the level at each of its inputs is defined. This search can be avoided by adding an extra field called, say, NumInpsReady, to the data structure of each circuit element. At the beginning of the procedure, NumInpsReady for each element can be initialized to zero. At each step of the algorithm, when the input level value is determined for a line c_i, the value of NumInpsReady can be incremented for the element with input c_i. At this time, the value of NumInpsReady for that circuit element can be compared with its fanin, i.e., the total number of inputs to the circuit element. If the two numbers are equal, then the output of the circuit element can be added to a list of lines whose input level values are ready for computation.)

(a) Write a step-by-step description of your procedure.

(b) What is the complexity of your procedure in terms of the number of lines in the circuit and any other parameter that you think affects the complexity significantly?

(c) Write a program that implements your procedure and print values of input level for every line in each of the following circuits: (i) c17, (ii) x3mult, and (iii) add2.

A skeleton program, description of data structures it uses, and a set of example circuit files including the three mentioned above are available at

http://books.cambridge.org/0521773563

to facilitate the development of this and many of the following programs.

3.16 Write programs that implement procedures that use *logic expressions* to compute the values implied at the outputs of following gates: (i) AND, (ii) NAND,

(iii) OR, (iv) NOR, (v) XOR, (vi) NOT, and (vii) BUF (non-inverting buffer). Use only logic AND, OR, and NOT operations provided by the programming language. Also, make sure that each of your programs can be used to evaluate the value at the output of a version of the gate with arbitrary number of inputs. Verify the correctness of your gate evaluation procedures using test cases.

Write a report describing the key design issues that you faced, key features of your procedures, and your verification methodology, including a description of the test cases used and results obtained.

3.17 Use the above procedures to obtain a program that implements Procedure SimpleLogicSimulation( ).

(a) Generate a set containing five randomly generated vectors suitable for application to each of the following circuits: (i) c17, (ii) x3mult, and (iii) add2. (See Problem 3.15 for location of these three circuit files.) Report the set of vectors.

(b) Use your implementation of Procedure SimpleLogicSimulation( ) to compute the values implied at primary outputs of each circuit for each of the corresponding vectors. Report for each circuit, each input vector and the values implied at the primary outputs.

3.18 Implement Procedure EventDrivenSimulation( ). Use the implemented procedure to repeat the simulations performed in the above problem using Procedure SimpleLogicSimulation( ). For each circuit, report the sequence of vectors applied, the response to each vector at the primary outputs of the circuit, and the number of events generated during the simulation of the vector.

3.19 Assuming that you are using a CPU that can execute bit-wise operations (AND, OR, and NOT) on $W = 2$ bit numbers, perform parallel simulation on circuit C_{17} shown in Figure 3.16 for vectors $P_1 = (0, 0, 0, 0, 0)$ and $P_2 = (1, 1, 1, 1, 1)$. Show the values $V(\ )$ implied by the above vectors at each line of the circuit.

3.20 If you had to repeat the parallel simulation for the circuits and vectors in the above problem using truth tables instead of logic expressions, what type of truth tables will be required? Comment on the feasibility of performing parallel simulation using truth tables for CPUs with $W = 32$ or 64.

3.21 Consider a two-input logic gate with inputs c_{i_1} and c_{i_2} and output c_j that implements the following function:

$$v(c_j) = v(c_{i_1})\ \overline{v(c_{i_2})}.$$

Identify equivalence and dominance relations between the six single SAFs associated with the inputs and outputs of the gate.

3.22 Perform fault collapsing by manually executing Procedure FaultCollapsing( ) on the circuit shown in Figure 3.1(a). Clearly describe the intermediate results and lists obtained at each step. Also, give the target fault list obtained after fault collapsing.

3.23 Prove that in a fanout-free circuit comprised only of primitive gates, fault collapsing can always be performed in such a manner that the only single SAFs remaining in the target fault list are at the primary inputs of the circuit. (Hint: use the equivalence and dominance relations between single SAFs associated with primitive gates and the way in which fault collapsing is performed by Procedure FaultCollapsing( ).)

3.24 Consider single SAFs f_i, f_j, and f_k in a combinational circuit C. Given that fault f_i dominates f_j, and that f_j dominates f_k, prove that f_i dominates f_k.

3.25 Fault simulation is to be performed on the circuit C_{17} shown in Figure 3.16. The fault list contains f_1: x_1 SA0, f_4: x_2 SA1, f_5: x_3 SA0, f_8: x_4 SA1, f_9: x_5 SA0, f_{12}: c_1 SA1, and f_{22}: c_6 SA1. The vectors to be simulated are $(0, 1, 0, 1, 1)$, $(1, 1, 0, 1, 0)$, $(1, 1, 1, 0, 1)$, and $(1, 0, 0, 0, 1)$.

Assuming that the programming language and the CPU can support bit-wise operations on four-bit integers, perform parallel fault simulation for the above scenario.

(a) How many simulation passes are required for the first vector?

(b) Show the fault mask for each of the passes for the first vector.

(c) Summarize the results of simulation for each vector, the number of passes required, the circuit versions simulated in each pass, the values implied at the primary outputs in each version of the circuit, and faults detected in each pass.

3.26 Perform parallel pattern single fault propagation (PPSFP) simulation for the scenario described in the above problem.

(a) In each pass in which fault f_{12} is simulated, show the manner in which the logic behavior of each line in the circuit is represented.

(b) How many simulation passes are required to complete the entire simulation?

(c) Summarize the results of simulation for each pass, the circuit version simulated, the values implied at the primary outputs for each vector, and faults detected in the pass.

3.27 Consider a two-input logic gate with inputs c_{i_1} and c_{i_2} and output c_j that implements the following function:

$$v(c_j) = v(c_{i_1})\,\overline{v(c_{i_2})}.$$

Derive a relation to compute the deductive fault list for output c_j, given (a) the deductive faults list for each input, and (b) the value implied by the current vector at each input in the fault-free version of the circuit.

3.28 A three-input logic gate implements the function given by the following logic equation

$$v(c_j) = \left[v(c_{i_1}) + v(c_{i_2})\right] v(c_{i_3}).$$

Derive a relation (or, a set of relations) to compute the deductive fault list for output c_j, given (a) the deductive faults list for each input, and (b) the value implied by the current vector at each input in the fault-free version of the circuit. Consider a circuit in which this gate is to be used in such a manner that the effect of any single SAF at any line in the transitive fanin of this gate may propagate to no more than one input of this gate. How can you simplify the relations to compute deductive fault lists derived above for use in such a circuit? What similarity do the simplified relations have to the rules for computing the sensitivity of each input of the gate for critical path tracing?

3.29 Identify the FFRs and the immediate dominator of each stem (if any) in the circuit shown in Figure 3.37.

Perform the critical path tracing for the vector (0, 1, 1, 1, 1) in such a manner so as to exploit the knowledge of criticality of the immediate dominator of a stem to reduce the complexity of explicit fault simulation performed to determine the criticality of the stem. Describe each major step of simulation, especially those that help reduce fault simulation complexity. Mark each line that is found to be critical and explicitly list all single SAFs detected by the vector.

3.30 Starting with a list containing each single SAF in the circuit shown in Figure 3.37, perform deductive fault simulation for the vector (1, 1, 0, 0, 0).

3.31 Repeat the above problem using the critical path tracing method. Perform explicit fault simulation for stem analysis.

3.32 In the above problem, the circuit under test has a single output. Hence, explicit fault simulation for analysis of a stem can be stopped at the immediate dominator of the stem.

Identify the immediate dominator of each stem of the above circuit. Quantify the reduction in the complexity of explicit fault simulation accrued when explicit fault simulation for analysis of a stem stops at its immediate dominator.

3.33 One of the approximate fault simulation techniques presented in this chapter uses the logical OR of the criticalities of the branches of a fanout system as the criticality of the stem. Repeat the fault simulation for the circuit, fault list, and vectors presented in Problem 3.25 using an approximate critical path tracing approach that uses the above approximation.

Compare the fault coverage obtained by the approximate method with that obtained by the exact fault simulation in Problem 3.25.

References

Abramovici, M., Menon, P. R., and Miller, D. T. (1984). Critical path tracing: an alternative to fault simulation. *IEEE Design & Test of Computer*, **1** (1), pp. 83–93.

Akers, S. B., Krishnamurthy, B., Park, S., and Swaminathan, A. (1990). Why is less information from logic simulation more useful in fault simulation? In *Proc. Int. Test Conference*, pp. 786–800.

Antreich, K. J. and Schulz, M. H. (1987). Accelerated fault simulation and fault grading in combinational circuits. *IEEE Trans. on Computer-Aided Design*, **6** (5), pp. 704–712.

Armstrong, D. B. (1972). A deductive method for simulating faults in logic circuits. *IEEE Trans. on Computers*, **21** (5), pp. 464–471.

Barzilai, Z., Carter, J. L., Rosen, B. K., and Rutledge, J. D. (1987). HSS-a high speed simulator. *IEEE Trans. on Computer-Aided Design*, **6** (4), pp. 601–616.

Becker, B., Hahn, R., Krieger, R., and Sparmann, U. (1991). Structure based methods for parallel pattern fault simulation in combinational circuits. In *Proc. European Design Automation Conference*, pp. 497–502.

Brglez, F. (1985). A fast fault grader: analysis and applications. In *Proc. Int. Test Conference*, pp. 785–794.

Brglez, F., Pownall, P., and Hum, R. (1984). Applications of testability analysis from ATPG to critical delay path tracing. In *Proc. Int. Test Conference*, pp. 705–712.

Goel, P. (1980). Test generation costs analysis and projections. In *Proc. Int. Symposium on Fault-Tolerant Computing*, pp. 77–84.

Goel, P. and Moorby, P. R. (1984). Fault simulation techniques for VLSI circuits. *VLSI Systems Design*, pp. 22–26.

Harel, D. and Krishnamurthy, B. (1987). Is there hope for linear time fault simulation? In *Proc. Int. Symposium on Fault-Tolerant Computing*, pp. 28–33.

Harel, D. and Krishnamurthy, B. (1988). A graph compaction approach to fault simulation. In *Proc. Design Automation Conference*, pp. 601–604.

Harel, D., Sheng, R., and Udell, J. (1987). Efficient single fault propagation in combinational circuits. In *Proc. Int. Conference on Computer-Aided Design*, pp. 2–5.

Hong, S. J. (1978). Fault simulation strategy for combinational logic networks. In *Proc. Int. Symposium on Fault-Tolerant Computing*, pp. 96–99.

Kirkland, T. and Mercer, M. R. (1987). A topological search algorithm for ATPG. In *Proc. Design Automation Conference*, pp. 502–508.

Lee, H. K. and Ha, D. S. (1991). An efficient, forward fault simulation algorithm based on the parallel pattern single fault propagation. In *Proc. Int. Test Conference*, pp. 946–955.

Maamari, F. and Rajski, J. (1988). A reconvergent fanout analysis for efficient exact fault simulation of combinational circuits. In *Proc. Int. Symposium on Fault-Tolerant Computing*, pp. 122–127.

Maamari, F. and Rajski, J. (1990a). A method of fault simulation based on stem regions. *IEEE Trans. on Computer-Aided Design*, **9** (2), pp. 212–220.

Maamari, F. and Rajski, J. (1990b). The dynamic reduction of fault simulation. In *Proc. Int. Test Conference*, pp. 801–808.

McCluskey, E. J. and Clegg, F. W. (1971). Fault equivalence in combinational circuits. *IEEE Trans. on Computers*, **20** (11), pp. 1286–1293.

Poage, J. F. and McCluskey, E. J. (1964). Derivation of optimal test sequences for sequential machines. In *Proc. Symposium on Switching Theory and Logic Design*, pp. 121–132.

Rogers, W. A., Guzolek, J. F., and Abraham, J. A. (1987). Concurrent hierarchical fault simulation: a performance model and two optimizations. *IEEE Trans. on Computers*, **6** (5), pp. 848–862.

Roth, J. P. (1966). Diagnosis of automata failures: a calculus and a method. *IBM Journal of Research & Development*, **10** (4), pp. 278–291.

Seshu, S. (1965). On an improved diagnosis program. *IEEE Trans. on Computers*, **14** (2), pp. 76–79.

Smith, S. P., Mercer, M. R., and Brock, B. (1987). Demand driven simulation: BACKSIM. In *Proc. Design Automation Conference*, pp. 181–187.

Smith, S. P., Mercer, M. R., and Underwood, B. (1988). D3FS: a demand driven deductive fault simulator. In *Proc. Int. Test Conference*, pp. 582–592.

Takahashi, N., Ishiura, N., and Yajima, S. (1991). Fault simulation for multiple faults using shared BDD representation of fault sets. In *Proc. Int. Conference on Computer-Aided Design*, pp. 550–553.

Tarjan, R. (1974). Finding dominators in directed graphs. *SIAM Journal of Computing*, **3**, pp. 62–89.

Ulrich, E. G. and Baker, T. (1973). The concurrent simulation of nearly identical digital networks. In *Proc. Design Automation Conference*, pp. 145–150.

Waicukauski, J. A., Eichelberger, E. B., Forlenza, D. O., Lindbloom, E., and McCarthy, T. (1985). Fault simulation for structured VLSI. *VLSI Systems Design*, pp. 20–32.

Wang, D. T. (1975). Properties of faults and criticalities of values under tests for combinational networks. *IEEE Trans. on Computers*, **24** (7), pp. 746–750.

4   Test generation for combinational circuits

In this chapter, we discuss automatic test pattern generation (ATPG) for combinational circuits. We begin by introducing preliminary concepts including circuit elements, ways of representing behaviors of their fault-free as well as faulty versions, and various value systems.

Next, we give an informal description of test generation algorithms to introduce some of the test generation terminology. We then describe direct as well as indirect implication techniques.

We discuss a generic structural test generation algorithm and some of its key components. We then describe specific structural test generation paradigms, followed by their comparison and techniques for improvement.

We proceed to some non-structural test generation algorithms. We describe test generation systems that use test generation algorithms in conjunction with other tools to efficiently generate tests.

Finally, we present ATPG techniques that reduce heat dissipated and noise during test application.

4.1   Introduction

While most practical circuits are sequential, they often incorporate the *full-scan* design for testability (DFT) feature (see Chapter 11). The use of full-scan enables tests to be generated using a combinational test generator. The input to the test generator is only the combinational part of the circuit under test (CUT), obtained by removing all the flip-flops and considering all the inputs and outputs of the combinational circuit as primary inputs and outputs, respectively. If the generated tests are applied using the full-scan DFT features and the test application scheme described in Chapter 11, the fault coverage reported by the combinational test generator is achieved. As will become clear in the following, when certain secondary objectives, such as minimization of test application time or minimization of heat dissipation during test application, are to be considered, it becomes necessary to distinguish a truly combinational circuit from a sequential circuit that uses full-scan.

The circuit model and notation introduced in Chapter 3 (Section 3.2) are also used in this chapter. The circuit under consideration, C, is a combinational logic circuit with

n inputs, $x_1, x_2, \ldots, x_n$, and m outputs, $z_1, z_2, \ldots, z_m$. We assume that the circuit is an interconnection of single-output gates where each gate implements an arbitrary logic function. The lines internal to the circuit are called $c_1, c_2, \ldots, c_k$, where c_j is either the output of a single-output gate, G_l, or a fanout of another line, c_i. The primary inputs, internal circuit lines, and primary outputs collectively constitute all circuit lines. Recall that, when there is no need to distinguish between internal lines and all circuit lines, we use symbols such as c and c_i to denote all circuit lines. In situations where it is necessary to distinguish between the internal lines and all circuit lines, symbols such as c and c_i are used to denote internal lines and symbols such as b and b_i are used to denote all circuit lines.

$F = \{f_1, f_2, \ldots, f_{N_f}\}$ denotes the fault list, i.e., the set of faults of interest. In this chapter, we will primarily consider fault lists that contain all possible single stuck-at faults (SAFs) in the given circuit, or some subset thereof. C^{f_i} denotes a copy of circuit C that has fault f_i. During a given execution of a test generation procedure, typically a single fault is considered. This fault will be called the **target fault** and the corresponding faulty version of the circuit will sometimes be referred to as C'.

Recall that a vector P is an n-tuple $(p_1, p_2, \ldots, p_n)$, where $p_j \in \{0, 1\}$, $1 \leq j \leq n$. Each p_j is called a component of the vector. Component values of vector P, $p_1, p_2, \ldots, p_n$, are applied, respectively, to primary inputs $x_1, x_2, \ldots, x_n$ of the circuit. The value implied by the vector at line c in the fault-free circuit C is denoted by $v(c)$. The value implied by the vector at line c in the faulty version with fault f_i, C^{f_i}, is denoted by $v^{f_i}(c)$. In particular, the value implied at line c in the faulty version with the target fault, C', is simply referred to as $v'(c)$.

A vector P is said to **detect**, **test**, or **cover** fault f_i in circuit C if the value implied at one or more of the outputs of C, due to the application of P at its inputs, is different from that obtained at the corresponding outputs of C^{f_i}. Any vector that detects a fault f_i in a circuit C is said to be a **test vector** or, simply, a **test**, for that fault in C. If no test vector exists for a fault, then the fault is said to be **untestable**.

In its simplest form, the objective of test generation is to generate a test vector for a given fault in a given circuit, or to declare it untestable. Typically, a set of tests must be generated within a given amount of computational effort. Hence, for some faults, test generation may be terminated before it is successful, i.e., before a test is generated or the fault proven untestable. Such a fault is sometimes called an **aborted** fault. A more practical version of test generation requires the generation of *a set of test vectors whose constituent vectors collectively detect all, or a maximal fraction of, the testable faults in the given fault list*. In other words, for a given amount of computational effort, the number of aborted faults must be minimized.

Another version of test generation requires the above objective to be satisfied using a *test set of minimum size*, i.e., one containing a minimum number of test vectors. Yet another version requires the detection of all, or a maximal fraction of, the faults in the target fault list using a *test sequence whose application to the circuit causes minimal heat dissipation*. In all the cases, the test generator also reports: (a) the fault coverage obtained by the generated tests, (b) a list containing each fault that was identified as being untestable, and (c) a list containing each aborted fault.

4.2   Composite circuit representation and value systems

In this chapter, we consider the class of combinational logic circuits described in Chapter 3 (Section 3.2). Next, we present some additional representations and value systems that are especially useful for test generation. Note that the information contained in the models presented in Section 3.2 is sufficient to completely define each of the following representations. In that sense, the value systems and other representations presented ahead should be viewed as alternative representations that are especially suitable for test generation.

Figures 4.1(a) and 4.1(b), respectively, show the fault-free and a faulty version of a circuit with a target fault. Vector $(0, 1)$ is applied to both versions of the circuit and simulation performed to determine the values at circuit lines. This indicates that $(0, 1)$ is a test for the given fault.

Figure 4.1(c) combines the above fault-free and faulty circuits into a single **composite circuit**, i.e., a combination of (a) the fault-free circuit that does not contain the fault shown, and (b) a faulty circuit that does. Associated with each line of the circuit are two logic values, $v(c)$ and $v'(c)$, which are the values implied by the input vector at the line in the fault-free and faulty circuits, respectively. These values are shown in the figure as $v(c)/v'(c)$. In most of the following discussion, we will use the composite circuit description.

4.2.1   Composite value system

Instead of using notation of the form $v(c)/v'(c)$ to denote the values implied at line c in the fault-free and faulty circuits, respectively, we will adopt composite value systems that are based on the D-notation (Roth, 1966).

4.2.1.1 Basic values of composite value systems

In the D-notation, $0/0$, $1/1$, $1/0$, and $0/1$ are denoted by $\underline{0}, \underline{1}, D$, and $\overline{D}$, respectively, and constitute the four **basic values** of all composite value systems presented ahead. Using this notation, the composite value at a line c will be represented by symbol $V(c)$, e.g., a line c with value $v(c)/v'(c) = 1/0$ will be labeled as $V(c) = D$. In Figure 4.1(d), the values $v(c)/v'(c)$ are shown as $V(c)$ using the above four values.

Since a line may also be assigned the unspecified value, x, in the fault-free and/or the faulty circuit, we also need to represent values such as $x/0$, $x/1$, $0/x$, $1/x$, and

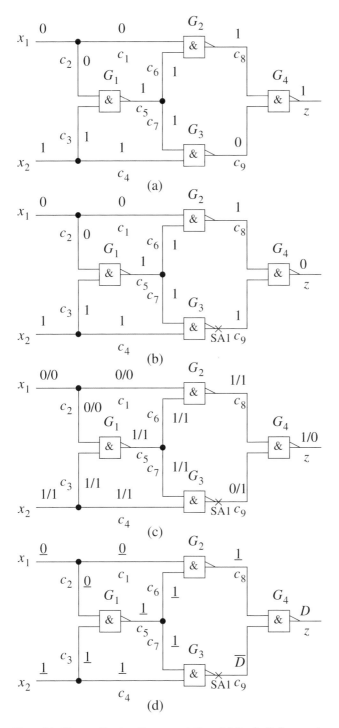

Figure 4.1 Composite circuit representation: (a) the fault-free version of an example circuit, (b) a faulty version of the circuit, (c) the composite circuit representation, and (d) the composite circuit representation with composite values

x/x. Note that these values can be represented using sets. We already have $0/0 = \{\underline{0}\}$, $1/1 = \{\underline{1}\}$, $1/0 = \{D\}$, and $0/1 = \{\overline{D}\}$. The above composite values with one or more unspecified values can be written as $x/1 = \{1/1, 0/1\} = \{\underline{1}, \overline{D}\}$, $1/x = \{1/1, 1/0\} = \{\underline{1}, D\}, \ldots, x/x = \{0/0, 1/1, 1/0, 0/1\} = \{\underline{0}, \underline{1}, D, \overline{D}\}$.

4.2.1.2 Possible sets of composite values

In general, a circuit line may be assigned any set that appears in the *power set* of the set $\{\underline{0}, \underline{1}, D, \overline{D}\}$, the set comprised of the four basic values (Cox and Rajski, 1994). (The **power set** of a set is the set of all subsets of the set, including the empty set.) Figure 4.2 shows the 16 possible sets of basic composite values. In addition to the sets of values, this figure also shows arrows between various sets, where an arrow from set S_i to S_j indicates that set S_j can be obtained by deleting one composite basic value from S_i. In general, if S_j is a proper subset of S_i, then S_j can be reached from S_i by following a sequence of one or more arrows.

4.2.1.3 Some composite value systems

Column 1 of Table 4.1 shows the 16 possible sets of values that may appear on a line. A value that is represented using a set that contains a single element is called a **completely specified** value. In contrast, a value represented by a set that contains multiple elements is said to be an **incompletely specified** value. Finally, the empty set indicates a **conflict** and its assignment to a circuit line indicates that no value can be assigned to the given line that is consistent with the values that have already been assigned to other circuit lines.

In addition, if a set of values S_j is a proper subset of another set of values S_i, then S_j is said to be **more specific** than S_i. A path exists from S_i to S_j in Figure 4.2 if S_j is more specific than S_i.

The table also shows four of the many possible value systems that may be used for test generation. The **five-valued system** (Roth, 1966) uses the symbols 0_5, 1_5, D_5, and $\overline{D}_5$ to, respectively, denote the completely specified values $\{\underline{0}\}$, $\{\underline{1}\}$, $\{D\}$, and $\{\overline{D}\}$. In addition, it uses a single symbol X_5 to denote each of the incompletely specified values. (This value system is the only one used in some of the following chapters. In those chapters, the values in the five-valued system are denoted simply as $0, 1, D, \overline{D}$, and X. In this chapter, we use the above symbols to be able to distinguish the symbols in the five-valued system from those of the *six-valued* and other systems discussed next.)

The **six-valued system** differs from the five-valued system in that it uses two symbols to denote incompletely specified values, namely χ_6 to denote the incompletely specified value $\{\underline{0}, \underline{1}\}$, and X_6 to denote each of the other incompletely specified values. The **nine-valued system** (Cha *et al.*, 1978) uses five different symbols to represent the incompletely specified values. Finally, the **16-valued system** (Cox and Rajski,

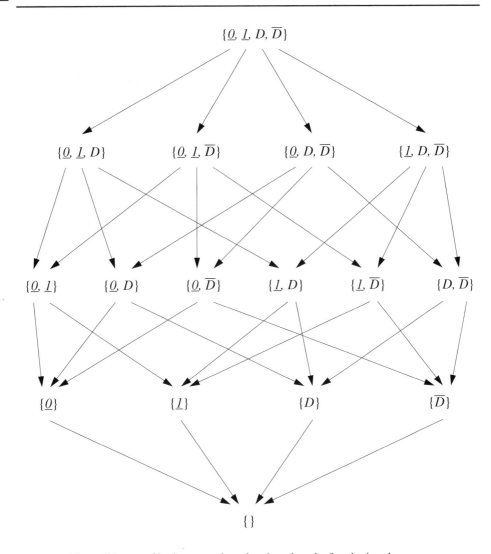

Figure 4.2 All possible sets of basic composite values based on the four basic values

1994) uses a distinct symbol for each of the incompletely specified values. Since the 16-valued system has a distinct symbol to describe each possible set of the four basic values of the composite system, instead of defining additional symbols, we will use each set as the corresponding symbol of the system.

4.2.1.4 Some properties of composite value systems

One important property is satisfied by all the above-mentioned value systems: among the multiple sets of values that are denoted by a single symbol, there always exists one set that is the superset of each of the other sets represented by that symbol. We will call this set the **leading value** of the corresponding symbol. For X_5, X_6, as well as

Table 4.1. *Some composite value systems (Cox and Rajski, 1994) (© 1994 IEEE)*

| 16-valued | Symbols in other value systems | | |
	Five-valued	Six-valued	Nine-valued
$\{\}$	—	—	—
$\{\underline{0}\}$	0_5	0_6	0_9
$\{\underline{1}\}$	1_5	1_6	1_9
$\{D\}$	D_5	D_6	D_9
$\{\overline{D}\}$	$\overline{D}_5$	$\overline{D}_6$	$\overline{D}_9$
$\{\underline{0}, \underline{1}\}$	X_5	χ_6	xx_9
$\{\underline{0}, D\}$	X_5	X_6	$x0_9$
$\{\underline{0}, \overline{D}\}$	X_5	X_6	$0x_9$
$\{\underline{1}, D\}$	X_5	X_6	$1x_9$
$\{\underline{1}, \overline{D}\}$	X_5	X_6	$x1_9$
$\{D, \overline{D}\}$	X_5	X_6	xx_9
$\{\underline{0}, \underline{1}, D\}$	X_5	X_6	xx_9
$\{\underline{0}, \underline{1}, \overline{D}\}$	X_5	X_6	xx_9
$\{\underline{0}, D, \overline{D}\}$	X_5	X_6	xx_9
$\{\underline{1}, D, \overline{D}\}$	X_5	X_6	xx_9
$\{\underline{0}, \underline{1}, D, \overline{D}\}$	X_5	X_6	xx_9

xx_9, the leading value is $\{\underline{0}, \underline{1}, D, \overline{D}\}$. When a line is assigned a symbol that denotes multiple sets of values, it is interpreted as being the leading value of the symbol.

In general, one symbol of a value system is said to be **more specific** than another, if the leading value of the former is a subset of the leading value of the latter. For example, in the six-valued system, each symbol that denotes a completely specified value, namely 0_6, 1_6, D_6, and $\overline{D}_6$, is said to be more specific than X_6. Furthermore, symbols 0_6 and 1_6 are said to be more specific than χ_6, which in turn is said to be more specific than X_6. These relations can be deduced by following the arrows in Figure 4.3.

The number of symbols (values) in a value system influences the complexity of representation of the symbols by only a small amount. A minimum of three bits is required to represent the symbols in the five-valued and six-valued systems; a minimum of four bits is required for the other two systems shown in Table 4.1. More importantly, the number of values in the system influences the size of the truth tables that need to be stored and/or the complexity of the implication operations (described ahead). However, arguably (e.g., see Cox and Rajski (1994)) such an increase in complexity is often off-set by the reduction in the complexity of search for the test.

In the following discussion, we will mostly use the six-valued system due to its simplicity and the 16-valued system due to its generality. However, the algorithms are

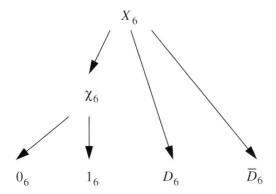

Figure 4.3   Values in the six-valued system

written such that they can be used in conjunction with any of the above value systems. In fact, the algorithms can be used with any value system which uses $\underline{0}, \underline{1}, D, \overline{D}$ as the basic values and in which any symbol denoting multiple values has a unique leading value.

4.2.2   Representation of behavior of circuit elements

The example composite circuit shown in Figure 4.1(d) illustrates the three scenarios under which the behavior of a gate or a fanout system must be described.

The first scenario, exemplified by gate G_4 in Figure 4.1(d), is that of a fault-free circuit element in the transitive fanout of the fault site. (We assume that the SA1 fault at c_9 is associated with the output of gate G_3.) For such an element, the values implied at the outputs, as well as those inputs that are in the transitive fanout of the fault site, may be affected by the presence of the fault. In other words, in the composite circuit description, the outputs and such inputs of such an element may take any subset of the values in set $\{\underline{0}, \underline{1}, D, \overline{D}\}$.

The second scenario is exemplified by gates G_1 and G_2. Not only are these gates fault-free, but they are not in the transitive fanout of the fault site, c_9. Note that the values at inputs and outputs of such a circuit element will always be subsets of set $\{\underline{0}, \underline{1}\}$.

The third scenario, namely a faulty element, is exemplified by gate G_3 whose output line c_9 has the SA1 fault. Recall that we consider single SAFs not only at the output lines of gates, but also at gate inputs, fanout stems, and fanout branches. Furthermore, sometimes we also consider a more general fault model.

In the following, we will discuss ways of describing circuit elements that belong to each of the above scenarios.

4.2.2.1 Description of a fault-free element

We first discuss the general case where all the inputs of a fault-free circuit element are in the transitive fanout of the fault site.

In this scenario, in the composite circuit, each input of the fault-free element may take any value from the set $\{\underline{0}, \underline{1}, D, \overline{D}\}$. Since the circuit element is itself fault-free, the output of the element for each of the possible combinations of these values can be determined by simulating two fault-free copies of the element with values $v(c)$ and $v'(c)$, respectively.

Example 4.1 Consider a two-input NAND gate with output c_j and inputs c_{i_1} and c_{i_2}. Let the values at the two inputs be given by $V(c_{i_1}) = \underline{1}$ and $V(c_{i_2}) = D$. $V(c_{i_1}) = \underline{1}$ can be decomposed into $v(c_{i_1}) = 1$ and $v'(c_{i_1}) = 1$. Similarly, $V(c_{i_2}) = D$ can be decomposed into $v(c_{i_2}) = 1$ and $v'(c_{i_2}) = 0$. The two-valued truth table of a fault-free two-input NAND gate can then be used (1) with input values $v(c_{i_1}) = 1$ and $v(c_{i_2}) = 1$ to obtain the output value, $v(c_j) = 0$, and (2) with input values $v'(c_{i_1}) = 1$ and $v'(c_{i_2}) = 0$ to obtain the output value, $v'(c_j) = 1$. The output values $v(c_j) = 0$ and $v'(c_j) = 1$ can then be combined to obtain $V(c_j) = \overline{D}$. $\qquad\square$

A similar procedure can be followed to obtain a four-valued truth table of any fault-free circuit element. Figure 4.4 shows the four-valued truth tables of fault-free versions of some circuit elements. In general, for a gate with output c_j and inputs $c_{i_1}, c_{i_2}, \ldots, c_{i_\alpha}$, such four-valued truth tables will be denoted by $\mathcal{G}_j^4(v_{i_1}, v_{i_2}, \ldots, v_{i_\alpha})$, where v_{i_l} is one of the basic composite values belonging to the set denoted by $V(c_{i_l})$. Similarly, the truth table of a branch c_{j_l} of a fanout system with stem c_i is described by $\mathcal{G}_{j_l}^4(v)$, where v is a basic composite value belonging to the set denoted by $V(c_i)$.

If any input of a fault-free circuit element is not in the transitive fanout of the fault site, then that input may only take values from set $\{\underline{0}, \underline{1}\}$, even in the composite circuit. The four-valued truth table of the element, along with this restriction, can be used to describe the behavior of such an element. For example, if input c_{i_1} of a two-input NAND gate is not in the transitive fanout of a fault site while its input c_{i_2} is, then its behavior is described by the first two rows of the truth table in Figure 4.4(a).

If none of the inputs of a fault-free gate are in the transitive fanout of the fault site, then the gate's inputs and its output can only take values from set $\{\underline{0}, \underline{1}\}$, even in the composite circuit. Such a circuit element can be described using a part of its four-valued truth table. For example, if neither of the inputs of a two-input NAND gate is in the transitive fanout of a fault site, then its behavior is described by the first two columns of the first two rows of the truth table in Figure 4.4(a) (i.e., by the top-left quarter of the truth table).

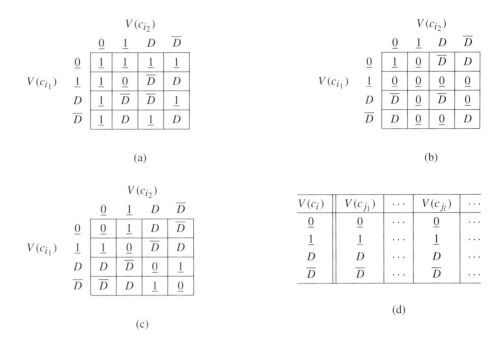

Figure 4.4 Four-valued truth tables of fault-free versions of some circuit elements: (a) two-input NAND gate, (b) two-input NOR gate, (c) two-input XOR gate, and (d) a fanout system

4.2.2.2 Description of a faulty circuit element

The above method can also be used to obtain four-valued truth tables for faulty circuit elements. First, the four-valued symbol at each input c, $V(c)$, is decomposed into two values $v(c)$ and $v'(c)$. The two-valued truth table of a fault-free circuit element is then used to compute the output value corresponding to the fault-free values v. The two-valued truth table of the faulty circuit element is used to compute the value corresponding to the faulty values v'. Finally, the two output values, v and v', are combined to obtain the corresponding four-valued output, V, of the faulty circuit element.

The focus of this chapter is on combinational circuits that contain no more than one faulty circuit element. More specifically, the faulty circuit element is assumed to have a single SAF at one of its input or output lines. In such cases, the inputs of the faulty circuit elements are always free of any fault effect, since all circuit elements in the transitive fanin of a faulty element must be fault-free. In other words, all inputs to the faulty circuit element may only take values that are more specific versions of the set of basic values $\{\underline{0}, \underline{1}\}$.

Figure 4.5 shows the four-valued truth tables for some faulty circuit elements for different combinations of input values. For each faulty circuit element in a composite circuit, a table similar to the one shown in this figure should be derived and used during

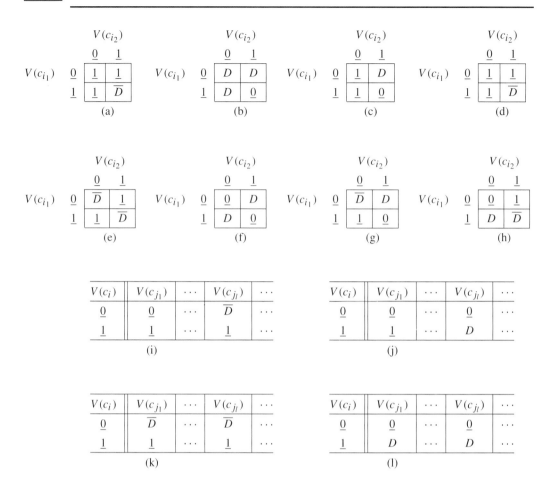

Figure 4.5 Four-valued truth tables of some faulty circuit elements. A two-input NAND gate with (a) output SA1, (b) output SA0, (c) input c_{i_1} SA1, and (d) input c_{i_1} SA0. A two-input XOR gate with (e) output SA1, (f) output SA0, (g) input c_{i_1} SA1, and (h) input c_{i_1} SA0. A fanout system with (i) branch c_{j_l} SA1, (j) branch c_{j_l} SA0, (k) stem c_i SA1, and (l) stem c_i SA0

test generation, as opposed to its counterpart for the fault-free version of the element. Again, for a gate with output c_j and a branch c_{j_l} of a fanout system, the tables will respectively be referred to as $\mathcal{G}_j^4()$ and $\mathcal{G}_{j_l}^4()$.

4.2.3   Representation of circuit element behavior for test generation

The four-valued truth tables for fault-free and each possible faulty version of a circuit element are sufficient to perform test generation. However, the information presented in these truth tables can be represented in other forms that are typically more compact and more suitable for most test generation operations. In particular, these representations reduce the complexity of many operations encountered during test generation, such as computation of the value at the output of a circuit element given a set of incompletely specified values at its inputs.

4.2.3.1 Compact representation of the behavior of fault-free circuit elements

The notion of *singular cover* of a fault-free circuit element can be generalized to obtain compact representations of four-valued truth tables of gates. First, all the input combinations that imply a particular value at the output are grouped. Second, whenever possible, multiple input combinations that imply identical values at the output are combined into cubes to obtain a more compact representation. Any method to generate prime implicants of a Boolean function can be generalized to identify such cubes.

Figure 4.6 shows all the six-valued cubes for the fault-free two-input NAND, AND, OR, and XOR gates. The cubes that imply values 1_6 and 0_6 at the gate output are listed first. These are followed by cubes that imply values D_6 or $\overline{D}_6$. The former set of cubes are referred to as the six-valued **singular cover** of the gate (Roth, 1966). The latter set of cubes describe the manner in which the fault effect at one or more inputs of the gate is propagated to its output. These sets of cubes are referred to as six-valued **propagation cubes** of the gate. Collectively, these two sets of cubes are referred to as the **cube cover** of the gate.

The truth table of a fault-free fanout element shown in Figure 4.4(d) can be easily rewritten in this form. The first two cubes shown in Figure 4.7 represent the six-valued singular cover of the fanout system; the last two cubes represent its six-valued propagation cubes.

Since the cubes are generated by combining truth table entries by using incompletely specified values, the number of cubes obtained depends on the incompletely specified values provided by the value system. The reader can verify that fewer 16-valued cubes can be found for a two-input NAND gate than the six-valued cubes shown in Figure 4.6.

4.2.3.2 Compact representation of the behavior of faulty circuit elements

The truth table of a faulty version of a circuit element can also be represented in a compact fashion using cubes. Again, the cubes for which all the outputs of the circuit element have fault-free values are listed first and constitute the six-valued **singular cover** of the faulty version of the gate. (Since we focus largely on combinational circuits with a single faulty circuit element, only those cubes that do not have any fault effect at the inputs are considered for a faulty circuit element.) The cubes for which at least one output of the circuit element has a fault effect are listed separately and constitute the six-valued **fault excitation cubes** (called *D-cubes of a failure* in Roth (1966)) of the faulty version of the gate. These cubes are named in this manner because it is *necessary* to imply at the inputs of the faulty circuit element a combination of input values contained in one of these cubes for the fault to be excited. Again, the

$V(c_{i_1})$	$V(c_{i_2})$	$V(c_j)$
0_6	X_6	1_6
X_6	0_6	1_6
D_6	$\overline{D}_6$	1_6
$\overline{D}_6$	D_6	1_6
1_6	1_6	0_6
$\overline{D}_6$	1_6	D_6
1_6	$\overline{D}_6$	D_6
$\overline{D}_6$	$\overline{D}_6$	D_6
D_6	1_6	$\overline{D}_6$
1_6	D_6	$\overline{D}_6$
D_6	D_6	$\overline{D}_6$

(a)

$V(c_{i_1})$	$V(c_{i_2})$	$V(c_j)$
1_6	1_6	1_6
0_6	X_6	0_6
X_6	0_6	0_6
D_6	$\overline{D}_6$	0_6
$\overline{D}_6$	D_6	0_6
D_6	1_6	D_6
1_6	D_6	D_6
D_6	D_6	D_6
$\overline{D}_6$	1_6	$\overline{D}_6$
1_6	$\overline{D}_6$	$\overline{D}_6$
$\overline{D}_6$	$\overline{D}_6$	$\overline{D}_6$

(b)

$V(c_{i_1})$	$V(c_{i_2})$	$V(c_j)$
1_6	X_6	1_6
X_6	1_6	1_6
D_6	$\overline{D}_6$	1_6
$\overline{D}_6$	D_6	1_6
0_6	0_6	0_6
D_6	0_6	D_6
0_6	D_6	D_6
D_6	D_6	D_6
$\overline{D}_6$	0_6	$\overline{D}_6$
0_6	$\overline{D}_6$	$\overline{D}_6$
$\overline{D}_6$	$\overline{D}_6$	$\overline{D}_6$

(c)

$V(c_{i_1})$	$V(c_{i_2})$	$V(c_j)$
1_6	0_6	1_6
0_6	1_6	1_6
D_6	$\overline{D}_6$	1_6
$\overline{D}_6$	D_6	1_6
0_6	0_6	0_6
1_6	1_6	0_6
D_6	D_6	0_6
$\overline{D}_6$	$\overline{D}_6$	0_6
0_6	D_6	D_6
D_6	0_6	D_6
1_6	$\overline{D}_6$	D_6
$\overline{D}_6$	1_6	D_6
1_6	D_6	$\overline{D}_6$
D_6	1_6	$\overline{D}_6$
0_6	$\overline{D}_6$	$\overline{D}_6$
$\overline{D}_6$	0_6	$\overline{D}_6$

(d)

Figure 4.6 Six-valued cube covers representing the behavior of the fault-free version of a two-input (a) NAND, (b) AND, (c) OR, and (d) XOR gate

cubes in these two sets will be collectively referred to as the cube cover of the faulty element.

$V(c_i)$	$V(c_{j_1})$	$\cdots$	$V(c_{j_l})$	$\cdots$
1_6	1_6	$\cdots$	1_6	$\cdots$
0_6	0_6	$\cdots$	0_6	$\cdots$
D_6	D_6	$\cdots$	D_6	$\cdots$
$\overline{D}_6$	$\overline{D}_6$	$\cdots$	$\overline{D}_6$	$\cdots$

Figure 4.7 Six-valued cube cover representing the behavior of a fault-free fanout system

A faulty gate: Since a gate is assumed to have a single output, in general, when the fault is excited, the output must take either value $\{D\}$ or $\{\overline{D}\}$. The cubes with either of these values as output are called fault excitation cubes of a faulty circuit. Consider the two-input NAND gate with inputs c_{i_1} and c_{i_2} with an SA1 fault at its output c_j. An examination of the corresponding four-valued truth table in Figure 4.5(a) shows that this fault can be excited only if values $V(c_{i_1}) = V(c_{i_2}) = \{\underline{1}\}$ are implied at the gate inputs. This implies $V(c_j) = \{\overline{D}\}$ at its output. This information is represented in the form of a fault-excitation cube of the corresponding faulty circuit. Figure 4.8(a) lists the fault excitation cube of a two-input NAND gate with an SA1 fault at its output.

Next, consider a two-input NAND gate with an SA0 fault at its output, c_j. In this case, the fault is excited and value $\{D\}$ implied at its output for the following three combinations of values implied at its inputs (a) $V(c_{i_1}) = \underline{0}$ and $V(c_{i_2}) = \underline{0}$, (b) $V(c_{i_1}) = \underline{0}$ and $V(c_{i_2}) = \underline{1}$, and (c) $V(c_{i_1}) = \underline{1}$ and $V(c_{i_2}) = \underline{0}$. In this case, the first two combinations of values may be combined and the first and third combinations may be combined to obtain the fault-excitation cubes of the faulty gate, as shown in Figure 4.8(b).

Figure 4.8 also shows the fault-excitation cubes of various faulty versions of a two-input XOR gate. Two main differences between the fault excitation cubes of NAND and XOR gates should be noted. First, for a given SAF at an input of a NAND gate, either D or $\overline{D}$, but not both, is implied at its output. In contrast, it is possible to imply either of these two values at the output of an XOR gate. Second, multiple input combinations may sometimes be combined into a single cube for a NAND gate; in contrast, no such combining is possible for an XOR gate.

A faulty fanout system: The four-valued truth table of a faulty fanout system can be organized in a similar manner to obtain its fault excitation cubes. Figure 4.8 also shows the fault excitation cubes of various faulty versions of a fanout system.

4.3   Test generation basics

Two key concepts, namely, *fault effect excitation* (FEE) and *fault effect propagation* (FEP), form the basis of all methods to search for test vectors. However, as will become

$V(c_{i_1})$	$V(c_{i_2})$	$V(c_j)$
1_6	1_6	$\overline{D}_6$

(a)

$V(c_{i_1})$	$V(c_{i_2})$	$V(c_j)$
0_6	χ_6	D_6
χ_6	0_6	D_6

(b)

$V(c_{i_1})$	$V(c_{i_2})$	$V(c_j)$
0_6	1_6	D_6

(c)

$V(c_{i_1})$	$V(c_{i_2})$	$V(c_j)$
1_6	1_6	$\overline{D}_6$

(d)

$V(c_{i_1})$	$V(c_{i_2})$	$V(c_j)$
0_6	0_6	$\overline{D}_6$
1_6	1_6	$\overline{D}_6$

(e)

$V(c_{i_1})$	$V(c_{i_2})$	$V(c_j)$
0_6	1_6	D_6
1_6	0_6	D_6

(f)

$V(c_{i_1})$	$V(c_{i_2})$	$V(c_j)$
0_6	1_6	D_6
0_6	0_6	$\overline{D}_6$

(g)

$V(c_{i_1})$	$V(c_{i_2})$	$V(c_j)$
1_6	0_6	D_6
1_6	1_6	$\overline{D}_6$

(h)

$V(c_i)$	$V(c_{j_1})$	$\cdots$	$V(c_{j_l})$	$\cdots$
0_6	0_6	$\cdots$	$\overline{D}_6$	$\cdots$

(i)

$V(c_i)$	$V(c_{j_1})$	$\cdots$	$V(c_{j_l})$	$\cdots$
1_6	1_6	$\cdots$	D_6	$\cdots$

(j)

$V(c_i)$	$V(c_{j_1})$	$\cdots$	$V(c_{j_l})$	$\cdots$
0_6	$\overline{D}_6$	$\cdots$	$\overline{D}_6$	$\cdots$

(k)

$V(c_i)$	$V(c_{j_1})$	$\cdots$	$V(c_{j_l})$	$\cdots$
1_6	D_6	$\cdots$	D_6	$\cdots$

(l)

Figure 4.8 Six-valued fault-excitation cubes of some faulty circuit elements. A two-input NAND gate with (a) output SA1, (b) output SA0, (c) input c_{i_1} SA1, and (d) input c_{i_1} SA0. A two-input XOR gate with (e) output SA1, (f) output SA0, (g) input c_{i_1} SA1, and (h) input c_{i_1} SA0. A fanout system with (i) branch c_{j_l} SA1, (j) branch c_{j_l} SA0, (k) stem c_i SA1, and (l) stem c_i SA0

clear in the following, the search for a test for a fault can – in the worst case – have exponential run-time complexity. Hence, a large number of additional concepts have been developed to make test generation practical.

In this section, we manually generate tests for selected target faults in example circuits to introduce some key concepts and procedures that are used to perform test generation.

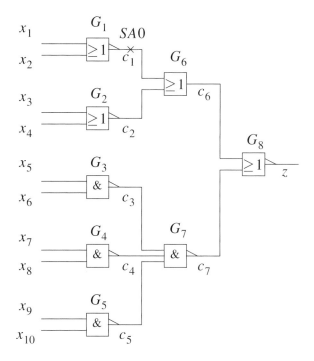

Figure 4.9 A 10-input fanout-free circuit.

Consider the 10-input fanout-free circuit shown in Figure 4.9 with primary inputs $x_1, x_2, \ldots, x_{10}$, internal lines $c_1, c_2, \ldots, c_7$, and a single primary output z. Let the target fault be the SA0 fault at line c_1. Let us begin by assigning value χ_6 to all lines not in the transitive fanout of the fault site and X_6 to the remaining lines.

Fault effect excitation is the process of creating a fault effect at one or more outputs of the faulty circuit element. The *local value assignments* for fault effect excitation are given by the fault excitation cubes. The reader can verify that the only combination of values that may be applied to the inputs of faulty gate G_1 to excite the above fault are 0_6 at x_1 as well as x_2. This results in a D_6 at the fault site, c_1.

Fault effect propagation is the process of assigning values to circuit lines such that a fault-effect propagates from an output of the faulty circuit element to a primary output of the circuit. One or more paths must exist between the fault site and a primary output such that each line along that path has D_6 or $\overline{D}_6$ for the fault effect to be propagated to that output (Roth, 1966).

The conditions for fault effect propagation require the propagation of D_6 at c_1 to the primary output of the circuit, z. In this fanout-free circuit, there exists only a single physical path from the fault site to the circuit output, namely the path $c_1 c_6 z$. The requirements for fault effect propagation can be dealt with – one gate at a time – considering the propagation of the fault effect (a) via G_6, from c_1 to c_6, and (b) via

G_8, to z. The six-valued cube cover of two-input OR gate G_6 provides the following propagation cubes.

c_1	c_2	c_6
D_6	0_6	D_6
0_6	D_6	D_6
D_6	D_6	D_6
$\overline{D}_6$	0_6	$\overline{D}_6$
0_6	$\overline{D}_6$	$\overline{D}_6$
$\overline{D}_6$	$\overline{D}_6$	$\overline{D}_6$

Note how a non-controlling value at one input helps propagate the fault effect from its other input to the gate output. Also note how multiple fault effects of the same polarity at the inputs of a primitive gate can also provide a fault effect at its output.

Since line c_2 does not belong to the transitive fanout of the fault site, it cannot carry a value D_6. Also, line c_1 already contains D_6. Hence, only the first of these alternatives can be achieved and any test for the target fault must imply 0_6 at c_2 to enable the propagation of D_6 via G_6 to c_6. Similarly, using the six-valued cube cover for a two-input NOR gate, one can determine that to propagate D_6 at c_6 via G_8 to obtain a $\overline{D}_6$ at z, a test vector must imply 0_6 at c_7. This process of propagating a fault effect from one or more inputs of a gate to its output – by assigning suitable values to the other inputs of the gate – is called D-**drive**.

The following table summarizes the above local value assignments that a vector must satisfy to detect the SA0 fault at line c_1.

Line	Required value
x_1	0_6
x_2	0_6
c_2	0_6
c_7	0_6

This part of the example also illustrates one of the key components of a test generation problem, namely, the process of translating the requirements for fault effect excitation and fault effect propagation into a set of logic value requirements at an appropriate set of circuit lines. It should be noted that the conditions in the above example are unique and easy to identify since the circuit is **fanout-free**. For a general circuit, these requirements are not unique, hence not so easy to identify.

The remainder of the example illustrates the process of finding a vector that will imply desired logic values at corresponding circuit lines. The logic value at a line is determined by the logic values applied at the inputs of its cone. Since this is a fanout-free circuit, the input cones of the lines at which specific values are required are independent, i.e., they do not share any gates or primary inputs.

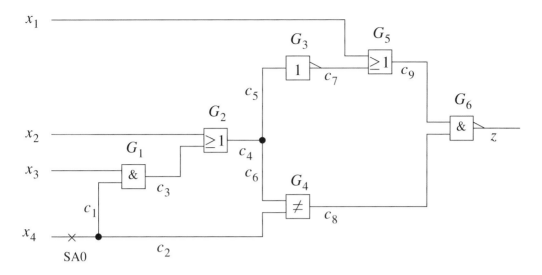

Figure 4.10 An example circuit for test generation (Roth, 1966) (© 1966 IBM).

Hence, the four values described above can be obtained independently without any conflict. Again, this is not true for a general circuit as will be seen in the next example.

The first two logic value requirements given in the above table are 0_6 at lines x_1 and x_2. Since both these lines are primary inputs, these requirements are satisfied by directly applying the desired values at the lines. Next, consider the third logic value requirement in the above table, namely, 0_6 at c_2. Analysis of the six-valued cube cover of two-input NOR G_2 shows that this requirement can be satisfied by any vector that assigns 1_6 to *either* x_3 or x_4. Next, consider the fourth logic value requirement, 0_6 at c_7. Using the six-valued cube cover of three-input NAND G_7, one can determine that this requirement can be satisfied by a vector only if it implies 1_6 at c_3, c_4, and c_5. Each of these requirements can then be analyzed in the same fashion. By continuing the analysis, it can be seen that a number of combinations of values to primary inputs $x_5, x_6, \ldots, x_{10}$ can satisfy these requirements. One of these combinations of values is 0_6 at inputs $x_5, x_6, \ldots, x_{10}$.

For our example circuit, one possible test vector is $(0_6, 0_6, 1_6, \chi_6, 0_6, 0_6, 0_6, 0_6, 0_6, 0_6)$. Note that when this vector is applied to the CUT, only the fault-free part of the composite value needs to be applied. Hence, the three-valued vector $(0, 0, 1, x, 0, 0, 0, 0, 0, 0)$ is a test for the SA0 fault at c_1.

The task of identifying values to be assigned to primary inputs such that a desired value is implied at an internal circuit line is called **justification**.

Next, consider the SA0 fault at primary input x_4 of the four-input circuit shown in Figure 4.10. Again, we initialize all the lines in the transitive fanout of the fault site to X_6 and the remaining lines to χ_6.

Since the target fault is at a primary input line, the fault can be excited merely by applying 1_6 at input x_4. Note that due to the presence of the fault, value D_6 appears at line c_1 as well as c_2. This process of computing the values at c_1 and c_2, once 1_6 is assigned to x_4, is a very simple example of *implication*, or more precisely, *forward implication*. Note that the values assigned at this stage of test generation to the inputs of XOR gate G_4 are X_6 at c_6 (assigned during initialization) and D_6 at c_2. The six-valued cover of a two-input XOR gate shows that the output of this XOR gate cannot be assigned any more specific value than X_6 at this time. Hence, the forward implication stops at line c_2. For to a similar reason, it also stops at c_1.

Now consider fault effect propagation. Note that a fault effect D_6 currently exists at an input of G_1 while value X_6 is assigned to its output. Since the value at the output, X_6, is less specific than D_6 or $\overline{D}_6$ while a fault effect exists at one or more of its inputs, gate G_1 is a candidate for fault effect propagation. For similar reasons, gate G_4 is also a candidate for fault effect propagation.

If the output of a gate is assigned an incompletely specified value whose set representation includes a D or $\overline{D}$ and a fault effect appears at one or more of its inputs, then the gate is added to a set $\mathcal{D}$, called the **D-frontier**. At any stage of test generation, the D-frontier contains all the gates that are candidates for fault effect propagation. Note that excluded from the D-frontier is any gate with fault effects at one or more of its inputs but whose output (a) already has a fault effect, or (b) has a value whose set representation does not include D or $\overline{D}$. A gate of the first type is excluded since the fault effect has already been propagated to its output; a gate of the latter type is excluded since a fault effect cannot be propagated to its output. At this stage of test generation for our example circuit, $\mathcal{D} = \{G_1, G_4\}$.

In this example, we can *choose* to propagate the fault effect either via G_1, G_4, or both. Let us first try to propagate the fault effect via G_1 and subsequently to the output of the circuit, z. The reader can perform D-drive one step at a time to find that assignments 1_6 at x_3 and 0_6 at x_2 propagate the fault effect via G_1 and G_2, respectively, to give a D_6 at c_4. Analysis of fanouts of c_4 determines that D_6 at c_4 implies that value D_6 will appear at both its fanouts, c_5 and c_6. Note that after this implication step, the fault effects at lines c_5 and c_6 must be considered (instead of the fault effect at c_4) during any future attempt to propagate the fault effect.

Forward implication of D_6 at line c_5 using the six-valued cube cover of NOT gate G_3 shows that a $\overline{D}_6$ must be assigned to its output, c_7. This $\overline{D}_6$ must now be considered for any future fault effect propagation, instead of the D_6 at c_5. Forward implication of D_6 at c_6 using the six-valued cubes of XOR gate G_4 shows that 0_6 is implied at c_8. This, in turn, can be further implied via NAND gate G_6 to obtain 1_6 at primary output z. The last implication, $z = 1_6$, demonstrates that **none** of the input combinations described by the current primary input assignments (χ_6 at x_1, 0_6 at x_2, 1_6 at x_3, and 1_6 at x_4) can detect the target fault. Hence, one must **backtrack**, i.e., review the most recent assignment that was made and replace it by an alternative assignment. Tracing our actions, we find that the last assignment that we had made was 0_6 at x_2 to propagate the fault effect via G_2. This assignment was followed by value assignments at lines c_4, c_5, c_6, c_7, c_8, and z. However, all those assignments were derived via implication as direct consequences of the assignment of 0_6 at x_2. Thus, we must erase all the assignments implied by this assignment.

Once all the above assignments are erased, one strategy is to assign the only other value that may be assigned to x_2, namely, 1_6. If we make this assignment, and perform implications, we get the following values: 1_6 at c_4, c_5, and c_6, 0_6 at c_7, $\overline{D}_6$ at c_8, and χ_6 at c_9. Note that the only gate in the D-frontier is G_6 since the fault effect at c_8 is now the only one that may be considered for propagation.

The six-valued cubes of NAND gate G_6 show that $\overline{D}_6$ at c_8 can only be propagated to z by assigning 1_6 at c_9. This assignment is then made to obtain a D_6 at z.

Note that the current values at the inputs of the circuit do not represent a test for the target fault since the value 1_6 assigned at c_9 is not consistent with that implied by the values assigned to the inputs of gate G_5.

If the value assigned at the output of a gate cannot be obtained by implication of the values at its inputs, then the gate is said to be **unjustified**. All unjustified lines in the circuit should be justified before test generation is complete. Hence, a **list of unjustified lines** $\mathcal{U}$ that contains all unjustified lines in the circuit is often maintained.

In our case, it can be determined that the value at c_9 can be justified only by assigning 1_6 to x_1. This is an example of *backward implication*, where the value at one input of an unjustified gate can be more completely specified by the values at its other inputs and output.

For our example circuit, we obtain vector $(1_6, 1_6, 1_6, 1_6)$. Hence, the corresponding three-valued vector $(1, 1, 1, 1)$ is a test for the SA0 fault at x_4 in our example circuit.

4.4 Implication

Logic implication, or simply **implication**, is the process of determining the logic values that appear at various lines of a circuit as a consequence of the values assigned to some of its lines.

Example 4.2 Consider the fault-free circuit shown in Figure 4.11. Assume that initially, the values at all the circuit lines are unspecified, as shown in Figure 4.11(a). Now consider a scenario where a test generator assigns a value 0 to line c_9, making $v(c_9) = 0$. Since c_9 is an input of two-input NAND gate G_4, $v(c_9) = 0 \Rightarrow v(z) = 1$. Also, c_9 is the output of two-input NAND gate G_3. Since the output of a NAND gate can be 0 if and only if each input of the NAND gate is 1, $v(c_9) = 0 \Rightarrow v(c_4) = 1$ and $v(c_7) = 1$. This process can be continued to finally obtain the logic values shown in Figure 4.11(b). ☐

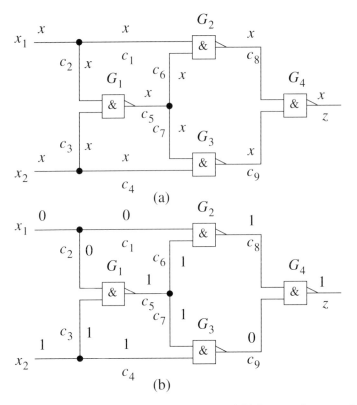

Figure 4.11 An example of logic implication: (a) initial status of an example fault-free circuit, and (b) logic values obtained via logic implication following the assignment $v(c_9) = 0$

The above process of starting with unspecified values at all the circuit lines except c_9, where 0 is assigned, to obtain all the values is an example of logic implication.

As will become apparent in the following, by identifying the values implied by the assigned values, a powerful implication procedure can decrease the amount of search required to find a test vector for a target fault and hence can decrease the time required for test generation. More importantly, when an untestable fault is targeted, a powerful implication procedure can decrease the time required to demonstrate that the fault is untestable.

In general, an *implication procedure* uses primitive *implication operations*. These operations can be classified into *direct* and *indirect* implication operations. A **direct implication operation** determines the value at an input or an output of a circuit element, given the values assigned to the element's inputs and outputs. In contrast, an **indirect implication operation** determines the value at a circuit line c_i given the values at lines other than the inputs and outputs of the circuit elements that are directly connected to c_i.

Implication operations are used, often in an event-driven fashion, by an implication procedure. An **implication procedure** determines logic values at circuit lines. It also

identifies and reports any inconsistency between the desired value at any line and
the value determined at the line via implication. In addition, a typical implication
procedure also updates various lists, such as the D-frontier, $\mathcal{D}$, and the list of
unjustified lines, $\mathcal{U}$.

4.4.1   Direct implication operations

As will be seen ahead, a test generation algorithm repeatedly executes its implication procedure, which in turn executes implication operations at appropriate circuit
elements. Each time an implication procedure is executed, values are assigned to one
or more circuit lines. Hence, during a typical execution of an implication procedure,
a value is already associated with each line of the circuit under consideration. The
execution of an implication operation on a circuit element can hence be viewed as an
update of the values associated with its outputs or inputs by more specific values. In the
following, we will refer to the values at the inputs and outputs of an element prior to
execution of an implication operation as **given** or **current** values; the values obtained
after the operation will be referred to as **new** or **updated** values.

An execution of a direct implication operation updates the value at the outputs or
inputs of a circuit element only by analyzing the values at its inputs and outputs. A
direct **forward implication operation** only proceeds from a line to lines directly in
its fanout and a direct **backward implication operation** only proceeds from a line
to lines directly in its fanin. In the scenario described in Example 4.2 and illustrated
in Figure 4.11, the update of the values at the output of G_4 and the inputs of G_3
are, respectively, examples of direct forward and backward implications of assignment
$v(c_9) = 0$.

At the core of implication operations is the set of functions describing the logic
behavior of various circuit elements. In the following, we will assume that the logic
behavior of a gate G with output c_j in the composite circuit is described using a
function $\mathcal{G}_j^4()$, which can be used to compute the value at the output of the gate for
any combination of the values $\underline{0}, \underline{1}, D$, and $\overline{D}$ applied at its inputs. Similarly, the logic
behavior of a fanout branch c_j in the composite circuit is described by a function
$\mathcal{G}_j^4()$ of the corresponding fanout system. In practice, functions $\mathcal{G}_j^4()$ are typically
represented in the form of four-valued truth tables or using higher-valued cube covers
of the corresponding circuit elements, as described earlier.

When implication is performed for each faulty circuit element in a composite
circuit, the corresponding function $\mathcal{G}_j^4()$ represents the four-valued truth table or
cube cover of the element with the corresponding fault. For example, consider
the composite circuit shown in Figure 4.1(d), which has an SA1 fault at line c_9.
When this circuit is being studied, $\mathcal{G}_9^4()$ represents the four-valued truth table for a
two-input NAND gate with output SA1 fault, shown in Figure 4.5(a). In contrast,
since gate G_1 with output c_5 is fault-free in the above scenario, $\mathcal{G}_5^4()$ represents

the four-valued truth table of a fault-free two-input NAND gate, shown in Figure 4.4(a).

In the following, the implication operations and procedure are described for the composite circuit and composite value systems that use $\underline{0}$, $\underline{1}$, D, and $\overline{D}$ as basic values. Also, each circuit element, fault-free or faulty, is described by its four-valued function.

The implication operations and procedure presented next can be adapted for the fault-free circuit by using (a) the two-valued function, $\mathcal{G}_j^2(\ )$, for each output c_j of each logic element, instead of its four-valued function, $\mathcal{G}_j^4(\ )$, and (b) a value system that is based on basic values 0 and 1, such as the three-valued system with symbols $0 = \{0\}$, $1 = \{1\}$, and $x = \{0, 1\}$. Furthermore, by merely using the two-valued function for each faulty circuit element, the operations and the procedure can be adapted to a faulty circuit.

In this chapter, we sometimes consider only the fault-free version of the circuit, sometimes one of its faulty versions, and, most of the time, the composite circuit, which is a combination of both these versions. Even though the circuit diagrams for these cases look very similar, confusion can be avoided by observing the values assigned to the circuit lines. Values from a composite value system are used whenever a composite circuit is considered. Values from the two- or three-valued system are used when either the fault-free or a faulty version of the circuit is considered. Between these two versions, the target fault is shown when a faulty version of the circuit is considered.

4.4.1.1 Direct forward implication operations

We will first describe the direct forward implication operation at a gate. This will be followed by its description at a fanout system.

Direct forward implication operation at a gate: Consider a gate G with output c_j, and inputs $c_{i_1}, c_{i_2}, \ldots, c_{i_\alpha}$. Let $V(c)$ denote the value associated with line c. (Recall that the value associated with each line in the composite circuit is a symbol from the composite value system in use and represents a set of basic composite values. Also, recall that whenever a single symbol is used to represent multiple sets of logic values, the symbol is deemed to denote its leading value.) The direct forward implication operation at gate G is comprised of two steps.

First, the value implied at the gate output by the given values at its inputs is determined as

$$V^* = \bigcup_{\forall v_l \in V(c_{i_l}),\ \forall l \in \{1,2,\ldots,\alpha\}} \mathcal{G}_j^4(v_1, v_2, \ldots, v_\alpha). \tag{4.1}$$

Note that value V^* given by the above operation is in the form of a set of basic logic values. In practice, the computed values are stored only as symbols of the logic system in use. Table 4.1 can be used to replace the set of basic values, V^*, by the leading value of the corresponding symbol of the value system. Note that the computed set of basic values should also be replaced by an appropriate symbol of the value system in use *after each of the following operations.*

Second, value $V(c_j)$ at the output of the gate is updated. In addition, the effect of the implication operation on the value at c_j is determined. If $V^* \supseteq V(c_j)$, then the implication operation leaves the value $V(c_j)$ **unchanged**. In all the other cases, the implication is said to have **changed** the value $V(c_j)$ and its new value is computed as

$$V(c_j) = V^* \bigcap V(c_j). \tag{4.2}$$

It is possible that the new value of $V(c_j)$ is equal to {}. This indicates that no value can be assigned to the output of G and hence the implication is said to result in a **conflict**.

The implication operation is also the natural place to update various lists such as $\mathcal{D}$, the D-frontier, and $\mathcal{U}$, the set of unjustified lines.

Whether or not c_j is an unjustified line can be determined by examining values V^* and $V(c_j)$. If the new value of $V(c_j)$ is identical to V^* computed above, the line c_j is said to be **justified**, since the values at the inputs of the circuit element imply the value at its output. Otherwise, i.e., when $V(c_j) \subset V^*$, line c_j as well as the corresponding circuit element are said to be **unjustified**.

A **gate belongs to the D-frontier** if at least one of the gate's inputs has the value that corresponds to $\{D\}$ or $\{\overline{D}\}$, and its incompletely specified output value corresponds to a set that contains multiple basic values including D, $\overline{D}$, or both. For example, when the six-valued system is used, this can occur only when at least one of the inputs has the value $D_6 = \{D\}$ or $\overline{D}_6 = \{\overline{D}\}$ and the output has the value $X_6 = \{\underline{0}, \underline{1}, D, \overline{D}\}$.

We will assume that the implication operation appropriately updates lists $\mathcal{D}$ and $\mathcal{U}$.

Example 4.3 Consider the two-input AND gate G with output c_j and inputs c_{i_1} and c_{i_2} shown in Figure 4.12(a). Let the initial values at the inputs and output be $V(c_{i_1}) = V(c_{i_2}) = V(c_j) = \{\underline{0}, \underline{1}, D, \overline{D}\}$. Assume that the value at input c_{i_2} is changed (say, by the execution of forward implication operation at the circuit element driving c_{i_2}) from $\{\underline{0}, \underline{1}, D, \overline{D}\}$ to $\{D\}$. First, consider how the implication operation will be performed for the 16-valued system, in which each possible set of values has a distinct symbol, denoted by the set itself. First, V^* is computed using the four-valued truth table of a fault-free two-input AND gate as $G_j^4()$ and Equation (4.1). This gives $V^* = \{\underline{0}, D\}$. Second, since the 16-valued system uses a distinct symbol to denote each possible set of values, V^* retains the above value even after it is replaced by the corresponding symbol. Third, a comparison of V^* with the given value at line c_j, $V(c_j)$, indicates that the value at c_j has changed. Hence, $V(c_j)$ is updated using Equation (4.2) which gives $V(c_j) = \{\underline{0}, D\}$, as shown in Figure 4.12(b).

Furthermore, the gate output is marked as having changed, the gate is said to be justified, and the gate is added to $\mathcal{D}$, the D-frontier. ☐

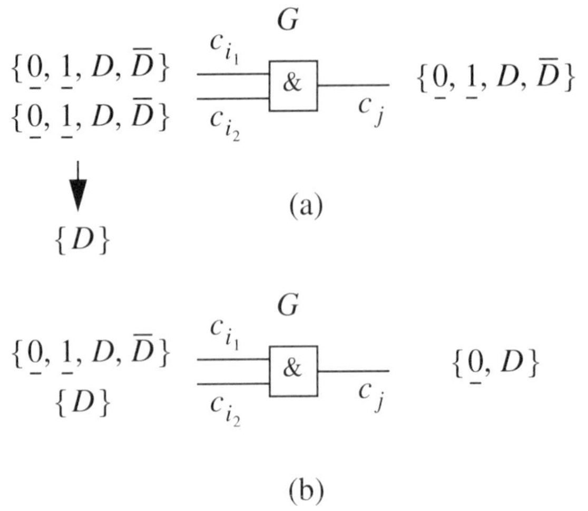

$$\{0, 1, D, \bar{D}\} \quad \overset{G}{\underset{c_{i_2}}{\overset{c_{i_1}}{\boxed{\&}}}} \quad \overset{}{\underset{c_j}{}} \{0, 1, D, \bar{D}\}$$

$\{0, 1, D, \bar{D}\}$

$\downarrow$

$\{D\}$

(a)

$$\{0, 1, D, \bar{D}\} \quad \overset{G}{\underset{c_{i_2}}{\overset{c_{i_1}}{\boxed{\&}}}} \quad \overset{}{\underset{c_j}{}} \{0, D\}$$

$\{D\}$

(b)

Figure 4.12 An example of forward implication operation at a gate using the 16-valued system: (a) initial status of the gate, and (b) the values determined by implication

$$X_6 = \{0, 1, D, \bar{D}\} \quad \overset{G}{\underset{c_{i_2}}{\overset{c_{i_1}}{\boxed{\&}}}} \quad \overset{}{\underset{c_j}{}} X_6 = \{0, 1, D, \bar{D}\}$$

$X_6 = \{0, 1, D, \bar{D}\}$

$\downarrow$

$D_6 = \{D\}$

(a)

$$X_6 = \{0, 1, D, \bar{D}\} \quad \overset{G}{\underset{c_{i_2}}{\overset{c_{i_1}}{\boxed{\&}}}} \quad \overset{}{\underset{c_j}{}} X_6 = \{0, 1, D, \bar{D}\}$$

$D_6 = \{D\}$

(b)

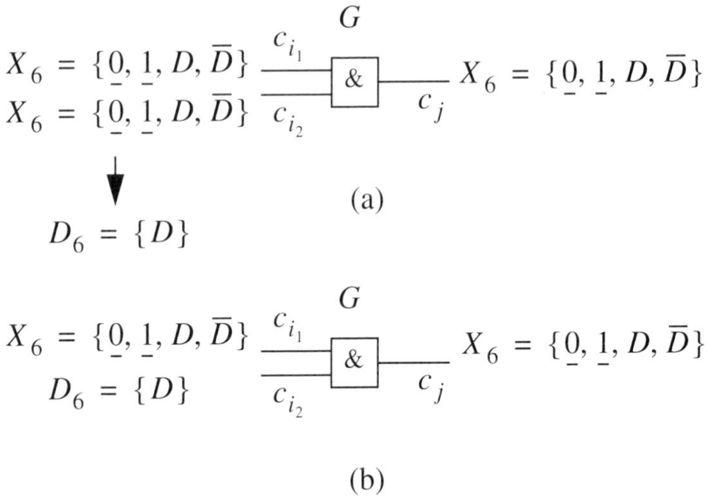

Figure 4.13 An example of forward implication operation at a gate using the six-valued system: (a) initial status of the gate, and (b) the values determined by implication

Example 4.4 Figures 4.13(a) and (b) show the implication operation described in Example 4.3, but assuming that the six-valued system is used. In this case, initially, the input values, old as well as new, are specified using appropriate symbols from the six-valued system. Figure 4.13(a) shows that initially $V(c_{i_1}) = V(c_{i_2}) = V(c_j) =$

X_6, and that $V(c_{i_2})$ is changed to D_6. In the six-valued system, X_6 is used to denote multiple values. Hence, as described above, before V^* is computed using Equation (4.1), X_6 is replaced by its leading value in the six-valued system, namely $\{\underline{0}, \underline{1}, D, \overline{D}\}$. Next, as in the above case, V^* is computed as $\{\underline{0}, D\}$. In this case, V^* is replaced by the corresponding symbol by using Table 4.1 to obtain $V^* = X_6$ (see Figure 4.13(b)). Hence, in this case, the value at c_j is not changed by the forward implication operation.

Next, the gate output is marked as unchanged, the gate is said to be justified, and the gate is added to $\mathcal{D}$, the D-frontier. □

The above example illustrates how the use of one value system may lead to loss of information that can be obtained by performing the same operation using a more powerful value system.

Direct forward implication at a fanout system: Consider a fanout system which has stem c_i and branches $c_{j_1}, c_{j_2}, \ldots, c_{j_\beta}$. In this case, for the fanout branch c_{j_l}, we have

$$V^* = \bigcup_{\forall v \in V(c_i)} \mathcal{G}^4_{j_l}(v), \tag{4.3}$$

and

$$V(c_{j_l}) = V(c_{j_l}) \bigcap V^*. \tag{4.4}$$

The above operations are repeated for each branch of the fanout system, i.e., $\forall l \in \{1, 2, \ldots, \beta\}$.

Note that whether the output value is changed, unchanged, or the forward implication operation results in a conflict can be determined in a manner similar to that for the forward implication operation at a gate.

An algorithm that implements the forward implication operation: Figure 4.14 shows an algorithm for forward implication operation, called ForwardImplicationOperation( ). The input to the operation is a line c which specifies the circuit element at which the operation is performed. c is either an input of a gate G_l or the stem of a fanout system. The operation is performed at the circuit element that has line c as an input. Note that the procedure returns to the calling procedure either the value SUCCESS or CONFLICT. A return value CONFLICT indicates a conflict has occured at one or more outputs of the circuit element; otherwise a return value SUCCESS is used. It also suitably updates flags at the outputs to indicate the effect of the implication operation on the values of the output lines. Finally, the procedure also updates lists $\mathcal{D}$ and $\mathcal{U}$.

ForwardImplicationOperation(c)

{

    if c is an input to gate G_l with output c_j

      then

      {

        compute set V^* using Equation (4.1)

        replace set V^* by an appropriate symbol of logic system in use

        if $V^* \supseteq V(c_j)$

          then changed(c_j)=0

          else changed(c_j)=1

        $V(c_j) = V(c_j) \cap V^*$

        if ($V(c_j) == \{\}$)

          then return(CONFLICT)

          else

          {

            update $\mathcal{D}$ and $\mathcal{U}$

            return(SUCCESS)

          }

      }

      else /* i.e., if c is the stem of a fanout system */

      {

        for each fanout branch c_{j_i} of the fanout system

        {

        compute set V^* using Equation (4.3)

        replace set V^* by an appropriate symbol of logic system in use

        if $V^* \supseteq V(c_{j_i})$

          then changed(c_{j_i})=0

          else changed(c_{j_i})=1

         $V(c_{j_i}) = V(c_{j_i}) \cap V^*$

        if ($V(c_{j_i}) == \{\}$)

          then return(CONFLICT)

          else

          {

            update $\mathcal{D}$ and $\mathcal{U}$

          }

        }

        return(SUCCESS)

      }

}

Figure 4.14 An algorithm for forward implication operation

4.4.1.2 Direct backward implication operation

The backward implication operation updates the value at one or more inputs of a gate or the stem of a fanout system. In the case of a gate, it uses the value at the gate's output as well as the values at the other inputs of the gate, if any. In the case of a fanout system, it uses the values at all branches of the fanout system.

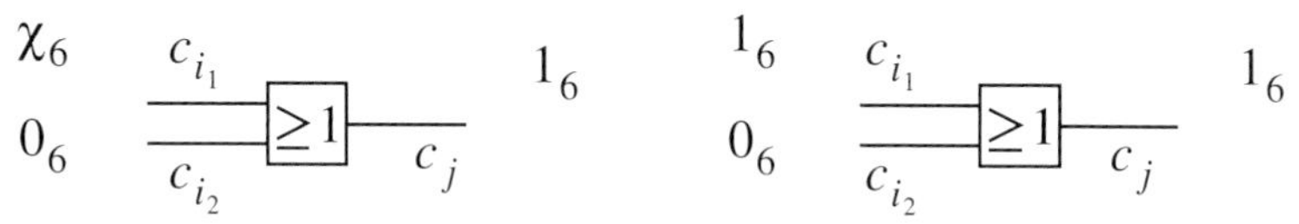

(a)   (b)

Figure 4.15 An example of backward implication: (a) initial values, and (b) values after backward implication

Direct backward implication operation at a gate: Consider gate G with output c_j and inputs $c_{i_1}, c_{i_2}, \ldots, c_{i_\alpha}$. Backward implication is performed one input at a time. Consider the qth input, c_{i_q}, where $q \in \{1, 2, \ldots, \alpha\}$. Let v'_q be one of the basic values contained in $V(c_{i_q})$. First, a value V^* is computed as follows:

$$V^* = \bigcup_{\substack{\forall v_l \in V(c_{i_l}); \\ \forall l \in \{1,2,\ldots,\alpha\}, l \neq q}} \mathcal{G}_j^4(v_1, v_2, \ldots, v_{q-1}, v'_q, v_{q+1}, \ldots, v_\alpha). \tag{4.5}$$

V^* is the set of all the possible basic values that appear at c_j, provided (a) value v'_q is the only basic value applied to the qth input of the gate, and (b) all possible combinations of basic values contained in the given input values are applied to all other gate inputs. If V^* does not contain any basic value contained in $V(c_j)$, then the presence of basic value v'_q in $V(c_{i_q})$ is no longer warranted, for the given values at the gate output and all its inputs other than c_{i_q}. Hence, the following operation is performed to update the value of $V(c_{i_q})$.

If $V^* \cap V(c_j) = \{\}$, then $V(c_{i_q}) = V(c_{i_q}) - \{v'_q\}$. \tag{4.6}

The steps given by Equations (4.5) and (4.6) are then repeated for each $v'_q \in V(c_{i_q})$. The above steps are then repeated for each input of the gate, i.e., $\forall q \in \{1, 2, \ldots, \alpha\}$.

Example 4.5 Consider the gate and values shown in Figure 4.15. First, consider input c_{i_1}, i.e., $q = 1$. Furthermore, since $V(c_{i_1}) = \chi_6$ contains basic values $\underline{0}$ as well as $\underline{1}$, first consider $v'_1 = \underline{0}$. For this case, Equation (4.5) can be used to obtain $V^* = \{\underline{0}\}$. Since $V^* \cap V(c_j) = \{\}$, it indicates that the presence of the corresponding basic value, $v'_1 = \underline{0}$, is not consistent with $V(c_j) = 1_6$ and $V(c_{i_2}) = 0_6$. Hence, the use of the step given by Equation (4.6) removes this value from $V(c_{i_1})$ and gives $V(c_{i_1}) = 1_6$.

The above steps are then repeated for $q = 1$ and $v'_1 = \underline{1}$, i.e., for the other basic value in $V(c_{i_1})$. Subsequently, all the above steps are repeated for the other input, i.e., for $q = 2$. Since these subsequent steps do not result in removal of any other value from $V(c_{i_1})$ or $V(c_{i_2})$, we finally obtain $V(c_{i_1}) = 1_6$, $V(c_{i_2}) = 0_6$, and $V(c_j) = 1_6$ as a result of the backward implication operation. $\square$

In the above example, the backward implication operation changes the value of $V(c_{i_1})$ and leaves the value of $V(c_{i_2})$ unchanged. These facts are reflected by appropriately setting the flags at lines c_{i_1} and c_{i_2}. Also note that the value at the output of the gate has been justified as a result of this implication operation. Line c_j is hence removed from set $\mathcal{U}$ to reflect this fact.

Direct backward implication operation at a fanout system: Consider the fanout system with stem c_i and branches $c_{j_1}, c_{j_2}, \ldots, c_{j_\beta}$. In this case, backward implication is performed to obtain the new value at the stem by considering one branch at a time. Consider branch c_{j_l} whose behavior is described by function $\mathcal{G}_{j_l}^4(\ )$. Let $v' \in V(c_i)$ be one of the basic values contained in $V(c_i)$. Value V^* is computed as follows:

$$V^* = \mathcal{G}_{j_l}^4(v').$$
(4.7)

V^* is the basic value that appears at c_{j_l}, provided value v' is the only basic value applied to stem c_i. If V^* does not contain any basic value contained in $V(c_{j_l})$, then the presence of basic value v' in $V(c_i)$ is no longer required for the given values at c_{j_l}. Hence, the following operation is performed to update the value of $V(c_i)$.

If $V^* \cap V(c_{j_l}) = \{\}$, then $V(c_i) = V(c_i) - \{v'\}$.
(4.8)

The above steps are then repeated for each branch of the fanout system, i.e., $\forall l \in \{1, 2, \ldots, \beta\}$. The entire process is then repeated for each of the other values $v' \in V(c_i)$.

An algorithm implementing backward implication operation, called BackwardImplicationOperation(c), can be developed using the above equations. In this case, the input parameter c is either the output of the gate or a branch of the fanout system at which backward implication operation needs to be performed. Again, assignment of value $\{\ \}$ at any line constitutes a conflict and is appropriately indicated by the return value, which is either SUCCESS or CONFLICT.

4.4.1.3 Reducing the time complexity of implication operations

During test generation, direct implication operations are performed repeatedly. Hence, Equations (4.1), (4.3), (4.5), and (4.7) are executed many times. Such repeated computation can contribute significantly to the time complexity of test generation.

Higher-valued truth tables: To reduce run-time complexity, the above equations are typically used to compute the value of V^* for each combination of input values, before any implication is performed on circuit lines. After each V^* obtained in this manner is replaced by the corresponding symbol of the value system in use, the results are stored in the form of multi-valued truth tables of gates.

For example, for forward implication, the six-valued truth tables obtained by using these equations are shown in Figure 4.16. (Alternatively, this could be achieved

$$v(c_{i_2})$$

	0_6	1_6	D_6	$\overline{D}_6$	χ_6	X_6
0_6	1_6	1_6	1_6	1_6	1_6	1_6
1_6	1_6	0_6	$\overline{D}_6$	D_6	χ_6	X_6
D_6	1_6	$\overline{D}_6$	$\overline{D}_6$	1_6	X_6	X_6
$\overline{D}_6$	1_6	D_6	1_6	D_6	X_6	X_6
χ_6	1_6	χ_6	X_6	X_6	X_6	X_6
X_6	1_6	X_6	X_6	X_6	X_6	X_6

$v(c_{i_1})$ labels the rows at left.

(a)

$v(c_i)$	$v(c_{j_1})$	$\cdots$	$v(c_{j_l})$	$\cdots$
0_6	0_6	$\cdots$	0_6	$\cdots$
1_6	1_6	$\cdots$	1_6	$\cdots$
D_6	D_6	$\cdots$	D_6	$\cdots$
$\overline{D}_6$	$\overline{D}_6$	$\cdots$	$\overline{D}_6$	$\cdots$
χ_6	χ_6	$\cdots$	χ_6	$\cdots$
X_6	X_6	$\cdots$	X_6	$\cdots$

(b)

Figure 4.16 Six-valued truth tables for fault-free versions of (a) a two-input NAND gate, and (b) a fanout system

by using Procedure ComputeHigherValuedTruthTable( ) given in Chapter 3.) Once obtained, these tables are used, instead of repeated executions of Equations (4.1) and (4.3), to obtain the value of V^* during the execution of implication operations. This helps reduce the run-time complexity of the implication operation.

Similar tables can also be derived for backward implication to avoid repeated computation of values using the above equations.

Higher-valued cube covers: As described in Section 3.2 and Section 4.2, for most commonly used gates, cube covers provide a compact representation of the behavior of the circuit element. Next, we will see how the above operations can be performed using cube covers. We will concentrate on the case of a fault-free gate. Other cases are similar and left as an exercise to the reader.

Forward implication using cube covers: Consider a gate G with output c_j and inputs $c_{i_1}, c_{i_2}, \ldots, c_{i_\alpha}$. Let $V(c_{i_1}), V(c_{i_2}), \ldots, V(c_{i_\alpha})$ be the given values at these inputs. These values can be combined into a cube with α components, $\mathcal{V} = (V(c_{i_1}), V(c_{i_2}), \ldots, V(c_{i_\alpha}))$. Let $\mathcal{C}^l = (\mathcal{C}_1^l, \mathcal{C}_2^l, \ldots, \mathcal{C}_\alpha^l)$ be the cube comprised of the input part of the lth row of the cube cover of the gate. The value of V^* is obtained as the union of the output part of each row l of the gate's cube cover for which cube $\mathcal{V}$

is compatible with cube C^l. Subsequently, Equation (4.2) can be used to determine the new value $V(c_j)$ at the output of the gate.

Alternatively, the given values at the inputs, $V(c_{i_1})$, $V(c_{i_2})$, ..., $V(c_{i_\alpha})$, and that at the output, $V(c_j)$, can be combined into a cube with $\alpha + 1$ components, namely $\mathcal{V}_{ex} = (V(c_{i_1}), V(c_{i_2}), \ldots, V(c_{i_\alpha}), V(c_j))$. Cube $C_{ex}^l = (C_1^l, C_2^l, \ldots, C_\alpha^l, C_{\alpha+1}^l)$ is then obtained by combining all the elements in the lth row of the cube cover of the gate. Cube $\mathcal{V}_{ex}$ is intersected with cube C_{ex}^l, for each row l, and all the values of l for which $\mathcal{V}_{ex}$ and C_{ex}^l are compatible are stored in set I.

The $(\alpha + 1)^{th}$ component of cube $\mathcal{V}_{ex}$, i.e., the given value of $V(c_j)$, is then intersected with the union of the output part of each cube C_{ex}^l, i.e., $C_{\alpha+1}^l$, to obtain the updated value of $V(c_j)$ as

$$V(c_j) = V(c_j) \cap \left[\bigcup_{\forall l \in I} C_{\alpha+1}^l \right]. \tag{4.9}$$

Note that if the given value at the output of the gate is partially specified, then the complexity of this approach can be lower than that of the previous approach where only the input parts of cubes were intersected.

Example 4.6 Consider the two-input AND gate, the scenario depicted in Example 4.4, and Figures 4.13(a) and (b). The given input values can be combined into a two-component cube (X_6, D_6). This can be intersected with the input part of each row of the gate's six-valued cube cover shown in Figure 4.6(b). The reader can verify that the above cube is compatible with rows 2, 5, 7, and 8 of the above cover. The output values of these rows, namely 0_6, 0_6, D_6, and D_6, are combined via a union operation to obtain $V^* = 0_6 \cup 0_6 \cup D_6 \cup D_6 = X_6$.

Alternatively, the other method described above can be used to first combine the current values at the inputs and output of the two-input AND gate into a three-component cube, (X_6, D_6, X_6). This cube can be intersected with the input as well as the output part of each row. The cube is found to be compatible with rows 2, 5, 7, and 8 and output value computed as described above.

Note that in this case the complexities of the two approaches are identical, since the given value at the gate output is X_6. For any other output value, the complexity of the second approach could have been lower.   □

Complete direct implication using cube covers: The second approach for forward implication described above can be extended to perform complete direct implication at a gate. Again, the given values at the inputs as well as the output can be combined into a cube $\mathcal{V}_{ex}$ with $\alpha + 1$ components. Intersection can be performed between this cube and each row of the gate's cube cover to determine set I, the rows of the cube cover that are compatible with the cube of the given input and output values, as described above. Subsequently, the updated value at input c_{i_1} is obtained as

$$V(c_{i_1}) = V(c_{i_1}) \bigcap \left[\bigcup_{\forall l \in I} C_1^l \right]. \tag{4.10}$$

The updated values at other inputs, $c_{i_2}, c_{i_3}, \ldots, c_{i_\alpha}$, as well as the output are obtained in a similar manner.

Example 4.7 Consider the gate used in Example 4.5 and shown in Figure 4.15(a). In this case, the given values at the inputs and output of the gate are combined into a three-component cube $(\chi_6, 0_6, 1_6)$. This cube is intersected with each row of the six-valued cube cover of the two-input OR gate shown in Figure 4.6(c). The reader can verify that this cube intersects only with the first row, giving the updated values as $V(c_{i_1}) = 1_6$, $V(c_{i_2}) = 0_6$, and $V(c_j) = 1_6$.   □

4.4.2   Implication procedure

An implication procedure executes the above implication operations on various elements of the given circuit, as necessary, to identify all the implications of the given logic values assigned to various circuit lines. We will first discuss the key concepts behind an implication procedure. This will be followed by a discussion of its variants.

4.4.2.1 Key concepts

In a typical test generation scenario, the test generation algorithm generally replaces values at one or more circuit lines by values that are more specific than those assigned at the end of the previous execution of the implication procedure. In the following discussion, we will assume that a list, called the **implication list** that contains these lines, is maintained. Each element in the list is called an **implication task** and takes the form (c_i, d_i), where c_i, called the **implication site**, is the circuit line at which the value was altered by the test generation algorithm and d_i is the direction in which implication of the new value is desired. If c_i is a primary input, then d_i is assigned direction FORWARD. If c_i is a primary output, then d_i is assigned direction BACKWARD. If c_i is a line internal to the circuit, then both forward and backward implications may be desired. In this case, two implication tasks at the line are added to the implication list, one with direction FORWARD and one with direction BACKWARD. Some test generation algorithms only use forward implication. In such algorithms, d_i = FORWARD for all the lines except primary outputs, which cannot be sites of forward implication.

Forward implication tasks generated by forward implication: Consider an implication site c_i where forward implication needs to be performed. If c_i is an input to gate

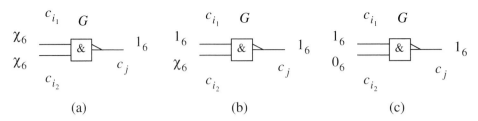

Figure 4.17 An example illustrating how backward implication at gate G following forward implication may help update additional values: (a) original state of gate inputs and output, (b) gate after $V(c_{i_1})$ is changed due to forward implication at the circuit element with output c_{i_1}, and (c) gate after backward implication

G with output line c_j, then a forward implication operation is performed on gate G. If c_i is the stem of a fanout system, then a forward implication operation is performed on that fanout system. In the former case, if the forward implication operation changes the value at the output c_j of G, then forward implication must subsequently be performed at c_j. In the latter case, forward implication must subsequently be performed at each branch of the fanout system where values are changed. In either case, this is accomplished by adding the corresponding forward implication tasks to the implication list.

Backward implication tasks created by forward implication: An interesting case arises when the forward implication operation is performed at line c_i which is an input of a multi-input gate G with output c_j. In this case, if c_j remains unjustified after the execution of the forward implication operation on gate G, then the new value at input c_i and the value at line c_j may make possible some backward implications at other gate inputs.

Example 4.8 Figure 4.17(a) depicts the status of a gate G prior to the execution of a forward implication operation at the gate. At this time, c_j is unjustified. However, the backward implication operation does not update any values, since the desired value at c_j, 1_6, can be obtained in more than one way, by making $V(c_{i_1}) = 0_6$ or $V(c_{i_2}) = 0_6$. Figure 4.17(b) shows a case where the value $V(c_{i_1})$ is changed due to a forward implication at the circuit element with output c_{i_1}. Subsequently, a forward implication operation is performed at gate G as shown in the figure. This operation does not alter the value at the gate output, c_j; it also leaves the gate output unjustified. However, now that $V(c_{i_1}) = 1_6$, the only way value $V(c_j) = 1_6$ can be justified is by replacing $V(c_{i_2})$ by 0_6. Figure 4.17(c) shows the gate after the backward implication is performed. Hence, execution of the backward implication operation at this gate will help identify implication $V(c_{i_2}) = 0_6$. Additional implications may then be caused in the circuit element with output c_{i_2}, and so on. $\square$

Implication tasks created by backward implication: An implication site c_i, where backward implication must be performed, is handled in a dual fashion to what was described above for a forward implication site. In this case, first, backward implication is performed at the circuit element whose output is c_i. A backward implication task is added to the implication list for each input of the circuit element where the value changes due to backward implication. Finally, in a dual fashion to the forward implication case – where forward implication at a gate that remains unjustified is followed by backward implication at the gate – backward implication at a fanout system is followed by forward implication at the system.

The implication procedure terminates after it finishes processing all the implication events in its list. An implication procedure that incorporates the concepts presented above is shown in Appendix 4.A. This procedure is called Implication(S), where input S is the implication list described earlier.

4.4.2.2 Special cases

We will discuss three main variations of the above approach to accommodate (a) test generation algorithms where only forward implication is performed, (b) adaptations to perform implications on the fault-free version of the circuit, and (c) adaptations to faulty circuits.

Forward implication only: Some test generation algorithms only perform forward implications. The above implication procedure can be simplified for this case. First, initially, the implication list only contains forward implication events. Second, during the processing of a forward implication event at a multi-input gate, a backward implication task is not added to the task list.

Typically, many test generation algorithms that only use forward implications assign values only to the primary inputs. In such cases, prior to the execution of the implication procedure, the value at each circuit line is initialized to the completely unspecified value of the value system in use, namely, X_5, X_6, or $\{\underline{0}, \underline{1}, D, \overline{D}\}$ for lines in the transitive fanout of a fault site and X_5, χ_6, or $\{\underline{0}, \underline{1}\}$ for all the other lines. All the primary input values are then changed to the given values, corresponding implication tasks added to the implication list, and the above version of the implication procedure executed. In such cases, there are no unjustified lines in the circuit and no conflict is ever reported.

Implications on a fault-free circuit only: This can be accomplished by using (a) for each circuit line c_j, the two-valued function for the fault-free version of the circuit, $\mathcal{G}_j^2()$, and (b) a value system that uses as basic values 0 and 1, such as the three-valued system with symbols $0 = \{0\}$, $1 = \{1\}$, and $x = \{0, 1\}$.

Implications on a faulty version of the circuit only: This can be accomplished by using the above version with the difference that for each output c_j of the faulty circuit element, function $\mathcal{G}_j^2()$ refers to the behavior described by the two-valued truth table of the particular faulty version of the element.

4.4.3 Indirect implications

The direct implication operations described above compute the values at one or more outputs or inputs of a circuit element, given the values at the inputs and/or outputs of that circuit element. An implication procedure iteratively executes direct implication operations to identify implications of the values assigned to some circuit lines at distant circuit lines in run time that is proportional to the number of lines in the circuit.

Any implication procedure that relies solely on direct implication operations is **incomplete**, i.e., it cannot always identify all the implications of the given values at various circuit lines. *Indirect implication operations* and techniques help make implication more complete.

4.4.3.1 Motivation for indirect implication

The following example illustrates that direct implication is incomplete.

Example 4.9 Consider a portion of the fault-free version of an example subcircuit shown in Figure 4.18. Assume that initially the value at each line is unspecified. Next, consider the scenario where the test generator makes the assignment $v(c_8) = 0$. This assignment may be processed by adding to implication list S the implication task $(c_8, \text{BACKWARD})$. It can be verified that when this implication task is processed by the direct implication procedure, it will not identify any additional implication.

Since the above subcircuit has only three inputs, c_1, c_2, and c_3, all possible combinations of values can be applied at these inputs to verify that $v(c_8) = 0$ only if $v(c_2) = 0$. In other words, $v(c_8) = 0$ *implies* $v(c_2) = 0$. Since the direct implication procedure cannot identify this implication, it is incomplete. □

As shall be seen ahead, identification of more complete implications can reduce the amount of search that a test generator may need to perform. Hence, if an implication procedure can be made more complete at low computational complexity, then the overall complexity of test generation may be reduced.

4.4.3.2 Static indirect implications

Many indirect implications, such as $v(c_8) = 0 \Rightarrow v(c_2) = 0$ in the above example scenario, can be identified in a pre-processing step preceeding test generation for individual target faults. Since these indirect implications are identified in a

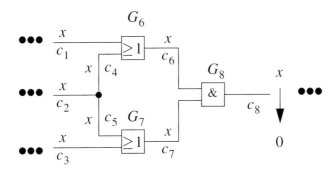

Figure 4.18 A fault-free subcircuit and an example scenario illustrating incompleteness of direct implications (Schulz *et al.*, 1988) (© 1988 IEEE)

pre-processing step, they do not take advantage of the values assigned to various circuit lines by the test generator. Hence, such indirect implications are called **static indirect implications** and are typically performed only on the fault-free version of the circuit. Indirect implications identified in the pre-processing step can be stored and used during each execution of the implication procedure to identify additional implications of the assigned values.

Types of implications identified: Two types of static indirect implications are commonly identified (Schulz *et al.*, 1988), namely

1 $v(c_j) = u \Rightarrow v(c_i) = w$, where $u, w \in \{0, 1\}$, and
2 $v(c_j) = K$, where K is a fixed logic value, independent of the input vector applied to the circuit.

The objective of the identification of indirect implication is to improve the quality of implication *at low computation cost*. Hence, a methodology to statically identify indirect implications should focus on those implications that cannot be identified by the direct implication procedure described earlier.

A methodology to identify static indirect implications: Consider the following methodology to identify static indirect implications (Schulz *et al.*, 1988). Start with the *fault-free version* of the given circuit. Second, initialize the value at each circuit line to the unspecified value, x. Recall that, in the set notation, $x = \{0, 1\}$. (Since this is only the fault-free version of the circuit, a composite value system need not be used.) Next, select a circuit line, say c_i, and assign it the value 0, i.e., make $v(c_i) = 0$, and perform direct implications using an appropriate version of the implication procedure given before. Direct implication may assign specific logic values, i.e., either 0 or 1, at other circuit lines. Consider one of these lines, say c_j, such that the procedure gives $v(c_j) = 0$. Since prior to the application of direct implication the only specified value in the circuit was $v(c_i) = 0$, direct implication demonstrates that

$$v(c_i) = 0 \quad \Rightarrow \quad v(c_j) = 0.$$

This result obtained via direct implication can be used to *deduce* an additional implication via the observation that if one desires $v(c_j) = 1$, then $v(c_i) \neq 0$. In other words, the new implication that can be deduced is

$$v(c_j) = 1 \quad \Rightarrow \quad v(c_i) = 1.$$

Once all the implications due to assignment $v(c_i) = 0$ are identified, and appropriate implications deduced therefrom, the circuit can be reinitialized, by assigning x to all the lines, and the above process repeated with $v(c_i) = 1$.

One noteworthy possibility is that there may exist a line c_l such that in one step the implication $v(c_l) = u \Rightarrow v(c_i) = 0$ (where u is some logic value, either 0 or 1) is deduced, and in another step the implication $v(c_l) = u \Rightarrow v(c_i) = 1$ is deduced. Together these two implications indicate a contradiction and show that line c_l cannot be assigned the value u, and that $v(c_l)$ is constant at logic value $\bar{u}$, independent of the values assigned to other circuit lines. Once this fact is deduced, any implications deduced at line c_l are eliminated and implication $v(c_l) = \bar{u}$ is recorded.

Another possibility is that the execution of the direct implication procedure following circuit initialization and assignment $v(c_i) = u$ returns a value CONFLICT. This demonstrates that u cannot be assigned to c_i. Hence, implication $v(c_i) = \bar{u}$ is recorded.

The above process should subsequently be repeated for each line in the circuit.

Identifying implications for storage: The objective of indirect implication is to identify implications that cannot be obtained via direct implication. This has three main consequences. First, since in the case described above, the implication $v(c_i) = 0 \Rightarrow v(c_j) = 0$ was identified using only direct implications, it need not be stored. Second, consider the case where direct implication gives implication $v(c_i) = w \Rightarrow v(c_j) = u$ from which implication $v(c_j) = \bar{u} \Rightarrow v(c_i) = \bar{w}$ may be deduced. If c_j is not in the transitive fanout of c_i, then the specific value u at c_j is obtained via a sequence of direct implications of the value w assigned to c_i. In many such cases, the deduced implication $v(c_j) = \bar{u} \Rightarrow v(c_i) = \bar{w}$ can be identified by direct implication following assignment $v(c_j) = \bar{u}$. Hence, implications that may be deduced at lines that are not in the transitive fanout of c_i are not recorded. Third, the implication of the form $v(c_j) = \bar{u} \Rightarrow v(c_i) = \bar{w}$ deduced from implication $v(c_i) = w \Rightarrow v(c_j) = u$, which was obtained via direct implication, may not be meaningful to store under some conditions. For example, if c_j is the output of a two-input AND gate with c_i as an input, then the implication $v(c_j) = 1 \Rightarrow v(c_i) = 1$ can also be obtained by using only the direct implication procedure and hence need not be stored.

In general, implications obtained by executing only the direct implication procedure need not be stored; only the implications deduced therefrom are candidates for storage.

A general method to identify which deduced implication at line c_j should be stored can be based on gate G that drives line c_j. A deduced implication at c_j must not be stored if the corresponding value associated with c_j is equal to $\overline{R(G)}$, the complement of the controlled response of gate G.

For example, the deduced indirect implication $v(c_j) = 1 \Rightarrow v(c_i) = 1$ should **not** be stored if gate G is an AND gate or a NOR gate, since the controlled response of each of these gates is 0, i.e., $\overline{R(G)} = 1$. On the other hand, this implication may be stored if G is a NAND, OR, or XOR gate; in the former two cases, since $\overline{R(G)} = 0$, and in the last case, since $R(G)$ as well as $\overline{R(G)}$ are undefined.

Algorithm to identify static indirect implications: An algorithm to identify static indirect implications, called IdentifyStaticIndirectImplications( ), is shown in Figure 4.19. For each line c_i and each logic value u, direct implications are performed following the initialization of each circuit line to $x = \{0, 1\}$. In each step, deduced implications, if any, are stored in set $GI(c_j, u)$, which is a set of two-tuples of form (c_i, w), where the presence of the two-tuple (c_i, w) in $GI(c_j, u)$ denotes that indirect implication $v(c_j) = u \Rightarrow v(c_i) = w$ was deduced. Implications of the form $v(c_j) = K$, where K is a constant logic value, are stored by adding a two-tuple of the form (c_j, K) to set FVL.

Example 4.10 The reader is invited to apply the above indirect implication algorithm to the fault-free subcircuit shown in Figure 4.18 to check that $\{(c_2, 0)\} \in GI(c_8, 0)$, i.e., it identifies the indirect implication $v(c_8) = 0 \Rightarrow v(c_2) = 0$.   □

Example 4.11 The reader is also invited to apply the algorithm to the fault-free circuit shown in Figure 4.20 and verify that $(c_8, 1) \in FVL$, i.e., it identifies that $v(c_8)$ is always 1, independent of the input vector.   □

Utilizing identified static indirect implications: The above algorithm to identify static indirect implications is executed in a pre-processing step, before any specific target fault is considered. Hence, during the test generation for a target fault, the implications identified in this manner are applicable only to the portion of the circuit that is not in the transitive fanout of any of the fault sites. That is, an implication $v(c_j) = K$, where K is a constant, indicated by $(c_j, u) \in FVL$, may be used only if line c_j is not in the transitive fanout of any fault site. Similarly, an indirect implication of form $v(c_j) = u \Rightarrow v(c_i) = w$, indicated by the fact that $(c_i, w) \in GI(c_j, u)$, is valid only if both c_i and c_j are not in the transitive fanout of any fault site. Furthermore, most static indirect implications are backward and hence useful only if the test generation algorithm utilizes backward implications.

The static indirect implications can be easily utilized by an implication procedure in the manner outlined by the following example. Consider a scenario where at some

IdentifyStaticIndirectImplications()
{

    $FVL = \{\}$ /* FVL: fixed value lines */

    $GI(c_i, u) = \{\}; \forall c_i, \forall u \in \{0, 1\}$ /* GI: Global Implications (Indirect) */

    for each circuit line c_i

        for each value $w \in \{0, 1\}$

        {

            for all lines c_j assign $v(c_j) = x$

            $v(c_i) = w$

            add implication task $(c_i, Forward)$ to implication list S

            if Implication(S) == CONFLICT

                then

                {

                    $FVL = FVL \cup \{(c_i, \overline{w})\}$

                    $GI(c_i, 0) = \{\}$

                    $GI(c_i, 1) = \{\}$

                }

                else

                {

                    for each line $c_j \neq c_i$ such that $v(c_j) \neq x$ **and** c_j is in

                            the transitive fanout of c_i **and** $(c_j, v(c_j)) \notin FVL$

                  if $(c_i, w) \in GI(c_j, \overline{v(c_j)})$

                      then

                      {

                          $GI(c_j, v(c_j)) = \{\}$

                          $GI(c_j, \overline{v(c_j)}) = \{\}$

                          $FVL = FVL \cup \{(c_j, v(c_j))\}$

                      }

                    else

                      $GI(c_j, \overline{v(c_j)}) = GI(c_j, \overline{v(c_j)}) \cup \{(c_i, \overline{w})\}$

                }

        }

    for each line c_j that is the output of gate G_l that has a controlling value

        $GI(c_j, \overline{R(G_l)}) = \{\}$ /* $R(G_l)$ is the controlled response of gate G_l */

}

Figure 4.19 A procedure to statically identify indirect implications

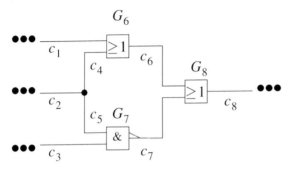

Figure 4.20 An example fault-free subcircuit to illustrate static identification of indirect implications

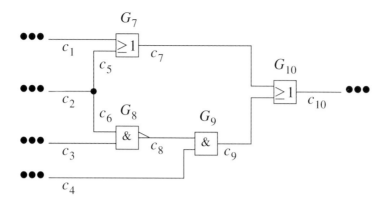

Figure 4.21 An example fault-free subcircuit to illustrate the need for dynamic indirect implications

stage during the execution of an implication procedure on a composite circuit, the value at line c_i, which is not in the transitive fanout of any fault site, is changed to 0_6. All implication operations normally executed by a direct implication procedure are executed in the manner described earlier. In addition, for each c_j that is not in the transitive fanout of the fault site and $(c_j, w) \in GI(c_i, 0)$, the value of c_j, $V(c_j)$, can be replaced by 0_6 if $w = 0$, or by 1_6 if $w = 1$. If this changes the value at line c_j, then appropriate implication tasks with implication site c_j are added to the implication list. The case in which the value at a line c_i is replaced by 1_6 is similar. Note that all the flags associated with c_j as well as $\mathcal{D}$ and $\mathcal{U}$ must be updated to accommodate any changes in $V(c_j)$.

4.4.3.3 Dynamic indirect implication

In the form described above, static indirect implication cannot identify all the implications. This is illustrated by the fault-free subcircuit shown in Figure 4.21. It can be verified via exhaustive simulation that $v(c_{10}) = 0$ only if $v(c_4) = 0$. However, the above procedure for static indirect implication fails to identify the corresponding indirect implication, namely $v(c_{10}) = 0 \Rightarrow v(c_4) = 0$.

Motivation for dynamic indirect implications: The procedure given earlier to identify indirect implications can be generalized in several ways to improve its completeness. First, it may be executed for each faulty version of the circuit. This can help identify additional implications that are useful only during the test generation for the particular target fault. Second, it identifies implications by assigning one specific value at one line. The process is repeated for each possible value at each line. The procedure may be generalized by repeating the process for all the combinations of values to each set of two lines, three lines, and so on. (See Problems 4.19 and 4.20 for hints on issues that must be addressed for each of the above generalizations.)

The above generalizations may identify additional implications. However, the time complexity of the procedure to identify indirect implications will significantly increase. Furthermore, in many cases, the additional indirect implications identified may be very specific and hence used rarely. For example, an indirect implication identified only for a faulty version of a circuit with a specific target fault is useful only during the generation of a test for that target fault.

Often it is computationally more efficient to keep low the complexity of the implication procedure by identifying static indirect implications that can be used throughout the entire test generation process. The completeness of implication can then be enhanced by performing *dynamic indirect implication* during various stages of test generation. **Dynamic indirect implications** utilize the knowledge of (a) the specific target fault under consideration, and (b) the values already assigned to various circuit lines by the test generator. Hence, dynamic indirect implications are only valid at the specific stage of test generation at which they are identified. Despite their specificity, such procedures can be quite efficient.

Types of implication identified: Dynamic indirection implications can identify at least three main types of implications.

1 *Conflict:* It is possible that the values already assigned to various circuit lines by the test generator are inconsistent and/or cannot lead to a test for the target fault. Due to incompleteness of implications, it is possible that this fact is not known prior to the execution of dynamic indirect implication. Execution of dynamic indirect implication may expose this fact. This indicates the need for backtracking.

2 *Constant logic values:* These implications take the form $V(c_i) = K$, where K is a constant composite value of the value system in use.

3 *Indirect implication relations:* These are of the form $V(c_j) = u \Rightarrow V(c_i) = w$, where u, w are composite values denoted by appropriate symbols of the composite value system in use.

A methodology for dynamic indirect implication: One possible way to dynamically identify indirect implications is to execute a modified version of the procedure to identify static indirect implications. The first modification that is required is generalization of the procedure to work with a composite value system and on a composite circuit with the given target fault. Second, the values already assigned to various circuit lines by the test generator must be saved. These values must be used to initialize the circuit prior to the execution of indirect implication at each line and for each possible value assignment. Also, after the completion of indirect implication, these values must be restored to circuit lines. Third, the indirect implication procedure must be modified to deal with the possibility of identification of inconsistencies between the values already assigned by the test generator.

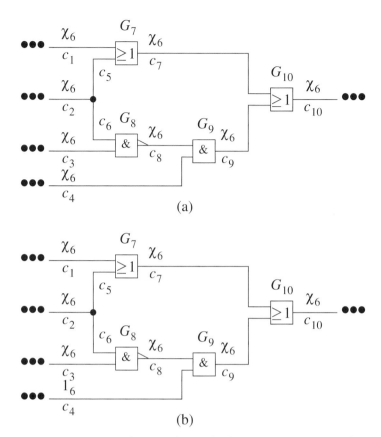

Figure 4.22 (a) An example composite subcircuit, (b) a value assignment following which dynamic indirect implications can be identified by a generalized version of IdentifyStaticIndirectImplications( )

Example 4.12 Consider as an example the part of the *composite circuit* shown in Figure 4.22(a) which corresponds to the circuit shown in Figure 4.21. Assume that the target fault in this circuit is located at some line (not shown in this figure) in the transitive fanout of the subcircuit shown. As described earlier, execution of Procedure IdentifyStaticIndirectImplications( ) on the fault-free circuit during pre-processing will not identify any indirect implications within this subcircuit.

Now consider a scenario where the test generator has assigned the values shown in Figure 4.22(b). This assignment of values will constitute an *initial assignment* for the following indirect implication identification. (Note that composite values are used since a composite circuit is being considered.)

As in Procedure IdentifyStaticIndirectImplications( ), in each step, first a line is selected. A fully specified value that is more specific than the current value at the line is then assigned. Direct implications are then performed and indirect implications are deduced. Consider the step in which line c_2 is selected. First, assignment $V(c_2) = 0_6$

is tried and direct implication performed to obtain $V(c_5) = V(c_6) = 0_6$ and $V(c_8) = V(c_9) = V(c_{10}) = 1_6$.

Subsequently, implications are deduced as follows. In the above scenario, $V(c_2) = 0_6 \Rightarrow V(c_{10}) = 1_6$. Hence, $V(c_{10}) \neq 1_6 \Rightarrow V(c_2) \neq 0_6$. Since the initial value at line c_{10} is χ_6, if $V(c_{10}) \neq 1_6$ then $V(c_{10}) = 0_6$. For the same reason, if $V(c_2) \neq 0_6$ then $V(c_2) = 1_6$. Hence, the above deduced implication can be rewritten as $V(c_{10}) = 0_6 \Rightarrow V(c_2) = 1_6$.

Next, the above process can be repeated by (a) restoring the values to those shown in Figure 4.22(b), and (b) trying the only other possible more specific value assignment at c_2, namely $V(c_2) = 1_6$. Following the same steps as above, this time implication $V(c_{10}) = 0_6 \Rightarrow V(c_2) = 0_6$ can be deduced. Since this contradicts the implication deduced earlier, it can be concluded that $V(c_{10})$ cannot take the value 0_6. Hence, the above two deductions collectively demonstrate that $V(c_{10}) = 1_6$, given the initial value assignment shown in Figure 4.22(b).   □

An example scenario where dynamic indirect implication can identify conflicts is described in Schulz and Auth (1989).

A complete implication methodology: We now illustrate another dynamic indirect implication methodology (Kunz and Pradhan, 1994), called *recursive learning*, via an example. In principle, if an unlimited amount of memory is available and unlimited run-time is allowed, this technique can identify all the implications of the existing value assignments.

Consider the circuit shown in Figure 4.23. Assume that at some stage of test generation, the values assigned to the circuit lines are as shown in Figure 4.23(a). Note that the value $V(c_8) = 0_6$ is unjustified. The next step is the identification of alternative value assignments at the inputs of this gate. Such value assignments at a gate can be enumerated by first considering the current values at the inputs and output of a gate as a cube. This cube can then be intersected with each cube in the cube cover of the gate. Each valid intersection obtained above represents an alternative value assignment at the gate.

In our example, the given values at the inputs and output of gate G_4, $(\chi_6, \chi_6, 0_6)$, can be intersected with each cube in the cube cover of the fault-free version of a two-input XOR gate (Figure 4.6(d)) to obtain cubes that are compatible with these values. In this case, the following compatible cubes are obtained.

c_4	c_5	c_8
0_6	0_6	0_6
1_6	1_6	0_6

Next, we try the first possible way of justifying $V(c_8) = 0_6$. We begin by assigning the values given by the cube at various lines in the gate, namely $V(c_4) = 0_6$ and

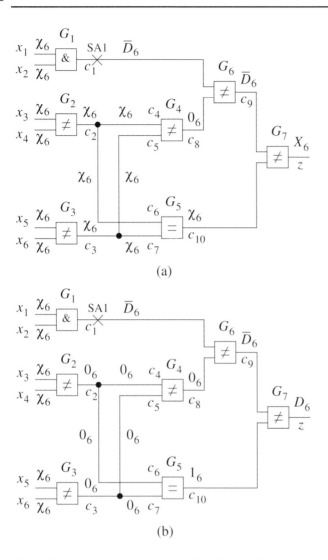

Figure 4.23 An example of dynamic indirect implication: (a) original status of the composite circuit, (b) value assignments obtained by processing one alternative value assignment at G_4, (c) value assignments for the other alternative, and (d) the circuit status after dynamic indirect implication. (This circuit is a modified version of a circuit in Goel (1981) (© 1981 IEEE)) (Continued)

$V(c_5) = 0_6$. This is followed by execution of the direct implication procedure, which gives the values shown in Figure 4.23(b). We then note the values that have become more specific as a result of the above assignment and implication. In this case, values $V(c_2) = 0_6$, $V(c_3) = 0_6$, $V(c_4) = 0_6$, $V(c_5) = 0_6$, $V(c_6) = 0_6$, $V(c_7) = 0_6$, $V(c_{10}) = 1_6$, and $V(z) = D_6$, are noted. The values at the various circuit lines are then restored to those shown in Figure 4.23(a).

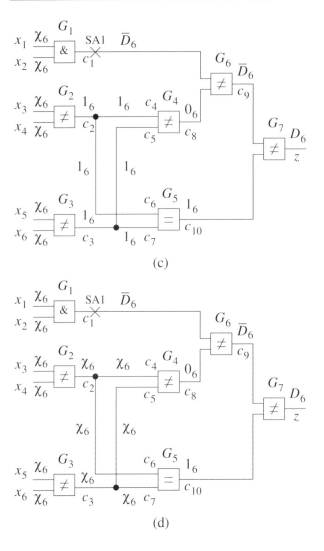

Figure 4.23 (Continued) An example of dynamic indirect implication: (a) original status of the composite circuit, (b) value assignments obtained by processing one alternative value assignment at G_4, (c) value assignments for the other alternative, and (d) the circuit status after dynamic indirect implication. (This circuit is a modified version of a circuit in Goel (1981) (© 1981 IEEE))

Next, we repeat the above process for the only other possible way of justifying value $V(c_8) = 0_6$, namely $V(c_4) = 1_6$ and $V(c_5) = 1_6$, as given by the other compatible cube listed above. The direct implication procedure is then executed, the more specific values noted, and the values in circuit lines restored to that in Figure 4.23(a). In this case, values $V(c_2) = 1_6$, $V(c_3) = 1_6$, $V(c_4) = 1_6$, $V(c_5) = 1_6$, $V(c_6) = 1_6$, $V(c_7) = 1_6$, $V(c_{10}) = 1_6$, and $V(z) = D_6$, are noted (see Figure 4.23(c)).

Once each of the alternative ways of justifying value $V(c_8) = 0_6$ has been tried and the values that become more specific in each case noted, the noted values are analyzed. Any implication that is common to all of the possibilities is identified as an indirect implication of the current values. In the above example, values $V(c_{10}) = 1_6$, and $V(z) = D_6$ are indirect implications identified by the procedure and are shown in Figure 4.23(d).

The above method can be applied recursively to identify all implications of the current values at circuit lines as outlined next in procedure RecursiveIndirectImplication( ).

Procedure 4.1   [RecursiveIndirectImplication( )]

1 If there exist one or more unjustified lines, then select one of these lines as the implication site. If no unjustified line exists, then select as the implication site one of the lines that has an incompletely specified value. If no line in the circuit is unjustified and no line has an incompletely specified value, then this procedure has successfully completed; transfer control to the calling procedure.

2 Form a cube with values at the inputs and output of the gate whose output is an implication site. Intersect this with each cube in the cube cover of the gate (singular cover and propagation cubes, if the gate is fault-free; singular cover and fault excitation cubes, if the gate has the target fault). Each cube in the intersection represents an alternative way in which values at the circuit lines may be further specified.

3 One at a time, process each alternative value assignment by assigning the values given by the corresponding cube.

At this stage, in addition to the execution of the direct implication procedure, the entire dynamic indirect implication may be *recursively* executed.

If this implication results in a conflict, then return the status CONFLICT. Otherwise return the values that become more specific along with the status SUCCESS.

4 If each of the alternatives returns CONFLICT, then an inconsistency is identified and, in turn, return CONFLICT to the calling procedure. Otherwise, analyze the line values that became more specific during various alternative assignments. Identify as indirect implication any implication that is common to all the sets of values returned by successful recursive calls. If this is not the top level of recursion, then return to the calling procedure all the indirect implications identified in this manner. Otherwise, assign these indirect implications to the circuit lines. The procedure may then be repeated.

4.5   Structural test generation algorithms: preliminaries

In this section, we discuss some test generation *algorithms* that analyze the *structure* of the given CUT to generate a test for a given target fault, or declare it untestable. A test generation *system* uses a test generation algorithm and other procedures, such as random vector generators, fault simulators, and test set compactors, to obtain a set of vectors that efficiently detect all target faults in a given fault list (or a large proportion thereof). The discussion of test generation systems is deferred to a later section.

Test generation for a given target fault is a search process that is said to be **successful** if either a test is generated for the fault or the fault is proven to be untestable. A test generation algorithm is said to be **complete** if it can guarantee that test generation will be successful for any given target fault in any given CUT, assuming sufficient memory is available and sufficient run-time is allowed. The **efficiency** of a test generation algorithm is loosely measured by the proportion of all target faults for which it is successful and the corresponding run-time.

All test generation algorithms discussed ahead are complete. These and any other test generation algorithms can guarantee completeness only by *enumerating all possibilities* in the search space. However, since the search space is typically very large even for circuits of moderate size, a test generation algorithm must *implicitly enumerate* the space of possibilities to achieve high efficiency.

We begin with a discussion of a generic test generation algorithm and the description of some of its generic components.

4.5.1   A generic structural test generation algorithm

A generic structural test generation algorithm is described next.

1 **Reading CUT:** The fault-free version of the CUT is read.
2 **Pre-processing:** Two operations are commonly performed on the fault-free version of the circuit in the pre-processing step.
   (a) *Computation of testability measures:* Controllability and observability measures, described ahead, are computed.
   (b) *Static indirect implication identification:* Static indirect implications are identified using algorithms described earlier and stored in lists $GI(\ )$ and FVL.
3 **Target fault insertion:** *This and subsequent steps are repeated for each target fault.*
   A target fault is selected and a composite circuit model is created for the corresponding CUT with this fault. Each fault-free circuit element is described by the cube cover of its fault-free version; each faulty circuit element is described by the cube

cover of its version with the appropriate fault. (Strategies for selecting the target fault will be discussed later when the complete test generation system is described.)

4 **Initialization:** Suitable initial values are assigned to each circuit line.

If the five-valued system is used, symbol X_5 is assigned to each line, i.e., $V(c) = X_5$, $\forall$ lines c in the CUT.

If the six-valued system is used, then the initial values are assigned as follows.

$$V(c) = \begin{cases} \chi_6, & \text{if line } c \text{ is not in the transitive fanout of the fault site,} \\ X_6, & \text{otherwise.} \end{cases} \tag{4.11}$$

The rationale behind the above initialization is that only the lines that are in the transitive fanout of one or more fault sites may take values $\{D\}$ or $\{\overline{D}\}$; all the other lines may only take values $\{\underline{0}\}$ or $\{\underline{1}\}$. Use of the six-valued system enables more precise initialization of circuit lines and can decrease the number of choices that need to be enumerated during test generation. It can hence decrease the overall complexity of test generation at a slight increase in the size of truth tables of fault-free and faulty circuit elements. Also, it helps distinguish lines that may have a fault effect from those that may not.

The use of the 16-valued system allows for even more precise initialization. Consider the case where only a single element in the circuit is assumed to be faulty. In this case, first the values at circuit lines are initialized as follows.

$$V(c) = \begin{cases} \{\underline{0}, \underline{1}\}, & \text{if line } c \text{ is not in the transitive fanout of the fault site,} \\ \{\underline{0}, \underline{1}, D, \overline{D}\}, & \text{otherwise.} \end{cases}$$

$$\tag{4.12}$$

Subsequently, the implication tasks $(c_j, \text{FORWARD})$, for each output c_j of the faulty circuit element, are added to the implication task list and forward implication performed as described in Section 4.4.2.2. In this case, for example, the output of a faulty NAND gate with an SA1 fault is initialized to $\{\underline{1}, \overline{D}\}$. Note that only D or $\overline{D}$, but not both, will appear in the set of values assigned to many lines in the circuit. Problem 4.22 illustrates the difference between the initial values at circuit lines for the above three value systems.

5 **Identification of test generation subtasks:** In this step, the following values are analyzed.

- The values at the outputs of the faulty circuit element.
- The values at the primary outputs of the CUT.
- The gates in the D-frontier, $\mathcal{D}$.
- The lines in the set of unjustified lines, $\mathcal{U}$.

The analysis, whose details will be described ahead, determines which of the following cases has occurred.

- *Successful test generation:* The given vector at the primary inputs (which may be partially specified) is a test vector.

  If this is found to be the case, test generation for the current target fault is complete. The generated test is reported and the fault removed from the fault list.

  If any more faults remain in the fault list, then the test generation procedure is repeated starting at Step 3.

- *Conflict:* The given partial vector is not a test vector, and the incompletely specified values at the inputs cannot be replaced by compatible more specific values to obtain a test. This signals the need to backtrack and Step 8 is executed.

- *Continuation of search:* The given partial vector is not a test, but the values at some inputs may be replaced by compatible more specific values to possibly lead to a test.

  In this case, possible test generation subtasks (TGSTs), such as fault excitation, fault-effect propagation, and/or line justification, are identified. In some cases, the completion of one or more of these TGSTs may be identified as being necessary to satisfy, if the given partial vector is to be further specified to become a test.

6 **Identification of value assignments that facilitate completion of TGSTs:** This step is executed if the need to continue search is identified in the above step.

   Various structural algorithms differ significantly in how they implement this step. The tasks performed in this step may include (a) for one or more TGSTs, enumeration of alternative local value assignments that help the completion of the TGST, (b) prioritization of the TGSTs and the corresponding alternative local value assignments, and (c) identification of one or more value assignments. Some test generation algorithms may not perform some of these tasks and different algorithms perform them differently. However, at the end of this step, a value assignment at a circuit line is identified that may facilitate the completion of one or more subtasks in a given set of TGSTs.

7 **Value assignment:** Prior to making the assignments identified in the above step, the current *status of the test generator* is saved to facilitate backtracking, if it becomes necessary. The status of the test generator may include the values at circuit lines and alternatives that have not yet been considered.

   The selected value assignments are then made.

   This is followed by execution of the implication procedure. If the implication procedure returns CONFLICT, then the program backtracks and executes Step 8. Otherwise, the algorithm continues by executing Step 5 to identify a new set of TGSTs for the new partial vector and new set of values at circuit lines.

8 **Backtracking:** As can be seen from the control flow described in the above steps, this step is executed only when it is determined that the partial vector generated thus far cannot be further specified to obtain a test vector for the target fault.

In this case, the effects of the last assignment on the status of the test generator are eliminated by restoring the status to that saved prior to the last assignment. The alternatives that had not been considered at that time are examined.

If no such alternatives are found then the program backtracks to the status prior to the previous value assignment, until untried alternatives are found. If there exists no previous value assignment at which untried alternatives exist, then the target fault has been demonstrated as being untestable. The target fault is moved from the fault list to a list of untestable faults and test generation for the given target fault is terminated. If any faults remain in the target fault list, then the procedure is repeated starting at Step 3.

On the other hand, if an alternative assignment exists at a prior stage, then one of these value assignments is selected and the algorithm continues by executing Step 7 of the algorithm to make the alternative assignments.

Steps 1–5 of the above algorithm are basic steps that are common to most structural test generation algorithms. Since the first four steps have already been described above, next we will describe Step 5. Steps 6–8 above constitute the core of a test generation algorithm. The details of these steps vary significantly from one algorithm to another. Hence, these steps will be detailed when specific structural test generation algorithms are described.

4.5.2   Identification of TGSTs

As described in the above generic test generation algorithm, the application of each logic value is followed by execution of the logic implication procedure. Recall that if any inconsistency is found between the applied values and the values implied by them, then the implication procedure returns CONFLICT and invokes backtracking. In contrast, successful completion of logic implication, i.e., when the implication returns the value SUCCESS, provides an updated (a) set of values at various circuit lines, (b) D-frontier $\mathcal{D}$, and (c) list of unjustified lines $\mathcal{U}$. These values are analyzed as follows to identify TGSTs.

4.5.2.1 Fault effect excitation

The values at the outputs of the faulty circuit element are the indicators of fault excitation. If each output of the faulty circuit element is already assigned $\{\underline{0}\}$ or $\{\underline{1}\}$, then the current vector has clearly failed to excite the target fault and cannot be further specified to obtain a test for a target fault. This is also the case if a value $\{\underline{0}, \underline{1}\}$ is assigned to each output of the faulty circuit element. In these cases, no TGST can be identified and backtracking is initiated. On the other hand, if among the outputs of the faulty circuit element (a) none have a fault effect, but (b) at least one has a value that may be further specified into $\{D\}$ or $\{\overline{D}\}$, then the sole TGST that is identified is fault

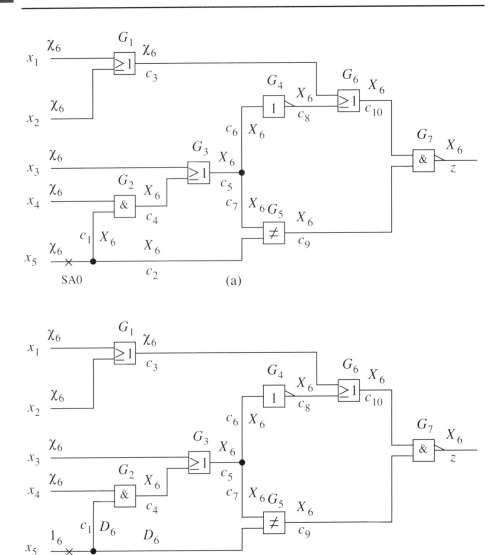

Figure 4.24 An example execution of structural test generation algorithms. (This circuit is a modified version of a circuit in Roth (1966)) (Continued)

excitation. For example, this is the case for the circuit, target fault, and values shown in Figure 4.24(a).

This subtask entails further specification of the current vector so as to obtain a $\{D\}$ or a $\{\overline{D}\}$, as appropriate, at one or more outputs of the faulty circuit element. Note that this is a necessary task. Furthermore, most test generation algorithms perform this subtask first, since unless the fault is excited, it is difficult to carry out fault effect

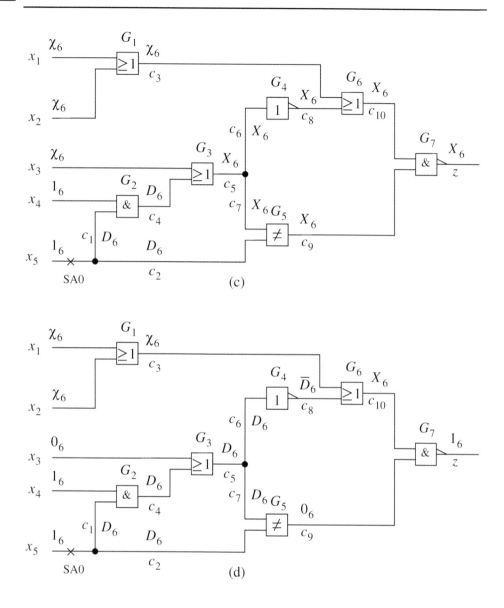

Figure 4.24 (Continued) An example execution of structural test generation algorithms. (This circuit is a modified version of a circuit in Roth (1966)) (Continued)

propagation. The third possibility is that the target fault is already excited, i.e., a $\{D\}$ or $\{\overline{D}\}$ already exists at one or more outputs of the faulty circuit element. When this is the case, e.g., in the scenario depicted in Figure 4.24(b), the identification of TGSTs proceeds as described next.

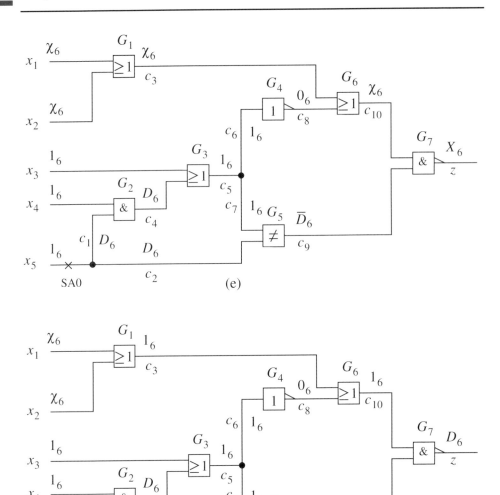

Figure 4.24 (Continued) An example execution of structural test generation algorithms. (This circuit is a modified version of a circuit in Roth (1966)) (Continued)

4.5.2.2 Fault effect propagation

Once the target fault is excited, the attention turns to the propagation of the fault effect to one or more primary outputs of the CUT. If the fault effect, i.e., $\{D\}$ or $\{\overline{D}\}$, is already at one or more of the CUT outputs, then fault effect propagation has been achieved and the attention is turned to justification, described in the next subsection. If

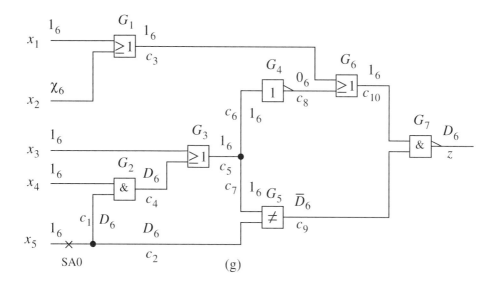

Figure 4.24 (Continued) An example execution of structural test generation algorithms. (This circuit is a modified version of a circuit in Roth (1966))

this is not the case, the D-frontier, $\mathcal{D}$, is examined. Each gate in $\mathcal{D}$ is a candidate for fault effect propagation. Furthermore, if $\mathcal{D}$ is empty, then it indicates that no candidates exist for fault effect propagation. In this case, unless the fault effect is already at one of the outputs of the CUT, backtracking must be initiated. On the other hand, i.e., when $\mathcal{D}$ is not empty, one option is to declare propagation of fault effect through each gate in $\mathcal{D}$ as a distinct TGST. However, in this case, additional analysis may be performed to provide a much more precise and comprehensive list of TGSTs.

First, for a gate to be a meaningful site for fault effect propagation, it must not only belong to $\mathcal{D}$ but there must exist at least one path from the output of the gate to a CUT output such that, at each line along the path, the incompletely specified value implied by the current vector contains basic value D or $\overline{D}$. We will call such a path a **potential propagation path**. All potential propagation paths for a gate in $\mathcal{D}$ can be obtained by tracing the paths from the gate to CUT outputs and checking the value at each line along the path. Clearly, any gate that is in $\mathcal{D}$ but for which no potential propagation path exists from the output of the gate to a CUT output can be removed from $\mathcal{D}$. This operation helps eliminate some spurious fault effect propagation subtasks from the list of possible TGSTs and hence makes the list of TGSTs more precise. This process is called **x-path check** (Goel, 1981). For example, in Figure 4.24(c), the D-frontier is comprised of gates G_3 and G_5. In this case, at least one potential propagation path exists from the output of each of these gates to the primary output. In the scenario depicted in Figure 4.24(d), the D frontier contains gate G_6. However, in this case, no potential propagation path exists from this gate to the primary output.

Next, consider scenarios where one or more fault effect propagation subtasks may be identified as being necessary for the current vector. One special case of the above occurs when $\mathcal{D}$ contains a single gate. In such a case, clearly the fault effect must be propagated through that gate. This situation is called **unique D-drive**. Furthermore, if all the potential propagation paths from all the gates in $\mathcal{D}$ pass through a particular gate, then fault effect propagation through that gate constitutes a necessary subtask. This more general case is called **future unique D-drive** and is identified as described next.

A necessary fault effect propagation subtask can be identified by first creating a **potential propagation graph**. This graph contains a node corresponding to each gate that falls along any potential propagation path from the output of any gate in $\mathcal{D}$. It also contains one node for each CUT primary output that lies on one or more potential propagation paths. A directed edge from one node to another exists if the output of the gate denoted by the former node drives the input of the gate denoted by the latter, through a circuit line that belongs to one or more potential propagation paths. Next, each node that corresponds to a gate in $\mathcal{D}$ is labeled as a source and each node that corresponds to a CUT output is labeled as a sink. If the removal of an edge from this graph will cause every source to become completely disconnected from every sink, then the gate corresponding to the graph node that is the source of the corresponding edge is a necessary site for fault effect propagation. Such edges are called **dominators** (Tarjan, 1974; Kirkland and Mercer, 1987) of a directed graph and can be identified using graph theory techniques.

Example 4.13 Figure 4.25 shows the potential propagation graph obtained by using the above procedure for the scenario depicted in Figure 4.24(c). The reader can verify that the only node whose removal from this graph will cause all the sources to be disconnected from the sink is G_7. Hence, the future-unique D-drive analysis shows that it is necessary to propagate the fault effect via G_7 for a test vector to be obtained by further specifying the components of the current partial vector shown in Figure 4.24(c).

□

Identification of a more comprehensive list of TGSTs through additional analysis, such as x-path check and future unique D-drive, can reduce the number of additional backtracks required before test generation for a target fault is completed. However, the complexity of additional analysis should be compared with the expected reduction in complexity due to fewer backtracks to determine how much of the above analysis should be performed.

4.5.2.3 Justification

Set $\mathcal{U}$ contains gates that are unjustified, i.e., the value at the output of each such gate is more specific than that obtained when a forward implication operation is performed at the gate. Hence, justification of the current value at the output of each gate in

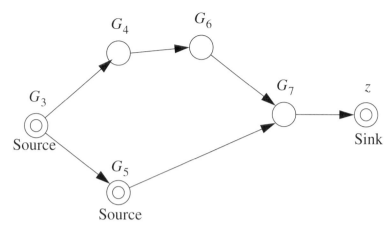

Figure 4.25 Potential propagation graph for the scenario in Figure 4.24(c). This graph provides an example of future unique D-drive

$\mathcal{U}$ constitutes a distinct justification task. Note that each justification task must be completed for the current vector to become a test vector.

If the value $\{D\}$ or $\{\overline{D}\}$ already exists at one or more outputs of the circuit and there are no gates in $\mathcal{U}$, then the current vector is a test for the target fault and test generation for the fault is completed.

Figure 4.26 summarizes the procedure to identify TGSTs. Each leaf of the tree shows the action taken for that case (such as backtrack) or the TGSTs that are identified.

4.5.3   Testability measures

Testability measures are used by test generation algorithms as heuristic measures to determine (a) the order in which TGSTs are to be considered, and (b) the order in which alternative value assignments for a TGST are tried. Two types of testability measures are commonly used, *controllability* and *observability*. The **controllability** of a line is a measure of the difficulty of obtaining a desired value at that line. Typically, two controllability measures are defined for each line c, **0-controllability**, $CC^0(c)$, and **1-controllability**, $CC^1(c)$. These are, respectively, the measures of difficulty of obtaining values 0 and 1 at the line. The **observability** of a line, $O(c)$, is a measure of the difficulty of observing a value, i.e., propagating the fault effect from the line to one or more CUT outputs.

We will discuss the testability measure framework presented in Goldstein and Thigpen (1980). Several other frameworks, including those presented in Abramovici *et al.* (1986a), Brglez *et al.* (1984), Grason (1979) and Jain and Agrawal (1985), can be found in the literature. Very often, the values of these measures are computed only for the fault-free version of the circuit. This enables their computation at a low complexity

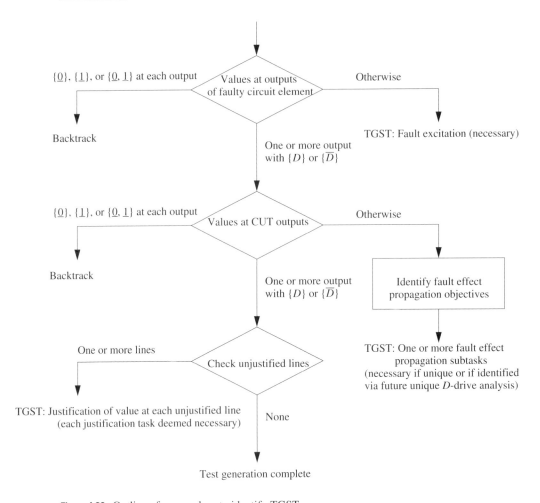

Figure 4.26 Outline of a procedure to identify TGSTs

during the pre-processing phase of a test generation algorithm. Consequently, such cost functions are limited in their ability to order alternative value assignments involving fault effects, i.e., $\{D\}$ or $\{\overline{D}\}$. Even though the measures are computed for the fault-free circuits, we use the six-valued singular covers and fault effect propagation cubes of each circuit element to compute controllability and observability values, respectively. Use of the composite circuit (with no fault!) and six-valued system is warranted since they provide the basis for estimating the difficulty of fault effect propagation.

When tests are to be generated for sequential circuits (Chapter 5), it is often worthwhile to compute these values for a given faulty circuit as well. In such a case, these values will be denoted by CC_g^0, CC_g^1, and O_g for the fault-free circuit; the corresponding values for the circuit with the target fault will be denoted by CC_f^0, CC_f^1, and O_f.

4.5.3.1 Components for computation of testability measures

Six major components characterize a given set of testability measures. First is the value CC_{pi} assigned to the 0- and 1-controllability of each primary input. Second is a set of values $CC_\kappa(c_j)$, where $CC_\kappa(c_j)$ is associated with each output c_j of each circuit element. Third is the **controllability transfer function** that enables computation of the value of the above measures at each output of a circuit element, given their values at each of its inputs. These components are used to compute the 0- and 1-controllability of each line in the circuit. The fourth component is value O_{po} assigned to the observability of each CUT output. The fifth component is a set of values $O_\kappa(c_j)$, where $O_\kappa(c_j)$ is associated with each input c_j of each logic element. The sixth component is the **observability transfer function**, which enables the computation of the observability of each input of a circuit element, given (a) the observabilities of each of its outputs, and (b) the 0- and 1-controllabilities of each of its inputs. These components, along with the 0- and 1-controllability values for all the circuit lines, are used to compute the observabilities of each line in the circuit.

4.5.3.2 Controllability transfer function

The form of controllability and observability cost functions is introduced next with the aid of a two-input NAND gate with inputs c_{i_1} and c_{i_2} and output c_j. Assume that the 0- and 1-controllability of each of its inputs are already known.

Consider the combination of input values that give a 0 at the output of the NAND gate in a fault-free circuit. The six-valued cube cover of the gate shows that this can occur if and only if the following combination of values is applied at its inputs.

c_{i_1}	c_{i_2}
1_6	1_6

The difficulty of obtaining 1 at these inputs is given by $CC^1(c_{i_1})$ and $CC^1(c_{i_2})$, respectively. These measures of difficulty need to be combined in some fashion to obtain a measure of difficulty of obtaining both these values simultaneously. One possible way of combining them is $CC^1(c_{i_1}) + CC^1(c_{i_2})$, where the *addition operation* is used whenever multiple value assignments must be satisfied simultaneously. (Some other operations, such as multiplication, may also be used – provided several other changes are made to the other components.) Since, for our example gate, this is the only combination of input values that gives 0_6 at the gate's output, the measure of difficulty of setting c_j to 0_6 can be computed by adding to the above expression the value $CC_\kappa(c_j)$, which is a constant value associated with each output of each circuit element. This gives $CC^0(c_j) = CC^1(c_{i_1}) + CC^1(c_{i_2}) + CC_\kappa(c_j)$.

Next, consider the combinations of input values that give a 1 at the output of a NAND gate in a fault-free circuit. The cube cover of the gate shows that this occurs when any of the following combinations of values is applied at its inputs.

c_{i_1}	c_{i_2}
0_6	X_6
X_6	0_6
D_6	$\overline{D}_6$
$\overline{D}_6$	D_6

Consider the first cube that gives 1 at the output. The difficulty of obtaining any one of the vectors contained in this cube at the inputs of the NAND gate is given by $CC^0(c_{i_1})$. Similarly, the difficulty of obtaining one of the vectors in the second cube can be measured by $CC^0(c_{i_2})$.

The difficulty of obtaining a 1 at c_j due to the application of the vector given by the third cube is a little tricky to compute, since the measures are computed only for the fault-free circuit and this cube contains fault effects. One approximation that can be used in this case is that the difficulty of obtaining a specific fault effect at a line can be measured by the difficulty of obtaining the corresponding fault-free value at the line. In other words, the difficulty of obtaining values D_6 and $\overline{D}_6$ at a line c can be, respectively, approximated as $CC^1(c)$ and $CC^0(c)$. Hence, the difficulty of obtaining the vector given by the third cube can be written as $CC^1(c_{i_1}) + CC^0(c_{i_2})$; that for the fourth cube can be written as $CC^0(c_{i_1}) + CC^1(c_{i_2})$.

Since a 1 at the output can be obtained by the application of a combination of values contained in any of the above four cubes, the above four measures should be combined in some way. One possible way to combine these values is by taking the minimum of the above values, since the difficulty of obtaining a vector in any of these cubes is no greater than the difficulty of obtaining a vector in any one of the cubes. Hence, we obtain $CC^1(c_j) = \min[CC^0(c_{i_1}), CC^0(c_{i_2}), CC^1(c_{i_1}) + CC^0(c_{i_2}), CC^0(c_{i_1}) + CC^1(c_{i_2})] + CC_\kappa(c_j)$.

In general, the 0-controllability at the output of an arbitrary gate with inputs $c_{i_1}, c_{i_2}, \ldots, c_{i_\alpha}$ and output c_j can be computed as follows. First, the cubes in its six-valued singular cover that imply value 0_6 at the gate output are enumerated. For each such cube $\mathcal{C} = (\mathcal{C}_1, \mathcal{C}_2, \ldots, \mathcal{C}_\alpha)$, the measure is computed as

$$Cost(\mathcal{C}) = \sum_{l=1,2,\ldots,\alpha} cost(\mathcal{C}_l), \tag{4.13}$$

where

$$cost(\mathcal{C}_l) = \begin{cases} CC^0(c_{i_l}), & \text{if } \mathcal{C}_l \text{ is } 0_6 \text{ or } \overline{D}_6, \\ CC^1(c_{i_l}), & \text{if } \mathcal{C}_l \text{ is } 1_6 \text{ or } D_6, \\ 0, & \text{otherwise.} \end{cases} \tag{4.14}$$

In the following, $Cost(\mathcal{C})$ will be referred to as the **cost of cube** $\mathcal{C}$ and $cost(\mathcal{C}_l)$ will be referred to as the **cost of a component** $\mathcal{C}_l$ of a cube $\mathcal{C}$.

The 0-controllability of the gate output is then computed by taking the minimum of all the above values and adding a fixed value associated with the output as

$$CC^0(c_j) = \min_{\forall \mathcal{C} \text{ for which } V(c_j) = 0_6} [Cost(\mathcal{C})] + CC_\kappa(c_j). \tag{4.15}$$

The 1-controllability at the output can be computed in a similar manner.

A similar approach may also be used to compute the controllability of each output c_{j_l} (where $l = 1, 2, \dots, \beta$) of a fanout system with stem c_i. In this case, however, the final result can be written simply as

$$CC^0(c_{j_l}) = CC^0(c_i) + CC_\kappa(c_{j_l}), \tag{4.16}$$

for each possible l. The expression for the corresponding 1-controllability is similar.

4.5.3.3 Observability transfer function

Next, assume that observability $O(c_j)$ of output c_j of a gate with inputs $c_{i_1}, c_{i_2}, \dots, c_{i_\alpha}$ has already been computed. The observability of input c_{i_ν}, where ν takes any value from the set $\{1, 2, \dots, \alpha\}$, can be computed by considering those fault effect propagation cubes of the gate that have a D_6 or $\overline{D}_6$ at input c_{i_ν}. For each such cube $\mathcal{C} = (\mathcal{C}_1, \mathcal{C}_2, \dots, \mathcal{C}_\alpha)$, first the following value is computed.

$$OCost_\nu(\mathcal{C}) = \sum_{l=1,2,\dots,\alpha;l\neq\nu} cost(\mathcal{C}_l), \tag{4.17}$$

where $cost(\mathcal{C}_l)$ is given by Equation (4.14). $OCost_\nu(\mathcal{C})$ will be referred to as the **observability cost of a cube** $\mathcal{C}$ **for input** c_{i_ν}.

The observability of input c_{i_ν} is then computed by taking the minimum of the above values for all cubes for which $V(c_j) = D_6$ or $\overline{D}_6$, adding the output observability, and adding a constant:

$$O(c_{i_\nu}) = \min_{\forall \mathcal{C}}[OCost_\nu(\mathcal{C})] + O(c_j) + O_\kappa(c_{i_\nu}) \tag{4.18}$$

The observability of stem c_i of a fanout system with branches $c_{j_1}, c_{j_2}, \dots c_{j_\beta}$ can be simply written as

$$O(c_i) = \min_{l=1,2,\dots,\beta}[O(c_{j_l})] + O_\kappa(c_i). \tag{4.19}$$

4.5.3.4 Procedure to compute testability measures

Testability measures are computed by a two-pass, breadth-first traversal of the circuit as described in the following procedure.

Procedure 4.2 [ComputeTestabilityMeasures( )]

1 The first traversal starts at the primary inputs of the CUT. For each primary input x_l, $CC^0(x_l)$ as well as $CC^1(x_l)$ are initialized to CC_{pi}.

2 Next, starting at the primary inputs, the circuit is traversed breadth-first. During this traversal, whenever an output of a circuit element is visited, its controllabilities are computed using the controllability transfer function described above.

3 The second breadth-first traversal starts at the primary outputs with the initialization of the observability of each output z_l, $O(z_l)$, to O_{po}.

4 The circuit lines are then traversed backward starting at the primary outputs in a breadth-first manner. Whenever an input of a circuit element is visited, its observability is computed using the observability transfer function described above.

4.5.3.5 Variations of testability measures

The main purpose of testability measures is to provide a mechanism to prioritize choices that may exist at any stage of the search for a test vector. Testability measures provide a *heuristic* basis for prioritization of alternatives. In the measures described above, it is deemed more difficult to imply a logic value, say 0, at a line c_i than to imply a logic value, say 1, at a line c_j, if $CC^0(c_i) > CC^1(c_j)$. Similarly, it is deemed more difficult to propagate the fault effect from a line c_i to any CUT output than from line c_j, if $O(c_i) > O(c_j)$. Next, we will discuss variations of the testability measures to illustrate various components that measure the difficulty of controlling or observing a value at a line.

Variation 1: First, consider a variation in which the above transfer functions are used with $CC_{\text{pi}} = 1$, $CC_\kappa(c_j) = 0$, $O_{\text{po}} = 0$, and $O_\kappa(c_j) = 0$. The reader can verify that in this case, if a CUT is comprised only of primitive logic gates and has no fanout, the 0-controllability of a line becomes equal to the minimum number of primary inputs at which specific values need to be assigned to imply 0 at that line. Furthermore, the observability of a line becomes equal to the minimum number of primary inputs at which specific values need to be assigned to propagate a fault effect from that line to the CUT output.

Variation 2: Next, consider another variation of the above transfer functions where $CC_{\text{pi}} = 0$, $CC_\kappa(c_j) = 1$, $O_{\text{po}} = 0$, $O_\kappa(c_j) = 1$, and the summation operations ($\sum$) in Equations (4.13) and (4.17) are replaced by maximum operations (max). In this case, for a CUT that is comprised only of primitive gates and has no fanout, the 0-controllability of a line captures some measure of the number of circuit elements that must be traversed from the line to primary inputs at which specific values are required to imply 0 at the line.

The above variations, which are only two of many, of the cost functions indicate that the difficulty of setting a line to a value or observing the value at a line can be measured in terms of many different parameters and their combinations. The above transfer functions can be viewed as providing a mechanism to create a set of testability measures that capture the combination of parameter values that are deemed most likely to accelerate the particular test generation algorithm for which they are developed.

Only CUTs comprised of primitive gates that do not have fanouts were used in the above discussion of variations. This is due to the fact that for an arbitrary circuit even such simple variations of these measures do not capture exactly the desired parameter values. Hence, for a general circuit, in the first variation, a controllability of a line captures some heuristic measure of the number of inputs that need to be set for the corresponding value to be implied at the line. In the second variation, a controllability of a line captures some heuristic measure of the distance of the line from primary inputs at which specific values are required to imply the corresponding value at the line. A suitable combination of values of CC_{pi}, O_{po}, CC_κ, and O_κ is used in practice.

4.5.3.6 Use of testability measures

In addition to the above limitations, these testability measures are limited in two ways due to the fact that they are computed in the pre-processing phase. First, all measures, including observability, are computed using the fault-free version of the circuit. Second, the measures are not updated to take advantage of the logic values that are implied by the partial vector already generated by the test generation algorithm. For example, consider a fault-free two-input XOR gate in a CUT under two possible scenarios. In the first scenario, the values at its inputs c_{i_1} and c_{i_2} as well as output c_j are all X_6. Assume that value 0_6 is desired at the output. The cube comprised of the current values at the inputs and output of the gate can be intersected with the rows of the six-valued cube cover of the XOR gate to obtain the following alternative value assignments that can give the desired value at the output.

c_{i_1}	c_{i_2}
0_6	0_6
1_6	1_6
D_6	D_6
$\overline{D}_6$	$\overline{D}_6$

The cost of each cube in the above set may be computed by using Equation (4.13), which is a part of the controllability transfer function. For example, the cost of the first cube can be computed as $CC^0(c_{i_1}) + CC^0(c_{i_2})$. Equation (4.15) can then be used to combine the costs of the cubes to compute the value of $CC^0(c_j)$, as

$$CC^0(c_j) = \min\left[CC^0(c_{i_1}) + CC^0(c_{i_2}), CC^1(c_{i_1}) + CC^1(c_{i_2}), CC^1(c_{i_1}) + CC^1(c_{i_2}),\right.$$
$$\left. CC^0(c_{i_1}) + CC^0(c_{i_2})\right] + CC_\kappa(c_j).$$

Now consider the second scenario in which the current values at inputs c_{i_1} and c_{i_2} are 0_6 and X_6, respectively, and the value at the output is X_6. If the value 0_6 is desired at c_j, then the intersection of the cube containing the current values with the six-valued cube cover of an XOR gate gives

c_{i_1}	c_{i_2}
0_6	0_6

First, consider the cost of this cube. If Equation (4.13) is used, it gives the cost as $CC^0(c_{i_1}) + CC^0(c_{i_2})$. This does not reflect the fact that the current vector already implies the value desired by the cube at input c_{i_1}.

In general, the cost of a cube $\mathcal{C} = (\mathcal{C}_1, \mathcal{C}_2, \dots, \mathcal{C}_\alpha)$ can be computed in order to take advantage of the current value assignment as

$$Cost(\mathcal{C}) = \sum_{l=1,2,\dots,\alpha} cost(\mathcal{C}_l), \tag{4.20}$$

where

$$cost(\mathcal{C}_l) = \begin{cases} 0, & \text{if the current value at } c_{i_l}, V(c_{i_l}), \text{ is identical to } \mathcal{C}_l, \\ CC^0(c_{i_l}), & \text{if } \mathcal{C}_l \text{ is } 0_6 \text{ or } \overline{D}_6, \\ CC^1(c_{i_l}), & \text{if } \mathcal{C}_l \text{ is } 1_6 \text{ or } D_6, \\ 0, & \text{otherwise.} \end{cases} \tag{4.21}$$

For the above example, this equation gives the cost of the cube as $CC^0(c_{i_2})$. Next, consider the fact that, for the given set of current values, only one cube can give the value 0_6 at the output of the cube. The overall difficulty of obtaining this value at the output is given merely by the cost of the single cube computed above plus the constant $CC_\kappa(c_j)$ as

$$CC^0(c_j) = CC^0(c_{i_2}) + CC_\kappa(c_j).$$

In general, each TGST alternative can be represented by a set of cubes where each cube is (a) compatible with the current values at the inputs and outputs of a circuit element, and (b) provides the desired value. Note that, due to the addition of condition (a) above, this set of cubes is limited to those that are relevant for the current values at the inputs and outputs of the circuit element. The cost of each cube in the above set can be computed using modified versions of the controllability transfer function as exemplified by Equations (4.20) and (4.21). The cost of the TGST alternative can then be computed by taking the minimum of the costs of each cube in the above set and adding appropriate constants. This step is similar to those described by Equations (4.15) and (4.18), with the difference that the minimum is taken over the limited set of cubes.

The cost functions computed above are used in conjunction with principles described in Section 4.6.2.2 to prioritize alternative TGSTs, compatible cubes, and value assignments. The manner in which they can be used is illustrated throughout Section 4.6.

4.6 Specific structural test generation paradigms

In this section, we discuss two specific structural test generation paradigms. In the first paradigm, which is exemplified by the D-algorithm (Roth, 1966), the search space is constituted of alternative TGSTs and alternative local value assignments for the TGSTs. In the second paradigm, exemplified by PODEM (Goel, 1981), the search space is constituted of alternative value assignments at circuit inputs and/or at selected circuit lines. We first describe each of the two algorithms. This is followed by a comparison of the two paradigms and additional techniques that may be used to enhance their performance.

4.6.1 Searching the space of TGSTs and local value assignments: D-algorithm

The D-algorithm is the first known complete test generation algorithm. This algorithm established the paradigm of searching the space of TGSTs and local value assignments. In the following, we describe a variant of the D-algorithm to illustrate the concepts behind this search paradigm.

The first four steps of this algorithm can be identical to those of the generic test generation algorithm discussed in Section 4.5. Consider an example circuit with an SA0 fault at input x_5, as shown in Figure 4.24. The status of the CUT after initialization using the six-valued system is shown in Figure 4.24(a).

After initialization, the CUT is analyzed and TGSTs identified using the approach described in Section 4.5.2. In the first step following initialization, the target fault is not yet excited. Hence, the first TGST that is identified is fault excitation. The circuit element under consideration for this TGST is the circuit element with the target fault. This TGST is considered necessary since most test generators in this category do not proceed unless this TGST is completed.

4.6.1.1 Identification of local value assignments for fault excitation

The objective of fault excitation is to ensure that one or more outputs of the faulty circuit element have the value D_6 or $\overline{D}_6$. *In this case, the local value assignments of interest are the values that may be assigned at the outputs of the faulty circuit element.*

The identification of local value assignments begins with the intersection of the current values at the inputs and outputs of the faulty circuit element with its fault excitation cubes. Each non-empty cube obtained as a result of these intersections is a **candidate** for local value assignment for fault excitation.

If no candidate cube is found, then there exists no value assignment that can excite the target fault and backtrack is initiated. Assuming that one or more candidate cubes are found, first we consider the case where the element under consideration is a faulty gate. This will be followed by a discussion of the case where the element is a fanout system.

A faulty gate: Fault effect excitation can only result in the implication of either the value D_6 or $\overline{D}_6$ at the output of the faulty gate. However, in some cases, it may be possible to imply only one of these values.

Consider as an example the case where the target fault is an SA0 fault at output c_j of a NAND gate with inputs c_{i_1} and c_{i_2}. Assume that the current values at its inputs and output are those obtained after initialization using the six-valued system, namely, $V(c_{i_1}) = \chi_6$, $V(c_{i_2}) = \chi_6$, and $V(c_j) = X_6$. Intersection of a cube containing these values, (χ_6, χ_6, X_6), with the *fault excitation cubes of the faulty version of the NAND gate* gives the following compatible cubes. (In most of the following sections, for each compatible cube, the output value will also be shown.)

c_{i_1}	c_{i_2}	c_j
0_6	χ_6	D_6
χ_6	0_6	D_6

The above example illustrates a case where all the candidate cubes for fault excitation have the same value in their output fields. In such cases, a unique local value assignment is identified for fault excitation. For the above example, the local assignment $V(c_j) = D_6$ is the unique assignment associated with fault excitation.

In other cases, some compatible cubes may have the value D_6 in their output fields while others have the value $\overline{D}_6$. For example, consider the case where the target fault is an SA0 fault at input c_{i_1} of an XOR gate with inputs c_{i_1} and c_{i_2} and output c_j. If the current values at the gate inputs and output are $V(c_{i_1}) = \chi_6$, $V(c_{i_2}) = \chi_6$ and $V(c_j) = X_6$, then the intersection gives the following compatible cubes.

c_{i_1}	c_{i_2}	c_j
1_6	0_6	D_6
1_6	1_6	$\overline{D}_6$

In all such cases, two alternative assignments, (a) $V(c_j) = D_6$, and (b) $V(c_j) = \overline{D}_6$, are identified. In this case, the cost of the first cube can be computed as $CC^1(c_{i_1}) + CC^0(c_{i_2})$ and that of the second cube can be obtained as $CC^1(c_{i_1}) + CC^1(c_{i_2})$. The cost of obtaining $\{D\}$ at c_j can be computed as the cost of the first cube (since this is the only compatible cube that gives this value at c_j) plus a constant as $CC^1(c_{i_1}) + CC^0(c_{i_2}) + CC_\kappa(c_j)$. Similarly, the cost of obtaining a $\overline{D}_6$ at c_j can be computed as $CC^1(c_{i_1}) + CC^1(c_{i_2}) + CC_\kappa(c_j)$. These costs can be used to prioritize two alternative ways of fault excitation, for example, in some cases, the alternative with lower cost may be selected.

A faulty fanout system: A fanout system has two or more outputs. However, for each faulty version of the fanout system with a single SAF, at most one compatible cube is found. For example, consider a fanout system with input c_i and outputs c_{j_1}, c_{j_2}, and c_{j_3} with an SA0 fault at input c_i. Assuming the current values $V(c_i) = \chi_6$, $V(c_{j_1}) = X_6$, $V(c_{j_2}) = X_6$, and $V(c_{j_3}) = X_6$, the intersection gives the following compatible cube.

c_i	c_{j_1}	c_{j_2}	c_{j_3}
1_6	D_6	D_6	D_6

In this case, the unique set of assignments, namely, $V(c_{j_1}) = D_6$, $V(c_{j_2}) = D_6$, and $V(c_{j_3}) = D_6$, is identified as being necessary for fault effect excitation.

Next, consider a single SA1 fault at output c_{j_1} of a fanout system, with initial values $V(c_i) = \chi_6$, $V(c_{j_1}) = X_6$, $V(c_{j_2}) = \chi_6$, and $V(c_{j_3}) = \chi_6$. In this case, the set of assignments $V(c_{j_1}) = \overline{D}_6$, $V(c_{j_2}) = 0_6$, and $V(c_{j_3}) = 0_6$, is found to be necessary for fault effect excitation.

For our running example circuit shown in Figure 4.24, the element under consideration is the fanout system with stem x_5. For the values obtained after initialization, shown in Figure 4.24(a), only one compatible cube for fault excitation is found.

x_5	c_1	c_2
1_6	D_6	D_6

Consequently, the necessary local value assignments for fault effect excitation are $V(c_1) = D_6$ and $V(c_2) = D_6$. The values at circuit lines after making this assignment, and performing direct implications, are shown in Figure 4.24(b).

4.6.1.2 Identification of local value assignments for fault effect propagation

Once the fault is excited, the TGST identification procedure identifies gates through which the fault effect may be propagated at a particular test generation step. This is repeated until the fault effect is propagated to one or more outputs of the CUT or no fault effect propagation alternative can be found.

For each gate that is identified as the site of a propagation subtask, the alternative local value assignments of interest are the possible faulty values that may be assigned to the output of the gate. These alternative value assignments can be enumerated by intersecting the cube comprised of the current values at the inputs and outputs of the element with the propagation cubes of the element. Each non-empty cube obtained in this manner denotes an alternative combination of value assignments that can help propagate the fault effect to the output of the corresponding gate.

For some gates, all alternative cubes may have the same faulty value at the output, e.g., $\overline{D}_6$. For example, consider a NAND gate with inputs c_{i_1} and c_{i_2} and output c_j that is identified as a site of fault propagation. If the current values at the inputs and output of the gate are $V(c_{i_1}) = D_6$, $V(c_{i_2}) = X_6$, and $V(c_j) = X_6$, then the following two compatible cubes are found for fault effect propagation through this gate.

c_{i_1}	c_{i_2}	c_j
D_6	1_6	$\overline{D}_6$
D_6	D_6	$\overline{D}_6$

Since the value at the output of this gate is $\overline{D}_6$ for both these cubes, a unique value assignment, namely $V(c_j) = \overline{D}_6$, is identified for fault effect propagation through this gate. (In particular, for propagation of a D_6 from one input of a primitive gate to its output, some combination of D_6 and the gate's non-controlling value must be applied to its other inputs.)

The cost of both the cubes can be computed as $CC^1(c_{i_2})$, using the method outlined in Section 4.5.3.6. The cost of propagating the fault effect through the gate to obtain $\overline{D}_6$ at c_j can therefore be computed as $\min[CC^1(c_{i_2}), CC^1(c_{i_2})] + CC_\kappa(c_j) = CC^1(c_{i_2}) + CC_\kappa(c_j)$.

For other gates, some compatible cubes have one faulty value, say D_6, at the output while others may have the other faulty value, $\overline{D}_6$. For example, consider an XOR gate with inputs c_{i_1} and c_{i_2} and output c_j that is identified as a site of fault propagation. If the current values at the inputs and output of the gate are $V(c_{i_1}) = D_6$, $V(c_{i_2}) = X_6$, and $V(c_j) = X_6$, then the following two compatible cubes are found for fault effect propagation through this gate.

c_{i_1}	c_{i_2}	c_j
D_6	0_6	D_6
D_6	1_6	$\overline{D}_6$

In such a case, two alternative local value assignments, namely $V(c_j) = D_6$ and $V(c_j) = \overline{D}_6$, are identified for fault effect propagation through this gate. The cost of the above two alternative local value assignments can be computed as $CC^0(c_{i_2}) + CC_\kappa(c_j)$ and $CC^1(c_{i_2}) + CC_\kappa(c_j)$, respectively.

When cases of unique D-drive and future unique D-drive are identified, some of these fault effect propagation subtasks are identified as being necessary. In such cases, the attention may first be limited only to the necessary propagation subtasks. If no propagation subtask is identified as being necessary, then all subtasks are considered. Typically, the observability of the gate output is used to select one of the gates for fault effect propagation. The above analysis is then performed on the selected gate to (a) identify alternative local value assignments to propagate the fault effect to the output of the gate; (b) enumerate all possible local value assignments at the inputs of the gate (in the form of compatible cubes), for each alternative faulty value at the output; and (c) order the alternative local value assignments at the outputs and the corresponding compatible cubes at the inputs.

For our example circuit, for the state shown in Figure 4.24(b), the analysis of the
D-frontier identifies the fault effect propagation subtasks: (a) propagate via G_2, and
(b) propagate via G_5. (An additional necessary subtask, namely, propagate via G_7,
would have been identified if a more complex procedure was used to identify a future
unique D-drive.) For the first alternative, the following compatible cube is found.

x_4	c_1	c_4
1_6	D_6	D_6

The cost of the cube can be computed as $CC^1(x_4)$ and the cost of obtaining D_6 at
c_4 can be computed as $CC^1(x_4) + CC_\kappa(c_4)$. For the second alternative, the following
compatible cubes can be found.

c_2	c_7	c_9
D_6	0_6	D_6
D_6	1_6	$\overline{D}_6$

Two alternative local value assignments are found for this fault effect propagation
TGST, namely $V(c_9) = D_6$ and $V(c_9) = \overline{D}_6$. The costs of the above cubes
are, respectively, $CC^0(c_7)$ and $CC^1(c_7)$. The costs of the alternative local value
assignments are $CC^0(c_7) + CC_\kappa(c_9)$ and $CC^1(c_7) + CC_\kappa(c_9)$, respectively.

The observabilities of c_4 and c_9 would typically be compared to decide whether to
propagate the fault effect via G_2 or G_5. If the second alternative is selected, the above
computed costs would be used to decide whether a D_6 or $\overline{D}_6$ is assigned to c_9.

4.6.1.3 Identification of local value assignments for line justification

Once the fault effect is propagated to one or more primary outputs of the CUT,
the attention is focused on line justification. *In this case, the alternative local value
assignments of interest are the values assigned to one or more inputs of the circuit
element whose output is the justification site.* Note that this is different from other
types of TGSTs where the value assignments of interest are the faulty values that may
be assigned at the output of the site under consideration.

The process of identifying alternative local value assignments for an unjustified gate
in the circuit starts by enumerating various compatible cubes. This is accomplished by
intersecting the current values at the gate inputs and outputs with the corresponding
values of each cube in the cube cover of the gate. (Recall that if the unjustified
gate is fault-free, then the cube cover is comprised of the gate's singular cover and
propagation cubes of the fault-free version. If the unjustified gate is faulty, then it is
comprised of the singular cover and fault excitation cubes of the faulty version of the
gate with the target fault. However, in this case, only the intersection with the fault
excitation cubes will yield the desired results.)

For example, consider a fault-free NAND gate with inputs c_{i_1} and c_{i_2} and output c_j with current values $V(c_{i_1}) = \chi_6$, $V(c_{i_2}) = \chi_6$, and $V(c_j) = 1_6$. The above values are intersected with each cube in the cube cover of the gate to identify the following compatible cubes.

c_{i_1}	c_{i_2}	c_j
0_6	χ_6	1_6
χ_6	0_6	1_6

Again, the cost of each combination of local value assignments can be computed as described in Section 4.5.3.6. In this case, the cost of the first compatible cube is $CC^0(c_{i_1})$; that of the second cube is $CC^0(c_{i_2})$.

Next, consider a version of the above NAND gate with an SA0 fault at its output. During the fault excitation phase, the D-algorithm will merely assign the value D_6 to its output and proceed with fault effect propagation. The value at each of its inputs may be left at its initial value, χ_6, leaving unjustified the values at this gate. If the values at this gate are to be justified later during the justification phase, then the following cubes are found by intersecting the current values at the inputs and output of the gate with the fault excitation cubes of the faulty version of the NAND gate with an SA0 fault at its output (see Figure 4.8(b)).

c_{i_1}	c_{i_2}	c_j
0_6	χ_6	D_6
χ_6	0_6	D_6

Each of the above cubes represents a possible local value assignment that can help justify the values at this gate. In this case, the cost of the first compatible cube is $CC^0(c_{i_1})$; that of the second cube is $CC^0(c_{i_2})$.

4.6.1.4 Remaining steps of the algorithm

During each **round** of execution of the test generation algorithm, the above concepts can be used to identify TGSTs and alternative local value assignments for each TGST. These may be prioritized using the costs computed above and the rules outlined in Section 4.6.2.2.

The next step is to make value assignments. First, the first untried local value assignment of the first untried TGST is selected and marked as having been tried. Second, the current state of the test generation algorithm is saved. In the D-algorithm, the state is comprised of *values at circuit lines, untried alternative TGSTs, and untried alternative local value assignments.* For convenience, the contents of $\mathcal{D}$ and $\mathcal{U}$ may also be saved along with the state. Third, the selected local value assignments are made by assigning the values to corresponding circuit lines. Finally, the implication procedure is executed.

If the implication procedure returns CONFLICT, then backtracking is initiated. Otherwise, the algorithm continues with the next round, beginning with the identification of new TGSTs.

The details of the backtracking steps are as described in Step 8 of the generic structural test generation algorithm.

4.6.1.5 An execution example of the D-algorithm

To keep the following description simple, we will skip a discussion of the cost computation for and prioritization of various TGSTs and their alternative local value assignments. Instead, we assume that alternatives are tried in the *default order*, i.e., in the order in which they are generated. We also assume that indirect implications are neither identified in the pre-processing step nor used during the implications performed during various steps of execution of test generation. Finally, we assume that during the identification of TGSTs, x-path check is performed but the sites of future unique D-drive are not identified.

The steps of the test generation algorithm are executed iteratively until test generation terminates. In the following, each such iteration is referred to as a **round** of execution of the algorithm. In Table 4.2, the acronyms FEE and FEP are used to respectively denote fault effect excitation and fault effect propagation. Finally, a mark "$\sqrt{}$" is used to mark alternatives that have already been tried.

Now consider the example circuit shown, along with the target fault, in Figure 4.24. First, the circuit lines are initialized using the six-valued system to obtain the values shown in Figure 4.24(a).

Subsequently, TGSTs are identified. The only TGST that is identified at this step is fault effect excitation at the fault site, the fanout system with stem x_5. The intersection of the current values with the fault excitation cubes of the faulty version of the fanout stem with the target fault gives only a single set of local value assignments, namely $V(c_1) = D_6$ and $V(c_2) = D_6$, as shown in row R1(b) of Table 4.2. This TGST as well as the corresponding combination of local value assignments are marked as tried. The state of the test generation algorithm is saved and this value assignment is made. This is followed by execution of the implication procedure. Figure 4.24(b) shows the values at various circuit lines. Note that the fanout stem is justified due to the execution of the implication procedure. As shown in row R1(b) of Table 4.2, the D-frontier, $\mathcal{D}$, contains gates G_2 and G_5, and the list of unjustified lines, $\mathcal{U}$, is empty.

The TGSTs identified in the second round include fault effect propagation through each of the gates in $\mathcal{D}$. The value assignments for propagation via G_2 are found by intersecting current values at the inputs and output of the gate with the propagation cubes of a two-input NAND gate. For this TGST, the only local value assignment found is $V(c_4) = D_6$. The alternative value assignments for the other TGST, propagation via G_5, are found using the propagation cubes of a two-input XOR gate. Two alternative value assignments (a) $V(c_9) = D_6$, and (b) $V(c_9) = \overline{D}_6$, are found. The first TGST is selected and the first alternative local value assignment, namely $V(c_4) = D_6$, is then selected and marked as tried. The state of the test generator is then saved and the values given by the selected local value assignment are actually assigned to the circuit lines. The implication procedure is then executed to obtain the values at the circuit lines shown in Figure 4.24(c). Note that the values at gate G_2 are justified during the execution of the implication procedure. The new values of $\mathcal{D}$ and $\mathcal{U}$ are as shown in row R2 of Table 4.2.

In the third round, the TGSTs identified and the alternative local value assignments are shown in row R3 of the table. The first alternative value assignment is selected, marked as tried, state of test generator saved, selected assignment made, and direct implication performed. The values at the circuit lines after implication are shown in Figure 4.24(d) and the values of $\mathcal{D}$ and $\mathcal{U}$ are shown in row R3 of the table. Note that the output of the CUT has value 1_6 while the D-frontier is not empty. This is a scenario in which x-path check, executed during TGST identification in the next step, can help identify the need to backtrack.

The fourth round of the algorithm begins by identification of TGSTs. During this process, x-path check eliminates all gates from $\mathcal{D}$. Since no primary output already has D_6 or $\overline{D}_6$, this indicates the need to backtrack. The state of the test generator is retrieved from the stack. The circuit state is restored to that at the beginning of the third round, shown in Figure 4.24(c). The alternative TGSTs obtained from the stack contain the TGSTs and alternative value assignments that were left untried at the end of the third round. These alternatives are copied in row R5 of the table.

Due to the backtrack in the previous round, TGST identification is not carried out in the fifth round. Instead, the first untried TGST in the restored list (propagate via G_5) is selected and the first untried alternative local value assignment, $V(c_9) = D_6$, is selected. The selected alternative is marked as tried, the state of the test generator saved, and the selected assignment made. Subsequently, the implication procedure is executed. In this case, the implication procedure returns CONFLICT. The reader can verify that the values $V(c_4) = D_6$, $V(x_3) = \chi_6$, and $V(c_5) = 0_6$ are incompatible. This is identified by the implication procedure and causes the value CONFLICT to be returned. This signals the need to backtrack. Again, the state is restored to that at the beginning of the fifth round (which, due to the backtrack in round 4, was identical to that shown in Figure 4.24(c)). The alternative TGSTs obtained from the stack are copied in row R6 of the table.

Again, due to backtrack in the previous step, TGST identification is not carried out in the sixth round. Instead, the TGSTs and alternative local value assignments restored from the stack are used. At this stage, only one TGST, propagation via G_5, and only

Table 4.2. *An example execution of the D-algorithm*

	Initial state	TGSTs identified	Alternative assignments	Values assigned	Implication results	$\mathcal{D}$	$\mathcal{U}$	Comments
R1(a)	—	—	—	—	Figure 4.24(a)	{}	{}	Initialization
R1(b)	Figure 4.24(a)	√(A) FEE at fanout x_5	√(i) $V(c_1)=D_6$; $V(c_2)=D_6$	$V(c_1)=D_6$; $V(c_2)=D_6$	Figure 4.24(b)	$\{G_2, G_5\}$	{}	
R2	Figure 4.24(b)	√(A) FEP via G_2 (B) FEP via G_5	√(i) $V(c_4)=D_6$ (i) $V(c_9)=D_6$ (ii) $V(c_9)=\overline{D}_6$	$V(c_4)=D_6$	Figure 4.24(c)	$\{G_3, G_5\}$	{}	
R3	Figure 4.24(c)	√(A) FEP via G_3 (B) FEP via G_5	√(i) $V(c_5)=D_6$ (i) $V(c_9)=D_6$ (ii) $V(c_9)=\overline{D}_6$	$V(c_5)=D_6$	Figure 4.24(d)	$\{G_6\}$	{}	
R4	Figure 4.24(d)	None (Backtrack)			Figure 4.24(c)*	$\{G_3, G_5\}$	{}	*State restored from R3
R5	Figure 4.24(c)*	√(A) FEP via G_3 (B) FEP via G_5	√(i) $V(c_5)=D_6$ √(i) $V(c_9)=D_6$ (ii) $V(c_9)=\overline{D}_6$	$V(c_9)=D_6$	CONFLICT (Backtrack) Figure 4.24(c)†	$\{G_3, G_5\}$	{}	TGSTs restored from R3 †State restored from R4
R6	Figure 4.24(c)†	√(A) FEP via G_3 √(B) FEP via G_5	√(i) $V(c_5)=D_6$ √(i) $V(c_9)=D_6$ √(ii) $V(c_9)=\overline{D}_6$	$V(c_9)=\overline{D}_6$	Figure 4.24(e)	$\{G_7\}$	{}	TGSTs restored from R5
R7	Figure 4.24(e)	√(A) FEP via G_7	√(i) $V(z)=D_6$	$V(z)=D_6$	Figure 4.24(f)	{}	$\{G_1\}$	
R8	Figure 4.24(f)	√(A) Justify G_1	√(i) $V(x_1)=1_6$ (ii) $V(x_2)=1_6$	$V(x_1)=1_6$	Figure 4.24(g)	{}	{}	

one local value assignment, $V(c_9) = \overline{D}_6$, remain. This value assignment is selected, marked as tried, the state of the test generation algorithm saved, and the selected value assignment is made. The implication procedure is then executed to obtain the values shown in Figure 4.24(e). Again, execution of the implication procedure justifies all values. The values of $\mathcal{D}$ and $\mathcal{U}$ are shown in row R6 of the table.

In the seventh round, a single TGST, propagate via G_7, and a single alternative local value assignment, $V(z) = D_6$, are identified. The state of the test generator is saved, assignment made, and implication performed to obtain the circuit state shown in Figure 4.24(f) and the values of $\mathcal{D}$ and $\mathcal{U}$ are shown in row R7 of the table.

In the eighth round, TGST identification finds D_6 at the output of the CUT and finds $\mathcal{U}$ non-empty. Hence, it identifies a single TGST, justify G_1, and two alternative local value assignments: (a) $V(x_1) = 1_6$, and (b) $V(x_2) = 1_6$. The first assignment is selected, marked as tried, the state saved, selected assignment made, and implication performed to obtain the values shown in Figure 4.24(g).

The values at the end of above round constitute a test for the target fault. This fact is actually identified when TGST identification is carried out in the ninth round. Note that the generated vector $(1_6, \chi_6, 1_6, 1_6, 1_6)$ will be written in the form $(1, x, 1, 1, 1)$ using the three-valued system for application to each manufactured copy of the CUT.

Figure 4.27 shows the structure of the search tree for the above example execution of the D-algorithm. Here, VA stands for value assignment.

4.6.2   Searching the space of input vectors: PODEM

In the path-oriented decision making (PODEM) algorithm, the search space is constituted of values assigned at primary inputs or, in some variations, a combination of a selected set of fanout stems and primary inputs. In the original PODEM, values are assigned only at the primary inputs and only the forward implication procedure is used. Some variations assign values at internal lines, especially fanout stems, and may use an implication procedure that performs forward as well as backward implications. In the following, we will describe the original PODEM algorithm.

The first consequence of this search strategy is that the search space becomes predictable in size, for an n-input circuit, the maximum number of backtracks required is 2^n. In contrast, the search space size is much more difficult to compute for the D-algorithm, since the number of TGSTs and alternative local value assignments cannot be computed readily in terms of circuit parameters, such as the number of primary inputs, number of gates, and so on.

The second consequence of assigning values only at the primary inputs is that no unjustified lines exist during any stage of execution of the test generation algorithm.

The third consequence is that only forward implication is required. This reduces the complexity of each implication step. It also renders useless static indirect implications, most of which are useful only if backward implications are performed.

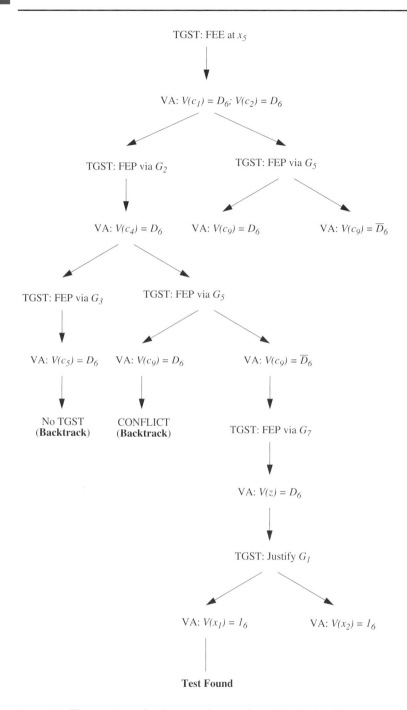

Figure 4.27 The search tree for the example execution of the D-algorithm

The fourth consequence is that at each step of execution of the test generation algorithm, only the values at the primary inputs, and the corresponding untried alternative values, need to be stored on the stack. Whenever backtracking becomes necessary, these values are assigned and forward implication procedure is executed to obtain the values at all circuit lines.

Next, we describe the PODEM algorithm. We will focus on some of the key mechanisms unique to PODEM. The first four steps of the generic structural test generation algorithm described in Section 4.5.1 are used with minor modifications. In particular, since backward implication is not used, static identification of indirect implications in the pre-processing step need not be performed.

4.6.2.1 Identification of an objective

TGSTs may also be identified using the procedure described in Section 4.5.2. The first key difference is that, since unjustified lines do not exist, line justification subtasks are never identified. In other words, only fault effect excitation and fault effect propagation subtasks are identified. Note that the above-mentioned procedure to identify TGSTs need not be modified, since it can naturally accommodate this fact.

Once the TGSTs and alternative local value assignments for each TGST are identified, they cannot be used directly, since most alternative local value assignments are typically at lines internal to the circuit. Instead, each alternative local value assignment is treated as a possible **objective** and is handled as described ahead.

4.6.2.2 Rules for prioritizing alternatives

The cost of a particular combination of values at the outputs of a circuit element under consideration can be computed as described in Section 4.5.3.6. Each alternative local value assignment at the inputs of the circuit element under study is typically represented as a cube. The cost of each cube can also be computed as described in Section 4.5.3.6. Finally, the cost of obtaining a specific value at an input of the circuit element under consideration can be computed using Equation (4.21). These cost functions are then used in conjunction with the rules below that aim to reduce the amount of space searched by the test generation algorithm before a test is found.

The following rules for prioritizing alternative TGSTs and the corresponding alternative value assignments are described in Goel (1981). In the following, **goal** is a generic term used to describe a subtask. For the circuit element under study, it also describes an alternative value assignment at its outputs, cubes describing alternative combinations of values at the inputs, and values at different inputs.

1 Among multiple goals, if some are necessary and others are not, then narrow the consideration to the necessary goals.

2 Among a set of goals where each goal is necessary, select the one with the highest difficulty, i.e., with the highest cost.

3 Among multiple goals where each goal may potentially be individually sufficient to

accomplish the overall task, select the one with the lowest difficulty, i.e., with the lowest cost.

For example, if during a particular stage of the test generation algorithm, three fault effect propagation subtasks are identified of which two are identified as being necessary, then the first rule suggests that the attention should be focused on the two necessary subtasks. Furthermore, the second rule suggests that the more difficult of these two subtasks should be assigned higher priority. On the other hand, if there exist three alternatives, none of which is identified as being necessary, then according to the third rule, the easiest one should be assigned the highest priority.

Similarly, when multiple cubes exist that can satisfy a given value assignment, then the cube with the lowest cost should be assigned the highest priority, since the existence of multiple cubes demonstrates the existence of choice. Furthermore, for a given cube, each of the specific input values must be implied for the cube to imply the desired value at the outputs of the circuit element under consideration. Hence, among these values, the value assignment with the highest cost should be selected.

These rules attempt to reduce the space searched during test generation. The most difficult of a number of necessary goals may not be satisfiable. If this goal is targeted first and is found to be unsatisfiable, then backtracking may be initiated without wasting any effort on searching for a solution for any of the other necessary goals. In contrast, when there exists a choice of goals, if a solution can be found for the easiest goal, then none of the other, more difficult, goals need to be satisfied. Since more search may have been required to satisfy one of the more difficult goals, trying the easiest choice first can help reduce the amount of search required.

The identified TGSTs, their alternative value assignments, the corresponding cubes, and the component values of each cube can all be prioritized using their costs and the above rules. The value assignment that is assigned the highest priority is selected as the objective of the given test generation step. For example, for the circuit shown in Figure 4.24, the first objective identified will be to imply value D_6 at line c_1 as well as c_2. In general, an objective is comprised of a set of lines that are the outputs of the circuit element under consideration and a set of desired values at these lines. For example, the above fault excitation objective can be written as (c_1, D_6) and (c_2, D_6).

4.6.2.3 Translating an objective to a primary input assignment

As seen above, an analysis of the circuit identifies TGSTs and alternative local value assignments that are, in general, at lines internal to the circuit. As the test generation objective is one of these value assignments, the objective is, in general, a value assignment at one or more internal circuit lines. However, in PODEM, value assignments are made only at the primary inputs. Hence, a procedure is required to obtain a primary input assignment that may facilitate the completion of the selected objective. PODEM uses a procedure called *backtrace*, which is one of its most used procedures.

The process of **backtrace** begins at the circuit element under consideration at whose output the objective values are given. The current values at the inputs of the element and the values desired at its outputs, as given by the objective, are intersected with the cubes in the cube cover of the appropriate version of the circuit element. When a non-empty intersection is found between a partially specified current value at one of its input components and the corresponding value in the gate's cube, the resulting cube may be saved *even if the intersection of another input component and the corresponding value in the cube from the gate's cover is empty.* Such **incompatible cubes** obtained during intersection may be used to pursue backtrace when no **compatible cube** is found during intersection.

In general, this process generates a number of alternative cubes. If multiple alternative cubes are found, then the cubes are prioritized and the one with the highest priority is selected.

The values given by the selected cube at the inputs of the circuit element are then analyzed. For some inputs, the values given by the selected cube are identical to the current values at the corresponding circuit lines. All such values are ignored. The values at the remaining inputs are prioritized using controllabilities and the above rules. Finally, the input value that is assigned the highest priority and the circuit line at which this value occurs are selected as the next objective. In the following, we say that the backtrace procedure **visited** this line with the selected value as the **desired value**.

The above process is repeated until the resulting objective is located at a primary input.

Consider the circuit shown in Figure 4.28. Assume the initial values given in the figure. Let the initial objective be the assignment of 1_6 at line c_{10}, i.e., $(c_{10}, 1_6)$. The intersection of current values at the inputs of G_6 and the desired value at its output with the cubes of G_6 gives the following cubes.

c_3	c_8	c_{10}
1_6	X_6	1_6
X_6	1_6	1_6

Let us assume that the first cube is selected based on cost and the above rules. In this cube, only the value at c_3 is more specific than the current value. Hence, obtaining 1_6 at c_3 becomes the next objective. As mentioned earlier, we say that the backtrace procedure visited c_3 with value 1_6. Intersection is then performed between the current values at the inputs of G_1 and the value desired by the objective at its output and the cubes of the two-input OR gate G_1. The following cubes are obtained.

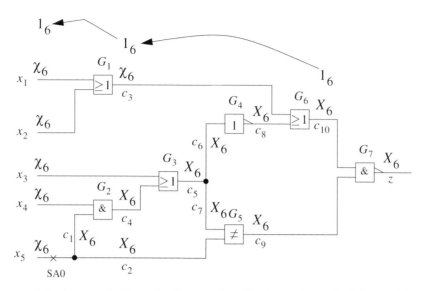

Figure 4.28 An example illustrating the execution of backtrace. (This circuit is a modified version of a circuit in Roth (1966))

x_1	x_2	c_3
1_6	χ_6	1_6
χ_6	1_6	1_6

Again, if the first cube is selected using the cost function and rules, then the next objective is obtained as assignment of 1_6 at x_1. Since this objective is located at a primary input, backtrace terminates by providing the assignment $V(x_1) = 1_6$ as the assignment that PODEM should make. The sequence of objectives obtained by the backtrace procedure are shown in Figure 4.28.

4.6.2.4 Remaining steps of the algorithm

The remaining steps of the algorithm are very similar to those described in the generic test generation algorithm. First, the primary input identified by and the value provided by backtrace are selected for assignment, the opposite value is noted as an untried alternative at the selected input, and the state of test generation is saved. Recall that in PODEM, the state of the test generator is comprised of the values applied at its inputs (since the remaining values can be derived by executing the implication procedure) and the set of untried alternative value assignments at primary inputs. This is followed by the assignment of the selected value at the selected primary input and the execution of the forward implication procedure. In this case, the implication procedure always returns SUCCESS, because no values are assigned to the internal circuit lines, hence no conflict can occur.

In the subsequent round, TGST identification is performed. This may (a) identify that the assignment made in the previous step cannot detect the target fault, (b) identify that assignments made in the previous steps constitute a test vector for the target fault, or (c) identify one or more TGSTs. In the first case, backtracking is initiated; in the second case, test generation for the target fault is complete; and in the third case, test generation continues and the above process is repeated.

4.6.2.5 An execution example of PODEM

Consider the circuit shown in Figure 4.24 with the SA0 fault at line x_5. (Coincidentally, the values obtained at circuit lines at the end of each round (except one) of PODEM are identical to those for the D-algorithm. Hence, we will use the same figures to save space. It should, however, be apparent from the following that the steps in the algorithm are quite different and that the similarity for this circuit and target fault is indeed highly coincidental.)

The steps taken by PODEM to find a test for the target fault in this CUT will now be outlined. To enable a direct comparison with the execution of the D-algorithm, we will use the topological order in the following. (The execution of both algorithms using testability measures and above rules is left as an exercise.)

The first round of execution of the PODEM algorithm begins with initialization, which results in the values shown in Figure 4.24(a). TGST identification is then performed and it identifies fault excitation as the only necessary TGST. Alternative value assignments for this are enumerated to obtain a single compatible cube.

x_5	c_1	c_2
1_6	D_6	D_6

Since only a single value at x_5, $V(x_5) = 1_6$, can excite the fault, it is selected as the objective for backtrace. This objective is sometimes written in the form $(x_5, 1_6)$. Since the objective is already specified at a primary input, backtrace terminates immediately giving the assignment $V(x_5) = 1_6$. This assignment is then selected and the other possible assignment at the selected input, 0_6 at x_5, is noted as an untried alternative at the input. The state of the test generator is then saved, the selected assignment made, and forward implication performed to obtain $V(c_1) = D_6$ and $V(c_2) = D_6$. The values at the circuit lines are shown in Figure 4.24(b) and the D-frontier, $\mathcal{D}$, and set of unjustified lines, $\mathcal{U}$, are shown in row R1(b) of Table 4.3.

The second round begins with the identification of TGSTs. TGSTs identified in this round include fault effect propagation through each of the gates in $\mathcal{D}$. The first TGST, propagation via G_2, is selected and the current values at the inputs and output of G_2 are intersected with the propagation D-cubes of a two-input AND gate to obtain the following cube.

x_4	c_1	c_4
1_6	D_6	D_6

The first value is selected as the objective, $(x_4, 1_6)$, and backtrace executed. Since x_4 is a primary input, backtrace terminates immediately giving the value assignment $V(x_4) = 1_6$. The other possible assignment at x_4, $V(x_4) = 0_6$, is noted as an untried alternative. The state of the test generator, which in this case is comprised of all the primary input values assigned prior to this step and untried alternative value assignments identified above, is saved. The selected assignment, $V(x_4) = 1_6$, is then made and forward implication performed to obtain the values shown in Figure 4.24(c) and the contents of $\mathcal{D}$ and $\mathcal{U}$ shown in row R2 of Table 4.3.

The TGSTs identified in the third round are (a) propagation via G_3, and (b) propagation via G_5. The former is selected and the following cube is obtained by intersecting the current values at the inputs and output of G_3 with the propagation cubes of a two-input OR gate.

x_3	c_4	c_5
0_6	D_6	D_6

The first value is selected to obtain the objective $(x_3, 0_6)$, which leads to the primary input assignment $V(x_3) = 0_6$. The other possible assignment at x_3 is noted as untried. The primary input values assigned until the end of previous round and the above untried assignment are saved. The selected assignment is made and forward implication performed to recreate all the values shown in Figure 4.24(d) and the contents of $\mathcal{D}$ and $\mathcal{U}$ shown in row R3 of Table 4.3. (Please note that each row of Table 4.3 summarizes the corresponding round of this example execution of PODEM.)

The fourth round of the algorithm begins by identification of TGSTs. During this process, a check of the potential propagation path eliminates all gates from $\mathcal{D}$, indicating the need to backtrack. The last saved state of the test generation algorithm is retrieved from the stack. The circuit state is restored by assigning the primary input values obtained from the stack and performing forward implication to obtain the values shown in Figure 4.24(c). The untried alternative assignment, $V(x_3) = 1_6$, is also obtained from the test generation state saved in the previous round.

Due to the backtrack in the previous step, TGST identification is not carried out in the fifth round. Instead, the untried value assignment, $V(x_3) = 1_6$, is selected. Note that no alternative value assignments remain at this input at this stage of execution of the algorithm. This fact as well as the previously made primary input assignments are saved as the state of the test generation algorithm. The selected assignment is then made and forward implication performed to obtain the values shown in Figure 4.24(e). Note that $\mathcal{D}$ only contains a single gate, G_7.

In the sixth round, a single objective, propagate via G_7, is identified. In the manner described in Section 4.6.2.3 and illustrated in Figure 4.28, backtrace terminates by

Table 4.3. *An example execution of PODEM*

	Initial state	TGST selected	Initial objective	Backtrace result	Untried alternative	Implication results	$\mathcal{D}$	$\mathcal{U}$	Comments
R1(a)	—	—	—	—	—	Figure 4.24(a)	{}	{}	Initialization
R1(b)	Figure 4.24(a)	FEE at fanout x_5	$(x_5, 1_6)$	$(x_5, 1_6)$	$(x_5, 0_6)$	Figure 4.24(b)	$\{G_2, G_5\}$	{}	
R2	Figure 4.24(b)	FEP via G_2	$(x_4, 1_6)$	$(x_4, 1_6)$	$(x_4, 0_6)$	Figure 4.24(c)	$\{G_3, G_5\}$	{}	
R3	Figure 4.24(c)	FEP via G_3	$(x_3, 0_6)$	$(x_3, 0_6)$	$(x_3, 1_6)$	Figure 4.24(d)	$\{G_6\}$	{}	
R4	Figure 4.24(d)	None (Backtrack)	—	—	$(x_3, 1_6)^*$	Figure 4.24(c)*	$\{G_3, G_5\}$	{}	*Restored from R3
R5	Figure 4.24(c)	—	$(x_3, 1_6)^{\dagger}$	$(x_3, 1_6)$	None	Figure 4.24(e)	$\{G_7\}$	{}	†Untried alternative used
R6	Figure 4.24(e)	FEP via G_7	$(c_{10}, 1_6)$	$(x_1, 1_6)$	$(x_1, 0_6)$	Figure 4.24(g)	{}	{}	

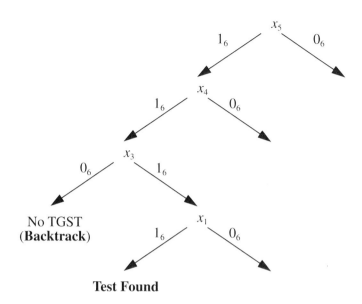

Figure 4.29 The search tree for the example execution of PODEM

providing the objective $(x_1, 1_6)$. Again, the other alternative assignment at this input, namely $(x_1, 0_6)$, is noted and all the primary input values assigned until the end of the previous round are saved. The selected assignment, $V(x_1) = 1_6$, is then made and forward implication performed to obtain the values shown in Figure 4.24(g). (Note that the state shown in Figure 4.24(f) is not encountered during the above execution of PODEM.)

The fact that the vector obtained at the end of the above round is a test is identified when TGST identification is carried out at the beginning of the next round.

The search tree traversed during the above execution of PODEM is shown in Figure 4.29.

4.6.3   Comparisons and enhancements

We will begin by a comparison of the D-algorithm and PODEM. This will be followed by a discussion of known techniques for their enhancement.

4.6.3.1 Comparison

The two algorithms differ primarily in their search strategies. As is clear from the above discussion and examples, this drastically changes the search tree. In fact, the maximum size of the search tree is difficult to predict for the D-algorithm.

In addition to the differences in the nature of the search tree, the two algorithms also differ in (a) the composition and size of state information that must be stored in the stack, and (b) how the knowledge of necessary conditions is utilized.

State of the test generation algorithm: For the D-algorithm, the state of the algorithm is described as being comprised of values at all circuit lines and all untried TGSTs and alternative local value assignments. In contrast, for PODEM, the state of the algorithm is described as the values assigned to primary input lines and the untried value assignment at each input where a value is assigned. Hence, as described above, the stack size for the D-algorithm may be much larger than that for PODEM. However, following a backtrack, PODEM needs to execute the forward implication procedure to recreate the values at all circuit lines. In contrast, as described above, in the D-algorithm, no computation needs to be performed to obtain the values at circuit lines or alternative TGSTs and value assignments. In other words, PODEM reduces stack size, i.e., storage complexity, through an increase in run-time complexity.

The above is an example of storage vs. computation trade-off that is common in computing. Such a trade-off can also be made to reduce the state information that needs to be stored in the stack by the D-algorithm. In this case, the state of the algorithm can be comprised of (a) the value at each primary input and at each unjustified line, and (b) two integers, respectively representing the sequence numbers of TGST alternatives and alternative value assignments that have been tried (e.g., the second TGST alternative, or the first alternative local value assignment). If the state information is stored in this form, then following a backtrack, first the values at circuit lines stored in the stack are assigned and the implication procedure executed. Next, the procedure to identify TGSTs and alternative value assignments is executed. The sequence numbers of the TGSTs and alternative value assignments can be used to then identify all the untried alternatives. Hence, even though in the form described above the stack sizes required for the two paradigms seem to be different, they can be made comparable by exploiting the above storage vs. run-time complexity trade-offs.

Utilization of known necessary conditions: In some rounds of execution of a test generation algorithm, necessary assignments may be identified at multiple CUT lines. The assignment of all these necessary values followed by execution of the implication procedure can help assign values to other circuit lines. Consequently, the search space is constrained. The D-algorithm can reduce search space in this manner, since it allows values to be assigned to lines internal to the circuit. In contrast, when necessary value assignments are identified at multiple internal circuit lines, in any step of PODEM only one of these value assignments can be selected as a backtrace objective. During backtrace, the knowledge of necessary value assignments at other lines is not utilized in any way to constrain the search space. This fact is identified in Fujiwara and Shimono (1983) and an algorithm called FAN, which allows assignment of identified necessary values to internal circuit lines but uses a search strategy that is closely related to that used in PODEM, is developed.

4.6.3.2 Techniques for accelerating test generation

The techniques for accelerating test generation can be classified into two main categories (Schulz *et al.*, 1988), deterministic and heuristic. We classify a technique as deterministic if it is guaranteed to reduce the amount of search required for the generation of a test for any given fault in any CUT. A technique is called a heuristic if it is expected to reduce the search in many cases, but universal reduction in the amount of search cannot be guaranteed. Note that even a deterministic technique to reduce search may not be attractive, if the increase in run-time complexity due to its use exceeds the typical reduction in complexity it provides by reducing the amount of search.

Deterministic techniques discussed below fall into three main categories: identification of alternative search spaces, use of powerful implication techniques, and reuse of results of prior search. In the following, we also discuss the multiple backtrace heuristic to reduce search. Several testability measures have been proposed to reduce search, but are not discussed here since they have been described in Section 4.5.3.

Static indirect implication: We have described earlier a methodology to identify static indirect implications. We have also discussed how these indirect implications can be used by an implication procedure. Additional implications of values assigned by the test generator can be found and search space reduced by using an implication procedure that exploits indirect implications identified during a pre-processing step of test generation.

Dynamic indirect implication: A dynamic indirect implication strategy, such as the ones discussed earlier, can also help reduce the search space by identifying additional value assignments implied by current value assignments.

A dynamic indirect implication procedure can be used during test generation in many different ways. One possible option is to use the procedure recursively to identify conditions necessary for fault effect propagation (Kunz and Pradhan, 1994). The procedure can be used to first identify necessary value assignments for propagation of fault effect through each gate in the D-frontier. In this process, the procedure can help identify the fact that fault effect propagation through some of these gates is infeasible. If the necessary conditions for fault effect propagation through each of the remaining gates in the D-frontier have any value assignments in common, those value assignments are identified as being necessary for fault effect propagation. Several other techniques to reduce the search space using dynamic indirect implication are described in Kunz and Pradhan (1994).

A technique that performs indirect implications using the 16-valued system and *reduction lists* is presented in Cox and Rajski (1994).

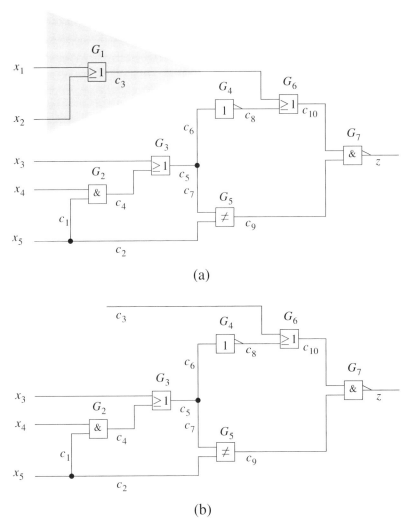

Figure 4.30 An example illustrating the use of head lines: (a) original circuit C with a cone of a head line, (b) reduced circuit C_{red} with head lines as inputs. (This circuit is a modified version of a circuit in Roth (1966))

Alternative search space: Consider the circuit shown in Figure 4.30(a) and the input cone of line c_3, which has two inputs, x_1 and x_2, and a single output, c_3. One special characteristic of this cone is that it has no fanouts and hence any line in the cone is connected to the rest of the circuit only via line c_3.

Now consider a target fault f_i at any line in the circuit that is not within the above cone. Test generation using a PODEM-like algorithm may require enumeration of all possible vectors that can be applied to primary inputs, $x_1, x_2, \ldots, x_5$, a total of 2^5 vectors. However, the values assigned to inputs x_1 and x_2 can influence the detection of f_i only through the value they imply at c_3. This fact, first described in the FAN algorithm (Fujiwara and Shimono, 1983), can be used in the following manner to reduce the search space.

First, this cone can be removed from the circuit to obtain a circuit C_{red} with primary inputs c_3, x_3, x_4, and x_5, as shown in Figure 4.30(b). Test generation can then be performed by enumerating the space of input vectors. In this case, the number of vectors enumerated will be no larger than 2^4, since C_{red} has four inputs. Once a vector that detects fault f_i in C_{red} is found, a test vector that detects the fault in the original circuit can be found by justifying the value assigned to line c_3 in the above vector through its input cone. Since this cone is fanout-free, this can be accomplished without any backtracks.

The above method can be applied to an arbitrary CUT. A large reduction in search space can be achieved if the number of inputs of C_{red}, obtained by removing cones as above, is much lower than the number of inputs of the original circuit.

In general, cones such as the one described above can be found by first identifying *head lines* which can be defined as follows. Any circuit line that does not lie in the transitive fanout of any fanout system is called a **free line**; all other lines, including every branch of every fanout system, are called **bound lines**. A free line that has a bound line in its fanout is called a **head line**. The cone feeding a head line is fanout-free and can be removed from the circuit during test generation for a fault in any bound line, in the manner illustrated above, as long as every gate in the circuit has a non-empty on-set as well as a non-empty off-set. The reader is invited to develop an efficient method to generate a test for an SAF at a free line that belongs to one of the cones that is removed.

In the above circuit, lines x_1, x_2, x_3, x_4, x_5, and c_3 are free lines, all other lines are bound lines. Lines x_3, x_4, x_5, and c_3 are head lines. In this circuit, the only head line that has multiple inputs in its input cone is c_3. Hence, this cone can be removed to obtain the circuit C_{red} as described above.

The above approach allows the removal of parts of a circuit that (a) have primary inputs as inputs, (b) have a single output, and (c) are fanout-free. In Kirkland and Mercer (1987), a technique has been presented that allows the removal of a larger class of subcircuits. Specifically, the subcircuit removed need only satisfy the first two of the above three requirements. Essentially, a subcircuit may be removed despite having fanout branches as long as each fanout branch reconverges before the output of the subcircuit.

For a given circuit C, let C_{rem}^j be a subcircuit with output line c_j that can be removed; let the circuit obtained after removing C_{rem}^j from C be called C_{red}^*. Note that c_j is a primary input of C_{red}^* while it is the output of subcircuit C_{rem}^j that has been removed.

As in the above approach, a test can be generated for any fault located at any circuit element in C_{red}^* by only considering that part of the circuit. If a specific value is assigned at the line c_j, it is justified using the subcircuit C_{rem}^j. This is similar to the above approach with one important difference: in this case, search may be required to justify the value at c_j, since the subcircuit C_{rem}^j is not fanout-free. However, it should be noted that when tests are generated for each target fault in C_{red}^*, only two specific values may be required at c_j by those vectors, namely 0_6 and 1_6. Hence, once the search has been performed to find a way to justify a 0_6 at c_j during the generation of test for one fault, the results of the search can be reused to justify the value 0_6 at c_j for any subsequent vector.

Non-conflicting assignments: As described earlier, during test generation, desired local value assignments are typically identified at internal circuit lines while, in many test generators, values are actually assigned at primary inputs or a handful of selected internal lines. The backtrace procedure translates the desired local value assignments at internal lines to values at lines where assignments can actually be made. The backtrace procedure has two main weaknesses. First, it only starts with a single objective, even when multiple objectives, i.e., multiple desired local value assignments, are known. Second, it only provides a heuristic value assignment, i.e., the value assignment identified by it is not guaranteed to facilitate the attainment of the original objective and may sometimes even prevent its attainment.

In Fujiwara and Shimono (1983), a **multiple backtrace** approach has been presented to remedy the first weakness of backtrace. In this approach, backtrace is performed starting at the site of each known desired local value assignment and it stops when it reaches a fanout branch. Once all backtrace operations that can reach the branches of a particular fanout system have reached the respective branches, the value desired by a majority of backtraces is selected as the next objective at the stem. All backtrace operations then continue their backward traversal. This process continues until a desired value is identified at a line where a value assignment can actually be made.

In Cox and Rajski (1994), a concept called *tendency list* has been developed. A **tendency list** for a value u_i at line c_i, $tlst(c_i, u_i)$, is the set of primary input assignments that cannot in any way preclude the possibility of obtaining the value u_i at line c_i. For example, in the circuit in Figure 4.28,

$$tlst(c_3, 0_6) = \{(x_1, 0_6), (x_2, 0_6)\},$$

since the assignment of 0_6 at either of these two inputs cannot preclude the possibility of obtaining a 0_6 at c_3. Similarly, $tlst(c_3, 1_6) = \{(x_1, 1_6), (x_2, 1_6)\}$.

If $(c_{i_1}, u_{i_1}), (c_{i_2}, u_{i_2}), \ldots, (c_{i_l}, u_{i_l})$ are all the justification objectives necessary to complete generation of a test for a fault, then the assignments in the set

$$tlst(c_{i_1}, u_{i_1}) \cap tlst(c_{i_2}, u_{i_2}) \cap \cdots \cap tlst(c_{i_l}, u_{i_l}) \tag{4.22}$$

are **non-conflicting**, i.e., each assignment in this set can only facilitate the attainment of one or more of the justification objectives without the risk of precluding the attainment of any of the objectives.

If one or more non-conflicting assignments are made, then the search space is reduced and a test is guaranteed to exist in the search space obtained after making the assignment, if a test existed in the search space prior to the assignment.

Reuse of search: The value assignments made during the generation of a test for a target fault can be represented as a tree of the type shown in Figure 4.29. During this process, the search problem at the current node in the tree may be equivalent to that encountered earlier. In such cases, the results of the previous search can be reused to reduce the complexity of search.

Consider a composite circuit with a target fault f at the end of some intermediate round of test generation. Let P_i be the partially specified vector at that test generation round and assume that the values implied by the vector have been identified. The *E-frontier* at the end of this round of test generation can be defined using the following procedure. Start at a primary output and mark it as visited. If the output has a partially specified value whose set representation contains a D, a $\overline{D}$, or both, mark each line in the fanin of the line as visited. For each internal circuit line c that is marked visited in this manner, mark each line in its fanin as visited, provided c has an incompletely specified value. Repeat the above process starting at each primary output until all the lines that can be visited are processed. The *E*-**frontier** can be defined as a set of two-tuples where each element is either (a) of the form $(x_l, V(x_l))$, where x_l is a primary input of the circuit marked as visited and $V(x_l)$ is the value implied at x_l, or (b) of the form $(c_l, V(c_l))$, where c_l is an internal circuit line that is marked as visited, and at which the current vector specifies a completely specified value $V(c_l)$.

Example 4.14 Consider the circuit shown in Figure 4.31 (Giraldi and Bushnell, 1990). First, consider some intermediate round during the generation of a test vector for fault f_4: x_2 SA1. The partial vector generated for this fault at this stage and the value implied by the vector are shown in Figure 4.31(a).

The procedure outlined above will mark lines z, c_{10}, x_1, c_8, c_6, c_9, c_5, c_2, and x_5 as visited. Among the internal lines marked visited, lines c_8, c_6, and c_5 have completely specified values D_6, 1_6, and 1_6, respectively. The primary inputs visited, namely x_1 and x_5, each have value χ_6. Hence, the *E*-frontier at the end of this round of test generation is $\{(x_1, \chi_6), (x_5, \chi_6), (c_5, 1_6), (c_6, 1_6), (c_8, D_6)\}$.   □

The significance of the *E*-frontier is illustrated by the above example and Figure 4.31(a). Simply, the lines in the *E*-frontier separate the circuit lines whose values can be more completely specified to help detect the target fault from those whose values are either already completely specified or cannot influence the detection of

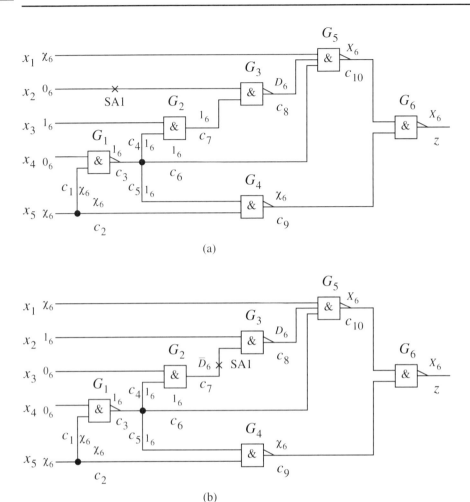

Figure 4.31 Examples of the E-frontier (Giraldi and Bushnell, 1990) (© 1990 IEEE)

the target fault. Hence, the state of the circuit at the end of the round of search is completely described by the E-frontier at that time.

Suppose that a particular E-frontier is encountered during a certain round of test generation for fault f_1 when the partial vector was P_i. First, consider the case where the same E-frontier is encountered during some round of test generation for the same fault when the partial vector is P_j. If the test generator was unsuccessful in finding a test vector for the fault by further specifying the values in the incompletely specified vector P_i, then the forward search is also futile at the current round of test generation. Backtracking should hence be initiated without wasting any time on such a search. On the other hand, if a test vector P_i^* was indeed found for the fault by further specifying the values of the partially specified vector P_i, then each additional value assignment that was made to P_i to obtain P_i^* can now be made to P_j to obtain a test vector P_j^*

for the fault. In either case, the results of the search conducted following the first time an E-frontier was encountered can help eliminate the need for search each subsequent time that the E-frontier is encountered.

With some minor modifications, the above reasoning also applies to the scenario where an E-frontier that was encountered during some round of test generation for a fault f_1 is encountered again during a round of test generation for another fault f_2. This case is illustrated in the following example; details of the methodology can be found in Giraldi and Bushnell (1990).

Example 4.15 Consider the example scenario described in Example 4.14 and depicted in Figure 4.31(a) along with the corresponding partial test vector $(\chi_6, 0_6, 1_6, 0_6, \chi_6)$. Assume that following the occurrence of the E-frontier $\{(x_1, \chi_6), (x_5, \chi_6), (c_5, 1_6), (c_6, 1_6), (c_8, D_6)\}$, the value assignments $(x_1, 1_6)$ and $(x_5, 0_6)$ were added onto the partial vector to obtain a test vector $(1_6, 0_6, 1_6, 0_6, 0_6)$ for fault f_4.

Now consider an intermediate round during the test generation for fault f_{24}: c_7 SA1, as depicted in Figure 4.31(b). At this stage, the partial vector for fault f_{24} is $(\chi_6, 1_6, 0_6, 0_6, \chi_6)$ and the E-frontier is identical to that in Example 4.14.

A vector for fault f_{24} can hence be obtained by adding to the above partial vector the value assignments that were made to the partial vector for fault f_4 following the occurrence of an identical E-frontier. Hence, without performing any search, the assignments $(x_1, 1_6)$ and $(x_5, 0_6)$ can be added onto the above partial vector for fault f_{24} to obtain a test vector $(1_6, 1_6, 0_6, 0_6, 0_6)$. $\qquad \square$

4.7   Non-structural test generation techniques

Structural algorithms for test generation directly and continually analyze the gate-level description of a circuit and implicitly enumerate all possible input combinations to find a test vector for a target fault. In contrast, an algebraic algorithm converts the test generation problem into an algebraic formula and applies algebraic techniques to simplify and then solve the formula to obtain a test. Early algebraic techniques were based on *Boolean difference* (Sellers *et al.*, 1968) and were not practical due to their high computational complexity. Recent developments in efficient circuit representation via *binary decision diagrams (BDDs)* (Bryant, 1986) have inspired many algebraic ATPG algorithms for SAFs (Gaede *et al.*, 1988; Srinivasan *et al.*, 1993; Stanion *et al.*, 1995). The memory requirements and computation time of these techniques are comparable (in some cases, superior) to those of the structural techniques. These techniques work well for circuits that can be efficiently represented by BDDs. For some circuits, however, the BDD representations are impractically large, making the BDD-based techniques inapplicable. In this section, we discuss some of these non-structural test generation techniques.

4.7.1   Test generation based on satisfiability

The satisfiability-based approach proposed in Larrabee (1992) translates the test generation problem into a formula in *conjunctive normal form (CNF)* (Cormen *et al.*, 1992). Any input combination (vector) which makes the CNF formula evaluate to 1, i.e., which **satisfies** the formula, is a test for the fault. This approach is not strictly algebraic because no algebraic manipulations are employed. However, a branch-and-bound strategy, similar to the one used in structural algorithms, is used to find an input combination that satisfies the formula. Still, it is not a structural algorithm because all the operations are performed on the formula in CNF instead of the gate-level circuit description. The performance of an efficient implementation of this technique (Stephan *et al.*, 1992) compares favorably to the best known structural algorithms.

4.7.1.1 Basic terminology

A **literal** is a Boolean variable or its complement (e.g., a or $\bar{a}$). An **OR-clause** is an OR of one or more literals. An **r-clause** is an OR-clause with exactly r distinct literals. A **formula** is composed of parentheses, literals, and Boolean operations (e.g., NOT, AND, OR, NAND, and NOR). A Boolean formula is in **conjunctive normal form (CNF)** if it is expressed as AND of OR-clauses. A Boolean formula is in r-CNF, if each OR-clause is an r-clause. An **assignment** for a Boolean formula is a set of Boolean input values (0 or 1). A **satisfying assignment** for a single-output Boolean formula y is an assignment such that y evaluates to 1. A Boolean formula is **satisfiable** if it has a satisfying assignment. The problem of **Boolean satisfiability** is to determine whether a Boolean formula is satisfiable and to find a satisfying assignment if it is. As an example, the Boolean formula $y = (a + b + \bar{d})(\bar{a} + \bar{c} + d)$ is in three-CNF. The formula is satisfiable with a satisfying assignment $\{a = 1, c = 0\}$, among others.

4.7.1.2 ATPG formulation

Test generation based on satisfiability can be divided into two independent steps: extraction of the CNF formula and identification of a satisfying assignment. Different fault models require different CNF formula extractors, but can use identical satisfiability solvers (though some heuristics may be more efficient for a specific fault model). This section describes CNF formula extraction for single SAFs (Larrabee, 1992; Stephan *et al.*, 1992).

Fault-free circuit: First, consider a two-input AND gate with inputs c_{i_1} and c_{i_2} and output c_j whose behavior is represented by the equation

$$v(c_j) = v(c_{i_1})v(c_{i_2}).$$

Table 4.4. *CNF formulae for fault-free primitive gates and fanout system*

Circuit element	CNF formula
NOT	$[v(c_{i_1}) + v(c_j)][\overline{v(c_{i_1})} + \overline{v(c_j)}]$
AND	$[v(c_{i_1}) + \overline{v(c_j)}][v(c_{i_2}) + \overline{v(c_j)}][\overline{v(c_{i_1})} + \overline{v(c_{i_2})} + v(c_j)]$
NAND	$[v(c_{i_1}) + v(c_j)][v(c_{i_2}) + v(c_j)][\overline{v(c_{i_1})} + \overline{v(c_{i_2})} + \overline{v(c_j)}]$
OR	$[\overline{v(c_{i_1})} + v(c_j)][\overline{v(c_{i_2})} + v(c_j)][v(c_{i_1}) + v(c_{i_2}) + \overline{v(c_j)}]$
NOR	$[\overline{v(c_{i_1})} + \overline{v(c_j)}][\overline{v(c_{i_2})} + \overline{v(c_j)}][v(c_{i_1}) + v(c_{i_2}) + v(c_j)]$
Fanout system	$[v(c_i) + \overline{v(c_{j_l})}][\overline{v(c_i)} + v(c_{j_l})]$, for each branch c_{j_l}

For clarity, the above representation is said to be in *equation form*. Alternatively, the equation can be written as

$$v(c_{i_1})v(c_{i_2}) \oplus v(c_j) = 0, \text{ or}$$

$$\overline{v(c_{i_1})}v(c_j) + \overline{v(c_{i_2})}v(c_j) + v(c_{i_1})v(c_{i_2})\overline{v(c_j)} = 0, \text{ or}$$

$$\left[v(c_{i_1}) + \overline{v(c_j)}\right]\left[v(c_{i_2}) + \overline{v(c_j)}\right]\left[\overline{v(c_{i_1})} + \overline{v(c_{i_2})} + v(c_j)\right] = 1.$$

Now the *left-hand side* of the equation is in CNF and will be called the **CNF of the gate**. Note that only the values in the truth table of an AND gate can satisfy the CNF formula. The CNF formulae for some commonly used gates are summarized in Table 4.4. For each gate, the formula can be easily extended to a version with multiple inputs.

A logic circuit consists of gates interconnected by wires and fanout systems. The circuit can be represented in CNF by simply concatenating the CNF formula for each individual gate. Consider the circuit shown in Figure 4.32. The value implied at the primary output, $v(z)$, can be written in terms of the values applied at the primary inputs in equation form as

$$v(z) = \overline{v(x_1)v(x_2)v(x_3) + \overline{v(x_3)}}.$$

$v(z)$ can also be written as the CNF formula

$$CNF_g = \underbrace{\left[v(x_3) + \overline{v(c_1)}\right]\left[\overline{v(x_3)} + v(c_1)\right]}_{\text{Fanout branch } c_1}\underbrace{\left[v(x_3) + \overline{v(c_2)}\right]\left[\left[\overline{v(x_3)} + v(c_2)\right]\right]}_{\text{Fanout branch } c_2}$$

$$\underbrace{\left[v(x_1) + \overline{v(c_3)}\right]\left[v(x_2) + \overline{v(c_3)}\right]\left[v(c_1) + \overline{v(c_3)}\right]\left[\overline{v(x_1)} + \overline{v(x_2)} + \overline{v(c_1)} + v(c_3)\right]}_{\text{AND } G_1}$$

$$\underbrace{\left[v(c_2) + v(c_4)\right]\left[\overline{v(c_2)} + \overline{v(c_4)}\right]}_{\text{NOT } G_2}\underbrace{\left[\overline{v(c_3)} + \overline{v(z)}\right]\left[\overline{v(c_4)} + \overline{v(z)}\right]\left[v(c_3) + v(c_4) + v(z)\right]}_{\text{NOR } G_3},$$

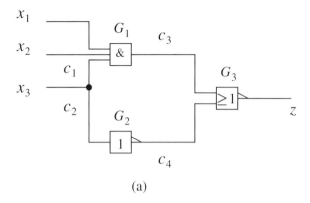

(a)

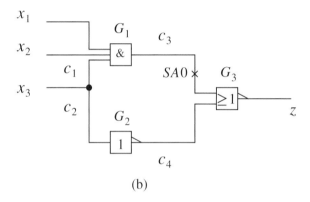

(b)

Figure 4.32   An example circuit: (a) fault-free version, (b) version with c_3 SA0 (Chen and Gupta, 1996) (© 1996 ACM)

where any satisfying assignment of CNF_g represents the behavior of the fault-free version of the example circuit. (The CNF for the AND gate shown above contains a four-clause. This clause can be replaced by a product of multiple three-clauses.)

Faulty circuit: To model a faulty version of the circuit, for each line c_j in the transitive fanout of the fault site, a new value $v'(c_j)$ is created to represent the value implied at the line in the faulty version of the circuit. As an example, the circuit shown in Figure 4.32(a) is shown in Figure 4.32(b) with an SA0 fault at c_3. The clauses for the CNF formula for the faulty circuit are given by

$$CNF_f = \left[\overline{v'(c_3)} + \overline{v'(z)}\right]\left[\overline{v(c_4)} + \overline{v'(z)}\right]\left[v'(c_3) + v(c_4) + v'(z)\right].\tag{4.23}$$

The new variables $v'(c_3)$ and $v'(z)$ (faulty values) are introduced to indicate that lines c_3 and z may have different values in the fault-free and faulty circuits.

Condition for fault detection: To derive a test for a target fault, it is insufficient to concatenate the CNF formulae for the fault-free and faulty circuits and to find a satisfying assignment for the resulting CNF formula. In fact, if a satisfying assignment can be found for the CNF formula for the fault-free circuit, it will always satisfy the CNF formula of the faulty circuit – by assigning the value implied at each line in the fault-free circuit to the corresponding line in the faulty circuit.

The creation of a fault effect at a line can be captured by introducing **active variables** and **active clauses**. For each line c_j in the transitive fanout of the fault site, an active variable $a(c_j)$ may be defined. If a fault effect appears at c_j, then c_j is said to be **active** and the value implied at c_j in the fault-free circuit is the complement of that in the faulty circuit. In other words,

$$a(c_j) \quad \Rightarrow \quad v(c_j) \neq v'(c_j).$$

This can be expressed in terms of satisfiability of the following CNF formula.

$$CNF_a(c_j) = \left[\overline{a(c_j)} + v(c_j) + v'(c_j)\right]\left[\overline{a(c_j)} + \overline{v(c_j)} + \overline{v'(c_j)}\right].$$

For a single-output circuit with output z, we define the CNF formula that enforces the conditions for fault detection as

$$CNF_d = CNF_a(z)a(z).$$

The first term is the active clause for output z and the second term denotes the condition that the output should be active. Together, the two terms of CNF_d denote the conditions under which the fault effect appears at z.

The CNF formulae corresponding to the fault-free circuit, faulty circuit, and conditions for fault detection in a single-output circuit can be concatenated to obtain the following CNF formula:

$$CNF_g CNF_f CNF_d. \tag{4.24}$$

Any assignment of primary input values that satisfies the above CNF formula is a test vector for the target fault in the single-output circuit.

In general, for a multi-output circuit with outputs $z_1, z_2, \ldots, z_m$, the third part of the above formula, CNF_d, must be replaced by a concatenation of active clauses for each of its primary outputs, $CNF_a(z_1), CNF_a(z_2), \ldots, CNF_a(z_n)$, along with one or more clauses indicating that one or more output is active. This can simply be written as

$$CNF_d = CNF_a(z_1)CNF_a(z_2) \cdots CNF_a(z_m)\left[a(z_1) + a(z_2) + \cdots + a(z_m)\right].$$

Note that the last clause in the above formula may have more than three terms. However, it can be manipulated and expressed using multiple clauses, each with three or fewer terms, as

$$CNF_d = CNF_a(z_1)CNF_a(z_2) \cdots CNF_a(z_m) [a(z_1) + a(z_2) + b_1]$$
$$\left[\overline{b_1} + a(z_3) + b_2\right]\left[\overline{b_2} + a(z_4) + b_3\right] \cdots \left[\overline{b_{m-2}} + a(z_m)\right].$$

Again, any assignment of values that satisfies the CNF formula

$$CNF_g CNF_f CNF_d, \tag{4.25}$$

corresponds to a test vector for the target fault in the multi-output circuit.

4.7.1.3 Techniques for accelerating test generation

The problem of finding a satisfying assignment to a three-CNF formula is known to belong to a class of problems believed to be difficult. Hence, techniques are required to reduce the complexity of finding a satisfying assignment.

Necessary conditions: If any local value assignment is known to be necessary for detection of the target fault, then additional clauses can be added to the CNF formula in Equation (4.25). For example, a necessary condition for the excitation of an SA0 fault at line c_j leads to the following formula

$$CNF_g CNF_f CNF_d v(c_j)\overline{v'(c_j)}.$$

Note that any vector that satisfies this formula also satisfies the formula in Equation (4.25). However, the additional clauses reduce the search space by assigning specific values to some of the literals in the original formula. Clauses can similarly be added to exploit other known necessary assignments, such as those arising out of unique D-drive and future unique D-drive.

Static indirect implications: If a static indirect implication of the form $v(c_i) = 1 \Rightarrow v(c_j) = 1$ is identified during a pre-processing phase, then an additional clause

$$\left[\overline{v(c_i)} + v(c_j)\right]$$

can be concatenated to the formula in Equation (4.25) to explicitly add this information. Similar clauses can be added for each of the other specific forms of indirect implications.

Two-SAT: In a typical CNF formula for test generation, a large fraction of clauses have only two terms. The problem of finding a satisfying solution for a two-CNF formula is known as two-SAT and polynomial-time algorithms exist for the problem.

In Larrabee (1992), an approach is presented to exploit the above fact to accelerate test generation via satisfiability. In this approach, clauses with two terms are separated from the remaining clauses in the CNF formula. Whenever possible, these clauses are partitioned into different CNF formulae and the two-SAT algorithm is used to find a satisfying assignment for each such formula. These assignments are then substituted into the original formula. If a satisfying assignment for the original formula can subsequently be found, then test generation is complete. Otherwise, an alternative set of satisfying assignments is found for each of the two-CNF formulae and the entire process is repeated. A similar approach has also been reported in Chakradhar *et al.* (1993) where global signal dependences are computed using *transitive closure*.

4.7.2   Test generation using binary decision diagrams

Binary decision diagrams (BDDs) are a compact way of representing the logic behavior of combinational circuits based on Shannon's expansion where a logic function $g(x_1, x_2, \ldots, x_n)$ is expressed in terms of its *cofactors* as

$$g(x_1, x_2, \ldots, x_{i-1}, x_i, x_{i+1}, \ldots, x_n) = x_i g(x_1, x_2, \ldots, x_{i-1}, 1, x_{i+1}, \ldots, x_n)$$
$$+ \overline{x_i} g(x_1, x_2, \ldots, x_{i-1}, 0, x_{i+1}, \ldots, x_n).$$

In the above expression, $g(x_1, x_2, \ldots, x_{i-1}, 1, x_{i+1}, \ldots, x_n)$ and $g(x_1, x_2, \ldots, x_{i-1}, 0, x_{i+1}, \ldots, x_n)$ are **cofactors** of the function g obtained by setting the value of input variable x_i to 1 and 0, respectively. A BDD for a given function g can be obtained by iteratively computing cofactors of the function in terms of its input variables.

4.7.2.1 The structure of a BDD

A BDD for a single-output function is an acyclic graph with two types of vertices, **terminal** and **non-terminal**. Associated with a terminal vertex a is an attribute **value**, $value(a)$, which takes either of the two logic values, 0 or 1. Associated with a non-terminal vertex a is (a) an attribute **index**, $index(a) = i$, which refers to the value of an input variable x_i, $v(x_i)$, and (b) two children vertices called $low(a)$ and $high(a)$. An edge with a label 0 exists between a vertex a_1 and a_2 if $low(a_1)$ is vertex a_2. Similarly, an edge with a label 1 exists between a vertex a_1 and a_2 if $high(a_1)$ is vertex a_2.

Four example BDDs representing the logic behavior of a two-input AND gate with inputs x_1 and x_2 and output c_j are shown in Figure 4.33. Each terminal vertex is denoted by a square with the associated attribute value shown within. Each non-terminal vertex is denoted by a circle with the variable shown within. Label c_j is sometimes written next to the vertex that is the root of the BDD representing the logic behavior of line c_j. For the above example, the root of each of the BDDs represents the logic behavior of the two-input AND gate. For a given combination of logic values, $v(x_1)$ and $v(x_2)$, the graph is traversed as follows. If the index associated with root a is i and value $v(x_i) = 0$, then vertex $low(a)$ is visited; otherwise, vertex $high(a)$ is visited. This process is repeated until a terminal vertex is reached. The

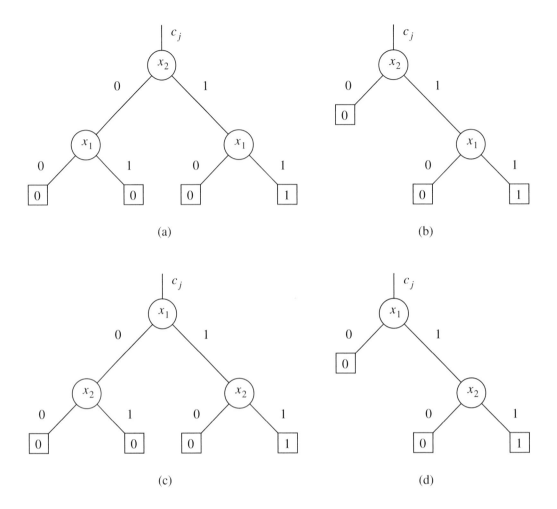

Figure 4.33 Four BDDs representing the behavior of a two-input AND gate.

attribute value of that terminal vertex gives the value $v(c_j)$ for the given values $v(x_1), v(x_2), \ldots, v(x_n)$. The reader can verify that each BDD in Figure 4.33 correctly represents the truth table of a two-input AND gate.

Ordered binary decision diagrams (OBDDs) (Bryant, 1986) are a special case of BDDs described above where the indices of non-terminal vertices are ordered in the following manner. If attribute $low(a_i)$ of a non-terminal vertex a_i refers to a non-terminal vertex a_j, then $index(a_i) < index(a_j)$. Furthermore, if attribute $high(a_i)$ of a non-terminal vertex a_i refers to a non-terminal vertex a_j, then $index(a_i) < index(a_j)$. The example BDDs in Figures 4.33(c) and (d) are ordered while those in Figures 4.33(a) and (b) are not. Alternative orders are typically considered, since the size of the BDD for a function may depend on the order of inputs. In such cases, the inputs of the function may be permuted via appropriate renaming.

Reduced OBDDs (Bryant, 1986) add a further restriction that for the given order of input variables each BDD is minimized. The BDD in Figure 4.33(d) is a reduced OBDD. Since most BDD-based test generators consider the reduced OBDDs, in the following they will simply be referred to as BDDs.

4.7.2.2 Some properties of reduced OBDDs

A detailed discussion of some of the basic properties of such BDDs can be found in Bryant (1986). We only describe those properties that are important for the use of BDDs for test generation.

Canonical representation: First, for the given function and ordering of input variables, a reduced OBDD is unique. Hence, it is a **canonical** representation of the function. An important consequence of this property is that algorithms to identify the equality of two different functions have complexity that is proportional to the number of vertices in each BDD.

Existence of efficient procedures to compute the BDD of a given circuit: Given the BDDs of two functions g_1 and g_2, the BDD of the function $g = g_1 \diamond g_2$, where $\diamond$ is any Boolean operation, can be obtained in asymptotic complexity that is no greater than the product of the numbers of vertices in the BDD representations of each of the two functions. A procedure called *Apply( )* is used to generate the BDD of a given circuit as follows. The circuit lines are levelized and a BDD is obtained for each line with input level one. The BDD of the output of a gate which has input level two is then computed by using the BDDs representing its inputs and Procedure Apply( ). This process is repeated until a BDD is obtained for each primary output.

Figure 4.34(a) shows an example circuit. The BDD for the function implemented by lines c_3 (the output of a three-input AND) and c_4 (the output of an inverter) are shown in Figures 4.34(b) and 4.34(c), respectively. The Apply( ) function is then used to combine the above two BDDs to obtain the BDD of the function implemented by primary output z, as shown in Figure 4.34(d). Of the three BDDs shown in this figure, only the BDD that represents the behavior of the primary output is used for test generation.

For a target single SAF, the BDD of the corresponding faulty version of the circuit can be obtained in a similar manner. The only difference is that, at the line that is the site of the target fault, the BDD of the line obtained for the fault-free circuit is replaced by a BDD comprised only of a terminal node with value equal to the stuck-at value of the target fault. The process is illustrated in Figure 4.35 for the above example circuit and a target single SA0 fault at c_3.

Finally, Procedure Apply( ) can be used to obtain the XOR of the behaviors of the primary output of the fault-free and faulty versions of the circuit. As described in

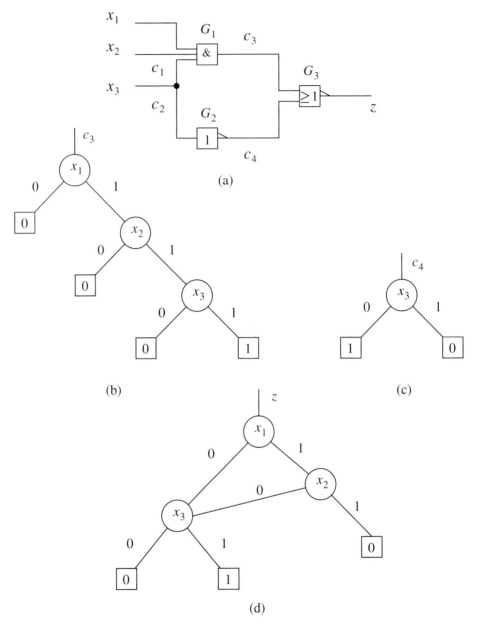

Figure 4.34 Creating a BDD for a circuit: (a) an example circuit (Chen and Gupta, 1996) (© 1996 ACM), (b) BDD for line c_3, (c) BDD for line c_4, and (d) BDD for output z.

Chapter 3 (Section 3.1, Figure 3.2), any vector that implies a logic 1 at the output of such a circuit and corresponding BDD is a test vector for the target fault. Furthermore, if no such vector exists, then the target fault is untestable.

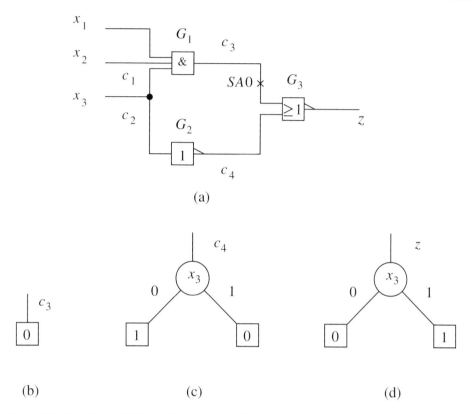

Figure 4.35 Creating a BDD for a faulty version of a circuit: (a) a faulty version of an example circuit (Chen and Gupta, 1996) (© 1996 ACM), (b) BDD for line c_3, (c) BDD for line c_4, and (d) BDD for output z.

Figure 4.36(c) shows the BDD obtained for the example circuit and target fault discussed above. The reader can verify that each vector that implies the value 1 at output Ω is indeed a test vector for the SA0 fault at c_3 in the above example circuit.

Identification of a vector that satisfies the function: The above steps have already reduced the problem of test generation for a target fault to that of identifying an input vector that satisfies the XOR function of the outputs of the fault-free and faulty circuits. One of the important properties of reduced OBDDs is that such a vector can be found with complexity that is proportional to the number of inputs in the circuit. This is accomplished by a depth-first traversal of the BDD along any sequence of vertices.

For our example circuit and target fault, the traversal starts at Ω of the BDD shown in Figure 4.36(c). First $v(x_1) = 0$ is tried. Since that leads to a terminal vertex with value 0, the other assignment for x_1, namely $v(x_1) = 1$, is then tried. This process is repeated for each of the other variables until a terminal vertex with value 1 is found for $v(x_1) = 1$, $v(x_2) = 1$, and $v(x_3) = 1$. Hence, the process terminates with a test vector $(1, 1, 1)$.

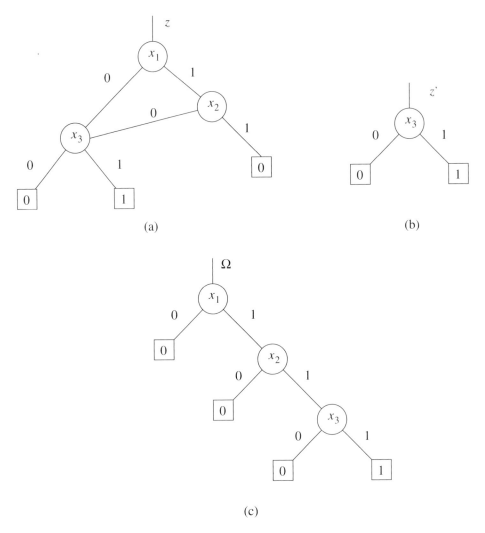

Figure 4.36 Combining the BDDs of the fault-free and faulty circuits to obtain a BDD that represents all tests for a fault: (a) BDD for output z of the fault-free circuit, (b) BDD for the output (labeled as z') of the version with the target fault, and (c) BDD representing the XOR of z and z'.

4.7.2.3 Generalizations and limitations

The above description is limited to single-output circuits. The approach can be easily modified for multi-output circuits. In this case, the BDD of line Ω of the auxiliary circuit C_{aux} shown in Figure 3.2 must be computed and a vector that implies 1 at Ω found.

More efficient versions of the generic BDD-based test generation approach described above are presented in Srinivasan *et al.* (1993) and Stanion *et al.* (1995).

Also, the BDD for line Ω implicitly represents all test vectors that detect the target fault. Relatively efficient procedures can be used to identify multiple (or all) test vectors for the target fault (Stanion *et al.*, 1995).

If the target fault is untestable, then no satisfying assignment exists for Ω. In such cases, the BDD for Ω would only contain a single vertex, namely a terminal vertex with value 0. Hence, in the BDD-based method, untestable faults can be easily identified.

The key limitation of the BDD-based method for test generation is that for some circuits, the size of BDDs is impractically large. The method cannot be used for such circuits.

4.8   Test generation systems

Given a CUT and a fault list, a set of test vectors T, also called a *test set*, may be generated by iteratively selecting one target fault from the list and using a test generation algorithm to either generate a test for the fault or to prove it untestable. Such an approach is impractical for most circuits. First, the worst-case run-time complexity of any complete test generation algorithm is large — test generation problem is NP-complete. This is especially important in the light of the fact that many circuits for which test sets need to be generated can be very large, in terms of the number of primary inputs, number of lines, and number of faults in the fault list. Second, cost and competitive reasons place constraints on computational resources — memory as well as run-time. This necessitates the imposition of a limit on the amount of run-time that a test generation algorithm may spend on each target fault. Such a limit is called the **backtrack limit**, since it is expressed in terms of the maximum number of backtracks that the algorithm is allowed before it must discontinue test generation for, or **abort**, a fault. Ideally, test generation should be successfully completed for *every* fault in the list, i.e., either a test must be generated for the fault or it must be proven untestable, within these constraints. In practice, however, it is desired that test generation be successful for a *vast majority* of the faults in the list. Third, the cost of applying the generated test set to manufactured chips should be low. Finally, in some cases, the generated test set may be required to satisfy additional constraints. For example, the heat dissipation during application of tests to a chip must be below a given safe level, otherwise the chip may be damaged during testing.

A **test generation system** works within the above constraints and attempts to satisfy the above objectives by using test generation algorithms in conjunction with other tools. A fault simulator, a tool for prioritization of faults in the list, and a set of tools for obtaining a compact test set, are some of the other key components of a test generation system. In the following, we will describe key concepts behind these tools and some typical ways in which they are used by test generation systems.

4.8.1   Use of a fault simulator

A fault simulator can be used in four main ways to improve the efficiency of a test generation algorithm.

Coverage of non-target faults: A test generation algorithm targets a specific fault and generates a test vector for it. Often, the generated vector is a test for many faults other than the one targeted. The generation of a test vector is hence followed by fault simulation. Any fault that still belongs to the fault list and is detected by the generated vector is identified as detected and removed from the fault list.

Identification of non-target faults that are detected in the above manner can help improve test generation efficiency in two main ways. First, the test generation time that would have otherwise been spent on each such non-target fault may be saved. This helps decrease the overall run-time. Second, there may exist faults for which the test generation algorithm may have failed to generate a test under the specified backtrack limit, but the fault is detected serendipitously by a vector generated for another target fault. Whenever this occurs, it increases the percentage of faults for which test generation is successful. These benefits are provided at the expense of the time spent on fault simulation. However, in practice the benefits outweigh this cost.

Increasing the fault coverage of a deterministic vector: Very often, a test generator generates an incompletely specified test vector for a target fault. In such cases, each incompletely specified value in the vector may be replaced by any compatible fully specified value without affecting the coverage of the target fault. Sometimes, these unspecified values are specified in several alternative ways and fault simulation performed. The fully specified version of the vector which detects the largest number of non-target faults in the fault list is then added to the set, T_d, of deterministically generated tests. Of course, the faults detected by the selected version of the vector are marked as tested and deleted from the fault list. Such an approach increases the test generation efficiency in the same ways as above. However, in this case the cost of fault simulation may be a little higher, especially if fault simulation is performed for multiple versions of the generated vector. Furthermore, this may sometimes reduce the effectiveness of some test compaction techniques discussed ahead, thereby increasing somewhat the test application cost.

Functional test vectors: Sometimes, in addition to the vectors generated by a test generation system, a set, T_f, of *functional vectors* must be applied during the testing of a given CUT. These vectors are typically developed in different phases of CUT design that precede test generation, such as design and/or verification, and correct operation of the circuit for these vectors is sometimes deemed crucial. In such cases, fault simulation is performed for the given vectors and the faults tested by any of these

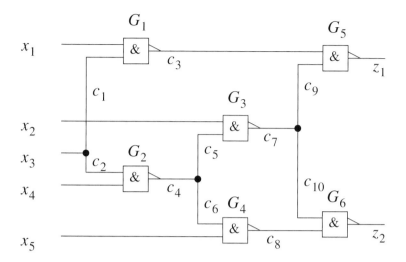

Figure 4.37 An example circuit

vectors are marked as detected and removed from the fault list. This helps enhance test generation efficiency as described above. Note that in this case, T_f is a part of T, the final test set to be applied to the CUT during testing.

Random test generation: For most combinational circuits, a small set of randomly generated test vectors can detect a substantial percentage of SAFs. For most combinational circuits, this percentage lies in the 60–90% range. In cases where functional vectors are provided for use during test, they achieve the same result.

In cases where functional vectors are not provided, before tests are generated deterministically, a small set, T_r, of vectors may be generated randomly and fault simulation performed. The faults detected by vectors in T_r are then marked as tested and removed from the fault list. Of course, the vectors in T_r must be added to T.

A set of randomly generated tests of a given size typically provides lower fault coverage compared to a set with an equal number of deterministically generated test vectors. In such an approach, test generation efficiency may be attained at the expense of an increase in test application cost. This increase in test application cost may be avoided if fault simulation of random vectors is performed to prioritize the faults in the target fault list, as described ahead. This additional cost may also be reduced if random test generation is followed by test compaction, also described ahead.

Example 4.16 Consider the circuit shown in Figure 4.37. Let

$\{x_1$ SA1, x_2 SA1, x_3 SA0, x_3 SA1, x_4 SA1, x_5 SA1, c_1 SA1, c_2 SA1,

c_3 SA1, c_4 SA1, c_4 SA0, c_5 SA1, c_6 SA1, c_7 SA0, c_7 SA1, c_8 SA1,

c_9 SA1, c_{10} SA1, z_1 SA0, z_1 SA1, z_2 SA0, z_2 SA1$\}$

be the fault list of interest. (This fault list is obtained by starting with all the single SAFs in the circuit and using the fault collapsing procedure in Lee and Ha (1991).) The most straightforward manner in which tests can be generated for these faults using an ATPG is by individually targeting each fault to generate a vector for each fault. One such ATPG run generates the following set of test vectors:

$\{(1, 0, 1, x, x), (1, 0, 0, x, x), (0, 1, 1, 0, x), (0, 1, x, 0, x), (0, 1, 1, 1, x),$

$(x, 1, 0, 1, x), (0, 0, x, 0, x), (0, 0, x, x, x), (0, 1, x, 0, x), (x, 1, 1, 1, x),$

$(0, 0, 1, x, x), (1, 0, 1, x, x), (1, 0, 0, x, x), (1, x, 1, x, x), (0, 0, x, x, x),$

$(0, 1, x, 0, x), (x, 0, x, 0, 0), (x, 0, x, 0, 1), (x, x, 1, 1, 1), (x, 1, x, 0, x),$

$(x, 0, x, x, 0), (x, 1, x, 0, 0)\}.$

In this simple test generation system, each of the 22 faults in the given fault list is individually targeted by the ATPG. Consequently, 22 incompletely specified vectors are generated. (In general, each vector may not be incompletely specified though a majority typically are.) An examination of the vectors generated reveals that some vectors, such as $(1, 0, 1, x, x)$, appear multiple times in the generated test set. Clearly, a maximum of one copy of a vector needs to be retained in the test set. Finally, some vectors as well as their more incompletely specified counterparts appear in the test set. For example, vector $(0, 1, 1, 0, x)$ as well as its more incompletely specified counterpart $(0, 1, x, 0, x)$ appear in the generated test set. Clearly, the former vector can detect any fault detected by the latter. Hence, if a vector P and one or more of its more incompletely specified counterparts exist in the generated test set, then each counterpart can be removed from the set. The test set obtained by removing duplicates and more incompletely specified counterparts from the above set of 22 vectors gives the following set of 11 vectors:

$\{(1, 0, 1, x, x), (1, 0, 0, x, x), (0, 1, 1, 0, x), (0, 1, 1, 1, x),$

$(x, 1, 0, 1, x), (0, 0, x, 0, x), (0, 0, 1, x, x), (x, 0, x, 0, 0),$

$(x, 0, x, 0, 1), (x, x, 1, 1, 1), (x, 1, x, 0, 0)\}.$

□

Example 4.17   Again, consider the circuit and the fault list described in Example 4.16. This time assume that first a sequence of four random vectors,

$\{(1, 1, 0, 1, 0), (0, 1, 1, 0, 1), (1, 0, 0, 1, 0), (0, 1, 0, 0, 1)\},$

is generated and fault simulation performed to identify faults in the given fault list that are detected by these vectors. Each such fault is then removed from the fault list and the following fault list with seven faults is obtained:

$\{x_1 \text{ SA1}, x_3 \text{ SA0}, c_3 \text{ SA1}, c_4 \text{ SA1}, c_5 \text{ SA1}, c_6 \text{ SA1}, c_8 \text{ SA1}\}.$

One of the faults in this fault list, say c_6 SA1, is then selected as a target fault and the incompletely specified vector $(x, x, 1, 1, 1)$ is generated.

One option would be to add this incompletely specified vector to the above test set, perform fault simulation for this vector, and delete all the faults detected by the vector from the fault list.

Alternatively, the incompletely specified components in the above vector can be specified in a number of different ways and fault simulation performed for each completely specified version of the vector. Since the above vector contains only two incompletely specified component values, we can generate all the four versions, namely $(0, 0, 1, 1, 1)$, $(0, 1, 1, 1, 1)$, $(1, 0, 1, 1, 1)$, and $(1, 1, 1, 1, 1)$, and perform fault simulation for each version. (For a large circuit, the vector may contain a large number of incompletely specified components and fault simulation may be performed for only a few completely specified counterparts of the generated vector.) The above four vectors respectively detect four, five, four, and five of the seven faults in the above fault list. We can select either of the two vectors that detect five faults from the above fault list. Assume that we select vector $(0, 1, 1, 1, 1)$ and add it to the test set to obtain an updated test set,

$$\{(1, 1, 0, 1, 0), (0, 1, 1, 0, 1), (1, 0, 0, 1, 0), (0, 1, 0, 0, 1), (0, 1, 1, 1, 1)\}.$$

The faults detected by the added vector can then be deleted from the fault list to obtain the fault list $\{c_3$ SA1, c_8 SA1$\}$.

The above process can be repeated by targeting c_3 SA1, generating an incompletely specified vector, performing fault simulation for each of its fully specified counterparts, and selecting vector $(1, 0, 1, 0, 1)$, which detects both the faults in the above fault list.

Since the fault list is empty, 100% fault coverage has been attained and test generation concludes with a test set with six vectors, namely

$$\{(1, 1, 0, 1, 0), (0, 1, 1, 0, 1), (1, 0, 0, 1, 0),$$
$$(0, 1, 0, 0, 1), (0, 1, 1, 1, 1), (1, 0, 1, 0, 1)\}.$$

In contrast to Example 4.16, in this case ATPG was used only for two target faults. Fault simulation was performed for completely specified counterparts of each generated vector and the one that provided highest fault coverage was added to the test set. ☐

4.8.2   Selection of target faults

Techniques for selecting target faults from the fault list can help increase test generation efficiency and/or decrease test application cost by generating more compact sets of test vectors.

Fault collapsing: The simplest technique for target fault selection is fault collapsing, described in Chapter 3. For typical circuits, a significant degree of fault collapsing can be accomplished at low computational complexity.

Since each test generation step is typically followed by fault simulation, equivalence fault collapsing does not reduce the number of target faults for which test generation must be performed. This is due to the fact that whenever any of the faults in an equivalence set of faults is selected as a target fault and a test generated for that fault, then subsequent fault simulation eliminates from the fault list all the faults that are equivalent to the target fault. Similarly, dominance fault collapsing may also not help reduce the number of faults that need to be targeted by test generation, provided that a fault f_i is selected as a target fault only after all the faults that it dominates are targeted. In such a scenario, if any of the faults that f_i dominates is detected, then f_i is anyway removed from the fault list before it is selected as a target fault. However, even in such a case, fault collapsing helps reduce the fault simulation effort expended during test generation.

Prioritizing faults by their random vector testabilities: In most circuits, some faults are tested by many vectors while others are tested by a few vectors. Most faults that are tested by a small set of randomly generated test vectors have a large number of deterministic test vectors. A small set, T_r, of randomly generated vectors may be added to the overall test set T as described before. However, since most of the faults tested by vectors in T_r have many test vectors, with a very high probability they are also tested by one or more of the test vectors generated subsequently for the remaining faults in the fault list. Hence, the addition of T_r to the overall test set T may unnecessarily increase the number of tests in T and hence increase the test application cost. This increase in test application cost can be avoided by generating a small set of random vectors, performing fault simulation to identify the faults tested by the vectors, and assigning lower priority to these faults. Subsequently, deterministic test generation is first performed on the faults in the fault list that have higher priorities.

By the time tests are generated by the deterministic test generation algorithm for the high-priority faults, most faults that are assigned lower priority in this manner are tested and removed from the fault list. However, some faults that were assigned low priorities may not be tested by the deterministically generated tests for other faults, since it is possible that any small set of randomly generated vectors may coincidentally test a fault that is really hard to test. At this stage, one possibility is to identify which vectors in set T_r test such faults and add only those vectors to the overall test set T. The other alternative is to continue deterministic test generation on the faults in the fault list until all the faults are removed from the list.

Prioritizing faults by their membership in independent fault sets: Two *faults* f_i and f_j in a circuit are said to be **independent** if there exists no vector that is a test for both the faults. Otherwise, the two faults are said to be **concurrently testable**. A set of faults is said to be an **independent fault set** if each pair of faults in the set is independent.

A precise identification of the independence relationship between any two faults in an arbitrary CUT is difficult, since it entails identification of all the necessary as well as all the possible sets of sufficient conditions for the detection of the two faults followed by a comparison of these conditions. However, by examining any set of necessary conditions for the detection of faults, the independence between many pairs of faults can be established, as shown in the procedure given next.

Procedure 4.3 [**IdentifyIndependentFaultSets( )**]

1 For each fault f_i in the fault list, perform the following steps.
   (a) Use the D-algorithm to identify necessary conditions for fault excitation and fault effect propagation for f_i.
       This can be achieved by only processing the TGSTs that are identified as necessary, only making those value assignments for which no alternative assignments exist, and stopping before any arbitrary assignment is made.
   (b) Record the values assigned to each circuit line by the D-algorithm. For each line c, record the fault-free value as $v(c)$ as well as $v^{f_i}(c)$. If the six-valued system is used, then the D-algorithm may assign one of the following values: $0_6, 1_6, D_6, \overline{D}_6, \chi_6$, or X_6, to $V(c)$. The values 0, 1, 1, 0, x, and x are, respectively, assigned to $v(c)$ as well as $v^{f_i}(c)$.
2 For each pair of faults f_i and f_j in the fault list and for each line c in the circuit, intersect value $v^{f_i}(c)$ with $v^{f_j}(c)$.
   If at least one value for f_i, say $v^{f_i}(c)$, has an empty intersection with the corresponding value for f_j, say $v^{f_j}(c)$, then faults f_i and f_j **are** independent. Otherwise, they **may be** concurrently testable.
3 Generate a graph where each node corresponds to a fault, such as f_i. Add an undirected edge between two nodes f_i and f_j if the corresponding faults are identified as being independent.
4 Identify all the *maximal cliques* in this graph. (A **clique** in a graph is a set of nodes such that an edge exists between each pair of nodes in the set. A **maximal clique** is a clique that is not contained in any clique with a larger number of nodes.) Each maximal clique in the graph corresponds to an independent fault set.

Powerful procedures to identify necessary assignments, such as analysis of potential propagation paths and future unique D-drive should be incorporated within the D-algorithm used by the above procedure. Their use may help identify additional necessary conditions. Also, powerful static and dynamic implication procedures should be used for the same reason.

In practice, identification of independent fault sets may be limited to faults within each fanout-free region within the circuit. Efficient procedures to accomplish this for circuits with primitive gates can be found in Pomeranz *et al.* (1991) and Akers *et al.* (1987).

Once the independent fault sets are identified, the faults in the largest identified independent fault set may be assigned the highest priority. This can help improve the test generation efficiency since the first few tests will be those generated for each of the faults in the largest independent fault set. Such tests may cover fairly distinct sets of faults, thereby reducing the number of subsequent test generation steps and the complexity of the fault simulation step that follows each test generation step.

4.8.3   Techniques to obtain a compact set of tests

Two types of technique are used to obtain a compact test set for a given CUT and fault list. In the first type, first a test generation algorithm is used to generate a set of vectors that detects all the faults in the list, except for faults that were proven to be untestable or for which test generation failed. From here onwards, a test generation algorithm is not used but the tests in the generated test set are manipulated to reduce its size. Such techniques are called **static test compaction** techniques, since the test set is manipulated without the benefit of generating new tests. (Fault simulation may however be performed.) The other class of techniques, called **dynamic test compaction**, may use a test generator to generate alternative tests with properties that facilitate compaction.

4.8.3.1   Static test compaction

Given a set of tests, without the further benefit of a test generator, its constituent tests may be manipulated in two main ways. First, a test vector that only detects faults that are detected by other tests in the set can be identified and *eliminated*. Second, two or more incompletely specified test vectors may be *combined* into a single test vector.

Static test compaction by test elimination: In any test generation system, fault simulation is performed following the generation of each test vector. Hence, when a test vector P_i is generated, it detects one or more faults that are not detected by any vector generated *prior* to the generation of P_i. However, vectors generated subsequently may detect all the faults that are detected by P_i and not by any of its preceeding vectors. Hence, it is often possible to eliminate some vectors from the set of vectors generated.

A simple approach to identify and eliminate some of the unwanted vectors from a test set is called **reverse order fault simulation** (e.g., see Schulz *et al.* (1988)). In this approach, starting with the complete fault list, test vectors are simulated in an order that is reverse of the order in which they were generated. Any vector that does not detect any fault not detected by the vectors simulated earlier can then be dropped. This method can sometimes reduce the test set size significantly. For example, in real-life circuits, reverse order simulation can sometimes help eliminate most of the vectors that were generated randomly prior to deterministic vector generation. However, such a method cannot guarantee that a *minimal test set* is found, where a **minimal test set** is a set of vectors from which no vector can be removed without decreasing the fault coverage provided by the set. For example, any vector P_i that tests (a) at least one fault not tested by vectors generated before P_i, and (b) at least one fault not tested by vectors generated afterwards, will be retained in the test set. However, if each fault in category (a) is tested by a vector generated after P_i and each fault in category (b) is tested by a vector generated before P_i, then vector P_i can be removed from the test set without decreasing its fault coverage.

Example 4.18   Again, consider the circuit and the fault list described in Example 4.16. Consider the set of six vectors,

$$\{(1, 1, 0, 1, 0), (0, 1, 1, 0, 1), (1, 0, 0, 1, 0),$$
$$(0, 1, 0, 0, 1), (0, 1, 1, 1, 1), (1, 0, 1, 0, 1)\},$$

generated in Example 4.17. Recall that four random vectors were generated first and fault simulation performed. This was followed by generation of two additional deterministic vectors.

Starting with the original fault list, the vectors in the above set are simulated in the reverse order of how they were originally generated. The reader can verify that the fourth and second vectors in the above set, i.e., $(0, 1, 0, 0, 1)$ and $(0, 1, 1, 0, 1)$, can be deleted from the above set. For each of these vectors, each fault that was detected for the first time by the vector was also detected by some other vector generated afterwards. Reverse order fault simulation helps obtain a set with four vectors,

$$\{(1, 1, 0, 1, 0), (1, 0, 0, 1, 0), (0, 1, 1, 1, 1), (1, 0, 1, 0, 1)\},$$

that provides 100% coverage of the faults in the given fault list.

□

Another approach called **double detection** (Kajihara *et al.*, 1993) helps identify a minimal set of tests. In this method, each fault is simulated until it is detected by two vectors before it is removed from the fault list used by fault simulation. (The fault may, however, be removed from the fault list used for test generation as soon as it is detected the first time.) When fault simulation is performed for a vector P_i during test generation, each fault detected by P_i that was not detected by any previous vector is marked as detected. The number of faults detected in this manner by P_i is stored in a variable $N_{f.det}(P_i)$. Also, for each fault f_j detected by P_i in this manner,

Table 4.5. *An example set of test vectors and faults detected by each vector*

Vector	Faults detected
P_1	f_1, f_3
P_2	f_1, f_2, f_3
P_3	f_1, f_4
P_4	f_2, f_4, f_5

the fact that the fault is detected is recorded by setting $ND(f_j) = 1$. In addition, a variable $FD(f_j)$ is used to identify P_i as the vector that first detected the fault. All the faults detected by P_i that were already detected by exactly one previous vector, i.e., all f_j such that $ND(f_j) = 1$ and P_i detects f_j, are processed differently. First, $ND(f_j)$ is incremented to two and f_j is dropped from the fault list used by the fault simulator. Second, vector P_j that had previously detected f_j is identified by examining $FD(f_j)$. $N_{f.det}(P_j)$ is then decremented to reflect the fact that one of the faults that P_j had detected for the first time has been detected by at least one of the subsequent vectors.

If the above process is followed to the end, then each vector P_i for which $ND(P_i) \geq 1$ detects at least one fault that is not detected by any other vector. None of these vectors can be eliminated from the test set without decreasing the fault coverage. Hence, each of them is retained. In contrast, if for a vector P_j, $ND(P_j) = 0$, then the vector is a candidate for elimination. The elimination is performed as follows. First, starting with the complete fault list, all the vectors in the original test set that are not to be eliminated are simulated and the faults detected by these vectors removed from the fault list. Next, the vectors that are candidates for elimination are simulated one-by-one in the order that is reverse of the order in which they were originally generated by the test generator. Any vector that does not detect at least one fault in the fault list is eliminated. The entire process outlined above can be repeated until no vector is eliminated in a pass. This method guarantees that the resulting test set is minimal. However, it does not guarantee that the resulting test set is of *minimum size* that may be obtained by elimination of vectors from the given test set.

Example 4.19 Consider a circuit and four vectors, where each vector detects faults as described in Table 4.5.

Now let us follow the steps of the double detection fault simulation methodology. The fault list is initialized to

$$\{f_1, f_2, f_3, f_4, f_5\}.$$

For each fault f_j, the values of $ND(f_j)$ and $FD(f_j)$ are initialized as shown in Figure 4.38(a). For each vector P_i, the value of $N_{f.det}(P_i)$ is initialized as shown in Figure 4.38(b).

The first vector, P_1, is then simulated and found to detect f_1 and f_3. The value of ND is incremented for each of these two faults. Since this is the first vector that detects these faults, the variable FD is updated for each of these faults. Also, since, thus far, vector P_1 has detected two faults that are detected for the first time, value $N_{f.det}$ is updated to reflect this fact. Figures 4.38(c) and 4.38(d) denote the updated values. Faults f_1 and f_3 are retained in the fault list despite being detected by P_1, since they have thus far been detected only by one vector. Hence, the fault list remains unchanged.

The second vector, P_2, is then simulated and identified as detecting f_1, f_2 and f_3. The value of ND is incremented for each of these faults. During this process, it is identified that this is the first vector to detect fault f_2 (since the value of $ND(f_2)$ was incremented from zero to one). It is also identified that faults f_1 and f_3 are now detected two times. Variable $FD(f_2)$ is hence updated to identify P_2 as the first vector to detect f_2. Also, since f_1 is now known to be detected by two vectors, the vector that first detected f_1 is determined by examining the value of $FD(f_1)$. Since $FD(f_1) = P_1$, we now know that P_1 is no longer the only vector that detects f_1 as was previously noted. Hence, the value of $N_{f.det}(P_1)$ is decremented. A similar process is followed for the other fault that is detected for the second time, f_3, further decrementing the value of $N_{f.det}(P_1)$ to zero. The updated values of the variables are shown in Figures 4.38(e) and 4.38(f). Faults f_1 and f_3 are now removed from the fault list.

The above process is repeated for the next two vectors to obtain the values shown in Figures 4.38(g) and 4.38(h). These figures demonstrate that other than P_4 which detects one fault not detected by any other vector (since $N_{f.det}(P_4) = 1$), each fault detected by each other vector is detected by at least one other vector. Hence, P_4 cannot be eliminated from the test set. □

Example 4.20 Consider the process of elimination of vectors from the set of four vectors identified in the above example. Since P_4 cannot be eliminated from the given set of vectors, starting with the complete list of faults,

$$\{f_1, f_2, f_3, f_4, f_5\},$$

fault simulation is performed for this vector to identify that it detects f_2, f_4, and f_5. These faults are deleted from the fault list to obtain the updated fault list,

$$\{f_1, f_3\}.$$

Simulation is then performed for each of the other vectors in reverse order. P_3 is identified as detecting f_1 and retained in the fault list. Subsequently, P_2 is identified as detecting f_3 and retained in the fault list. Since all the faults are detected by this time, P_1 can be eliminated from the fault list. Hence a test set with three vectors,

$$\{P_2, P_3, P_4\},$$

is obtained.

Fault f_j	$ND(f_j)$	$FD(f_j)$
f_1	0	
f_2	0	
f_3	0	
f_4	0	
f_5	0	

(a)

Vector P_i	$N_{\text{f.det}}(P_i)$
P_1	
P_2	
P_3	
P_4	

(b)

Fault f_j	$ND(f_j)$	$FD(f_j)$
f_1	1	P_1
f_2	0	
f_3	1	P_1
f_4	0	
f_5	0	

(c)

Vector P_i	$N_{\text{f.det}}(P_i)$
P_1	2
P_2	
P_3	
P_4	

(d)

Fault f_j	$ND(f_j)$	$FD(f_j)$
f_1	2	P_1
f_2	1	P_2
f_3	2	P_1
f_4	0	
f_5	0	

(e)

Vector P_i	$N_{\text{f.det}}(P_i)$
P_1	0
P_2	1
P_3	
P_4	

(f)

Fault f_j	$ND(f_j)$	$FD(f_j)$
f_1	2	P_1
f_2	2	P_2
f_3	2	P_1
f_4	2	P_3
f_5	1	P_4

(g)

Vector P_i	$N_{\text{f.det}}(P_i)$
P_1	0
P_2	0
P_3	0
P_4	1

(h)

Figure 4.38 Values of variables stored for double detection. (a) and (b) initial values; (c) and (d) after simulation of P_1; (e) and (f) after simulation of P_2; and (g) and (h) after simulation of P_4.

The entire double detection fault simulation methodology can now be repeated for this set containing three vectors. The reader can verify that this time P_3 can be eliminated to find a test set with only two vectors. □

The problem of finding a test set of minimum size via elimination of vectors from a given set can be formulated as a *covering problem*. This approach begins by recording all the faults detected by each vector in the set. Next, a minimum set of vectors that detects all the faults is found using a covering method similar to that used to obtain a minimum sum-of-product representation of a truth table. The complexity of obtaining and storing the fault coverage information required by the method can itself be high; finding a minimum cover is also known to be a hard problem. Hence, such a method is not used very often.

Static test compaction by test combining: A vector generated by a test generator for a target fault is, in general, partially specified. Sometimes, as described before, the partially specified input components of such a vector may be specified immediately after its generation, in an attempt to maximize its fault coverage. If this is the case, then no two vectors may be combined. If this is not the case, and there exist more than one vector in the test set that have incompletely specified components, then it is possible to obtain a compact test set by combining two or more vectors into a single vector.

Two vectors are said to be compatible if their intersection is non-empty. For example, the three-bit vectors, $(0, x, 1)$ and $(0, 1, x)$, are compatible, since their intersection is the vector $(0, 1, 1)$. On the other hand, the vectors $(0, x, 1)$ and $(1, x, 1)$ are not compatible. A set of vectors is compatible if and only if each pair of vectors in the set is compatible.

The problem of finding a test set of minimum size that can be obtained by combining vectors in a given test set can also be formulated as a covering problem. First, a graph that contains a node corresponding to each vector in the test set is created. An edge is added between two nodes if the corresponding vectors are compatible. A compatible set of vectors form a clique in this graph. A test set of minimum size can be obtained by finding a minimum clique cover of the graph, where a minimum clique cover is comprised of a minimum number of cliques that cover all the vectors in the original test set. Finally, one vector is obtained per clique by intersecting all the vectors in the clique.

The complexity of finding a minimum test set in this manner is high. Hence, in practice, the partially specified component values of each test vector are specified immediately following its generation, either as described before or as described ahead in dynamic test compaction.

Example 4.21 Consider the circuit shown in Figure 4.37 and the following set of 11 incompletely specified vectors derived in Example 4.16:

$\{(1, 0, 1, x, x), (1, 0, 0, x, x), (0, 1, 1, 0, x), (0, 1, 1, 1, x),$
$(x, 1, 0, 1, x), (0, 0, x, 0, x), (0, 0, 1, x, x), (x, 0, x, 0, 0),$
$(x, 0, x, 0, 1), (x, x, 1, 1, 1), (x, 1, x, 0, 0)\}.$

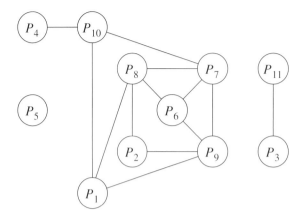

Figure 4.39 Compatibilities between incompletely specified vectors in an example test set

Let the above vectors be called P_1, P_2, ..., P_{11}. Figure 4.39 shows the compatibility relation between various vectors – each node represents a vector and an edge exists between nodes representing two vectors if they are compatible.

This graph contains two cliques of size three. First is the clique comprised of vectors P_6, P_7, and P_8; second is that comprised of vectors P_6, P_7, and P_9. Several cliques of size two exist, such as one with P_1 and P_{10}. Only one clique of size one exists, namely P_5 by itself.

Let us use a greedy approach for clique covering. First select the first clique with three vectors, P_6, P_7, and P_8. The existence of such a clique implies that the three constituent vectors are compatible and can be combined into one vector. Specifically, these incompletely specified vectors, i.e., (0, 0, x, 0, x), (0, 0, 1, x, x), and (x, 0, x, 0, 0), can be combined into one vector, (0, 0, 1, 0, 0). Note that P_9 is the only vector that would be covered if the second clique of size three is selected. Since cliques of size two, such as the one with vectors P_2 and P_9, can cover more than one vector, the second clique of size three is not selected. The selection process can continue via combination of P_1 and P_{10} into (1, 0, 1, 1, 1), P_2 and P_9 into (1, 0, 0, 0, 1), P_3 and P_{11} into (0, 1, 1, 0, 0). Finally, the incompletely specified vectors P_4, (0, 1, 1, 1, x), and P_5, (x, 1, 0, 1, x), are added to the set.

This gives the following test set with six vectors,

$$\{(0, 0, 1, 0, 0), (1, 0, 1, 1, 1), (1, 0, 0, 0, 1),$$
$$(0, 1, 1, 0, 0), (0, 1, 1, 1, x), (x, 1, 0, 1, x)\}.$$

Note that the vectors are combined without any information about the actual fault coverage. For example, it is possible that the vector (1, 0, 1, 1, 1), obtained by combining P_1 and P_{10}, may detect faults that were not detected by either P_1 or P_{10}. For this reason, it may be possible to delete one or more vectors from the above set without reducing fault coverage. (The reader is advised to perform fault simulation to see if this is indeed possible.) ☐

4.8.3.2 Dynamic test compaction

A very large range of techniques have been developed for dynamic compaction. Here, we describe the most common. This involves generation of test vectors that detect a large number of single SAFs in a given fault list. Two main approaches are taken for test generation, as discussed next.

In the first approach (e.g., Goel and Rosales (1979); Pomeranz *et al.* (1991)), a test vector is first generated for a target fault. If the generated vector contains any partially specified component values, then another fault, sometimes called a *surrogate fault*, is selected. A **surrogate fault** is a fault that belongs to the fault list for which it seems possible to obtain a test by specifying the given partially specified vector. Whether or not a fault may be detectable in this manner can be determined by creating a circuit model with the fault as the target fault, assigning the given partially specified vector at the inputs of the circuit model, and executing the step of a structural test generation algorithm that identifies TGSTs. If this step identifies that the given vector may not be further specified to obtain a test for the surrogate fault, then the surrogate fault may be replaced by another fault from the fault list. On the other hand, if TGSTs are indeed identified for the selected surrogate, then other steps of the test generator can be used to generate a test. The only modification that is made is that the values in the partially specified vector that were assigned prior to test generation for the surrogate fault are not altered during surrogate test generation.

In the second approach, a group of target faults is selected from the fault list and generation of a test vector that can detect each of these faults is attempted. If it is found during any round of test generation that one or more of the target faults cannot be detected by further specifying the given partial vector, then a decision is made to either backtrack or to drop from the target fault list those faults that cannot be tested by further specifying the current vector. This process is continued until a test vector is generated for some subset of the selected set of target faults.

Example 4.22 In Example 4.17, we specified various incompletely specified values in a generated vector in multiple ways, performed fault simulation for each vector obtained in this manner, and selected the one that provided the largest increase in fault coverage. While for a small example circuit, it is practical to try all possible combinations of assignments to the incompletely specified values, for circuits with a large number of inputs, only a small fraction of all possible combinations can be tried. Typically, a few randomly generated combinations of values are assigned to the incompletely specified values.

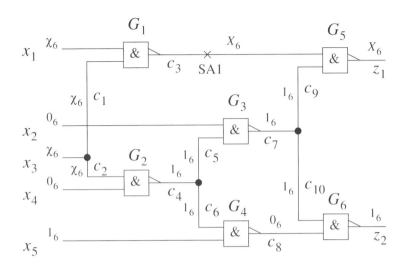

Figure 4.40 Targeting c_3 SA1 starting with an incompletely specified vector for c_8 SA1

Higher fault coverage can often be obtained by selecting another fault as a target fault that may be detected by further specifying the values of the given incompletely specified vector and performing test generation for that fault.

Consider an intermediate stage during test generation for the circuit shown in Figure 4.40. Assume that previously generated vectors have detected all the faults except $\{c_3$ SA1, c_8 SA1$\}$. Assume that at this stage, fault c_8 SA1 is targeted and the incompletely specified vector $(x, 0, x, 0, 1)$ is generated. Now, we can select the only remaining fault in the fault list, c_3 SA1, as the surrogate target fault and attempt test generation starting with the above vector. Figure 4.40 shows the circuit with this target fault and the above vector. PODEM can now be applied to further specify the components of the above vector to generate a test for the secondary target fault. The reader can verify that this test generation will be successful and generate a fully specified vector, $(1, 0, 1, 0, 1)$ which will detect c_3 SA1 in addition to c_8 SA1. $\square$

4.9   Test generation for reduced heat and noise during test

Attainment of high fault coverage has traditionally been the main objective of test generation techniques. In recent years, reducing the heat dissipated during test application has become an important auxiliary objective.

The importance of this problem is illustrated by the example circuit reported in Zorian (1993), where heat dissipation during test application was significantly higher (100–200%) than during the circuit's normal operation. This increase can be attributed to several factors. First, significant correlation exists between successive vectors applied to a block of combinational logic during its normal operation. In contrast, the correlation between consecutive test vectors generated by a test generator is very low, since a test is generated for a given target fault without any consideration of the previous vector in the test sequence. Second, the use of DFT techniques can further decrease the correlation between successive test vectors. For example, this is true for the *scan* DFT technique (see Chapter 11) since the values applied during successive clock cycles at the state inputs of a logic block represent shifted values of scan test vectors and circuit responses. This is even more true for built-in self-test techniques which typically employ random pattern generators to generate test sequences, where the correlation between consecutive vectors is provably small (see Chapter 12; Section 12.2). Finally, circuit miniaturization requires the package size to be reduced to aggressively match the average heat dissipation during the circuit's normal operation.

It is important to ensure that heat dissipation during test application does not destroy a chip during test. It is also important to reduce switching noise during test application, since excessive switching noise may cause a logical error in a fault-free chip leading to an unnecessary loss of yield.

In this section, we discuss a technique to generate sequences of test vectors that provide high fault coverage while minimizing heat dissipation and switching noise during test application (Wang and Gupta, 1998). The technique discussed here focuses on combinational circuits; techniques for scan circuits are discussed in Chapter 11 (Section 11.6.2).

4.9.1   Preliminaries

In the following, the CUT is assumed to be a CMOS digital circuit. Heat dissipation in CMOS can be divided into static and dynamic. Static dissipation is due to leakage currents that currently have small magnitudes in digital CMOS circuits. Hence, for such circuits, the dynamic dissipation is the dominant term. Dynamic dissipation occurs at a node when it switches from $0 \rightarrow 1$ (or from $1 \rightarrow 0$) and is typically divided into two components caused by *short-circuit current* and *charge/discharge current*, respectively. The former is caused by a current pulse that flows from power supply to ground when both nMOS and pMOS transistors are simultaneously on during switching. The charge/discharge current is the current that charges and discharges the output capacitive load and typically dominates dynamic dissipation. Under the zero-delay model (i.e., when the gates are assumed to switch with zero delay) and ignoring logic hazards, the dynamic power dissipation in the circuit can be approximated as described next.

Let P^{pr} and P be the previous and the current test vectors applied to the CUT. Let $v^{\mathrm{pr}}(c_i)$ and $v(c_i)$, respectively, be the values implied at a line c_i by these two vectors. The amount of heat dissipated due to the application of test vector P is approximately proportional to

$$\sum_{i=1}^{L}\{v(c_i) \oplus v^{\mathrm{pr}}(c_i)\}C_{\mathrm{L}}(c_i), \tag{4.26}$$

where $C_{\mathrm{L}}(c_i)$ is the load capacitance at line c_i.

4.9.2   Transition testability measures

The notion of controllability can be modified to obtain the notion of **transition 0-controllability**, $CC_{\mathrm{tr}}^{0}(c_i)$, which is a heuristic estimate of the amount of heat dissipated in the circuit if a value 0 is to be implied at line c_i by a vector P, assuming that the vector P^{pr} was applied before P. Transition 1-controllability, $CC_{\mathrm{tr}}^{1}(c_i)$, is similarly defined.

The testability measures described earlier can be modified in the following key ways. First, the constant CC_{pi} is assigned a distinct value for each primary input x_i, as $CC_{\mathrm{pi}}(x_i) = C_{\mathrm{L}}(x_i)$. Second, the constant $CC_{\kappa}(c_i)$ associated with each line c_i is defined as $CC_{\kappa}(c_i) = C_{\mathrm{L}}(c_i)$. Third, three-valued singular covers are used for each circuit element rather than five- or six-valued cube covers. Fourth, the costs associated with cubes in the singular cover of a gate with inputs $c_{i_1}, c_{i_2}, \ldots, c_{i_\alpha}$ and output c_j can be computed as described next. Let $\mathcal{C} = (\mathcal{C}_1, \mathcal{C}_2, \ldots, \mathcal{C}_\alpha)$ be a cube in the singular cover of the gate. The *transition cost of a component $\mathcal{C}_l$ of a cube $\mathcal{C}$*, $cost_{\mathrm{tr}}(\mathcal{C}_l)$, can be defined as

$$cost_{\mathrm{tr}}(\mathcal{C}_l) = \begin{cases} 0, & \text{if the previous value at } c_{i_l}, v^{\mathrm{pr}}(c_{i_l}), \text{ is identical to } \mathcal{C}_l, \\ CC_{\mathrm{tr}}^{0}(c_{i_l}), & \text{if } \mathcal{C}_l \text{ is } 0, \\ CC_{\mathrm{tr}}^{1}(c_{i_l}), & \text{if } \mathcal{C}_l \text{ is } 1, \\ 0, & \text{otherwise.} \end{cases} \tag{4.27}$$

The **transition cost of cube** $\mathcal{C}$, $Cost_{\mathrm{tr}}(\mathcal{C})$, can then be defined as

$$Cost_{\mathrm{tr}}(\mathcal{C}) = \sum_{l=1,2,\ldots,\alpha} cost_{\mathrm{tr}}(\mathcal{C}_l). \tag{4.28}$$

For a primary input x_i, the transition 0-controllability is given by

$$CC_{\mathrm{tr}}^{0}(x_i) = \begin{cases} 0, & \text{if } v^{\mathrm{pr}}(c_j) = 0, \\ CC_{\mathrm{pi}}(x_i), & \text{otherwise.} \end{cases} \tag{4.29}$$

$CC_{\mathrm{tr}}^{1}(x_i)$ can be defined in a similar manner.

The transition 0-controllability of the gate output can be computed as

$$CC_{\mathrm{tr}}^{0}(c_j) = \begin{cases} 0, & \text{if } v^{\mathrm{pr}}(c_j) = 0, \\ \min_{\forall \mathcal{C} \text{ for which } v(c_j)=0}[Cost_{\mathrm{tr}}(\mathcal{C})] + CC_{\kappa}(c_j), & \text{otherwise.} \end{cases} \tag{4.30}$$

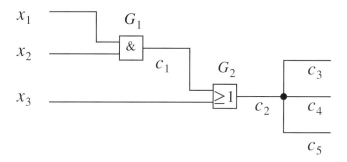

Figure 4.41   An example circuit to illustrate the computation of transition testability measures (Wang and Gupta, 1998) (© 1998 IEEE)

The transition 1-controllability at the gate output, $CC_{tr}^1(c_j)$, can be computed in a similar manner. The transition observability of a line can be obtained by making similar changes to the notion of observability defined earlier.

The following example illustrates how transition controllabilities are computed.

Example 4.23   Figure 4.41 shows a part of an example circuit. Assume that the previous vector applied to the circuit was $P^{pr} = (1, 1, 0)$.

Assume that the capacitive load of a line is equal to its fanout. Hence, we get $C_L(x_1) = 1, C_L(x_2) = 1, C_L(x_3) = 1, C_L(c_1) = 1$, and $C_L(c_2) = 3$.

These values can be used to define, $CC_{pi}(x_1) = 1, CC_{pi}(x_2) = 1$, and $CC_{pi}(x_3) = 1$. We can also define, $CC_\kappa(c_1) = 1$ and $CC_\kappa(c_2) = 3$.

For primary input x_1, $CC_{tr}^0(x_1) = 1, CC_{tr}^1(x_1) = 0$. Similarly, $CC_{tr}^0(x_2) = 1$, $CC_{tr}^1(x_2) = 0$. For x_3, $CC_{tr}^0(x_3) = 0, CC_{tr}^1(x_3) = 1$.

The value of $CC_{tr}^0(c_1)$ can be computed by first enumerating the cubes in the off-set of the gate with output c_1, namely cubes $(0, x)$ and $(x, 0)$. Consider the cost computation for the first component of the first cube. Since the value implied at the line by the previous vector was different than the value of this component, the cost of this component is given by the transition 0-controllability of the line, i.e., by $CC_{tr}^0(x_1) = 1$. The cost of this cube is given as $Cost_{tr}(0, x) = 1$. The cost of the other cube can also be computed to be 1. The transition 0-controllability of c_1 can now be computed as $CC_{tr}^0(c_1) = \min(1, 1) + 1 = 2$. The reader can follow the above steps to verify that $CC_{tr}^0(c_2) = 2 + 3 = 5$.   □

A transition observability measure, $O_{tr}(\ )$, can be obtained via similar modification of a classical observability measure.

It should be noted that the above calculations depend on the values implied by the previous vector P^{pr}. Hence, the transition testability measures must be recomputed prior to generation of each vector.

4.9.3 Test generation algorithm

First, the test generation algorithm selects a target fault by using the transition testability measures. For an SAw fault at line c_i, the transition detection cost is computed as $CC_{tr}^{\overline{w}}(c_i) + O_{tr}(c_i)$. The transition detection cost is computed for each remaining fault in the target fault list and the fault with the least cost is selected as the next target fault. Second, a test is generated for the target fault using any of the test generation algorithms described earlier that has been modified to use the above transition testability measures instead of the usual testability measures. Finally, if the test vector generated, P, has any incompletely specified values, the number of transitions depends on the assignment of 0 or 1 to the incompletely specified value. If 0 is the controlling value of more gates in the fanout of that primary input, then the value 0 is assigned; otherwise, the value 1 is assigned.

Subsequently, fault simulation is performed for the vector, the faults detected by the vector are dropped, and the above process is repeated.

Summary

- The composite values $\underline{0}$, $\underline{1}$, D, and $\overline{D}$ can be used to represent the values implied by a vector at a circuit line in the fault-free circuit as well as the version with the target fault. Composite value systems used in this chapter use the above four values as basic values, and truth tables for fault-free and faulty versions of circuit elements can be represented using these values.
- The use of a composite value system with a larger number of values increases the complexity of each implication step and/or the size of each truth table. However, its use may decrease the number of backtracks required to generate a test vector for a target fault. Also, if cube covers are stored for every circuit element, then use of a composite value system with a larger number of values may even reduce the size of the representation of the behavior of many circuit elements.
- Test generation can be divided into two main parts, fault effect excitation and fault effect propagation.
- Fault effect excitation requires assignment of values at lines in the transitive fanin of the fault site. Fault effect propagation via a gate G_i requires assigning values at lines in the transitive fanin of those inputs of G_i that do not already have a fault effect.
- Circuit gates that must be considered for fault-effect propagation during any intermediate round of test generation are stored in a set called the D-frontier, $\mathcal{D}$.
- A circuit line that has been assigned a value that is not justified by the values assigned at the line's fanins is said to be unjustified. All unjustified lines in a circuit during any intermediate round of test generation are stored in a list $\mathcal{U}$.

- Implication of values currently assigned to circuit lines can help reduce the amount of search required to generate a test vector for a fault or to prove it to be untestable.
- Direct implication operations are iteratively used by a direct implication procedure. Such a procedure can identify inconsistencies between values assigned to various circuit lines.
- Implication operations may be accelerated using truth tables or cube covers for the symbols of the value system in use.
- In general, direct implication may be incomplete for any vector with one or more components with incompletely specified values.
- Static indirect implication may identify implications that cannot be identified by a direct implication procedure. Such implications are generated for the fault-free version of the circuit, starting with a completely unspecified vector.
- Dynamic indirect implications are identified for a composite circuit with a given target fault and a partially specified vector.
- If the target fault is not excited during a particular round of test generation, then the TGST identification procedure lists alternative ways of exciting the fault.
- If the target fault is already excited, then TGST identification produces alternative ways of propagating the fault effect. At increasingly higher complexities, the unique D-drive, x-path check, and future unique D-drive provide increasingly more precise set of fault effect propagation alternatives.
- Controllability measures provide heuristic measures of the difficulties of obtaining logic values 0 and 1 at a line. Observability measure provides heuristic measure of the difficulty of propagating a fault effect from a line to one or more of the primary outputs of the circuit. Typically, controllability measures are computed via a breadth-first traversal of the CUT starting at primary inputs. Subsequently, the observability measures are computed via a backward breadth-first traversal of the circuit starting at the primary outputs.
- In test generation algorithms like the D-algorithm, values may be assigned at internal circuit lines as well as primary inputs. The search space is comprised of alternative TGSTs and its size is unpredictable. Since values may be assigned at internal lines, both forward and backward implication operations are performed.
- In test generation algorithms like PODEM, values are assigned only at primary inputs, only forward implication is performed, and the search space is comprised of the space of all possible vectors that can be applied to the circuit.
- Test generation techniques based on satisfiability formulate the test generation problem as a Boolean satisfiability problem.
- Test generation algorithms based on BDDs are suitable for generating all the test vectors for a fault. However, they cannot be applied to circuits for which the size of the BDD becomes impractically large.
- Fault simulation is typically performed for functional vectors and/or randomly generated test vectors prior to deterministic test pattern generation. This helps

reduce the complexity of test generation by reducing the number of faults that must be targeted.

- Fault simulation is typically performed following the generation of a vector for a target fault. This helps reduce the overall complexity of test generation.
- If a vector generated for a target fault is incompletely specified, then the values at the incompletely specified components may be specified in alternative ways to obtain different versions of the test vector. Fault simulation can then be performed for each version and the version that provides the highest fault coverage may be added to the test set.
- Alternatively, the incompletely-specified vector may be added to the test vector set. After the completion of test generation, compatible vectors may be combined to reduce the size of the test set.
- Dynamic test compaction can typically provide more compact test sets.
- Test vectors may sometimes be generated to reduce the switching activity in the circuit to minimize the heat dissipation and switching noise during test application.

Additional reading

A neural network based ATPG formulation has been presented in Chakradhar *et al.* (1993). Genetic algorithm has been applied to generate tests (Corno *et al.*, 1996; Rudnick *et al.*, 1997) (also see Chapter 5).

An ATPG can be used to identify whether or not a single SAF in a combinational circuit is testable. The ATPG techniques presented in Kunz and Pradhan (1994) and Schulz *et al.* (1988) are demonstrated as being effective for this purpose. In addition, techniques such as the one in Iyer and Abramovici (1996) have been developed specifically for this purpose. Demonstrating that an SAF in a combinational circuit is untestable can help simplify the circuit. Techniques to optimize logic, such as Entrena and Cheng (1993), have been designed to exploit this fact. Also based on this principle, ATPG techniques, such as Kunz and Pradhan (1994), have been applied for logic verification. (See Chapter 5 for a discussion of redundancy identification in sequential circuits.)

Exercises

4.1 Derive the two-valued truth table of the faulty versions of the following gates.
    (a) A two-input NOR gate, with inputs c_{i_1} and c_{i_2} and output c_j, with an SA0 fault at c_{i_2}.
    (b) A two-input AND gate with an SA1 fault at input c_{i_2}.
    (c) A four-input gate with inputs c_{i_1}, c_{i_2}, c_{i_3}, and c_{i_4} that implements the function $\overline{v(c_{i_1})\, v(c_{i_2}) + v(c_{i_3})\, v(c_{i_4})}$ with an SA0 fault at input c_{i_3}.

(d) A gate that implements the function $v(c_j) = v(c_{i_1}) + \overline{v(c_{i_2})}$ with an SA1 fault at input c_{i_1}.

(e) A gate that implements the function $v(c_j) = v(c_{i_1}) + \overline{v(c_{i_2})}$ with an SA1 fault at input c_{i_2}.

4.2 Derive the four-valued truth table of fault-free versions of each of the following gates.

  (a) A two-input AND.

  (b) A two-input NOR.

  (c) A two-input XNOR.

4.3 Derive the four-valued truth table for the faulty versions of the gates described in Problem 4.1. (Assume that no fault effect can appear at any of the gate inputs.)

4.4 If we assume that multiple faulty circuit elements may exist in a given circuit, then a fault effect may appear at any of the inputs of a faulty gate. In such cases, it becomes necessary to expand the four-valued truth tables for faulty gates to include combinations of input values that contain one or more D and/or $\overline{D}$. Derive four-valued truth tables for faulty versions of a two-input NAND gate for the following faults: (a) an SA1 fault at one of its inputs, (b) an SA0 fault at one of its inputs, (c) an SA1 fault at its output, and (d) an SA0 fault at its output.

4.5 Repeat the above problem for a two-input XOR gate.

4.6 Obtain six-valued singular covers of fault-free versions of the following gates.

  (a) A two-input AND.

  (b) A two-input NOR.

  (c) A two-input XNOR.

  (d) A two-input gate that implements the function $v(c_{i_1}) \, \overline{v(c_{i_2})}$.

4.7 Obtain 16-valued singular covers of the fault-free versions of the following gates.

  (a) A two-input NAND.

  (b) A two-input XOR.

  Compare the size of each 16-valued cover with the corresponding six-valued cover.

4.8 Perform the direct forward implication operation on the following gates.

  (a) A two-input NAND gate with values 1_6, X_6, and X_6 at its inputs and output, respectively.

  (b) A two-input NAND gate with values 0_6, X_6, and X_6 at its inputs and output, respectively.

  (c) A two-input NAND gate with values 0_6, X_6, and χ_6 at its inputs and output, respectively.

  (d) A two-input NAND gate with values 0_6, X_6, and 0_6 at its inputs and output, respectively.

  (e) A two-input NAND gate with values 1_6, D_6, and X_6 at its inputs and output, respectively.

(f) A two-input NAND gate with values 1_6, D_6, and χ_6 at its inputs and output, respectively.

(g) A three-input NAND gate with values 1_6, χ_6, D_6, and X_6 at its inputs and output, respectively.

(h) A three-input NAND gate with values 1_6, χ_6, X_6, and χ_6 at its inputs and output, respectively.

4.9 Derive five-valued truth tables for (a) a two-input NAND gate, (b) a two-input XOR gate, and (c) a fanout system with four branches.

4.10 Perform the direct forward implication operation on a two-input XOR gate with values $\{\underline{0}, \underline{1}\}$, $\{D, \overline{D}\}$, and $\{\underline{0}, \underline{1}, D, \overline{D}\}$, at its inputs and its output, respectively.

4.11 Repeat the above problem using the cube cover of the gate.

4.12 Perform the direct backward implication operation at each of the following gates.

(a) A two-input NAND gate with values $\{\underline{0}, \underline{1}\}$, $\{\underline{0}, \underline{1}, D, \overline{D}\}$, and $\{D, \overline{D}\}$ at its inputs and output, respectively.

(b) A two-input XOR gate with values $\{D\}$, $\{\underline{0}, \underline{1}, D, \overline{D}\}$, and $\{D, \overline{D}\}$ at its inputs and output, respectively.

(c) A two-input NAND gate with values $\{\underline{0}\}$, $\{\underline{0}, \underline{1}, D, \overline{D}\}$, and $\{D, \overline{D}\}$ at its inputs and output, respectively.

4.13 Repeat the above problem using the cube cover of the gate.

4.14 Assume that the six-valued system was used instead of the 16-valued system in each of the scenarios in Problem 4.12.

(a) In each scenario, how would the initial values at inputs and output be represented?

(b) Starting with these initial values, what would be the final values at gate inputs and output if direct backward implication was performed using the six-valued system?

4.15 Identify static indirect implications in fault-free subcircuits shown in (a) Figure 4.18, and (b) Figure 4.20.

4.16 If static indirect implication $v(c_i) = K$, where K is a Boolean constant, is derived by considering a subcircuit, would this indirect implication also be identified if static indirect implications were identified using the complete circuit that includes the subcircuit?

4.17 If static indirect implication $v(c_i) = u \Rightarrow v(c_j) = w$, where $u, w \in \{0, 1\}$, is derived by considering a subcircuit, would this indirect implication also be identified if static indirect implications were identified for the complete circuit that includes the subcircuit?

4.18 Would any static indirect implication be identified in a fanout-free circuit comprised of (multiple) primitive gates?

4.19 Modify Procedure IdentifyStaticIndirectImplications() to identify indirect implications for a faulty version of the circuit with a given target fault.

4.20 Procedure IdentifyStaticIndirectImplications() may be generalized to identify

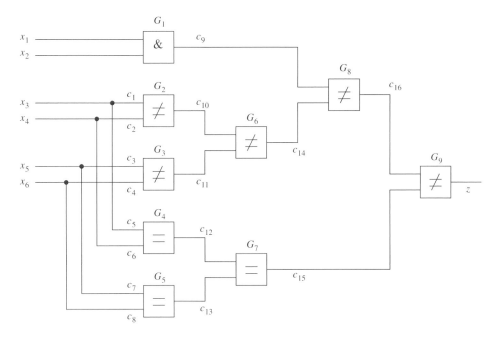

Figure 4.42   An example circuit (Goel, 1981) (© 1981 IEEE)

indirect implications by performing direct implications following the assignment
of each possible combination of values to each pair of lines, c_i and c_j.

(a) What are the types of implications that may be deduced by this procedure?

(b) What modifications need to be made to Procedure IdentifyStaticIndirectIm-
plications( ) to identify static indirect implications in this manner?

(c) Apply your procedure to the fault-free subcircuit shown in Figure 4.21. Does
your procedure identify implication $v(c_{10}) = 0 \Rightarrow v(c_4) = 0$?

4.21 If static indirect implications are to be identified on a composite circuit with a
given target fault, then one of the generalizations that will be required is the
use of a composite value system. In such a generalized algorithm that uses
the 16-valued system, direct implication may identify the implication $V(c_i) =
\{0\} \Rightarrow V(c_j) = \{1\}$. What indirect implication may be deduced from this direct
implication?

4.22 Consider the circuit shown in Figure 4.42 with an SA0 fault at line c_9. Assume
that initialization is performed using values from the six-valued system. Subsequently, assignments $V(c_9) = D_6$ and $V(c_{14}) = 0_6$ are made and the direct
implication procedure executed. Execute Procedure RecursiveIndirectImplica-
tion( ) recursively on this circuit. Limit the depth of recursion to two, report the
results at intermediate stages, and the final indirect implications identified.

4.23 Perform initialization on the circuit in Figure 4.43 with an SA0 fault at x_5 using
the (a) five-valued system, (b) six-valued system, and (c) 16-valued system.

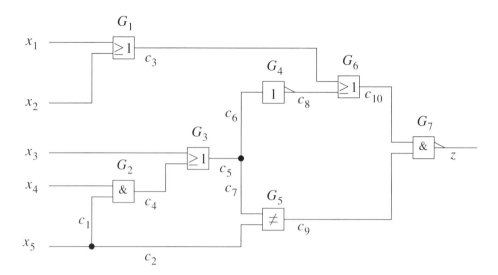

Figure 4.43 An example circuit C_{16}. (This circuit is a modified version of a circuit in Roth (1966))

Comment on the differences in the values assigned to lines.

4.24 Using Variation 1 described in Section 4.5.3.5, compute the testability measures for each line in the circuit in Figure 4.43.

4.25 Using Variation 2 described in Section 4.5.3.5, compute the testability measures for each line in the circuit in Figure 4.9. Analyze the values of $CC^0(c)$ and $CC^1(c)$ at each line c and describe what parameter is captured by these values.

4.26 Generate a test for fault x_5 SA0 in the circuit shown in Figure 4.43 using (a) the D-algorithm, (b) the six-valued system, and (c) the testability measures computed in Problem 4.24 and the rules presented in Section 4.6.2.2.

4.27 Repeat Problem 4.26 using PODEM instead of the D-algorithm.

4.28 Repeat Problem 4.26 using more powerful techniques to identify TGSTs, including future unique D-drive.

4.29 Use the six-valued system and the D-algorithm to generate a test for fault c_9 SA0 in the circuit shown in Figure 4.42.

4.30 Repeat Problem 4.29 using PODEM. Compare the time complexities (in terms of number of backtracks) of the two algorithms for this case.

4.31 Use the 16-valued system and the D-algorithm to generate a test for fault c_9 SA0 in the circuit shown in Figure 4.42.

4.32 In the above problem, what is the value obtained at the circuit's primary output after implication is performed following fault effect excitation? Are all lines justified at this stage? Is the partial vector at this stage a test for the target fault? If not, can it be modified in any simple manner to obtain a test?

4.33 Write a test generation procedure that generates test vectors for single SAFs in fanout-free circuits.

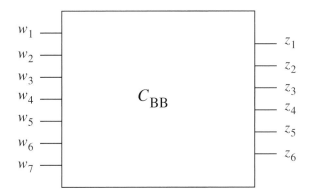

Figure 4.44 A combinational black-box

4.34 A complete set of test vectors is given for the seven-input six-output combinational circuit block, C_{BB}, shown in Figure 4.44. However, the internal structure of this circuit is not known. This circuit is combined with the circuit shown in Figure 4.43 to obtain a new circuit. In this new combinational circuit, output z of the circuit shown in Figure 4.43 drives input w_1 of C_{BB}. How can a complete test set be obtained for the new combinational circuit?

4.35 Two combinational blocks, C_1 with n_1 inputs and one output and C_2 with n_2 inputs and an arbitrary number of outputs, are given. A combinational circuit C with $n_1 + n_2 - 1$ inputs is obtained by connecting the output of C_1 to one of the inputs of C_2. Given that the upper bound on the search required to generate a vector for an SAF in a general n-input combinational circuit is 2^n, formulate a test generation strategy for circuit C for which the upper bound on the search required is $2^{n_1} + 2^{n_2}$.

4.36 Write the CNF for the fault-free version of circuit in Figure 4.1(a). Write the CNF for the version of the circuit with a SA1 fault at line c_9 shown in Figure 4.1(b). Generate a test vector for the above fault.

4.37 Generate a test vector for the circuit and fault shown in Figure 4.1(b) using the BDD approach.

4.38 Consider the circuit shown in Figure 4.10 and a fault list comprised of each possible single SAF.

(a) Perform fault simulation for a set of vectors containing the following four randomly generated vectors: $P_1 = (1, 1, 0, 0)$, $P_2 = (0, 1, 1, 0)$, $P_3 = (0, 0, 1, 1)$, and $P_4 = (1, 1, 0, 1)$. Drop from the fault list all faults detected by the above set of vectors.

(b) Generate a test vector for a fault that still remains undetected in the circuit. If the vector is incompletely specified, randomly assign values to its unspecified bits. Perform fault simulation and drop from the fault list all faults detected by this vector. Add the vector to the set of vectors. Repeat test generation and fault simulation until all the target faults are detected.

(c) Starting with the original fault list, simulate each vector in the above test set in reverse order to that in which the vector was simulated/generated. Can any of the vectors in the set be deleted?

4.39 Consider the circuit, target fault list, and the final set of vectors obtained in Example 4.16. Use the double detection method to obtain a minimal subset of the above set of vectors whose constituent vectors detect all the above target faults.

4.40 Consider the circuit in Figure 4.1(a) and a target fault list obtained by performing fault collapsing starting with the set containing all single SAFs. Generate test vectors using the technique described in Section 4.9.

Appendix 4.A Implication procedure

Procedure 4.4 [Implication(S)]

Implication(S)

{

/* $S = \{S_1, S_2, \ldots\}$ is the implication list */
/* S_i are the implication tasks to be performed */
/* $S_i = (l_i, d_i)$, where l_i is a line at which implication must be performed in direction d_i */
/* d_i is either FORWARD or BACKWARD */

while $S \neq \{\}$
{
  select an $S_i \in S$, where $S_i = (l_i, d_i)$
  $S = S - S_i$
  case(d_i)
    FORWARD:
    {
      if ForwardImplicationOperation(l_i) $==$ CONFLICT
        return(CONFLICT)
      if l_i is an input to a gate G_j with output l_j
        then
        {
          if $l_j \in \mathcal{U}$ and G_j has multiple inputs
            $S = S \cup \{(l_j, \text{BACKWARD})\}$
          if changed(l_j) and l_j is not a PO / the value at l_j has changed */*
            $S = S \cup \{(l_j, \text{FORWARD})\}$
        }
      else / i.e., if l_i is the stem of a fanout system with branches $l_{j_0}, l_{j_1}, \ldots$ */*

>   *for each branch l_{j_k} of the fanout system*
>   *if changed(l_{j_k})* **and** *l_{j_k} is* **not** *a PO*
> $S = S \cup \{l_{j_k},$ FORWARD)
>   *break*
>   }
> BACKWARD:
> {
>   *if BackwardImplicationOperation(l_i) ==* CONFLICT
>   *return(*CONFLICT*)*
>   *if l_i is an output of a gate G_i with inputs $l_{j_0}, l_{j_1}, \ldots$*
>   *then*
>   *for each input l_{j_k} of gate G_i*
>   *if changed(l_{j_k})* **and** *l_{j_k} is* **not** *a PI*
> $S = S \cup \{l_{j_k},$ BACKWARD)
>   *else /* i.e., if l_i is a branch of a fanout system with stem l_j */*
>   {
> $S = S \cup \{(l_j,$ FORWARD)}
>   *if changed(l_j)* **and** *l_j is* **not** *a PI /* the value at l_j has changed */*
> $S = S \cup \{(l_j,$ BACKWARD)}
>   }
>   *break*
>   }
> }
> *return(*SUCCESS*)*
> }

References

Abramovici, M., Kulikowski, J. J., Menon, P. R., and Miller, D. T. (1986a). SMART and FAST: test generation for VLSI scan-design circuits. *IEEE Design and Test of Computers*, pp. 43–54.

Abramovici, M., Menon, P. R., and Miller, D. T. (1986b). Checkpoint faults are not sufficient target faults for test generation. *IEEE Trans. on Computers*, **C-35** (8), pp. 769–771.

Abramovici, M., Miller, D., and Henning, R. (1990). Global cost functions for test generation. In *Proc. Int. Test Conference*, pp. 35–43.

Akers, S. B., Joseph, C., and Krishnamurthy, B. (1987). On the role of independent fault sets in the generation of minimal test sets. In *Proc. Int. Test Conference*, pp. 1100–1107.

Brglez, F., Pownall, P., and Hum, R. (1984). Applications of testability analysis: from ATPG to critical delay path tracing. In *Proc. Int. Test Conference*, pp. 705–712.

Bryant, R. E. (1986). Graph-based algorithms for Boolean function manipulation. *IEEE Trans. on Computers*, **C-35** (8), pp. 677–691.

Cha, C., Donath, W., and Ozguner, F. (1978). 9-V algorithm for test pattern generation of combinational digital circuits. *IEEE Trans. on Computers*, **C-27** (3), pp. 193–200.

Chakradhar, S. T., Agrawal, V. D., and Rothweiler, S. G. (1993). A transitive closure algorithm for test generation. *IEEE Trans. on Computer-Aided Design*, **11** (7), pp. 1015–1028.

Chandra, S. J. and Patel, J. H. (1989). Experimental evaluation of testability measures for test generation. *IEEE Trans. on Computer-Aided Design*, **8** (1), pp. 93–97.

Chen, C.-A. (1996). *Test generation and embedding for built-in self-test*. Ph.D. thesis, University of Southern California, Los Angeles, CA.

Chen, C.-A. and Gupta, S. K. (1996). A satisfiability-based test generator for path delay faults in combinational circuits. In *Proc. Design Automation Conference*, pp. 209–214.

Cormen, T. H., Leiserson, C. E., and Rivest, R. L. (1992). *Introduction to Algorithms*. The MIT Press, New York, NY.

Corno, F., Prinetto, P., Rebaudengo, M., and Reorda, M. S. (1996). GATTO: a genetic algorithm for automatic test pattern generation for large synchronous sequential circuits. *IEEE Trans. on Computer-Aided Design*, **15** (8), pp. 991–1000.

Cox, H. and Rajski, J. (1994). On necessary and nonconflicting assignments in algorithmic test pattern generation. *IEEE Trans. on Computer-Aided Design*, **13** (4), pp. 515–230.

Entrena, L. and Cheng, K.-T. (1993). Sequential logic optimization by redundancy addition and removal. In *Proc. Int. Conference on Computer-Aided Design*, pp. 301–315.

Fujiwara, H. and Shimono, T. (1983). On the acceleration of test generation algorithms. *IEEE Trans. on Computers*, **C-32** (12), pp. 1137–1144.

Gaede, R. K., Mercer, M. R., Butler, K. M., and Ross, D. E. (1988). CATAPULT: concurrent automatic testing allowing parallelization and using limited topology. In *Proc. Design Automation Conference*, pp. 597–600.

Giraldi, J. and Bushnell, M. L. (1990). EST: the new frontier in automatic test-pattern generation. In *Proc. Design Automation Conference*, pp. 667–672.

Goel, P. (1981). An implicit enumeration algorithm to generate tests for combinational logic circuits. *IEEE Trans. on Computers*, **C-30** (3), pp. 215–222.

Goel, P. and Rosales, B. C. (1979). Test generation and dynamic compaction of tests. In *Proc. Int. Test Conference*, pp. 189–192.

Goel, P. and Rosales, B. C. (1981). PODEM-X: an automatic test generation system for VLSI logic structures. In *Proc. Design Automation Conference*, pp. 260–268.

Goldstein, L. H. (1978). *Controllability/Observability Analysis of Digital Circuits*. Technical Report 78-1895, Sandia Laboratories.

Goldstein, L. H. and Thigpen, E. L. (1980). SCOAP: Sandia controllability/observability analysis program. In *Proc. Design Automation Conference*, pp. 190–196.

Grason, J. (1979). TMEAS, a testability measurement program. In *Proc. Design Automation Conference*, pp. 156–161.

Ivanov, A. and Agarwal, V. K. (1986). Testability measures – what do they do for ATPG? In *Proc. Int. Test Conference*, pp. 129–138.

Iyer, M. A. and Abramovici, M. (1996). FIRE: a fault-independent combinational redundancy identification algorithm. *IEEE Trans. on VLSI Systems*, **4** (2), pp. 295–301.

Jain, S. K. and Agrawal, V. D. (1985). Statistical fault analysis. *IEEE Design & Test of Computers*, **2** (1) pp. 38–44.

Kajihara, S., Pomeranz, I., Kinoshita, K., and Reddy, S. M. (1993). Cost-effective generation of minimal test sets for stuck-at faults in combinational logic circuits. In *Proc. Design Automation Conference*, pp. 102–106.

Kajihara, S., Pomeranz, I., Kinoshita, K., and Reddy, S. M. (1994). On compacting test sets by addition and removal of test vectors. In *Proc. VLSI Test Symposium*, pp. 202–207.

Kirkland, T. and Mercer, M. R. (1987). A topological search algorithm for ATPG. In *Proc. Design Automation Conference*, pp. 502–508.

Kunz, W. and Pradhan, D. K. (1993). Accelerated dynamic learning for test pattern generation in combinational circuits. *IEEE Trans. on Computer-Aided Design*, **12** (5), pp. 684–694.

Kunz, W. and Pradhan, D. K. (1994). Recursive learning: a new implication technique for efficient solutions to CAD problems. *IEEE Trans. on Computer-Aided Design*, **13** (9), pp. 1143–1157.

Larrabee, T. (1989). Efficient generation of test patterns using Boolean difference. In *Proc. Int. Test Conference*, pp. 795–801.

Larrabee, T. (1992). Test pattern generation using Boolean satisfiability. *IEEE Trans. on Computer-Aided Design*, **11** (1), pp. 4–15.

Lee, H. K. and Ha, D. S. (1991). An efficient, forward fault simulation algorithm based on the parallel pattern single fault propagation. In *Proc. Int. Test Conference*, pp. 946–955.

McDonald, J. F. and Benmehrez, C. (1983). Test set reduction using the subscripted *D*-algorithm. In *Proc. Int. Test Conference*, pp. 115–121.

Patel, S. and Patel, J. (1986). Effectiveness of heuristics measures for automatic test pattern generation. In *Proc. Design Automation Conference*, pp. 547–552.

Pomeranz, I., Reddy, L. N., and Reddy, S. M. (1991). COMPACTEST: a method to generate compact test sets for combinational circuits. In *Proc. Int. Test Conference*, pp. 194–203.

Roth, J. P. (1966). Diagnosis of automata failures: a calculus and a method. *IBM Journal of Research and Development*, **10** (4), pp. 278–291.

Rudnick, E. M., Patel, J. H., Greenstein, G. S., and Niermann, T. M. (1997). A genetic algorithm framework for test generation. *IEEE Trans. on Computer-Aided Design*, **16** (9), pp. 1034–1044.

Savir, J. (1983). Good controllability and observability do not guarantee good testability. *IEEE Trans. on Computers*, **C-32** (12), pp. 1198–1200.

Schulz, M. H. and Auth, E. (1989). Improved deterministic test pattern generation with applications to redundancy identification. *IEEE Trans. on Computer-Aided Design*, **8** (7), pp. 811–816.

Schulz, M. H., Trischler, E., and Sarfert, T. M. (1988). SOCRATES: a highly efficient automatic test pattern generation system. *IEEE Trans. on Computer-Aided Design*, **7** (1), pp. 126–137.

Sellers, F. F., Hsiao, M. Y., and Bearnson, L. W. (1968). Analyzing errors with the Boolean difference. *IEEE Trans. on Computers*, **C-17** (7), pp. 676–683.

Srinivasan, S., Swaminathan, G., Aylor, J. H., and Mercer, M. R. (1993). Combinational circuit ATPG using binary decision diagrams. In *Proc. VLSI Test Symposium*, pp. 251–258.

Stanion, R. T., Bhattacharya, D., and Sechen, C. (1995). An efficient method for generating exhaustive test sets. *IEEE Trans. on Computer-Aided Design*, **14** (12), pp. 1516–1525.

Stephan, P. R., Brayton, R. K., and Sangiovanni-Vincentelli, A. L. (1992). *Combinational Test Generation using Satisfiability*. Technical Report UCB/ERL M92/112, Univ. of California, Berkeley.

Tarjan, R. (1974). Finding dominators in directed graphs. *SIAM Journal of Computing*, **3**, pp. 62–89.

Trischler, E. and Schulz, M. (1985). Applications of testability analysis to ATG: methods and experimental results. In *Proc. Int. Symposium on Circuits and Systems*, pp. 691–694.

Tromp, G.-J. (1991). Minimal test sets for combinational circuits. In *Proc. Int. Test Conference*, pp. 204–209.

Wang, S. and Gupta, S. K. (1998). ATPG for heat dissipation minimization during test application. *IEEE Trans. on Computers*, **C-47** (2), pp. 256–262.

Wunderlich, H.-J. (1985). PROTEST: a tool for probabilistic testability analysis. In *Proc. Design Automation Conference*, pp. 204–211.

Zorian, Y. (1993). A distributed BIST control scheme for complex VLSI devices. In *Proc. VLSI Test Symposium*, pp. 4–9.

5   Sequential ATPG

In this chapter, we first discuss the challenges we face in test generation and fault simulation of sequential circuits. We then discuss classification of the fault simulation methods, test generation methods, and different types of faults.

Next, we discuss how the fault list can be collapsed in a sequential circuit. We show that the concept of fault dominance is only selectively applicable to such circuits.

Under fault simulation, we discuss a method which combines the best of different conventional fault simulation methods and tries to avoid their pitfalls.

Under test generation, we discuss three types of methods: those which derive tests from the state table, those which assume full reset capability, and those which do not assume any reset capability. We show how test generation techniques obtained for synchronous sequential circuits can be extended to asynchronous sequential circuits. Then we discuss methods for compacting the test set.

5.1   Classification of sequential ATPG methods and faults

Sequential automatic test pattern generation (ATPG) is a difficult problem. The many challenges we face in this area include reduction in the time and memory required to generate the tests, reduction in the number of cycles needed to apply the tests to the circuit, and obtaining a high fault coverage. Adding to the complexity of this problem is that, unlike a combinational circuit where an untestable fault is also redundant, an untestable fault is not necessarily redundant in a sequential circuit. In spite of the excellent progress made in the last decade in this area, it is usually not feasible to satisfactorily test an entire chip using sequential ATPG. It needs to be aided in this process through testability insertion. Testability insertion is covered in Chapters 11 and 13.

In this section, we discuss the different categories in which the fault simulation and test generation methods for sequential circuits can be classified. We then show how faults in sequential circuits can be classified.

Table 5.1. *The fault simulation task (Niermann et al. 1992) (© 1992 IEEE)*

		Test vectors		
Circuit	P_1	$\cdots$ P_j	$\cdots$	P_s
Good	G_1	$\cdots$ G_j	$\cdots$	G_s
Faulty$_1$	$F_{1,1}$	$\cdots$ $F_{1,j}$	$\cdots$	$F_{1,s}$
Faulty$_2$	$F_{2,1}$	$\cdots$ $F_{2,j}$	$\cdots$	$F_{2,s}$
.	.	$\cdots$.	$\cdots$	.
Faulty$_i$	$F_{i,1}$	$\cdots$ $F_{i,j}$	$\cdots$	$F_{i,s}$
Faulty$_{i+1}$	$F_{i+1,1}$	$\cdots$ $F_{i+1,j}$	$\cdots$	$F_{i+1,s}$
.	.	$\cdots$.	$\cdots$	.
Faulty$_r$	$F_{r,1}$	$\cdots$ $F_{r,j}$	$\cdots$	$F_{r,s}$

5.1.1   Classification of fault simulation techniques

In fault simulation, each test vector from a test sequence is applied to the good (fault-free) sequential circuit as well as to every faulty sequential circuit. If the output response of any faulty circuit differs from the response of the good circuit, the corresponding fault is said to be detected.

Table 5.1 shows the tasks that need to be performed in fault simulation of sequential circuits (Niermann *et al.*, 1992). Each column corresponds to a test vector and each row corresponds to a circuit. There are s test vectors and $r+1$ circuits, one good circuit and r faulty circuits. In the case of the single stuck-at fault (SAF) model, the faulty circuits would correspond to different single SAFs. Thus, there are $s(r+1)$ circuit states represented in the table. The task of fault simulation is to find all primary output vectors for the $s(r+1)$ states and determine which faulty circuits have output vectors different from the good circuit. Fault simulation techniques can be classified based on the order in which they fill the table.

In **concurrent and deductive** fault simulation, the table is filled from left to right. The values of all faulty circuit states are concurrently computed for a vector using previously computed information. For example, to compute $F_{i,j}$, they use values from $F_{i,j-1}$, which are different from G_{j-1}, and the current good circuit state G_j. This is shown in Figure 5.1(a). This approach results in a small number of simulation events. However, since a large list of active faulty circuit values is associated with each line in the circuit, the memory requirement is large. In addition, in deductive fault simulation, set intersection and set union operations have to be performed on fault lists. These are expensive operations.

In **differential** fault simulation, $F_{i,j}$ is exclusively determined from $F_{i-1,j}$, as shown in Figure 5.1(b). To determine if a fault is detected, a count is kept of the number of times a primary output has changed. Differential fault simulation requires very little memory because it stores only one copy of all the line values of the circuit and the differences between adjacent circuits. However, it cannot drop detected faults

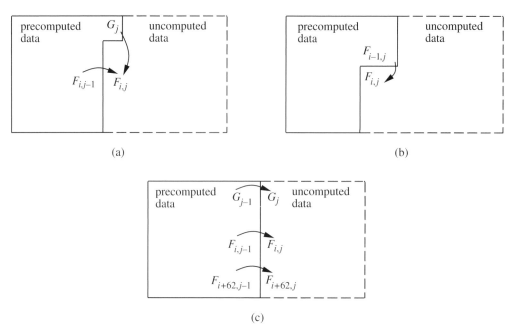

Figure 5.1   Table filling: (a) concurrent and deductive, (b) differential, (c) parallel (Niermann et al. 1992) (© 1992 IEEE)

easily because subsequent faulty circuits rely on differences from previous faulty circuits.

In **parallel** fault simulation, $F_{i,j}$ is determined exclusively from $F_{i,j-1}$. It takes advantage of the word-level parallelism of the computer to simulate one good circuit and 63 faulty circuits (assuming a 64-bit machine) in one pass, as shown in Figure 5.1(c). This speeds up fault simulation. However, it suffers from the repetition of the good circuit simulation in every pass and an inability to drop detected faults. After a fault is detected, one bit space is wasted for the remainder of that simulation pass.

In a fault simulation technique called **single-fault propagation**, each circuit state is generated from the good circuit state in the same column. For example, $F_{1,j}$ is generated from G_j by simulating the fault site of $Fault y_1$ as the initial fault source. In this technique, restoring the state of the good circuit before every faulty circuit simulation results in a performance penalty. In **parallel-pattern single-fault propagation**, the computer word parallelism is taken advantage of for parallel simulation of several faulty circuit states in the same row of the table. In **parallel-pattern parallel-fault propagation**, the computer word parallelism is used for parallel simulation of several faulty circuit states in multiple rows of the table.

5.1.2   Classification of test generation techniques

The taxonomy of various sequential test generation approaches is given in Figure 5.2 (Cheng, 1996). Most of these approaches are limited to test generation for synchronous

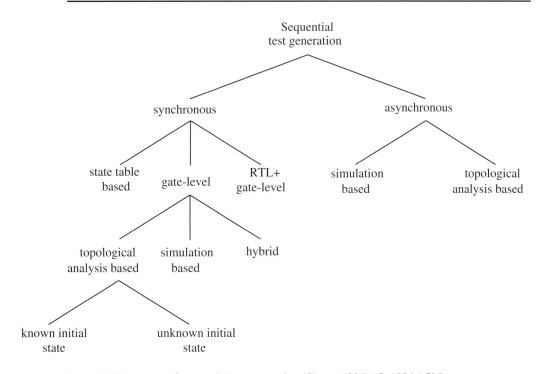

Figure 5.2 Taxonomy of sequential test generation (Cheng, 1996) (© 1996 ACM)

sequential circuits which may have some asynchronous reset/clear circuitry. However, some advances have also been made in test generation for asynchronous sequential circuits. Under the synchronous category, the three main approaches are based on the state table, gate-level circuits, and both register-transfer level (RTL) and gate-level circuits. The state table based approach is only useful for pure controllers for which a state table is either available or easily extractable from its lower-level description. The gate-level approach can itself be divided up into *topological analysis based, simulation based*, and *hybrid*. The **topological analysis** based approach uses an *iterative array model* of the sequential circuit. Most methods in this category assume that the initial state of the sequential circuit is unknown, while a few others assume that the initial state is known to avoid the problem of initializing the memory elements. The topological analysis based approach assuming an unknown initial state is by far the most common approach among all sequential test generation approaches. The simulation based approach uses an enhanced logic or fault simulator. The hybrid approach combines the topological analysis based and simulation based approaches. When the RTL description of the circuit is available in addition to its gate-level one, test generation can be significantly sped up since it becomes easier to justify a state and propagate the error to an observable output.

For asynchronous circuits, the two main approaches are simulation-based and topological analysis based. In the first approach, a potential test sequence is first derived by ignoring the circuit delays, and then simulated using the appropriate delay models to make sure it does not get invalidated by races, hazards and oscillations. If invalidation does occur, another potential test sequence is derived and the process repeated. In the second approach, synchronous topological analysis based test generation is adapted to asynchronous circuits by only allowing the use of stable states.

5.1.3   Classification of faults

In the case of a combinational circuit, an undetectable fault is also redundant, and can be removed. However, this is not true in the case of a sequential circuit. In this section, we analyze the relationships among detectable faults, undetectable faults, redundant faults, test strategy, and operation mode of the circuit under test (Pomeranz and Reddy, 1993).

Sequential circuits generally have two modes of operation: *synchronization mode* and *free mode*. In the **synchronization mode**, the operation always starts with a specified input sequence, denoted by R. When power-up reset exists, $|R| = 0$ ($|R|$ is the cardinality of R), and the circuit operation begins from a reset state. If hardware reset is available as a special input, which is always employed to reset the circuit at the beginning of operation, then $|R| = 1$. For a general *synchronizing sequence*, $|R| \geq 1$. Such a sequence can synchronize the circuit to a single state or to a state within a subset of states. Under the **free mode** of operation, no synchronization is done, and the sequential circuit starts operating from whatever state it happens to be in at the time.

We can have two test strategies, corresponding to the two operation modes defined above: *restricted* and *unrestricted*. Under the **restricted test strategy**, all test sequences start with sequence R. Under the **unrestricted test strategy**, any sequence can be generated as a test sequence.

Both the above test strategies can be used under one of two test generation approaches: *single observation time* and *multiple observation time*. Under the **single observation time** (SOT) approach, a fault f is said to be detectable if there exists an input sequence I such that for every pair of initial states S and S_f of the fault-free and faulty circuits, respectively, the response $z(I, S)$ of the fault-free circuit to I is different from the response $z_f(I, S_f)$ of the faulty circuit at a specific time unit j. To conform to existing test generation methods, this definition is further restricted to require that the fault-free response be different from the faulty response at time unit j on a specific primary output. Under the **multiple observation time** (MOT) approach, a fault f is said to be detectable if there exists an input sequence I such that for every pair of initial states S and S_f of the fault-free and faulty circuits, respectively, $z(I, S)$ is different from $z_f(I, S_f)$ at *some* time unit. A fault is said to be undetectable if it is not detectable under the specified test approach.

Every fault that is detectable under the SOT approach is also detectable under the MOT approach. In fact, a fault may be undetectable under the SOT approach, but

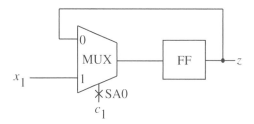

Figure 5.3 A faulty circuit illustrating the difference between the SOT and MOT approaches

may be detectable under the MOT approach, as the following example shows (Cheng, 1996).

Example 5.1 Consider the circuit in Figure 5.3. The stuck-at 0 (SA0) fault on select line c_1 of the multiplexer prevents the flip-flop from getting initialized. Since no input sequence can initialize the flip-flop in the faulty circuit to a known binary value, no input sequence can produce a D or $\overline{D}$ at output z. Thus, this fault is undetectable under the SOT approach. However, let us next consider the following test sequence on (x_1, c_1): $\{(1, 1), (0, 1)\}$. The fault-free response at z is $\{1, 0\}$ (all flip-flops in this chapter are assumed to be D flip-flops). If the initial state pair is $x/1$ (i.e., the initial state of the fault-free circuit is unknown x and that of the faulty circuit is 1), then the response of the faulty circuit is $\{1, 1\}$ and the fault is detected on the second vector. Similarly, if the initial state pair is $x/0$, the response of the faulty circuit is $\{0, 0\}$, and the fault is detected on the first vector. Since the union of $x/1$ and $x/0$ covers all possible combinations of initial state pairs, this test sequence detects the fault under the MOT approach. Since it cannot be predicted beforehand which vector will produce an error at z, z must be observed for both the vectors.   □

Another advantage of the MOT approach over the SOT approach is that the length of test sequences derived under MOT may be smaller than those derived under SOT. The reason is that in the SOT approach, a synchronization sequence is needed which could be long, whereas in the MOT approach, full synchronization may not be needed. The disadvantage of the MOT approach is that test generation is more complex. Also, although in Example 5.1, for the given test sequence, there was only one possible fault-free response, in general, there may be different fault-free responses from different initial states. Thus, under the MOT approach, the tester would need to store several fault-free responses for the observation time units under question and compare the response from the actual circuit with all these responses. Existing testers do not provide this feature yet.

We consider redundant faults next. A fault is redundant under a given operation mode (synchronization or free) and test strategy (restricted or unrestricted) if, under all possible input sequences applicable under that operation mode and test strategy,

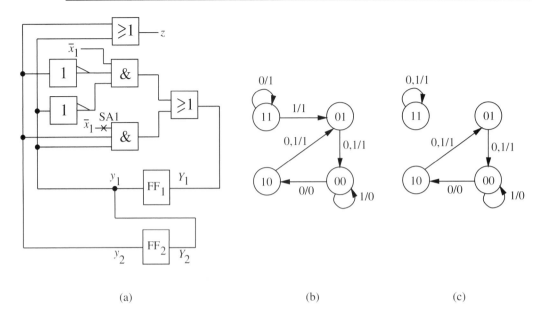

Figure 5.4 Illustration of a partially detectable fault

the fault does not affect the expected output responses. The concept of *partial test* is important to understand the difference between an undetectable and a redundant fault. A **partial test** under the free operation mode and unrestricted test strategy is an input sequence I such that for at least one initial state S_f of the faulty circuit, the output response, $z_f(I, S_f)$, is different from the fault-free response, $z(I, S)$, for every fault-free initial state S. For the synchronization mode and restricted test strategy, the only difference in this definition is that each such input sequence must start with the synchronization sequence R. A fault cannot be guaranteed to be detected by a partial test, but is detected if the circuit starts in a state for which the partial test constitutes a test. A fault is said to be **partially detectable** if at least one partial test exists for it. A fault is redundant if and only if it is not partially detectable, since such a fault is not manifested at a primary output under any input sequence.

Example 5.2 Consider the sequential circuit shown in Figure 5.4(a), where y_1, y_2 denote present state variables and Y_1, Y_2 denote next state variables. The target fault is the SA1 fault shown in the figure. The state diagrams of the fault-free and faulty circuits are shown in Figures 5.4(b) and 5.4(c), respectively. The input sequence $x_1 = \{0, 1, 1\}$ will initialize the fault-free circuit to state $(y_1, y_2) = (0, 0)$. However, the faulty circuit cannot be initialized since the state $(y_1, y_2) = (1, 1)$ is disconnected from the other states in the state diagram. The fault-free response to the input sequence $x_1 = \{0, 1, 1, 0\}$ is $z = \{x, x, x, 0\}$, where x denotes an unknown value. If the initial state of the faulty circuit is $(y_1, y_2) = (1, 1)$, then the faulty response for this input

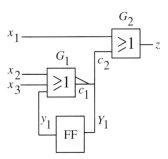

Figure 5.5 A circuit illustrating the self-hiding effect

Table 5.2. *Test sequence for the circuit in Figure 5.5*

Test			fault-free					x_2 SA0					c_1 SA1				
x_1	x_2	x_3	y_1	c_1	c_2	Y_1	z	y_1	c_1	c_2	Y_1	z	y_1	c_1	c_2	Y_1	z
1	1	1	x	0	0	0	1	x	0	0	0	1	x	1	1	1	1
1	1	0	0	0	0	0	1	0	1	1	1	1	1	1	1	1	1
0	0	0	0	1	1	1	1	1	0	0	0	0	1	1	1	1	1

sequence is $z_f = \{1, 1, 1, 1\}$, and the fault is detected. However, if the faulty circuit is in any of the other three states, then the faulty response is $z_f = \{x, x, x, 0\}$, and the fault is not detected. Therefore, this SA1 fault is detectable if the power-up initial state of the faulty circuit is $(1, 1)$, and undetectable otherwise. In other words, it is a partially detectable fault. □

5.2   Fault collapsing

We saw earlier in Chapter 3 (Section 3.4.1) how fault collapsing can be employed to reduce the burden of test generation and fault simulation in combinational circuits. The two methods that were used for this purpose were *fault equivalence* and *fault dominance*. Fault propagation through sequential circuits, however, is more complicated because of the presence of memory elements and feedback paths. Because of this complication, the concept of fault dominance may not always hold in a sequential circuit, as explained through an example next.

Consider the sequential circuit given in Figure 5.5. Fault c_1 SA1 dominates fault x_2 SA0 if only the combinational portion of this circuit is considered. However, if the whole sequential circuit is considered, the following test sequence on (x_1, x_2, x_3): $\{(1, 1, 1), (1, 1, 0), (0, 0, 0)\}$ detects x_2 SA0, but does not detect c_1 SA1, as shown in Table 5.2. This happens because of a phenomenon called *self-hiding* which is explained ahead (Chen *et al.*, 1991).

From here on, the concepts of fault equivalence and dominance in a combinational circuit are referred to as *c-equivalence* and *c-dominance*, respectively. The corresponding extensions of these concepts to a sequential circuit are referred to as *s-equivalence* and *s-dominance*. These extended concepts are defined as follows. Two faults α and β are said to be **s-equivalent** in a sequential circuit if and only if the function under fault α is equal to the function under fault β for every input sequence applied to the circuit. A fault β is said to **s-dominate** another fault α in a sequential circuit if and only if every test sequence for α is also a test sequence for β.

There are two phenomena: *self-hiding* and *delayed reconvergence*, which prevent a c-dominant fault from also being s-dominant in some cases. These are discussed next.

5.2.1 Self-hiding

When a faulty signal propagates through a sequential circuit, it may pass through a feedback path and propagate back to the node it originated from. Consider the sequential circuit in Figure 5.5 and its test sequence given in Table 5.2 again. The first vector $(1, 1, 1)$ initializes the D flip-flop. The second vector $(1, 1, 0)$ activates both the faults x_2 SA0 and c_1 SA1. After the third vector $(0, 0, 0)$ is applied, the faulty signal at Y_1 propagates to y_1. Under this vector, in the presence of the c_1 SA1 fault, further propagation of the faulty signal is masked by the fault itself. This phenomenon is called **self-hiding**. However, in the presence of the x_2 SA0 fault, the faulty signal propagates to circuit output z. This example shows that even though c_1 SA1 c-dominates x_2 SA0, c_1 SA1 does not s-dominate x_2 SA0.

Define the inversion parity of a path in a circuit to be the number of inversions along the path modulo 2. A closed path in a sequential circuit is said to be an **O-path** with respect to an input of a gate if and only if the path originates from and terminates at the input of the gate and has an odd inversion parity. In Figure 5.5, gate G_1 has an O-path, (y_1, c_1, Y_1, y_1), with respect to its input y_1. It should be noted that an O-path must pass through at least one memory element and a gate may have more than one O-path with respect to one of its inputs. A gate in a sequential circuit is called **non-self-hiding** if and only if it has O-paths with respect to at most one of its inputs. For example, since gate G_1 has only one O-path given above, it is non-self-hiding. If there is no O-path in a non-self-hiding gate, then all its gate inputs are said to be **s-dominatable**. If it has one O-path, then the gate input from which the O-path originates is said to be s-dominatable. The term s-dominatable implies that the fault at this gate input may be s-dominated by the corresponding c-dominant fault at the gate output. This concept will be used later for dominance fault collapsing after we discuss the concept of delayed reconvergence.

Table 5.3. *Test sequence for the circuit in Figure 5.6*

Test			fault-free						x_1 SA0						c_1 SA1					
x_1	x_2	x_3	y_1	c_1	c_2	c_3	Y_1	z	y_1	c_1	c_2	c_3	Y_1	z	y_1	c_1	c_2	c_3	Y_1	z
1	1	0	x	0	0	0	1	x	x	0	0	0	1	x	x	1	1	1	0	1
1	0	0	1	0	0	0	1	1	1	1	1	1	0	1	0	1	1	1	0	1
1	1	1	1	0	0	0	0	1	0	0	0	0	0	0	0	1	1	1	0	1

5.2.2 Delayed reconvergence

Another phenomenon that invalidates the fault dominance relationship in sequential circuits is called *delayed reconvergence*. This is a result of the fact that faulty signals may be stored in memory elements and then reconverge with themselves. As an example, consider the faults x_1 SA0 and c_1 SA1 in the sequential circuit given in Figure 5.6. Even though c_1 SA1 c-dominates x_1 SA0, it can be seen that the test sequence given in Table 5.3 detects x_1 SA0, but not c_1 SA1. In other words, c_1 SA1 does not s-dominate x_1 SA0. In this test sequence, $(1, 1, 0)$ initializes the D flip-flop. The next vector $(1, 0, 0)$ activates both these faults. The faulty signals are propagated to line Y_1. With the application of the last vector $(1, 1, 1)$, the faulty signal passes through the flip-flop and propagates to line y_1. This vector also activates c_1 SA1 whose effect propagates to line c_2. The two faulty signals at the input of gate G_2 kill each other's effect and a fault-free value is obtained at output z. In this situation, only x_1 SA0 is detected.

The above phenomenon is called **delayed reconvergence**. This phenomenon arises when the faulty signals originating from c-dominant and c-dominated faults of a gate, some passing through memory elements, reconverge at some gates with different inversion parities. For example, gate G_1 has two propagation paths to G_2. The propagation path including lines c_1 and c_2 does not pass through the flip-flop and has even inversion parity, whereas the other propagation path including lines c_1, c_3, Y_1 and y_1, passes through the flip-flop and has odd inversion parity. A gate for which this phenomenon occurs, such as G_1 in this case, is called a delayed-reconvergent gate. On the other hand, a gate is said to be a non-delayed-reconvergent gate if and only if the gate is on paths which (a) do not reconverge at any gate, or (b) reconverge at gates, but every path passing through each of these gates does not pass through any memory element, or (c) pass through different number of memory elements and reconverge at some gates, and the paths which pass through any of these gates have the same inversion parity. Since the above condition is satisfied for gates G_2 and G_3, they are non-delayed-reconvergent gates.

For an irredundant sequential circuit, self-hiding and delayed reconvergence are the only two phenomena that can prevent a c-dominant and c-dominated pair of faults related to a gate from also being s-dominant and s-dominated.

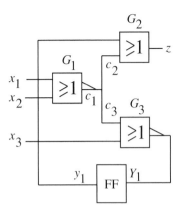

Figure 5.6   A circuit illustrating the delayed reconvergence effect

5.2.3   Collapsing and analysis of faults

A gate in a sequential circuit is said to be **non-SD** if and only if it is both non-self-hiding and non-delayed-reconvergent. For a non-SD gate G in an irredundant sequential circuit, the c-dominant fault at the output of G s-dominates each s-dominatable input of G. This is because the test sequence for the dominated fault also activates the dominant fault. Since the gate is non-SD, the effect of the dominant fault cannot be masked. Thus, dominance fault collapsing can be used for all non-SD gates in the circuit.

The equivalence relationship in a combinational circuit remains valid for a sequential circuit. In other words, c-equivalence implies s-equivalence.

Further fault collapsing is possible through an analysis of fanouts and memory elements. A fanout branch in a sequential circuit is said to be **prime** if and only if all the propagation paths from other fanout branches of its fanout stem to the primary output pass through this fanout branch. For example, in Figure 5.5, for the fanout branches c_2 and Y_1 of fanout stem c_1, since the path from Y_1 to z goes through c_2, c_2 is prime. The faults at a prime fanout branch and at its corresponding fanout stem are s-equivalent. This is because the function of the circuit under an SAF on a prime fanout branch cannot be influenced by the values of other fanout branches.

Exactly analogous to the definition of a non-delayed-reconvergent gate, one can define a non-delayed-reconvergent D flip-flop. Since a D flip-flop has only one input, no self-hiding phenomenon can occur for such a flip-flop. Thus, a non-delayed-reconvergent D flip-flop is also a non-SD flip-flop. In an irredundant sequential circuit, the output SAF of a non-SD D flip-flop s-dominates the corresponding input SAF of the flip-flop. This is again owing to the fact that once such faults are activated, their effects cannot be masked by the corresponding input SAF.

Let us apply the above techniques to the sequential circuit in Figure 5.5. There are eight lines in this circuit. Hence, there are 16 possible SAFs. Gates G_1 and G_2 as well as the D flip-flop are non-SD, and c_2 is a prime fanout branch. Through equivalence fault collapsing, we can remove the following faults: x_1 SA1, x_2 SA1, x_3 SA1, z SA1, c_1 SA0. Through dominance fault collapsing in the two gates, we can remove c_1 SA1 and z SA0. Through prime fanout branch analysis, we can drop c_2 SA0 and c_2 SA1. Finally, because the D flip-flop is non-SD, we can drop y_1 SA0 and y_1 SA1. After dropping these 11 SAFs, only five SAFs remain. Testing those five SAFs will test the whole circuit.

The above techniques are applicable to irredundant sequential circuits. However, many sequential circuits encountered in practice are not irredundant. In such circuits, it is possible that irredundant faults get collapsed to redundant ones for which no test exists. Still, the fault list derived from the above techniques can be useful as an initial fault list.

5.3   Fault simulation

In this section, we will describe a fault simulation algorithm for synchronous sequential circuits which is a hybrid of the concurrent, differential, and parallel fault simulation algorithms (Niermann *et al.*, 1992; Lee and Ha, 1996). It retains the advantage of fault dropping that is possible with the concurrent algorithm, while exploiting the word-level parallelism of the parallel algorithm, and retaining the low memory requirement of the differential algorithm.

We first give some background material.

5.3.1   Background

Most commonly used sequential test generators and fault simulators are based on a combinational iterative array model of the sequential circuit. Such a model is shown in Figure 5.7, where the feedback signals are generated from the copies of the combinational logic of the sequential circuit in the previous **time frames**. A time frame is depicted by a rectangle in the figure. The inputs of a time frame are the primary inputs $(x_1, x_2, \ldots, x_n)$, and the present state lines $(y_1, y_2, \ldots, y_p)$ which are the outputs of the flip-flops. The outputs of a time frame are the primary outputs $(z_1, z_2, \ldots, z_m)$, and the next state lines $(Y_1, Y_2, \ldots, Y_p)$ which are the data inputs of the flip-flops. A clock pulse needs to be applied between each pair of successive time frames in order to update the logic values at the present state lines of a time frame from the next state lines of the previous time frame. A single SAF in a sequential circuit corresponds to a multiple SAF in the above model where each time frame contains that single SAF.

Consider the combinational logic of the sequential circuit. If the fanout stems of the combinational logic are removed, the logic is partitioned into fanout-free regions

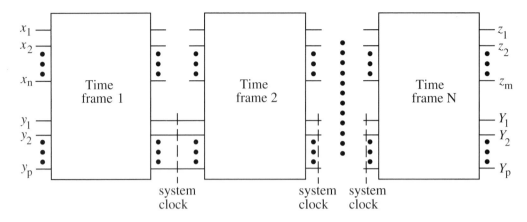

Figure 5.7   An iterative array model

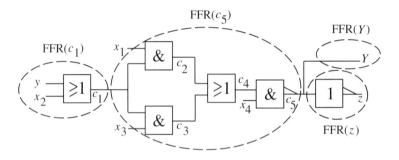

Figure 5.8   A partition of the combinational logic of a sequential circuit

(FFRs). Let FFR(i) denote the FFR whose output is i. The output of an FFR can be a stem, a primary output or a next state line. In the combinational logic, if all the paths from a line r to primary outputs and next state lines go through line q, then line q is said to be the **dominator** of line r. If there is no other dominator between a signal and its dominator, the dominator is said to be an **immediate dominator**. Stem i is a dominator of all lines within FFR(i). A stem may or may not have a dominator.

Example 5.3   Consider the combinational logic of a sequential circuit shown in Figure 5.8. This logic corresponds to one time frame in the iterative array model. The logic can be partitioned into four FFRs as shown. Since all paths from stem c_1 to primary output z and next state line Y pass through lines c_4 and c_5, lines c_4 and c_5 are dominators of line c_1, and line c_4 is the immediate dominator. However, even though line c_5 is a stem, it does not have a dominator.   □

The behavior of a sequential circuit can be simulated by repeating the simulation of its combinational logic in each time frame. If the effect of fault f does not propagate to a next state line in a particular time frame, the fault-free and faulty values of the

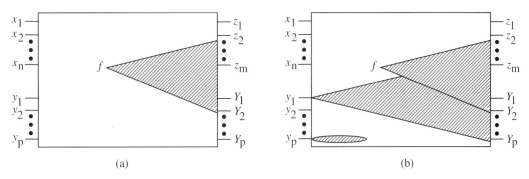

Figure 5.9 Propagation of fault effect: (a) single event fault, and (b) multiple event fault (Lee and Ha, 1996) (© 1996 IEEE)

corresponding present state line are the same in the following time frame. If the fault-free and faulty values of each present state line are identical in a time frame, the fault is said to be a **single event fault** in that time frame. If there exists at least one present state line whose fault-free and faulty values are different, the fault is said to be a **multiple event fault**. The propagation of the fault effect for a single event fault as well as a multiple event fault is illustrated in Figure 5.9. The shaded areas depict the fault effect propagation zone. As shown in Figure 5.9(a), the fault effect only originates at the fault site for a single event fault. However, the fault effect originates at multiple sites for a multiple event fault, as shown in Figure 5.9(b). A single event fault at one time frame may become a multiple event fault at another time frame, and vice versa.

Assuming a computer word size of 64, the algorithm selects 64 faulty sequential circuits to be simulated in parallel. These 64 circuits constitute a fault group.

The line value data structures are shown in Figure 5.10. A one-dimensional array is used to store the fault-free circuit value of each line in the circuit. Another array stores the value of each line in the faulty circuit. Each faulty circuit line has a *group id* to indicate if the line values of the faulty circuits in the current group are different from the fault-free circuit value. Each value consists of 64-bit words, V_0 and V_1, where each bit is used to store a different faulty circuit's value. A four-valued logic (0, 1, x, and Z) is used. x denotes an unknown logic value and Z denotes a tristate value. In order to code the four values, two bits are employed, one in V_0 and one in V_1: 0 is coded as $(1, 0)$, 1 as $(0, 1)$, x as $(0, 0)$, and Z as $(1, 1)$.

Table 5.4 shows the logic used to evaluate different gates, 64 faulty circuits at a time. Each gate is assumed to have two inputs, A and B, and an output, C. A_0 and A_1 are bits that correspond to A in V_0 and V_1. B_0, B_1, C_0, and C_1 are similarly defined. The tristate gate has an input, A, and an enable input, E. The BUS element has two inputs, A and B, wired to a bus. Tristate gate outputs are assumed to only be connected to a BUS element. The fault-free circuit values are also represented as a 64-bit word to allow easy comparison with faulty circuit values.

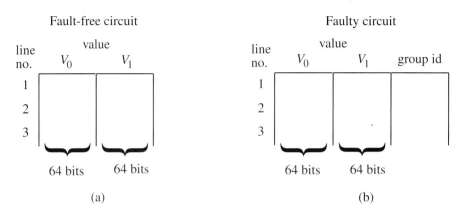

Figure 5.10 Line value storage: (a) fault-free circuit, and (b) faulty circuit (Niermann *et al.*, 1992) (© 1992 IEEE)

Table 5.4. *Gate evaluations (Niermann et al., 1992) (© 1992 IEEE)*

Gate	C_0	C_1
AND	$A_0 + B_0$	$A_1 B_1$
OR	$A_0 B_0$	$A_1 + B_1$
INV	A_1	A_0
XOR	$A_0 B_0 + A_1 B_1$	$A_0 B_1 + A_1 B_0$
TRISTATE	$E_0 + A_0 E_1$	$E_0 + A_1 E_1$
BUS	$A_0 B_0 B_1 + A_0 A_1 B_0$	$A_1 B_0 B_1 + A_0 A_1 B_1$

5.3.2   The fault simulation method

During fault simulation, each test vector from the test sequence is considered one at a time. The fault-free circuit is simulated first for the test vector in question. This is based on standard event-driven logic simulation. Then a fault group of 64 faults is selected for evaluation. To evaluate a fault group, the *group id* is first incremented so that the faulty values of other faulty circuits are not used in the evaluation of these faulty circuits. The faults from the selected fault group are injected into the circuit, and the values for the state obtained from the previous input vector are inserted into the faulty circuits. After evaluating the faulty circuits in this fault group, the state values are stored for the next vector. Thereafter, another group of 64 faults from the fault list is selected and evaluated. This process repeats until all faults in the fault list are evaluated for the test vector.

Several heuristics can be employed to speed up fault simulation by reducing the number of faults to be simulated in parallel, making fault injection efficient, and deriving a fault ordering to reduce the number of gate evaluations. These heuristics and various other steps of the fault simulation algorithm are described next.

5.3.2.1 Reduction of faults to be simulated in parallel

Before faults from the fault list are grouped into fault groups, it is advisable to reduce
the number of faults that need to be simulated in a given time frame. The number of
multiple event faults cannot be reduced using this approach. However, the number of
non-stem single event faults can be reduced. This can be done in two phases. In the
first phase, all single event non-stem faults inside FFRs are mapped to corresponding
stem faults through local fault simulation of the non-stem faults. In the second phase,
mapped stem faults that have a faulty value are examined for possible elimination.

Let us denote line p stuck-at α fault, $\alpha \in \{0, 1\}$, as p SAα. Suppose that fault p
SAα is a single event fault inside an FFR and it propagates to the corresponding output
stem of the FFR, say q. Let the fault-free value of line q and its value in the presence
of p SAα be β and β^*, respectively, where $\beta, \beta^* \in \{0, 1, x\}$. Consider the pseudofault
that changes the value of line q from β to β^*. If β^* is the same as β, fault p SAα does
not propagate to line q. In this case, we say that the equivalent pseudofault q SAβ^*
is insensitive (note that this equivalence is only for the given time frame). Otherwise,
the pseudofault is said to be sensitive. If the pseudofault is insensitive, fault p SAα
need not be considered any further since it cannot result in an error at a primary output
or next state line. If the pseudofault is sensitive, the replacement of p SAα with the
pseudofault cannot change the behavior of the faulty circuit for this time frame. Thus,
for example, if the fault-free value at stem q is 0, all non-stem faults inside the FFR
can be mapped to at most two stem faults, q SA1 or q SAx, since q SA0 is insensitive.
This completes the first phase.

Some of the mapped stem faults can be further eliminated in the second phase. If
the stem is a primary output or a next state line, the simulation of the corresponding
stem fault is trivial. Otherwise, the stem fault is examined to see if it propagates to a
certain gate(s) by local fault simulation. This gate(s) is chosen depending on whether
the stem has a dominator or not.

If stem q has a dominator, there is a unique stem r such that FFR(r) includes the
immediate dominator of q. If local fault simulation indicates that the fault at q can
propagate to r and the corresponding fault at r is sensitive, then the fault at q can
be dropped from further consideration in this time frame, provided that r is neither
a primary output nor a next state line, and the fault at r has not been simulated in a
previous pass for the current test vector. The last condition ensures that a stem fault
is simulated at most once for each test vector. Consider the circuit in Figure 5.11.
Stem q has an immediate dominator s which is included in FFR(r). The fault q SA1
propagates to r, where fault r SA0 is sensitive. If r SA0 has not already been simulated,
it is included in a fault group and simulated in parallel.

If stem q does not have a dominator, then it is examined if the corresponding fault
at q propagates through any of the gates it feeds. If so, it is included in a fault group.
Otherwise, it is dropped from further simulation in this time frame. For example,

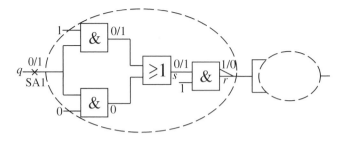

Figure 5.11 Simulation of a stem fault with a dominator

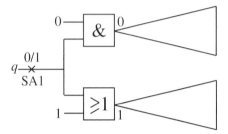

Figure 5.12 Simulation of a stem fault without a dominator

consider the circuit in Figure 5.12. Since stem q does not have a dominator, the gates that q feeds are examined. Since q SA1 does not propagate through either gate that q feeds, q SA1 is dropped from further consideration.

The above fault reduction strategy can substantially reduce the number of faults that need to be simulated in a given time frame. For some circuits, a reduction ratio of 15 has been reported.

5.3.2.2 Group id

Before the evaluation of any fault group, the *group id* is incremented. The *group id* helps distinguish between faulty circuit values from different fault groups. For the fault-free circuit, the value of every line is kept. However, for a faulty circuit, every line has a value and a *group id*, as shown earlier in Figure 5.10. If the *group id* associated with a line value does not match the current *group id*, then the value is a residual value from a previous fault group propagation, and the fault-free circuit value is, therefore, the correct value for the current fault group. For example, consider the gate evaluation shown in Figure 5.13. For simplicity, four-bit parallelism is used in this example, as opposed to 64-bit parallelism. Suppose the current *group id* is 20. Since the *group id* of the lower input is not 20, the fault-free circuit value of the lower input is used for the gate evaluation. The second row in Table 5.4 can be used for this gate evaluation. Without this feature, the faulty value array would need to be cleaned up by copying the fault-free circuit values after the simulation of each fault group.

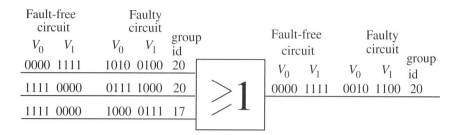

Figure 5.13 Gate evaluation

5.3.2.3 Fault injection

Each fault in the fault group must be injected into the circuit. Traditionally, faults are injected with the help of a **bit-mask** associated with each gate. This mask acts like a flag associated with the gate inputs and outputs. It indicates if the values used during gate evaluation are those produced by predecessor gates or the stuck-at value associated with the gate. This requires that flags be examined for every gate even though only a few gates have faults. An efficient alternative is to use functional fault injection, as described next.

When a fault is present at the input or output of a gate, it changes the function of the gate. This implies that faults can be injected into the circuit by introducing a new type of gate whose function reflects the behavior of the fault. For example, consider a circuit with an XOR gate whose function is $z(x_1, x_2) = \bar{x}_1 x_2 + x_1 \bar{x}_2$. Suppose that an SA0 fault is injected at input x_1 of the XOR gate. This fault effectively results in a new type of gate whose function is $z(x_1, x_2) = x_2$. If we replace the XOR gate with this new type of gate, the fault is effectively injected into the circuit.

5.3.2.4 Fault dropping and storing state values

A fault is detected if there is a faulty value at a primary output and the fault-free circuit has a known value at that output. When a fault is detected, it is removed from the fault list. If a fault is not detected, all the next state lines with values different from the fault-free values are stored. These values are inserted into the circuit when the next test vector is considered for a fault group containing this fault.

5.3.2.5 Fault ordering

In parallel fault simulation, a proper fault grouping is crucial for exploiting the 64-bit word-level parallelism of the computer. If faults that cause the same events are in the same fault group, the number of events required for the simulation of faulty circuits is reduced. The fault list constantly changes due to the dropping of detected faults. Therefore, it is important to consider the order of the complete fault list. During fault

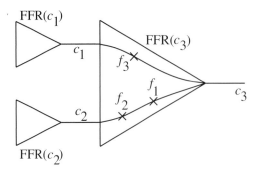

Figure 5.14 Static fault ordering (Lee and Ha, 1996) (© 1996 IEEE)

simulation, a group of 64 faults is selected by traversing the ordered list from beginning to end. Two fault ordering methods can be employed to make this process efficient, a static method used in a pre-processing step followed by a dynamic method used during fault simulation.

In static fault ordering, the non-stem faults inside each FFR are grouped together first, and the resulting groups of faults are ordered by traversing FFRs in a depth-first manner from the primary outputs. This fault ordering is performed once during pre-processing. Consider the circuit in Figure 5.14. Let signals c_1, c_2 and c_3 be stem lines. Suppose that there are three multiple event faults, f_1, f_2, and f_3, in FFR(c_3). Since the effect of these three faults always passes through stem c_3, putting them in the same fault group is likely to reduce the number of events. Let $list(c_1)$ and $list(c_2)$ represent the ordered fault lists within FFR(c_1) and FFR(c_2), respectively. Using the above method, the faults are arranged in the order f_1, f_2, f_3, $list(c_2)$, and $list(c_1)$. This method only affects multiple event non-stem faults (which are always put into fault groups for parallel fault simulation), but not single event non-stem faults which are mapped to stem faults, as explained before.

The dynamic fault ordering method is based on the assumption that the activity of some faults is much higher than that of other faults during fault simulation. If high activity faults are grouped together, it is likely the propagation paths of the faults will overlap with each other, thus reducing overall gate evaluation time.

In fault simulation, all flip-flops are initially set to x. With the progress of simulation, most flip-flops are initialized to 0 or 1. However, in the presence of a fault, a flip-flop may be prevented from being initialized, i.e., it may remain at x. Consider a fault situation where the x value propagates from the flip-flop to a primary output. If the fault-free circuit has a 0 or 1 at this output, then such a fault is called **potentially detected** since the unknown value at this output may attain a different value than the fault-free value in the actual circuit. The activity of potentially detected faults is high compared to other faults since the value x propagates all the way from a present state line to a primary output. To exploit this fact, the fault list obtained through static fault ordering can be divided up into two lists, list *A* and list *B*. List *A* contains faults which have not been potentially detected so far, and list *B* contains faults which have been potentially detected at least once. A fault in list *A* is moved to list *B* whenever it is potentially detected. Fault groups are now formed by traversing lists *A* and *B* separately.

The above fault ordering methods typically reduce the number of simulation events by about 20%.

5.4   Test generation for synchronous circuits

In this section, we describe test generation methods based on the state table and the gate-level implementation. The latter includes methods based on topological analysis, simulation and a hybrid of these two.

5.4.1   Test generation using a state table

For a controller-type of circuit for which a state table exists or can be easily extracted from its gate-level implementation, one possible test generation approach is to utilize a functional fault model based on the state table (Cheng and Jou, 1992). Such an approach can significantly improve the test generation time. The simplest functional fault model is the *single-state-transition* (SST) fault model. This model assumes that the fault causes the destination state of one state transition to be faulty while retaining its correct input/output label. Test sets derived from the SST fault model usually provide a very high coverage of single SAFs and other non-modeled faults in the corresponding gate-level implementation.

Most faults in lower-level implementations of the circuit actually map to *multiple-state-transition* (MST) faults in a state table. Such a fault corrupts the destination states of multiple state transitions. However, a test set for SST faults also detects almost all MST faults in practical finite-state machines (FSMs). Therefore, we concentrate on the SST fault model only. Also, we make the assumption that the fault does not increase the number of states in the state table.

Each state transition in an FSM is associated with a four-tuple: ⟨input, source state, destination state, output⟩. A state transition is corrupted if its destination state, its output label, or both are faulty. However, a test sequence that detects a fault that corrupts the destination state also detects the fault that corrupts either the output label or both the destination state and the output label of the same transition. This can be seen as follows. A test sequence for detecting a corrupt destination state consists of three subsequences: the **initialization sequence** that forces the FSM to the source state of the faulty transition, the input label of the transition that activates the

Table 5.5. *A state table*

| PS | NS, z | |
	$x_1 = 0$	$x_1 = 1$
A	B, 1	C, 0
B	C, 0	D, 1
C	A, 0	A, 0
D	A, 1	D, 1

fault, and the **state-pair differentiating sequence** (SPDS) that differentiates between the good and faulty destination states, i.e., produces different output sequences starting from these states. If the output label is faulty then the initialization sequence and the activation vector already detect the fault. Thus, we need to consider only those faults that result in a faulty destination state. Also, we will derive the above subsequences from only the fault-free FSM, whereas, strictly speaking, they should be derived from both the fault-free and faulty machines. This significantly speeds up test generation without much compromise in accuracy. In the small number of cases when accuracy is compromised, we will show later how the situation can be remedied.

The number of SST faults in an N-state M-transition machine is $M(N - 1)$. This number may be too large for large FSMs. Fortunately, we can further reduce the number of SST faults through fault collapsing. For each state transition, instead of targeting $N - 1$ possible faulty destination states, we usually need to target a small subset of these faulty states. Suppose an SST fault f_1 corrupts the transition, depicted by its four-tuple, $\langle x_i, S_i, S_j, z_i \rangle$, to $\langle x_i, S_i, S_k, z_i \rangle$, and fault f_2 corrupts this transition to $\langle x_i, S_i, S_l, z_i \rangle$. If the SPDS of S_j and S_k also differentiates between S_j and S_l, then fault f_2 dominates fault f_1, and f_2 can be removed from the fault list.

Example 5.4 Consider the state table of an FSM shown in Table 5.5, where PS denotes the present state and NS denotes the next state. Since input 0 differentiates between states A and B as well as A and C, SPDS(A, B) = SPDS(A, C) = 0. Similarly, SPDS(A, D) = 1. Next, consider the state transition $\langle 0, D, A, 1 \rangle$ in the fourth row and $x_1 = 0$ column. The destination state of this transition can be corrupted in three ways to give the following faulty transitions: $\langle 0, D, B, 1 \rangle$, $\langle 0, D, C, 1 \rangle$, $\langle 0, D, D, 1 \rangle$. However, since SPDS($A$, B) is also SPDS(A, C), the first two of these faulty transitions can be collapsed into just the first one.   □

Another reasonable heuristic for further fault collapsing, which reduces the test length while still resulting in a high fault coverage, is as follows. If there are two transitions in the state table having identical source state, destination state and output label, then they are collapsed into a single transition in the specification. For example, in the state table in Table 5.5, the two transitions from state C satisfy this condition. Hence, the faulty transition for only one of these two transitions needs to be considered, not for both.

In the pre-processing phase of test generation, the *transfer sequences* between every state pair is first computed. A **transfer sequence** from state S_i to state S_j is the shortest input sequence that takes the machine from S_i to S_j. Then, the relevant SPDSs are computed. The test generation procedure consists of three steps: initialization, excitation and state differentiation. In the initialization step, the machine is brought from the current state to the source state of the faulty transition using an appropriate transfer sequence. In the excitation step, the faulty transition is executed. In the state differentiation step, the corresponding SPDS is applied to differentiate between the good and faulty states.

The initialization and state differentiating sequences for different SST faults can be overlapped by dynamically ordering the fault list. This can be done with the help of a fault simulator. The fault simulator records the current state of each faulty machine and drops the detected faults from the fault list. In order to determine the next target fault, the information on the current state of each faulty machine is used. A fault for which the current state of the faulty machine differs from the current state of the good machine is the preferred next target fault. The reason is that such a fault is already activated and does not require the initialization sequence and activation vector. If such a fault does not exist, we choose a fault which is easiest to activate. For example, if the source state of the faulty transition of a fault is the same as the current state of the good machine, then this fault will be selected since it will not need an initialization sequence.

Example 5.5 Consider the SST fault which corrupts the transition $\langle 0, D, A, 1 \rangle$ to $\langle 0, D, B, 1 \rangle$ in the state table in Table 5.5. In order to derive a test sequence for this fault, we first derive a transfer sequence from the initial state A to D. Such a sequence is $\{0, 1\}$. Then the activation vector $x_1 = 0$ is applied. Finally, SPDS$(A, B) = 0$ is applied. Hence, the complete test sequence is $\{0, 1, 0, 0\}$.   □

Since the SPDSs are obtained from the state table of the fault-free machine, the derived test sequence is not guaranteed to detect the targeted SST fault. The fault simulator can be used to verify whether or not the pre-generated SPDS is valid. If not, an alternative SPDS is generated. Since the modeled fault consists of only one faulty state transition, the pre-generated SPDS works in most cases. Empirically, it has been determined that less than 1% of the SSTs require the generation of an alternative sequence. An accurate algorithm which generates test sequences from both the fault-free and faulty machines is given in Pomeranz and Reddy (1994). However, the price paid for guaranteed accuracy is an increase in the test generation time.

5.4.2   Test generation using a gate-level implementation

From the earlier classification of test generation algorithms, we know that gate-level test generation can be either topological analysis based, simulation based, or a hybrid of the two. Next, we describe algorithms from each of these categories.

5.4.2.1 Test generation based on topological analysis

In this section, we show how sequential test generation can be done with the help of the iterative array model (Cheng, 1996). Topological analysis based algorithms generate a test sequence to activate the fault and propagate its effect to a primary output by finding a sensitized path through multiple time frames.

In the sequel, we first assume that the initial state of the circuit is unknown, and later give algorithms to take advantage of the situation when the initial state is known.

Extended D-algorithm: One possible way to generate a test sequence for a fault in a sequential circuit is to extend the D-algorithm. We can start with some time frame, say time frame 0, and use the D-algorithm to generate a test vector for this time frame. If the error propagates to at least one primary output, no further error propagation is needed. However, if it propagates to only the next state lines, a new time frame is added as the next time frame to further propagate the error. This process continues until the error reaches a primary output. If the test vector requires particular logic values at the present state lines in time frame 0, a new time frame is added as the previous time frame. State justification is then performed backwards through the previous time frame. This process continues until no specific logic values are required at the present state lines.

Example 5.6 Consider the iterative array shown in Figure 5.15. In time frame i, the signals are superscripted with i. First, time frame 0 is obtained for the shown SA0 fault. After applying the D-algorithm to this time frame, the fault effect D is obtained at the next state line Y and a value 0 needs to be justified at the present state line y. For further error propagation, time frame 1 is added to the right of time frame 0. The fault effect can be successfully propagated to primary output z in this time frame. Therefore, we next justify the required value at y in time frame 0. This is done by adding another time frame -1 to the left of time frame 0. $x_1 = 0$ at the input of this time frame is sufficient to obtain $y = 0$ in time frame 0. Note that since the SAF is present in each time frame, we need to make sure that the fault-free and stuck-at values are the same in this state justification step. There is no need to add any further time frames to the left since no specific value is required at line y in time frame -1. Thus, the test sequence for the above fault given at (x_1, x_2) is $\{(0, x), (1, 1), (x, 1)\}$.   □

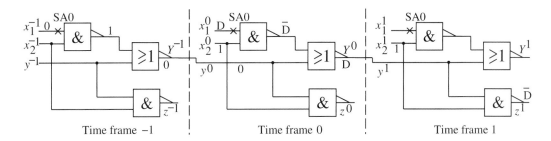

Figure 5.15 Application of the extended D-algorithm

The nine-valued logic: Although the above approach is straightforward and intuitive, it was pointed out in Muth (1976) that the five-valued logic $(0, 1, x, D, \overline{D})$ used in the D-algorithm is not appropriate for sequential circuits. The reason is that the five-valued model overspecifies the value requirements at some nodes in the circuit, which sometimes prevents it from arriving at a test sequence even when one exists for the fault. Muth suggested a nine-valued logic to take care of this problem. This logic properly accounts for the repeated effects of the fault in each time frame. Each of the nine values represents an ordered pair of ternary values: 0, 1 or x (don't care). The first value of the pair represents the ternary value of the fault-free circuit and the second value represents the ternary value of the faulty circuit. Thus, the nine ordered pairs are as follows: $0/0, 0/1, 0/x, 1/0, 1/1, 1/x, x/0, x/1, x/x$.

Example 5.7 Consider the iterative array shown in Figure 5.16. Using the shown values, the fault effect is propagated to primary output z in time frame 0. Hence, no further error propagation is needed. To justify the required value 1 at present state signal y in time frame 0, we need to add time frame -1 to the left. State justification in time frame -1 results in a conflict at primary input x_1 on which a 1 is required while it has an SA0 fault. Thus, the algorithm mistakenly concludes that no two-vector test sequence exists to detect this fault. Next, consider the iterative array shown in Figure 5.17 which uses nine-valued logic. In order to propagate the error from x_1 to the output of gate G_1 in time frame 0, the other input of G_1 must have a 1 for the fault-free circuit, but has no value requirement for the faulty circuit. In nine-valued logic, this requirement is denoted as $1/x$. Eventually, we have a value requirement of $1/x$ at line y. Due to this relaxed requirement, there is no conflict at line x_1 in time frame -1. The corresponding test sequence for this fault is, thus, $\{(1, 1), (1, 0)\}$. ☐

Reverse time processing: The extended D-algorithm and the nine-valued algorithm combine *forward and reverse time processing* to obtain a test sequence. **Forward time processing** consists of error propagation from the next state lines to a primary output through forward time frames. **Reverse time processing** consists of state justification through backward time frames. Such a mixed time processing approach

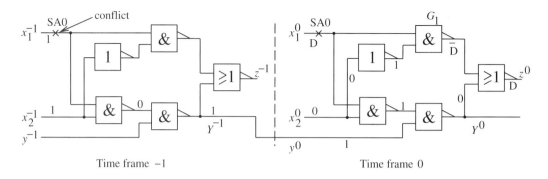

Figure 5.16 Test generation with five-valued logic

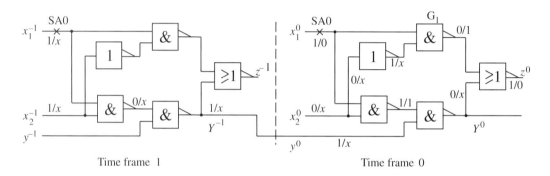

Figure 5.17 Test generation with nine-valued logic

has the disadvantage that a large number of time frames needs to be maintained during test generation since the time frames are only partially processed. The reverse time processing approach, which was introduced in the extended backtrace algorithm (Marlett, 1978), overcomes this disadvantage. This approach works backwards in time from the last time frame to the first. It pre-selects a path, possibly consisting of many time frames, from the fault site to a primary output. An attempt is then made to sensitize this path backwards in time from the primary output. If the attempt succeeds, backward justification is continued for the required value at the fault site. Else, another path is selected. The advantage of such an approach is that at any time only the current and previous time frames need to be maintained. The required values in time frame k are completely justified before starting the justification of required values in time frame $k - 1$. Hence, the values justified at the internal nodes of time frame k can be discarded when the justification starts in time frame $k - 1$. This means that the memory usage is low and the implementation is straightforward. Of course, the decision points and their corresponding circuit status still need to be stacked for backtracking later, if necessary. Another advantage of reverse time processing is that it is easy to identify *state repetition*. **State repetition** is a situation where a state visited earlier is again required to be justified. Since further justification would just be a waste

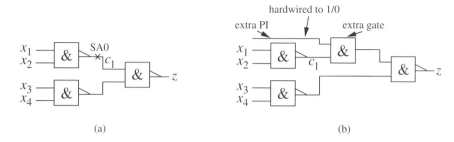

Figure 5.18 Fault injection

of time, backtracking should occur immediately. Since justification in time frame k is completed before justification begins in time frame $k - 1$, state repetition can be easily identified by comparing the newly required state with the list of previously encountered states. Similarly, a list of illegal states, which have previously been found to be unjustifiable, can also be maintained. If the newly required state belongs to this list, then further search in that direction can be terminated.

The disadvantage of the above testing approach is that it considers only one path at a time. When the number of paths from the fault site to a primary output is very large, trying each path individually may be impractical. Another disadvantage is that faults which require simultaneous sensitization of multiple paths may not be detected.

The disadvantages of the extended backtrace algorithm can be overcome by another reverse time processing algorithm called the BACK algorithm (Cheng, 1988). Instead of pre-selecting a path, this algorithm pre-selects a primary output. It assigns a 1/0 or 0/1 from the nine-valued logic to the selected primary output and justifies the value backwards. A testability measure called *drivability* is used to guide the backward justification process from the primary output to the fault site. The **drivability** of a signal measures the difficulty associated with propagating a 1/0 or 0/1 from the fault site to the signal. The drivability measure is precomputed for both the fault-free and faulty circuits based on the controllability measures (see Chapter 4; Section 4.5.3).

Before the drivability computation begins, an extra two-input gate is inserted into the circuit to inject the targeted fault. An SA0 fault can be injected with the help of a two-input AND gate, as shown in Figure 5.18. One input of the extra gate is made a primary input (PI) and hardwired to 1/0 to obtain the correct functionality in the fault-free and faulty circuits. Similarly, an SA1 fault can be injected with the help of a two-input OR gate, whose extra input is hardwired to 0/1. The advantage of this fault injection technique is that it enables equal treatment of all the gates. Note that the extra gate is only added to facilitate test generation; the actual circuit implementation does not have the extra gate.

To obtain drivability, the controllability of the different nodes in both the fault-free and faulty circuits need to be first obtained. Let us denote the 0-controllability and 1-controllability of a signal c in the good circuit as $CC_g^0(c)$ and $CC_g^1(c)$, respectively.

In the faulty circuit, which is obtained after modifying the fault-free circuit with the extra two-input gate, let us denote the 0-controllability and 1-controllability of signal c as $CC_f^0(c)$ and $CC_f^1(c)$, respectively. These controllabilities can be computed in the pre-processing phase. Consider the faulty circuit in which an extra AND gate is injected with the added primary input c_1. Since this input is hardwired to $1/0$, we set $CC_g^0(c_1) = \infty$, $CC_g^1(c_1) = 0$, $CC_f^0(c_1) = 0$, $CC_f^1(c_1) = \infty$. The controllabilities of all the signals reachable from the faulty signal (i.e., those in its fanout cone) are updated based on the fault-free signal controllabilities and the controllabilities of the extra input c_1. Thereafter, two drivability measures, $1/0\_drive$ and $0/1\_drive$, are derived for each signal. These measures indicate the relative difficulty in propagating $1/0$ and $0/1$ from the fault site to the signal under question. If a signal c is not reachable from the fault site, both $1/0\_drive(c)$ and $0/1\_drive(c)$ are set to ∞. If the hardwired value of the extra primary input of the injected gate is $1/0$ $(0/1)$, its $1/0\_drive$ $(0/1\_drive)$ is set to 0 and its $0/1\_drive$ $(1/0\_drive)$ is set to ∞. This enables the computation of the drivabilities of signals in the fanout of the injected gate. Consider one such AND gate with output c_2 and inputs w_1 and w_2 in the fanout cone of the injected gate. To obtain $1/0$ at c_2 requires that one of its inputs have $1/0$ and the other input have logic 1 in the fault-free circuit. Thus, $1/0\_drive(c_2)$ can be derived as the minimum of $\{1/0\_drive(w_1), CC_g^1(w_2)\}$ and $\{1/0\_drive(w_2), CC_g^1(w_1)\}$. Since the input with $1/0$ has a controlling value 0 in the faulty circuit, there is no value requirement for the other input in the faulty circuit. Similarly, to obtain $0/1$ at c_2 requires that one of its inputs have $0/1$ and the other input have logic 1 in the faulty circuit. Thus, $0/1\_drive(c_2)$ can be derived as the minimum of $\{0/1\_drive(w_1), CC_f^1(w_2)\}$ and $\{0/1\_drive(w_2), CC_f^1(w_1)\}$. In this case, since the input with $0/1$ has a controlling value 0 in the fault-free circuit, there is no value requirement for the other input in the fault-free circuit.

In order to justify $1/0$ $(0/1)$ at a gate's output during backward justification, the method selects the input with the smallest $1/0\_drive$ $(0/1\_drive)$ as the signal to be sensitized. This enables efficient and implicit sensitization of paths through the circuit.

After deriving the drivability measures, the method first selects a primary output with the smallest drivability measure and assigns it $1/0$ or $0/1$. If all the primary outputs with finite drivability have been considered, the fault is declared to be undetectable. All the required values are justified in the current time frame. If a decision is made, it is stored in the decision stack along with the circuit status. In the case of a conflict, backtracking is done to the last decision point. If no value is required at the present state lines, a test sequence is found. The obtained test sequence can be fault simulated to detect other faults. If some value is required at the present state lines, it is checked if the state is a repeated or an illegal one. If true, backtracking is done. If false, the state requirement is used as the next state of a previous time frame, and the above process is repeated.

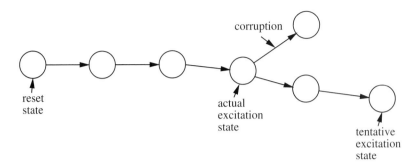

Figure 5.19 Corruption of a justification sequence

Some of the more advanced test generation techniques, employing both forward time and reverse time processing, can be found in Niermann and Patel (1991), Lee and Reddy (1991), Kelsey *et al.* (1993), Chen and Bushnell (1994) and Glaeser and Vierhaus (1995).

Test generation from a known initial state: Test generation algorithms which start from a known initial state, called the reset state, generally make use of the state transition graph (STG) of the sequential circuit (Cho *et al.*, 1993). The assumption of a known initial state is valid for controllers with an external reset signal which can reset its flip-flops. For such circuits, the STG is manageable. Implicit state enumeration techniques, based on binary decision diagrams (BDDs), allow one to deal with a large number of states. However, test generation for large sequential circuits, consisting of hundreds of flip-flops, is still beyond the scope of such methods.

The reset state is assumed to be valid for both fault-free and faulty circuits. In other words, faults affecting the reset mechanism are not tackled. The STG is implicitly traversed in a pre-processing step to determine which states are reachable from the reset state, and what input sequence is required to reach them. For example, in Figure 5.4(b), if state $(0, 0)$ is assumed to be the reset state, then states $(0, 1)$ and $(1, 0)$ are reachable from it, but state $(1, 1)$ is not. The reachability information is stored in the form of BDDs.

Test generation is divided up into three phases. In the first phase, a combinational test vector, called the **excitation vector**, is generated which either propagates the fault effect to a primary output or to a next state variable. The reachability information is exploited to make sure that the present state part of the excitation vector, called the **excitation state**, is reachable from the reset state. In the second phase, the justification sequence is derived. Owing to the pre-stored reachability information, no backtracking is required in this phase. A third phase is required when the fault effects propagate to the next state outputs, but not the primary outputs. In this phase, a state differentiating sequence is derived. This sequence differentiates between the fault-free and faulty machines.

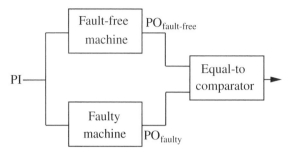

Figure 5.20　A product machine

Since the justification sequence is derived from the reachability information obtained from the fault-free sequential circuit, it may not be valid in the presence of the fault. However, a prefix of the invalid justification sequence is a valid justification in this case, though for a different excitation state. This is illustrated in Figure 5.19. This prefix can be obtained by fault simulating the initially derived justification sequence. If the prefix detects the fault at a primary output, a valid test sequence has been obtained. Otherwise, the next state variables to which the fault effects are propagated are marked as needing a differentiating sequence.

To collect the reachability information, the STG is implicitly traversed. This implies exercising of all the transitions from valid (i.e., reachable) states. Implicit enumeration avoids to some extent the problems that arise when the STG has a very large number of states. This technique exercises multiple state transitions simultaneously and stores sets of states visited so far in the form of **characteristic functions**. These functions evaluate to 1 if the states represent set members. For example, suppose the state variables of a circuit are $y_1, y_2, \ldots, y_{50}$. Then the characteristic function $y_1 + y_{40}$ represents all states whose first or 40th bits are equal to 1. There are 3×2^{48} such states. Yet the representation is quite compact. In most cases, this is more efficient than storing the entire set. However, in the worst case, the size of a characteristic function can be exponential in the number of variables. Given a set of present states, the set of states reachable with one state transition is computed using breadth-first traversal. The process is repeated with the newly reached states until no new states are reached. To obtain the initial justification sequence, a characteristic function closest to the reset state is first found which evaluates to 1 for the given excitation state. Then the path from the reset state to this state, represented by the characteristic function, is traversed backwards.

To obtain a differentiating sequence, the fault simulator is initialized with the pair of states reached in the fault-free and faulty machines at the end of the justification sequence and excitation vector. Then state differentiation is tried by simulation with randomly generated input sequences. A limit is imposed on the number of such random sequences and their lengths.

If the above approach is not successful, a product machine is formed, as shown in Figure 5.20. Here PI and PO denote primary inputs and primary outputs, respectively.

The product machine is traversed to see if a reachable state can be obtained for which the comparator output indicates a mismatch between its inputs. If so, a test sequence is found. To make this process more efficient, only the primary outputs that the fault under question can affect are used in this fashion. For the other primary outputs, only the fault-free machine is sufficient. Still, for large sequential circuits, this approach can be prohibitively expensive.

5.4.2.2 Simulation-based test generation

Simulation-based test generation methods are enhancements from logic simulation or fault simulation. The basic approach is as follows. To generate a test for a given fault or set of faults, a candidate test vector or test sequence is generated. The fitness of the vector or sequence is evaluated through logic or fault simulation. The vector or sequence with the highest fitness value, based on some specified cost function, is selected and the others are discarded. This process continues until some pre-specified halting condition is met. The methods under this category differ with each other in how they generate new vectors or sequences and what their cost function is. Even within a single method, generally, multiple cost functions are used for different phases of test generation.

The advantage of simulation-based approaches is that they, in general, require less run-time than array-model based approaches. Their disadvantages are that they cannot identify undetectable faults, tend to generate larger test sequences, and have difficulty in detecting hard-to-activate faults. However, methods have been devised to overcome the last two disadvantages.

In this section, we will concentrate on two promising approaches. The first one is based on logic simulation and fault activation, and the second one is based on a genetic algorithm.

Test generation based on logic simulation and fault activation: This approach avoids fault-oriented test generation and uses randomized search to obtain test sequences at low computational cost (Pomeranz and Reddy, 1997). It is based on the following observation. Consider an input sequence $T_1 T_2$ such that T_1 brings the fault-free circuit to state S_0 and T_2 takes the fault-free circuit through a cycle consisting of states $S_0 \rightarrow S_1 \rightarrow \cdots \rightarrow S_j = S_0$. Next, consider the test sequence $T = T_1 T_2 \cdots T_2$ obtained by concatenating T_1 with k repetitions of T_2. Suppose that in the presence of the fault, the circuit goes through the sequence of states $S_0' \rightarrow S_1' \rightarrow \cdots \rightarrow S_j'$ during the application of T_2. If this sequence does not create a cycle, or creates a different cycle than the one traversed by the fault-free circuit, then the fault effects are repeatedly propagated by T to the next state variables, and are, therefore, likely to be detected. Generally, only a small number of repetitions of T_2 is required to derive an effective test sequence. The computation time involved in

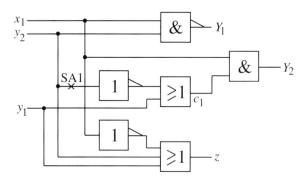

Figure 5.21   Fault detection using repeated sub-sequences

Table 5.6. *State and output sequences*

		0		1		1		1		1		1		1		1	
Fault-free:	xx	$\rightarrow$	10	$\rightarrow$	11	$\rightarrow$	01	$\rightarrow$	00	$\rightarrow$	11	$\rightarrow$	01	$\rightarrow$	00	$\rightarrow$	11
			1		1		1		1		0		1		1		0	

		0		1		1		1		1		1		1		1	
Faulty:	xx	$\rightarrow$	10	$\rightarrow$	11	$\rightarrow$	01	$\rightarrow$	00	$\rightarrow$	10	$\rightarrow$	11	$\rightarrow$	01	$\rightarrow$	00
			1		1		1		1		0		1		1		1	

fault activation is kept low by considering just one fault with every cycle and always simulating the fault starting from the fault-free state.

To motivate the method, let us consider an example first.

Example 5.8   Consider the combinational logic of the sequential circuit shown in Figure 5.21. The fault under consideration is the SA1 fault shown in the figure. The input sequence $\{0, 1, 1, 1, 1\}$ takes the fault-free circuit through the following state sequence: $xx \xrightarrow{0} 10 \xrightarrow{1} 11 \xrightarrow{1} 01 \xrightarrow{1} 00 \xrightarrow{1} 11$, where the values above the arrow are the input values. Since a cycle is created with the application of the last three input bits, we divide the input sequence into two parts: $T_1 = \{0, 1\}$ and $T_2 = \{1, 1, 1\}$. Let us consider the test sequence $T = \{0, 1, \ 1, 1, 1, \ 1, 1, 1\}$. The corresponding state and output sequences are shown in Table 5.6 (the values below the arrow are the output values). The fault is activated at time frame 5. Thereafter, the fault-free and faulty circuits reach different states in each time frame. T detects the fault at time frame 8. □

In the above example, the fault was activated after T_2 was applied once. In general as well, the fault is required to be activated during the first application of T_2 to keep the total test sequence length as small as possible.

This method generates a sequence T of fixed length L by selecting the input vectors of T one at a time, starting from the input vector at time frame 1. At every time frame, it generates N random input vectors and simulates them on the fault-free circuit. One of these vectors is chosen based on as close a satisfaction of certain requirements as possible. These requirements are based on generating cycles which are effective in detecting faults.

The method maintains the present state of the fault-free circuit in variable S. Initially, S is the all-x state. It also maintains the set of all states traversed by the fault-free circuit so far and the time frames when they were first reached. For every random input vector, the fault-free and faulty circuits are placed in state S and this vector is simulated in both circuits. If one or more input vectors, among the N vectors simulated, terminate a cycle, the method chooses the one which maximizes the length of the periodic part of the cycle, and stops. However, to avoid trivial cycles, the cycle length is required to be greater than or equal to a pre-specified minimum length. Also, it is required that the targeted fault be activated at least once during the periodic part of the cycle. At each step, that input vector is selected which activates the fault at as many primary outputs (primary objective) and next state variables (secondary objective) as possible. A tertiary objective in choosing a vector is that it specify as many next state variables from x to 0 or 1. For circuits with long synchronizing sequences, this objective is useful in ensuring that the fault-free circuit is synchronized by the derived sequence to a particular state.

If a cycle cannot be closed, the method stops after reaching a pre-specified maximum sequence length. As long as a cycle cannot be closed, it prefers to traverse states that have not been traversed before. However, even in some such cases, one could use a synchronizing sequence to close the cycle. For example, consider the input sequence $\alpha_1\alpha_2\alpha_3\alpha_4\alpha_5$. Suppose that the fault-free circuit gets synchronized by $\alpha_1\alpha_2$ and the targeted fault gets activated when α_4 is applied. Then the sequence $\alpha_1\alpha_2\alpha_3\alpha_4\alpha_1\alpha_2$ is guaranteed to create a cycle because the synchronized state will be visited twice.

Once the basic sequence is derived as above, the test sequence T is derived by duplicating the periodic part of the basic sequence. However, in the process of duplication, if a pre-specified maximum test sequence length is reached, T is terminated. Once T is generated, it is simulated and the faults it detects are dropped from further consideration.

The above process is repeated with different upper bounds on the cycle length. The reason is that shorter cycles potentially result in shorter test sequences and are, therefore, preferable. However, short test sequences may not be able to detect all the faults. Therefore, the upper bound on the cycle length is increased until a pre-specified limit, in order to improve the likelihood of detecting hard-to-detect faults.

When the above sequences are concatenated, they may detect additional faults not detected by the individual test sequences. Also, after concatenation it may be possible to remove the suffix of the test sequence that does not detect any new faults, thus reducing test length. It has been observed that concatenating the test sequences in the reverse order in which they were generated is, in general, more effective. The following values give good results for the benchmark circuits: $N = 100$, upper bounds on cycle length = 100, 200 and 400, lower bound on cycle length = 10, and upper bound on test sequence length = 500.

Test generation based on the genetic algorithm: The genetic algorithm has been popular for solving many optimization problems. It will be used in the subsequent chapters for solving other testing-related problems too. It has also found popularity for sequential test generation (Saab *et al.*, 1992; Hsiao *et al.*, 1996a, 1997). The last two references contain an algorithm which can detect more faults, frequently with smaller test sequences, than even deterministic test generators. Such an algorithm starts with a population of individuals. Each individual is a sequence of test vectors. Each such vector is applied to the primary inputs of the circuit in a particular time frame. The test sequence length depends on the *sequential depth*. **Sequential depth** is defined as the minimum number of flip-flops in a path between the primary inputs and the farthest gate in the combinational logic.

In the first stage of test generation, the test sequence length is made equal to the sequential depth. The test sequence length is doubled in the second stage, and doubled again in the third stage, since hard-to-detect faults may require longer sequences to activate and/or propagate. The population size (i.e., number of individuals) is set equal to $4 \times \sqrt{\text{sequence length}}$ when the number of primary inputs is 15 or fewer, and $16 \times \sqrt{\text{sequence length}}$ when the number of primary inputs is 16 or more. Larger populations are needed to accommodate longer individual test sequences in the interest of maintaining diversity. The quality of each individual is judged based on its value of a *fitness function*. The quality is measured in terms of the fault detection capability, controllability/observability measures, etc., and is explained in detail later. A fault simulator is used to evaluate the fitness. The population is initialized with some random as well as specific test sequences depending on the particular phase of test generation. Then the evolutionary processes of *selection, crossover* and *mutation* are used to generate a new population from the current one. Two individuals are selected from the existing population, with selection biased towards more highly fit individuals. The two individuals are crossed by randomly swapping bits between them to create two new individuals. The probability of swapping bits is set at $1/2$. Each bit in the two new individuals is mutated with a small probability of 0.01. This just involves complementing the bit. The two new individuals are then placed in the new population. This process continues until the new generation is completely filled. Evolution from one generation to another continues until a sequence is found to activate the targeted fault or propagate its effects to the primary outputs or a pre-specified maximum number of generations is reached. Since selection is biased towards more highly fit individuals, the average fitness is expected to increase from one generation to the next. However, the best individual may appear in any generation.

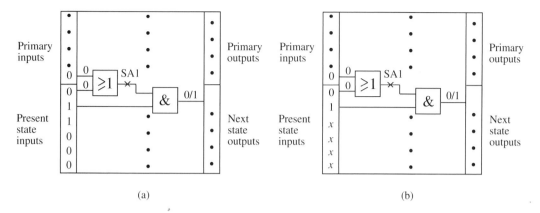

Figure 5.22   Fault activation and state relaxation

The test generation strategy uses several passes through the fault list, and targets faults individually in two phases. The first phase is geared towards activating the fault and the second phase towards propagating its effects from the flip-flops to primary outputs. During the fault activation phase, if no activation sequence can be found directly from the state where the previous sequence left off, then the *single-time-frame mode* is entered. In this mode, the aim is to find a test vector, consisting of primary input and flip-flop values, which can activate the fault in a single time frame. This is illustrated in Figure 5.22(a). Once such a vector is obtained, its flip-flop values are relaxed to a state which has as many don't cares as possible and can still activate the fault, as shown in Figure 5.22(b). This improves the chances of the state justification step that follows.

State justification is performed using a genetic algorithm with an initial population consisting of random sequences and any useful state-transfer sequences. If the relaxed state S_i matches a previously visited state S_j, and a state-transfer sequence exists that takes the circuit from current state S_k to S_j, then this sequence is added to the population. However, if S_i does not match any of the previously visited states, several sequences are genetically engineered to try to justify the target state. Several candidate states are selected from the set of previously visited states which closely match S_i. This is based on the number of matching flip-flop values in the states. The set of sequences that justify these candidate states from S_k are added to the population. The sequences in the population are simulated from S_k. Consider the scenario in Figure 5.23, in which an attempt is made to justify state $01xxxx$. The only bit that sequence T_1 cannot justify is the first bit. Similarly, the only bit that sequence T_2 cannot justify is the second bit. T_1 and T_2 are included in the population, and are used to genetically engineer the desired sequence T_3. Presence of the fault may invalidate a sequence that was previously used for traversing a set of states. However, since the genetic algorithm performs simulation in the presence of the fault to derive a sequence for the current situation, any such derived sequence will be a valid one.

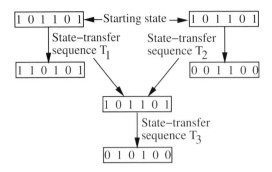

Figure 5.23 Genetic state justification

If the required state justification sequence is found, it is appended to the test set. A fault simulator is invoked to remove any additional faults detected by the sequence. The test generator then proceeds to the fault propagation phase. Otherwise, the current target fault is aborted, and test generation continues for the next fault in the fault list.

In the fault propagation phase, three types of *distinguishing sequences* are considered. A **distinguishing sequence** of type A for flip-flop j is a sequence which produces two distinct output responses in a fault-free machine for two initial states which differ in the jth position and are independent of all other flip-flop values. An example of two such states is $xx0xxx$ and $xx1xxx$. A distinguishing sequence of type B for flip-flop j is a sequence which, when applied to the fault-free machine with j^{th} flip-flop $= 0$ (or 1) and applied to the faulty machine with the same flip-flop $= 1$ (or 0), produces two distinct output responses independent of the values of all other flip-flops. A distinguishing sequence of type C is similar to type B except that a subset of flip-flops are assigned specific logic values.

The distinguishing sequences of type A are pre-generated statically for the fault-free machine. Sequences of type B and C are dynamically generated for both fault-free and faulty machines during test generation. The population is initialized with random sequences. Distinguishing sequences of the flip-flops to which fault effects have been propagated in the last phase are used as seeds in place of some of the random sequences. Since small population sizes are used in order to reduce execution time, the number of distinguishing sequences stored per flip-flop is limited to five.

A term called *distinguishing power* is associated with each distinguishing sequence. It indicates how well the sequence can propagate a fault effect from the corresponding flip-flop to the primary outputs. If a smaller subset of the flip-flops needs to be specified for this purpose, it indicates a higher distinguishing power. Furthermore, a distinguishing sequence capable of propagating a fault effect under a different faulty machine is given a higher power than a sequence which is not.

For flip-flops without an associated distinguishing sequence of any type, an observability value is used to indicate how observable is the flip-flop in the genetic algorithm framework. Initially, all flip-flops in the circuit are set to a certain observability value. As time progresses, the observability of a flip-flop is decreased if no distinguishing sequence can be found for it. The lower the observability value, the harder it is to generate a test sequence for that flip-flop. This value helps avoid propagation of fault effects to hard-to-observe flip-flops.

If a sequence that propagates the fault effects to the primary outputs is successfully found through genetic manipulation over successive generations, the sequence is added to the test set. The fault simulator is invoked to identify and drop other detected faults. Test generation continues with the next fault.

Since the fault activation and fault propagation phases have different aims, their fitness functions differ. The parameters that are used to judge the fitness of an individual in the population are as follows:

$P1$:   Fault detection
$P2$:   Sum of dynamic controllabilities
$P3$:   Matches of flip-flop values
$P4$:   Sum of distinguishing powers
$P5$:   Induced faulty circuit activity
$P6$:   Number of new states visited.

Parameter $P1$ is self-explanatory for the fault propagation phase. It is included in the fault activation phase to account for fault effects that directly propagate to the primary outputs in the time frame in which they are activated. To improve state justification and fault propagation, hard-to-control or hard-to-observe flip-flops are identified dynamically and are given lower controllability values. Using these values, justification of difficult states in the first phase and the propagation of fault effects to hard-to-observe flip-flops in the first and second phases can be avoided. The quality of the state to be justified is indicated by $P2$. Maximizing $P2$ during single-time-frame fault activation makes the state more easily justifiable and also avoids unjustifiable states. On the other hand, minimizing $P2$ during state justification expands the search space by forcing visits to hard-to-justify states. $P3$ guides the genetic algorithm to match the required flip-flop values in the state to be justified during state justification, from the least controllable to the most controllable flip-flop value. $P4$ is an indicator of the quality of the set of flip-flops reached by the fault effects. Maximizing $P4$ improves the probability of fault effects reaching flip-flops with more powerful distinguishing sequences. $P5$ is a measure of the number of events generated in the faulty circuit, with events on more observable gates weighted higher. For example, events are weighted higher for logic cones of primary outputs or flip-flops with more powerful distinguishing sequences. $P6$ is used to expand the search space. $P6$ is considered in the final stage only when the fault detection count drops to a very low value. The fitness function is defined for different phases as follows:

Fault activation phase:

Multiple time frame:

    1 $fitness = 0.2P1 + 0.8P4.$

Single time frame:

    2 $fitness = 0.1P1 + 0.5P2 + 0.2(P4 + P5 + P6).$

State justification:

    3 $fitness = 0.1P1 + 0.2(k - P2) + 0.5P3 + 0.2(P5 + P6)$, where k is a constant.

Fault propagation phase:

    4 $fitness = 0.8P1 + 0.2(P4 + P5 + P6).$

In the fault activation phase, the fitness function places a heavier weight on the quality of flip-flops reached by the fault effects. If no sequence is obtained to activate the targeted fault, single-time-frame activation and state justification are used in a second attempt to activate the fault. In this case, the fitness function favors states that can be easily justified in single-time-frame activation, while it favors hard-to-reach states during state justification since such states may be necessary to reach the target state. In the fault propagation phase, the aim is to find a sequence which propagates the fault effects to a primary output. Hence, the emphasis is on fault detection.

Since the genetic algorithm is not capable of declaring a fault untestable, targeting such faults is a wastage of effort. Therefore, after the first stage of the genetic algorithm, a sequential test generator is used to identify and remove from consideration many of the untestable faults. A small time limit per fault is used to minimize execution time.

5.4.2.3 Hybrid test generation

Some hybrid methods have also been proposed which combine the topological analysis based methods with simulation based methods in various ways (Saab *et al.*, 1994; Rudnick and Patel, 1995; Hsiao *et al.*, 1996b). For example, after a fixed number of generations without any improvement in fault coverage, an array-model based test generator can be used to derive test sequences for hard-to-detect faults, as well as to seed the current population. A fast run of a simulation-based test generator followed by a slow, but more successful, run of a deterministic test generator can provide better results in a shorter time than obtained by the deterministic test generator alone. The two test generation approaches can be repeated many times in each stage.

Another way to combine the two approaches is to use deterministic algorithms for fault excitation and propagation and genetic algorithms for state justification. If the genetic algorithm is not successful in state justification, the deterministic algorithm can be used for it.

5.5   Test generation for asynchronous circuits

There is a revived interest in asynchronous sequential circuits because of their potential for high performance and low power consumption. However, lack of high testability has been one of the hindrances in their widespread use. Since most asynchronous circuits are delay-dependent, care must be taken to avoid unpredictable behavior because of hazards and races. When an input vector is applied to it, an asynchronous circuit may cycle through many intermediate, *unstable* states before settling down in a *stable* state. This cycling occurs through the combinational logic of the asynchronous circuit over multiple time frames. It is also possible that the circuit never settles down even when no new input vector is applied. This phenomenon is called *oscillation*. It is important to take the above factors into account in the test generation process.

Since explicit consideration of delay-dependent behavior and unstable states is impractical during test generation for most designs, different models are generally used for test generation and fault simulation. In simulation-based methods, the test generator derives a potential test while ignoring circuit delays. The potential test is then simulated using proper delay models to check its validity. However, since many potential tests are invalidated in the simulation phase, this approach represents much wasted effort, and usually does not result in high fault coverage. It also necessitates the development of new software tools for asynchronous circuit test generation.

An alternative and promising approach is to adapt synchronous circuit test generation methods to asynchronous circuit test generation (Banerjee *et al.*, 1996). This approach also enables test generation for a circuit in which an asynchronous part is embedded in a larger synchronous part. A synchronous test model (STM) for the asynchronous circuit is used for this purpose. Since this is just a model, there is no actual modification needed in the hardware implementation of the asynchronous circuit. The main feature of the model is that it imposes the *fundamental mode* of operation during test generation. This means that a new input vector cannot be applied until the circuit has stabilized. The tests generated for the STM can be easily translated into tests for the asynchronous circuit.

A fault in an asynchronous circuit is detectable if the fault effect can be observed at a primary output and the circuit is in a stable state. If observation of a fault effect can only occur in an unstable state in both the good and faulty circuits, then the fault is undetectable. Therefore, a test for a target fault is invalid if the test assumes observation of the fault effect when the circuit (both good and faulty) is in an unstable state.

The cycle length of an asynchronous circuit is the maximum number of passes through the feedback path that is necessary for the circuit to stabilize, for any input change. This is equal to u, where u is the maximum number of unstable states between any two stable states. In the STM, both cycling and unstable states are taken into account. The model approximates the continuous-time behavior of the circuit

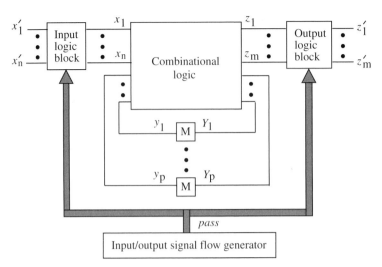

Figure 5.24 The synchronous test model for asynchronous circuits (Banerjee *et al.*, 1996) (© 1996 IEEE)

through division of time into fine units. It replaces the asynchronous latches on the feedback path with clocked flip-flops, and makes the clock period equal to the critical path delay of the combinational logic. Thus, each update of the state of the clocked flip-flops simulates a pass through the asynchronous latch. To properly simulate the cycling behavior, we need to maintain the primary input values for the duration of the cycle. In addition, the primary outputs should be observed only when the cycling is complete.

The STM is shown in Figure 5.24. Here, x_i' and z_i' denote the primary inputs and outputs, respectively, of the STM. The model flip-flops, which replace the asynchronous latches, are denoted by M. The STM also includes three additional blocks: an input logic block (ILB), an output logic block (OLB), and an input/output signal flow generator (IOSFG). The IOSFG generates a *pass* signal to control the application of new input vectors as well as to observe the primary outputs of the asynchronous circuit. It ensures that the circuit operates in the fundamental mode. It assumes that the maximum number of unstable states between any two stable states is known for both the fault-free and faulty circuits. Possible implementations of the ILB, OLB and IOSFG are shown in Figure 5.25. In the ILB implementation, when *pass* is low, x_i maintains its previous value. When *pass* goes high, x_i takes the value of the new input line x_i'. In the OLB implementation, when *pass* is high, $z_i' = z_i$, else z_i' remains at steady 0. The IOSFG implementation consists of a modulo-u counter, where u is the cycle length, and an AND gate. Thus, *pass* is asserted once every u cycles. For test generation, only the faults that correspond to those in the asynchronous circuit are targeted. Faults in the ILB, OLB and IOSFG need not be targeted as they are just part of the model, not the actual circuit.

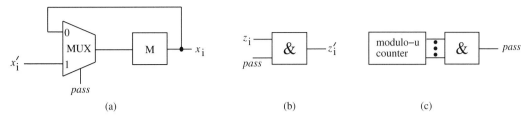

Figure 5.25 Implementations of (a) ILB, (b) OLB, and (c) IOSFG (Banerjee *et al.*, 1996) (© 1996 IEEE)

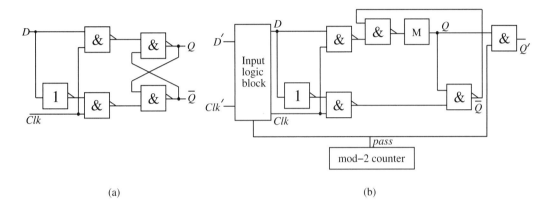

Figure 5.26 An asynchronous circuit and its STM (Banerjee *et al.*, 1996) (© 1996 IEEE)

Example 5.9 Consider the asynchronous circuit shown in Figure 5.26(a). Its STM is shown in Figure 5.26(b). D' and CLK' are the primary inputs of the STM and Q' its primary output. In the STM, one of the feedback loops is broken with the introduction of the clocked flip-flop M as shown. It is easy to check that this circuit requires only two cycles to stabilize. Thus, IOSFG can be implemented using a modulo-2 counter. Hence, the signal *pass* is asserted once every two clock cycles.

Tests for targeted faults can be derived using any synchronous sequential test generator. For example, consider fault Q SA1 in the asynchronous circuit. The test sequence which detects this fault in the STM specified at $(D', CLK', pass)$ is $\{(0, 0, 0), (0, 1, 1)\}$. The test for the asynchronous circuit is obtained by extracting every vector for which $pass = 1$. Thus, the translated test sequence at (D, CLK) is $\{(0, 1)\}$.   □

Even when an asynchronous circuit is embedded in a synchronous circuit, using the STM for the former allows one to obtain a synchronous model for the complete circuit. Thus, any synchronous sequential test generator can handle such circuits as well.

5.6 Test compaction

Shorter test sets result in smaller test application times. This has a strong influence on the testing costs since more chips can be tested in a given amount of time and fewer expensive testers are needed. In addition, if the test set size exceeds the amount of memory available in a tester for storing tests, it necessitates a memory reloading operation, which increases testing time.

There are two methods for obtaining compact test sets: *static* and *dynamic*. A **static test compaction** method tries to reduce the size, without sacrificing fault coverage, of a test set obtained through any sequential test generator. A **dynamic test compaction** method tries to reduce test set size concurrently with test generation. We describe both types of methods next.

5.6.1 Static test compaction

Three useful methods for static test compaction are based on the concepts of vector insertion, vector omission and vector selection (Pomeranz and Reddy, 1996). Of these, the method of reducing the length of a test sequence through vector omission is the most powerful, and is described next.

Suppose a test sequence $\{P_1, P_2, \ldots, P_l\}$ has been given, and the vectors are applied to the circuit at time units $u_1, u_2, \ldots, u_l$, respectively. Suppose the set of faults detected by this test sequence is F. The omission of vector P_i can only affect the detection of faults from F which are detected at or after time u_i. It may also cause a fault which is undetected by the test sequence to be detected after P_i is omitted from it. These effects can be taken into account through fault simulation.

The test vectors are considered for omission in the order in which they appear in the test sequence. For $i = 1, 2, \ldots, l$, P_i is omitted and the fault coverage of the reduced test sequence is recomputed by simulating only those faults from F that were detected at or after time u_i as well as the undetected faults. If the fault coverage after omission is not lower than the fault coverage before omission, then the omission is accepted. Else, P_i is restored. If a vector P_i cannot be omitted, then after omitting a subsequent vector, it may now become possible to omit P_i. Therefore, after a vector omission, the reduced test sequence is completely considered once again, until no more vectors can be omitted.

Example 5.10   Consider the circuit given in Figure 5.27. Suppose the test sequence, $\{(1, 0), (1, 1), (0, 0), (0, 1), (0, 1), (0, 0)\}$, has been derived for primary inputs (x_1, x_2). This test sequence, upon fault simulation, is found to detect all single SAFs in the circuit except x_2 SA1. If we omit the first vector from this test sequence, the fault coverage goes down because x_1 SA0 is no longer detected, neither is x_2

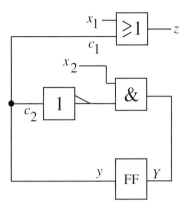

Figure 5.27   A circuit to illustrate static compaction

SA1. Thus, this omission is not accepted. Exactly the same situation occurs if the second vector is omitted. However, when the third vector is omitted, obtaining the reduced test sequence, $\{(1, 0), (1, 1), (0, 1), (0, 1), (0, 0)\}$, the fault coverage remains the same as before. Thus, this omission is accepted. We next omit vectors from the reduced test sequence from the beginning. Again, we find that omitting the first two vectors reduces the fault coverage. However, if the third vector is omitted, obtaining the test sequence, $\{(1, 0), (1, 1), (0, 1), (0, 0)\}$, the fault coverage remains the same. Therefore, this omission is accepted. If we try to omit any more vectors, we find that the fault coverage reduces. Therefore, we have found a test sequence of length four which has the same fault coverage as the initial test sequence of length six. In fact, they detect exactly the same set of faults.   □

5.6.2   Dynamic test compaction

As mentioned earlier, genetic algorithms can be used for efficient test generation. Together with fault simulation, genetic algorithms can also be used for efficient dynamic test compaction (Rudnick and Patel, 1996). Test generators frequently assume that the circuit starts from an unknown initial state. However, if the circuit state is known, some of the vectors at the beginning and end of a partially-specified test sequence may not be needed to detect the targeted fault. In addition, the unspecified bits in such a test sequence are usually filled randomly with 1 or 0. Using a genetic algorithm to fill these bits allows more faults to be detected when fault simulation is done with the derived test sequence. Since increasing the number of faults detected by a test sequence reduces the number of faults that the test generator has to target, an improvement in test generation time is also made possible.

Even if the test generator assumes an unknown initial state, the fault simulator may have information on the circuit state if any test vectors were included in the test set

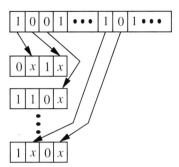

Figure 5.28 Genetic filling of unspecified bits

before. Therefore, for a generated test sequence of length L, fault simulations are performed for the targeted fault with the first k vectors removed from the test sequence, $1 \leq k \leq L - 1$. The shortest subsequence that detects the fault is selected. Thereafter, any vectors at the end of the sequence, which do not contribute to the detection of the targeted fault, are removed.

The unspecified bits in the resulting subsequence are filled with a genetic algorithm, as illustrated in Figure 5.28. In this example, the two unspecified bits in the first vector of the test sequence are filled with the first two bits of the genetic algorithm individual, the only unspecified bit in the second vector is filled with the third bit of the individual, and so on, until all unspecified bits are filled. The objective of this filling procedure is to maximize the number of secondary (i.e., non-targeted) faults detected by the test sequence. All the genetic algorithm individuals are initialized with random values, and unspecified bits in the test sequence are filled with successive bits from an individual, as shown above. The fully specified test sequence is fault simulated to determine its fitness value. This value reflects the quality of the solution, primarily in terms of the fault coverage. All the individuals of a genetic population are similarly evaluated. A new generation is then evolved using the usual genetic operators: selection, crossover and mutation. Mutation is done by complementing a single bit in the individual. This process continues for a fixed number of generations. The best test sequence found in any generation is added to the test set.

The number of new faults detected is the primary metric in the fitness function. However, to differentiate among test sequences that detect the same number of new faults, the number of fault effects propagated to the next state variables is also included in the fitness function. The number of propagated fault effects is divided by the number of faults simulated and the number of flip-flops to ensure that the number of detected faults is the dominant metric. Thus,

$$fitness = \#faults\ detected + \frac{\#faults\ propagated\ to\ flip\text{-}flops}{(\#faults\ simulated)(\#flip\text{-}flops)}.$$

To avoid the high computational cost of fault simulation for large circuits, the fitness of an individual can be estimated by using a random sample of faults. A random sample size of 100 is adequate in most cases. To avoid wasting time, the faults identified as untestable by the test generator are not included in the sample.

Good results have been obtained for a large class of circuits by using population sizes between 8 and 64, the number of generations limited to 8, a crossover probability of 1 and a mutation probability of 1/64.

Summary

- Fault simulation techniques for sequential circuits can be classified under various categories such as deductive, concurrent, differential, single-pattern parallel-fault, parallel-pattern single-fault and parallel-pattern parallel-fault.
- Test generation for synchronous sequential circuits can be done using its state table, purely from its gate-level implementation, purely from its higher level description such as register-transfer level, or hierarchically using more than one level of the design hierarchy.
- Gate-level test generation for synchronous sequential circuits can be based on topological analysis or simulation or a hybrid of these two approaches.
- Gate-level topological analysis methods either start from a known initial state called the reset state, or they assume an unknown initial state.
- Test generation for asynchronous sequential circuits can be based on simulation or topological analysis.
- Sequential circuits generally have two modes of operation: synchronization mode and free mode. Correspondingly, there are two test strategies: restricted and unrestricted. These test strategies can be used under one of two test generation approaches: single observation time and multiple observation time.
- A fault in a sequential circuit can be testable, untestable or redundant under a given operation mode and test strategy. An untestable fault is not necessarily redundant.
- A fault, which dominates another fault when the combinational logic of the sequential circuit is considered separately, may not dominate it in the sequential circuit. This can be due to two phenomena: self-hiding and delayed reconvergence.
- An efficient fault simulation method can be based on a hybrid of concurrent, differential and parallel fault simulation algorithms.
- Tests for controller-type sequential circuits can be efficiently derived using its state table and single-state-transition fault model.
- The most common method for sequential ATPG is based on the iterative array model derived from its gate-level implementation.
- Genetic algorithms, which generate tests based on simulations, have shown a lot of promise, as have hybrid gate-level test generation methods.
- Any test generator for synchronous sequential circuits can be adapted to test

asynchronous sequential circuits by using a synchronous test model which imposes the fundamental mode operation during test generation.

- There are two ways to compact a test set of a sequential circuit: static and dynamic. Static methods post-process the test set derived by a sequential test generator to try to drop vectors without sacrificing fault coverage. Dynamic methods are employed during test generation.

Additional reading

In Kung and Lin (1994), an extension of the fault simulation algorithm presented in this chapter is given which performs a parallel version of serial fault simulation on hypertrophic faults in parallel with fault-free simulation. Such faults prevent a flip-flop from being set to logic 0 or 1.

A dynamic fault grouping method is presented in Graham *et al.* (1997) to be used in conjunction with fault simulation.

A parallel pattern parallel fault sequential circuit fault simulator has been given in Amin and Vinnakota (1996). This is specially useful when the fault list is partitioned for multiprocessor simulation.

In Abramovici *et al.* (1992), a sequential circuit testing approach is given which makes the clock inactive and applies a group of vectors at the primary inputs in each state. This allows many faults to be combinationally detected.

In El-Maleh *et al.* (1995), it is shown that retiming preserves testability with respect to a single SAF test set by adding a prefix sequence of a pre-determined number of arbitrary input vectors. Retiming also allows a significant speed-up in test generation time in some cases.

Exercises

5.1 The circuit in Figure 5.6 has nine lines and, hence, 18 SAFs. Show that after equivalence and dominance fault collapsing and D flip-flop analysis, a set of only eight SAFs can be obtained.

5.2 Evaluate the fault-free and faulty circuit output values of the three-input AND gate shown in Figure 5.29.

5.3 Evaluate the fault-free and faulty circuit output values of the three-input XNOR gate shown in Figure 5.30.

5.4 Obtain the set of collapsed SST faults of minimum cardinality for the state table given in Table 5.5.

5.5 Derive a test sequence for the SST fault that corrupts the transition $\langle 0, C, D, 0 \rangle$

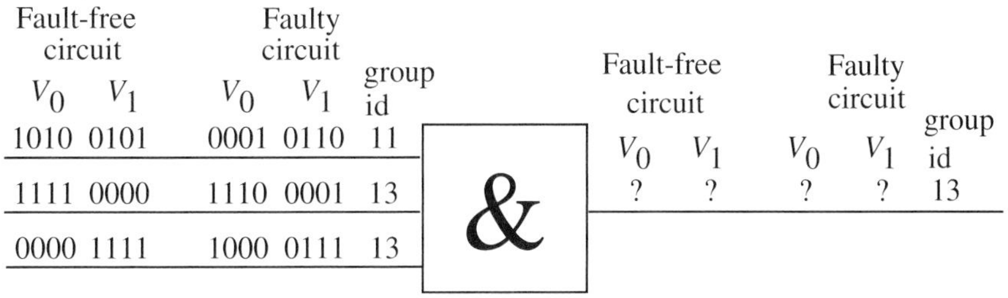

Fault-free circuit V_0 V_1	Faulty circuit V_0 V_1	group id
1010 0101	0001 0110	11
1111 0000	1110 0001	13
0000 1111	1000 0111	13

$\&$

Fault-free circuit V_0 V_1	Faulty circuit V_0 V_1	group id
? ?	? ?	13

Figure 5.29

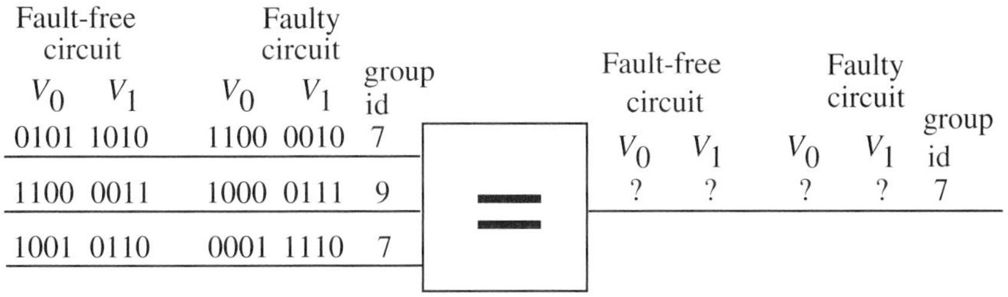

Fault-free circuit V_0 V_1	Faulty circuit V_0 V_1	group id
0101 1010	1100 0010	7
1100 0011	1000 0111	9
1001 0110	0001 1110	7

$=$

Fault-free circuit V_0 V_1	Faulty circuit V_0 V_1	group id
? ?	? ?	7

Figure 5.30

Table 5.7. *A state table*

	NS, z	
PS	$x_1 = 0$	$x_1 = 1$
A	B, 1	A, 1
B	C, 1	D, 0
C	D, 0	A, 1
D	B, 0	B, 0

to $\langle 0, C, C, 0 \rangle$ in Table 5.7. Which fault would be the best to target next from the point of view of reducing the test length?

5.6 For the x_2 SA0 fault in the circuit in Figure 5.27, derive a test sequence using the extended D-algorithm.

5.7 In the circuit in Figure 5.6, derive a test sequence for the c_3 SA1 fault using the nine-valued logic. Repeat for the c_2 SA0 fault.

5.8 For the sequential circuit given in Figure 5.4, suppose that the two flip-flops are resettable with an external reset signal, and $(0, 0)$ is the specified reset state. From the state reachability analysis, determine which faults in the circuit are undetectable.

5.9 Consider the input sequence $\{0, 0, 1\}$ for the circuit given in Figure 5.31. Identify which portion of the sequence is the periodic part, i.e., closes a cycle of states. Show that this periodic part activates the fault shown in the figure. Derive a test

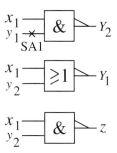

Figure 5.31

sequence T from the above sequence, assuming the allowed upper bound on test sequence length is five. Show that T also detects the shown fault.

5.10 For the sequential circuit given in Figure 5.27, derive a test sequence of length at most five which detects all single SAFs in it.

5.11 In a combinational circuit, it is never possible for any input vector, applied twice in consecutive clock cycles, to detect both the SA0 and SA1 faults on any line c in the circuit. However, give an example to show that it is possible for a particular input vector, applied twice in consecutive clock cycles, in a test sequence for a sequential circuit to detect both the SA0 and SA1 faults on some line in the combinational logic of the circuit.

References

Abramovici, M., Rajan, K.B., and Miller, D.T. (1992). FREEZE!: a new approach for testing sequential circuits. In *Proc. Design Automation Conference*, pp. 22–25.

Amin, M.B. and Vinnakota, B. (1996). ZAMBEZI – a parallel pattern parallel fault simulator for sequential circuits. In *Proc. VLSI Test Symposium*, pp. 438–443.

Banerjee, S., Chakradhar, S.T., and Roy, R.K. (1996). Synchronous test generation model for asynchronous circuits. In *Proc. Int. Conference on VLSI Design*, pp. 178–185.

Chen, J.E., Lee, C.L., and Shen, W.J. (1991). Single-fault fault-collapsing analysis in sequential logic circuits. *IEEE Trans. on Computer-Aided Design*, **10** (12), pp. 1559–1568.

Chen, X. and Bushnell, M.L. (1994). Dynamic state and objective learning for sequential circuit test generation using decomposition equivalence. In *Proc. Int. Symposium on Fault-Tolerant Computing*, pp. 446–455.

Cheng, W.-T. (1988). The BACK algorithm for sequential test generation. In *Proc. Int. Conference on Computer Design*, pp. 66–69.

Cheng, K.-T. and Jou, J.-Y. (1992). A functional fault model for finite state machines. *IEEE Trans. on Computer-Aided Design*, **11** (9), pp. 1065–1073.

Cheng, K.-T. (1996). Gate-level test generation for sequential circuits: a survey. *ACM Trans. on Design Automation of Electronic Systems*, **1** (3), pp. 405–442.

Cho, H., Hachtel, G.D., and Somenzi, F. (1993). Redundancy identification/removal and test generation for sequential circuits using implicit state enumeration. *IEEE Trans. on Computer-Aided Design*, **12** (7), pp. 935–945.

El-Maleh, A., Marchok, T., Rajski, J., and Maly, W. (1995). On test set preservation of retimed circuits. In *Proc. Design Automation Conference*, pp. 176–182.

Glaeser, U. and Vierhaus, H.T. (1995). FOGBUSTER: an efficient algorithm for sequential test generation. In *Proc. European Design Automation Conference*, pp. 230–235.

Graham, C. R., Rudnick, E. M., and Patel, J. H. (1997). Dynamic fault grouping for PROOFS: a win for large sequential circuits. In *Proc. Int. Conference on VLSI Design*, pp. 542–544.

Hsiao, M.H., Rudnick, E.M., and Patel, J.H. (1996a). Automatic test generation using genetically-engineered distinguishing sequences. In *Proc. VLSI Test Symposium*, pp. 216–223.

Hsiao, M.H., Rudnick, E.M., and Patel, J.H. (1996b). Alternating strategy for sequential circuit ATPG. In *Proc. European Design & Test Conference*, pp. 368–374.

Hsiao, M.H., Rudnick, E.M., and Patel, J.H. (1997). Sequential circuit test generation using dynamic state traversal. In *Proc. European Design & Test Conference*, pp. 22–28.

Kelsey, T.P., Saluja, K.K., and Lee, S.Y. (1993). An efficient algorithm for sequential circuit test generation. *IEEE Trans. on Computers*, **42** (11), pp. 1361–1371.

Kung, C.-P. and Lin, C.-S. (1994). HyHOPE: a fast fault simulator with efficient simulation of hypertrophic faults. In *Proc. Int. Conference on Computer-Aided Design*, pp. 714–718.

Lee, D.H. and Reddy, S.M. (1991). A new test generation method for sequential circuits. In *Proc. Int. Conference on Computer-Aided Design*, pp. 446–449.

Lee, H.K. and Ha, D.S. (1996). HOPE: an efficient parallel fault simulator for synchronous sequential circuits. *IEEE Trans. on Computer-Aided Design*, **15** (9), pp. 1048–1058.

Marlett, R. (1978). EBT: a comprehensive test generation technique for highly sequential circuits. In *Proc. Design Automation Conference*, pp. 332–339.

Muth, P. (1976). A nine-valued circuit model for test generation. *IEEE Trans. on Computers*, **C-25** (6), pp. 630–636.

Niermann, T. and Patel, J.H. (1991). HITEC: a test generation package for sequential circuits. In *Proc. European Design Automation Conference*, pp. 214–218.

Niermann, T., Cheng, W.-T., and Patel, J.H. (1992). PROOFS: a fast, memory-efficient sequential circuit fault simulator. *IEEE Trans. on Computer-Aided Design*, **11** (2), pp. 198–207.

Pomeranz, I. and Reddy, S.M. (1993). Classification of faults in synchronous sequential circuits. *IEEE Trans. on Computers*, **42** (9), pp. 1066–1077.

Pomeranz, I. and Reddy, S.M. (1994). On achieving complete fault coverage for sequential machines. *IEEE Trans. on Computer-Aided Design*, **13** (3), pp. 378–386.

Pomeranz, I. and Reddy, S.M. (1996). On static compaction of test sequences for synchronous sequential circuits. In *Proc. Design Automation Conference*, pp. 215–220.

Pomeranz, I. and Reddy, S.M. (1997). ACTIV-LOCSTEP: a test generation procedure based on logic simulation and fault activation. In *Proc. Int. Symposium on Fault-Tolerant Computing*, pp. 144–151.

Rudnick, E.M. and Patel, J.H. (1995). Combining deterministic and genetic approaches for sequential circuit test generation. In *Proc. Design Automation Conference*, pp. 183–188.

Rudnick, E.M. and Patel, J.H. (1996). Simulation-based techniques for dynamic test sequence compaction. In *Proc. Int. Conference on Computer-Aided Design*, pp. 67–73.

Saab, D.G., Saab, Y.G., and Abraham, J.A. (1992). CRIS: a test cultivation program for sequential VLSI circuits. In *Proc. Int. Conference on Computer-Aided Design*, pp. 216–219.

Saab, D.G., Saab, Y.G., and Abraham, J.A. (1994). Iterative [simulation-based genetics + deterministic techniques] = complete ATPG. In *Proc. Int. Conference on Computer-Aided Design*, pp. 40–43.

6 I_{DDQ} testing

I_{DDQ} testing refers to detection of defects in integrated circuits through the use of supply current monitoring. This is specially suited to CMOS circuits in which the quiescent supply current is normally very low. Therefore, an abnormally high current indicates the presence of a defect. In order to achieve high quality, it is now well-established that integrated circuits need to be tested with logic, delay as well as I_{DDQ} tests.

In this chapter, we first give an introduction to the types of fault models that I_{DDQ} testing is applicable to, and the advantages and disadvantages of this type of testing. We then present test generation and fault simulation methods for detecting such faults in combinational as well as sequential circuits. We also show how the I_{DDQ} test sets can be compacted.

We look at techniques for I_{DDQ} measurement based fault diagnosis. We derive diagnostic test sets, give methods for diagnosis and evaluate the diagnostic capability of given test sets.

In order to speed up and facilitate I_{DDQ} testing, various built-in current sensor designs have been presented. We look at one of these designs.

We next present some interesting variants of current sensing techniques that hold promise.

Finally, we discuss the economics of I_{DDQ} testing.

6.1   Introduction

In the quiescent state, CMOS circuits just draw leakage current. Therefore, if a fault results in a drastic increase in the current drawn by the circuit, it can be detected through the monitoring of the quiescent power supply current, I_{DDQ}. In such a testing method, the error effects of the fault no longer have to be propagated to circuit outputs. The faults just have to be activated. Because observability is no longer a problem, I_{DDQ}-testable faults are easier to derive tests for.

Consider the inverter shown in Figure 6.1(a). Suppose it has a defect d consisting of a short across the source and drain nodes of its pMOS transistor. We saw in Chapter 2 (Section 2.2.4.2) that such a defect is modeled as a stuck-on fault. When the input changes to logic 1, i.e., input voltage V_{IN} becomes high, the defect provides a direct

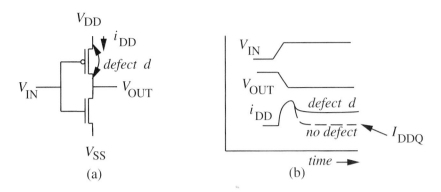

Figure 6.1 A defective inverter and its voltage/current waveforms

conduction path from V_{DD} to V_{SS}. This results in orders of magnitude increase in I_{DDQ}. Thus, I_{DDQ} testing can detect such a defect. The voltage and current waveforms are shown in Figure 6.1(b).

From the above example, we can deduce that I_{DDQ} testing will be successful for faults through which a conduction path can be activated from V_{DD} to V_{SS}. Bridging faults (BFs) are an important class of faults for which this is possible. Inductive fault analysis (see Chapter 2; Section 2.3) shows that up to 50% of the defects in manufactured chips may map to BFs. With shrinking geometries, this percentage is likely to become even higher in the coming years.

An **unrestricted BF** is a BF which results in unwanted conduction between any arbitrary pair of nodes in the circuit. Examples of nodes are wires connecting gates, points internal to a gate, V_{DD} and V_{SS}. If the shorted nodes (including internal, input and output) belong to the same gate then it is said to be an **intra-gate BF**. If the shorted nodes involve two distinct gates then it is said to be an **inter-gate BF**. BFs are usually defined at the gate level or switch level (Midkiff and Bollinger, 1993). At the gate level, a BF may be a short between a wire and V_{DD} or V_{SS}, or a short between two distinct wires. The former maps to stuck-at faults (SAFs). The latter can be classified under intra-gate/inter-gate as well as non-feedback/feedback. A BF between nodes c_i and c_j is said to be a **non-feedback BF** (NFBF) only if there does not exist a path from c_i to c_j or c_j to c_i in the circuit. Otherwise, it is called a **feedback BF** (FBF).

BFs defined at the switch level involve at least one interior node of the gate. An important fault model defined at this level is called the *leakage fault model*. In a fault-free MOS transistor, there should not be any current between gate and source, gate and drain, or gate and bulk (Mao and Gulati, 1992). The current flowing between source and bulk or drain and bulk should be less than a specified value if the bulk is connected to V_{DD} (V_{SS}) for a pMOS (nMOS) transistor. In addition, the current between source and drain should be less than a specified value when the transistor is off. However, certain defects, such as gate oxide shorts, can result in a substantial leakage current between certain terminals of the transistor. Six **leakage faults** are defined as follows for shorts between two terminals of the transistor: gate–source, gate–drain, source–drain, bulk–source, bulk–drain and bulk–gate. A stuck-on fault can be modeled by a leakage fault between source and drain. Hence, stuck-on faults need not be separately targeted.

When inductive fault analysis was performed on the layouts of many combinational benchmark circuits, it was found that about 65% to 85% of the BFs are gate-level BFs and about 15% to 35% of the BFs are switch-level BFs (Midkiff and Bollinger, 1993). Most of the switch-level BFs can be detected by targeting leakage faults.

For gate-level BFs, test generation can be done with the gate-level model of the circuit. This is substantially faster than using the switch-level model. In fact, leakage faults can also be detected using the gate-level model. For detecting unrestricted BFs, both the gate-level and switch-level models are required.

An obvious advantage of I_{DDQ} testing is that it allows us to detect many faults which may be undetectable by just monitoring the logic values at the circuit outputs. It is important to screen out chips with such defects because they are likely to fail early in the field. Thus, for obtaining very low defect levels, we must have I_{DDQ} testing in our arsenal of techniques that we use to attack the fault detection problem. The disadvantage of I_{DDQ} testing is that it is very slow. It may be possible to apply I_{DDQ} test vectors only at a frequency of 10 to 100 KHz. This means that it is important to obtain compact I_{DDQ} test sets. However, this problem can be substantially alleviated by including built-in current sensors on the chip.

6.2   Combinational ATPG

In this section, we will first discuss how we can detect leakage faults in combinational circuits. Then we will consider the general case of unrestricted BFs, and finally discuss test compaction.

6.2.1   Leakage fault detection

I_{DDQ} test generation can be of two types: *direct* and *indirect*. In **direct test generation**, the test set is derived by directly targeting the fault model. In **indirect test generation**, the test set is first derived for some other fault model (such as stuck-at) and a subset of this test set is then selected to detect the I_{DDQ}-testable faults. We will describe both such approaches for detecting leakage faults next.

6.2.1.1   Direct method

In order to motivate the direct method, let us first look at an example. Consider the NOR gate in Figure 6.2(a). Denote the gate, source, drain and bulk terminals of a

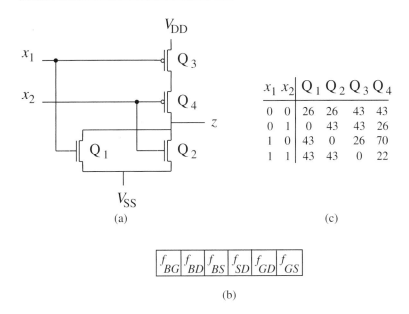

x_1 x_2	Q_1	Q_2	Q_3	Q_4
0 0	26	26	43	43
0 1	0	43	43	26
1 0	43	0	26	70
1 1	43	43	0	22

(a)   (c)

f_{BG}	f_{BD}	f_{BS}	f_{SD}	f_{GD}	f_{GS}

(b)

Figure 6.2 A NOR gate, leakage fault representation, and leakage fault table

transistor by G, S, D and B, respectively. Similarly, denote a bulk–gate leakage fault by f_{BG}, a source–drain leakage fault by f_{SD}, and so on. Suppose that the exhaustive set of two-bit vectors is fed to the NOR gate. A *leakage fault table* can then be obtained by observing the logic states at the terminals of each transistor. For example, if $(G, S, D, B) = (0, 0, 1, 0)$ for an nMOS transistor then faults f_{GD}, f_{SD} and f_{BD} will be detected. The reason is that the corresponding two terminals of the nMOS transistor will be activated to opposite logic values, thus creating a path from V_{DD} to V_{SS} in the presence of the fault. We will assume from here on that the bulk terminal of an nMOS (pMOS) transistor is connected to V_{SS} (V_{DD}).

The **leakage fault table** is a $2^j \times k$ matrix where j is the number of inputs of the gate and k is the number of transistors in it. Each non-zero entry of the matrix has two integers, which correspond to a six-bit binary number where each bit represents a leakage fault, as shown in Figure 6.2(b). A bit is set to 1 (0) if the corresponding leakage fault is detected (not detected) by I_{DDQ} testing. The first (second) integer in each leakage fault table entry represents the octal value of the first (last) three bits in the six-bit number. For example, the leakage fault table for the NOR gate is given in Figure 6.2(c). The entry corresponding to input vector $(0, 0)$ and nMOS transistor Q_1 is 26 whose two integers correspond to $(0, 1, 0, 1, 1, 0)$. This means that $(0, 0)$ will detect the following leakage faults: f_{BD}, f_{SD}, f_{GD}. This is derived from the fact that for $(0, 0)$, the (G, S, D, B) terminals of Q_1 will have the values $(0, 0, 1, 0)$. Other entries are similarly derived. Note that in the NOR gate figure, the upper (lower) terminal of an nMOS transistor is

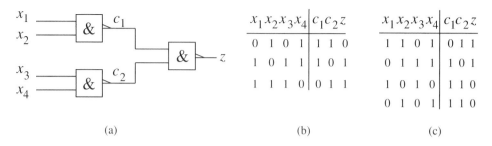

(a)   (b)   (c)

Figure 6.3 A gate-level model and its direct and indirect test sets

the drain (source) terminal, whereas for a pMOS transistor it is the source (drain) terminal.

There are three redundant leakage faults in the NOR gate: f_{BS} for transistors Q_1, Q_2 and Q_3 since the bulk and source terminals of these transistors always assume the same logic value. The minimal test set for detecting all the remaining detectable faults is $\{(0, 0), (0, 1), (1, 0)\}$. Readers may recall that this is precisely the single SAF test set of the NOR gate. A similar experiment with a two-input NAND gate shows that $\{(0, 1), (1, 0), (1, 1)\}$ is a minimal test set for detecting all detectable leakage faults in it. Again, this test set is the single SAF test set for the NAND gate.

The above observations can actually be generalized, as follows. Suppose that a test set is derived by targeting all single SAFs at gate inputs and listing a fault as detected if its effects can merely be propagated to the gate's output, rather than a primary output of the circuit. Such a test set is called a **pseudo-SAF test set** and is usually much smaller than the conventional SAF test set for large circuits. This test set can detect all detectable leakage faults (Malaiya and Su, 1982; Aitken, 1991). In order to reduce the size of the test set, when an SAF is activated by a vector, one can perform logic simulation of the circuit with this vector to determine which other SAFs also got activated. These faults can then be dropped from the fault list.

Example 6.1 Consider the gate-level model of a static CMOS circuit shown in Figure 6.3(a). Its minimal pseudo-SAF test set is given in Figure 6.3(b). This test set provides vectors $(0, 1)$, $(1, 0)$ and $(1, 1)$ to each of the three NAND gates. However, when these vectors are applied to a particular gate, its output value is not necessarily propagated to circuit output z. □

Most multiple leakage faults will also be detected by a pseudo-SAF test set since all it takes to detect a multiple fault with I_{DDQ} testing in fully complementary MOS circuits is that at least one of the constituent single faults activate a path from V_{DD} to V_{SS}.

6.2.1.2 Indirect method

An indirect method for detecting leakage faults can be obtained for standard cell library based designs as follows. First, the standard cells are characterized at the switch level by exhaustively feeding all possible input vectors to them and obtaining leakage fault tables. This characterization is a one-time cost incurred for the cell library. In the indirect method, the assumption is that a test set targeting some other fault model has been given to us. We take one vector at a time from this test set and perform gate-level logic simulation to determine what the logic values are at the inputs of each gate in the circuit. The fault tables generated earlier are then used to infer what faults are detected in the entire circuit with this vector. If a vector detects at least one new leakage fault, it is greedily selected (Mao and Gulati, 1992).

Example 6.2   Consider the circuit in Figure 6.3(a) once again. Its minimal single SAF test set is given in Figure 6.3(c). If this is the test set given to us, then we find that each of the first three vectors detects a new leakage fault. However, the fourth vector does not detect a new leakage fault. Hence, the leakage fault test set consists of just the first three vectors.   □

Although for the small circuit used in the above examples, it was not obvious, in the experiments reported in Mao and Gulati (1992), the size of the obtained indirect leakage fault test set was less than 1% of the size of the production test set that was used as the starting point. The advantage of the indirect test set is that the testing speed needs to be slowed down for I_{DDQ} testing only for vectors from this test set. For these and the remaining vectors in the production test set, logic monitoring can be done as usual.

6.2.2   ATPG for unrestricted BFs

In this section, we will first discuss conditions for the detection of BFs. Then we will consider fault collapsing methods for BFs. Next, we will present a method for representing fault lists which makes test generation and fault simulation of BFs time- and memory-efficient. We will look at a genetic algorithm for test generation. Thereafter, we will present a fault simulation method. We will limit ourselves to the consideration of a BF between two lines only. This is owing to the fact that if a BF between multiple nodes is present and we activate a path from V_{DD} to V_{SS} through any two nodes involved in the BF, I_{DDQ} testing will detect the multi-node BF as well. Also, we will consider all two-node BFs in the circuit from here on. It becomes necessary to do this in the absence of layout information. However, if layout information is available, then inductive fault analysis can help reduce the candidate set of two-node BFs.

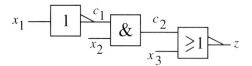

Figure 6.4 Detection of FBFs

6.2.2.1 Conditions for detecting BFs

Let $P(r)$ denote the logic value of node r on the application of vector P to the fault-free circuit. For the NFBF $\langle r_1, r_2 \rangle$, as we discussed earlier, the only requirement for detection is that $P(r_1)$ and $P(r_2)$ assume opposite logic values. However, this represents an **optimistic condition** for the detection of FBFs. A **pessimistic condition** for detecting FBFs can be derived as follows. Consider an FBF $\langle r_1, r_2 \rangle$, in which there is a path from r_1 to r_2. Let $P'(r_2)$ be the logic value at r_2 on the application of P obtained by setting r_1 to x (don't care) and using three-valued simulation. Then P is a test for this FBF if and only if $P(r_1)$ and $P'(r_2)$ assume opposite logic values and $P'(r_2) \neq x$ (Chakravarty and Thadikaran, 1996).

The need for the pessimistic condition can be illustrated through the circuit in Figure 6.4. Suppose that this circuit is a gate-level model of a static CMOS circuit. Consider the input vector P for which $(x_1, x_2, x_3) = (0, 1, 0)$. In the fault-free circuit, $P(c_1) = 1$ and $P(z) = 0$. There is a conducting path from node c_1 to V_{DD} and node z to V_{SS}. In the faulty circuit which contains the FBF $\langle c_1, z \rangle$, the values on nodes c_1 and z will be ambiguous. Since the values on inputs x_2 and x_3 are unaffected by the FBF, the value on node c_2 will be ambiguous too. Thus, there may or may not be a conduction path from z to V_{SS}. Hence, P may not activate a low resistance conduction path from V_{DD} to V_{SS}. Therefore, $(0, 1, 0)$ may not be a valid test for the FBF $\langle c_1, z \rangle$ even though $P(c_1) \neq P(z)$. Next, consider vector $(0, 1, 1)$. For this vector, even if c_1 is set to x, the value on z, i.e., $P'(z)$, is 0. Thus, there is guaranteed to be a conduction path between V_{DD} and V_{SS}, and the fault is detected.

For simplicity of exposition, henceforth we will assume that fault detection and diagnosis are based on the optimistic condition.

6.2.2.2 Fault collapsing

In order to reduce the test generation and fault simulation effort, we need to collapse the initial list of BFs. Suppose two nodes r_1 and r_2 exist in the circuit such that for every input vector P, $P(r_1) = P(r_2)$. Then the BF $\langle r_1, r_2 \rangle$ is redundant. Furthermore, if a set of vectors T is such that it detects the BFs between node r_1 and nodes in set R, then T will also detect the BFs between node r_2 and nodes in R. Hence, every BF involving node r_2 can be replaced with a corresponding fault involving node r_1 (Reddy et al., 1995; Thadikaran and Chakravarty, 1996).

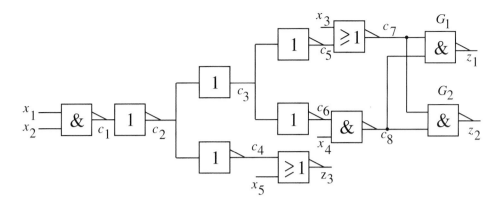

Figure 6.5 Example of fault collapsing

The first method for fault collapsing that uses the above arguments involves the identification of trees containing inverters and/or buffers in the circuit. Consider the root c_i of such a tree. If there exists at least one inverter with output c_j in this tree such that the path from c_i to c_j in the tree does not contain any other inverters, then we need to consider only those BFs for which one of the nodes is c_i or c_j. The BFs involving the other nodes in the tree can be ignored. If many inverters such as c_j exist, one can be picked randomly. If no such inverter exists (i.e., the tree consists of only buffers) then only node c_i from the tree needs to be considered for the BFs in the circuit.

Consider the gate-level model shown in Figure 6.5. Node c_1 is the root of a tree of inverters and buffers. The path from c_1 to c_2 does not contain any other inverters. Therefore, the BFs involving nodes c_3, c_4, c_5, c_6, among themselves or with other nodes, need not be considered.

The second method for fault collapsing involves fanout nodes. Consider a set of nodes S such that each node in S has the same fanin nodes and realizes the same function. Then a BF between any pair of nodes in S is redundant, and the above arguments can again be applied. For example, in Figure 6.5, the fanins of nodes z_1 and z_2 are the same and these two nodes realize the same function. Therefore, only the BFs involving either z_1 or z_2, but not both, need to be considered. This argument can be extended to the internal nodes of gates G_1 and G_2 as well, i.e., only those BFs need to be considered which involve the internal nodes of either gate G_1 or G_2, but not both.

Simple fault collapsing techniques, such as above, result in up to 90% reduction in the number of BFs that need to be considered in the benchmark circuits.

6.2.2.3 Fault list representation

An efficient way to represent the fault list is to use the **set of ordered pairs of sets** (SOPS) of the form $\{\langle R_1, S_1 \rangle, \langle R_2, S_2 \rangle, \ldots, \langle R_k, S_k \rangle\}$, where R_i and S_i

are a set of nodes in the circuit. The ordered pair of sets $\langle R_i, S_i \rangle$ denotes the set of BFs $\langle r, s \rangle$, where $r \in R_i$, $s \in S_i$ (Thadikaran *et al.*, 1995). For example, the SOPS $\{\langle \{c_1, c_2\}, \{c_3, c_4\} \rangle, \langle \{c_1, c_2\}, \{c_1, c_2\} \rangle\}$ denotes the set of BFs $\{\langle c_1, c_3 \rangle, \langle c_1, c_4 \rangle, \langle c_2, c_3 \rangle, \langle c_2, c_4 \rangle, \langle c_1, c_2 \rangle\}$.

In the above example, c_1 and c_2 are repeated three times. If a large number of nodes are repeated many times, the space used by the representation increases. This can be contained through the use of pointers. For example, if $A_0 = \{c_1, c_2\}$ and $A_1 = \{c_3, c_4\}$, then the above SOPS can be simply represented as $\{\langle A_0, A_1 \rangle, \langle A_0, A_0 \rangle\}$.

6.2.2.4 Test generation

An efficient way to generate tests for BFs is through the genetic algorithm (Thadikaran and Chakravarty, 1996). The few faults that remain undetected through this algorithm can be detected by applying an SAF test generator to a transformed circuit.

The genetic algorithm has been used for many optimization problems. It is specially attractive in the context of test generation when there is a large number of faults to be considered, as is the case for BFs. The BF test generation problem can be formulated as an optimization problem as follows. We compute one vector at a time. Suppose we have computed a set of vectors $\{P_1, P_2, \ldots, P_{i-1}\}$ and the list of detected faults with respect to these vectors is $F(P_{i-1})$. The next vector P_i is chosen such that the number of newly detected faults in $F(P_i)$ is maximized.

We first start with a population of vectors where the population size is fixed. Then we evolve the population over a fixed number of generations based on the genetic algorithm concepts of selection, crossover and mutation. Once a new generation of vectors is created, the best vector of the generation is computed based on a fitness function. A useful fitness function, $FF(v)$ for vector v, that we can use is to make it equal to the number of newly detected BFs in $F(v)$. Higher the fitness function, better the vector. After evolving through all generations, the best vector generated during the process is selected as the new test vector. The various concepts are described below.

Selection: To generate the initial population for the next generation of vectors, the roulette wheel method is used. The wheel is divided up into *population_size* (i.e., the size of the population) sectors and each sector in the wheel is assigned to one vector. The sectors are of unequal size. The sector size assigned to vector v is proportional to $FF(v)$. The wheel is turned *population_size* times. For each spin of the wheel, if it stops in the sector for vector w then w is selected for the next generation. A vector can be selected many times.

Crossover: This is the process of producing two offsprings from two parents. If vectors P_1 and P_2 are the two parent vectors, a bit position r is randomly chosen for crossover. Two new vectors, W_1 and W_2, are created as follows. Bit positions 1 through $r - 1$ of W_1 (W_2) are the same as those of P_1 (P_2). The remaining bits of P_1 (P_2) are copied into positions $r, r + 1, \ldots$ of W_2 (W_1). One possible way to choose the vectors for crossover is to mate the first vector with the second, the third vector with the fourth, and so on.

Mutation: Each bit of every vector in the population is scanned and flipped with some fixed mutation probability.

A population size of 25, number of generations of 10 and a mutation probability of 0.001 have been found to give good results.

The best vector, P_i, found in the ith step is used to compute $F(P_i)$ from $F(P_{i-1})$ through incremental simulation. A method for doing this is given in the next sub-section. After computing a number of vectors, the best vector selected may have the same fitness value as the previous vector included in the test set. If the fitness value of the best selected vector does not change for some fixed number (typically five) of consecutive vectors, then this test generation process can be terminated.

Next, the BFs that are not detected by the test set generated in the genetic algorithm phase have to be detected or proved to be redundant. This is accomplished through an SAF test generator. However, to use such a test generator, the circuit needs to be modified as follows.

Gate-level BFs: For a BF $\langle c_1, c_2 \rangle$ where both c_1 and c_2 are gate outputs, we insert an XOR gate G with inputs c_1 and c_2. The target fault given to the SAF test generator is a stuck-at 0 (SA0) fault at the output of G. If a test is found for this fault then it would drive c_1 and c_2 to opposite logic values in the fault-free case and, hence, be a test for the BF. Else, the BF is redundant.

BFs between two internal nodes: For a BF $\langle c_1, c_2 \rangle$ involving two internal nodes, we use the detection criterion $\langle c_1 = 0, c_2 = 1 \rangle$ or $\langle c_1 = 1, c_2 = 0 \rangle$, as before. This is illustrated through the CMOS circuit in Figure 6.6. Consider the BF shown in the figure. There are two ways to detect this BF.

(i) $c_3 = 0$ and $c_5 = 1$: This requires ($c_1 = 1$) and ($x_4 = 0, x_5 = 0$). We can introduce gate G_1 in a circuit model such that the output of G_1 is logic 1 if this condition holds, as shown in Figure 6.7.

(ii) $c_3 = 1$ and $c_5 = 0$: This requires ($x_1 = 1, c_1 = 0$) and ($x_4 = 1$ or $x_5 = 1$). As before, we can introduce gate G_2 in a circuit model such that the output of G_2 is 1 if this condition holds, as shown in Figure 6.7.

Using G_1 and G_2, we can obtain a test for the BF $\langle c_3, c_5 \rangle$ if either the output of G_1 or G_2 is 1. This is accomplished by adding a two-input OR gate G, as shown in the figure. The target SAF is an SA0 at the output of G. This approach is, of course, also applicable when one of the shorted nodes is a gate-level node and the other is an internal node.

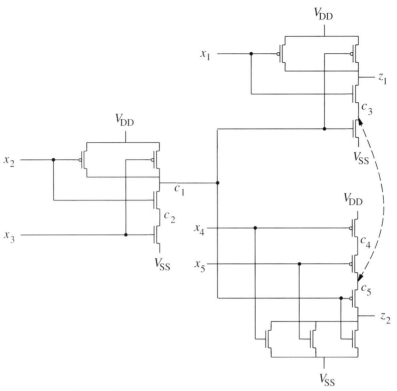

Figure 6.6   BFs between internal nodes

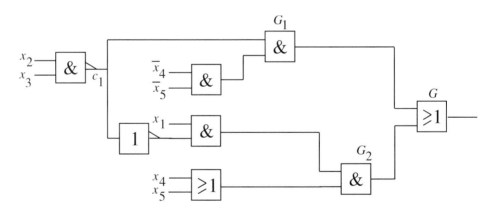

Figure 6.7   Modeling of BFs between internal nodes

6.2.2.5 Fault simulation

In fault simulation of BFs, we want to find out what BFs are detected by a given test set and what the resulting fault coverage is. We will use the SOPS fault list representation introduced earlier for this purpose. We need to first introduce the two concepts of *node partitions* and *indistinguishable pairs* (Thadikaran *et al.*, 1995).

Table 6.1. *Test set for the circuit in Figure 6.6*

Test	x_1	x_2	x_3	x_4	x_5	c_1	c_2	c_3	c_4	c_5	z_1	z_2
P_1	0	1	1	1	1	0	0	f	f	0	1	0
P_2	1	0	1	0	0	1	0	0	1	1	0	0
P_3	0	1	0	0	0	1	1	0	1	1	1	0

Consider the circuit in Figure 6.6 once again (ignore the dashed BF). Suppose a test set $T = \{P_1, P_2, P_3\}$ shown in Table 6.1 is given for this circuit. We obtain the values in the table after fault-free logic simulation. Here f denotes the value of a floating node which is connected to neither V_{DD} nor V_{SS} for the given vector. We can see that nodes x_2 and z_1 have the same logic values for all three tests. Such a set of nodes is considered equivalent with respect to T. Such equivalence classes of nodes are referred to as **node partitions**. For our example, the node partitions with respect to T are $\{\{x_1\}, \{x_2, z_1\}, \{x_3\}, \{x_4, x_5\}, \{c_1, c_5\}, \{c_2\}, \{c_3\}, \{c_4\}, \{z_2\}\}$.

Consider nodes c_3 and z_2 in the above example. For P_2 and P_3, the logic values of these nodes are the same. Since it is possible that even for P_1, the floating node c_3 could have the same value as z_2, the BF $\langle c_3, z_2 \rangle$ is not guaranteed to be detected. Such a fault can only be called a **potentially detected fault**. However, for our analysis we will be pessimistic and count such a fault under the undetected category. Since c_3 and z_2 are in distinct node partitions, we introduce the notion of indistinguishable pairs to keep track of such undetected BFs, as follows. Consider two node partitions A_i and A_j with respect to a test set T. We say that $\langle A_i, A_j \rangle$ forms an **indistinguishable pair** if for each vector of the test set, they either have the same logic value from the set $\{0, 1\}$ or one of them has the value f. For our example, $\langle \{c_1, c_5\}, \{c_4\} \rangle$ is an indistinguishable pair.

The above concepts can be utilized in a list-based fault simulation scheme. In this scheme, we maintain a list of node partitions and another list of indistinguishable pairs. For every test $P_i \in T$, both lists are updated.

We will first show how node partitions are derived for every test vector. A straightforward way to compute node partitions is to compute a table like Table 6.1 and then do a pairwise comparison of values assigned to each node. This is an expensive algorithm. A faster method is as follows. Let NP_0 be the initial set of node partitions and NP_i be the set of node partitions obtained after processing the set of vectors $\{P_1, \ldots, P_i\}$. For our running example, we can see that:

$NP_0 = \{\{x_1, x_2, x_3, x_4, x_5, c_1, c_2, c_3, c_4, c_5, z_1, z_2\}\}$

$NP_1 = \{\{x_1, c_1, c_2, c_5, z_2\}, \{x_2, x_3, x_4, x_5, z_1\}, \{c_3, c_4\}\}$

$NP_2 = \{\{c_2, z_2\}, \{x_1, c_1, c_5\}, \{x_2, x_4, x_5, z_1\}, \{x_3\}, \{c_3\}, \{c_4\}\}$

$NP_3 = \{\{z_2\}, \{c_2\}, \{x_1\}, \{c_1, c_5\}, \{x_4, x_5\}, \{x_2, z_1\}, \{x_3\}, \{c_3\}, \{c_4\}\}$

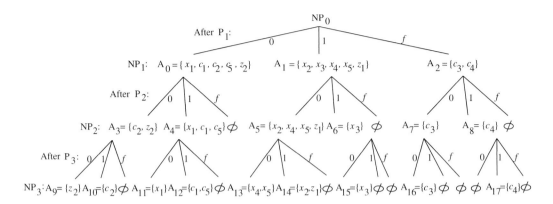

Figure 6.8 Node partition tree

We start with just one node partition. Then each partition is replaced with its 0, 1 and f subsets. This process is illustrated for our example through the node partition tree shown in Figure 6.8. In this tree, ϕ denotes a null set.

To understand the algorithm for computing the indistinguishable pairs, we continue with our example. We first initialize the initial set of indistinguishable pairs IP_0 to ϕ. On the application of P_1, NP_0 is split and $NP_1 = \{A_0, A_1, A_2\}$ is obtained, as shown in Figure 6.8. Using the SOPS notation, we know that BFs in $\langle A_0, A_2 \rangle$ and $\langle A_1, A_2 \rangle$ are not detected by P_1. These indistinguishable pairs belong to IP_1, as shown in Figure 6.9. Splitting an NP_i, in general, gives rise to two new indistinguishable pairs, such as above, corresponding to the $0/f$ and $1/f$ arcs. However, for each of A_0, A_1 and A_2, the f arc points to ϕ. Therefore, splitting these members of NP_1 to obtain NP_2 does not result in any indistinguishable pair. The other entries in IP_2 are a result of entries in IP_1. For example, after the application of P_2, A_0 gives rise to A_3, A_4, and A_2 gives rise to A_7, A_8. Since A_3, A_7 correspond to logic 0, and A_4, A_8 correspond to logic 1, $\langle A_3, A_7 \rangle$ and $\langle A_4, A_8 \rangle$ are indistinguishable pairs. This is depicted by arcs in Figure 6.9 from $\langle A_0, A_2 \rangle$ to $\langle A_3, A_7 \rangle$ and $\langle A_4, A_8 \rangle$. The other nodes in the indistinguishable pairs tree in Figure 6.9 are similarly derived. The two rules for deriving the set of indistinguishable pairs, IP_i, from members of NP_i and IP_{i-1} are summarized in Figure 6.10.

The BF coverage can be obtained from the node partitions and indistinguishable pairs as follows. After the application of P_3 in our example, we obtain $NP_3 = \{\{z_2\}, \{c_2\}, \{x_1\}, \{c_1, c_5\}, \{x_4, x_5\}, \{x_2, z_1\}, \{x_3\}, \{c_3\}, \{c_4\}\}$, and $IP_3 = \{\langle\{z_2\}, \{c_3\}\rangle, \langle\{c_1, c_5\}, \{c_4\}\rangle, \langle\{x_4, x_5\}, \{c_3\}\rangle\}$. The total number of undetected BFs is the number of undetected BFs in NP_3 and IP_3. The nine members of NP_3 contribute undetected BFs as follows: $0+0+0+1+1+1+0+0+0 = 3$. The three members of IP_3 contribute undetected BFs as follows: $1 + 2 + 2 = 5$. Hence, the total number of undetected BFs is eight. Since there are 12 nodes in the circuit, there are 66 (i.e., 12 choose 2) possible two-node BFs. Therefore, the coverage of test set T is 58/66.

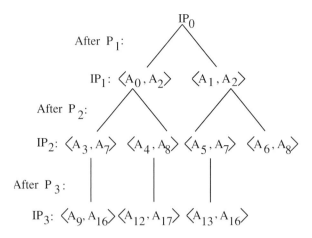

After P_1:
IP$_0$
IP$_1$: $\langle A_0, A_2 \rangle$ $\langle A_1, A_2 \rangle$

After P_2:
IP$_2$: $\langle A_3, A_7 \rangle$ $\langle A_4, A_8 \rangle$ $\langle A_5, A_7 \rangle$ $\langle A_6, A_8 \rangle$

After P_3:
IP$_3$: $\langle A_9, A_{16} \rangle$ $\langle A_{12}, A_{17} \rangle$ $\langle A_{13}, A_{16} \rangle$

Figure 6.9 Indistinguishable pairs tree

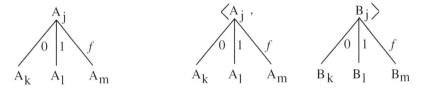

Rule 1: indistinguishable pairs
are $\langle A_k, A_m \rangle$, $\langle A_l, A_m \rangle$

Rule 2: indistinguishable pairs are
$\langle A_k, B_k \rangle$, $\langle A_k, B_m \rangle$, $\langle A_l, B_l \rangle$, $\langle A_l, B_m \rangle$,
$\langle A_m, B_k \rangle$, $\langle A_m, B_l \rangle$, $\langle A_m, B_m \rangle$

Figure 6.10 Rules for obtaining indistinguishable pairs (Thadikaran *et al.*, 1995) (© 1995 IEEE)

Formally, after the application of tests $P_1, P_2, \ldots, P_i$, let $NP_i = \{D_1, \ldots, D_k\}$ and $IP_i = \{\langle A_1, B_1 \rangle, \ldots, \langle A_m, B_m \rangle\}$. Then the total number of undetected BFs after applying these i vectors is equal to $NP\_size(NP_i) + IP\_size(IP_i)$, where

$$NP\_size(NP_i) = \sum_{j=1}^{k} \frac{|D_j| \times (|D_j| - 1)}{2}$$

$$IP\_size(IP_i) = \sum_{j=1}^{m} |A_j| \times |B_j|$$

and $|Z|$ denotes the cardinality of set Z.

In the list-based scheme presented above, we explicitly enumerated the indistinguishable pairs. A refinement of this technique under a *tree-based scheme* has also been presented in Thadikaran *et al.* (1995), which uses an implicit representation of indistinguishable pairs. This refinement can significantly reduce the time and space requirements for the algorithm.

6.2.3   Test compaction

Test compaction for I_{DDQ} tests is very important because of the slow speed at which such tests are applied. This can be accomplished through a *direct* scheme or an *indirect* scheme.

In the **direct scheme**, the aim to be ideally achieved is to halve the number of undetected faults with the addition of each new test vector to the test set. If the number of targeted BFs is α, and in each step such a test vector can be found, then the size of the minimum test set is $\lceil \log_2 \alpha \rceil$. However, finding such a test set, even when one exists, is computationally expensive. Still, this number provides a lower bound on achievable test set size. A method, which uses a combination of random pattern test generation, deterministic test generation and reverse-order fault simulation, to try to achieve the above aim is given in Reddy *et al.* (1995). In this approach, a pre-defined number, R_t, of random test vectors is first generated and fault simulated. Among these, the vector that detects the maximum number of BFs is selected. Empirically, $R_t = 10\,000$ has been found to be useful. After a desired BF coverage is reached, the method shifts to deterministic test generation which tries to halve the number of undetected BFs, whenever possible. After a test set is obtained in the above fashion, reverse-order fault simulation is used to simulate vectors in the reverse order in which they were generated. A vector which does not detect any new faults is discarded.

In the **indirect scheme**, a test set T is assumed to be given and the aim is to select a subset of T which achieves the same fault coverage as T (Chakravarty and Thadikaran, 1994). This problem can be formulated as follows. We can derive a table with the test vectors from the test set given as rows and all the targeted BFs given as columns. There is a 1 at the intersection of vector P and BF β if P detects β. A minimum subset of T is chosen such that the vectors in it together have 1s for all the BFs. An exact solution to this problem is very time- and memory-consuming. The SOPS notation can be used to reduce the memory requirement. To reduce the time requirement, various heuristics can be used. A greedy heuristic is to traverse the vectors in T in a fixed order and select a vector only if it detects a new BF. Reverse-order fault simulation is, in fact, one such method. A more time-consuming heuristic is to first obtain the list of faults detected by each vector and at each step pick the vector that detects the maximum number of BFs. The detected BFs are removed, and this process is repeated until no BF remains unconsidered.

6.3   Sequential ATPG

In this section, we will extend some of the previously presented methods to sequential circuits. Some background information is presented first.

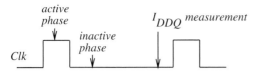

Figure 6.11　Clock phases and I_{DDQ} measurement

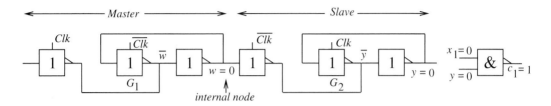

Figure 6.12　A master–slave flip-flop and a combinational block

6.3.1　The state change problem

We had earlier stated that under the optimistic condition of detecting a BF $\langle r_1, r_2 \rangle$, all we need to do is to drive r_1 and r_2 to opposite logic values in the fault-free circuit. This condition can also be used to detect BFs in sequential circuits except for BFs involving internal nodes of flip-flops. This is illustrated next. We assume that the I_{DDQ} measurement is made during the inactive phase of the clock, as shown in Figure 6.11. In the circuit in Figure 6.12, c_1 represents a node in the combinational logic of the sequential circuit and y, w represent the output of a master–slave flip-flop and one of its internal nodes, respectively. On the application of some test vector P, let the logic values of the different nodes be as shown. Consider the BF $\langle c_1, w \rangle$. In the fault-free case, $P(c_1) \neq P(w)$. Yet P is not a test for this BF due to the *state change problem* described next. Assume that node c_1 is stronger than node w. This may pull w up to logic 1. Since I_{DDQ} is measured during the inactive phase of the clock, inverter G_1 is enabled. Therefore, in the presence of the above BF, the faulty value at w can feed back through the loop $w - \bar{w} - w$. Hence, $\bar{w}$ is set to 0 and w to 1. This faulty value propagates through $\bar{y}$ to y. The new value of y will not propagate to c_1 because $x_1 = 0$ kills the propagation. This means that the fault will not be detected in the current cycle. In addition, the incorrect value at y may propagate to change the value of the next state outputs. This is called the **state change problem** (Lee and Breuer, 1992). Next, consider the BF $\langle c_1, y \rangle$. Since $P(c_1) \neq P(y)$, this BF is detected for the following reason. The faulty value at node y cannot feed back through the loop $y - \bar{y} - y$ because during the inactive phase of the clock, inverter G_2 is disabled and hence this loop is open. For the above reasons, shorts involving internal nodes of a flip-flop are usually not considered during test generation or fault simulation, but its output node is considered.

6.3.2   Potentially detected faults

In fault simulation of sequential circuits, we are given an I_{DDQ} test sequence T, and we are asked to compute the BF coverage of T. Generally, the initial state of the circuit is assumed to be unknown. One source of difficulty in fault simulation is the presence of potentially detected faults. For simulating the circuit, we set the initial value of all flip-flop outputs to the unknown value X. Now consider a BF $\langle c_1, c_2 \rangle$ such that on the application of a test vector P, $P(c_1) = 1$ and $P(c_2) = X$. During testing, $P(c_2)$ could be either 0 or 1. If it is 1, the BF is not detected. However, if it is 0, the BF is detected. Thus, the BF may or may not be detected depending on the initial state. We will be pessimistic and assume that such faults are not detected.

6.3.3   Next state computation

In order to simulate BFs in sequential circuits, we use logic simulation and vector P_i to compute the good (fault-free) next state. The next state computed in this fashion is used in the logic simulation of the following vector P_{i+1} in the sequence (Thadikaran *et al.*, 1995). The use of the good next state is valid for the following reason. Suppose that L_0 is the original BF list and suppose after processing vectors $P_1, \ldots, P_{i-1}$ and dropping the detected BFs from the fault list at each step, the reduced fault list L_{i-1} is obtained. Assume that present state PS_i is obtained after fault-free logic simulation with the above test sequence. After processing P_i, all BFs $\langle r_1, r_2 \rangle$, such that r_1 and r_2 assume opposite logic values, are dropped from L_{i-1} to obtain L_i. Thus, all faults in L_i are such that the present state PS_{i+1} obtained by simulating $P_1, \ldots, P_i$ in the presence of BFs in L_i is the same as the present state obtained by fault-free logic simulation of $P_1, \ldots, P_i$.

    The above assumption is also valid in the presence of potentially detected faults for the following reason. Consider such a fault $\langle r_1, r_2 \rangle$ in a sequential circuit where on application of vector P_i, $P_i(r_1) = 1$ and $P_i(r_2) = X$. Let NS_i be the fault-free next state and NS_i' be the next state with this fault in the circuit. During testing, $P_i(r_2)$ is either 0 or 1. If it is 0, the fault is detected. If it is 1, the BF is not activated and $P_i(r_1) = P_i(r_2)$. In this case, $NS_i' = NS_i$.

6.3.4   Test generation

The fault collapsing methods given earlier for combinational circuits are also valid for sequential circuits. In addition, a genetic algorithm can be used to do test generation for sequential circuits as well. One method which is based purely on a genetic algorithm is given in Lee *et al.* (1996). Although the first few test sequences derived through this genetic algorithm can result in 98% or higher BF coverage, detection of some of the remaining hard-to-detect BFs may require a large number of extra test sequences.

Also, the few BFs that remain undetected cannot be proved to be redundant through the use of this method. Therefore, as in the case of combinational circuits, one can derive a few test sequences through a genetic algorithm to achieve a high BF coverage, and then switch to deterministic test generation.

For test generation based on a genetic algorithm, we start with some randomly generated test sequences. A fitness function is used to evaluate each test sequence and a population of test sequences evolves over a couple of generations in order to improve the fitness of individual test sequences in the population. A fitness function $FF(S)$ which maximizes the BF coverage while minimizing test length for a test sequence S is given below:

$$FF(S) = \sum_{i=1}^{s}(s - i + 1) \cdot R_i,$$

where s is the length of the test sequence S (i.e., the number of vectors in S) and R_i is the fraction of remaining faults first detected by the ith vector of S. Because the fault coverage is weighted, the genetic algorithm tries to maximize the fault coverage of the early vectors. This makes the overall test set more compact.

After the fitness function of each test sequence is computed, the best test sequences are selected. For each selected pair of test sequences, a crossover operation is performed to produce a pair of offspring test sequences. In addition, mutations are occasionally introduced in some randomly chosen bits to enhance the diversity of the sequences. After these operations, a new generation of potentially more highly fit test sequences are produced which replaces the previous generation. The best test sequences from a population may be selected directly from one generation to the next. This is determined by a parameter called **generation gap** which determines the fraction of test sequences that are replaced in evolving the next generation. This process is repeated until some termination condition is met. For example, the genetic algorithm could be stopped when five consecutive generations fail to achieve any progress. Then the best test sequence from all generations is selected and added to the test set. Finally, test generation is itself stopped when a new test sequence fails to detect any new faults.

In order to target the remaining undetected BFs, we can use deterministic test generation as follows (Jha *et al.*, 1992). The gate-level or internal node BFs are modeled at the gate level in exactly the same fashion as required for combinational test generation earlier. If for the corresponding SAF, a test vector cannot be found by a combinational test generator then the BF is combinationally redundant. If a test vector can be found, then we need a state justification step. This means that the part of the test vector that corresponds to the present state needs to be reached from the all-X initial state (or the reset state as the case may be). If this is not possible then another test vector is generated through the combinational test generator. If for each such test vector, state justification is not possible, then the BF is sequentially untestable.

Table 6.2. *Test set for the circuit in Figure 6.13*

Test	x_1	x_2	c_1	c_2	c_3	c_4	c_5	c_6	z
P_1	0	1	X	1	0	X	X	1	1
P_2	0	0	0	1	1	f	1	0	0
P_3	0	0	1	1	0	0	1	1	1

As explained earlier, state justification can be done through the fault-free sequential circuit.

6.3.5 Fault simulation

A simple modification of the fault simulation algorithm presented earlier for combinational circuits also makes it applicable to sequential circuits. In sequential circuits, in addition to 0, 1, and f, a node can also take on the value X. The concept of node partitions remains the same as before. However, the concept of indistinguishable pairs has to be modified as follows.

Consider two node partitions A_i and A_j with respect to a test set T. We say that $\langle A_i, A_j \rangle$ forms an **indistinguishable pair** if for each vector of the test set, A_i and A_j either have the same logic value from the set $\{0, 1\}$ or one of them has the value f or X. The extension of the previous method can be illustrated with the following example.

Example 6.3 Consider the sequential circuit shown in Figure 6.13. Suppose the test set consisting of the sequence $T = \{P_1, P_2, P_3\}$ shown in Table 6.2 is given for this circuit. We obtain the values in the table after fault-free logic simulation. The node partitions with respect to T can be seen to be $\{\{x_1\}, \{x_2\}, \{c_1\}, \{c_2\}, \{c_3\}, \{c_4\}, \{c_5\}, \{c_6, z\}\}$. Similarly, the indistinguishable pairs can be seen to be $\langle \{x_1\}, \{c_4\} \rangle$, $\langle \{x_2\}, \{c_4\} \rangle$, $\langle \{c_1\}, \{c_6, z\} \rangle$, $\langle \{c_2\}, \{c_5\} \rangle$, $\langle \{c_3\}, \{c_4\} \rangle$. □

The list-based fault simulation method given earlier for combinational circuits is directly applicable with the modification that f and X have to be treated similarly. This is made possible because we are allowed to do fault-free logic simulation to get the next state. The method for determining fault coverage also remains the same.

6.3.6 Test compaction

We present an indirect scheme for compaction of test sets derived for sequential circuits. Recall that in such a scheme, a test set T is assumed to be given and the aim is to select a subset of this test set for I_{DDQ} testing such that the same fault coverage is obtained as the original test set.

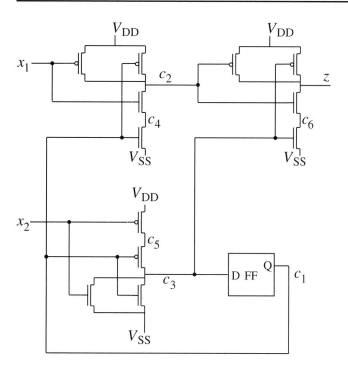

Figure 6.13 A sequential circuit

A greedy heuristic would be to traverse the test set in the given order and select a vector only if it detects a new BF (Chakravarty and Thadikaran, 1994). However, care must be taken regarding potentially detected faults. If such a fault is not activated then our use of fault-free logic simulation to get the next state is correct. However, if it is activated then the next state obtained can be different from the fault-free next state used in fault simulation. This implies that if a BF exists, which is a potentially detected fault for vector P_i in the test set, then P_i must be selected for I_{DDQ} testing. More sophisticated heuristics, such as those used for combinational circuits, can also be used for compacting test sets for sequential circuits, as long the above caveat about potentially detected faults is kept in mind.

6.4   Fault diagnosis of combinational circuits

In this section, we will discuss fault diagnosis issues related to I_{DDQ} testing. The first problem that we will address is that of analysis, defined as follows. Given a combinational circuit C, determined by an I_{DDQ} test set T to be faulty, determine which BF is present in C (Burns, 1989; Aitken, 1991; Chakravarty and Liu, 1992). The second problem we will consider is that of test set evaluation. For this problem, various diagnostic measures will be defined, on the basis of which we can determine

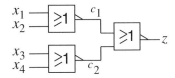

Figure 6.14   A gate-level circuit to illustrate the analysis algorithm

Table 6.3. *Test set for the circuit in Figure 6.14*

Test	x_1	x_2	x_3	x_4	c_1	c_2	z	E
P_1	0	0	0	0	1	1	0	0
P_2	0	1	0	1	0	0	1	1
P_3	0	0	1	1	1	0	0	0

the diagnostic quality of a given test set (Chakravarty, 1996). The third problem is of diagnostic test generation, where an I_{DDQ} test set, which has good diagnostic measures, is generated (Chakravarty *et al.*, 1995).

6.4.1   Analysis

For analysis, we will limit ourselves to gate-level BFs (Chakravarty and Liu, 1992). However, the method we describe next can be easily extended to unrestricted BFs. For the given test set T, we assume that the information is given for each test vector in T as to whether the circuit gave a faulty response for that vector or not. Also, we assume that only single two-line BFs are present.

An analysis algorithm starts with an initial set of all possible faults. Then it drops faults from this set by processing each input vector from the test set and the circuit response to this vector. This process continues until the test set is exhausted. To make this process efficient, we will use the SOPS representation defined earlier. The algorithm we describe here uses only logic simulation, and no fault simulation.

The algorithm is best illustrated through an example first.

Example 6.4   Consider a CMOS circuit gate-level model shown in Figure 6.14. Suppose the test set $T = \{P_1, P_2, P_3\}$ shown in Table 6.3 is given for this circuit. An entry 1 in the last column E indicates that there was a large I_{DDQ} on the application of the corresponding test vector, else the entry is 0. We obtain the values on lines c_1, c_2 and z through fault-free logic simulation. The node partition tree for this example is shown in Figure 6.15. Let K_i be the SOPS denoting the possible set of BFs after processing vector P_i. Since P_1 did not give a faulty response, the fault must be between lines in A_0 or between lines in A_1, but not between a line in A_0 and a line in A_1. Thus,

$$K_1 = \{\langle A_0, A_0 \rangle, \langle A_1, A_1 \rangle\}$$

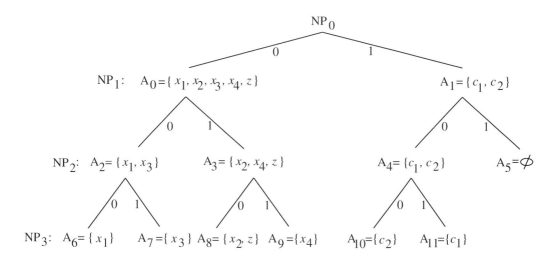

Figure 6.15 Node partition tree

Next, we process P_2. Since there was a faulty response for this vector, the BF must be between lines in A_2 and A_3 or between lines in A_4 and A_5. However, since A_5 is null, K_2 reduces to

$$K_2 = \{\langle A_2, A_3 \rangle\}$$

Finally, we process P_3. Since there was no faulty response for this vector, the BF must be between lines in A_6 and A_8 or between lines in A_7 and A_9. Thus,

$$
\begin{aligned}
K_3 &= \{\langle A_6, A_8 \rangle, \langle A_7, A_9 \rangle\} \\
&= \{\langle x_1, x_2 \rangle, \langle x_1, z \rangle, \langle x_3, x_4 \rangle\}
\end{aligned}
$$

K_3 contains the three BFs which are consistent with the test set and the observed response. One can check that for each of these three BFs, the logic values of the two lines in the BF only differ for P_2, which is the vector that yielded the faulty response. $\square$

Formally, let $K_{i-1} = \{\langle R_1, S_1 \rangle, \langle R_2, S_2 \rangle, \ldots, \langle R_k, S_k \rangle\}$, and let the ith value of E, corresponding to vector P_i, be e_i. For $1 \leq j \leq k$, we process $\langle R_j, S_j \rangle$ to form new ordered pairs. These new ordered pairs are added to K_i, and $\langle R_j, S_j \rangle$ is deleted from K_{i-1}. Let $R_j^0(P_i)$ $(R_j^1(P_i))$ be the subset of R_j whose member lines are at logic 0 (1) when P_i is applied. Similarly, define subsets of S_j. We need to consider two cases.

Case 1: $R_j = S_j$. If $e_i = 0$, then the new ordered pairs are $\langle R_j^0(P_i), R_j^0(P_i) \rangle$ and $\langle R_j^1(P_i), R_j^1(P_i) \rangle$. If $e_i = 1$, then the new ordered pair is $\langle R_j^0(P_i), R_j^1(P_i) \rangle$.

Case 2: $R_j \neq S_j$. If $e_i = 0$, then the new ordered pairs are $\langle R_j^0(P_i), S_j^0(P_i) \rangle$ and $\langle R_j^1(P_i), S_j^1(P_i) \rangle$. If $e_i = 1$, then the new ordered pairs are $\langle R_j^0(P_i), S_j^1(P_i) \rangle$ and $\langle R_j^1(P_i), S_j^0(P_i) \rangle$.

Note that if either R or S is null, then $\langle R, S \rangle$ denotes a null set of faults and is, hence, not added to K_i. Similarly, if R is null, then $\langle R, R \rangle$ denotes a null set and is not added to K_i. If there are N vectors in the test set, then the union of the set of BFs represented by the ordered pairs in K_N denotes the set of possible BFs.

6.4.2   Diagnostic fault simulation

Given a test set, we are frequently interested in finding out how good are its diagnostic capabilities. In order to answer this question, we have to first define some quantitative measures, called **diagnostic measures**, as follows.

Two faults, f_1 and f_2, are said to be **equivalent** with respect to a test set T if and only if for each $P_i \in T$, the responses of the two faulty circuits with f_1, f_2 in them, respectively, are the same. This concept can be used to partition the set of BFs into **equivalence classes**. Let $E_1, E_2, \ldots, E_m$ be the equivalence classes of the set of two-node BFs with respect to test set T, L be the number of two-node BFs, and P be the number of fault equivalence classes of size 1. Let **residual set** denote the set of candidate BFs that remain after the analysis of the test set is done with the obtained response from the circuit (in this section, we will consider unrestricted two-node BFs). Then we can define the following diagnostic measures.

Diagnostic resolution (DR) of a test set T is the percentage of BFs that can be completely distinguished from all other BFs using T. In other words, $DR = \frac{P}{L} \times 100$.

Diagnostic power (DP) of a test set T is the percentage of pairs of BFs that are not equivalent with respect to T.

Expected residual set (ERS) of a test set T is the expected value of the size of the residual set computed using T. In other words, $ERS = \sum_{i=1}^{m} \frac{|E_i| \times |E_i|}{L}$.

The process of computing DR, DP and ERS is referred to as **diagnostic fault simulation**. For simplicity, we will use DR and ERS as the two diagnostic measures from here on.

Diagnostic simulation consists of two steps: (a) updating of node partitions, and (b) updating of equivalence classes. The algorithm iterates through these two steps for each input vector. We have already discussed updating of node partitions earlier. Therefore, we will concentrate on updating of equivalence classes here.

We first illustrate the process of diagnostic fault simulation through an example.

Example 6.5   Consider Figure 6.6, its test set given in Table 6.1, and its node partition tree given in Figure 6.8 once again. Suppose this test set is the one we want to evaluate to determine its diagnostic capabilities. Based on the information

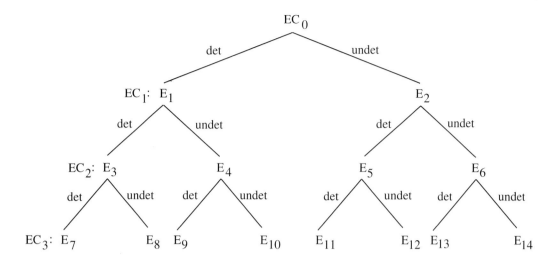

Figure 6.16 Equivalence class tree

in the node partition tree, we can form an **equivalence class tree** as follows. We first initialize the initial set of equivalence classes EC_0 to all the two-node BFs in the circuit. Then we process vector P_1. From the node partition tree, we know that the BFs denoted by $\{\langle A_0, A_1 \rangle\}$ are detected by P_1, whereas the BFs denoted by $\{\langle A_0, A_0 \rangle, \langle A_1, A_1 \rangle, \langle A_2, A_2 \rangle, \langle A_0, A_2 \rangle, \langle A_1, A_2 \rangle\}$ are not detected. We denote these two equivalence classes as E_1 and E_2, respectively. The new set of equivalence classes becomes $EC_1 = \{E_1, E_2\}$, as shown in Figure 6.16. Vectors P_2 and P_3 can be similarly processed. The expanded form of the different equivalence classes in the tree is given in Table 6.4, along with the number of BFs in each equivalence class. We need to keep in mind that an ordered pair of the type $\langle A_i, A_i \rangle$, where A_i contains just one node of the circuit, can be deleted, as it does not represent a BF.

After processing P_1, P_2, P_3, we obtain $EC_3 = \{E_7, E_8, E_9, E_{10}, E_{11}, E_{12}, E_{13}, E_{14}\}$, which can be used to obtain the diagnostic measures. Since there are 12 nodes in the circuit, there are 66 possible two-node BFs. Hence, $L = 66$. Since there is no equivalence class in EC_3 which contains only one BF, $DR = \frac{0}{45} \times 100 = 0\%$. ERS can be determined as follows.

$$ERS = \frac{(49 + 49 + 36 + 36 + 121 + 100 + 144 + 64)}{66}$$
$$= 9.08 \qquad \qquad \qquad \qquad \Box$$

Formally, the rules for computing EC_i from EC_{i-1} are as follows. Let $EC_{i-1} = \{E_1, E_2, \ldots, E_m\}$. First, NP_i is computed from NP_{i-1}. Next, each $E_j \in EC_{i-1}$ is split into two set of BFs, E_j^{det} and E_j^{undet}, where E_j^{det} (E_j^{undet}) is the subset of E_j that is detected (not detected) by P_i.

Let E_j be represented by the SOPS $\{\langle R_1, S_1 \rangle, \langle R_2, S_2 \rangle, \ldots, \langle R_l, S_l \rangle\}$. To compute E_j^{det}, for each $\langle R_k, S_k \rangle \in E_j$ we compute det_k, the subset of BFs represented by this

Table 6.4. *Expanded equivalence classes for the tree in Figure 6.16*

Equivalence class	Number of BFs
$E_1 = \{\langle A_0, A_1 \rangle\}$	25
$E_2 = \{\langle A_0, A_0 \rangle, \langle A_1, A_1 \rangle, \langle A_2, A_2 \rangle, \langle A_0, A_2 \rangle,$	$10 + 10 + 1 + 10$
$\quad \langle A_1, A_2 \rangle\}$	$+ 10 = 41$
$E_3 = \{\langle A_4, A_5 \rangle, \langle A_3, A_6 \rangle\}$	$12 + 2 = 14$
$E_4 = \{\langle A_3, A_5 \rangle, \langle A_4, A_6 \rangle\}$	$8 + 3 = 11$
$E_5 = \{\langle A_3, A_4 \rangle, \langle A_5, A_6 \rangle, \langle A_7, A_8 \rangle, \langle A_3, A_8 \rangle,$	$6 + 4 + 1 + 2$
$\quad \langle A_4, A_7 \rangle, \langle A_5, A_8 \rangle, \langle A_6, A_7 \rangle\}$	$+ 3 + 4 + 1 = 21$
$E_6 = \{\langle A_3, A_3 \rangle, \langle A_4, A_4 \rangle, \langle A_5, A_5 \rangle, \langle A_3, A_7 \rangle,$	$1 + 3 + 6 + 2$
$\quad \langle A_4, A_8 \rangle, \langle A_5, A_7 \rangle, \langle A_6, A_8 \rangle\}$	$+ 3 + 4 + 1 = 20$
$E_7 = \{\langle A_{11}, A_{14} \rangle, \langle A_{12}, A_{13} \rangle, \langle A_{10}, A_{15} \rangle\}$	$2 + 4 + 1 = 7$
$E_8 = \{\langle A_{11}, A_{13} \rangle, \langle A_{12}, A_{14} \rangle, \langle A_9, A_{15} \rangle\}$	$2 + 4 + 1 = 7$
$E_9 = \{\langle A_9, A_{14} \rangle, \langle A_{10}, A_{13} \rangle, \langle A_{12}, A_{15} \rangle\}$	$2 + 2 + 2 = 6$
$E_{10} = \{\langle A_9, A_{13} \rangle, \langle A_{10}, A_{14} \rangle, \langle A_{11}, A_{15} \rangle\}$	$2 + 2 + 2 = 6$
$E_{11} = \{\langle A_9, A_{12} \rangle, \langle A_{10}, A_{11} \rangle, \langle A_{14}, A_{15} \rangle,$	$2 + 1 + 2$
$\quad \langle A_{16}, A_{17} \rangle, \langle A_9, A_{17} \rangle, \langle A_{12}, A_{16} \rangle,$	$+ 1 + 1 + 2$
$\quad \langle A_{13}, A_{17} \rangle\}$	$+ 2 = 11$
$E_{12} = \{\langle A_9, A_{11} \rangle, \langle A_{10}, A_{12} \rangle, \langle A_{13}, A_{15} \rangle,$	$1 + 2 + 2$
$\quad \langle A_{10}, A_{17} \rangle, \langle A_{11}, A_{16} \rangle, \langle A_{14}, A_{17} \rangle,$	$+ 1 + 1 + 2$
$\quad \langle A_{15}, A_{16} \rangle\}$	$+ 1 = 10$
$E_{13} = \{\langle A_9, A_{10} \rangle, \langle A_{11}, A_{12} \rangle, \langle A_{13}, A_{14} \rangle,$	$1 + 2 + 4$
$\quad \langle A_{10}, A_{16} \rangle, \langle A_{11}, A_{17} \rangle, \langle A_{14}, A_{16} \rangle,$	$+ 1 + 1 + 2$
$\quad \langle A_{15}, A_{17} \rangle\}$	$+ 1 = 12$
$E_{14} = \{\langle A_{12}, A_{12} \rangle, \langle A_{13}, A_{13} \rangle, \langle A_{14}, A_{14} \rangle,$	$1 + 1 + 1$
$\quad \langle A_9, A_{16} \rangle, \langle A_{12}, A_{17} \rangle, \langle A_{13}, A_{16} \rangle\}$	$+ 1 + 2 + 2 = 8$

pair of sets that are detected by P_i, as follows. Let R_k^0, R_k^1, R_k^f be pointers to partitions in NP_i which are subsets of R_k set to 0, 1, f, respectively, by P_i. S_k^0, S_k^1, S_k^f are similarly defined. Then,

$$det_k = \{\langle R_k^0, S_k^1 \rangle, \langle R_k^1, S_k^0 \rangle\}$$

$$E_j^{det} = \bigcup_{k=1}^{l} det_k$$

Let $undet_k$ be the subset of BFs represented by $\langle R_k, S_k \rangle$ which are not detected by P_i. Then,

$$undet_k = \{\langle R_k^0, S_k^0 \rangle, \langle R_k^1, S_k^1 \rangle, \langle R_k^f, S_k^f \rangle, \langle R_k^0, S_k^f \rangle,$$
$$\langle R_k^1, S_k^f \rangle, \langle R_k^f, S_k^0 \rangle, \langle R_k^f, S_k^1 \rangle\}$$

$$E_j^{undet} = \bigcup_{k=1}^{l} undet_k$$

In det_k or $undet_k$, if any ordered pair denotes a null set of BFs, it is deleted. Also, if $R_k = S_k$, then the two ordered pairs in det_k denote the same set of BFs. Hence, one of them is deleted. Similarly, if $R_k = S_k$, then in $undet_k$, $\langle R_k^0, S_k^f \rangle$ is the same as $\langle R_k^f, S_k^0 \rangle$, and $\langle R_k^1, S_k^f \rangle$ is the same as $\langle R_k^f, S_k^1 \rangle$. After deleting the redundant ordered pairs, we are guaranteed that the remaining ordered pairs represent disjoint sets of BFs. Because of this property, the number of BFs represented by an equivalence class E_j can be simply determined by summing the number of BFs represented by each ordered pair in E_j.

6.4.3 Diagnostic test generation

In this section, we will discuss how to obtain test sets which have good diagnostic measures. We will describe a simulation-based test generator which uses a genetic algorithm. We may recall the use of a genetic algorithm for deriving fault detection test sets earlier.

We can formulate the diagnostic test generation problem as an optimization problem as follows. We will compute one vector at a time. Suppose we have computed a set of vectors $\{P_1, P_2, \ldots, P_{i-1}\}$ and the set of equivalence classes with respect to these vectors is EC_{i-1}. The next vector P_i is chosen such that the diagnostic measures with respect to the new set of equivalence classes EC_i is optimized. In other words, DR is maximized, ERS is minimized, etc.

We start with a population of vectors with a fixed population size. We then evolve the population over a fixed number of generations based on the genetic algorithm concepts of selection, crossover and mutation. Once a new generation of vectors is created, the best vector of the generation is derived based on a fitness function. A useful fitness function, $FF(v)$, which returns a real number for vector v, can be derived as follows. Let $E_1, E_2, \ldots, E_p$ be the p largest equivalence classes in EC_{i-1}. As mentioned before, for all $1 \leq j \leq p$, let $E_j^{det}(v)$ $(E_j^{undet}(v))$ be the subset of E_j that is detected (not detected) by vector v, and let q be the number of such non-null subsets. Then

$$FF(v) = \frac{1}{q} \times \sum_{j=1}^{p} \min\{|E_j^{det}(v)|, |E_j^{undet}(v)|\}$$

The higher the value of the fitness function, the better the vector. This is because closer the values of $E_j^{det}(v)$ and $E_j^{undet}(v)$, the higher the value of $\min\{|E_j^{det}(v)|, |E_j^{undet}(v)|\}$ and lower the combined contribution of $E_j^{det}(v)$ and $E_j^{undet}(v)$ towards ERS. The selection, crossover and mutation steps are exactly the same as those used in the fault detection algorithm given earlier.

A population size of 100, number of generations of 10, a mutation probability of 0.001 and $p = 10$ have been found to give good results.

After evolving through all generations, the best vector generated during the process is selected as the new test vector. This vector is used to compute EC_i by splitting all

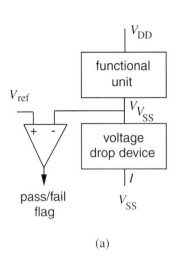
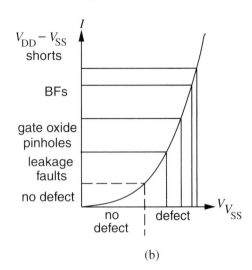

(a)   (b)

Figure 6.17 Built-in current sensor

the equivalence classes in EC_{i-1}, as explained earlier, through incremental simulation. After computing a number of vectors, the best vector selected may have the same fitness value as the previous vector included in the test set. If the fitness value of the best selected vector does not change for some fixed number (typically five) of consecutive vectors, then the test generation process can be terminated.

6.5   Built-in current sensors

Implementing I_{DDQ} testing with off-chip current sensing has two problems: its inherent resolution limit, and the need to modify test equipment to measure currents. The I/O drivers on the chips consume a lot of current, and the fluctuation in their current can overshadow some of the I_{DDQ} abnormalities in the internal logic blocks. Also, even with a modified testing equipment, the current cannot be measured at the maximum speed of the tester. Some of these problems can be overcome with the use of built-in current sensors in the chips (Maly and Nigh, 1988; Hsue and Lin, 1993; Hurst and Singh, 1997).

The basic structure of the built-in current sensor is given in Figure 6.17(a) (Hsue and Lin, 1993). It is composed of a voltage drop device and a voltage comparator. At the end of each clock cycle, the virtual V_{SS} voltage, $V_{V_{SS}}$, is compared with the reference voltage, V_{ref}. The value of V_{ref} is chosen such that $V_{V_{SS}} \leq V_{ref}$ for a fault-free functional unit, and $V_{V_{SS}} > V_{ref}$ for a faulty one. It is also possible to locate the voltage drop device between V_{DD} and the functional unit.

A built-in current sensor must satisfy certain objectives. It should be able to detect abnormal I_{DDQ} currents within a relatively large range (typically, several

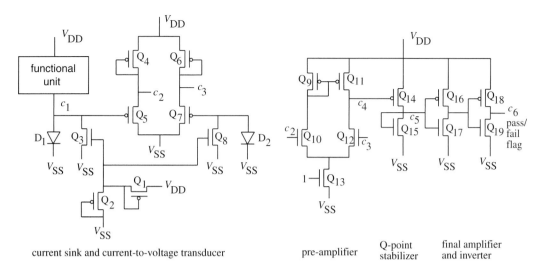

current sink and current-to-voltage transducer   pre-amplifier   Q-point stabilizer   final amplifier and inverter

Figure 6.18 A built-in current sensor implementation (Hsue and Lin, 1993) (© 1993 IEEE)

microamperes or higher). However, the voltage drop across the current sensor due to peak transient currents (which could be in the ampere range) should be minimized. In addition, the sensor should be as small as possible to reduce its area overhead and reduce the probability of malfunction in the sensor itself. In order to achieve the above objectives, the voltage drop device must have non-linear $I-V$ characteristics, as shown in Figure 6.17(b).

A detailed implementation of a built-in current sensor is given in Figure 6.18. It consists of the following subcircuits: current sink, current-to-voltage transducer, voltage comparator, voltage amplifier, and voltage quiescent-point (Q-point) stabilizer. In this sensor, diode D_1 is used as the voltage drop device since a diode has the desired $I-V$ characteristics. The other diode D_2 in the current sink and current-to-voltage transducer parts of the sensor provides a reference current (voltage) in order to detect the voltage difference when a current is injected into D_1 from the functional unit. Transistors Q_3 and Q_8 determine the current sensitivity of the sensor. An I_{DDQ} current, which distinguishes a good functional unit from a bad one, depends on the size of the functional unit and the wafer processing method used to fabricate it. For example, when the I_{DDQ} *threshold* is set to 5 μA, the drain current flowing through Q_3 and Q_8 should be of the order of 5 μA. If the I_{DDQ} current is above this threshold, the chip is assumed to be defective, else not.

When a current (voltage) is detected at node c_1, the sensor translates it into an appropriate voltage between V_{SS} and V_{DD}. Therefore, the sensor acts like a DC amplifier. The Q-point of this amplifier dictates the functionality of the sensor. A variation in V_{DD} can change the Q-point of the sensor and affect its stability. Thus, a Q-point stabilizer consisting of transistors Q_{14} and Q_{15} is introduced. This stabilizes the quiescent voltage at node c_5 irrespective of the variation in V_{DD}. Transistors Q_9, Q_{10}, Q_{11} and Q_{12} form a differential amplifier. The drain voltage of Q_{11} at node c_4 follows the variation in V_{DD}. This means that the gate-to-source voltage of transistor Q_{14} remains stable irrespective of the change in V_{DD}. The output at node c_6 gives the pass/fail value of the sensor (logic 1 for pass and logic 0 for fail).

For $V_{DD} = 5$ V and an I_{DDQ} threshold of 25 μA, the operating speed of such a built-in current sensor is about 10 MHz. Because of the voltage drop across the diodes of about 0.5 V, the available voltage across the functional unit is only $(V_{DD} - 0.5)$ V. This can significantly slow down the functional unit. To avoid this situation, the diodes can by connected to -0.5 V instead of V_{SS}. However, the disadvantage of this solution is that a dual polarity of supply voltage is required during I_{DDQ} testing. The area of the sensor is dominated by the size of the two diodes, since they need to sink large transient currents while performing I_{DDQ} testing.

In general, when the functional unit is very large, it may have to be partitioned into smaller subcircuits so that a built-in current sensor can be provided for each subcircuit.

In order to obtain higher operating speeds of up to 30 MHz, a differential built-in current sensor design presented in Hurst and Singh (1997) can be used.

6.6   Advanced concepts in current sensing based testing

In this section, we will briefly discuss five advanced concepts which seem to hold promise for the development of more sophisticated testing methods based on current sensing. These concepts are (a) i_{DD} pulse response testing (Beasley *et al.*, 1993), (b) dynamic current testing (Segura *et al.*, 1995), (c) testing through depowering (Rius and Figueras, 1995), (d) current signatures (Gattiker and Maly, 1996), and (e) techniques for maintaining the applicability of I_{DDQ} testing to deep-submicron CMOS (Soden *et al.*, 1996).

6.6.1 i_{DD} **pulse response testing**

It is possible to detect many faults by pulsing both the V_{DD} and V_{SS} voltage rails (V_{CC} and V_{EE} for bipolar circuits) while applying a fixed bias voltage to the primary inputs of the circuit. Fault detection is done through temporal and spectral analysis of the resulting transient rail current (i_{DD}). The bias voltage of the inputs is set to the midpoint in the power supply rail voltages. For example, if V_{DD} and V_{SS} are 5 V and 0 V, respectively, then the inputs are set to 2.5 V. This generally enables all the transistors inside the circuit to turn on to some mode of operation at some point in the pulsing process. The power supply rails are pulsed from the midpoint voltage to their normal operating voltage. In doing so, the CMOS devices in the circuit enter one or more of the following states: sub-threshold (weak inversion), saturation or linear (ohmic). This technique makes the current characteristics of each device observable by momentarily turning it on. Thus, for testing purposes, pulsing the supply voltage rails and a fixed bias voltage at the primary inputs provide the controllability needed to excite the fault, and the analysis of i_{DD} provides the desired observability.

The defects typically detected by this method are gate–oxide shorts, opens, polysilicon BFs and metal BFs. This method is applicable to both digital and analog circuits. It can also detect changes in fabrication parameters of analog devices which can significantly affect their performance.

6.6.2   Dynamic current testing

Although I_{DDQ} monitoring is a powerful technique for testing integrated circuits, many opens and parametric defects escape detection because they prevent quiescent power supply current elevation. Many of these defects can be detected if the dynamic current $(i_{DD}(t))$ is monitored instead of the static current. This technique may be useful in detecting faults which escape detection through both I_{DDQ} and logic testing.

The power supply current waveform consists of a series of sharp peaks corresponding to input/output transitions. During the quiescent period, the current drawn is negligible. The sharp peaks occur for two reasons: (a) simultaneous conduction of nMOS and pMOS transistors during the transition, and (b) the charge/discharge of internal capacitances. Defects in CMOS circuits may change the current waveform by eliminating or suppressing the sharp peaks or causing elevated current to appear during the quiescent periods. If the quiescent current is not affected by the defect, I_{DDQ} testing cannot detect it. However, since $i_{DD}(t)$ testing monitors the current waveform, it can detect these defects.

The effectiveness of $i_{DD}(t)$ testing is directly dependent on quantifying the changes in the current peaks owing to defects in the circuit. This can be done by monitoring the amount of charge flowing through the circuit during and after the transition. The value of this charge, Q, is given by:

$$Q = \int_{clk} i_{DD}(t)dt,$$

where clk denotes the clock period. For a non-defective circuit, the value of the charge, $Q_{non-def}$, will be within the operating margins of the circuit such that $Q_{non-def}^{min} < Q_{non-def} < Q_{non-def}^{max}$. $Q_{non-def}^{min}$ is the minimum charge that can flow through the circuit during any transition and $Q_{non-def}^{max}$ is the maximum charge. $i_{DD}(t)$ testing is able to detect a defect if the charge flowing through the defective circuit is outside the above range.

A built-in current sensor to implement the $i_{DD}(t)$ testing scheme is shown in Figure 6.19. The sensor provides a voltage value proportional to the charge flowing through

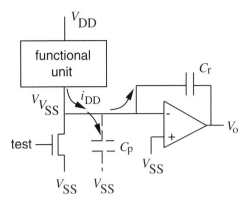

Figure 6.19 A built-in current sensor for $i_{DD}(t)$ testing (Segura *et al.*, 1995) (© 1995 IEEE)

the circuit in a given period of time. The sensor operation is based on the integration of the current through a capacitor. If the functional unit parasitic capacitor, C_p, is used for this purpose, then the variation of its capacitance with process deviations and the electrical state of the circuit has to be taken into account. To avoid this scenario, the functional unit current is integrated into a separate smaller capacitor C_r. The value of this capacitance is multiplied by a suitable factor using the Miller effect to take into account the fact that most of the current flows into the small capacitor and not C_p. During testing, the *test* control signal is 0, otherwise it is 1.

The maximum voltage, $V_{V_{SS}}$, at the virtual V_{SS} node in a defect-free circuit during an input/output transition is

$$V_{V_{SS}} = \frac{Q_{\text{non-def}}}{(1+A)C_r}.$$

Assuming that the gain A of the amplifier is high, i.e., $A \gg 1$, the sensor output V_o ranges between:

$$V_{DD} - \frac{Q_{\text{non-def}}^{\text{max}}}{C_r} < V_o < V_{DD} - \frac{Q_{\text{non-def}}^{\text{min}}}{C_r}.$$

A defect is detected when the output voltage of the sensor is outside this range.

6.6.3   Depowering

The concept of CMOS circuit testing through depowering is based on disconnecting the power supply line in the circuit's quiescent state. In the non-defective case, the capacitances of different nodes in the circuit discharge very slowly due to very small I_{DDQ} currents. In I_{DDQ}-defective circuits, the discharge is much faster because I_{DDQ} is orders of magnitude larger. Some time after the discharge starts, the logic states of some nodes begin changing. When this change propagates to a primary output, the defect is detected by observing the time at which the change occurs.

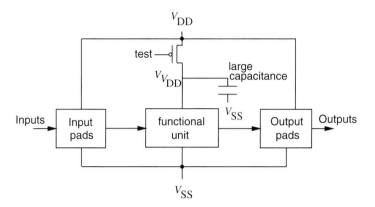

Figure 6.20   Implementation of depowering (Rius and Figueras, 1995) (© 1995 IEEE)

To implement this concept, there needs to be two sets of power supply lines: the peripheral power line connected to the circuit input/output pads, and the core power line connected to the internal CMOS circuitry. The core power line is connected to the power supply through a switch, which could be implemented as a pMOS transistor fed by a *test* signal. This is shown in Figure 6.20.

The testing strategy is as follows. An input vector is applied to the integrated circuit to excite the targeted defect, such that one or more of the circuit nodes directly connected to an output pad is high. After the transients die down, the switch is opened by making the *test* signal logic 1. The voltage of the circuit outputs, which are at logic 1, begins to decay following the decay in $V_{V_{DD}}$. After some time t_s, these output voltages become logic 0, and the change becomes observable at the output pins. Simulations for medium-sized circuits, such as adders, indicate that t_s is in the 10 to 100 nanosecond range. However, for large CMOS circuits, t_s could be in the 100 microsecond range. One possible way to reduce t_s is to partition the functional unit under test into sub-units and provide a small switch for each sub-unit.

An advantage of depowering over other off-chip current sensing approaches is that while the latter require changes in the testing equipment, depowering does not, since ultimately only the logic values at the output pins are monitored.

6.6.4   Current signatures

Conventional I_{DDQ} testing methods use a single I_{DDQ} threshold value. I_{DDQ} is measured for a given test set and the maximum measurement is compared with the threshold value. If the maximum measurement is higher than the threshold, the test fails, else it passes. Thus, the choice of the current threshold is quite critical in this approach. If the threshold is set too low, then many good dies may get rejected. If it is set too high, then bad dies may escape detection. Ideally, we need to develop an approach which can identify dies that have harmful defects, even if they only result

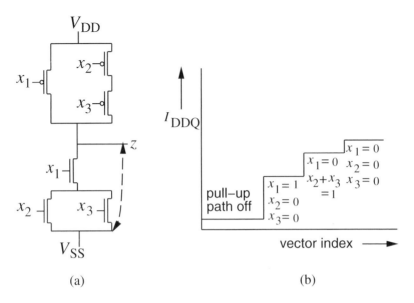

Figure 6.21   A defective circuit and its current signature

in a slight increase in I_{DDQ}, without having to lower the I_{DDQ} threshold so much that good dies get rejected. This is the idea behind *current signatures*.

A **current signature** of a circuit consists of grouping the current measurements into different levels, where in a given level the measurements are of the same magnitude, and then ordering the levels by magnitude. The following example illustrates this concept.

Example 6.6   Consider the circuit shown in Figure 6.21(a). It has a BF as shown between nodes z and V_{SS}. This BF does not conduct I_{DDQ} for any vector for which $x_1(x_2 + x_3) = 1$. It produces three levels of I_{DDQ} measurements for the following three sets of vectors on (x_1, x_2, x_3): (a) $(1, 0, 0)$, (b) $\{(0, 0, 1), (0, 1, 0), (0, 1, 1)\}$, and (c) $(0, 0, 0)$, as shown in Figure 6.21(b).

If the BF had not been present, only one level of low I_{DDQ} would have been obtained for all the vectors. On the other hand, if the BF had been between V_{DD} and V_{SS}, only one level of high I_{DDQ} would have been obtained for all the vectors.   □

The above example shows that the key information about the defect in the circuit is contained in the number and values of all current measurements obtained for it.

In order to exploit current signatures for fault detection, let us first define two types of I_{DDQ} defects: passive and active. A **passive defect** is one which involves only non-switching nodes of the circuit. Such a defect provides a direct and static path between V_{DD} and V_{SS}, and produces a constant level of I_{DDQ} for all test vectors. Such defects are activated regardless of the circuit state or the input vector. An example is a leaky junction between the well and the substrate. An **active defect** is one which involves switching nodes of the circuit. The BF in the circuit in Figure 6.21(a) is an example. Such defects produce abnormal I_{DDQ} only when a test vector activates them. In other words, they result in multi-level current signatures.

Since passive defects do not directly impact the quality of information-carrying signals in the circuit, they are unlikely to cause a fault. On the other hand, since active defects affect switching nodes, they are likely to cause a fault. A vast majority of active defects may be detected simply by determining if the current signature has more than one level. A multi-level current signature usually indicates the presence of a harmful defect even if all the current levels in the signature are very small.

If a current signature with just one high level is obtained for the die, and the die passes other tests such as logic tests, then it is probable that the die just has a passive defect. Such a die need not be rejected based on I_{DDQ} testing.

6.6.5   Applicability to deep-submicron CMOS

The effectiveness of I_{DDQ} testing is threatened by the rush of semiconductor technology to smaller feature sizes and faster, denser circuits. In the quiescent state of a CMOS circuit, about half the transistors are off. Each of the off transistors contributes a small leakage current (I_{off}) owing to junction leakage. In a circuit with millions of transistors, the total I_{DDQ}, which is the sum of all transistor I_{off} currents, grows.

Another, perhaps more significant, cause of higher off-state leakage is a lowering of the transistor threshold voltage V_T. For 5 V technologies, V_T is in the 0.7–1.0 V range. As V_{DD} drops to the 1–2 V range, V_T is projected to reduce to 0.3 V. If V_T is not reduced as V_{DD} drops, the switching speed of the transistor suffers. However, as V_T is reduced, the leakage current increases drastically. Therefore, as technology progresses, I_{DDQ} of a defect-free circuit could rise from a few nanoamperes to tens of milliamperes or more, decreasing the effectiveness of I_{DDQ} testing.

To combat the above problem, some possible solutions are as follows. I_{DDQ} can be lowered by three to four orders of magnitude for $V_T = 0.3V$ if the measurement temperature is lowered from $22\,°C$ to $-55\,°C$. With a factor of 10^4 reduction, a defect-free I_{DDQ} of 10 mA will reduce to 1 µA, making I_{DDQ} testing effective again.

Another approach is to increase V_T temporarily by increasing the source-to-substrate bias voltage. In this approach, all sources, wells and substrate regions are connected to separate package pins. During I_{DDQ} testing, substrate and well back-biases are driven at the pin level. This may enable four orders of magnitude reduction in I_{DDQ}.

The third approach is to have built-in current sensors for a fixed number of transistors. Thus, larger the circuit, larger the number of sensors.

The fourth approach is to use transistors with different thresholds on the same chip. Transistors on off-critical paths are doped for higher thresholds, thus reducing their I_{off}.

Finally, processing techniques based on channel doping can be used to significantly reduce I_{off}.

6.7   Economics of I_{DDQ} testing

In this section, we will discuss when I_{DDQ} testing is economically viable.

A major problem associated with I_{DDQ} testing is that it results in about 1 to 3% *yield loss* (Davidson, 1994). **Yield loss** is based on chips that fail the I_{DDQ} test, but pass all other tests. Some of this yield loss is real in the sense that some "good" chips get termed as bad. Here a "good" chip is defined as one which has a defect that does not affect its normal operation for a particular application. Two examples of such defects are those resulting in a redundant fault and those resulting in an increase in off-critical path delays. For example, consider a logically redundant SA0 fault in a circuit. Suppose this fault is caused by a BF to V_{SS}. I_{DDQ} testing will detect this fault. Such a fault represents a real circuit defect. However, the circuit may not fail until other defects occur in it, which may not happen until much later, if ever. Therefore, discarding a chip with a redundant fault in it usually represents an yield loss. Note that in some circuits, redundancies are deliberately added for circuit performance or noise immunity. Detecting such redundant faults should not be considered to be contributing to yield loss. Similarly, if sufficient slack exists on an off-critical path in a circuit such that an I_{DDQ}-testable fault on it increases the delay of the path, but not by enough to affect circuit performance, then rejection of such a chip represents an yield loss.

Of course, there may be many defects that actually change the logical behavior of the circuit, but are missed by other tests, while they are caught through I_{DDQ} testing. Also, some I_{DDQ}-testable defects that do not change the logical behavior of the circuit could still prevent it from meeting its functional specifications. Suppose a battery-powered application, such as a pacemaker, requires very small stand-by power consumption. In this case, high I_{DDQ} should always lead to rejection of the chip.

Based on the above discussions, one can propose a simple economic model for I_{DDQ} testing. Figure 6.22(a) shows how the universe of chips can be divided into various regions based on the outcomes of logic and I_{DDQ} testing. Region V represents defect-free chips which pass both the tests. Region III represent chips which fail both tests and are, hence, bad. Region IV represents chips which have failed the logic test, but passed the I_{DDQ} test. This may happen, for example, for some stuck-open faults, or when the fault coverage of the I_{DDQ} test is not that high. Region I represents the most interesting case where an I_{DDQ} test may be of benefit. In Figure 6.22(b), we have blown up this region and divided it into two parts. Region Ia (Ib) represents "customer bad (good) chips" which will (will not) fail in the customer application. The size of Region I depends on the fault coverages of the logic and I_{DDQ} tests, as well as the number of

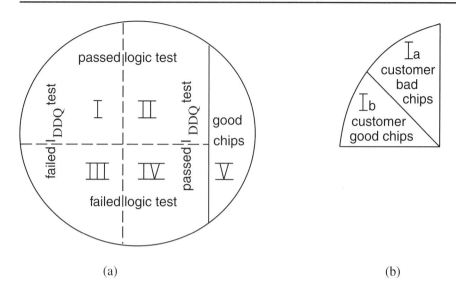

Figure 6.22 Chip categories and regions of yield loss (Davidson, 1994) (© 1994 IEEE)

defective chips manufactured. The relative sizes of Regions Ia and Ib depend on the customer quality requirements.

Let N_{Ia} (N_{Ib}) represent the number of chips in Regions Ia (Ib). Let C_α be the cost of sending a "customer bad chip" to a customer. This includes warranty repairs, lost business, etc. Let C_β be the cost of failing a "customer good chip." This is the cost of the chip. Let C_γ be the incremental cost of I_{DDQ} testing, which is incurred by additional test development, test hardware, test time, etc. This cost reduces the benefit of I_{DDQ} testing. Let C_δ be the cost of enhancing a logic test to the level required if I_{DDQ} testing were not used. Avoiding this cost is a benefit of I_{DDQ} testing. Then, the cost benefit of using I_{DDQ} testing is given by:

$$benefit_{I_{DDQ}} = (N_{Ia} \times C_\alpha + C_\delta) - (N_{Ib} \times C_\beta + C_\gamma).$$

The following two examples illustrate how this economic model can be used.

Example 6.7 Let us consider a high-reliability, low-volume defense application. For such an application, the cost of failure is very high. The cost of lost yield is small since only a handful of chips is needed. The cost of test generation is high since the cost is amortized over a small number of chips. In this case, improved quality and test generation times enabled by I_{DDQ} testing should more than make up for the increased cost due to yield loss.

Suppose the manufacturer manufactures 20 000 chips. We assume that $N_{Ib} = 0$ since it is not acceptable to ship any defective chip. Suppose $N_{Ia} = 3\%$. This number is obtained assuming that the logic test coverage is very high and 3% of defective chips passing the logic test are detectable only through I_{DDQ} testing. This means that there are 600 such chips. C_{α} is very high for this application. Suppose it is $10 000. The actual cost may be higher. However, since some defective chips may never be used, the average cost of failure is reduced. Assume C_{δ} is $100 000. Reducing this cost would result in an increase in N_{Ia} which increases the benefits of I_{DDQ} testing. Next, suppose that C_{γ} is $20 000 because of the low-volume requirement. Then the benefit of I_{DDQ} testing is given by:

$$benefit_{I_{DDQ}} = 600 \times \$10\,000 + \$100\,000 - \$20\,000 = \$6.08 \text{ million.} \qquad \Box$$

Example 6.8 Let us next consider a high-volume, low-cost application for electronic toys and games. For this application, the cost of failure is low to medium. The cost of yield loss is high due to the low-margin nature of the business. The cost of test generation is small since manual or automatic test generation is amortized over millions of chips. For such applications, the loss of yield due to I_{DDQ} testing may not be tenable.

In this application, the size of Region Ib is relatively large, C_{α} is relatively low, and C_{β} is relatively high (because of low margins and price pressures). Suppose that two million chips are manufactured. Let N_{Ia} be 0.5% of 2 million, or 10 000 chips, since these are probably simple, low technology chips. Let N_{Ib} be 2% of 2 million, or 40 000 chips. Let $C_{\alpha} = \$10$ since many defects will not be exercised and the warranties are short. Let $C_{\beta} = \$3$. Let $C_{\gamma} = \$100\,000$ because of the high volume. Finally, assume that C_{δ} is 0 since a logic test would probably be adequate for simple chips. Thus, the benefit of I_{DDQ} testing is given by:

$$benefit_{I_{DDQ}} = 10\,000 \times \$10 - 40\,000 \times \$3 - \$100\,000 = -\$120\,000.$$

Since $benefit_{I_{DDQ}}$ is negative in this case, it is not advantageous to do I_{DDQ} testing. $\Box$

Summary

- I_{DDQ} is the quiescent power supply current which gets drastically elevated when certain faults, such as BFs and leakage faults, are activated. Such faults provide a conducting path between V_{DD} and V_{SS}.
- About 50% of the defects in manufactured chips may map to BFs. With shrinking geometries, this percentage is likely to become even higher in the coming years.
- BFs may be categorized in various ways: restricted vs unrestricted, intra-gate vs inter-gate, feedback vs non-feedback, gate-level vs switch-level.
- Switch-level leakage faults can be detected by pseudo-SAF test sets.
- Test compaction of an I_{DDQ} test set is very important because of its slow test application time.
- Feedback BFs may not be detected by just providing complementary logic values to the two lines in question. Three-valued simulation can be used to detect such

faults. However, to reduce the test generation complexity, it is usually optimistically assumed that providing complementary values to the two lines is enough.

- The set of ordered pairs of sets representation of BFs is a very efficient way of representing a very large number of BFs.
- Genetic algorithms, which employ the concepts of selection, crossover and mutation, are very useful for test generation for fault detection and diagnosis purposes.
- The concepts of node partition tree and indistinguishable pairs tree make fault simulation of BFs very efficient.
- Fault diagnosis and diagnostic fault simulation of a circuit can be done with the help of a node partition tree and equivalence class tree.
- The quality of a diagnostic test set can be evaluated on the basis of diagnostic measures such as diagnostic resolution, diagnostic power and expected residual set.
- The problems associated with the test application time, resolution limit and need for special test equipment of off-chip current sensing can be alleviated by using built-in current sensors.
- Extensions of the current sensing method, such as i_{DD} pulse response testing, dynamic current testing, depowering, and current signatures, hold future promise.
- In order to maintain the applicability of I_{DDQ} testing to deep-submicron CMOS, some special steps need to be taken. These steps may be one or more of the following: lowering the temperature while testing, temporarily increasing the transistor threshold voltage, using built-in current sensors with each partition in a partitioned circuit, using transistors with different thresholds, and using processing techniques based on channel doping.
- While I_{DDQ} testing may be very important in high reliability applications, it may not be that useful in high-volume, low-cost applications, from the economic point of view.

Additional reading

In Dalpasso *et al.* (1995), a test generation technique is presented which targets the highest values of current during fault activation such that either a higher fault coverage is obtained or a less accurate sensor is needed.

The effectiveness of using a combination of I_{DDQ}, functional and scan tests in attaining a high defect coverage of integrated circuits has been established in Hawkins *et al.* (1989) and Maxwell *et al.* (1992).

A description of an I_{DDQ} monitor for test fixtures based on the Quality Test Action Group (QTAG) standard is given in Baker *et al.* (1994).

In Williams *et al.* (1996), the sensitivity of I_{DDQ} testing to individual device parameters is studied and it is shown how device scaling affects I_{DDQ} testing.

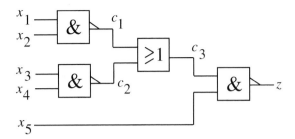

Figure 6.23

In Thibeault (1999), differential I_{DDQ} and I_{DDQ} testing methods are compared and the superiority of the former is established in reducing the probability of false test decisions.

In Jiang and Vinnakota (2000), a new metric, called energy consumption ratio, is defined. It improves the effectiveness of dynamic current testing by reducing the impact of process variations.

Exercises

6.1 Obtain the gate-level model of a NAND–NAND static CMOS implementation of a full-adder. Use it to derive a minimal leakage fault test set which detects all detectable leakage faults. Compare the size of this test set to the minimal single SAF test set for this circuit.

6.2 For the circuit shown in Figure 6.23:
   (a) Derive a test vector for the FBF $\langle c_1, z \rangle$ under the pessimistic condition of detection.
   (b) What are the gate-level FBFs detected by vector $(x_1, x_2, x_3, x_4, x_5) = (0, 0, 0, 1, 1)$ under this condition.

6.3 In the circuit in Figure 6.5:
   (a) How many gate-level two-node FBFs and NFBFs are there?
   (b) How many of the gate-level two-node BFs remain after fault collapsing?
   (c) Of the collapsed set of BFs, how many are detected by the following test set T applied to $(x_1, x_2, x_3, x_4, x_5)$: $\{(1, 0, 0, 1, 0), (1, 1, 0, 0, 0), (0, 1, 1, 1, 1)\}$.
   (d) Obtain another test set T' from T by deriving two new vectors from the first two vectors in T by crossover at bit position 3, and mutating bit position 5 in the third vector in T. What is the BF coverage of T' for the collapsed set of faults?

6.4 For the circuit shown in Figure 6.24, derive a gate-level model for detecting the BF $\langle c_1, c_2 \rangle$. Obtain all possible test vectors for this BF by targeting the appropriate SAF in the gate-level model.

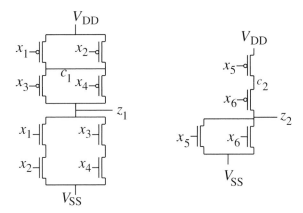

Figure 6.24

6.5 For the circuit shown in Figure 6.23, a test set that detects all gate-level two-node BFs is $\{(0, 0, 0, 0, 1), (1, 1, 0, 0, 1), (1, 0, 1, 0, 0), (1, 1, 1, 1, 0), (0, 0, 0, 0, 0), (0, 1, 1, 1, 0)\}$. Obtain a minimum subset of this test set which also detects all such BFs.

6.6 For the circuit shown in Figure 6.13, determine the node partitions, indistinguishable pairs and bridging fault coverage of the following test sequence: $\{(1, 0), (0, 1), (1, 1)\}$.

6.7 For the circuit shown in Figure 6.13, a test sequence that detects all gate-level two-node BFs is $\{(0, 1), (1, 1), (1, 0), (0, 0), (0, 1), (0, 0)\}$. Find a minimal subset of this test sequence for which I_{DDQ} testing will also detect all such BFs.

6.8 For the circuit shown in Figure 6.14, consider the following test set: $\{P_1, P_2, P_3\} = \{(0, 0, 0, 0), (0, 0, 1, 1), (1, 0, 1, 0)\}$. Suppose that vector E, which indicates faulty or fault-free response, is $(1, 0, 0)$. Determine one possible additional test vector, P_4, which can be added to the end of the test set, such that K_4 is guaranteed to have at most one BF irrespective of whether I_{DDQ} measurement indicates a fault-free or faulty response for P_4.

6.9 For the gate-level BFs in the circuit in Figure 6.14:
(a) Obtain the diagnostic measures, DP and ERS, for the test set T given in Table 6.3.
(b) Obtain another test set T' from T by deriving two new vectors from the first two vectors in T by crossover at bit position 2, and mutating bit position 1 in the third vector in T. What are DP and ERS of T'?

6.10 Obtain a CMOS complex gate to implement the function $z = \bar{x}_1(\bar{x}_2 + \bar{x}_3)$. What type of current signature would the different input vectors produce for a BF between V_{DD} and output z? Associate respective vectors with each I_{DDQ} measurement level in the current signature.

References

Aitken, R.C. (1991). Fault location with current monitoring. In *Proc. Int. Test Conference*, pp. 623–632.

Baker, K., Bratt, A., Richardson, A. and Welbers, A. (1994). Development of a class 1 QTAG monitor. In *Proc. Int. Test Conference*, pp. 213–222.

Beasley, J.S., Ramamurthy, H., Ramirez-Angulo, J. and DeYong, M. (1993). i_{DD} pulse response testing of analog and digital CMOS circuits. In *Proc. Int. Test Conference*, pp. 626–634.

Burns, D.J. (1989). Locating high resistance shorts in CMOS circuits by analyzing supply current measurement vectors. In *Proc. Int. Symposium for Testing and Failure Analysis*, pp. 231–237.

Chakravarty, S. and Liu, M. (1992). Algorithms for I_{DDQ} measurement based diagnosis of bridging faults. *J. of Electronic Testing: Theory & Applications*, **3** (4), pp. 377–385.

Chakravarty, S. and Thadikaran, P. (1994). A study of I_{DDQ} subset selection algorithms for bridging faults. In *Proc. Int. Test Conference*, pp. 403–411.

Chakravarty, S., Fuchs, K. and Patel, J. (1995). Evaluation and generation of I_{DDQ} diagnostic test sets for bridging faults in combinational circuits. Technical Report No. 95-11, Dept. of Computer Science, S.U.N.Y. at Buffalo.

Chakravarty, S. and Thadikaran, P. (1996). Simulation and generation of I_{DDQ} tests for bridging faults in combinational circuits. *IEEE Trans. on Computers*, **45** (10), pp. 1131–1140.

Chakravarty, S. (1996). A sampling technique for diagnostic fault simulation. In *Proc. VLSI Test Symposium*, pp. 192–197.

Dalpasso, M., Favalli, M. and Olivo, P. (1995). Test pattern generation for I_{DDQ}: increasing test quality. In *Proc. VLSI Test Symposium*, pp. 304–309.

Davidson, S. (1994). Is I_{DDQ} yield loss inevitable? In *Proc. Int. Test Conference*, pp. 572–579.

Gattiker, A.E. and Maly, W. (1996). Current signatures. In *Proc. VLSI Test Symposium*, pp. 112–117.

Hawkins, C.F., Soden, J.M., Fritzmeier, R.R. and Horning, L.K. (1989). Quiescent power supply current measurement for CMOS IC defect detection. *IEEE Trans. on Industrial Electronics*, **36** (5), pp. 211–218.

Hsue, C.-W. and Lin, C.-J. (1993). Built-in current sensor for I_{DDQ} test in CMOS. In *Proc. Int. Test Conference*, pp. 635–641.

Hurst, J.P. and Singh, A.D. (1997). A differential built-in current sensor for high speed IDDQ testing. *IEEE J. of Solid State Circuits*, **32** (1), pp. 122–125.

Jha, N.K., Wang, S.-J. and Gripka, P.C. (1992). Multiple input bridging fault detection in CMOS sequential circuits. In *Proc. Int. Conference on Computer Design*, pp. 369–372.

Jiang, W. and Vinnakota, B. (2000). IC test using the energy consumption ratio. *IEEE Trans. on Computer-Aided Design*, **19** (1), pp. 129–141.

Lee, K.J. and Breuer, M.A. (1992). Design and test rules for CMOS circuits to facilitate I_{DDQ} test of bridging faults. *IEEE Trans. on Computer-Aided Design*, **11** (5), pp. 659–669.

Lee, T., Hajj, I.N., Rudnick, E.M. and Patel, J.H. (1996). Genetic-algorithm-based test generation for current testing of bridging faults in CMOS VLSI circuits. In *Proc. VLSI Test Symposium*, pp. 456–462.

Malaiya, Y.K. and Su, S.Y.H. (1982). A new fault model and testing technique for CMOS devices. In *Proc. Int. Test Conference*, pp. 25–34.

Maly, W. and Nigh, P. (1988). Built-in current testing – feasibility study. In *Proc. Int. Conference on Computer-Aided Design*, pp. 340–343.

Mao, W. and Gulati, R.K. (1992). QUIETEST: a methodology for selecting I_{DDQ} test vectors. *J. of Electronic Testing: Theory & Applications*, **3** (4), pp. 349–357.

Maxwell, P., Aitken, R., Johansen, V., and Chiang, I. (1992). The effectiveness of I_{DDQ}, functional and scan tests: how many fault coverages do we need? In *Proc. Int. Test Conference*, pp. 168–177.

Midkiff, S.F. and Bollinger, S.W. (1993). Classification of bridging faults in CMOS circuits: experimental results and implications for test. In *Proc. VLSI Test Symposium*, pp. 112–115.

Reddy, R.S., Pomeranz, I., Reddy, S.M. and Kajihara, S. (1995). Compact test generation for bridging faults under I_{DDQ} testing. In *Proc. VLSI Test Symposium*, pp. 310–316.

Rius, J. and Figueras, J. (1995). Detecting I_{DDQ} defective CMOS circuits by depowering. In *Proc. VLSI Test Symposium*, pp. 324–329.

Segura, J.A., Roca, M., Mateo, D. and Rubio, A. (1995). An approach to dynamic power consumption current testing of CMOS ICs. In *Proc. VLSI Test Symposium*, pp. 95–100.

Soden, J.M., Hawkins, C.F. and Miller, A.C. (1996). Identifying defects in deep-submicron CMOS ICs. *IEEE Spectrum*, **33** (9), pp. 66–71.

Thadikaran, P., Chakravarty, S. and Patel, J. (1995). Fault simulation of I_{DDQ} tests for bridging faults in sequential circuits. In *Proc. Int. Symposium on Fault-Tolerant Computing*, pp. 340–349.

Thadikaran, P. and Chakravarty, S. (1996). Fast algorithms for computing I_{DDQ} tests for combinational circuits. In *Proc. Int. Conference on VLSI Design*, pp. 103–106.

Thibeault, C. (1999). On the comparison of ΔI_{DDQ} and I_{DDQ} testing. In *Proc. VLSI Test Symposium*, pp. 143–150.

Williams, T.W, Dennard, R.H., and Kapur, R. (1996). I_{DDQ} test: sensitivity analysis of scaling. In *Proc. Int. Test Conference*, pp. 786–792.

7 Functional testing

In this chapter, we describe functional testing methods which start with a functional description of the circuit and make sure that the circuit's operation corresponds to its description. Since functional testing is not always based on a detailed structural description of the circuit, the test generation complexity can, in general, be substantially reduced. Functional tests can also detect design errors, which testing methods based on the structural fault model cannot.

We first describe methods for deriving universal test sets from the functional description. These test sets are applicable to any implementation of the function from a restricted class of networks.

We then discuss pseudoexhaustive testing of circuits where cones or segments of logic are tested by the set of all possible input vectors for that cone or segment.

Finally, we see how iterative logic arrays can be tested, and how simple design for testability schemes can make such testing easy. We introduce a graph labeling method for this purpose and apply it to adders, multipliers and dividers.

7.1   Universal test sets

Suppose the description of a function is given in some form, say a truth table. Consider the case where a fault in the circuit can change the truth table in an arbitrary way. How do we detect all such faults? One obvious way is to apply all 2^n vectors to it, where n is the number of inputs. The only faults not guaranteed to be detected this way are those that convert the behavior of the combinational circuit into a sequential one, e.g., stuck-open or feedback bridging faults. If we do not want to pay the price of **exhaustive testing** with all the input vectors, we need to impose some restrictions on the way a circuit is synthesized. Suppose the circuit is implemented as shown in Figure 7.1, where inversions, if any, are all at the primary inputs.

There may be any number of gate levels in the circuit. Any other circuit which can be reduced to a circuit of this type using De Morgan's theorem, is also acceptable. Then, indeed, one can derive a *universal test set* from the functional description, applicable to any restricted implementation of this type (Betancourt, 1971; Akers, 1973; Reddy, 1973; Pitchumani and Soman, 1986; Gupta and Jha, 1988; Chen and Lee, 1994). Such a test set will detect any change in the truth table.

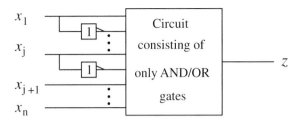

Figure 7.1 A restricted implementation

In order to understand the concept of universal test sets, we first need a framework of definitions.

Definition 7.1 A vector $X = (x_1, x_2, \ldots, x_n)$ is said to **cover** another vector $Y = (y_1, y_2, \ldots, y_n)$ if for all i, whenever $y_i = 1$, then $x_i = 1$ too.

Definition 7.2 A function $z(x_1, x_2, \ldots, x_n)$ is **positive (negative) unate** in variable x_i if for all $x_1, \ldots, x_{i-1}, x_{i+1}, \ldots, x_n$, $z(x_1, \ldots, x_{i-1}, 0, x_{i+1}, \ldots, x_n)$ is covered by (covers) $z(x_1, \ldots, x_{i-1}, 1, x_{i+1}, \ldots, x_n)$. If the function is not positive or negative unate in x_i, it is said to be **binate** in x_i.

The function is said to be unate if it is either positive or negative unate in each of its variables, else it is called binate. For example, $z = x_1 x_2 + \bar{x}_3$ is unate since it is positive unate in x_1 and x_2, and negative unate in x_3, whereas $z = x_1 x_2 + \bar{x}_2 \bar{x}_3$ is binate since it is not unate in x_2.

Definition 7.3 An input vector which makes the function 1 (0) is called a **true (false) vector**. A true vector which does not cover any other true vector is called a **minimal true vector**. Similarly, a false vector which is not covered by any other false vector is called a **maximal false vector**.

Suppose we form an expanded truth table for the given function which has a column for each literal that it depends on (not just each variable). For example, for $z = x_1 x_2 + \bar{x}_2 \bar{x}_3$, the expanded truth table is given in Table 7.1.

Consider all the minimal expanded true vectors and maximal expanded false vectors derived from such an expanded truth table. The union of all such vectors is the **universal test set**. In Table 7.1, the minimal true and maximal false vectors are indicated in the respective columns. In this table, the vector $(1, 0, 1, 1)$ is not a minimal true vector, for example, because it covers $(0, 0, 1, 1)$; however, $(0, 0, 1, 1)$ is a minimal true vector because it does not cover any other true vector. Similarly, $(0, 0, 1, 0)$ is not a maximal false vector since it is covered by $(1, 0, 1, 0)$, whereas $(1, 0, 1, 0)$ is such a vector because it is not covered by any other false vector. The universal test set with respect to inputs (x_1, x_2, x_3) for this example is $\{(0, 0, 0), (0, 1, 0), (1, 0, 1), (1, 1, 1)\}$.

Table 7.1. *Expanded truth table*

x_1	x_2	$\bar{x}_2$	$\bar{x}_3$	z	minimal true vector	maximal false vector
0	0	1	0	0		
0	0	1	1	1	✓	
0	1	0	0	0		
0	1	0	1	0		✓
1	0	1	0	0		✓
1	0	1	1	1		
1	1	0	0	1	✓	
1	1	0	1	1		

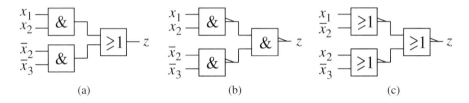

(a)   (b)   (c)

Figure 7.2   Some example circuits to which the universal test set is applicable

This test set is applicable to any circuit implementation of the given function which is of the form shown in Figure 7.1, or can be reduced to such a form through the use of De Morgan's theorem. Some such example circuits are shown in Figure 7.2.

Note that the circuit in Figure 7.2(b) can be transformed into the circuit in Figure 7.2(a) using De Morgan's theorem. The circuit in Figure 7.2(c) can be similarly transformed into a two-level OR–AND circuit which would be of the desired form. Readers can check that the universal test set derived above detects all single stuck-at faults (SAFs) in each of these implementations. In fact, the universal test set is much more powerful, as we will see next.

A nice property that a minimal true vector has is that it is guaranteed to detect the faults detected by any other true vector that covers it in the expanded truth table. The proof of this statement is based on the **monotone property** of an AND/OR circuit such as the one in Figure 7.1, which says that for two input vectors X and Y, if X covers Y then the output $z(X)$ covers $z(Y)$. For most typical fault models, such as single or multiple stuck-at, the monotone property is maintained even in the faulty circuit. Suppose that X is an expanded true vector and Y an expanded minimal true vector. Let the corresponding outputs of the faulty circuit be $z_f(X)$ and $z_f(Y)$, respectively. If X detects the fault then $z_f(X)$ must be logic 0. Since $z_f(X)$ covers $z_f(Y)$, from the monotone property it means that $z_f(Y)$ must be logic 0 too. This implies that Y is also a test for the fault. Similarly, a maximal false vector is guaranteed to detect any fault detected by any other false vector that is covered by it (the covering relationship is always evaluated in the expanded form of these vectors).

Consider our running example again. The above property suggests that the expanded minimal true vector $(0, 0, 1, 1)$ will detect all faults detected by $(1, 0, 1, 1)$. Similarly, the expanded maximal false vector $(1, 0, 1, 0)$ will detect all faults detected by $(0, 0, 1, 0)$.

Since we have not assumed that the circuit implementation of the given function, as long as it is of the form in Figure 7.1, is necessarily irredundant, the universal test set concept is also applicable in the presence of redundancies. In redundant circuits, it will detect all detectable single or multiple faults. The above arguments imply that the universal test is as powerful as the exhaustive test set. However, the disadvantage of a universal test set is that it reduces to the exhaustive test set when the function is binate in all its input variables. For such a function, no expanded true (false) vector would cover any other expanded true (false) vector. Therefore, no vector can be dropped.

For large functions, deriving the universal test set from the truth table would be too compute- and memory-intensive. Also, most of the time we are concerned with multi-output circuits, not single-output ones. An efficient algebraic method, based on Shannon's decomposition, to get a compacted universal test for any restricted multi-output circuit for the given set of functions, is given in Chen and Lee (1994).

7.2   Pseudoexhaustive testing

When exhaustive testing with all the 2^n vectors for an n-input circuit is not feasible, we can try to retain most of its advantages, while requiring much fewer vectors, through a technique called **pseudoexhaustive testing**. In this technique, only different segments of the circuit are exhaustively tested. **Segments** consist of logic which are not necessarily disjoint. Depending on the segmentation method, different types of pseudoexhaustive testing methods can be developed, as follows.

1 Verification testing
2 Hardware segmentation
3 Sensitized segmentation

We discuss these methods in detail next.

7.2.1   Verification testing

In many multi-output combinational circuits, each output depends on only a subset of primary inputs. A circuit in which no output depends on all the inputs is called a **partial dependence** circuit (McCluskey, 1984). Suppose we think of the cone of logic feeding any output as a segment. A testing approach in which each of these segments is simultaneously tested in an exhaustive fashion is called **verification testing**. If two

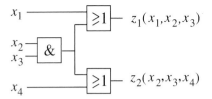

Figure 7.3   A circuit for illustrating pseudoexhaustive testing

inputs never appear in the same output function, then the same test signal can be applied to both the inputs. Thus, the number of test signals, p, can be less than the number of inputs, n.

The first step in the derivation of a pseudoexhaustive test set is to obtain a dependence matrix. For an n-input, m-output circuit, the **dependence matrix** has m rows, one for each output, and n columns, one for each input. An entry (i, j) in this matrix is 1 if and only if output i depends on input j. The other entries are 0. For example, for the circuit given in Figure 7.3, the dependence matrix is given below.

$$
\begin{array}{c}
\quad\quad x_1 \; x_2 \; x_3 \; x_4 \\
\begin{array}{c} z_1 \\ z_2 \end{array}
\left[
\begin{array}{cccc}
1 & 1 & 1 & 0 \\
0 & 1 & 1 & 1
\end{array}
\right]
\end{array}
$$

We next partition the columns in the dependence matrix into different sets such that each row of a set has at most one 1-entry, and the number of such sets p is a minimum. The matrix so obtained is called the **partitioned dependence matrix**. In general, such a partitioning may not be unique. For our example, this matrix is given below.

$$
\begin{array}{c}
\quad\quad x_1 \; x_4 \;\; x_2 \;\; x_3 \\
\begin{array}{c} z_1 \\ z_2 \end{array}
\left[
\begin{array}{cc|c|c}
1 & 0 & 1 & 1 \\
0 & 1 & 1 & 1
\end{array}
\right]
\end{array}
$$

A **reduced partitioned dependence matrix** is obtained from a partitioned dependence matrix by replacing each row within each partition by the logical OR of the entries in it. This is an $m \times p$ matrix. This matrix is shown below for our running example.

$$
\begin{array}{c}
\quad\quad\quad x_1 \\
\quad x_4 \; x_2 \; x_3 \\
\begin{array}{c} z_1 \\ z_2 \end{array}
\left[
\begin{array}{ccc}
1 & 1 & 1 \\
1 & 1 & 1
\end{array}
\right]
\end{array}
$$

For each column in the reduced partitioned dependence matrix, one test signal is required. The test signal feeds all the inputs corresponding to that column. Therefore, for our example, three test signals are required; the first one feeds x_1 and x_4, the second one feeds x_2 and the third one feeds x_3. A **reduced verification test set** (RVTS)

Table 7.2. $U(3, 3)$ for
the circuit in Figure 7.3

x_1		
x_4	x_2	x_3
0	0	0
0	0	1
0	1	0
0	1	1
1	0	0
1	0	1
1	1	0
1	1	1

specifies the set of input vectors that needs to be applied through the test signals to the circuit.

Let w be the maximum number of 1s in any row of the partitioned (or reduced partitioned) dependence matrix. This denotes the maximum number of inputs that any output depends on. The number of 1s in a vector is frequently referred to as its **weight**. A universal RVTS $U(p, w)$ is a p-column matrix in which all submatrices of w columns and all rows contain all possible vectors of w bits.

When $p = w$, $U(p, w)$ just consists of all possible 2^w vectors of w bits. Such a circuit is called a **maximal test concurrency** (MTC) circuit. For our example, $p = w = 3$. Hence, the circuit in Figure 7.3 is indeed an MTC circuit. The RVTS $U(3, 3)$ for this circuit is shown in Table 7.2. This test set exhaustively tests both the logic cones feeding outputs z_1 and z_2, since all eight vectors are applied to both sets of inputs (x_1, x_2, x_3) and (x_2, x_3, x_4). The exhaustive testing for the whole circuit, on the other hand, would have required 16 test vectors.

Next, consider a circuit for which $p = w + 1$. The block diagram of one such circuit is shown in Figure 7.4. The reduced partitioned dependence matrix (which is non-unique in this case) is shown below.

$$
\begin{array}{c}
\begin{array}{ccc} & x_3 & \\ x_1 & x_2 & x_4 \end{array} \\
\begin{array}{c} z_1 \\ z_2 \\ z_3 \\ z_4 \end{array}
\begin{bmatrix} 1 & 1 & 0 \\ 1 & 0 & 1 \\ 1 & 0 & 1 \\ 0 & 1 & 1 \end{bmatrix}
\end{array}
$$

The RVTS $U(p, w)$ for $p = w + 1$ is simply the set of all even (or all odd) parity vectors of length p. For the above example, the test set based on even parity is shown in Table 7.3. This test set exhaustively tests each of the four output logic cones. This is a substantial reduction from the exhaustive test set of size 16 for the whole circuit.

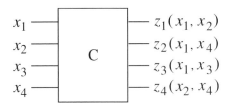

Figure 7.4 Circuit C for which $p = 3$, $w = 2$

Table 7.3. $U(3, 2)$ *for*
the circuit in Figure 7.4

x_1	x_2	x_3 x_4
0	0	0
0	1	1
1	0	1
1	1	0

Finally, consider the case when $p > w + 1$. For this case, $U(p, w)$ consists of different sets of input vectors, each set containing all possible vectors of a given weight. Table 7.4 gives $U(p, w)$ for various values of p and w. Here $(a_1, a_2, \ldots)$ under column $U(p, w)$ refers to a set of p-bit vectors consisting of all vectors of weight a_1 and all vectors of weight a_2, and so on. For the (p, w) pairs for which an entry does not exist in the table, one can increase the value of w to obtain another RVTS, which may, however, not be minimal. For example, there is no entry for $(p, w) = (8, 7)$. However, by increasing the value of w to eight, one can obtain an RVTS consisting of all 256 vectors of eight bits.

7.2.2   Hardware segmentation

Verification testing is not feasible when one or more outputs of the circuit depend on too many primary inputs. To test such circuits pseudoexhaustively, we need a more general segmentation technique which allows one to control and observe some internal lines of the circuit (McCluskey and Bozorgui-Nesbat, 1981; Patashnik, 1983; Roberts and Lala, 1984; Archambeau, 1985; Min and Li, 1986; Shperling and McCluskey, 1987; Udell, 1989). One way to do this is through the use of extra hardware, such as multiplexers. This technique is called **hardware segmentation**. Another technique, called **sensitized segmentation**, applies vectors to the primary inputs in such a fashion that the inputs and outputs of internal segments become indirectly controllable and observable, respectively. This technique will be discussed later.

Table 7.4. $U(p, w)$ for various p and w (McCluskey, 1984) (© 1984 IEEE)

w	range for p	$U(p, w)$
2	$p > 3$	$(0, p - 1)$ or $(1, p)$
3	$p > 4$	$(1, p - 1)$
4	$p > 5$	$(1, p - 2)$ or $(2, p - 1)$
5	$p > 6$	$(2, p - 2)$
6	$p = 8$	$(1, 4, 7)$
6	$p > 8$	$(2, p - 3)$ or $(3, p - 2)$
7	$p = 9$	$(0, 3, 6, 9)$
7	$p > 9$	$(3, p - 3)$
8	$p = 10$	$(0, 3, 6, 9)$ or $(1, 4, 7, 10)$
8	$p = 11$	$(0, 4, 8)$ or $(3, 7, 9)$
8	$p > 11$	$(3, p - 4)$ or $(4, p - 3)$
9	$p = 11$	$(1, 4, 7, 10)$
9	$p = 12$	$(0, 4, 8, 12)$
9	$p > 12$	$(4, p - 4)$
10	$p = 12$	$(1, 4, 7, 10)$ or $(2, 5, 8, 11)$
10	$p = 13$	$(0, 4, 8, 12)$ or $(1, 5, 9, 13)$
10	$p = 14$	$(0, 5, 10)$ or $(4, 9, 14)$
10	$p > 14$	$(4, p - 5)$ or $(5, p - 4)$

In the most general form of hardware segmentation, the circuit is segmented into $k \geq 2$ possibly overlapping segments S_i, which are tested exhaustively. Each segment S_i has $n(S_i) = n_i + p_i$ inputs, where n_i inputs come from the primary inputs and p_i new inputs come from the segmentation cuts. The cost of pseudoexhaustive testing is defined in terms of its test length, as follows.

$$Cost = \sum_{segments\ S_i} 2^{n(S_i)} = \sum_{i=1}^{k} 2^{n_i + p_i}$$

Finding k segments of the circuit that yield the smallest test length is known to be NP-complete. Therefore, various heuristics have been presented in the above-mentioned references to solve this problem.

Consider the segmentation of a circuit into two segments (i.e., $k = 2$), as shown in Figure 7.5. Access to the internal inputs and outputs of the two segments can be provided by using multiplexers as shown in Figure 7.6. By controlling the multiplexers, all the inputs and outputs of the segments can be accessed through primary inputs and outputs. For example, in order to test segment S_1, the multiplexers can be controlled as

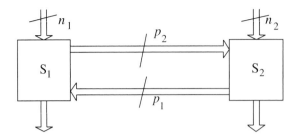

Figure 7.5   Circuit segmentation for $k = 2$

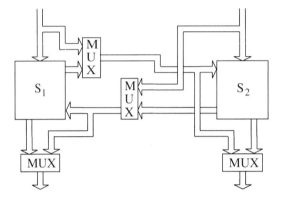

Figure 7.6   Hardware segmentation using multiplexers

shown in Figure 7.7. We can see that the intersegment inputs of S_1 are now completely controllable, and its intersegment outputs are completely observable. Segment S_2 can be similarly tested.

Let us apply the above segmentation technique to the circuit shown in Figure 7.8. Consider segments S_1 and S_2. There is only one intersegment connection c_1 in this example. In order to test S_1, we need to apply all possible input vectors to x_1, x_2 and x_3, and make c_1 observable. This can be done by inserting a multiplexer before output z_2 to select between that output and c_1. In order to test S_2, we need to apply all possible vectors to c_1, x_4 and x_5, and observe z_2. This can be done by inserting a multiplexer on line c_1 to select between that line and one of the inputs x_1, x_2 or x_3.

The disadvantage of hardware segmentation is that a considerable portion of the overall circuitry is comprised of extra multiplexers. This adversely affects both the area and performance of the circuit. However, it is frequently possible to avoid the addition of extra test circuitry by a technique called *sensitized segmentation*, which is discussed next.

7.2.3   Sensitized segmentation

In sensitized segmentation, testing of each segment is made possible by sensitizing paths from primary inputs to segment inputs, and from segment outputs to circuit

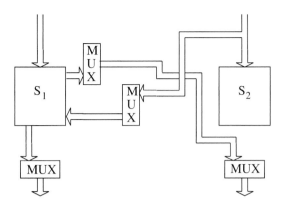

Figure 7.7 Configuration to test segment S_1

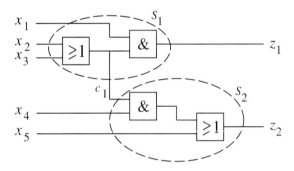

Figure 7.8 Hardware segmentation example

outputs. In many cases, each segment can still be exhaustively tested. However, there may exist cases where path sensitization may not be enough to supply all possible input vectors to the embedded segment and propagate the errors to the circuit outputs. In such cases, segments are tested to the maximum extent possible. Not being able to guarantee exhaustive testing of all segments is the only disadvantage this method has over hardware segmentation. However, its advantage is that it does not incur any area or performance penalties.

Let us consider the circuit in Figure 7.8 once again. However, let us look at another segmentation with four segments, as shown in Figure 7.9. To do sensitized pseudoexhaustive testing of segment S_1, we have to feed all four two-bit vectors to x_2 and x_3, while sensitizing paths from the two outputs of this segment to the circuit outputs. This can be done by making $x_1 = 1$, $x_4 = 1$, and $x_5 = 0$. To test segment S_2, we need to feed the AND gate its exhaustive test set and observe z_1. To test segment S_3, we need to exhaustively test the AND gate and sensitize a path from its output to z_2 by making $x_5 = 0$. Finally, to test segment S_4, we need to exhaustively test the OR gate and observe z_2. All the above conditions can be simultaneously satisfied by the pseudoexhaustive test set given in Table 7.5. The length of the test has been

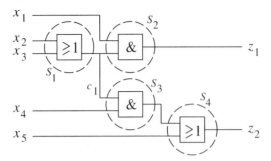

Figure 7.9 Sensitized circuit segmentation

Table 7.5. *A pseudoexhaustive test set for the circuit in Figure 7.9*

x_1	x_2	x_3	x_4	x_5	S_1	S_2	S_3	S_4
1	0	0	1	0	✓	✓	✓	✓
1	0	1	1	0	✓	✓	✓	✓
1	1	0	1	0	✓			
1	1	1	1	0	✓			
0	0	1	0	0		✓	✓	
0	0	0	0	0		✓	✓	
0	0	0	0	1				✓
0	0	1	1	1				✓

minimized by overlapping the test sets for the different segments. The test vectors for the corresponding segments have been indicated under the respective segment column. Note that sensitized segmentation lets us reduce the test set size to eight, whereas for verification testing the size of the test set in this case would be 16 (because z_2 depends on four of the five primary inputs).

7.3   Iterative logic array testing

An **iterative logic array** (ILA) consists of repeated interconnection of identical logic cells, as shown in Figure 7.10. For a typical cell i, the external *cell inputs* are denoted by $x_{i1}, x_{i2}, \ldots, x_{ip}$, and the *cell outputs* by $z_{i1}, z_{i2}, \ldots, z_{ir}$. In addition, it receives inputs $y_{i1}, y_{i2}, \ldots, y_{iq}$, called *input carries*, from the previous cell, and sends $Y_{i1}, Y_{i2}, \ldots, Y_{iq}$, called *output carries*, to the next cell. The input carries to cell 1 and all the external cell inputs are assumed to be directly controllable, and the output carries of the last (leftmost) cell and all the cell outputs are assumed to be directly observable. In some applications, only the outputs of the leftmost cell are of interest,

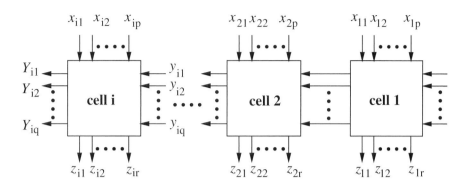

Figure 7.10   An iterative logic array

in which case the other cells do not have any cell outputs, and the output carries of the leftmost cell are the only observable outputs.

The ILA described above is an example of a one-dimensional unidirectional ILA. It is called **unidirectional** because the interconnections between cells are in one direction only. We will primarily be concentrating on one- and two-dimensional unidirectional ILAs here.

For analyzing the testability of ILAs, it is frequently useful to describe its operation with a **flow table** or a **flow diagram**. This specifies the cell outputs and output carries in terms of the cell inputs and input carries. This is analogous to a state table for a sequential circuit whose combinational logic is the same as the cell logic of the ILA. In this analogy, the input (output) carries correspond to present (next) states. Consider a ripple-carry adder. Its ith full adder (FA) cell, flow table and flow diagram are shown in Figure 7.11(a), (b) and (c), respectively. In the flow diagram, the entries within the nodes correspond to carries and the arcs correspond to $x_{i1}x_{i2}/z_i$.

While it is usually possible to derive a test set for an ILA which is linear in its size (Kautz, 1967; Menon and Friedman, 1971), we can frequently exploit the repetitive nature of the hardware to obtain a much smaller test set. A very useful concept in this regard, called *C-testability* (C stands for constant), was presented in Friedman (1973). A **C-testable** ILA can be tested with a constant number of test vectors irrespective of the size (i.e., the number of cells) of the ILA. Since then, many works have focussed on C-testability and easy testability of ILAs (Dias, 1976; Parthasarthy and Reddy, 1981; Sridhar and Hayes, 1981a, 1981b; Elhuni *et al.*, 1986; Wu and Capello, 1990; Chatterjee and Abraham, 1991a, 1991b).

The fault model generally used for ILAs is the **single cell fault model**, where the truth table of any one cell is assumed to be altered in an arbitrary fashion by the fault. To test such faults, each cell is fed its exhaustive test set. We will refer to this fault model as the **general fault model**. Sometimes, the fault model is restricted to arbitrary alteration of subcells of the cell. In this case, only the subcells are exhaustively tested, not the whole cell. We will refer to this fault model as the **restricted fault model**.

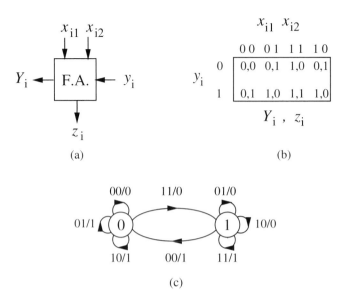

Figure 7.11 A ripple-carry adder cell, its flow table and flow diagram

Under the above fault models, ILA testing is equivalent to sensitized segmentation based pseudoexhaustive testing. Here, the segments are the cells or subcells.

We next consider conditions under which an ILA is C-testable.

7.3.1   C-testable ILAs

Consider a transition in the flow table of the ILA. In order to C-test the ILA for this transition, we must be able to feed the cell input pattern and input carries corresponding to this transition to different cells at regular intervals, say s, along the array. If this is possible, then the transition is called a **repeatable transition**. For such a transition it is possible to drive the cell back from the next state reached after the transition to its initial state. Then by shifting the test so obtained, we can test other cells for this transition. In all, this transition would need s test vectors. This insight leads us to the following necessary and sufficient conditions for C-testability of ILAs (Friedman, 1973; Dias, 1976; Sridhar and Hayes, 1981a, 1981b):

- the flow table for the ILA must be reduced (i.e., no redundant states), and
- every transition in the flow table must be repeatable.

The flow table needs to be reduced in order that errors on output carries of any cell can be propagated to an observable output. The flow diagram of a flow table which satisfies the above conditions can only have one or more strongly connected components (in a strongly connected component each node is reachable from each other node of the component). If these conditions are not satisfied, then a unidirectional ILA can be made C-testable by adding just one extra column to the flow table (Dias,

Table 7.6. *Test set for a ripple-carry adder*

x_{41}	x_{42}	x_{31}	x_{32}	x_{21}	x_{22}	x_{11}	x_{12}	y_1
0	0	0	0	0	0	0	0	0
0	1	0	1	0	1	0	1	0
1	0	1	0	1	0	1	0	0
0	0	1	1	0	0	1	1	0
1	1	0	0	1	1	0	0	1
0	1	0	1	0	1	0	1	1
1	0	1	0	1	0	1	0	1
1	1	1	1	1	1	1	1	1

1976; Sridhar and Hayes, 1981a). The idea is to link all the states of the flow table into a single loop by appropriately choosing the next state (output carries) entries in this extra column. This makes the flow diagram strongly connected. The cell output entries in this column can be chosen to make the flow table reduced, if it is not already so.

Consider the ripple-carry adder again. Its flow table is reduced and its flow diagram is strongly connected. Therefore, the above conditions are met and it is C-testable. Six of the transitions in the flow table take the state back to itself, and hence are trivially repeatable. For example, one could apply logic 0 to the input carry of the rightmost cell, and apply either only 00 or 01 or 10 repeatedly to the various cells to obtain three of the test vectors. Similarly, one could apply logic 1 to the input carry of the rightmost cell, and apply either only 01 or 10 or 11 repeatedly to the various cells to obtain the remaining three test vectors. For all these six test vectors, $s = 1$. For the transition with cell inputs 11 (00) from state 0 (1), one could apply 00 (11) to bring it back to its initial state. Thus, these are repeatable transitions with $s = 2$. This means that for the first of these two test vectors, we apply logic 0 to the input carry of the rightmost cell and 11 (00) to odd (even) numbered cells, and for the second vector we apply logic 1 to the input carry of the rightmost cell and 00 (11) to odd (even) numbered cells. Any error on the output carries of any cell gets propagated to the cell output of the next cell to its left. The reason is that the cell output of a full adder implements a three-input XOR function, which will always indicate error if only one of its inputs is in error. This shows that the ripple-carry adder is C-testable under the general fault model with eight test vectors since each of its cells is exhaustively tested. The complete test set for a ripple-carry adder with four cells is shown in Table 7.6.

7.3.2　Testing of ILAs by graph labeling

A powerful graph labeling technique was presented in Chatterjee and Abraham (1991a) for testing commonly-occurring ILAs that perform arithmetic functions. This technique can also automatically point out where small amounts of hard-

Table 7.7. *Basic bit vectors*

V_1	V_2	V_3	V_4	C_1	C_2	C_3	C_4
0	0	0	0	0	0	0	0
0	0	1	1	1	1	0	0
0	1	0	1	1	0	1	0
0	1	1	0	1	0	0	1
1	0	0	1	0	1	1	0
1	0	1	0	0	1	0	1
1	1	0	0	0	0	1	1
1	1	1	1	1	1	1	1

ware can be added to the ILA to make it C-testable. We discuss this technique next.

This technique is based on a data flow graph representation of the circuit where the nodes represent logical entities connected together to realize the desired function. An ILA cell can contain more than one node from the data flow graph. For such cases, although results for both general and restricted fault models are presented in Chatterjee and Abraham (1991a), we will restrict ourselves to the latter. In other words, the nodes (logical entities) will be tested in a pseudoexhaustive manner, not necessarily the complete cell. The branches of the data flow graph of the ILA to be tested are labeled with symbols representing bit vectors. These bit vectors exploit the properties of full addition which is a basic building block of arithmetic ILAs. The advantage of this method comes from its ease of application as well as its applicability to a wide class of ILAs for arithmetic functions, even those consisting of different types of cells.

Consider the set of bit vectors $V_1 - V_4$ and $C_1 - C_4$ shown in Table 7.7. Let function $g(\,)$ be the bitwise modulo-2 summation of vectors V_i, V_j and V_k, and function $f(\,)$ be the bitwise carry produced in this summation. Thus, these functions can be defined as follows:

$$g(V_i, V_j, V_k) = V_i \oplus V_j \oplus V_k \tag{7.1}$$

$$f(V_i, V_j, V_k) = V_i \cdot V_j + V_i \cdot V_k + V_j \cdot V_k. \tag{7.2}$$

We also consider eight other bit vectors $\bar{V}_1 - \bar{V}_4$ and $\bar{C}_1 - \bar{C}_4$, where $\bar{V}_i$ ($\bar{C}_i$) is the bit-by-bit complement of V_i (C_i), $1 \le i \le 4$. Let S be the set of all these sixteen bit vectors. Then the following relations can be obtained for members of S:

$$g(V_i, V_j, V_k) = V_l \tag{7.3}$$

$$f(V_i, V_j, V_k) = C_l \tag{7.4}$$

$$g(C_i, C_j, C_k) = C_l \tag{7.5}$$

$$f(C_i, C_j, C_k) = V_l \tag{7.6}$$

$$g(V_i, V_j, \bar{V}_k) = \bar{V}_l \tag{7.7}$$

$$f(V_i, V_j, \bar{V}_k) = C_k \tag{7.8}$$

$$g(V_i, \bar{V}_j, \bar{V}_k) = V_l \tag{7.9}$$

$$f(V_i, \bar{V}_j, \bar{V}_k) = \bar{C}_i \tag{7.10}$$

$$g(\bar{V}_i, \bar{V}_j, \bar{V}_k) = \bar{V}_l \tag{7.11}$$

$$f(\bar{V}_i, \bar{V}_j, \bar{V}_k) = \bar{C}_l \tag{7.12}$$

$$g(V_i, C_i, C_j) = V_j \tag{7.13}$$

$$f(V_i, C_i, C_j) = C_j \tag{7.14}$$

$$g(V_i, C_j, C_k) = V_l \tag{7.15}$$

$$f(V_i, C_j, C_k) = V_i \tag{7.16}$$

where $i \neq j \neq k \neq l$ and $1 \leq i, j, k, l \leq 4$, in each of these equations.

The following properties can also be derived for these vectors and functions (Chatterjee and Abraham, 1987, 1991a):

Property 7.1 The bitwise sum and carry obtained from any three vectors in S is also a vector in S.

Property 7.2 Any combination of three vectors with indices i, j, k from the set $\{V_i, \bar{V}_i, V_j, \bar{V}_j, V_k, \bar{V}_k\}$, where $i \neq j \neq k$ and $1 \leq i, j, k \leq 4$, contains all possible three-bit patterns, as do combinations of vectors from $\{C_i, \bar{C}_i, C_j, \bar{C}_j, C_k, \bar{C}_k\}$.

Property 7.3 Equations (7.3)–(7.16) can be written with the V and C variables interchanged.

Property 7.4 Equations (7.3)–(7.16) can be written with the variables on the left- and right-hand sides of the equations replaced with their complemented values.

To test the ILA which realizes an arithmetic function, a data flow graph of the ILA needs to be constructed. In this graph, full adders are represented as nodes, and the remaining circuitry is partitioned for testing purposes with each partition also represented as a node. Thus, one cell of the ILA may have more than one node. The interconnection among nodes is based on the array schematic. Each node is exhaustively tested. The regular nature of ILAs is exploited to find vectors that can be repetitively applied to the ILA cells in various directions of signal flow.

To test the ILA, we assign vectors from set S to the different branches of the corresponding data flow graph as follows (this is called **branch labeling**). Each input

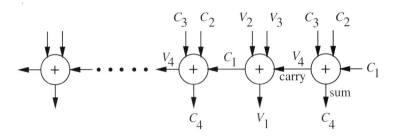

Figure 7.12 Labeling of the data flow graph of a ripple-carry adder

and output of a full adder is assigned a unique label consistent with the full addition operation, such that these labels can be applied repetitively in various signal flow directions. Then logic values at the external inputs of the ILA are specified so that values indicated by branch labeling can be applied to the full adders. Usually, this also leads to exhaustive testing of the other nodes in the data flow graph. If not, additional vectors are added to make this possible. If a branch labeling satisfying the above conditions exists, it is called a **consistent labeling**. Due to discontinuities in the regular cell interconnection structure at the ILA boundaries, a consistent labeling may not always exist. In such cases, design for testability techniques can be used by adding extra logic (such as XOR gates) and test inputs at the boundaries in order to accommodate graph labeling. Care must be taken to test the extra logic also in a pseudoexhaustive manner.

    We next look at some applications of graph labeling.

7.3.2.1 Ripple-carry adder

Let us revisit the problem of testing a ripple-carry adder. Its data flow graph is shown in Figure 7.12. Here, each node denotes a full addition operation. A consistent labeling is also shown in the figure. Equations (7.5) and (7.6) have been used to derive the labeling of the rightmost node, and Equations (7.3) and (7.4) for the next-to-rightmost node. Thereafter, the labeling repeats. From Property 7.2, each node gets its exhaustive test set. Also, we saw in the previous section that any error at the carry output of a node gets propagated to the sum output of the node to its left. This shows that the ripple-carry adder is C-testable with only eight test vectors. However, this test set is slightly different from the type of test set we had derived earlier in Table 7.6.

7.3.2.2 Carry-save array multiplier

To multiply two operands $A = (a_{n-1}, a_{n-2}, \ldots, a_0)$ and $B = (b_{n-1}, b_{n-2}, \ldots, b_0)$, a carry-save array multiplier requires an $n \times n$ array of multiplier cells and a row consisting of a ripple-carry adder. A 3×3 carry-save array multiplier is shown in Figure 7.13.

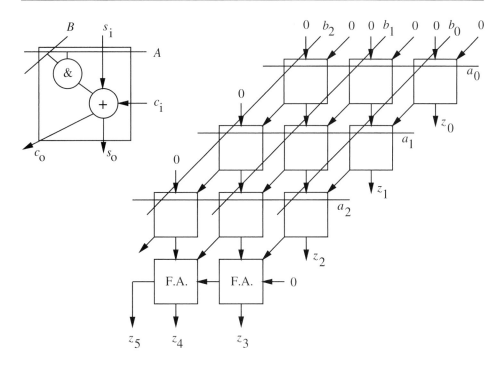

Figure 7.13   A 3×3 carry-save array multiplier and its cell

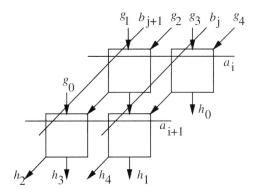

Figure 7.14   Block of four cells of the carry-save array multiplier

Each cell of the array consists of an AND gate and a full adder as shown. The outputs of the cell are: $c_o = (AB)(s_i + c_i) + s_i c_i$ and $s_o = (AB) \oplus s_i \oplus c_i$. Array inputs marked logic 0 are fed 0 during normal operation. However, for testing purposes, they will be assumed to be controllable. A 2×2 block of cells of the multiplier is shown in Figure 7.14 with external inputs a_i, a_{i+1} and b_j, b_{j+1}.

A labeling of the full adders of the cells based on Equations (7.3)–(7.6) is shown in the data flow graph in Figure 7.15(a) (notice the similarity of this labeling

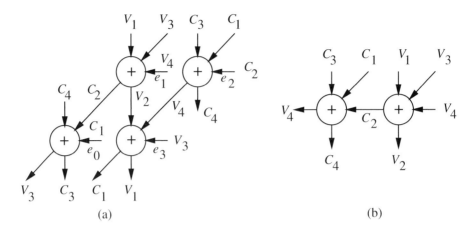

Figure 7.15 (a) Labeling of full adders of the block of cells, (b) labeling for carry propagate row

Table 7.8. *Assignment of external inputs to generate given labeling*

C_1	V_4	C_2	V_3	a_i	a_{i+1}	b_j	b_{j+1}
0	0	0	0	0	0	0	0
1	1	1	1	1	1	1	1
1	1	0	0	1	1	0	1
1	0	0	1	0	1	1	1
0	1	1	0	1	0	1	1
0	0	1	1	1	1	1	0
0	0	0	0	0	0	0	0
1	1	1	1	1	1	1	1

along the diagonals with the labeling used for the ripple-carry adder in Figure 7.12).

Observe that the label on both the elements in each pair, (g_0, h_0), (g_1, h_1), (g_2, h_2), (g_3, h_3) and (g_4, h_4), in Figure 7.14 is identical. This means that the respective cell input patterns can be applied repetitively to the array cells if we can ensure that the labels corresponding to e_0, e_1, e_2 and e_3 in Figure 7.15(a) can be generated through external inputs. Table 7.8 shows the assignment of values to the external inputs of the block of cells to generate the corresponding bit vectors at e_0–e_3 (given in the first four columns) through the AND gates. The labeling applies their respective exhaustive test sets to all full adders and AND gates.

To test the last carry-propagate row, consider the labeling in the data flow graph in Figure 7.15(b) for a block of two cells. This labeling has been derived from the labeling of the outputs of the block of cells in Figure 7.15(a). Again, the labels repeat every two cells. Thus, the labelings in Figures 7.15(a) and (b) can be used to test a carry-save array multiplier of any size. Any error on the sum or carry lines of any

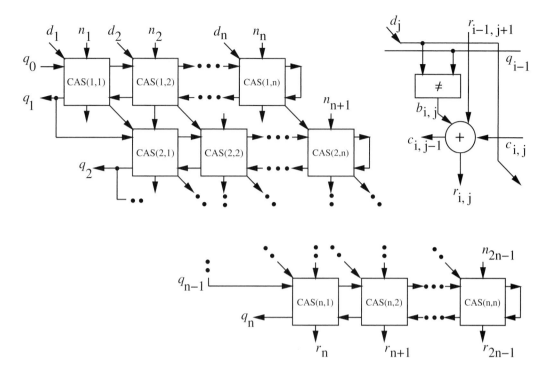

Figure 7.16 An $n \times n$ non-restoring array divider and its cell

internal cell is guaranteed to propagate to the observable outputs through a chain of XOR gates. Therefore, the carry-save array multiplier is C-testable with eight input vectors under the restricted fault model. In practice, the leftmost diagonal of the array in Figure 7.13 can be omitted.

7.3.2.3 Non-restoring array divider

A non-restoring array divider can be implemented as an ILA. The basic cell in such an array is a *controlled adder/subtracter* (CAS) cell. Each cell has an XOR gate and a full adder as shown in Figure 7.16. The logic of the CAS(i, j) cell is specified by the following equations:

$$r_{i,j} = r_{i-1,j+1} \oplus c_{i,j} \oplus b_{i,j}$$

$$c_{i,j-1} = c_{i,j} r_{i-1,j+1} + b_{i,j} c_{i,j} + b_{i,j} r_{i-1,j+1}$$

where $b_{i,j} = q_{i-1} \oplus d_j$. The cell interconnection for an $n \times n$ array is shown in the same figure. The dividend and divisor are represented by $N = (n_1, n_2, \ldots, n_{2n-1})$ and $D = (d_1, d_2, \ldots, d_n)$, respectively, whereas the quotient is represented by $Q = (q_1, q_2, \ldots, q_n)$, and the remainder by $R = (r_n, r_{n+1}, \ldots, r_{2n-1})$. n_1, d_1, q_1 and r_n are the respective sign bits. The q_{i-1} input is used to control the add ($q_{i-1} = 0$)

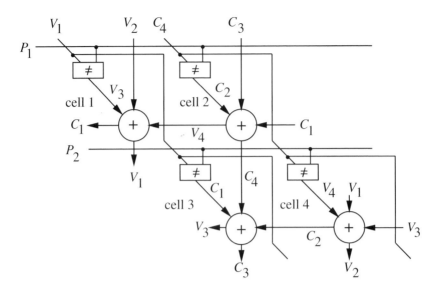

Figure 7.17 Labeling of a block of four cells

or subtract ($q_{i-1} = 1$) operations in the ith row. q_0 is held at logic 1 permanently during normal operation, since the initial operation performed in the top row is always subtraction.

Consider the data flow graph for a 2×2 block of cells shown in Figure 7.17 and the corresponding labeling. Notice the similarity with the labeling used earlier for the carry-save array multiplier in Figure 7.15. This labeling exhaustively tests the full adders in cells 1, 2, 3 and 4. Also, the $r_{i-1,j+1}$ input labels of cells 1 and 2 are the same as the $r_{i,j}$ output labels of cells 4 and 3, respectively. Thus, the labeling can be repetitively applied to the rows of the array. Similarly, the $c_{i,j}$ input labels of cells 2 and 4 are the same as $c_{i,j-1}$ output labels of cells 1 and 3, respectively. Therefore, the labeling can be repetitively applied to the columns of the array as well. The external input d_j of cells 1 and 3 is labeled V_1 and that of cells 2 and 4 is labeled C_4. We need to control the external inputs P_1 and P_2 in such a fashion that we obtain the labels shown at the output of each of the XOR gates. This means that the label for P_1, denoted as L_{P_1}, is given by $V_1 \oplus V_3 = C_2 \oplus C_4 = 01011010$. Similarly, the label for P_2, denoted as L_{P_2}, is given by $V_1 \oplus C_1 = V_4 \oplus C_4 = 01111110$. With these labels, all the XOR gates in the cells are also tested exhaustively.

The application of the above labeling to a 3×3 non-restoring array divider is shown in Figure 7.18. In order to repetitively apply the labeling scheme developed in Figure 7.17, we need some extra logic (two extra XOR gates for most rows, and two extra inputs T_1 and T_2) at the left and right boundaries, as shown. For the left boundary, output labels C_1 and V_3 need to be transformed to L_{P_2} and L_{P_1}, respectively. Since $C_1 \oplus L_{P_2} = V_3 \oplus L_{P_1} = V_1$, this means that V_1 should be applied to test input T_1

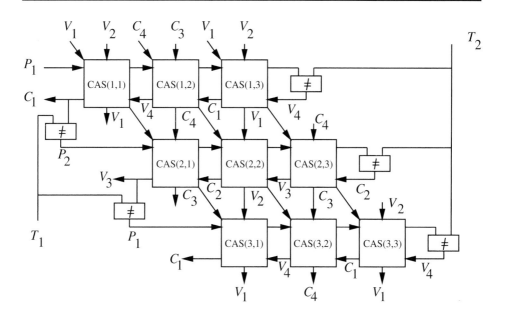

Figure 7.18 Labeling of the 3 × 3 non-restoring array divider

in the test mode. Similarly, at the right boundary, input labels V_4 and C_2 have to be obtained from L_{P_1} and L_{P_2}, respectively. Since $V_4 \oplus L_{P_1} = C_2 \oplus L_{P_2} = V_2$, T_2 is fed V_2 in the test mode. In the normal mode, both T_1 and T_2 are fed constant 0.

Note that the extra XOR gates are also exhaustively tested. Any error at the $r_{i,j}$ output of any cell is propagated to the output of the respective column, whereas any error at the $c_{i,j-1}$ output of the cell is propagated to the outputs of the successive columns or to the $c_{i,j-1}$ output of the leftmost cell of the row. Therefore, the modified non-restoring array divider is C-testable with just eight vectors under the restricted fault model.

Summary

- A universal test set consisting of minimal true vectors and maximal false vectors can be derived for restricted implementations of logic functions which can detect all detectable single and multiple SAFs in it.
- Pseudoexhaustive testing is a technique through which different segments of the circuit are exhaustively tested.
- Based on the segmentation method used, pseudoexhaustive testing methods can be divided into three categories: verification testing, hardware segmentation, and sensitized segmentation.

- In verification testing, each output cone of a multi-output combinational circuit is exhaustively tested.
- When one or more outputs of the circuit depend on too many primary inputs, verification testing is not possible. For such cases, controllability and observability of some internal lines of the circuit can be provided with the help of some extra hardware, such as multiplexers. This is called hardware segmentation.
- In sensitized segmentation, testing of each circuit segment is made possible by sensitizing paths from primary inputs to segment inputs, and from segment outputs to circuit outputs.
- An ILA consists of repeated interconnection of identical logic cells.
- An ILA is frequently analyzed for its testability based on its flow table.
- The most frequently used fault model for an ILA is the single cell fault model which assumes that the truth table of one cell in the ILA changes in an arbitrary fashion.
- An ILA which is testable with the same number of vectors irrespective of its size is called C-testable.
- For an ILA to be C-testable, its flow table must be reduced and every transition in the flow table must be repeatable.
- Commonly-occurring ILAs, which perform arithmetic functions, can be tested with a graph labeling technique.
- A ripple-carry adder and carry-save array multiplier can be shown to be C-testable with just eight test vectors using the graph labeling technique.
- A non-restoring array divider is C-testable with eight vectors when two extra inputs and some extra logic are added to the array.

Additional reading

In Cheng and Patel (1986), testing of two-dimensional ILAs is discussed.

In Bhattacharya and Hayes (1986), easily testable arithmetic ILAs are presented.

In Shen and Ferguson (1984), Takach and Jha (1991) and Jha and Ahuja (1993), designs of easily testable multipliers and dividers are given.

In Blanton and Hayes (1996), the testing properties of a class of regular circuits called convergent trees are investigated. Comparators, multiplexers and carry-lookahead adders are examples of such trees.

Exercises

7.1 Prove that for any circuit implementation of the form shown in Figure 7.1 for the given function, a maximal expanded false vector detects all detectable single and multiple SAFs that any expanded false vector covered by it detects.

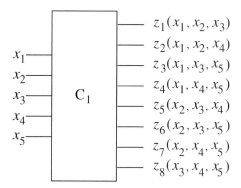

$$z_1(x_1, x_2, x_3)$$
$$z_2(x_1, x_2, x_4)$$
$$z_3(x_1, x_3, x_5)$$
$$z_4(x_1, x_4, x_5)$$
$$z_5(x_2, x_3, x_4)$$
$$z_6(x_2, x_3, x_5)$$
$$z_7(x_2, x_4, x_5)$$
$$z_8(x_3, x_4, x_5)$$

Figure 7.19 Circuit C_1

7.2   Show that the partitioned dependence matrix of the dependence matrix given below is not unique. Does the size of the verification test set depend on which partitioned dependence matrix is chosen? Derive the verification test set from one such matrix.

$$
\begin{array}{c}
 \\
z_1 \\
z_2 \\
z_3
\end{array}
\begin{array}{cccccc}
x_1 & x_2 & x_3 & x_4 & x_5 & x_6 \\
0 & 0 & 0 & 1 & 1 & 1 \\
1 & 0 & 0 & 0 & 1 & 0 \\
0 & 1 & 1 & 0 & 0 & 0
\end{array}
$$

7.3   Given a dependence matrix, give a method for obtaining a partitioned dependence matrix with a minimum number of partitions.

7.4   For circuit C_1 given in Figure 7.19, find a minimal verification test set which tests each of the output logic cones exhaustively (hint: the universal RVTS in this case does not lead to a minimal verification test set).

7.5   Show that any universal RVTS $U(p, w)$ is also a $U(p, j)$ for all j less than w.

7.6   Show that for an MTC circuit, the minimum-sized RVTS has 2^w vectors.

7.7   Consider the set of all even parity vectors of length p. Prove that this set constitutes the universal RVTS $U(p, p - 1)$.

7.8   For the circuit in Figure 7.8, derive a pseudoexhaustive test set based on sensitized segmentation, assuming that the circuit is divided up into two segments as shown.

7.9   Present a heuristic segmentation method for minimizing the size of the sensitized segmentation pseudoexhaustive test set.

7.10 Give a flow table which is reduced but in which all transitions are not repeatable (hence the corresponding ILA is not C-testable). Add an extra column to the flow table and specify entries in it to make all transitions repeatable.

7.11 Prove Property 7.1.

7.12 Give a graph labeling for testing the carry save array multiplier based on

Equations (7.7), (7.8) and the corresponding equations derived from Property 7.3. Show the actual test set for a 3×3 array multiplier.

7.13 Give a graph labeling for testing the non-restoring array divider based on Equations (7.7), (7.8) and the corresponding equations derived from Property 7.3. Show the actual test set for an augmented 4×4 array divider.

7.14 For your favorite ILA of an arithmetic function which is not considered in this chapter, use the graph labeling technique to either show that the ILA is already C-testable or add extra logic to the array to make it C-testable (some examples of ILAs you could consider are carry-propagate array multiplier, Pezaris array multiplier, and restoring array divider).

References

Akers, S.B. (1973). Universal test sets for logic networks. *IEEE Trans. on Computers*, **C-22** (9), pp. 835–839.

Archambeau, E.C. (1985). *Network Segmentation for Pseudoexhaustive Testing*. CRC Tech. Report, No. 85-10, Stanford University.

Betancourt, R. (1971). Derivation of minimum test sets for unate logical circuits. *IEEE Trans. on Computers*, **C-20** (11), pp. 1264–1269.

Bhattacharya, D. and Hayes, J.P. (1986). Fast and easily testable implementations of arithmetic functions. In *Proc. Int. Symp. on Fault-Tolerant Computing*, pp. 324–329.

Blanton, R.D. and Hayes, J.P. (1996). Testability of convergent tree circuits. *IEEE Trans. on Computers*, **45** (8), pp. 950–963.

Chatterjee, A. and Abraham, J.A. (1987). On the C-testability of generalized counters. *IEEE Trans. on Computer-Aided Design*, **CAD-6** (5), pp. 713–726.

Chatterjee, A. and Abraham, J.A. (1991a). Test generation, design for testability and built-in self-test for arithmetic units based on graph labeling. *J. of Electronic Testing: Theory & Applications*, **2**, pp. 351–372.

Chatterjee, A. and Abraham, J.A. (1991b). Test generation for iterative logic arrays based on an N-cube of cell states model. *IEEE Trans. on Computers*, **40** (10), pp. 1133–1148.

Chen, B. and Lee, C.L. (1994). A complement-based fast algorithm to generate universal test sets for multi-output functions. *IEEE Trans. on Computer-Aided Design*, **13** (3), pp. 370–377.

Cheng, W.T. and Patel, J.H. (1986). Testing in two-dimensional iterative logic arrays. In *Proc. Int. Symp. on Fault-Tolerant Computing*, pp. 76–81.

Dias, F.J.O. (1976). Truth table verification of an iterative logic array. *IEEE Trans. on Computers*, **C-25** (6), pp. 605–612.

Elhuni, H., Vergis, A. and Kinney, L. (1986). C-testability of two-dimensional iterative arrays. *IEEE Trans. on Computer-Aided Design*, **CAD-5** (4), pp. 573–581.

Friedman, A.D. (1973). Easily testable iterative systems. *IEEE Trans. on Computers*, **C-22** (12), pp. 1061–1064.

Gupta, G. and Jha, N.K. (1988). A universal test set for CMOS circuits. *IEEE Trans. on Computer-Aided Design*, **7** (5), pp. 590–597.

Jha, N.K. and Ahuja, A. (1993). Easily testable non-restoring and restoring gate-level cellular array dividers. *IEEE Trans. on Computer-Aided Design*, **12** (1), pp. 114–123.

Kautz, W.H. (1967). Testing for faults in combinational cellular logic arrays. In *Proc. 8th Annual Symp. on Switching & Automata Theory*, pp. 161–174.

McCluskey, E.J. and Bozorgui-Nesbat, S. (1981). Design for autonomous test. *IEEE Trans. on Computers*, **C-30** (11), pp. 866–875.

McCluskey, E.J. (1984). Verification testing – a pseudoexhaustive test technique. *IEEE Trans. on Computers*, **C-33** (6), pp. 541–546.

Menon, P.R. and Friedman, A.D. (1971). Fault detection in iterative logic arrays. *IEEE Trans. on Computers*, **C-20** (5), pp. 524–535.

Min, Y. and Li, Z. (1986). Pseudoexhaustive testing strategy for large combinational circuits. *Computer System Science & Engg.*, **1** (4), pp. 213–220.

Parthasarthy, R. and Reddy, S.M. (1981). A testable design of iterative logic arrays. *IEEE Trans. on Computers*, **C-30** (11), pp. 833–841.

Patashnik, O. (1983). *Circuit Segmentation for Pseudoexhaustive Testing*. Center for Reliable Computing Tech. Report No. 83-14, Stanford University.

Pitchumani, V. and Soman, S. (1986). An application of unate function theory to functional testing. In *Proc. Int. Symposium on Fault-Tolerant Computing*, pp. 70–75.

Reddy, S.M. (1973). Complete test sets for logic functions. *IEEE Trans. on Computers*, **C-22** (11), pp. 1016–1020.

Roberts, M.W. and Lala, P.K. (1984). An algorithm for the partitioning of logic circuits. *IEE Proc.*, **131** (4), Pt. E, pp. 113–118.

Shen, J.P. and Ferguson, F.J. (1984). The design of easily testable array multipliers. *IEEE Trans. on Computers*, **C-33** (6), pp. 554–560.

Shperling, I. and McCluskey, E.J. (1987). Circuit segmentation for pseudoexhaustive testing via simulated annealing. In *Proc. Int. Test Conference*, pp. 112–124.

Sridhar, T. and Hayes, J.P. (1981a). A functional approach to testing bit-sliced microprocessors. *IEEE Trans. on Computers*, **C-30** (8), pp. 563–571.

Sridhar, T. and Hayes, J.P. (1981b). Design of easily testable bit-sliced systems. *IEEE Trans. on Computers*, **C-30** (11), pp. 842–854.

Takach, A.R. and Jha, N.K. (1991). Easily testable gate-level and DCVS multipliers. *IEEE Trans. on Computer-Aided Design*, **10** (7), pp. 932–942.

Udell Jr., J.G. (1989). Pseudoexhaustive test and segmentation: formal definitions and extended fault coverage results. In *Proc. Int. Symposium on Fault-Tolerant Computing*, pp. 292–298.

Wu, C.W. and Capello, P.R. (1990). Easily testable iterative logic arrays. *IEEE Trans. on Computers*, **39** (5), pp. 640–652.

8 Delay fault testing

Delay fault testing exposes temporal defects in an integrated circuit. Even when a circuit performs its logic operations correctly, it may be too slow to propagate signals through certain paths or gates. In such cases, incorrect logic values may get latched at the circuit outputs.

In this chapter, we first describe the basic concepts in delay fault testing, such as clocking schemes, testability classification, and delay fault coverage.

Next, we present test generation and fault simulation methods for path, gate and segment delay faults in combinational as well as sequential circuits. We also cover test compaction and fault diagnosis methods for combinational circuits. Under sequential test generation, we look at non-scan designs. Scan designs are addressed in Chapter 11.

We then discuss some pitfalls that have been pointed out in delay fault testing, and some initial attempts to correct these problems.

Finally, we discuss some unconventional delay fault testing methods, which include waveform analysis and digital oscillation testing.

8.1 Introduction

Delay fault (DF) testing determines the operational correctness of a circuit at its specified speed. Even if the steady-state behavior of a circuit is correct, it may not be reached in the allotted time. DF testing exposes such circuit malfunctions. In Chapter 2 (Section 2.2.6), we presented various DF models, testing for which can ensure that a circuit is free of DFs. These fault models include the gate delay fault (GDF) model and the path delay fault (PDF) model. GDFs can themselves be divided into gross GDFs (G-GDFs) and small GDFs (S-GDFs). G-GDFs are also called transition faults (TFs). We also presented the segment delay fault (SDF) model, which is intermediate to the GDF and PDF models. We classified the two-pattern tests (which consist of an initialization vector and a test vector) required for DFs into robust and non-robust. Robust tests were further divided into hazard-free robust and general robust. A special type of non-robust test, called validatable non-robust, was also defined. In this chapter, we will be targeting these different types of fault models and tests for combinational and sequential circuits.

An underlying assumption behind the fault models and most of the testing methods presented here is that the gate propagation delays are fixed and independent of the input logic values. Therefore, it is assumed that if a circuit passes a test for a given fault, then the fault will not cause an incorrect circuit operation for any other sequence of input patterns. This does not strictly correspond to what happens in actual circuits. Thus, this assumption can lead to some pitfalls in DF testing, which is discussed in detail in Section 8.7. However, making this assumption keeps the DF testing problem tractable.

In this section, we present some basic background material. We first describe two clocking schemes used in DF testing based on *variable clock* and *rated clock*. We then present various testability classifications of PDFs. We finally consider how realistic DF coverages can be obtained.

8.1.1 Clocking schemes for delay fault testing

DF testing techniques are either based on a variable clock scheme or a rated clock scheme (Bose *et al.*, 1998a). The most commonly used is the variable clock scheme. Consider the combinational circuit shown in Figure 8.1(a). In the **variable clock** scheme, two clocks are required to separately strobe the primary inputs and outputs in such a circuit, as shown in Figure 8.1(b). In a two-pattern test (P_1, P_2), the first pattern (or vector) P_1 is applied to the primary inputs at time t_1 and the second pattern P_2 at time t_2. The shaded area represents the amount of time required for the faulty circuit to stabilize after the application of P_1. The circuit response is observed at time t_3. This two-pattern test is used to determine if the propagation delay of a path, through which a GDF, SDF or PDF is being tested, exceeds the time interval $(t_3 - t_2)$. Due to the skewed input/output strobing, interval $(t_3 - t_2)$ is less than the interval $(t_2 - t_1)$. $(t_2 - t_1)$ is the time allowed for signal values to stabilize in the faulty circuit. $(t_3 - t_2)$ is the maximum allowable path delay for the rated frequency of operation. If we assume that no path delay in a faulty circuit exceeds twice the clock period, then $(t_2 - t_1)$ should be at least twice the interval $(t_3 - t_2)$. This DF testing methodology increases the test application time and makes the hardware required for controlling the clock or the test application software more complex. However, it makes test generation and fault simulation easier.

In the **rated clock** scheme, all input vectors are applied at the rated circuit speed using the same strobe for the primary inputs and outputs, as shown in Figure 8.1(c). All the path delays in the fault-free circuit are assumed to be less than the interval $(t_2 - t_1)$. However, paths in the faulty circuit may have delays exceeding this interval. Therefore, logic transitions and hazards that arise at time instance t_1 due to the vector pair (P_0, P_1) may still be propagating through the circuit during the time interval $(t_3 - t_2)$. This is shown in Figure 8.1(c). In addition, other transitions may originate at time instance t_2 due to the vector pair (P_1, P_2). If we assume as before that no path delay in the faulty

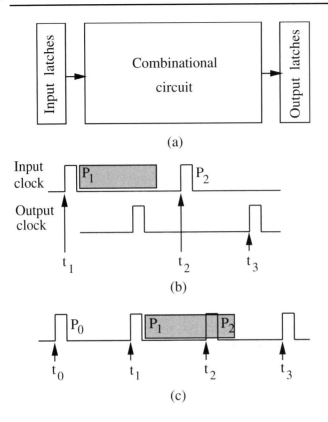

Figure 8.1 Different clocking schemes for combinational circuit testing

circuit exceeds twice the clock period, then signal conditions during interval $(t_3 - t_2)$ depend on three vectors (P_0, P_1, P_2). This, in general, makes test generation and fault simulation more complex. However, this is the type of scenario one encounters more often in the industry. Contrast the above situation with the one in Figure 8.1(b) for the variable clock scheme, where signal conditions during the interval $(t_3 - t_2)$ depend on the vector pair (P_1, P_2) only.

In sequential circuits, paths may also start and end at the internal latches. In the *slow–fast variable clocking scheme*, the circuit is brought to the state required for testing using a **slow clock**, such as the one shown in Figure 8.1(b). This is shown in Figure 8.2 using the iterative array model for sequential circuits (see Chapter 5, Figure 5.7). Here, TF_i denotes the ith time frame. The fault is activated by using a **fast clock** (or rated clock) for one cycle. At this point, the faulty logic value may be stored in an internal latch. This fault is propagated to a primary output using a slow clock again. Using a slow clock for state justification as well as error propagation ensures that they are not invalidated by the DF. This considerably simplifies test generation and fault simulation. However, since a slow clock is used in the state justification and error propagation phases, no other DF can be activated. Hence, the test sequence for

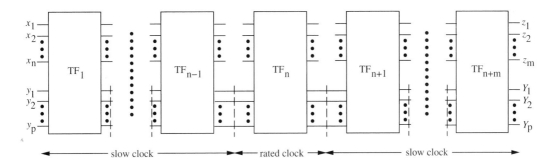

Figure 8.2 Slow–fast clocking scheme for sequential circuit testing

the circuit under test is extremely long. If all vectors are applied at the rated speed, they can activate many DFs with the same test sequence, thereby reducing test length. The price paid is, of course, in terms of much more complex test generation and fault simulation methods.

In this chapter, we will discuss both variable clock and rated clock methods.

8.1.2   Testability classification of PDFs

There are various ways to classify PDFs in combinational circuits. We will consider four main ways of doing it: Cheng's classification (Cheng and Chen, 1993), Lam's classification (Lam *et al.*, 1995), Gharaybeh *et al.*'s classification (1995), and Sivaraman's classification (Sivaraman and Strojwas, 1997). These classifications are based on the variable clock scheme. We present some background material first.

An input of a gate is said to have a **controlling value** if it uniquely determines the output of the gate independently of its other inputs. Otherwise, the value is said to be **non-controlling**. For example, 1 is the controlling value of an OR or NOR gate, and 0 the non-controlling value.

A path R in a circuit is a sequence $(g_0 g_1 \cdots g_r)$, where g_0 is a primary input, $g_1 \cdots g_{p-1}$ are gate outputs, and g_r is a primary output of the circuit. A segment is a sub-sequence of a path. Let the gate with output g_j be denoted as G_j. The **inversion parity** of R is the number of inversions along R modulo 2. If the inversion parity is 0 (or even), then the value of the associated literal, l, is g_0, else it is $\bar{g}_0$. An **on-input** of R is a connection between two gates along R. A **side-input** of R is any connection to a gate along R other than its on-input. Let the inputs of gate G_j be ordered as $i_1, \ldots, i_n$. For an input i_k, where $1 \le k \le n$, its low-order side-inputs are $i_1, \ldots, i_{k-1}$. A path that starts at a primary input and ends at a side-input of R is called a **side-path** of R. The logic value of line g_j, when vector P_i is applied to the primary inputs, is denoted as $g_j(P_i)$.

There are two PDFs (or logical paths) for each physical path R depending on the direction of signal transition along R. There are two equivalent ways of depicting a

PDF: by considering the transition at its input or its output. If the desired transition at input g_0 of R is a rising (falling) one, the PDF is denoted by $\uparrow R$ ($\downarrow R$). Alternatively, if the desired transition at output g_r of R is a rising (falling) one, the PDF is denoted by $R \uparrow$ ($R \downarrow$). Similar notations are used for the two SDFs for each physical segment. The term logical path is used to refer to both PDFs and SDFs.

Delay assignment refers to the assignment of a delay (a number or a range) to each connection and each transition (rising or falling) in a circuit. Given a design, there is a delay assignment for the fault-free circuit which represents the desired timing behavior. We call it the fault-free delay assignment. However, the presence of different delay defects results in different delay assignments.

Consider a two-pattern test (P_1, P_2) derived for a PDF. Suppose P_2 becomes stable at time $t = 0$. Under a delay assignment, the logic values at each gate output will ultimately stabilize. The logic value at which a gate output g_i stabilizes is called its **stable value** under the two-pattern test. The time at which g_i stabilizes under P_2 is called the **stable time** of g_i under P_2.

Under a delay assignment, the stable value and stable time at each input of a gate, say G, is known under (P_1, P_2). Suppose line g_i feeds G. Then g_i is said to **dominate** G if the stable value and stable time at G's output are determined by those at g_i. A path is considered to be **sensitized** under a given delay assignment and two-pattern test if each on-input of the path dominates the gate it is connected to. If such a two-pattern test exists for the path for a given delay assignment M, then the path is said to be **sensitizable under** M. If a path cannot be sensitized under any two-pattern test, it is called a **false path**. It is quite possible that a path is sensitizable under one delay assignment and false under another one.

An input g_i of gate G is said to be the earliest controlling input for a two-pattern test, T, if the stable value at g_i under T is a controlling value and its stable time is earlier than other inputs of G which also stabilize at the controlling value. T sensitizes a path if and only if each on-input of the path is either the earliest controlling input or the latest non-controlling input with all its side-inputs being non-controlling inputs too. In other words, under the above condition, each on-input of the path dominates the gate it is connected to.

8.1.2.1 Cheng's classification

Cheng and Chen classify the PDFs under *robustly testable*, *non-robustly testable*, *functionally sensitizable* and *functionally redundant*. This is shown in Figure 8.3. We briefly described robustly and non-robustly testable faults, with examples, in Chapter 2 (Section 2.2.6). However, we define these faults and their corresponding tests more rigorously here.

A two-pattern test (P_1, P_2) is a **non-robust test** for a PDF if and only if it satisfies the following conditions: (i) it launches the desired logic transition at the primary input of the targeted path, and (ii) all side-inputs of the targeted path settle to non-controlling

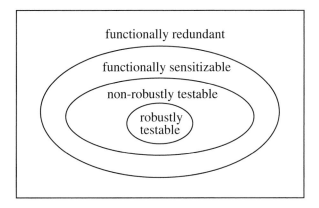

Figure 8.3 Testability classification of PDFs

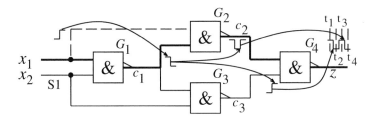

Figure 8.4 An XOR gate implementation

values under P_2. For example, consider the XOR gate implementation shown in Figure 8.4. It shows the signals at each node when the two-pattern test $\{(0, 1), (1, 1)\}$ is applied (Gharaybeh *et al.*, 1995). The arrows show how the signal transitions propagate. Signal value $S1$ denotes a steady logic 1 at input x_2. This is a non-robust test for PDF $x_1 c_1 c_2 z \downarrow$, which is shown in bold. This is because if the observation point is t_2 then this test is invalidated since we obtain the correct output 0 for the second vector even though the above PDF is present. This would happen if PDF $x_1 c_2 z \uparrow$ is also present in the circuit. Thus, a non-robust test cannot guarantee the detection of the targeted PDF.

A **robust test** can detect the targeted PDF independent of the delays in the rest of the circuit. It must satisfy: (i) the conditions of non-robust tests, and (ii) whenever the logic transition at an on-input is from a non-controlling to a controlling value, each corresponding side-input should maintain a steady non-controlling value. For example, consider the XOR gate implementation of Figure 8.4 once again. $\{(0, 0), (1, 0)\}$ is a robust test for PDF $x_1 c_2 z \uparrow$, which is shown in dashed/bold lines. Such tests were referred to as general robust tests in Section 2.2.6.

We saw above that a non-robust test for a targeted PDF may be invalidated by other PDFs. However, given the presence of tests in the test set that robustly test for the invalidating PDFs, a non-robust test is called validatable. For example, in Figure 8.4, the rising transition at z just after t_2 corresponds to $x_1c_2z \uparrow$, which was shown to have a robust test $\{(0, 0), (1, 0)\}$ above. Suppose this test is in the test set. If the circuit passed this test, then the observation time can only be t_3 or t_4 when $\{(0, 1), (1, 1)\}$ is applied. In both cases, this test is valid. Thus, $\{(0, 1), (1, 1)\}$ is a validatable non-robust test for $x_1c_1c_2z \downarrow$. This is because either the PDF is caught if the observation time is t_3 or the circuit is free of this fault if the observation time is t_4.

If there exists an input vector P_2 such that all the side-inputs of connection g_i along the PDF settle to non-controlling values on P_2 when the on-input g_i has a non-controlling value (there is no requirement for the side-inputs of g_i when the on-input g_i has a controlling value), then the PDF is said to be **functionally sensitizable**. Otherwise, the PDF is said to be functionally unsensitizable. This definition is independent of the delay assignment because there is no requirement for the stable times.

A PDF is called **functionally redundant** if the logical path is a false path under all delay assignments. Otherwise, it is called functionally irredundant. It can be shown that if a PDF is functionally unsensitizable, it is functionally redundant. Note that this is just a sufficient condition, not a necessary one. In other words, a PDF may be functionally sensitizable, yet it may be functionally redundant. This is because functional sensitizability is defined with respect to only a single vector, whereas a PDF requires a two-pattern test. Therefore, even if a vector P_2 has been found to functionally sensitize a path, it does not mean that an initialization vector P_1 can always be found such that the two-pattern test (P_1, P_2) can sensitize the logical path corresponding to the PDF.

A functionally redundant fault can never be sensitized under any possible delay defect. Therefore, the existence of such faults will not affect the performance of the circuit. For a functionally irredundant PDF, there exists at least one delay assignment under which the path is sensitizable. Since we do not know beforehand what may go wrong with the semiconductor manufacturing process, we have to make a conservative assumption that any arbitrary delay assignment may occur. Therefore, we need to worry about all functionally irredundant PDFs.

For example, consider the logical path (or PDF) $x_1c_1c_2z \uparrow$ in the circuit in Figure 8.4. To propagate the desired transition through this path, we need a test vector P_2 that sets x_1 and c_2 to logic 0 and c_1 and z to logic 1. To functionally sensitize this logical path, since the on-input c_1 has a non-controlling value, its side-input x_1 should also have a non-controlling value, i.e., P_2 must set x_1 to 1. However, this leads to a contradiction. Therefore, this PDF is functionally redundant.

8.1.2.2 Lam's classification

Lam *et al.* classify PDFs under *robust-dependent* (RD) and non-RD. They show that there exist PDFs which can never impact circuit delay unless some other PDFs are also present. These are termed RD faults. They need not be considered in DF testing. This leads to a more accurate and increased DF coverage.

More formally, let D be the set of all PDFs in circuit C and R be a subset of D. If for all τ, the absence of PDFs (with delays greater than τ) in $D - R$ implies that the delay of C is $\leq \tau$, then R is said to be **robust-dependent**.

A PDF is non-RD if and only if there exists a two-pattern test (P_1, P_2) that satisfies the following conditions: (i) it launches the desired logic transition at the primary input of the targeted path, (ii) if an on-input of the targeted path has a non-controlling value under P_2, the corresponding side-inputs have non-controlling values under P_2, and (iii) if an on-input of the targeted path has a controlling value under P_2, the corresponding low-order side-inputs have non-controlling values under P_2 (Sparmann *et al.*, 1995).

The set of RD PDFs is independent of the delay assignment, i.e., R can be eliminated from consideration in DF testing under every delay assignment. This is illustrated through the following example.

Example 8.1 Consider the circuit in Figure 8.5. In this circuit, PDFs $x_2 z \downarrow$ and $x_2 c_1 z \downarrow$ are robustly tested by the two-pattern test $\{(1, 1), (1, 0)\}$ and PDF $x_2 z \uparrow$ is robustly tested by the two-pattern test $\{(0, 0), (0, 1)\}$. The remaining three PDFs, $x_1 c_1 z \downarrow$, $x_1 c_1 z \uparrow$, and $x_2 c_1 z \uparrow$, are not robustly or non-robustly testable. However, they are not functionally redundant.

Let us first consider PDF $x_1 c_1 z \downarrow$. We argue that PDF $x_2 c_1 z \downarrow$ has a longer delay than $x_1 c_1 z \downarrow$. For a falling transition to propagate through the path $x_1 c_1 z$, the final value of x_2 must be 0 (otherwise the transition will be blocked in gate G_2). The output of gate G_1 changes (after some delay) to logic 0 when any input changes to logic 0. Therefore, x_2 must initially be at logic 1 at the moment the falling transition on x_1 reaches G_1, in order for this transition to propagate through G_1. Since x_2 eventually changes to logic 0, the delay of PDF $x_2 c_1 z \downarrow$ is longer than the delay of PDF $x_1 c_1 z \downarrow$. Since a robust test exists for $x_2 c_1 z \downarrow$, the existence of this longer fault can be detected. Hence, $x_1 c_1 z \downarrow$ need not be tested if $x_2 c_1 z \downarrow$ is tested. In other words, if the circuit passes the test for $x_2 c_1 z \downarrow$ at a given clock period then the delay of $x_1 c_1 z \downarrow$ is guaranteed to be within the specified clock period. Thus, $x_1 c_1 z \downarrow$ is RD. Similarly, it can be shown that $x_1 c_1 z \uparrow$ and $x_2 c_1 z \uparrow$ are also RD. Since all the three non-RD faults are robustly testable, this circuit can be said to have 100% robust delay fault coverage, even though only 50% of the PDFs are testable. $\square$

8.1.2.3 Gharaybeh's classification

Gharaybeh *et al.* classify PDFs into three categories: *singly-testable* (ST), *multiply-testable* (MT), and *singly-testable-dependent* (STD).

A PDF is said to be **singly-testable** if and only if there exists a test that guarantees its detection when it is the only PDF in the circuit. This test is either robust or validatable

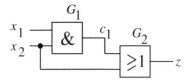

Figure 8.5   Circuit illustrating RD PDFs

non-robust or non-robust. For example, in the XOR gate implementation in Figure 8.4, all PDFs are ST except $x_1c_1c_2z \uparrow$ and $x_2c_1c_3z \uparrow$. In this case, each ST PDF has a robust or validatable non-robust test.

A PDF is said to be **singly-testable-dependent** if and only if it cannot propagate a late transition given the absence of certain ST faults. Since STD faults cannot propagate late transitions, they do not affect the circuit speed and need not be tested. This is illustrated next.

Example 8.2   Consider PDF $x_1c_1c_2z \uparrow$ in the XOR gate implementation given in Figure 8.4. For testing this PDF, the corresponding transition at the on-input of gate G_2 must be $0 \rightarrow 1$. Also, whenever this transition arrives, there is a $1 \rightarrow 0$ transition on the side-input of G_2. Let us temporarily switch our attention to PDF $x_1c_2z \downarrow$. This is ST because it has a robust test $\{(1,0),(0,0)\}$. Suppose the circuit passes this test. This means that $x_1c_2z \downarrow$ is not present. In order for $x_1c_1c_2z \uparrow$ to propagate a late transition, the $0 \rightarrow 1$ transition at the on-input of G_2 must arrive before the $1 \rightarrow 0$ transition at its side-input. However, given that $x_1c_2z \downarrow$ is absent, this is not possible. Therefore, $x_1c_1c_2z \uparrow$ cannot propagate a late transition if an ST fault is absent. This implies that $x_1c_1c_2z \uparrow$ is STD.   ☐

A PDF is **multiply-testable** if and only if it is neither ST nor STD. MT PDFs can affect the circuit speed. This is illustrated next.

Example 8.3   Consider the circuit in Figure 8.6. PDFs $x_1c_1z \downarrow$ and $x_3c_1z \downarrow$ are neither ST nor STD. Therefore, they are MT. Now consider $x_1c_2z \downarrow$. This has a robust two-pattern test $\{(1,x,0),(0,x,0)\}$. Similarly, $x_3c_3z \downarrow$ can be seen to have a robust two-pattern test $\{(0,x,1),(0,x,0)\}$. Suppose the circuit passes these tests. Next, consider the two-pattern test $\{(1,1,1),(0,1,0)\}$. The corresponding waveforms are shown in the figure. Since the robust tests for $x_1c_2z \downarrow$ and $x_3c_3z \downarrow$ were passed, the signals at lines c_2 and c_3 are known to settle at logic 0 before the transition arrives late at line c_1. Therefore, this transition will propagate to z and will follow the faster of $x_1c_1z \downarrow$ or $x_3c_1z \downarrow$ transitions. If either $x_1c_1z \downarrow$ or $x_3c_1z \downarrow$ is not present, then the signal at z will settle before the observation time t_1. However, if both PDFs are present, a faulty response of logic 1 will be observed at t_1. Thus, this multiple PDF can adversely affect the circuit operation. However, the absence of MT faults cannot

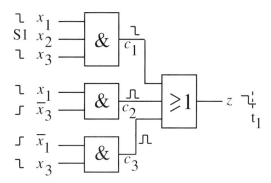

Figure 8.6   Circuit illustrating MT PDFs

be guaranteed. For example, as mentioned above, PDF $x_3 c_1 z \downarrow$ is MT. However, we cannot be sure that it is absent. It manifests itself only if $x_1 c_1 z \downarrow$ is also present.   □

8.1.2.4 Sivaraman's classification

Sivaraman *et al.* classify PDFs into three categories: *primitive single path delay fault* (SPDF), *primitive multiple path delay fault* (MPDF), and *primitive-dependent*.

An MPDF $\Pi \downarrow (\Pi \uparrow)$ consists of a set of SPDFs such that each path $R \in \Pi$ shares the same output and has a fault $R \downarrow (R \uparrow)$. The MT PDF classification introduced by Gharaybeh does not show us which MPDFs need to be targeted for testing. Identifying all MPDFs in a circuit is infeasible since one would need to target all path pairs, path triples, and so on. In this context, *primitive* PDFs were defined in Ke and Menon (1995). The ST PDFs belong to the set of primitive PDFs. Also, MPDFs whose constituent SPDFs are *jointly non-robustly testable*, and which are *minimal*, belong to the set of primitive PDFs. Joint non-robust testability of an MPDF implies that the desired transitions occur at the primary inputs of the SPDFs in the MPDF, and all side-inputs of the SPDFs settle to non-controlling values under the second vector of the two-pattern test. Minimality of the MPDF implies that none of its subsets is jointly non-robustly testable. It has been shown that if all primitive PDFs are tested and shown to be absent then the circuit's correct timing behavior can be guaranteed at the tested speed as well as any slower speed.

In Example 8.3, $\{x_1 c_1 z \downarrow, x_3 c_1 z \downarrow\}$ is an example of a primitive MPDF.

Based on the concept of primitive PDFs, Gharaybeh's classification of PDFs can be further refined as follows. Primitive SPDFs have the same meaning as Gharaybeh's ST PDFs. Primitive MPDFs are a subset of MT PDFs since some MT PDFs may not be jointly non-robustly testable and minimal. Primitive-dependent PDFs are those that need not be tested if all primitive PDFs are guaranteed to be absent. This is a superset of the STD PDFs defined by Gharaybeh.

Lam's classification	non-RD		RD
Gharaybeh's classification	ST	MT	STD
Cheng's classification	robustly or non-robustly testable	functionally sensitizable	functionally redundant
Sivaraman's classification	primitive SPDF	primitive MPDF	primitive-dependent

Figure 8.7 Comparison of classifications (Sivaraman and Strojwas, 1997) (© 1997 IEEE)

8.1.2.5 Comparison of different PDF testability classifications

Figure 8.7 compares the different PDF testability classifications we have presented so far. The STD PDF set is a subset of the RD set, but may be a superset of the functionally redundant set.

Some RD PDFs are MT, and not STD. For example, in the circuit in Figure 8.6, the RD set consists of either $x_1c_1z \downarrow$ or $x_3c_1z \downarrow$, but not both. However, in the circuits in Figures 8.4 and 8.5, the RD and STD sets are equal.

Some STD PDFs are not functionally redundant. For example, in the circuit in Figure 8.5, the STD set is $\{x_1c_1z \downarrow, x_1c_1z \uparrow, x_2c_1z \uparrow\}$, while the functionally redundant set is empty. However, in the circuits of Figures 8.4 and 8.6, the STD and functionally redundant sets are equal.

8.1.3   Delay fault coverage

The most conventional way of deriving the percentage DF coverage of a given test set is to take the ratio of the number of detected DFs and the total number of possible DFs under the given fault model and multiply by 100. However, this generally leads to very low DF coverage for many circuits.

We saw in the previous section that RD, STD, functionally redundant or primitive-dependent PDFs need not be targeted. In fact, it has been shown that up to 81% of the PDFs in ISCAS '85 benchmark circuits are functionally redundant (Cheng and Chen, 1993). Recall from Figure 8.7 that RD and STD sets can be larger than the functionally redundant set. Therefore, if a test set has been derived for PDFs, it makes sense to remove such faults from consideration before determining the PDF coverage. Thus, the percentage PDF coverage under this modified scenario would be the ratio of the number of detected PDFs and the total number of PDFs that need to be targeted, multiplied by 100.

The above definitions of DF coverage, however, do not take into account the probability distribution of delay fault sizes caused by fabrication process defects. What we really want to know is the percentage of fabricated faulty chips which can be detected by a given test set. Such a realistic DF coverage metric is presented in

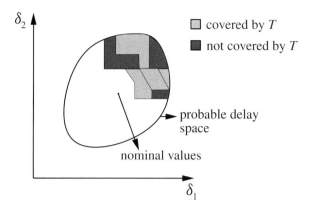

Figure 8.8 Coverage with respect to the delay space

Sivaraman and Strojwas (1996). It is independent of the DF model, i.e., it is equally applicable to PDFs, GDFs and SDFs. It takes into account delay fault size distributions caused by fabrication defects. It is not limited in scope to single DFs. Finally, it can handle DFs which are detected only for certain delay assignments. This percentage DF coverage metric is statistical and is defined as follows for a given test set T.

Statistical coverage$(T) = 100 \times$ Probability$(T$ detects DF | fabricated chip has a DF$)$.

Consider a circuit with a delay assignment Δ consisting of delays $\delta_1, \delta_2, \ldots, \delta_p$, which correspond to the delays of various connections and rising/falling delays in the circuit. Δ is a set of random variables, and can be expressed as a joint probability density function, $f(\Delta)$, which is a consequence of the fabrication process effects. For the sake of illustration, consider a hypothetical case in which $\Delta = (\delta_1, \delta_2)$, as shown in Figure 8.8.

The space of probable delay assignments is represented by the bounded region. In this region, the joint probability density function of delay value combinations is greater (or smaller) than some given value. Within this space, the region representing those delay value combinations for which the circuit will have a delay greater than the system clock period is shown shaded. The test set T may detect some DFs for some delay assignments, not others. The sub-regions covered by T are shown lightly shaded, and the ones not covered by T are shown darkly shaded. The probability that a fabricated chip has a DF is equal to $\int_R f(\Delta)d\Delta$, where the volume integral (for a p-dimensional delay space) is taken over R, the region where the chip has a DF; i.e., over all the shaded regions in the figure. The probability that the chip has a DF and the DF is detected by T is equal to $\int_{R_1} f(\Delta)d\Delta$, where the volume integral is taken over R_1, the sub-regions covered by T, i.e., over the lightly shaded regions in the figure. The ratio of the two probabilities, multiplied by 100, gives the statistical DF coverage of T.

It has been observed that there can be a very wide variation in the results reported by the conventional DF coverage metrics and the statistical DF coverage metric.

8.2   Combinational test generation

In this section, we discuss methods to detect PDFs, GDFs and SDFs in combinational circuits. Since a large number of PDFs do not need to be detected (see Section 8.1.2) or may be untestable, it would be a good idea to quickly identify such faults, and present only the remaining faults to the test generator. Therefore, we present such preprocessing methods first. A large number of test generation methods have been presented in the literature. We look at some representative ones next. We finally look at some test compaction methods. We assume a variable clock scheme, unless otherwise stated.

8.2.1   Untestability analysis

A very fast method has been given in Heragu *et al.* (1997a) to identify a large number of, though not all, untestable SDFs and PDFs under various sensitization conditions, using precomputed static logic implications.

We say that a logical path is **sensitized** by the two-pattern test (P_1, P_2) if and only if the corresponding PDF is either functionally sensitized, robustly tested, or non-robustly tested by (P_1, P_2). We will make use of the following intuitive result in this section.

Lemma 8.1 Consider a path $R = g_0 g_1 \cdots g_r$, and its corresponding logical paths $R \downarrow$ and $R \uparrow$. If $R \downarrow (R \uparrow)$ is sensitized by a two-pattern test (P_1, P_2), then the value of g_j on P_2 is logic 0 (1) if the inversion parity of the segment between g_0 and g_j is even, and logic 1 (0) if the inversion parity is odd.

Let $S_R(x^\alpha, y^\beta)$, where $\alpha, \beta \in \{0, 1\}$, denote the set of logical paths passing through lines x and y such that for every path $R \downarrow (R \uparrow)$ in $S_R(x^\alpha, y^\beta)$, $[x(P_2) = \alpha]$ and $[y(P_2) = \beta]$ are necessary conditions for $R \downarrow (R \uparrow)$ to be sensitized by the two-pattern test (P_1, P_2), based on Lemma 8.1.

If $x = \alpha$ implies $y = \beta$, we denote this fact by $(x = \alpha) \Rightarrow (y = \beta)$.

We next present various lemmas which form the basis for untestability analysis under different sensitization conditions.

8.2.1.1   Robust untestability analysis

If all SDFs or PDFs in the set $S_P(x^\alpha, y^\beta)$ are robustly untestable, we denote this fact by $[S_P(x^\alpha, y^\beta) = RU]$.

Lemma 8.2 For lines x and y, if $(x = \alpha) \Rightarrow (y = \beta)$, and β is the controlling value of gate G that y feeds, then for all other inputs z of G and for all $\gamma \in \{0, 1\}$, $[S_P(x^\alpha, z^\gamma) = RU]$.

This lemma follows from the following arguments. Consider a logical path in $S_P(x^\alpha, z^\gamma)$. For this path, y is a side-input. Therefore, from the definition of robust testability, for a two-pattern test (P_1, P_2) for the corresponding PDF or SDF, the following conditions are necessary: (i) $x(P_2) = \alpha$, and (ii) $y(P_2) = \bar{\beta}$, the non-controlling value of G. However, since $(x = \alpha) \Rightarrow (y = \beta)$, these two conditions cannot be simultaneously satisfied. The following lemmas similarly follow from the definition of robust testability.

Lemma 8.3 For lines x and y, if $(x = \alpha) \Rightarrow (y = \beta)$, where α is the non-controlling value of gate G_1 that x feeds, and β is the controlling value of gate G_2 that y feeds, then for all other inputs u of G_1, for all other inputs v of G_2, and for all $\gamma, \delta \in \{0, 1\}$, $[S_P(u^\gamma, v^\delta) = RU]$.

Lemma 8.4 For lines x and y, if $(x = \alpha) \Rightarrow (y = \beta)$, and β is the controlling value of gate G that y feeds, then for all other inputs z of G, $[S_P(x^{\bar{\alpha}}, z^\beta) = RU]$.

Lemma 8.5 For lines x and y, if $(x = \alpha) \Rightarrow (y = \beta)$, then $[S_P(x^{\bar{\alpha}}, y^\beta) = RU]$ and $[S_P(x^\alpha, y^{\bar{\beta}}) = RU]$.

Lemma 8.6 If a line x is identified as having a constant value assignment α, then for all $\gamma \in \{0, 1\}$, $[S_P(x^\gamma, x^\gamma) = RU]$.

The above lemma follows from the fact that there can be no logic transition on x under any two-pattern test (P_1, P_2). Thus, all logical paths going through x are robustly untestable.

Information on stuck-at fault (SAF) redundancies can also be used in this analysis, as follows (Park and Mercer, 1987).

Lemma 8.7 If the stuck-at α fault on line x is untestable, where $\alpha \in \{0, 1\}$, then $[S_P(x^{\bar{\alpha}}, x^{\bar{\alpha}}) = RU]$.

Note that the difference between Lemmas 8.6 and 8.7 is that an SAF may be untestable even if the corresponding line does not assume a constant value assignment. This would happen if the logic value, opposite to the stuck value, is controllable, but not observable.

Example 8.4 Consider the circuit in Figure 8.4 once again. Let us see how we can exploit the logic implication $(x_1 = 0) \Rightarrow (c_2 = 1)$. From Lemma 8.5,

$[S_P(x_1^1, c_2^1) = RU]$ and $[S_P(x_1^0, c_2^0) = RU]$. Under a two-pattern test (P_1, P_2), $x_1(P_2) = 1$ and $c_2(P_2) = 1$ are necessary conditions for sensitizing segment $x_1 c_1 c_2$ for a rising transition at x_1, according to Lemma 8.1. Thus, $[S_P(x_1^1, c_2^1) = RU]$ implies that this SDF is not robustly testable. Since this SDF is a part of PDF $x_1 c_1 c_2 z \downarrow$, this PDF is not robustly testable either. Similarly, $[S_P(x_1^0, c_2^0) = RU]$ implies that PDF $x_1 c_1 c_2 z \uparrow$ is not robustly testable.

The logic implication $(x_2 = 0) \Rightarrow (c_3 = 1)$ implies in a similar fashion that PDFs $x_2 c_1 c_3 z \downarrow$ and $x_2 c_1 c_3 z \uparrow$ are also not robustly testable. □

8.2.1.2 Functional unsensitizability analysis

If all SDFs or PDFs in set $S_P(x^\alpha, y^\beta)$ are functionally unsensitizable, we denote this fact by $[S_P(x^\alpha, y^\beta) = FU]$. The following lemmas can be used to identify functionally unsensitizable faults.

Lemma 8.8 For lines x and y, if $(x = \alpha) \Rightarrow (y = \beta)$, then $[S_P(x^\alpha, y^{\bar{\beta}}) = FU]$.

Lemma 8.9 If a line x is identified as having a constant value assignment α, then $[S_P(x^{\bar{\alpha}}, x^{\bar{\alpha}}) = FU]$.

Other lemmas similar to those presented for robust untestability analysis can be derived for identifying functionally unsensitizable faults. However, they have not been found to be very useful.

Note that the condition in Lemma 8.8 forms a subset of the one in Lemma 8.5. Thus, from Example 8.4, we can see that PDFs $x_1 c_1 c_2 z \uparrow$ and $x_2 c_1 c_3 z \uparrow$ are not only robustly untestable, but also functionally unsensitizable.

8.2.1.3 Non-robust untestability analysis

If all SDFs or PDFs in set $S_P(x^\alpha, y^\beta)$ are non-robustly untestable, we denote this fact by $[S_P(x^\alpha, y^\beta) = NU]$. To identify non-robustly untestable faults, Lemmas 8.2 and 8.3 can be used by replacing RU by NU, and Lemmas 8.8 and 8.9 can be used by replacing FU by NU.

8.2.1.4 Robust dependability analysis

To identify RD faults, modified versions of Lemmas 8.2, 8.3, 8.8 and 8.9 can be used. This modification is needed based on the fact that identifying RD faults depends on a given ordering of side-inputs (see the definition of such faults).

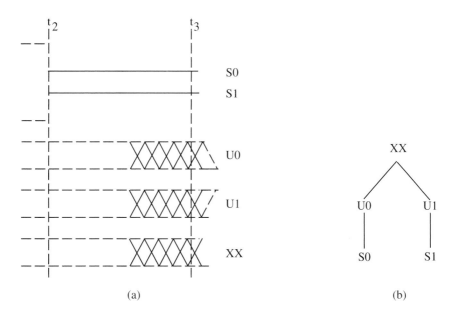

Figure 8.9 Five-valued logic system and the covering relationship

8.2.2   Test generation for PDFs based on a five-valued logic system

In this section, we present a test generation method for PDFs based on a simple five-valued logic system (Lin and Reddy, 1987; Cheng *et al.*, 1996). The number of logic values determines the time and memory complexity of the methods based on them. Fewer the logic values, less complex the implementation. However, fewer logic values, in general, also imply less efficiency (i.e., test generation takes longer). We will consider other logic systems later.

The five logic values we consider here are $\{S0, S1, U0, U1, XX\}$. They are depicted in Figure 8.9(a). Each logic value represents a type of signal on the lines in the circuit in the time interval (t_2, t_3). Let the two-pattern test be (P_1, P_2). Let the initial value (final value) of a line in the circuit be the binary logic value on the line after P_1 (P_2) has been applied and the circuit has stabilized. Under the variable clock scheme, recall from Figure 8.1(b) that at time t_2, any signal on a line in the circuit has stabilized to the initial value of that line. However, at time t_3, a signal in a faulty circuit may not have stabilized to the final value of the line. The purpose of DF testing is to make sure that signals do stabilize by time t_3. $S0$ $(S1)$ represents signals on lines whose initial and final values are 0 (1). Furthermore, the line remains free of hazards, i.e., there are no momentary transitions to the opposite logic value. $U0$ $(U1)$ represents signals on lines whose final value is 0 (1). The initial values of these lines could either be 0 or 1. In addition, in the time interval (t_2, t_3), the lines could have hazards. Obviously, signal $U0$ $(U1)$ includes signal $S0$ $(S1)$. XX represents signals whose initial and final values are unspecified. The covering relationship among the five logic values is shown

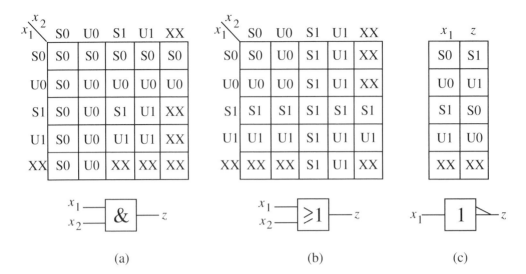

Figure 8.10 Implication tables

in Figure 8.9(b). Logic value $U0$ covers $S0$, $U1$ covers $S1$, and XX covers both $U0$ and $U1$.

In Figure 8.10, implication tables for the five-valued logic system is given for AND, OR and NOT gates. From these tables and the associative law of Boolean algebra, one can determine the outputs of multiple-input AND, OR, NAND and NOR gates, given their inputs.

8.2.2.1 Test generation for robustly testable PDFs

In deriving two-pattern tests for PDFs, signals $U0$ and $U1$ are interpreted in two different ways. The $U0$ ($U1$) signal on an on-input is interpreted as a $1 \rightarrow 0$ ($0 \rightarrow 1$) transition. The $U0$ and $U1$ signals on lines in the circuit, other than on-inputs, are interpreted according to Figure 8.9(a), i.e., with final values of 0 and 1, respectively.

The following key result is used for robust test generation of PDFs. A two-pattern test (P_1, P_2) robustly tests a PDF if and only if: (i) it launches the desired transition at the input of the targeted path, and (ii) the side-inputs have logic values that are covered by the values indicated in Figure 8.11. Such side-inputs are called robust. This result follows directly from the definition of a robust test. Figure 8.12(a) shows examples of robust side-inputs for an AND gate, where the on-input is shown in bold.

Example 8.5 Consider the bold path in the circuit in Figure 8.13. Suppose the PDF with a rising transition at input x_3 of this path needs to be tested. Therefore, we place signal $U1$ at x_3. The current side-input is x_4. From Figure 8.11, we find that signal $S0$ must be placed on x_4. From the implication table in Figure 8.10(b), we find that the

gate type on-input transition	AND or NAND	OR or NOR
Rising(U1)	U1	S0
Falling(U0)	S1	U0

Figure 8.11 Robustly sensitizing input values

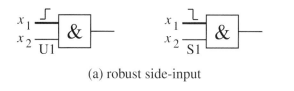

(a) robust side-input

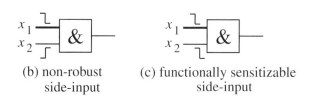

(b) non-robust   (c) functionally sensitizable
side-input   side-input

Figure 8.12 Different types of side-inputs for an AND gate

signal on line c_2 is, therefore, $U1$. The side-input of c_2 is c_1. From Figure 8.11, we should place $S0$ at c_1. This results in $U0$ at line c_3 (from the tables in Figures 8.10(b) and 8.10(c)). The side-input now is c_4. From Figure 8.11, we should place $U0$ at c_4, which allows propagation of $U0$ to primary output z. At this point, the sensitization of the path under test is complete. Next, signals $S0$ at line c_1 and $U0$ at line c_4 need to be justified at the primary inputs. This is accomplished as shown in Figure 8.13. In general, this step may need backtracking, just as in the case of SAF test generation. The corresponding two-pattern test is $\{(0, x, 0, 0, x), (0, x, 1, 0, 0)\}$. Note that for on-input x_3, $U1$ was interpreted as a $0 \rightarrow 1$ transition, whereas $U0$ on x_5 was interpreted as just a signal with a final value 0. However, if the unknown value for x_5 in the first vector is chosen to be 1, obtaining the two-pattern test $\{(0, x, 0, 0, 1), (0, x, 1, 0, 0)\}$, then readers can check that this two-pattern test also robustly tests for PDF $x_5c_4z \downarrow$. $\square$

8.2.2.2 Test generation for non-robustly testable PDFs

There may be PDFs in a circuit that are not robustly testable. However, they may be non-robustly testable. The method presented above can be very easily modified to perform non-robust test generation. The only modification that is needed is to relax the conditions for side-input values, as shown in Figure 8.14. This directly follows from

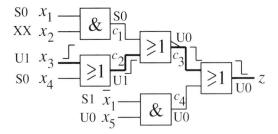

Figure 8.13 Circuit illustrating robust test generation

gate type on-input transition	AND or NAND	OR or NOR
Rising(U1)	U1	U0
Falling(U0)	U1	U0

Figure 8.14 Non-robustly sensitizing input values

the definition of a non-robust two-pattern test. Figure 8.12(b) shows an example of a non-robust side-input for an AND gate.

When the on-input has a transition from a non-controlling value to a controlling value, non-robust tests simply require that the side-inputs settle to a non-controlling value on the second vector, as opposed to the requirement of a steady non-controlling value for robust tests. The former side-inputs are called non-robust. Thus, a non-robustly testable PDF has at least one non-robust side-input with the other side-inputs being robust for any two-pattern test. The number of non-robust side-inputs can be different for different two-pattern tests. To reduce the chance of test invalidation, we should try to reduce the number of non-robust side-inputs. The amount of time by which the non-controlling value on the side-input becomes stable before the on-input becomes stable is called the **slack** of the side-input. If the slacks of all the side-inputs for the PDF under test are positive, then there can be no test invalidation. Thus, another objective of test generation is also to maximize the slack of the non-robust side-inputs. Such two-pattern tests can tolerate larger timing variations at these side-inputs.

The quality of non-robust tests can be improved further by converting them into validatable non-robust tests, if possible. A two-pattern test, T, obtained as above with a minimal number of non-robust side-inputs and maximal slack can be processed further as follows. If T has don't cares at some primary inputs, we should specify the don't cares in such a fashion that the number of transitions at the primary inputs is minimized (i.e., $U1$ is specified as 11, $U0$ is specified as 00, and XX as 00 or 11). After performing the implications of the new two-pattern test, T', we examine the

gate type / on-input transition	AND or NAND	OR or NOR
Rising(U1)	U1	XX
Falling(U0)	XX	U0

Figure 8.15 Functionally sensitizing input values

non-robust side-inputs and identify the PDFs that need to be robustly tested to validate T'. If these identified paths are indeed robustly testable, the non-robust two-pattern test in question is validatable.

8.2.2.3 Test generation for functionally sensitizable PDFs

The conditions for side-input values can be further relaxed for functional sensitizability, as shown in Figure 8.15. These conditions also follow from the definition of functional sensitizability. However, as mentioned before, even when these conditions are satisfied, the PDF may still be functionally redundant. For example, if the side-input of a gate has a stable controlling value then the on-input fault effect cannot be propagated further. Figure 8.12(c) shows an example of a functionally sensitized side-input (these are the side-inputs with an assignment of XX) for an AND gate.

We can improve the quality of the two-pattern test for functionally sensitizable PDFs in a manner analogous to how the quality of a non-robust two-pattern test is improved. First, we should try to minimize the number of functionally sensitized side-inputs. This can be done by maximizing the number of side-inputs which are either robust or non-robust. Second, we should try to make sure that the arrival times of functionally sensitizable side-inputs are as late as possible relative to the on-input transition. This increases the probability that the on-input transition will propagate through the gate in question. Third, as before, we should try to minimize the number of transitions at the primary inputs.

8.2.3   Test generation for PDFs based on stuck-at fault testing

In this section, we establish a relationship between robust PDF and SAF test generation (Saldanha *et al.*, 1992). This relationship is based on a circuit called *leaf-dag*, which is functionally equivalent to the given circuit. A **leaf-dag** is a circuit composed of AND and OR gates with fanout and inverters only permitted at the primary inputs. An inverter is not allowed fanout. Every circuit consisting of AND, OR, NAND, NOR and NOT gates can be converted to a leaf-dag. However, in the worst case, this may take an exponential number of gate duplications.

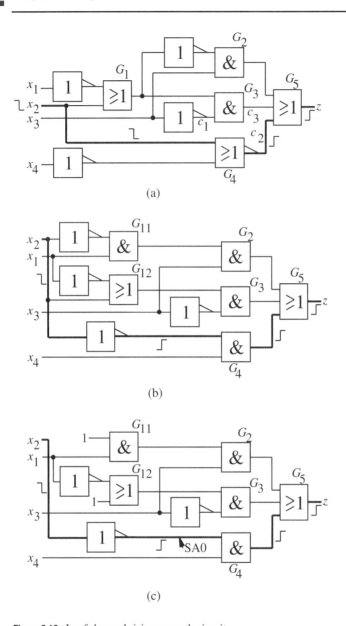

Figure 8.16 Leaf-dag and rising-smooth circuit

As an example, consider the circuit in Figure 8.16(a). Its leaf-dag is shown in Figure 8.16(b). This is obtained by pushing all inverters to the primary inputs. Gate G_1 is first duplicated to gates G_{11} and G_{12}. Gate G_{11} becomes an AND gate when the inverter is pushed from its output to its inputs (using De Morgan's theorem). Similarly, OR gate G_4 becomes an AND gate.

The **I-edge** of path R with primary input x_i in a leaf-dag refers to the first connection of R (this connection is after the inverter if one exists). The *I-edge* is said to be associated with input x_i. Let R_r be the set of side-paths to all the OR gates along R, such that the *I-edge* of each $Q \in R_r$ is associated with x_i. The **rising-smooth circuit** for path R is obtained by replacing the *I-edge* of each $Q \in R_r$ by 1. A **failing-smooth circuit** is defined analogously, by considering AND gates instead of OR gates, and 0 instead of 1. For example, consider the bold path R in the leaf-dag in Figure 8.16(b). The *I-edge* of R is the connection from the inverter to gate G_4. The rising-smooth circuit for R is shown in Figure 8.16(c). This is obtained from the leaf-dag by replacing the *I-edges*, associated with x_2, of side-paths to all the OR gates along R by 1.

For every logical path in the original circuit, there is a corresponding logical path in the leaf-dag. Therefore, testing of PDFs in the former can actually be done with the help of the latter. Also, the test for a rising-transition PDF on an *I-edge* in a leaf-dag can be related to the test for a stuck-at 0 (SA0) fault on the corresponding *I-edge* in the rising-smooth circuit. For example, consider the SA0 fault on the *I-edge* of the bold path in the rising-smooth circuit in Figure 8.16(c). The only test for this fault is $(0, 0, 1, 1)$. This forms the test vector of the two-pattern test for the rising-transition PDF on the *I-edge* in the leaf-dag in Figure 8.16(b). In other words, it forms the test vector of the two-pattern test for the falling-transition PDF on the bold path starting at x_2 in the original circuit as well as the leaf-dag. The corresponding initialization vector of the two-pattern test is simply obtained by complementing the logic value assigned to x_2. Thus, $\{(0, 1, 1, 1), (0, 0, 1, 1)\}$ is the desired two-pattern test for this PDF. Similarly, the test for a falling-transition PDF on an *I-edge* in a leaf-dag can be related to the test for a stuck-at 1 (SA1) fault on the corresponding *I-edge* in the failing-smooth circuit.

The above method allows an SAF test generator to be used for PDF testing. However, as mentioned before, the disadvantage of a leaf-dag is that it may require an exponential number of gate duplications. An extension of this method is given in Saldanha *et al.* (1992) to avoid this problem by using another functionally equivalent circuit instead of a leaf-dag. This equivalent circuit is guaranteed to be no larger than four times the size of the original circuit. In practice, it only requires an increase of a few gates.

8.2.4   Test generation for primitive PDFs

Primitive faults were defined earlier in Section 8.1.2.4. It is necessary and sufficient to test for all primitive PDFs in order to guarantee timing correctness of the circuit. Recall that some primitive PDFs are *multi-path* faults. In this section, we consider robust and non-robust testing of primitive PDFs based on the concept of *sensitizing cubes* (Tekumalla and Menon, 1997). These cubes specify all mandatory assignments for testing a PDF, and reduce the search space for test generation. We consider the subcircuit feeding each primary output separately.

First, some definitions are in order. A value assignment to a subset of inputs of a circuit that produces a 0 or 1 at the primary output is called a **cube**. A cube that produces a 0 (1) is referred to as a 0-cube (1-cube). A cube may also be presented as a product of input literals. For example, for a circuit with inputs x_1, x_2, x_3, $(1, 1, x)$ represents a cube which can also be represented as x_1x_2. This cube contains two vectors: $(1, 1, 0)$ and $(1, 1, 1)$.

A **multi-path** M consists of a set of single paths $\{R_1, R_2, \ldots, R_n\}$ to the same primary output. Every on-input of every gate G on $R_i \in M$ is an on-input of M. Every other input of G is a side-input of M. An MPDF $M \uparrow (M \downarrow)$ depicts a situation in which every single path in M has a rising (falling) PDF at its output. A multi-path M is **statically sensitized** to 1 (0) by a vector P if it sets the multi-path's output to 1 (0) and sets all side-inputs to their respective non-controlling values. A fault $M \uparrow (M \downarrow)$ is primitive if (i) M is statically sensitizable to a 1 (0) at its output, and (ii) there exists no multi-path $M_s \subset M$ that is statically sensitizable to a 1 (0) at its output. A cube q is a **sensitizing cube** of a multi-path M if (i) it sets every side-input to the non-controlling value, and (ii) there exists no $q' \subset q$ (i.e., with fewer literals) that satisfies (i).

We obtain the sensitizing 1-cubes of a multi-level circuit from a two-level sum-of-products expression, called the collapsed form of the circuit. The collapsed form is obtained by expanding the factored expression representing the circuit. All the rules of Boolean algebra are used to reduce the expression. The sensitizing 0-cubes are obtained from the collapsed form of the circuit in which the primary output of the original circuit is complemented.

In the following discussion, we will give results based on the set of sensitizing 1-cubes only. Similar results can be obtained based on the set of sensitizing 0-cubes. An essential vector of a sensitizing 1-cube is one which is not included in any other sensitizing 1-cube. In order to identify primitive faults, only essential vectors and vectors common to sensitizing cubes, which do not have essential vectors, need to be considered. Cubes containing essential vectors are processed first. The path sensitized by an essential vector is determined by tracing the path backwards from the primary output to the primary inputs. Such a path corresponds to a primitive PDF.

In the next phase, sensitizing cubes without essential vectors are processed. Let S be the set of sensitizing cubes without essential vectors. The vectors common to proper subsets of S are determined. Let P_j be the common vector of a set of cubes $S_j \subset S$. The MPDF statically sensitized by P_j is primitive if no $S_i \supset S_j$ has a common vector.

The sensitizing 1-cubes correspond to rising-transition PDFs at the primary output.

Example 8.6 Consider the circuit in Figure 8.17. The set of sensitizing 1-cubes are $(0, x, 1)$, $(x, 0, 1)$, $(1, 0, x)$ and $(1, x, 0)$. First, essential vectors $(0, 1, 1)$ in $(0, x, 1)$ and $(1, 1, 0)$ in $(1, x, 0)$ are obtained. Vectors $(0, 1, 1)$ and $(1, 1, 0)$ statically sensitize the paths $x_1c_1c_3z$ and $x_3c_2c_4z$, respectively. Cubes $(x, 0, 1)$ and $(1, 0, x)$ remain for further processing. The intersection of these two cubes is $(1, 0, 1)$. This statically

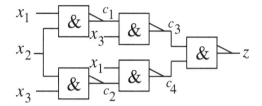

Figure 8.17 Circuit with a primitive MPDF

sensitizes multi-path $(x_2c_1c_3z, x_2c_2c_4z)$. Thus, $(0, 1, 1)$, $(1, 1, 0)$, and $(1, 0, 1)$ are the three static sensitization vectors. The corresponding primitive PDFs are $x_1c_1c_3z \uparrow$, $x_3c_2c_4z \uparrow$, and $(x_2c_1c_3z, x_2c_2c_4z)\uparrow$, respectively. The primitive PDFs for the falling transitions can be similarly identified from the set of sensitizing 0-cubes.   □

The static sensitization vectors obtained during primitive PDF identification are used as test vectors of the two-pattern tests to test these faults. The initialization vector is obtained by complementing the values at the start of the targeted multi-path. If all the remaining primary input values are kept constant in the two-pattern test, all side-inputs to the multi-path will either have a stable non-controlling value or a final non-controlling value. The two-pattern test is robust if all gates with a final controlling value on on-inputs have stable non-controlling values at side-inputs. Since the test vector in the two-pattern test is a static sensitization vector, the two-pattern test is still guaranteed to be non-robust if the above condition is not satisfied. As an example, $\{(1, 1, 1), (1, 0, 1)\}$ is a robust two-pattern test for the MPDF $(x_2c_1c_3z, x_2c_2c_4z)\uparrow$ in the circuit in Figure 8.17.

The practical limitation of this method is that it is not possible to obtain the collapsed two-level form for many multi-level circuits.

8.2.5   Test compaction for PDFs

In PDF testing, the number of faults to be tested is inherently very large. Hence, deriving compact tests for PDFs is an important issue. In this section, we present a method to derive such test sets based on the concept of compatible faults (Saxena and Pradhan, 1993). This method is based on identifying necessary signal values on lines in a circuit along with values a line cannot take in order to test a given path. It is dynamic in nature in the sense that it tries to generate a two-pattern test that simultaneously detects as many PDFs as possible. Static compaction methods try to compact test sets only after test generation is complete.

We first give some definitions. A set of faults is said to be **independent** if no two faults in the set can be detected by the same test. This set is said to be **maximal** if no other fault can be added to the set while still retaining the independence property of the set. Two faults f_1 and f_2 are **potentially compatible** if the necessary conditions

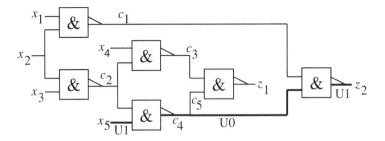

Figure 8.18 Circuit illustrating test compaction

for signal values in the circuit to detect f_1 do not contradict those required to detect f_2. The necessary conditions to test a fault are a set of constraints on signal values that must be satisfied in order to detect the fault. It is not always possible to generate a common test for potentially compatible faults. Such faults are truly compatible when they have a common test. If two faults are not potentially compatible, they are independent. If two faults are potentially compatible, they may still be independent. However, since the maximal set of potentially compatible faults is much easier to derive than the maximal independent fault set, they serve as a useful starting point.

Most PDFs that can be detected simultaneously are basically of two types: (i) PDFs that share many lines in common, for which most of the necessary conditions for testing the PDFs are the same, and (ii) PDFs that are totally disjoint, for which the necessary conditions imposed by one PDF have very little overlap with the necessary conditions imposed by the other.

There are three basic steps in the test compaction method: path activation, derivation of sensitizing values for side-inputs, and constraint propagation. These are illustrated through an example next.

Consider the circuit in Figure 8.18. This circuit has 11 physical paths, and, hence, 22 logical paths. Suppose the PDF to be tested is $x_5 c_4 z_2 \uparrow$, as shown in bold in the figure. The objective now is to derive a list of other paths in the circuit that may be simultaneously tested with this PDF. The compatible fault set contains only this PDF at this point. First, we concern ourselves with only robust testing. We perform the following steps.

Path activation: This step involves placing appropriate signal values on the path being tested. We will use the five-valued logic system given in Section 8.2.2 for this purpose. This gives the following signal values: $x_5 = U1$, $c_4 = U0$, and $z_2 = U1$.

Next, in order to determine which PDFs can be tested simultaneously with the above PDF, we first determine which PDFs cannot be so tested based on the above necessary assignments. It is observed that any PDF requiring $x_5 = U0$ or $c_4 = U1$ or $z_2 = U0$ cannot be tested. Thus, this rules out PDFs $x_5 c_4 z_1 \downarrow$, $x_5 c_4 z_2 \downarrow$, $x_2 c_2 c_4 z_2 \downarrow$, $x_3 c_2 c_4 z_2 \downarrow$, $x_2 c_2 c_4 z_1 \downarrow$, and $x_3 c_2 c_4 z_1 \downarrow$.

Derivation of sensitizing values for side-inputs: For the PDF being tested, the side-inputs are c_1 and c_2. Based on the propagation rules for robust testing (see Figure 8.11), one can observe that the required assignments for these side-inputs are: $c_1 = S1$ and $c_2 = U1$. Since line c_1 assumes a steady value, no PDF can be tested through it. This rules out $x_1c_1z_2 \uparrow$, $x_1c_1z_2 \downarrow$, $x_2c_1z_2 \uparrow$, and $x_2c_1z_2 \downarrow$. Since $c_2 = U1$, any PDF that requires c_2 to be $U0$ cannot be tested. This rules out the following PDFs, which have not already been ruled out: $x_2c_2c_3z_1 \downarrow$ and $x_3c_2c_3z_1 \downarrow$.

Constraint propagation: We observe that line c_5 assumes value $U0$ due to fanout. Thus, no PDF through line c_3 can be tested because its robust propagation to output z_1 is prevented by $c_5 = U0$. The newly ruled out PDFs are therefore $x_2c_2c_3z_1 \uparrow$, $x_3c_2c_3z_1 \uparrow$, $x_4c_3z_1 \uparrow$, and $x_4c_3z_1 \downarrow$.

Based on the above reasoning, out of the 21 remaining PDFs, we have identified 16 as being incompatible with $x_5c_4z_2 \uparrow$. Hence, the search is now restricted to only five PDFs: $x_5c_4z_1 \uparrow$, $x_2c_2c_4z_1 \uparrow$, $x_2c_2c_4z_2 \uparrow$, $x_3c_2c_4z_1 \uparrow$, and $x_3c_2c_4z_2 \uparrow$. If we now pick $x_5c_4z_1 \uparrow$ to be in the compatible set C, then $C = \{x_5c_4z_2 \uparrow, x_5c_4z_1 \uparrow\}$. We now superimpose the necessary conditions for testing $x_5c_4z_1 \uparrow$ over the conditions for testing $x_5c_4z_2 \uparrow$, and repeat this process until no more paths can be considered. Finally, $C = \{x_5c_4z_2 \uparrow, x_5c_4z_1 \uparrow, x_2c_2c_4z_2 \uparrow, x_2c_2c_4z_1 \uparrow\}$. This is the maximal potentially compatible fault set. Test generation can now be performed to satisfy necessary conditions imposed by all these paths. In this case, such a test does exist, and is given by $(x_1, x_2, x_3, x_4, x_5) = (S0, U0, S1, S0, U1)$.

We can now start with the next undetected PDF and repeat the process until all PDFs have been considered for robust testing. Among the PDFs that remain, similar test generation should be done except that now rules for non-robust propagation (see Figure 8.14) can be used. Finally, for the PDFs that still remain, rules for functional sensitization (see Figure 8.15) can be used.

Further test compaction can be achieved by overlapping two-pattern tests when the test vector of one two-pattern test can serve as the initialization vector of another.

8.2.6   Test generation for GDFs

As mentioned in Chapter 2 (Section 2.2.6) and the introduction of this chapter, GDFs can be divided into two categories: G-GDFs and S-GDFs. For a G-GDF, the delay defect size is assumed to be larger than the system clock period. For an S-GDF, this is not true. We will consider both types of faults in this section (Park and Mercer, 1992; Mahlstedt, 1993).

8.2.6.1 Test generation for G-GDFs

In G-GDF testing, since the delay defect size is assumed to be larger than the system clock period, the delay fault can be exposed by appropriately sensitizing an arbitrary

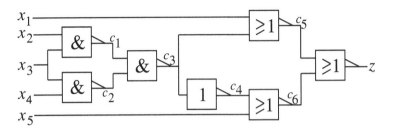

Figure 8.19 G-GDF testing

path through the faulty gate. Consider a two-pattern test (P_1, P_2) for a slow-to-rise G-GDF at the output, g_i, of some gate in a circuit. The two-pattern test needs to satisfy two conditions: (i) $g_i(P_1) = 0$, and (ii) $g_i(P_2) = 1$ and a path can be sensitized from g_i to some primary output under P_2. Thus, the two-pattern test launches a rising transition at the fault site and makes the effect observable at a primary output for the second vector. In the presence of the G-GDF, both the fault site and the primary output under question will have an error for vector P_2. Note that P_2 is simply a test for the SA0 fault at g_i. For testing a slow-to-fall fault at g_i, the conditions are similar except that $g_i(P_1) = 1$ and $g_i(P_2) = 0$. In this case, P_2 is an SA1 fault test for g_i. In this method, test invalidation due to hazards at the primary output under question is ignored. However, the possibility of such test invalidation can be reduced by choosing two-pattern tests in which P_1 and P_2 differ in only one bit, whenever possible.

Example 8.7 Consider the circuit in Figure 8.19. Suppose there is a slow-to-rise G-GDF at line c_3. First, we need to derive a vector P_1 which makes $c_3 = 0$. Such a vector is $(x, x, 0, x, x)$. Then, we need to derive a vector P_2 which makes $c_3 = 1$ and sensitizes any path from c_3 to z. $(0, x, 1, 1, 1)$ is one such vector. Thus, one possible two-pattern test is $\{(0, 0, 0, 1, 1), (0, 0, 1, 1, 1)\}$, which reduces the number of bits in which the two vectors differ to just one. □

8.2.6.2 Test generation for S-GDFs

In S-GDF testing, the fault is tested along the longest functional path (i.e., a path that can be sensitized) through the fault site so that timing failures caused by even the smallest delay defects can be detected. The longest functional path delay is the largest fault-free circuit propagation delay of a path which passes through the fault site and which can be sensitized in the normal circuit's operation. The delay fault will be detected if the delay defect size is greater than the *slack* of the functional path. The **slack** of a functional path is defined as the difference between the clock period and its fault-free path delay.

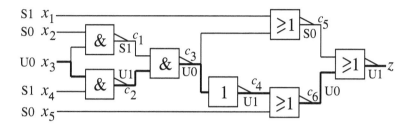

Figure 8.20　S-GDF testing

The longest structural path delay for an S-GDF is the largest propagation delay of a path passing through the fault site based on a structural analysis of the circuit without consideration of the functionality of the path.

Determining the longest functional path through a fault site is a difficult problem. If the longest structural path is not sensitizable, then the next longest structural path must be evaluated, and so on, until the actual longest functional path is identified. Path sensitization can be done based on various multi-valued logic systems. We use the five-valued logic system introduced earlier in Section 8.2.2 to illustrate the method through the following example.

Example 8.8　Consider the circuit in Figure 8.19 once again. Suppose the slow-to-fall S-GDF at c_3 is under consideration. This circuit is reproduced in Figure 8.20 where a possible longest structural path through the fault site is shown in bold. This path is actually functional and can be sensitized by the signal values shown in the figure. This corresponds to the two-pattern test $\{(1, 0, 1, 1, 0), (1, 0, 0, 1, 0)\}$. Here robust propagation rules from Figure 8.11 have been used.　□

8.2.7　Test generation for SDFs

As mentioned before, SDFs bridge the gap between PDFs and GDFs (Heragu *et al.*, 1996a). These are faults that affect the delay of a segment of L lines, where L can be chosen based on available statistics about the types of manufacturing defects. Once L is chosen, the fault list comprises of all segments of length L and all paths whose entire length is less than L. It is assumed that an SDF results in a large enough increase in delay so as to cause a DF in all the paths that include the segment. Thus, when $L = 1$, an SDF reduces to a G-GDF. When L has as many lines as a complete path from a primary input to a primary output, an SDF reduces to a PDF.

There are three types of SDF tests: robust, transition and non-robust. Each of these is discussed next.

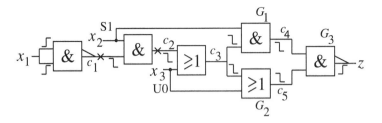

Figure 8.21 SDF testing

8.2.7.1 Robust SDF test

As in the case of PDFs, a test for an SDF is robust if it guarantees detection of the fault irrespective of the delays of all the other signals in the circuit. Generating robust tests consists of three phases: transition launching, fault excitation and fault effect propagation. Consider a segment $c_1 c_2 c_3$ of lines c_1, c_2 and c_3. To test an SDF on this segment, the desired transition (rising or falling) has to be launched at c_1 with or without a hazard. Then the SDF is excited by propagating the transition robustly along the segment. This is done in a manner analogous to PDFs. Finally, the fault effect is propagated from c_3 to a primary output. The propagation conditions for this part are more relaxed than those required for PDFs. This is illustrated by the following example.

Example 8.9 Consider the two-line segment $c_1 c_2$ in the circuit in Figure 8.21. Suppose the SDF $\downarrow$ $c_1 c_2$ needs to be tested. The figure shows a test for this fault using the five-valued logic system introduced earlier. The falling transition is first launched at line c_1 by feeding a rising transition at x_1. Then this transition is robustly propagated to c_2 by making $x_2 = S1$. Finally, the transition at c_2 is propagated to output z using relaxed propagation rules. From the assumption that the SDF affects all paths passing through it, the falling transitions at the inputs of both gates G_1 and G_2 are delayed. Therefore, the falling transitions at both the inputs of gate G_3 are also delayed implying that the rising transition at z is delayed. Thus, the desired two-pattern test is $\{(0, 1, x), (1, 1, 0)\}$. The SDF propagation rules are considered relaxed because the PDF propagation rules do not allow propagation of two falling transitions through a NAND gate in either the robust or non-robust case (see Figures 8.11 and 8.14). □

The above example illustrates the use of relaxed propagation rules when dealing with reconvergent paths that have passed through the fault site. However, in other cases, the conventional rules of robust propagation apply.

8.2.7.2 Transition SDF test

Transition tests for SDFs are similar to those of G-GDFs (or TFs). A two-pattern test (P_1, P_2) is a transition test for an SDF if the desired transition is launched at the origin O of the segment and the SA0 (SA1) fault at O is detected through the segment by P_2 for a rising (falling) transition at O.

Consider the SDF $\downarrow c_1 c_2$ in Figure 8.21 once again. A transition test for this fault is simply $\{(0, x, x), (1, 1, 0)\}$. The rising transition at x_1 results in the desired falling transition at c_1. Then the second vector just sensitizes an SA1 fault on c_1 to output z through segment $c_1 c_2$.

8.2.7.3 Non-robust SDF test

Analogous to robust testing of SDFs, non-robust testing of SDFs also consists of three phases: transition launching, fault excitation and fault effect propagation. The first phase is the same as before. The next two phases are also the same except that a restricted non-robust propagation rule is used. Contrary to general non-robust propagation rules, the restricted rule requires different initial and final values on all the on-path signals. In the case of fault effect propagation, different propagation rules are used as before when dealing with reconvergent paths passing through the fault site, else the restricted non-robust propagation rule is used.

Consider the SDF $\downarrow c_1 c_2$ in Figure 8.21 once again. A test satisfying the above conditions is $\{(0, 1, x), (1, 1, 0)\}$. The rising transition at x_1 results in the desired falling transition at c_1. x_2 cannot be made $U1$, unlike in the case of general non-robust propagation, since this will not make the initial and final values on the on-path line c_2 different. However, x_3 can be made $U0$ without violating the restricted non-robust propagation rule. This two-pattern test is the same as the robust test derived earlier for this SDF. However, in general, the robust and non-robust tests may be different.

8.2.8   At-speed test generation

All the test generation methods presented so far have been under the variable clock scheme. In this section, we wish to consider test generation under the rated clock scheme. Rated clock tests are also called at-speed tests.

Suppose we make the assumption that the DF does not cause the delay of the path it is on to exceed two clock cycles. Then a trivial way of obtaining an at-speed test from a two-pattern test, (P_1, P_2), derived under a variable clock scheme, is to use the three-pattern test (P_1, P_1, P_2) (Bose *et al.*, 1998b). By doing this, the signal values in the circuit are guaranteed to stabilize when the first vector is left unchanged for two clock cycles. This method is applicable to two-pattern variable clock tests derived for any fault model: PDF, GDF or SDF. If we relax our assumption and say that the DF does not cause the delay of the path it is on to exceed n clock cycles, then we can simply derive an $(n + 1)$-pattern test where the first vector P_1 is replicated n times.

Although the above method of obtaining at-speed tests is quite straightforward, the obtained tests may not be the most efficient. An eight-valued logic has been presented in Bose *et al.* (1998b) to derive more compact test sets.

8.3   Combinational fault simulation

Fault simulation helps determine the fault coverage of a specified test set. A fault simulator can also be integrated into an ATPG system to drop detected faults from a fault list. In this section, we will discuss methods for simulating PDFs, GDFs and SDFs under the variable clock scheme. We will also discuss an at-speed fault simulation method.

8.3.1   PDF simulation

Most existing PDF simulators can be classified under: (i) exact methods using path enumeration, or (ii) non-enumerative approximate fault coverage estimators. Simulators in the first category provide an exact fault coverage, but are unable to efficiently tackle circuits which contain a large number of paths. Simulators in the second category do not enumerate the paths, and hence are efficient. However, their fault coverage estimate is only approximate. A non-enumerative exact PDF simulator has been presented in Gharaybeh *et al.* (1998), which overcomes the disadvantages of the above-mentioned methods. We discuss this method next.

The key to exact non-enumerative PDF simulation is an efficient data structure, called the **path status graph** (PSG). This holds the detection status of PDFs. Initially, the PSG contains all the fanout structures of the circuit, and there is a one-to-one correspondence between paths in the PSG and in the original circuit. The status of PDFs is distributed in the PSG using flags attached to graph edges. The status of a PDF is true (i.e., detected) if all the edges on the path are true. All PDFs with at least one false edge have a false (i.e., undetected) status. Thus, the status of a PDF can be extracted from the PSG by performing an AND operation on the status of the corresponding edges. The efficiency of this data structure comes from the fact that the status flags for one edge can potentially be shared by an exponential number of PDFs. For example, this makes it possible for this simulator to simulate all of the approximately 10^{20} PDFs in the combinational ISCAS'85 benchmark c6288.

Figure 8.22 illustrates the extraction of the PSG from a circuit. For the circuit in Figure 8.22(a), the PSG is given in Figure 8.22(b). For each primary input and output of the circuit, there is a node in the PSG. Other vertices of the PSG correspond to multi-fanout gates. Therefore, only gate G_1 has a corresponding vertex in the PSG.

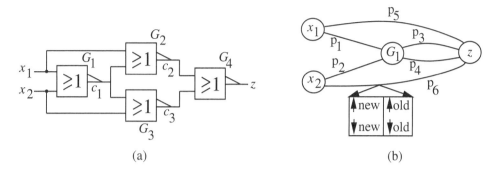

Figure 8.22 Extracting the PSG from a circuit

Edges in the PSG correspond to fanout-free subpaths between nodes in the circuit that are represented in the PSG. For example, there are two edges $\{p_3, p_4\}$ between vertices G_1 and z corresponding to paths c_1c_2z and c_1c_3z.

The status of a PDF can be further classified as: (i) PDF never detected, (ii) PDF detected by previous two-pattern tests only, (iii) PDF detected by the current two-pattern test only, and (iv) PDF detected by the current two-pattern test and previous ones. To represent these, each edge has two flags: (i) *old*: the edge belongs to at least one PDF detected previously, and (ii) *new*: the edge belongs to at least one PDF detected currently. Thus, for example, a PDF detected by previous and current tests has *old* = true and *new* = true for all its edges. Since each path has two corresponding PDFs, there are two sets of status flags for each edge. $\{\uparrow old, \uparrow new\}$ and $\{\downarrow old, \downarrow new\}$. These correspond to rising and falling PDFs with respect to the direction of transition at the source of each subpath.

In Figure 8.22, if the PDF $\downarrow x_1c_1c_3z$ is detected by the current two-pattern test, then flags $\downarrow new(p_1)$ and $\uparrow new(p_4)$ become true. For all the subsequent vectors, the following status flags are true: $\downarrow old(p_1)$ and $\uparrow old(p_4)$. This is because a PDF detected currently is detected by previous vectors with respect to all the subsequent vectors.

For fault simulation, the *new* and *old* flags are initially made false for all the edges in the PSG. Then multiple iterations are performed, each for the simulation of a two-pattern test. In each iteration, logic simulation is performed using a logic system suitable for the selected PDF detection criterion, e.g., robust or non-robust. The *new* flags are set to true for each edge in the PSG through which at least one PDF is detected by the current two-pattern test. Then, selected vertices in the PSG are split to avoid simulation error. This step prevents overestimation of fault coverage. Since faults detected in the current iteration are "detected by previous tests" for subsequent iterations, the *old* and *new* flags are readjusted for reuse. This is done by setting the *old* flag true for each edge having the *new* flag true, and then setting the *new* flag false. Vertices having input edges with *old* flags set true in the current iteration are marked as

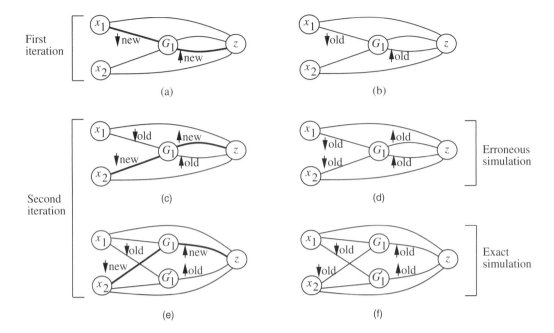

Figure 8.23 Fault simulation

candidates for recombination. Recombination is the opposite of splitting, and reduces the size of the PSG. At the end of simulation, the detected PDFs are those that have the *old* flag true for all of their edges.

Example 8.10 Consider the circuit in Figure 8.22(a) and its PSG once again. Suppose we need to robustly simulate two two-pattern tests: $\{(1, 0), (0, 0)\}$ and $\{(0, 1), (0, 0)\}$. If we use the five-valued logic system introduced earlier, the first two-pattern test corresponds to $x_1 = U0$, $x_2 = S0$. On simulating, we get $c_1 = U1$, $c_2 = c_3 = U0$, $z = U1$. This detects the PDF $\downarrow x_1 c_1 c_3 z$. The corresponding flags that are set to true are shown in Figure 8.23(a). At the end of the first iteration, the flags that are set to true are shown in Figure 8.23(b). Then the second two-pattern test is simulated. This corresponds to $x_1 = S0$, $x_2 = U0$. On simulating, we get $c_1 = U1$, $c_2 = c_3 = U0$, $z = U1$. This detects the PDF $\downarrow x_2 c_1 c_2 z$. The corresponding flags that are set to true are shown in Figure 8.23(c), which are then converted into the true flags shown in Figure 8.23(d). The PSG in Figure 8.23(d) yields four detected PDFs, as can be verified by tracing the paths that have the *old* flag true on all corresponding edges. The correct number is two because only two two-pattern tests were simulated, and each detected one PDF. Figure 8.23(e) shows the correct PSG after the execution of the function *split*. At the end of the second iteration, the flags that are set to true are shown in Figure 8.23(f). This tells us that so far two PDFs have been robustly detected. □

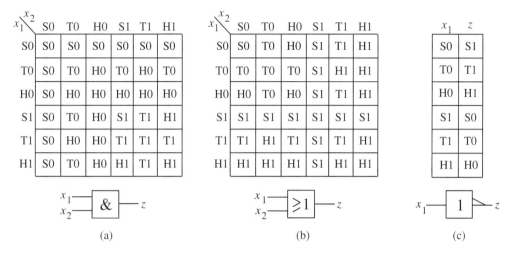

Figure 8.24   Implication tables (Dumas *et al.*, 1993) (© 1993 IEEE)

8.3.2   GDF simulation

As mentioned before, there are two types of GDFs: G-GDFs and S-GDFs. Simulation of G-GDFs can be done with a minor extension of SAF simulation (Waicukauski and Lindbloom, 1986). For two consecutive vectors in a test set, it just needs to be ascertained through logic simulation which nodes in the circuit have a transition, and if the second vector detects the relevant SAF at the node. Thus, in this section, we will concentrate on S-GDFs.

The S-GDF simulation method that we discuss next is based on a critical path tracing method (Dumas *et al.*, 1993). It avoids explicit simulation of each S-GDF in the circuit. In the case of S-GDFs, the quality of test set is not only determined by the number of S-GDFs detected, but also by the detected size of the faults. The fault size indicates the minimum increase in delay the S-GDF must cause in order to be detected. Such a minimum fault size is called the **detection threshold**. The fault simulation method determines an approximate detection threshold.

A six-valued logic is used to perform simulation of the fault-free circuit for the given two-pattern test. The six values are $S0$, $S1$, $T0$, $T1$, $H0$, and $H1$. $S0$ ($S1$) represents a stable 0 (1). A falling (rising) transition with or without a hazard is denoted by $T0$ ($T1$). Finally, signals with the initial and final values 0 (1), which may or may not be affected by a hazard, are denoted by $H0$ ($H1$). Initially, the primary inputs of the circuit are initialized to their signal values based on the two-pattern test. Primary inputs with the same initial and final values are encoded by symbols $S0$ or $S1$, while those with different initial and final values are encoded by symbols $T0$ or $T1$. Next, the logic values of all the other signals in the circuit are determined based on implication tables. These tables provide the output symbol of a gate based on its input symbols, and are shown in Figure 8.24 for some primitive gates.

In order to determine the size of the detected faults, the earliest arrival time (EAT) and latest stabilization time (LST) of all the signals in the circuit have to be established. This can be done simultaneously with the simulation process. The EAT of a signal on line l, denote by $EAT(l)$, is the earliest time at which the signal on l may assume a value different from its initial value $IV(l)$. The LST of a signal on line l, denoted by $LST(l)$, is the time after which the signal on l always stays at its final value $FV(l)$. For primary input signals encoded with $T0$ or $T1$, we set $EAT = LST = 0$. For primary input signals encoded with $S0$ or $S1$, we set $EAT = +\infty$ and $LST = -\infty$. To compute the EAT and LST values, consider an AND/NAND gate G with output y, where $SI(0)$ ($SF(0)$) is the set of all input signals to the gate with an IV (FV) of 0, and $SI(1)$ ($SF(1)$) is the set of all input signals to the gate with an IV (FV) of 1 (Pramanick and Reddy, 1997). Let t_G^r (t_G^f) be the rising (falling) propagation delay of gate G. The preliminary estimates of $EAT(y)$ and $LST(y)$ are given by

$$PreEAT(y) = \begin{cases} \max[EAT(l)|l \in SI(0)] + t_G^r & \text{if } SI(0) \neq \phi \\ \min[EAT(l)|l \in SI(1)] + t_G^f & \text{if } SI(0) = \phi \end{cases}$$

$$PreLST(y) = \begin{cases} \min[LST(l)|l \in SF(0)] + t_G^f & \text{if } SF(0) \neq \phi \\ \max[LST(l)|l \in SF(1)] + t_G^r & \text{if } SF(0) = \phi \end{cases}$$

These estimates are improved by detecting certain cases with steady signal values on y, as follows.

$$EAT(y) = \begin{cases} PreEAT(y) & \text{if } PreEAT(y) \leq PreLST(y) \\ +\infty & \text{otherwise} \end{cases}$$

$$LST(y) = \begin{cases} PreLST(y) & \text{if } PreEAT(y) \leq PreLST(y) \\ -\infty & \text{otherwise} \end{cases}$$

Similar computations hold for OR/NOR gates through the Principle of Duality.

For signals encoded with $H0$ and $H1$, there is a need to preserve the preliminary value of EAT, as we will see later. This is done with the help of a parameter called $PreVEAT$ as follows.

If (line l is encoded by $H0$ or $H1$ and $EAT(l) = +\infty$ and $LST(l) = -\infty$) then

$$PreVEAT(l) = \begin{cases} PreEAT(l) & \text{if } PreEAT(l) \neq +\infty \\ 0 & \text{otherwise} \end{cases}$$

After the simulation with the six-valued logic and EAT/LST computations are done in a forward phase, different lines in the circuit are characterized as sensitive or non-sensitive based on the following propagation rules. If only one input of a gate has a controlling final logic value and if the signal on this line is $T0$, $T1$, $H0$ or $H1$, then the line is sensitive. If all inputs of a gate have a non-controlling final logic value,

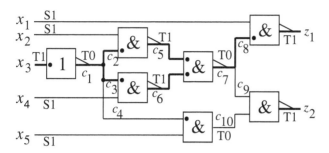

Figure 8.25 S-GDF simulation

then each input with a $T0$, $T1$, $H0$ or $H1$ signal is sensitive. Otherwise, no input is sensitive. Note that in the case of XOR and XNOR gates, there is no controlling value. Therefore, all inputs of such gates with the value $T0$, $T1$, $H0$ or $H1$ are sensitive.

The next phase of the method is a backward phase which makes use of the information on sensitive lines to perform critical path tracing. This phase provides a list of S-GDFs. It starts with all sensitized primary outputs (i.e., those with signals $T0$, $T1$, $H0$ or $H1$) and propagates their critical states towards primary inputs based on which lines are sensitive. This is illustrated by the following example.

Example 8.11   Consider the circuit in Figure 8.25. It shows the signal values on various lines based on six-valued logic simulation for the two-pattern test $\{(1, 1, 0, 1, 1), (1, 1, 1, 1, 1)\}$. In critical path tracing, we start from all sensitized primary outputs, and recursively mark as critical every sensitive input of gates which have a critical output. Critical lines in the circuit are shown in bold and sensitive inputs are marked by dots. The list of detectable slow-to-rise S-GDFs obtained after the backward phase are on the following inputs and outputs of different gates: $\{x_3, c_5, c_6, z_1, z_2\}$. Similarly, the list of detectable slow-to-fall S-GDFs are on $\{c_1, c_2, c_3, c_7, c_8\}$.   ☐

In order to compute the detection threshold of an S-GDF, we define **reference time** T_{ref}. It represents the projection of the observation time at the primary outputs to the internal lines of the circuit. If the delay fault size on a critical line is such that the logic values of the signal in the fault-free and faulty cases are different at time T_{ref}, then the corresponding S-GDF on this line will always be detected by the two-pattern test.

Suppose T_{obs} is the time at which the primary outputs are observed. In the beginning, we set $T_{\text{ref}} = T_{\text{obs}}$ for each sensitized primary output, and $T_{\text{ref}} = 0$ for all other primary outputs. The detection threshold $\Delta(z)$ of an S-GDF on a sensitized primary output z is then given by:

$$\Delta(z) = T_{\text{ref}}(z) - LST(z)$$

If a line l fans out, let y be its fanout branch. Otherwise, if l feeds some gate G, let the gate's output be y. Let t_G^{tr} represent either t_G^{r} or t_G^{f} depending on what the final value on y is. The detection threshold, $\Delta(l)$, of an S-GDF on line l is then derived as follows.

If (l is a fanout stem) then

$$T_{\text{ref}}(l) = \begin{cases} \min\{T_{\text{ref}}(y)\} & \text{if } y \text{ is critical} \\ \max\{T_{\text{ref}}(y)\} & \text{if no } y \text{ is critical} \end{cases}$$

Else $T_{\text{ref}}(l) = T_{\text{ref}}(y) - t_G^{\text{tr}}$
If (l is critical) then
$\Delta(l) = T_{\text{ref}}(l) - LST(l)$

The above procedure may not produce the correct results if line l is encoded by $H0$ or $H1$. For such cases, after running the above procedure, the procedure given next is run.

If (l is encoded by $H0$ or $H1$ and $EAT(l) = +\infty$ and
$\quad LST(l) = -\infty$) then
$\quad T_{\text{ref}}^*(l) = \max\{T_{\text{ref}}(l), PreVEAT(l)\}$
Else $T_{\text{ref}}^*(l) = T_{\text{ref}}(l)$
If (l is critical) then
$\quad \Delta(l) = T_{\text{ref}}^*(l) - LST(l)$

Consider the circuit in Figure 8.25 again and all the S-GDFs detected by the two-pattern test shown in the figure (see Example 8.11). Let both the rising and falling propagation delay of an AND/NAND gate be two time units, and correspondingly for an inverter let it be one time unit. Suppose that $T_{\text{obs}} = 8$. Then the above procedures indicate that the detection threshold for all the detected S-GDFs is one time unit. For example, the LST of output z_1 is 7 whereas its $T_{\text{ref}} = T_{\text{obs}} = 8$. The difference gives the detection threshold. Similarly, one can find that the difference between T_{ref} and LST of all bold lines from x_3 to z_1 is one.

8.3.3   SDF simulation

We saw in Section 8.2.7 that there are three types of SDF tests: robust, transition and non-robust. Conditions for each type of SDF test were presented there. We will illustrate the use of these conditions through examples here for SDF simulation.

Consider the two-pattern test shown in the circuit in Figure 8.25. This test robustly detects DFs along the bold paths. Suppose we are interested in robustly detecting SDFs of length $L = 2$. Then the SDFs that are robustly detected can be seen to be: $\uparrow x_3 c_1$, $\downarrow c_1 c_5$, $\downarrow c_1 c_6$, $\uparrow c_5 c_7$, $\uparrow c_6 c_7$, and $\downarrow c_7 z_1$ (note that we do not make a distinction between a fanout stem and its branches for fault detection). Thus, segments of robustly detected PDFs are also robustly detected.

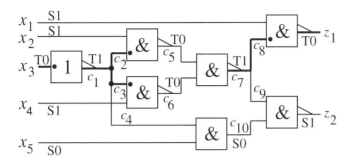

Figure 8.26 SDF simulation

Robust detection is also possible under the relaxed propagation rules which come into play when there is a reconvergent fanout. For this case, we first identify lines which have a transition propagated to a primary output (not necessarily robustly). SDFs which end on these lines are possible candidates for detection. For example, consider the two-pattern test shown in Figure 8.26. Using six-valued logic simulation, the identified lines are x_3, c_1, c_7, and z_1. Suppose we are interested in segments with $L = 2$ again. Then the SDFs that we need to investigate are $\downarrow x_3 c_1$ and $\uparrow c_7 z_1$. In the presence of $\downarrow x_3 c_1$, the falling transitions on c_5 and c_6 are both delayed (faulty). Thus, we need to distinguish between various types of falling transitions: (i) those that have passed through the fault site and are carrying the fault effect, (ii) those that are robustly propagated, and (iii) those that are not robustly propagated. Similarly, three types of rising transitions need to be considered. The two extra types of falling transitions and the two extra types of rising transitions can be included in a ten-valued logic system (Heragu *et al.*, 1996b). This would show both the above SDFs to be detected by this two-pattern test.

Transition SDF simulation is very similar to G-GDF simulation. In other words, it can be based on a minor extension of SAF simulation. For a two-pattern test, one can first determine through logic simulation if the desired transition was launched at the origin of the segment, and if the corresponding SAF at the origin was detected through the segment.

Non-robust SDF simulation is similar to robust SDF simulation except that restricted non-robust propagation rules are used instead of the robust propagation rules.

8.3.4   At-speed fault simulation

In this section, we discuss a method for robust fault simulation of PDFs when vectors are applied at the rated clock (or at-speed) (Hsu and Gupta, 1996).

Assume that a test sequence $P_1, P_2, \ldots, P_{t-1}, P_t$ is applied at-speed to the circuit under test. There are two conditions that must be satisfied by this sequence for robust detection of a PDF R by P_t. First, (P_{t-1}, P_t) must be a robust two-pattern test as

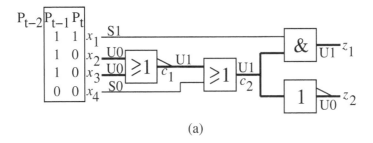

(a)

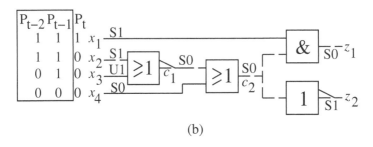

(b)

Figure 8.27 At-speed PDF simulation

defined by the variable clock scheme. This can be checked with any variable clock PDF simulation method, such as the one presented in Section 8.3.1. Second, all on-path lines must be stable at their steady values implied by P_{t-1}. To simplify the analysis, we will assume that the worst-case delay along any path in the circuit is bounded by twice the circuit's clock period.

To make sure all on-inputs are stable before P_t is applied, we need to consider the two-pattern test (P_{t-2}, P_{t-1}). We will use the familiar five-valued logic system, consisting of $S0$, $S1$, $U0$, $U1$ and XX, for this purpose. A simple way to check the stability of the on-inputs is to perform circuit simulation with (P_{t-2}, P_{t-1}). The lines in the circuit that are assigned values $S0$ ($S1$) are guaranteed to be stable at 0 (1) before the application of P_t. In addition, if the circuit passes some two-pattern tests, then during the simulation of subsequent vectors, the stability checking method can take advantage of the fact that the circuit is free from some PDFs. This is illustrated next.

Example 8.12 Consider the circuit in Figure 8.27. Figure 8.27(a) shows the circuit simulation with the two-pattern test (P_{t-1}, P_t) to identify PDFs that are robustly detected under the variable clock scheme. These PDFs are indicated in bold and are $\downarrow x_2 c_1 c_2 z_1$, $\downarrow x_2 c_1 c_2 z_2$, $\downarrow x_3 c_1 c_2 z_1$, and $\downarrow x_3 c_1 c_2 z_2$. Figure 8.27(b) shows the circuit simulation with the two-pattern test (P_{t-2}, P_{t-1}). Simple stability checking shows that the dashed lines are stable. Since the on-inputs of PDFs $\downarrow x_2 c_1 c_2 z_1$ and $\downarrow x_2 c_1 c_2 z_2$ are stable at the values implied by P_{t-1}, these faults are indeed robustly detected at-speed.

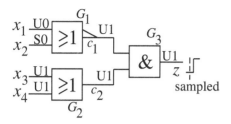

Figure 8.28   Effect–cause fault diagnosis

Next, suppose that PDF $\uparrow x_3c_1c_2z_1$ has already been robustly tested by a previous test. If the circuit passed this test, then the rising transition is guaranteed to propagate through the path $x_3c_1c_2z_1$ in one clock cycle. This means that line x_3 is stable at logic 1 before P_t is applied. Thus, PDFs $\downarrow x_3c_1c_2z_1$ and $\downarrow x_3c_1c_2z_2$ are also robustly detected at-speed by the vectors shown in the figure.   □

8.4   Combinational delay fault diagnosis

In this section, we discuss methods to diagnose both PDFs and GDFs.

8.4.1   PDF diagnosis

We discuss here a path-oriented, effect–cause methodology for diagnosing PDFs (Hsu and Gupta, 1998). This assumes that the two-pattern tests have been applied to manufactured circuits under the variable clock scheme, and data on the pass/fail status for each test is known. We first illustrate this methodology through an example using the familiar five-valued logic system.

Example 8.13   Consider the circuit in Figure 8.28. Suppose the two-pattern test is as shown. If the circuit fails this test, i.e., a 0 is captured at z, then one or more DFs exist in the circuit. Let us first look at gate G_3. A slow-to-rise problem at either line c_1 or c_2 can cause the slow-to-rise effect at z. Thus, an OR condition can be used to describe the relationship between the two causes.

A similar analysis can lead to the causes for the slow-to-rise transitions at c_1 and c_2. At gate G_1, input x_2 has a steady non-controlling value. Therefore, it cannot contribute to the slow-to-rise problem at c_1. Hence, input x_1 is the only suspect at this gate. At gate G_2, the time at which the rising transition occurs at line c_2 depends on the earlier of the rising transitions at inputs x_3 and x_4. Therefore, the transition at line c_2 can be delayed only if the transitions at both these inputs are delayed. An AND condition can be used to describe the relationship between these two causes.

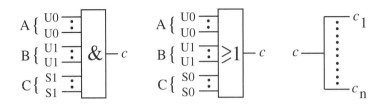

Figure 8.29 Set propagation

The above analysis shows that the possible PDFs are $\downarrow x_1 c_1 z$ (R_1), $\uparrow x_3 c_2 z$ (R_2), and $\uparrow x_4 c_2 z$ (R_3). PDF R_1 can alone explain the error at the primary output. However, neither R_2 nor R_3 alone can explain the error, only their simultaneous presence can. Thus, the result of the diagnosis can be explained by the set $\{R_1, R_2 R_3\}$. □

In the above example, tracing of PDFs was started at the primary output for ease of exposition. However, we next use breadth-first search from primary inputs to primary outputs to generalize this tracing method.

The **Cartesian product** of n sets $A_1, A_2, \ldots, A_n$ is defined as $\{(a_1, a_2, \ldots, a_n) \mid a_i \in A_i, i = 1, 2, \ldots, n\}$. In the **AND-product** of n sets $A_1, A_2, \ldots, A_n$, denoted by $\coprod_{i=1}^{n} A_i$, each $(a_1, a_2, \ldots, a_n)$ in the Cartesian product is replaced by $a_1 \cdot a_2 \cdots a_n$.

Let (P_1, P_2) be a two-pattern test for which the circuit failed. Let $S_r(x_i)$ be the set of all PDFs starting at primary input x_i with a rising transition. Set $S_f(x_i)$ is defined similarly for a falling transition. In Figure 8.29, two primitive gates are shown. A is the set of inputs that have been assigned $U0$, B is the set of inputs that have been assigned $U1$, C is the remaining inputs with a steady non-controlling value.

Under the above scenario, $S_r(c)$ or $S_f(c)$ can be derived using the following rules. For an AND gate:

$$S_f(c) = \coprod_{c_i \in A} S_f(c_i) \text{ if } A \neq \phi \text{ or } S_r(c) = \bigcup_{c_j \in B} S_r(c_j) \text{ if } A = \phi$$

For a NAND gate:

$$S_r(c) = \coprod_{c_i \in A} S_f(c_i) \text{ if } A \neq \phi \text{ or } S_f(c) = \bigcup_{c_j \in B} S_r(c_j) \text{ if } A = \phi$$

For an OR gate:

$$S_r(c) = \coprod_{c_j \in B} S_r(c_j) \text{ if } B \neq \phi \text{ or } S_f(c) = \bigcup_{c_i \in A} S_f(c_i) \text{ if } B = \phi$$

For a NOR gate:

$$S_f(c) = \coprod_{c_j \in B} S_r(c_j) \text{ if } B \neq \phi \text{ or } S_r(c) = \bigcup_{c_i \in A} S_f(c_i) \text{ if } B = \phi$$

Finally, for an inverter:

$$S_f(c) = S_r(d) \text{ and } S_r(c) = S_f(d)$$

where d is the input of the inverter.

Next, consider a fanout stem c with fanouts $c_1, \ldots, c_n$. $S_f(c_i)$ contains those elements of $S_f(c)$ in which all the ANDed PDFs contain c_i. $S_r(c_i)$ is similarly derived.

If a slow-to-rise (slow-to-fall) fault is detected at a primary output z, then set $S_r(z)$ $(S_f(z))$ will contain one or more product terms, any of which can explain the fault. Thus, this set will be referred to as the **suspect set**. Each product term is the logical AND of one or more PDFs, whose collective presence can explain the observed error due to the fault.

Let us revisit the circuit in Figure 8.28. For the given two-pattern test, we obtain $S_r(c_1) = S_f(x_1) = \{R_1\}$, where $R_1 = \downarrow x_1 c_1 z$. Similarly, $S_r(c_2) = \bigsqcup \{S_r(x_3), S_r(x_4)\} = \{R_2 R_3\}$, where $S_r(x_3) = \{R_2\}$, $S_r(x_4) = \{R_3\}$, $R_2 = \uparrow x_3 c_2 z$, and $R_3 = \uparrow x_4 c_2 z$. These sets are obtained by breadth-first search from the primary inputs. At the next logic level, we obtain $S_r(z) = S_r(c_1) \bigcup S_r(c_2) = \{R_1, R_2 R_3\}$. This is the suspect set we had obtained in Example 8.13.

We construct a suspect set for each failing two-pattern test. However, typically, the circuit will pass a majority of two-pattern tests. This means that PDFs robustly detected by these passing tests cannot be present in the circuit. Results of these tests can be used to whittle down the size of the suspect set by removing those product terms which contain PDFs robustly detected by the passing tests. For example, in Example 8.13, if the circuit passed a robust two-pattern test for PDF $R_2 = \uparrow x_3 c_2 z$, then the suspect set reduces from $\{R_1, R_2 R_3\}$ to just $\{R_1\}$.

8.4.2   GDF diagnosis

In this section, we will discuss a method for diagnosing GDFs (both G-GDFs and S-GDFs) based on critical path tracing and the six-valued logic system which were presented in Section 8.3.2 (Girard *et al.*, 1992). The only information that is assumed to be available is for which two-pattern tests the circuit fails and at which primary outputs.

For a failing two-pattern test, six-valued logic simulation of the circuit is first performed. Then all lines which are sensitive are determined. The effect of a GDF can propagate through a gate with no sensitive inputs only if two or more of its inputs are simultaneously affected by the fault. Recall that the reason a gate becomes insensitive is if it has two or more inputs with a final controlling value (i.e., $T0/H0$ for an AND/NAND gate or $T1/H1$ for an OR/NOR gate). If we assume only single GDFs are present in the circuit, the above scenario is only possible if a fanout stem exists which has reconvergent paths to the inputs of the gate in question which have controlling final values. The search for potential GDFs proceeds towards the primary inputs from the fanout stem thus found. The above information is used to determine which GDFs are detected by the two-pattern test using the method illustrated in Example 8.11. The set of suspect GDFs can be found by taking the intersection of the sets of GDFs detected at the different failing primary outputs. The suspect set can be further reduced by intersecting with the suspect sets obtained for other failing two-pattern tests.

Example 8.14 Consider the two-pattern test shown in the circuit in Figure 8.25 once again. The signal values obtained through six-valued simulation and the sensitive inputs of different gates are also shown in the circuit. Suppose that the circuit fails at both the primary outputs for this two-pattern test. In other words, both z_1 and z_2 assume the faulty value of 0 after the second vector is applied. The failure at z_1 can be explained with the slow-to-rise or slow-to-fall GDFs on the lines shown in bold that lead to z_1. However, since the NAND gate feeding z_2 does not have any sensitive inputs, the failure at z_2 can only be explained by a GDF at the fanout stem c_1 which has reconvergent paths to this gate or by a GDF on a line feeding c_1. Thus, the only possible suspect GDFs are a slow-to-rise GDF on x_3 or a slow-to-fall GDF on c_1.   □

8.5   Sequential test generation

In this section, we discuss various DF test generation methods for sequential circuits. We first look at methods which can quickly identify untestable faults. We then discuss test generation methods for non-scan sequential circuits (scan circuits are discussed in detail in Chapter 11). We assume a variable clock scheme, as illustrated in Figure 8.2.

8.5.1   Untestability analysis

In Section 8.2.1, we discussed methods for quickly identifying untestable DFs in combinational circuits. We discuss extensions of these methods to sequential circuits here (Heragu, 1998). We use the same notations.

We assume that logic implications of both logic 0 and 1 for every line in the circuit are available over two time frames of the sequential circuit. For lines x and y, an implication $(x = \alpha) \Rightarrow (y_{+1} = \beta)$, where $\alpha, \beta \in \{0, 1\}$, indicates that if the signal value on x is α in a particular time frame, then the signal value on y is β in the next time frame. Similarly, an implication $(x = \alpha) \Rightarrow (y_{-1} = \beta)$ indicates that if the signal value on x is α in a particular time frame, then the signal value on y is β in the previous time frame.

We next present some lemmas which form the basis for untestability analysis. All these lemmas follow from the basic definitions of DF testability.

Lemma 8.10 For lines x and y, if $(x = \alpha) \Rightarrow (y_{+1} = \beta)$, and β is the controlling value of gate G that y feeds, then for all other inputs z of G and for all $\gamma \in \{0, 1\}$, $[S_P(x^{\bar{\alpha}}, z^{\gamma}) = RU]$.

Lemma 8.11 For lines x and y, if $(x = \alpha) \Rightarrow (y_{+1} = \beta)$, then $[S_P(x^{\bar{\alpha}}, y^{\bar{\beta}}) = RU]$.

Lemma 8.12 For lines x and y, if $(x = \alpha) \Rightarrow (y_{-1} = \beta)$, then $[S_P(x^{\alpha}, y^{\beta}) = RU]$.

Lemma 8.13 For lines x and y, if $(x = \alpha) \Rightarrow (y_{-1} = \beta)$, and β is the controlling value of gate G that y feeds, then for all other inputs z of G, $[S_P(x^{\alpha}, z^{\beta}) = RU]$.

Lemma 8.14 For lines x and y, if $(x = \alpha) \Rightarrow (y_{+1} = \beta)$, where x is a primary input or flip-flop output, then $[S_P(x^{\bar{\alpha}}, y^{\bar{\beta}}) = FU]$.

Lemma 8.15 For a flip-flop with input x and output y, if the stuck-at α fault on y is combinationally untestable, then $[S_P(x^{\bar{\alpha}}, x^{\bar{\alpha}}) = RU]$.

Other lemmas similar to those presented above can be derived for identifying non-robustly untestable and robust-dependent PDFs.

Example 8.15 Consider the sequential circuit in Figure 8.30(a). Its iterative array model with an expansion over two time frames is shown in Figure 8.30(b). For a two-pattern test (P_1, P_2), vector P_1 will be applied in the first time frame using a slow clock whereas vector P_2 will be applied in the second time frame using a rated clock. In general, of course, more time frames may be needed to the left and right for state justification and error propagation, respectively. Let us look at the following sequential implication: $(x_2(P_1) = 0) \Rightarrow (c_2(P_2) = 1)$. From Lemma 8.11, $[S_P(x_2^1, c_2^0) = RU]$. Let us now look at PDF $\uparrow x_2c_1c_2z$ in the sequential circuit in Figure 8.30(a). From Lemma 8.1 (see Section 8.2.1), $x_2(P_2) = 1$ and $c_2(P_2) = 0$ are necessary conditions for sensitizing this PDF. Therefore, $[S_P(x_2^1, c_2^0) = RU]$ implies that this PDF is not robustly testable. To test this PDF, we need the conditions $x_2(P_1) = 0$, $x_2(P_2) = 1$ and $y(P_2) = 1$, which are not simultaneously satisfiable.   □

8.5.2   Test generation

In this section, we discuss variable clock testing methods based on the iterative array model (Chakraborty and Agrawal, 1997). We will first target testing of PDFs, and discuss extensions to other types of DFs later.

The test for a PDF should: (a) produce a signal transition at the origin of the path, (b) propagate the transition through the path, and (c) allow observation of the value at the destination flip-flop of the path (if the destination is a primary output, then this

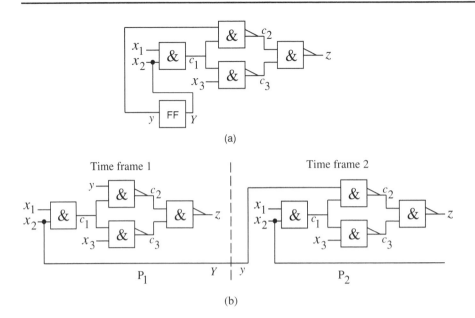

Figure 8.30 Untestability analysis

step is not needed). Steps (a) and (b) are called the **initialization** and **path activation** phases, respectively, and step (c) is called the **error propagation** phase.

Suppose we want to test a PDF on a path from flip-flop FF_1 to flip-flop FF_2. We first need a vector sequence T_1 to initiate a signal transition at FF_1 and propagate it to FF_2. Then we need a vector sequence T_2 to propagate the captured logic state of FF_2 to the primary output. T_1 and T_2 together form the desired test sequence.

The last two vectors of T_1 are needed for the activation of the selected path. The preceding vectors of T_1 create the initialization condition for path activation. Only the last vector of T_1 is applied at the rated clock. The reason the other vectors are applied using a sufficiently slower clock is that we need to guarantee that the required signal transition actually occurs at the origin of the path irrespective of arbitrary delays in the circuit. Similarly, T_2 is applied using a slower clock to avoid interference from other DFs in error propagation.

Test generation is done as follows. Consider a PDF on a path from flip-flop FF_1 to flip-flop FF_2. The desired transition ($0 \rightarrow 1$ or $1 \rightarrow 0$) is selected at the output of the source flip-flop. The signal values are derived at the inputs of the gates on the selected path to propagate the transition through the path using an appropriate logic system. We will use the usual five-valued logic system ($S0$, $S1$, $U0$, $U1$, XX) for this purpose. Backward justification of these signal values is then performed to the primary inputs of the circuit to generate the input vector sequence T_1. Because of the signal representation, two time frames are simultaneously considered. The logic value latched at FF_2 can be considered as D or $\overline{D}$, where the fault-free and faulty circuit values are 1 and 0 or vice versa, respectively. If $S0$ ($S1$) propagates to a flip-flop other than the destination flip-flop, then that flip-flop assumes the value 0 (1) in the subsequent vector. If $U0$ or $U1$ propagates to the other flip-flop, then that flip-flop is assigned the fault-free value in the subsequent vector provided it is the path destination of at least one robustly activated fault. This is called the **optimistic update rule**. This rule guarantees that the targeted PDF cannot be masked by the presence of the other PDFs in the circuit (Bose *et al.*, 1993a). Otherwise, that flip-flop assumes the unknown logic value x. Next, sequence T_2 is derived to propagate the D or $\overline{D}$ captured in FF_2 to a primary output. T_2 distinguishes between the fault-free and faulty state values. If conflicts arise during backward justification or path activation, they are resolved using all available choices for robust or non-robust tests. If no choices are left, the fault is declared untestable.

PDFs which start at primary inputs have only two vectors in T_1, and those that end at primary outputs do not require a propagation sequence T_2. Otherwise, they are treated in the same fashion as above.

Example 8.16 Consider the sequential circuit in Figure 8.31(a). Suppose we are interested in deriving a test sequence to detect a PDF on the path from flip-flop output y to its input Y, as shown in bold, for a falling transition at the origin of the path. This PDF can be robustly sensitized using the five-valued logic system with the signal values shown in this figure. These signal values correspond to time frames 2 and 3 in the iterative array model shown in Figure 8.31(b). The flip-flop value $y = 1$ in time frame 2 needs to be justified in time frame 1, as shown in the figure.

In the fault-free circuit, we expect $Y = 1$ in time frame 3. However, when the targeted PDF is present, Y will become 0 in this time frame. Thus, this value can be represented as D. We need time frame 4 to propagate it to primary output z, as shown in the figure. A rated clock is needed only in time frame 3. Slower clocks are needed for the other time frames. Thus, the desired test sequence at (x_1, x_2) is $\{(1, 1), (0, x), (0, x), (1, 0)\}$. □

It is easy to extend the above test generation method to most DFs other than PDFs. Using an appropriate logic system, we can first derive a two-pattern test for the fault in the combinational logic of the circuit. This two-pattern test is applied over two time frames. If a logic 0 or 1 needs to be justified on the present state lines in the time frame corresponding to the first vector of the two-pattern test, then time frames are added to the left for this purpose until an unknown initial state is reached. Similarly, time frames are added to the right to propagate errors from next state lines to a primary output. State justification and error propagation can be done with methods similar to those used for SAFs in Chapter 5 (Section 5.4.2).

The shortcoming of the variable clock scheme is that since a rated clock is used in only one time frame, it precludes activation of faults in consecutive clock cycles. This

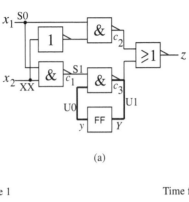

(a)

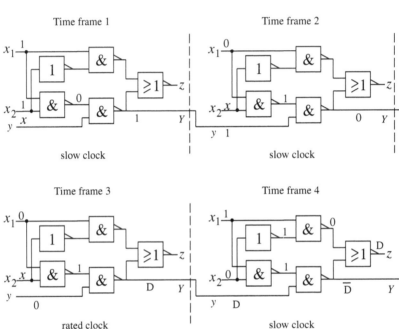

(b)

Figure 8.31 Sequential test generation

results in a large number of test sequences. This problem can be somewhat alleviated if a rated clock is used in multiple consecutive time frames while a slow clock is used only when needed (Pomeranz and Reddy, 1992).

8.6　Sequential fault simulation

In this section, we discuss sequential fault simulation methods for both PDFs and GDFs.

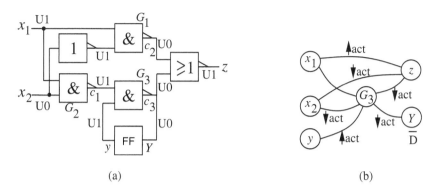

(a)   (b)

Figure 8.32   Sequential PSG

8.6.1   PDF simulation

We had discussed a PDF simulation method for combinational circuits based on an efficient data structure called the path status graph (PSG) in Section 8.3.1. This method is applicable under the variable clock scheme. We use an extension of this method for sequential fault simulation (Parodi *et al.*, 1998). The non-enumerative nature of the simulation algorithm is maintained by using a layered PSG and restricting the number of time frames over which a fault must be detected.

In a sequential circuit, many paths terminate at flip-flops which are not directly observable. The logic cones of all activated paths that terminate at the same flip-flop are recorded as a single fault effect. The cone of paths is saved in the PSG format for the complete fault description. The fault record just consists of the destination flip-flop, the time frame when the fault originated, and the true value that would be latched if the circuit were not faulty.

In order to avoids cycles in the PSG, flip-flop nodes are considered as *pseudo primary input/output* pairs. Consider the sequential circuit shown in Figure 8.31(a) again. It has been reproduced in Figure 8.32(a). Its sequential PSG is shown in Figure 8.32(b) (ignore the signal values in Figure 8.32(a) and the flags on the edges in Figure 8.32(b) for now).

To facilitate fault storage, the PSG data structure is modified as follows. Instead of recording the activated paths in the same graph, layers are added to the graph. In the first (bottom) layer, all the activated paths in the current time frame are tagged. These are the paths that are either directly observable at the primary outputs or whose effects might be observable in future time frames. This layer also keeps a tab on all the paths that have been tested. Higher layers contain flags corresponding to the previously activated, but yet unobserved, paths. Each layer is a snapshot of the PSG in one time frame. As the time frame advances, the PSGs are moved one layer up. Using this new data structure, stored faults are identified by noting the affected flip-flop and the time frame of fault occurrence.

Non-enumerative PDF simulation consists of two parts: simulation of the combinational logic and stored fault list management. The first part consists of event-driven simulation of the active portion of the circuit using the chosen logic system, and recording the circuit activity in the current time frame layer of the PSG. This part is similar to the PSG-based combinational fault simulation method presented in Section 8.3.1. The possible faults in the activated paths terminating at primary outputs are considered tested since they are directly observable. The possible faults in activated paths terminating at flip-flops give rise to a D or $\overline{D}$ at the flip-flop.

The second part of the simulation determines which of the captured faults become observable at the primary outputs, which get propagated to the other flip-flops, and which should be eliminated. This part also consists of event-driven simulation. However, in this case, the faulty circuit behavior is simulated separately. Each stored fault gives rise to one faulty circuit behavior. Using machine word parallelism, 32 or 64 faulty circuits can be simulated at a time.

Each active fault has a list of flip-flops at which its effect is present. When a fault is simulated, the value complementary to that of the fault-free circuit is placed in the flip-flop where the fault effect was. This is done for groups of 32 or 64 active faults. Once a set of faults is simulated, the effect of each fault is checked at all primary outputs and flip-flops. If the fault becomes observable at a primary output, the logic cone of paths that gave rise to the fault (obtained from the fault's initial time frame and the original path destination flip-flop) is flagged as tested in the present layer of the PSG, and the fault is eliminated. If the fault only becomes observable at other flip-flops, the list of flip-flops where the fault effect is presented is updated. If the fault does not become observable at any primary output or flip-flop, it is eliminated.

If an active fault is detected after a few time frames, the cone of paths that gave rise to that fault is copied from a higher level layer to the bottom layer of the PSG. This is because the bottom layer must maintain a record of all tested paths. Before the simulation of the next time frame is initiated, the current bottom layer produces a new bottom layer consisting of all previously tested paths, and the next level layer consisting of paths activated, but not detected in the current time frame. All other layers are moved one level higher.

Since the number of prior time frames in the graph is fixed, if a fault remains active for those many time frames and is still not detected, then it is dropped. In the rare cases when this happens and the fault is actually detectable at a later time frame, the simulator will report a slightly lower fault coverage than the actual fault coverage.

The above method can be easily extended to perform at-speed (rated clock) PDF simulation by using a vector of flags to record whether the signal was stable in the previous N time frames (N can be user-selectable), and adjusting the signal values accordingly.

Example 8.17 Suppose the following input sequence is fed to inputs (x_1, x_2) of the circuit in Figure 8.32(a): $\{(1, 1), (0, 1), (0, 1), (1, 0), (1, 0)\}$. Also, suppose that the first three vectors are fed using a slow clock, the fourth one using the rated clock, and the fifth one using a slow clock. Thus, PDFs can only be detected in time frame 4. Using five-valued logic, the signal values for time frames 3 and 4 are as shown in Figure 8.32(a). The corresponding edges in the PSG where falling (rising) transitions are propagated are depicted by $\downarrow act$ ($\uparrow act$). The effects of the following PDFs are propagated to primary output z: $\uparrow x_1c_2z$, $\downarrow x_2c_2z$, $\downarrow x_2c_1c_3z$, and $\uparrow yc_3z$. Paths are also activated to the next state line Y. In the fault-free case, Y will be 0 in time frame 4, whereas in the faulty case, it will be 1. Thus, it has the error signal $\overline{D}$. This information is kept in a higher layer PSG. When the fifth vector is applied, this error signal propagates to output z. Thus, two additional PDFs are detected: $\downarrow x_2c_1c_3Y$ and $\uparrow yc_3Y$. This information is made available to the bottom PSG layer. □

8.6.2   GDF simulation

Simulation of G-GDFs can be done in a straightforward manner under the variable clock scheme as follows. For the given input sequence, if in the time frame preceding the one with the rated clock and the time frame with the rated clock, a circuit node has a $0 \rightarrow 1$ ($1 \rightarrow 0$) transition and the effect of an SA0 (SA1) fault at the node is propagated to the next state line, then the G-GDF is activated (if the effect is propagated to a primary output, it is already detected). After that, all that needs to be ascertained is that an error signal D or $\overline{D}$ at the next state lines gets propagated to a primary output in the subsequent time frames using a slow clock. A method for at-speed G-GDF simulation of sequential circuits can be found in Cheng (1993).

S-GDFs can be similarly detected based on a direct extension (Cavallera *et al.*, 1996) of the method we discussed in Section 8.3.2 for S-GDF simulation of combinational circuits. We will assume a variable clock scheme.

We will use a six-valued logic system, and consider the size of S-GDFs implicitly through detection threshold computation, as before. Recall that the detection threshold Δ represents the minimum fault size for a fault to be observed at a given output.

The simulation process consists of initialization, fault activation and error propagation. In the initialization phase, logic simulation is performed in the initial time frames, where a slow clock is used, to arrive at the state from which the fault is activated using the rated clock in the next time frame. When the rated clock is applied, faults whose effects reach the primary outputs are detected with respect to their detection threshold. This is determined based on the method discussed in Section 8.3.2. Also, lists of faults are produced whose effects reach the next state lines. These faults result in a D or $\overline{D}$ at the next state lines when the fault size exceeds the detection threshold. In the error propagation phase, where a slow clock is used again, faults whose effects reach the primary outputs are declared detected with respect to their detection threshold.

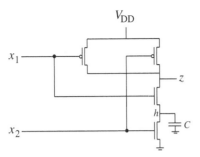

Figure 8.33 A two-input NAND gate

Error propagation rules for at-speed S-GDF simulation of sequential circuits have also been presented in Cavallera *et al.* (1996).

8.7   Pitfalls in delay fault testing and some remedies

DF testing can only be as good as the delay models it is based on. As mentioned in the introduction section, an underlying assumption behind the testing methods presented so far is that the gate propagation delays are fixed and independent of the input logic values. However, this is not true in practice, and can lead to many pitfalls (Franco and McCluskey, 1994; Pierzynska and Pilarski, 1997). We discuss these pitfalls here and an initial test generation attempt that has been presented to avoid them.

8.7.1   Pitfalls

To consider the pitfalls of the assumptions made above, let us consider the two-input NAND gate shown in Figure 8.33, which also shows the hidden node h and the capacitance, C, associated with it. When both primary inputs assume the controlling value 0, both the nMOS transistors are off. Therefore, the charge stored in C is determined by the last input vector for which the primary inputs had at most one controlling value. This charge can affect the gate delay significantly.

Nodes considered unrelated to the path under test in a gate-level circuit can also affect the path delay. This may be due to capacitive coupling between these nodes and the path. Such nodes may also affect the capacitances of the path nodes. Examples are side-inputs of the path under test and non-path inputs of fanout gates not included in the path. The latter are called side-fanout inputs. These nodes are depicted in Figure 8.34.

The fact that different gate inputs are not equivalent from the electrical point of view is generally disregarded in DF testing. However, simulations for the NAND gate show that compared to the gate delay for the two-pattern sequence $\{(1, 1), (0, 0)\}$, the

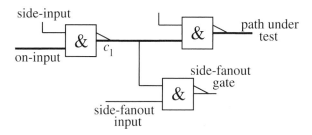

Figure 8.34 Side-inputs and side-fanout inputs

gate delays for the two-pattern sequences $\{(1, 1), (1, 0)\}$ and $\{(1, 1), (0, 1)\}$ are higher by 62% and 90%, respectively, in a particular CMOS process. Thus, the gate delay is substantially different depending on whether the $1 \rightarrow 0$ transition is at x_1 or x_2 or both. However, since both $\{(1, 1), (1, 0)\}$ and $\{(1, 1), (0, 1)\}$ are required robust two-pattern tests for the NAND gate, this phenomenon does not lead to any pitfalls.

Consider four scenarios under which two-pattern sequences cause a falling transition at output z of the NAND gate: (a) $\{(0, 1), (1, 1)\}$, (b) $\{(1, 0), (1, 1)\}$, (c) $\{(0, 0), (1, 1)\}$ with the previous value at the hidden node h being 0, and (d) $\{(0, 0), (1, 1)\}$ with the previous value at h being 1. Simulations show that the normalized gate delays for these scenarios are 1.0, 1.06, 1.26 and 1.33, respectively. Even though all these tests are robust, the best test is given by the fourth scenario since it provokes the longest gate delay by causing a transition at the side-input from the controlling to the non-controlling value. Since this scenario requires that h be at 1, a pre-initialization vector $(1, 0)$ has to be added, thus resulting in a three-pattern test $\{(1, 0), (0, 0), (1, 1)\}$. Choosing robust two-pattern tests in the first two scenarios leads to the pitfall that the longest delay is not exercised through the gate and, therefore, the quality of DF testing suffers.

Let us now switch our attention to fanout effects. Consider the circuit in Figure 8.34 once again. It has been observed through simulations that the longest on-path delay occurs when line c_1 feeds the x_1 input (i.e., the one that feeds the nMOS transistor closest to the output) of the NAND gate on the on-path as well as the side-fanout gate, and the side-fanout input is at the non-controlling value 1. The on-path delay is not affected much by the value at the hidden nodes in the side-fanout gates. Of course, as the number of side-fanout gates satisfying the above conditions increases, so does the on-path delay. Most DF testing methods do not take this effect into account.

Another effect that has been observed is that transitions at side-fanout inputs may result in considerably longer or considerably shorter on-path delays than the case when the side-fanout inputs are stable.

The above observations indicate that different robust tests for a DF may result in very different delays along the path. In fact, it is possible that the path delay under a non-robust test is significantly longer than the path delay under any robust test. This is true even for completely robustly testable circuits. This implies that to obtain high-quality DF tests, these effects need to be targeted during test generation. Some initial attempts in this regard are discussed next.

8.7.2   High-quality test generation

Work in trying to target the above-mentioned effects in test generation is currently in its infancy. We discuss here a method which targets the side-input and pre-initialization effects for robustly detecting PDFs in combinational circuits under a variable clock scheme (Chen *et al.*, 1997). The aim is to find a robust test for a PDF which causes the worst-case delay through the path under test. This method works at the gate level. The only switch-level information assumed to be available for the different gates in the circuit is which gate input feeds the transistor closest to the gate output, which input feeds the next transistor, and so on.

To derive a high-quality robust two-pattern test, (P_1, P_2), one of the conditions we should try to satisfy is as follows. When the on-path has a controlling-to-non-controlling transition, e.g., $0 \rightarrow 1$ for an AND/NAND gate and $1 \rightarrow 0$ for an OR/NOR gate, the side-inputs should also have a controlling-to-non-controlling transition. Note that the classical robust tests just require the side-inputs to have a non-controlling value for the second vector in this case (see Figure 8.11). When the on-path has a non-controlling-to-controlling transition, the requirements are the same as those for classical tests, i.e., the side-inputs should have a steady non-controlling value.

Pre-initialization is needed for an on-path gate when the transition at its on-input as well as at least one side-input is controlling-to-non-controlling. For gates that require pre-initialization, it is desirable to pre-charge the capacitances at the hidden nodes to the non-controlling value of the gate. This is done using a pre-initialization vector P_0 in such a manner that the hidden nodes hold the charge until the application of P_2. This is done to maximize delay. Since the desired P_0 for a gate can conflict with that for another gate, as a compromise, pre-initialization could use a multi-vector sequence or attempt to pre-initialize fewer hidden nodes. Alternatively, another two-pattern test for the PDF in question can be used and the above process of trying to find a pre-initialization vector repeated.

It is possible that a high-quality robust two-pattern test cannot be found for some PDFs in the first step. For such a PDF, we attempt to find, among the robust two-pattern tests for the PDF, one that excites the maximum propagation delay. This can be done by maximizing the number of side-inputs with a controlling-to-non-controlling transition when the on-input has a similar transition.

This method does not target another effect that was mentioned in the previous section, that of values at the side-fanout inputs.

Example 8.18   Consider the circuit given in Figure 8.13 once again, where the two-

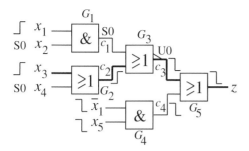

Figure 8.35 High-quality robust test generation

pattern test $\{(0, x, 0, 0, x), (0, x, 1, 0, 0)\}$ was derived for PDF $\uparrow x_3c_2c_3z$. This circuit has been reproduced in Figure 8.35. For line c_4, the above two-pattern test merely requires that the logic value be 0 for the second vector. However, since the on-input of gate G_5 has a controlling-to-non-controlling transition, so should c_4. A two-pattern test that satisfies this condition is shown in Figure 8.35. In addition, to maximize the delay through gate G_5, its hidden node must be pre-initialized to logic 0. Assuming that c_4 is the input of G_5 that feeds the pMOS transistor closest to its output, pre-initialization can be done with the vector $(x, 0, 0, 0, 0)$. Thus, the complete three-pattern test is $\{(x, 0, 0, 0, 0), (0, 0, 0, 0, 1), (1, 0, 1, 0, 0)\}$. □

8.8   Unconventional delay fault testing techniques

In this section, we will discuss two unconventional methods for detecting DFs. The first one is based on output waveform analysis (Franco and McCluskey, 1991) and the second one is based on creating oscillations in the circuit (Arabi *et al.*, 1998).

8.8.1   Output waveform analysis

DFs change the shape of output waveforms by moving the signal transitions in time. Hence, since the output waveforms contain information about circuit delays, DFs can be detected by analyzing them, instead of simply latching the output values at sampling time. We next discuss a method based on post-sampling output waveform analysis in which the output waveforms are observed after sampling time. In the fault-free circuits, the output waveforms should be stable after sampling time, whereas in the faulty circuit there will be transitions which will be caught.

Consider the two-pattern test (P_1, P_2) shown in Figure 8.36, where T_c denotes the system clock period. If an incorrect value is latched at sampling time, then the DF is already detected. When hazards are present at the output, as may be the case for non-robust testing, a correct value may be latched and the conventional method will be

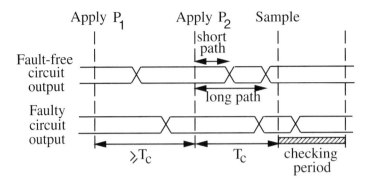

Figure 8.36 Post-sampling waveform analysis

unable to detect the DF. However, if we check for the stability of the output waveform in the *checking period*, then the DF can still be detected if it causes any transitions in this period. Compact stability checkers can be used for this purpose. They can be used in conjunction with both deterministic and random-pattern testing of digital circuits.

The above method can also be extended to pre-sampling waveform analysis, where output waveforms are checked to see if they are stable even before sampling time. This is aimed at detecting GDFs which exceed their specifications, but do not result in an incorrect value at a path output. The reason one may be interested in detecting such faults is that they may portend reliability problems.

8.8.2   Digital oscillation testing

The oscillation testing method consists of sensitizing a path in the circuit under test and incorporating it in a ring oscillator to test for DFs and SAFs. To obtain oscillations, the method makes sure that there are an odd number of inverters in the loop. Faults manifest themselves by causing a deviation from the fault-free oscillation frequency. This concept is illustrated with the following example.

Example 8.19   Consider the circuit in Figure 8.37(a). The path in bold is sensitized and converted to an oscillator by applying the vector $(x_1, x_2, x_3, x_4) = (0, z, 0, 1)$. This vector implies that output z is connected to input x_2. Since there are an odd number of inversions on this path, oscillations are created at the output. By measuring the oscillation frequency, a PDF or GDF on this path for both the rising and falling transitions can be detected. Thus, there is no need for a two-pattern test, nor is there a need to separately test DFs for falling and rising transitions. In addition, both SA0 and SA1 faults on all the lines on the bold path are also detected. Such faults will not cause any oscillation at all. Some other SAFs, such as x_1 SA1 and x_3 SA1, are detected as well. The vector $(1, z, 0, 0)$ can detect DFs and SAFs on path $x_2 c_1 c_3 z$ in the same configuration.

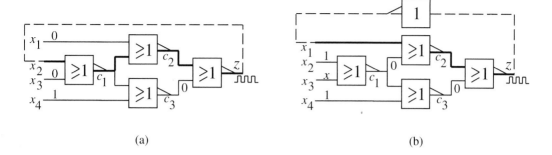

(a)   (b)

Figure 8.37 Oscillation testing

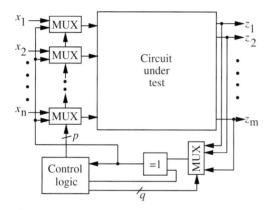

Figure 8.38 Implementation (Arabi *et al.*, 1998) (© 1998 IEEE)

To test the bold path in Figure 8.37(b), we first note that it has an even number of inversions. Thus, in this case, output z is connected to input x_1 via an inverter. The vector that sensitizes this path is $(z, 1, x, 1)$. □

The block diagram of a typical implementation of this method is shown in Figure 8.38. The output multiplexer is used to select the output and establish the feedback loop. Input multiplexers allow input values to be fed directly or through a feedback loop. The XOR gate establishes an inverting (non-inverting) loop when its control input is made 1 (0). The control logic controls the multiplexers and XOR gate, and measures the oscillation frequency. This method can also be used for on-chip testing by letting the control logic feed the required vectors to the circuit in the test mode.

Summary

- DF testing determines if the circuit is operating properly at the designated speed.
- DF models deal with temporal faults in paths, segments or gates for rising or falling transitions.

- DFs in combinational circuits are generally detected by two-pattern tests which consist of an initialization vector followed by a test vector.
- Two types of clocking schemes are used for DF testing: variable clock and rated clock.
- PDFs can be classified in various ways: (a) robustly testable, non-robustly testable, functionally sensitizable and functionally redundant, (b) robust-dependent and non-robust-dependent, (c) singly-testable, multiply-testable and singly-testable-dependent, and (d) primitive single PDF, primitive multiple PDF, and primitive-dependent.
- A large number of untestable SDFs and PDFs can be identified under various sensitization conditions using precomputed static logic implications.
- A commonly used logic system for DF testing consists of the following five values: $S0$, $S1$, $U0$, $U1$ and XX.
- A leaf-dag can be used to establish a relationship between DFs and SAFs.
- Testing for all primitive PDFs guarantees timing correctness.
- The concept of compatible faults can be used to compact a PDF test set.
- GDFs can be categorized under G-GDFs (or TFs) and S-GDFs.
- Testing of G-GDFs is similar to SAF testing.
- S-GDFs are tested along the longest functional path through the fault site so that system timing failures caused by even the smallest delay defects can be detected.
- Exact non-enumerative PDF simulation is possible for most circuits through an efficient data structure called the path status graph.
- S-GDF simulation can be done based on the concept of critical path tracing.
- For at-speed fault simulation, it is also necessary to check the on-path lines for stability.
- PDFs can be diagnosed based on effect–cause analysis.
- Critical path tracing can not only be used to detect GDFs, but to also diagnose them.
- Sequential logic implications can be used to quickly identify a large number of untestable DFs in sequential circuits.
- DFs in sequential circuits can be tested using time-frame expansion.
- PDF simulation of sequential circuits is also possible through the use of the path status graph data structure.
- An underlying assumption behind most of the DF testing methods is that the gate propagation delays are fixed and independent of the input logic values. This is not strictly true and can lead to pitfalls in testing. Methods to remedy this situation are still in their infancy.
- Two unconventional techniques for DF testing are based on waveform analysis and oscillation creation.

Additional reading

A static implication based method to quickly identify non-robustly untestable PDFs in combinational circuits can be found in Li *et al.* (1997).

A test generation method for combinational circuits, which targets PDF coverage under parametric fabrication process variations, is discussed in Sivaraman and Strojwas (1995).

A functional test generation method, which is suitable in the case when a gate-level implementation of a combinational circuit is not available, is given in Pomeranz and Reddy (1995a).

Generation of compact robust two-pattern tests for combinational circuits based on a 23-valued logic system is presented in Bose *et al.* (1993b).

Non-enumerative test generation and PDF coverage estimation methods for combinational circuits can be found in Pomeranz and Reddy (1995b), Kagaris *et al.* (1997), and Heragu *et al.* (1997b).

A method to derive accurate fault coverage for S-GDFs is given in Pramanick and Reddy (1997).

A method to identify the set of PDFs that do not need to be targeted in sequential circuits is given in Krstić *et al.* (1996).

A statistical DF coverage estimation for sequential circuits is given in Pappu *et al.* (1998).

Detection of delay flaws using a very low supply voltage is addressed in Chang and McCluskey (1996). Such flaws cause local timing failures which are not severe enough to cause malfunction at normal operating conditions. However, they indicate reliability problems in the circuit.

Exercises

8.1 The XOR gate implementation shown in Figure 8.4 has six physical paths, hence 12 logical paths. Six of the logical paths are robustly testable, four are validatable non-robustly testable, and two are functionally redundant. Put each of the 12 logical paths in one of these categories, and derive two-pattern tests for all the testable PDFs.

8.2 For the circuit in Figure 8.5, show the following.
   (a) PDFs $x_1c_1z \downarrow$, $x_1c_1z \uparrow$, and $x_2c_1z \uparrow$ are functionally sensitizable.
   (b) PDFs $x_1c_1z \uparrow$ and $x_2c_1z \uparrow$ are also RD.

8.3 In the circuit in Figure 8.4, show that PDF $x_2c_1c_3z \uparrow$ is STD.

8.4 Prove Lemma 8.4.

8.5 Identify all the robustly untestable, non-robustly untestable and functionally unsensitizable PDFs in circuit C_1 shown in Figure 8.39, based on Lemmas 8.1 through 8.9.

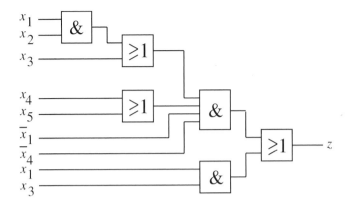

Figure 8.39 Circuit C_1

8.6 Consider PDF $x_3 c_1 c_3 z \uparrow$ in the circuit in Figure 8.16(a). Obtain the corresponding rising-smooth circuit, and use it to obtain the two-pattern test for this PDF.

8.7 Obtain the set of sensitizing 0-cubes for the circuit in Figure 8.17. Identify the corresponding primitive PDFs and obtain two-pattern tests for these PDFs.

8.8 Consider PDF $x_1 c_1 z_2 \uparrow$ in the circuit in Figure 8.18.
(a) Find a maximal set of PDFs which are potentially compatible with the above PDF for robust test generation.
(b) Find a maximal set of PDFs which are potentially compatible with the above PDF for non-robust test generation.

8.9 Derive a two-pattern test for the slow-to-fall S-GDF on line c_1 in the circuit in Figure 8.19. Is this a two-pattern test for the slow-to-fall G-GDF on this line as well?

8.10 Derive the robust and transition two-pattern tests for the SDF $\uparrow c_1 c_2 c_3$ in the circuit in Figure 8.21.

8.11 Use the five-valued logic system to perform non-robust PDF simulation of the circuit in Figure 8.22(a) with the following two two-pattern tests with the help of a PSG: $\{(0, 1), (1, 0)\}$, $\{(0, 0), (0, 1)\}$. Show the status of the PSG at each step.

8.12 Determine which S-GDFs are detected in the circuit in Figure 8.25 by the two-pattern test $\{(0, 0, 0, 0, 0), (1, 1, 0, 0, 1)\}$. Let both the rising and falling propagation delay of an AND/NAND gate be two time units, and correspondingly for an inverter let it be one time unit. Suppose that $T_{obs} = 9$. Find the detection threshold of each detected S-GDF.

8.13 What are the SDFs of length two that are robustly detected by the two-pattern test $\{(1, 1, 1, 1, 0), (1, 1, 0, 1, 1)\}$ in the circuit in Figure 8.26?

8.14 Show which PDFs in the circuit of Figure 8.27 are robustly detected at-speed with the following sequence of vectors: $\{(0, 1, 0, 0), (0, 0, 1, 0), (1, 1, 1, 1)\}$. Assume that the circuit has previously passed a test for PDF $\uparrow x_2 c_1 c_2 z_1$.

8.15 Suppose the following two-pattern test is fed to inputs (x_1, x_2, x_3, x_4) of the

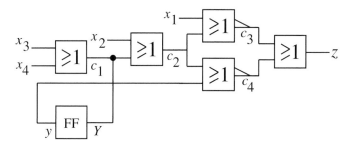

Figure 8.40 Circuit C_2

circuit in Figure 8.28: $\{(0, 0, 1, 1), (1, 1, 0, 0)\}$, and the circuit fails this test.

(a) What is the suspect set of PDFs that can explain this failure?

(b) Suppose that the circuit passes the two-pattern test $\{(0, 0, 1, 0), (0, 0, 0, 0)\}$. How does this reduce the suspect set derived above?

8.16 Identify all the PDFs which are robustly untestable in the circuit in Figure 8.40 by looking at logic implications over two time frames.

8.17 For the sequential circuit in Figure 8.31, derive separate robust test sequences based on the variable clock scheme for the following PDFs: $\downarrow x_1 c_1 Y$, $\uparrow y c_3 z$, and $\downarrow x_2 c_2 z$.

8.18 Suppose the following input sequence is fed to inputs (x_1, x_2) of the circuit in Figure 8.32(a): $\{(1, 1), (1, 0), (1, 1)\}$. Also, suppose that the first and third vectors are fed using a slow clock, whereas the second vector is fed using the rated clock. Find the PDFs robustly detected by this input sequence.

8.19 For PDF $\downarrow x_3 c_2 c_3 z$ in the circuit in Figure 8.35, derive a high-quality robust three-pattern test. Assume that x_4 (c_1) is the input of gate G_2 (G_3) that feeds the pMOS transistor closest to that gate's output.

8.20 Derive a minimal set of test vectors which can detect all DFs and SAFs in the circuit in Figure 8.37 using oscillation testing.

References

Arabi, K., Ihs, H., Dufaza, C. and Kaminska, B. (1998). Digital oscillation-test method for delay and stuck-at fault testing of digital circuits. In *Proc. Int. Test Conference*, pp. 91–100.

Bose, S., Agrawal, P. and Agrawal, V.D. (1993a). The optimistic update theorem for path delay fault testing. *J. of Electronic Testing: Theory & Applications*, **4**, pp. 285–290.

Bose, S., Agrawal, P. and Agrawal, V.D. (1993b). Generation of compact delay tests by multiple path activation. In *Proc. Int. Test Conference*, pp. 714–723.

Bose, S., Agrawal, P. and Agrawal, V.D. (1998a). A rated-clock test method for path delay faults. *IEEE Trans. on Very Large Scale Integration Systems*, **6** (2), pp. 323–342.

Bose, S., Agrawal, P. and Agrawal, V.D. (1998b). Deriving logic systems for path delay test generation. *IEEE Trans. on Computers*, **47** (8), pp. 829–846.

Cavallera, P., Girard, P., Landrault, C. and Pravossoudovitch, S. (1996). DFSIM: a gate-delay fault simulator for sequential circuits. In *Proc. European Design & Test Conf.*, pp. 79–87.

Chakraborty, T.J. and Agrawal, V.D. (1997). On variable clock methods for path delay testing of sequential circuits. *IEEE Trans. on Computer-Aided Design*, **11** (11), pp. 1237–1249.

Chang, J.T.-Y. and McCluskey, E.J. (1996). Detecting delay flaws by very-low-voltage testing. In *Proc. Int. Test Conference*, pp. 367–376.

Chen, L.-C., Gupta, S.K. and Breuer, M.A. (1997). High-quality robust tests for path delay faults. In *Proc. VLSI Test Symposium*, pp. 88–93.

Cheng, K.-T. and Chen, H.-C. (1993). Delay testing for non-robust untestable circuits. In *Proc. Int. Test Conference*, pp. 954–961.

Cheng, K.-T., Devadas, S. and Keutzer, K. (1993). Delay-fault test generation and synthesis for testability under a standard scan design methodology. *IEEE Trans. on Computer-Aided Design*, **12** (8), pp. 1217–1231.

Cheng, K.-T. (1993). Transition fault testing for sequential circuits. *IEEE Trans. on Computer-Aided Design*, **12** (12), pp. 1971–1983.

Cheng, K.-T., Krstić, A. and Chen, H.-C. (1996). Generation of high quality tests for robustly untestable path delay faults. *IEEE Trans. on Computers*, **45** (12), pp. 1379–1392.

Dumas, D., Girard, P., Landrault, C. and Pravossoudovitch, S. (1993). An implicit delay fault simulation method with approximate detection threshold calculation. In *Proc. Int. Test Conference*, pp. 705–713.

Franco, P. and McCluskey, E. (1991). Delay testing of digital circuits by output waveform analysis. In *Proc. Int. Test Conference*, pp. 798–807.

Franco, P. and McCluskey, E. (1994). Three-pattern tests for delay faults. In *Proc. VLSI Test Symposium*, pp. 452–456.

Gharaybeh, M.A., Bushnell, M.L. and Agrawal, V.D. (1995). Classification and test generation for path-delay faults using single stuck-fault tests. In *Proc. Int. Test Conference*, pp. 139–148.

Gharaybeh, M.A., Bushnell, M.L. and Agrawal, V.D. (1998). The path-status graph with application to delay fault simulation. *IEEE Trans. on Computer-Aided Design*, **17** (4), pp. 324–332.

Girard, P., Landrault, C. and Pravossoudovitch, S. (1992). A novel approach to delay-fault diagnosis. In *Proc. Design Automation Conf.*, pp. 357–360.

Heragu, K., Patel, J.H. and Agrawal, V.D. (1996a). Segment delay faults: a new fault model. In *Proc. VLSI Test Symposium*, pp. 32–39.

Heragu, K., Patel, J.H. and Agrawal, V.D. (1996b). SIGMA: a simulator for segment delay faults. In *Proc. Int. Conference on Computer-Aided Design*, pp. 502–508.

Heragu, K., Patel, J.H. and Agrawal, V.D. (1997a). Fast identification of untestable delay faults using implications. In *Proc. Int. Conference on Computer-Aided Design*, pp. 642–647.

Heragu, K., Agrawal, V.D., Bushnell, M.L. and Patel, J.H. (1997b). Improving a nonenumerative method to estimate path delay fault coverage. *IEEE Trans. on Computer-Aided Design*, **16** (7), pp. 759–762.

Heragu, K. (1998). *New Techniques to Verify Timing Correctness of Integrated Circuits*. Ph.D. thesis, Univ. of Illinois at Urbana-Champaign, 1998.

Hsu, Y.-C. and Gupta, S.K. (1996). A simulator for at-speed robust testing of path delay faults in combinational circuits. *IEEE Trans. on Computers*, **45** (11), pp. 1312–1318.

Hsu, Y.-C. and Gupta, S.K. (1998). A new path-oriented effect–cause methodology to diagnose delay failures. In *Proc. Int. Test Conference*, pp. 758–767.

Kagaris, D., Tragoudas, S. and Karayiannis, D. (1997). Improved nonenumerative path-delay fault-coverage estimation based on optimal polynomial-time algorithms. *IEEE Trans. on Computer-Aided Design*, **16** (3), pp. 309–315.

Ke, W. and Menon, P.R. (1995). Synthesis of delay-verifiable combinational circuits. *IEEE Trans. on Computers*, **44** (2), pp. 213–222.

Krstić, A., Chakradhar, S.T. and Cheng, K.-T. (1996). Testable path delay fault cover for sequential circuits. In *Proc. European Design Automation Conference*, pp. 220–226.

Lam, W.K., Saldanha, A., Brayton, R.K. and Sangiovanni-Vincentelli, A.L. (1995). Delay fault coverage, test set size, and performance trade-offs. *IEEE Trans. on Computer-Aided Design*, **14** (1), pp. 32–44.

Li, Z., Min, Y. and Brayton, R.K. (1997). Efficient identification of non-robustly untestable path delay faults. In *Proc. Int. Test Conference*, pp. 992–997.

Lin, C.J. and Reddy, S.M. (1987). On delay fault testing in logic circuits. *IEEE Trans. on Computer-Aided Design*, **CAD-6** (5), pp. 694–703.

Mahlstedt, U. (1993). DELTEST: deterministic test generation for gate delay faults. In *Proc. Int. Test Conference*, pp. 972–980.

Pappu, L., Bushnell, M.L., Agrawal, V.D. and Mandyam-Komar, S. (1998). Statistical delay fault coverage estimation for synchronous sequential circuits. *J. of Electronic Testing: Theory & Applications*, **12** (3), pp. 239–254.

Park, E.S. and Mercer, M.R. (1987). Robust and nonrobust tests for path delay faults in a combinational circuit. In *Proc. Int. Test Conference*, pp. 1027–1034.

Park, E.S. and Mercer, M.R. (1992). An efficient delay test generation system for combinational logic circuits. *IEEE Trans. on Computer-Aided Design*, **11** (7), pp. 926–938.

Parodi, C.S., Agrawal, V.D., Bushnell, M.L. and Wu, S. (1998). A non-enumerative path delay fault simulator for sequential circuits. In *Proc. Int. Test Conference*, pp. 934–943.

Pierzynska, A. and Pilarski, S. (1997). Pitfalls in delay fault testing. *IEEE Trans. on Computer-Aided Design*, **16** (3), pp. 321–329.

Pomeranz, I. and Reddy, S.M. (1992). At-speed delay testing of synchronous sequential circuits. In *Proc. Design Automation Conference*, pp. 177–181.

Pomeranz, I. and Reddy, S.M. (1995a). Functional test generation for delay faults in combinational circuits. In *Proc. Int. Conference on Computer-Aided Design*, pp. 687–694.

Pomeranz, I. and Reddy, S.M. (1995b). NEST: a nonenumerative test generation method for path delay faults in combinational circuits. *IEEE Trans. on Computer-Aided Design*, **14** (12), pp. 1505–1515.

Pramanick, A.K. and Reddy, S.M. (1997). On the fault detection coverage of gate delay fault detecting tests. *IEEE Trans. on Computer-Aided Design*, **16** (1), pp. 78–94.

Saldanha, A., Brayton, R.K. and Sangiovanni-Vincentelli (1992). Equivalence of robust delay-fault and single stuck-fault test generation. In *Proc. Design Automation Conference*, pp. 173–176.

Saxena, J. and Pradhan, D.K. (1993). A method to derive compact test sets for path delay faults in combinational circuits. In *Proc. Int. Test Conference*, pp. 724–733.

Sivaraman, M. and Strojwas, A.J. (1995). Test vector generation for parametric path delay faults. In *Proc. Int. Test Conference*, pp. 132–138.

Sivaraman, M. and Strojwas, A.J. (1996). Delay fault coverage: a realistic metric and an estimation technique for distributed path delay faults. In *Proc. Int. Conference on Computer-Aided Design*, pp. 494–501.

Sivaraman, M. and Strojwas, A.J. (1997). Primitive path delay fault identification. In *Proc. Int. Conference on VLSI Design*, pp. 95–100.

Sparmann, U., Luxenberger, D., Cheng, K.-T. and Reddy, S.M. (1995). Fast identification of robust dependent path delay faults. In *Proc. Design Automation Conference*, pp. 119–125.

Tekumalla, R. and Menon, P.R. (1997). Test generation for primitive path delay faults in combinational circuits. In *Proc. Int. Conference on Computer-Aided Design*, pp. 636–641.

Waicukauski, J.A. and Lindbloom, E. (1986). Transition fault simulation by parallel pattern single fault propagation. In *Proc. Int. Test Conference*, pp. 542–549.

9 CMOS testing

In this chapter, we discuss how CMOS circuits can be tested under various fault models, such as stuck-at, stuck-open and stuck-on. We consider both dynamic and static CMOS circuits. We present test generation techniques based on the gate-level model of CMOS circuits, as well as the switch-level implementation.

Under dynamic CMOS circuits, we look at two popular techniques: domino CMOS and differential cascode voltage switch (DCVS) logic. We consider both single and multiple fault testability of domino CMOS circuits. For DCVS circuits, we also present an error-checker based scheme which facilitates testing.

Under static CMOS circuits, we consider both robust and non-robust test generation. A robust test is one which is not invalidated by arbitrary delays and timing skews. We first show how test invalidation can occur. We then discuss fault collapsing techniques and test generation techniques at the gate level and switch level.

Finally, we show how robustly testable static CMOS designs can be obtained.

9.1 Testing of dynamic CMOS circuits

Dynamic CMOS circuits form an important class of CMOS circuits. A **dynamic CMOS** circuit is distinguished from a *static CMOS* circuit by the fact that each dynamic CMOS gate is fed by a clock which determines whether it operates in the *precharge phase* or the *evaluation phase*. There are two basic types of dynamic CMOS circuits: *domino CMOS* (Krambeck *et al.*, 1982) and DCVS logic (Heller *et al.*, 1984). Many variations of these circuits have also been presented in the literature (Goncalves and de Man, 1983; Friedman and Liu, 1984; Pretorius *et al.*, 1986; Lee and Szeto, 1986; Hwang and Fisher, 1989). However, from a testing point of view, these variations can be treated similarly. We first consider domino CMOS circuits.

9.1.1 Testing of domino CMOS circuits

Each **domino CMOS** gate in a domino CMOS circuit is fed by a clock and each such gate is always followed by an inverter. For example, consider a circuit with only one such gate and inverter, as shown in Figure 9.1. This circuit realizes the function

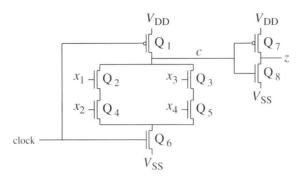

Figure 9.1 A domino CMOS circuit

$z = x_1x_2 + x_3x_4$. During the **precharge phase**, *clock* is made 0. This allows the clocked pMOS transistor Q_1 to charge node c to logic 1, and hence node z discharges to logic 0. During the **evaluation phase**, *clock* is made 1. This makes the clocked nMOS transistor Q_6 conduct and turns off Q_1. Whether node c is then pulled down to logic 0 or remains at logic 1 is determined by which input vector is applied to the circuit. For example, $(x_1, x_2, x_3, x_4) = (1, 1, 0, 0)$ will make $c = 0$, while $(1, 0, 1, 0)$ will let it remain at logic 1.

Every domino CMOS gate has to be followed by an inverter in order to ensure glitch-free operation of the circuit. During the precharge phase, the outputs of all the inverters are 0. So any nMOS transistors that they may feed in other domino CMOS gates are turned off at the beginning of the evaluation phase. This prevents accidental discharging of the precharged node in those gates. In addition, the primary input vectors are only changed during the precharge phase. The above two facts ensure correct operation.

Domino CMOS circuits enjoy area, delay and testability advantages over static CMOS circuits. The area advantage comes from the fact that the pMOS network of a domino CMOS gate consists of only one transistor. This also results in a reduction in the capacitive load at the output node, which is the basis for the delay advantage. The testability advantage will become obvious as we go through this section. The disadvantage of domino CMOS over static CMOS comes from the fact that it requires more power, since every domino CMOS gate needs to be precharged in every cycle, even if its output is to continue to be at logic 0. Also, because of the need to feed all domino gates with the clock, one needs to be much more careful in making sure that the circuit works properly. In addition, a gate-level circuit which has an inverter somewhere in the middle, not just at the primary inputs, cannot be directly implemented in domino CMOS. On the other hand, for any given Boolean function, it is always possible to obtain gate-level circuits which have inverters only at the primary inputs. Any such circuit can be directly implemented in domino CMOS.

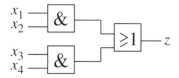

Figure 9.2 The gate-level model of the circuit in Figure 9.1

9.1.1.1 The gate-level model

In order to test a domino CMOS circuit, it is useful to derive its gate-level model first. This allows efficient gate-level test generation techniques to become applicable to it. A **gate-level model** for a domino CMOS gate is obtained from its nMOS network simply by replacing a parallel connection of transistors by an OR gate and a series connection by an AND gate. For example, the gate-level model for the domino CMOS gate and inverter given in Figure 9.1 is shown in Figure 9.2. The inversion introduced by the domino CMOS gate is canceled by the inverter, which is the reason no inverter appears at the output of this model. The clocked transistors need not be explicitly modeled at the gate level for testing purposes. For a more general domino CMOS circuit, similar gate-level models can be obtained for each domino CMOS gate/inverter pair and these gate-level models can be interconnected appropriately. Such a technique was first presented in Reddy *et al.* (1984a) for static CMOS circuits. However, it is also directly applicable to domino CMOS.

9.1.1.2 Test generation

We will see next how single stuck-open faults (SOpFs), stuck-on faults (SOnFs) and stuck-at faults (SAFs) in a domino CMOS circuit can be tested by simply deriving a test set for all single SAFs in its gate-level model, and appropriately arranging the sequence in which these vectors are applied. Single fault detection in domino CMOS circuits was considered in Oklobdzija and Kovijanic (1984), Manthani and Reddy (1984), Jha and Abraham (1984) and Barzilai *et al.* (1984).

Let us first look at a single domino CMOS gate, such as the one shown in Figure 9.1. If there is a stuck-at 0 (SA0) fault at the input of any transistor in the nMOS network, it will result in that transistor becoming stuck-open. Similarly, a stuck-at 1 (SA1) fault will lead to a stuck-on transistor. Therefore, if we were to derive a test set for all single SAFs from the gate-level model, it would detect all single SAFs, SOpFs and SOnFs in the nMOS network in the evaluation phase. One such test set is given in Table 9.1. From Chapter 4, we already know how to derive the SAF test set for any combinational gate-level circuit. Next, let us look at SOpFs in the clocked and inverter transistors. If the clocked nMOS transistor Q_6 is stuck-open, the equivalent fault will be an SA1 at node c or an SA0 at node z. Similarly, an SOpF in the inverter pMOS transistor Q_7 will result in an SA0 fault at z. Thus, the above two faults are obviously tested by the

Table 9.1. *Test set derived from the gate-level model of Figure 9.2*

x_1	x_2	x_3	x_4	z
1	1	0	0	1
1	0	1	0	0
0	0	1	1	1
0	1	0	1	0

test set in Table 9.1, specifically by vectors (1, 1, 0, 0) and (0, 0, 1, 1). An SOpF in the inverter nMOS transistor Q_8, however, requires a **two-pattern test**, one that will result in a 1 → 0 transition at z. Both {(1, 1, 0, 0), (1, 0, 1, 0)} and {(0, 0, 1, 1), (0, 1, 0, 1)} serve this purpose. These two-pattern tests also detect the SOpF in the clocked pMOS transistor Q_1.

Consider now the SOnFs in the clocked and inverter transistors. If Q_1 is stuck-on and a vector is applied which results in a conduction path through the nMOS network, such as the vector (1, 1, 0, 0) or (0, 0, 1, 1), then in the evaluation phase there will be a path activated from V_{DD} to V_{SS}. The resultant drastic increase in current can be detected through the current monitoring technique. If only logic monitoring is done, then this fault can only be detected if the resultant logic value at node c is 1. Similarly, an SOnF in the inverter transistor Q_7 (Q_8) can be detected through current monitoring when a vector is applied which normally makes $z = 0$ ($z = 1$). Detection of an SOnF in the clocked nMOS transistor Q_6 is not that straightforward. Suppose in some evaluation phase, a vector is applied which activates a conduction path through the nMOS network, such as the vector (1, 1, 0, 0) or (0, 0, 1, 1). Then when the clock becomes 0 to start the precharge phase, there is a path between V_{DD} to V_{SS}. If the vector is kept fixed for enough time to allow an appreciable increase in the current drawn by the circuit, then the fault is detected. However, a note of caution regarding such faults is that they may become undetectable if the domino CMOS gate is embedded in a domino CMOS circuit (i.e., when it is not directly fed by primary inputs). This is owing to the fact that the inputs of an embedded domino CMOS gate start becoming 0 as soon as the precharge phase starts. Thus, a conduction path from V_{DD} to V_{SS} may not be activated for enough time for current monitoring to be successful.

When current monitoring is not possible, the notion of n-dominance and p-dominance may be useful. A CMOS gate is called n-**dominant** (p-**dominant**) if its output becomes 0 (1) when both its nMOS and pMOS networks conduct due to a fault (Oklobdzija and Kovijanic, 1984). In order to ensure n-dominance or p-dominance, we need to adjust the resistances of the two networks accordingly. For example, to make a domino CMOS gate p-dominant, we can decrease the resistance of the clocked pMOS transistor by increasing its width slightly. Similarly, the inverter can be made n-dominant by increasing the width of its nMOS transistor. In a p-dominant domino

CMOS gate, an SOnF in the clocked pMOS transistor will be detectable through logic monitoring, but an SOnF in the clocked nMOS transistor will not be, even when this gate is not embedded. Similarly, if the inverter is n-dominant, then an SOnF in its nMOS transistor will be detectable, but not in its pMOS transistor.

An SAF at node c or z is obviously detectable. An SAF at a branch of the clock line results in an SOpF or SOnF in the corresponding clocked transistor, which has already been covered above. Thus, the only remaining fault that we need to consider is an SAF at the stem of the clock line. An SA0 fault on this line makes the clocked nMOS (pMOS) transistor stuck-open (stuck-on). This results in an SA1 (SA0) behavior at node c (z), which is detectable. An SA1 fault on the clock stem makes the clocked nMOS (pMOS) transistor stuck-on (stuck-open). This is detected by the two-pattern test which causes a $1 \rightarrow 0$ transition at z (or equivalently, a $0 \rightarrow 1$ transition at c).

The key point we need to remember from all the above discussions is that a test set derived for all single SAFs in the gate-level model will also detect all detectable single SOpFs, SOnFs and SAFs in the domino CMOS gate and inverter, as long as the test vectors are arranged to allow at least one $1 \rightarrow 0$ transition at the output. Generalizing this method to arbitrary domino CMOS circuits is quite straightforward, as we will see next.

Consider the circuit shown in Figure 9.3. Its gate-level model is shown in Figure 9.4. As before, we can first derive a test set for all single SAFs in the gate-level model. The vectors in this test set should be arranged appropriately since we know that each of the three internal inverter nMOS transistors would require a two-pattern test. This two-pattern test should satisfy two conditions:

- It should create a $1 \rightarrow 0$ transition at the corresponding node, c_3, c_4 or z.
- The test vector (i.e., the second vector) of the two-pattern test should sensitize a path from the inverter output to the circuit output.

One way of obtaining the above two-pattern tests would be to ensure that for each inverter output, a vector which tests for an SA0 fault on it is followed by a vector which detects an SA1 fault on it. We must not forget the inverter at the primary input which generates $\bar{x}_1$ from x_1. This would require two two-pattern tests, one for each of its transistors. One of these two-pattern tests would create a $1 \rightarrow 0$ transition at $\bar{x}_1$ and the other one a $0 \rightarrow 1$ transition, and the test vectors of both the two-pattern tests would sensitize a path from $\bar{x}_1$ to circuit output z. A test set with vectors ordered to satisfy all these conditions is given in Table 9.2. The first and second vectors form the required two-pattern test for both c_3 and z, and the third and fourth vectors form the required two-pattern test for c_4. The two two-pattern tests for the primary input inverter are formed by the first–second and second–third vectors respectively. In the worst case, some vectors may have to be repeated to satisfy all the required conditions, although for this example this was not necessary.

As before, we can see that the test set derived in the above fashion detects all detectable single SOpFs and SOnFs, as well as SAFs affecting nMOS networks or

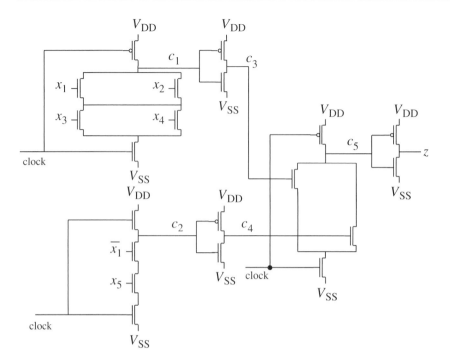

Figure 9.3 A domino CMOS circuit

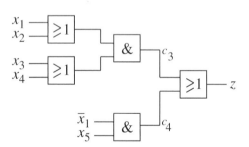

Figure 9.4 A gate-level model of the circuit in Figure 9.3

nodes such as c_i or z. The only new fault we need to consider is an SAF on the clock stem or one of its branches which affects more than one domino CMOS gate. For example, if the main clock stem has an SA0 fault it will make all the three clocked nMOS transistors stuck-open and the three clocked pMOS transistors stuck-on. This will result in a unidirectional SA0 fault on nodes c_3, c_4 and z. Is such a fault guaranteed to be detected by the test set we have derived? The answer is, fortunately, yes. This unidirectional fault will be detected by any vector which tests for a single SA0 fault on either c_3 or c_4 or z. The reason is that there are no inversions inside the gate-level model (the only inversions can be at the primary inputs). Consider the test for an SA0 fault on c_3. This test should normally result in $c_3 = 1$, and also $z = 1$, since there

Table 9.2. *Test set derived from the gate-level model of Figure 9.4*

x_1	x_2	x_3	x_4	x_5	c_3	c_4	z
0	1	0	1	0	1	0	1
1	1	0	0	1	0	0	0
0	0	0	0	1	0	1	1
0	0	1	0	0	0	0	0
1	0	1	0	0	1	0	1

are no inversions from c_3 to z. The presence of the SA0 fault on c_3 will make both these nodes 0. If, in addition, SA0 faults are also present on c_4 or z or both, they can only further aid in testing, since they cannot cause the correct value of 1 to appear on z.

Next, consider an SA1 fault on the clock stem or one of its branches which affects more than one domino CMOS gate. This would result in an SOpF (SOnF) in each of the affected clocked pMOS (nMOS) transistors. We know that such a double fault within a single domino CMOS gate is detected by the standard two-pattern test we derived for each such gate. However, when such a double fault is present in more than one domino CMOS gate, would the two-pattern test derived for any of these gates detect this multiple fault also? The answer again is yes (this is left as an exercise to the reader, see Exercise 9.1).

The above discussions imply that to test a general domino CMOS circuit for all detectable single SOpFs, SOnFs and SAFs, all we need to do is first derive an SAF test set from its gate-level model, and then order the tests so that the required two-pattern tests for all the inverters in the circuit are present. Current monitoring can be employed to detect some of the faults not detectable through logic monitoring. A test set derived for a domino CMOS circuit is inherently **robust** for detecting SOpFs (the word "robust" was first coined in connection with static CMOS testing (Reddy *et al.*, 1984b)). This means that timing skews among the primary inputs or arbitrary delays in the circuit cannot invalidate the tests. The reason is that the primary input vector is only changed during the precharge phase, and a transistor, which is not fed by a primary input, in any nMOS network is non-conducting at the beginning of the evaluation phase. Thus, accidental conduction of a path in parallel to the path containing the transistor with an SOpF, which would result in test invalidation, is not possible. A more detailed discussion of robust tests is given in Section 9.2.1 ahead.

9.1.1.3 Multiple fault testability

We will show next that domino CMOS circuits can be relatively easily synthesized to guarantee the detection of most of their multiple faults whose constituent faults consist of SOpFs, SOnFs and SAFs. We will have to combine results from different areas of testing to demonstrate this, and thus a bit of a digression is required here.

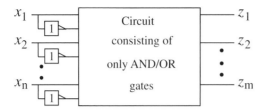

Figure 9.5 An algebraically factored multi-level circuit

In Dandapani and Reddy (1974), Hachtel *et al.* (1992), and Devadas *et al.* (1993), logic synthesis methods are given to ensure detection of all multiple SAFs in a multi-level logic circuit. Suppose the starting point of synthesis is a multi-output two-level circuit, in which an SAF on any given line is testable at each output that line feeds (such circuits are sometimes called **single-output minimized** two-level circuits). Then if a logic synthesis method called *algebraic factorization* (without the use of the complement) is applied to the two-level circuit, the resultant multi-level circuit can be tested for all multiple SAFs using a test set which detects all single SAFs in the two-level circuit (Hachtel *et al.*, 1992). Note that the two-level test set should be derived such that it tests for each SAF at each output that line feeds, or else the method given in Jacob and Agrawal (1992) could also be used (see Chapter 13; Section 13.1.1). A smaller test set with the same properties can also be derived (Devadas *et al.*, 1993).

Suppose a function z is expressed as $qg + p$. If q and g have no inputs in common, then both q and g are said to be **algebraic divisors** of z, where p is the remainder (Brayton *et al.*, 1987). For example, $g = x_1 x_2 + x_3$ is an algebraic divisor of $z = x_1 x_2 x_4 + x_3 x_4 + x_5$. If we express z as $gx_4 + x_5$ in this example, g is said to be **algebraically resubstituted** in z. By identifying algebraic divisors common to two or more expressions and resubstituting them, one can convert a two-level circuit into a multi-level circuit. This process is what is referred to as **algebraic factorization**. If the complement of the algebraic divisor is not used in this factorization, it is said to be algebraic factorization without the use of the complement. This kind of factorization leads to a multi-level circuit of the form shown in Figure 9.5.

At this point, we realize that a gate-level model of any domino circuit would be exactly of the above form. Consider any such domino CMOS circuit whose gate-level model corresponds to the algebraically factored multi-level circuit. From previous discussions, it is clear that if there is a multiple fault consisting of SAFs, SOpFs and SOnFs affecting only the various nMOS networks in the domino CMOS circuit then an equivalent multiple SAF can be found in the gate-level model. Therefore, any such fault will be tested by the test set mentioned above. However, multiple faults in domino CMOS circuits can also affect the other parts in these circuits. This is discussed next.

In Jha (1988), it was shown that any multiple fault consisting of SOpFs and SOnFs (not necessarily limited to the nMOS networks), which are individually detectable, can be detected by a test set in which a set of initialization vectors (called the *initialization set*) is followed by a set of vectors which detects all multiple SAFs in the gate-level model. This result can be shown to be extensible to multiple faults which have SAFs as well as SOpFs and SOnFs as constituents (the SAFs can even be on the clock lines). An assumption behind this result is that when both the nMOS and pMOS networks of a gate become non-conducting through faults, then any charge at the output node is eventually lost through stray capacitances. The **initialization set** is a set of vectors that can activate every conduction path in every nMOS logic network in the domino CMOS circuit. Consider the circuit shown in Figure 9.3 once again. There are four conduction paths in the nMOS network of the domino CMOS gate with output c_1: x_1x_3, x_1x_4, x_2x_3, and x_2x_4. Similarly, there are one and two conduction paths, respectively, in the gates with outputs c_2 and c_5. All these paths can be activated by the initialization set $\{(1, 1, 1, 1, x), (0, x, x, x, 1)\}$, where x denotes a *don't care*. If both a primary input and its complement are used, different vectors always exist in the initialization set which will provide both a 0 and a 1 to it. The application of the initialization set ensures that the mapping of multiple faults in the domino CMOS circuit to multiple SAFs in the gate-level model is valid for the duration when the multiple SAF test set is applied next. If this test set is very large, then in order to ensure the continued validity of the above mapping, it can be broken up into smaller parts and the initialization set applied before each such part.

For our running example, we can check that the gate-level model given in Figure 9.4 can be derived through algebraic factorization from the two-level AND–OR circuit which implements $z = x_1x_3 + x_1x_4 + x_2x_3 + x_2x_4 + \bar{x}_1x_5$. A test set which detects all single SAFs in this AND–OR circuit is $\{(1, 0, 1, 0, 0), (1, 1, 0, 0, 1), (0, 0, 1, 1, 0), (0, 0, 0, 0, 1), (0, 1, 0, 1, 0), (1, 0, 0, 1, 0), (0, 1, 1, 0, 0)\}$. Thus, if the two vectors from the initialization set, such as $\{(1, 1, 1, 1, 1), (0, 1, 1, 1, 1)\}$, precede the above test set, then the corresponding set of nine vectors will detect all the aforementioned multiple faults in the domino CMOS circuit shown in Figure 9.3. The sequence, in which vectors within the initialization set or vectors within the test set are applied, is no longer relevant.

9.1.2   Testing of DCVS circuits

The **DCVS** logic circuit implementation technique is an extension of the domino CMOS technique. However, the DCVS technique can be used to implement both inverting and non-inverting functions. A DCVS circuit can potentially be faster and more area-efficient compared to a static CMOS circuit. DCVS testing has been considered in Montoye (1985), Barzilai *et al.* (1985), and Kanopoulos and Vasanthavada (1990).

Consider the DCVS gate in Figure 9.6 which implements the carry function of a full adder: $z = x_1x_2 + x_1x_3 + x_2x_3$. Each DCVS gate always produces the desired function and its complement. Just like domino CMOS, the circuit has a precharge phase and an

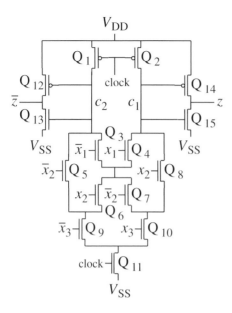

Figure 9.6 A DCVS circuit for the carry function

evaluation phase depending on whether the clock is low or high. Because of sharing of transistors between the nMOS logic required for implementing the function and its complement, usually the nMOS network of a DCVS gate is non-series-parallel in nature. This makes it inefficient to do testing from a gate-level model. It is better to do test generation directly from the switch-level description. For the DCVS gate in Figure 9.6, one possible test set is given in Table 9.3. The first vector $(1, 0, 0)$ detects single SOpFs in transistors Q_5, Q_9, Q_{11} and Q_{12} at output $\bar{z}$, single SOnFs in transistors Q_6 and Q_{10} at output z through logic monitoring, and SOnFs in transistors Q_1, Q_{13} and Q_{14} through current monitoring (c.m.). Similarly, for other vectors.

In general, the set of vectors should be such that every transistor in the nMOS network should have at least one conduction path activated through it, in order to detect all SOpFs. Additional checking may be necessary to ascertain if all the SOnFs are also detected. When an SOpF is detected, the two outputs become $(0, 0)$, and when an SOnF is detected through logic monitoring, the two outputs become $(1, 1)$. A two-pattern test is not required for testing the SOpFs in the nMOS network transistors or the two inverter pMOS transistors because the precharge phase automatically provides initialization for this purpose. However, SOpFs in the two inverter nMOS transistors do require two-pattern tests. To detect these faults, all we need to do is to provide a $1 \rightarrow 0$ transition at both z and $\bar{z}$, or in other words, both $1 \rightarrow 0$ and $0 \rightarrow 1$ transitions at z. By properly arranging the sequence of vectors, as in Table 9.3, this is easily accomplished. These two-pattern tests also detect the SOpFs in the two clocked pMOS transistors. As in the case of domino CMOS, an SOnF in the clocked

Table 9.3. *Test set for the DCVS gate*

Test				Stuck-open	at	Stuck-on	at	Stuck-on (c.m.)
x_1	x_2	x_3	z					
1	0	0	0	Q_5,Q_9,Q_{11},Q_{12}	$\bar{z}$	Q_6,Q_{10}	z	Q_1,Q_{13},Q_{14}
0	1	1	1	Q_8,Q_{10},Q_{11},Q_{14}	z	Q_7,Q_9	$\bar{z}$	Q_2,Q_{12},Q_{15}
0	0	1	0	$Q_3,Q_7,Q_{10},Q_{11},Q_{12}$	$\bar{z}$	Q_4,Q_8	z	Q_1,Q_{13},Q_{14}
1	1	0	1	$Q_4,Q_6,Q_9,Q_{11},Q_{14}$	z	Q_3,Q_5	$\bar{z}$	Q_2,Q_{12},Q_{15}

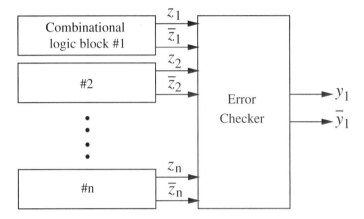

Figure 9.7 Error checking scheme for DCVS circuits

nMOS transistor can be detected if the DCVS gate gets fed directly from the primary inputs, but may be undetectable if the gate is embedded in a larger DCVS circuit. The arguments for detecting SAFs are also similar to those used for domino CMOS circuits.

Let us now suppose that a DCVS gate with a single fault is embedded in a larger DCVS circuit, and that this fault requires logic monitoring. When the fault is activated, it will result in a $(0, 0)$ or $(1, 1)$ at the output of the faulty DCVS gate. When $(0, 0)$ gets fed to the next DCVS gate and is sensitized to its two outputs, it can again only result in a $(0, 0)$ at the corresponding inverter outputs. The reason is that both the precharged nodes will not be able to find a path to ground and will remain at 1. Similarly, a $(1, 1)$, when sensitized, can only propagate as $(1, 1)$ at the inverter outputs of the next DCVS gate.

The above arguments suggest a straightforward method for introducing testability into DCVS circuits. Consider the scheme shown in Figure 9.7 (Kanopoulos and Vasanthavada, 1990). The n combinational logic blocks can each have many DCVS gates in them, and may also share DCVS gates among themselves. The **error checker** can be a simple DCVS XOR gate, as shown in Figure 9.8 for the case $n = 3$. Every time the value of n is increased by 1, we just need to add four more transistors to the

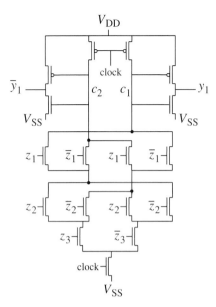

Figure 9.8 Error checker for $n = 3$

nMOS network of the DCVS XOR gate. When the value of n is large, we can use a tree of DCVS XOR gates as our error checker. A very helpful property that a DCVS XOR checker has is that if any one or more of its $(z_i, \bar{z}_i)$ inputs have $(0, 0)$ $((1, 1))$ then the checker outputs $(y_1, \bar{y}_1)$ become $(0, 0)$ $((1, 1))$ (see Exercise 9.2). Therefore, the errors from the combinational logic blocks can trivially propagate through the error checker. It is very easy to detect all single faults in the error checker itself (see Exercise 9.3). The checker usually results in a very small area overhead. In addition, it can be used to do both off-line and on-line testing. In order to test single faults in the combinational logic blocks, which may have many DCVS gates, pseudorandom test sets have been found to be quite effective (Barzilai *et al.*, 1985). This error-checking scheme can be trivially extended to sequential DCVS circuits as well, where the checker can check both the primary outputs and the next state outputs.

9.2 Testing of static CMOS circuits

In this section, we will discuss the test invalidation problem with static CMOS circuits, and both non-robust and robust test generation.

9.2.1 The test invalidation problem

A test for a static CMOS circuit can get invalidated, if not properly derived, because of two reasons: (a) primary input timing skews and arbitrary circuit delays, and (b) charge sharing. We discuss these next.

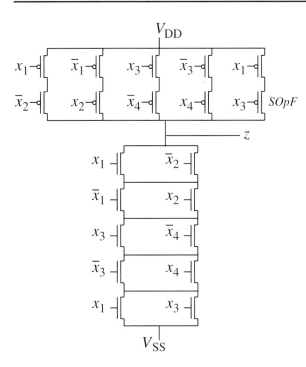

Figure 9.9 An AND–OR CMOS complex gate

9.2.1.1 Test invalidation due to timing skews and arbitrary delays

A two-pattern test derived for an SOpF may get invalidated due to different arrival times of logic values at the primary inputs and arbitrary delays through the circuit (Jain and Agrawal, 1983; Reddy *et al.*, 1983). A test which cannot be invalidated is called robust (Reddy *et al.*, 1984b).

Consider the CMOS complex gate shown in Figure 9.9 which implements the function $z = \bar{x}_1 x_2 + x_1 \bar{x}_2 + \bar{x}_3 x_4 + x_3 \bar{x}_4 + \bar{x}_1 \bar{x}_3$. It is called an **AND–OR complex gate** since it is based on a sum-of-products expression. Suppose the pMOS transistor fed by x_3 has an SOpF, as indicated. The initialization vector to test this fault needs to make $z = 0$. There are only three input vectors which can do this: $(1, 1, 0, 0)$, $(0, 0, 1, 1)$, $(1, 1, 1, 1)$. The test vector should try to activate a conduction path through the faulty transistor without activating any parallel conduction path in the pMOS network. The only candidate for this purpose is $(0, 0, 0, 0)$. Thus, we have three possible two-pattern tests: $\{(1, 1, 0, 0), (0, 0, 0, 0)\}$, $\{(0, 0, 1, 1), (0, 0, 0, 0)\}$, $\{(1, 1, 1, 1), (0, 0, 0, 0)\}$. For the first two-pattern test, depending on input timing skews, $(1, 0, 0, 0)$ or $(0, 1, 0, 0)$ could occur as intermediate vectors. Both these vectors would activate a conduction path in parallel to the one containing the faulty transistor, thus invalidating the two-pattern test. Similarly, the other two-pattern tests can also get invalidated. Therefore, for this SOpF, a robust test does not exist, and we

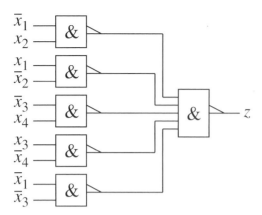

Figure 9.10 A NAND–NAND implementation

say that the complex gate is not fully robustly testable. This problem does not go away if we derive a NAND–NAND static CMOS implementation from the above function (a gate-level model for such an implementation is shown in Figure 9.10).

An SOpF in the pMOS transistor of the output NAND gate which is fed by the NAND gate realizing $\bar{x}_1\bar{x}_3$ is not robustly testable for the same reason as above.

Sometimes, even when a robust two-pattern test exists for an SOpF, another two-pattern test may be carelessly derived which can get invalidated. For example, in the AND–OR complex gate shown in Figure 9.9, $\{(1, 1, 0, 0), (0, 1, 1, 1)\}$ is a two-pattern test for an SOpF in either transistor in the pMOS network conduction path $x_1\bar{x}_2$, which can get invalidated. However, $\{(1, 1, 1, 1), (0, 1, 1, 1)\}$ is another two-pattern test for the same SOpFs, which is robust.

9.2.1.2 Test invalidation due to charge sharing

Test set invalidation can also occur due to the sharing of charge between the output node and the internal nodes in a CMOS gate (Reddy and Reddy, 1986). Consider the complex gate shown in Figure 9.11. For testing the SOpF in transistor Q_{15}, one possible two-pattern test is $\{(1, 0, 0, 1), (0, 0, 0, 1)\}$. The initialization vector $(1, 0, 0, 1)$ charges nodes 3, 4, 5, 6, 9 and 10 to logic 1. The charge at the other nodes is not determined by this vector, but depends on the preceding vector. Let the preceding vector be $(0, 1, 0, 1)$ (this could be the test vector of another two-pattern test applied previous to this one). This vector discharges nodes 2, 7, 8, 11 and 12 to logic 0. Since the initialization vector $(1, 0, 0, 1)$ does not affect these nodes, their logic 0 value is maintained. Next, when the test vector $(0, 0, 0, 1)$ is applied in the presence of the SOpF in Q_{15}, there is no conduction path from node 9 to V_{SS}. However, since transistors Q_2, Q_3, Q_{11} and Q_{12} now conduct, node 9 will have to share its charge with nodes 1, 2, 7 and 8. This depletion of the charge at node 9 may be sufficient so that its logic value is no longer recognized as logic 1, but becomes logic 0. This is the

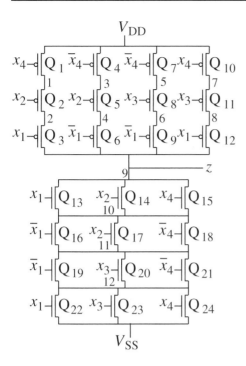

Figure 9.11   An AND–OR CMOS complex gate

logic value we would expect in the fault-free case. Thus, charge sharing has invalidated the two-pattern test.

In reality, test invalidation due to charge sharing is not a major problem because

- the capacitance associated with the output node of the CMOS gate is usually much larger compared to the other intermediate nodes, and
- most CMOS gates used in practice are not large enough for the possibility of extensive charge sharing leading to test invalidation to arise.

9.2.2   Fault collapsing

In order to reduce the test generation time, one can resort to *fault collapsing*, as explained in Chapter 3 (Section 3.4.1). The **checkpoints** for an arbitrary irredundant static CMOS combinational circuit for the SOpF model are as follows (Chandramouli, 1983; Shih and Abraham, 1986a):

- all transistors associated with the primary inputs which do not fan out, and
- all transistors associated with the fanout branches.

In other words, if one tests for all SOpFs at the above checkpoints, one also detects all other SOpFs in the circuit, under a given condition. This condition requires that the initialization vector of each two-pattern test derived for the checkpoint transistors also

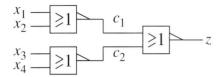

Figure 9.12 A NOR–NOR implementation

has its effects propagated from the fault site to the primary outputs. Another restriction is that the above checkpoints are only applicable to non-robust testing. This means that even if all the SOpFs in the checkpoint transistors are detected robustly by a test set, it is not guaranteed to detect all other SOpFs in the circuit robustly.

Consider a NOR–NOR static CMOS implementation whose gate-level model is shown in Figure 9.12. Here, none of the primary inputs fan out. Hence, they form the checkpoints. Each input feeds two transistors: one nMOS and one pMOS. Therefore, we need to derive a test set which detects SOpFs in the eight transistors fed by the checkpoints. To show the importance of the condition which requires that the effects of the initialization vector of each two-pattern test also be propagated to the primary output, consider the following test set, where the vectors are applied in the given sequence: $\{(1, 0, 0, 0),\ (0, 0, 1, 0),\ (0, 1, 1, 0),\ (0, 0, 0, 0),\ (1, 0, 1, 0),$ $(0, 0, 1, 0),\ (1, 0, 0, 0),\ (1, 0, 0, 1)\}$. Even though this test set detects all SOpFs in the eight checkpoint transistors, it does not detect the SOpF in the nMOS transistor fed by c_2. The reason is that the above condition is not met by this test set.

Let us consider another test set for the NOR–NOR circuit: $\{(0, 0, 1, 0),\ (1, 0, 1, 0),$ $(0, 0, 0, 1),\ (0, 1, 0, 1),\ (1, 0, 0, 0),\ (1, 0, 1, 0),\ (0, 1, 0, 0),\ (0, 1, 0, 1)\}$. This test set not only satisfies the above condition, but also tests the SOpFs in all the checkpoint transistors robustly. However, it still does not robustly detect the SOpFs in the nMOS transistors fed by c_1 and c_2 (it does obviously detect these faults non-robustly).

Another important result that has been shown in Chiang and Vranesic (1983) is that the test vector of a two-pattern test for an SOpF in an nMOS (pMOS) transistor of a fully complementary CMOS gate also detects an SOnF in the pMOS (nMOS) transistor fed by the other branch of the same input, assuming current monitoring is done. This can be seen from the fact that when this input changes values, then in the fault-free circuit, the output changes values. Thus, in the presence of the SOnF, both the networks will conduct, resulting in a very large increase in the current drawn by the circuit. This result implies that SOnFs need not be specifically targeted for test generation, if SOpF testing is already being done.

9.2.3   Test generation from the gate-level model

Since robust testing is more complex than non-robust testing, we will first concentrate on the latter. Non-robust testing can be done both from the gate-level model of the

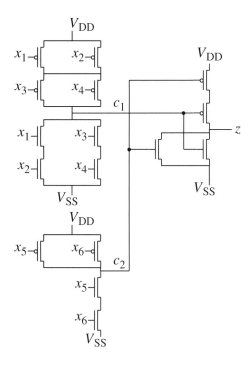

Figure 9.13   A static CMOS circuit

CMOS circuit and its switch-level model. This section deals with test generation from the gate-level model (Chandramouli, 1983; Jain and Agrawal, 1983; Reddy *et al.*, 1984a).

The basic idea behind the methods discussed here is to relate SOpF testing of CMOS circuits to SAF testing in the gate-level model. Since SAF test generation for combinational gate-level circuits is already a mature area, the additional complexity for SOpF test generation done this way is not much. The gate-level model of a static CMOS circuit can be derived from either the nMOS or the pMOS network of its constituent CMOS gates. We will assume that the nMOS network has been used for this purpose. The method for deriving the model is the same as that used for domino CMOS circuits earlier.

It is best to illustrate the SOpF test generation approach through an example first.

Example 9.1   Consider the static CMOS circuit given in Figure 9.13 and its corresponding gate-level model given in Figure 9.14. This circuit has six checkpoints (the six primary inputs), and hence twelve checkpoint transistors whose SOpFs need to be detected in order to detect all SOpFs in the circuit. Suppose we want to derive a two-pattern test for the SOpF in the nMOS transistor fed by input x_1. Since an SA0 fault at the immediate gate input of an nMOS transistor would equivalently make the transistor stuck-open, we map this fault to an SA0 fault at input x_1 in the gate-level

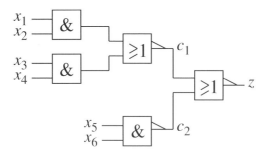

Figure 9.14 The gate-level model

model. Using any of the methods we have learned about in Chapter 4, we can derive a test for this SAF in the gate-level model. One such test is $(1, 1, 0, 1, 1, 1)$. This vector can now be used as the test vector of the two-pattern test for the SOpF in question. The initialization vector has to make $c_1 = 1$ and its effect has to be propagated to circuit output z. This can be trivially accomplished by the vector which detects an SA1 fault at x_1, such as $(0, 1, 0, 1, 1, 1)$. Thus, $\{(0, 1, 0, 1, 1, 1), (1, 1, 0, 1, 1, 1)\}$ forms a two-pattern test for the above SOpF. Similarly, in a two-pattern test for detecting the SOpF in the pMOS transistor fed by x_1, the test (initialization) vector should detect the SA1 (SA0) fault at x_1 in the gate-level model. Thus, $\{(1, 1, 0, 1, 1, 1), (0, 1, 0, 1, 1, 1)\}$ forms a two-pattern test for this SOpF. The above two two-pattern tests can actually be merged into a three-pattern test: $\{(1, 1, 0, 1, 1, 1), (0, 1, 0, 1, 1, 1), (1, 1, 0, 1, 1, 1)\}$. Such three-pattern tests can be derived for the SOpFs in the transistors fed by the other five checkpoint lines as well. □

Instead of deriving three-pattern tests, as above, for every checkpoint, one can do even better. For example, when the SOpF in the nMOS transistor fed by x_1 is detected, the SOpF in the nMOS transistor fed by x_2 is also automatically detected. Thus, for checkpoint x_2, only a two-pattern test to detect the SOpF in the pMOS transistor fed by it is needed. One such two-pattern test is $\{(1, 1, 0, 1, 1, 1), (1, 0, 0, 1, 1, 1)\}$. This two-pattern test can be merged with the three-pattern test derived above for checkpoint x_1 to give the following four-pattern test: $\{(1, 1, 0, 1, 1, 1), (0, 1, 0, 1, 1, 1), (1, 1, 0, 1, 1, 1), (1, 0, 0, 1, 1, 1)\}$. This gives us an idea as to how the above method can be generalized.

Consider a general static CMOS circuit and its checkpoints. Suppose for the line corresponding to checkpoint i in the gate-level model, the SA0 (SA1) test is t_i^0 (t_i^1). If checkpoint i is an input to an AND or NAND gate, and checkpoints $k_1, k_2, \ldots, k_r$ are also inputs of this gate, then derive the following test sequence for this gate: $\{t_i^0, t_i^1, t_i^0, t_{k_1}^1, t_i^0, t_{k_2}^1, \ldots, t_i^0, t_{k_r}^1\}$. Otherwise, if checkpoint i is an input to an OR or NOR gate, and checkpoints $k_1, k_2, \ldots, k_r$ are also inputs of this gate, then derive the following test sequence for this gate: $\{t_i^1, t_i^0, t_i^1, t_{k_1}^0, t_i^1, t_{k_2}^0, \ldots, t_i^1, t_{k_r}^0\}$. If no other checkpoint feeds the gate fed by checkpoint i, then the test sequence just consists of

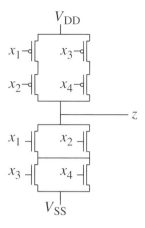

Figure 9.15 A static CMOS complex gate

the first three vectors from either of the above two sequences. To derive the test set for the whole circuit, we just need to concatenate all the test sequences derived in the above fashion.

Revisiting Example 9.1 again, we find that there are three gates in the gate-level model which are fed by checkpoints. Each of these gates has two inputs. Thus, a test sequence of length four is needed for each gate, for a total test set size of twelve for the whole circuit.

9.2.4   Test generation at the switch level

When the CMOS circuit contains pass transistors, it may be better to do test generation at the switch level. Also, for some opens and shorts, a switch-level model may allow a test to be found, whereas a gate-level model may not be able to yield a test (Galiay *et al.*, 1980). Some switch-level test generation methods have been presented in Chiang and Vranesic (1983), Agrawal and Reddy (1984), and Shih and Abraham (1986b).

The method given in Agrawal and Reddy (1984) uses a **tree representation** of a CMOS gate, which was first introduced in Lo *et al.* (1983). The nodes in the tree can be of the following types: T to denote a transistor, $(+)$ to denote a parallel connection of transistors, $(*)$ to denote a series connection of transistors, $(=)$ to combine the outputs of the trees which represent the nMOS and pMOS networks, and P to denote a pass transistor. For modeling a CMOS circuit, tree representations are obtained for every gate and are then interconnected together. Consider, for example, the CMOS gate in Figure 9.15. Its tree representation is given in Figure 9.16. The left (right) part of the tree represents the pMOS (nMOS) network. The "bubbles" at the inputs of the T nodes in the left half denote inversions.

The D-algorithm (see Chapter 4; Section 4.6.1) can be adapted to the tree representation. Corresponding to signals 0, 1, x, D and $\overline{D}$ used in the D-algorithm, we use

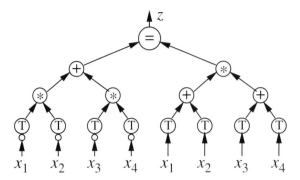

Figure 9.16 The tree representation

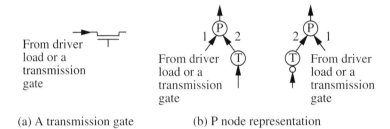

(a) A transmission gate   (b) P node representation

Figure 9.17 Representation of nMOS and pMOS pass transistors

Table 9.4. *Evaluation table for the T node*

Input	Output
0	OFF
1	ON
D	SD
$\overline{D}$	$\overline{SD}$
x	x

OFF, ON, x, SD and $\overline{SD}$, respectively. These signals are employed for the branches of the tree. SD denotes the case where the branch is ON (OFF) in the fault-free (faulty) circuit. Similarly, $\overline{SD}$ denotes the case where the branch is OFF (ON) in the fault-free (faulty) circuit. An SOpF is denoted by SD and an SOnF by $\overline{SD}$. A D ($\overline{D}$) at the input of an nMOS transistor implies SD ($\overline{SD}$) as the branch status, whereas for a pMOS transistor it implies $\overline{SD}$ (SD) as the branch status. The evaluation tables for the various types of nodes are given in Tables 9.4–9.8. In Tables 9.5–9.7, the first row and first column denote the two inputs to the node, and the output of the node is given at the intersection of the corresponding row and column. The transmission gate and its P node representation for an nMOS and pMOS transistor are given in Figure 9.17.

Table 9.5. *Evaluation table for the ($*$) node*

$*$	OFF	ON	SD	$\overline{SD}$	x
OFF	OFF	OFF	OFF	OFF	OFF
ON	OFF	ON	SD	$\overline{SD}$	x
SD	OFF	SD	SD	OFF	x
$\overline{SD}$	OFF	$\overline{SD}$	OFF	$\overline{SD}$	x
x	OFF	x	x	x	x

Table 9.6. *Evaluation table for the ($+$) node*

$+$	OFF	ON	SD	$\overline{SD}$	x
OFF	OFF	ON	SD	$\overline{SD}$	x
ON	ON	ON	ON	ON	ON
SD	SD	ON	SD	ON	x
$\overline{SD}$	$\overline{SD}$	ON	ON	$\overline{SD}$	x
x	x	ON	x	x	x

Table 9.7. *Evaluation table for the ($=$) node, nMOS (pMOS) tree output given horizontally (vertically)*

$=$	OFF	ON	SD	$\overline{SD}$	x
OFF	M	0	$\overline{D}(1)$	$D(1)$	x
ON	1	0	$\overline{D}$	D	x
SD	$D(0)$	0	$\overline{D}(1)$	D	x
$\overline{SD}$	$\overline{D}(0)$	0	$\overline{D}$	$D(1)$	x
x	x	0	x	x	x

If an SOpF causes both the nMOS and pMOS networks to be OFF, then the previous logic value is retained at the gate output. Thus, it enters a memory state labeled as M in Table 9.7. In this table, the logic value in parenthesis for some entries denote the value of M to which the CMOS gate output should be initialized in order to detect the

Table 9.8. *Evaluation table for the P node*

Input 2 →	Previous output = 0			Previous output = 1		
Input 1 ↓	OFF	ON	x	OFF	ON	x
0	0^*	0	0^*	1^*	0	x
1	0^*	1	x	1^*	1	1^*
0^*	0^*	0^*	0^*	1^*	x	x
1^*	0^*	x	x	1^*	1^*	1^*
x	0^*	x	x	1^*	x	x

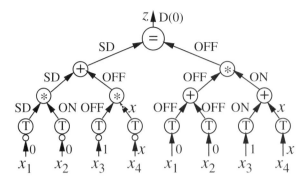

Figure 9.18 Test generation example

fault in question. For example, if the nMOS (pMOS) tree output is SD (OFF) then the output is $\overline{D}(1)$. This implies that the output would be $\overline{D}$ only if the output was initialized to 1. In order to simplify the model, in Table 9.7 it has been assumed that when both the nMOS and pMOS networks are ON due to a fault then the output is 0.

In Table 9.8, 0^* (1^*) represents the 0 (1) value at high impedance nodes. This table is used once for the fault-free circuit and then for the faulty circuit to determine the combined output value, i.e., to see if it is D or $\overline{D}$. When the input to a pass transistor is ON, its source and drain can interact with each other. To accurately model this phenomenon, one should keep in mind what the previous values on both the source and drain were. However, in order to reduce the complexity of test generation, the somewhat simplified model given in Table 9.8 is used.

The test generation method is geared towards SAF, SOpF and SOnF detection. An SD or $\overline{SD}$ is injected at the fault site and is propagated to the circuit output through what could be termed as **SD-drive** (analogous to D-drive in the D-algorithm). **Line justification** can then be done to justify at the primary inputs the different values specified in the SD-drive. Whenever needed, backtracking can be done to consider other choices.

Example 9.2 Suppose that we want to detect an SOpF in the pMOS transistor fed by x_1 in Figure 9.15. Let us first derive the test vector of the two-pattern test. Since x_1 must be 0 to activate this fault, the resultant branch signal at the output of the corresponding T node becomes SD, as shown in Figure 9.18. To propagate SD through the ($*$) node, the other input of this node should be ON. To further propagate SD through the ($+$) node, the other input of this node should be OFF. SD has now reached the pMOS tree output. The above values can be justified at the primary inputs as $(0, 0, 1, x)$. When this vector is applied to the nMOS tree, its output is OFF. Therefore, Table 9.7 gives the value at z to be $D(0)$. In order to initialize z to 0 we could use the vector $(1, x, 1, x)$. Thus, $\{(1, x, 1, x), (0, 0, 1, x)\}$ is a two-pattern test for the SOpF in question. $\square$

In order to speed up test generation, it was recognized in Shih and Abraham (1986b) that the complete tree representation for both the nMOS and pMOS networks is required only for the faulty CMOS gate. For other gates in the circuit, it is enough to obtain a tree representation for just one of the two networks, since both networks contain the same functional information. This also reduces the memory requirements for storing these representations.

9.2.5   Robust test generation

We have seen earlier that robustly detecting all the SOpFs at the checkpoints is not guaranteed to robustly detect all the internal SOpFs in an irredundant CMOS circuit. This seems to imply that we have to target each SOpF individually for robust test generation. However, a more efficient strategy would be to first derive the robust test set for the checkpoints, and then determine what other SOpFs also got robustly detected. Only the remaining SOpFs need to be targeted for further robust test generation. It is worth emphasizing at the outset, to avoid any confusion later, the assumption behind robust testing that primary inputs of the circuit do not have any glitches when the initialization vector is changed to the test vector. However, the inputs can have timing skews with respect to each other, and the delays within the circuit can be arbitrary.

Robust testing has been considered in Reddy *et al.* (1984b, 1985), Weiwei and Xieting (1986), and Jha (1989). We will discuss a method which is a simpler version of the method given in Reddy *et al.* (1984b). Reddy's method relies on a gate-level model presented in Jain and Agrawal (1983) which models both the nMOS and pMOS networks at the gate level. However, one can actually use the simplified gate-level model (Reddy *et al.*, 1984a) which models either the nMOS or the pMOS network of the CMOS gate. As we have done earlier in this chapter for the sake of simplicity, we will again assume that all gate-level modeling is done from only the nMOS networks of the CMOS gates in the circuit.

Let the two-pattern test be denoted as $\{P_1, P_2\}$. Assume that P_1 is fed at time t_1 and P_2 is fed at time t_2. Our approach would be to derive P_1, given P_2, such that the two-pattern test becomes robust. For robust testing, we have to be ever vigilant about static hazards inside the circuit. A **static 0 (1) hazard** is said to be present on a line when an input vector is changed to another, if the expected logic value on the line is 0 (1) for both the vectors, however, the line assumes the value 1 (0) momentarily.

We have seen earlier in Section 9.2.3 that for testing purposes we can model an SOpF in an nMOS (pMOS) transistor of a CMOS gate by an SA0 (SA1) fault in its gate-level model. Keeping this in mind, we can obtain the following theorem:

Theorem 9.1 If $\{P_1, P_2\}$ is a robust two-pattern test to detect an nMOS (pMOS) transistor SOpF in gate G of a CMOS circuit, then it is necessary and sufficient that the output of the faulty gate in the gate-level model be free of a static 1 (0) hazard.

Table 9.9. *Satisfying hazard-free requirements (Reddy* et al., *1984b) (© 1984 IEEE)*

Gate	Requirement at the output	Hazard status of the inputs
NOT	0-hazard-free	1-hazard-free
NOT	1-hazard-free	0-hazard-free
AND	0-hazard-free	at least one input 0-hazard-free
AND	1-hazard-free	all inputs 1-hazard-free
OR	0-hazard-free	all inputs 0-hazard-free
OR	1-hazard-free	at least one input 1-hazard-free

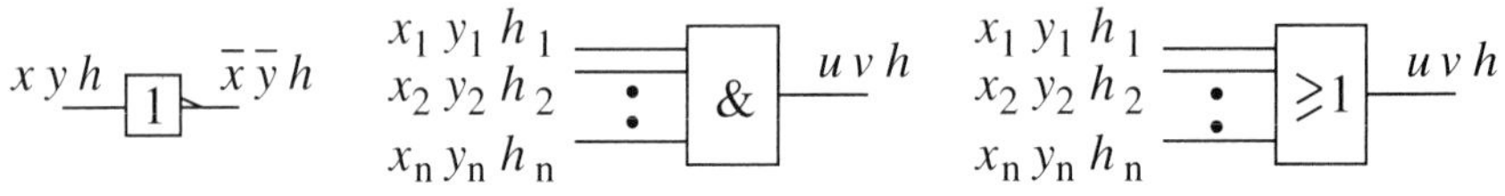

Figure 9.19 Hazard status calculation

This theorem indicates that if want to derive P_1, given P_2, then we need to avoid the corresponding static hazard. At times t_1 and t_2, a line in the gate-level model is assumed to take values from the set $\{0, 1, x\}$. A **hazard status** from the set $\{hp, hu, hf\}$ is also associated with a line, where the elements of this set denote *hazard present, hazard unknown*, and *hazard-free*, respectively. HFR denotes the set of lines which are to be made hazard-free at a given step in the test generation process. Table 9.9 shows how hazard-free requirements can be satisfied at the output of a NOT, AND and OR gate.

For finding the hazard status of the output of a gate, given the hazard status of its inputs, consider the primitive gates in Figure 9.19. For the NOT gate, x and y denote the input logic values at times t_1 and t_2 respectively, and its output hazard status is the same as its input hazard status. For the n-input AND and OR gates, x_i, y_i and h_i denote the logic values on input i at t_1 and t_2 and its hazard status, respectively. The corresponding output values and hazard status are u, v, and h. The output hazard status h can be computed as follows for the AND gate:

1 If $u = v = 0$ then

$$h = \begin{cases} hf & \text{if and only if at least one input is } 00hf \\ hu & \text{if and only if no input is } 00hf \text{ and at least} \\ & \text{one is } 00hu, 0xhu, x0hu \text{ or } xxhu \\ hp & \text{otherwise} \end{cases}$$

2 If $u = v = 1$ then

$$h = \begin{cases} hf & \text{if and only if } h_1 = h_2 = \cdots = h_n = hf \\ hp & \text{if and only if at least one of } \{h_1, h_2, \ldots, h_n\} \text{ is } hp \\ hu & \text{otherwise} \end{cases}$$

3 If $u = v = x$ or $u \neq v$ then $h = hu$.

For an OR gate, h can be computed as follows:

1 If $u = v = 1$ then

$$h = \begin{cases} hf & \text{if and only if at least one input is } 11hf \\ hu & \text{if and only if no input is } 11hf \text{ and at least} \\ & \text{one is } 11hu, 1xhu, x1hu \text{ or } xxhu \\ hp & \text{otherwise} \end{cases}$$

2 If $u = v = 0$ then

$$h = \begin{cases} hf & \text{if and only if } h_1 = h_2 = \cdots = h_n = hf \\ hp & \text{if and only if at least one of } \{h_1, h_2, \ldots, h_n\} \text{ is } hp \\ hu & \text{otherwise} \end{cases}$$

3 If $u = v = x$ or $u \neq v$ then $h = hu$.

With the above background, let us now see how we can derive P_1, given P_2. The method for doing this is very similar to the line justification step in the D-algorithm (see Chapter 4; Section 4.6.1). We assume that the hazard-free requirement is automatically satisfied on reaching the primary inputs, and hence the primary inputs do not need to be assigned the hf status beforehand. Initially, we set HFR to null. The faulty line in the gate-level model is assigned its stuck value and hazard status hf. All the other lines in the model are assigned x at t_1 and hazard status hu. The line in the model corresponding to the output of the faulty CMOS gate is added to HFR, and assigned $11hu$ ($00hu$) if we are testing for an SOpF in an nMOS (pMOS) transistor. If HFR is not empty, we pick a line, say line r, from it, and try to specify the inputs of the gate whose output is r such that the hazard status of r becomes hf. If this is possible, we delete r from HFR, else we backtrack to consider other choices and update HFR. Line justification is needed to satisfy the new requirements. Both forward and backward implication are also done. If this makes the hazard status of some line in HFR equal to hf, then this line can be dropped from HFR. Any conflict in logic values or hazard status during implication also triggers a backtrack to the last choice. If HFR becomes empty and no more line justifications remain to be done, then we obtain initialization vector P_1. Otherwise, we repeat the above process.

Example 9.3 Consider the CMOS circuit given in Figure 9.13 once again. Suppose we want to derive a two-pattern test for an SOpF in the nMOS transistor fed by x_1. In the gate-level model, the corresponding fault is SA0 on line c_3, as shown in Figure 9.20. We can use any of the previous methods to derive P_2 for this fault, one such vector being $(1, 1, 0, x, 1, 1)$. c_3 is now isolated from its input and is set to $00hf$.

The different steps involved in deriving P_1 for the above P_2 are shown in Table 9.10, where "$-$" indicates that the assignment to the line in the previous step remains valid. From the last step, P_1 can be seen to be $(0, x, 0, x, x, x)$. Although this is adequate, we see that the effect of this initialization vector is not propagated to the circuit output

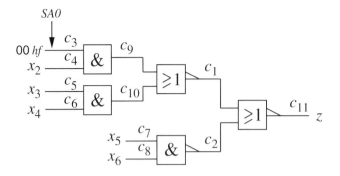

Figure 9.20 The gate-level model with the fault

Table 9.10. *Derivation of* P_1

Line label	Initial value	Step 1	Step 2	Step 3
c_1	$11hu$	$11hf$	—	—
c_2	$x0hu$	—	—	—
c_3	$00hf$	—	—	—
c_4	$x1hu$	—	—	—
c_5	$x0hu$	—	—	$00hf$
c_6	$xxhu$	—	—	—
c_7	$x1hu$	—	—	—
c_8	$x1hu$	—	—	—
c_9	$x0hu$	$00hu$	$00hf$	—
c_{10}	$x0hu$	$00hu$	—	$00hf$
c_{11}	$x0hu$	—	—	—
HFR	null	$\{c_9, c_{10}\}$	$\{c_{10}\}$	

z. However, doing so may, in general, let us robustly test other SOpFs along the path. Here, this means that we should specify x_5 and x_6 to be 1, instead of a *don't care*. ☐

9.3   Design for robust testability

Since many circuits are not completely robustly testable, we have to come up with methods which either start from the functional specification and generate a robustly testable circuit, or take an existing circuit and add extra testability logic to it to make it testable (Reddy *et al.*, 1983; Jha and Abraham, 1985; Reddy and Reddy, 1986; Liu and McCluskey, 1986; Sherlekar and Subramanian, 1988; Kundu and Reddy, 1988; Kundu, 1989; Kundu *et al.* 1991). We discuss some of these methods in this section.

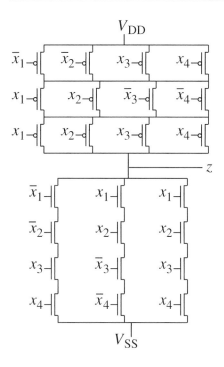

Figure 9.21   An OR–AND CMOS complex gate

9.3.1   Robustly testable CMOS complex gates

We have seen in Section 9.2.1 earlier that even irredundant CMOS complex gates are sometimes not robustly testable for all SOpFs. For example, the AND–OR CMOS complex gate shown in Figure 9.9, which implements the sum-of-products expression $z = \bar{x}_1 x_2 + x_1 \bar{x}_2 + \bar{x}_3 x_4 + x_3 \bar{x}_4 + \bar{x}_1 \bar{x}_3$, is not robustly testable. For the same function, the product-of-sums expression is $z = (x_1 + x_2 + \bar{x}_3 + \bar{x}_4)(\bar{x}_1 + \bar{x}_2 + x_3 + x_4)(\bar{x}_1 + \bar{x}_2 + \bar{x}_3 + \bar{x}_4)$. The corresponding complex gate, called the **OR–AND complex gate**, is given in Figure 9.21. This gate is completely robustly testable. However, in general, functions exist none of whose AND–OR or OR–AND complex gates are robustly testable (see Exercise 9.7). We consider necessary and sufficient conditions for robust testability next.

A vector (sometimes also called a vertex) is called *true (false)* if it results in a 1 (0) at the output. In a given irredundant sum-of-products (product-of-sums) expression, if a true (false) vertex of a prime implicant (implicate) is not a part of any other prime implicant (implicate), it is called distinguished. The following theorem gives necessary and sufficient conditions for robust testability of an AND–OR or OR–AND complex gate (Jha and Abraham, 1985). Not surprisingly, similar conditions are also applicable to NAND–NAND or NOR–NOR CMOS implementations as well, as independently shown in Reddy and Reddy (1986).

Theorem 9.2 A robust test set for an irredundant AND–OR (OR–AND) CMOS complex gate exists if and only if at least one distinguished true (false) vertex of every prime implicant (implicate) in the realization is only one bit different from some false (true) vertex.

The reason the AND–OR complex gate in Figure 9.9 is not robustly testable is that the only distinguished true vertex $(0, 0, 0, 0)$ of the prime implicant $\bar{x}_1\bar{x}_3$ is more than one bit different from all of the false vertices. Given that even irredundant AND–OR and OR–AND complex gates may have robust testability problems, we focus next on complex gates which can be guaranteed to be free of this problem.

If the nMOS network of the AND–OR complex gate is attached to the pMOS network of the OR–AND complex gate, we arrive at what is called a **hybrid CMOS complex gate** (Jha and Abraham, 1985). This is also called a PS-PS complex gate in Reddy and Reddy (1986). The hybrid CMOS complex gate for our example function is given in Figure 9.22. The unusual feature of such gates is that a series connection of transistors in the nMOS network does not imply a parallel connection of transistors fed by the same inputs in the pMOS network, and vice versa. However, this is not a necessary condition for a valid fully complementary CMOS gate realization, only a sufficient one. What is necessary for such a realization is that the two networks never simultaneously conduct for any input vector. This condition is satisfied by the hybrid gate.

The hybrid gate is guaranteed to be robustly testable. This guarantee comes from the structure of the two networks: in both, transistors are first connected in parallel, and then such structures are themselves connected in series. Consider the nMOS network of a hybrid gate. For an input vector which has 1s corresponding to literals in a prime implicant, the corresponding parallel connection of transistors in the nMOS network will be deactivated. For example, the parallel connection of transistors fed by x_1 and $\bar{x}_2$ at the top of the nMOS network of the hybrid gate in Figure 9.22 is derived from the prime implicant $\bar{x}_1 x_2$. This parallel connection will be deactivated if the input vector has $x_1 = 0$ and $x_2 = 1$. Since there is no conduction path from output z to V_{SS}, there must be one from V_{DD} to z, resulting in $z = 1$. This vector serves as the initialization vector to test for an SOpF in any of the transistors in the parallel connection. The test vector is derived by simply complementing the input in the initialization vector which corresponds to the transistor in question, and setting the other inputs such that a conduction path in the nMOS network can be activated. This is always possible to do in an irredundant complex gate because otherwise the transistor in question must be redundant. For our example, to test for an SOpF in the transistor fed by x_1 in the above parallel connection, one such test vector is $(1, 1, 0, 0)$. Thus, the two-pattern test becomes $\{(0, 1, x, x), (1, 1, 0, 0)\}$. Any such two-pattern test is robust because the transistors in parallel to the one being tested have a hazard-free 0 value. Similarly, to test a pMOS transistor in a hybrid gate for an SOpF, one can start with a prime

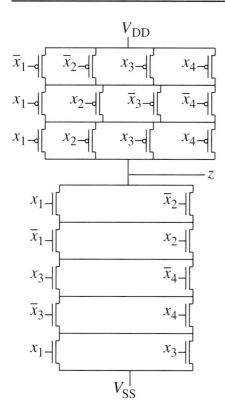

Figure 9.22 A hybrid CMOS complex gate

implicate to derive the initialization vector. This will deactivate the corresponding parallel connection of pMOS transistors. Then the test vector can be derived by just complementing the input of the transistor being tested, and setting the other inputs so that there is a conduction path through the pMOS network.

The guarantee of robust testability can still be given if the distributive law $xy + xz = x(y + z)$ or its dual $(x + y)(x + z) = (x + yz)$, where x, y and z are variables or sub-expressions, is used to simplify the two networks of the hybrid gate. For example, for the function considered above, we could have derived the nMOS network from the reduced expression $z = \bar{x}_1(x_2 + \bar{x}_3) + x_1\bar{x}_2 + \bar{x}_3x_4 + x_3\bar{x}_4$, and the pMOS network from the reduced expression $z = ((x_1+x_2)(\bar{x}_1+\bar{x}_2)+\bar{x}_3+\bar{x}_4)(\bar{x}_1+\bar{x}_2+x_3+x_4)$. This would save us one transistor in the nMOS network and two transistors in the pMOS network. Such a reduced hybrid gate is still robustly testable (see Exercise 9.17).

9.3.2 Robustly testable circuits using extra inputs

One common method for imparting testability to an untestable circuit is to add extra controllable inputs. This is the approach taken in Reddy *et al.* (1983), Reddy and

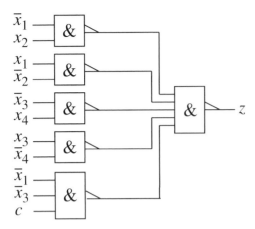

Figure 9.23 A robustly testable NAND–NAND implementation

Reddy (1986) and Liu and McCluskey (1986) to solve the problem of robust testability. We discuss next how an extra input can be used to make a two-level circuit robustly testable.

Consider an irredundant NAND–NAND implementation of the given function. In such an implementation, all SOpFs in the first-level NAND gates are robustly testable. The reason is that when an SOpF in the pMOS network of such a NAND gate is being tested, the transistors in parallel can be provided a hazard-free 1 as an input. The nMOS network of a NAND gate is always robustly testable because there are no conduction paths in parallel that can be spuriously activated in the transition from the initialization vector to the test vector. Thus, SOpFs which are not robustly testable, if any, can only be present in the pMOS network of the output NAND gate. Consider the NAND–NAND implementation of our favorite function $z = \bar{x}_1 x_2 + x_1 \bar{x}_2 + \bar{x}_3 x_4 + x_3 \bar{x}_4 + \bar{x}_1 \bar{x}_3$. As mentioned in Section 9.2.1, an SOpF in the pMOS transistor of the output NAND gate which is fed by the NAND gate realizing $\bar{x}_1 \bar{x}_3$ is not robustly testable. This is because the condition given in Theorem 9.2 is not met. Such an implementation can be made robustly testable with the addition of just one extra input c (Reddy et al., 1983). This just involves adding a fanout of c to every first-level NAND gate whose output feeds a pMOS transistor in the output gate whose SOpF is not robustly testable. For the above example, this is shown in Figure 9.23. Consider the NAND gate fed with the extra input c. To robustly test the SOpF in the pMOS transistor fed by this NAND gate, one could now use the following two-pattern test: $(x_1, x_2, x_3, x_4, c) = \{(0, 0, 0, 0, 0), (0, 0, 0, 0, 1)\}$. Here the first four bits in each vector correspond to the distinguished true vertex of the corresponding prime implicant. Therefore, the outputs of the other first-level NAND gates are guaranteed to be hazard-free, and test invalidation cannot occur. In the general case, to robustly test the SOpF in the pMOS transistor of the output gate which is fed by a NAND gate with the added control input c, one can use the distinguished true vertex of the corresponding prime implicant, and just change c from 0 to 1. From duality, a similar method is, of course, also applicable to NOR–NOR two-level circuits.

In the above discussions, we have assumed that for robust testability the side inputs of only the CMOS gate being tested should be hazard-free (hazards are allowed elsewhere in the circuit). However, in Bryan *et al.* (1992) it was assumed that hazards elsewhere are also not allowed. Under this stringent model, it was shown that algebraic factorization preserves SOpF testability. It is known that algebraic factorization does not preserve SOpF testability under the less stringent model which has been traditionally used, and assumed in this chapter (Reddy, 1991). Therefore, for obtaining testable multi-level circuits, either one can start with a two-level circuit testable under the more stringent model, or use the traditional model, and verify that the final multi-level circuit was actually completely or at least highly robustly testable.

Summary

- There are two main types of dynamic CMOS circuits: domino CMOS and DCVS logic. They enjoy area, performance and testability advantages over static CMOS circuits, but disadvantage in terms of power.
- A domino CMOS circuit can be tested for single SAFs, SOpFs and SOnFs from its gate-level model.
- A domino CMOS circuit can be easily synthesized to be testable for most multiple faults consisting of SAFs, SOpFs and SOnFs, using an initialization set and a single fault test set derived from the corresponding two-level logic circuit.
- For DCVS circuits, it is best to do test generation at the switch level. Propagation of errors in DCVS circuits is relatively simple. An error checker with low overhead can be derived for DCVS circuits to do both on-line and off-line testing.
- Test sets of static CMOS circuits can get invalidated by timing skews and arbitrary delays, as well as charge sharing.
- An SOpF test set for the checkpoint transistors in an irredundant fully complementary static CMOS circuit also detects (possibly non-robustly) the other SOpFs in the circuit if the effects of the initialization vector of each two-pattern test are also propagated to the circuit output.
- An SOpF test set for an irredundant static CMOS circuit also detects all SOnFs in the circuit if current monitoring is done.
- SOpFs can be detected by mapping them to SAFs in the gate-level model.
- A tree representation of the CMOS gates in the circuit is useful for test generation when there are pass transistors present in it.

- For robust test generation, there should be no hazard at the output of the faulty gate. Hazards elsewhere in the circuit are allowed.
- A hybrid CMOS complex gate realization of any function is always robustly testable.
- Any irredundant two-level circuit can be made robustly testable, if not already so, with the addition of just one extra input.

Additional reading

In Ling and Bayoumi (1987), Jha and Tong (1990) and Tong and Jha (1990), testing of different variants of dynamic CMOS circuits, such as NORA CMOS, multi-output domino logic and zipper CMOS, is discussed.

In Jha (1990) and Jha (1993), testing of DCVS parity circuits, ripple-carry adders, and one-count generators are discussed.

In Rajski and Cox (1986), SOpF testing is based on three-pattern tests and tracing dynamic paths from primary inputs to outputs.

A method for transistor-level test generation is presented in Lee *et al.* (1992).

Robust testing of multiple faults in static CMOS circuits is accomplished in Shen and Lombardi (1992) with a combination of two-pattern test sequences and universal test sets.

Robust tests for SOpFs are derived from tests for SAFs in Chakravarty (1992).

In Reddy *et al.* (1992), a method is given to compact the two-pattern tests into a small test set.

In Greenstein and Patel (1992), a method is presented for addressing the bridging fault simulation problem under the voltage testing environment (as opposed to I_{DDQ} testing environment).

A fault coverage analysis and diagnosis method for SOpFs is discussed in Cox and Rajski (1988).

In Liu and McCluskey (1987), a specially designed shift-register latch facilitates the application of two-pattern tests to the circuit to test for SOpFs.

Exercises

9.1 Show that an SA1 fault on any branch of the clock line in a domino CMOS circuit is detected by the test set derived from its gate-level model, in which the vectors have been arranged such that each domino CMOS gate in the circuit gets fed its required two-pattern test.

9.2 Consider an n-input DCVS XOR gate with outputs $(y_1, \bar{y}_1)$.

    (a) Show that if any one or more of its inputs $(z_i, \bar{z}_i)$ have $(0, 0)$ due to an error, then $(y_1, \bar{y}_1)$ will also have $(0, 0)$, irrespective of the logic values that the other inputs carry.

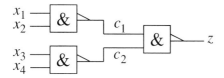

Figure 9.24 A NAND–NAND circuit

(b) If none of the inputs have $(0, 0)$, but one or more inputs have $(1, 1)$ due to an error, then show that $(y_1, \bar{y}_1)$ will also have $(1, 1)$, irrespective of the logic values that the other inputs carry.

9.3 Show that all detectable single SOpFs, SOnFs and SAFs in a DCVS XOR gate with any number of inputs is detected by a test set of size four. Extend this result to a DCVS parity tree (i.e., a tree of DCVS XOR gates) by showing that the size of the test set need be at most five, irrespective of the number of inputs the parity tree has, or the number of inputs any of the constituent XOR gates have.

9.4 Show that a DCVS full adder made up of carry and sum circuits similar to those given in Figures 9.6 and 9.8 is tested for all detectable single SOpFs, SOnFs and SAFs by a test set of size four. Extend this result to a DCVS ripple-carry adder by showing that the test set size remains four no matter how many stages the ripple-carry adder has.

9.5 Find a function of four variables none of whose OR–AND static CMOS complex gate realizations based on irredundant product-of-sums expressions is completely robustly testable for all SOpFs in the presence of arbitrary delays and timing skews. Are the corresponding NOR–NOR CMOS implementations of this function robustly testable for all SOpFs?

9.6 How many functions of four variables are there, none of whose irredundant AND–OR static CMOS complex gate realizations are robustly testable?

9.7 Find a function of five or more variables whose irredundant AND–OR and OR–AND complex gate realizations are both not robustly testable.

9.8 Find a two-pattern test for the SOpF in transistor Q_{16} in the AND–OR complex gate given in Figure 9.11 which can get invalidated by charge sharing. Is there a particular vector that should precede the two-pattern test for invalidation to occur? Assume that the output node has to share its charge with at least four other internal nodes before invalidation can take place.

9.9 For a NAND–NAND static CMOS circuit, whose gate-level model is shown in Figure 9.24, do the following:
   (a) Find a test set for all single SOpFs in the checkpoint transistors, which does not also detect all other SOpFs because the condition requiring propagation of the effects of the initialization vector of each two-pattern test to the primary output is not met.
   (b) Find a test set that robustly detects all SOpFs in the checkpoint transistors, yet does not robustly detect all the other SOpFs.

9.10 Prove that a test set which detects all single SOpFs in an irredundant fully complementary static CMOS circuit also detects all the single SOnFs in it if current monitoring is done.

9.11 Prove that any test set that detects all single SOpFs in transistors fed by the primary inputs in a fanout-free static CMOS circuit also detects (possibly non-robustly) all single SOpFs in the circuit, provided the initialization vector of each two-pattern test in the test set also has its effects propagated to the primary outputs.

9.12 For a static CMOS NOR–NOR circuit which implements the product-of-sums $(x_1 + x_2)(\bar{x}_1 + x_3)(x_1 + \bar{x}_3)$, derive a test set from its gate-level model which detects all SOpFs in it.

9.13 Prove Theorem 9.1.

9.14 Derive a robust test for an SOpF in the nMOS transistor fed by line c_1 in the CMOS circuit in Figure 9.13, based on the method given in Section 9.2.5.

9.15 Prove Theorem 9.2.

9.16 Derive a hybrid complex gate realization and its robust test set for the function $z = x_1 x_2 + \bar{x}_1 \bar{x}_2 + x_3 x_4 + \bar{x}_3 \bar{x}_4 + x_1 x_3$.

9.17 Prove that the reduced hybrid complex gate realization of any function is robustly testable.

9.18 Suppose we form a complex CMOS gate of a given function by attaching the nMOS network of its OR–AND complex gate with the pMOS network of its AND–OR complex gate. State the necessary and sufficient conditions on the Karnaugh map for such a complex gate to be robustly testable for all SOpFs. Give a proof for sufficiency.

9.19 Prove that just one extra control input is sufficient to provide robust testability to any irredundant NAND–NAND or NOR–NOR CMOS realization which is not robustly testable.

9.20 Either prove or give a counter-example to the following statement: if a combinational static CMOS circuit is robustly testable for all SOpFs then it is also robustly testable for all path delay faults.

References

Agrawal, P. and Reddy, S.M. (1984). Test generation at MOS level. In *Proc. Int. Conference on Computers, Systems & Signal Proc.*, pp. 1116–1119.

Barzilai, Z., Iyengar, V.S., Rosen, B.K. and Silberman, G.M. (1984). Fault modeling and simulation of SCVS circuits. In *Proc. Int. Conference on Computer Design*, pp. 42–47.

Barzilai, Z., Iyengar, V.S., Rosen, B.K. and Silberman, G.M. (1985). Accurate fault modeling and efficient simulation of differential CVS circuits. In *Proc. Int. Test Conference*, pp. 722–729.

Brayton, B., Rudell, R., Sangiovanni-Vincentelli, A. and Wang, A.R. (1987). MIS: a multiple-level logic optimization system. *IEEE Trans. on Computer-Aided Design*, **CAD-6** (11), pp. 1062–1081.

Bryan, M.J., Devadas, S. and Keutzer, K. (1992). Necessary and sufficient conditions for hazard-free robust transistor stuck-open testability in multilevel networks. *IEEE Trans. on Computer-Aided Design*, **11** (6), pp. 800–803.

Chakravarty, S. (1992). Heuristics for computing robust tests for stuck-open faults from stuck-at test sets. In *Proc. The European Conference on Design Automation*, pp. 416–420.

Chandramouli, R. (1983). On testing stuck-open faults. In *Proc. Int. Symposium on Fault-Tolerant Computing*, pp. 258–265.

Chiang, K.W. and Vranesic, Z.G. (1983). On fault detection in CMOS logic networks. In *Proc. Design Automation Conference*, pp. 50–56.

Cox, H. and Rajski, J. (1988). A method of fault analysis for test generation and fault diagnosis. *IEEE Trans. on Computer-Aided Design*, **7** (7), pp. 813–833.

Dandapani, R. and Reddy, S.M. (1974). On the design of logic networks with redundancy and testability considerations. *IEEE Trans. on Computers*, **C-23** (11), pp. 1139–1149.

Devadas, S., Keutzer, K. and Malik, S. (1993). A synthesis-based test generation and compaction algorithm for multifaults. *J. of Electronic Testing: Theory and Applications*, **4**, pp. 91–104.

Friedman, V. and Liu, S. (1984). Dynamic logic CMOS circuits. *IEEE J. of Solid-State Circuits*, **SC-19** (2), pp. 263–266.

Galiay, J., Crouzet, Y. and Vergniault, M. (1980). Physical versus logical fault models in MOS LSI circuits: impact on their testability. *IEEE Trans. on Computers*, **C-29** (6), pp. 527–531.

Goncalves, N.F. and de Man, H.J. (1983). NORA: a racefree dynamic CMOS technique for pipelined logic structures. *IEEE J. of Solid-State Circuits*, **SC-18** (3), pp. 261–266.

Greenstein, G.S. and Patel, J.H. (1992). E-PROOFS: a CMOS bridging fault simulator. In *Proc. Int. Conference on Computer-Aided Design*, pp. 268–271.

Hachtel, G.D., Jacoby, R.M., Keutzer, K. and Morrison, C.R. (1992). On properties of algebraic transformations and the synthesis of multifault-irredundant circuits. *IEEE Trans. on Computer-Aided Design*, **11** (3), pp. 313–321.

Heller, L.G., Griffin, W.R., Davis, J.W. and Thoma, N.G. (1984). Cascode voltage switch logic: a differential CMOS logic family. In *Proc. Int. Solid-State Circuits Conference*, pp. 16–17.

Hwang, I.S. and Fisher, A.L. (1989). Ultrafast compact 32-bit CMOS adders in multiple-output domino logic. *IEEE J. of Solid-State Circuits*, **24** (2), pp. 358–369.

Jacob, J. and Agrawal, V.D. (1992). Multiple fault detection in two-level multi-output circuits. *J. of Electronic Testing: Theory & Applications*, **3**, pp. 171–173.

Jain, S.K. and Agrawal, V.D. (1983). Test generation for MOS circuits using D-algorithm. In *Proc. Design Automation Conference*, pp. 64–70.

Jha, N.K. and Abraham, J.A. (1984). Totally self-checking MOS circuits under realistic physical failures. In *Proc. Int. Conference on Computer Design*, pp. 665–670.

Jha, N.K. and Abraham, J.A. (1985). Design of testable CMOS circuits under arbitrary delays. *IEEE Trans. on Computer-Aided Design*, **CAD-4** (7), pp. 264–269.

Jha, N.K. (1988). Testing for multiple faults in domino-CMOS logic circuits. *IEEE Trans. on Computer-Aided Design*, **7** (1), pp. 109–116.

Jha, N.K. (1989). Robust testing of CMOS logic circuits. *Int. J. of Computers & Electrical Engg.*, **15** (1), pp. 19–28.

Jha, N.K. (1990). Testing of differential cascode voltage switch one-count generators. *IEEE J. of Solid-State Circuits*, **25** (1), pp. 246–253.

Jha, N.K. and Tong, Q. (1990). Testing of multiple-output domino logic (MODL) CMOS circuits. *IEEE J. of Solid-State Circuits*, **25** (3), pp. 800–805.

Jha, N.K. (1993). Fault detection in CVS parity trees with application to strongly self-checking parity and two-rail checkers. *IEEE Trans. on Computers*, **42** (2), pp. 179–189.

Kanopoulos, N. and Vasanthavada, N. (1990). Testing of differential cascode voltage switch circuits. *IEEE J. of Solid-State Circuits*, **25** (3), pp. 806–813.

Krambeck, R.H., Lee, C.M. and Law, H.-F.S. (1982). High-speed compact circuits with CMOS. *IEEE J. of Solid-State Circuits*, **SC-17** (3), pp. 614–619.

Kundu, S. and Reddy, S.M. (1988). On the design of robust testable combinational logic circuits. In *Proc. Int. Symposium on Fault-Tolerant Computing*, pp. 220–225.

Kundu, S. (1989). Design of multioutput CMOS combinational logic circuits for robust testability. *IEEE Trans. on Computer-Aided Design*, **8** (11), pp. 1222–1226.

Kundu, S., Reddy, S.M. and Jha, N.K. (1991). Design of robustly testable combinational logic circuits. *IEEE Trans. on Computer-Aided Design*, **10** (8), pp. 1036–1048.

Lee, C.M. and Szeto, E.W. (1986). Zipper CMOS. *IEEE Circuits & Devices*, (5), pp. 10–17.

Lee, K.J., Njinda, C.A., and Breuer, M.A. (1992). SWiTEST: a switch level test generation system for CMOS combinational circuits. In *Proc. Design Automation Conference*, pp. 26–29.

Ling, N. and Bayoumi, M.A. (1987). An efficient technique to improve NORA CMOS testing. *IEEE Trans. on Circuits & Systems*, **CAS-34** (12), pp. 1609–1611.

Liu, D.L. and McCluskey, E.J. (1986). Design of CMOS VLSI circuits for testability. In *Proc. Custom Integrated Circuits Conf.*, pp. 421–424.

Liu, D.L. and McCluskey, E.J. (1987). CMOS scan-path IC design for stuck-open fault testability. *IEEE J. of Solid-State Circuits*, **SC-22** (5), pp. 880–885.

Lo, C.Y., Nham, H.N. and Bose, A.K. (1983). A data structure for MOS circuits. In *Proc. Design Automation Conf.*, pp. 619–624.

Manthani, S.R. and Reddy, S.M. (1984). On CMOS totally self-checking circuits. In *Proc. Int. Test Conference*, pp. 866–877.

Montoye, B. (1985). Testing scheme for differential cascode voltage switch circuits. *IBM Tech. Disclosure Bulletin*, **27** (10B), pp. 6148–6152.

Oklobdzija, V.G. and Kovijanic, P.G. (1984). On testability of CMOS-domino logic. In *Proc. Int. Symposium on Fault-Tolerant Computing*, pp. 50–55.

Pretorius, J.A., Shubat, A.S. and Salama, C.A.T. (1986). Latched domino CMOS logic. *IEEE J. of Solid-State Circuits*, **SC-21** (4), pp. 514–522.

Rajski, J. and Cox, H. (1986). Stuck-open fault testing in large CMOS networks by dynamic path tracing. In *Proc. Int. Conference on Computer Design*, pp. 252–255.

Reddy, S.M., Reddy, M.K. and Kuhl, J.G. (1983). On testable design for CMOS logic circuits. In *Proc. Int. Test Conference*, pp. 435–445.

Reddy, S.M., Agrawal, V.D. and Jain, S.K. (1984a). A gate-level model for CMOS combinational logic circuits with application to fault detection. In *Proc. Design Automation Conference*, pp. 504–509.

Reddy, S.M., Reddy, M.K. and Agrawal, V.D. (1984b). Robust tests for stuck-open faults in CMOS combinational logic circuits. In *Proc. Int. Symposium on Fault-Tolerant Computing*, pp. 44–49.

Reddy, M.K., Reddy, S.M. and Agrawal, P. (1985). Transistor level test generation for MOS circuits. In *Proc. Design Automation Conference*, pp. 825–828.

Reddy, S.M. and Reddy, M.K. (1986). Testable realizations for FET stuck-open faults in CMOS combinational logic circuits. *IEEE Trans. on Computers*, **C-35** (8), pp. 742–754.

Reddy, M.K. (1991). *Testable CMOS digital designs and switch-level test generation for MOS digital circuits*. Ph.D. Thesis, Dept. of Electrical & Computer Engineering, Univ. of Iowa, Iowa City.

Reddy, L.N., Pomeranz, I. and Reddy, S.M. (1992). COMPACTEST-II: a method to generate compact two-pattern test sets for combinational logic circuits. In *Proc. Int. Conference on Computer-Aided Design*, pp. 568–574.

Shen, Y.-N., and Lombardi, F. (1992). Detection of multiple faults in CMOS circuits using a behavioral approach. In *Proc. VLSI Test Symposium*, pp. 188–193.

Sherlekar, S.D. and Subramanian, P.S. (1988). Conditionally robust two-pattern tests and CMOS design for testability. *IEEE Trans. on Computer-Aided Design*, **7** (3), pp. 325–332.

Shih, H.-C. and Abraham, J.A. (1986a). Fault collapsing techniques for MOS VLSI circuits. In *Proc. Int. Symposium on Fault-Tolerant Computing*, pp. 370–375.

Shih, H.-C. and Abraham, J.A. (1986b). Transistor-level test generation for physical failures in CMOS circuits. In *Proc. Design Automation Conference*, pp. 243–249.

Tong, Q. and Jha, N.K. (1990). Testing of zipper CMOS logic circuits. *IEEE J. of Solid-State Circuits*, **25** (3), pp. 877–880.

Weiwei, M. and Xieting, L. (1986). Robust test generation algorithm for stuck-open fault in CMOS circuits. In *Proc. Design Automation Conference*, pp. 236–242.

10 Fault diagnosis

In this chapter, we discuss methods for diagnosing digital circuits. We begin by identifying the main objectives of diagnosis and defining the notions of response, error response, and failing vectors for a circuit under test (CUT), the fault-free version of the circuit, and each circuit version with a distinct target fault.

We then describe the purpose of fault models for diagnosis and describe the fault models considered in this chapter. The cause–effect diagnosis methodologies follow. In these methodologies, each faulty version of the circuit is simulated, implicitly or explicitly, and its response determined and compared with that of the CUT being diagnosed. We first describe post-test diagnostic fault simulation approaches where fault simulation is performed after the CUT response to the given vectors is captured. Subsequently, we describe fault-dictionary approaches where fault simulation is performed and the response of each faulty version stored in the form of a fault dictionary, before diagnosis is performed for any CUT.

Next, we present effect–cause approaches for diagnosis which start with the CUT response and deduce the presence or absence of a fault at each circuit line.

Finally, we present methods for generating test vectors for diagnosis.

10.1 Introduction

Diagnosis is the process of locating the faults present within a given fabricated copy of a circuit. For some digital systems, each fabricated copy is diagnosed to identify the faults so as to make decisions about repair. This is the case for a system such as a printed circuit board, where chips identified as being faulty can be replaced and the opens at or shorts between pins of a chip may be repaired via re-soldering. In such cases, diagnosis must be performed on **each** faulty copy of the system. Hence, the cost of diagnosis should be low.

Digital VLSI chips are, by and large, un-repairable and faulty chips must be discarded. However, diagnosis is still performed on a **sample** of faulty chips, especially whenever chip yield is low or the performance of a large proportion of fabricated chips is unacceptable. In such a scenario, the objective of diagnosis is to identify the root cause behind the common failures or performance problems to provide insights on how to improve the chip yield and/or performance. The insights provided are used in a variety of ways: (a) to change the design of the chip, (b) to change one or more design rules, (c) to change one or more steps of the design methodology, such as extraction, simulation, validation, and test generation, (d) to change the fabrication process, and so on. Relatively larger amounts of effort and resources may be expended on diagnosis in this scenario, since the accrued costs can be amortized over a large volume of chips produced. The ability to perform diagnosis on a sample of failing chips allows the cost of diagnosis to be viewed as fixed. This is unlike the other diagnosis scenario where each faulty copy of the system must be diagnosed to make decisions about repair. Even though most techniques described ahead are applicable to both these scenarios of diagnosis, the emphasis is entirely on the chip diagnosis scenario.

A fabricated copy of the circuit, referred to as a circuit under test (CUT), can be diagnosed by applying tests to the inputs of the circuit, capturing the response at its outputs, and analyzing the captured response. (During this process, lines made controllable and/or observable by design-for-testability circuitry, such as scan, are considered as primary inputs and outputs, respectively.) The following discussion is limited to this type of diagnosis, also called **logic diagnosis**. Logic diagnosis is typically supplemented by more invasive **physical diagnosis**. If the CUT is already packaged, then the packaging is removed. In addition, layers of material may be removed at specific areas of the chip. VLSI features may then be examined using microscopes. Probes, such as electron-beam probes, may also be used to make measurements at internal circuit lines.

Physical diagnosis is expensive and destructive. Hence, logic diagnosis is typically used to first identify the likely faults. Physical diagnosis is then used to verify the results of logic diagnosis, a process that is sometimes also called **root-cause analysis**. Since physical diagnosis is expensive as well as destructive, it is important for logic diagnosis to provide high *diagnostic accuracy* and *resolution*. A CUT is said to be accurately diagnosed if the set of faults identified by logic diagnosis as the possible causes of the CUT's failure includes the real cause of failure. **Diagnostic accuracy** is defined as the proportion of all CUTs that are diagnosed accurately. **Diagnostic resolution** can be defined as the average number of faults that are identified by logic diagnosis as possible causes of CUT failure.

In this chapter, we discuss techniques for logic diagnosis for digital VLSI chips. The objective is to describe the basic techniques. We hence limit ourselves to full-scan circuits and to the most commonly used fault models, namely single and multiple stuck-at faults (SAFs), two-line bridging faults (BFs), and transition faults (TFs). In two-line BFs, we further limit ourselves to *non-feedback bridging faults* (NFBFs) and to certain types of bridging behavior. An extensive reading list is provided for readers interested in the diagnosis of sequential circuits and for other fault models, including *feedback* BFs, *Byzantine* BFs, and path delay faults.

10.2   Notation and basic definitions

Let C be a circuit design and C^* a fabricated copy of the design on which diagnosis is being performed. In the following discussion, C^* is referred to as the CUT. Let C (also, C^* as well as each faulty version of C) have n inputs, $x_1, x_2, \ldots, x_n$, m outputs, $z_1, z_2, \ldots, z_m$, and k internal lines, $c_1, c_2, \ldots, c_k$. The primary inputs, internal lines, and primary outputs of each of these circuits are collectively referred to as **circuit lines** and denoted using symbols such as b and b_l. Let L denote the total number of lines in the circuit, i.e., $L = n + k + m$.

In most of this chapter, we assume that a set of N_v vectors, $P_1, P_2, \ldots, P_{N_v}$, is provided for diagnosis. (The set of vectors is often referred to as a **sequence of vectors**, since typically the vectors are applied to a CUT in the given order.) Recall that each vector P_i is an n-tuple $(p_{1,i}, p_{2,i}, \ldots, p_{n,i})$, where $p_{j,i}$ is called a *component* of the vector that is applied to input x_j. Since these tests are applied to a fabricated version of the circuit, C^*, each component of the vector is fully specified, i.e., it is assigned a logic 0 or 1.

Let the **fault model** describe a list of N_f faults, $f_1, f_2, \ldots, f_{N_f}$, and let C^{f_j} denote a faulty version of the circuit with fault f_j. Circuit C is also referred to as the fault-free circuit.

In each type of diagnosis described ahead, CUT C^* is mounted on an automatic test equipment (ATE) and the given set of vectors applied. Following the application of each vector, P_i, the response at each CUT output is measured. Let $v_i^*(z_l)$ denote the value captured at output z_l of the CUT in response to the application of vector P_i.

The corresponding response for the fault-free version of the circuit, C, and any faulty version C^{f_j}, is determined via simulation. In some techniques, simulation is performed (implicitly or explicitly) after a CUT is tested and responses captured; in other techniques, simulation is performed before any CUT is tested.

Let $v_i(z_l)$ denote the response expected at output z_l of the fault-free version of the circuit, C, due to the application of vector P_i. Finally, let $v_i^{f_j}(z_l)$ denote the corresponding response for C^{f_j}.

Example 10.1 Consider circuit C_7 and the vectors shown in Figure 10.1. The vectors applied to the circuit are $P_1 = (1, 1, 1)$, $P_2 = (1, 0, 1)$, $P_3 = (0, 1, 0)$, and $P_4 = (1, 1, 0)$. The response of the fault-free version of the circuit is shown in Figure 10.1 and also summarized in the first row of Table 10.1. Some of the response values can be written as $v_1(z_1) = 0$, $v_1(z_2) = 0$, $v_4(z_1) = 0$, and $v_4(z_2) = 1$. The reader can simulate each faulty version of this circuit with a distinct single stuck-at fault (SAF) to determine the faulty circuit responses shown in Table 10.1. Consider the fault x_2 SA0, which will be referred to as fault f_3. Some of the response values for the version of

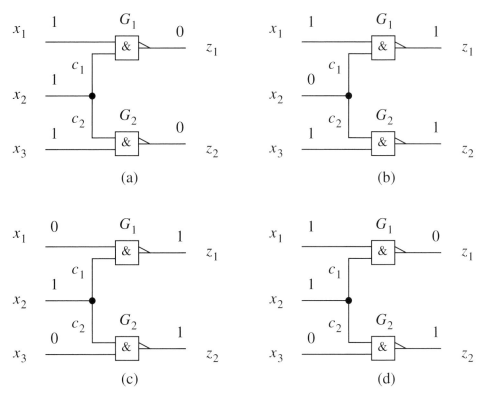

Figure 10.1 An example circuit C_7 with fault-free responses for a sequence of four vectors,
(a) $P_1 = (1, 1, 1)$, (b) $P_2 = (1, 0, 1)$, (c) $P_3 = (0, 1, 0)$, and (d) $P_4 = (1, 1, 0)$

the circuit with fault f_3, $C_7^{f_3}$, can be written as $v_1^{f_3}(z_1) = 1$, $v_1^{f_3}(z_2) = 1$, $v_4^{f_3}(z_1) = 1$, and $v_4^{f_3}(z_2) = 1$. □

The values $v_i^*(z_l)$, for all vectors P_i and all outputs $z_1, z_2, \ldots, z_m$, are collectively referred to as **response** Res^* of CUT C^*, where

$$Res^* = ((v_1^*(z_1), v_1^*(z_2), \ldots, v_1^*(z_m)), (v_2^*(z_1), v_2^*(z_2), \ldots, v_2^*(z_m)), \ldots,$$
$$(v_{N_v}^*(z_1), v_{N_v}^*(z_2), \ldots, v_{N_v}^*(z_m))).$$

Responses Res and Res^{f_j} can be defined in a similar manner for circuit C and C^{f_j}, respectively.

Example 10.2 Consider the fabricated copy C_7^*. Assume that C_7^* is tested using the above sequence of vectors and the response, $v_1^*(z_1) = 0$, $v_1^*(z_2) = 0$, $v_2^*(z_1) = 0$, $v_2^*(z_2) = 1$, $v_3^*(z_1) = 1$, $v_3^*(z_2) = 1$, $v_4^*(z_1) = 0$, and $v_4^*(z_2) = 1$, is captured. This can be collectively denoted as $Res^* = ((0, 0), (0, 1), (1, 1), (0, 1))$. □

Table 10.1. *A summary of fault-free and faulty responses for the circuit and vectors shown in Figure 10.1*

Circuit with fault	Response for vector P_i at output z_l							
	P_1		P_2		P_3		P_4	
	z_1	z_2	z_1	z_2	z_1	z_2	z_1	z_2
None	0	0	1	1	1	1	0	1
x_1 SA0								
c_1 SA0								
z_1 SA1	1	0	1	1	1	1	1	1
x_1 SA1	0	0	1	1	0	1	0	1
x_2 SA0	1	1	1	1	1	1	1	1
x_2 SA1	0	0	0	0	1	1	0	1
x_3 SA0								
c_2 SA0								
z_2 SA1	0	1	1	1	1	1	0	1
x_3 SA1	0	0	1	1	1	0	0	0
c_1 SA1	0	0	0	1	1	1	0	1
c_2 SA1	0	0	1	0	1	1	0	1
z_1 SA0	0	0	0	1	0	1	0	1
z_2 SA0	0	0	1	0	1	0	0	0

10.2.1 Errors

The captured CUT response $v_i^*(z_l)$ is said to be **erroneous** if its value is complementary to the corresponding response for the fault-free circuit, $v_i(z_l)$. Otherwise, the response is said to be **error-free**. In the former case, an **error is said to have occurred at CUT output** z_l **for vector** P_i. Occurrence or non-occurrence of an error at CUT output z_l in response to P_i is denoted by $e_i^*(z_l)$, which is assigned a value as

$$e_i^*(z_l) = \begin{cases} 0, \text{ if } v_i^*(z_l) \text{ is error-free,} \\ 1, \text{ if } v_i^*(z_l) \text{ is erroneous.} \end{cases}$$

If an error occurs at one or more CUT outputs for vector P_i, then the CUT is said to **fail** vector P_i and an **error is said to have occurred** for the vector. This is denoted by e_i^*, which is assigned a value as

$$e_i^* = e_i^*(z_1) + e_i^*(z_2) + \cdots + e_i^*(z_m),$$

where '+' denotes the logic OR operation.

Similarly, fault f_j is said to **cause an error at output** z_l following the application of vector P_i, if response $v_i^{f_j}(z_l)$ is complementary to the corresponding fault-free response, $v_i(z_l)$. This is denoted by $e_i^{f_j}(z_l)$, which is assigned a value as

$$e_i^{f_j}(z_l) = \begin{cases} 0, \text{ if } v_i^{f_j}(z_l) \text{ is error-free,} \\ 1, \text{ if } v_i^{f_j}(z_l) \text{ is erroneous.} \end{cases}$$

Table 10.2. *Error responses for the circuit and vectors shown in Figure 10.1*

Circuit with fault	Error for vector P_i at output z_l							
	P_1		P_2		P_3		P_4	
	z_1	z_2	z_1	z_2	z_1	z_2	z_1	z_2
None								
x_1 SA0								
c_1 SA0								
z_1 SA1	1				1			
x_1 SA1			1					
x_2 SA0	1	1			1			
x_2 SA1			1	1				
x_3 SA0								
c_2 SA0								
z_2 SA1		1						
x_3 SA1					1		1	
c_1 SA1			1					
c_2 SA1				1				
z_1 SA0			1		1			
z_2 SA0				1	1		1	

Finally, fault f_j is said to **cause an error** following the application of vector P_i if it causes an error at one or more outputs following the application of P_i. Note that if f_j causes an error for vector P_i, then C^{f_j} is said to **fail** that vector; equivalently, vector P_i is said to **detect** fault f_j. This is denoted by variable $e_i^{f_j}$, which is assigned a value as

$$e_i^{f_j} = e_i^{f_j}(z_1) + e_i^{f_j}(z_2) + \cdots + e_i^{f_j}(z_m).$$

The $e_i^{f_j}$ values for all the vectors are collectively denoted by **failing vectors of circuit with fault** f_j, FV^{f_j}, which can be written as

$$FV^{f_j} = (e_1^{f_j}, e_2^{f_j}, \ldots, e_{N_v}^{f_j}).$$

The failing vectors, FV^*, can be similarly defined for CUT C^*.

Values $e_i^*(z_l)$, for all vectors P_i and all outputs $z_1, z_2, \ldots, z_m$, are collectively referred to as the **error response** E^* of CUT C^*. These values are denoted as

$$E^* = ((e_1^*(z_1), e_1^*(z_2), \ldots, e_1^*(z_m)), (e_2^*(z_1), e_2^*(z_2), \ldots, e_2^*(z_m)), \ldots,$$
$$(e_{N_v}^*(z_1), e_{N_v}^*(z_2), \ldots, e_{N_v}^*(z_m))).$$

The error response E^{f_j} can be defined in a similar manner for each faulty circuit version C^{f_j}.

Example 10.3 Consider circuit C_7 and the sequence of test vectors shown in Figure 10.1 once again. The response of the fault-free circuit and that of each of its versions with a distinct single SAF are shown in Table 10.1. The error response for each faulty version of the circuit is summarized in Table 10.2. Note that the error value for the circuit with fault f_3, x_2 SA0, can be computed by comparing the value of $v_i^{f_3}(z_l)$ with that of $v_i(z_l)$ for each vector P_i and each output z_l. For example, since $v_1^{f_3}(z_1)$ and $v_1(z_1)$ are complementary, $e_1^{f_3}(z_1) = 1$. On the other hand, since $v_2^{f_3}(z_1)$ and $v_2(z_1)$ are not complementary, $e_2^{f_3}(z_1) = 0$. Collectively, the response values can be written as $Res^{f_3} = ((1, 1), (1, 1), (1, 1), (1, 1))$ and the error response values as $E^{f_3} = ((1, 1), (0, 0), (0, 0), (1, 0))$. The corresponding failing vectors can be written as $FV^{f_3} = (1, 0, 0, 1)$.

Consider the case where the response of a fabricated copy of this circuit, say C_7^*, is $Res^* = ((0, 0), (0, 1), (1, 1), (0, 1))$. The error response for C_7^* can similarly be computed as $E^* = ((0, 0), (1, 0), (0, 0), (0, 0))$. The corresponding failing vectors can be written as $FV^* = (0, 1, 0, 0)$. □

10.2.2 Comparing the CUT and faulty circuit responses

The central step in cause–effect diagnosis is a comparison of the CUT response with the response of each faulty version of the circuit.

Real fabrication defects in a CUT may not be modeled accurately by any fault in the fault model used for diagnosis. It is, therefore, necessary to define several types of matches between the error response of a CUT and that for C^{f_j}, a faulty version of the given circuit design. A range of such definitions have been proposed (e.g., Millman *et al.* (1990)). Here, we present the variations that are used ahead.

A **failing vector match** is said to occur between the response of a CUT and that of a circuit version with fault f_j if C^* and C^{f_j} fail identical vectors, i.e., $FV^* = FV^{f_j}$.

An **error response match** is said to occur between the response of a CUT and that of a circuit version with fault f_j, if C^* and C^{f_j} fail identical vectors and, for each failing vector, erroneous values are captured at identical sets of outputs of the two versions of the circuit, i.e., $E^* = E^{f_j}$.

The **failing vectors** of C^{f_j} are said to **cover** the failing vectors of a CUT if C^{f_j} fails a superset of vectors that C^* fails. Equivalently, $FV^{f_j} \sqsupseteq FV^*$, i.e., $e_i^* = 1$ only if $e_i^{f_j} = 1$.

The **error response** of C^{f_j} is said to **cover** the error response of C^* if (a) the failing vectors of C^{f_j} cover those of C^*, and (b) for each vector that C^{f_j} as well as C^* fail, the error responses of the two circuit versions are identical. Equivalently, (a) $FV^{f_j} \sqsupseteq FV^*$, and (b) for each vector P_i for which $e_i^* = e_i^{f_j} = 1$, $e_i^{f_j}(z_l) = e_i^*(z_l)$ for each output z_l. This covering is denoted as $E^{f_j} \sqsupseteq E^*$.

Example 10.4 For our running example scenario, suppose the following error response is obtained: $E^* = ((0, 0), (1, 0), (0, 0), (0, 0))$. A failing vector match occurs between this CUT and circuit versions with single SAFs x_2 SA1, c_1 SA1, and c_2 SA1. Of these, an error response match occurs between the CUT and the circuit version with fault c_1 SA1. In contrast, the failing vectors of circuit versions with single SAFs x_2 SA1, c_1 SA1, c_2 SA1, z_1 SA0, and z_2 SA0 all cover the failing vector of this CUT. Of these, only the error response of the circuit versions with single SAFs c_1 SA1 and z_1 SA0 cover the error response of this CUT. □

10.3 Fault models for diagnosis

Fault models play a significantly different role during diagnosis than during **test development**, i.e., test generation and fault simulation.

10.3.1 The role of a fault model during diagnosis

A fault model is considered useful for test generation as long as any set of test vectors that detects a large fraction of the faults in the corresponding fault list also detects a large proportion of real fabrication defects. Faults in the fault model need not accurately characterize the behavior of circuits with defects that really occur during fabrication. However, during diagnosis, faults in the model are required to characterize the behavior of circuits with realistic fabrication defects. In other words, the responses of one or more circuit versions, each with a distinct fault from the model, are expected to match responses of faulty CUTs with real fabrication defects. Hence, a fault model that is considered useful for test development may be inadequate for diagnosis.

For example, the single SAF model may be used for test development, even in a case where multiple SAFs are deemed likely to occur during fabrication. This is due to the fact that test sets with high SAF coverage typically provide high coverage of multiple SAFs. Now consider a scenario where a test set derived for single SAFs is used for diagnosis of a faulty CUT. If a multiple SAF exists in the fabricated CUT, the response of the faulty CUT may not match the response of any circuit version with any single SAF and an erroneous diagnosis may occur.

10.3.2 Fault models commonly used for diagnosis

While the single SAF model continues to be the most commonly used model for test development, diagnosis is performed using one or more of a range of fault models that are believed to characterize more accurately the behavior of defects that occur during VLSI manufacturing.

Single SAFs: Due to its continued popularity for test development, a wide range of fault simulators use the single SAF model. These tools can provide information required for diagnosis, such as the response for each faulty circuit version, if this model is used for diagnosis. Furthermore, conditions that a vector must satisfy to detect an SAu fault at a line b are subsets of the conditions that the vector must satisfy to detect other types of faults, including many types of BFs and some types of delay faults. (These relationships are described more precisely in the following discussion.) For these reasons, the single SAF model is commonly used for diagnosis.

Multiple SAFs: Realistic fabrication defects are sometimes assumed to cause multiple SAFs. The number of possible multiple SAFs in a circuit with L lines is $3^L - 1$. Since this number is very large for any circuit of realistic size, special effect–cause techniques have been developed to diagnose circuits under the multiple SAF model.

Two-line BFs: In Chapter 2, we discussed various two-line BFs, such as feedback, non-feedback, wired-AND, and wired-OR. For simplicity, we do not consider the feedback BFs and concentrate on non-feedback BFs (NFBFs).

In both wired-AND and wired-OR BF models, the fault effect appears on only one of the two sites of the fault. However, the site at which the fault effect appears depends on the values implied by the vector at the fault sites in the fault-free version of the circuit. One way in which a vector can detect a wired-AND BF between lines b_1 and b_2 (such a fault will be denoted as AB b_1b_2) is if it (a) implies logic 1 at b_1 in the fault-free circuit, (b) implies a 0 at b_2, and (c) propagates the fault effect from b_1 to one or more CUT outputs. Note that conditions (a) and (c) are collectively equivalent to the conditions for detection of an SA0 fault at b_1. Condition (b) is an additional condition that the vector must satisfy to detect the wired-AND BF. The only other way in which a vector can detect an NFBF AB b_1b_2 is if it is a test for an SA0 fault at b_2 that implies logic 0 at b_1.

As also pointed out in Chapter 2, the wired-AND and wired-OR models are, in general, not applicable to CMOS circuits, since both the strengths of the gates driving the sites of the BF, b_1 and b_2, and the resistance of the bridge determine how the presence of the fault alters the values at the two lines. This relationship is further complicated by the fact that, in general, the strength of a CMOS gate varies with the combination of logic values applied at its inputs. A **voting model** was proposed in Millman *et al.* (1990) and Acken and Millman (1992) to characterize the behavior of a BF in CMOS circuits. In this method, the logic values applied at the inputs of gates driving lines b_1 and b_2 are used to determine the transistors in each of these gates that are on. A resistance value is assigned to the channel of each transistor that is on and the equivalent resistance of each gate output is computed. The resistance of the short is assumed to be significantly lower than the equivalent resistance of each of the drivers, and is ignored. Figure 10.2(a) shows an example short between lines c_2 and

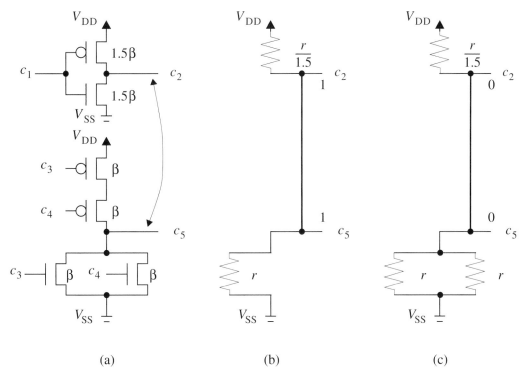

      (a)   (b)   (c)

Figure 10.2 A voting model to determine the behavior of a BF in a CMOS circuit: (a) an example short, (b) the voting model and output values for a vector P_i that implies $v_i(c_1) = 0$, $v_i(c_3) = 1$, and $v_i(c_4) = 0$, and (c) the output values for a vector P_j that implies $v_j(c_1) = 0$, $v_j(c_3) = 1$, and $v_j(c_4) = 1$

c_5 in a CMOS circuit. Here, β denotes the gain of a minimum-size nMOS transistor. Figures 10.2(b) and 10.2(c) show the voting models for two vectors, P_i and P_j, each of which implies 1 and 0 at lines c_2 and c_5, respectively, in the fault-free circuit. These models are used to determine the voltages obtained at the sites of the BFs for each combination of input values. In this model, it is further assumed that the voltages at the sites are typically interpreted as a logic 0 or a logic 1 by the gates in the fanout of each of these lines. These examples demonstrate how the values of the resistances, and hence the logic values assigned to the lines involved in the bridge, change with the combination of logic values applied to the gate inputs. This is true even for vectors that imply the same combination of values at the fault sites.

A BF in a CMOS circuit whose behavior is characterized by a voting model (either the one presented above or an extension thereof) is called a **driver-strength** BF. Under this model, the fault effect may originate at the site at which the vector implies 0 in the fault-free version of the circuit or at the site at which it implies 1, depending on the logic values applied to the inputs of the gates driving these lines and the values of the gains of their constituent transistors. This is illustrated by the example in Figure 10.2.

Table 10.3. *Conditions that a vector P_i, or a sequence of vectors (P_{i-1}, P_i), must satisfy to detect some types of faults, expressed in terms of the vector's ability to detect some single SAF*

Fault model	Fault type	Conditions
NFBF between b_1 and b_2	Wired-AND	(a) P_i implies 0 at b_1 and detects b_2 SA0, or (b) P_i implies 0 at b_2 and detects b_1 SA0
	Wired-OR	(a) P_i implies 1 at b_1 and detects b_2 SA1, or (b) P_i implies 1 at b_2 and detects b_1 SA1
	Driver-strength*	(a) P_i implies 0 at b_1 and detects b_2 SA0, or (b) P_i implies 1 at b_1 and detects b_2 SA1, or (c) P_i implies 0 at b_2 and detects b_1 SA0, or (d) P_i implies 1 at b_2 and detects b_1 SA1
TF at b	STR	P_{i-1} implies 0 at b and P_i detects b SA0
	STF	P_{i-1} implies 1 at b and P_i detects b SA1

*Only a subset of vectors satisfying these conditions will detect the fault.

However, for any given vector, the presence of a short between lines b_1 and b_2 is assumed to affect the logic value only at one of the two lines. Hence, the conditions for detection of a BF can again be described in terms of the conditions for detection of an SAu fault (where the value of u depends on the equivalent resistances of the gates driving the sites) at the line at which the fault effect originates, with the added condition that the vector imply a logic value u at the other line.

Table 10.3 shows all possible combinations of necessary conditions that a vector must satisfy to detect an NFBF of each of the above types.

Finally, the **Byzantine model** of a BF assumes a more general behavior of the fault sites. It assumes that the resistance of the short may be non-negligible and the voltage at each of the sites, b_1 and b_2, may fall between the logic values. The resulting intermediate voltage at any one of these lines may be interpreted differently by different gates in that line's fanout. Under this model, a fault effect may appear at one or both the lines involved in a bridge. Furthermore, the fault effect may propagate to the inputs of some arbitrary subset of gates driven by each fault site at which a fault effect appears. Under this model, no simple relationship exists between the conditions for detection of an SAF at one of the fault sites and the conditions for detection of the BF.

The TF model: The TF model is the simplest of the delay fault models. It is also referred to as the gross gate delay fault. (Chapters 2 and 8 describe this and other delay fault models.) Under this model, it is assumed that a single line in the circuit has a defect that causes the delay of either the rising or the falling transition at the line to increase by a sufficient amount to cause the delay of every path passing through the line to exceed the clock period. If the rising delay of line b is increased, the fault is called a slow-to-rise fault at line b, denoted as b STR. The fault b STF (slow-to-fall) is defined in a similar manner.

The conditions for detecting fault b STR are satisfied by any sequence of two vectors, (P_{i-1}, P_i), where P_{i-1} implies logic 0 at line b and P_i is a test for an SA0 fault at b.

Other delay fault models: The small gate delay fault model also assumes that a defect only alters the rise/fall time of a single line in a circuit. However, unlike the TF model, in this case the increase in delay may be small. The path delay fault model assumes that one or more defects may increase or decrease the transition time at any number of circuit lines on a path from a primary input to a primary output. The segment delay fault model considers paths which do not necessarily start at a primary input or end at a primary output.

In this chapter, we limit ourselves to single SAFs, multiple SAFs, TFs, and two-line NFBFs of types wired-AND, wired-OR, and driver-strength.

10.3.3   Modeled and dependent faults

As illustrated earlier, if the defect within a CUT is not characterized precisely by a modeled fault, then diagnosis may be unsuccessful. In some cases, diagnosis may be misleading, such as a TF at one line may be mis-diagnosed as an SAF at some distant circuit line. High costs are associated with such a misleading diagnosis. For example, during subsequent physical diagnosis, a significant amount of effort may be wasted examining defect-free areas of the CUT. Hence, it is crucial to consider all types of faults whose probability of occurrence is high. Some commonly studied combinations of fault types are (a) single SAFs and TFs, (b) single SAFs and BFs (Chess *et al.*, 1995; Lavo *et al.*, 1997; Millman *et al.*, 1990; Waicukauski and Lindbloom, 1989), and (c) single SAFs, TFs, and BFs. In many practical diagnosis scenarios, e.g., Waicukauski and Lindbloom (1989), a simultaneous presence of multiple faults is assumed likely; however, in most of this chapter, we assume that any given defective CUT has only one fault.

The most direct approach for diagnosis is to explicitly consider each type of fault that is likely to occur. In such a case, each likely fault is a **modeled fault**. If multiple fault types are considered, the fault model is said to be heterogeneous. In practice, two main reasons may prevent the use of such a direct approach. First, tools for simulation of some types of faults may be unavailable. Second, the run-time complexity of fault simulation and/or the storage space complexity of *fault dictionaries* (see Section 10.4.2) can be impractically high if the number of faults in the fault model is high. For example, for a circuit with L lines, only $2L$ faults need to be considered if a single SAF model is used. In contrast, $2L$ single SAFs and $L(L-1)/2$ two-line BFs may need to be considered when both types of faults are considered likely.

For above reasons, many diagnosis approaches explicitly model only some of the simplest types of faults, such as single SAFs. Each fault of any other type that is likely to occur is considered implicitly and is called a **dependent fault**. The conditions of detection for each dependent fault are expressed in terms of those for one or more of the corresponding modeled faults. Table 10.3 shows how the conditions for detection of TFs and NFBFs of some types can be expressed in terms of the conditions for detecting the corresponding SAFs. In this case, the STR fault at line b is considered as being dependent on fault b SA0. Similarly, the two-line, wired-AND BF between b_1 and b_2 is considered as being dependent on faults b_1 SA0 and b_2 SA0.

In an approach where some likely faults are modeled while others are considered as dependent, the failing vectors (FV^{f_j}) and error response (E^{f_j}) of each modeled fault f_j are determined by using a fault simulation tool for the fault. The relationships of the type shown in Table 10.3 are then used to obtain *approximate values* of the failing vectors and error response of each dependent fault. This process is illustrated for two common scenarios.

TFs as dependent on SAFs: From Table 10.3, any vector P_i in the given test sequence that detects the corresponding modeled fault, SA0 at line b, detects the dependent STR fault at line b, only if it is preceded by a vector that implies logic 0 at b. Hence, a CUT with an STR at line b fails a subset of vectors that a circuit with an SA0 fault at line b fails. Furthermore, for any vector that a circuit with an STR fault at line b fails, the error response is identical to that for a circuit with the single fault b SA0. In other words, $FV^{b\,\mathrm{SA0}} \sqsupseteq FV^{b\,\mathrm{STR}}$. Furthermore, $E^{b\,\mathrm{SA0}} \sqsupseteq E^{b\,\mathrm{STR}}$. The relationship between the dependent fault b STF and the corresponding modeled fault, b SA1, is similar.

Example 10.5 For the circuit and vectors shown in Figure 10.1, two vectors, P_3 and P_4, detect the SA1 fault at line x_3. Of these two vectors, only P_3 is preceded by a vector that implies logic 1 at x_3. Hence, only P_3 detects the dependent fault x_3 STF that corresponds to the above SAF. The error response of fault x_3 STF is hence $E^{x_3\,\mathrm{STF}} = ((0,0),(0,0),(0,1),(0,0))$, while that of the corresponding modeled fault, x_3 SA1, is $E^{x_3\,\mathrm{SA1}} = ((0,0),(0,0),(0,1),(0,1))$. Note that $E^{x_3\,\mathrm{SA1}} \sqsupseteq E^{x_3\,\mathrm{STF}}$. The reader can also verify that $FV^{x_3\,\mathrm{SA1}} \sqsupseteq FV^{x_3\,\mathrm{STF}}$. □

Two-line NFBFs as dependent on SAFs: Approximations to the failing vectors and error responses can again be obtained for each dependent fault as illustrated next. Consider the dependent two-line wired-AND BF between lines b_1 and b_2 (AB b_1b_2). Any vector that implies a 0 at line b_2 (in the fault-free circuit) **and** is a test for the single SA0 fault at line b_1 detects this BF. Similarly, any vector that implies 0 at line b_1 (in the fault-free circuit) **and** is a test for the single SA0 fault at line b_2 also detects this BF. Only a subset of vectors that detect b_1 SA0 imply a 0 at b_2. Similarly, only a

Table 10.4. *Approximations to failing vectors and error responses for some types of dependent faults in terms of single SAFs*

Dependent fault f_{dep}		$FV^{f_{dep}}$	$E^{f_{dep}}$
Fault model	Fault type		
NFBF	Wired-AND	$FV^{b_1\ \text{SA0}} + FV^{b_2\ \text{SA0}}$	$E^{b_1\ \text{SA0}} + E^{b_2\ \text{SA0}}$
between	Wired-OR	$FV^{b_1\ \text{SA1}} + FV^{b_2\ \text{SA1}}$	$E^{b_1\ \text{SA1}} + E^{b_2\ \text{SA1}}$
b_1 and b_2	Driver strength	$FV^{b_1\ \text{SA0}} + FV^{b_1\ \text{SA1}} +$ $FV^{b_2\ \text{SA0}} + FV^{b_2\ \text{SA1}}$	$E^{b_1\ \text{SA0}} + E^{b_1\ \text{SA1}} +$ $E^{b_2\ \text{SA0}} + E^{b_2\ \text{SA1}}$
TF at b	STR	$FV^{b\ \text{SA0}}$	$E^{b\ \text{SA0}}$
	STF	$FV^{b\ \text{SA1}}$	$E^{b\ \text{SA1}}$

subset of vectors that detect b_2 SA0 imply 0 at b_1. The above facts lead to the following relationships:

$$FV^{b_1\ \text{SA0}} + FV^{b_2\ \text{SA0}} \sqsupseteq FV^{\text{AB}\ b_1 b_2},$$

where '+' denotes a component-by-component logic OR operation, and

$$E^{b_1\ \text{SA0}} + E^{b_2\ \text{SA0}} \sqsupseteq E^{\text{AB}\ b_1 b_2}.$$

Table 10.4 summarizes such relationships for different types of two-line BFs and TFs, where each of these faults is treated as a dependent fault and each SAF as a modeled fault. In each case, the table entry covers the corresponding $FV^{f_{dep}}$ or $E^{f_{dep}}$ as appropriate.

Example 10.6 Consider the circuit and vectors shown in Figure 10.1 once again. The error response for the wired-AND BF between z_1 and z_2 (AB $z_1 z_2$) is $E^{\text{AB}\ z_1 z_2} = ((0, 0), (0, 0), (0, 0), (0, 1))$, while that of the corresponding modeled faults, z_1 SA0 and z_2 SA0, are $E^{z_1\ \text{SA0}} = ((0, 0), (1, 0), (1, 0), (0, 0))$ and $E^{z_2\ \text{SA0}} = ((0, 0), (0, 1), (0, 1), (0, 1))$. The component-by-component logic OR of the above two error responses is $((0, 0), (1, 1), (1, 1), (0, 1))$, which covers the error response of the corresponding dependent fault. □

10.4 Cause–effect diagnosis

Cause–effect diagnosis strategies simulate, explicitly or implicitly, the faults that are deemed likely to occur. (Modeled faults are simulated explicitly while un-modeled faults are simulated implicitly.) Simulation of a fault (the cause) provides the error response (the effect) of a CUT that has that fault. Error responses obtained in this manner are then compared with those obtained for the CUT.

In one variation, fault simulation is performed after the response of the CUT is obtained. This variation, called **post-test diagnostic fault simulation**, is suitable for scenarios where diagnosis is performed for a small number of CUTs. In another variation, extensive fault simulation is performed prior to the diagnosis of any CUT. The error responses obtained via fault simulation are stored in the form of a *fault dictionary*. When the response of a CUT is obtained, the fault within the CUT is identified by comparing the CUT response with the error responses stored in the fault dictionary.

10.4.1   Post-test diagnostic fault simulation

In this paradigm, diagnosis begins with mounting the CUT, C^*, on an automatic test equipment (ATE). The given set of patterns, $P_1, P_2, \ldots, P_{N_v}$, are then successively applied and the CUT's response, $v_i^*(z_l)$, is measured at each output z_l, for each vector P_i. These measurements collectively constitute the response Res^* of the CUT. Next, the fault-free version of the circuit, C, is simulated and the corresponding values of response, namely $v_i(z_l)$, are computed. These responses collectively constitute response Res of the fault-free circuit version. Res^* is then compared with Res to identify the failing vectors, FV^*, as well as the error response, E^*, of the CUT.

Next, each faulty version of the circuit, C^{f_j}, where f_j is a fault that belongs to the fault list, is simulated for vector P_i and the response values $v_i^{f_j}(z_l)$ are computed at each output z_l. For each fault f_j, the computed response values, Res^{f_j}, are compared with those for the fault-free circuit to determine the error response of C^{f_j}, E^{f_j}, and its failing vector response, FV^{f_j}. This process is then repeated for each subsequent vector.

10.4.1.1 A simple approach

First, consider an ideal scenario where the defect within the CUT (if any) is assumed to be accurately characterized by one of the modeled faults. In other words, the behavior of C^* is assumed to be identical to that of the fault-free circuit C or to that of one of the faulty versions of the circuit, namely $C^{f_1}, C^{f_2}, \ldots, C^{f_{N_f}}$. In such a scenario, almost any fault simulation approach described in Chapter 3 can be used to perform fault simulation.

The first vector is applied to the CUT and the response at its outputs is captured. A fault is retained in the fault list if the error response of the corresponding faulty circuit version matches that of the CUT; otherwise the fault is discarded. The above process is repeated for each subsequent vector. If faults $f_{j_1}, f_{j_2}, \ldots, f_{j_q}$ remain in the fault list after the last vector in the sequence, P_{N_v}, is processed, then the CUT is diagnosed as having **one** of these faults.

Example 10.7 Consider the circuit and vectors shown in Figure 10.1. If the response, $Res^* = ((0, 0), (1, 0), (1, 1), (0, 1))$, is captured for a CUT, then its error response is

$E^* = ((0, 0), (0, 1), (0, 0), (0, 0))$. Simulation is performed for each single SAF for vector P_1. Single SAFs x_1 SA0, x_2 SA0, x_3 SA0, c_1 SA0, c_2 SA0, z_1 SA1, and z_2 SA1 each cause errors for P_1, while the CUT response for P_1 is error-free. Hence, each of these faults is dropped from the fault list. For the second vector, P_2, only the remaining faults are simulated. Faults x_1 SA1 and x_3 SA1 are eliminated, since they do not cause any errors for this vector while the CUT response has an error at output z_2; faults x_2 SA1, c_1 SA1, and z_1 SA0 are eliminated, since they cause errors at output z_1 where an error is not captured for the CUT for this vector. After this step, only two faults, c_2 SA1 and z_2 SA0, are retained in the fault list. After the simulation of P_3, fault z_2 SA0 is also eliminated, leaving c_2 SA1 as the result of diagnosis. □

10.4.1.2 The need for a more sophisticated approach

Any fault simulation algorithm can be used to implement the above approach for post-test simulation-based diagnosis. Two factors motivate the development of more advanced approaches for diagnosis. First, the run-time complexity of the above approach can be high. Second, in real-life scenarios, the CUT often has one or more defects that are not precisely characterized by any of the modeled faults. In the following discussion, we address these two factors.

10.4.1.3 Acceleration of diagnostic fault simulation: only modeled faults

Consider the case in which the defect in any given CUT is assumed to be accurately characterized by a modeled fault. The following discussion is limited to single SAFs, TFs, and two-line NFBFs. Furthermore, only wired-AND, wired-OR, or driver-strength types of BF are considered.

One key way in which the run-time complexity of fault simulation, and hence that of diagnosis, can be reduced is by reducing the number of faults that need to be simulated for each vector. The following concepts are employed to accomplish this task.

Equivalent faults: Consider two faulty circuit versions C^{f_1} and C^{f_2}, which have the single SAF f_1 and f_2, respectively. If faults f_1 and f_2 are *equivalent* (see Chapter 3; Section 3.4.1), the response of C^{f_1} and C^{f_2} will be identical for any set of test vectors. Hence, from any set of equivalent faults, only one fault needs to be retained in the fault list used for simulation. However, it is generally possible to distinguish between faults f_1 and f_2 if f_1 dominates f_2. Hence, only equivalent faults should be collapsed during diagnosis.

A suitable procedure for fault collapsing can be obtained by modifying Procedure FaultCollapsing( ) described in Chapter 3 (Section 3.4.1.2) to (a) consider other types of target faults, such as TFs and two-line NFBFs of the above types, and (b) collapse only equivalent faults.

Faults not detected by the given test set: Consider a fault f_j that is not detected by the set of test vectors used for diagnosis. Clearly, a faulty version of the circuit with such a single fault f_j cannot be distinguished from the fault-free circuit by the given test vectors. Neither can such a fault be distinguished from any other fault that is also not detected by the test vectors. Finally, each such fault is definitely distinguished from any other that is detected by the test vectors. Hence, each such fault can be removed from the fault list used for fault simulation during diagnosis (Kautz, 1968). The information required for this operation can be easily obtained, especially if the faults that are not detected by the given test vectors are identified when the test vectors are generated. (When such faults are deleted from the fault list, it does affect the results of diagnosis for a CUT that does not fail any of the given test vectors.)

The use of a fault list obtained after the execution of the above operations can significantly reduce the complexity of fault simulation during diagnosis.

Existence of a path from a fault site to outputs with an erroneous response: Further reduction in fault simulation complexity can be obtained by utilizing the information contained in the responses measured by the ATE at the outputs of the CUT (Savir and Roth, 1982; Waicukauski and Lindbloom, 1989). Let $z_{l_1}, z_{l_2}, \ldots, z_{l_q}$ be the CUT outputs at which an erroneous response is captured for a particular failing vector. A single SAF can cause an erroneous response at output z_{l_1} only if the fault is located at a line in the transitive fanin of z_{l_1}. This condition should hold for each circuit output at which an erroneous response has been captured for the failing vector being analyzed. Hence, only those single SAFs whose fault sites belong to the transitive fanin of **each** of the outputs $z_{l_1}, z_{l_2}, \ldots, z_{l_q}$ need to be simulated for the particular failing vector. The same is true of a single TF.

An NFBF between two lines b_1 and b_2 can cause an erroneous response captured at CUT outputs only if b_1 and/or b_2 are in the transitive fanin of **each** output at which erroneous response has been captured for the CUT.

The above conditions can be applied to each successive vector that the CUT fails in order to further reduce the number of faults simulated for that and each subsequent vector.

Conditions for circuits containing primitive gates only (Waicukauski and Lindbloom, 1989): Consider a circuit that is comprised only of primitive gates, namely AND, OR, NAND, NOR, and inverter. Let the current vector imply a 0 at output z_l in the fault-free circuit. An SA0 fault at a line b can cause an error at z_l for the vector only if there exists one or more paths from b to z_l that contain an **odd** number of inverting gates, namely, NAND, NOR, and inverter. This is due to the fact that an SA0 fault at line b can only cause the value at the line to change from a 1 to a 0. This can cause the value at line z_l to change from 0 to 1 only if there exists such a path. Similarly, an SA1 fault at line b can cause an error at z_l only if there exists one or more paths from line b to z_l that contain an **even** number of inverting gates. Similar conditions also hold for cases where the current vector implies a 1 at z_l in the fault-free circuit. The conditions for faults b STR and b STF are identical to those for b SA0 and b SA1, respectively.

Consider a wired-AND BF between lines b_1 and b_2. Such a fault can cause an error at output z_l at which the vector implies logic 0 in the fault-free circuit only if there exists at least one path from either b_1 or b_2 to z_l that has an odd number of inverting gates. The conditions for the other output value and for other types of BFs are similar. The reader is advised to derive such conditions for each such scenario (see Problem 10.3).

Example 10.8   Consider vector P_2 and the example circuit shown in Figure 10.1(b). Since the value at output z_1 is 1 and one inverting gate separates x_1 and z_1, the SA0 fault at line x_1 cannot cause an error at output z_1 for this vector.   □

Each of the above conditions can be used if the values implied by a vector at the outputs of a fault-free circuit version are known; no fault simulation needs to be performed.

Fault-free value at each line for a given vector: In many fault simulation algorithms, including deductive, parallel pattern single fault propagation, and critical path tracing (see Chapter 3; Section 3.5), simulation of a particular vector begins with simulation of the fault-free version of the circuit. Let P_i be the current vector under consideration that the CUT has failed. Once the value implied by P_i at each line b, $v_i(b)$, is computed, the SAu fault at line b can be dropped from the fault list, if $u = v_i(b)$. This is due to the fact that any such fault is not excited by P_i and hence cannot be the cause of errors at any output of the CUT. Similarly, if $v_i(b) = 0$, then b STR can be deleted; otherwise, b STF can be deleted (see Problem 10.5).

Consider a wired-AND NFBF between lines b_1 and b_2. If a vector P_i implies 0 (or 1) at both lines b_1 and b_2 in the fault-free version of the circuit, then the fault is not excited. If P_i is a vector that the CUT has failed, the above BF can be dropped from the fault list prior to fault simulation for the vector. On the other hand, if vector P_i implies 1 and 0 at lines b_1 and b_2, respectively, in the fault-free version of the circuit, then the presence of the fault would cause an error only at line b_1 and only by changing its value from 1 to 0. In such a case, the BF needs to be retained in the fault list only if b_1 is in the transitive fanin of each of the outputs at which erroneous responses are captured for the CUT for P_i. Similar conditions can be derived for the other combination of values implied at the lines and for other types of BFs.

Fault dropping: In traditional fault simulation, a fault may be dropped from the fault list as soon as it is detected for the first time. In diagnostic fault simulation the fault dropping strategy is different. A fault may be dropped even before the first vector that detects it is simulated. This occurs when the CUT fails a vector that does not detect that fault. On the other hand, a fault may not be dropped even after it is simulated for one or more vectors that detect the fault. In general, a fault is retained in the fault list as long as the error response captured at CUT outputs continues to match the error response obtained via simulation of the corresponding faulty version of the circuit.

The above conditions can be used to accelerate post-test diagnostic fault simulation as described next. First, consider the case where only the single SAF model is used for diagnosis. Nearly any fault simulation algorithm described in Chapter 3 can be modified easily to incorporate the above conditions, since they require manipulation of the fault list using information that is readily available, namely circuit structure, values implied by a vector at lines of a fault-free version of the circuit, and errors observed at the outputs for a particular fault. In Waicukauski and Lindbloom (1989), the parallel-pattern single fault propagation fault simulator has been modified to obtain such a diagnostic fault simulator.

Next, consider the case where the defect in a CUT is assumed to be precisely modeled either by a single SAF or by a single two-line NFBF. In such a scenario, at the beginning of diagnosis of a CUT, the fault list is initialized to contain each single SAF as well as each such NFBF. Some faults are eliminated from the fault list using the above techniques. Subsequently, fault simulation is performed for the next vector, say, P_i. One alternative is to use a fault simulator that simulates both type of faults. If such a simulator is not available, then two simulators can be used, one that can simulate single SAFs and another that can simulate the NFBFs. Only those faults whose error responses match that of the CUT are then retained in the fault list for the analysis of subsequent vectors.

Example 10.9 The entire process can be illustrated for the single SAF model using the example circuit and vectors shown in Figure 10.1. Let $Res^* = ((0, 0), (1, 0),$ $(1, 1), (0, 1))$ be the response captured for a CUT. The error response of this CUT is $E^* = ((0, 0), (0, 1), (0, 0), (0, 0))$.

First, equivalence fault collapsing is performed to identify two sets of equivalent faults, $\{x_1 \text{ SA0}, c_1 \text{ SA0}, z_1 \text{ SA1}\}$ and $\{x_3 \text{ SA0}, c_2 \text{ SA0}, z_2 \text{ SA1}\}$. Only one fault is retained for each equivalence class, say $z_1 \text{ SA1}$ and $z_2 \text{ SA1}$. Next, faults $x_1 \text{ SA1}$, $c_1 \text{ SA1}$, $z_1 \text{ SA0}$, and $z_1 \text{ SA1}$ are eliminated from the fault list since no path exists from the sites of each of these faults to z_2, the only output at which an erroneous response is captured for the CUT. Since the failing vector P_2 implies a 1 at output z_2 in the fault-free circuit, faults $z_2 \text{ SA1}$ and $x_2 \text{ SA0}$ are eliminated based on the number of inverting gates between these lines and z_2 and the stuck-at values associated with these faults. Finally, if the logic values implied by the failing vector P_2 at each line in the fault-free version of the circuit are known, then fault $x_3 \text{ SA1}$ is eliminated, since it cannot be excited by the failing vector. Hence, only three faults, namely $x_2 \text{ SA1}$, $c_2 \text{ SA1}$, and $z_2 \text{ SA0}$, remain in the fault list, prior to fault simulation of even one fault. Fault simulation can then be performed using the reduced fault

list in a manner similar to that described earlier to determine that fault c_2 SA1
is the one that causes an error response that is identical to that observed for the
CUT. ☐

10.4.1.4 Defects not precisely described by modeled faults

Now consider the scenario where some likely fabrication defects are believed to be
not precisely characterized by any of the modeled faults. This scenario can be divided
into two main categories. The first one is where realistic defects are assumed to be
characterized accurately by either a modeled fault or a dependent fault. The second
category is where no particular relationship may exist between a likely defect and a
modeled fault, other than that the defects characterized by modeled faults are among
the most likely.

Realistic defects characterized by modeled or dependent faults: In such a case,
the previously described methodology is applied with some key modifications. First,
simulation is performed only for modeled faults while diagnosis is performed for
modeled as well as dependent faults. Fault simulation is used to determine the response
of each version of the circuit with a distinct modeled fault. Subsequently, relations
such as the ones described in Table 10.4 are used to obtain the approximate value
of the error response of each dependent fault. Second, different matching criteria are
used for modeled and dependent faults. While the error response of a modeled fault
must match that of a faulty CUT's response for the fault to be declared present within
that CUT, the approximate value of the error response of a dependent fault need only
cover the error response of the CUT. Some details of such a diagnostic procedure are
described next. The procedure will subsequently be illustrated for a diagnostic case
where single SAFs are considered as modeled faults and two-line wired-AND NFBFs
are considered as dependent faults.

1 The fault list used for fault simulation, F_s, is comprised of modeled faults only. In
   addition, a separate fault list F_d, which contains each modeled fault as well as each
   dependent fault, is used for diagnosis.

2 Prior to the simulation of a failing vector, the conditions described before are
   used to eliminate from F_d those modeled and dependent faults that cannot be the
   cause of errors captured at CUT outputs for that vector. A modeled fault f that is
   deleted from F_d may be deleted from F_s only if no dependent fault of f remains in
   F_d.

   For example, fault b_1 SA0 may be retained in F_s even after it is deleted from F_d, if
   one of the corresponding dependent faults, say the wired-AND BF between b_1 and
   b_2, still remains in F_d.

   This is due to the fact that the approximate error response of a dependent fault
   is computed using the error response of the corresponding modeled fault(s). A

modeled fault must hence be simulated as long as even one of its dependent faults remains in F_d.

3 Fault simulation is performed for the next vector, say P_i, and the error response for each modeled fault in F_s is obtained. Relations of the type shown in Table 10.4 are then used to obtain the approximate value of the error response of each dependent fault in F_d.

4 The error response of each modeled fault in F_d is compared with that of the CUT. If the two error responses do not match, then the modeled fault is deleted from F_d. If the CUT does not fail the vector, then each dependent fault that currently belongs to F_d is retained. (Why? See Problem 10.9.) On the other hand, if the CUT fails the vector, then any dependent fault whose error response does not match that of the CUT is removed from F_d. F_s is then updated by removing any modeled fault (a) that no longer belongs to F_d, **and** (b) none of whose corresponding dependent faults belong to F_d.

5 The above steps are repeated for each vector in the given test sequence. The faults that remain in F_d at the end of this process are declared as possible faults within the given CUT.

Example 10.10   Consider the example circuit and vectors shown in Figure 10.1. Consider single SAFs as modeled faults and two-line wired-AND NFBFs as dependent faults.

In the above diagnostic simulation methodology, F_s is initialized to contain each single SAF in the circuit and F_d is initialized to contain each single SAF and each two-line wired-AND NFBF.

Consider a CUT with response $Res^* = ((1, 0), (1, 1), (1, 0), (0, 1))$. The error response for such a CUT is given by $E^* = ((1, 0), (0, 0), (0, 1), (0, 0))$.

Since the fault effect appears at the first and the second outputs of the circuit for the first and third vectors, respectively, all single SAFs at lines that only fan out to one of the outputs, namely, x_1, x_3, c_1, c_2, z_1, and z_2, are removed from fault list F_d. All NFBFs between any pair of lines that fan out to only one output, namely AB $x_1 c_1$ and AB $c_2 x_3$, are also removed from F_d. All single SAFs are detected by this test set, hence no fault can be removed from the lists on the basis of the fact that it is not covered by the given test set. For vector P_1, the response at output z_1 of the fault-free circuit is 0. Hence, an error at this output cannot be caused due to the SA1 fault at x_2. This fault is deleted from F_d. A similar analysis of vector P_3 eliminates x_2 SA0, AB $x_1 x_2$, AB $x_1 x_3$, AB $x_2 x_3$, and AB $x_3 z_1$ from F_d. After the removal of all the above faults, $F_d = \{AB\ x_1 z_2, AB\ z_1 z_2\}$. Note that none of the modeled faults remain in F_d. However, the modeled faults x_1 SA0 and z_2 SA0 have the first fault in the above F_d as their dependent fault. Similarly, z_1 SA0 and z_2 SA0 have the second fault in the above F_d as their dependent fault. Hence, these faults remain in F_s, which is of the form $\{x_1$ SA0, z_1 SA0, z_2 SA0$\}$.

Fault simulation is then performed for each of the three faults in F_s and the error responses obtained for vector P_1. Fault x_1 SA0 is detected and its error response for this vector is $e_1^{x_1\ \text{SA0}}(z_1) = 1$, and $e_1^{x_1\ \text{SA0}}(z_2) = 0$. Equivalently, $E_1^{x_1\ \text{SA0}} = (1, 0)$. The other two faults are not detected and hence each of their error responses for this vector is $(0, 0)$. These error responses for the modeled faults are used to derive the error response of each of the two dependent faults in F_d. Using the relations in Table 10.4, we get

$$E_1^{\text{AB}\ x_1 z_2} = E_1^{x_1\ \text{SA0}} + E_1^{z_2\ \text{SA0}} = (1, 0),\ \text{and}$$

$$E_1^{\text{AB}\ z_1 z_2} = E_1^{z_1\ \text{SA0}} + E_1^{z_2\ \text{SA0}} = (0, 0).$$

Since the CUT fails vector P_1, and the error response of fault AB $z_1 z_2$ does not match that of the CUT, that fault is deleted from F_d. The error response of the other fault in F_d, namely AB $x_1 z_2$, matches that of the CUT and hence that fault is retained in F_d. This gives $F_d = \{ABx_1 z_2\}$. The modeled fault z_1 SA0 no longer has any dependent faults in F_d. Hence, $F_s = \{x_1\ \text{SA0}, z_2\ \text{SA0}\}$.

The above process is then repeated for the next vector, P_2. The reader can verify that this gives $E_2^{x_1\ \text{SA0}} = (0, 0)$ and $E_2^{z_2\ \text{SA0}} = (0, 1)$. The error response of the circuit with fault AB $x_1 z_2$ is computed as

$$E_2^{\text{AB}\ x_1 z_2} = E_2^{x_1\ \text{SA0}} + E_2^{z_2\ \text{SA0}} = (0, 1).$$

However, for this vector the CUT response is error-free. Hence, fault AB $x_1 z_2$ is retained in F_d. The process is then repeated for each of the subsequent vectors. The process terminates by providing fault AB $x_1 z_2$ as the only possible fault within the given CUT.   □

Other un-modeled faults: In some cases, realistic defects are believed to behave like multiple SAFs. We will discuss special effect–cause techniques for diagnosis of faults of arbitrary multiplicities in Section 10.5.

Finally, in some cases, no relationship can be assumed between the behavior of defects in some CUTs and the faults in the fault model. In such cases, no systematic technique can be developed to aid diagnosis and the probability of mis-diagnosis is high. In all such cases, typically fault simulation is performed for each modeled fault and the faults whose error responses are most similar to those of the CUT are reported as the candidate faults within the CUT.

10.4.2 Fault dictionary based diagnosis

This method is useful when diagnosis must be performed for a large number of fabricated copies of a circuit. Once the test sequence to be used for diagnosis is obtained, fault simulation is performed to determine the response of every faulty version of the circuit, each with a distinct target fault. In some cases, a fault may be dropped from the fault list after it is detected by κ vectors, while in other cases a fault is never dropped from the fault list. For each fault, the error response and failing vectors are determined. This information is collectively called a **fault dictionary** and is stored in one of many formats.

Diagnosis of CUTs can begin following the creation of the fault dictionary for the circuit. The error response of the CUT is compared with those stored in the fault dictionary for each fault in the fault list. If the error response obtained for the CUT matches (using an appropriate matching criterion) that for one or more faults, then the CUT is diagnosed as having one of those faults.

10.4.2.1 Complete fault dictionary

A **complete fault dictionary** contains the complete response of each faulty version of the circuit for each test vector. The response may be stored as is or in the form of an error response. In the first form, the **dictionary entry** for each fault f_j, $\mathcal{E}^{f_j}$, is the response of the circuit with that fault, i.e., $\mathcal{E}^{f_j} = Res^{f_j}$.

Example 10.11   For the circuit and test vectors shown in Figure 10.1 and the single SAF model, the complete fault dictionary contains the information shown in Table 10.1. □

The storage required to store the complete dictionary in this form is $N_v \times N_f \times m$ bits, where N_v, N_f, and m are the number of vectors in the given test sequence, the number of faults in the model, and the number of circuit outputs, respectively. This can be quite large. It can be reduced by replacing each response value by the corresponding error value. In other words, the response to vector P_i of the circuit with fault f_j at output z_l, $v_i^{f_j}(z_l)$, is replaced by the corresponding error value, $e_i^{f_j}(z_l)$, which is computed as $e_i^{f_j}(z_l) = v_i(z_l) \oplus v_i^{f_j}(z_l)$. In this form, the dictionary entry for fault f_j, $\mathcal{E}^{f_j} = E^{f_j}$. A complete fault dictionary in this form for our running example is shown in Table 10.5. In this format, the number of 1s in the dictionary is typically small, because of the fact that each fault is typically detected by a few vectors and each such vector causes erroneous response at a fraction of the outputs. Hence, this table is typically sparse.

Any representation used to minimize space required to store a sparse matrix (see any textbook on data structures) can be used to obtain a compact representation of the second version of the complete fault dictionary. One common approach represents, for each fault, each vector that detects the fault and the outputs at which it causes an erroneous response.

Example 10.12   The above representation for the complete fault dictionary in Table 10.5 is shown in Table 10.6 using the vector-output two-tuple. □

Table 10.5. *A complete fault dictionary describing error responses in a tabular form for the example circuit and vectors shown in Figure 10.1*

Fault	Error for vector P_i at output z_l							
	P_1		P_2		P_3		P_4	
	z_1	z_2	z_1	z_2	z_1	z_2	z_1	z_2
x_1 SA0								
c_1 SA0								
z_1 SA1	1						1	
x_1 SA1					1			
x_2 SA0	1	1					1	
x_2 SA1			1	1				
x_3 SA0								
c_2 SA0								
z_2 SA1		1						
x_3 SA1						1		1
c_1 SA1			1					
c_2 SA1				1				
z_1 SA0			1		1			
z_2 SA0				1		1		1

Each vector-output two-tuple is called a **point of detection** (Tulloss, 1980). Since there are N_v vectors and m circuit outputs, each point of detection requires $\lceil \log_2 N_v \rceil$ + $\lceil \log_2 m \rceil$ bits of storage. The storage space required to store the complete dictionary in this list representation is $M \times (\lceil \log_2 N_v \rceil + \lceil \log_2 m \rceil) + N_f \times \lceil \log_2 N_f \rceil$ bits, where M is the total number of points of detection in the fault dictionary.

Other list representations can be obtained by enumerating different combinations of faults, vectors and outputs (Boppana *et al.*, 1996). In one such representation, the vectors are enumerated. For each vector, a list of two-tuples, namely a fault detected by the vector and an output at which the fault causes an error for the vector, is enumerated. Another list representation can be obtained by enumerating all combinations of vectors and outputs. In such a list, each entry is a fault that causes an error at that output in response to that vector. A total of eight such representations and their storage-space complexities can be found in Boppana *et al.* (1996).

The run-time complexity of fault simulation required to generate a complete fault dictionary is high, since no faults can be dropped and each fault must be simulated for each vector in the test set. The storage space complexity can also be very high.

10.4.2.2 Reduced fault dictionaries

The high run-time complexity for generating and high space complexity for storing a complete fault dictionary has motivated the development of several smaller fault dictionaries.

Table 10.6. *A list representation of the complete fault dictionary shown in Table 10.5*

Fault	Points of detection
x_1 SA0	
c_1 SA0	
z_1 SA1	$P_1{:}z_1$; $P_4{:}z_1$
x_1 SA1	$P_3{:}z_1$
x_2 SA0	$P_1{:}z_1$; $P_1{:}z_2$; $P_4{:}z_1$
x_2 SA1	$P_2{:}z_1$; $P_2{:}z_2$
x_3 SA0	
c_2 SA0	
z_2 SA1	$P_1{:}z_2$
x_3 SA1	$P_3{:}z_2$; $P_4{:}z_2$
c_1 SA1	$P_2{:}z_1$
c_2 SA1	$P_2{:}z_2$
z_1 SA0	$P_2{:}z_1$; $P_3{:}z_1$
z_2 SA0	$P_2{:}z_2$; $P_3{:}z_2$; $P_4{:}z_2$

Table 10.7. *A pass–fail dictionary corresponding to the dictionary shown in Table 10.5*

Fault	Vector P_i pass/fail			
	P_1	P_2	P_3	P_4
x_1 SA0				
c_1 SA0				
z_1 SA1	1			1
x_1 SA1			1	
x_2 SA0	1			1
x_2 SA1		1		
x_3 SA0				
c_2 SA0				
z_2 SA1	1			
x_3 SA1			1	1
c_1 SA1		1		
c_2 SA1		1		
z_1 SA0		1	1	
z_2 SA0		1	1	1

Table 10.8. *List format of a pass–fail dictionary corresponding to the dictionary shown in Table 10.5*

Fault	Failing vectors
x_1 SA0	
c_1 SA0	
z_1 SA1	P_1; P_4
x_1 SA1	P_3
x_2 SA0	P_1; P_4
x_2 SA1	P_2
x_3 SA0	
c_2 SA0	
z_2 SA1	P_1
x_3 SA1	P_3; P_4
c_1 SA1	P_2
c_2 SA1	P_2
z_1 SA0	P_2; P_3
z_2 SA0	P_2; P_3; P_4

Pass–fail fault dictionary: For each faulty circuit version, a **pass–fail fault dictionary** contains a list of the vectors that detect the fault.

Example 10.13 For the circuit and test vectors shown in Figure 10.1 and the single SAF model, the pass–fail fault dictionary contains the information shown in Table 10.7. □

The space required to store the pass–fail dictionary in this form is $N_v \times N_f$ bits, which is smaller by a factor of m compared to the space required for the full fault dictionary. This dictionary can also be stored in various list formats.

Consider a list format for a pass–fail dictionary that is analogous to that described in Table 10.6 for the complete fault dictionary. In this format, each set of equivalent faults is enumerated and for each representative equivalent fault, the vectors that detect the fault are listed.

Example 10.14 Table 10.8 shows in the above format a pass–fail dictionary corresponding to the complete fault dictionary shown in Table 10.5. □

The storage space required for the list form of a pass–fail dictionary is lower than that for the corresponding complete dictionary by a factor approximately equal to the average number of outputs at which any fault causes errors. The fault simulation complexity is only a little lower than that required to generate a complete fault dictionary for the following reason. Since each vector that a fault fails must be identified, a fault cannot be dropped from the simulation. However, since the development of such a dictionary does not require information about the outputs at which a fault causes errors, faster fault-simulation techniques, such as those based on critical path tracing (see Chapter 3; Section 3.5.5), can be used to somewhat reduce the fault simulation complexity.

The above benefits are at a cost of reduced diagnostic resolution.

Example 10.15 For our example, in the pass–fail dictionary, single SAFs x_2 SA0 and z_1 SA1 have identical dictionary entries. In contrast, in the complete fault dictionary, the two faults have distinct entries.   □

Experimental results in Aitken (1995) report a scenario where the pass–fail dictionary was found to sometimes provide more accurate diagnosis of BFs. One possible explanation is based on the conjecture that while a BF simulator models a BF accurately enough to correctly identify all the vectors that the fault fails, it does not model it accurately enough to correctly identify the outputs at which the fault causes errors for each failing vector. Hence, the information contained in the pass–fail dictionary is quite accurate, while the additional information in a complete fault dictionary may be inaccurate and sometimes lead to incorrect diagnosis.

κ-detection fault dictionary: A **κ-detection fault dictionary** contains a part of the error response of each faulty circuit version, starting with the error response for the first vector and ending with that for the κ-th vector that detects the fault (Abramovici *et al.*, 1991). For any vector that follows the vector that detects a fault for the κ-th time, the response of the circuit version with that fault is assumed to be error-free. A **κ-detection pass–fail fault dictionary** contains information that identifies, for each fault, the first κ vectors in the given test sequence that detect the fault. A κ-detection dictionary with $\kappa = 1$ is called a **stop on first error (SOFE) dictionary** (Abramovici *et al.*, 1991).

Example 10.16 Tabular as well as list representations for a 1-detection fault dictionary are shown in Table 10.9 for the complete fault dictionary shown in Table 10.5.   □

The resolution of a κ-detection fault dictionary is typically lower than that of the corresponding complete dictionary. In the above example, the 1-detection dictionary contains identical entries for single SAFs c_1 SA1 and z_1 SA0, while the complete fault dictionary contains distinct entries for these faults.

The storage space complexity to store the list version of a κ-detection dictionary is always smaller than that for the complete dictionary and is *upper bounded* by $\kappa \times m \times N_f \times (\lceil \log_2 N_v \rceil + \lceil \log_2 m \rceil) + N_f \times \lceil \log_2 N_f \rceil$ bits.

The run-time complexity of fault simulation required to generate a κ-detection dictionary is lower than that required to generate a complete dictionary, since in the

Table 10.9. *A κ-detection dictionary ($\kappa = 1$) corresponding to the dictionary shown in Table 10.5: (a) tabular, and (b) list representation*

(a)

Fault	Error for vector P_i at output z_l							
	P_1		P_2		P_3		P_4	
	z_1	z_2	z_1	z_2	z_1	z_2	z_1	z_2
x_1 SA0								
c_1 SA0								
z_1 SA1	1							
x_1 SA1					1			
x_2 SA0	1	1						
x_2 SA1			1	1				
x_3 SA0								
c_2 SA0								
z_2 SA1		1						
x_3 SA1						1		
c_1 SA1			1					
c_2 SA1				1				
z_1 SA0			1					
z_2 SA0				1				

(b)

Fault	Points of detection
x_1 SA0	
c_1 SA0	
z_1 SA1	$P_1{:}z_1$
x_1 SA1	$P_3{:}z_1$
x_2 SA0	$P_1{:}z_1$; $P_1{:}z_2$
x_2 SA1	$P_2{:}z_1$; $P_2{:}z_2$
x_3 SA0	
c_2 SA0	
z_2 SA1	$P_1{:}z_2$
x_3 SA1	$P_3{:}z_2$
c_1 SA1	$P_2{:}z_1$
c_2 SA1	$P_2{:}z_2$
z_1 SA0	$P_2{:}z_1$
z_2 SA0	$P_2{:}z_2$

former a fault can be dropped from the fault list after it is detected κ times. At one extreme, when $\kappa = 1$, the fault simulation complexity becomes similar to that required to determine fault coverage.

A complete dictionary and the corresponding κ-detection dictionary differ signifi-cantly in the *negative information* they provide (Millman *et al.*, 1990), where **negative information** is the information about vectors that do not detect a fault. In a complete dictionary, any vector that is reported as having an error-free response for a fault is known to not detect the fault. In contrast, for a κ-detection fault dictionary, the same can be assumed only of those vectors that are applied before the vector that detects the fault for the κth time. No such assumption can be made about any vector applied after the vector that detects the fault for the κth time, since all such vectors are assumed to have an error-free response for the fault, independent of whether or not they really detect the fault.

Example 10.17 The absence of vector P_1 from the entry for z_1 SA0 as well as that for z_2 SA0 in the 1-detection dictionary in Table 10.9(b) demonstrates that P_1 does not detect either of these two single SAFs. However, any such inference drawn on the basis of the absence of P_3 from the entry of each of these faults will be erroneous. This fact that can be confirmed by examining the corresponding complete dictionary in Table 10.5. □

Hybrid dictionary: A **hybrid fault dictionary** (Pomeranz and Reddy, 1997) is one in which for each vector P_i, an error response is reported at selected outputs, while the remaining outputs are combined into a single group and the error responses for these outputs are combined into a single pass–fail value.

Let g_i be the set of outputs whose error responses are combined into a single pass–fail value for vector P_i. The error response is retained in the dictionary for each output that does not belong to set g_i. Note that the complete and pass–fail dictionaries can each be seen as an extreme special case of such a dictionary. In a complete dictionary, g_i is empty for each vector, P_i, and the error response is provided for each vector and each output. In a pass–fail dictionary, g_i is the set of all circuit outputs for each vector P_i; consequently only pass–fail information is provided for each vector.

Example 10.18   Table 10.10 shows an example complete fault dictionary, along with a hybrid dictionary, and its pass–fail version. For the hybrid dictionary shown in this table, outputs z_1, z_2, and z_4 are grouped into group g_1 for vector P_1, while output z_3 is left out. For vector P_2, outputs z_1 and z_4 are grouped into group g_2, while outputs z_2 and z_3 are each left out. Using the above notation, the structure of this hybrid dictionary can be described by its groups $g_1 = \{z_1, z_2, z_4\}$ and $g_2 = \{z_1, z_4\}$.   □

Note that, for each vector P_i, the structure of the complete fault dictionary can be described by $g_i = \{\}$, while that of the pass–fail dictionary can be described by $g_i = \{z_1, z_2, \ldots, z_m\}$.

A heuristic procedure to obtain a hybrid dictionary is described in Pomeranz and Reddy (1997). First, both a complete and a pass–fail dictionary are created for the circuit. The procedure begins by considering the hybrid dictionary as being identical to the pass–fail dictionary, i.e., with $g_i = \{z_1, z_2, \ldots, z_m\}$, for each vector P_i. Pairs of faults that cannot be distinguished by the pass–fail dictionary but can be distinguished by the complete fault dictionary are identified and stored in a set π. For each vector P_i, and for each output z_l that belongs to g_i, a new version of the hybrid dictionary is temporarily created in which output z_l is taken out of g_i. The number of pairs of faults in set π that can be distinguished is counted to determine the increase in diagnostic resolution provided by the temporary hybrid dictionary. Of all the temporary hybrid dictionaries, one that provides the highest increase in diagnostic resolution is selected as the next hybrid dictionary. Set π is updated to contain only the pair of faults that cannot be distinguished by this hybrid dictionary but can be distinguished by the complete dictionary. The above process of enumerating temporary hybrid dictionaries and selecting the one that provides the highest increase in resolution is repeated until set π becomes empty. Note that this procedure is always guaranteed to terminate. At worst, the final hybrid dictionary will become identical to the complete fault dictionary; typically, its size is much smaller.

Table 10.10. *An example of a hybrid dictionary: (a) a complete dictionary, (b) one of the corresponding hybrid dictionaries, and (c) the corresponding pass–fail dictionary*

(a) A complete dictionary

Fault	Error for vector P_i at output z_l							
	P_1				P_2			
	z_1	z_2	z_3	z_4	z_1	z_2	z_3	z_4
f_1	1	1						
f_2	1	1	1					
f_3	1	1			1		1	1
f_4	1			1	1	1		1

(b) A hybrid dictionary

Fault	Error for vector P_i for groups of outputs				
	P_1			P_2	
	$\{z_3\}$	$\{z_1, z_2, z_4\}$	$\{z_2\}$	$\{z_3\}$	$\{z_1, z_4\}$
f_1		1			
f_2	1	1			
f_3		1		1	1
f_4		1	1		1

(c) A pass–fail dictionary

Fault	Pass–fail for vector P_i	
	P_1	P_2
f_1	1	
f_2	1	
f_3	1	1
f_4	1	1

Example 10.19 Consider the complete and pass–fail dictionaries shown in Table 10.10(a) and Table 10.10(c), respectively. The pairs of faults that can be diagnosed when the complete fault dictionary is used but cannot be diagnosed using the pass–fail dictionary are stored as $\pi = \{(f_1, f_2), (f_3, f_4)\}$. The pass–fail dictionary in Table 10.10(c) is considered the first hybrid dictionary. Next, for each vector and each output, the output is removed from the group and an alternative hybrid dictionary is created. Some of the alternative dictionaries created in this manner are shown in Table 10.11(a), (b), and (c). Of these, several alternatives help distinguish one out of two pairs of faults in set π. One of these alternatives, say the one shown in Table 10.11(b), is designated as the next hybrid dictionary. This helps distinguish the pair of faults (f_3, f_4). Set π is hence updated to $\pi = \{(f_1, f_2)\}$. The process is then repeated to enumerate the next round of alternative hybrid dictionaries and one, say

Table 10.11. *Some temporary hybrid dictionaries enumerated during the search*

(a)

Fault	Error for vector P_i		
	P_1		P_2
	$\{z_1\}$	$\{z_2, z_3, z_4\}$	$\{z_1, z_2, z_3, z_4\}$
f_1	1	1	
f_2	1	1	
f_3	1	1	1
f_4	1	1	1

(b)

Fault	Error for vector P_i		
	P_1		P_2
	$\{z_2\}$	$\{z_1, z_3, z_4\}$	$\{z_1, z_2, z_3, z_4\}$
f_1	1	1	
f_2	1	1	
f_3	1	1	1
f_4		1	1

(c)

Fault	Error for vector P_i		
	P_1	P_2	
	$\{z_1, z_2, z_3, z_4\}$	$\{z_4\}$	$\{z_1, z_2, z_3\}$
f_1	1		
f_2	1		
f_3	1	1	1
f_4	1	1	1

(d)

Fault	Error for vector P_i			
	P_1		P_2	
	$\{z_2\}$	$\{z_3\}$	$\{z_1, z_4\}$	$\{z_1, z_2, z_3, z_4\}$
f_1	1		1	
f_2	1	1	1	
f_3	1		1	1
f_4			1	1

the one shown in Table 10.11(d), is designated as the next hybrid dictionary. Since this hybrid dictionary can distinguish every pair of faults that can be distinguished by the complete fault dictionary, the procedure terminates and provides this as the final hybrid dictionary. ◻

In addition to the dictionaries described above, several others have been proposed. Some of these methods use space compaction, where some vector-output pairs are removed or combined, as in some of the above approaches. Others perform time compression, i.e., use a linear compressor (such as a linear feedback shift register) to compress the error response of each faulty version of the circuit, at each output, for each vector, into a shorter signature.

10.4.2.3 Use of a fault dictionary for diagnosis

The manner in which a dictionary is used for diagnosis depends mainly on what can be assumed about the defect in the CUT. Three main scenarios are considered.

1 *Modeled faults:* In this scenario, one of the faults in the fault model used to prepare the fault dictionary is assumed to accurately characterize the behavior of the defect in any CUT.

2 *Modeled or dependent faults:* In this scenario, the defect in any given CUT is assumed to be characterized accurately either by a modeled fault or by a dependent fault. In this case, the dictionary for the faults in the model can be interpreted in a more general manner to permit diagnosis.

3 *Arbitrary defects:* In both the above scenarios, it is generally assumed that there exists a small probability that the defect in a CUT may not be accurately characterized by any of the modeled faults or dependent faults. In contrast, in this scenario, diagnosis is performed assuming that this probability is relatively large. In this case, the reasons for using the particular fault model may be (a) non-availability of a better fault model, or (b) if a better model exists, then impractically high complexity of dealing with that model or unavailability of appropriate tools for the model.

Modeled faults: Diagnosis is straightforward in this case. The error response of the CUT is transformed into the format of the entry of the given dictionary. For example, if a complete dictionary is used for diagnosis, then the error response of the CUT is not transformed in any manner. On the other hand, if a pass–fail dictionary is provided, then for each vector the errors observed at all the outputs of the CUT are combined into a single pass–fail value. In other words, error response E^* is transformed into failing vectors FV^*. Similar transformations can be identified for other types of dictionaries described before, namely the κ-detection and hybrid dictionaries.

Once the error response of the CUT is transformed in the above manner, it is compared with the dictionary entry for each set of equivalent faults. The sets of faults whose entries match the transformed CUT error response are reported as the possible fault sets. For a complete dictionary, the above match corresponds to an error response match; for a pass–fail dictionary, it corresponds to a failing vector match.

Modeled or dependent faults: In all such scenarios, a higher diagnostic resolution can be obtained by creating directly a fault dictionary for all the faults that are deemed likely to occur. That is, ideally the set of modeled faults should be expanded to include the other faults that are also believed as being likely to occur. If this approach is taken, then all the likely defects in the CUT are accurately characterized by one or more modeled faults, and the procedure described above can be used for diagnosis.

Sometimes, such a direct approach cannot be taken due to the unavailability of fault simulators for some types of faults, or due to the impractically high complexity of dealing explicitly with those faults. In such cases, the dictionary is created for a subset of faults, which are considered modeled faults; the remainder of the faults are considered as dependent faults. Whenever a combination of modeled and related faults are considered for diagnosis, two changes are made. First, the dictionary is augmented, either explicitly or implicitly, to include an entry for each dependent fault. Second, the matching criterion used for comparing the CUT error response to the dictionary entries is changed.

Commonly, the dictionary entry for a dependent fault is implicitly represented in terms of the entries for one or more modeled faults to reduce dictionary size. In such cases, typically the dictionary entry for the dependent fault is not completely accurate, and is hence referred to as an **approximate entry**. For example, when the complete dictionary representation is used, the error response for the dependent fault obtained in this manner is referred to as **approximate error response**.

Let us consider two specific examples of such a diagnosis scenario. As the first scenario, consider a case where the single SAF model is used to generate the dictionary, but each likely defect is assumed to behave either as a single SAF or a TF. As the second scenario, consider a case where the single SAF model is used to generate the dictionary, but each likely defect is assumed to behave either as a single SAF or a two-line NFBF.

TFs as dependent faults on single SAFs: As described in Section 10.3 and Table 10.4, the error response for the SA0 fault at line b covers the error response of fault b STR, i.e., $E^{b\ SA0} \sqsupseteq E^{b\ STR}$. Due to the above relationship, it is possible to use a fault dictionary created for single SAFs to diagnose circuits where a defect may be assumed to behave as a single SAF or a TF. In such a scenario, first the dictionary created for single SAFs is augmented to include an entry for each TF. Then the manner in which the dictionary is used is modified.

First, consider the case where a complete fault dictionary for single SAFs is provided. In the light of the above relationship between the conditions for detection of an SAF and the corresponding TF, the error response associated with an SA0 (SA1) fault on line b is now also associated with the STR (STF) fault at b. The matching criterion is then specified as described next to take into account that, depending on the value implied at line b by the previous vector in the test sequence, a vector that detects an SA0 fault at line b *may* or *may not* detect the STR fault at that line. Under the new matching criteria, any TF whose approximate error response *covers* the error response of the CUT is declared as a possible fault within the CUT. Furthermore, any SAF whose *error response matches* that of the CUT is also declared as a possible fault. A field can be added to each row of the dictionary to indicate the matching criterion used for the corresponding fault(s).

Next, consider a scenario where a pass–fail dictionary is provided for single SAFs. In this case, a new entry is created for an STR fault at a line b by using the pass–fail information for b SA0 fault. Similarly, an entry is created for each STF fault. An SAF is declared as a possible fault if the corresponding failing vectors *match* those of the CUT. In contrast, a TF is declared as a possible fault if the corresponding approximate failing vectors *cover* those of the CUT.

Example 10.20 Consider the circuit and the sequence of test vectors shown in Figure 10.1 once again. First, the single SAF dictionary shown in Table 10.5 is modified to list each TF. In addition, each entry of the table is augmented with the corresponding matching criterion. For each entry corresponding to an SAF, the criterion is 'match'; for each entry corresponding to a TF, the criterion is 'cover'. The dictionary obtained in this manner is shown in Table 10.12. □

Table 10.12. *A fault dictionary for SAFs and TFs for the circuit and vectors in Figure 10.1, derived from the SAF dictionary in Table 10.5*

Fault	Comparison criterion	Error for vector P_i at output z_l							
		P_1		P_2		P_3		P_4	
		z_1	z_2	z_1	z_2	z_1	z_2	z_1	z_2
x_1 SA0	Match								
c_1 SA0	Match								
z_1 SA1	Match	1						1	
x_1 SA1	Match					1			
x_2 SA0	Match	1	1					1	
x_2 SA1	Match			1	1				
x_3 SA0	Match								
c_2 SA0	Match								
z_2 SA1	Match		1						
x_3 SA1	Match						1		1
c_1 SA1	Match			1					
c_2 SA1	Match				1				
z_1 SA0	Match			1		1			
z_2 SA0	Match				1		1		1
x_1 STR	Cover								
c_1 STR	Cover								
z_1 STF	Cover	1						1	
x_1 STF	Cover					1			
x_2 STR	Cover	1	1					1	
x_2 STF	Cover			1	1				
x_3 STR	Cover								
c_2 STR	Cover								
z_2 STF	Cover		1						
x_3 STF	Cover						1		1
c_1 STF	Cover			1					
c_2 STF	Cover				1				
z_1 STR	Cover			1		1			
z_2 STR	Cover				1		1		1

Note that the changes to the dictionary are so systematic that it is really not necessary to explicitly change the dictionary. Typically, to save storage space, the program that matches the CUT error responses to dictionary entries is altered to take into account such changes, while working off the single SAF dictionary.

Example 10.21 Assume that error response E^* for a given CUT for the circuit considered in the above example is $((0, 0), (0, 0), (0, 1), (0, 0))$. This is compared with each row in Table 10.12; for each row marked 'match' a match is sought, and for each row marked 'cover' a cover is sought. Faults x_3 STF and z_2 STR are declared as the two faults (among all the SAFs and TFs) that may be present within the CUT.   □

Two-line NFBFs as dependent on single SAFs: An approach similar to the one above can also be used for two-line NFBFs of type wired-AND, wired-OR, and driver-strength (Chess *et al.*, 1995; Lavo *et al.*, 1997; Millman *et al.*, 1990; Narayanan *et al.*, 1997; Waicukauski and Lindbloom, 1989).

A dictionary entry for a wired-AND NFBF between lines b_1 and b_2 (say, fault f_1) can be constructed using the error responses for fault b_1 SA0 (say, fault f_2) and fault b_2 SA0 (say, fault f_3), as described in Table 10.4. Second, since a vector that detects one of these SA0 faults may or may not detect the BF (depending on whether it also implies logic 0 at the other line), the response of a CUT with fault f_1 may be covered by the above dictionary entry. This is reflected by marking the matching criterion on the dictionary entry for fault f_1 as 'cover'.

Example 10.22   The above method can be employed to augment the SAF dictionary shown in Table 10.5 to obtain the dictionary for SAFs (modeled) and two-line wired-AND NFBFs (dependent) as shown in Table 10.13.   □

The approximate values of failing vectors and error responses for a BF described above can be further refined using additional information provided by the SAF dictionary (Lavo *et al.*, 1998a). As described in Table 10.3, only those vectors which detect fault b_1 SA0 and imply logic 0 at line b_2 or vice versa detect the wired-AND BF between b_1 and b_2. For some vectors, the SAF dictionary provides the information whether the second condition is satisfied.

Example 10.23   In the dictionary shown in Table 10.13, the dictionary entries for faults x_1 SA0 and z_2 SA0 are

$$E^{x_1 \text{SA0}} = ((1, 0), (0, 0), (0, 0), (1, 0)), \text{ and}$$
$$E^{z_2 \text{SA0}} = ((0, 0), (0, 1), (0, 1), (0, 1)).$$

An examination of the above error responses for vector P_4 indicates that P_4 detects the fault x_1 SA0 as well as z_2 SA0. This clearly indicates that P_4 implies logic 1 at x_1 as well as at z_2 (in the fault-free version of the circuit). Hence, P_4 cannot detect the wired-AND BF between x_1 and z_2. This fact can be used to obtain the following approximate value of the error response for the wired-AND BF between x_1 and z_2:

$$((1, 0), (0, 1), (0, 1), (0, 0)).$$

This should be compared with the approximate value of the same error response shown in Table 10.13, namely

$$((1, 0), (0, 1), (0, 1), (1, 1)).$$

Table 10.13. *A fault dictionary for SAFs and two-line wired-AND BFs for the circuit and vectors in Figure 10.1, derived from the SAF dictionary in Table 10.5*

Fault	Comparison criterion	Error for vector P_i at output z_l							
		P_1		P_2		P_3		P_4	
		z_1	z_2	z_1	z_2	z_1	z_2	z_1	z_2
x_1 SA0	Match								
c_1 SA0	Match								
z_1 SA1	Match	1						1	
x_1 SA1	Match					1			
x_2 SA0	Match	1	1					1	
x_2 SA1	Match			1	1				
x_3 SA0	Match								
c_2 SA0	Match								
z_2 SA1	Match		1						
x_3 SA1	Match						1		1
c_1 SA1	Match			1					
c_2 SA1	Match				1				
z_1 SA0	Match			1		1			
z_2 SA0	Match				1		1		1
$x_1 x_2$ AB	Cover								
$x_1 x_3$ AB	Cover								
$x_2 x_3$ AB	Cover	1	1					1	
$x_1 z_2$ AB	Cover	1			1		1	1	1
$x_3 z_1$ AB	Cover			1	1	1			
$z_1 z_2$ AB	Cover			1	1	1	1		1

Note that the new approximate value of the error response of this BF is closer to the actual value of its error response which the reader can derive by explicit simulation of the fault as

$$((1, 0), (0, 0), (0, 1), (0, 0)).$$

□

In general, for a wired-AND bridge between b_1 and b_2, the approximate value of the error response given in Table 10.13 is modified by removing all the error values for any vector P_i for which an error appears in the dictionary entry of b_1 SA0 as well as in the entry for b_2 SA0. Similar modifications can be derived for each of the other types of BFs. The reader is invited to make all such changes to the dictionary shown in Table 10.13.

Once the dictionary is obtained for the SAFs and BFs, the procedure described before can be used to perform diagnosis assuming that a defect may be characterized either by a single SAF or by a two-line BF of the above type.

10.5   Effect–cause diagnosis

The diagnostic fault simulation and fault dictionary approaches both require simulation of each possible fault in a fault model. Fault simulation can be considered as a process that starts with each possible *cause* of error, namely a modeled fault, and computes the corresponding error response, i.e., its *effect*. Such diagnostic approaches were collectively referred to as cause–effect. The run-time complexity of a cause–effect approach is reasonable for the single SAF model, where the number of faults, N_f, is proportional to the number of lines in the circuit, L. It may even be considered practical for circuits with a moderate number of lines for a two-line BF model, where N_f is proportional to L^2. In contrast, in the multiple SAF model, N_f is proportional to 3^L. For such a model, it is clearly impossible to perform fault simulation for each modeled fault, even for a circuit with a small number of lines.

Effect–cause diagnostic approaches have been developed to enable diagnosis for fault models that consider faults with arbitrary multiplicities for which the run-time complexity of any cause–effect approach is impractically high. In **effect–cause** approaches to diagnosis (Abramovici and Breuer, 1980; Cox and Rajski, 1988), the response of the CUT to each vector, i.e., the effect, is captured and the circuit analyzed to eliminate those faults, i.e., the causes, whose absence from the circuit can be inferred from the response. For example, for a multiple SAF model, the response captured for the CUT is used to analyze the circuit to determine which lines do not have either the SA0 or the SA1 fault. In this manner, an effect–cause approach eliminates the need to explicitly consider each multiple SAF and hence makes run-time complexity of diagnosis practical. In this section, we limit the discussion to multiple SAFs.

10.5.1   Basics of effect–cause diagnosis

When diagnosis of a particular CUT begins, it is assumed that any line in the CUT may either be fault-free (ff), have an SA0 fault, or have an SA1 fault. The **fault status** of each line b is denoted by $fs(b)$ and is initialized as $fs(b) = \{ff, \text{SA0}, \text{SA1}\}$. A line that is known to have neither of the SAFs is said to be **fault-free** and has a fault status $\{ff\}$.

Next, the given vectors are applied to the CUT using an ATE and the response captured. These values become the starting point of a deduction process, where the actual values that must appear at internal lines of the CUT to cause the captured response are deduced. Let $v_i^*(b_l)$ denote the **actual value**, i.e., the value deduced at line b_l of the CUT when vector P_i is applied. At a CUT output, the captured response is the actual value. (Recall that b_l denotes a primary input, an internal line, or a primary

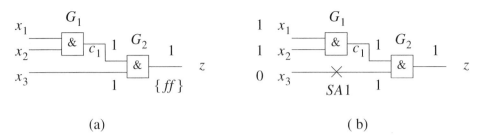

Figure 10.3 The meaning of the value deduced at a line: (a) the value deduced at a gate input based on the value deduced at its output, and (b) the presence of a fault at a line causes the value deduced at a line to be different from that implied at the line

output.) At an input b_l of a gate G_j, the deduced value for a vector P_i, $v_i^*(b_l)$, denotes how the value at b_l is interpreted by gate G_j.

Example 10.24 Consider the deduced values shown in Figure 10.3(a). Since logic 1 is deduced at output z of gate G_2, whose output is known to be fault-free, logic 1 is deduced at both the inputs of gate G_2, namely c_1 and x_3.

Consider the scenario depicted in Figure 10.3(b) where the fault status of each line is $\{ff\}$, except x_3, for which $fs(x_3) = \{SA1\}$. Due to the presence of an SA1 fault at x_3, gate G_2 interprets the value at the line as a 1, while logic 0 is applied to the line. Hence, logic 1 is deduced at line x_3 for this scenario. ◻

Fault-status update: A line at which a logic value u is deduced for some vector is free of an SA$\bar{u}$ fault. This is due to the fact that if a line b has an SA$\bar{u}$ fault, then only the value $\bar{u}$ can be deduced at that line. Hence, each time a value is deduced at a line, the fault-status of the line may potentially be updated.

A line b in the CUT is declared fault-free if logic 0 is deduced at the line for one vector, say P_i, and logic 1 is deduced for another, say P_j. That is, a line is deduced as being fault-free if there exist two vectors P_i and P_j such that $v_i^*(b)$ and $v_j^*(b)$ are complementary.

Constraints on backward implications: Backward implications of logic values deduced at an output b of a circuit element are not valid if b may be faulty. For example, if logic 1 is deduced at output b_l of an AND gate and fault status $fs(b_l)$ is such that $SA1 \in fs(b_l)$, then no value can be deduced at its inputs, since the value at the output may be caused due to the presence of the SA1 fault. In contrast, if the output of a logic gate is known to have neither of the SAFs, then the logic behavior of a fault-free version of the gate can be used to deduce values at its inputs. Similarly, if a fanout branch of a fanout system is known to be free of either of the SAFs, then the value at the branch is identical to the value at the stem of the fanout system. In general, the value deduced at an output of a circuit element can be implied

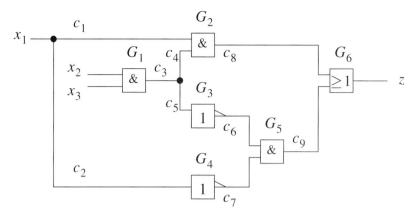

Figure 10.4 An example circuit used to illustrate effect–cause diagnosis (Abramovici and Breuer, 1980) (© 1980 IEEE)

Table 10.14. *The values deduced using the simple deduction procedure for the example circuit in Figure 10.4 and CUT responses shown in the last column of the table*

					Value deduced at line								
Vector	x_1	x_2	x_3	c_1	c_2	c_3	c_4	c_5	c_6	c_7	c_8	c_9	z
$P_1 = (0, 1, 1)$											0	0	0
$P_2 = (1, 1, 0)$													1
$P_3 = (1, 0, 1)$											0	0	0
$P_4 = (1, 1, 1)$													1
$P_5 = (0, 0, 1)$											0	0	0

backward via the element only if that output is known to be free of either of the two SAFs.

The above concepts can be integrated into any implication procedure (see Chapter 4; Section 4.4.2) to obtain a simple procedure to deduce the values at lines within the given CUT for each vector. Such a procedure is illustrated via the following example.

Example 10.25 Consider the circuit in Figure 10.4 and vectors shown in Table 10.14. Assume that response $Res^* = (0, 1, 0, 1, 0)$ is captured at the CUT output for vectors P_1, P_2, P_3, P_4, and P_5, respectively.

The output responses to the first two vectors show that the CUT output, z, does not have any SAF. Since z has been declared as being fault-free, backward implication can be performed for the values observed at z. This process is successful only for vectors P_1, P_3, and P_5, for which a response 0 is observed at z. In each of these cases, logic 0 is deduced at lines c_8 and c_9. These two lines are identified as being free of an SA1 fault, but neither can be identified as being fault-free. Hence, this simple version of the deduction procedure stops at this point providing the results shown in Table 10.14. □

10.5.2 Enhancements to effect–cause deduction

In the following discussion, we describe two sets of concepts proposed for enhancing effect–cause deduction. These two sets are presented separately here. However, in practice, it is possible to incorporate some of the concepts from one to further improve the other.

10.5.2.1 Forced values and enumerative search

One enhancement to the deduction process is based on the concept of *forced value of a line* (Abramovici and Breuer, 1980). A line b is said to have a **forced value** u for a vector P_i, denoted as $\Phi_i(b) = u$, if

$$\begin{cases} \text{either } v_i^*(b) = u, \\ \text{or } v_j^*(b) = \bar{u} \text{ for each vector } P_j. \end{cases}$$

If no such relation can be inferred, then the line is said to not have a forced value for that vector. This is denoted by the notation $\Phi_i(b) = $ undefined.

Properties of forced values: The following properties of forced values are due to Abramovici and Breuer (1980). Stronger versions of these properties are reported in Abramovici and Breuer (1980); here we only present those aspects of these results that are used by the deduction algorithm.

1 For every primary input x_l and vector $P_i = (p_{1,i}, p_{2,i}, \ldots, p_{n,i})$, $\Phi_i(x_l) = p_{l,i}$, where $p_{l,i}$ is the component of the vector P_i that is assigned to input x_l. This is due to the fact that the only fault whose presence can alter the value at x_l is an SAF at that input. If an SAu fault is present at input x_l, then $v_i^*(x_l) = u$ for all the vectors. Otherwise, i.e., if neither of the SAFs is present at x_l, then $v_i^*(x_l)$ is equal to the value applied to the line by vector P_i.

2 Let b_j be the stem of a fanout system and b_l one of its branches. If $\Phi_i(b_j) = u$, where u is a logic value, then $\Phi_i(b_l) = u$. This statement follows from the following two facts. First, if b_l does not have an SAF, then the value at b_l is identical to that at b_j, namely, either u for P_i or $\bar{u}$ for all the vectors. Second, if b_l has an SAw fault, then $v_i^*(b_l) = w$ for all vectors P_i. If $w = u$, then again the value at b_l for P_i is u. On the other hand, if $w = \bar{u}$, then the value at b_l is $\bar{u}$ for all the vectors.

3 If b_l is the output of a non-inverting primitive gate (i.e., an AND or OR gate), with inputs $b_{j_1}, b_{j_2}, \ldots, b_{j_\alpha}$, then $\Phi_i(b_l) = u$ if $\Phi_i(b_{j_q}) = u$ for **each** input b_{j_q} of the gate. Consider an AND gate whose output b_l is fault-free. If $u = 1$, then each input has the forced value 1. This implies that either every input has logic 1 for vector P_i or one or more of the inputs have a logic 0 for all the vectors. In the former case, b_l takes logic 1 in P_i. In the latter case, b_l takes logic 0 for all the vectors. In other words, $\Phi_i(b_l) = 1$ when $\Phi_i(b_{j_q}) = 1$ for **all** inputs b_{j_q}. Next, consider the case when $u = 0$. In this case, either at least one input has logic 0 for vector P_i or every

input has logic 1 for all the vectors. Clearly, fault-free b_l takes logic 0 in the former case; in the latter case it takes logic 1 for all the vectors. In other words, $\Phi_i(b_l) = 0$ when $\Phi_i(b_{j_q}) = 0$ for **all** inputs b_{j_q}. The case when output b_l has an SAF is similar to the case of the fanout branch.

4 If b_l is the output of an inverting primitive gate (i.e., an inverter, a NAND gate, or a NOR gate), with inputs $b_{j_1}, b_{j_2}, \ldots, b_{j_\alpha}$, then $\Phi_i(b_l) = u$ if $\Phi_i(b_{j_q}) = \bar{u}$ for **each** input b_{j_q} of the gate. The proof for this case is similar to that for the above case.

Computation of forced values: The above conditions can be used to compute the forced value of each line for each vector by using a variant of a forward implication procedure that we will call Compute-$\Phi()$. The main difference between a forward implication procedure and Compute-$\Phi()$ lies in the rules used to compute the values of Φ_i at outputs of a circuit element, given their values at the element's inputs. The rules used by Compute-$\Phi()$ are as follows. For a fanout system with stem b_j and branches $b_{l_1}, b_{l_2}, \ldots, b_{l_\beta}$, $\Phi_i(b_{l_q}) = \Phi_i(b_j)$ for each branch b_{l_q}. For a non-inverting primitive gate with inputs $b_{j_1}, b_{j_2}, \ldots, b_{j_\alpha}$ and output b_l,

$$\Phi_i(b_l) = \begin{cases} u, \text{ if } \Phi_i(b_{j_q}) = u \text{ for each input } b_{j_q} \text{ of the gate, and} \\ \text{undefined, otherwise.} \end{cases}$$

The rule for inverting primitive gates is similar.

For a vector $P_i = (p_{1,i}, p_{2,i}, \ldots, p_{n,i})$, Compute-$\Phi()$ starts by initializing $\Phi_i(x_l) = p_{l,i}$, for each primary input x_l, and $\Phi_i(b_j) = $ undefined, for every other line b_j in the circuit. The procedure then uses the above rules to compute values of Φ_i of each line in a manner similar to how a forward implication procedure computes logic values.

Note that the above conditions are only provided for primitive gates. How the above approach can be extended to compute forced values at the output of some types of non-primitive gates (e.g., complex CMOS gates) is the subject of Problem 10.22.

Conditional forced values: The above rules for computing forced values can be generalized in the following manner based on the notion of **conditional forced values**. Consider a fanout system with stem b_j and branches $b_{l_1}, b_{l_2}, \ldots, b_{l_\beta}$. If a fully specified value $v_i^*(b_j)$ has been deduced at b_j, then for each branch b_{l_q}, $\Phi_i(b_{l_q}) = v_i^*(b_j)$, otherwise $\Phi_i(b_{l_q}) = \Phi_i(b_j)$. For a non-inverting primitive gate with inputs $b_{j_1}, b_{j_2}, \ldots, b_{j_\alpha}$ and output b_l,

$$\Phi_i(b_l) = \begin{cases} u, \text{ if for each input } b_{j_q} \text{ of the gate, either } v_i^*(b_{j_q}) = u \text{ or } \Phi_i(b_{j_q}) = u, \text{ and} \\ \text{undefined, otherwise.} \end{cases}$$

The rules for inverting primitive gates can be modified in a similar manner.

As the subsequent deduction process deduces values at circuit lines, Procedure Compute-$\Phi()$ can be invoked to update the forced values at those lines.

Table 10.15. *Forced values for the circuit in Figure 10.4 for the vectors shown in the first three columns of Table 10.14 (Abramovici and Breuer, 1980) (© 1980 IEEE)*

Vector	x_1	x_2	x_3	c_1	c_2	c_3	c_4	c_5	c_6	c_7	c_8	c_9	z
					Forced value for line								
P_1	0	1	1	0	0	1	1	1	0	1			
P_2	1	1	0	1	1					0			
P_3	1	0	1	1	1					0			
P_4	1	1	1	1	1	1	1	1	0	0	1	0	
P_5	0	0	1	0	0					1			

Example 10.26 The forced values computed for the example circuit shown in Figure 10.4 are shown in Table 10.15 for the vectors shown in the first three columns of the table. □

Using forced values: Consider a line b at which value u is deduced for vector P_i. If a forced value $\bar{u}$ is computed at that line for vector P_i, i.e., $\Phi_i(b) = \bar{u}$, then u can be deduced at the line for all the vectors. Otherwise, for each other vector P_j for which a forced value u was computed for that line, i.e., $\Phi_j(b) = u$, value u can be deduced at the line, i.e., $v_j^*(b) = u$.

A deduction procedure that utilizes forced values: The simple deduction procedure described before can be enhanced to utilize the concept of forced value. We first identify the key concepts behind such a procedure *for a single-output circuit*. (The generalization to a multi-output circuit is relatively straightforward.) Then we describe its execution for some example cases.

In a pre-processing phase, Procedure Compute-$\Phi()$ is used to compute the forced values of lines in the given circuit.

The deduction process begins by examining the response captured at the output, say z. If the response value is u for a vector, fault SA$\bar{u}$ can be eliminated from the fault status, $fs(z)$, of the output. Once $fs(z) = \{ff\}$, backward implication can be performed for obtaining the response value at the input of the circuit element driving z for each vector.

As described earlier, the deduced value at a line b is implied backwards via the circuit element driving b only if b is known to be fault-free, i.e., $fs(b) = \{ff\}$. Similarly, the value deduced at a line b is implied forward to another line in its fanout only if that line is known to be fault-free. The above conditions can be easily incorporated in the implication procedures presented in Chapter 4 (Section 4.4.2) to obtain a procedure for deducing values, called Procedure Deduce-Values( ). In addition to the above changes, Procedure Deduce-Values( ) must also incorporate the following features.

First, when a value is deduced at a line, the *fault status* of the line is *updated*. Second, such an update may enable the *implication of values deduced* at the line for one or more vectors. As an example, consider a line b that is the output of an AND gate. Assume that in an earlier implication step, logic 1 was deduced at the line for a vector P_i. Assume that at that time, b was not known to be fault-free. Further, assume that in the current implication step, logic 0 is deduced at the line for a different vector P_j. This implication will cause b to be declared fault-free, i.e., $fs(b) = \{ff\}$. This in turn would enable the implication of deduced value 1 for P_i backward through the AND gate. This will lead to the deduction of logic 1 at each input of the AND gate for vector P_i and may lead to additional implications.

The third additional task that needs to be performed when logic u is deduced at a line b for vector P_i deals with appropriate *utilization of forced values*. If $\Phi_i(b) = \bar{u}$, then logic u is deduced at the line for each vector P_j. Otherwise, logic u is deduced at the line for each vector P_j for which a forced value u exists at that line, i.e., for each vector P_j such that $\Phi_j(b) = u$. If a value is deduced at line b in the above step for any vector, then all the tasks that are normally performed at a line when a new value is deduced are also performed.

The fourth, and final, task that is performed when a value is deduced at a line is the *computation of conditional forced values*. This is simply accomplished by executing Procedure Compute-$\Phi(\ )$ again. The complexity of this process can be decreased by developing an event-driven version (see Chapter 3; Section 3.3.3.1) of Procedure Compute-$\Phi(\ )$.

The following example illustrates the above deduction process.

Example 10.27 Consider the three-input circuit shown in Figure 10.4 once again. Consider a diagnosis scenario where the five vectors, $P_1, P_2, \ldots, P_5$, shown in the first column of Table 10.14 are used for diagnosis. Consider a fabricated copy of this circuit for which the responses to the five vectors are 0, 1, 0, 1, and 0, respectively.

The diagnosis process begins by computing the forced values for each vector. These were derived earlier in Table 10.15. Table 10.16 and Table 10.17 show a detailed trace of the deduction process. (Table 10.17 is just the continuation of Table 10.16.)

Assume that deduction begins with the response at the CUT output for P_1. All values that can possibly be deduced based on this information are then deduced. The process is then repeated for the response for each subsequent vector. (The exact order in which operations are performed is not important. We have adopted the above order for ease of illustration.) In Step 1 shown in the table, the response at output z to vector P_1 is processed. This is depicted as $PR(z, P_1, 0)$ in the table. First, the response value, 0, is assigned to the deduced value at z for P_1, shown in the table as $v_1^*(z) = 0$. This assignment indicates that z does not have an SA1 fault. The fault status of z, $fs(z)$ is hence updated and shown as $fs(z) = \{ff, \text{SA0}\}$. Since z is a primary output, the deduced value cannot be implied forward; since line z is not yet known to be fault-free,

Table 10.16. *Deduction for the circuit in Figure 10.4 for a CUT with response* $\{0, 1, 0, 1, 0\}$

	Task	Assignment	fs update	Implication	Φ use	Φ update
1	$PR(z, P_1, 0)$	$v_1^*(z) = 0$	$fs(z) = \{ff, \text{SA0}\}$			
2	$PR(z, P_2, 1)$	$v_2^*(z) = 1$	$fs(z) = \{ff\}$	$BI(z, P_1)$		
3	$BI(z, P_1)$	$v_1^*(c_8) = 0$	$fs(c_8) = \{ff, \text{SA0}\}$			$A\Phi(c_8, P_1, 0)$
		$v_1^*(c_9) = 0$	$fs(c_9) = \{ff, \text{SA0}\}$		$U\Phi(c_9, P_1, 0)$	$A\Phi(c_9, P_1, 0)$
4	$U\Phi(c_9, P_1, 0)$	$v_4^*(c_9) = 0$				
5	$PR(z, P_3, 0)$	$v_3^*(z) = 0$		$BI(z, P_3)$		
6	$BI(z, P_3)$	$v_3^*(c_8) = 0$				$A\Phi(c_8, P_3, 0)$
		$v_3^*(c_9) = 0$				$A\Phi(c_9, P_3, 0)$
7	$PR(z, P_4, 1)$	$v_4^*(z) = 1$		$BI(z, P_4)$		
8	$BI(z, P_4)$	$v_4^*(c_8) = 1$	$fs(c_8) = \{ff\}$	$BI(c_8, P_4)$		
9	$BI(c_8, P_4)$	$v_4^*(c_1) = 1$	$fs(c_1) = \{ff, \text{SA1}\}$		$U\Phi(c_1, P_4, 1)$	
		$v_4^*(c_4) = 1$	$fs(c_4) = \{ff, \text{SA1}\}$		$U\Phi(c_4, P_4, 1)$	
10	$U\Phi(c_1, P_4, 1)$	$v_2^*(c_1) = 1$				
		$v_3^*(c_1) = 1$		$FI(c_1, P_3)$		
11	$FI(c_1, P_3)$	$v_3^*(c_4) = 0$	$fs(c_4) = \{ff\}$	$BI(c_4, P_3)$		$A\Phi(c_4, P_3, 0)$
				$BI(c_4, P_4)$		
12	$BI(c_4, P_3)$	$v_3^*(c_3) = 0$	$fs(c_3) = \{ff, \text{SA0}\}$			$A\Phi(c_3, P_3, 0)$
						$R\Phi(c_3, P_3)$
13	$R\Phi(c_3, P_3)$					$A\Phi(c_5, P_3, 0)$
						$A\Phi(c_6, P_3, 0)$
14	$BI(c_4, P_4)$	$v_4^*(c_3) = 1$	$fs(c_3) = \{ff\}$	$BI(c_3, P_4)$	$U\Phi(c_3, P_4, 1)$	
15	$BI(c_3, P_4)$	$v_4^*(x_2) = 1$	$fs(x_2) = \{ff, \text{SA1}\}$		$U\Phi(x_2, P_4, 1)$	
		$v_4^*(x_3) = 1$	$fs(x_3) = \{ff, \text{SA1}\}$		$U\Phi(x_3, P_4, 1)$	
16	$U\Phi(x_3, P_4, 1)$	$v_1^*(x_3) = 1$				
		$v_3^*(x_3) = 1$		$FI(x_3, P_3)$		
		$v_5^*(x_3) = 1$				
17	$FI(x_3, P_3)$	$v_3^*(x_2) = 0$	$fs(x_2) = \{ff\}$		$U\Phi(x_2, P_3, 0)$	
18	$U\Phi(x_2, P_3, 0)$	$v_5^*(x_2) = 0$		$FI(x_2, P_5)$		
19	$FI(x_2, P_5)$	$v_5^*(c_3) = 0$		$FI(c_3, P_5)$		$A\Phi(c_3, P_5, 0)$
						$R\Phi(c_3, P_5)$
20	$R\Phi(c_3, P_5)$					$A\Phi(c_4, P_5, 0)$
						$A\Phi(c_5, P_5, 0)$
						$A\Phi(c_6, P_5, 1)$
						$A\Phi(c_9, P_5, 1)$

the value cannot be implied backward. Since no forced value exists for z for any of the vectors, the forced value cannot be used. Finally, adding the value deduced at an output to the forced value table does not serve any purpose.

Next, the response 1 at output z for vector P_2 is processed. First, the value is assigned to $v_2^*(z)$. Second, the line is found to be fault-free and its fault status updated.

Table 10.17. *Continuation of the deduction for the circuit in Figure 10.4 for a CUT with response* $\{0, 1, 0, 1, 0\}$

	Task	Assignment	*fs* update	Implication	Φ use	Φ update
21	$FI(c_3, P_5)$	$v_5^*(c_4) = 0$		$FI(c_4, P_5)$		
22	$FI(c_4, P_5)$	$v_5^*(c_8) = 0$				$A\Phi(c_8, P_5, 0)$
23	$U\Phi(x_2, P_4, 1)$	$v_1^*(x_2) = 1$		$FI(x_2, P_1)$		
		$v_2^*(x_2) = 1$				
24	$FI(x_2, P_1)$	$v_1^*(c_3) = 1$		$FI(c_3, P_1)$		
25	$FI(c_3, P_1)$	$v_1^*(c_1) = 0$	$fs(c_1) = \{ff\}$	$BI(c_1, P_1)$	$U\Phi(c_1, P_1, 0)$	
				$BI(c_1, P_2)$		
				$BI(c_1, P_3)$		
				$BI(c_1, P_4)$		
26	$BI(c_1, P_1)$	$v_1^*(x_1) = 0$	$fs(x_1) = \{ff, \text{SA0}\}$		$U\Phi(x_1, P_1, 0)$	
27	$U\Phi(x_1, P_1, 0)$	$v_5^*(x_1) = 0$		$FI(x_1, P_5)$		
28	$FI(x_1, P_5)$	$v_5^*(c_1) = 0$				
29	$BI(c_1, P_2)$	$v_2^*(x_1) = 1$	$fs(x_1) = \{ff\}$		$U\Phi(x_1, P_2, 1)$	
30	$U\Phi(x_1, P_2, 1)$	$v_3^*(x_1) = 1$				
		$v_4^*(x_1) = 1$				
31	$BI(c_1, P_3)$					
32	$BI(c_1, P_4)$					
33	$U\Phi(c_1, P_1, 0)$					
34	$U\Phi(c_3, P_4, 1)$					
35	$U\Phi(c_4, P_4, 1)$	$v_1^*(c_4) = 1$				
36	$PR(z, P_5, 0)$	$v_5^*(z) = 0$		$BI(z, P_5)$		
37	$BI(z, P_5)$	$v_5^*(c_9) = 0$			$U\Phi(c_9, P_5, 0)$	
38	$U\Phi(c_9, P_5, 0)$	$v_2^*(c_9) = 0$	$fs(c_9) = \{\text{SA0}\}$	$FI(c_9, P_2)$		$A\Phi(c_9, P_2, 0)$
39	$FI(c_9, P_2)$	$v_2^*(c_8) = 1$		$BI(c_8, P_2)$		$A\Phi(c_8, P_2, 1)$
40	$BI(c_8, P_2)$	$v_2^*(c_4) = 1$		$BI(c_4, P_2)$		$A\Phi(c_4, P_2, 1)$
41	$BI(c_4, P_2)$	$v_2^*(c_3) = 1$		$BI(c_3, P_2)$		$A\Phi(c_3, P_2, 1)$
						$R\Phi(c_3, P_2)$
42	$R\Phi(c_3, P_2)$					$A\Phi(c_5, P_2, 1)$
						$A\Phi(c_6, P_2, 0)$
43	$BI(c_3, P_2)$	$v_2^*(x_3) = 1$			$U\Phi(x_3, P_2, 1)$	
44	$U\Phi(x_3, P_2, 1)$		$fs(x_3) = \{\text{SA1}\}$			

Third, this makes possible backward implication at z for each vector for which a value has been deduced at z, namely, vectors P_1 and P_2. In the table, we show only the former, since the latter does not lead to any additional deductions at this time. (To reduce the size of the table, only operations that result in additional deductions of values, update of fault status, or a change in forced values are shown.) This pending task is depicted as $BI(z, P_1)$ in Step 2. Again, forced values are neither used nor updated.

The pending backward implication task, $BI(z, P_1)$, is then executed in Step 3. This leads to the deduction of values at lines c_8 and c_9. The fault status of each line is updated. Since c_9 has the forced value 0 for one of the vectors (P_4), the value deduced at c_9 can be used to deduce values at c_9 for other vectors. (Forced values are shown in Table 10.18.) The task that leads to the use of the deduced value in this manner is hence scheduled, depicted as $U\Phi(c_9, P_1, 0)$. Finally, the value deduced at each line is added to the forced value table, depicted as $A\Phi(c_8, P_1, 0)$ and $A\Phi(c_9, P_1, 0)$. This, however, does not result in re-computation of any other forced value.

Next, in Step 4, the pending task of using the forced value at c_9 is performed. At this time, since logic 0 was deduced at c_9 in the previous step, logic 0 can be deduced at this line for each vector for which it has a forced value 0. In this case, this results in the assignment $v_4^*(c_9) = 0$.

The values deduced thus far are shown in Table 10.19(a). The above process is carried out for the response for each successive vector and the deduced values at the end of completion of deduction for each response are shown in Tables 10.19(b), 10.19(c) and 10.19(d), respectively.

Some of the highlights of the deduction process are now detailed. In Step 8, c_8 is deduced to be fault-free. This causes backward implication to be scheduled at c_8 for every vector for which a value is deduced at c_8. In Step 11, a fanout branch c_4 is deduced as being fault-free. Again, the values are implied backwards and eventually lead to the deduction that the corresponding stem, c_3, is also fault-free.

In Step 12, the forced value at c_3 for P_3 is updated. This makes possible re-computation of other forced values. Hence, a task is scheduled to recompute forced values for P_3 starting at line c_3. This task is depicted as $R\Phi(c_3, P_3)$ in this row. It is subsequently executed in Step 13 and leads to the computation of forced values at lines c_5 and c_6. The updated forced values at the end of Step 13 are shown in Table 10.18(a). The forced values are continually updated and eventually additional forced values are derived at Step 20. The forced values at the end of Step 22 are shown in Table 10.18(b).

In Step 17, forward implication is performed at line x_3 for vector P_3. This triggers backward implication which ends with the deduction of a value at x_2, which is the other input of the gate driven by x_3.

In Step 38, the forced value 1 at line c_9 for vector P_5 (see Table 10.18(b)) is compared with the logic 0 deduced at the line. Since for this vector the deduced value at line c_9 is different from the forced value, 1, the complementary value, 0, is deduced at the line for each vector. In this case, this leads to the assignment $v_2^*(c_9) = 0$. This is also equivalent to the deduction that line c_9 is SA0.

The diagnosis of the above CUT finishes with the deduction that lines x_3 and c_9 have SA1 and SA0 faults, respectively, while lines $x_1, x_2, c_1, c_3, c_4, c_8$, and z are fault-free. Note that since c_9 is deduced as being SA0, the status of lines c_2, c_5, c_6, and c_7 cannot be determined. The deduction for this CUT is hence complete. (Final forced values are shown in Table 10.18(c) and final deduced values are shown in Table 10.19(d).)   □

Table 10.18. *Forced values for the circuit in Figure 10.4 for the CUT deduction detailed in Tables 10.16 and 10.17*

(a) Forced values after first round of recomputations (end of Step 13)

Vector	x_1	x_2	x_3	c_1	c_2	c_3	c_4	c_5	c_6	c_7	c_8	c_9	z
P_1	0	1	1	0	0	1	1	1	0	1	0	0	
P_2	1	1	0	1	1					0			
P_3	1	0	1	1	1	0	0	0	1	0	0	0	
P_4	1	1	1	1	1	1	1	1	0	0	1	0	
P_5	0	0	1	0	0					1			

(b) Forced values after second round of recomputations (end of Step 22)

Vector	x_1	x_2	x_3	c_1	c_2	c_3	c_4	c_5	c_6	c_7	c_8	c_9	z
P_1	0	1	1	0	0	1	1	1	0	1	0	0	
P_2	1	1	0	1	1					0			
P_3	1	0	1	1	1	0	0	0	1	0	0	0	
P_4	1	1	1	1	1	1	1	1	0	0	1	0	
P_5	0	0	1	0	0	0	0	0	1	1	0	1	

(c) Final forced values

Vector	x_1	x_2	x_3	c_1	c_2	c_3	c_4	c_5	c_6	c_7	c_8	c_9	z
P_1	0	1	1	0	0	1	1	1	0	1	0	0	
P_2	1	1	0	1	1	1	1	1	0	0	1	0	
P_3	1	0	1	1	1	0	0	0	1	0	0	0	
P_4	1	1	1	1	1	1	1	1	0	0	1	0	
P_5	0	0	1	0	0	0	0	0	1	1	0	1	

Continuation of deduction via search: In general, the above deduction process is not guaranteed to be complete.

Example 10.28 For the circuit shown in Figure 10.4 and a CUT with response $\{0, 1, 1, 1, 0\}$, the above deduction process stops with the deduced values shown in Table 10.20(a). At this stage lines x_1, c_1, c_8, and z are found to be fault-free but the status of each of the other lines is unknown. $\square$

The process of deduction can be continued via search. First, the values deduced thus far are saved. One simple strategy for search is to first assume logic u at a line c_j for some vector P_i and continue deduction. This may lead to a (a) contradiction, (b) a complete deduction, or (c) an incomplete deduction. In the first case, the assignment

Table 10.19. *Values deduced for the circuit in Figure 10.4 for the CUT deduction detailed in Tables 10.16 and 10.17 (Abramovici and Breuer, 1980) (© 1980 IEEE)*

(a) Deduced values after processing the response for P_2 (end of Step 4)

						Value deduced at line						
Vector	x_1	x_2	x_3	c_1	c_2	c_3	c_4	c_5	c_6	c_7	c_8	c_9	z
P_1												0	0	0
P_2														1
P_3														
P_4													0	
P_5														

(b) Deduced values after processing the response for P_3

						Value deduced at line						
Vector	x_1	x_2	x_3	c_1	c_2	c_3	c_4	c_5	c_6	c_7	c_8	c_9	z
P_1												0	0	0
P_2														1
P_3												0	0	0
P_4													0	
P_5														

(c) Deduced values after processing the response for P_4

						Value deduced at line						
Vector	x_1	x_2	x_3	c_1	c_2	c_3	c_4	c_5	c_6	c_7	c_8	c_9	z
P_1	0	1	1	0		1	1				0	0	0
P_2	1	1		1									1
P_3	1	0	1	1		0	0				0	0	0
P_4	1	1	1	1		1	1				1	0	1
P_5	0	0	1	0		0	0				0		

(d) Deduced values after processing the response for P_5

						Value deduced at line						
Vector	x_1	x_2	x_3	c_1	c_2	c_3	c_4	c_5	c_6	c_7	c_8	c_9	z
P_1	0	1	1	0		1	1				0	0	0
P_2	1	1	1	1		1	1				1	0	1
P_3	1	0	1	1		0	0				0	0	0
P_4	1	1	1	1		1	1				1	0	1
P_5	0	0	1	0		0	0				0	0	0

Table 10.20. *Values deduced for the circuit in Figure 10.4 for the vectors shown in Table 10.14 and response {0, 1, 1, 1, 0}. (a) Values deduced without search (Abramovici and Breuer, 1980); (b) values deduced after assuming logic 0 at line c_9 for vector P_3; (c) values deduced after assuming logic 0 and 1 at c_9 for vectors P_3 and P_2, respectively; and (d) values deduced after assuming logic 0 at c_9 for vectors P_3 and P_2*

(a)

Vector	x_1	x_2	x_3	c_1	c_2	c_3	c_4	c_5	c_6	c_7	c_8	c_9	z
						Values deduced at line							
P_1	0			0			1				0	0	0
P_2	1			1									1
P_3	1			1									1
P_4	1			1			1				1	0	1
P_5	0			0							0	0	0

(b)

Vector	x_1	x_2	x_3	c_1	c_2	c_3	c_4	c_5	c_6	c_7	c_8	c_9	z
						Values deduced at line							
P_1	0			0			1				0	0	0
P_2	1			1									1
P_3	1			1			1				1	0	1
P_4	1			1			1				1	0	1
P_5	0			0							0	0	0

(c)

Vector	x_1	x_2	x_3	c_1	c_2	c_3	c_4	c_5	c_6	c_7	c_8	c_9	z
						Values deduced at line							
P_1	0	1	1	0		1	1	1	0	1	0	0	0
P_2	1	1	0	1		0		0	1	1		1	1
P_3	1	1	1	1		1	1	1	0	1	1	0	1
P_4	1	1	1	1		1	1	1	0	1	1	0	1
P_5	0	1	1	0		1	1	1	0	1	0	0	0

(d)

Vector	x_1	x_2	x_3	c_1	c_2	c_3	c_4	c_5	c_6	c_7	c_8	c_9	z
						Values deduced at line							
P_1	0			0			1				0	0	0
P_2	1			1			1				1	0	1
P_3	1			1			1				1	0	1
P_4	1			1			1				1	0	1
P_5	0			0							0	0	0

u at line c_j for vector P_i was erroneous. Hence, this process can be terminated. In the second case, the process terminates with a possible diagnosis which is reported. In the third case, deduction can be continued by iteratively using the same search strategy. In any event, the assignment of u to line c_j for vector P_i leads to either no diagnosis or one or more possible diagnoses which can be reported. Subsequently, the deductions prior to the assignment of logic u to line c_j for vector P_i are restored and the alternative assignment, namely logic $\bar{u}$ at line c_j for vector P_i, is explored and deduction carried out. This can also lead to either no diagnosis or one or more diagnoses that can be reported. Iterative application of the above search strategy is guaranteed to provide all possible fault scenarios for any given CUT.

Example 10.29 Consider the example CUT again. In this case, the search process can begin with the saving of deduced values shown in Table 10.20(a). Subsequently, logic 0 is assumed at line c_9 for vector P_3 and deduction continued. This leads to an incomplete deduction shown in Table 10.20(b). The above search technique can again be applied by saving these deductions and assuming a value, say 1, at c_9 for vector P_2. This leads to the complete deduction shown in Table 10.20(c) and provides a diagnosis in which lines x_2, c_4, and c_7 are found to have SA1 faults while the other lines (with the exception of c_2 whose status cannot be determined since c_7 has an SA1 fault) are identified as being fault-free. Note that this diagnosis is the result of assuming logic 0 and 1 at line c_9 for vectors P_3 and P_2, respectively. The deduced values corresponding to the above incomplete deduction shown in Table 10.20(b) are then restored and the alternative to the last arbitrary assignment is explored by assigning 0 to line c_9 for vector P_2. This leads to the deduction of values shown in Table 10.20(d), which reports a diagnosis where line c_4 has an SA1 fault and line c_9 has an SA0 fault.

The deduction process then returns to the first arbitrary assignment, namely 0 at line c_9 for vector P_3. The deduced values shown in Table 10.20(a) are restored and the alternative assignment, namely 1 at line c_9 for vector P_3, is tried. In this case, this leads to a contradiction. Hence, the process terminates with the two diagnoses reported above.   □

10.5.2.2 Processing groups of vectors using a composite value system

In another approach (Cox and Rajski, 1988), vectors are processed collectively in groups of size two. (Larger groups can be considered to provide higher diagnostic resolution at increased complexity.) Assume that vectors P_i and P_j in the given set of vectors are processed collectively as a group. The deduction process has two main components, namely forward implication and backward implication. In forward implication, all possible values that may be implied at each line by P_i and P_j are computed. This computation takes into account faults that may be present at each circuit line. The response captured at the CUT outputs by the ATE is then used to perform backward implication. In this process, at each line, the values that may be implied at the line as well as the fault status of the line are updated.

Next, we describe the details of each of the above operations as well as the overall deduction algorithm.

Forward implication: A forward implication procedure (similar to that described in Chapter 4; Section 4.4.2) is extended to (a) collectively consider a pair of vectors, and (b) consider the effect of each *possible combination of faults* in the transitive fanin of a line.

Combinations of faults: Recall that the fault status of each line b is maintained in set $fs(b)$. This set can contain one or more of the following values: ff, SA0, and SA1. If the set contains only one of these three elements, then the exact status of the line is known. In contrast, when $fs(b)$ has more than one element, then the exact status of the fault is not known. For example, $fs(b) = \{ff, SA0\}$ indicates that line b may either be fault-free or have an SA0 fault, but it does not have an SA1 fault.

In a circuit with lines $b_1, b_2, \ldots, b_L$, **all possible combinations of faults** are given by $fs(b_1) \times fs(b_2) \times \cdots \times fs(b_L)$, where '$\times$' denotes the cross-product operation. For example, for a circuit with three lines, b_1, b_2, and b_3, with $fs(b_1) = \{ff\}$, $fs(b_2) = \{ff, SA0\}$, and $fs(b_3) = \{SA0, SA1\}$, all possible combinations of faults are given by $\{(ff, ff, SA0), (ff, SA0, SA0), (ff, ff, SA1), (ff, SA0, SA1)\}$, where each element of the set is an ordered list of faults at lines b_1, b_2, and b_3, respectively.

Collective consideration of two vectors: Collective consideration of a pair of vectors necessitates an extension of the *value system*. (See Chapter 4 (Section 4.2.1) for a detailed discussion of value systems, including concepts of *basic values*, *composite values*, *forward and backward implication operations*, *implication procedures*, and *indirect implications*.) For a single vector, two *basic values*, 0 and 1, are required. When incompletely specified values need to be represented, the three-valued *composite value system* with symbols 0, 1, and x, where $0 = \{0\}$, $1 = \{1\}$, and $x = \{0, 1\}$, are used. Note that *each composite value is described as a set of basic values*.

When two vectors, $P_i = (p_{1,i}, p_{2,i}, \ldots, p_{n,i})$ and $P_j = (p_{1,j}, p_{2,j}, \ldots, p_{n,j})$, are considered collectively, **four basic values**, 00, 01, 10, and 11, are required. (Note that these values are different from the four basic values used in Chapter 4.) Each basic value denotes a distinct combination of fully specified values, where the first (second) symbol corresponds to the value for P_i (P_j). Hence, if the lth components of the two vectors are $p_{l,i} = 0$ and $p_{l,j} = 1$, value 01 is assigned to input x_l of the circuit during the analysis of the pair of vectors P_i and P_j.

The use of a new **four-valued system** necessitates the development of new descriptions of the input–output logic behavior of each circuit element. Figure 10.5(a) shows the truth table of the fault-free version of a two-input AND gate for this value system.

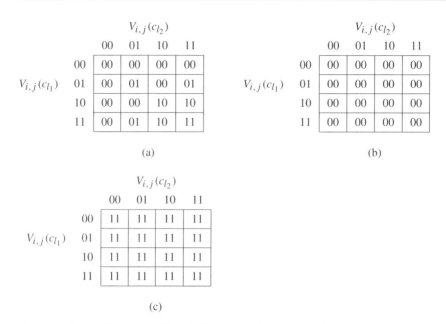

Figure 10.5 Four-valued truth tables of different versions of a two-input AND gate: (a) fault-free, (b) output SA0, and (c) output SA1

This table can be obtained by separating the first and the second symbols of the basic value at each gate input, using the two-valued truth table of the gate to compute the output value corresponding to the combination of first symbols, computing the output value for the combination of second symbols in a similar manner, and combining the two output values obtained above into a single two-symbol value. For example, if the input values for the gate are 01 and 11, then the first symbols of each input value give the combination (0, 1) and the second symbols give the combination (1, 1). The two-valued truth table of the gate gives the values 0 and 1 as outputs for the above two combinations, respectively. These are combined as 01 to obtain the value shown in the truth table for input values 01 and 11. Figure 10.5(b) and Figure 10.5(c) respectively show the four-valued truth tables for the versions of the AND gate with an SA0 and SA1 fault at the gate output, respectively. These tables are obtained following the above procedure using the two-valued truth tables of the corresponding faulty versions of the gate.

The above approach can be used to derive four-valued truth tables for the fault-free as well as each faulty version of any logic gate.

Consider now the case of a primary input x_l at which vectors P_i and P_j apply logic 0 and 1, respectively. These values can be collectively represented using the basic value 01. Initially, before deduction of values and fault status commences, the fault status of each line is initialized to $\{ff, \text{SA0}, \text{SA1}\}$. Hence, consider the case where $fs(x_l) = \{ff, \text{SA0}, \text{SA1}\}$. In this case, even though value 01 is applied to input x_l by the two vectors, the actual value at the line may be 01 if x_l is fault-free, 00 if x_l has an SA0 fault, and 11 if x_l has an SA1 fault. Hence, considering all possible faults at x_l, the application of the given pair of vectors may imply a value at line x_l that belongs to set $\{01, 00, 11\}$. This demonstrates that even if each vector in the given pair of vectors is fully specified, the value implied at a line may be one of a set of the four basic values. This is the case since the precise status of each circuit line is not known during the deduction process. This process is deemed complete whenever the precise status of each line becomes known.

Next, consider the case where vectors P_i and P_j apply logic 0 and x, respectively, to x_l. These values can be collectively represented using the set of basic values $\{00, 01\}$. In this case, even if x_l is fault-free, i.e., if $fs(x_l) = \{ff\}$, the actual value at x_l may still be 00 or 01, depending on how the incompletely specified value is set. Furthermore, if $fs(x_l) = \{ff, SA0, SA1\}$, the application of the given pair of vectors may imply a value at x_l that belongs to set $\{01, 00, 11\}$.

The above discussion demonstrates the need for a composite value system that is based on the basic values of the four-valued system. There are 16 composite values, represented by sets $\{\}$, $\{00\}$, $\{01\}$, $\{10\}$, $\{11\}$, $\{00, 01\}$, $\{00, 10\}$, $\{00, 11\}$, $\{01, 10\}$, $\{01, 11\}$, $\{10, 11\}$, $\{00, 01, 10\}$, $\{00, 01, 11\}$, $\{00, 10, 11\}$, $\{01, 10, 11\}$, and $\{00, 01, 10, 11\}$. The set of possible values that can appear at a line b in response to vectors P_i and P_j is denoted by $V_{i,j}(b)$ and takes one of the above values. Consider a case where $V_{i,j}(b)$ is assigned the composite value $\{01, 11\}$. This value can be interpreted in terms of the actual values implied at the line by the two vectors, namely $v_i^*(b)$ and $v_j^*(b)$, as follows: either $v_i^*(b) = 0$ and $v_j^*(b) = 1$, or $v_i^*(b) = 1$ and $v_j^*(b) = 1$. In general, multiple combinations of $v_i^*(b)$ and $v_j^*(b)$ values exist due to two factors. First, either one or both of the given vectors may contain one or more components that are incompletely specified. Second, the exact fault status may not be known for one or more lines in the transitive fanin of b, i.e., there may exist one or more lines in the transitive fanin of b whose fault status, $fs(b)$, may have more than one element.

The forward implication operation: The forward implication operation for an output b of a circuit element G can be viewed as the process of computing value $V_{i,j}(b)$ given (a) the $V_{i,j}$ value for each of the element's inputs, and (b) the fault status of line b. The forward implication procedure described in Chapter 4 (Section 4.4.1) can be used to compute the value of $V_{i,j}(b)$ for each of the three cases, namely, (i) line b is fault-free, (ii) line b has an SA0 fault, and (iii) line b has an SA1 fault. In the first case, the value, say $V_{i,j}^{ff}(b)$, can be computed using the four-valued truth table (or a function) representing the behavior of a fault-free version of element G. The same operation can be repeated using the truth table representing the behavior of the version of G with an SA0 fault at its output b to compute the value $V_{i,j}^{SA0}(b)$. The value $V_{i,j}^{SA1}(b)$ can be defined and computed in a similar manner. Once the three values described above are obtained, the value of $V_{i,j}(b)$ for a given fault status $fs(b)$ can be computed as follows.

$$V_{i,j}(b) = \bigcup_{\forall f \in fs(b)} V_{i,j}^f(b).$$

Table 10.21. $V_{i,j}(b)$ values for different values of $fs(b)$

Fault status $fs(b)$	$V_{i,j}(b)$
$\{ff\}$	$V_{i,j}^{ff}(b)$
$\{\text{SA0}\}$	$V_{i,j}^{\text{SA0}}(b)$
$\{\text{SA1}\}$	$V_{i,j}^{\text{SA1}}(b)$
$\{ff, \text{SA0}\}$	$V_{i,j}^{ff}(b) \cup V_{i,j}^{\text{SA0}}(b)$
$\{ff, \text{SA1}\}$	$V_{i,j}^{ff}(b) \cup V_{i,j}^{\text{SA1}}(b)$
$\{\text{SA0}, \text{SA1}\}$	$V_{i,j}^{\text{SA0}}(b) \cup V_{i,j}^{\text{SA1}}(b)$
$\{ff, \text{SA0}, \text{SA1}\}$	$V_{i,j}^{ff}(b) \cup V_{i,j}^{\text{SA0}}(b) \cup V_{i,j}^{\text{SA1}}(b)$

Table 10.21 shows the value of $V_{i,j}(b)$ for various values of $fs(b)$.

Example 10.30 Figure 10.6 shows an example circuit where the fault status of each line is $\{ff, \text{SA0}, \text{SA1}\}$. Consider vectors $P_1 = (0, 1, 0)$ and $P_2 = (1, 1, 0)$. The values applied at primary inputs x_1, x_2, and x_3 are $\{01\}$, $\{11\}$, and $\{00\}$, respectively. Since x_1 may be fault-free or have either of the two SAFs, $V_{1,2}(x_1) = \{01, 00, 11\}$. (In this and all the following composite values, the first element in the set corresponds to the values implied at the line in the fault-free circuit, assuming that the given vectors are completely specified.) Next, consider the computation of the value at line c_3. Given the values at the inputs of gate G_1, the forward implication operation in Chapter 4 (Section 4.4.1) and the truth table of the fault-free version of a two-input AND gate (see Figure 10.5(a)) can be used to compute $V_{1,2}^{ff}(c_3) = \{01, 00, 11\}$. Similarly, using the truth tables of each faulty version of the AND gate (Figures 10.5(b) and 10.5(c)), values $V_{1,2}^{\text{SA0}}(b)$ and $V_{1,2}^{\text{SA1}}(b)$ can be computed as $\{00\}$ and $\{11\}$, respectively. Since $fs(c_3) = \{ff, \text{SA0}, \text{SA1}\}$, $V_{1,2}(b)$ can be computed as the union of the above three values to obtain $\{01, 00, 11\}$.

The circuit lines can be traversed in a breadth-first manner and the above operation repeatedly performed to compute values $V_{i,j}(b)$ for each line. These are shown in Figure 10.6. $\square$

Backward implication: Once the values are computed for each line, backward implication is carried out. The process starts by first analyzing each primary output. Consider primary output z_l. Let the values captured at z_l in response to vectors P_i and P_j be $v_i^*(z_l)$ and $v_j^*(z_l)$, respectively.

First, the above values can be used to update the composite value assigned to z_l, $V_{i,j}(z_l)$. Among the set of basic values that appear in $V_{i,j}(z_l)$, only the basic value $v_i^*(z_l)v_j^*(z_l)$ needs to be retained and the other values can be deleted. At this stage, if basic value $v_i^*(z_l)v_j^*(z_l)$ does not appear in composite value $V_{i,j}(z_l)$, then the fault in

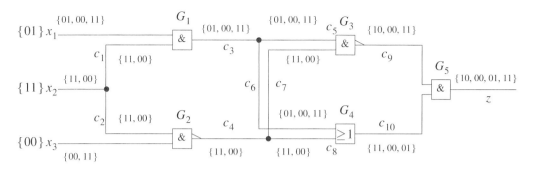

Figure 10.6 An example circuit and forward implication for vector pair $P_1 P_2$ (Cox and Rajski, 1988) (© 1988 IEEE)

the given CUT is not a multiple SAF. In that case, the diagnosis process for multiple SAFs can be terminated and that for other likely faults started.

Second, the new value of $V_{i,j}(z_l)$ is used to update the fault status of line z_l, $fs(z_l)$. If $V_{i,j}(z_l)$ does not contain basic value 00, then the SA0 fault is removed from $fs(z_l)$; if it does not contain 11, then the SA1 fault is removed. In this process, some primary outputs may be identified as fault-free, i.e., their fault status would be identified as being $\{ff\}$.

The above operations are performed at each primary output of the circuit. Then, backward implication is performed at each primary output that is identified as fault-free. The backward implication operation is identical to that described in Chapter 4 (Section 4.4.1). This operation updates values $V_{i,j}(b)$ at each line b in the fanin of the element at which it was executed. If $V_{i,j}(b)$ changes as a result of this operation, the following three tasks are performed. First, the circuit elements in the fanout of line b are scheduled for forward implication. Second, the fault status of line b is updated. This process is identical to that performed at each primary output. Finally, if line b is fault-free, the circuit element in the fanin of the line is scheduled for backward implication.

Example 10.31 Figure 10.7(a) shows the circuit and the vector pair shown in Figure 10.6. If the response of a CUT to vector pair $P_1 P_2$ is 10, then backward implication begins by deleting all the values except 10 from the composite value at primary output z. (This is depicted by crossing out from the set all the values except 10.) This shows that z is free of either of the single SAFs, which is reflected by updating its fault-status as $fs(z) = \{ff\}$. Since z is fault-free and the value at z has been updated, backward implication is performed at G_5, the gate driving z. The procedure described in Chapter 4 (Section 4.4.1.2) is used to update the values at c_{10} and c_9 as shown in the figure. Line c_{10} is deduced as being free of the SA0 fault, but may still have an SA1 fault; in contrast, c_9 is deduced as being fault-free. Hence, backward implication continues only at G_3. This process continues until no more values can be updated. Figure 10.7(a) shows the results obtained after the completion

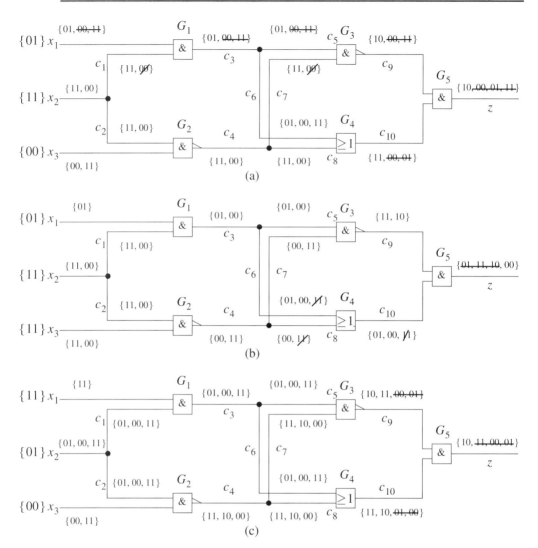

Figure 10.7 An example of deduction via forward and backward simulation of pairs of vectors (Cox and Rajski, 1988): (a) vector pair $P_1 P_2$, (b) $P_3 P_4$, and (c) $P_5 P_2$ (© 1988 IEEE)

of backward implication for this pair of vectors. Column 2 of Table 10.22 shows the fault status of circuit lines at the end of this implication.

Figure 10.7(b) shows the results of the forward and backward implication for vector pair $P_3 P_4$ for CUT response 00. The only difference from the forward implication for vector pair $P_1 P_2$ is that in this case initially the fault status of each line is assumed to be that at the end of processing of the first pair of vectors, $P_1 P_2$. Column 3 of Table 10.22 shows the fault status of circuit lines after the processing is completed for this pair of vectors. Figure 10.7(c) and Column 4 of Table 10.22, respectively, show the values deduced for vector pair $P_5 P_2$ and the fault status at the end of processing for CUT response 10.

Enhancements to the above approach: Given a set of vectors $P_1, P_2, \ldots, P_{N_v}$, the above process can be repeated for each pair of vectors in the set. In general, this may cause a vector to be simulated many times, each time in conjunction with a different vector. In the above example, vector P_2 has been simulated as a part of the first pair, $P_1 P_2$, and then as a part of the third pair, $P_5 P_2$.

When a vector is simulated multiple times, different values may be inferred at a line for the vector. Note that the value at any circuit line during forward as well backward implication is, in general, incompletely specified. This is true even if each vector in the pair being simulated is completely specified, since the circuit is being simulated for multiple combinations of faults. This causes the implication to be incomplete in the sense that the values that appear in the composite value at a line are, in general, a superset of the union of values that can actually appear at that line for any of the faulty versions of the circuit (as given by the fault status of each line).

Example 10.32 As an evidence of the above phenomenon, for the above example examine the composite values $V_{1,2}(c_3)$ and $V_{5,2}(c_3)$ at line c_3 for vector pairs $P_1 P_2$ and $P_5 P_2$, respectively. (See Figure 10.7(a) and Figure 10.7(c), respectively.) $V_{1,2}(c_3) = \{01\}$ while $V_{5,2}(c_3) = \{01, 00, 11\}$. Note that the former composite value indicates that the value deduced at line c_3 for vector P_2 is 1, under the combinations of faults implied by the fault status values shown in Column 2 of Table 10.22. In contrast, the latter composite value reports that the value at line c_3 due to vector P_2 may be either 0 or 1 for the combinations of faults implied by the fault status values shown in Column 4 of Table 10.22. □

Of special note in the above example is the fact that the latter case considers only a subset of combinations of faults considered in the former case. This shows that depending on which vector P_i is paired with, the implication may provide either more or less precise information about logic values that may appear at a line when that vector is applied.

The concept of *adjacency* has been developed in Cox and Rajski (1988) to refine the value assignments obtained by considering vectors in pairs. In the following, we describe the special case of adjacency, called *input adjacency*. Consider a scenario where a pair of vectors $P_i P_j$ is simulated and a second scenario where pair $P_i P_l$ is simulated. At each line b in the circuit, the composite value $V_{i,j}(b)$ is computed in the first case and value $V_{i,l}(b)$ in the second case. Clearly, the first symbol of each basic value in $V_{i,j}(b)$ and $V_{i,l}(b)$ represents the value for the same input vector. Such values are said to be **input adjacent**.

Input adjacency can be utilized to improve the quality of deductions, as illustrated by the following example.

Example 10.33 Consider the scenario in the above two examples. We found earlier that $V_{1,2}(c_3) = \{01\}$ and $V_{5,2}(c_3) = \{01, 00, 11\}$. Since the second values for vector

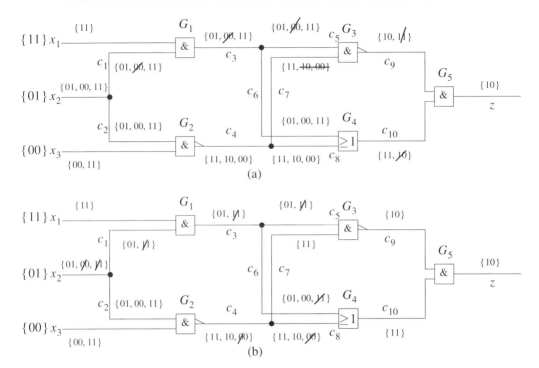

Figure 10.8 Additional deduction via adjacency for vector pair $P_5 P_2$: (a) values obtained after utilizing adjacencies between the second vectors of the first pair $(P_1 P_2)$ and the third pair $(P_5 P_2)$, and (b) results after performing implications

pairs $P_1 P_2$ and $P_5 P_2$ are input adjacent, one can conclude that $V_{5,2}(c_3)$ should not contain value 00. This can be used to update $V_{5,2}(c_3)$ as shown in Figure 10.8(a). Similarly, the above input adjacency can be used to update the values in Figure 10.7(c) at lines c_1, c_5, c_7, c_9, and c_{10}. The values obtained after the update are shown in Figure 10.8(a). The above update of values enables additional implications. For example, value 11 is deleted from $V_{5,2}(c_9)$ in the above step. Since c_9 is known to be free of either single SAF, this enables backward implication, which leads to the deletion of value 11 from c_5. Several other such implications can be performed. The values obtained at circuit lines for this pair of vectors after the completion of all such implications are shown in Figure 10.8(b). The additional implications resulting from processing of input adjacency between $P_1 P_2$ and $P_5 P_2$ also update the fault status of several lines. The new fault status obtained after processing all the input adjacencies for this vector pair are shown in Column 5 of Table 10.22. □

The above methodology can then be applied to other pairs of these five vectors as well as other pairs containing additional vectors, if any. The fault status obtained at the end of the process is the final result of diagnosis. If we stop this deduction after processing the input adjacency for pair $P_5 P_2$ as described above, then as reported in

Table 10.22. *The fault status of each line in the circuit, vector pairs, and responses shown in Figure 10.7. Column 5 shows the fault status after the input adjacency between vector pair $P_1 P_2$ and $P_5 P_2$ is utilized as shown in Figure 10.8*

| Line | Fault status after processing vector pair | | | |
	$P_1 P_2$	$P_3 P_4$	$P_5 P_2$	$P_5 P_2^A$
x_1	$\{ff\}$	$\{ff\}$	$\{ff\}$	$\{ff\}$
x_2	$\{ff, \text{SA0}, \text{SA1}\}$	$\{ff, \text{SA0}, \text{SA1}\}$	$\{ff, \text{SA0}, \text{SA1}\}$	$\{ff\}$
x_3	$\{ff, \text{SA0}, \text{SA1}\}$	$\{ff, \text{SA0}, \text{SA1}\}$	$\{ff, \text{SA0}, \text{SA1}\}$	$\{ff, \text{SA0}, \text{SA1}\}$
c_1	$\{ff, \text{SA1}\}$	$\{ff, \text{SA1}\}$	$\{ff, \text{SA1}\}$	$\{ff\}$
c_2	$\{ff, \text{SA0}, \text{SA1}\}$	$\{ff, \text{SA0}, \text{SA1}\}$	$\{ff, \text{SA0}, \text{SA1}\}$	$\{ff, \text{SA0}, \text{SA1}\}$
c_3	$\{ff\}$	$\{ff\}$	$\{ff\}$	$\{ff\}$
c_4	$\{ff, \text{SA0}, \text{SA1}\}$	$\{ff, \text{SA0}, \text{SA1}\}$	$\{ff, \text{SA0}, \text{SA1}\}$	$\{ff, \text{SA1}\}$
c_5	$\{ff\}$	$\{ff\}$	$\{ff\}$	$\{ff\}$
c_6	$\{ff, \text{SA0}, \text{SA1}\}$	$\{ff, \text{SA0}\}$	$\{ff, \text{SA0}\}$	$\{ff, \text{SA0}\}$
c_7	$\{ff, \text{SA1}\}$	$\{ff, \text{SA1}\}$	$\{ff, \text{SA1}\}$	$\{ff, \text{SA1}\}$
c_8	$\{ff, \text{SA0}, \text{SA1}\}$	$\{ff, \text{SA0}\}$	$\{ff, \text{SA0}\}$	$\{ff\}$
c_9	$\{ff\}$	$\{ff\}$	$\{ff\}$	$\{ff\}$
c_{10}	$\{ff, \text{SA1}\}$	$\{ff\}$	$\{ff\}$	$\{ff\}$
z	$\{ff\}$	$\{ff\}$	$\{ff\}$	$\{ff\}$

Column 5 of Table 10.22, the given CUT may have any combination of faults including SA0 or SA1 at x_3, SA0 or SA1 at c_2, SA1 at c_4, SA0 at c_6, and SA1 at c_7.

10.6   Generation of vectors for diagnosis

So far in this chapter, the test vectors used for diagnosis were assumed to be given. In this section, we describe techniques to generate such vector sets/sequences.

In its simplest form, the problem of test generation for diagnosis can be defined as the problem of generating a vector that distinguishes between two faults, say f_1 and f_2. In the following discussion, a test generator that can solve the above problem is called a **diagnostic test generator**. In contrast, a test generator that generates a set of vectors to detect faults in a circuit is referred to as a **traditional test generator**.

Given a pair of faults f_1 and f_2, a diagnostic test generator must provide a vector P that causes distinct error responses for f_1 and f_2. A complete set of diagnostic vectors can be obtained by repeatedly using a diagnostic test generator. Each time, the diagnostic test generator can be used to generate a vector to distinguish a pair of faults that the current set of test vectors cannot distinguish. Such a test generation strategy may not lead to a very compact set of diagnostic tests. Alternative formulations of the diagnostic test generation problem can include such additional objectives as minimization of (i) the size of the test set, (ii) the amount of fault simulation effort required for simulation-based diagnosis of CUTs using the sequence of vectors, and

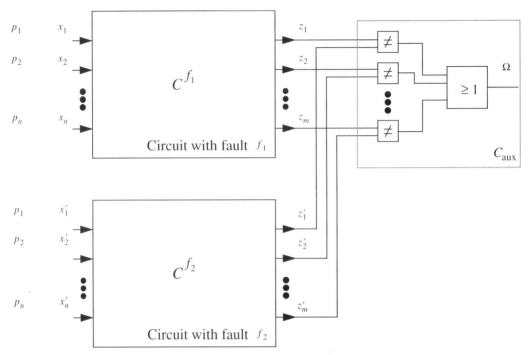

Figure 10.9 Representation of the conditions under which a vector P distinguishes between faults f_1 and f_2 in a circuit C

(iii) the size of fault dictionary generated. Most of the developments in diagnostic test generation have focused on the simpler formulation of the problem. We assume that the diagnostic test generator is given a pair of faults f_1 and f_2, whose sites are lines b_1 and b_2, respectively.

10.6.1   Conditions for a vector to distinguish between a pair of faults

A fully specified vector P is said to distinguish between two faults f_1 and f_2 in an m-output circuit if there exists at least one output z_l, such that the response of the vector at z_l for circuit version C^{f_1} is complementary to that for circuit version C^{f_2}. Let $v^{f_1}(z_l)$ and $v^{f_2}(z_l)$, respectively, denote these responses to vector P. The above condition is depicted in Figure 10.9, where $\Omega = 1$ indicates that vector $P = (p_1, p_2, \ldots, p_n)$ distinguishes between faults f_1 and f_2.

One key consequence of the above condition is that a vector that distinguishes between faults f_1 and f_2 must detect at least one of the two faults. This is due to the fact that the response of each faulty version of the circuit to a vector that detects neither of the two faults will be identical to that of the fault-free circuit version. Consequently, the responses of both the faulty circuit versions will be identical. The above condition leads to the following properties regarding untestable faults. In the next section, we describe how these properties enable any pair of faults that contain one or more untestable faults from being dropped from consideration by the diagnostic test pattern generator.

Pr 1: If faults f_1 and f_2 are both untestable, then the two faults are indistinguishable.

Pr 2: If fault f_1 is untestable and fault f_2 is testable, then any vector that detects f_2 distinguishes between f_1 and f_2.

The following property regarding a pair of testable faults f_1 and f_2 can also be deduced from the conditions for a vector to distinguish between a pair of faults. This property is used by one of the test generation approaches described ahead.

Pr 3: If a vector P does not distinguish between two faults f_1 and f_2 and it detects fault f_1, then it also detects fault f_2.

The following properties can also be inferred from the above conditions and demonstrate that a large proportion of all possible pairs of faults are distinguished by any complete set of vectors generated for fault detection.

Pr 4: If f_1 is an SA0 fault at line b_1 and f_2 an SA1 fault at the same line, then any vector that detects either fault distinguishes between the two faults.

Pr 5: If f_1 is a fault at line b_1 and f_2 is an SAu fault at a different line b_2, then any vector that detects f_1 but does not distinguish f_1 from f_2, distinguishes f_1 and f_3, where f_3 is the SA$\bar{u}$ fault at line b_2.

Pr 6: If the sites of f_1 and f_2, namely lines b_1 and b_2, do not have any common output in their transitive fanout, then any vector that detects either (or both) of these faults distinguishes between the two faults.

Pr 7: If the site of f_1 has one output, say z_l, in its transitive fanout that is not in the transitive fanout of the site for f_2, then any vector that detects f_1 by creating an error at output z_l distinguishes between the two faults.

The last two properties are also used by some of the diagnostic test generators that are primarily based on modification of traditional test generators.

10.6.2 A diagnostic test generation system

A circuit with L lines has $L(L-1)/2$ pairs of lines, each with two possible faults. Even for a moderate-sized circuit, targeting each pair of faults individually is impractical. Fortunately, it is also unnecessary to take such an approach, since the number of pairs of faults that must be explicitly targeted can be significantly reduced through two main steps.

First, fault collapsing can be performed to decrease the number of faults to be considered. Clearly, any two faults f_1 and f_2 that are *equivalent* (see Chapter 3 (Section 3.4.1) for definitions of this and other related concepts) are indistinguishable. In contrast, if f_1 *dominates* f_2, then it is generally possible to distinguish between the two faults. Hence, fault collapsing should be limited to collapsing only equivalent faults. The procedure for fault collapsing in Chapter 3 (Section 3.4.1.2) can be easily modified to perform such collapsing.

Second, diagnostic test generation is typically preceded by conventional test generation to generate a **detection test set**, i.e., a set of vectors that detects a large fraction of (or, ideally, all) testable faults in the circuit. A detection test set is often used as a part of the diagnostic test set. The use of such a test set drastically reduces the number of pairs of faults for which diagnostic test generation must be carried out. Assume for the sake of the following discussion that the detection test set covers all testable faults in the circuit.

Consider the faults that are identified by the conventional test generator as being untestable. All such faults can be removed from consideration if a complete detection test set is used as a part of the diagnostic test set. This follows from properties Pr 1 and Pr 2 described in the previous section.

Elimination of pairs of equivalent faults (via fault collapsing) and pairs of faults containing one or more untestable faults (via the use of a complete detection test set as a part of the diagnostic test set) can help significantly reduce the complexity of diagnostic test pattern generation. This is due to the fact that a diagnostic test generator is sure to be unable to generate a distinguishing vector for either of these two fault pair types. In fact, if a diagnostic test pattern generator targets such a pair of faults, then it may waste a significant amount of effort trying to generate a vector.

In addition to allowing elimination of pairs of faults with one or more untestable faults, the use of a complete detection test set as a part of diagnostic test set also helps reduce the number of pairs of faults for which diagnostic test generation must be explicitly performed. Properties Pr 4 through Pr 7 in Section 10.6.1 identify several fault pair types which any complete detection set is guaranteed to distinguish. Typically, even an incomplete detection test set that covers a large proportion of testable faults can greatly reduce the number of pairs of faults for which diagnostic test pattern generation must be explicitly performed.

10.6.3   Diagnostic test generation approaches

Techniques for generating diagnostic tests can be classified into two main categories. In the first category are techniques that use test generation techniques for fault detection (i.e., a traditional test generation technique) as a primary vehicle to obtain a vector that distinguishes between the given pair of faults. In the second category are techniques that focus directly on distinguishing between faults f_1 and f_2.

10.6.3.1 Diagnostic test generation based on fault detection

Deadening: One of the earliest approaches to diagnostic test pattern generation was

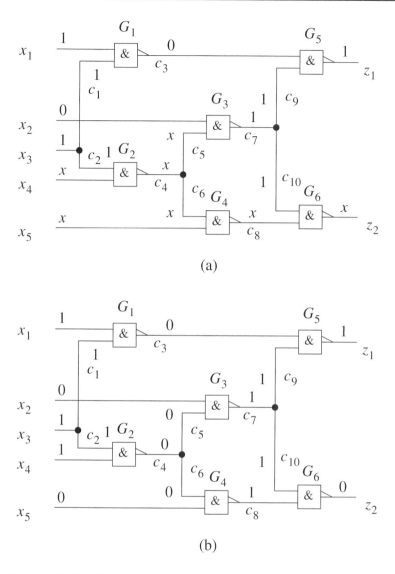

Figure 10.10 Specification of partially specified input values to increase the diagnostic power of a vector. (a) A vector generated by a conventional test generator, and (b) a completely-specified version of the vector used for diagnosis

presented in Roth *et al.* (1967). In this approach, a traditional test generator is first used to generate a test for a given target fault. Typically, the generated vector is partially specified. **Deadening** is the process of assigning a value to each partially specified component of a vector such that, for a large number of primitive gates, the gate's controlling value is implied at two or more of its inputs. This typically reduces the number of non-target faults detected by the generated vector, thereby increasing the number of faults from which the target fault can be distinguished.

Example 10.34 Consider for example circuit C_{17} shown in Figure 10.10(a) with a target fault c_1 SA0. For this target fault, a test generator may generate a vector $P = (p_1, p_2, p_3, p_4, p_5) = (1, 0, 1, x, x)$. The components p_4 and p_5 may be assigned logic 1 and 0, respectively, to deaden gates G_3 and G_4, as shown in Figure 10.10(b). The reader can verify that this vector, $(1, 0, 1, 1, 0)$, helps distinguish the target fault, c_1 SA0, from faults x_1 SA1, x_2 SA0, x_2 SA1, x_3 SA1, x_4 SA0, x_4 SA1, x_5 SA0, x_5 SA1, c_1 SA1, c_2 SA0, c_2 SA1, c_3 SA0, c_4 SA0, c_4 SA1, c_5 SA0, c_5 SA1, c_6 SA0, c_6 SA1, c_7 SA0, c_7 SA1, c_8 SA0, c_8 SA1, c_9 SA0, c_9 SA1, c_{10} SA0, c_{10} SA1, z_1 SA1, z_2 SA0, and z_2 SA1. □

Modification of conventional test generation: The above approach for distinguishing between the given target fault and other faults relies on prevention of detection of non-target faults. This can result in an unnecessary increase in the size of the detection as well as diagnostic test sets.

Another approach to generate a test to distinguish between a pair of faults, f_1 and f_2, using a test generator is based on the following two principles (Savir and Roth, 1982).

1 If there exists one or more outputs that are in the transitive fanout of one of these faults and not of the other, then generate a test vector that detects that fault at such an output.

2 If the above approach is not successful, then select an output that is in the transitive fanout of both the faults and generate a test that propagates the effect of one fault to that output but does not propagate the effect of the other fault. Such a test can be generated by initializing the fault site of fault f_1 with a fault effect, while initializing the site of f_2 with the corresponding stuck-at value. A traditional test generation procedure can then be used to generate a test for fault f_1 that produces a fault effect at that output. If the above approach is unsuccessful, then the reverse possibility is tried.
If the above steps are unsuccessful, then another output is selected and the steps are repeated.

Example 10.35 Consider the circuit shown in Figure 10.11. If the two given faults are x_3 SA1 (f_1) and c_1 SA1 (f_2), then this approach first determines that both the outputs are in the transitive fanout of x_3 while only output z_1 is in the transitive fanout of c_1. Hence, the approach first tries to generate a test for fault f_1 that will cause a fault effect at output z_2, which is not in the transitive fanout of the site of the other fault. Next, consider the pair of faults x_1 SA0 and c_3 SA0. The sites of these two faults have the same output in their transitive fanout, namely, z_1. If this pair of faults is given, then the above procedure tries to generate a test for one of these faults at that output while not detecting the other. This can be achieved by assigning logic 0 to line c_3 in the fault-free circuit to prevent fault c_3 SA0 from excitation and then generating a test for the other fault, x_1 SA0. □

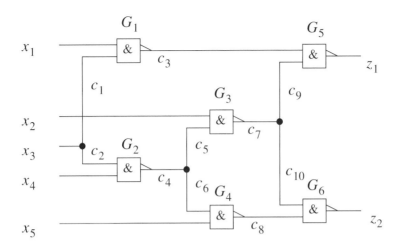

Figure 10.11   An example circuit C_{17} to illustrate diagnostic test generation

Modification of an existing vector that detects both faults: A third approach (Pomeranz and Fuchs, 1998) starts with a complete detection test set for the given circuit. For any pair of faults, say f_1 and f_2, that are not distinguished by the vectors in this test set, there exists a test vector in the set that detects both the faults but causes errors at an identical set of outputs for each of the two faulty circuit versions (see Pr 3 in Section 10.6.1). One of the components of such a vector P is selected randomly, and complemented to obtain a new vector, P'. Fault simulation is performed for vector P', for faults f_1 and f_2. If P' does not distinguish between the two faults and does not detect either of the faults, then it is discarded and the above steps are repeated for P. On the other hand, if P' does not distinguish between the two faults but detects both of them, then it is used instead of P and the above steps are repeated. Finally, if the vector distinguishes between the two faults, then it is added to the original set of vectors.

Example 10.36   Consider as an example the circuit shown in Figure 10.12(a). Assume that a set of vectors is given that does not distinguish between faults x_1 SA1 and z_1 SA1. Furthermore, assume that only one vector in the given test set, $P = (0, 0, 1, 1, 1)$, detects this pair of faults. One of the components of P, say, the first, is complemented to obtain a new vector $P' = (1, 0, 1, 1, 1)$. Fault simulation for these two faults shows that P' does not distinguish between the two faults. Furthermore, it does not detect either of these faults. Hence, P' is discarded. Again, another randomly selected component of P, say the fourth, is complemented to obtain $P'' = (0, 0, 1, 0, 1)$, as shown in Figure 10.12(b). Fault simulation reveals that this vector does not distinguish between the two faults but it detects both faults. P is replaced by P'' and the entire process is repeated. A randomly selected component, say the third, is complemented to obtain $P''' = (0, 0, 0, 0, 1)$, as shown in Figure 10.12(c).

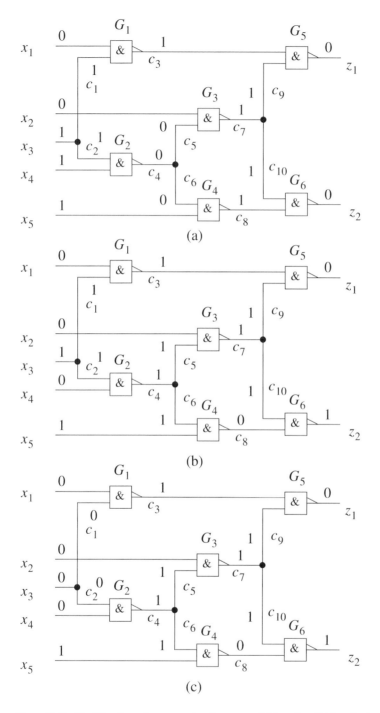

Figure 10.12 Modification of a vector that detects x_1 SA1 and z_1 SA1 to find a vector that distinguishes between the two faults. (a) A vector P that detects both faults, (b) a modified version, P'', and (c) a vector P''' that distinguishes between the two faults

Fault simulation determines that this vector distinguishes between the two faults. This vector is then added to the diagnostic test set (which originally was identical to the detection test set). □

10.6.3.2 Test generation approaches that directly focus on distinguishing between faults

Generating a diagnostic test using three circuit versions: The method presented in Camurati *et al.* (1990) generates a vector that distinguishes between a pair of faults f_1 and f_2 by considering three copies of the circuit, fault-free (C), that with fault f_1 (C^{f_1}), and that with fault f_2 (C^{f_2}). In the first phase, C and C^{f_1} are considered and a traditional test generator is used to generate a vector that detects f_1.

Example 10.37 For circuit C_{17} and a pair of faults f_1: x_3 SA0 and f_2: c_1 SA0, an incompletely-specified vector that detects f_1 is shown in Figure 10.13(a). This vector is simulated for each of the three circuit versions. In Figure 10.13(a), these three values are shown at each line and correspond to the values implied at that line in circuit versions C, C^{f_1}, and C^{f_2}, respectively. □

Any line at which complementary values are implied for circuit versions C^{f_1} and C^{f_2} is said to have a Δ. For the above example and vector, lines x_3 and c_2 have Δ. If further specification of one or more components of the given vector may lead to complementary values at a line, then the line is said to have a δ. In the above example, lines x_4, x_5, c_4, c_5, c_6, c_8, and z_2 are said to have δ. A δ-path is a path between a line that has a Δ and a primary output, where every line in the path has a δ.

Note that even though x_4 and x_5 each have a δ, they will not be included in any δ-path in any subsequent step of test generation. Such lines can be eliminated from consideration by assigning δ only to the lines in the transitive fanout of the sites of either of the target faults.

If any output has a Δ, then the vector generated in the first phase already distinguishes between the two faults. At the other extreme, if no δ-path exists in the circuit, then the components of the current vector cannot be specified to distinguish between the two faults. In such a case, the test generator in the first phase is used to generate another vector for fault f_1 and the entire process is repeated. A third possibility is that there exists one or more δ-paths. In this case, the vector generated in the first phase becomes the starting point for the second phase of the algorithm. In this phase, circuit versions C^{f_1} and C^{f_2} are considered and the unspecified values in the given vector specified until a vector is found for which a Δ appears at one or more outputs. This process is similar to a test generation process with three notable differences. First, each of the circuit versions considered has one faulty circuit element, the first one has f_1 and the second has f_2. Second, test generation starts with the partially specified vector generated in the first phase, instead of starting with a completely unspecified vector. Finally, the goal here is to specify values such that a Δ appears at one or more

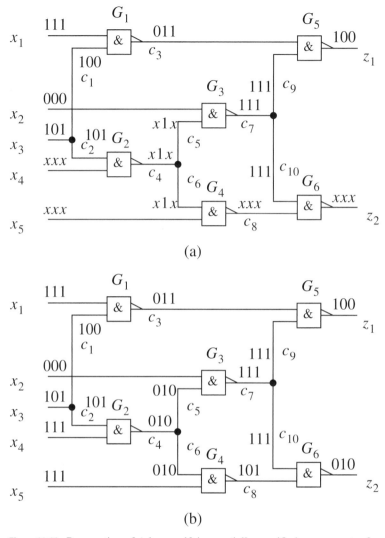

Figure 10.13 Propagation of Δ by specifying partially specified components of a vector using a systematic search. (a) An incompletely-specified vector that detects x_3 SA0, and (b) a vector that distinguishes between x_3 SA0 and c_1 SA0

outputs. An algorithm for this phase can be obtained by making changes to any test generation algorithm for combinational circuits. If this phase succeeds, then a vector that distinguishes the two faults is found; otherwise, the entire process is repeated starting at the first phase.

Example 10.38   For the initial incompletely-specified vector shown in Figure 10.13(a), the Δ at line c_2 can be propagated to line c_4 by assigning a 1 at x_4; subsequently, assignment of a 1 at x_5 propagates the Δ to z_2 via c_8. The resulting vector that distinguishes between faults x_3 SA0 and c_1 SA0, and the values implied at the lines in the three versions of the circuit are shown in Figure 10.13(b). □

Generating a diagnostic test using two circuit versions: A vector to distinguish between two faults, f_1 and f_2, can be generated by considering circuit versions C^{f_1} and C^{f_2}, without considering the fault-free version of the circuit C (Gruning *et al.*, 1991). In such a case, each version of the circuit has one of the given faults. The objective of diagnostic test generation is to generate a vector that implies different values at one or more outputs of the circuit for the two versions of the circuit. Such a difference in values can only be created at the sites of the two faults. Once such a difference is created, then it must be propagated to one or more primary outputs. This process is very similar to that of traditional test generation with the three differences noted for the previous approach.

Let fault f_1 be b_1 SAu and f_2 be b_2 SAw. Recall that a Δ can originate only at b_1 and/or b_2. The algorithm in Gruning *et al.* (1991) takes special steps to excite the Δ, if one line, say b_1, is in the transitive fanin of the other. In this case, this algorithm first assigns logic $\bar{u}$ at line b_1 to excite a Δ there and leaves the value at all other lines, including b_2, unspecified. A modified version of any test generator is then used to propagate the Δ to a circuit output and justify values at each line. If this process is unsuccessful, then the above process is repeated by assigning logic u to line b_1, logic $\bar{w}$ to line b_2 to create a Δ at b_2, and the unspecified value at each remaining line. At the end of this process, either a test that distinguishes f_1 from f_2 is found, or, given sufficient time, the two faults are declared indistinguishable.

Example 10.39 Consider the example circuit shown in Figure 10.14(a). Let f_1 be the SA0 fault at line x_3 and f_2 be the SA0 fault at line c_1. Test generation begins with assigning logic 1 to the site of the first fault, i.e., line x_3. This excites fault f_1 and hence logic 0 appears at the line in circuit version C^{f_1}. Furthermore, since f_2 is located at a line not in the transitive fanin of the site of f_1, the fault-free value 1 appears at x_3 in circuit C^{f_2}. These two values are shown as 01 at line x_3 in Figure 10.14(a), where the first value, 0, corresponds to the value at x_3 in C^{f_1} and the second value, 1, to that in C^{f_2}. Unspecified values are applied at each of the remaining lines, including c_1, the site of fault f_2, and implication is performed. The values obtained at the end of implication are shown in Figure 10.14(a). The implication process ends up justifying the values assigned to line x_3 by assigning logic 1 at primary input x_3. Note that a Δ appears at c_2 and can potentially be propagated via paths $(c_4, c_5, c_7, c_9, z_1)$, $(c_4, c_5, c_7, c_{10}, z_2)$, or (c_4, c_6, c_8, z_2).

For each of the above cases, the Δ must be propagated to c_4, hence logic 1 is applied at x_4. The values obtained after the completion of implication are shown in Figure 10.14(b). This process continues until the vector shown in Figure 10.14(c)

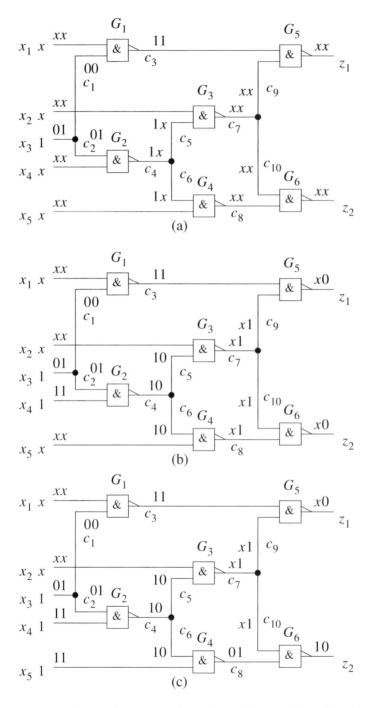

Figure 10.14 Diagnostic test generation using two faulty versions of the circuit for f_1: x_3 SA0 and f_2: c_1 SA0. (a) Excitation of Δ, (b) propagation of Δ, and (c) a test vector that distinguishes between f_1 and f_2

is found that distinguishes between these two faults. Thus, vector $(x, x, 1, 1, 1)$ distinguishes between faults x_3 SA0 and c_1 SA0.   □

Summary

- Diagnosis is the process of identifying faults that are the likely causes of failure of a fabricated copy of a CUT.
- For some digital systems, diagnosis may be performed to facilitate repairs; for VLSI chips, diagnosis is performed mainly to improve the design of the chip, the design methodology, or the fabrication process, so as to increase yield and enhance performance for chips to be fabricated in the future.
- The quality of diagnosis is measured by diagnostic accuracy and diagnostic resolution.
- The error response of the CUT and that of each faulty version of the circuit with a distinct fault is the primary information used for logic diagnosis.
- In different diagnosis scenarios, different types of matches are sought between the error response of the CUT and of each faulty version.
- The fault model plays a very different role during diagnosis than during test development. To be useful for diagnosis, modeled faults should accurately characterize the behavior of most faulty CUTs.
- During diagnosis, some faults may be explicitly modeled while additional faults may be considered implicitly as dependent faults. The response of each faulty version with a modeled fault is determined via fault simulation; an approximate value of the response of each faulty version with a dependent fault may be deduced using the responses of one or more faulty versions with modeled faults.
- Post-test diagnostic fault simulation complexity can be reduced by using information such as the outputs at which an erroneous response is observed, fault-free response value at those outputs, and numbers of inverting gates along various paths from the site of a fault to such an output.
- Fault dictionaries can be compressed by storing the error response instead of the response, combining error response values at multiple outputs into a single pass/fail value, storing a selected subset of error response values, and so on. Also, representations used to store sparse matrices and those used to store trees can be used to reduce storage complexity.
- Post-test diagnostic simulation as well as fault-dictionary based diagnosis can be performed considering some faults as modeled and others as dependent. Matching criteria used for comparing the CUT response with responses of each faulty circuit version should be modified to deal with the approximate nature of error response of each dependent fault.

- Effect–cause methodologies enable diagnosis for scenarios where faults with arbitrary multiplicities must be considered.
- Effect–cause methodologies combine information obtained from different vectors to obtain high diagnostic resolution. In the approach based on forced values, these values provide such a linkage; in another approach, this linkage is provided via concurrent simulation of pairs of vectors and the notion of adjacency.
- The conditions for a vector to distinguish between a pair of faults can be expressed using the corresponding faulty versions.
- Traditional test generation approaches can be modified to generate a vector that distinguishes between a pair of faults.
- Test generation techniques that create differences between the values at the corresponding lines of two faulty circuit versions and propagate the difference to an output provide a direct mechanism to generate a vector that distinguishes between a given pair of faults.

Additional reading

Fault models play a crucial role in diagnosis. Accurate fault models are discussed in Hawkins *et al.* (1994), Maly (1987), and Mei (1974). In general, a diagnosis approach must either use a very accurate fault model or use a flexible matching criterion (Lavo *et al.*, 1996). Some diagnosis approaches that use advanced fault models are presented in Maxwell and Aitken (1993) and Chakravarty and Gong (1993). Techniques to model realistic faults as dependent faults have been presented in Lavo *et al.* (1996). Additional flexible matching approaches have been presented in Lavo *et al.* (1996), Richman and Bowden (1985), and Sheppard and Simpson (1996). A probabilistic approach that improves diagnosis by combining information from fault models as well as other statistical data has been presented in Lavo *et al.* (1998b). Finally, an approach to perform diagnosis independent of any fault model has been presented in Boppana and Fujita (1998).

Post-test fault simulation approaches have been described in Arzoumanian and Waicukauski (1981) and Waicukauski and Lindbloom (1989). Both of these techniques have been used for diagnosis of large circuits.

Interesting fault dictionary approaches can be found in Richman and Bowden (1985) and Tulloss (1980). A hybrid approach that uses a combination of dictionary and fault simulation has been presented in Ryan *et al.* (1991).

An industrial diagnostic experiment is described in Saxena *et al.* (1998).

Most of the above approaches focus mainly on combinational (or full-scan) circuits. Numerous approaches have been developed for sequential circuits. Approaches have also been developed for diagnosis during built-in self-test where complete CUT

Table 10.23. *Two sets of test vectors for diagnosis of the circuit C_{17} shown in Figure 10.11*

TS1	TS2
$P_1 = (1, 0, 0, 1, 0)$	$P_1 = (1, 0, 0, 1, 0)$
$P_2 = (0, 1, 1, 1, 1)$	$P_2 = (0, 1, 1, 1, 1)$
$P_3 = (1, 1, 0, 1, 0)$	$P_3 = (1, 1, 0, 1, 0)$
$P_4 = (1, 0, 1, 0, 1)$	$P_4 = (1, 0, 1, 0, 1)$
	$P_5 = (0, 0, 0, 0, 0)$

response is typically not available. Finally, several diagnosis approaches have been presented for diagnosing delay faults.

Exercises

10.1 Consider circuit C_{17} shown in Figure 10.11 and test set TS1 shown in Table 10.23.

Perform fault simulation to determine the response, error response, and failing vectors for

(a) each single SAF,

(b) each TF, and

(c) each two-line NFBF of type

　i.　wired-AND,

　ii.　wired-OR, and

　iii.　driver-strength.

For the driver-strength model, assume that the gain of each nMOS transistor is three times that of each pMOS transistor. Furthermore, assume that each primary input is driven by a flip-flop, whose output has an inverter whose transistors are sized as those in the other gates.

10.2 For circuit C_{17} (Figure 10.11) and test set TS1 in Table 10.23, consider a CUT with response $Res^* = ((0, 0), (0, 0), (0, 0), (1, 1))$.

(a) Which single SAFs can be eliminated from post-test diagnostic fault simulation due to the fact that their sites are not in the transitive fanin of every output at which an erroneous response is captured?

(b) Which TFs can be eliminated from simulation?

(c) Which two-line NFBFs of each of the following types

　i.　wired-AND,

　ii.　wired-OR, and

　iii.　driver-strength (make the same assumption as in Problem 10.1(c)),

can be eliminated?

10.3 Consider a circuit comprised of primitive CMOS gates (AND, OR, NAND, NOR, and inverter). Let z_l be a primary output of the circuit at which the current vector P implies a value $u \in \{0, 1\}$ in the fault-free version of the circuit. (Assume that only the values implied by the current vector at primary outputs is known, the values implied at internal circuit lines are not known.)

Consider the two-line NFBF of types wired-AND, wired-OR, and driver-strength. For each of the above type of NFBFs, what type of path must exist from each/either site of an NFBF and output z_l for the fault to be considered for post-test diagnostic fault simulation?

Compare the types of paths required for each of the above types of NFBFs.

10.4 Consider a family of complex CMOS gates called AOI(γ_1, γ_2) with inputs $b_1, b_2, \ldots, b_{\gamma_1+\gamma_2}$, and output b_j. Each gate in the family implements a function of the form

$$v(b_j) = \overline{v(b_1)v(b_2)\ldots v(b_{\gamma_1}) + v(b_{\gamma_1+1})v(b_{\gamma_1+2})\ldots v(b_{\gamma_1+\gamma_2})}.$$

Consider a circuit that is comprised only of inverting primitive gates (NAND, NOR, and inverter) and AOI(γ_1, γ_2). What types of paths must exist between an output z_l at which current vector P implies a value $u \in \{0, 1\}$ (in the fault-free circuit version) and the site of an SAw fault for the SAF to be a possible cause of an erroneous response captured at output z_l for vector P?

Repeat for TFs, and NFBFs of types wired-AND, wired-OR, and driver-strength.

10.5 Assume that the value implied by current vector P at each circuit line and the circuit outputs at which erroneous responses are observed for a given CUT are known. Derive a complete set of conditions that can be used to eliminate from post-test diagnostic fault simulation each fault of each of the following types.

(a) SAF.

(b) TF.

(c) Two-line NFBF of type

    i.   wired-AND,

    ii.   wired-OR, and

    iii. driver-strength.

10.6 Perform post-test diagnostic fault simulation for circuit C_{17} (Figure 10.11), test set TS1 shown in Table 10.23, and a CUT with response $((0, 0), (0, 0), (0, 0), (1, 1))$, for each of the following combinations of faults. In this problem, assume that each fault is treated as a modeled fault.

(a) Single SAFs.

(b) Single SAFs and TFs.

(c) Single SAFs and two-line NFBFs of type wired-AND.

(d) Single SAFs and two-line NFBFs of type driver-strength (make the same assumption as in Problem 10.1(c)).

(e) Single SAFs, TFs, and two-line NFBFs of type wired-AND.

Tabulate the number of faults for which fault simulation was performed for each vector and the results of diagnosis for each case.

10.7 Repeat the above problem assuming that in each case, only the single SAFs are explicitly modeled and the other faults are considered as dependent on the single SAFs.

In this case, tabulate the number of faults simulated for each vector, the faults in fault lists F_s and F_d after processing of each vector, and the results of diagnosis.

For each target fault set, compare the complexity and diagnostic resolution for the approach followed in this problem with those for the approach followed in the previous problem.

10.8 Repeat the above two problems for other CUTs with responses
   (a) $((1, 1), (1, 1), (1, 1), (1, 1))$,
   (b) $((1, 0), (1, 0), (1, 1), (1, 1))$, and
   (c) $((0, 0), (0, 0), (1, 0), (1, 0))$.

10.9 Consider a scenario where the SAFs are explicitly modeled while TFs and NFBFs of type wired-AND, wired-OR, and driver-strength are considered as dependent faults. Prove that in such a scenario, if a CUT does not fail a vector P_i, then none of the dependent faults can be deleted from fault list F_d when P_i is processed.

10.10 For circuit C_{17} (Figure 10.11) and test set TS2 shown in Table 10.23, obtain a complete fault dictionary for each single SAF. Compare the size of the tabular representation of the dictionary (where the dictionary entry for each fault is the error response of the corresponding faulty version) and that of a list representation of the dictionary.

10.11 Use the procedure described in Section 10.4.2.2 to create a minimal hybrid dictionary for the complete dictionary shown in Table 10.5. Compare the space required for storing the hybrid dictionary with that for the complete dictionary.

10.12 Obtain pass–fail, 1-detect, and 2-detect dictionaries for the complete fault dictionary obtained in Problem 10.10. Identify all pairs of non-equivalent single SAFs that cannot be distinguished by each of these dictionaries. Repeat using the complete dictionary.

Compare each of the above dictionaries and the complete dictionary in terms of the storage space required and the numbers of pairs of non-equivalent faults that are not distinguished.

10.13 For circuit C_{17} (Figure 10.11), you are given four CUTs with responses
   (a) CUT1: $((0, 0), (0, 0), (0, 0), (1, 1), (0, 0))$,
   (b) CUT2: $((1, 1), (1, 1), (1, 1), (1, 1), (1, 1))$,
   (c) CUT3: $((1, 0), (1, 0), (1, 1), (1, 1), (1, 0))$, and
   (d) CUT4: $((0, 0), (0, 0), (1, 0), (1, 0), (0, 0))$.

Use the complete fault dictionary obtained in Problem 10.10 to identify the single SAFs that may be present within each of these CUTs.

10.14 Repeat the above problem using the pass–fail dictionary obtained in Problem 10.12.

Compare the diagnostic resolution obtained with that obtained using the complete dictionary in the previous problem.

10.15 Given the CUT versions for C_{17} with responses shown in Problem 10.13, how would the 1-detect dictionary derived in Problem 10.12 be used to perform diagnosis? Clearly detail the matching criterion that would be used when comparing the error response of a CUT with a dictionary entry and explain the new matching criterion using illustrative examples.

Compare the diagnostic resolution obtained for these four CUTs for the 1-detection, 2-detection, and complete dictionaries.

10.16 For the example circuit and vectors shown in Figure 10.1, obtain a complete dictionary considering SAFs and TFs as modeled faults. Compare this dictionary with that obtained considering SAFs and TFs as modeled and dependent faults, respectively.

Consider a CUT with response $((0, 0), (1, 1), (1, 0), (0, 1))$. Use the above two dictionaries to perform diagnosis. Do the two dictionaries provide different results? Repeat for another CUT with response $((0, 0), (1, 1), (1, 0), (0, 0))$.

Do there exist other CUT responses for which the two dictionaries provide different diagnosis results? If so, provide such examples.

10.17 For the example circuit and vectors shown in Figure 10.1, obtain a complete dictionary considering SAFs and two-line wired-AND NFBFs as modeled faults. Compare this dictionary with that obtained considering SAFs and NFBFs as modeled and dependent faults, respectively.

Consider a CUT with response $((1, 1), (1, 1), (1, 1), (1, 1))$. Use the above two dictionaries to perform diagnosis. Do the two dictionaries provide different results? Repeat for another CUT with response $((1, 0), (1, 1), (1, 1), (1, 1))$.

What does this illustrate about the trade-off between considering each fault as a modeled fault and considering a subset of faults as dependent faults?

Do there exist other CUT responses for which the two dictionaries provide different diagnosis results? If so, provide such examples.

10.18 Consider the hybrid dictionary obtained in Problem 10.11 for the circuit and vectors shown in Figure 10.1.

(a) Using this dictionary as a starting point, create a dictionary that considers single SAFs and two-line wired-AND NFBFs as modeled and dependent faults, respectively.

(b) Use the procedure described in Section 10.4.2.2 to create a minimal hybrid dictionary for the dictionary obtained in Problem 10.17.

(c) List the pairs of faults that cannot be distinguished using the hybrid dictionary obtained in Part 10.18(a) of this problem but can be distinguished by the dictionary obtained in Part 10.18(b).

10.19 Consider the 1-detection dictionary in Table 10.9 for the circuit and vectors in Figure 10.1. Using this dictionary, obtain a dictionary considering SAFs and two-line wired-AND NFBFs as modeled and dependent faults, respectively. What is the maximum number of vectors for which an error would appear in the dictionary entry for an NFBF ? What would be the consequence if only the first of these erroneous values is retained in the dictionary? Use example CUT responses to illustrate your answer.

10.20 Compute the forced values for the circuit shown in Figure 10.6 for vectors $P_1, P_2, \ldots, P_5$ shown in Figure 10.7.

10.21 Perform effect–cause diagnosis using forced values for the circuit and vectors shown in the previous problem. If the process terminates without providing a complete diagnosis, then use a search process until alternative complete diagnoses are obtained.

10.22 Consider the CMOS complex gate AOI(γ_1, γ_2) described in Problem 10.4. Derive a relationship between the forced values at the inputs of this gate and that at its output.

10.23 Obtain four-valued truth tables of the type described in Section 10.5.2.2 for (a) a two-input NAND gate, (b) a two-input NOR gate, and (c) a two-input XOR gate.

10.24 Perform effect–cause diagnosis considering pairs of vectors for the circuit in Figure 10.4 and vectors shown in Table 10.14.

10.25 Consider circuit C_{17} shown in Figure 10.11. Use properties Pr 1, Pr 2, Pr 4, and Pr 6 from Section 10.6.1 to identify the pairs of single SAFs that will be distinguished by any complete set of detection test vectors.
Identify the pairs of non-equivalent faults in this circuit that are not distinguished by the set of vectors in TS1 in Table 10.23.

10.26 Consider circuit C_{17} shown in Figure 10.11 and test set TS1 shown in Table 10.23. Modify a vector in TS1 that detects x_3 SA1 as well as c_1 SA1 faults but does not distinguish between the two faults, using the approach described in Section 10.6.3.1.

10.27 Consider circuit C_{17} shown in Figure 10.11. Generate a vector that distinguishes between faults x_3 SA1 and c_1 SA1 using the diagnostic test generation approach that uses three copies of the circuit.

10.28 Repeat the above problem using the diagnostic test generation approach that uses two copies of the circuit.

References

Abramovici, M. and Breuer, M. A. (1980). Multiple fault diagnosis in combinational circuits based on an effect–cause analysis. *IEEE Trans. on Computers*, **C-29** (6), pp. 451–460.

Abramovici, M., Breuer, M. A., and Friedman, A. D. (1991). *Digital Systems Testing and Testable Design*. IEEE Press, Piscataway, NJ, USA.

Acken, J. M. and Millman, S. D. (1992). Fault model evolution for diagnosis: accuracy vs precision. In *Proc. Custom Integrated Circuits Conference*, pp. 13.4.1–13.4.4.

Aitken, R. (1995). Finding defects with fault models. In *Proc. Int. Test Conference*, pp. 498–505.

Arzoumanian, Y. and Waicukauski, J. (1981). Fault diagnosis in LSSD environment. In *Proc. Int. Test Conference*, pp. 86–88.

Boppana, V. and Fujita, M. (1998). Modeling the unknown! Towards model-independent fault and error diagnosis. In *Proc. Int. Test Conference*, pp. 1094–1011.

Boppana, V., Hartanto, I., and Fuchs, W. K. (1996). Full fault dictionary storage based on labeled tree encoding. In *Proc. VLSI Test Symposium*, pp. 174–179.

Camurati, P., Medina, D., Prinetto, P., and Reorda, M. S. (1990). A diagnostic test pattern generation algorithm. In *Proc. Int. Test Conference*, pp. 52–58.

Chakravarty, S. and Gong, Y. (1993). An algorithm for diagnosing two-line bridging faults in combinational circuits. In *Proc. Design Automation Conference*, pp. 520–524.

Chess, B., Lavo, D. B., Ferguson, F. J., and Larrabee, T. (1995). Diagnosis of realistic bridging faults with single stuck-at information. In *Proc. Int. Conference on Computer-Aided Design*, pp. 185–192.

Cox, H. and Rajski, J. (1988). A method of fault analysis for test generation and fault diagnosis. *IEEE Trans. on Computer-Aided Design*, **7** (7), pp. 813–833.

Gruning, T., Mahlstedt, U., and Koopmeiners, H. (1991). DIATEST: a fast diagnostic test pattern generator for combinational circuits. In *Proc. Int. Conference on Computer-Aided Design*, pp. 194–197.

Hawkins, C. F., Soden, J. M., Righter, A. W., and Ferguson, F. J. (1994). Defect classes: an overdue paradigm for CMOS IC testing. In *Proc. Int. Test Conference*, pp. 413–425.

Henderson, C. L. and Soden, J. M. (1997). Signature analysis for IC diagnosis and failure analysis. In *Proc. Int. Test Conference*, pp. 310–318.

Kautz, W. H. (1968). Fault testing and diagnosis in combinational digital circuits. *IEEE Trans. on Computers*, **C-17** (4), pp. 352–366.

Lavo, D. B., Larrabee, T., and Chess, B. (1996). Beyond the Byzantine generals: unexpected behavior and bridging fault diagnosis. In *Proc. Int. Test Conference*, pp. 611–619.

Lavo, D. B., Chess, B., Larrabee, T., Ferguson, F. J., Saxena, J., and Butler, K. M. (1997). Bridging fault diagnosis in the absence of physical information. In *Proc. Int. Test Conference*, pp. 887–893.

Lavo, D. B., Chess, B., Larrabee, T., and Ferguson, F. J. (1998a). Diagnosis of realistic bridging faults with single stuck-at information. *IEEE Trans. on Computer-Aided Design*, **17** (3), pp. 255–268.

Lavo, D. B., Chess, B., Larrabee, T., and Hartanto, I. (1998b). Probabilistic mixed-model fault diagnosis. In *Proc. Int. Test Conference*, pp. 1084–1093.

Maly, W. (1987). Realistic fault modeling for VLSI testing. In *Proc. Design Automation Conference*, pp. 173–180.

Maxwell, P. C. and Aitken, R. C. (1993). Biased voting: a method for simulating CMOS bridging faults in the presence of variable gate logic thresholds. In *Proc. Int. Test Conference*, pp. 63–72.

Mei, K. (1974). Bridging and stuck-at faults. *IEEE Trans. on Computers*, **C-23** (7), pp. 720–727.

Millman, S. D., McCluskey, E. J., and Acken, J. M. (1990). Diagnosing CMOS bridging faults with stuck-at fault dictionaries. In *Proc. Int. Test Conference*, pp. 860–870.

Narayanan, S., Srinivasan, R., Kunda, R. P., Levitt, M. E., and Bozorgui-Nesbat, S. (1997). A fault dictionary methodology for UltraSPARC-I microprocessor. In *Proc. European Design and Test Conference*, pp. 494–500.

Pomeranz, I. and Reddy, S. M. (1997). On dictionary-based fault location in digital circuits. *IEEE Trans. on Computers*, **C-46** (1), pp. 48–59.

Pomeranz, I. and Fuchs, W. K. (1998). A diagnostic test generation procedure for combinational circuits based on test elimination. In *Proc. Asia Test Symposium*, pp. 486–491.

Richman, J. and Bowden, K. R. (1985). The modern fault dictionary. In *Proc. Int. Test Conference*, pp. 696–698.

Roth, J. P., Bouricius, W. G., and Schneider, P. R. (1967). Programmed algorithms to compute tests to detect and distinguish between failures in logic circuits. *IEEE Trans. on Computers*, **C-16** (5), pp. 567–580.

Ryan, P. G., Rawat, S., and Fuchs, W. K. (1991). Two-stage fault location. In *Proc. Int. Test Conference*, pp. 963–968.

Savir, J. and Roth, J. P. (1982). Testing for, and distinguishing between failures. In *Proc. Int. Test Conference*, pp. 165–172.

Saxena, J., Butler, K. M., Balachandran, H., Lavo, D. B., Chess, B., Larrabee, T., and Ferguson, F. J. (1998). On applying non-classical defect models to automated diagnosis. In *Proc. Int. Test Conference*, pp. 748–757.

Sheppard, J. W. and Simpson, W. R. (1996). Improving the accuracy of diagnostics provided by fault dictionaries. In *Proc. VLSI Test Symposium*, pp. 180–185.

Soden, J. M., Anderson, R. E., and Henderson, C. L. (1997). IC failure analysis: magic mystery and science. *IEEE Design & Test of Computer*, pp. 59–69.

Tulloss, R. E. (1980). Fault dictionary compression: recognizing when a fault may be unambiguously represented by a single failure detection. In *Proc. Int. Test Conference*, pp. 368–370.

Waicukauski, J. A. and Lindbloom, E. (1989). Failure diagnosis of structured VLSI. *IEEE Design & Test of Computer*, pp. 49–60.

11 Design for testability

In this chapter, first we describe the full-scan methodology, including example designs of scan flip-flops and latches, organization of scan chains, generation of test vectors for full-scan circuits, application of vectors via scan, and the costs and benefits of scan. This is followed by a description of partial scan techniques that can provide many of the benefits of full scan at lower costs. Techniques to design scan chains and generate and apply vectors so as to reduce the high cost of test application are then presented.

We then present the boundary scan architecture for testing and diagnosis of inter-chip interconnects on printed circuit boards and multi-chip modules.

Finally, we present design for testability techniques that facilitate delay fault testing as well as techniques to generate and apply tests via scan that minimize switching activity in the circuit during test application.

11.1 Introduction

The difficulty of testing a digital circuit can be quantified in terms of cost of *test development*, cost of *test application*, and costs associated with *test escapes*. **Test development** spans circuit modeling, test generation (automatic and/or manual), and fault simulation. Upon completion, test development provides test vectors to be applied to the circuit and the corresponding fault coverage. **Test application** includes the process of accessing appropriate circuit lines, pads, or pins, followed by application of test vectors and comparison of the captured responses with those expected. The cost associated with a high **test escape**, i.e., when many actual faults are not detected by the derived tests, is often reflected in terms of loss of customers. Even though this cost is often difficult to quantify, it influences the above two costs by imposing a suitably high fault coverage requirement to ensure that test escape is below an acceptable threshold.

Design for testability (DFT) can loosely be defined as changes to a given circuit design that help decrease the overall difficulty of testing. The changes to the design typically involve addition or modification of circuitry such that one or more new modes of circuit operation are provided. Each new mode of operation is called a **test mode** in which the circuit is configured only for testing. During normal use, the circuit is configured in the **normal mode** and has identical input-output logic behavior as the original circuit design. However, the timing of the circuit may be affected by the presence of the DFT circuitry.

The key objective of DFT is to reduce the difficulty of testing. As seen in Chapter 5, ATPG techniques for sequential circuits are expensive and often fail to achieve high fault coverage. The main difficulty arises from the fact that the state inputs and state outputs cannot be directly controlled and observed, respectively. In such cases, a circuit may be modified using the *scan* design methodology, which creates one or more modes of operation which can be used to control and observe the values at some or all flip-flops. This can help reduce the cost of test development. In fact, in many cases, use of some scan is the only way by which the desired fault coverage target may be achieved. However, in most cases, the use of scan can increase the test application time and hence the test application cost. In a slightly different scenario, DFT circuitry may be used to control and observe values at selected lines within the circuit using *test points* so as to reduce the number of test vectors required. In such a scenario, use of DFT may help reduce the test application cost.

Next, consider a typical board manufacturing scenario where pretested chips are assembled onto a printed circuit board. Assuming that the chips were not damaged during assembly, assembled boards must be tested to ensure correctness of interconnects between chips. One possible way to test these interconnects is via application of tests and observation of responses at, respectively, the primary inputs and outputs of the board, commonly referred to as a board's **connector**. In this method, the circuit under test (CUT) is the aggregate of the circuitry in each chip on the board and all inter-chip interconnections. Clearly, in such a scenario, test development is prohibitively expensive. Test development may in fact be impossible due to the unavailability of the details of the circuit within each chip. Another alternative is to use physical *probes* to obtain direct access to the input and output pins of chips to test the interconnect. The first difficulty in such an approach is that a value must be applied to a pin via a probe in such a manner that it does not destroy the driver of the pin. Second, the cost of such a physical probing can be very high, especially since in packaging technologies, device pins are very close to each other. Finally, in many new packaging technologies, such as *ball-grid array*, many of the 'pins' are located across the entire bottom surface of the chip and cannot even be probed. The *boundary scan* DFT technique provides a mechanism to control and observe the pins of a chip directly and safely. Once again, DFT not only reduces costs of test development and application, it makes high test quality achievable.

Finally, consider other test objectives, such as delay fault testing and managing switching activity during test application. A two-pattern test is required to detect a delay fault in a combinational logic block. Special DFT features are hence required when comprehensive delay fault testing is desired. The switching activity during test application via scan can be significantly higher than during a circuit's normal

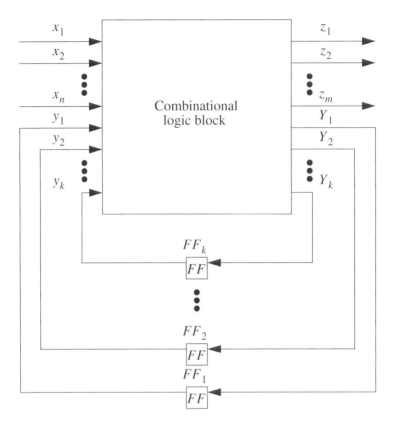

Figure 11.1 A general model of a sequential circuit

operation. Hence, to avoid damaging the CUT, special care must be taken during test development and application to reduce switching activity during scan testing.

11.2 Scan design

Consider the sequential circuit shown in Figure 11.1, comprised of a block of combinational circuit and a set of k flip-flops. The primary inputs and outputs of the sequential circuit are $x_1, x_2, \ldots, x_n$, and $z_1, z_2, \ldots, z_m$, respectively. The present state variables, $y_1, y_2, \ldots, y_k$, constitute the **state inputs** of the combinational circuit. The next state variables, $Y_1, Y_2, \ldots, Y_k$, constitute the **state outputs** of the combinational circuit. Y_l and y_l are, respectively, the input and the output of flip-flop FF_l, $1 \leq l \leq k$.

Test development for sequential circuits is difficult mainly due to the inability to control and observe the state inputs and outputs, respectively. In fact, for many sequential circuits, the cost of test development is so high that the fault coverage attained in practice is unacceptably low.

The **scan** DFT methodology (Williams and Angell, 1973; Eichelberger and Williams, 1977) helps reduce the test development cost and enables attainment of high fault coverage. In this methodology, the flip-flops (or latches) are designed and connected in a manner that enables two modes of operation. In the **normal mode**, the flip-flops are connected as shown in Figure 11.1. In the **test mode**, the flip-flops are reconfigured as one or more shift-registers, called **scan registers** or **scan chains**. During the normal mode, the response at the state outputs is captured in the flip-flops. These values can be observed by switching the circuit to test mode, clocking the scan register, and observing the output of the last flip-flop in each scan chain. Furthermore, values to be applied at the state inputs in the subsequent test may be simultaneously shifted into the flip-flops. Hence, for the purposes of test development, the state inputs and outputs can be treated as being similar to primary inputs and outputs, respectively. In other words, only the combinational logic block with inputs $x_1, x_2, \ldots, x_n$ **and** $y_1, y_2, \ldots, y_k$ and outputs $z_1, z_2, \ldots, z_m$ **and** $Y_1, Y_2, \ldots, Y_k$ need to be considered for the purposes of test development. By reducing the problem of generating tests for a sequential circuit to one of generating tests for a combinational circuit, scan DFT methodology reduces the test development cost. Furthermore, in many cases, it enables attainment of acceptable fault coverage. (To distinguish from the scan designs discussed in subsequent sections, whenever all flip-flops (or latches) are connected in this manner, the design is referred to as **full scan**.)

Next, we discuss the details of scan methodology using *multiplexed-input scan* flip-flops followed by those using *level-sensitive scan design* elements.

11.2.1 Multiplexed-input scan flip-flops

An ideal flip-flop based sequential circuit is shown in Figure 11.1, where a sequential circuit is comprised of a combinational logic block and edge-triggered D flip-flops. Furthermore, the flip-flops are clocked using a clock input to the chip, which is not *gated*, i.e., combined with any logic value in any manner. We first describe the details of scan design for such an ideal circuit. This is followed by a description of methodologies for test development and application.

11.2.1.1 Scan chain design

Scan DFT methodology requires modification of the circuit to add a test mode of operation in which the flip-flops may be configured as one or more scan chains.

First, the flip-flop design must be modified. In the normal mode, the value at state output Y_i is latched into flip-flop FF_i. In the test mode, the value at the output of the previous flip-flop in the scan chain is latched. One way to add this functionality into the flip-flop is through the addition of a multiplexer, as shown in Figure 11.2(a). In this design, the control input of the multiplexer is controlled by the mode control

Figure 11.2 A multiplexed-input scan flip-flop (Williams and Angell, 1973) (© 1973 IEEE). (a) An example implementation, and (b) its symbol

input, mc. Let the two data inputs of the multiplexer be D and D_s, respectively. When $mc = normal$, the value applied at input D is latched into the flip-flop. Alternatively, when $mc = test$, the value applied at input D_s is latched. This design is referred to as a **multiplexed-input scan** flip-flop and represented using the symbol shown in Figure 11.2(b). Each D flip-flop in a circuit may be replaced by such a flip-flop and input D of flip-flop FF_i connected to state output Y_i.

Second, the order in which flip-flops are connected to form a scan chain is determined and input D_s of flip-flop FF_i is connected to the output of flip-flop FF_j, where FF_j is designated as the predecessor of FF_i in the scan chain. Input D_s of the first flip-flop in a chain is the scan chain input and is denoted as *ScanIn*, while the output of the last flip-flop in a chain is the output of the scan chain and denoted as *ScanOut*. The input and output of a chain are, respectively, connected to an input and output pin of the chip, either directly or via boundary scan circuitry (see Section 11.5).

Figure 11.3(a) shows details of scan chain design. This figure assumes that flip-flops are configured as a single scan chain and ordered in a particular way. Figure 11.3(b) and Figure 11.3(c) show the effective connections when the flip-flops are configured in the normal mode ($mc = normal$) and test mode ($mc = test$), respectively. In the normal mode, the input to flip-flop FF_i is state output Y_i and its output drives state input y_i. (Actually, it also drives the D_s input of the next flip-flop in the scan chain, but that does not influence the value latched in that flip-flop, since the multiplexer of that scan flip-flop is also set to read data from the corresponding state output of the circuit.) In the test mode, the input to flip-flop FF_i is the output of the previous flip-flop in the scan chain. The only exception is the first flip-flop in the chain, where the value applied at *ScanIn* of the chain is input. The output of flip-flop FF_i drives input D_s of the next flip-flop in the scan chain *as well as* state input y_i of the combinational circuit.

The above-mentioned figure can be easily generalized by the reader to depict the more general scenario where flip-flops are configured into multiple scan chains and ordered in an alternative fashion. Note that in such designs, multiple scan chain inputs and outputs are required, one for each scan chain.

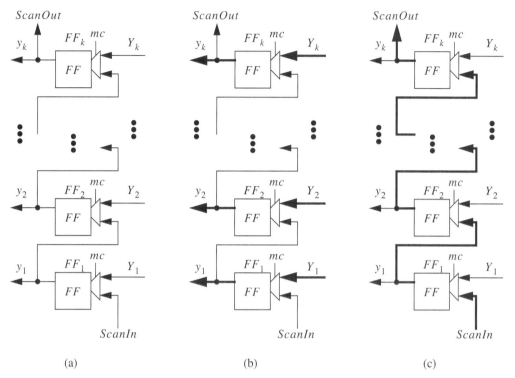

Figure 11.3 A generic scan chain: (a) structure, (b) configuration in normal mode, and (c) configuration in test mode

11.2.1.2 Test generation

The use of scan allows the desired value to be shifted into each flip-flop, or **scanned in**, using the test mode and scan chains. Each state input of the combinational circuit may hence be viewed as being directly controllable. After the application of a test, the values at state outputs may be captured into the flip-flops by configuring them in their normal mode. The values thus captured may be shifted out, or **scanned out**, using the test mode and observed at the corresponding scan output pin, *ScanOut*. The state outputs of the combinational circuit may hence be viewed as being directly observable. Hence, only the combinational logic block of the sequential circuit need be considered for test generation, as shown in Figure 11.4. Consequently, a test generator for combinational circuits may be used to generate a set of tests for the faults in the sequential circuit. Assume that test vectors $P_1, P_2, \ldots, P_{N_v}$ are generated. Note that the combinational logic block of the sequential circuit has a total of $n + k$ inputs, since the circuit has n primary inputs and k state inputs. Hence, a vector P_i has the form $(p_{1,i}, p_{2,i}, \ldots, p_{n,i}, p_{n+1,i}, p_{n+2,i} + \cdots, p_{n+k,i})$. Of these, values $p_{1,i}, p_{2,i}, \ldots, p_{n,i}$ are applied to primary inputs $x_1, x_2, \ldots, x_n$, respectively, and will collectively be referred to as the **primary input part** of test vector P_i. Values

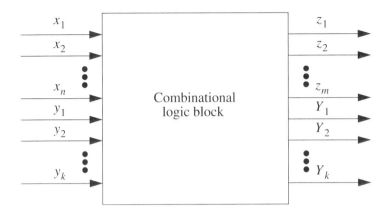

Figure 11.4 Circuit model used for test generation for the scan version of the sequential circuit in Figure 11.1

$p_{n+1,i}$, $p_{n+2,i}$, ..., $p_{n+k,i}$ are applied to state inputs y_1, y_2, ..., y_k, respectively, and will be referred to as the **state input part** of the vector.

11.2.1.3 Test application

The primary input part of each vector P_i generated by the combinational test generator can be applied directly to the primary inputs of the sequential circuit. In contrast, the state input part of each vector can be applied to the respective state inputs only via the scan chain, as described next.

Assume that the flip-flops are configured as a single scan chain. Also assume that, in the test mode, the *ScanIn* input is connected to the D_s input of FF_1, whose output drives the D_s input of FF_2, and so on. Finally, the output of FF_k is connected to the *ScanOut* line. In such a circuit, the following steps are used to apply vector P_i.

1 The circuit is set into test mode by setting $mc = test$.

2 For the next k clock cycles, the bits of the state input part of the vector are applied in the order $p_{n+k,i}$, $p_{n+k-1,i}$, ..., $p_{n+2,i}$, $p_{n+1,i}$, at the *ScanIn* pin.
   Arbitrary values may be applied to the primary inputs during the above clock cycles. At the end of these cycles, values $p_{n+1,i}$, $p_{n+2,i}$, ..., $p_{n+k,i}$ are shifted into flip-flops FF_1, FF_2, ..., FF_k, respectively, and are hence applied to state inputs y_1, y_2, ..., y_k.

3 Values $p_{1,i}$, $p_{2,i}$, ..., $p_{n,i}$ are then applied to primary inputs x_1, x_2, ..., x_n, respectively.
   At the end of this step, all bits of vector P_i are applied to the corresponding inputs (primary or state, as appropriate) of the combinational logic block.

4 The circuit is configured in its normal mode by setting $mc = normal$ and one clock pulse is applied.

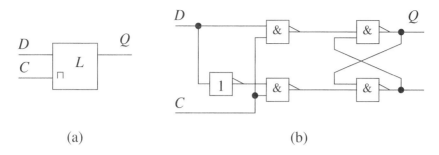

Figure 11.5 A positive latch: (a) symbol, (b) an implementation

This causes the response at the state outputs of the combinational logic block, $Y_1, Y_2, \ldots, Y_k$, to be captured in flip-flops $FF_1, FF_2, \ldots, FF_k$, respectively. The response at primary outputs $z_1, z_2, \ldots, z_m$ are observed directly at this time.

5 The response captured in the scan flip-flops is scanned out and observed at the *ScanOut* pin in Steps 1 and 2 when the above procedure is repeated to apply the next test vector, P_{i+1}.

k clock cycles are required to scan the state input part of each test vector into the flip-flops. One clock cycle is then required to apply the vector. Finally, $k - 1$ clock cycles are required to scan out the response captured at the flip-flops for the vector. Since the response for vector P_i is scanned out at the same time as when the state input part of the next vector, P_{i+1}, is scanned in, a total of $N_v(k + 1) + k - 1$ clock cycles are required to apply N_v test vectors.

11.2.2 Level-sensitive scan design

A level-sensitive *D*-**latch**, or, simply, a latch, is a storage device with inputs D and clock, and outputs Q and its complement, $\overline{Q}$, where output Q is also the state of the latch. A **positive latch** holds its previous state (and hence its outputs) constant, when the clock is low. When the clock is high, the state of the latch as well as its output Q assume the value at its input D. Any changes in the value applied at D during this period are reflected, after some delay, at its outputs. The symbol for a positive latch is shown in Figure 11.5(a), along with an example implementation in Figure 11.5(b). A negative latch holds its state and outputs constant when the clock is high and its state and outputs follow changes in the value at its D input when the clock is low. The positive pulse within the block depicting the latch indicates that this latch is a positive latch; a negative latch can be denoted by a symbol obtained by replacing the positive pulse by a negative pulse in this symbol.

A flip-flop is often obtained by using two latches in a master–slave configuration. One example of such an implementation is shown in Figure 11.6. In this design, two clocks, C and B, are used. These clocks must be **non-overlapping**, i.e., they must

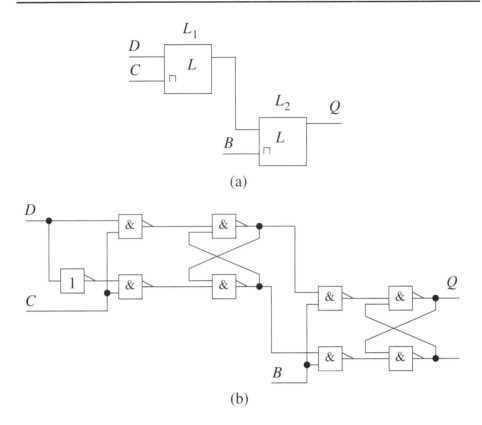

Figure 11.6 A master–slave flip-flop: (a) symbol, (b) an implementation

not assume the value 1 at the same time. In particular, B can be complement of C. In general, the value at the D input of latch L_1 is latched into the flip-flop at the falling edge of clock C and appears at the Q output of L_2 when clock B goes high.

Two main types of latch-based designs are commonly used. First is a variation of the circuit shown in Figure 11.1, where each flip-flop is implemented using a pair of latches in a master–slave configuration as described above. Such a circuit is depicted in Figure 11.7. Latches are always used in pairs in master–slave configuration in such a design and hence it is called a **double-latch** design (Eichelberger and Williams, 1977). In contrast is the **single-latch** design where single latches are used, as illustrated in Figure 11.8. In such a design, if there exists a combinational path from the output of a latch clocked with clock C_1 to the input of another latch, then that latch must be clocked by clock C_2, where clocks C_1 and C_2 are non-overlapping. Such a style is commonly used in certain types of circuits.

Next, we discuss methodologies to incorporate scan into both types of *ideal latch-based circuits*. In **ideal latch-based circuits**, we assume that a set of clocks are inputs to the chip. These clocks are used only to drive clock inputs of latches and are not

Primary
inputs

Primary
outputs

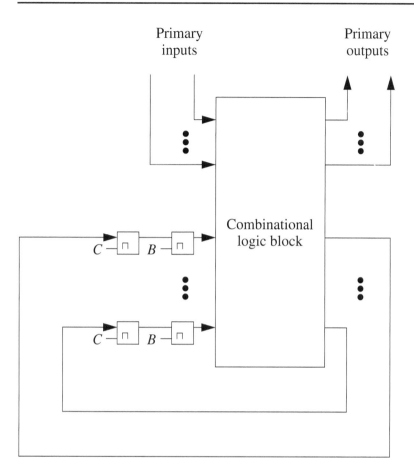

Combinational
logic block

Figure 11.7 A double-latch sequential circuit

combined in any fashion with any other signal in the circuit. Finally, the clock inputs
of latches are driven only by such primary input clocks.

11.2.2.1 Level-sensitive scan element

As described above, a memory element in a scan chain must be capable of selecting
the value from one of two inputs, namely, the state output in the normal mode and
either the *ScanIn* pin or the scan output of the previous element in the chain in the test
mode. Furthermore, since multiple scan elements must be connected as a shift-register,
each scan element must have a functionality that is equivalent to that of a flip-flop or a
master–slave latch configuration. One design of a scan element that uses a combination
of a multiplexer and a flip-flop was discussed earlier. Next, we discuss an alternative
design that uses test clocks to distinguish between the normal and test modes. This
design was presented in Eichelberger and Williams (1977) and is especially suited for
latch-based circuits.

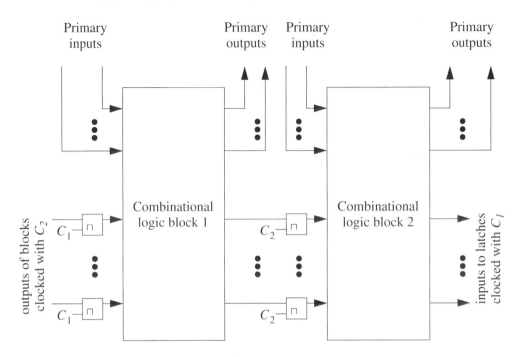

Primary
inputs

Primary
outputs

Primary
inputs

Primary
outputs

outputs of blocks
clocked with C_2

Combinational
logic block 1

C_1

C_1

C_2

C_2

Combinational
logic block 2

inputs to latches
clocked with C_1

Figure 11.8 A single-latch sequential circuit

Figure 11.9 shows an alternative design of a scan element. This design, called **level-sensitive (LS) scan** element, uses three clocks and is obtained by replacing the first latch, L_1, in the master–slave design shown in Figure 11.6 by a **two-port latch** L^* called L_1^*. An L^* latch has two inputs and two clocks. In the normal mode, clock C is used and clock A is kept low and the value at input D is latched. In the test mode, clock A is used, clock C is kept low, and the value at input D_s is latched.

For the master–slave design, in the normal mode, non-overlapping clocks C and B are used while clock A is held low. In this mode, the value at input D of latch L_1^* is input to the flip-flop. In the test mode, clock C is kept low and non-overlapping clocks A and B are used to input a value at input D_s of latch L_1^* into the element.

Now consider the double-latch circuit shown in Figure 11.7. Each pair of L_1 and L_2 latches in this design can be replaced by the above scan element with latches L_1^* and L_2. Input D, clocks C and B, and output Q of L_2 are connected as for the original pair of latches. In addition, input D_s of latch L_1^* of the first scan element is connected to the input *ScanIn* pin. The D_s input of L_1^* of the next scan element in the scan chain is connected to the Q output of the L_2 latch of the previous element, and so on. Finally, the Q output of the L_2 latch of the last element is also connected to the *ScanOut* pin. In addition to the *ScanIn* and *ScanOut* pins, an additional pin is also provided for clock A, which is connected to the clock A input of each element in the scan chain. Figure 11.10 shows the circuit obtained after the above changes are made to the circuit in Figure 11.7.

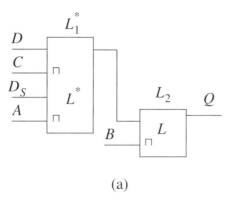

(a)

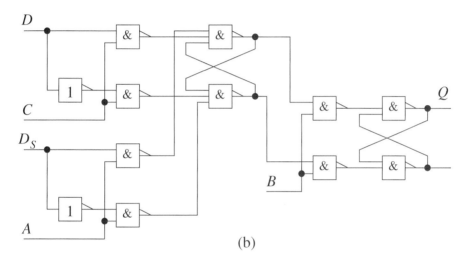

(b)

Figure 11.9 An LS scan element for a double-latch design (Eichelberger and Williams, 1977) (©
1977 IEEE)

Next, consider the single-latch circuit shown in Figure 11.8. In this design, some
latches are clocked using clock C_1 and others with clock C_2. In such a case, each
latch is replaced with the scan element shown in Figure 11.11, which is similar to
that described above (and shown in Figure 11.9), except that the output of latch L_1^*
is called Q, while the output of L_2 is called Q_S. The output Q of a scan element
is used to drive the input of the combinational logic block that was driven in the
original circuit version by the latch that the scan element replaces. Clock C of the
element is connected to the clock that was connected to the clock of the corresponding
latch in the original circuit version, i.e., C_1 or C_2. Clock input A of each L_1^* latch
and B of each L_2 latch of the scan element are driven by non-overlapping clocks
A and B, respectively. Input D_s and output Q_S of L_1^* and L_2 latches, respectively,
are used to connect the scan elements into a scan chain, as shown in Figure 11.12.
During the normal mode, clocks A and B are held low, and clocks C_1 and C_2 are

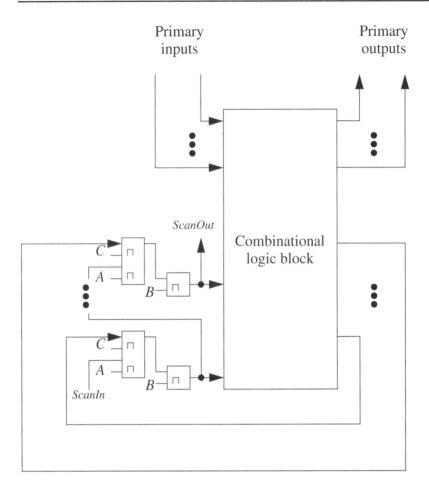

Figure 11.10 Scan version of a double-latch sequential circuit using LS scan elements (Eichelberger and Williams, 1977) (© 1977 IEEE)

used. During the test mode, clocks C_1 and C_2 are held low, and clocks A and B are used.

11.2.2.2 Test generation and application

As in the case of flip-flop based circuits, the use of full scan allows tests to be generated by considering only the combinational blocks of logic in a sequential circuit. As above, for purpose of test generation, the latch outputs and latch inputs are viewed as primary inputs and outputs, respectively.

Test application is also similar to that outlined in Section 11.2.1.3, with the difference that when alternative scan designs are used, clocks are used to control the mode of operation of the scan elements (in multiplexed-input scan flip-flop design, the control input mc is used for this purpose).

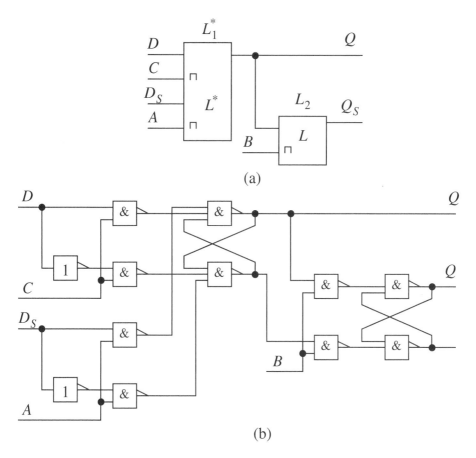

Figure 11.11   An LS scan element for a single-latch design (Eichelberger and Williams, 1977) (©
1977 IEEE)

The use of clocks can be exemplified by considering how tests are applied to the
scan version of the double-latch design shown in Figure 11.10. The application of
a test vector $P_i = (p_{1,i}, p_{2,i}, \ldots, p_{n,i}, p_{n+1,i}, p_{n+2,i}, \ldots, p_{n+k,i})$ is accomplished
as follows. First, bits of the state input part of the vector are sequentially applied
to the *ScanIn* in appropriate order, and non-overlapping clocks A and B are used to
shift these bits into the scan elements while clock C is held low. This process, which
takes k clock cycles, is followed by the application of the primary input part of the
test vector to the corresponding primary inputs of the combinational logic block. A
combination of clocks C and B is then used to apply vector P_i to the circuit and
capture the response at the state outputs into the scan elements. During this period,
clock A is held low. Note that due to the use of clocks C and B in this step, each
pair of latches operates in its normal mode. The response at the primary outputs are
observed and the response captured in the scan elements is scanned out in the next
$k - 1$ clock cycles.

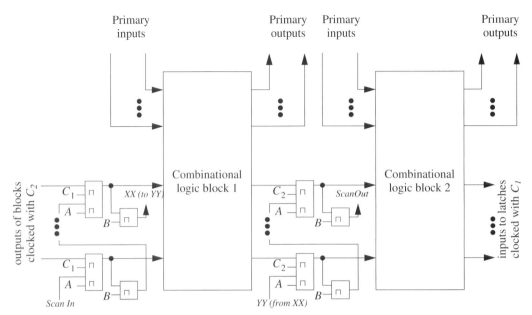

Figure 11.12 Scan version of a single-latch sequential circuit using LS scan elements (Eichelberger and Williams, 1977) (© 1977 IEEE)

The test application methodology for the single-latch design is similar and is left as an exercise to the reader.

11.2.3 Generalizations

In this section, we discuss (1) alternative scan chain configurations, (2) application of scan to non-ideal circuits, and (3) testing of the scan circuit itself.

11.2.3.1 Alternative scan chain configurations

The above discussion was limited to cases where all the scan elements in a circuit are configured as a single scan chain. In a circuit with a large number of flip-flops, a single scan chain is long and hence a large number of scan clock cycles are required for the application of each test vector. In such cases, the test application time is large.

To decrease the test application time, scan elements are often configured as **multiple scan chains**. The scan elements are partitioned into a number of sets and elements in each set are configured as a separate scan chain with its own *ScanIn* and *ScanOut* pins. To reduce the number of additional pins required to implement scan, each *ScanIn* and *ScanOut* pin may be multiplexed with a suitable primary input and output, respectively. This is possible since during test application via scan, the primary input part of a vector is applied to the primary inputs only when the circuit is configured in the normal mode, i.e., when the scan elements do not need to shift data through the *ScanIn* and *ScanOut* pins.

Two main variations are possible in a multiple scan chain configuration. In the first variation, all scan chains are controlled by a single mode control signal *mc*. In the second variation, **multiple mode control** lines are used. In this variation, each mode control line controls the mode of a distinct set of chains. Since mode control pins cannot be easily multiplexed with logic inputs, this can increase the number of additional pins required. As will become clear later, the need for additional pins can be avoided for chips with boundary scan. Methodologies that use multiple mode control lines do so to further reduce the test application time, e.g., Narayanan and Breuer (1995a).

Another variation of scan chain that is also motivated by the need to reduce the test application time is **reconfigurable scan chain**. In this design, the scan elements are configured as a single scan chain. However, the outputs of multiple scan elements are connected via a multiplexer to the *ScanOut* pin.

11.2.3.2 Application of scan to non-ideal circuits

The above discussion has been limited to circuits that are ideal in terms of how they are clocked. First, it was assumed that each latch or flip-flop was clocked directly by a **primary input clock**, i.e., a clock signal that is input to the chip and is not altered in any manner via combination with other logic signals. Second, it was assumed that each clock signal drives only the clock inputs of latches or flip-flops. However, many practical designs do not satisfy these conditions. For example, **gated clocks**, i.e., clocks derived by combining primary input clocks with (non-clock) primary inputs or logic values within the circuit, are commonly used. In some circuits, such as an asynchronous counter, the clock input of a flip-flop may be connected to the data output of another flip-flop. For such a circuit, modifications required to obtain a scan version are not as straightforward as described above. The modifications depend on the specific manner in which the above ideal conditions are violated. However, the modified circuit must, *when configured in the test mode*, satisfy the following conditions (adapted from Eichelberger and Williams (1977)) to be able to use the above scan methodology for test generation and application.

1 Each latch or flip-flop must be suitably modified such that it may be configured as a scan element and each scan element must be part of a scan chain.

2 There must exist a set of control inputs, i.e., control pins (such as one or more mode control pins) and clocks, such that application of suitable values or transitions to these control inputs enables each scan chain to work as a shift-register. This can be achieved if the following two requirements are satisfied in the test mode.

   (a) The value to be latched into each scan element should depend only on the value at the output of the preceding scan element in the chain (or, for the first element in a chain, on the value at the corresponding *ScanIn* pin).

   (b) It should be possible to clock each scan element in a manner that causes the element to shift data at the output of its predecessor in the chain.

Specific versions of these rules for circuits with LS scan elements are presented in Eichelberger and Williams (1977).

The reader is invited to design a flip-flop based design of an asynchronous counter and try to make it testable using the scan methodology.

11.2.3.3 Testing of the scan circuitry

Tests for a scan circuit must include tests for the logic inserted in the circuit to implement scan. Some faults, such as a stuck-at fault (SAF) at the D_s input of the scan flip-flop design in Figure 11.2, can only be detected when the flip-flops are configured in the test mode. Other faults, such as the SAF at the D input of the above multiplexer, can be tested only by applying specific values to the inputs of the combinational logic and capturing the response of the circuit during a test application clock. Tests for the faults in the former category are generated by considering the scan elements configured in test mode as the circuit under test (CUT). Typically, tests for these faults are generated manually. These tests generally contain a short sequence of 1s and 0s applied prior to the application of tests to the combinational logic block.

The additional circuitry required to implement scan, such as the multiplexer in the scan flip-flop, can be combined with the combinational logic block and the faults in the second category discussed above added to the target fault list for the combinational logic block. The test set generated for the combinational logic block hence includes test vectors that detect the faults in the additional scan logic that belong to the second category.

11.2.4 Costs and benefits of scan

Additional logic and routing are required to implement scan. Hence, the layout area of the scan version of a circuit is higher than that of its non-scan counterpart. This increase in area, typically called **area overhead**, increases the cost of fabrication of the circuit in two main ways. First, higher layout area means fewer copies of a chip can be manufactured on a given semiconductor wafer. Second, an increase in area is accompanied with a decrease in yield. Hence, a smaller percentage of manufactured chips work.

If additional logic required for a scan circuit is incorporated in a straightforward manner into the circuit, then additional delay is introduced into the circuit. This can sometimes necessitate a reduction in the rate at which the circuit may be clocked. Consider as an example the multiplexed-input scan flip-flop. In this design, one multiplexer is added on many combinational paths in the circuit. The impact on clock rate of the circuit can be significant for high-speed circuits, where a combinational logic block contains only a few levels of logic gates. The performance penalty of the LS scan element design is due to the replacement of a pair of two-input AND gates with three-input AND gates. Any decrease in the speed of the circuit during normal operation is often referred to as **performance penalty**.

Two main approaches are used to minimize the performance penalty of scan. In the first approach, the size of the transistors in the scan element are appropriately increased to reduce delays. This approach helps reduce performance penalty of scan but further increases the area overhead. In the second approach, the additional logic incorporated for scan, such as the multiplexer added in the multiplexed-input scan flip-flop design, is combined with the combinational logic block and the entire logic is re-synthesized to reduce delay. This approach is practical only for those cases where the combinational logic block was synthesized using automatic tools in the first place. Even in those cases, this approach often reduces performance penalty at the expense of further increase in area overhead.

The third main cost of scan is the need for extra pins required for signals such as *ScanIn*, *ScanOut*, mode control, and test clocks. Some of these signals, especially *ScanIn* and *ScanOut*, can be multiplexed with existing primary input and output pins. In the normal mode, these pins function as originally intended; in the test mode, they function as *ScanIn* and *ScanOut*. As will be seen ahead, scan design may be integrated within the boundary scan architecture, reducing the number of additional pins required.

The fourth cost of scan is that, for most circuits, test application time increases. The use of scan decreases the number of vectors required to test a circuit, but several clock cycles are required to apply each test vector via scan. In most circuits, this leads to an increase in the overall test application time. Some exceptions exist, however. This cost can be mitigated by using multiple scan chains and other techniques described ahead.

Incorporation of scan increases the design complexity to some extent. For example, in a design methodology where standard cells are used, the cell library must be extended to include cells for scan versions of latches and flip-flops.

One key benefit of scan is that for many sequential circuits, the use of some amount of scan is the only way to attain an acceptable fault coverage. This is one of the key reasons behind acceptance of scan. However, increased controllability and observability provided by scan are also useful for purposes other than test generation and application. Scan is used extensively to locate design errors and weaknesses during debugging of *first silicon*, i.e., the first batch of chips fabricated for a new design. Also, scan is used to locate failing components when an operational system fails. In such scenarios, scan is used to reduce the time required to repair a system, hence to increase system availability.

11.3 Partial scan

The area overhead associated with full scan can be reduced by replacing only a subset of the flip-flops in a circuit by scan flip-flops. Such an approach is called **partial scan**.

It has the potential for reducing the performance penalty of scan (Trischler, 1980). The key questions in the partial-scan methodology are (a) which flip-flops to select for replacement with scan flip-flops, (b) how to generate tests for the partial-scan circuit, and (c) how to apply tests using partial scan.

Flip-flops should be selected for partial scan in a manner that minimizes costs of scan while preserving its benefits. To minimize area overhead, the number of flip-flops selected must be minimized. To reduce performance penalty, selection of flip-flops that are in critical paths of circuits (i.e., paths whose delays are equal, or close, to the clock period) should be avoided. At the same time, flip-flops should be selected to ensure that the controllability and observability enhancements enable successful generation of tests that provide an acceptably high fault coverage.

Once one or more flip-flops in a given circuit are selected for partial scan, a circuit model may be obtained for test generation. Such a circuit model, called a **kernel** (Gupta *et al.*, 1990), can be obtained by removing each selected flip-flop FF_i from the given circuit, considering the state output Y_i that drives the D input of the flip-flop in the given circuit as a primary output of the kernel, and considering the state input y_i that is driven by the Q output of the flip-flop as a primary input. (If Q as well as $\overline{Q}$ outputs of a flip-flop drive lines in the original circuit, then the line driven by the Q output may be considered as an input to the kernel. This primary input may then be used to drive via an inverter the lines in the original circuit driven by the $\overline{Q}$ output of the flip-flop. A similar approach may be employed when only the $\overline{Q}$ output of a flip-flop drives a line in the original circuit.)

In the following discussion, we divide the approaches to partial scan into two categories. In the first category are the **structural** approaches for partial scan. In these approaches, flip-flops for partial scan are selected to ensure that the kernel obtained satisfies certain structural properties that are known to simplify test generation. In the second category are approaches that select flip-flops to scan by estimating the ease of test generation (using testability measures, a sequential test generator, or some other technique) to identify flip-flops to control or observe.

11.3.1   Structural approaches

Most sequential test generators cannot guarantee high fault coverage for an arbitrary sequential circuit. However, there exist special classes of sequential circuits for which test generation is much more predictable in the sense that attainment of high fault coverage is feasible in practice. These classes of circuits are often characterized in terms of their **sequential structure**, which captures how the values at flip-flop inputs (state outputs) as well as primary outputs of a circuit depend on the values at flip-flop outputs (state inputs) and primary inputs.

We begin with some basic definitions followed by a description of some subclasses of sequential circuits. Then we discuss the methodologies for selection of scan flip-flops followed by the techniques for test generation and application.

11.3.1.1 Some definitions and a representation

A **path** from a line c_i to a line c_j in its transitive fanout in a sequential circuit is the sequence of lines starting with c_i and ending with c_j, where each line in the sequence is a fanout of the previous line in the sequence. The lines, gates, fanout systems, and flip-flops via which a path passes are also said to belong to the path. This is a generalization of the definition of a path for a combinational circuit, since such a path may pass via one or more flip-flops. A **combinational path** is a path that does not pass via any flip-flop.

A path is said to be **cyclic** if it contains multiple instances of one or more lines. A path in which no line appears more than once is called an **acyclic** path. A sequential circuit is said to be **acyclic** if all the paths in the circuit are acyclic. The **sequential depth of an acyclic path** is the number of flip-flops in the path. Note that, a combinational path is a path with a sequential depth of zero. The **sequential depth of an acyclic sequential circuit** is the maximum of the sequential depths of the paths in the circuit.

The main properties of the structure of the sequential circuit that are of interest in the following discussion are how the values at flip-flop inputs (state outputs) and primary outputs depend on the values at the flip-flop outputs (state inputs) and primary inputs. These dependencies can be represented using a directed graph called an **s+graph**. This graph contains a node $v(x_i)$ for each primary input x_i, a node $v(FF_i)$ for each flip-flop FF_i, and a node $v(z_i)$ for each primary output z_i. A directed edge exists from node $v(x_i)$ to node $v(z_j)$, if a combinational path exists from x_i to z_j. An edge exists from node $v(x_i)$ to node $v(FF_j)$, if a combinational path exists from x_i to the D input of FF_j. An edge exists from node $v(FF_i)$ to node $v(z_j)$, if a combinational path exists from the output (Q and/or $\overline{Q}$) of FF_i to z_j. Finally, an edge exists from node $v(FF_i)$ to node $v(FF_j)$, if a combinational path exists from the output of FF_i to the D input of FF_j. (The s+graph is an extension of the s graph described in Cheng and Agrawal (1990).) Figure 11.13 shows an example circuit, where CLB denotes a combinational logic block, and the corresponding s+graph.

11.3.1.2 Subclasses of sequential circuits

We next discuss some subclasses of sequential circuits.

Linear pipeline kernels: Let S be the set of all the flip-flops in a circuit. If S can be partitioned into sets $S_1, S_2, \ldots, S_l$ such that (a) the value at the input of any flip-flop in set S_1 depends only on the values at the primary inputs (and does not depend on the value at the output of any flip-flop in the circuit), (b) for each i, where $2 \le i \le l$, the value at the input of any flip-flop in set S_i depends only on the values at the outputs of flip-flops in set S_{i-1}, and (c) the value at each primary output depends only on the values at the outputs of the flip-flops in set S_l, then the circuit is said to be a **linear pipeline**. The lengths of all the sequential paths in a linear pipeline

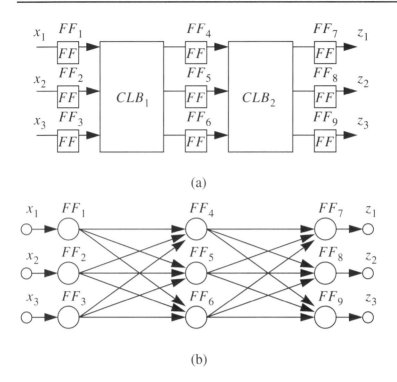

Figure 11.13 (a) An example linear pipeline kernel, and (b) corresponding s+graph

circuit are identical. This length is also called the **depth** of the linear pipeline. One can verify that in the circuit in Figure 11.13(a), flip-flops can be partitioned into sets $S_1 = \{FF_1, FF_2, FF_3\}$, $S_2 = \{FF_4, FF_5, FF_6\}$, and $S_3 = \{FF_7, FF_8, FF_9\}$, which satisfy the above conditions.

A linear pipeline is one of the simplest subclasses of sequential circuits. Even though such a circuit is sequential, a combinational test generator may be used to generate tests as follows. First, the given linear pipeline circuit C^{lp} is transformed into a combinational circuit C^{lp*} by (a) removing each flip-flop in the circuit, (b) connecting directly the line driving its D input to the line driven by its Q output, and (c) connecting via an inverter the line driving its D input to the line driven by its $\overline{Q}$ output. Second, a set of tests can be generated for SAFs in circuit C^{lp*}. It can be demonstrated that application of vectors in the above test set to circuit C^{lp} provides SAF coverage that is identical to that reported by the combinational test generator for circuit C^{lp*}. This coverage can be obtained by applying the generated vectors to the primary inputs of the linear pipeline in successive clock cycles. After the application of the last vector in the set, a number of clock cycles equal to the sequential depth of the circuit must be applied to the circuit and responses at its outputs observed.

The definition of a more general type of pipeline can be found in Kunzman and Wunderlich (1990).

Balanced kernels: *Balanced kernels* (Gupta *et al.*, 1990) form a subclass of sequential circuits that includes pipeline circuits and can be defined as follows. A sequential circuit is said to be **balanced** if (a) the circuit is acyclic, and (b) all the paths between any given primary input x_i and primary output z_j have equal sequential depths. An example balanced circuit is shown in Figure 11.14. The s+graph for this circuit is also shown in the figure. The above two conditions have their counterparts in the s+graph, namely (a) the s+graph must be acyclic, and (b) all the paths between any given input node and output node in the s+graph must pass via an equal number of flip-flop nodes.

As with a pipeline kernel, a combinational test generator can be used to generate tests for a balanced sequential kernel (Gupta *et al.*, 1990). Given a balanced kernel C^b, a combinational circuit model C^{b^*} is obtained by replacing each flip-flop in the kernel by a wire and/or an inverter as described above for the pipeline circuit. The circuit C^{b^*} obtained in this manner for the example circuit in Figure 11.14 is shown in Figure 11.15. A combinational test generator can then be used to generate a test for each SAF in circuit C^{b^*}.

The different bits of a test generated by a combinational test generator for circuit C^{b^*} may need to be applied to the primary inputs at different times. For example, assume that a vector $P_i = (p_{1,i}, p_{2,i}, p_{3,i}, p_{4,i}, p_{5,i})$ is generated for a particular SAF in circuit C^{b^*} shown in Figure 11.15. Recall that bit $p_{j,i}$ of the vector is to be applied to input x_j. The value applied to x_1, $p_{1,i}$, appears at the inputs of combinational logic block CLB_4 immediately. In contrast, the value implied at the inputs of CLB_4 by $p_{3,i}$, the value applied at x_3, appears after two clock cycles. (The first clock cycle is required for the implied values to be latched into $FF_3, FF_4, \ldots, FF_6$; the second clock cycle is required for the implied values to propagate via CLB_2 and CLB_3 and the response to be latched into $FF_7, FF_8, \ldots, FF_{10}$, whose outputs drive the inputs of CLB_4.) Hence, if bits $p_{1,i}$, $p_{2,i}$, and $p_{5,i}$ are applied in clock cycle $i + 2$, then $p_{3,i}$ and $p_{4,i}$ must be applied in clock cycle i.

The following labeling procedure can be used to determine the relative times at which the bits of a vector generated for circuit model C^{b^*} must be applied at the corresponding inputs of circuit C^b. Assume that the s+graph for circuit C^b is connected. Associate with each node a in the s+graph a label $\eta(a)$. The label associated with a flip-flop node is assumed to be associated with its input D. Initialize each label $\eta(a)$ as `unlabeled`. Select a primary input x_j and assign $\eta(x_j) = 0$. Assign $\eta(l) = 0$ to each successor node l of x_j. Process each node (primary input, flip-flop, or primary output) that is assigned a label in the above step, by assigning suitable labels to each of its predecessor and successor nodes that does not have a numerical label. While processing successor nodes of node a, if a flip-flop is traversed from node a to its successor b, then $\eta(b) = \eta(a) + 1$; otherwise, $\eta(b) = \eta(a)$. Similarly, while processing predecessor nodes of node a, if a flip-flop is traversed from node a to its predecessor b, then $\eta(b) = \eta(a) - 1$; otherwise, $\eta(b) = \eta(a)$. Repeat the above process until a numerical label is assigned to each node. If the s+graph is

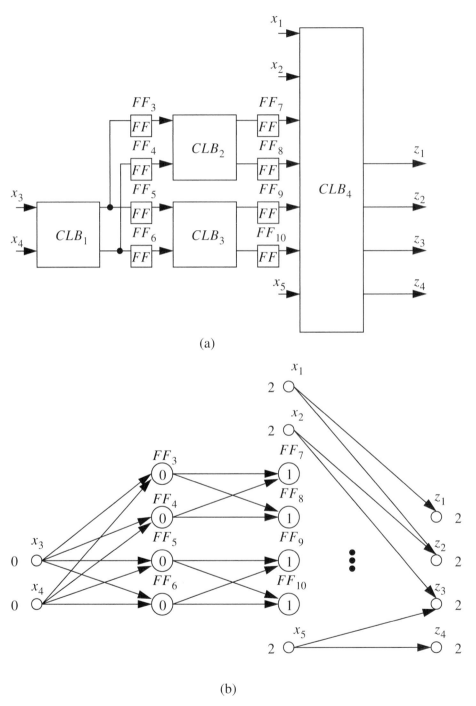

(a)

(b)

Figure 11.14 (a) An example balanced kernel (Gupta *et al.*, 1990) (© 1990 IEEE), and (b) corresponding s+graph (Note: Only some of the edges are shown.)

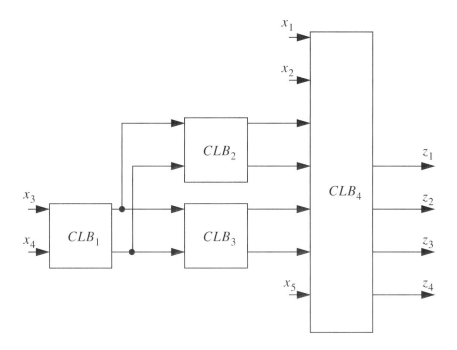

Figure 11.15 Combinational circuit model used to generate tests for the balanced kernel in Figure 11.14 (Gupta *et al.*, 1990) (© 1990 IEEE)

not connected, then the above procedure must be repeated for each of its connected components.

Example 11.1 Consider the s+graph shown in Figure 11.14(b). We begin at primary input x_3 by assigning $\eta(x_3) = 0$. Since no flip-flop is traversed between x_3 and its successor nodes, FF_3, FF_4, FF_5, and FF_6, each of the corresponding nodes is assigned the same label as x_3. Each of the newly labeled nodes is processed next beginning with, say, FF_3. Since no flip-flop is traversed between FF_3 and its only unlabeled predecessor node, x_4, it is assigned a label as $\eta(x_4) = \eta(FF_3) = 0$. Next, successor nodes of FF_3 are considered, beginning with, say FF_7. Since one flip-flop, namely FF_3, must be traversed to reach the input of FF_7 from the input of FF_3, the assignment is $\eta(FF_7) = \eta(FF_3) + 1 = 1$. (Recall that for a flip-flop node, label η is associated with the input of the corresponding flip-flop. Hence, a primary input predecessor node of a flip-flop can be reached without traversal of any flip-flop while a flip-flop predecessor node can be reached only via the traversal of that flip-flop. Also, any successor node (a primary output or flip-flop) can only be reached via traversal of the flip-flop under consideration.) The above process is repeated until numerical labels are assigned to each node. The results of the above labeling process for this example s+graph are shown in Figure 11.14(b).   □

Table 11.1. *An example of reformatting of test vectors generated for a combinational circuit model C^{b^*} of a balanced kernel C^b for application to the kernel: (a) Test vectors generated for C^{b^*} in Figure 11.15, and (b) vectors reformatted for application to kernel C^b in Figure 11.14(a)*

(a)

Vector	Bits of vector				
P_1	$p_{1,1}$	$p_{2,1}$	$p_{3,1}$	$p_{4,1}$	$p_{5,1}$
P_2	$p_{1,2}$	$p_{2,2}$	$p_{3,2}$	$p_{4,2}$	$p_{5,2}$
P_3	$p_{1,3}$	$p_{2,3}$	$p_{3,3}$	$p_{4,3}$	$p_{5,3}$
P_4	$p_{1,4}$	$p_{2,4}$	$p_{3,4}$	$p_{4,4}$	$p_{5,4}$
$\vdots$	$\vdots$	$\vdots$	$\vdots$	$\vdots$	$\vdots$

(b)

Normal clock	Values applied at input				
	x_1	x_2	x_3	x_4	x_5
1			$p_{3,1}$	$p_{4,1}$	
2			$p_{3,2}$	$p_{4,2}$	
3	$p_{1,1}$	$p_{2,1}$	$p_{3,3}$	$p_{4,3}$	$p_{5,1}$
4	$p_{1,2}$	$p_{2,2}$	$p_{3,4}$	$p_{4,4}$	$p_{5,2}$
5	$p_{1,3}$	$p_{2,3}$	$\vdots$	$\vdots$	$p_{5,3}$
6	$p_{1,4}$	$p_{2,4}$			$p_{5,4}$
$\vdots$	$\vdots$	$\vdots$			$\vdots$

The above labels can be used to determine the clock cycle in which each bit of a vector generated by the combinational test generator must be applied. Let $P_1, P_2, \ldots, P_{N_v}$ be the test vectors generated by the combinational test generator. Assuming that the balanced kernel has l inputs, bit $p_{j,i}$ of vector $P_i = (p_{1,i}, p_{2,i}, \ldots, p_{l,i})$ should be applied at x_j in clock cycle $i + \eta(x_j)$.

Example 11.2 For the above example, the bits of a set of tests $P_1, P_2, \ldots, P_{N_v}$ generated for circuit model C^{b^*} (Figure 11.15) can be *reformatted* for application to kernel C^b (Figure 11.14(a)) as shown in Table 11.1. Table 11.1(a) shows the test vector bits as generated for C^{b^*}. Table 11.1(b) shows the clock cycles at which each vector bit must be applied to C^b. $\square$

Acyclic kernels: *Acyclic kernels* form a more general class of sequential circuits. An **acyclic kernel** is one in which all the paths are acyclic. The s+graph of an acyclic graph contains no cycles. Recall that the sequential depth of an acyclic kernel is the maximum of the sequential depth of all its paths.

While for an arbitrary sequential circuit, a sequential test generator may need to consider sequences with a large number of vectors, for an acyclic kernel with sequential depth d, the test generator needs to consider only sequences with a maximum of d vectors. Intuitively, this follows from the following observations. First, in an acyclic kernel, the value at a flip-flop in a given clock cycle cannot affect the value implied at that flip-flop in any future clock cycle. In fact, the flip-flops in the circuit can be levelized as follows. If all the predecessor nodes of a flip-flop in the s+graph are primary inputs, then the flip-flop is assigned **sequential level** one. If all the predecessor nodes of a flip-flop in the s+graph are primary inputs or flip-flops with sequential level one, then the flip-flop is assigned sequential level two, and so on. The highest sequential level that may be assigned to any flip-flop in an acyclic circuit is equal to its sequential depth, d.

Second, it can be shown that the value at a flip-flop at sequential level one is completely determined by the values applied at the primary inputs. The value at a flip-flop at sequential level two is completely determined by the values at the primary inputs and the values at the flip-flops at sequential level one. Hence, the value at a flip-flop at sequential level two is completely determined by the values applied at the circuit inputs over two consecutive clock cycles. In this manner, it can be demonstrated that the value at any flip-flop in an acyclic circuit with sequential depth d is determined uniquely by the values applied at primary inputs over d consecutive clock cycles.

Due to the above fact, typically, the complexity of sequential ATPG for an acyclic kernel is lower than that for an arbitrary sequential circuit. Second, for any given target SAF in an acyclic kernel with sequential depth d, a test sequence contains no more than d vectors. If no such sequence can be found, then the target fault is untestable and search for a longer test sequence is unwarranted.

The knowledge of above facts can be easily incorporated into a general sequential test generator to tailor it to efficiently generate tests for an acyclic kernel.

Kernels with only self-loops: The final type of kernels that we consider are those whose s+graphs contain only one type of cycles, *self-loops*. A **self-loop** in an s+graph is a cycle formed by the presence of an edge whose source and destination nodes are identical. In the s+graph shown in Figure 11.16, a self-loop exists around FF_4. The s+graph also contains two other cycles that are not self-loops.

Consider an example s+graph comprised of a single flip-flop node with a self-loop that has additional edges from primary inputs and to primary outputs. A sequential circuit with such an s+graph can only have a single flip-flop. The self-loop indicates that the value at the input of the flip-flop depends on the value at the output of the flip-flop. Hence, the corresponding circuit is a two-state finite-state machine (FSM). One of the important properties of a two-state FSM is that its state can be set to any desired value through the application of a single vector (Cheng and Agrawal, 1990).

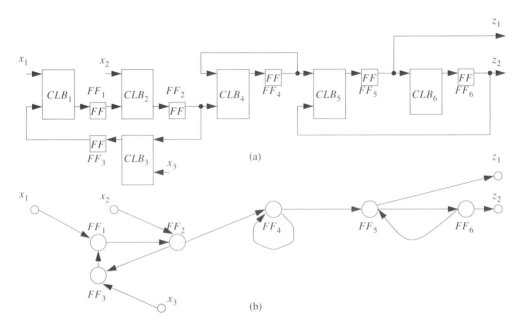

Figure 11.16 (a) An example sequential circuit, and (b) s+graph for the circuit (Cheng and Agrawal, 1990) (ⓒ 1990 IEEE)

Now consider the s+graph shown in Figure 11.17(a). The only type of cycle this s+graph contains is one self-loop. Hence, the corresponding sequential circuit has one two-state FSM embedded in an otherwise acyclic sequential circuit. Due to the fact that it is relatively easy to set the state of a two-state FSM, test generation for such a sequential kernel is typically easier than for an arbitrary sequential circuit.

In general, a sequential circuit corresponding to an s+graph that contains multiple self-loops but no other cycles can be viewed as a set of two-state FSMs that interact in a non-cyclic manner. Test generation for such circuits is hence typically easier than for arbitrary sequential circuits (Cheng and Agrawal, 1990).

11.3.1.3 Selection of flip-flops to scan

Structural approaches to partial scan start by analyzing a given sequential circuit and creating a graph model like the s+graph. Next, the methodology specifies the type of kernel. The methodologies in Kunzman and Wunderlich (1990) and Gupta *et al.* (1990) target balanced kernels, the methodology in Gupta and Breuer (1992) targets acyclic kernels, while the methodology in Cheng and Agrawal (1990) targets kernels whose s+graphs contain self-loops but are otherwise acyclic.

The s+graph can be used to select flip-flops to scan for each of the above methodologies. When a flip-flop FF_i is selected for scan, the node corresponding to FF_i is removed from the s+graph. One input node, called FF_i^I, and one output node,

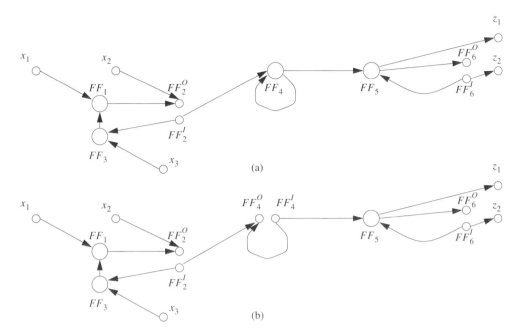

Figure 11.17 s+graph obtained after selection of flip-flops to scan from the circuit in Figure 11.16: (a) FF_2 and FF_6 selected for scan, and (b) FF_2, FF_4, and FF_6 selected for scan (Cheng and Agrawal, 1990) (© 1990 IEEE)

called FF_i^O, are added to the s+graph. All edges that had FF_i as their destinations in the original s+graph now have output node FF_i^O as their destination; all the edges that originally had FF_i as the source, now have input node FF_i^I as their source. Scan flip-flops are selected and s+graph modified after each selection, until the resulting s+graph satisfies all the properties required by the partial-scan methodology being implemented.

Example 11.3 Figure 11.17 shows the s+graphs obtained after two different selections of scan flip-flops for the s+graph shown in Figure 11.16. In the first scenario, FF_2 and FF_6 are selected. The s+graph obtained by removal of the corresponding nodes and all incoming and outgoing edges is shown in Figure 11.17(a). This s+graph is not acyclic but it only has a self-loop. Hence, it corresponds to a kernel that has self-loops but is otherwise acyclic. In the second scenario, FF_2, FF_4, and FF_6 are selected. The resulting s+graph is shown in Figure 11.17(b) and is acyclic and all the paths between any pair of input and output nodes have equal sequential depth. This graph hence corresponds to a balanced kernel.   □

Since all the properties of the types of kernels required by each methodology can be expressed in terms of the properties of the s+graph of the given sequential circuit, the problem of selecting flip-flops to scan for each of the above methodologies can be mapped into a graph problem. One key graph problem that is considered is called the **minimum feedback vertex set** (MFVS), where the objective is to find a minimum number of nodes in a given graph, whose removal from the graph (along with all the incoming and outgoing edges) will make the graph acyclic. Procedures that identify flip-flops to scan by solving the MFVS problem have been proposed in Lee and Reddy (1990), Chakradhar *et al.* (1994), and Ashar and Malik (1994).

To obtain kernels that have self-loops but are otherwise acyclic, one can proceed as follows. First, an s+graph is created for the given sequential circuit. Then all the edges that constitute self-loops are simply removed from the s+graph. Any of the above procedures to identify scan flip-flops can then be applied and flip-flops selected until an acyclic s+graph is obtained. All the self-loops that were removed from the s+graph for the given circuit in the first step are added as follows. If a flip-flop FF_i with self-loop was not selected for scan, then the self-loop is added around the corresponding node $v(FF_i)$ in the s+graph. Otherwise, an edge is added between nodes $v(FF_i^I)$ and $v(FF_i^O)$. The kernel corresponding to the resulting s+graph may contain self-loops but will otherwise be acyclic.

Balanced kernels may be obtained by modifying any of procedures to solve the MFVS problem, or using the procedure outlined in Gupta *et al.* (1990).

11.3.2   Testability and test generation based approaches

Consider the following greedy heuristic for selection of scan flip-flops. The heuristic begins by considering the given non-scan circuit as the *initial design*. A number of *alternative designs* are then generated, where each alternative design is obtained by selecting one non-scan flip-flop in the initial design for scan. A figure-of-merit is then computed for each alternative design. (Different figures-of-merit will be discussed later.) The alternative design with the most attractive figure-of-merit is then selected as the next initial design. The above procedure is repeated until either a design is obtained for which satisfactory fault coverage is achieved or a specified maximum number of flip-flops is selected for scan.

We now describe three different ways of computing the figures-of-merit for a given design.

11.3.2.1 Testability measures

In its simplest form, the classical testability measures, namely the 0-controllability (CC^0), 1-controllability (CC^1), and observability (O) can be computed for each line c_i in a given circuit, as described in Chapters 4 and 5 (e.g., see Sections 4.5.3 and 5.4.2.1). Recall that specific values are assigned to CC^0 and CC^1 at each primary input and the values of these measures are then computed for each internal line in the circuit, including the state outputs and inputs. Subsequently, specific values are assigned to O at each primary output and the value of this measure is then computed for each line in the circuit, including state outputs and inputs.

The difficulty of detecting a fault f can be computed in terms of these measures as

$$TCost(f) = \begin{cases} CC^1(c_i) + O(c_i), \text{ if } f \text{ is an SA0 fault at line } c_i, \text{ or} \\ CC^0(c_i) + O(c_i), \text{ if } f \text{ is an SA1 fault at line } c_i. \end{cases} \qquad (11.1)$$

The figure-of-merit for an alternative design can then be computed simply as the summation of $TCost(f)$ for all the faults of interest. This figure-of-merit can then be used in the greedy heuristic outlined above to select flip-flops for partial scan.

11.3.2.2 Empirical testability measures

In Kim and Kime (1990), the run-time required for a sequential test generator to generate a sequence of vectors that assigns a value 0 at a line c_i is used as the *empirical* measure of 0-controllability $CC^0(c_i)$. The 1-controllability, $CC^1(c_i)$, and observability, $O(c_i)$, are computed in a similar manner. To reduce the run-time complexity required to compute the testability measures, the values of controllabilities and observabilities are computed empirically only for the state inputs and state outputs, respectively. For a circuit with k flip-flops, this is accomplished by running the test generator to generate a total of $3k$ sequences; one sequence is generated to imply a 0 at a state input y_i, another to imply a 1, and a third to sensitize the corresponding state output, Y_i, to a primary output. An appropriate initial value is then assigned to each controllability measure at each primary input and to the observability at each primary output. This helps assign a value to each controllability measure at each input (primary as well as state) and to observability at each output (primary and state). The approach described in Chapter 4 (Section 4.5.3.4) can then be used to compute the values of these measures for each line within the combinational logic block. (In some cases, the sequential test generator may abort before it can generate a sequence of vectors that assigns a desired logic value at a state input or sensitizes a state output to a primary output. In such cases, the computation of values of these measures for each circuit line is a little more involved (Kim and Kime, 1990).)

The value of $TCost(f)$ can be computed for each fault f of interest using Equation (11.1), and used in the greedy heuristic for selection of partial-scan flip-flops.

11.3.2.3 Uncontrollability analysis

The methodology in Abramovici *et al.* (1991) assumes that functional test vectors are provided for the original circuit and used during manufacturing testing in a non-scan test phase. The test development process must hence target only those faults that are not detected by the functional tests. Furthermore, it assumes that only a combinational test generator is used to generate tests for each target fault. Thus, if flip-flop FF_i is not replaced by a scan flip-flop, the combinational test generator cannot assign either

(a)

	$V(c_{i_2})$			
	{}	{0_u}	{1_u}	{0_u, 1_u}

$V(c_{i_1})$

{}	{}
{0_u}	{1_u}
{1_u}	{0_u}
{0_u, 1_u}	{0_u, 1_u}

(a)

(b) $V(c_{i_2})$

$V(c_{i_1})$	{}	{0$_u$}	{1$_u$}	{0$_u$, 1$_u$}
{}	{}	{}	{0$_u$}	{0$_u$}
{0$_u$}	{}	{1$_u$}	{0$_u$}	{0$_u$, 1$_u$}
{1$_u$}	{0$_u$}	{0$_u$}	{0$_u$}	{0$_u$}
{0$_u$, 1$_u$}	{0$_u$}	{0$_u$, 1$_u$}	{0$_u$}	{0$_u$, 1$_u$}

(b)

$V(c_i)$	$V(c_{j_1})$	$V(c_{j_2})$	...
{}	{}	{}	...
{0$_u$}	{0$_u$}	{0$_u$}	...
{1$_u$}	{1$_u$}	{1$_u$}	...
{0$_u$, 1$_u$}	{0$_u$, 1$_u$}	{0$_u$, 1$_u$}	...

(c)

Figure 11.18 Truth tables for computing the uncontrollability status of circuit lines. (a) NOT, (b) two-input NAND gate, and (c) a fanout system

of the specific logic values to the corresponding state input, y_i; neither can it observe the corresponding state output, Y_i.

Uncontrollability status: Two logic values, namely, 0_u and 1_u, are defined to determine the effects of making some state inputs uncontrollable. 0_u indicates that value 0 cannot be implied at the line; 1_u is similarly defined. These values are used to obtain a composite value system that contains four composite values, {}, {0$_u$}, {1$_u$}, and {0$_u$, 1$_u$}. The composite value at a line indicates the logic values that cannot be implied at the line, called its **uncontrollability status**. A line to which composite value {0$_u$} is assigned is said to be **0-uncontrollable**. Similarly, a line with value {1$_u$} is said to be **1-uncontrollable** while a line with value {0$_u$, 1$_u$} is said to be **uncontrollable**. Figure 11.18 shows the truth tables for some primitive gates and a fanout system. The uncontrollability status of the lines in a circuit can be determined by assigning the desired uncontrollability status to each of its inputs and computing the uncontrollability of each line. The process of computing the uncontrollability status is similar to logic simulation (see Chapter 3; Section 3.3.2) with the difference that the above composite values and corresponding truth tables are used.

Example 11.4 Figure 11.19(a) shows the uncontrollability status of each line in an example circuit obtained starting with uncontrollability value {0$_u$, 1$_u$} at inputs x_4 and x_5. □

Unobservability status: Uncontrollability analysis is followed by identification of lines in the given circuit that are **unobservable**. A line may be unobservable because

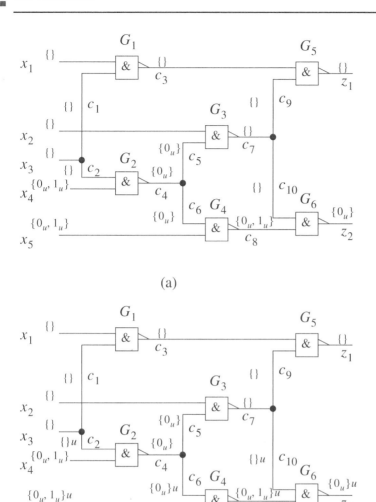

Figure 11.19 (a) Uncontrollability status of each line obtained starting with status $\{0_u, 1_u\}$ at inputs x_4 and x_5 and $\{\}$ at the other inputs, and (b) unobservability status starting with unobservable z_2

some state output, say Y_i, of the circuit is unobservable if the corresponding flip-flop is not replaced by a scan flip-flop. Alternatively, a line that is an input to a primitive gate may be unobservable because some other input of that gate is $\bar{a}$-uncontrollable, where a is the controlling value of that gate. The unobservable lines can be identified via a backward traversal of the circuit lines. First, the state outputs that are unobservable are marked. The *output level* of each circuit line (see Procedure 3.6 in Section 3.4.1.2) is determined and the circuit lines are traversed in a non-decreasing order of their output level values. When a line is visited, its unobservability is determined as follows. If the line is a fanout stem, then it is marked unobservable if each of its fanout branches is unobservable. If the line is an input to a primitive gate with controlling value a then it is marked unobservable if (a) the output of the gate is unobservable, or (b) any of its other inputs is $\bar{a}$-uncontrollable. (An inverter can be viewed as a primitive gate with one input. In this case, its input is unobservable if its output is unobservable.) The rules for non-primitive gates are more complicated and left as an exercise to the reader.

Example 11.5 Consider the uncontrollability values computed in Example 11.4 and shown in Figure 11.19(a). Assume that output z_2 is unobservable (denoted by u). Figure 11.19(b) shows the unobservability of different lines. c_8 is unobservable since the output of G_6, namely z_2, is unobservable. c_6 is unobservable since it satisfies both the above conditions for unobservability. Finally, fanout branch c_2 is unobservable because the other input of AND gate G_2 is 1-uncontrollable.   □

Identifying faults that cannot be detected for a given partial-scan design: Assume that a list of faults of interest as well as an alternative partial-scan design are given. The latter is comprised of a sequential circuit and a list of flip-flops that are already selected for scan. First, each state input that is driven by the output of a non-scan flip-flop is marked uncontrollable, i.e., value $\{0_u, 1_u\}$ is assigned at that input. Each state output that drives the data input of a non-scan flip-flop is marked unobservable, i.e., labeled u. The uncontrollability and unobservability status of each line in the circuit is determined using the above procedure.

An SA0 fault at line c_i cannot be detected if (a) the line c_i is 1-uncontrollable, or (b) line c_i is unobservable. The conditions under which an SA1 fault at a line cannot be detected are similar. All the faults in the target fault list that cannot be detected under the above partial-scan design can be identified using these conditions.

Example 11.6 Consider the circuit, uncontrollability status, and unobservability values shown in Figure 11.19(b). Faults x_4 SA0, x_4 SA1, x_5 SA0, x_5 SA1, c_2 SA0, c_2 SA1, c_4 SA1, c_5 SA1, c_6 SA0, c_6 SA1, c_8 SA0, c_8 SA1, c_{10} SA0, c_{10} SA1, z_2 SA0, and z_2 SA1, are identified as being undetectable by this analysis. Some faults, such as x_4 SA0, are undetectable due to uncontrollability; others, such as c_{10} SA0, due to unobservability. Some other faults, such as x_5 SA0, are undetectable due to uncontrollability as well as unobservability.   □

Selection of flip-flops to scan: The above mechanism to compute the number of faults of interest that are not detectable in a target fault list can be used as a figure-of-merit (or, more precisely, a figure-of-demerit). This figure-of-merit can be used to select scan flip-flops in conjunction with the greedy heuristic described above.

11.3.2.4 Generation and application of tests to a partial-scan circuit

We next discuss how a partial-scan circuit can be tested.

Creation of kernels: Once partial-scan flip-flops are identified for a given sequential circuit C, a kernel C^{ker} is obtained for test generation by (a) removing each flip-flop selected for partial scan from the sequential circuit, (b) relabeling the input to the flip-flop (one of the state outputs) as a primary output of the kernel, and (c) relabeling the line driven by the Q output of the flip-flop (one of the state inputs) as a primary input of the kernel. (If the $\overline{Q}$ output of the flip-flop also drives lines in the combinational circuit, then those lines are driven via an inverter by the primary input corresponding to the line driven by the Q output of that flip-flop.)

Test generation: A modified version of a sequential test generator that takes advantage of the special properties of the kernel is typically used for acyclic kernels as well as for kernels that have self-loops but are otherwise acyclic. A similar approach can be used for balanced kernels. Alternatively, a combinational circuit model C^{b*} may be derived for a balanced kernel C^b, followed by generation of a set of tests using a combinational test generator. The resulting tests can be reformatted using the labeling procedure described earlier.

In any case, a set of tests to be applied to the circuit are obtained for the given kernel.

Modification of sequential circuit: Two main modifications are made to the given sequential circuit C. First, the flip-flops selected for scan are replaced by scan flip-flops as described earlier. The mode control signals mc of these flip-flops are connected together to a new pin of the chip called, say, mc. The D_s inputs and Q outputs of these flip-flops are connected to form a scan chain as before.

Second, the clock distribution circuitry is modified such that, in the test mode, only the flip-flops in the scan chain load data from their scan input, D_s, when an appropriate clock edge is applied to the clock pin of the chip. At this time, the value in each non-scan flip-flop is held constant. In the normal mode, all the flip-flops load data from their normal inputs, D, whenever an appropriate clock edge is applied to the clock pin.

Application of tests: Test application is very similar to the methodology described in Section 11.2.1.3. Prior to the ith test application cycle, the circuit is configured in the test mode by setting $mc = test$. Recall that in this mode each non-scan flip-flop holds its state value. The desired values to be applied to the scan flip-flops are fed to the *ScanIn* pin in an appropriate order over the next SCL clock cycles, where SCL is the length of the scan chain. (During these clock cycles, the *ScanOut* pin of the chip is monitored to observe and compare the response of the test applied in the $(i-1)^{th}$

test application cycle.) Next, the desired values are applied to the primary inputs in the ith test application cycle. The circuit is then configured in the normal mode of operation by assigning $mc = normal$ and one clock pulse is applied. In this mode, all the flip-flops are clocked. A test vector is hence applied and the response captured in the flip-flops as well as observed at primary outputs. The above process is repeated, until all the tests are applied and all the responses observed.

11.4   Organization and use of scan chains

One key issue that must be addressed is the high cost of test application via scan. Scan provides serial access to the state inputs and outputs. Hence, a large number of clock cycles is required to apply each test vector. In this section, we describe techniques that can be used to reduce the test application time for a scan circuit, while preserving its ability to provide high fault coverage at reduced test generation complexity.

11.4.1   Circuit model

We begin by describing the models and terminology used ahead to represent large circuits. Most of this model has been borrowed from Gupta *et al.* (1991), Narayanan *et al.* (1992, 1993) and Narayanan (1994).

11.4.1.1 Maximal combinational blocks and registers

A sequential circuit is assumed to be comprised of combinational circuit elements and flip-flops. For ease of representation, combinational circuit elements may often be combined into a number of **maximal combinational blocks** using the following procedure (Gupta *et al.*, 1991). Initially, each combinational circuit element is considered an individual block. If any circuit element in one block is connected *via a combinational connection* to any element in another block, then the two are combined into a single block. This process is repeated until no two blocks can be combined, i.e., blocks of maximal size are obtained. Let $CLB_1, CLB_2, \ldots, CLB_{N_b}$ be the blocks obtained in this manner. Subsequently, the flip-flops may be combined into multi-bit parallel-load *registers*. A **register** is the set of *all* flip-flops whose normal data inputs are driven by the outputs of logic block CLB_i and whose normal outputs drive the inputs of logic block CLB_j. Such a register is said to be a **driver** for block CLB_j and a **receiver** for block CLB_i. (Note that any register that provides feedback connections between some outputs and some inputs of a block is a driver and receiver for the same block. For such a register, CLB_i and CLB_j are identical.) Let $Lreg_i$ denote the **length of register** R_i, i.e., the number of flip-flops in R_i.

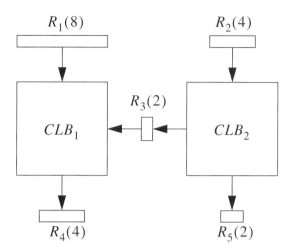

Figure 11.20   An example circuit with two blocks and five registers (Narayanan *et al.*, 1992) (© 1992 IEEE)

11.4.1.2 Kernels and their drivers and receivers

In a full-scan circuit, each flip-flop is replaced by a scan flip-flop to provide direct control and observation of the corresponding state input and output, respectively. Hence, each combinational logic block can be tested independently of any other block. Therefore, each combinational block constitutes a distinct kernel. In other words, in a full-scan circuit, kernel K_i can be considered synonymous with combinational block CLB_i and the number of kernels, N_k, is equal to the number of blocks, N_b. In a partial-scan circuit, the set of kernels constitutes a partition of the set of combinational blocks in the circuit. In general, $N_k \leq N_b$. Furthermore, a kernel may contain one or more non-scan registers in addition to one or more combinational blocks. The constitution of each kernel is determined by the specifics of the given circuit and the registers selected for scan.

A register whose normal outputs are inputs to a kernel is said to be a **driver of that kernel**. A register whose normal data inputs are driven by outputs of a kernel is said to be a **receiver of that kernel**. Finally, a register is said to be a **pure driver** if it is not a receiver for any kernel. The notion of a **pure receiver** can be defined in a similar manner.

Example 11.7   Consider the circuit shown in Figure 11.20, which contains two combinational logic blocks CLB_1 and CLB_2. Let $R_1, R_2, \ldots, R_5$ denote parallel-load registers or, simply, registers, containing 8, 4, 2, 4, and 2 flip-flops, respectively.

In this circuit, registers R_1 and R_2 are pure drivers that are drivers of blocks CLB_1 and CLB_2, respectively. Registers R_4 and R_5 are pure receivers that are receivers of blocks CLB_1 and CLB_2, respectively. In contrast, R_3 is neither a pure driver nor a pure receiver since it is a driver for CLB_1 and a receiver for CLB_2.   □

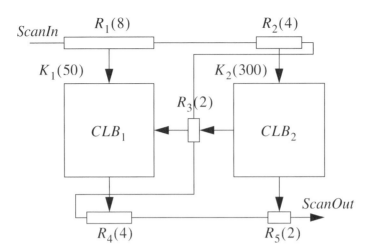

Figure 11.21 An example circuit whose registers are configured as a single chain (Narayanan *et al.*, 1992) (© 1992 IEEE)

Example 11.8 Consider a full-scan version of the example circuit in Figure 11.20 that contains a single scan chain as shown in Figure 11.21. In this case, kernel K_1 is identical to block CLB_1 and kernel K_2 is identical to CLB_2.

Registers R_1 and R_2 are pure drivers that are drivers for kernels K_1 and K_2, respectively. Registers R_4 and R_5 are pure receivers that are receivers for K_1 and K_2, respectively. In contrast, R_3 is a driver for K_1 and a receiver for K_2. □

11.4.1.3 Scan chain and test vectors

In the following discussion, we consider circuits with a single scan chain as well as those with multiple scan chains. In a circuit with a single scan chain, we assume that the chain has length SCL and the chain's scan input and output are called *ScanIn* and *ScanOut*, respectively. The chain is configured into normal and test modes when values *normal* and *test* are, respectively, applied to the mode control input *mc*.

In circuits with multiple chains, we denote the number of chains by N_{sc}. The ith chain, *ScanChain$_i$*, is assumed to have SCL_i flip-flops, scan input *ScanIn$_i$*, and scan output *ScanOut$_i$*. We also assume that a single mode control signal *mc* is used to either configure every chain in the normal mode or every chain in the test mode. Such a scan design only permits **synchronous** operation of scan chains where at any given time all the chains are configured in the same mode.

Let NT_i be the number of vectors to be applied to kernel K_i. Without loss of generality, assume that $NT_1 \leq NT_2 \leq \cdots \leq NT_{N_k}$.

The **driving weight of a register** R_i, DRW_i, is the number of vectors for which the register is used as a driver. The driving weight of a pure receiver is defined as zero. The **receiving weight of a register** R_i, RCW_i, can be defined similarly. If NT_j tests are required to test kernel K_j, then any register R_l that is a driver for K_j has a driving weight $DRW_l = NT_j$. Also, any register R_i that is a receiver for K_j has a receiving weight $RCW_i = NT_j$.

Example 11.9 Consider the full-scan circuit shown in Figure 11.21. Assume that kernel K_1 requires 50 test vectors while kernel K_2 requires 300 test vectors, i.e., $NT_1 = 50$ and $NT_2 = 300$.

The driving weight of R_1, DRW_1, is 50. Since R_1 is a pure driver, its receiving weight, RCW_1, is zero. Similarly, for pure receiver R_5, $DRW_5 = 0$ and $RCW_5 = 300$. For driver/receiver R_3, $DRW_3 = 50$ and $RCW_3 = 300$. ☐

11.4.2 Characteristics of scan vectors

The techniques described in the remainder of this section exploit the following characteristics of scan vectors.

Incompletely specified test vectors: As seen in Chapter 4 (Section 4.8), most test vectors for a combinational block of logic are incompletely specified. Hence, for most vectors, it is necessary to scan in desired values only into a subset of flip-flops in the scan chain.

Need to observe a fraction of scan flip-flops: For many faults detected by a vector, fault effects propagate to one or more primary outputs of the circuit. The effects of the remaining faults detected by the vector often propagate to a handful of scan flip-flops. The faults in the former category can be detected by observing the response at the primary outputs, without the need to scan the response captured in the scan flip-flops. The faults in the latter category can typically be detected by scanning out the response captured in a subset of flip-flops.

Relative use of different flip-flops: Desired values need to be scanned into some flip-flops for a much larger number of vectors than into others (Morley and Marlett, 1991). Also, the response captured into some flip-flops need to be scanned out for a much larger number of vectors than others. One reason why this occurs is due to the characteristics of logic blocks. For example, for most combinational circuit blocks, the fault effects of many faults propagate to some outputs (e.g., the output of a cone with a large number of gates), while effects of a few faults propagate to some other outputs.

A second, and more common, reason lies in the structure of large circuits. Typically, such circuits can be viewed as being composed of a large number of combinational blocks that are connected to each other via flip-flops as described above. When full scan is used, each combinational logic block can be viewed as an independent kernel. Typically, some kernels require a larger number of test vectors than others. Hence, the desired values must be scanned into registers that drive some kernels for more vectors than into registers that drive some other kernels. Similarly, the response captured in some registers must be scanned out for many more vectors than the response captured in some other registers.

Example 11.10 Consider the circuit in Figure 11.21 once again. The desired values must be scanned into each flip-flop in R_1 and R_3 (drivers of K_1) for 50 vectors. In contrast, the desired values must be scanned into R_2 (the driver of K_2) for 300 vectors. Similarly, the response captured in R_4 (the receiver of K_1) must be scanned out for 50 vectors, while those in R_3 and R_5 (receivers of K_2) must be scanned out for 300 vectors. □

Trade-offs between scan and non-scan testing: For a sequential circuit with full scan, a single vector can be generated to detect any given testable fault in the circuit. This vector can be applied to the circuit using the scan chains. Application of this vector may, however, require many clock cycles to scan in the vector and to scan out the response. On the other hand, it may be possible to use a sequential test generator to generate a sequence of vectors to detect the same fault without the use of scan. In this case, the number of vectors is typically larger than one, but each vector in the sequence can be applied in one clock cycle.

11.4.3 Test sessions and test application schemes

Testing of a large scan circuit can be divided into a number of *test sessions*. In each **test session**, the desired values are shifted into some subset of drivers and responses captured at some subset of receivers are observed. Multiple sessions are often possible for two main reasons. First, when a circuit is partitioned into multiple kernels by the scan chains, a different combination of kernels may be tested in each session. Second, some test vectors for a kernel may be incompletely specified and hence may require application of specific values only to a subset of drivers of the kernel. Similarly, for some vectors, the effects of the faults detected in a kernel may only propagate to a subset of the kernel's receivers. In such cases, tests that require specific data to be shifted into every driver and observation of the response at every receiver may be applied in one session. In each subsequent session, vectors that require the control of a particular subset of drivers and observation of a particular subset of receivers can be applied. In the following discussion, we use examples from the former scenario to illustrate the notion of test sessions and associated concepts.

Any circuit may be tested in one test session in which tests are applied to all the kernels simultaneously. The number of vectors applied during this test session is the maximum of the number of vectors required for any kernel. Such a scheme is referred to as the **combined** test application scheme (Narayanan *et al.*, 1992). This scheme does not take advantage of the fact that different kernels require different number of test vectors. The number of test vectors applied in the session is

$$NTS^{\text{Com}} = \max(NT_1, NT_2, \dots, NT_{N_k}). \tag{11.2}$$

Example 11.11 For the circuit shown in Figure 11.21, one test session with 300 vectors is required. The first 50 vectors are obtained by concatenating one vector for the first kernel with one vector for the second. The remaining 250 vectors are essentially the remaining vectors for the second kernel concatenated with arbitrary values (since the first kernel has already been completely tested).   □

In contrast, in an **overlapped** test application scheme (Narayanan *et al.*, 1992), in the first session, tests vectors are applied to all the kernels. Each test vector in this session is obtained by concatenating one test vector for each kernel. A sufficient number of vectors is used until all the test vectors required for one or more kernels are applied. In the next session, tests are applied only to those kernels for which some test vectors remain. Again, each vector is obtained by concatenating a vector for each kernel that is being tested during the session. This process is continued until testing of one or more of these kernels is completed. A sufficient number of such test sessions is used to complete testing of every kernel.

The overlapped scheme requires a maximum of N_k test sessions. Consider the case where the numbers of test vectors for the kernels satisfy the relations

$$NT_1 < NT_2 < \cdots < NT_{N_k}.$$

In this case, $N_{\text{Ses}} = N_k$ sessions are required. In the first session, $TSes_1$, the number of vectors required is

$$NTS^{\text{Ovr}}(TSes_1) = NT_1. \tag{11.3}$$

For each of the subsequent sessions, $TSes_i$, where $2 \le i \le N_{\text{Ses}}$, the number of vectors required is

$$NTS^{\text{Ovr}}(TSes_i) = NT_i - NT_{i-1}. \tag{11.4}$$

The test lengths for various kernels may not satisfy the above relations. However, the kernels can always be numbered to satisfy the general conditions, namely

$$NT_1 \le NT_2 \le \cdots \le NT_{N_k}.$$

In such cases, the relations derived for the number of vectors applied in each session for the above case can still be used. However, if any two or more kernels have identical test lengths, then the above equation will indicate that zero vectors are required for some sessions. In practice, all such *trivial* sessions can be simply ignored. Furthermore, Equations (11.3) and (11.4) can be used to compute test lengths for such cases.

Example 11.12 In the circuit considered in Figure 11.21, two sessions are required under the overlapped test application scheme. In the first session, $TSes_1$, $NTS(TSes_1) = 50$ vectors are applied to both the kernels where each test vector is obtained by concatenating one vector for each kernel. In the second session, $TSes_2$, $NTS(TSes_2) = 250$ vectors are applied to kernel K_2 only. Each of these vectors is a vector for K_2 that is not applied in the first session. ☐

Note that the combined test application scheme can be viewed as a special case of the overlapped scheme where all the test vectors are applied in a single session and the number of vectors is given by Equation (11.2).

In the **isolated** test application scheme, in each session, test vectors are applied to a distinct kernel. The number of vectors applied in a session simply equals the number of vectors required for the kernel being tested in that session.

Example 11.13 In the circuit shown in Figure 11.21, two sessions are required under the isolated test application scheme. In the first session, 50 vectors are applied to kernel K_1 only. In the second session, 300 vectors are applied to kernel K_2 only. ☐

The set of **active drivers for session** $TSes_j$, denoted as $AD(TSes_j)$, is the set of drivers into which desired values are shifted during the session. Similarly, the set of **active receivers for session** $TSes_j$, $AR(TSes_j)$, is the set of receivers observed following the application of each vector during the session.

Again, the combined scheme can be viewed as a special case of the overlapped scheme in the following sense. In the combined scheme, all the vectors are applied in one session in which all the drivers and receivers are active. Since all the drivers and receivers are also active in the first session under the overlapped scheme, the combined scheme can be viewed as a variant of the overlapped scheme where all the kernels are completely tested in the first session.

The following example illustrates these concepts.

Example 11.14 Consider the circuit in Figure 11.21 once again. Recall that under the overlapped test application scheme, in the first test session, $TSes_1$, 50 vectors are applied to both the kernels. In this session, the active drivers are

$$AD(TSes_1) = \{R_1, R_2, R_3\},$$

and the active receivers are

$$AR(TSes_1) = \{R_3, R_4, R_5\}.$$

At the end of this session, testing of kernel K_1 is complete. In the next session, $TSes_2$, only kernel K_2 is tested. In this session, the active driver is

$$AD(TSes_2) = \{R_2\},$$

and the active receivers are

$$AR(TSes_2) = \{R_3, R_5\}.$$

Note that in session $TSes_1$, register R_3 is used as a driver as well as a receiver. In contrast, in session $TSes_2$, it is used only as a receiver.

As described above, the combined scheme has only one session. Active drivers and receivers for this session are identical to those for the first session of the overlapped scheme.   □

11.4.4 Scan-shift policies

The purpose of scan is two-fold. First, the response to the last vector captured in the scan flip-flops are shifted out and observed at the output of the scan chains. Second, bits of the next test vector are shifted in.

11.4.4.1 Flush policy

The most straightforward scan policy is called **flush**, where the contents of every flip-flop in each scan chain are shifted out and new test data are shifted into each flip-flop. If the circuit contains a single scan chain of length SCL, then the flip-flops are configured in the test mode and a minimum of SCL clock pulses are applied to shift out the content of each flip-flop in the chain and observe it at the scan chain output, $ScanOut$, while, simultaneously, suitable values are applied at the scan chain input, $ScanIn$, and shifted into these flip-flops.

If the circuit contains N_{sc} scan chains of lengths $SCL_1, SCL_2, \ldots, SCL_{N_{sc}}$, then for the application of each test vector, all the chains are configured in the test mode, and

$$SC^{Flsh} = \max(SCL_1, SCL_2, \ldots, SCL_{N_{sc}}) \tag{11.5}$$

clock pulses are applied, where SC is referred to as the **shift cycle** and SC^{Flsh} denotes the shift cycle under the flush policy. During this period, values at outputs $ScanOut_i$ of scan chains are observed and appropriate values applied to $ScanIn_i$ inputs, where $i = 1, 2, \ldots, N_{sc}$.

Test application time under the flush policy: First, consider the combined test application scheme. SC^{Flsh} clock cycles are required to shift each vector. One more clock cycle is then required to apply the vector and capture the responses. The responses for one vector, say P_i, are scanned out for observation during the scanning in of the next vector, P_{i+1}. Finally, additional $SC^{Flsh} - 1$ clock cycles are required to scan out the response captured for the last vector. In this test application scheme, the total number of vectors applied is NTS^{Com} given by Equation (11.2). Hence, the test application time under this scheme and flush policy is

$$NTS^{Com}(SC^{Flsh} + 1) + SC^{Flsh} - 1 \tag{11.6}$$

clock cycles.

Next, consider the overlapped test application scheme. In this case, testing is conducted in N_{Ses} sessions where the number of vectors applied in session $TSes_i$ is given either by Equation (11.3) or Equation (11.4). In general, the test application time can be written as

$$\sum_{i=1}^{N_{\text{Ses}}} NTS^{\text{Ovr}}(TSes_i) \left[SC^{\text{Flsh}}(TSes_i) + 1 \right] + SC^{\text{Flsh}}(TSes_1) - 1, \tag{11.7}$$

where $SC^{\text{Flsh}}(TSes_i)$ is the shift cycle for session $TSes_i$. However, as shown in Equation (11.5), under the flush policy the shift cycle is identical for each session. Hence, the above equation for test application time can be rewritten as

$$\sum_{i=1}^{N_{\text{Ses}}} NTS^{\text{Ovr}}(TSes_i) \left[SC^{\text{Flsh}} + 1 \right] + SC^{\text{Flsh}} - 1, \tag{11.8}$$

where SC^{Flsh} is given by Equation (11.5). Since, $\sum_{i=1}^{N_{\text{Ses}}} NTS^{\text{Ovr}}(TSes_i) = NT_{N_k}$, the test length is

$$NT_{N_k} \left[SC^{\text{Flsh}} + 1 \right] + SC^{\text{Flsh}} - 1. \tag{11.9}$$

Note that the expression for test application time for the overlapped scheme under the flush policy is identical to that given for the combined scheme in Equation (11.6). This indicates that under the flush policy, the overlapped scheme does not provide any improvement over the combined test application scheme.

Example 11.15 Consider a single scan configuration of the circuit shown in Figure 11.21. Under the flush policy, test application begins with the configuration of the scan chain in the test mode. For the next $SC^{\text{Flsh}} = 20$ clock cycles, bits of the first vector, P_1, are successively applied at input *ScanIn*. Once P_1 is completely scanned in, the flip-flops in the scan chain are configured in their normal mode for one clock cycle and the response of the kernels to P_1 is captured into the flip-flops. The above process is repeated for each vector with one minor difference, namely, the value appearing at output *ScanOut* is observed during each scan clock cycle.

Since kernels K_1 and K_2, respectively, require $NT_1 = 50$ and $NT_2 = 300$ vectors, under the combined test application methodology, $NTS^{\text{Com}} = \max(50\,300) = 300$. Hence, the test application time under the combined scheme and flush policy is $300(20 + 1) + 20 - 1 = 6319$.

One can verify that in the overlapped scheme, two sessions are used with $NTS^{\text{Ovr}}(TSes_1) = 50$, and $NTS^{\text{Ovr}}(TSes_2) = 250$. However, for each session, the shift cycle under the flush policy is 20. Hence, the test application time under the overlapped scheme under the flush policy is also 6319. □

11.4.4.2 Active-flush policy

This is a variation of the flush policy where only the *active chains* are flushed. A **chain** is said to be **active** during a test session, $TSes_i$, if any register in the chain is either an active driver or an active receiver.

Clearly, the active-flush policy becomes identical to the flush policy for a circuit with a single chain. Now consider a circuit with multiple chains. In any session in which all the kernels are being tested, all the chains are active and the active-flush policy is identical to the flush policy described above. Next, consider a session, $TSes_j$, in which some proper subset of all the kernels are being tested and scan chains $ScanChain_{i_1}$, $ScanChain_{i_2}$, ..., $ScanChain_{i_\alpha}$ are active. During the application of any vector in this session, all the chains are configured in the test mode, and

$$SC^{\text{AFlsh}}(TSes_j) = \max(SCL_{i_1}, SCL_{i_2}, \ldots, SCL_{i_\alpha}) \tag{11.10}$$

clock pulses are applied, where $SC(TSes_j)$ is called the **shift cycle for session** $TSes_j$ and $SC^{\text{AFlsh}}(TSes_j)$ denotes this cycle under the active flush policy. During each of these clock cycles, values at outputs $ScanOut_{i_l}$ of scan chains are observed and appropriate values applied to $ScanIn_{i_l}$ inputs, where $l = 1, 2, \ldots, \alpha$.

Test application time under the active flush policy: First, consider the combined test application scheme. In this scheme, all the vectors are applied in a single session in which all the drivers and receivers are assumed to be active. Hence, the test application time for the active flush policy is no different from that for the flush policy as given by Equation (11.6).

Next, consider the overlapped test application scheme. In this case, testing is conducted in N_{Ses} sessions where the number of vectors applied in session $TSes_i$, $NTS^{\text{Ovr}}(TSes_i)$, is given either by Equation (11.3) or Equation (11.4). The test application time for the overlapped scheme under the active flush policy can be written as

$$\sum_{i=1}^{N_{\text{Ses}}} NTS^{\text{Ovr}}(TSes_i) \left[SC^{\text{AFlsh}}(TSes_i) + 1 \right] + SC^{\text{AFlsh}}(TSes_1) - 1, \tag{11.11}$$

where $SC^{\text{AFlsh}}(TSes_i)$ is given by Equation (11.10). *Note that the test application time may be a few clocks lower than that given by the above equation. However, we will use this version due to its simplicity.*

Example 11.16 Consider a configuration with two scan chains for the example circuit in Figure 11.20, as shown in Figure 11.22. The first scan chain, $ScanChain_1$, is comprised of register R_1 followed by register R_4 and has length $SCL_1 = 12$. Chain $ScanChain_2$ is comprised of R_2, R_3, and R_5 in that order and has length $SCL_2 = 8$.

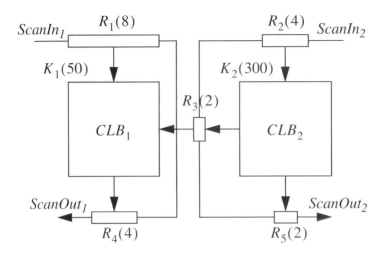

Figure 11.22 An example circuit whose registers are configured as two scan chains

Under the overlapped scheme, $NTS^{Ovr}(TSes_1) = 50$ vectors are applied to both the kernels in session $TSes_1$ and $NTS^{Ovr}(TSes_2) = 250$ vectors are applied to kernel K_2 in session $TSes_2$.

In the first session, all the registers are active. Hence, both chains are active and the shift cycle for the session is $SC^{AFlsh}(TSes_1) = \max(SCL_1, SCL_2) = 12$.

In the second session, only $ScanChain_2$ is active and $SC^{AFlsh}(TSes_2) = SCL_2 = 8$. For this session, the number of clock cycles needed during scan operation for the active-flush policy is lower than that for the flush policy. The test application time for the overlapped scheme under the active flush policy is 2911. This should be compared with the 3911 clock cycles for the overlapped scheme under the flush policy.   □

11.4.4.3 Minimum-shift policy

Often, appropriate values can be shifted into all the drivers and responses captured in the receivers observed in fewer scan cycles than the above flush policies. This will be illustrated by using an example circuit with a single scan chain.

Example 11.17   Consider the circuit with the single scan chain shown in Figure 11.21. Note that registers R_1, R_2, and R_3 are drivers while R_3, R_4, and R_5 are receivers. Hence, even in a session in which both kernels are being tested simultaneously, test data need to be shifted only into the first three registers of the scan chain. Also, only the response in the last three registers needs to be shifted out for observation. For the given order of registers in the scan chain, 14 scan clock cycles are sufficient to control all the drivers and observe the response captured in all the receivers. This is significantly lower than 20 scan clock cycles required per vector under the flush policy.   □

The observations made in the above example can be generalized as follows. For a single-chain circuit and the minimum-shift policy, let $SCin$ denote the minimum number of scan clock cycles required to shift in bits of any desired test vector into all the drivers. Let $SCout$ denote the minimum number of scan clock cycles required to observe at $ScanOut$ the response captured in all the receivers. Then the minimum-shift policy only requires the chain to be clocked

$$SC = \max(SCin, SCout)$$

or fewer times during scanning of each vector. (More precisely, when the bits of the first vector are scanned in, then only $SCin$ clock cycles are required. Only $SCout - 1$ clock cycles are required when the response to the last vector is scanned out. SC clock cycles are required during the application of each of the other vectors.)

In a circuit with multiple scan chains, let $SCin_i$ be the number of scan clock cycles required to shift in the desired values into each driver of the ith scan chain. Let $SCout_i$ be the number of clock cycles required to shift out the response captured in each of the receivers of the chain. Then the shift cycle of the ith chain is

$$SC_i = \max(SCin_i, SCout_i).$$

The shift cycle of the entire multiple scan chain design is

$$SC^{MS} = \max_i SC_i. \tag{11.12}$$

11.4.4.4 Active minimum-shift policy

This is a variation of the minimum-shift policy where only scan chains that are active during a particular session, say $TSes_j$, are considered. Furthermore, for each chain, $ScanChain_{i_l}$, that is active during session $TSes_j$, the shift cycle of the i_l^{th} chain in session $TSes_j$, $SCin_{i_l}(TSes_j)$, is defined as the minimum number of scan clock cycles required to shift in the desired values into each active driver of the i_l^{th} scan chain. Similarly, $SCout_{i_l}(TSes_j)$ is defined as the minimum number of clock cycles required to shift out the response captured in each of the active receivers of the chain. Then the shift cycle of the i_l^{th} chain for session $TSes_j$ is

$$SC_{i_l}(TSes_j) = \max(SCin_{i_l}(TSes_j), SCout_{i_l}(TSes_j)).$$

The shift cycle of the entire multiple scan chain design for session $TSes_j$ is

$$SC^{AMS}(TSes_j) = \max\left[SC_{i_1}(TSes_j), SC_{i_2}(TSes_j), \ldots, SC_{i_\alpha}(TSes_j)\right], \tag{11.13}$$

where $ScanChain_{i_1}, ScanChain_{i_2}, \ldots, ScanChain_{i_\alpha}$ are the chains that are active in session $TSes_j$.

Test application time under the active minimum-shift policy: First, consider the combined test application scheme. Under this scheme, all the tests are applied in one session and $SC^{AMS}(TSes_1)$ clock cycles are required to apply each vector. The total number of vectors applied is NTS^{Com} given by Equation (11.2). Hence the test application time under the combined test application scheme and active minimum-shift policy is

$$NTS^{Com}\left[SC^{AMS}(TSes_1) + 1\right] + SC^{AMS}(TSes_1) - 1 \qquad (11.14)$$

clock cycles.

Next, consider the overlapped test application scheme. In this case, testing is conducted in N_{Ses} sessions where the number of vectors applied in session $TSes_i$ is given either by Equation (11.3) or (11.4). The test application time can be written as

$$\sum_{i=1}^{N_{Ses}} NTS^{Ovr}(TSes_i)\left[SC^{AMS}(TSes_i) + 1\right] + SC^{AMS}(TSes_1) - 1, \qquad (11.15)$$

where $SC^{AMS}(TSes_i)$ is the shift cycle for session $TSes_i$ given by Equation (11.13). *Note that the test application time may be a few clock cycles smaller than that given by the above equation. However, we will use this version due to its simplicity.*

Example 11.18 Consider a configuration with two scan chains for the example circuit in Figure 11.20, as shown in Figure 11.22. Recall that in session $TSes_1$ both kernels are tested, and in session $TSes_2$ only kernel K_2 is tested.

In $TSes_1$, all the registers are active. For the first chain, the shift cycle can be computed as $SC_1(TSes_1) = \max(8, 3) = 8$. Similarly, $SC_2(TSes_1) = \max(6, 3) = 6$. Hence, the shift cycle for the entire circuit for session $TSes_1$ is $SC^{AMS}(TSes_1) = \max(SC_1(TSes_1), SC_2(TSes_1)) = \max(8, 6) = 8$.

In $TSes_2$, only R_2, R_3, and R_5 are active. Consequently, only $ScanChain_2$ is active. Also, note that in this session, R_3 is used only as a receiver. The shift cycle for this chain for session $TSes_2$ can hence be computed as $SC_2(TSes_2) = \max(4, 3) = 4$. Since this is the only active chain during this session, the shift cycle for the entire circuit for the session is $SC^{AMS}(TSes_2) = SC_2(TSes_2) = 4$.

The total number of clock cycles required to apply 50 vectors in the first session and 250 vectors in the second is $50(8 + 1) + 250(4 + 1) + 8 - 1 = 1707$ cycles. □

11.4.5 Scan chain organization

Three main operations are commonly considered during optimization of scan chain design. First, the flip-flops/registers are connected in an appropriate order in the scan chain. Second, in designs with multiple scan chains, flip-flops/registers are assigned to different chains. Third, additional control circuitry is sometimes added to scan chains to enable reconfiguration of chains in some test sessions.

11.4.5.1 Organization of a single scan chain

The key operation for the optimization of a single scan chain design is the ordering of registers. The problem of identifying an optimal order under the most general of the above test application schemes and shift policies has been shown to be computationally complex (NP-complete) (Narayanan *et al.*, 1992). In the following discussion, we illustrate the key factors that should be considered during ordering of registers. These factors provide insights into the general problem of scan chain ordering and can be used to develop good heuristics for this purpose. They can also help generate optimal orderings for some special cases.

It should be noted that an optimal ordering of registers depends on the test application scheme and shift policy employed. In addition, it depends on the characteristics of the design and the number of test vectors required for each kernel.

Benefits of ordering: In a single chain design, the order of registers in the scan chain influences the test application time by affecting the shift cycles for one or more test sessions. One important implication of this fact is that the test application time cannot be reduced under the flush or active flush policies. Also, it has been demonstrated in Narayanan (1994) that the overlapped test application scheme is optimal for such designs. Furthermore, it can be shown that the active minimum-shift policy always leads to a lower test application time than the minimum-shift policy. Hence, we will limit the following discussion to the overlapped test application scheme and active minimum-shift policy, as illustrated next.

Example 11.19 Consider the single scan chain design in Figure 11.21. In this design, the registers are ordered as R_1, R_2, R_3, R_4, and R_5. Under the overlapped scheme, two test sessions are required, $TSes_1$ in which $SC(TSes_1) = 50$ vectors are applied to both the kernels and $TSes_2$ in which $SC(TSes_2) = 250$ vectors are applied to kernel K_2 only. The active drivers and receiver for the two sessions are as identified in Example 11.14:

$AD(TSes_1) = \{R_1, R_2, R_3\}$,

$AR(TSes_1) = \{R_3, R_4, R_5\}$,

$AD(TSes_2) = \{R_2\}$,

$AR(TSes_2) = \{R_3, R_5\}$.

Under the active minimum-shift policy, the shift cycles for the two sessions can be computed as

$SC(TSes_1) = 14$,

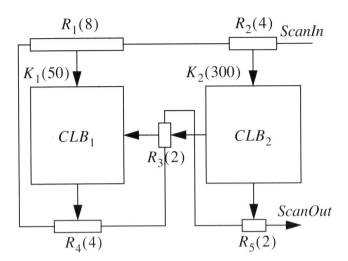

Figure 11.23 An alternative single scan chain design for the circuit in Figure 11.20 (Narayanan *et al.*, 1992) (© 1992 IEEE)

$SC(TSes_2) = 12$.

Hence, the test application time for this chain order is $50(14 + 1) + 250(12 + 1) + 14 - 1 = 4013$ cycles.

Now consider an alternative single chain design shown in Figure 11.23 where the registers are ordered as R_2, R_1, R_4, R_3, and R_5. The shift cycles for this design are

$SC(TSes_1) = 18$,

$SC(TSes_2) = 4$.

Hence, the test application time for this chain order is $50(18+1)+250(4+1)+18-1 = 2217$ cycles.   □

The above example illustrates that the order of registers in the scan chain can reduce the test application time under the active minimum-shift policy. The key idea is that the order of registers in the chain determines the shift cycle for each session. Hence, reordering registers in the chain can change the shift cycle and therefore the test application time.

Let $(R_{i_1}, R_{i_2}, \ldots, R_{i_{N_t}})$ be the sequence in which registers appear in the scan chain. For a session $TSes_j$, the values of $SC(TSes_j)$ can be determined as follows. First, the driver that is active in session $TSes_j$, say R_{i_γ}, and is farthest from *ScanIn* is identified. Then the value of $SCin(TSes_j)$ is computed as

$$SCin(TSes_j) = \sum_{l=1}^{\gamma} Lreg_{i_l},$$

where $Lreg_{i_l}$ is the length of (or the number of flip-flops in) R_{i_l}. The value of $SCout(TSes_j)$ is computed in an analogous manner. The value of $SC(TSes_j)$ is computed as

$$SC(TSes_j) = \max \left[SCin(TSes_j), SCout(TSes_j) \right].$$

The approach described earlier can then be used to compute the overall test application time.

Some simple observations can be made about the impact of register ordering in the scan chain. First, placement of drivers closer to *ScanIn* reduces *SCin*. Second, placement of receivers closer to *ScanOut* reduces *SCout*. Third, placement of drivers that have higher weights closer to *ScanIn* helps reduce the value of *SCin* for later sessions. Similarly, placement of receivers with higher weights closer to *ScanOut* helps reduce the value of *SCout* for later sessions.

In a general circuit, the above simple rules cannot be applied directly to obtain an optimal chain ordering. This is due to the fact that some registers are drivers as well as receivers. Second, the placement suggested by the first rule only facilitates the reduction in the value of *SCin* and may in fact increase the value of *SCout*. The second rule above suffers from a similar pitfall. These factors make the problem of ordering registers in a scan chain in a general circuit computationally complex (Narayanan *et al.*, 1992). In general, various permutations of registers need to be tried in some systematic manner to find the optimal order of registers. The following result helps reduce the number of permutations that need to be tried.

The **interchange property** (Narayanan *et al.*, 1992) states that if any scan chain contains register R_i closer to *ScanIn* than register R_j where $DRW_i \leq DRW_j$ and $RCW_i \geq RCW_j$, then the positions of the two registers can be interchanged to obtain a scan chain that will require a lower test application time, provided either (a) R_i and R_j are adjacent to each other in the given chain, or (b) the lengths of the two registers, namely $Lreg_i$ and $Lreg_j$, are equal.

The above result can be used to derive an optimal scan chain order for circuits with some special characteristics.

1 In a circuit where all the registers are either pure drivers or pure receivers, an optimal scan chain order is one in which the pure drivers appear closer to *ScanIn*, sorted with decreasing driver weights, followed by pure receivers sorted with increasing receiver weights.

2 In a circuit where all the kernels require an equal number of vectors, an optimal scan chain contains three parts. The first part contains all the pure drivers, the second part contains all the driver/receivers, and the third part contains all the pure receivers. Within each part, the registers can appear in an arbitrary order.

For a more general scenario, the above orders may not be optimal. This was illustrated in Example 11.19. One approach for obtaining an optimal solution is to try various orders in a systematic manner, such as using a branch-and-bound search, where the interchange property is used to reduce the search space (Narayanan *et al.*, 1992). Another approach is to start with several random orders and use the interchange property iteratively to reorder registers. Several other approaches have also been presented in the literature.

11.4.5.2 Organization of multiple scan chains

The key operations for the optimization of a multiple scan chain design are (a) assigning registers to different chains, and (b) ordering registers within each chain. The general problem of designing optimal multiple chains is known to be computationally complex (NP-complete) (Narayanan *et al.*, 1993). Next, we first discuss scan chain design for the flush and active flush shift policies. Subsequently, we discuss issues in scan chain design under the minimum-shift and active minimum-shift policies.

Organization of scan chains under flush policies: Under either of the flush policies, the order of registers within a scan chain does not matter; only how the registers are assigned to chains matters.

The following example illustrates the effects of assigning registers to different chains.

Example 11.20 Suppose the registers in the circuit shown in Figure 11.20 need to be assigned to two scan chains.

First, consider the overlapped test application scheme and flush policy. In such a case, during each session each chain is flushed during the application of each vector. Hence, the shift cycle is minimum when two chains of equal length are used. This can be obtained in many ways, one of which is shown in Figure 11.24. Each of the two chains has length 10, where $ScanChain_1$ has registers R_1 and R_3, while $ScanChain_2$ has registers R_2, R_4, and R_5. The shift cycle for each test is $\max(SCL_1, SCL_2) = 10$ and the overall test application time is $50(10 + 1) + 250(10 + 1) + 10 - 1 = 3309$ cycles.

Next, consider the overlapped test application scheme and active flush policy. For the above two-chain configuration, both chains are active in both test sessions. Hence, even under the active flush policy, the shift cycle for each session is 10 and the overall test application time remains 3309 cycles. In contrast, consider the two-chain design shown in Figure 11.22. $ScanChain_1$ contains R_1 and R_4 while $ScanChain_2$ contains R_2, R_3, and R_5. The lengths of the two scan chains are 12 and 8, respectively. In this design, $ScanChain_1$ is active during the first test session in which 50 vectors are applied and $ScanChain_2$ is active in both test sessions. Consequently, the shift cycles in the first and second sessions are 12 and 8, respectively. The test application time can now be calculated as $50(12 + 1) + 250(8 + 1) + 12 - 1 = 2911$ cycles. □

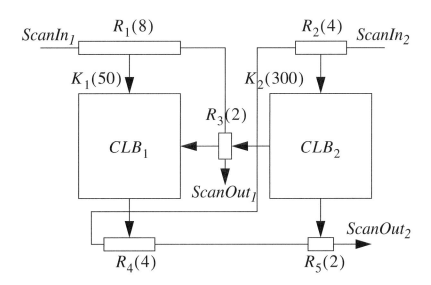

Figure 11.24   An alternative two scan chain design for the circuit in Figure 11.20

The above example provides an intuitive understanding of the following observations, whose proofs can be found in Narayanan (1994).

First, under the flush policy, equal length scan chains are optimal. If individual flip-flops of a register can be assigned to different scan chains, then the above objective can be attained using a straightforward algorithm. If flip-flops of a register cannot be assigned to different chains, then the problem of scan chain design under this policy reduces to an assignment of registers into a given number of chains such that the length of the longest chain is minimized.

Second, under the active flush policy, the following key observation can be made. Registers may be assigned to chains in such a manner that longer chains become inactive after the first few sessions to reduce the shift cycle for subsequent sessions. Under the assumption that individual flip-flops of registers can be assigned to different chains, the above observation can be used to establish the following methodology. First, the flip-flops can be ordered in terms of decreasing weights, where the weight of a flip-flop is the maximum of its driver and receiver weights. Second, in the above order, flip-flops can be assigned to chains of increasing lengths. The only question that needs to be answered is when to add a flip-flop to an existing chain and when to start a new chain. Clearly, the length of the shortest chain can be no smaller than one and no larger than $1/N_{sc}$ times the total number of flip-flops, where N_{sc} is the number of chains. We can try each possible length in this range. For each alternative design of the shortest chain, the next chain would need to be no shorter than the previous chain and no longer than $1/(N_{sc} - 1)$ times the number of remaining flip-flops. Again, each possible length in this range can be tried. This process can be repeated for each of the N_{sc} chains. A dynamic programming approach for efficiently carrying out the above steps to obtain optimal chain designs is given in Narayanan *et al.* (1993). The following example illustrates this approach.

Example 11.21 Consider the circuit shown in Figure 11.20 for which an optimal three-chain design is desired. First, the flip-flops are ordered in terms of decreasing weights as $R_2^1, R_2^2, R_2^3, R_2^4, R_3^1, R_3^2, R_5^1, R_5^2, R_1^1, R_1^2, \ldots, R_1^8, R_4^1, R_4^2, R_4^3, R_4^4$, where R_i^j denotes the jth flip-flop of register R_i.

Let $\mathcal{A}(\alpha, \beta, \gamma)$ denote an algorithm that optimally organizes the last α flip-flops in the above ordered list into β chains under the constraint that the shortest chain has no fewer than γ flip-flops and every subsequent chain has no fewer flip-flops than the previous chain.

In terms of the above notation, the problem at hand is solved by making the call $\mathcal{A}(20, 3, 1)$. The shortest chain can be no shorter than one and no longer than $\lfloor 20/3 \rfloor = 6$. These six alternatives configurations for the shortest chain will include the first, or the first two, or the first three, . . ., or the first six of the flip-flops in the above ordered list. For each of these alternatives, the other two scan chains can be designed by invoking $\mathcal{A}(19, 2, 1), \mathcal{A}(18, 2, 2), \mathcal{A}(17, 2, 3), \mathcal{A}(16, 2, 4), \mathcal{A}(15, 2, 5), \mathcal{A}(14, 2, 6)$, respectively. For each of the invocations, the above process can be repeated. For example, the invocation $\mathcal{A}(16, 2, 4)$ aims to organize into two chains the last 16 flip-flops in the above list, i.e., the fifth to the 20th flip-flop in the original ordered list of flip-flops. Also, the shorter chain must have no fewer than four flip-flops. The shorter chain may have the first four of these flip-flops, or the first five, . . ., or the first eight. For each of these alternatives, the above algorithm can be invoked again with appropriate parameter values.

The above process can be repeated systematically to obtain the optimal chain design. In this case, the optimal chain design contains three chains that, respectively, contain the first four, the next four, and the remaining 12 flip-flops in the above ordered list, as shown in Figure 11.25. The corresponding test application time is 1911 cycles. □

Organization of scan chains under minimum-shift policies: The general problem of designing optimal multiple scan chains under the minimum-shift and active minimum-shift policies is shown to be computationally complex (NP-complete) (Narayanan *et al.*, 1993). Heuristics have been presented only for some special cases, such as one where all the kernels require an equal numbers of test vectors. Under this condition, the active minimum-shift policy becomes identical to the minimum-shift policy. Also, the overlapped scheme becomes identical to the combined scheme.

For the above case, first consider a scenario where the individual flip-flops of registers can be separated and assigned to different scan chains. If N_{sc} chains are desired, the test application time is minimized by the following scan chains. (Note that this is one of many optimal designs.) An equal number of pure driver, pure receiver,

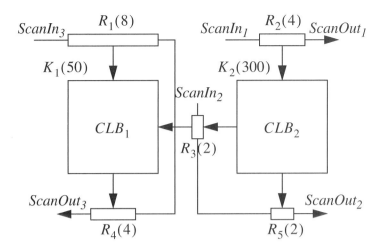

Figure 11.25 An example circuit with two blocks and five registers whose registers are configured as three scan chains

and driver/receiver flip-flops can be assigned to each chain. Furthermore, within each chain, the pure driver flip-flops appear first, followed by driver/receivers, and finally the pure receivers.

In the case where the flip-flops within a single register cannot be distributed over multiple chains, the registers need to be distributed across chains so as to minimize over all the chains the maximum of the sum of the numbers of pure driver and driver/receiver flip-flops and the sum of the numbers of driver/receiver and pure receiver flip-flops in each chain. Once the registers are assigned to the chains, within each register, the pure drivers should be followed by driver/receivers which should be followed by pure receivers.

11.4.6 Reconfigurable chains

Many of the previous examples illustrate that, even under the active minimum-shift policy, shift cycles for many sessions are influenced by registers that are not active in that session. For example, of the two single scan chain configurations described in Example 11.19, the first fails to achieve the minimum possible shift cycle for the second session while the second fails to achieve it for the first session. The first chain configuration is further analyzed in the following example.

Example 11.22 For the circuit in Figure 11.20, consider the single chain configuration shown in Figure 11.26(a). This configuration achieves the minimum possible shift cycle for the first session. However, in the second session only receivers R_3 and R_5 are active. While the total number of active driver flip-flops is four **and** the number of

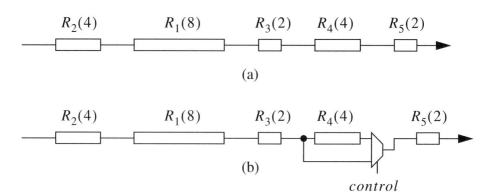

Figure 11.26 Example illustrating advantages of a reconfigurable chain. (a) A classical chain, and (b) a reconfigurable version

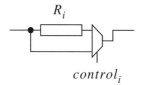

Figure 11.27 A generic reconfigurable scan version of a register

active receiver flip-flops is four, the shift cycle is seven, since R_4 is placed between the two active receivers. (As illustrated in Example 11.19, an alternative location of R_4 prevents the attainment of a minimum-shift cycle in the first test session.)

One solution is to add a 2-to-1 multiplexer to the above chain, as shown in Figure 11.26(b). In the first session, the multiplexer control input is set such that register R_4 is not bypassed. In the second session, the multiplexer can be set to bypass this register and allow a shift cycle of four. ☐

The use of multiplexers to reconfigure scan chains has been proposed in Morley and Marlett (1991) and Narayanan and Breuer (1995b). In its simplest form, each scan register in the circuit can be accompanied by a multiplexer, as shown in Figure 11.27. The scan inputs and outputs of the register can be connected as usual when the register is incorporated into a scan chain. The control input of the multiplexer forms an additional control signal that must be used during scan testing. In each session in which the register is active, the control signal is set such that the register is a part of the scan chain. In any session in which the register is inactive, the control signal is set to bypass the register.

The overheads of the above approach are threefold. First, the use of a multiplexer increases area overhead. Also, additional routing is required for the bypass wire. However, since this wire is confined to the local region around the register, its contribution to the area overhead may be small. Second, an additional area overhead is incurred due to the routing required for control signals. Third, additional pins are required for the control signals to the multiplexers. This can be mitigated by creating an FSM to drive these control signals. The complexity of the FSM can be reduced by taking advantage of the following characteristic of these control signals. Register R_i is active for the first l_i test sessions, after which it becomes inactive and can be bypassed.

The above overheads are affected by the order of registers in the scan chain. For example, if two registers that are active in an identical set of sessions are assigned adjacent locations in the chain, then a single multiplexer can be used to bypass the two registers. If no constraints are placed on the overheads, all possible register orders can be tried to identify the one that minimizes the test application time at minimum overheads. When the number of bypass multiplexers must be limited, then it may become impossible to achieve the above minimum test application time. In such cases, minimization of the test application time becomes the objective. Approaches to solve the latter version of this problem have been presented in Morley and Marlett (1991).

11.4.7   Use of scan chains

The above discussion assumes that each test vector is applied via scan. In this scenario, for each vector, bits of the vector are scanned into scan chains while bits of the response captured for the previous vector are scanned out. One way to reduce the test application time is to allow other types of test sessions. Next, we discuss two such alternative approaches that can be used in conjunction with the above approach.

11.4.7.1 Reusing scanned state for multiple vectors

In Abramovici et al. (1992), an approach has been presented to reduce the number of vectors for which bits are shifted into and out of scan flip-flops. This approach takes advantage of the fact that a large number of possible faults in a circuit can be grouped into sets, where each fault in the set has one or more test vectors (a) that propagate the effect of the fault to a primary output, and (b) whose state input parts are compatible with those of the vectors for other faults in the set.

The faults in any such set can be tested in the following manner. First, test vectors can be generated for each fault in the set that have compatible values at state inputs (but may have different values at primary inputs), and propagate fault effects to primary outputs. Next, the values desired at the state inputs by each of these vectors are first shifted into the scan flip-flops. The scan flip-flops are then put into a hold mode, where their contents are held constant. The primary input part of each vector is then applied and the response at primary outputs observed. Faults in another such set are then processed and the above steps are repeated to test each fault in that set.

Example 11.23 Consider a circuit with three primary inputs, two state inputs, two state outputs, and two primary outputs. Consider two sets of faults that satisfy the above requirements. Assume that the faults in the first set can be detected by vectors $P_1 = (1, 1, 1, 1, x)$, $P_2 = (0, 1, 1, 1, x)$, and $P_3 = (1, 0, 0, 1, 0)$, where the first three and the last two bits of each vector are to be applied to the primary and state inputs, respectively. Furthermore, assume that each of these vectors propagates the effects of the corresponding faults in the first set to one or more of the primary outputs. Let $P_4 = (1, 0, 1, 0, 0)$ and $P_5 = (0, 0, 1, 0, x)$ be similar vectors for the faults in the second set.

The test application for these two sets of faults under the proposed methodology begins with configuring the two flip-flops in the test mode and scanning in the values $(1, 0)$ into the respective flip-flops. When the second bit is being scanned in, the primary input part of P_1, $(1, 1, 1)$, is applied to the circuit and the response at primary outputs observed. The flip-flops are then put in the hold mode for the next two clock cycles while primary input parts of P_2 and P_3, i.e., $(0, 1, 1)$ and $(1, 0, 0)$, are applied and responses observed at the primary outputs.

The flip-flops are then configured in the test mode and $(0, 0)$ is scanned into appropriate flip-flops. Again, the primary input part of P_4, $(1, 0, 1)$, is applied when the second bit is being scanned in, and the response obseved. The flip-flops are then put into the hold mode for the next clock cycle while the primary input part of P_5, $(0, 0, 1)$, is applied and the response observed at the primary outputs.

If the above five vectors were applied using a normal scan methodology, a total of $5(2 + 1) + 2 - 1 = 16$ clock cycles would be required. In contrast, by virtue of grouping the vectors with compatible state input values, a total of seven clock cycles are required to detect these faults that do not require observation of the response at the state outputs.   □

The above methodology is not complete and must be used in conjunction with the classical scan test application methodologies for the detection of faults that do not have tests with compatible values at the state inputs or require observation of responses at the state outputs.

11.4.7.2 Combining scan and non-scan testing

If a fault f in a sequential circuit is targeted by a sequential test generator, then two possible scenarios may result. First, the test generator may fail to generate a test for the fault, either because it is sequentially untestable or because of the inadequacies of the test generation algorithm. Second, it may generate a sequence of test vectors that can be applied to the sequential circuit to detect the target fault. Under the second scenario, assume that the number of vectors in the sequence generated is α.

If the sequential circuit has full scan, the combinational blocks in the circuit can be input to a combinational test generator and a single scan vector generated for target fault f. Assume that β clock cycles are required to apply this vector. If $\alpha < \beta$, then it may be more beneficial to test fault f using the sequence of vectors generated by the sequential test generator. This is due to the fact that α vectors generated by the sequential test generator can be applied in the normal mode of the circuit. Hence, each vector can be applied in one clock cycle, leading to a saving of $\beta - \alpha$ clock cycles over scan testing. On the other hand, if $\alpha > \beta$ **or** the sequential test generator is unable to generate a test sequence for f, then the test vector for the fault should be generated using a test generator for combinational circuits and applied via scan.

Clearly, allowing a combination of scan and non-scan testing can only lead to more efficient testing than testing exclusively via scan. Also, it can achieve a high fault coverage that can be typically obtained only by using some scan vectors.

In practice, the simple comparison performed above is complicated by the fact that any sequence of vectors generated for f using a sequential test generator as well as any scan vector generated by a combinational test generator typically detects many other faults. In such cases, the comparison of the non-scan and scan testing should be generalized to take into account the numbers of non-target faults detected by the two tests. Methodologies that combine non-scan and scan testing to obtain more efficient and yet complete testing of full-scan circuits can be found in Pradhan and Saxena (1992) and Lee and Saluja (1992).

11.5 Boundary scan

In the previous sections, the focus was on enhancing the testability of a digital circuit that is implemented on a silicon die and may be packaged to obtain a chip. At the next level of integration, many such digital circuits are mounted and interconnected on a substrate. In a chips-on-board (COB) system, digital circuits are packaged to obtain chips and mounted on a printed-circuit-board (PCB). In a multi-chip module (MCM), unpackaged digital circuits, also called bare die, are integrated on a silicon or PCB-like substrate. In this section, we describe the *boundary scan architecture* that simplifies testing of COBs and MCMs.

To simplify the following discussion, we discuss chips, PCBs, and COB systems. However, nearly all the discussions are also applicable to MCMs.

11.5.1 Objectives of board-level DFT

A COB system is obtained by integrating chips on a PCB using some bonding process, such as soldering. A PCB physically supports the chips and **traces**, i.e., metal connectors on the PCB that constitute inter-chip interconnections.

Each chip used during board assembly is pretested by the chip's vendor and declared fault-free. The following types of faults are commonly targeted during board testing.

1 Inter-chip interconnect faults. These types of faults may occur due to defects introduced during PCB manufacturing and when chips are bonded to the PCB. First, a chip pin may not be properly bonded to the PCB, leading to an open circuit. Second, extra solder may flow between pins and PCB traces and cause shorts between interconnects. Third, some PCB traces may have opens or shorts due to defects that occur during PCB manufacturing.

2 Faults internal to the chip. Even though each chip is tested prior to board assembly, two main types of on-chip faults are sometimes targeted by board-level tests. In the first category are faults that may be caused due to improper handling during board assembly, such as shock or excessive heat during soldering. Second, certain types of faulty behavior may become apparent only after a chip is integrated into the board, such as its ability to drive a large load at chip pins. Note that both these types of faults are much more likely to occur in the **pad drivers** and **receivers**, which drive output pins of the chip and/or are driven by its input pins. As seen ahead, due to the location of *boundary scan cells* within a chip, these drivers and receivers are tested along with the inter-chip interconnects. In addition to the above faults, there exists a small likelihood that improper handling of a chip causes a fault in the **system logic**, i.e., the on-chip logic circuit. When that is a concern, following board assembly, boundary scan can be used to re-test the system logic within a chip.

Many faulty PCBs can be repaired by re-soldering, replacing faulty chips, or replacing faulty boards. Hence, diagnosis must be performed to identify the types and locations of faults to make decisions about repair. Therefore, board-level DFT mechanisms must provide support for the following.

1 Testing and diagnosis of faults in inter-chip interconnections.
2 In-situ re-testing of on-chip system logic.
3 Debugging of the system design.

As will be seen ahead, the DFT circuitry that is commonly used to achieve the first two objectives achieves the third without any significant modifications.

11.5.2   Basics of boundary scan

One simple way to develop tests for a board is to view the entire board as a single digital circuit and use techniques employed for digital chips. This approach is not practical for two main reasons. First, the entire circuit is too large for most test generators. More importantly, the net-list of the circuitry within each chip is not available. To circumvent this problem, **in-circuit tests** were used in the past. In this test methodology, physical probes were used to access the input/output pins of chips to test inter-chip interconnects as well as on-chip circuitry. Care had to be taken, however, to design pad drivers of chips as well as the circuitry driving the probes to

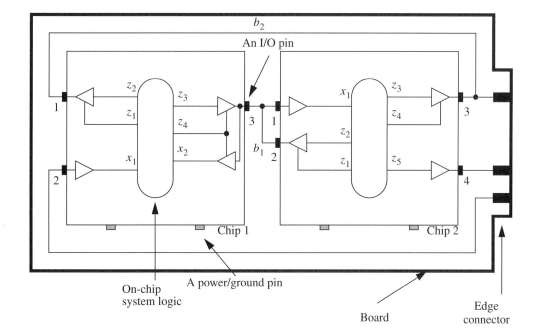

Figure 11.28 An example board with two chips

ensure that such tests could be performed without damaging the chips. Two main facts made such probing nearly impossible and provided the motivation for the development and wide adoption of *boundary scan*. First, the decrease in the inter-pin separation necessitated the use of finer, and hence more expensive, probes. Second, it became increasingly common for many chips to have pins distributed across their entire base. Since most such pins cannot be accessed via probes in multi-layered PCBs, in-circuit testing became very ineffective.

The key idea behind **boundary scan** is to incorporate DFT circuitry in each chip that provides direct access to the chip's input and output pins via scan chains. This access can then be used to achieve each of the above objectives as illustrated by an example. Figure 11.28 shows an example board with two chips, with five and six pins, respectively. In each chip, one pair of pins is used for power and ground connections. In in-circuit testing, an open on net b_1 is tested by probing Pin 3 of Chip 1 and Pins 1 and 2 of Chip 2. Note that Pin 3 of Chip 1 is a *bi-directional driver/receiver*, Pin 2 of Chip 2 is a *tri-state driver*, and Pin 1 of Chip 2 is a *simple receiver*. Unless otherwise specified, in all the examples in this section we assume that if a receiver is disconnected from its drivers due to opens, then it interprets the value as a 1. We also assume that a control value 1 enables a tri-state driver and configures a bi-directional driver/receiver as a driver.

To test for an open at the simple receiver of net b_1, one probe can be used to apply a value 0 at Pin 2 of Chip 2 while other probes are used to observe the values at the

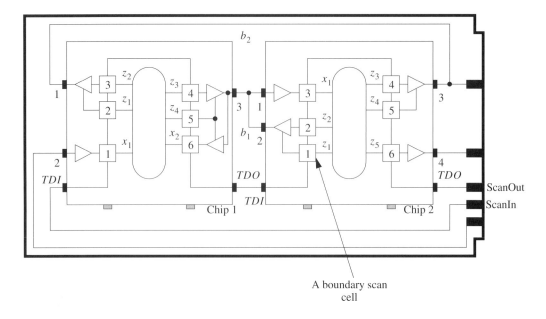

A boundary scan
cell

Figure 11.29   Boundary scan version of the example board with two chips

other two pins of the net. Similarly, a short between nets b_1 and b_2 can be tested by probing an appropriate subset of the above pins as well as Pin 1 of Chip 1 and Pin 3 of Chip 2.

Figure 11.29 shows a conceptual view of a version of the above board where boundary scan is incorporated in each chip. In this case, a **boundary scan register (BSR)** is formed using **boundary scan cells (BSCs)** to provide access to the input and output pins of the two chips. By serially scanning in appropriate values at the scan-in input of the board, values can be applied to, and responses captured from, each pin. The BSR can hence be used to test for the faults discussed above, and physical probing is no longer necessary. For example, an open at the simple receiver of net b_1 can be tested by (a) shifting the sequence of values $(x, x, x, x, 0, 1, x, 0, x, x, x, x)$ at the scan-in input of the board (where the leftmost value is scanned in first), (b) applying the values shifted into the BSCs onto the corresponding pins, (c) capturing responses from the input pins, and (d) scanning the response out and observing it at the scan-out output of the board. In the above example, if there is no open at the simple receiver of net b_1, the fourth bit of the response at the scan-out output is 0; otherwise this bit is 1.

11.5.3   Boundary scan architecture

A typical board contains chips from multiple manufacturers. Hence, the boundary scan circuitry in various chips must be inter-operable. IEEE Std 1149.1 (IEEE, 1994) is a standard that has been adopted to ensure such inter-operability. In this section, we describe some of the salient features of this standard. Nearly all the optional features of the standard and most details have been omitted and can be found in IEEE (1994).

11.5.3.1 The overall architecture

Boundary scan is a serial architecture in the sense that both test data and control signals are applied serially. The first key component of the architecture is the **test access port (TAP)**, which has a minimum of four pins that provide access to the boundary scan circuitry within a chip. The second component of the architecture is an FSM, called the **TAP controller**, which interprets the bits of the control sequence applied at an appropriate pin and generates some of the signals to control the on-chip test circuitry. The third component is an **instruction register (IR)** into which any one of the several instructions supported by a particular chip is loaded. The instruction at the outputs of the IR is decoded to produce additional signals to control the on-chip test logic. The final component of the architecture is a set of **test data registers (TDRs)**, including the *BSR*, a *bypass register (BypR)*, and other optional registers, that are used to shift test data in and out of chips and (in most cases) to apply tests and capture responses. IR contains a shift-register and circuitry that enables a new instruction to be scanned into the shift-register without changing the values that appear at its parallel outputs. This is also the case for some TDRs, especially the BSR.

Figure 11.30 illustrates the overall architecture.

11.5.3.2 The test access port (TAP)

The test access port must include four pins, **test clock (TCK)**, **test mode select (TMS)**, **test data input (TDI)**, and **test data output (TDO)**.

Test clock: A separate test clock, TCK, is used to control all the operations of the test circuitry. This allows the test circuitry to operate independently of the system clocks, which can differ significantly from one chip to another. The use of a separate test clock permits the test circuitry in different chips to operate with each other. In some test modes (such as the SAMPLE instruction), this also enables the test circuitry to operate without disturbing the operation of the on-chip system logic. However, as seen ahead, it becomes necessary to synchronize TCK with one or more system clocks for certain types of tests.

All the input data and non-clock control signals are sampled and all the output data and non-clock control signals updated in such a manner that race-free inter-operability of chips with boundary scan is ensured (IEEE, 1994).

Test mode select: TMS is an input at which a control value is applied in each cycle of TCK. In conjunction with the instruction at the outputs of IR, these control values

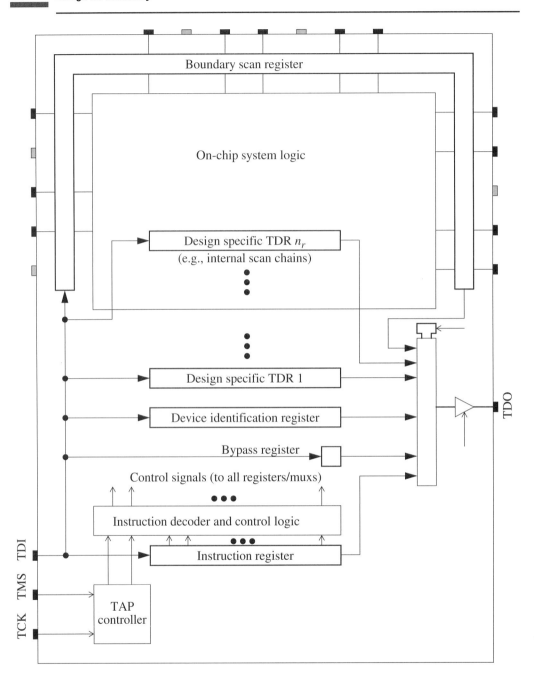

Figure 11.30 Overall architecture of boundary scan within a chip

provide clock-by-clock control of the test circuitry. These values also provide the control required to replace the instruction in the IR.

Test data input: TDI is the serial input at which bits of the instruction are applied when a new instruction is being loaded into the IR. At some other times, this works as the scan input, and bits of the next test vector are applied serially. In other clock cycles, no specific data need to be applied.

Test data output: TDO is the serial output where the bits of the contents of the shift-register within the IR appear when a new instruction is being loaded. These response bits may be analyzed to identify certain types of faults in the IR. In the clock cycles in which bits of the test vector are scanned into one or more TDRs, this pin functions as a scan-out pin and the bits captured in those TDRs in response to the last test vector appear serially for comparison. In all the other clock cycles, this output is tri-stated.

In addition, the TAP may contain a fifth optional input, called **test reset (TRST*)**. Application of a 0 at this input can asynchronously reset all the test logic.

11.5.3.3 The TAP controller

The TAP controller is an FSM that interprets the control inputs applied serially to the TMS input. It has the state transition graph (STG) shown in Figure 11.31.

At each rising edge of TCK, the value applied at the TMS input is sampled, and the state of the controller has a transition following the next rising edge of TCK. The state transitions are shown in Figure 11.31.

The controller state is reset to the *Test-Logic-Reset* state when (a) the chip is powered up, (b) five or more consecutive 1s are applied at TMS, or (c) 0 is applied at the optional TAP input, TRST*. When the TAP enters this state, the instruction at the output of the IR is initialized to the BYPASS instruction, whose binary code is 11 . . . 1. In this state, test logic is disabled and the system logic is allowed to perform its normal operation, unaffected in any manner by the test logic.

For most instructions, in the *Run-Test/Idle* state the test logic is idle and the system logic operates normally. For some instructions, such as RUNBIST, self-test is performed on the system logic while a pre-specified number of consecutive 0s are applied to TMS to retain this TAP controller state.

Select-DR-Scan, *Exit1-DR*, *Exit2-DR*, *Select-IR-Scan*, *Exit1-IR*, and *Exit2-IR* are temporary states. These states are required because the control values are applied serially and a sequence of multiple values are required at states with more than two successor states. In each of these states, the IR as well as all the TDRs hold their states and outputs. Furthermore, the TDO output is tri-stated.

The states in the rightmost column of the STG in Figure 11.31 serve to manipulate the state of the IR and its output. In the *Capture-IR* state, 1 and 0 are loaded into the least significant bit (LSB) of the shift-register contained in the IR and its adjacent bit,

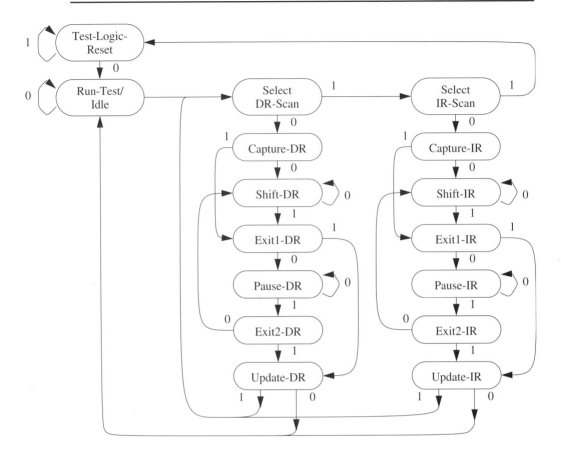

Figure 11.31 The STG of the TAP controller (IEEE, 1994) (© 1994 IEEE)

respectively, where LSB is the bit closest to TDO. (Specific values may also be loaded into the other bits of the register, if any.) For each clock cycle when the controller is in the *Shift-IR* state, the shift-register within the IR is connected between TDI and TDO and bits of an instruction applied serially at TDI are scanned in. At the same time, the bits loaded into the shift-register when the TAP was in the *Capture-IR* state are scanned out and observed at TDO. Since a 1 and a 0 are loaded into two least significant stages of the shift-register within the IR, an SA1 or SA0 fault in the register can be detected by observing the values appearing at TDO in these clock cycles. The *Pause-IR* state is used to temporarily suspend scanning in of the bits of the instruction. This is useful if the automatic test equipment (ATE), or any other controller that controls boundary scan test, needs to suspend scanning while it fetches additional data from its hard disk. The process of scanning (and, if necessary, pausing) is repeated until bits of a desired instruction are shifted into appropriate bits of the shift-register within the IR. Finally, in the *Update-IR* state, the state of the shift-register within the IR is clocked at the falling edge of TCK into its outputs and a new instruction is put in place.

Table 11.2. *A summary of major control signals*

Control input	Description
*Reset**	0 during *Test-Logic-Reset* TAP state
Select	1 during *Test-Logic-Reset*, *Run-Test/Idle*, *Exit1-IR*, *Exit2-IR*, *Pause-IR*, *Capture-IR*, *Update-IR* TAP states (changes following rising edge of TCK)
Enable	1 during *Shift-DR* and *Shift-IR* TAP states
ShIR	1 during *Shift-IR* TAP state
ShDR	1 during *Shift-DR* TAP state
ClkIR	TCK gated in *Shift-IR* and *Capture-IR* TAP states
UpIR	TCK inverted and gated in *Update-IR* TAP state
ClkDR	TCK gated in *Shift-DR* and *Capture-DR* TAP states
UpDR	TCK inverted and gated in *Update-DR* TAP state

Other than the *Test-Logic-Reset* state, this is the only state in which the instruction changes.

The states in the middle column of the STG of the TAP controller are similar, except that analogous operations are performed on the TDRs selected by the current instruction.

The TAP controller states are used to generate two types of signals. In the first category are **control signals** *Reset**, *Select*, *Enable*, *ShIR*, and *ShDR*. In the second category are **gated clocks** *ClkIR*, *UpIR*, *ClkDR*, and *UpDR*. Each of these signals is obtained by gating clock TCK. Table 11.2 describes the nature of these control signals. All the control signals, except *Select*, change during the first falling edge of TCK after which TAP first enters the particular state. Gated clocks *UpIR* and *UpDR* are inverted. Hence, a falling edge of TCK is passed on as a rising edge during the *Update-IR* and *Update-DR* TAP states, respectively.

11.5.3.4 The instruction register

An example cell that may be used to implement an IR is shown in Figure 11.32, where *M* is a multiplexer. If the instructions for the given chip are n_{IR}-bits long, then n_{IR} such cells are connected in the manner shown in Figure 11.33 to obtain an IR. The *SerialIn* input of one cell is connected to the *SerialOut* output of the previous cell. Hence, the D_1 flip-flops of the cells in the IR form a shift-register. The *SerialIn* input of the first cell is connected to TDI, while the *SerialOut* output of the last cell is connected to one of the inputs of the multiplexer whose output drives TDO (see Figure 11.30). The cell of the IR closest to TDO and TDI are considered the least and the most significant bits of the register, respectively.

If *ShIR* is held high, the desired value is shifted into the shift-register from TDI, and its previous contents shifted out at TDO at each rising edge of the gated clock

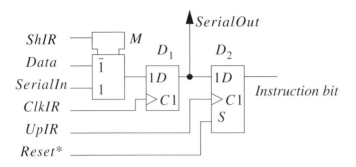

Figure 11.32 An example implementation of an IR cell (IEEE, 1994) (© 1994 IEEE)

ClkIR. This combination of control signal values occurs in the *Shift-IR* state of the TAP controller.

If *ShIR* is held low, the value at the *Data* input of each cell is loaded into its D_1 flip-flop at the rising edge of gated clock *ClkIR*. The standard requires that values 0 and 1 be loaded into the least significant stage of the shift-register within the IR and the adjacent stage, respectively. This combination of control signal values occurs in the *Capture-IR* state of the TAP controller.

The D_2 flip-flop in each cell constitutes its output stage. The values at the outputs of the D_2 flip-flops constitute the current instruction. The presence of an output stage allows the current instruction to remain in place while the next instruction is scanned into the shift-register. The intermediate values that are present in the shift-register during scanning are hence prevented from being considered as instructions. Once a valid instruction is completely shifted in, it is applied to the output of the IR at the rising edge of the *UpIR* gated clock and becomes the new instruction. This occurs at the falling edge of TCK when the TAP is in the *Update-IR* state.

11.5.3.5 Bypass register

The bypass register is a mandatory one-stage register. Figure 11.34 shows a design that meets all the requirements of the standard.

Consider an instruction in which this register is selected. When *ShBypR* is high, the value at TDI is shifted to TDO at each rising edge of the gated clock *ClkBypR*. This combination of control signal values occurs during the *Shift-DR* state of the TAP controller.

As described in Section 11.5.4, in conjunction with the mandatory instruction BYPASS, this register enables efficient testing through reduction in the length of the overall scan chain by selecting a one-stage bypass register within each chip that is not being tested at the given time.

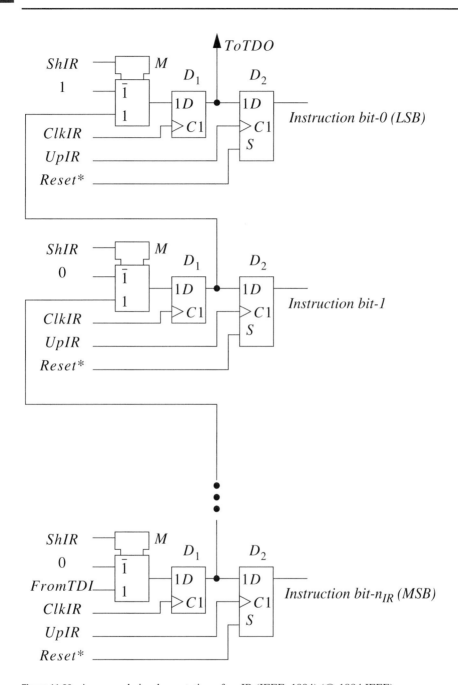

Figure 11.33　An example implementation of an IR (IEEE, 1994) (© 1994 IEEE)

11.5.3.6 Boundary scan register

The BSR is the most important TDR and one of two TDRs required by the standard. As shown in Figure 11.30, This register is inserted between the on-chip system logic

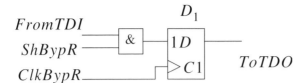

Figure 11.34 An example implementation of a bypass register cell (IEEE, 1994) (© 1994 IEEE)

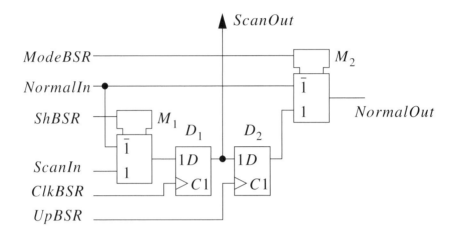

Figure 11.35 An example implementation of a BSR cell (IEEE, 1994) (© 1994 IEEE)

and pad-logic (driver or receiver) at chip pins and is used to support the mandatory EXTEST and SAMPLE/PRELOAD instructions as well as many non-mandatory instructions.

Cell design: The cell shown in Figure 11.35 (IEEE, 1994) meets all the requirements of the standard for all types of pins. Alternative cell designs that also meet the standard for all types of pins can be found in the literature. In addition, it is possible to obtain smaller versions of the above cell that are suitable for each specific type of pin (see Problem 11.22).

The D_1 flip-flops in such BSCs are configured as a shift-register. The D_2 flip-flops hold the test data applied by the register. As in the case of the IR, this prevents the data in the shift-register during intermediate stages of scan-in and scan-out from being applied as test data. The M_2 multiplexers determine whether the values at the *NormalIn* inputs or the test data in the register are applied to the *NormalOut* outputs of the cells. The M_1 multiplexers determine whether the D_1 flip-flops operate as a shift-register for scan-in and scan-out or capture the values appearing at the *NormalIn* inputs.

Cell placement: A chip has logic input, logic output, and power and ground pins. It may also have one or more input pins that bring in external clocks. BSCs are placed at all the logic input and output pins; they are not placed at clock, power, and ground pins.

At a logic input x_i of the system that is driven by Pin j, a cell is used in the following manner. The *NormalIn* input of the cell is connected to the output of the input receiver driven by Pin j and the *NormalOut* output of the cell is connected to x_i. Consider the board design in Figure 11.28 whose boundary scan version is shown in Figure 11.29. The details of BSC 1 of Chip 1 in Figure 11.29 at input Pin 2 of the chip are shown in Figure 11.36.

Logic output pins can be classified into three main types, two-state, three-state, and bi-directional. Pin 4 of Chip 2, Pin 3 of Chip 2, and Pin 3 of Chip 1 in Figure 11.29 are, respectively, examples of these three types of pins.

At a two-state output pin that is driven by output z_i of the system logic and drives Pin j of the chip, the *NormalIn* input and *NormalOut* output of the cell are connected to z_i and the input of the pad driver that drives Pin j, respectively. For a three-state output that drives Pin j and whose data and control inputs are driven, respectively, by outputs z_{i_d} and z_{i_c} of the system logic, two cells are used. The *NormalIn* input and *NormalOut* output of the first cell are connected to z_{i_d} and the data input of the tri-state pad driver whose output drives Pin j. The *NormalIn* input and the *NormalOut* output of the second cell are connected to z_{i_c} and the control input of the tri-state pad driver whose output drives Pin j. The details of how Cells 2 and 3 of Chip 1 in Figure 11.29 are configured at the tri-state output Pin 1 of the chip are also shown in Figure 11.36. Consider a bi-directional Pin j which is either driven by output z_{i_d} of the system logic or drives its input x_{i_δ}. Depending on the value at control output z_{i_c}, three cells may be used as shown in Figure 11.29. Two of these cells are configured as the corresponding cells for a tri-state output pin; the third cell is configured as the cell at a logic input pin.

Formation of the boundary scan register: After the BSCs are placed at the locations described above and their *NormalIn* inputs and *NormalOut* outputs connected, the *ScanIn* input of the first cell is connected to TDI. The *ScanIn* input of each subsequent cell is connected to the *ScanOut* output of the previous cell in the chain. The *ScanOut* output of the last cell is connected to an input of the multiplexer whose output drives the TDO pin (see Figure 11.30).

Key modes of operation of the BSR: In the **normal** mode of operation, a 0 is applied at *ModeBSR*, the gated clocks *ClkBSR* and *UpBSR* are both disabled, and the value at *ShBSR* does not affect the operation of the register. In this case, the value applied at the *NormalIn* input of each cell appears at its *NormalOut* output; the remainder of the logic is disabled. The data flow during this mode of operation is shown in Figure 11.37.

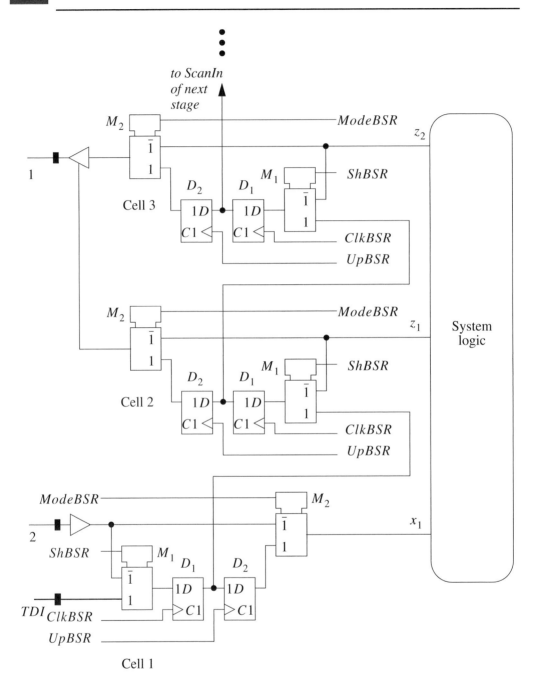

Figure 11.36 Details of the designs of a part of the BSR for the example board in Figure 11.29

In the **normal-sample** mode, a 0 is again applied at *ModeBSR* and the value applied at the *NormalIn* input of a cell appears at its *NormalOut* output. The system logic hence continues to operate as in the normal mode. Also, the gated clock *UpBSR* is disabled.

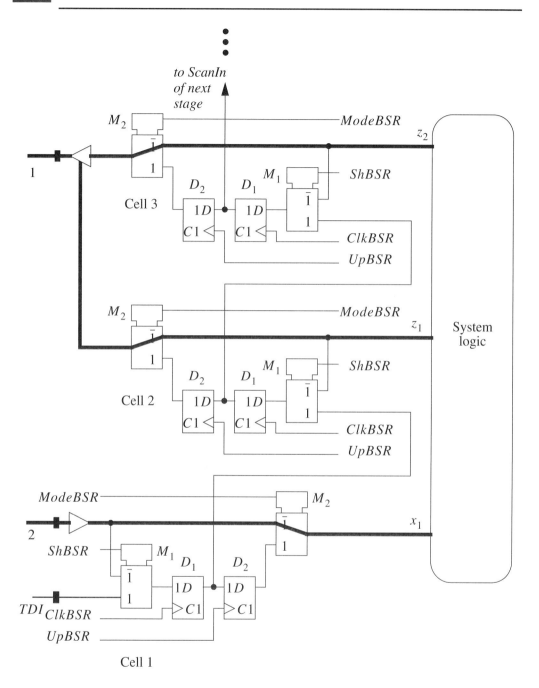

Figure 11.37 Normal mode of operation of the BSR

However, in this case, the value 0 is applied at *ShBSR* and one clock pulse is applied at *ClkBSR*. The edge of this gated clock is properly synchronized with the system clock that controls the value applied at the *NormalIn* input of the cell and the value

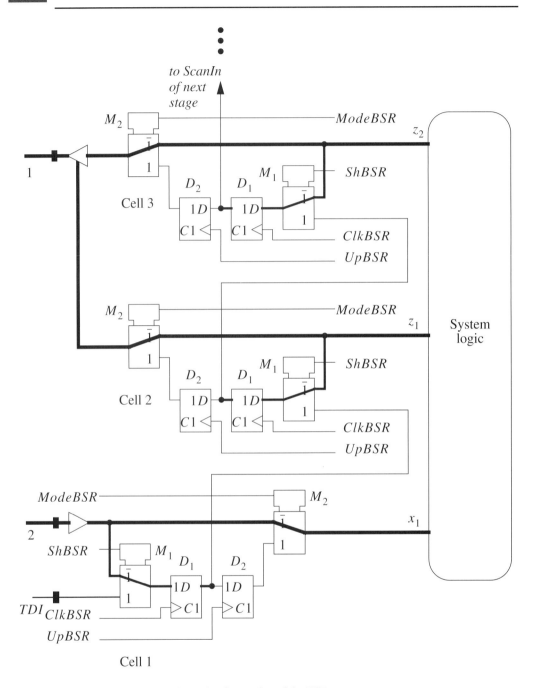

Figure 11.38 Normal-sample mode of operation of the BSR

passing from the cell's *NormalIn* input to its *NormalOut* pin is captured in the cell's D_1 flip-flop. *This mode occurs only when the TAP controller is in the Capture-DR state.* The data flow during this mode of operation is shown in Figure 11.38.

In the **normal-scan** mode, a 0 is again applied at *ModeBSR* and the value applied at the *NormalIn* input of each cell appears at the cell's *NormalOut* output and the system logic continues to operate as in the normal mode. The gated clock *UpBSR* is kept inactive and a 1 is applied at the *ShBSR* inputs of the cells while a sequence of clock pulses is applied at the *ClkBSR* inputs. This shifts the data into the D_1 flip-flops of the cells in the BSR and accomplishes one or both of the following purposes. First, any value captured in the D_1 flip-flops during the normal-sample mode is scanned out and can be observed at the TDO output. Second, new test data values to be loaded into the BSR cells can be scanned in from the TDI input. Note that the system logic continues to operate in the normal mode during the entire process. *This mode occurs only when the TAP controller is in the Shift-DR state.* The data flow during this mode of operation is shown in Figure 11.39.

In the **normal-preload** mode, a 0 is again applied at *ModeBSR* and the value applied at the *NormalIn* input of each cell appears at the cell's *NormalOut* output and the system logic continues to operate as in the normal mode. In addition, one clock pulse is applied at *UpBSR* while *ClkBSR* is disabled. This loads the data in the D_1 flip-flop of each cell in the BSR into its D_2 flip-flop. *This mode occurs only when the TAP controller is in the Update-DR state.* The data flow during this mode of operation is shown in Figure 11.40.

In the **test-capture** mode, a 1 is applied at *ModeBSR* and the value loaded in the D_2 flip-flop of a cell appears at its *NormalOut* output. At the same time, *UpBSR* is held inactive but a 0 is applied at *ShBSR* and one clock pulse is applied at *ClkBSR*. In this case, the value appearing at *NormalIn* in each cell in the BSR is captured in its D_1 flip-flop. The value appearing at *NormalOut* of the cell remains unchanged. Hence, the system logic and interconnect continue to operate under the same vector. *This mode occurs only when the TAP controller is in the Capture-DR state.* The data flow during this mode of operation is shown in Figure 11.41.

In the **test-scan** mode, a 1 is again applied at *ModeBSR* and the value in the D_2 flip-flop of each BSR cell appears at its *NormalOut* output and the system logic and interconnect continue to operate in the test mode. In addition, value 1 is applied at *ShBSR* and a sequence of clock pulses is applied at *ClkBSR*. This shifts the data in the D_1 flip-flops of the cells in the BSR and accomplishes one or both of the following purposes. First, any value captured in the D_1 flip-flops in the test-capture mode is scanned out and can be observed at the TDO output. Second, new test data values to be loaded into the BSR cells can be scanned in from the TDI input. Furthermore, *UpBSR* is held inactive; hence, the test data applied at *NormalOut* are held constant and the circuit continues to operate under the same vector while data are scanned into and out of the shift-register contained in the BSR. *This mode occurs only when the TAP controller is in the Shift-DR state.* The data flow during this mode of operation is shown in Figure 11.42.

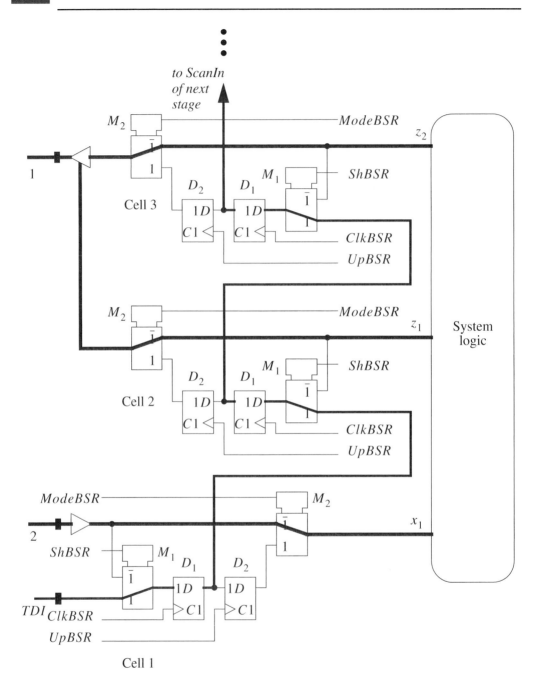

Figure 11.39 Normal-scan mode of operation of the BSR

In the **test-update** mode, a 1 is again applied at *ModeBSR*. In addition, a clock is applied at *UpBSR* while *ClkBSR* is held inactive. This loads the data in the D_1 flip-flop of each cell in the BSR into its D_2 flip-flop, which then appears at the *NormalOut*

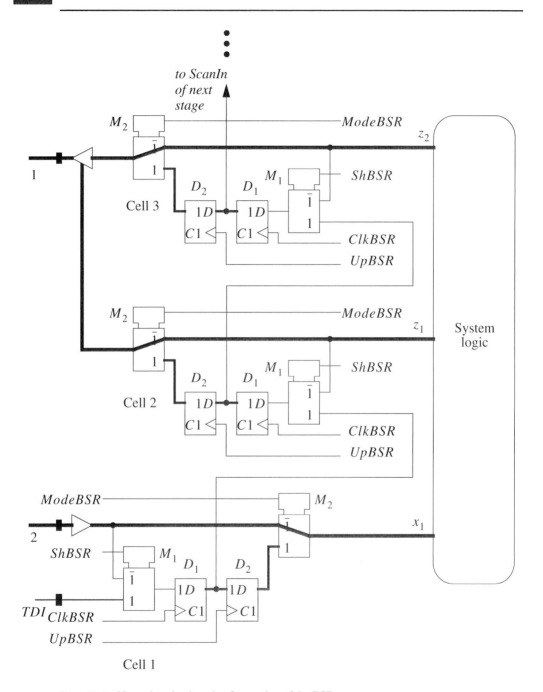

Figure 11.40 Normal-preload mode of operation of the BSR

output of the cell. *This mode occurs only when the TAP controller is in the Update-DR state.* The data flow during this mode of operation is shown in Figure 11.43.

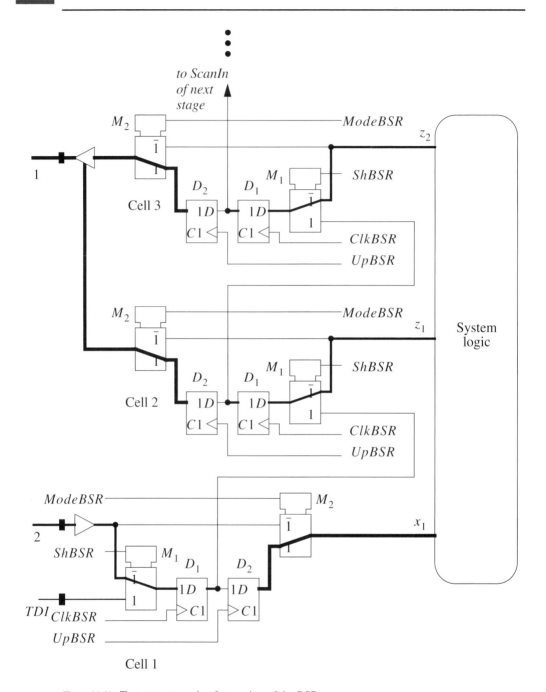

Figure 11.41   Test-capture mode of operation of the BSR

11.5.4 Instructions: testing via boundary scan

Each boundary scan implementation must support the **mandatory** instructions BY-PASS, SAMPLE/PRELOAD, and EXTEST. In addition, most implementations also

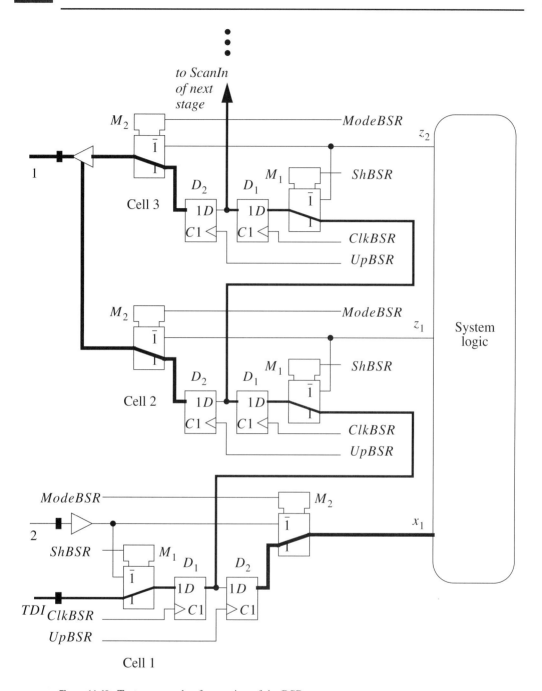

Figure 11.42   Test-scan mode of operation of the BSR

support INTEST and RUNBIST. We next describe these instructions. The reader is advised to refer to IEEE (1994) for other instructions that are sometimes supported.

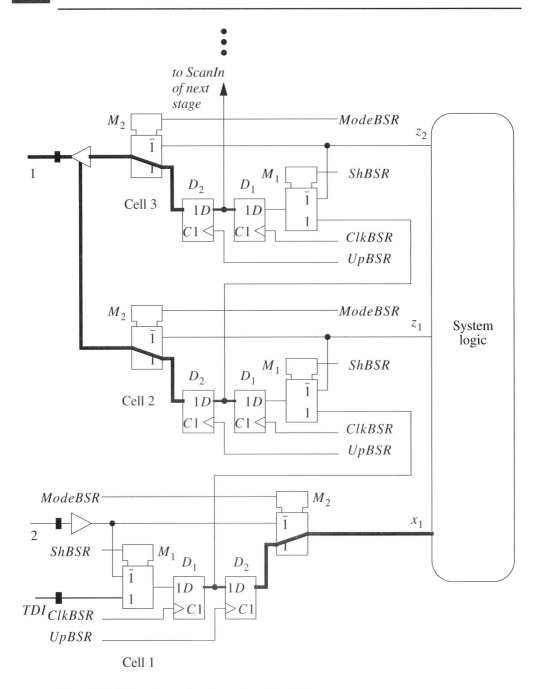

Figure 11.43 Test-update mode of operation of the BSR

Each instruction provides a different test capability. It does so (a) by selecting the TDR that is activated, and/or (b) by determining the values of the control signals and gated clocks during each state of the TAP controller. The selection of the TDR that is activated influences the operation of the boundary scan circuitry in two main ways. First, the selected TDR is connected between TDI and TDO by appropriately configuring the multiplexer controlling TDO (see Figure 11.30). Second, various control signals and gated clocks are applied to the selected register. For example, if the BSR is selected during a particular instruction, then control signals *ModeDR* and *ShDR* and gated clocks *ClkDR* and *UpDR* are passed on to the corresponding inputs of the BSR. In contrast, the corresponding inputs of other TDRs are held inactive while this instruction remains in place. Each of these registers that performs some function during normal operation is configured to operate in its normal mode. For the above example instruction, signals *ModeBypR*, *ShBypR*, *ClkBypR*, and *UpBypR* of the bypass register are held at appropriate values to hold the bypass register inactive.

Next, we describe the purpose of some of the commonly used instructions along with some details of their operation. A complete description can be found in IEEE (1994).

BYPASS: This is a mandatory instruction that must have the op-code $11 \ldots 1$. We illustrate how this instruction enables efficient testing after we describe the operation of the boundary scan circuitry in the presence of this instruction.

The presence of the BYPASS instruction at the outputs of the IR of a chip selects that chip's BYPASS register. As described above, this causes the BYPASS register of the chip to be connected between its TDI and TDO pins. It also causes control signal *ShDR* to be passed onto signal *ShBypR* and gated clock *ClkDR* to be passed onto clock input *ClkBypR*. (Each of the other TDRs in the chip is configured in its normal mode via application of suitable values to its control signals and gated clocks.) In the presence of this instruction, a rising edge of TCK appears at gated clock *ClkDR*, and hence at gated clock *ClkBypR*, during each cycle when its TAP controller is in the *Shift-DR* state. Furthermore, the *ShDR*, and hence the *ShBypR*, signal is high during each of the above clock cycles. (The value at this signal for the other clock cycles is not important.) Finally, the rest of the test logic is deactivated and the system logic is configured to perform its normal operation.

The selection of the BYPASS register between the TDI and TDO pins of a chip helps maintain the continuity between its TDI and TDO. This helps maintain the integrity of any scan chain that passes via multiple chips, while reducing the length of the chain by replacing a longer TDR by a one-bit BYPASS register. This enables efficient testing, as illustrated by the following example.

Example 11.24 Consider a scenario where both the chips on a board need to be re-tested after board assembly where one chip requires 1000 vectors while the other requires only 100 vectors. Assume that the lengths of the appropriate TDRs within these chips are n_1 and n_2, respectively. In the first session, the scan chain is configured to include the appropriate TDR within each chip. 100 tests are then applied to each chip using a scan chain of length $n_1 + n_2$. Subsequently, in the second session, the BYPASS instruction can be used to bypass the TDR of the second chip, by placing that chip's BYPASS register between its TDI and TDO. The remaining 900 tests can then be applied to the first chip using a scan chain whose total length is $n_1 + 1$ bits. □

The BYPASS instruction provides similar benefits during other test modes as well as during the debugging of system design using the SAMPLE instruction.

SAMPLE/PRELOAD: This is a mandatory instruction that serves dual purposes of (a) sampling the data being transferred at the chip inputs and outputs in a particular clock cycle during its normal operation (SAMPLE), and (b) pre-loading of suitable data into the D_2 flip-flops of the BSR to be applied as the first vector when subsequently an instruction that selects BSR as the TDR is applied. This instruction achieves the above objectives without affecting the normal system operation in any manner.

This instruction selects only the BSR. As described above, this has two main consequences. First, the BSR is connected between TDI and TDO of the chip. Second, control signals *ModeDR* and *ShDR*, as well as gated clocks *ClkDR* and *UpDR*, are passed on to the corresponding control inputs of the BSR, namely *ModeBSR*, *ShBSR*, *ClkBSR*, and *UpBSR*. In addition, appropriate values are applied to the corresponding signals of every other data register (if any) to configure it in its normal mode of operation.

Values of control signals and gated clocks: As long as this instruction is active, control signal *ModeBSR* is held low, allowing the chip inputs and outputs (which pass via the cells of BSR; see Figure 11.36) to carry their normal values. Also, the system logic is allowed to operate in its normal mode. Hence, in the presence of this instruction at the outputs of its IR, a chip performs its normal operation.

Next, consider the remaining control inputs to the BSR, namely control signal *ShBSR* and gated clocks *ClkBSR* and *UpBSR*, when the TAP controller traverses the states in the second column of Figure 11.31, i.e., states between *Select DR-scan* and *Update-DR*. In the TAP controller state *Capture-DR*, these control inputs take a combination of values that configures the cells of the BSR in the *normal-sample* mode. TCK and the system clocks that control the value of data at the inputs and outputs of the system logic are appropriately synchronized and the value at each input/output is captured in the D_1 flip-flop of the corresponding cell of the BSR. Subsequently, appropriate values are applied at TMS to cause the TAP controller to visit the TAP state *Shift-DR* an appropriate number of times. Each time this state is visited, the BSR operates in the *normal-scan* mode and the contents of the D_1 flip-flops of each BSR cell are shifted towards TDO. The values captured in these flip-flops during the *normal-sample* mode are observed at the output of the scan chain to which this particular BSR belongs. The above process can be repeated to sample and observe the values at chip inputs/outputs, i.e., to implement the functionality required by the SAMPLE instruction.

While the TAP is in the *Shift-DR* state, appropriate values can be applied at the scan input of the chain to which this BSR belongs to scan in the desired values into the D_1 flip-flops of the BSR. Subsequently, when the TAP reaches the *Update-DR* state, values at the control inputs cause the BSR to be in the *normal-preload* mode. The value scanned into the D_1 flip-flop of each BSR cell is then transferred to the corresponding D_2 flip-flop. This makes the desired values, available at an input of the M_2 multiplexer of each cell in the BSR, ready for application to the corresponding input/output of the chip when an appropriate instruction is subsequently applied. The combination of the *normal-scan* mode each time the TAP controller is in the *Shift-DR* state followed by the *normal-preload* mode when the TAP controller is in the *Update-DR* state helps implement the functionality desired of the PRELOAD instruction.

The purpose of the SAMPLE instruction is to facilitate debugging of system design. In contrast, the purpose of the PRELOAD instruction is to facilitate initialization of the value in the D_2 flip-flop of each BSR cell such that safe values are applied at the chip inputs and outputs when a new instruction that activates the BSR is applied. This is illustrated in some detail next.

EXTEST: The purpose of this mandatory instruction is to facilitate testing of inter-chip interconnects. The op-code of this instruction is required to be $00\ldots0$. The presence of this instruction causes only the BSR to be selected between TDI and TDO of a chip, as in the case of the above instruction.

Two key differences with the above instruction should be noted. First, in the presence of this instruction, value 1 is applied to signal *ModeBSR*. This causes the value at the output of the D_2 flip-flop of the BSR cell to be applied to the chip output. In other words, values scanned into the BSR cells and clocked into their D_2 flip-flops are applied to the inter-chip interconnects instead of the normal values. The second main difference with the above instruction is how the system logic operates in the presence of this instruction. During the execution of EXTEST, the system logic may be disabled to ensure that no damage is caused to the circuit due to the application of any value loaded by the user into the D_2 flip-flop of each BSR cell. Alternatively, the on-chip system logic may be designed in such a way that application of any arbitrary combination of values at its inputs will not cause the logic any damage.

When the EXTEST instruction is applied at the outputs of a chip's IR, the data in the D_2 flip-flops of the BSR cells are applied to the inter-chip interconnects. In other words, in the presence of this instruction, the values in the D_2 flip-flops of the BSR cells are applied to the inter-chip interconnects. If the inter-chip interconnect contains multi-driver tri-state nets or bi-directional nets, then the application of some combinations of values can cause multiple drivers to drive opposite values on a net, thereby causing damage to drivers.

Example 11.25   Consider the example design shown in Figure 11.29. Assume that each tri-state driver is enabled when 1 is applied at its control input. Also assume that each bi-directional driver is configured in the output mode if a value 1 is applied to its control input.

If 1, 1, 0, and 1 are applied to Cell 2 of Chip 1, Cell 3 of Chip 1, Cell 4 of Chip 2 and Cell 5 of Chip 2, then both the tri-state drivers in this example circuit will be enabled and drive opposite values on a net. This may cause excessive current to flow and damage one or more drivers.   □

To avoid the possibility of such damage, the PRELOAD instruction should be used to preload a legal combination of values into the D_2 flip-flops of BSR cells *prior* to the application of the EXTEST instruction at the outputs of the IR. Furthermore, each set of values scanned into the BSR during subsequent testing should also constitute a legal combination of values.

We assume that the preload instruction is used as described above to avoid the possibility of circuit damage. We also assume that the system logic is either disabled or designed in such a manner that the application of any combination of values cannot cause damage. Having precluded the possibility of circuit damage, we now describe the operation of the BSR during EXTEST.

The values applied to the other control inputs of the BSR are such that it operates as follows. First, when the TAP controller is in the *Capture-DR* state, the BSR is configured in the *test-capture* mode. At this time, the response at the inter-chip interconnects to the previously applied vector is captured in the D_1 flip-flops of the BSR cells. The values applied to the interconnect remain unchanged. Subsequently, a sequence of values is applied to TMS to ensure that the *Shift-DR* state is visited an appropriate number of times. Each time the TAP is in the *Shift-DR* state, the BSR is configured in the *test-scan* mode and the response captured into the D_1 flip-flop of each BSR cell is scanned out for observation at the output of the scan chain. At the same time, bits of the next vector to be applied to the interconnect are applied to the input of the scan chain and values scanned into the D_1 flip-flops of the BSR cells. Finally, after the bits of the captured response are scanned out and the desired values scanned into the D_1 flip-flops of the BSR cells, the TAP enters the *Update-DR* state. In this state, the BSR operates in its *test-update* mode and the value in the D_1 flip-flop of each BSR cell is loaded into the corresponding D_2 flip-flop and applied to the inter-chip interconnect.

If different types of cells are used at different types of pins, the response may be captured only at the BSR cells at the input pins.

Example 11.26   Consider the example boundary scan circuit shown in Figure 11.29.

Assume that it is desired to apply the following sequence of vectors to the interconnect:

$$P_1 = (x, 1, 1, 0, 1, x, 0, x, x, x, 0, 0),$$
$$P_2 = (x, 1, 0, 1, 1, x, 0, x, x, x, 0, 0),$$

and so on, where the first bit of the vector is applied to Cell 1 of Chip 1, the second bit to Cell 2 of Chip 1, . . ., the seventh bit to Cell 1 of Chip 2, the eighth bit to Cell 2 of Chip 2, and so on.

As described above, before the EXTEST instruction is applied, the SAMPLE/PRELOAD instruction is applied. The bits of the first vector are scanned into the cells of the BSR and finally loaded into the D_2 flip-flop of each cell. The EXTEST instruction is then applied. When the TAP reaches the *Capture-DR* state for the first time after the application of the EXTEST instruction, the response of the interconnect to P_1 is captured into the D_1 flip-flops of the BSR cells. Subsequently, this response is scanned out and bits of the next vector scanned in. The bits scanned in are then loaded into the D_2 flip-flops of the BSR cells. When the TAP reaches the *Capture-DR* state, the response of the interconnect to P_2 are captured. The above process is repeated until all the desired vectors are applied to the interconnect.　　　　□

INTEST: The purpose of this instruction is to enable static, i.e., slow speed, re-testing of the system logic within a chip. The operation of this instruction is very similar to that for the EXTEST instruction, with two notable exceptions. First, in this instruction, all the output pins are configured in inactive modes. For example, tri-state drivers are disabled and bi-directional driver/receivers configured as receivers. Second, the system logic is not disabled. Instead, it is configured in such a manner that it is clocked once between the *Update-DR* and *Capture-DR* TAP states to obtain its response to the vector applied at its inputs after the *Update-DR* state. This response is then captured in the *Capture-DR* state. (If different cells are used at different types of pins, then the responses may be captured only at the output pins.) Subsequently, the response is scanned out and bits of the next vector scanned in as in the case of the above instruction.

RUNBIST: The RUNBIST instruction allows the user to access BIST resources on a chip even after the chip is mounted on a board. This can enable at-speed testing of on-chip logic without the use of scan in each vector.

Details of the implementation of this instruction vary significantly from one chip to another. However, typically, the internal registers which constitute the linear-feedback shift-register (LFSR) random-pattern generators and response analyzers (RAs) are configured as a single TDR and selected during this operation. Values are scanned into these registers and constitute seeds of LFSRs and initial values of signatures. (See Chapter 12 for details of BIST techniques.) The TAP is then taken to the *Run-Test/Idle* state and held there for a prescribed number of clock cycles. During this period, the

Table 11.3. *A summary of operation for some boundary scan instructions.*

	BYPASS	SAMPLE/ PRELOAD	Instruction EXTEST	INTEST	RUNBIST
Register selected	BypR	BSR	BSR	BSR	Internal chain
Value of *ModeDR*	–	0	1	1	n/a
Mode of TDR during *Update-DR*	–	Normal-sample	Test-capture	Test-capture	–
Mode of TDR during *Shift-DR*	Scan	Normal-scan	Test-scan	Test-scan	Scan
Mode of TDR during *Update-DR*	–	Normal-preload	Test-update	Test-update	Update
Mode of TDR during *Run-Test/Idle*	Idle	Idle	Idle	Idle	Run test
System logic	Normal	Normal	Disabled*	Single-step[†]	BIST[‡]
Comments			Outputs disabled		

* Disabled or designed to operate safely for any combination of input values.
[†] Single-step between the *Update-DR* and *Capture-DR* TAP states.
[‡] System logic run under the control of the BIST controller.

on-chip BIST controller executes self-test. BIST is designed in such a manner that the values applied at the inputs and outputs of the chip do not affect the tests applied and responses captured by the BIST circuitry.

Finally, the above chain is accessed and signatures captured in the RAs (or even one-bit overall pass/fail value) are scanned out.

Table 11.3 provides a summary of operation for the above instructions.

11.5.5   Testing and diagnosing interconnect faults via boundary scan

The boundary scan architecture described before provides the resources to apply tests to and observe responses from the inter-chip interconnects. In this section, we describe how these resources can be used to test and diagnose faults in the interconnects. We begin by describing a model of interconnects. This is followed by a discussion of the type of faults that commonly occur in the interconnects. We then describe the conditions that should be satisfied by the tests applied to detect these faults. Finally, we describe different sets of vectors for test and diagnosis of interconnect faults.

11.5.5.1 A model of the inter-chip interconnect

The interconnect between a set of chips with boundary scan is comprised of all the circuitry and interconnections between the BSCs of any pair of chips. Consider a

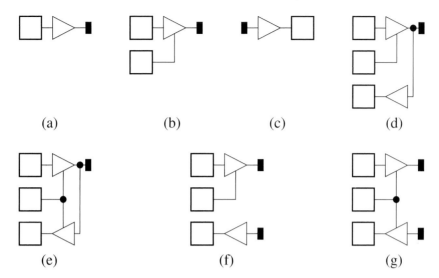

Figure 11.44 Structure of different types of drivers, receivers, and driver/receivers: (a) a simple driver, (b) a tri-state driver, (c) a simple receiver, (d) and (e) single-pin versions of a bi-directional driver/receiver, and (f) and (g) two-pin versions of a bi-directional driver/receiver

simple net in a COB system. Such an interconnect typically begins at the BSC of one chip. The cell drives a pad-driver whose output drives a pad. The pad is connected to a pin of the chip which in turn is soldered to the trace on the PCB. The PCB trace connects this pin of the chip with a pin of another chip. At that point, the trace is soldered to the pin of the other chip. That pin is connected to a BSC via a pad and pad-receiver. The above description can be generalized in a straightforward manner to describe other types of nets.

Most of the physical details of the interconnect can be abstracted in such a manner that each net can be viewed in terms of a number of *drivers*, *receivers*, and wires.

Drivers and receivers: A driver represents the circuitry comprised of one or more BSCs, a pad-driver, pad, and the pin. A receiver represents a pin, pad, pad-receiver, and one or more BSCs.

A **simple driver**, which is also called a **two-state driver**, contains a single BSC, as shown in Figure 11.44(a). In contrast, a **tri-state driver** contains two BSCs, as shown in Figure 11.44(b). The first cell is called a data cell and determines the value at the data input of the pad-driver. The second is called a control cell and determines whether the pad-driver is enabled or disabled. If a value that enables the pad-driver is applied to the control cell, the output of the pad-driver is the value in the data cell. Otherwise, the output of the pad-driver is tri-stated and conceptually has zero drive capacity and infinite output impedance. (For practical reasons, the output impedance of a disabled driver may not be infinite. However, it must be significantly larger than that of an enabled driver.) A **simple receiver**, simply called a **receiver**, contains a single BSC, as shown in Figure 11.44(c). One type of **bi-directional driver/receiver** can be viewed as a combination of a tri-state driver and receiver, as shown in Figure 11.44(d). Note that this design contains one pad-driver and one pad-receiver, and the output of the former and the input of the latter are connected to the same pad and hence the same pin. Another version of bi-directional driver/receiver is shown in Figure 11.44(e). Either of the above versions of the driver/receiver is said to be enabled when it is configured as a driver. When the driver/receiver is disabled, it does not drive any value on the pin; instead it captures the value at the pin implied by some other driver. Versions of bi-directional driver/receivers that use two pins are shown in Figures 11.44(f) and (g). In these versions, when the driver/receiver is enabled, the top pin is driven by the driver; when it is disabled, the top pin is tri-stated and the value applied at the bottom pin is captured.

Types of nets: A **simple net** is driven by a single two-state driver and has one or more receivers. The **in-degree** of a net is the number of drivers of the net. The **out-degree** of a net is the number of receivers of the net. For a simple net, the in-degree is always one, while its out-degree may take any value greater than or equal to one. A **tri-state net** has one or more tri-state drivers and one or more receivers. The in-degree and out-degree of a tri-state net are the number of drivers and receivers, respectively; each value is greater than or equal to one. A **bi-directional net** is one that is connected to one or more bi-directional driver/receivers. In addition, it may contain a number of tri-state drivers and simple receivers. A **non-trivial bi-directional net** should either contain a minimum of two bi-directional driver/receivers or a minimum of one bi-directional driver, one tri-state driver, and one receiver. Also note that non-trivial tri-state and bi-directional nets do not contain any two-state drivers.

For practical reasons, tri-state and bi-directional nets are typically designed in such a manner that if none of the tri-state drivers or bi-directional driver/receivers to the net is enabled, then the net is driven to one of the logic values (logic 0 or logic 1; determined by the design of the circuit) via a weak **keeper** whose output impedance is significantly higher than that of any enabled driver. Hence, when any tri-state driver or bi-directional driver/receiver of a net is enabled, it can easily overpower the keeper and drive the net to the desired value. In many logic families, it is quite common for each receiver to have such a keeper. In a PCB that contains chips from such a family, the receivers of the net capture the value driven by the keeper if the receiver is not driven by any driver. Such a situation arises either because the drivers are all disabled or because a receiver is disconnected from the drivers due to one or more opens in the PCB trace.

11.5.5.2 Fault models for nets

During interconnect testing and diagnosis, the fault is assumed to be confined to parts of the interconnects that are outside the chip. More specifically, the faults are assumed

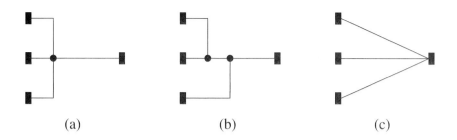

(a)   (b)   (c)

Figure 11.45 Some example structures of a net with four pins

to occur due to faulty soldering of chip pins to the PCB or due to faulty PCB traces. First, consider the soldering between the chip pin and PCB. If an insufficient amount of solder is deposited at the pin, the connection may have significantly higher resistance and may even be open. Oxidation of solder can cause a similar problem. On the other hand, deposition of too much solder at a pin may cause a short circuit between the pin and its neighbors. Next, consider the PCB traces. Defects during PCB manufacture can cause any PCB trace to be open or may cause a short between two PCB traces. It should be noted that the drivers and receivers may be connected using varied patterns of PCB traces. Figure 11.45 shows some example PCB traces that may be used to connect a net connected to four pins.

Based on such physical reasoning, the following types of faults are commonly considered between the inter-chip interconnect.

Stuck-at faults (SAFs): A net may be SA0 or SA1. In the former case, each receiver of the net captures value 0, irrespective of the value driven on the line by any driver driving the net. The behavior of a net with an SA1 fault can be described in an analogous manner.

Opens: An **open** can occur in any PCB trace that connects the pins of a net. The behavior of a net with an open depends on the layout of the PCB traces of the net, location of the open, and characteristics of the receivers. First, consider the case where each receiver has a keeper that drives the input of the receiver to logic 1 whenever the input is not driven by any driver. In such a scenario, consider a driver that is on one side of an open and a receiver that is on its other side. When only this driver is enabled and value 0 driven on the net, value 1 is captured at the particular receiver.

Example 11.27 Consider the net with two drivers and two receivers shown in Figure 11.46(a). For this net and location of open shown, a faulty value is captured at each of the receivers when the driver driving Pin 4 is enabled and drives value 0 on the net while the other driver is disabled. Next, consider an alternative fault location shown in Figure 11.46(b). In this case, the receiver connected to Pin 1 captures an

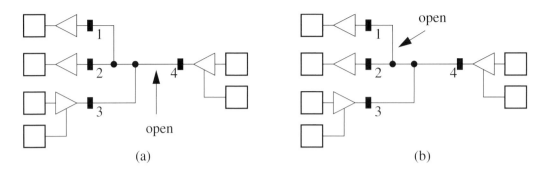

Figure 11.46 Some alternative locations of opens on an example net

erroneous value when either or both drivers are enabled and drive value 0 onto the
net.   □

The case where each receiver has a keeper that drives its input to 0 when no driver
is driving the input is analogous to the above case. For some types of receivers, the
behavior of an open net is analogous to one of the above cases even when no keeper
is used. For example, the behavior of a TTL receiver is similar to the case described
above with a keeper that drives value 1. For some other types of receivers, the value
captured at the receiver may be indeterminate when its input is not driven. In such
cases, it is difficult to test for some open faults. In practice, such cases do not occur
very often and are not considered here.

Shorts: Consider a **short** between two nets. The values captured at the receivers of
the two nets depend upon the type of drivers driving the nets and their strengths. For
example, if the drivers driving the two nets are TTL and if a driver of one net drives
value 0 and that of the other drives value 1, then the receivers of the nets will capture
value 0. In other words, a short between nets in a PCB whose constituent chips use
TTL drivers can be modeled as a **wired-AND**. A **wired-OR** model can be defined in
an analogous manner. The following models of shorts are generalizations of the above
models of shorts. For example, a net i is said to **0-dominate** a net j in the presence
of a short between the two nets, if when a 0 is driven on net i then all receivers of
the two nets capture value 0. The notion of net i **1-dominating** net j can be defined
in a similar manner. Note that a short between two nets can be defined as wired-AND
if each net 0-dominates the other. Finally, in some cases, the value captured at the
receivers of two nets i and j in the presence of a short is always the value driven by
the drivers of net i. Such a short is said to be a **strong driver** short, where net i is said
to have stronger drivers. Consider as an example the special case of a short between
net i and a tri-state net j under the condition that all the drivers of net j are disabled.
Under this condition, the receivers of both the nets will capture the value driven on
net i.

In the presence of shorts, the value at receiver inputs may have some intermediate value in some cases that is neither logic 0 nor 1. Under such cases, testing of shorts is difficult. Such cases are typically avoided via appropriate design of drivers, receivers, and keepers and are not discussed ahead.

11.5.5.3 Vectors for testing

In this section, we describe the tests for detection and diagnosis of faults in interconnects. We begin by describing the conditions to be satisfied by vectors that detect each type of fault.

Single SAFs: To excite an SA0 fault on a net, at least one driver of the net should be enabled and a 1 applied at its data input to drive a 1 onto the net. At least one receiver should then capture the response at its input to detect the fault. The conditions for the detection of an SA1 fault on a net are analogous.

Opens: Consider a net where each receiver has an internal keeper that causes its input to be 1 whenever its input is not driven. In such a case, any open in a simple net can be detected by applying a 0 at the input of the driver to the net to drive a 0 onto the net and observing each of the receivers of the net. Any one of the opens in a tri-state net with l drivers can be detected by applying the following set of l vectors. In the ith vector, the ith driver of the net is enabled, a 0 is applied at this driver's data input, and each of the other drivers is disabled. The response is then captured at each receiver of the net. The l vectors are obtained by enumerating the vectors of the above type for each driver, i.e., for all values of $i = 1, 2, \ldots, l$. The vectors for testing all the opens in a bi-directional net are very similar. In this case, each vector should in turn enable one bi-directional or one tri-state driver and drive value 0 on the net. Furthermore, values should be observed at each receiver as well as each bi-directional driver/receiver that is not enabled.

Vectors to detect each open in a net where each receiver has a keeper that causes its input to be 0 whenever its input is not driven can be obtained by simply replacing value 0 in each of the above vectors by 1.

Shorts: A single SAF as well as a single open fault only affects a single net. In contrast, a short affects multiple nets. Consider a short between a pair of two-state nets, net i and net j. First, assume that the short is of the wired-AND type. Such a short can be detected by applying a 0 and 1 to the drivers of nets i and j, respectively, and observing the response at any of the receivers of net j. It can also be detected by applying a 1 and 0 to the drivers of nets i and j, respectively, and observing the response at any of the receivers of net i. If the short is assumed to be of the type wired-OR, then each of the above conditions can be modified as follows. Instead of observing receivers of the net on which 1 is driven, in this case one or more receivers of the other net are observed. If net i 1-dominates net j, then a 1 and 0 should be driven on net i and j, respectively, and the response observed at one or more receivers of net j. The reader can similarly identify vectors that detect each of the other type of shorts.

Next, consider a wired-AND short between a pair of three-state nets i and j. Assume that each receiver of each net has a keeper that implies value 1 at the receiver's input if the net is not driven. The wired-AND short between these two nets can be detected by enabling any one of the drivers of net i and driving value 0, enabling any of the drivers of net j and driving value 1, and observing the response at one or more receivers of net j. If the in-degree of net i is l_i and that of net j is l_j, then $l_i l_j$ distinct vectors of the above type can be applied to the pair of nets. In addition to vectors of the above type, this short can also be detected by a vector where one driver of net i is enabled and value 0 is driven on the net, all drivers of net j are disabled, and the response observed at any one or more of the receivers of net j. Note that a vector of this type can detect the fault independently of whether the short between the two nets is wired-AND, wired-OR, or one of the dominant types. This is due to the fact that the receiver of a tri-stated net is driven only by a keeper that is much weaker than any enabled driver. Hence, if an opposite value is driven on the other net, that driver can always overpower the keeper via the short. Clearly, the fault can also be detected by a similar vector where net i is disabled and net j is enabled.

A complete set of vectors for detection – two-state nets only: Consider a set of N_{nets} two-state nets. Assume that each receiver has a keeper that assigns value 1 to its own input when the net is not driven by any driver. Also assume wired-AND shorts.

A set of vectors must contain at least one vector that drives value 0 on net i to detect the SA1 fault on the net, for each $i = 1, 2, \ldots, N_{nets}$. Note that for the types of nets under consideration, such a vector also detects any opens in net i. It must also contain at least one vector that drives value 1 at net i to detect an SA0 fault at that net, for each $i = 1, 2, \ldots, N_{nets}$. Furthermore, it must contain at least one vector that applies opposite logic values to nets i and j, for each pair of values of $i = 1, 2, \ldots, N_{nets}$ and $j = 1, 2, \ldots, N_{nets}$, where $i \neq j$.

A minimal set of vectors that satisfies the above conditions for N_{nets} nets is called a **modified counting sequence** (Goel and McMahon, 1982) and contains $N_v^{count} = \lceil \log_2(N_{nets} + 2) \rceil$ vectors of the form $P_1 = (1, 0, 1, 0, 1, 0, 1, 0, 1, 0, \ldots, 1, 0)$, $P_2 = (0, 1, 1, 0, 0, 1, 1, 0, 0, 1, \ldots, 0, 1)$, $P_3 = (0, 0, 0, 1, 1, 1, 1, 0, 0, 0, \ldots, 1, 0)$, and so on. Here P_j has the following form. It contains one run of 0s of length $2^{j-1} - 1$, followed by a run of 1s of length 2^{j-1}, followed by a run of 0s of length 2^{j-1}, and so on. Figure 11.47 shows the $N_v^{count} = 4$ vectors required for $N_{nets} = 8$ nets.

The values applied to various nets by any single vector is sometimes called a **parallel test vector (PTV)**. In contrast, the sequence of values applied to one net

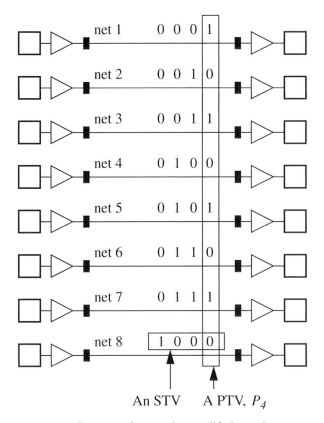

Figure 11.47 Sequence of vectors in a modified counting sequence for an example circuit with eight simple nets

by a sequence of vectors is sometimes called a **serial test vector (STV)** (Jarwala and Yau, 1989). Note that in the order in which the vectors are shown in Figure 11.47, the STV for net i is a binary representation of number i. Also, note that no net is assigned an STV $00\ldots0$ or $11\ldots1$. Hence, an SA0 as well as an SA1 fault at each net is covered. Also, either combination $(0, 1)$ or $(1, 0)$ is applied by some vector to every pair of nets. Hence, wired-AND shorts between any pair of nets are also covered.

One can verify that the modified counting sequence contains $N_v^{\text{modcount}} = \lceil \log_2(N_{\text{nets}} + 2) \rceil$ vectors and covers any combination of SAFs and open faults in any number of nets and wired-AND shorts between any number of nets.

To ensure coverage of more general forms of shorts, it is desired that both the combinations, $(0, 1)$ as well as $(1, 0)$, are applied to each pair of nets. A set of vectors that contains each vector in the modified counting sequence as well as the complement of each vector accomplishes this task. Clearly, such a sequence contains $N_v^{\text{tccount}} = 2\lceil \log_2(N_{\text{nets}} + 2) \rceil$ vectors. We refer to this set of test vectors as the **true/complement counting sequence**.

Table 11.4. *Application of modified counting sequence to an example circuit shown in Figure 11.48.*

Vector	Value applied to cell													
	3	4	5	6	9	10	11	12	13	15	17	19	21	23
P_1	1	e	x	d	0	e	x	d	1	0	1	0	1	0
P_2	0	e	x	d	1	e	x	d	1	0	0	1	1	0
P_3	x	d	0	e	x	d	0	e	0	1	1	1	1	0
P_4	0	e	x	d	0	e	x	d	0	0	0	0	0	1

One can verify that the true/complement counting sequence covers any combination of SAFs, open faults in any number of nets, and shorts of arbitrary type between any number of nets.

A complete set of vectors for detection – general nets: Now consider a general interconnect that includes two-state, three-state as well as bi-directional nets. Again, assume that each receiver has a keeper that assigns value 1 to its input when the net is not driven by any driver. Also, assume general shorts.

The true/complement counting sequence described above can be used to detect each SAF, open, and pair-wise short of general type if it is applied in the following manner. Each time value 0 is applied to a tri-state or bi-directional net i, a different driver of the net is enabled and value 0 implied on the net; at the same time, each of the other drivers of net i is disabled. If the vectors in the above sequence are exhausted before each driver of net i is enabled to drive logic 0 on the net, then some of the vectors in the set can be repeated. Alternatively, a sufficient number of all-zero vectors can be added to the above set of vectors.

Example 11.28 Consider an example interconnect with tri-state nets shown in Figure 11.48 which contains eight nets. The four PTVs of a modified counting sequence can be applied as shown in Table 11.4. For each vector, values applied at each driver and control cell are shown. d and e, respectively, denote the value that disables or enables the corresponding tri-state driver. Following the application of each vector, the response is captured at each receiver. □

11.5.5.4 Vectors for diagnosis

While the above set of vectors is sufficient for detecting faults that are of interest, they do not provide high diagnostic resolution.

Example 11.29 Consider a pair-wise wired-AND short between nets 3 and 5 of the example interconnect in Figure 11.47. In the presence of this fault, the sequence of responses captured at the receivers of net 1 and net 3, as well as net 5, will be identical,

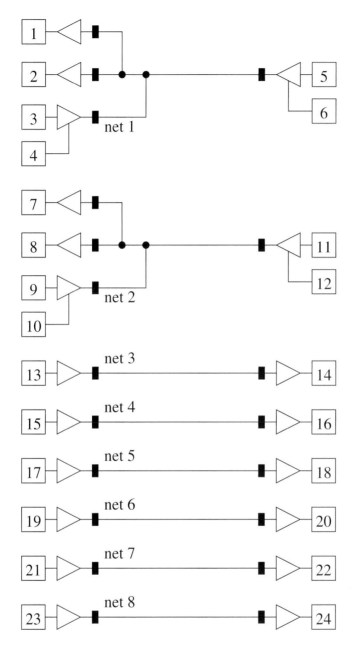

Figure 11.48 An example interconnect used to illustrate how a modified counting sequence is applied to a circuit with a mix of simple and tri-state nets where each receiver has a keeper that drives its input to 1 when it is not driven by any driver

namely 0001. At the receiver of each of the remaining nets, a fault-free response will be captured. The same responses will be captured for a circuit which contains a wired-AND short between nets 1, 3, and 5. Hence, this test is unable to determine which of these shorts has occurred in a PCB with such a faulty response. As shall be seen next, tests that provide higher diagnostic resolution can be found.   □

Next, we discuss a few of the many tests proposed to improve diagnostic resolution. Again, consider a set of N_{nets} two-state nets. Assume that each receiver has a keeper that assigns value 1 to its own input when the net is not driven by any driver. Also, assume wired-AND shorts.

Walking-one test: A walking-one (Hassan *et al.*, 1988) test set contains N_{nets} vectors. Without loss of generality, the vectors are of the form $P_1 = (1, 0, 0, 0, \ldots, 0)$, $P_2 = (0, 1, 0, 0, \ldots, 0)$, $P_3 = (0, 0, 1, 0, \ldots, 0)$, $\ldots$, $P_{N_{nets}} = (0, 0, 0, 0, \ldots, 1)$.

For the above type of nets, this test provides a significantly higher diagnostic resolution. One can verify that for the short in the above example, this test clearly identifies the faulty nets.

True/complement walking test: Again, consider a set of N_{nets} two-state nets and assume that each receiver has a keeper that assigns value 1 to its own input when the net is not driven by any driver. Also, assume shorts of the more general type. As with the counting sequence, the walking-one test can be augmented by adding N_{nets} vectors, each of which is a complement of each of the original walking-one vectors.

Now consider a general set of N_{nets} nets, including tri-state and bi-directional nets. The true/complement walking test can be used to detect each SAF, open, and pair-wise short of a general type if it is applied in the following manner. Each time value 0 is applied to the tri-state or bi-directional net i, a different driver of the net is enabled and value 0 implied; at the same time, each of the other drivers of net i is disabled. In the very unlikely scenario where the maximum in-degree of net i is greater than the number of nets, the vectors in the above test set are exhausted before each driver of net i is enabled to drive logic 0 on the net. In this case, some of the vectors in the set can be repeated. Alternatively, a sufficient number of all-zero vectors can be added to the above set of vectors.

11.6   DFT for other test objectives

The objective of each DFT technique presented thus far in this chapter is to efficiently achieve coverage of static faults, such as SAFs and shorts. In this section, we consider DFT for two alternative objectives. First, we describe DFT techniques for delay fault testing. This is followed by DFT and test development techniques to reduce switching activity during testing.

11.6.1   DFT for delay fault testing

Detection of a delay fault in a combinational circuit requires the application of a **two-pattern test**, say (P_i, P_{i+1}) (see Chapter 8). In a typical delay fault test protocol, the **initialization vector** of the two-pattern test, P_i in this case, is applied and circuit lines are allowed to stabilize. The **test vector**, P_{i+1}, is then applied and the circuit response to P_{i+1} is captured at the system clock speed. To ensure that the initialization attained due to the application of the initialization vector is not disturbed in unexpected ways, no other value is allowed to appear at any input of the combinational logic block between application of the two vectors. Due to the need for consecutive application of **two-pattern tests**, many DFT designs for SAFs need to be revamped for delay fault testing.

11.6.1.1 Delay fault testing using classical scan

The primary inputs of a circuit are completely controllable in the sense that any pair of sub-vectors can be applied in consecutive clock cycles to the primary inputs. In contrast, only a fraction of all possible pairs of sub-vectors can be applied to the state inputs, even in a full-scan design.

Next, we discuss the three main ways in which pairs of vectors can be applied to a combinational logic block in a classical full-scan design.

Application via scan shifting: In the **scan shifting** approach (Cheng *et al.*, 1991b; Patil and Savir, 1992), the flip-flops are configured in the test mode and the state input part of the first vector of the pair is scanned in by applying appropriate values at the *ScanIn* input. The primary input part of the first vector is applied at this time. The flip-flops are then kept in the test mode, an appropriate value is applied at *ScanIn*, and the flip-flops clocked once to apply the next combination of values at the state inputs of the circuit. Simultaneously, the desired values are applied to the primary inputs. The above combination of values at the circuit's primary and state inputs constitutes the next vector applied to the combinational logic.

Only a fraction of all possible pairs of vectors can be applied in this manner, as illustrated next.

Example 11.30 Consider the circuit shown in Figure 11.49, where the combinational logic has two primary inputs, x_1 and x_2, and three state inputs, y_1, y_2, and y_3. Note that the flip-flop driving y_1 is closest to *ScanIn*, followed by those driving y_2 and y_3. Consider a pair of vectors $P_1 = (0, 1, 1, 0, 1)$ and $P_2 = (1, 1, 1, 0, 0)$, where the first two bits of each vector constitute its primary input part to be applied to x_1 and x_2, respectively, and the last three bits constitute its state input part to be applied to y_1, y_2, and y_3, respectively.

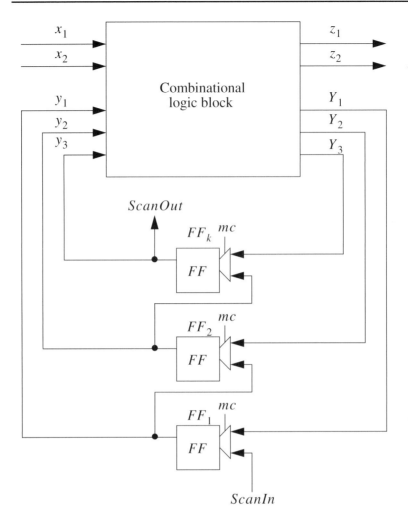

Figure 11.49 An example full-scan circuit

A delay fault test comprised of P_1 followed by P_2 cannot be applied via scan shifting. This is due to the fact that one clock pulse only allows the contents of the scan chain to be shifted by one position and a new value to be scanned into its first stage. Hence, if the flip-flops are configured in their test mode as shown in the figure, then the values $(1, 0, 1)$ at y_1, y_2, and y_3, respectively, can be followed by either $(0, 1, 0)$ or $(1, 1, 0)$. Therefore, only the following vectors can be applied to the circuit after P_1 via scan shifting: $(0, 0, 0, 1, 0)$, $(0, 0, 1, 1, 0)$, $(0, 1, 0, 1, 0)$, $(0, 1, 1, 1, 0)$, $(1, 0, 0, 1, 0)$, $(1, 0, 1, 1, 0)$, $(1, 1, 0, 1, 0)$, and $(1, 1, 1, 1, 0)$. □

Due to the constraints imposed by scan testing, any given combination of k-bit values at the state inputs can only be followed by two out of 2^k possible combination of values. In all, under the scan shifting approach, any given completely specified $(n + k)$-bit vector can be followed by only 2×2^n out of a possible 2^{n+k} vectors.

Application via functional justification: In the **functional justification** approach (Cheng *et al.*, 1991b; Savir and Patil, 1994), the flip-flops are configured in the test mode and the state input part of the first vector of a pair is scanned in by applying an appropriate sequence of values at *ScanIn*. The primary input part of the first vector is then applied. Subsequently, flip-flops are configured in the normal mode and clocked once. At the same time, the primary input part of the second vector in the pair is applied at the primary inputs. Hence, the primary input part of the second vector and state outputs for the first vector are applied to the combinational logic as the second vector of the pair.

Again, only a fraction of all possible pairs of vectors can be applied in this manner as illustrated next.

Example 11.31 Consider the circuit and the pair of vectors described in the previous example. The first vector of the pair, $P_1 = (0, 1, 1, 0, 1)$, can be first applied as described above. Next, the flip-flops are configured in their normal mode and the response at the next state outputs of the combinational logic block is captured in the flip-flops. Assume that the response to P_1 at the state inputs is $(1, 1, 1)$. In this case, the following vectors can be applied following the application of P_1: $(0, 0, 1, 1, 1)$, $(0, 1, 1, 1, 1)$, $(1, 0, 1, 1, 1)$, and $(1, 1, 1, 1, 1)$.

Again, in this case the desired pair of vectors P_1 and P_2 cannot be applied.   □

In general, in the functional justification approach, any completely specified $(n + k)$-bit vector can be followed by 2^n out of 2^{n+k} possible vectors.

Application via hold: In the **hold** approach (Fang and Gupta, 1994), the flip-flops are configured in the test mode and the first vector of a pair is applied using scan. The flip-flops are then configured in a hold mode and the primary input part of the second vector in the pair is applied to the primary inputs. (The flip-flops may be configured in the hold mode either if they are designed to support such a mode or through suppression of the clock using added clock gating circuitry.) Hence, the state input part of the subsequent vector is identical to that for the first vector.

Again, only a fraction of all possible pairs of vectors can be applied in this manner, as illustrated next.

Example 11.32 Consider the circuit and the pair of vectors described in the previous example. The first vector of the pair, $P_1 = (0, 1, 1, 0, 1)$, can be first applied via scan. Next, the flip-flops are configured in their hold mode and, depending on the values applied to the primary inputs, the application of P_1 may be followed by the application of: $(0, 0, 1, 0, 1)$, $(0, 1, 1, 0, 1)$, $(1, 0, 1, 0, 1)$, and $(1, 1, 1, 0, 1)$.

Again, in this case the desired pair of vectors P_1 and P_2 cannot be applied.   □

In general, in the hold approach, any completely specified $(n + k)$-bit vector can be followed by 2^n out of a possible 2^{n+k} vectors.

11.6.1.2 Summary of test application methodologies

In general, in the best case, assuming that each of the three test application methodologies can be used, any completely specified $(n + k)$-bit vector can be followed by no more than 4×2^n out of a possible 2^{n+k} vectors.

Despite the ability to apply only a fraction of all possible pairs of vectors, typically moderate to high delay fault coverage can be attained using these methodologies. This is due to the fact that the vectors for most target faults are incompletely specified. Often, bits of an incompletely specified vector P_1 can be specified in suitable ways to enable the application of the desired P_2. First, each unspecified value, x, in the state input part of vector P_1 may sometimes be specified so as to enable the application of the desired P_2 via scan shifting or hold. Second, each x in the primary and state input parts of P_1 may sometimes be specified to facilitate the application of the desired P_2 via functional justification.

The following example illustrates how many of the bits of a two-pattern test for a given target delay fault may be incompletely specified.

Example 11.33   A pipeline kernel is an extreme case where a large number of values are left unspecified.

Consider the combinational logic block CLB_2 in the circuit shown in Figure 11.13(a). To test a path within CLB_2, only the values at the inputs of the block are specified in the initialization vector, say P_1, as well as in the test vector, say P_2.

P_1 requires specific values to be scanned into FF_4, FF_5, and FF_6 and leaves the values at FF_1, FF_2, and FF_3 unspecified. When possible, values can be assigned to these flip-flops in such a manner that P_1 produces at the outputs of CLB_1 a response that loads into FF_4, FF_5, and FF_6 the values required by the test vector, P_2, for the target delay fault within CLB_2. $\square$

11.6.1.3 Enhanced scan for delay fault testing

A combination of the above test application methodologies may fail to provide the desired coverage of delay faults despite the exploitation of the incompletely specified vectors in the manner described above. This is especially true when outputs of the combinational logic under test feed back to its inputs and a more stringent delay fault model, such as the *path delay fault model* (see Chapter 8), is used. In such cases, the scan chain design may be changed in the following ways to achieve the desired coverage of delay faults.

Enhanced scan cells: One alternative is to use an **enhanced scan cell** instead of the scan flip-flops described earlier. A conceptual schematic of one such design is shown in Figure 11.50. (The actual implementation can be very different so as to minimize area overhead and performance penalty.) This scan cell design is similar to the cell

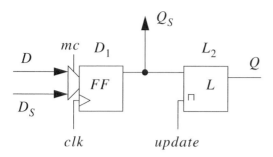

Figure 11.50   An example circuit implementing the functionality desired of an enhanced scan cell

design for a boundary scan register and is used in a similar manner. A state output Y_i is connected to the D input of the cell while the cell's output Q is connected to the corresponding state input, y_i. The D_1 flip-flops within the enhanced scan cells are connected to form a scan chain by connecting the D_S input of one cell to the Q_S output of the previous cell in the scan chain. The mode control signal mc and the normal circuit clock control the operation of the D_1 flip-flops. The new control signal *update* determines whether latch L_2 is in the hold or transparent mode.

During normal circuit operation, value *normal* is applied to the control signal mc and *update* is held high. In the test mode, if value *test* is applied to mc while *update* is held low, then the clock can be used to scan in values into the scan chain formed by D_1 flip-flops while holding the state input part of the last vector at the state inputs of the circuit. Once the state input part of the next vector is scanned in, it can be applied to the circuit's state inputs by making *update* high.

Clearly, the use of an enhanced scan cell enables consecutive application of any pair of vectors. Hence, use of enhanced scan guarantees the coverage of all testable delay faults in the circuit. The cost of such a design can be high, however. In the example cell design shown in Figure 11.50, each normal flip-flop in the circuit is supplemented by a latch and some additional control circuitry. This causes area overhead as well as performance penalty. These can be reduced somewhat by using a more intricate cell design (e.g., Dervisoglu and Stong (1991)). However, the overheads due to an enhanced scan cell are higher than those for a conventional scan cell. Also, additional routing area is required for the *update* signal.

Chains with additional flip-flops: An alternative scan chain design that can guarantee the coverage of all testable delay faults in a circuit can be obtained using standard scan cells, as shown in Figure 11.51 (Patil and Savir, 1992). In this approach, additional flip-flops can be inserted between any two scan flip-flops whose outputs drive state inputs of the combinational logic block. Let such a scan flip-flop be called an **original flip-flop**, say FF_i. Let the flip-flop inserted before FF_i in the scan chain be called the corresponding **dummy flip-flop**, FF_i^*.

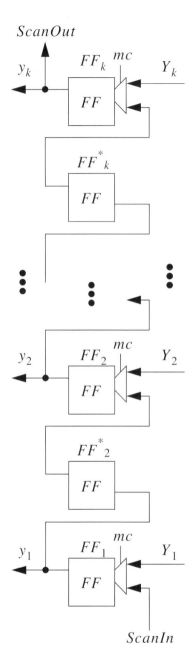

Figure 11.51 A scan chain with dummy flip-flops

The above scan chain design can be used for consecutive application of a pair of vectors, P_1 and P_2, to the combinational logic block, as follows. Configure the original as well as dummy flip-flops in the test mode. Apply corresponding bits of the state input parts of the two vectors in an interleaved manner until bit i of the state input part of P_1 is shifted into the original flip-flop FF_i and the corresponding bit of the state input part of P_2 is shifted into the dummy flip-flop FF_i^*. Apply the primary input part of P_1 to the primary inputs of the circuit. This results in the application of bits of P_1 to the inputs of the combinational logic block. Subsequently, apply the value to be applied at y_1 for P_2 at ScanIn and apply one clock pulse to the scan chain to shift its contents by one step. This results in the application of the state part of P_2 to the state inputs of the combinational logic. Simultaneous application of the primary input part of P_2 completes the application of the entire P_2 to the inputs of the combinational logic block. Such a scan chain guarantees that all testable delay faults in the combinational circuit can be tested. The extra area overhead over the classical scan approach is due to the addition of dummy flip-flops.

The original flip-flops in the chain can be ordered and some dummy flip-flops eliminated while ensuring that the two-pattern tests for the target delay faults can be applied to the combinational logic block. This can help reduce the area overhead of this approach (Cheng *et al.* 1991a,b; Patil and Savir 1992; Chen and Gupta, 1998).

11.6.2 Reduction of switching activity during scan testing

As described in Chapter 4 (Section 4.9), the switching activity in a circuit during test application can be significantly higher than that during the circuit's normal operation. This can sometimes cause heat dissipation that is severe enough to destroy some fault-free chips. It can also cause power/ground noise which may cause some fault-free chips to fail the test.

Next, we describe two approaches for reducing the switching activity during test application. We begin with a discussion of the main causes of high switching activity.

11.6.2.1 Switching activity during scan testing

Consider a full-scan version of the sequential circuit shown in Figure 11.1. Assume that its flip-flops are organized into a single scan chain, as shown in Figure 11.3(a).

Consider the application of vector $P_i = (p_{1,i}, p_{2,i}, \ldots, p_{n,i}, p_{n+1,i}, p_{n+2,i}, \ldots, p_{n+k,i})$ following the application of vector P_{i-1}. Recall that values $p_{1,i}, p_{2,i}, \ldots, p_{n,i}$ constitute the primary input part of test vector P_i, and values $p_{n+1,i}, p_{n+2,i}, \ldots, p_{n+k,i}$ constitute the state input part of the vector.

As described in Section 11.2.1.3, once P_{i-1} is scanned in, it is applied to the combinational logic. Subsequently, the response of the combinational logic to P_{i-1} is captured. In the next k clock cycles, the response to P_{i-1} is scanned out while the bits of the state input part of P_i are scanned in.

Example 11.34 Consider a full-scan version of a circuit with three primary inputs and four state inputs whose flip-flops are configured into a single chain as shown in

Table 11.5. *Example illustrating application of vectors via scan*

	Value at input							Action
	x_1	x_2	x_3	y_1	y_2	y_3	y_4	
1	0	0	1	0	0	1	0	P_{i-1} applied
2	x	x	x	0	1	0	1	Response captured
3	x	x	x	x	0	1	0	Scan shift
4	x	x	x	1	x	0	1	Scan shift
5	x	x	x	1	1	x	0	Scan shift
6	0	x	1	0	1	1	x	P_i applied

Figure 11.3(a). Assume that $P_{i-1} = (0, 0, 1, 0, 0, 1, 0)$ and $P_i = (0, x, 1, 0, 1, 1, x)$. Also, assume that $(0, 1, 0, 1)$ is produced at the state outputs of a fault-free circuit in response to the application of P_{i-1}.

Table 11.5 summarizes the process of applying the above sequence of vectors. The first row of the table depicts the application of P_{i-1}. The second row shows the response capture cycle. This is followed by $k - 1$ clock cycles in which the bits of the response for P_{i-1} are scanned out of, and the bits of the state input part of P_i are scanned into, the flip-flops. In the kth cycle, the last bit of the state input part of P_i is scanned in and the primary input part of P_i is applied. This leads to application of P_i to the inputs of the combinational logic. The above process is then repeated.   □

The above example illustrates the following general characteristics of test application to a full-scan circuit.

Importance of scan shifting cycles: k out of $k + 1$ clock cycles required for the application of a test vector are spent scanning out the response to one vector and scanning in the state input part of the next. Consequently, a large fraction of switching activity during the application of a vector occurs during scan shifting.

Unspecified values: As the table illustrates, a large number of input values are unspecified. First, the values at the primary inputs during every cycle other than the test application cycles are completely unspecified and not determined in any manner by P_i. Second, an ATPG algorithm may generate a vector P_i for a target fault with some bits left unspecified. An unspecified value at a primary input (such as that at x_2 for P_i in the above example) appears only for one clock cycle. In contrast, an unspecified value at a state input (such as that at y_4 for P_i) appears during multiple clock cycles, each time at a different state input. As will be seen later, these unspecified values can be replaced by specific values to reduce the switching activity during scan testing.

Table 11.6. *Example illustrating application of vectors via enhanced scan*

	x_1	x_2	x_3	FF_1	FF_2	FF_3	FF_4	y_1	y_2	y_3	y_4
					Value at						
1	0	0	1	0	0	1	0	0	0	1	0
2	0	0	1	0	1	0	1	0	0	1	0
3	0	0	1	x	0	1	0	0	0	1	0
4	0	0	1	1	x	0	1	0	0	1	0
5	0	0	1	1	1	x	0	0	0	1	0
6	0	x	1	0	1	1	x	0	1	1	x

11.6.2.2 Reduced switching activity using enhanced scan cells

As indicated above, a large fraction of switching activity occurs during the k cycles used primarily for scan shifting. This fact can be exploited to significantly reduce switching activity during scan testing using two key ideas (Gerstendorfer and Wunderlich, 1999).

First, each scan cell can be replaced by an enhanced scan cell of the type shown in Figure 11.50. The use of such a scan cell enables the response to a vector, say P_{i-1}, at the state outputs to be captured while holding the values at the state inputs of the combinational circuit. This is also the case when the captured response is scanned out and the state input part of the next vector, P_i, is scanned in. Only during the test application cycle do the values at the state inputs need to be changed, via application of logic 1 at the *update* control inputs of the cells.

Second, the primary input part of P_{i-1} can be applied to the primary inputs until the cycle in which the primary input part of P_i needs to be applied.

Example 11.35 Consider the circuit, vectors, and the response described in Example 11.34. Assume that enhanced scan cells are used instead of the conventional scan cells. Also, assume that the approach described above is used to apply test vectors.

Table 11.6 shows the details of test application. Note that the values at the inputs of the combinational logic block change only during each test application cycle. Hence, all switching activity in the circuit during the scan shifting cycles is eliminated.   □

The above approach reduces the average switching activity during scan testing by eliminating the switching activity during all the cycles except the test application cycles. The switching activity during the test application cycles can be reduced by using the ATPG technique described in Chapter 4 (Section 4.9) to generate test vectors.

The above benefits are accrued through the use of enhanced scan cells instead of classical scan cells. As described earlier, the use of these cells causes higher area overhead and performance penalty.

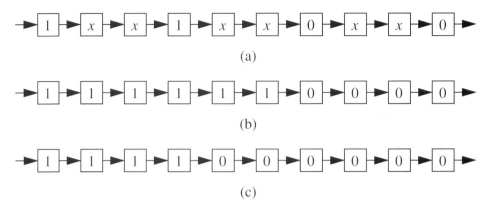

(a)

(b)

(c)

Figure 11.52 Assignment of unspecified values in the state input part of a vector: (a) original vector, (b) and (c) two alternative assignments

11.6.2.3 Reducing switching activity using classical scan cells

In this approach, reduction in switching activity is accomplished through generation of certain types of test vectors for target faults and assigning suitable values to the unspecified values described in Section 11.6.2.1 (Wang and Gupta, 1997).

As described earlier, during scan testing, the primary input part of a vector needs to be applied to the circuit inputs only during one clock cycle, namely the test application cycle. In contrast, bits of the state input part of the same vector must be scanned in over several clock cycles. Hence, an undesirable value in the state input part of a vector generated by an ATPG algorithm has the potential to cause switching activity for many clock cycles while that in the primary input part need only cause activity during one clock cycle.

Based on the above fact, it is desirable to generate a test vector for a target fault that assigns specific values, i.e., 0 or 1, to a minimum number of state inputs. Furthermore, if a choice exists between assigning a specific value at two state inputs, then the one closer to the *ScanIn* input should be chosen, since the value to be applied to this input can cause switching activity for fewer scan shifting clock cycles. This can be accomplished by using a controllability cost function that assigns low controllability values to the primary inputs and progressively increasing controllability values to the state inputs in the order in which the corresponding flip-flops are placed in the scan chain. The details of such an ATPG algorithm can be found in Wang and Gupta (1997).

Once a test vector is obtained, the unspecified bits in its state input part are specified in a manner illustrated in Figure 11.52 so as to assign identical values to the state inputs driven by adjacent flip-flops in the scan chain. Clearly, this helps reduce the switching activity at the state inputs during scan shifting cycles.

Further reduction in switching activity is obtained by assigning specific values to the primary inputs during all cycles other than the scan shifting cycles. These values can be assigned so as to *block* the transitions caused at the state inputs from propagating

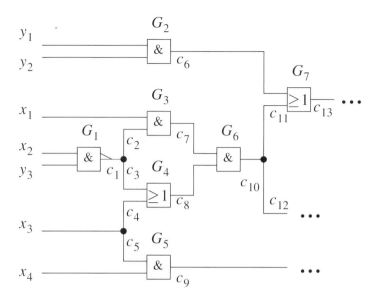

Figure 11.53 An example circuit to illustrate blocking objectives

well into the combinational logic block. The first key concept in this approach is that of a *blocking objective*. A gate G_j is said to be a **blocking objective** if one or more of its inputs are in the transitive fanout of the primary inputs while one or more of its inputs are not.

Example 11.36 Consider the sub-circuit shown in Figure 11.53 which has three state inputs, y_1, y_2, and y_3, and four primary inputs, x_1, x_2, x_3, and x_4.
    In this circuit, gates G_1 and G_7 are the two blocking objectives.   □

    Collectively, blocking objectives are gates that are closest to the state inputs at which a transition caused at those inputs can possibly be blocked from propagating further by applying fixed values at the primary inputs. Note that in the above example, gate G_2 is not a blocking objective since it is not possible to assign values to the primary inputs to block the transitions caused at this gate due to shifting of the contents of the scan chain. Detailed definitions of the concept of blocking objectives and its generalizations can be found in Wang and Gupta (1997).
    An ideal combination of values at the primary inputs during scan shifting is one that blocks each gate that is a blocking objective by implying the gate's controlling value at one or more of its inputs, independent of the value applied to the state inputs.

Example 11.37 In the sub-circuit shown in Figure 11.53, any transitions that may occur at the state inputs during scan shifting can be blocked by assigning logic 1 at

x_1 and logic 0 at x_2. Note that assignment of logic 0 at x_2 blocks any transition at y_3 from propagating past blocking objective G_1. Also, note that collectively the above two assignments imply logic 1 at input c_{11} of blocking objective G_7. If these values are applied at the primary inputs during scan shifting clock cycles, then any transition caused at the state inputs cannot create transitions at the gates in the transitive fanouts of the gates that are blocking objectives. □

Clearly, assignment of a combination of values to the primary inputs during scan shifting cycles blocks any switching activity at one or more state inputs as early as possible. Procedures to find good assignments are described in Wang and Gupta (1997).

The following example illustrates the entire methodology.

Example 11.38 Consider the sub-circuit shown in Figure 11.53 and discussed in the previous examples.

First, the test generator outlined above is used to generate a vector that specifies values at a minimum number of state inputs. For example, consider a target fault in the part of the circuit not shown in this figure whose detection requires logic 1 at lines c_{12} and c_{13}. A vector $(1, 0, x, x, x, x, x)$, where the value at each state input is left unspecified, may be obtained to detect this fault.

Second, the unspecified values in the state input part may be specified. In this example, depending on the response captured in the scan flip-flops for the previous vector, every x in the state input part is assigned logic 0 or logic 1.

Finally, during each scan shifting cycle, the sub-vector $(1, 0, x, x)$ is applied at primary inputs to block any transitions caused at the state inputs. □

Summary

- DFT can informally be defined as changes to a given circuit design that help decrease the overall difficulty of testing. In some cases, DFT may be essential to achieve a desired fault coverage.
- Scan design enables direct control of state inputs and direct observation of state outputs. Partial scan enables this for selected flip-flops while full scan enables this for all the flip-flops. Due to this fact, full scan enables tests to be generated considering only the combinational part of the given circuit and using a test generator for combinational circuits.
- In an ideal flip-flop-based design, each flip-flop may be replaced by a multiplexed-input scan flip-flop and appropriate connections made to configure the flip-flops into one or more scan chains. A control input mc can be used to configure the flip-flops in the normal mode ($mc = normal$) or as scan chains ($mc = test$).

- $N_v(k + 1) + k - 1$ clock cycles are required to apply N_v test vectors via scan to a circuit with k state inputs that has full scan with a single chain.
- In an ideal level-sensitive double-latch design, an LS scan element that uses an additional test clock instead of a mode control input can be used. In this case, the LS scan element replaces the master latch of each double-latch by a two-port latch that uses two clocks.
- In a single-latch design, the LS element requires significantly higher overhead. This is because an LS element must be added to each latch in the original design.
- In non-ideal circuits, clocks and logic values may interact in many ways. In such circuits, scan and functional circuitry must satisfy additional conditions to enable scan testing.
- Scan helps attain higher fault coverage at the expense of area overhead and performance penalty. Furthermore, test application time may increase. Area overhead may lead to lower yield; performance penalty may be reduced at the expense of a further increase in area overhead.
- Scan can help not only during high-volume manufacturing testing but also for debugging first-silicon for design errors.
- Linear pipelines, general pipelines, and balanced circuits represent sequential kernels for which vectors can be generated using test generators for combinational circuits. Generated vectors must, however, be applied to any such kernel in a specific manner.
- Acyclic kernels require sequential test generators. However, the length of a sequence of vectors required to detect a given target fault is bounded by the sequential depth of the circuit.
- The s+graph of a sequential circuit describes the sequential nature of the circuit and hence can be used to identify flip-flops to scan under partial-scan methodologies that create kernels with specific structural properties.
- Testability measures, such as controllabilities and observability, can be used to estimate the degree of difficulty of detecting each target fault in a given partial-scan design.
  1 The above estimates can be used to compute a figure-of-merit that may be used to compare alternative partial-scan designs.
  2 Classical testability measures or empirical testability measures may be used for computing this figure-of-merit.
  3 The above figure-of-merit may be used to identify flip-flops for scan. It can be used in conjunction with an iterative greedy approach to select flip-flops.
- If only a combinational test generator is available, the output and input of each flip-flop can be assumed to be uncontrollable and unobservable, respectively. For a given partial-scan design, the number of target faults that can be detected using a combinational test generator can be estimated using uncontrollability and unobservability analysis.

- In a partial-scan design, scan and non-scan flip-flops are typically designed in such a manner that when values are scanned into scan flip-flops, the non-scan flip-flops hold their state values.
- In an ideal flip-flop-based circuit, the flip-flops can be grouped into registers and gates into combinational logic blocks.
- A register may be a pure driver, a pure receiver, or a driver/receiver.
- Selection of registers for scan helps partition the circuit into kernels. Each kernel contains one or more combinational logic blocks. A kernel may also contain one or more registers.
- The kernels in a circuit can be tested over a number of test sessions. Combined, overlapped, and isolated test application schemes may be used. The overlapped scheme is optimal for many scan shift policies.
- Scan shift policies include flush, active flush, minimum shift, and active minimum shift.
- Registers can be ordered into a single scan chain for the overlapped test application scheme under the minimum shift policy. They can also be assigned to multiple chains and ordered within each chain for the overlapped test application scheme under the active flush policy.
- Multiplexers can be used in conjunction with registers to obtain reconfigurable scan chains.
- In some cases, a combination of non-scan and scan testing leads to a lower test application time compared to testing exclusively via scan.
- The objectives of board-level DFT are to test inter-chip interconnects, retest chips in-situ, and to provide support for debugging system design.
- The boundary scan standard uses a four(optionally, five)-pin port called test access port (TAP) to input/output control and test data.
- In conjunction with the instruction register, TAP controller, and control logic, the control and test data inputs generate control signals — control inputs and gated clocks — for all the boundary scan test logic within a chip.
- The instruction register contains a shift register and an output stage. This enables a new instruction to be completely shifted in place before it is applied.
- A boundary scan cell is placed at each two-state logic input and each two-state logic output pin of a chip. At tri-state and bidirectional pins, cells are not only placed at the logic pins but also at the control signals.
- The boundary scan register can be configured in normal, normal-sample, normal-scan, normal-preload, test-capture, test-scan, and test-update modes. Most key instructions involving boundary scan use some combination of these modes.
- Mandatory instructions and several optimal instructions provide mechanisms to apply tests and capture responses. They also provide mechanisms to control DFT and BIST circuitry within a chip.

- The modified counting sequence provides an effective way to detect faults in inter-chip interconnects. Modified walking tests provide comprehensive diagnosis of interconnect faults.
- Delay fault testing in a full-scan circuit requires consecutive application of a sequence of two vectors to the combinational logic. In a standard full-scan design, scan shifting, functional justification, and hold approaches can be used. Cumulatively, these approaches enable the application of no more than 4×2^n out of a possible 2^{n+k} fully-specified vectors following the application of another fully-specified vector.
- If many bits of two-pattern tests for target delay faults are unspecified, then high delay fault coverage may be achieved using standard full-scan design. This may not be the case if the outputs of a combinational logic block under test feed back to its inputs.
- Enhanced scan cells enable the application of any two-pattern test. Alternatively, dummy flip-flops may be inserted in a standard scan chain to enable the application of any two-pattern test.
- Much of the switching activity during test application via scan occurs during scan shifting when the values at the primary inputs are unspecified.
- Switching activity during scan shifting can be reduced by using enhanced scan cells. Alternatively, switching activity can be reduced under the standard scan design by (a) generating vectors for target faults with a maximum number of unspecified values in their state input parts, (b) specifying these values so as to reduce the number of transitions during scan shifting, and (c) applying values to the primary inputs of the circuit during scan shifting to further reduce the number of transitions.

Additional reading

Heuristics/techniques to find minimum feeback vertex sets are described in Ashar and Malik (1994), Chakradhar *et al.* (1994), and Lee and Reddy (1990). Many other approaches for partial scan have been reported. In Min and Rogers (1990), another approach based on pipeline kernels is presented. In Park and Akers (1992), kernels with cycles of length K or less are considered, while in Bhawmik *et al.* (1991), partial-scan flip-flops are selected to eliminate strongly connected components in the s+graph. Re-timing techniques can be used to change the number and location of flip-flops in a sequential circuit. This changes the s+graph of the circuit and often decreases the number of flip-flops that must be selected for partial scan to obtain kernels with a desired structure. A combination of re-timing and elimination of all cycles other than self-loops is used in Kagaris and Tragoudas (1993), Dey and Chakradhar (1994), and Chakradhar and Dey (1994) to reduce the overheads of partial scan.

A partial-scan design where the scan as well as non-scan flip-flops use the same clock is presented in Cheng (1995). Efficient scan chain designs are presented in Kim and Schultz (1994) for circuits that use multiple clocks with different frequencies. A partial-scan approach that minimizes the performance overhead of scan is presented in Lin *et al.* (1996).

Several other DFT approaches to reduce test application time have been reported. In Higami *et al.* (1994), techniques are presented to order scan flip-flops and minimize the shift-cycle by generating vectors that minimize the number of flip-flops that are active for detecting specific target faults. In Kanjilal *et al.* (1993), Hsu and Patel (1995), and Kuo *et al.* (1995), techniques are presented to analyze the state transition graph (STG) of an FSM to improve the test generation and test application costs by reducing some measure of distance between states in the FSM's STG. Techniques to reduce test application time by taking advantage of asynchronous multiple scan chains are proposed in Narayanan and Breuer (1995a). An approach where a parity tree is added at the state outputs to enhance observability of state outputs and reduce cost of test application via scan is presented in Fujiwara and Yamamoto (1992).

Many non-scan DFT techniques have been proposed. The technique in Chickermane *et al.* (1993) adds test points, the technique in Abramovici *et al.* (1993) adds the capability to reset subsets of flip-flops to facilitate traversal of the state space of the sequential circuit, the technique in Rajan *et al.* (1996) transforms the clock, and that in Baeg and Rogers (1994) uses scan as well as clock line control.

Designs of controllers that can be added to a PCB to facilitate testing via boundary scan are presented in Yau and Jarwala (1990) and Lien and Breuer (1989). Approaches for integrating and re-using boundary scan at the system level are discussed in IEEE (1995), Whetsel (1992), and Bhavsar (1991).

Other tests for boundary scan interconnects are presented in Cheng *et al.* (1990). Approaches for diagnosis of interconnect faults are presented in Yau and Jarwala (1989), Lien and Breuer (1991), and Shi and Fuchs (1995).

Techniques to obtain high delay fault coverage while using a minimum number of dummy flip-flops are presented in Cheng *et al.* (1991a,b), Patil and Savir (1992), and Chen and Gupta (1998).

Exercises

11.1 A general model of a sequential circuit is shown in Figure 11.1. Consider a specific circuit that has $n = 4$ primary inputs, $m = 3$ primary outputs, and $k = 3$ state inputs and outputs. Assume that test generation has been performed on the combinational logic block of this circuit to obtain the following test vectors and responses.

Show the details of the scan chain design using a multiplexed-input scan flip-flop. Also, describe how the above vectors are applied and the corresponding response observed on a cycle-by-cycle basis.

Vectors							Responses					
x_1	x_2	x_3	x_4	y_1	y_2	y_3	z_1	z_2	z_3	Y_1	Y_2	Y_3
1	0	1	1	0	1	1	0	0	1	1	1	0
0	0	0	1	0	1	0	1	1	1	0	1	0
0	1	1	1	1	0	1	0	1	1	0	0	1

11.2 Consider the sequential circuit shown in Figure 5.4.

(a) Generate a set of vectors to detect all single SAFs in a non-scan version of this circuit.

(b) Show the details of a full-scan version of this circuit obtained using the multiplexed-input scan flip-flop.

(c) Generate a set of vectors that can be used to detect all single SAFs in the full-scan version of this circuit.

(d) Compare the test application for the non-scan version of the circuit with that for its full-scan version.

11.3 Repeat above two problems using the LS scan element. In each case, start by first modifying the original design to use the master–slave flip-flop shown in Figure 11.6.

11.4 Design a three-bit, modulo-eight asynchronous counter using master–slave D-flip-flops shown in Figure 11.6 and logic gates.

Discuss the issues faced in making this circuit testable using the LS scan elements. Show the details of your scan design.

11.5 Consider the scan design obtained in Problem 11.2. Assume that each gate with α inputs has a delay of α units.

(a) Show a detailed gate-level design (including gate-level details of the scan flip-flop) of the full-scan version. What is the minimum clock period at which the full-scan version of this circuit can be operated? What is the performance penalty of using full-scan DFT in this manner?

(b) Redesign the combinational logic along with the multiplexers used in the scan flip-flops to reduce the clock period of the full-scan version of this design.

(c) Compare the area overheads (approximately measured as the number of gates) and performance penalties of the two full-scan versions described above.

11.6 Consider the circuit in Figure 11.54 where each cell called FA is a one-bit full-adder whose scan output is a primary output and carry output drives the carry input of the next stage (if any).

(a) Is this circuit a linear pipeline kernel? Is this a balanced kernel?

(b) Design a full-adder cell using AND, OR, NAND, NOR, and XOR gates and inverters. Generate a set of vectors that detect all single SAFs in a single full-adder cell, assuming that all its inputs and outputs are controllable and observable, respectively.

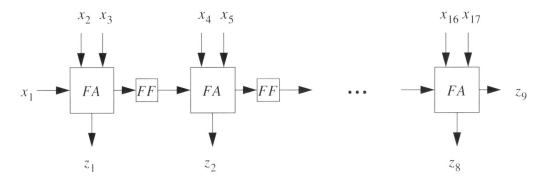

Figure 11.54 An example sequential circuit

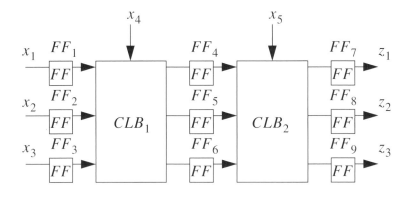

Figure 11.55 An example kernel

(c) Use the above full-adder cell in the given circuit to obtain a detailed description of the circuit. How can the test vectors generated for each full-adder cell be applied to the non-scan version of the given design?

11.7 Consider the example kernel shown in Figure 11.55.

(a) Draw an s+graph for this kernel.

(b) Is this kernel a pipeline? Is this kernel balanced?

(c) Show the circuit model that can be used to generate tests for this kernel using a combinational ATPG algorithm.

(d) Assume that a combinational ATPG algorithm gives you vectors (1, 1, 1, 0, 0) and (0, 0, 0, 1, 1). Reformat these vectors, clearly showing the test application clock cycle in which each bit of the above vectors is applied.

11.8 Consider the example circuit shown in Figure 11.56.

(a) Select a minimum number of flip-flops to scan that would give a balanced kernel. (You may use a trial-and-error approach for this step.)

(b) Show the detailed design of a partial-scan version of the circuit obtained by replacing the flip-flops selected in the above step by scan flip-flops.

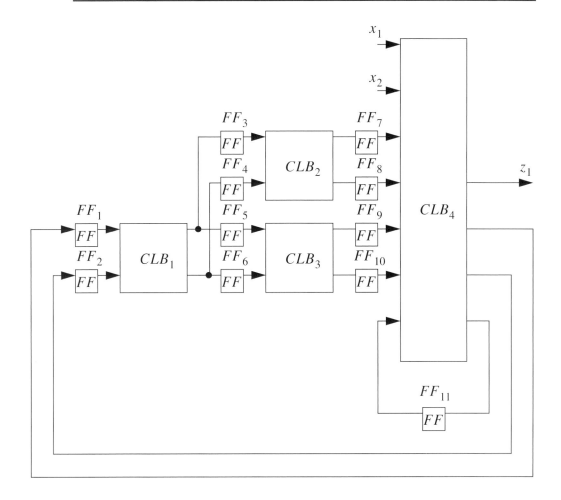

Figure 11.56 An example sequential circuit

    (c) Clearly describe how a set of vectors generated using a combinational ATPG algorithm can be applied to the partial-scan version of the circuit.

11.9   Select a minimum number of flip-flops for scan in the circuit in the above problem to obtain one or more kernels that are acyclic with the exception of self-loops. Show the s+graph of the original circuit as well as that of the partial-scan version obtained.

    How does the overhead of this partial-scan design compare with that obtained in the above problem?

11.10 A sequence of test vectors must be generated to detect a single SAF in combinational logic block CLB_3 of the circuit shown in Figure 11.57.

    (a) For each primary input, identify clock cycles (e.g., t, $t + 1$, and so on) at which specific values may be required at that input.

    (b) Relative to the above clock cycles, identify the clock cycles at which an

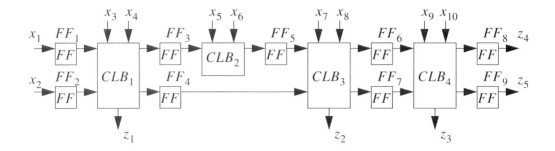

Figure 11.57 An example acyclic kernel

> erroneous response (if any) due to the target fault would be expected at each primary output.
>
> (c) What is the maximum number of distinct clock cycles for which specific values may be required at any single input to detect this fault? Is this number smaller than the sequential depth of the circuit? Explain your answer.

11.11 Generalize the labeling procedure described in Section 11.3.1.2 for formatting vectors for balanced kernels to perform the task described in part (a) of the above problem.

11.12 Derive a truth table that gives the uncontrollability status of the output of a two-input XOR gate, given the uncontrollability status of each of its inputs.

11.13 Consider the circuit shown in Figure 11.58 and a target fault list containing each single SAF in its combinational logic. Use uncontrollability and unobservability analysis to select one flip-flop for partial scan. Repeat the above analysis and select another flip-flop for partial scan. Show details of your analyses and the final partial-scan design.

11.14 Consider the circuit shown in Figure 11.59 along with the number of flip-flops in each register and the number of test vectors for each kernel.
Organize the registers into a single scan chain that minimizes the test application time under the overlapped test application scheme and minimum-shift policy.

11.15 Repeat the above problem assuming that kernel K_1 requires 1000 test vectors instead of 30.

11.16 Design a single reconfigurable scan chain that uses a single control signal and minimizes the test application time under the overlapped test application scheme and minimum-shift policy for the circuit shown in Figure 11.59.

11.17 Assuming that individual flip-flops of a register can be separated and assigned to different scan chains, design two scan chains that minimize test application time for the circuit in Figure 11.59 under the overlapped test application scheme and active flush policy.

11.18 Repeat the above problem for active minimum shift policy.

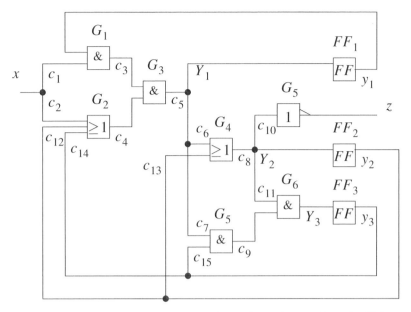

Figure 11.58 An example sequential circuit (Chakradhar and Dey, 1994) (© 1994 ACM)

11.19 Assume that a tree of XOR gates is added to the circuit to observe the response at the kernel outputs for each vector, and scan flip-flops are no longer required to capture and observe the responses to test vectors. Repeat the scan chain designs in Problems 11.14, 11.15, 11.17, and 11.18.

11.20 For each of the scan chain designs obtained in Problem 11.19, is it possible to use a smaller XOR tree that observes only the response at some kernel outputs while the scan flip-flops are used to observe the response at the remaining outputs? Comment on the implications for these examples.

11.21 Assuming that a chip has a four-bit instruction register, identify the values applied at the TMS, TDI, and TCK inputs to load and apply the BYPASS instruction stipulated by the boundary scan standard.
Identify the clock cycles in which TDO is **not** tri-stated. What values appear at TDO during each of these clock cycles?

11.22 Show versions of the BSR cell shown in Figure 11.35 that are suitable for (a) a two-state logic input pin, and (b) a two-state logic output pin. Assume only mandatory instructions need to be supported.

11.23 Use the version of the BSR cell described in the text along with those obtained in the above problem to obtain a detailed design of the BSR for the chips shown in Figure 11.29. Again assume that only mandatory instructions need to be supported.

11.24 Consider a board that contains three chips that are to be re-tested in-situ using the INTEST instruction. Assume that each chip has a two-bit instruction register and uses 01 as the op-code for INTEST. Further assume that Chip 1, Chip 2,

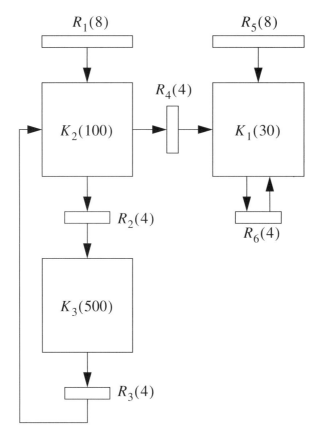

Figure 11.59   A sequential circuit with three combinational kernels (Narayanan *et al.*, 1993)

and Chip 3 have 40, 60, and 60 cells, respectively, and require 100, 500, and 700 vectors, respectively, for re-test.

Show how a combination of BYPASS and INTEST instructions can be used to re-test the three chips. How many clock cycles are required to shift in all the instructions, test vectors, and test responses? How many clock cycles would be required if the BYPASS instruction were not provided?

11.25 Show details of how the modified counting sequence can be applied to the example board design with two chips shown in Figure 11.29.

11.26 Repeat the above problem for the modified walking test.

11.27 Obtain a full-scan version of the circuit shown in Figure 11.58. Identify all the fully specified two-pattern tests that can be applied to your design via (a) scan shifting, (b) functional justification, and (c) hold.

11.28 Identify the blocking objectives for the circuit shown in Figure 11.58. Does there exist a primary input value that implies a controlling value at an input of every blocking objective? What primary input value would you use during scan shifting cycles to minimize switching activity during scan shifting?

11.29 Generate all possible test vectors for the c_4 SA0 fault in the combinational logic of the full-scan version of the circuit in Figure 11.58. Which of these vectors would you select to minimize the switching activity during scan testing?

References

Abramovici, M., Kulikowski, J. J., and Roy, R. R. (1991). The best flip-flops to scan. In *Proc. Int. Test Conference*, pp. 166–173.

Abramovici, M., Rajan, K. B., and Miller, D. T. (1992). Freeze: a new approach for testing sequential circuits. In *Proc. Design Automation Conference*, pp. 22–25.

Abramovici, M., Parikh, P., Mathew, B., and Saab, D. (1993). On selecting flip-flops partial reset. In *Proc. Int. Test Conference*, pp. 1008–1012.

Ashar, P. and Malik, S. (1994). Implicit computation of minimum-cost feedback-vertex sets for partial scan and other applications. In *Proc. Design Automation Conference*, pp. 77–80.

Baeg, S. and Rogers, W. (1994). Hybrid design for testability combining scan and clock line control and method for test generation. In *Proc. Int. Test Conference*, pp. 340–349.

Bhavsar, D. (1991). An architecture for extending the IEEE Standard 1149.1 test access port to system backplanes. In *Proc. Int. Test Conference*, pp. 768–776.

Bhawmik, S., Lin, C.-J., Cheng, K. T., and Agrawal, V. D. (1991). PASCANT: a partial scan and test generation system. In *Proc. Int. Custom Integrated Circuits Conference*, pp. 17.3.1–17.3.4.

Chakradhar, S. T., Balakrishnan, A., and Agrawal, V. D. (1994). An exact algorithm for selecting partial scan flip-flops. In *Proc. Design Automation Conference*, pp. 81–86.

Chakradhar, S. T. and Dey, S. (1994). Synthesis and retiming for optimum partial scan. In *Proc. Design Automation Conference*, pp. 87–93.

Chen, C. and Gupta, S. (1998). Efficient BIST TPG design and test set compaction for delay testing via input reduction. In *Proc. Int. Conference on Computer Design*, pp. 32–39.

Cheng, K.-T. and Agrawal, V. D. (1990). A partial scan method for sequential circuits with feedback. *IEEE Trans. on Computers*, **39** (4), pp. 544–548.

Cheng, K.-T., Devadas, S., and Keutzer, K. (1991a). A partial enhanced-scan approach to robust delay-fault test generation for sequential circuits. In *Proc. Int. Test Conference*, pp. 403–410.

Cheng, K.-T., Devadas, S., and Keutzer, K. (1991b). Robust delay-fault test generation and synthesis for testability under a standard scan design methodology. In *Proc. Design Automation Conference*, pp. 80–86.

Cheng, K. T. (1995). Partial scan designs without using a separate scan clock. In *Proc. VLSI Test Symposium*, pp. 277–282.

Cheng, W.-T., Lewandowski, J. L., and Wu, E. (1990). Diagnosis for wiring interconnects. In *Proc. Int. Test Conference*, pp. 565–571.

Chickermane, V., Rudnick, E., Banerjee, P., and Patel, J. (1993). Non-scan design-for-testability techniques for sequential circuits. In *Proc. Design Automation Conference*, pp. 236–241.

Dervisoglu, B. and Stong, G. E. (1991). Design for testability: using scanpath techniques for path-delay test and measurement. In *Proc. Int. Test Conference*, pp. 365–374.

Dey, S. and Chakradhar, S. T. (1994). Retiming sequential circuits to enhance testability. In *Proc. VLSI Test Symposium*, pp. 28–33.

Eichelberger, E. B. and Williams, T. W. (1977). A logic design structure for design for testability. In *Proc. Design Automation Conference*, pp. 462–468.

Fang, W.-C. and Gupta, S. K. (1994). Clock grouping: a low cost DFT methodology for delay fault testing. In *Proc. Design Automation Conference*, pp. 94–99.

Fujiwara, H. and Yamamoto, A. (1992). Parity-scan design to reduce the cost of test application. In *Proc. Int. Test Conference*, pp. 283–292.

Gerstendorfer, S. and Wunderlich, H.-J. (1999). Minimized power consumption for scan-based BIST. In *Proc. Int. Test Conference*, pp. 77–84.

Goel, P. and McMahon, M. T. (1982). Electronic chip-in-place test. In *Proc. Int. Test Conference*, pp. 83–90.

Gupta, R., Gupta, R., and Breuer, M. A. (1990). The BALLAST methodology for structured partial scan design. *IEEE Trans. on Computers*, **39** (4), pp. 538–543.

Gupta, R., Srinivasan, R., and Breuer, M. A. (1991). Reorganizing circuits to aid testability. *IEEE Design and Test of Computers*, April, pp. 49–57.

Gupta, R. and Breuer, M. A. (1992). Testability properties of acyclic structures and applications to partial scan design. In *Proc. VLSI Test Symposium*, pp. 49–54.

Hassan, A., Rajski, J., and Agarwal, V. K. (1988). Testing and diagnosis of interconnects using boundary scan architecture. In *Proc. Int. Test Conference*, pp. 126–137.

Higami, Y., Kajihara, S., and Kinoshita, K. (1994). Reduced scan shift: a new testing method for sequential circuits. In *Proc. Int. Test Conference*, pp. 624–630.

Hsu, F. and Patel, J. (1995). A distance reduction approach to design for testability. In *Proc. VLSI Test Symposium*, pp. 158–386.

IEEE (1994). *Supplement(B) to IEEE standard test access port and boundary-scan architecture.*

IEEE (1995). *IEEE Standard Module Test and Maintenance (MTM) Bus Protocol.*

Jarwala, N. and Yau, C. W. (1989). A New framework for analyzing test generation and diagnosis algorithms for wiring interconnects. In *Proc. Int. Test Conference*, pp. 63–70.

Kagaris, D. and Tragoudas, S. (1993). Partial scan with retiming. In *Proc. Design Automation Conference*, pp. 249–254.

Kanjilal, S., Chakradhar, S., and Agrawal, V. D. (1993). A synthesis approach to design for testability. In *Proc. Int. Test Conference*, pp. 754–763.

Kim, K. S. and Kime, C. R. (1990). Partial scan by use of empirical testability. In *Proc. Int. Conference on Computer-Aided Design*, pp. 314–317.

Kim, K. S. and Schultz, L. (1994). Multi-frequency, multi-phase scan chain. In *Proc. Int. Test Conference*, pp. 323–330.

Kunzman, A. and Wunderlich, H. J. (1990). An analytical approach to the partial scan problem. *Journal of Electronic Testing: Theory and Applications*, **1** (1), pp. 163–174.

Kuo, T.-Y., Liu, C.-Y., and Saluja, K. (1995). An optimized testable architecture for finite state machines. In *Proc. VLSI Test Symposium*, pp. 164–169.

Lee, D. H. and Reddy, S. M. (1990). On determining scan flip-flops in partial scan designs. In *Proc. Int. Conference on Computer-Aided Design*, pp. 322–325.

Lee, S. Y. and Saluja, K. K. (1992). An algorithm to reduce test application time in full scan designs. In *Proc. Int. Conference on Computer-Aided Design*, pp. 17–20.

Lien, J.-C. and Breuer, M. A. (1989). A universal test and maintenance controller for modules and boards. *IEEE Trans. on Industrial Electronics*, **36**, pp. 231–240.

Lien, J.-C. and Breuer, M. A. (1991). Maximal diagnosis for wiring networks. In *Proc. Int. Test Conference*, pp. 96–105.

Lin, C.-C., Marek-Sadowska, M., Cheng, K.-T., and Lee, M. (1996). Test point insertion: scan paths through combinational logic. In *Proc. Design Automation Conference*, pp. 268–273.

Min, H. and Rogers, W. (1990). A test methodology for finite state machines using partial scan design. *Journal of Electronic Testing: Theory and Applications*, **3**, pp. 127–137.

Morley, S. P. and Marlett, R. A. (1991). Selectable length partial scan: a method to reduce vector length. In *Proc. Int. Test Conference*, pp. 385–392.

Narayanan, S., Njinda, C., and Breuer, M. A. (1992). Optimal sequencing of scan registers. In *Proc. Int. Test Conference*, pp. 293–302.

Narayanan, S., Gupta, R., and Breuer, M. A. (1993). Optimal configuring of multiple scan chains. *IEEE Trans. on Computers*, **42** (9), pp. 1121–1131.

Narayanan, S. (1994). *Scan Chaining and Test Scheduling in an Integrated Scan Design System.* PhD thesis, University of Southern California, Los Angeles, CA.

Narayanan, S. and Breuer, M. A. (1995a). Asynchronous multiple scan chains. In *Proc. VLSI Test Symposium*, pp. 270–276.

Narayanan, S. and Breuer, M. A. (1995b). Reconfiguration techniques for a single scan chain. *IEEE Trans. on Computer-Aided Design*, **14** (6), pp. 750–765.

Park, S. J. and Akers, S. B. (1992). A graph theoretic approach to partial scan design by K-cycle elimination. In *Proc. Int. Test Conference*, pp. 303–311.

Patil, S. and Savir, J. (1992). Skewed-load transition test: part II, coverage. In *Proc. Int. Test Conference*, pp. 714–722.

Pradhan, D. K. and Saxena, J. (1992). A design for testability scheme to reduce test application in full scan. In *Proc. VLSI Test Symposium*, pp. 55–60.

Rajan, K., Long, D., and Abramovici, M. (1996). Increasing testability by clock transformation. In *Proc. VLSI Test Symposium*, pp. 224–230.

Savir, J. and Patil, S. (1994). Broad-side delay test. *IEEE Trans. on Computer-Aided Design*, **13** (8), pp. 1057–1064.

Shi, W. and Fuchs, W. K. (1995). Optimal interconnect diagnosis of wiring networks. *IEEE Trans. on VLSI Systems*, **3** (3), pp. 430–436.

Trischler, E. (1980). Incomplete scan path with an automatic test generation methodology. In *Proc. Int. Test Conference*, pp. 153–162.

Wang, S. and Gupta, S. (1997). ATPG for heat dissipation minimization during scan testing. In *Proc. Design Automation Conference*, pp. 614–619.

Whetsel, L. (1992). A proposed method of accessing 1149.1 in a backplane environment. In *Proc. Int. Test Conference*, pp. 206–216.

Williams, M. and Angell, J. (1973). Enhancing testability of large-scale integrated circuits via test points and additional logic. *IEEE Trans. on Computers*, **C-22**, pp. 46–60.

Yau, C. W. and Jarwala, N. (1989). A unified theory for designing optimal test generation and diagnosis algorithms for board interconnects. In *Proc. Int. Test Conference*, pp. 318–324.

Yau, C. W. and Jarwala, N. (1990). The boundary-scan master: target applications and functional requirements. In *Proc. Int. Test Conference*, pp. 311–315.

12 Built-in self-test

In this chapter, we discuss built-in self-test (BIST) of digital circuits. We begin with a description of the commonly used test pattern generators, namely linear feedback shift-registers and cellular automata, and the properties of sequences they generate. This is followed by an analysis of test length vs. fault coverage for testing using random and pseudo-random sequences. Two alternative approaches are then presented to achieve the desired fault coverage for circuits under test (CUTs) for which the above test pattern generators fail to provide adequate coverage under given constraints on test length. We then discuss various test response compression techniques, followed by an analysis of the effectiveness of commonly used linear compression techniques.

The second part of the chapter focuses on issues involved in making a large digital circuit self-testable. We begin with a discussion of some of the key issues and descriptions of reconfigurable circuitry used to make circuits self-testable in an economical fashion. We then discuss two main types of self-test methodologies, in-situ BIST and scan-based BIST, followed by more detailed descriptions of the two methodologies in the last two sections.

The third part of the chapter contains description of BIST techniques for delay fault testing as well as for testing with reduced switching activity.

12.1   Introduction

Built-in self-test refers to techniques and circuit configurations that enable a chip to test itself. In this methodology, test patterns are generated and test responses are analyzed on-chip. Hence, the simplest of BIST designs has a pattern generator (PG) and a response analyzer (RA) as shown in Figure 12.1.

BIST offers several advantages over testing using automatic test equipment (ATE). First, in BIST the test circuitry is incorporated on-chip and no external tester is required, provided power and clocks are supplied to the chip. This is especially attractive for high-speed circuits which otherwise require expensive testers.

Second, self-test can be performed at the circuit's normal clock rate. In the conventional paradigm, this is always challenging for leading-edge chip designs, since ATE must keep pace with the increasing circuit speeds. This can provide significant

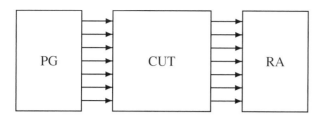

Figure 12.1 A general self-test configuration

advantages in terms of coverage of faults that may otherwise be detected only after a chip is integrated into the system.

Third, a self-testable chip has the ability to perform self-test even after it is incorporated into a system. This can be used either for periodic testing or to diagnose system failures. This aspect of self-test has made it an integral part of high reliability systems or systems where high availability, hence rapid failure isolation, is desired.

To be sure, there are several costs associated with BIST. First, incorporation of the self-test capability requires addition of hardware to the chip. This increases the silicon area required to implement the chip, which increases the cost of manufacturing chips. The increase in manufacturing cost occurs not only due to the fact that larger silicon die must be used, but also because an increase in die area decreases die yield. Second, the hardware added to the circuit can increase delays of normal circuit paths, decreasing the speed at which the circuit can be used during its normal operation. The above two costs of BIST are commonly referred to as **area overhead** and **performance penalty**, respectively.

Third, design errors in a self-testable circuit may be harder to identify than in a circuit with DFT features such as scan. This is due to the fact that while a self-testable circuit is easy to test, its internal lines are not necessarily easy to control and observe from the outside; instead, they are controlled and observed by on-chip test circuitry. Furthermore, for a failing circuit, typically self-test only reports a failure without providing much information about the failure. These factors can make it difficult to debug design errors in a self-testable design. However, once all the errors in the design have been removed and the chip is produced in large quantities, the above advantages of BIST can be accrued.

Figure 12.1 shows two key components of a self-testable design, namely, a PG and an RA. However, its simplicity hides some of the key issues in the design of self-testable circuitry. In the following, we begin with a description of commonly used PGs. Since these PGs generate pseudo-random sequences, the estimation of the quality of testing obtained becomes a central aspect of BIST design. We then discuss two approaches to deal with CUTs for which pseudo-random PGs fail to provide the desired fault coverage under given constraints on test length. RAs used in BIST are discussed next, along with the question of their effectiveness. Then we discuss different ways in which self-test is implemented for large circuits.

12.2   Pattern generators

A PG is an autonomous finite-state machine (FSM), i.e., it has no external input other than clock(s), that is typically configured as a shift register with additional feedback connections. A PG is said to be **linear** if its feedback circuitry can be implemented using only exclusive-or (XOR) gates; otherwise, it is said to be **non-linear**. In the following, we describe two types of linear PGs, namely, linear feedback shift registers (LFSRs) and cellular automata (CA).

Since a PG is an autonomous FSM, when its state values, i.e., the contents of its flip-flops, are initialized to a known initial state, called the **seed**, it has a repeatable sequence of state transitions. The values of the state variables are used as test vectors (patterns) in one of a variety of ways. In the **parallel mode**, the values appearing at v-out-of-n, where $v \leq n$, PG flip-flops at a given clock cycle are used as a test vector. In the **serial mode**, the values appearing at the output of a specific PG flip-flop over v consecutive clock cycles are used as a v-bit test vector. In general, in the **parallel–serial mode**, a test vector may comprise of values appearing at a set of PG flip-flops over several consecutive clock cycles. In the following, we present those properties of state transitions of PGs that are relevant to common modes in which a PG may be used.

12.2.1   Linear feedback shift registers (LFSRs)

An n-stage LFSR is obtained by adding linear feedback to a shift register with flip-flops $D_0, D_1, \ldots, D_{n-1}$. Two types of LFSRs are commonly employed in practice, **internal-XOR** (Figure 12.2) and **external-XOR** (Figure 12.3).

12.2.1.1   Structure of LFSRs – feedback polynomial

In an internal-XOR LFSR, the output of the last flip-flop in the shift register, D_{n-1}, is fed back to the inputs of a selected subset of flip-flops, as shown in Figure 12.2(a). A **feedback polynomial**,

$$\phi(x) = \phi_n x^n + \cdots + \phi_1 x + \phi_0,$$

is used to describe the feedback connections. First, the fact that the output of D_{n-1} (which can be viewed as the input of non-existent D_n) is used as the feedback signal is indicated by always assigning $\phi_n = 1$. For $i = 0, 1, \ldots, n - 1$, $\phi_i = 1$ indicates that feedback occurs at the input of D_i, and $\phi_i = 0$ indicates that no feedback occurs there. Finally, a **non-trivial** n-stage internal-XOR LFSR has $\phi_0 = \phi_n = 1$. One can verify that if $\phi_0 = 0$, then the n-stage LFSR can be viewed as combination of a shift register and an LFSR with fewer stages. Also, $\phi_n = 0$ eliminates the feedback and the entire LFSR can be viewed as an n-stage shift register.

Feedback to the input of flip-flop D_i, $i = 1, 2, \ldots, n-1$, is achieved by adding a two-input XOR gate between D_{i-1} and D_i, connecting the feedback signal and the output of D_{i-1} to its inputs, and connecting its output to the input of D_i. No such XOR gate is added if $\phi_i = 0$, since in that case no feedback occurs at the input of D_i.

Example 12.1 Figure 12.2(b) shows an example four-stage internal-XOR LFSR with $\phi(x) = x^4 + x + 1$. Note that since $\phi_3 = \phi_2 = 0$, no feedback occurs at the inputs of D_2 and D_3. Also note how the XOR gate is used to implement the feedback connection at the input of D_1 and no XOR gate is required for the feedback at the input of D_0. □

The flip-flops of an external-XOR LFSR are named in the opposite order of how they are named for an internal-XOR LFSR. In this case, the outputs of a selected subset of flip-flops are combined into a single signal, which is fed back to the input of the first flip-flop in the shift register, D_{n-1}. Again, a feedback polynomial is used to characterize the feedback connections. The fact that the signal is fed back to the input of D_{n-1} (which can be viewed as being driven by non-existent D_n) is indicated by always assigning $\phi_n = 1$. For $i = 0, 1, \ldots, n-1$, $\phi_i = 1$ indicates that the output of D_i is selected for feedback. Once again, $\phi_0 = \phi_n = 1$ for a non-trivial n-stage external-XOR LFSR.

An array of two-input XOR gates is used to combine the outputs of the selected shift register flip-flops into a single feedback signal.

Example 12.2 Figure 12.3(b) shows an example four-stage external-XOR LFSR with $\phi(x) = x^4 + x + 1$. □

12.2.1.2 Relationship between the LFSR sequence and feedback polynomial

Pattern generation using either type of LFSR starts with initialization of the flip-flops to an initial state, called the **seed**. The flip-flops are then clocked to cause LFSR state transitions, whose exact nature depends on the feedback connections and hence the feedback polynomial.

The feedback polynomial determines not only the exact sequence of state transitions generated by the LFSR but also strongly influences its basic properties. For example, starting with any **non-zero** seed, i.e., any seed other than $00\ldots0$, a four-stage internal-XOR LFSR with feedback polynomial $\phi(x) = x^4 + x + 1$ generates a sequence of 15 states before the initial state, and hence the sequence, repeats. The same is true for the LFSR with $\phi(x) = x^4 + x^3 + 1$. In contrast, a four-stage internal-XOR LFSR with feedback polynomial $\phi(x) = x^4 + 1$ generates much shorter sequences,

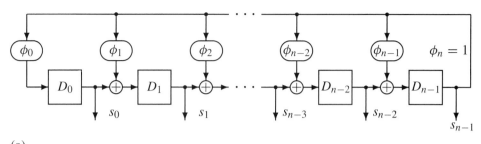

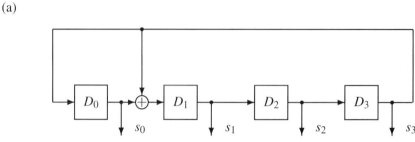

(a)

(b)

Figure 12.2 Internal-XOR LFSR: (a) the general structure with $\phi(x) = \phi_n x^n + \cdots + \phi_1 x + \phi_0$, and (b) an example with $\phi(x) = x^4 + x + 1$

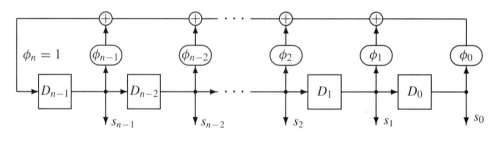

(a)

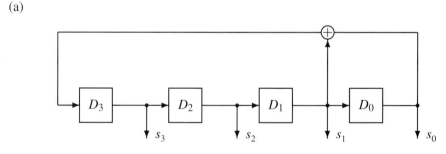

(b)

Figure 12.3 External-XOR LFSR: (a) the general structure with $\phi(x) = \phi_n x^n + \cdots + \phi_1 x + \phi_0$, and (b) an example with $\phi(x) = x^4 + x + 1$

whose lengths depend on the seed. For example, for seed 1111, this LFSR generates a sequence of length one; for seed 1000, it generates a sequence of length four.

All coefficients of the feedback polynomial, ϕ_i, are Boolean, i.e., $\phi_i \in \{0, 1\}$. Due to the fact that all logic elements used are linear, addition and multiplication operations are implemented as XOR ($0 + 0 = 0$, $0 + 1 = 1$, $1 + 0 = 1$, and $1 + 1 = 0$) and AND ($0.0 = 0$, $0.1 = 0$, $1.0 = 0$, and $1.1 = 1$), respectively. In other words, the computation is carried out using *modulo-2* arithmetic. An important consequence of using modulo-2 arithmetic is that subtraction is identical to addition.

The relationship between the sequence generated at any stage of a non-trivial LFSR and the feedback polynomial is now described (Bardell *et al.*, 1987). Let $s_i(\tau)$ denote the state of, and hence the value appearing at the output of, flip-flop D_i after the application of the τth clock. The sequence appearing at the output of D_{n-1} can be represented as $s_{n-1}(0), s_{n-1}(1), \ldots, s_{n-1}(j)$, and so on. We can use a **generating function**, $G_{n-1}(x)$, to represent this sequence using a parameter x to denote time shift, where

$$G_{n-1}(x) = s_{n-1}(0) + s_{n-1}(1)x + \cdots + s_{n-1}(j)x^j + \cdots$$
$$= \sum_{j=0}^{\infty} s_{n-1}(j)x^j. \tag{12.1}$$

Values at the outputs of different stages of an internal-XOR LFSR with $\phi_n = 1$ satisfy the following relations.

$$s_{n-1}(j) = \phi_{n-1}s_{n-1}(j - 1) + s_{n-2}(j - 1),$$
$$s_{n-2}(j - 1) = \phi_{n-2}s_{n-1}(j - 2) + s_{n-3}(j - 2),$$
$$\vdots$$
$$s_1(j - n + 2) = \phi_1 s_{n-1}(j - n + 1) + s_0(j - n + 1), \text{ and}$$
$$s_0(j - n + 1) = \phi_0 s_{n-1}(j - n).$$

These relations can be used to derive the following **recurrence relation**:

$$s_{n-1}(j) = \sum_{i=0}^{n-1} \phi_i s_{n-1}(j - n + i). \tag{12.2}$$

The above recurrence relation describes the values appearing at s_{n-1} at the jth clock cycle in terms of the values appearing at s_{n-1} at some previous clock cycles.

Using the above recurrence relation in Equation (12.1), we get

$$G_{n-1}(x) = \sum_{j=0}^{\infty} \left[\sum_{i=0}^{n-1} \phi_i s_{n-1}(j - n + i) \right] x^j,$$
$$= \sum_{i=0}^{n-1} \phi_i x^{n-i} \sum_{j=0}^{\infty} s_{n-1}(j - n + i)x^{j-n+i},$$

$$= \sum_{i=0}^{n-1} \phi_i x^{n-i} \left[s_{n-1}(-n+i)x^{-n+i} + \cdots + s_{n-1}(-1)x^{-1} + s_{n-1}(0) \right.$$
$$\left. + s_{n-1}(1)x + \cdots \right],$$

$$= \sum_{i=0}^{n-1} \phi_i x^{n-i} \left[s_{n-1}(-n+i)x^{-n+i} + \cdots + s_{n-1}(-1)x^{-1} + G_{n-1}(x) \right].$$

(12.3)

Collecting all the terms containing $G_{n-1}(x)$ on the left hand side, we obtain,

$$\left[1 + \sum_{i=0}^{n-1} \phi_i x^{n-i} \right] G_{n-1}(x)$$

$$= \sum_{i=0}^{n-1} \phi_i x^{n-i} \left[s_{n-1}(-n+i)x^{-n+i} + \cdots + s_{n-1}(-1)x^{-1} \right],$$

which gives

$$G_{n-1}(x) = \frac{\sum_{i=0}^{n-1} \phi_i x^{n-i} \left[s_{n-1}(-n+i)x^{-n+i} + \cdots + s_{n-1}(-1)x^{-1} \right]}{1 + \sum_{i=0}^{n-1} \phi_i x^{n-i}}.$$

(12.4)

The numerator of the above expression is a function of the initial states of the LFSR flip-flops represented in terms of $s_{n-1}(-n)$, $s_{n-1}(-n+1)$, ..., $s_{n-1}(-1)$, i.e., the values appearing at the output of D_{n-1} in the n clock cycles preceding the zeroth clock cycle. Hence, the numerator captures the effects of the seed on the sequence of values appearing at the output of D_{n-1}. If the initial state of the LFSR is equivalent to the appearance of the sequence $s_{n-1}(-n) = 1$, $s_{n-1}(-n+1) = 0$, ..., $s_{n-1}(-1) = 0$ at the output of D_{n-1} prior to the zeroth clock cycle, then for an LFSR with $\phi_0 = 1$ the above expression can be rewritten as

$$G_{n-1}(x) = \frac{1}{1 + \sum_{i=0}^{n-1} \phi_i x^{n-i}}.$$

(12.5)

The denominator of the above expression,

$$1 + \sum_{i=0}^{n-1} \phi_i x^{n-i},$$

can be re-written in terms of the feedback polynomial as $x^n \phi(1/x)$, since $\phi_n = 1$ for an n-stage LFSR. Using this in the above equation, we finally obtain

$$G_{n-1}(x) = \frac{1}{x^n \phi(1/x)}.$$

(12.6)

The above equation shows how the sequence generated at the output of D_{n-1} depends on the feedback polynomial of a non-trivial LFSR.

$$\begin{pmatrix} 0 & 0 & 0 & \cdots & 0 & \phi_0 \\ 1 & 0 & 0 & \cdots & 0 & \phi_1 \\ 0 & 1 & 0 & \cdots & 0 & \phi_2 \\ 0 & 0 & 1 & \cdots & 0 & \phi_3 \\ \vdots & \vdots & \vdots & \ddots & \vdots & \vdots \\ 0 & 0 & 0 & \cdots & 1 & \phi_{n-1} \end{pmatrix}$$

(a)

$$\begin{pmatrix} 0 & 1 & 0 & \cdots & 0 & 0 \\ 0 & 0 & 1 & \cdots & 0 & 0 \\ \vdots & \vdots & \vdots & \ddots & \vdots & \vdots \\ 0 & 0 & 0 & \cdots & 1 & 0 \\ 0 & 0 & 0 & \cdots & 0 & 1 \\ \phi_0 & \phi_1 & \phi_2 & \cdots & \phi_{n-2} & \phi_{n-1} \end{pmatrix}$$

(b)

Figure 12.4 State transition matrix, A, for an n-stage LFSR: (a) internal-XOR, and (b) external-XOR

Starting with an appropriate recurrence relation, the above derivation can be repeated to obtain a similar generating function for an external-XOR LFSR (Bardell *et al.*, 1987).

Alternatively, an LFSR, which is an autonomous FSM, can be described using a **state transition relation**. Let

$$\sigma(\tau) = \begin{bmatrix} s_0(\tau) \\ s_1(\tau) \\ \vdots \\ s_{n-1}(\tau) \end{bmatrix} \tag{12.7}$$

be the n-bit column vector representing the state of the LFSR flip-flops in the τth clock cycle. (Note that if the LFSR is used in parallel mode, then $\sigma(\tau)$ is the n-bit test vector that is applied to the inputs of the CUT in the τth clock cycle.) The state transition relation for the LFSR can be written in the form

$$\sigma(\tau + 1) = A \cdot \sigma(\tau), \tag{12.8}$$

where A is an $n \times n$ binary **state transition matrix** and '$\cdot$' represents matrix multiplication that uses modulo-2 arithmetic. Figure 12.4 shows the general form of matrix A for non-trivial LFSRs of both types, where $\phi_n = 1$.

The **characteristic equation** of an FSM is given by

$$M(\lambda) = \det(A - \lambda I) = 0, \tag{12.9}$$

where I is an identity matrix of the same size as A. For both types of LFSRs, the **characteristic polynomial** of an LFSR, $M(\lambda)$, satisfies the relation

$$M(\lambda) = \phi(\lambda). \tag{12.10}$$

Hence, the feedback polynomial of an LFSR is also its characteristic polynomial.

Example 12.3 One can verify that for the four-stage LFSRs described in the previous examples, the characteristic polynomial is indeed $M(\lambda) = \lambda^4 + \lambda + 1$. $\qquad\square$

12.2.1.3 Properties of LFSR generated sequences

The above analysis shows that the feedback polynomial of an LFSR determines the basic properties of the sequences generated at its outputs. Hence, the feedback polynomial should be selected to ensure that the LFSR generates sequences with desirable characteristics. An LFSR may be used in parallel, serial, or some intermediate parallel–serial mode. In each of these modes, an LFSR may be used to apply a few or a large number of test vectors.

The mode and manner in which an LFSR is used as a PG determines the desirable characteristics of the generated sequence. For example, when an LFSR is used as a parallel PG to apply a very long sequence of test vectors, it is desirable that the LFSR generate all possible n-bit vectors before repeating any vector. This property, called **completeness**, is desirable since it ensures that all possible vectors will be applied to the CUT. Furthermore, **uniqueness** of vectors ensures that test time is not wasted by repeated application of some vectors. On the other hand, if an LFSR is used as a parallel PG but a short sequence of vectors is applied, then any short subsequence generated by the PG should contain vectors sampled uniformly from the space of all possible 2^n vectors. This is desirable to ensure high coverage of faults, since detection of different faults may require the application of vectors from different parts of the space of all possible 2^n vectors. The desirable characteristics of the sequences can also be identified for cases where the LFSR is used as a serial PG.

Due to its linearity, any LFSR with an **all-zero**, i.e., $00\ldots0$, state will remain in that state. Furthermore, it can also be shown that starting at any non-zero initial state, any non-trivial LFSR can never reach the all-zero state. Hence, for an n-stage LFSR, the maximal number of state transitions that can occur before the initial state repeats is $2^n - 1$. The sequence generated at the output of any flip-flop of such an LFSR is called a **maximal length sequence (MLS)**.

Coincidentally, sequences of vectors generated by LFSRs that generate MLS, which will be referred to as **ML-LFSRs,** possess the characteristics desired for most common modes and manners in which an LFSR is used as a PG. To generate an MLS, an LFSR must use a feedback polynomial drawn from the set of *primitive polynomials*. (The property of primitive polynomials that is of interest in the following is that an LFSR is an ML-LFSR if and only if its feedback polynomial is primitive. Formal definitions and properties of primitive polynomials can be found in Peterson and Weldon (1972) and Lin and Costello (1983).) Table 12.1 shows primitive polynomials of various degrees. Extensive tables of primitive polynomials can be found in Bardell *et al.* (1987). Next, we present some key properties of sequences generated by ML-LFSRs.

Near-completeness: Let $S(\tau) = (s_0(\tau), s_1(\tau), \ldots, s_{n-1}(\tau))$ be an n-bit **parallel test vector (PTV)** generated at the outputs of an n-stage LFSR at the τth clock cycle.

Table 12.1. *A table of primitive polynomials*

Degree	Primitive polynomial
2	$x^2 + x + 1$
3	$x^3 + x + 1$
3	$x^3 + x^2 + 1$
4	$x^4 + x + 1$
4	$x^4 + x^3 + 1$
5	$x^5 + x^2 + 1$
6	$x^6 + x + 1$
7	$x^7 + x + 1$
8	$x^8 + x^4 + x^3 + x^2 + 1$
9	$x^9 + x^4 + 1$
10	$x^{10} + x^3 + 1$

Since, starting with any non-zero seed, an ML-LFSR generates a sequence of $2^n - 1$ state transitions before repeating the initial state, it generates a sequence containing all the possible PTVs, with the sole exception of the all-zero PTV, before repeating any PTV. Due to this fact, an ML-LFSR is said to be a **near-complete** parallel PG.

Now consider the case where an ML-LFSR is used as a serial PG. Further assume that the values appearing at the output of the ith shift register flip-flop over n consecutive clock cycles are used as an n-bit **serial test vector (STV)**. In other words, successive test vectors applied to the CUT are $S_i(\tau) = (s_i(\tau), s_i(\tau + 1), \ldots, s_i(\tau + n - 1))$; $S_i(\tau + 1) = (s_i(\tau + 1), s_i(\tau + 2), \ldots, s_i(\tau + n))$; and so on. It can be shown that even in this case, a near-complete set of test vectors will be applied to the n inputs of the CUT. Hence, an n-stage ML-LFSR is also a near-complete generator of n-bit serial vectors.

Note that the above approach for obtaining n-bit serial vectors is called **overlapped sampling**. Another approach, called *disjoint sampling*, is described in Section 12.10.2 and Problem 12.5. As will be seen in Section 12.10.2, additional conditions need to be satisfied to obtain a near-complete set of STVs when serial vectors are generated using disjoint sampling.

An interesting case arises when an n-stage ML-LFSR PG is used to generate v-bit vectors, where $v < n$. In this case, completeness can be guaranteed, but only when $2^n - 1$ vectors are applied. Furthermore, in a sequence of this length, each v-bit vector will appear 2^{n-v} times, except for the all-zero vector, which will appear $2^{n-v} - 1$ times.

Uniqueness: Near-completeness automatically implies that an n-stage ML-LFSR generates $2^n - 1$ unique parallel vectors before the first vector is generated again. Similarly, an n-stage ML-LFSR generates $2^n - 1$ unique serial vectors of the type described above.

As discussed above, if a ν-bit PTV or STV is generated, where $\nu < n$, then the uniqueness of vectors is not guaranteed.

Randomness: At each output of an ML-LFSR, a sequence that is repeatable, but satisfies many empirical criteria for randomness, is generated (Golomb, 1967).

1 The number of 1s and number of 0s in the sequence differ by one, i.e., the sequence contains almost equal numbers of 1s and 0s.

   Due to this property, the probability of getting a 1 (or a 0) at any specific bit position of a PTV or STV is close to 0.5.

2 Define a **run of 0s of length** j as any subsequence of a sequence that contains j consecutive 0s and is preceded and succeeded by 1s; a **run of 1s of length** j is defined in a similar fashion. A **run of length** j is a run of 0s or 1s of length j.

   The total number of runs in one period of an MLS generated by an n-bit ML-LFSR is 2^{n-1}. Half of these runs are of length one, a quarter are of length two, an eighth are of length three, and so on. The longest run of 0s has length $n - 1$, while the longest run of 1s has length n. The total numbers of runs of 0s and 1s are equal. Additionally, for any run length $\nu \in \{1, 2, \ldots, n - 2\}$, the numbers of runs of 0s and 1s are almost equal.

   This property is useful in ensuring randomness for PTVs as well as STVs. For PTVs, it guarantees that none of the bit positions of the vector has a fixed value for many consecutive clock cycles, as is the case for a binary counter. This helps ensure that PTVs from different parts of the space of all possible 2^n vectors appear in relatively short sequences. In the serial mode, this ensures that STVs from different parts of the space of 2^n vectors appear. This is in sharp contrast with a binary counter, where STVs at any flip-flop output contain only a small number of distinct vectors.

3 Let MLS^j be the sequence obtained by cyclically shifting one period of an MLS by j steps. In other words, if $MLS = s_i(0), s_i(1), \ldots, s_i(2^n - 2)$, then $MLS^j = s_i(j), s_i(j+1), \ldots, s_i(2^n - 2), s_i(0), s_i(1), \ldots, s_i(j-1)$. For any $j \neq 0$, a bit-by-bit comparison of MLS and MLS^j results in a number of matches that exceeds the number of mismatches by one. Clearly, for $j = 0$, all the bits of MLS and MLS^j match.

   This property ensures that when MLS and MLS^j are applied to two CUT inputs (which occurs, for example, when these two CUT inputs are connected to outputs of stages i and $i + j$ of an external-XOR LFSR), the probability of an identical value being applied to the inputs is approximately the same as that of opposite values being applied. Note that if this were not true, then the correlation between the values applied to the two inputs could compromise the quality of the LFSR as a parallel PG.

In the following, any sequence that has the above properties will be called **pseudo-random** and PGs that generate such sequences will be called **pseudo-random pattern generators (PRPGs)**.

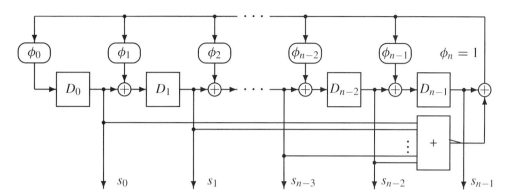

Figure 12.5 An LFSR that generates complete sequences

Despite exhibiting pseudo-randomness, the sequence generated at the output of the flip-flop D_i of an external-XOR LFSR is identical to the sequence obtained by shifting by one clock cycle the sequence generated at the output of D_{i-1}. The sequences generated at the consecutive stages of an external-XOR LFSR are hence said to have a **phase shift** of one. Such **low phase shift** between the sequences generated at the consecutive stages of an LFSR can lead to low fault coverage when the LFSR is used as a parallel PG and only a small number of vectors are used.

12.2.1.4 LFSRs that generate complete sequences

An LFSR with a primitive feedback polynomial can be modified, via the addition of some non-linear circuitry, to generate a **complete sequence** that contains the MLS generated by the original LFSR as well as the all-zero state (McCluskey and Bozorgui-Nesbat, 1981).

Consider the internal-XOR LFSR shown in Figure 12.2(a). If the feedback polynomial is primitive, this LFSR generates an MLS. Now consider the case where the state of this LFSR is $\sigma_i = (0, \ldots, 0, 1)$, i.e., the flip-flop D_{n-1} contains logic 1 and all the other flip-flops contain 0s. Due to the presence of the feedback, in the next clock cycle, the LFSR will proceed to another non-zero state, say σ_j, whose exact value depends on the feedback polynomial.

The above LFSR can be modified by the addition of an $(n-1)$-input NOR gate and a two-input XOR gate, as shown in Figure 12.5. When the state of the LFSR is $\sigma_i = (0, \ldots, 0, 1)$, the output of NOR gate is 1, which, in turn, alters the value of the feedback signal to 0. Hence, in the subsequent cycle, the LFSR state becomes all-zero. When the LFSR is in the all-zero state, the output of the NOR gate is again 1. This time, since the content of D_{n-1} is 0, the feedback signal becomes 1, and the next state of the LFSR is σ_j. Hence, the addition of the NOR and XOR gates modifies the state transition of the LFSR to include the all-zero state between $\sigma_i = (0, \ldots, 0, 1)$ and its successor state, σ_j. Furthermore, it can be shown that all the other state transitions of the LFSR shown in Figure 12.5 are identical to those of the LFSR shown in Figure 12.2(a). Hence, the LFSR shown in Figure 12.5 generates a complete sequence.

12.2.1.5 LFSRs that generate reverse-order sequences

In some self-test scenarios, a pair of test sequences are required, where one contains vectors in the reverse order of how they appear in the other. If the feedback polynomial of the original LFSR is $\phi(x)$, then the corresponding **reverse-order sequence** is generated by an LFSR with feedback polynomial $\phi^*(x) = x^n \phi(1/x)$. This can be verified easily by writing the generating functions of the two LFSRs,

$$G_{n-1}(x) = \frac{1}{x^n \phi(1/x)}, \text{ and}$$

$$G^*_{n-1}(x) = \frac{1}{x^n \phi^*(1/x)} = \frac{1}{\phi(x)}.$$

One can verify that the sequence represented by the first generating function is the reverse-order sequence of that represented by the second function.

The above observation has two main implications. First, the STV generated at any output of one of these LFSRs is the reverse of that generated at the corresponding output of the other. Second, if the values at outputs $s_0, s_1, \ldots, s_{n-1}$ are considered to constitute an n-bit PTV for one of these LFSRs and the values at outputs $s_{n-1}, \ldots, s_1, s_0$ are considered a PTV for the other, then the sequence of PTVs generated by one LFSR is the reverse of that generated by the other.

Example 12.4   Table 12.2 illustrates that the above facts hold for the sequences generated by internal-XOR LFSRs with feedback polynomials $\phi(x) = x^5 + x^2 + 1$ and $\phi^*(x) = x^5 + x^3 + 1$.   □

12.2.2 Cellular automata

Cellular automata (CA) (Wolfram, 1984) are another class of autonomous FSMs which have been studied as PRPGs for BIST. Cellular automata that have been studied extensively as PRPGs are *one-dimensional* arrays of *binary* flip-flops, say $D_0, D_1, \ldots, D_{n-1}$. The next state of flip-flop D_i, $s_i(\tau + 1)$, is a *linear* function of its own current state, $s_i(\tau)$, and those of the flip-flops that are its immediate neighbors to the left and right, i.e., $s_{i-1}(\tau)$ and $s_{i+1}(\tau)$. In a CA with **cyclic boundary conditions**, the first and the last stages are considered to be each other's neighbors. In contrast, in a CA with **null boundary conditions**, the first and last stages are assumed to have no neighbors to their left and right, respectively. In this case, the current states of the left neighbor of the first stage and the right neighbor of the last stage are assumed to be always 0.

Table 12.2. *Sequence generated by a five-stage, internal-XOR LFSR with feedback polynomials (a)* $x^5 + x^2 + 1$, *and (b)* $x^5 + x^3 + 1$

	(a)						(b)				
i	i^{th} PTV					i	i^{th} PTV				
0	1	0	0	0	0	0	1	0	0	0	0
1	0	1	0	0	0	1	0	1	0	0	0
2	0	0	1	0	0	2	0	0	1	0	0
3	0	0	0	1	0	3	0	0	0	1	0
4	0	0	0	0	1	4	0	0	0	0	1
5	1	0	1	0	0	5	1	0	0	1	0
6	0	1	0	1	0	6	0	1	0	0	1
7	0	0	1	0	1	7	1	0	1	1	0
8	1	0	1	1	0	8	0	1	0	1	1
9	0	1	0	1	1	9	1	0	1	1	1
10	1	0	0	0	1	10	1	1	0	0	1
11	1	1	1	0	0	11	1	1	1	1	0
12	0	1	1	1	0	12	0	1	1	1	1
13	0	0	1	1	1	13	1	0	1	0	1
14	1	0	1	1	1	14	1	1	0	0	0
15	1	1	1	1	1	15	0	1	1	0	0
16	1	1	0	1	1	16	0	0	1	1	0
17	1	1	0	0	1	17	0	0	0	1	1
18	1	1	0	0	0	18	1	0	0	1	1
19	0	1	1	0	0	19	1	1	0	1	1
20	0	0	1	1	0	20	1	1	1	1	1
21	0	0	0	1	1	21	1	1	1	0	1
22	1	0	1	0	1	22	1	1	1	0	0
23	1	1	1	1	0	23	0	1	1	1	0
24	0	1	1	1	1	24	0	0	1	1	1
25	1	0	0	1	1	25	1	0	0	0	1
26	1	1	1	0	1	26	1	1	0	1	0
27	1	1	0	1	0	27	0	1	1	0	1
28	0	1	1	0	1	28	1	0	1	0	0
29	1	0	0	1	0	29	0	1	0	1	0
30	0	1	0	0	1	30	0	0	1	0	1
$\vdots$	$\vdots$					$\vdots$	$\vdots$				

Exclusive use of near-neighbor connections makes CA amenable to compact layout by avoiding the long feedback connections required to implement LFSRs. However, it will be seen in the following that the number of XOR gates required to implement CA that generate an MLS is typically large. Hence, the area required to implement a CA PRPG can be significantly higher than an LFSR (Hortensius *et al.*, 1989). The main advantages of near-neighbor connections are: (a) due to the absence of long feedback connections, a CA can be operated at a high clock rate, and (b) the modular

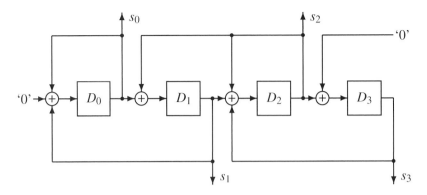

Figure 12.6 An example four-stage cellular automaton

connections make it easier to obtain a CA with larger number of stages by cascading CA containing fewer stages. Figure 12.6 shows an example of a four-stage CA.

The main feature of CA that makes them attractive as PRPGs is that the sequences generated by CA pass more empirical tests for randomness than the sequences generated by an LFSR (Hortensius *et al.*, 1989). The key issues that will be discussed in the following is the design of CA that generate MLS, called **ML-CA**, and the properties of the ML sequences generated by CA.

A state transition relation can be used to describe the state transitions of a one-dimensional, binary, linear CA. Recall that $\sigma(\tau) = [s_0(\tau), s_1(\tau), \ldots, s_{n-1}(\tau)]^T$, where T denotes a matrix transpose operation, is an n-bit column vector representing the state values, i.e., the contents of the flip-flops of the CA at time τ. The state transition relation for the CA can be written in terms of $\sigma(\tau)$ as

$$\sigma(\tau + 1) = A \cdot \sigma(\tau), \tag{12.11}$$

where A is an $n \times n$ binary state transition matrix and '.' represents matrix multiplication that uses modulo-2 arithmetic. The characteristic equation of the CA can be written in terms of the state transition matrix, A, as

$$M(\lambda) = \det(A - \lambda I) = 0, \tag{12.12}$$

where I is an identity matrix of the same size as A. Finally, the characteristic polynomial of a CA, $M(\lambda)$, is given by

$$M(\lambda) = \det(A - \lambda I). \tag{12.13}$$

The characteristic polynomial of any FSM defines many key characteristics of its state transition.

Example 12.5 The state transition matrix, A, for the four-stage CA shown in Figure 12.6 is shown in Figure 12.8(a). Given this value of A, it is easy to verify that $M(\lambda)$, which can be computed as $\det(M - \lambda \cdot I)$, is $\lambda^4 + \lambda + 1$. $\qquad \square$

12.2.2.1 CA that generate MLS

If the characteristic polynomial of a CA is *irreducible*, then the CA is similar to an LFSR with the same characteristic polynomial (Serra *et al.*, 1990). (A polynomial with binary coefficients is said to be **irreducible** if it cannot be factorized into lower degree polynomials with binary coefficients. The set of irreducible polynomials includes all the primitive polynomials.) Similarity of two FSMs implies that the state transition graphs of the two FSMs are identical, provided that the states are appropriately renamed. This result implies that a CA with a primitive characteristic polynomial generates an MLS. This follows from the facts that such a CA is similar to an LFSR with the same characteristic polynomial, and that an LFSR with a primitive characteristic (feedback) polynomial generates an MLS. Furthermore, one period of the MLS appearing at the output of any stage of an ML-CA, which has a primitive characteristic polynomial, is identical to an MLS obtained by appropriately shifting the sequence generated at any stage of an LFSR with the same characteristic polynomial (Bardell, 1990a).

Only CA with null-boundary conditions can have primitive characteristic polynomials. Furthermore, only CA that use combinations of two linear rules, namely *rule-90* and *rule-150*, can have primitive characteristic polynomials (Serra *et al.*, 1990). These rules can be written as

rule-90: $s_i(\tau + 1) = s_{i-1}(\tau) + s_{i+1}(\tau)$, and

rule-150: $s_i(\tau + 1) = s_{i-1}(\tau) + s_i(\tau) + s_{i+1}(\tau)$.

Since we are interested primarily in ML-CA, the remainder of this discussion will be confined to rule-90/150 CA with null boundary conditions.

Any n-stage CA of the above type can be described uniquely using a **rule polynomial**, $\rho(x) = \rho_0 + \rho_1 x + \cdots + \rho_{n-1} x^{n-1}$, where $\rho_i = 1$ if stage i of the CA uses rule-150, otherwise, $\rho_i = 0$. The general structure of such a CA is shown in Figure 12.7 and its state transition matrix is shown in Figure 12.8(b). Due to null boundary conditions and exclusive use of rules-90/150: (a) the upper and lower diagonal of this matrix contain all 1s, (b) all the elements of the matrix above the upper diagonal and below the lower diagonal are 0s, and (c) the entries on the main diagonal are ρ_i.

Since the state transition matrix of a CA can be defined using the coefficients of its rule polynomial, i.e., ρ_i, the only variables contained in its characteristic polynomial, $M(\lambda) = \det(A - \lambda \cdot I)$, are ρ_i. Hence, based on the results presented above, an n-stage ML-CA can be obtained by finding the values of ρ_i, $0 \le i \le n - 1$, such that its characteristic polynomial is primitive. For example, the characteristic polynomial of a three-stage CA can be written in terms of the coefficients of its rule polynomial $\rho(x)$, i.e., ρ_0, ρ_1, and ρ_2, as

$$M(\lambda) = \lambda^3 + (\rho_0 + \rho_1 + \rho_2)\lambda^2 + (\rho_0\rho_1 + \rho_0\rho_2 + \rho_1\rho_2)\lambda + (\rho_0\rho_1\rho_2 + \rho_0 + \rho_2).$$

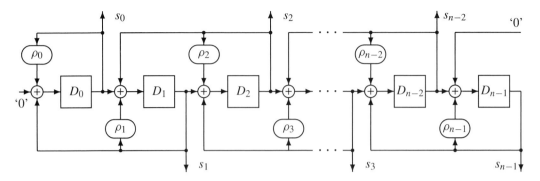

Figure 12.7 General structure of an n-stage CA with $\rho(x) = \rho_0 + \rho_1 x + \cdots + \rho_{n-1} x^{n-1}$

$$
\begin{pmatrix}
1 & 1 & 0 & 0 \\
1 & 0 & 1 & 0 \\
0 & 1 & 1 & 1 \\
0 & 0 & 1 & 0
\end{pmatrix}
$$
(a)

$$
\begin{pmatrix}
\rho_0 & 1 & 0 & 0 & \cdots & 0 & 0 & 0 \\
1 & \rho_1 & 1 & 0 & \cdots & 0 & 0 & 0 \\
0 & 1 & \rho_2 & 1 & \cdots & 0 & 0 & 0 \\
0 & 0 & 1 & \rho_3 & \cdots & 0 & 0 & 0 \\
\vdots & \vdots & \vdots & \vdots & \ddots & \vdots & \vdots & \vdots \\
0 & 0 & 0 & 0 & \cdots & \rho_{n-3} & 1 & 0 \\
0 & 0 & 0 & 0 & \cdots & 1 & \rho_{n-2} & 1 \\
0 & 0 & 0 & 0 & \cdots & 0 & 1 & \rho_{n-1}
\end{pmatrix}
$$
(b)

Figure 12.8 State transition matrices of (a) a four-stage CA with $\rho(x) = 1 + x^2$, and (b) an n-stage CA with $\rho(x) = \rho_0 + \rho_1 x + \cdots + \rho_{n-1} x^{n-1}$

Equating each coefficient of the above polynomial to the corresponding coefficients of a primitive polynomial, say, $\lambda^3 + \lambda^2 + 1$, one can obtain the rule polynomial $\rho(x) = x^2$. One can verify that a CA with this rule polynomial generates maximal length sequences.

The complexity of the above straightforward method to find a rule polynomial that describes an ML-CA grows rapidly with n, the number of stages in the CA. A computationally efficient method to find the rule polynomials $\rho(x)$ and an extensive set of rule polynomials describing CA that generate MLS are presented in Slater and Serra (1990). Table 12.3 shows rule polynomials corresponding to some primitive characteristic polynomials.

12.2.2.2 Properties of CA-generated MLS

The MLS generated at any output of a CA with primitive characteristic polynomial $M(\lambda)$ is identical (with appropriate shift) to that generated at any stage of an LFSR with an identical characteristic polynomial. This implies that an MLS generated by a CA has the same properties as an LFSR-generated MLS. Hence, the properties of

Table 12.3. *A table of ML-CA rule polynomials (Slater and Serra, 1990)*

Degree	Characteristic polynomial	ML-CA rule polynomial
2	$x^2 + x + 1$	1
3	$x^3 + x + 1$	$1 + x$
3	$x^3 + x^2 + 1$	x^2
4	$x^4 + x + 1$	$1 + x^2$
4	$x^4 + x^3 + 1$	$1 + x + x^3$
5	$x^5 + x^2 + 1$	$1 + x + x^2 + x^3$
6	$x^6 + x + 1$	$x + x^2$
7	$x^7 + x + 1$	$1 + x^2 + x^3 + x^6$
8	$x^8 + x^4 + x^3 + x^2 + 1$	$x + x^2$

STVs generated by ML-CA are identical to those of STVs generated by ML-LFSRs. Furthermore, an n-stage ML-CA generates a complete set of n-bit PTVs and generates $2^n - 1$ unique PTVs before repeating the first PTV.

The observed difference in the randomness of PTVs generated by ML-CA and ML-LFSRs arise only due to the phase difference between the sequences generated at consecutive PG stages.

Example 12.6 Table 12.4 shows the sequences generated by a four-stage ML-CA and a four-stage internal-XOR ML-LFSR, both with characteristic polynomial $M(\lambda) = \lambda^4 + \lambda + 1$. (The four-stage LFSR is shown in Figure 12.2(b) and the four-stage CA is shown in Figure 12.6.) Note that due to the similarity, the MLS at all the outputs of the CA and LFSR are identical, after they are adjusted using an appropriate phase shift. However, the phase shift between each pair of adjacent stages of the LFSR and CA are $(-3, 1, 1)$ and $(-4, -6, 1)$, respectively. □

In general, for PGs with larger number of stages, the phase difference between most pairs of adjacent LFSR stages remains one, while that between adjacent CA stages is large.

12.3　Estimation of test length

Determination of the relationship between *test sequence length* and *test quality* is central to the development of random or pseudo-random tests. **Test quality** is measured either by *fault coverage* or by **test confidence**, i.e., the probability of detection of all target faults (or, some subset thereof). In scenarios where a desired test length is given, fault coverage, or its expected value, must be determined for that length. In other scenarios, it may be required to determine the length of a (pseudo-)random sequence that will achieve a given fault coverage or will guarantee the detection of faults with a given level of confidence.

Table 12.4. *Sequences generated by four-stage PGs with* $M(\lambda) = \lambda^4 + \lambda + 1$: *(a) internal-XOR LFSR, and (b) CA*

(a)

i	i^{th} PTV			
0	1	0	0	0
1	0	1	0	0
2	0	0	1	0
3	0	0	0	1
4	1	1	0	0
5	0	1	1	0
6	0	0	1	1
7	1	1	0	1
8	1	0	1	0
9	0	1	0	1
10	1	1	1	0
11	0	1	1	1
12	1	1	1	1
13	1	0	1	1
14	1	0	0	1

(b)

i	i^{th} PTV			
0	1	0	0	0
1	1	1	0	0
2	0	1	1	0
3	1	1	0	1
4	0	1	0	0
5	1	0	1	0
6	1	0	1	1
7	1	0	0	1
8	1	1	1	0
9	0	0	0	1
10	0	0	1	0
11	0	1	1	1
12	1	1	1	1
13	0	0	1	1
14	0	1	0	1

Fault simulation can be used to perform the above-mentioned tasks. However, while self-test sequences containing billions of vectors can be applied in a few seconds at the clock rates of today's circuits, the run-time complexity of fault simulation is too high to simulate sequences of such lengths. Hence, it is often necessary to *estimate* the fault coverage for a given test length or the test length for a desired fault coverage or test confidence. We will discuss estimation for random as well as pseudo-random sequences. We begin with a discussion of the detectabilities of faults in a given CUT.

12.3.1   Detectabilities of faults in the CUT

The relationship between test length and test quality depends on the nature of the CUT. If all the faults in the CUT have a large number of tests, then a short sequence can provide high fault coverage. In contrast, if many faults in the CUT have very few tests, then a long sequence is required to obtain high fault coverage.

The **detectability** of a fault f in an n-input CUT, κ_f, is the number of vectors that can detect the fault.

Example 12.7 Let f be the stuck-at 1 (SA1) fault at the output of a circuit comprised of a single two-input AND gate. Since f has three tests, {00, 01, 10}, the detectability of f, κ_f, is three. □

The **detectability profile** of a circuit, H, is the frequency-distribution of the detectabilities of all the faults within the circuit. H is a vector $\{h_{\kappa_{min}}, h_{\kappa_{min}+1}, \dots, h_{\kappa_{max}}\}$, where h_κ is the number of faults in the circuit that have detectability κ. Note that the detectability of an untestable fault is zero. We will assume that untestable faults are not included in the set of faults being *targeted* by the analysis. Hence, in our analysis, $1 \leq \kappa_{min}$ and $\kappa_{max} \leq 2^n$. Faults with detectabilities close to κ_{min} are called **hard-to-detect** faults. Finally, we will denote the total number of target faults by n_f, which can be computed as

$$n_f = \sum_{i=\kappa_{min}}^{\kappa_{max}} h_i. \tag{12.14}$$

Example 12.8 If all the six single stuck-at faults (SAFs) at the inputs and output of the above-mentioned circuit, namely a single two-input AND gate, are targeted, then $n_f = 6$ and the detectability profile of the circuit is given by $\{h_1 = 5, h_3 = 1\}$. □

Determination of the detectability profile of a circuit requires an estimation of the number of tests for each target fault. Accurate computation of the detectability profile is hence computationally complex. However, the detectability profile can be estimated by using the techniques described next to estimate fault detectabilities. That will be followed by techniques to estimate the fault coverage vs. test length relationship for random and pseudo-random testing.

12.3.2 Estimation of the detectability profile of a circuit

The following analysis for estimating fault coverage for a given set of randomly selected vectors requires as an input the detectability profile of the circuit. In practice, a probabilistic measure of detectability of each fault f, $D_p(f)$, is first estimated. **Probabilistic detectability** for a fault f in an n-input combinational circuit, $D_p(f)$, is the probability that the fault is detected by a vector P selected randomly from the set of all possible 2^n vectors. The detectability of each fault f can then be estimated as $D_p(f) \times 2^n$.

12.3.2.1 Probabilistic controllability

Central to the estimation of $D_p(f)$ are *probabilistic controllability* values. These values can be viewed as a variation of the controllability values defined in Chapter 4 (Section 4.5.3). Consider an n-input combinational circuit. The **probabilistic 0-controllability** of a line c_i, $CC_p^0(c_i)$, is the probability that logic 0 is implied at line c_i by a vector selected randomly from the set of all possible 2^n n-bit vectors. The **probabilistic 1-controllability** of a line c_i, $CC_p^1(c_i)$, can be defined in an analogous manner.

The probabilistic controllability values have two notable characteristics. First, these measures describe the ease of setting a line to logic 0 or 1 in a probabilistic sense. For example, $CC_p^0(c_i)$ denotes the ease of obtaining logic 0 at line c_i by vectors that are randomly selected. This is useful in BIST since many BIST PGs generate pseudo-random vectors. Note that this is in contrast with the variants of controllability described in Chapter 4 (Section 4.2.3), which provide estimates of the difficulty with which a test generation algorithm can construct a vector that implies the corresponding value at c_i. Second, in this case, a high value of $CC_p^0(c_i)$ indicates that it is easier to set the line c_i to logic 0. This is also in contrast with the variations presented in Chapter 4.

Next, we describe two approaches to compute the values of the above controllabilities.

Circuit traversal approaches: Despite the above differences, the overall framework for computation of controllability values in Chapter 4 (Section 4.2.3) can be used to compute the probabilistic controllability values. In this case, for each primary input x_i, $CC_p^0(x_i)$ and $CC_p^1(x_i)$ are initialized to 0.5. This captures the fact that a PRPG generates logic 0 and logic 1 at each input with equal probabilities. Recall that in the framework in Chapter 4, the *controllability transfer function* provides the rule for computing the controllability values at a gate's output given their values at its inputs. This function can be derived for each type of gate, as illustrated next for a NAND gate with inputs c_1 and c_2 and output c_3. A logic 0 is implied at the output only if logic 1 is implied at both of its inputs. Assuming that the logic values at the two inputs of the NAND gate are *independent*, the following expressions give the probabilistic controllability values at the gate's output in terms of their values at its inputs.

$$CC_p^0(c_3) = CC_p^1(c_1)CC_p^1(c_2), \tag{12.15}$$
$$CC_p^1(c_3) = 1 - CC_p^1(c_1)CC_p^1(c_2). \tag{12.16}$$

Once similar expressions are derived for other types of gates, the breadth-first approach for computing the controllability values described in Chapter 4 can be used to compute the probabilistic controllability values for each line in the circuit.

The main limitation of the above approach is that it assumes independence between the logic values at the inputs of a gate. While this is true for any fanout-free circuit as well as any circuit that does not have any reconvergent fanouts, it is clearly not true for a circuit with reconvergent fanouts. This is illustrated by the following example.

Example 12.9 Consider the circuit shown in Figure 12.9. This circuit has a fanout at x_2 that reconverges at G_4. The above approach is used to compute the probabilistic controllability values for circuit lines, and the probabilistic 0-controllability values, CC_p^0, obtained in this manner are shown in the figure. Note that for any line c_i, $CC_p^1(c_i) = 1 - CC_p^0(c_i)$.

Simulation of all possible vectors shows that the actual probabilistic 0-controllability at the output z is 0.75. In contrast, the above approach estimates this

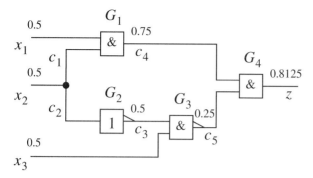

Figure 12.9 Computation of $CC_p^0()$ values via circuit traversal (circuit obtained from Briers and Totton (1986) (© 1986 IEEE))

value to be 0.8125, since it assumes that the logic values at c_4 and c_5 are independent.

□

Several approaches have been proposed to take into account the effects of correlation caused due to reconvergent fanouts. First, in the approach proposed in Parker and McCluskey (1975), equations of the above type are used to obtain *symbolic expressions* of probabilistic controllability values for a line in terms of their values at primary inputs. This approach provides accurate values of signal probability, but its time and memory complexities are too high for it to be practical for any circuit of realistic size. Second, the approach presented in Bardell *et al.* (1987) computes a lower and upper bound on each controllability value at each line that takes into account the effects of possible correlation. Finally, when binary decision diagrams are constructed for a circuit as outlined in Chapter 4, the number of vectors that assign a particular logic value at a line can be computed.

Statistical simulation approaches: Many recent approaches statistically estimate the values of probabilistic controllabilities. In these approaches, a number of vectors are selected randomly from the set of all possible vectors. Fault-free circuit simulation is performed for each selected vector and the number of vectors that assign logic 0 as well as the number of vectors that assign logic 1 at a line c_i are counted. $CC_p^0(c_i)$ and $CC_p^1(c_i)$ are then computed by dividing these counts by the number of vectors simulated. The main weakness of this method is that a large number of vectors may be required to obtain meaningful estimates of CC_p^0 and CC_p^1, especially for lines that have extreme controllability values.

The main advantage of this method is that it naturally captures the effects of correlations caused due to reconvergent fanouts. In some applications, such as *test point insertion* (see Section 12.4), it is desirable to identify gates whose output controllability values are significantly affected by such correlations. Such gates can be identified by using the statistically obtained controllability values in conjunction with the expressions similar to those described in Equations (12.15) and (12.16) as follows. To distinguish between the values computed statistically and those computed at a gate output using equations of the above type, let $CC^0_{p,s}(c_i)$ and $CC^1_{p,s}(c_i)$ denote the probabilistic controllability values computed statistically and let the corresponding values at a line using the above equations be denoted by $CC^0_{p,c}(c_i)$ and $CC^1_{p,c}(c_i)$. Consider a NAND gate G with inputs c_1 and c_2 and output c_3. Assume that the statistical simulation approach has already been used to compute the values of $CC^0_{p,s}$ and $CC^1_{p,s}$ for each of these circuit lines. Next, assuming $CC^0_{p,c}(c_1) = CC^0_{p,s}(c_1)$, $CC^1_{p,c}(c_1) = CC^1_{p,s}(c_1)$, $CC^0_{p,c}(c_2) = CC^0_{p,s}(c_2)$, and $CC^1_{p,c}(c_2) = CC^1_{p,s}(c_2)$, compute the values of $CC^0_{p,c}(c_3)$ and $CC^1_{p,c}(c_3)$ using Equations (12.15) and (12.16). Any significant difference between the values of, say $CC^0_{p,s}(c_3)$ and $CC^0_{p,c}(c_3)$, indicates the presence of correlation between the logic values implied at the inputs of the gate.

Example 12.10   In the previous example, statistical simulation will give, among others, the values, $CC^0_{p,s}(c_4) = 0.75$, $CC^0_{p,s}(c_5) = 0.25$, and $CC^0_{p,s}(z) = 0.75$. $CC^0_{p,c}(z)$ can then be computed using $CC^0_{p,s}(c_4)$ and $CC^0_{p,s}(c_5)$ as

$$CC^0_{p,c}(z) = 1 - \left[1 - CC^0_{p,s}(c_4)\right]\left[1 - CC^0_{p,s}(c_5)\right] = 0.8125.$$

(The above equation is a variation of Equation (12.15).) Since the value of $CC^0_{p,c}(z)$ computed above is different from $CC^0_{p,s}(z)$, a correlation between the logic values at lines c_4 and c_5 is discovered.   □

Statistical simulation is sometimes extended to provide more specific information about the correlation between the logic values at the inputs of various gates. For example, in Briers and Totton (1986) and Tamarapalli and Rajski (1996), during simulation of randomly selected vectors the number of times each distinct *combination* of values appears at the inputs of each gate is also recorded. For example, for the above two-input NAND gate, the number of vectors which imply combinations $(0, 0)$, $(0, 1)$, $(1, 0)$, and $(1, 1)$ at the gate inputs are recorded separately. If one or more of these combinations do not occur in a simulation where adequate number of randomly selected vectors are simulated, then that not only indicates some fairly significant correlation between the values but also provides additional information about the type of correlation. Furthermore, availability of this type of information makes possible the use of more accurate versions of Equations (12.15) and (12.16) that consider correlations between the logic values at the gate inputs (Tamarapalli and Rajski, 1996).

12.3.2.2 Probabilistic observability

The other component that is required to compute the detectability of a particular SAF is the value of observability of a line. Again, we define the **probabilistic observability** of a line c_i in an n-input combinational circuit as the probability that a randomly selected vector will sensitize one or more paths from line c_i to one or more of the circuit outputs. This value can be computed by using the general approach described in Chapter 4 (Section 4.2.3). In this case, we first initialize the observability of each primary output z_i as $O_p(z_i) = 1$. The probabilistic observability is computed for the other circuit lines via a breadth-first, backward traversal of the circuit lines starting at the primary output. Consider a two-input NAND gate with inputs c_1 and c_2 and output c_3. Assume that the probabilistic controllability values are known for each of the above lines and the probabilistic observability is known only for the gate output, c_3. The probabilistic observability of input c_1 can be computed as follows:

$$O_p(c_1) = CC_p^1(c_2)O_p(c_3). \tag{12.17}$$

Similar expressions can be derived for other types of gates.

At a fanout system with stem c_1 and branches $c_2, c_3, \ldots, c_{\beta+1}$, observability of the stem can be computed by using the following equation.

$$O_p(c_1) = \max \left[O_p(c_2), O_p(c_3), \ldots, O_p(c_{\beta+1}) \right]. \tag{12.18}$$

This equation typically underestimates the probabilistic observability of the stem (Jain and Agrawal, 1985); however, in some cases it can provide an overestimate.

Alternatively, especially if statistical simulation is used to compute the controllability values, the **probabilistic sensitivity** of each gate input c_i, $Sens_p(c_i)$, is computed by counting the number of vectors simulated for which the gate input is sensitive to its output, c_j, i.e., a change in the value at the gate input will change the value at its output. Then, the probabilistic sensitivity values can be used during the backward, breadth-first computation of probabilistic observability values instead of Equation (12.17). For example, for a gate with inputs c_1 and c_2 and output c_3, $O_p(c_1)$ can be computed given $O_p(c_3)$ as

$$O_p(c_1) = Sens_p(c_1)O_p(c_3). \tag{12.19}$$

Again, Equation (12.18) can be used at each fanout system.

12.3.2.3 Computation of detectability

Once any combination of the above approaches is used to estimate the probabilistic 0- and 1-controllability and observability of each circuit line, the probabilistic detectability of a fault f that is an SA0 or SA1 fault at line c_i can be computed as

$$D_p(f) = \begin{cases} CC_p^1(c_i)O_p(c_i), & \text{if } f \text{ is an SA0 fault at } c_i, \\ CC_p^0(c_i)O_p(c_i), & \text{if } f \text{ is an SA1 fault at } c_i. \end{cases} \tag{12.20}$$

The detectability profile for the circuit can then be computed by dividing the interval from 0 to 1 into ranges of appropriate widths and merely tabulating the number of faults whose probabilistic detectabilities are within each range.

Next, we study random as well pseudo-random testing scenarios in which a PRPG is used as a parallel test pattern generator. In the latter case, we assume that the PRPG is capable of generating complete sequences. An interesting case arises when the number of stages in a PRPG, v, is greater than n, the number of circuit inputs. In such a scenario, each n-bit test for a fault f in the circuit may be applied up to 2^{v-n} times and the effective detectability of fault f is given by $2^{v-n}\kappa_f$. In the following, we will consider only the case where the PRPG has n stages. The analysis can be repeated using the modified values of fault detectabilities to study the case where the PRPG has a larger number of stages than the number of CUT inputs (McCluskey *et al.*, 1988). The following analysis of random testing can, however, be directly applied to a case in which a v-stage random PG, where $v > n$, is used.

12.3.3 Random testing

Patterns generated by a random pattern generator (RPG) can be viewed as being obtained via *random sampling with replacement* from the set of all possible n-bit vectors, since a vector may be generated more than once.

12.3.3.1 Escape probability of a fault

The probability that a fault f with detectability κ will escape detection after the application of a single vector is given by

$$Q_\kappa^r(1) = \frac{N - \kappa}{N},$$

where $N = 2^n$, since $N - \kappa$ out of N vectors do not detect the fault f. The escape probability of f after the application of l vectors can then be computed as

$$Q_\kappa^r(l) = \left(\frac{N - \kappa}{N}\right)^l. \tag{12.21}$$

Substituting $\frac{N-\kappa}{N}$ for a in the known inequality $\ln(a) \le a - 1$, one can obtain

$$\left(\frac{N - \kappa}{N}\right)^l \le e^{-\kappa l/N}. \tag{12.22}$$

Substitution of the above relation in Equation (12.21) gives

$$Q_\kappa^r(l) \le e^{-\kappa l/N}. \tag{12.23}$$

When $\kappa/N \le 0.1$, the bound in Equation (12.22) is very tight and the two sides can be assumed to be approximately equal (McCluskey *et al.*, 1988). Note that $\kappa/N \le 0.1$ is precisely the case where probabilistic analysis is required since otherwise a very short sequence will detect all the faults. Hence, we can use the following approximation,

$$Q_\kappa^r(l) \approx e^{-\kappa l/N}. \tag{12.24}$$

For a given l, the escape probability of a fault decreases rapidly with increases in κ. The above approximation for $Q_\kappa^r(l)$ illustrates this relationship very clearly by showing that $Q_\kappa^r(l)$ decreases exponentially with an increase in the value of κ. This indicates that hard-to-detect faults play a crucial role in determining the quality of random testing. This will become more clear in the following.

12.3.3.2 Fault coverage estimation

The expected value of fault coverage, $C^r(N_v)$, obtained by the application of a random sequence of length N_v can be computed by using the value of $Q_\kappa^r(l)$ from Equation (12.21). Let $U^r(N_v) = 1 - C^r(N_v)$ be the **uncoverage** for the given random sequence. The expected value of uncoverage can be estimated as a weighted sum of the escape probabilities of all the target faults as

$$E(U^r(N_v)) = \frac{1}{n_f} \sum_{\kappa=\kappa_{min}}^{\kappa_{max}} h_\kappa \left(\frac{N-\kappa}{N}\right)^{N_v}. \tag{12.25}$$

Alternatively, the approximation in Equation (12.24) can be used to obtain

$$E(U^r(N_v)) \approx \frac{1}{n_f} \sum_{\kappa=\kappa_{min}}^{\kappa_{max}} h_\kappa e^{-\kappa N_v/N}. \tag{12.26}$$

Note that since the value of Q_κ^r decreases rapidly with an increase in the value of κ, typically only the first few terms of the summation in Equation (12.25) are numerically significant. However, it should be noted that each term in the above summation has the form $h_\kappa \left(\frac{N-\kappa}{N}\right)^{N_v}$. Hence, in circuits where h_κ increases rapidly from $\kappa = \kappa_{min}$ to some value, κ^*, even faults with higher detectabilities may contribute to the uncoverage.

The expected value of fault coverage can be simply obtained as $E(C^r(N_v)) = 1 - E(U^r(N_v))$.

12.3.3.3 Estimation of test length

The random test length is typically computed to ensure that each target fault is detected with a probability exceeding $1 - \epsilon$, where ϵ is the desired level of confidence. Since the hardest-to-detect faults, i.e., faults with detectabilities κ_{min}, have the highest escape probability, a test sequence length that guarantees the detection of a fault with detectability κ_{min} with the desired level of confidence will achieve the above objective. In other words, N_v must be computed such that $Q_{\kappa_{min}}^r(N_v) \le \epsilon$.

Using the expressions for $Q_\kappa^r(l)$ from Equations (12.21) and (12.24), we get the minimum test length required to achieve this level of confidence as

$$N_v = \frac{\ln(\epsilon)}{\ln\left(\frac{N-\kappa_{min}}{N}\right)}, \text{ and} \tag{12.27}$$

$$N_v = -\frac{\ln(\epsilon)N}{\kappa_{min}}. \tag{12.28}$$

12.3.4 Pseudo-random testing

PGs that generate pseudo-random sequences, such as ML-LFSRs, are used extensively in BIST. Next, we study PRPGs that generate complete sequences. Recall that such PGs generate sequences that exhibit randomness and can generate complete sequences without repeating any vectors. Hence, vectors generated by a PRPG can be viewed as being obtained via *random sampling without replacement* from the set of all possible N vectors. Also, recall that each vector is generated with equal probability.

12.3.4.1 Escape probability of a fault

Consider a fault f with detectability κ. The probability that the fault will escape detection by the first vector, $Q_\kappa^{pr}(1)$, is given by

$$Q_\kappa^{pr}(1) = \frac{N - \kappa}{N},$$

since $N - \kappa$ out of N vectors do not detect the fault. If the application of the first vector does not detect f, in the next step, one of the remaining $N - 1$ vectors is applied; of this set of vectors, $N - \kappa - 1$ are not tests for f. Using this reasoning in an iterative fashion, the following expression can be obtained for the escape probability of the fault f after the application of l vectors (Malaiya and Yang, 1984; McCluskey *et al.*, 1988).

$$Q_\kappa^{pr}(l) = \left(\frac{N - \kappa}{N}\right)\left(\frac{N - \kappa - 1}{N - 1}\right) \cdots \left(\frac{N - \kappa - l + 1}{N - l + 1}\right)$$

$$= \prod_{i=1}^{l}\left(\frac{N - \kappa - i + 1}{N - i + 1}\right). \tag{12.29}$$

The following inequalities can be derived using the result that $\frac{a}{b} > \frac{a-1}{b-1}$, if $a < b$.

$$\frac{N - \kappa - j + 1}{N - j + 1} \le \frac{N - \kappa}{N}, \quad \text{for all } j \in \{1, \ldots, l\}.$$

$$\frac{N - \kappa - l + 1}{N - l + 1} \le \frac{N - \kappa - j + 1}{N - j + 1}, \quad \text{for all } j \in \{1, \ldots, l\}.$$

Substitution of the above inequalities in Equation (12.29) helps bound the value of $Q_\kappa^{pr}(l)$ from below and above as

$$\left(\frac{N - \kappa - l + 1}{N - l + 1}\right)^l \le Q_\kappa^{pr}(l) \le \left(\frac{N - \kappa}{N}\right)^l. \tag{12.30}$$

Note that the above inequality demonstrates that $Q_\kappa^{pr}(l) \le Q_\kappa^r(l)$. This is due to the fact that vectors are sampled without replacement in pseudo-random testing. The above upper bound is tight when $l \ll N$ and $l \ll N - \kappa$. Typically, $l \ll N$ in most pseudo-random testing scenarios. Also, if $l \ll N$, then $l \ll N - \kappa$ for hard-to-detect faults, since for such faults, $\kappa \ll N$.

Using the bound given in Equation (12.23) in the above upper bound gives

$$Q_\kappa^{\text{pr}}(l) \leq e^{-\kappa l/N}. \tag{12.31}$$

12.3.4.2 Fault coverage estimation

The expected value of fault coverage, $C^{\text{pr}}(N_\text{v})$, obtained by application of a pseudo-random sequence of length N_v can be computed using the value of $Q_\kappa^{\text{pr}}(l)$ from Equation (12.29). Let $U^{\text{pr}}(N_\text{v}) = 1 - C^{\text{pr}}(N_\text{v})$ be the uncoverage for the given pseudo-random sequence. The expected value of uncoverage can be computed as a weighted sum of the escape probabilities of all the target faults as

$$
\begin{aligned}
E(U^{\text{pr}}(N_\text{v})) &= \frac{1}{n_\text{f}} \sum_{j=\kappa_{\min}}^{\kappa_{\max}} h_j Q_j^{\text{pr}}(N_\text{v}) \\
&= \frac{1}{n_\text{f}} \sum_{j=\kappa_{\min}}^{\kappa_{\max}} h_j \prod_{i=1}^{N_\text{v}} \left(\frac{N - j - i + 1}{N - i + 1} \right).
\end{aligned} \tag{12.32}
$$

Recall that n_f is the number of target faults.

Alternatively, the bound in Equation (12.31) can be used to obtain

$$E(U^{\text{pr}}(N_\text{v})) \leq \frac{1}{n_\text{f}} \sum_{j=\kappa_{\min}}^{\kappa_{\max}} h_j e^{-j N_\text{v}/N}. \tag{12.33}$$

12.3.4.3 Estimation of test length

Since $Q_\kappa^{\text{pr}}(l) \leq Q_\kappa^{\text{r}}(l)$, the values of N_v given in Equations (12.27) and (12.28) are clearly sufficient to achieve a test confidence ϵ. Hence, these values are upper bounds on the test length that provides this test confidence.

The lower bound given in Equation (12.30) can be used to show that the smallest value of N_v that satisfies the relation

$$\left(\frac{N - \kappa_{\min} - N_\text{v} + 1}{N - N_\text{v} + 1} \right)^{N_\text{v}} \leq \epsilon \tag{12.34}$$

is a lower bound on the value of N_v that will provide the desired level of test confidence.

In this manner, a range can be computed within which lies the minimum N_v value that can achieve the given level of test confidence. If the upper bound on N_v (given by Equation (12.27) or (12.28)) is much smaller than N, then it is also very close to the lower bound described above.

12.4   Test points to improve testability

In scenarios where expected fault coverage for a given CUT is less than that desired under the given constraint on test length, one is faced with two alternatives. Either the CUT can be made more testable via insertion of *test points* to improve the detectabilities of hard-to-detect faults, or the PG can be customized to generate vectors that are more suitable for the given CUT. These two alternatives are, respectively, discussed in this and the next section.

Practical tools for BIST commonly use test points to enhance the testability of a CUT for which commonly used BIST PGs fail to provide the desired fault coverage under the given constraints on test length. The popularity of such methods stems from the fact that they guarantee achievement of any desired testability target, provided a sufficient number of test points is used. The costs of attaining this guarantee are area overhead and potential performance penalty. The area overhead is due to the additional logic and routing required to add the test points to the circuit. Performance penalty is sustained if the additional circuitry increases delays of critical paths.

12.4.1   Types of test points

Test points can be divided into two broad categories. The first type is a **control point**, where circuitry is added to a help control the value at a line, called the **site** of the control point. By enabling control of the value at its site, a control point also alters controllabilities of the lines in its fanout. As will be seen later, a control point also alters the observabilities of lines. The second type of test point is an **observation point**, which enhances the observability of its site by enabling observation of the value at the line. In this manner, an observation point also enhances observabilities of lines in the fanin of its site.

Figure 12.10 shows some example implementations of control points. Figure 12.10(a) shows part of a circuit where line c_i is the output of gate G_1 as well as one of the inputs of G_2. In the normal mode, the value at line c_i is determined by the values applied at the inputs of G_1 and the function implemented by G_1. The values implied by different vectors at c_i can be changed to facilitate testing by adding a control point with c_i as its site. Figure 12.10(b)–(e) show some common control points. The one shown in Figure 12.10(b) is sometimes called a **zero control point**, since application of logic 0 at y_i causes the value at c_i to become logic 0, irrespective of the value implied by a vector at the output of G_1. Note that when logic 1 is applied to y_i, then this control point is **inactive**, i.e., the logic operation of the circuit is not affected by the presence of this control point. Similarly, the one shown in Figure 12.10(c) is called a **one control point** which is inactive when logic 0 is applied to y_i. The alternative shown in Figure 12.10(d) can be called an **inversion control point**. This control point

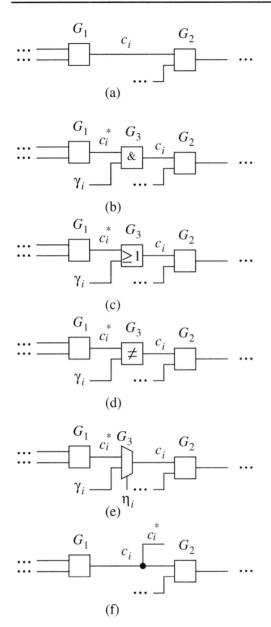

Figure 12.10 Types of test points: (a) original circuit, (b) a zero control point, (c) a one control point, (d) an inversion control point, (e) a complete control point, and (f) an observation point

is inactive when a logic 0 is applied at γ_i; when logic 1 is applied at γ_i, the value at c_i is complement of that implied by the vector at the output of G_1. Finally, the alternative in Figure 12.10(e) provides the greatest degree of control and will be called a **complete control point**. Application of suitable values at inputs γ_i and η_i of this design can make it operate in the normal mode or as a zero control point or as a one control point.

Note that even though in Figure 12.10 each type of control point is depicted as a separate gate, it may actually be implemented as a part of one of the gates in the original circuit. For example, a zero control point shown in Figure 12.10(b) may be implemented by adding transistors to the pull-up and pull-down networks of gate G_1, as described in Savaria and Kaminska (1988). In a custom VLSI design, most of the control points may be implemented in this manner to reduce their performance penalty. In designs obtained using standard cell libraries, a control point is typically implemented as a separate gate, however.

An observation point is considerably simpler to implement. As shown in Figure 12.10(f), all that is required is creation of a fanout of line c_i, say c_i^*, that is observed as a primary output of the CUT.

12.4.2 Effects of test points

An appropriate type of control point at line c_i can improve the 0-controllability and/or 1-controllability of line c_i. It can also change the controllabilities at the lines in the transitive fanout of c_i.

12.4.2.1 Causes of poor controllability

A circuit line may have a poor controllability for two main reasons. First, in some instances, a line c_i may have an extreme controllability value because of the inherent nature of the gates in its transitive fanin. For example, the output of a 16-input NAND gate shown in Figure 12.11(a) has low 0-controllability, despite the fact that each input of the gate is a primary input and hence has good 0- as well as 1-controllability. Note that this would also be case if the 16-input NAND was implemented as a tree of two-input AND and NAND gates. However, in that case, the controllability values will take on extreme values only at lines closer to the output of the tree. Second, in some instances, correlation between logic values implied at the input of a gate may cause the output of have poor 0- or 1-controllability, despite the fact that each of the inputs of the gate has a good controllability value. Such a correlation is often caused by the presence of reconvergent fanouts in the sub-circuit driving line c_i. An example of such a case is shown in Figure 12.11(b). In this circuit, the 1-controllability of line z is zero, despite the fact that each of the inputs of the AND gate G_3 that drives z has a non-zero value for 1-controllability. (Recall that in this chapter, we are using probabilistic versions of controllabilities and observability.) This indicates that due to the fanouts driving lines c_5 and c_6, the event of obtaining logic 1 at c_5 is not independent of that of obtaining logic 1 at c_6.

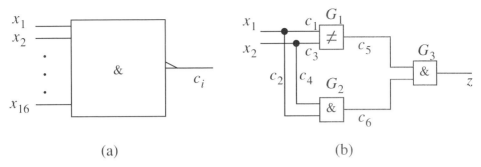

Figure 12.11 Causes of poor controllability: (a) high fan-in, and (b) correlation between logic values at a gate's inputs

12.4.2.2 Effects of control points

The manner in which a control point affects controllabilities of lines mirrors the two cases described above.

In the first case discussed above, depicted in Figure 12.11(a), the 0-controllability of the NAND gate's output can be enhanced by adding at c_i a control point of any type **other than** a one control point. For example, if a zero control point is used at c_i, then logic 0 is obtained whenever a 0 is applied at the corresponding control point input, y_1. In contrast, when a logic 1 is applied at the control point input, the value at the output is unaltered by the presence of the control point. Note that when such a control point is active, the effects of faults in the transitive fanin of the control point site cannot propagate via its site. This is only one example of how the activation of a control point may affect observabilities of other lines.

Alternatively, an inversion control point may be used. In this case, when logic 1 is applied at the control point input, the complement of the value implied at the output of the gate is applied at c_i. Application of logic 0 at the control point input inactivates such a control point.

For either of the above alternatives, the desired 0-controllability at line c_i can be obtained by activating the control point an appropriate number of times. In contrast to the one control point, an inversion control point does not prevent the propagation, via its site, of the effects of faults in its site's transitive fanin.

The other alternative of adding a one control point at each of its inputs may also enhance the 0-controllability at the gate's output. However, in the case of a 16-input gate, it is not possible to improve the 0-controllability of c_i unless the 1-controllability of each input of the NAND gate is significantly increased. In a more general scenario, where some or all the inputs of this 16-input NAND may be driven by cones of logic, this may adversely effect the detectabilities of faults in the fanin cones of their sites.

A third alternative is to first decompose the 16-input NAND gate into a tree of gates with a smaller number of inputs. For example, the 16-input NAND may be replaced by four four-input AND gates whose outputs drive the inputs of a four-input NAND gate. Once the gate is decomposed in this manner, appropriate control points can be inserted at some of the lines internal to the tree of gates to enhance the controllabilities as desired.

In the second case, which is depicted in Figure 12.11(b), the 1-controllability of z can be enhanced by adding a one control point, inversion control point, or complete control point at z. Alternatively, a one, inversion, or complete control point can be added at c_5 or c_6. In some of these cases, activation of the added control point can completely eliminate the harmful correlation between the values at c_5 and c_6. Addition and activation of an appropriate control point at c_1, c_2, c_3, or c_4 can help reduce, but not completely eliminate, the detrimental effects of this correlation on the 1-controllability at z.

The following general observations can be inferred from the above examples. First, addition of a control point can help enhance the 0- or 1-controllability of a line that is inherently uncontrollable due to the nature of the function implemented by the fanin cone of the logic. Control points can also be used to reduce or eliminate the effects of correlations that might be the cause of low 0- or 1-controllability at some circuit lines.

In addition to the above effects on controllability values, when activated a control point also affects the observability values of lines in the circuit. The above examples have already illustrated some ways in which the observabilities of lines in the fanin of the site of a control point are affected by activation of the control point. An inversion control point typically causes lower reduction in observabilities of the lines in the fanin of its site.

Activation of a control point may also affect the observabilities of lines that are in the transitive fanin of lines in its site's transitive fanout. The following example illustrates this fact.

Example 12.11 Consider the part of a circuit shown in Figure 12.12, where the 1-controllability of line c_1 is low. A one control point can be added at, say, c_1 to improve the 1-controllability of c_1. Assuming that the circuit lines are independent, this will increase the 1-controllabilities of the lines in the transitive fanout of c_1 including c_3, c_7, c_4, and c_8.

Let us examine the effects on the observabilities of lines in various parts of the circuit. Clearly, a higher 1-controllability of line c_3 can potentially enhance the observabilities of the lines in the fanin cone of c_2. Similarly, the increase in 1-controllability of c_8 can potentially enhance the observabilities of the lines in the fanin cone of c_9. In contrast, the increase in 1-controllability of c_4 can potentially decrease the observabilities of the lines in the transitive fanin of c_5. The observabilities of the lines in the transitive fanin of c_6 may also be adversely affected. □

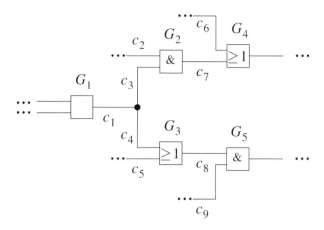

Figure 12.12   An example scenario illustrating the effects of a control point

12.4.2.3 Observation points

The causes of low observability mirror those for low controllability, namely (a) inherently low observability due to long propagation paths with unfavorable controllabilities at the inputs of gates along the paths, and (b) adverse correlation because of reconvergence.

An observation point at line c_i enhances the observability of lines in the fanin cone of c_i. It has no impact on the controllability of any line in the circuit. Hence, addition of an observation point can only enhance the detectability of faults in a circuit.

12.4.3   Selection of test points

Some approaches for test point selection aim to improve the testability of the circuit for any set of pseudo-random (or random) vectors while others try to improve the testability for a given sequence of vectors. The approaches of the first type often use testability measures to characterize the testability of the circuit. Approaches of the second type rely on some combination of logic simulation, fault simulation, statistical estimation of controllability and observability values, path tracing, and using probabilistic expressions of the type described in Section 12.3.2 or their advanced versions. In the following, we describe the key steps that are common to the approaches of the second type. In this discussion, we provide the details of how each step is carried out in one of the approaches (Briers and Totton, 1986).

Simulation of given vectors: The first key task in dealing with test point insertion for a given set of vectors is logic and fault simulation. This can help identify the faults that are detected by the given set of vectors. The remaining faults can be further classified into (i) faults that were not excited, (ii) faults at gate inputs that were excited but not propagated even to the output of the gate with the fault site, and (iii) faults that were excited and propagated to some gate output.

The faults in the first two categories require control points for their detection while those in the third category may be detected by adding observation points only. However, since addition of control points may help detect the faults in the third category, control point selection may be carried out before observation point selection.

At each line, the number of vectors that imply a logic 0 as well as the number of vectors that imply a logic 1 at the line are recorded. In some approaches, distinct combinations of patterns that appear at the inputs of each gate are also recorded.

Identification of candidate control points: The second key task is enumeration of candidate control points. At each line, a zero control point and a one control point may be considered. The list of candidates must include those that facilitate detection of faults not detected by the given vectors. Special consideration may be given to the faults in the first two categories (see (i) and (ii) above). Faults in the first category are due to extreme controllability at the fault site. For example, a fault f that is an SA0 fault at line c_i belongs to this category if the 1-controllability of c_i is low. One candidate test point that can facilitate excitation of f is a one control point at c_i. In addition, control points in the fanin of c_i may also help accomplish this. Low 1-controllability at c_i may be caused due to high fanin of the input cone of c_i or due to correlations. The presence of correlation can be identified by comparing the 0-controllability of c_i as determined via counting during the statistical simulation step and those recomputed using the appropriate controllability values at the gate inputs in the probabilistic expressions (see Example 12.10 in Section 12.3.2).

If the controllability at site c_i of fault f is low due to high fanin, then candidate control points are lines in the fanin of c_i with controllability values that may be changed to obtain a desired change in the controllability values at c_i. If the controllability at c_i is low due to a correlation, then fanout stems whose branches reconverge at the gate with output c_i can be identified via backward traversal of circuit paths starting at c_i. Of these stems, the stem(s) that cause the desire correlation can then be identified. One way in which this can be accomplished is via re-simulation of the given vectors with the following change. In each re-simulation run, one of the branches of one of these stems is assigned a value that is complementary to the value implied by the vector at its stem. The fanout stem is declared the cause of undesirable controllability at c_i if the above re-simulation at the stem improves the controllability at c_i. In such a case, all the lines along the path between the fanout stem and c_i are considered as sites for candidate control points. For example, based on the above analysis of the circuit shown in Figure 12.11(b), lines c_1, c_2, c_3, c_4, c_5, and c_6, will be declared as potential sites for control points for improving the 1-controllability at z.

Next, consider a fault in the second category that is excited but whose effect is not propagated even to the output of the gate at whose input the fault is located. In such a case, the above re-simulation approach can be used to identify the fanout stems which cause the undesirable correlation at the inputs of the gate with the target fault site.

Selection of control points: Once control point candidates are identified for each fault in the first two categories, a minimum number of control points can be identified. This can be formulated approximately as a minimum covering problem where a minimum number of control points are selected such that at least one control point candidate is included for each fault in the above two categories. Covering is often performed heuristically, since the number of candidates in a typical circuit can be large.

Observation points: Once the control points are selected, the given vectors are augmented to accommodate the additional inputs associated with the control points and fault simulation is performed. Hopefully, each fault not detected by the original vectors is now either detected or falls into the third category. For each fault of the latter type, each circuit line to which the effect of the fault propagates is identified as a potential observation point candidate. A covering problem is then formulated and solved to identify observation points to be added.

12.5 Custom pattern generators for a given circuit

A PRPG is versatile in the sense that it can generate vectors that are likely to provide reasonable coverage for a variety of different CUTs. However, if a PRPG fails to provide the desired fault coverage for a particular CUT within a given test length, a custom PG may be designed to provide a higher fault coverage for the CUT.

Two considerations are central to the design of a custom PG for a given CUT. First, the PG should be able to generate vectors that detect faults in the given CUT. Second, the PG should be inexpensive to implement in hardware.

A large number of approaches have been proposed to achieve the above objectives. We discuss techniques that can be broadly classified into one of three categories, namely (1) embedding vectors, (2) embedding statistics of vectors, and (3) embedding subspaces of vectors.

12.5.1 Embedding vectors

In this category are approaches which first generate test vectors for the target faults in the CUT and subsequently design the PG such that it generates a sequence of vectors that cover the test vectors for the CUT. At one extreme, this involves the selection of an RPG or PRPG, such as an LFSR with an appropriate feedback polynomial, and a seed (initial state). At the other extreme, this may involve major redesign of a standard RPG. (In the following discussion, we will use RPG to refer to RPG as well as PRPG.)

12.5.1.1 Selection of an LFSR and seed

An n-stage LFSR with a primitive feedback polynomial generates an MLS containing $2^n - 1$ parallel test vectors in some order that satisfies many criteria for randomness. The problem of finding the optimal seed for the LFSR can be formulated as follows.

Assume that a test set containing test vectors $P_1, P_2, \ldots, P_{N_v}$ is generated for the given circuit using a deterministic ATPG system. An optimal seed is an initial state of the LFSR starting at which the shortest sequence of PTVs covers each of the above test vectors. The following example illustrates this process.

Example 12.12   Consider a five-input combinational CUT where five faults need to be targeted. Consider a scenario where a combinational ATPG generates the following set of test vectors

$$\{P_1 = (0, 1, 1, 0, 1), P_2 = (1, 0, 0, 1, 0), P_3 = (0, 1, 1, 1, 1), P_4 = (1, 0, 1, 0, 1)\}.$$

Table 12.2(a) shows the sequence of PTVs generated by a five-stage, internal-XOR LFSR with feedback polynomial $\phi(x) = x^5 + x^2 + 1$. If this LFSR is used with a default seed value, say, $(1, 0, 0, 0, 0)$, then 30 vectors must be applied before each of the above four test vectors is covered by the LFSR generated sequence.

In contrast, one can verify that if the same LFSR is used in conjunction with the seed $(1, 0, 1, 0, 1)$, then the four test vectors in the above set are covered by a sequence of eight vectors generated by the LFSR.   □

Suppose an LFSR with a particular feedback polynomial is given. The position of a fully specified vector in the sequence of PTVs generated by the LFSR, starting with the initial state $(1, 0, \ldots, 0)$, is known as the vector's **discrete logarithm** (e.g., see Lempel *et al.* (1995)). In the above example, for an LFSR with $\phi(x) = x^5 + x^2 + 1$, the discrete logarithm of vector $P_1 = (0, 1, 1, 0, 1)$ is 28 and that of vector $P_2 = (1, 0, 0, 1, 0)$ is 29. Once the discrete logarithm of each test vector for the CUT is computed, the optimal seed can be identified by finding the seed that minimizes the maximum difference (taken modulo-$(2^n - 1)$) between the discrete logarithms of any pair of vectors in the given set. A heuristic methodology for solving this problem is presented in Lempel *et al.* (1995).

By repeating the above process for various primitive polynomials of degree n, a combination of an LFSR and seed that minimizes the test application time can be found.

It should be noted that the above approach may provide a combination of an LFSR and seed with lower test length if multiple vectors are provided for each target fault. The following example illustrates this process.

Example 12.13 Consider the five-input combinational CUT and five target faults discussed in the previous example. Consider a scenario where a combinational ATPG generates an incompletely-specified test vector for each fault:

$$\{(0, 1, 1, 0, x), (0, 1, 1, 1, x), (1, x, 1, x, x), (x, 1, x, 0, x), (x, 0, x, x, 0)\}.$$

Using Table 12.2(a), one can verify that for the LFSR with primitive feedback polynomial $\phi(x) = x^5 + x^2 + 1$ and seed $(0, 1, 1, 0, 0)$, the same five target faults considered in the above example are covered using a sequence of six vectors.   □

The above example illustrates the advantages of generating multiple test vectors for each target fault. In the limit, every test vector for each target fault may be generated. However, the complexity of such a direct approach is impractically high. First, generation of each possible test vector for a target fault may itself have impractically high run-time complexity. Subsequently, the discrete logarithm must be computed for each generated vector and the covering problem solved.

In practice, the benefits of considering alternative vectors for each target fault can be obtained at significantly lower complexity than the direct approach described above. The key idea is to generate several test vectors (all, if possible) for each hard-to-detect fault for use during the selection of LFSR feedback polynomial and seed. By using a suitable criteria for determining which faults should be considered as hard-to-detect in the above process, the coverage of remaining faults can be assured with a high level of confidence (see Lempel *et al.* (1995) for one such criterion).

12.5.1.2 LFSR re-seeding

In the above approach, starting with a seed, the sequence of vectors generated by an LFSR is applied to the circuit until the last of the desired test vectors is applied to the given CUT. During this process, many vectors that are not required to test faults in the CUT must be applied. Clearly, the test length can be decreased if N_{v1} LFSR generated vectors are applied starting with one seed, followed by application of N_{v2} vectors starting with a second seed, and so on. In the limit, each desired vector can be used as a seed. While such an extreme approach minimizes test application time, the area overhead of BIST is often impractical since each vector must be stored on chip for use as a seed. In practice, use of a small number of seeds can yield a PG design where the test length is significantly smaller than that for a single seed while the area overhead of BIST PG is acceptably small.

Example 12.14 Consider the five-input combinational CUT and five target faults discussed in the previous example. Consider the scenario where a combinational ATPG provides an incompletely-specified test vector for each fault:

$$\{(0, 1, 1, 0, x), (0, 1, 1, 1, x), (1, x, 1, x, x), (x, 1, x, 0, x), (x, 0, x, x, 0)\}.$$

Using Table 12.2(a), one can verify that each of the five target faults is detected using an LFSR with primitive feedback polynomial $\phi(x) = x^5 + x^2 + 1$ and (1) applying two vectors starting with seed $(0, 1, 1, 0, 0)$, and (2) applying two vectors starting with seed $(1, 1, 1, 1, 0)$. □

Several heuristic approaches have been proposed to identify multiple seeds and number of vectors applied starting at the seed so as to minimize the overall test application time under a given constraint on the maximum number of seeds. A greedy heuristic starts with a vector for the fault with the minimum detectability as a seed and applies a specific number of vectors. If some target faults remain undetected, then the above process is repeated using a vector for that fault that has the minimum detectability among faults that remain undetected.

12.5.1.3 Modifying the sequence generated by an LFSR

The test application time using an LFSR is typically significantly larger than what is required for applying the test set generated using a deterministic ATPG. This is due to the fact that in an LFSR-generated sequence, vectors that detect faults in a given CUT (*useful vectors*) are separated by many other vectors that do not detect any of the remaining target faults in the CUT (*non-useful vectors*). The test application time using an LFSR can hence be reduced if non-useful vectors can be replaced by useful vectors that occur much later in the sequence. This would increase the frequency with which useful vectors are applied to the circuit and hence reduce the test application time.

The approach is illustrated in the following example.

Example 12.15 Once again, consider the five-input combinational CUT and five target faults discussed in the previous example, where a combinational ATPG provides an incompletely-specified test vector for each fault:

$$\{(0, 1, 1, 0, x), (0, 1, 1, 1, x), (1, x, 1, x, x), (x, 1, x, 0, x), (x, 0, x, x, 0)\}.$$

Table 12.2(b) shows that if an LFSR with primitive feedback polynomial $x^5 + x^3 + 1$ is used in conjunction with seed $(0, 1, 1, 1, 1)$ to apply four vectors, then the sequence of vectors $(0, 1, 1, 1, 1)$, $(1, 0, 1, 0, 1)$, $(1, 1, 0, 0, 0)$, and $(0, 1, 1, 0, 0)$ is applied to the circuit.

The only incompletely-specified vector in the set of test vectors for the CUT that is not covered by the above set of vectors is $(x, 0, x, x, 0)$. Furthermore, each vector in the sequence generated by the LFSR, **except** $(1, 1, 0, 0, 0)$, covers one or more of the test vectors for the CUT that is not covered by any other vector in the sequence.

Consider a five-input, five-output combinational circuit that implements the following set of incompletely specified logic functions.

Input values	Output values
$(0, 1, 1, 1, 1)$	$(0, 1, 1, 1, x)$
$(1, 0, 1, 0, 1)$	$(1, x, 1, x, x)$
$(1, 1, 0, 0, 0)$	$(x, 0, x, x, 0)$
$(0, 1, 1, 0, 0)$	$(0, 1, 1, 0, x)$

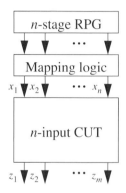

Figure 12.13 General architecture of remapping PG (Chatterjee and Pradhan, 1995; Touba and McCluskey, 1995) (© 1995 IEEE)

If any such circuit is added between the outputs of the LFSR and the inputs of the CUT, then all the five target faults in the CUT can be tested using four vectors generated by the above LFSR and using a single seed, $(0, 1, 1, 1, 1)$. □

The general architecture of a PG for this approach is depicted in Figure 12.13. The outputs of the n-stage RPG are input to a mapping logic and the outputs of the mapping logic drive the inputs of the CUT. The mapping logic is so called because it maps vectors generated by the RPG to vectors to be applied at CUT inputs. Any RPG generated vector that is deemed necessary for detection of one or more target faults in the CUT passes unchanged via the mapping logic. A subset of the other vectors generated by the RPG is mapped to test vectors for faults that are not detected by the RPG vectors that pass unchanged via the mapping logic. The remainder of the RPG generated vectors can be mapped to any vector.

The area overhead of this technique depends on the complexity of the mapping logic. The complexity of this logic depends on the choice of the RPG vectors that are mapped to the desired test vectors for the CUT as well as the exact nature of this mapping. Procedures to design low-complexity mapping logic are based on approaches in logic synthesis and their details can be found in Chatterjee and Pradhan (1995) and Touba and McCluskey (1995).

12.5.2   Embedding statistics of vectors

Weighted-random testing is an alternative approach for obtaining high fault coverage for a circuit for which an RPG fails to provide the desired fault coverage at acceptable test length. In this approach, the characteristics of test vectors for hard-to-detect faults are captured using a number of n-tuples of probabilities that are used to determine the frequency with which logic 1 is applied at the inputs of an n-input CUT.

Each n-tuple of probabilities, $w_i = (w_{i,1}, w_{i,2}, \ldots, w_{i,n})$, is called a **weight set**. When weight set w_i is used to generate weighted random vectors, the **weighted-random pattern generator (WRPG)** applies logic 1 to the jth CUT input with probability $w_{i,j}$. For example, when the weight set $(0.5, 0.25, 0.75, 0, 1)$ is used to generate patterns for a five-input CUT, logic 1 is applied to the first CUT input with probability 0.5 as by a classical, non-weighted RPG. In contrast, logic 1 is not applied to the fourth CUT input and is always applied to the fifth input. Finally, logic 1 is applied to the second and third CUT inputs with probabilities 0.25 and 0.75, respectively. Note that a classical RPG, such as an ML-LFSR, can be viewed as a special case given by a weight set $w_i = (0.5, 0.5, \ldots, 0.5)$.

The motivation behind weighted-random testing can be illustrated using example scenarios. Consider a scenario where an SAF in an n-input CUT requires the vector $(1, 1, \ldots, 1)$. A classical RPG that applies a logic 1 (and hence also logic 0) at each input with probability 0.5 generates this vector with a probability 0.5^n, which is very small for large n. A weighted RPG can enhance this probability by, say, using a weight set $(0.75, 0.75, \ldots, 0.75)$, to 0.75^n.

12.5.2.1 Computation of weight sets

The purpose of weight set computation is to capture the test vector requirements of the target faults in the given CUT. Consider a weight set $w_i = (w_{i,1}, w_{i,2}, \ldots, w_{i,n})$ and a vector $P_j = (p_{j,1}, p_{j,2}, \ldots, p_{j,n})$, where $p_{j,l} \in \{0, 1, x\}$. The probability that a given test vector P_j is generated by a WRPG using the weight set w_i is (Hartmann and Kemnitz, 1993)

$$pr(j, i) = \prod_{\forall l \mid p_{j,l} \neq x} \left[w_{i,l} p_{j,l} + (1 - w_{i,l})(1 - p_{j,l}) \right]. \tag{12.35}$$

Now consider a scenario where one is given a set containing N_{tv} test vectors whose application to the CUT is deemed necessary for the coverage of target faults. Let $Tses_i$ be a **test session** in which WRPG uses weight set w_i to generate and apply $N_v^{ses_i}$ vectors to the CUT. The probability that vector P_j is **not** applied to the CUT in session $Tses_i$ can be computed as (Hartmann and Kemnitz, 1993)

$$U_v(j, i) = [1 - pr(j, i)]^{N_v^{ses_i}}. \tag{12.36}$$

In $Tses_i$ in which $N_v^{ses_i}$ vectors are generated using weight set w_i, the expected fraction of the given set of N_{tv} vectors for the CUT that are **not** applied is given by

$$U(i) = \frac{1}{N_{tv}} \sum_{j=1}^{N_{tv}} U_v(j, i).$$

(12.37)

This quantity is called the **expected uncoverage** of the ith test session. The ith weight set should be selected to minimize the expected uncoverage of the session.

Given a set of test vectors for the CUT, the following intuitive heuristic may be used to compute a weight set. The weight corresponding to the lth input, $w_{i,l}$, can be computed as the ratio of the number vectors in the given set that contain a 1 in their l^{th} component to the number of vectors in the given test set that contain a completely specified value (i.e., 0 or 1) in that component.

Example 12.16   Consider the five-input combinational CUT and five target faults for which a combinational ATPG has generated the following set of incompletely-specified vectors.

$$\{(0, 1, 1, 0, x), (0, 1, 1, 1, x), (1, x, 1, x, x), (x, 1, x, 0, x), (x, 0, x, x, 0)\}.$$

The weight for the first input, $w_{i,1}$, can be computed as 1/3, since one vector requires logic 1 at the first CUT input and a total of three vectors require completely specified values at that input. Similarly, one can compute $w_{i,2} = 3/4$, $w_{i,3} = 1$, $w_{i,4} = 1/3$, and $w_{i,5} = 0$.

The resulting weight set is $w_i = (1/3, 3/4, 1, 1/3, 0)$.   □

An approach to compute optimal weights by using numerical downhill-climbing to minimize the value of the expected uncoverage of a weight set is presented in Hartmann and Kemnitz (1993). It takes advantage of the fact that for the inputs for which the above heuristic provides an extreme weight, i.e., 0 or 1, that weight value is optimal. Hence, the numerical approach needs to only search for better values for the remaining weights.

12.5.2.2 Multiple weight sets

As seen above, each test vector for the CUT affects the values of weights within a weight set. Often a weight set that is optimal for one subset of vectors in the set of test vectors for a given CUT is unsuitable for the remainder of the vectors in the test set. In some cases, different CUT test vectors may pose contradictory requirements on the weight set in such a manner that the optimal set of weights for the complete set of test vectors for the CUT takes the form $(0.5, 0.5, \ldots, 0.5)$. The following example illustrates such a scenario.

Example 12.17   Consider a CUT that requires the following set of test vectors for detection of target faults (Waicukauski *et al.*, 1989; Wunderlich, 1990):

$$\{(1, 1, 1), (0, 1, 1), (1, 0, 1), (1, 1, 0), (0, 0, 0), (1, 0, 0), (0, 1, 0), (0, 0, 1)\}.$$

The above heuristic for computing a weight set gives the set $(0.5, 0.5, 0.5)$. For this set of test vectors, this happens to be the optimal weight set (Wunderlich, 1990).   □

When taken in its entirety, the above set of test vectors leads to a weight set that corresponds to classical unweighted testing due to the fact that the requirements of one subset of test vectors in the set (say, the first four test vectors) are diametrically opposite of those of another subset (the last four test vectors). Even in cases where the requirements of different partitions of the test vector set for the CUT are not completely contradictory and the optimal weight set for the entire set of test vectors is different from $(0.5, 0.5, \ldots, 0.5)$, the resulting weight set may fail to provide the desired fault coverage within the desired test length.

One approach that is commonly used to improve the effectiveness of weighted random testing in the presence of such contradictory requirements is to use multiple weight sets. Conceptually, each weight set represents optimally the requirements of one subset of test vectors for the CUT. The computation of weight sets is typically performed as described below.

The first test session is usually an un-weighted session, i.e., a weight set $w_1 = (0.5, 0.5, \ldots, 0.5)$ is used. A desired number of test vectors, say N_v^{ses1}, is applied. All the faults in the original target fault list that are detected in this session are dropped.

Next, a deterministic ATPG is used to generate a set of test vectors for the faults that remain in the target fault list. The above heuristic or the optimal approach presented in Hartmann and Kemnitz (1993) can then be used to compute a weight set for the entire set of test vectors. The quality of the weight set can be evaluated by computing the expected uncoverage. If this quality is satisfactory, then the weight set is used for the second session. On the other hand, if the expected uncoverage is high, a subset of the test vectors for the CUT are identified and a weight set computed for the vectors in the subset. Various heuristics can be employed to identify this subset of vectors. The key idea is to identify vectors that contain compatible values at most of their bit positions (e.g., see Pomeranz and Reddy (1993)). (For example, vectors $(x, 0, 1, x, 0, 1, x, 0, 1)$ and $(x, x, x, 0, 0, 0, 1, !, 1)$ are compatible in all the bit positions **except** the sixth and eighth.) In either case, a weight set is obtained for the next session. This weight set is used to generate and apply N_v^{ses2} vectors and the faults detected in the session are dropped from the target fault list.

If the desired fault coverage is not achieved and the area overhead limit is not reached, then the process described for the second session can be repeated to compute a weight set for the third session, and so on.

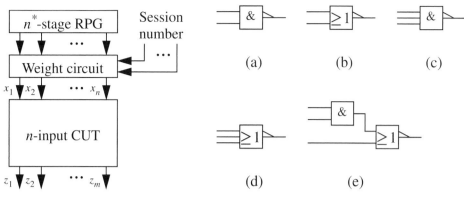

Figure 12.14 General architecture of WRPG

Figure 12.15 Example weight circuits: (a) 0.75, (b) 0.25, (c) 0.875, (d) 0.125, (e) 0.375

12.5.2.3 Implementation of WRPG

The general architecture of a WRPG is depicted in Figure 12.14. The RPG can be an ML-LFSR or ML-CA. The number of RPG stages, n^*, may be equal to or greater than the number of CUT inputs, n. (The reason for this will become clear later.)

The **weight circuit** can be implemented as a combinational circuit that takes as inputs (a) an n^*-bit vector, and (b) a control value describing the session number. Its output is an n-bit vector to be applied to the CUT. Essentially, this circuitry uses the weight set for the particular session number and the n^*-bit random pattern to obtain a weighted random pattern, as described next.

Note that each RPG output can be viewed as having a weight ≈ 0.5. Multiple signals can be combined via logic gates to implement other weight values. Figure 12.15 shows some example circuits for implementing weights. Each input to each of these circuits is an RPG output. The circuit in Figure 12.15(a) implements a weight of 0.75. The other gates in this figure implement weights of 0.25, 0.875, 0.125, and 0.375, respectively. One can verify that weights that are multiples of $1/2^l$ can be implemented using l or fewer RPG inputs and using a CMOS complex gate with $2l$ or fewer transistors.

Example 12.18 Consider the weight set generated in Example 12.16, namely

$(1/3, 3/4, 1, 1/3, 0)$.

These weights can be quantized to reduce the complexity of weight circuitry as

$(3/8, 3/4, 7/8, 3/8, 1/8)$.

The corresponding weight set can be implemented as shown in Figure 12.16.   □

The inputs $a_1, a_2, \ldots, a_{14}$ of the weight computation circuit should be connected to the outputs of an RPG. One choice is to use a five-stage RPG. This would necessitate some RPG outputs to fan out to multiple inputs of the weight circuitry. This can create

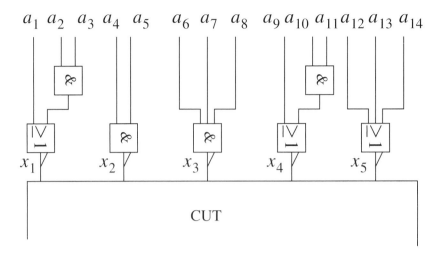

a_1 a_2 a_3 a_4 a_5 a_6 a_7 a_8 a_9 a_{10} a_{11} a_{12} a_{13} a_{14}

x_1 x_2 x_3 x_4 x_5

CUT

Figure 12.16 Weight circuit for a five-input circuit implementing weight set $(3/8, 3/4, 7/8, 3/8, 1/8)$.

correlations between the values applied to some CUT inputs. For example, if a_4 and a_5 are connected to the same RPG output as a_6 and a_7, respectively, then if logic 1 is applied to CUT input x_2 then only logic 1 can be applied to x_3. In other words, any vector containing the combination of values $(v(x_2), v(x_3)) = (1, 0)$ will never be applied to the CUT. Alternatively, an RPG with n^* stages, where n^* is the total number of inputs to the weight circuitry, can be used and such correlations avoided.

If multiple weight sets are used, the weight circuitry needs to include not only the weight generators but also some control circuitry to configure them for different weight sets in different sessions.

12.5.3 Embedding subspaces of vectors

All possible vectors for an n-input combinational circuit can be collectively viewed as a space of 2^n n-bit vectors. Typically, all commonly targeted static faults (mainly, stuck-at) in any CUT can be detected by vectors belonging to certain subspace within this space. The objective of the approaches described next is to identify subspaces that (a) contain at least one vector for each target fault in the given CUT, (b) contain a small number of vectors, and (c) can be generated using an RPG or a counter and a small amount of additional circuitry. The above characteristics guarantee complete fault coverage, low test length, and low area overhead, respectively.

Pseudo-exhaustive testing can be viewed as a classical example of the subspace embedding approach. Consider an n-input combinational CUT. Recall that a **cone** of a CUT is the entire subcircuit in the transitive fanin of one of its outputs. Also, recall from Chapter 7 that a CUT is said to be tested **pseudo-exhaustively** if all the possible combinations of values are applied to each of its cones.

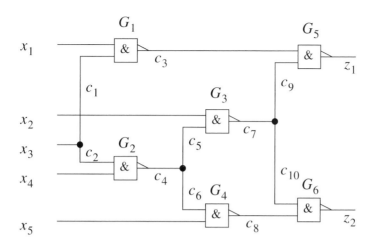

Figure 12.17　A five-input, two-output example circuit

Example 12.19　Consider the five-input, two-output combinational circuit shown in Figure 12.17. Corresponding to output z_1 is cone $Cone_1$ with inputs x_1, x_2, x_3, and x_4; corresponding to output z_2 is $Cone_2$ with inputs x_2, x_3, x_4, and x_5.

The exhaustive test set for this CUT is the space containing all possible (32) five-bit vectors. Classical pseudo-exhaustive testing techniques (McCluskey, 1984) can help identify a subspace containing the following 16 vectors and capable of detecting all the static faults within each cone.

$$
\begin{array}{llllll}
\{ & (0, & 0, & 0, & 0, & 0) \\
 & (0, & 0, & 0, & 1, & 0) \\
 & (0, & 0, & 1, & 0, & 0) \\
 & (0, & 0, & 1, & 1, & 0) \\
 & (0, & 1, & 0, & 0, & 0) \\
 & (0, & 1, & 0, & 1, & 0) \\
 & (0, & 1, & 1, & 0, & 0) \\
 & (0, & 1, & 1, & 1, & 0) \\
 & (1, & 0, & 0, & 0, & 1) \\
 & (1, & 0, & 0, & 1, & 1) \\
 & (1, & 0, & 1, & 0, & 1) \\
 & (1, & 0, & 1, & 1, & 1) \\
 & (1, & 1, & 0, & 0, & 1) \\
 & (1, & 1, & 0, & 1, & 1) \\
 & (1, & 1, & 1, & 0, & 1) \\
 & (1, & 1, & 1, & 1, & 1) & \} \\
\end{array}
$$

In this four-dimensional subspace, x_1, x_2, x_3, and x_4 can be viewed as **independent inputs** at which all possible combination of values are applied. x_5 can be viewed

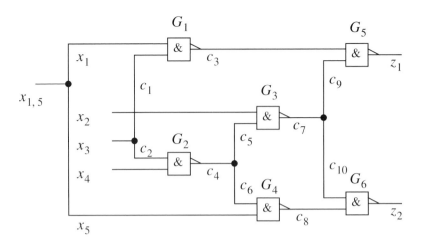

Figure 12.18 A five-input, two-output example circuit after x_1 and x_5 are combined into a test signal

as a dependent input at which the value applied at x_1 is applied. In other words, all combinations of $v(x_1)$, $v(x_2)$, $v(x_3)$, and $v(x_4)$ are enumerated and $v(x_5) = v(x_1)$. $\square$

Subspaces of the type described in the above example are identified using the following notion. Two inputs x_i and x_j of a circuit are said to be **PE-compatible** if there does not exist any cone to which x_i and x_j both belong (McCluskey, 1984). In the test mode, a pair of compatible inputs can be combined into a single pseudo-input, called a **test signal**, without losing coverage of any static fault within any cone. In the above example, x_1 and x_5 are PE-compatible and can be combined, in the test mode, into a single pseudo-input, say $x_{1,5}$ which fans out to x_1 and x_5. As above, this can also be denoted as $v(x_5) = v(x_1)$.

The notion of PE-compatibility can be used in an iterative manner. First, one pair of PE-compatible inputs can be combined into a test signal. The circuit can be viewed as having as primary inputs the test signal and other inputs that are not combined into test signals. The above procedure of identifying compatible inputs and combining them into test signals can be repeated.

Example 12.20 In the scenario in the above example, after x_1 and x_5 are combined, the CUT can be viewed as having as its inputs the test signal $x_{1,5}$ and inputs x_2, x_3, and x_4. This is depicted in Figure 12.18. In this case, no pair of PE-compatible inputs exist in this version of the CUT. In this case, the test signals identified above represent an optimal solution if only PE-compatibility is considered. $\square$

Figure 12.19(a) shows the generic architecture of the test pattern generator for pseudo-exhaustive testing. If the CUT has n_I independent inputs, then an n_I-stage exhaustive pattern generator, such as a complete LFSR or a binary counter, is used as the base pattern generator. Each output of the base pattern generator drives a CUT

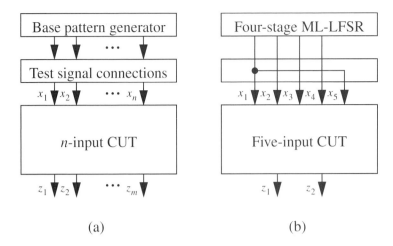

Figure 12.19 A PG that exploits PE-compatibility: (a) general architecture, (b) for the example five-input, two-output circuit (Chen and Gupta, 1998a) (© 1998 IEEE)

input that is considered independent. Some of these inputs also fan out to the other dependent inputs, as given by the final test signals.

Example 12.21 Figure 12.19(b) shows the details of the design of the test pattern generator for pseudo-exhaustive testing of the CUT described in the above examples. Since x_1, x_2, x_3, and x_4 can be viewed as independent inputs, a four-stage complete LFSR can be used as a base pattern generator. The four outputs of the LFSR can be connected to the above four CUT inputs. Since x_5 is a dependent input to which the value applied at x_1 is applied, the output of the base PG that drives x_1 is used to also drive x_5. (Note that in the above case, inputs x_2, x_3, x_4, and x_5 could have been considered independent inputs and x_1 considered dependent on x_5.) □

The above iterative approach does not guarantee the identification of an optimal solution, i.e., the smallest subspace. The covering approach described in McCluskey (1984) can be used to find optimal solutions to the pseudo-exhaustive problem using the notion of PE-compatibility. In Akers (1985), a generalization of the type of test signals is presented which helps obtain subspaces for pseudo-exhaustive testing that are, in general, smaller than those obtained by limiting oneself to test signals of the above type.

12.5.3.1 Additional types of compatibilities

The limitation of the pseudo-exhaustive approach is that the number of independent inputs for a CUT is always greater than or equal to the number of inputs in any of its cones. This is especially severe since many circuits contain one or more cones that depend on all the inputs. In such cases, PE-compatibility cannot identify any inputs that can be combined into test signals to reduce the test application time. One solution that has been posed to this problem is to use control and observation points within the CUT to obtain smaller cones in the test mode. Alternatively, more powerful notions of compatibility can accomplish the same objective without partitioning the circuit lines.

Inputs x_i and x_j of a CUT are said to be **D-compatible** if x_i and x_j can be combined into a test signal $x_{i,j}$ without reducing the coverage of target faults (Chen and Gupta, 1998a). In other words, if x_i and x_j are D-compatible, then in the test mode x_i can be viewed as an independent signal while x_j can be viewed as being dependent on x_i in the manner $v(x_j) = v(x_i)$.

Example 12.22   Consider the circuit in Figure 12.17. Assume that all the single SAFs in this circuit are to be targeted. One can verify via exhaustive fault simulation that all the target faults in the circuit can be tested even if x_2 and x_4 are combined into a test signal, i.e., x_2 and x_4 are D-compatible. Also, x_1 and x_5 are D-compatible. Finally, x_3 and x_5 are D-compatible. Let us pick one of these pairs of compatible inputs, say, x_2 and x_4, and combine them into a test signal, say $x_{2,4}$.

After the above combination, we can view the circuit as having inputs x_1, $x_{2,4}$, x_3, and x_5, where $x_{2,4}$ fans out to x_2 and x_4. In this version of the circuit, inputs x_3 and x_5 can be verified as being D-compatible and hence can be combined into a test signal, $x_{3,5}$.

Since no more D-compatibilities can be found, the iterative approach terminates by determining that the CUT inputs can be combined into three test signals, namely x_1, $x_{2,4}$, and $x_{3,5}$ while guaranteeing the detection of all the single SAFs in this circuit.

In other words, all the single SAFs in this circuit can be tested using vectors from a three-dimensional space in which all combinations of values $v(x_1)$, $v(x_2)$, and $v(x_3)$ are applied, and $v(x_4) = v(x_2)$, $v(x_5) = v(x_3)$.

Figure 12.20(a) shows the details of the design of the above PG.   □

The following notions can be used in conjunction with those presented above to obtain smaller subspaces of vectors that guarantee the coverage of all the target faults in a given CUT.

Inputs x_i and x_j of a CUT are said to be **I-compatible** if x_i and x_j can be combined into a test signal $x_{i,j}$, where the test signal $x_{i,j}$ directly drives x_i and drives x_j via an inverter, without reducing the coverage of target faults (Chen and Gupta, 1998a). In other words, if x_i and x_j are I-compatible, then in the test mode x_i can be viewed as an independent signal while x_j can be viewed as being dependent on x_i in the manner $v(x_j) = \overline{v(x_i)}$.

Inputs x_i and x_j of a CUT are said to be **dec-compatible** if all the target faults can be detected without applying any vector where logic 1 is applied to x_i as well as x_j (Chakrabarty *et al.*, 1997). This definition can be extended to any number of inputs $x_{i_1}, x_{i_2}, \ldots, x_{i_l}$, which are said to be dec-compatible if all the target faults in the CUT

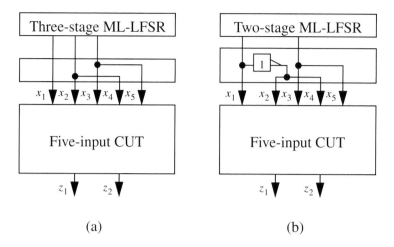

Figure 12.20 PGs for the five-input, two-output example circuit using: (a) D-compatibility only, and
(b) D-compatibility and I-compatibility (Chen and Gupta, 1998a) (© 1998 IEEE)

can be detected using a set of test vectors where, in any single vector, logic 1 is applied
to no more than one of these inputs. In this case, the values for these l inputs can be
generated using a decoder and a binary counter with $\lceil \log_2 l \rceil$ stages.

Inputs $x_{i_1}, x_{i_2}, \ldots, x_{i_l}$ of a CUT are said to be **C(l)-compatible** to another input x_j
if all the target faults can be detected when x_j can be driven by the output of an l-input
gate whose inputs are driven by $x_{i_1}, x_{i_2}, \ldots, x_{i_l}$ (Hamzaoglu and Patel, 2000). In other
words, in the test mode $v(x_j) = \psi[v(x_{i_1}), v(x_{i_2}), \ldots, v(x_{i_l})]$, where ψ is some logic
function.

Example 12.23 Figure 12.20(b) shows the details of the PG for the circuit in
Figure 12.17 obtained using the notions of D-compatibility and I-compatibility. In
this case, x_1 and x_3 can be viewed as independent test signals. In the test mode all
combinations of values $v(x_1)$ and $v(x_3)$ are applied. Values at the other inputs are
obtained as $v(x_2) = \overline{v(x_1)}$, $v(x_4) = \overline{v(x_1)}$, and $v(x_5) = v(x_3)$.   □

12.6   Response compression

If a PG is used to generate and apply a test sequence with N_v vectors to an m-output
CUT, mN_v bits of response are generated at the CUT outputs. On-chip, bit-by-bit
comparison of the CUT response with that expected of the fault-free version of the
CUT (referred to as the **fault-free response**) would require a large on-chip memory
to store the complete fault-free response. To avoid the use of a large on-chip memory,
the CUT response is compressed using one of the compressors described ahead which
compress the response by orders of magnitude. The compressed value of the CUT

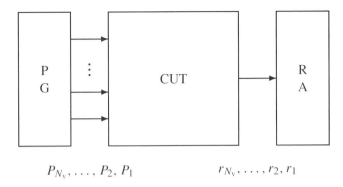

Figure 12.21   A generic view of self-test

response is called a **signature** which is compared with the compressed value of the fault-free response, called the **fault-free signature**. Self-test using compression decreases the on-chip memory required, since only the fault-free signature needs to be stored instead of the complete fault-free response. Hence, compression techniques play a crucial role in decreasing the silicon area required to implement BIST.

Compression techniques used in BIST are lossy, i.e., it is possible for two different CUT responses to be compressed into the same signature. The phenomenon where a faulty CUT's response is different from the fault-free response and yet is compressed into the fault-free signature is called **aliasing**. Occurrence of aliasing causes loss of fault coverage, since a match between the CUT signature and fault-free circuit signature causes the faulty CUT to be declared fault-free. Minimizing the probability of aliasing and minimizing the silicon area required to implement a compressor are the two main objectives of BIST compressor design. In the following, we discuss two main classes of compressors, *count-based* and *linear*.

A generic block diagram of BIST circuitry is shown in Figure 12.21, assuming a single-output CUT. The PG generates a sequence of N_v tests, $P_{N_v}, \ldots, P_2, P_1$, where each test is an n-bit vector that is applied in parallel to the inputs of the n-input CUT. Upon the application of test P_i, response r_i^* appears at the output of the fault-free circuit. The response to the same vector at the output of a CUT is denoted by r_i. The response sequences for the fault-free and the faulty circuits are denoted by $Res^* = r_{N_v}^*, \ldots, r_2^*, r_1^*$ and $Res = r_{N_v}, \ldots, r_2, r_1$, respectively. We assume that the vectors are applied in the order $P_1, P_2, \ldots, P_{N_v}$ and the response bits appear in the corresponding order. The CUT response is compressed using a response analyzer, which is typically one of the following compressors.

12.6.1   Count-based compressors

We first discuss two types of count-based compressors, *one's count* and *transition count*.

12.6.1.1 One's count compressors

As the sequence of tests is applied to the CUT, a response sequence, $Res = r_{N_v}, \ldots, r_2, r_1$, containing 0s and 1s appears at the CUT output, starting with r_1. **One's count compression** technique is based on counting the number of 1s in this response sequence.

Example 12.24   One's count for a CUT response sequence $Res = 1, 1, 1, 0, 1, 0, 0$ is given by, $1C(Res) = 4$. □

One's count can be computed using a counter to count the number of clock cycles for which the output of the CUT is logic 1. Since the minimum and maximum values of $1C(Res)$ for an N_v-bit response sequence Res are zero and N_v, respectively, a modulo-(N_v+1) counter, which can be implemented using $\lceil \log_2(N_v+1) \rceil$ flip-flops, is required. Hence, using a small amount of hardware, one's counting can help compress an N_v-bit response into a $\lceil \log_2(N_v + 1) \rceil$-bit signature. However, this compression is lossy and can cause aliasing, i.e., the one's count for some faulty CUTs may be the same as that for the fault-free circuit *despite the fact that the response of the faulty circuit, Res, is different from Res**.

Example 12.25   If the response of a CUT is $Res = 1, 1, 1, 0, 1, 0, 0$ and the corresponding fault-free circuit response is $Res^* = 1, 0, 1, 1, 1, 0, 0$, then the faulty CUT will be declared fault-free by one's counting. Aliasing occurs due to the fact that only the one's counts of the CUT and the fault-free circuit, $1C(Res)$ and $1C(Res^*)$, respectively, are compared and, for the above values of Res and Res^*, $1C(Res) = 1C(Res^*) = 4$. □

The number of possible response sequences which are different from Res^* but have one's count identical to $1C(Res^*)$ is one measure of aliasing. Sometimes called **aliasing volume**, it is given by:

$$\binom{N_v}{1C(Res^*)} - 1,$$

where $\binom{N_v}{1C(Res^*)}$ is the number of N_v-bit sequences with $1C(Res^*)$ 1s, one of which is Res^*.

12.6.1.2 Transition count compressors

In transition count compression, the number of $1 \rightarrow 0$ and $0 \rightarrow 1$ transitions in the CUT response sequence is used as the signature. Since $r_i \oplus r_{i+1} = 1$ if and only if a

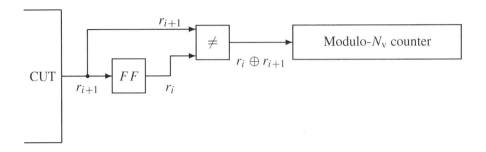

Figure 12.22 A transition count compressor

transition occurs between response bits r_i and r_{i+1}, the transition count of a response sequence Res can be defined as

$$TC(Res) = \sum_{i=1}^{N_v-1} r_i \oplus r_{i+1}.$$

Example 12.26 A CUT response sequence $Res = 1, 1, 1, 0, 1, 0, 0$ has two $0 \to 1$ transitions and one $1 \to 0$ transition (recall that we need to look at bits from right to left). Hence, $TC(Res) = 3$. $\quad\square$

The above equation for $TC(Res)$ clearly shows the key components required for the hardware implementation of a transition count compressor. A flip-flop is required to store response bit r_i for one clock cycle until r_{i+1} becomes available. A two-input XOR gate is required to compute the value of $r_i \oplus r_{i+1}$. Finally, a counter is required to count the number of clock cycles for which the XOR gate output is logic 1. Since the minimum and maximum transition counts for an N_v-bit response are zero and $N_v - 1$, respectively, a modulo-N_v counter, which can be implemented using $\lceil \log_2 N_v \rceil$ flip-flops, is required. Figure 12.22 shows a complete implementation of the transition count compressor. Note that to correctly compute the signature, the D flip-flop in Figure 12.22 should be initialized to r_1 before self-test starts.

Like one's count compression, transition counting is also lossy and can cause aliasing.

Example 12.27 If the response of a CUT is $Res = 1, 1, 1, 0, 1, 0, 0$ and the corresponding fault-free circuit response is $Res^* = 1, 1, 0, 0, 1, 0, 0$, then the faulty CUT will be declared fault-free by transition counting, since $TC(Res) = TC(Res^*) = 3$. $\quad\square$

The aliasing volume for transition counting, i.e., the number of possible response sequences Res which are not equal to Res^* but have a transition count equal to

$TC(Res^*)$, can be computed as follows. A transition is said to occur between successive pairs of bits r_i and r_{i+1} of a sequence Res if $r_i \neq r_{i+1}$. Since an N_v-bit sequence Res that starts with $r_1 = 0$ has $N_v - 1$ such pairs, the number of sequences with k transitions can be given by $\binom{N_v-1}{k}$. Considering the facts that the response sequence can also start with $r_1 = 1$ and that the fault-free circuit response, Res^*, is one of the sequences with transition count $TC(Res^*)$, the aliasing volume is given by

$$2\binom{N_v - 1}{TC(Res^*)} - 1.$$

12.6.1.3 Aliasing characteristics

As seen above, aliasing volumes for one's and transition count compressors are dependent on the values of their respective fault-free circuit signatures, $1C(Res^*)$ and $TC(Res^*)$. Furthermore, for both cases, the expression for aliasing volume is of such a form that its value is minimum when the fault-free circuit signature is either at its minimum or maximum value. For example, the aliasing volume for transition count, $2\binom{N_v-1}{TC(Res^*)} - 1$, attains its minimum value when $TC(Res^*)$ is either zero or $N_v - 1$. Only two sequences, namely, $0, 1, 0, 1, \ldots$ and $1, 0, 1, 0, \ldots$, have transition count of $N_v - 1$; since one of these two must be the fault-free circuit response, the other is the only erroneous response whose occurrence can cause aliasing. In contrast, aliasing volume is maximum when $TC(Res^*) \approx N_v/2$.

The number of 1s in the response of a circuit can be changed by using a different test set. The number of transitions can be changed, even for a given test set, by changing the order in which tests are applied. For example, consider two partitions of the given test set $\mathcal{T}$, $\mathcal{T}_0$ and $\mathcal{T}_1$, which contain all the test vectors for which the fault-free circuit response is a 0 and 1, respectively. Let $\mathcal{T}'$ be a test sequence obtained by selecting tests alternately from $\mathcal{T}_0$ and $\mathcal{T}_1$, until each test is selected once (note that if $\mathcal{T}_0$ and $\mathcal{T}_1$ have unequal number of vectors, then some vectors will appear in the sequence more than once). Clearly, $\mathcal{T}'$ will provide the same SAF coverage as $\mathcal{T}$; it will simultaneously minimize the aliasing volume by maximizing the value of $TC(Res^*)$ (Reddy, 1977). In fact, if this test sequence starts with a vector for which the fault-free circuit response is a 1 (0), aliasing can be completely eliminated by modifying the compressor to count only the $1 \to 0$ ($0 \to 1$) transitions. Other ways to order tests to decrease or eliminate aliasing are presented in Hayes (1976).

While tests can be reordered to decrease aliasing in transition counting and special test sets can be generated to minimize aliasing for one's counting, the application of the resulting test sequences in a BIST scheme is limited. This is due to the fact that a large amount of hardware may be required to implement a circuit that can generate most such sequences.

Compressor
Input

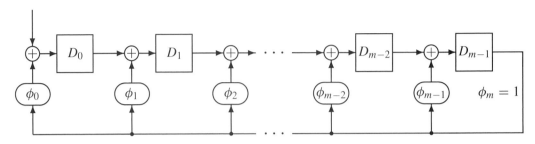

Figure 12.23 An m-stage LFSR compressor with $\phi(x) = x^m + \phi_{m-1}x^{m-1} + \cdots + \phi_0$

12.6.2 Linear compressors – LFSRs as compressors

An LFSR was first used as a compressor in a logic analyzer to reduce the amount of storage required while capturing circuit responses (Frohwerk, 1977). Since then it has increasingly become a method of choice for BIST compression, partly due to the fact that using a small amount of additional hardware, a normal register in the CUT can be reconfigured as an LFSR compressor and a pattern generator. An internal-XOR LFSR compressor can be obtained by simply adding an XOR gate to the corresponding PG. The resulting LFSR compressor is shown in Figure 12.23 which is also characterized by its feedback polynomial. As tests are applied, the CUT response bits are sequentially applied to the input of the compressor, starting with r_1. The contents of the LFSR flip-flops, $D_0, D_1, \ldots, D_{m-1}$ after the application of the last response bit, r_{N_v}, is the signature of the CUT. The fault-free circuit signature is the signature obtained by compressing the response of the fault-free circuit. In the following, we analyze internal-XOR LFSR response analyzers; the final conclusions about the aliasing volume and probabilities hold for external-XOR LFSRs as well.

12.6.2.1 Computation of the signature

Compression using an internal-XOR LFSR is often described using division of polynomials in x with modulo-2 coefficients (Peterson and Weldon, 1972). Note that the feedback connections of an LFSR compressor are already represented by its feedback polynomial. The CUT response, $Res = r_{N_v}, \ldots, r_2, r_1$, can also be represented as

$$Res(x) = r_1 x^{N_v-1} + \cdots + r_{N_v-1}x + r_{N_v}.$$

Note that r_1, the response bit to the first test, is assigned the highest power of x, $N_v - 1$. The fault-free circuit response, Res^*, can be similarly represented as

$$Res^*(x) = r_1^* x^{N_v-1} + \cdots + r_{N_v-1}^* x + r_{N_v}^*.$$

The process of LFSR compression is equivalent to the division of the CUT response $Res(x)$ by the LFSR feedback polynomial, $\phi(x)$, *where modulo-2 operations are used on the coefficients of the polynomials.* The remainder obtained as a result of this polynomial division, $S(x)$, is the polynomial representation of the contents of the LFSR flip-flops, i.e., the signature, assuming that the LFSR initial state was $00 \ldots 0$. In other words, the signature is given by the equation

$$Res(x) = Q(x)\phi(x) + S(x),$$

where $Q(x)$ and $S(x)$ are respectively the quotient and remainder of the division of the response $Res(x)$ by the feedback polynomial $\phi(x)$. Note that since $S(x)$ is the remainder of division of $Res(x)$ by $\phi(x)$, the degree of $S(x)$ must be less than the degree of $\phi(x)$.

Example 12.28 Consider an LFSR compressor with feedback polynomial $\phi(x) = x^4 + x + 1$ that is used to compress CUT response $Res = r_7, r_6, r_5, r_4, r_3, r_2, r_1 = 1, 1, 1, 0, 1, 0, 1$ or $Res(x) = x^6 + x^4 + x^2 + x + 1$. Polynomial division gives $Q(x) = x^2 + 1$ and signature $S(x) = x^3$, which implies that at the end of compression, the contents of the LFSR compressor are given by $D_3 = 1$, and $D_2 = D_1 = D_0 = 0$. Figure 12.24 shows the steps of the above polynomial division as well as the simulation of the LFSR for the above case. □

Similarly, the fault-free circuit signature, $S^*(x)$, is the remainder of the division of the fault-free circuit response by the LFSR feedback polynomial and is given by

$$Res^*(x) = Q^*(x)\phi(x) + S^*(x).$$

12.6.2.2 Condition for aliasing and aliasing volume

Since aliasing occurs when $S(x) = S^*(x)$ while $Res(x) \neq Res^*(x)$, the condition for aliasing to occur can be derived by adding the above two equations. This gives

$$Res(x) + Res^*(x) = \left[Q(x) + Q^*(x)\right]\phi(x) + S(x) + S^*(x).$$

Since $S(x) = S^*(x)$, *i.e.*, $S(x) + S^*(x) = 0$, the above relation becomes

$$Res(x) + Res^*(x) = \left[Q(x) + Q^*(x)\right]\phi(x).$$

Hence, aliasing occurs during compression using an LFSR if $Res(x) + Res^*(x)$ is divisible by the LFSR's feedback polynomial, $\phi(x)$.

Define an **error sequence**, $E = e_{N_v}, \ldots, e_2, e_1$, as the sequence obtained by bit-by-bit XOR of the CUT response sequence, Res, and the fault-free circuit response Res^*, i.e., $e_i = r_i + r_i^*$, for $i \in [1, N_v]$. Note that the corresponding polynomial $E(x)$, called the **error polynomial**, is equal to $Res(x) + Res^*(x)$. Furthermore, $E(x) \neq 0$ if and only if $Res(x) \neq Res^*(x)$. Hence, the above condition can be restated as: aliasing

(a)

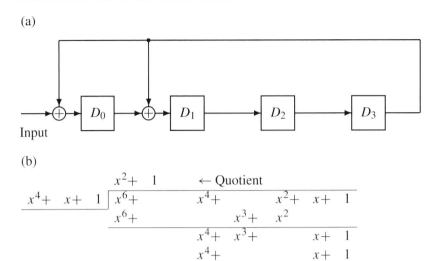

(b)

$$
\begin{array}{r}
x^2+ \quad 1 \quad\quad \leftarrow \text{Quotient} \\
x^4+ \quad x+ \quad 1 \,\big)\, \overline{x^6+ \qquad\qquad x^4+ \qquad\qquad x^2+ \quad x+ \quad 1} \\
x^6+ \qquad\qquad x^3+ \quad x^2 \\
\hline
x^4+ \quad x^3+ \qquad\qquad x+ \quad 1 \\
x^4+ \qquad\qquad\qquad x+ \quad 1 \\
\hline
\text{Signature} \rightarrow \qquad x^3
\end{array}
$$

(c)

Time	Input	LFSR state				Output
–	–	0	0	0	0	–
0	$r_1 = 1$	1	0	0	0	0
1	$r_2 = 0$	0	1	0	0	0
2	$r_3 = 1$	1	0	1	0	0
3	$r_4 = 0$	0	1	0	1	0
4	$r_5 = 1$	0	1	1	0	1
5	$r_6 = 1$	1	0	1	1	0
6	$r_7 = 1$	0	0	0	1	1

Figure 12.24 Compression using an example LFSR compressor: (a) LFSR compressor with $\phi(x) = x^4 + x + 1$, (b) computation of signature via polynomial division, and (c) simulation of LFSR compression

occurs during LFSR compression if the non-zero error polynomial is divisible by the LFSR's feedback polynomial, i.e., when $E(x) \neq 0$ and

$$E(x) = \left[Q(x) + Q^*(x) \right] \phi(x). \tag{12.38}$$

The above restatement demonstrates that the condition for aliasing for LFSR compression is independent of the exact value of the fault-free circuit response, $Res^*(x)$, and depends only on the errors, i.e., the bits where Res differs from Res^*. Note that this is different from count-based compression, where the number of 1s or the number of transitions in the fault-free circuit response sequence has significant impact on the aliasing volume.

The aliasing volume can be computed by examining the condition for aliasing given in Equation (12.38). The maximum degree of $E(x)$ is $N_v - 1$ and the degree of $\phi(x)$ is

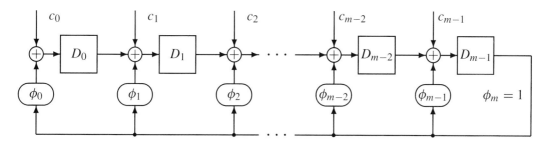

Figure 12.25 A multiple-input signature register

m. Hence the maximum degree of the term $\left[Q(x) + Q^*(x)\right]$ can be $N_v - 1 - m$. For a given $\phi(x)$, any polynomial with degree $\leq N_v - 1 - m$ can be used as $\left[Q(x) + Q^*(x)\right]$ in the above expression to obtain an error polynomial that will result in a fault-free circuit signature. Since there exist $2^{N_v - m}$ polynomials of degree $N_v - 1 - m$, the number of possible error polynomials that result in fault-free circuit signature is $2^{N_v - m}$. Eliminating the all-zero error polynomial from the above number, the number of error polynomials that cause aliasing is

$$2^{N_v - m} - 1.$$

12.6.2.3 Multiple-input signature register (MISR)

An LFSR compressor with m stages can be modified to compress responses of a circuit with m outputs, as shown in Figure 12.25. Note that this multiple-input signature register (MISR) is obtained by replacing the two-input XOR gates in the corresponding LFSR (see Figure 12.2(a)) by three-input gates. The m outputs of the CUT can be connected to the inputs of this MISR.

Due to the linearity of an MISR, the condition under which aliasing occurs in MISR compression is independent of the fault-free circuit response and depends solely on the error sequence and the feedback polynomial. However, unlike LFSR compression, where the error associated with the application of a test vector P_i, e_i, has only a single bit, in an m-output circuit, each e_i has m bits. The m bits of error e_i can be denoted as $e_{i,1}, e_{i,2}, \ldots, e_{i,m}$. We assume that error $e_{i,j}$ appears in response to test vector P_i at the jth output of the CUT and is input to the jth input of the MISR.

For an internal-XOR MISR in the all-zero state, the effect of applying a value $e_{i,j}$ at the jth input of the MISR at time τ is identical to its application at the $(j-1)^{\text{th}}$ MISR input at time $\tau - 1$. This property can be used to demonstrate that MISR compression of the multi-output error sequence $(e_{1,1}, e_{1,2}, \ldots, e_{1,m})$, $(e_{2,1}, e_{2,2}, \ldots, e_{2,m})$, $\ldots$, $(e_{N_v,1}, e_{N_v,2}, \ldots, e_{N_v,m})$ can be reduced to the compression of an equivalent sequence by an LFSR compressor with the same $\phi(x)$. The above MISR input sequence can be written as the following $N_v + m - 1$ bit **equivalent LFSR error sequence**: $(e_{1,m})$, $(e_{1,m-1} \oplus e_{2,m})$, $(e_{1,m-2} \oplus e_{2,m-1} \oplus e_{3,m})$, $\ldots$, $(e_{N_v-1,1} \oplus e_{N_v,2})$, $(e_{N_v,1})$.

Two facts should be noted about the above computation of the equivalent LFSR error sequence.

1 A number of error combinations may already be masked during the computation of the equivalent LFSR sequence. For example, consider the case where a fault in a CUT creates errors only for vectors P_1 and P_2 at the outputs $m - 1$ and m, respectively. The above reduction shows that such an erroneous response will result in an all-zero equivalent LFSR error sequence. Such multi-output error sequences will clearly cause aliasing *independent of the feedback polynomial of the compressor.*

2 Once the equivalent LFSR error sequence is obtained, the conditions for LFSR aliasing can be used to determine whether MISR compression of a given multi-output sequence causes aliasing.

In the next section, we will analyze in greater detail the aliasing probability for LFSR and MISR compressions.

12.7 Analysis of aliasing in linear compression

In the previous section, aliasing volume was used as a measure of compression quality. If we assume that all the possible erroneous CUT responses are equally likely, then the **aliasing probability** can be determined by computing the ratio of the aliasing volume to the number of possible circuit responses. However, in practice, all erroneous responses are not equally likely. We now describe **error models**, i.e., models that help estimate the probability of occurrence of any given erroneous CUT response. We then quantify the aliasing probability for LFSR and MISR compression.

Two main techniques for determining the aliasing probability are discussed. First, we discuss the techniques where compression is modeled as a Markov chain. This is followed by a discussion of coding theory methods to compute aliasing probability.

12.7.1 Error model

The most commonly used error model for multiple-output circuits is the independent error model (Williams *et al.*, 1988).

12.7.1.1 Independent error model

In this case, it is assumed that the errors at the various outputs of the circuit are independent, as shown in Figure 12.26. Here, q represents the **detection probability**, i.e., the probability that the response at any output of the CUT is in error. The probability of the response to a test vector at a particular CUT output being in error is assumed to be independent of (a) the errors at the other CUT outputs

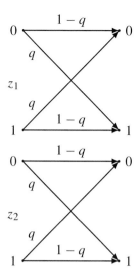

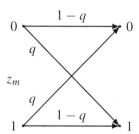

Figure 12.26   Errors at the outputs are totally independent (Pradhan and Gupta, 1991) (© 1991 IEEE)

(*spatial independence*), and (b) the errors caused in response to other tests (*temporal independence*).

The independence of errors in time is realistic, since in most BIST methodologies, a combinational CUT is tested using pseudo-random vectors where consecutive vectors have low correlation. However, the independence in space may not be realistic for many circuits.

12.7.1.2 Symmetric error model

The CUT is assumed to have m outputs, hence the response at the outputs of the CUT for one vector can be represented as an m-bit symbol.

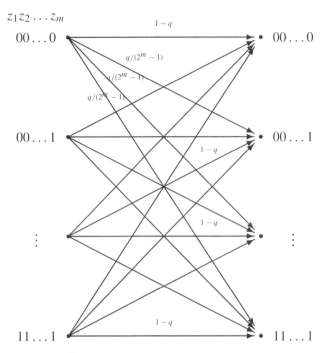

$z_1 z_2 \dots z_m$

00...0

00...1

11...1

$1-q$

$q/(2^m - 1)$

$q/(2^m - 1)$

$q/(2^m - 1)$

00...0

$1-q$

00...1

$1-q$

$1-q$

$1-q$

11...1

Figure 12.27 2^m-ary symmetric model for multiple output circuits (Pradhan and Gupta, 1991) (© 1991 IEEE)

Table 12.5. *Parameters for the two error models for a three-output circuit (Pradhan and Gupta, 1991) (© 1991 IEEE)*

Error	Independent	Symmetric
000	$(1-q)^3$	$1-q$
001	$q(1-q)^2$	$q/7$
010	$q(1-q)^2$	$q/7$
100	$q(1-q)^2$	$q/7$
011	$q^2(1-q)$	$q/7$
110	$q^2(1-q)$	$q/7$
101	$q^2(1-q)$	$q/7$
111	q^3	$q/7$

Under this model, the errors at the outputs of the CUT are represented as shown in Figure 12.27. It is assumed that all the $2^m - 1$ error values can occur with an equal probability, $q/(2^m - 1)$. Hence, the probability that a given output combination is in error is q. The probability that the output combination is error-free is $(1 - q)$.

Table 12.5 illustrates the probabilities of all the possible error values for these two different error models for a three-output circuit. It can be seen that the probabilities of occurrence of various error values are different under the two models.

To be precise, for a CUT, the value of q under the independent error model is different from the value of q under the symmetric error model. For accurate estimation of aliasing, it is crucial to have the correct value of q. One way to determine such a value is through fault simulation.

Example 12.29 Consider ALU 181, a 14-input and eight-output circuit. Through exhaustive simulation, the value of q under the independent error model can be determined as being 0.0356 and under the symmetric error model as being 0.251 35 (Pradhan and Gupta, 1991). □

In general, the knowledge of the CUT structure can be used to more accurately model the errors. Statistical fault simulation may be used instead of exhaustive simulation to obtain an estimate of the value of q under the chosen error model.

12.7.2 Markov modeling of linear compression

Consider the m-stage LFSR compressor with feedback polynomial $\phi(x)$ shown in Figure 12.23. The state transitions of the LFSR compressor can be obtained by modifying the state transition relation given in Equation (12.8) to accommodate the input, c, to the compressor as

$$
\begin{bmatrix} s_0(\tau+1) \\ s_1(\tau+1) \\ \vdots \\ s_{m-1}(\tau+1) \end{bmatrix} = A \cdot \begin{bmatrix} s_0(\tau) \\ s_1(\tau) \\ \vdots \\ s_{m-1}(\tau) \end{bmatrix} + \begin{bmatrix} 1 \\ 0 \\ \vdots \\ 0 \end{bmatrix} \cdot c(\tau),
\tag{12.39}
$$

where A is the state transition matrix of the LFSR, similar to that shown in Figure 12.4(a).

While the above relation can be used to derive the state transitions for a given input sequence, in practice, the analysis must only be carried out for an arbitrary error sequence whose statistics are known (e.g., given by the error model). For example, for a single-output circuit, both the above error models reduce to a scenario where each bit of the input is in error with a probability q and is error-free with a probability $1 - q$. Since the aliasing analysis of linear compression can be performed using the error sequence, this implies that each input bit is logic 1 with probability q. Also, the temporal independence assumption implies that the bits are independent in time.

Example 12.30 Figure 12.28 shows a probabilistic version of the state transition graph (STG) of a two-stage LFSR compressor with $\phi(x) = x^2 + x + 1$. The solid and dotted arrows, respectively, indicate transitions that occur when input c is 0 and 1. Given the above error model, at any time, the probabilities of the transition out of the current state occurring along the solid and dotted edge are $1 - q$ and q, respectively. Such a state transition can be modeled as a time-stationary Markov chain. □

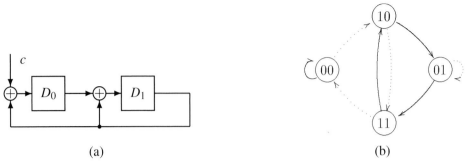

Figure 12.28 (a) A two-stage LFSR compressor, and (b) its probabilistic STG (Williams *et al.*, 1988) (© 1988 IEEE)

Let $\pi_i(\tau)$, $i = 0, 1, \ldots, 2^m - 1$, be the probability that the LFSR is in state i, where state i is an m-bit state vector, $[s_0(\tau), s_1(\tau), \ldots, s_{m-1}(\tau)]^T$, that is the binary representation of integer i (assuming $s_0(\tau)$ to be the least significant bit). The vector $[\pi_0, \pi_1, \ldots, \pi_{2^m-1}]^T$ will be called the **state probability vector**. The Markov chain can then be written as

$$
\begin{bmatrix}
\pi_0(\tau + 1) \\
\pi_1(\tau + 1) \\
\vdots \\
\pi_{2^m-1}(\tau + 1)
\end{bmatrix}
= M \cdot
\begin{bmatrix}
\pi_0(\tau) \\
\pi_1(\tau) \\
\vdots \\
\pi_{2^m-1}(\tau)
\end{bmatrix}.
\tag{12.40}
$$

Note that since we assume that the compressor starts in the all-zero state, $\pi_0(0) = 1, \pi_1(0) = 0, \ldots, \pi_{2^m-1}(0) = 0$.

Example 12.31 For the above two-bit LFSR compressor, this relation can be written as

$$
\begin{bmatrix}
\pi_0(\tau + 1) \\
\pi_1(\tau + 1) \\
\pi_2(\tau + 1) \\
\pi_{2^2-1}(\tau + 1)
\end{bmatrix}
=
\begin{bmatrix}
1 - q & 0 & 0 & q \\
q & 0 & 0 & 1 - q \\
0 & 1 - q & q & 0 \\
0 & q & 1 - q & 0
\end{bmatrix}
\cdot
\begin{bmatrix}
\pi_0(\tau) \\
\pi_1(\tau) \\
\pi_2(\tau) \\
\pi_{2^2-1}(\tau)
\end{bmatrix}.
\tag{12.41}
$$

□

In general, the entries in matrix M have two important properties. First, the entries in each column sum to one, since in each time step, one of the outgoing edges must be traversed for any state. Second, for LFSR compression, the entries in each row also sum to one. Such an M is said to be **doubly-stochastic** and its properties ensure that the state probability vector will eventually stabilize at the value $[1/2^m, 1/2^m, \ldots, 1/2^m]$.

Since we assume that the LFSR starts in the all-zero initial state, aliasing occurs if the LFSR ends in the all-zero final state for a non-zero error input. Hence, if the test length N_v is large then the aliasing probability for LFSR compression is $1/2^m - (1 - q)^{N_v}$, where the first term is the probability that, at the end of test application, the

LFSR is in the all-zero state and the second term is the probability of the occurrence of the all-zero error input. Note that this result holds for any non-trivial m-stage LFSR.

The Markov chain given above has been analyzed to derive the following key conclusions.

1 For short test sequence lengths, i.e., before the state probability vector of the circuit stabilizes to the above *asymptotic* state probability values, the aliasing probability can exceed $1/2^m - (1 - q)^{N_v}$. Furthermore, for short test lengths, the aliasing probability may be higher for an LFSR with a non-primitive $\phi(x)$ than one with primitive $\phi(x)$ (Williams *et al.*, 1988).

2 Let the STG of the given LFSR compressor, assuming an all-zero input, be comprised of b cycles of lengths $l_1, l_2, \ldots, l_b$. Also, let d_j be the number of 1s appearing at the output of any LFSR stage during its traversal of the jth cycle. The aliasing probability for the m-stage LFSR for test length N_v is bounded by (Damiani *et al.*, 1989)

$$ALP \leq \frac{1}{2^m} \sum_{j=1}^{b} l_j |1 - 2q|^{d_j N_v} - (1 - q)^{N_v}. \tag{12.42}$$

Since an LFSR with a primitive feedback polynomial has only two cycles, one with the all-zero state and the other with all the non-zero states, the above expression can be simplified to obtain the following expression for the aliasing probability for such an LFSR:

$$ALP \leq \frac{1}{2^m} + \left(1 - \frac{1}{2^m}\right) |1 - 2q|^{[2^m - 1]N_v} - (1 - q)^{N_v}. \tag{12.43}$$

12.7.3 Coding theory approach

This approach can be illustrated with the following example.

Example 12.32 Consider an LFSR compression scenario where the test response from a single-output circuit is compressed using the three-stage LFSR with primitive feedback polynomial $\phi(x) = x^3 + x + 1$, as shown in Figure 12.29. Assume that $N_v = 7$ tests are applied to the CUT. One can verify that the following error polynomials will result in an all-zero signature:

$$\{0, x^3 + x + 1, x^4 + x^2 + x, x^4 + x^3 + x^2 + 1, x^5 + x^2 + x + 1,$$
$$x^5 + x^3 + x^2, x^5 + x^4 + 1, x^5 + x^4 + x^3 + x, x^6 + x^2 + 1,$$
$$x^6 + x^3 + x^2 + x, x^6 + x^4 + x + 1, x^6 + x^4 + x^3, x^6 + x^5 + x,$$
$$x^6 + x^5 + x^3 + 1, x^6 + x^5 + x^4 + x^3 + x^2 + x + 1\}. \tag{12.44}$$

□

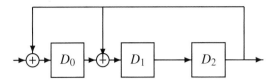

Figure 12.29 LFSR compressor with $\phi(x) = x^3 + x + 1$

When data are communicated over a noisy channel, or stored in a disk/memory which may cause errors, they are often protected by an error detecting/correcting code. In such cases, redundant bits are added to the original data. One of the common ways to determine whether data received from the channel, disk, or memory are error-free is by shifting them into an LFSR with appropriate feedback polynomial; the data are declared a valid codeword if the final value in the LFSR is all-zero. Consider a scenario where seven-bit codewords are transmitted and the code used is such that it must be decoded using the above LFSR. It is easy to see that in such a scenario, the polynomials shown in Equation (12.44) are the ones that will be declared as codewords.

Note how the LFSR is used in both cases. In the former, the CUT is declared fault-free when an all-zero remainder is obtained. In the latter, an all-zero remainder indicates that the input is a codeword; hence it is assumed to be correct. However, in both cases, the conclusion can be erroneous. In LFSR compression, any of the non-zero error polynomials listed above will lead to a faulty CUT being declared good. Analogously, in the latter case, noise could have corrupted the original data in a manner that it was transformed to a different codeword. In that case, the receiver's conclusion that it had received the correct data would be erroneous. In a linear code, the addition of a non-codeword to a codeword always gives a non-codeword; the addition of two codewords always gives a codeword. Hence, the original codeword can be modified into another codeword by a noise that corrupts a set of bits, represented by a polynomial H, only if H is a codeword. In the above example, the corrupted bits must correspond to one of the polynomials in the given set.

Finally, in self-test, the probability of occurrence of an error in a bit is q. Analogously, the probability of noise corrupting one bit of transmitted data can also be represented by q. Hence, it can be seen that the probability of aliasing during LFSR compression is identical to the probability of an incorrect codeword being received in the analogous coding scenario. The latter problem has been studied extensively and the approach and results developed there have been used to derive the following results (Pradhan and Gupta, 1991).

1 If a sequence of $N_{\mathrm{v}} = 2^m - 1$ binary responses at the output of a single-output circuit is being compressed by an LFSR with a primitive polynomial of degree m, the aliasing probability is given by

$$ALP = 2^{-m}\{1 + (2^m - 1)(1 - 2q)^{2^{m-1}}\} - (1 - q)^{2^m - 1}, \tag{12.45}$$

where q is the probability of any bit of the response being erroneous.

2 If the response of an m-output circuit is compressed using an MISR with degree-m primitive feedback polynomial, under the symmetric error model the aliasing probability for **any** test length N_v is given by

$$ALP = 2^{-m} \left[1 - 2^m (1-q)^{N_v} + (2^m - 1) \left(1 - \frac{2^m q}{2^m - 1} \right)^{N_v} \right]. \qquad (12.46)$$

3 If $N_v = v(2^m - 1)$ tests, where $v \geq 1$ is an integer, are applied to an m-output CUT whose response is compressed using an MISR with a degree-m primitive feedback polynomial, under the independent error model the aliasing probability is given by

$$ALP = \frac{1}{2^m} \left[1 + (2^m - 1)(1 - 2q)^{\frac{vm(2^m)}{2}} \right] - (1-q)^{vm(2^m - 1)}. \qquad (12.47)$$

12.8 BIST methodologies

The PGs and RAs described above can be used in many ways to self-test a circuit. The term **BIST methodology** is used to describe the key characteristics of how self-test is accomplished for a circuit. One set of key characteristics consists of descriptions of all the circuits used to implement self-test, including PGs and RAs, any circuitry used to connect their outputs and inputs to the inputs and outputs of the CUT, respectively, and any controller used to control the execution of self-test. Other characteristics include details of test application, such as the speed at which the tests are applied, how the tests are generated, and the class of circuit structures considered as CUTs. As will become clear in the following, these and other characteristics of a BIST methodology are often closely related. Furthermore, it will also become clear that an infinite number of BIST methodologies are possible. In the following, we begin with a description of the general model of the circuit to be tested.

12.8.1 Model of the circuit

The circuit to be made self-testable is assumed to be a synchronous sequential circuit containing logic gates and D flip-flops. All primary inputs and outputs of the circuit are assumed to be latched using flip-flops in sets D^I and D^O, respectively. Internal lines of the circuit that are made controllable or observable (see Secion 12.4) are assumed to be also included in set D^I and D^O, respectively. D^S is the set of state flip-flops of the circuit. The normal input of state flip-flop $D_i^S \in D^S$ is driven by the **state output**, Y_i, of the circuit and its output drives the circuit's **state input** y_i. Hence, state flip-flop D_i^S can be used to capture responses observed at state output Y_i and to apply tests to state input y_i.

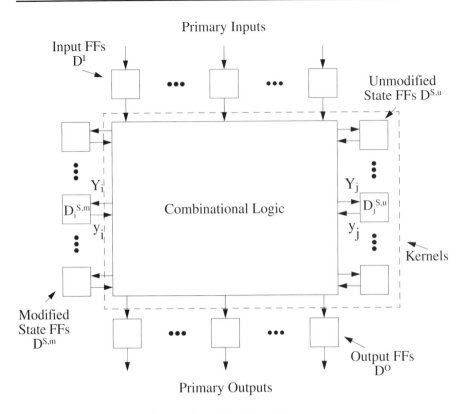

Figure 12.30 A model of the circuit to be made self-testable

We assume that a subset of the state flip-flops, $D^{S,m}$, are modified to enable reconfiguration in various **test modes** to apply desired tests to, and capture responses observed at, the corresponding state inputs and outputs. The remaining state flip-flops, $D^{S,u}$, are unmodified and always operate in the normal mode. This circuit model is illustrated in Figure 12.30.

In general, either $D^{S,m}$ or $D^{S,u}$, or both, may be empty sets. In the first case, only flip-flops at the primary inputs and outputs of the circuit are modified to enable reconfiguration in one or more test modes during self-test. The second case corresponds to **full intrusion BIST** where all the CUT flip-flops are modified.

12.8.2 Kernels

To simplify the development and application of self-test, a large circuit may be divided into a number of sub-circuits, each of which is treated as an individual CUT during self-test. In the following discussion, each such sub-circuit is called a **kernel**. All test circuitry required to apply vectors and analyze responses, including PGs and RAs, is considered as being external to the kernel. For example, consider the example circuit shown in Figure 12.31(a). (In this and many subsequent figures, sets of flip-flops will be shown grouped together as registers, such as R_1, R_2, and so on, and logic gates that are directly connected to each other will be collectively shown as combinational logic blocks, such as C_1, C_2, and so on. A technique for obtaining such a model of a circuit starting with its gate-level description is described in Chapter 11 (Section 11.4.1.1).) If this circuit is tested by configuring R_1 as a PG and R_2 as an RA, then kernel K_1 is merely the logic block C_1.

12.8.2.1 Types of kernels

The types of kernels considered by various BIST methodologies closely mirror the types of circuit structures allowed by various structural methodologies for partial scan flip-flop selection (see Chapter 11; Section 11.3.1). Most BIST methodologies consider only *combinational kernels*, such as the one obtained by replacing the registers at the inputs and outputs of a block of logic by PGs and RAs, respectively, such as K_1 in Figure 12.31(a). Other methodologies allow kernels that contain a number of combinational logic blocks and registers connected as linear *pipelines* of combinational blocks of logic, e.g., K_2 in Figure 12.31(b). Some methodologies allow *balanced kernels*, such as K_3 in Figure 12.31(c). (Recall that a kernel is said to be balanced if any two paths between any pair of registers in a circuit pass via an equal number of registers.) At least one methodology allows the use of kernels whose flip-flops may have self-loops but are otherwise acyclic (see Chapter 11; Section 11.3.1).

The type and size of a kernel can have a significant impact on the complexity of the self-test circuitry required for its testing, test application time, and quality of testing. For example, any pseudo-random sequence may provide high SAF coverage for a small combinational kernel, such as K_1. In contrast, a sequential kernel may require a PG that generates a sequence that contains a rich set of sub-sequences of ω vectors, where $\omega \geq 2$. A pipeline kernel, such as K_2, lies somewhere between these two extremes, since it may be tested by a simple PRPG, but the clocks of the PG and RA need to be controlled to take into account the fact that the response to a vector applied at the inputs of a pipeline kernel appears at its outputs a few clock cycles later (see Chapter 11; Section 11.3.1).

Kernel K_3 shown in Figure 12.31(c) is one-vector testable. This implies that the detection of an SAF in C_7 requires the application of a single vector at the kernel inputs, R_6 and R_9. However, due to the difference between the sequential depths of the paths from these registers to C_7, the corresponding values must be applied to R_6 and R_9 at times τ and $\tau + 1$, respectively (see Chapter 11; Section 11.3.1).

PRPGs used in BIST typically fail to apply a rich set of vectors if values are required at different times. For example, an LFSR PG formed by combining the bits of R_6 and R_9 would typically fail to generate a rich combination of vectors required to test C_7. Assume, for example, that R_6 and R_9 are each two-bits wide

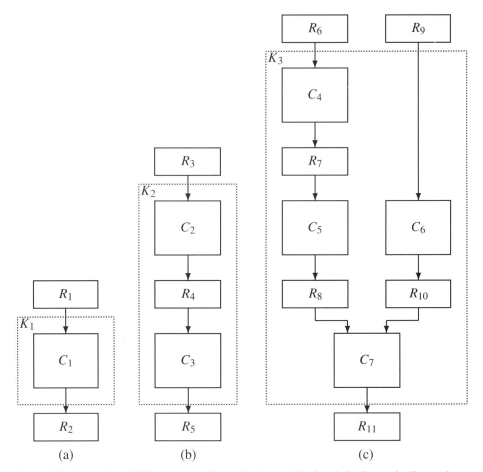

Figure 12.31 Examples of different types of kernels: (a) combinational, (b) linear pipeline, and (c) balanced

and are configured as a four-stage LFSR with $\phi(x) = x^4 + x + 1$. One can verify that such an LFSR will generate much fewer than $2^4 - 1 = 15$ vectors with the type of time-skew mentioned above. Hence, if such PRPGs are to be used, then it may be necessary to further restrict the types of kernels. This may be achieved by further partitioning kernel K_3 to obtain, say, a mix of combinational and pipeline kernels.

The size of a kernel can also affect the detectabilities of faults within a logic block. For example, the circuit shown in Figure 12.31(b) can be tested as a single pipeline or as two combinational kernels. While in the latter case, any D or $\overline{D}$ observed at the output of C_2 is directly captured by an RA, in the former case, the D or $\overline{D}$ must be propagated to the RA via C_3. Hence, typically the detectabilities of faults in a combinational logic block decrease as the block is incorporated into larger kernels.

12.8.2.2 Identification of kernels

Identification of appropriate kernels is one of the important aspects of BIST design. In many BIST methodologies, this task is made simple by allowing only the simplest types of kernels, e.g., combinational kernels or pipeline kernels with a bound on sequential depth, i.e., the number of registers along any path from a kernel input to a kernel output. Other methodologies use techniques for selecting scan flip-flops (see Chapter 11; Section 11.3.1.3) and control/observation points to implicitly define kernels.

Structural techniques to identify partial scan flip-flops can be used to identify kernels. For example, if balanced kernels are allowed, then the appropriate methodology referred to in Section 11.3.1.3 can be used. In the BIST design, the flip-flops identified by the partial scan methodology are replaced by appropriate test registers. Hence, the kernels are obtained by deleting these flip-flops and considering their inputs and outputs as kernel outputs and inputs, respectively. Each connected component of the resulting circuit constitutes a kernel.

Consider the example circuit shown in Figure 12.32(a). Assume that the BIST methodology allows balanced kernels, and hence the methodology referred to at the end of Section 11.3.1.3 can be used to identify flip-flops to be replaced by test registers. One possible solution is to replace R_1 and R_2 by scan registers. The two connected circuit components obtained after the removal of registers R_1 and R_2 from the circuit in this figure constitute kernels K_1 and K_2 and are shown in Figure 12.32(b).

12.8.3 Test mode configurations and reconfiguration circuitry

Let $D_1, D_2, \ldots, D_l$ be a set of D flip-flops of the CUT. Let $Y_1, Y_2, \ldots, Y_l$ be the state outputs of the circuit that drive the normal inputs of these flip-flops. Let $y_1, y_2, \ldots, y_l$ be the state inputs of the circuit driven by the outputs of these flip-flops. Figure 12.33(a) shows how a register R, with flip-flops $D_1, D_2, \ldots, D_l$, is configured during the normal operation of the circuit. At every clock cycle, the values appearing at $Y_1, Y_2, \ldots, Y_l$ are captured by the register and applied to $y_1, y_2, \ldots, y_l$. We will refer to such an operation as the **normal** mode operation of register R.

The simplest of the various test modes in which a register R may be configured is the **scan** mode shown in Figure 12.33(b). In this mode, the flip-flops do not capture the values from the lines $Y_1, Y_2, \ldots, Y_l$. Instead, the flip-flops are configured as a scan chain where the output of flip-flop D_i is clocked into flip-flop D_{i+1}. The output of the last flip-flop, D_l, and the input of the first flip-flop, D_1, are, respectively, the scan output and scan input of the register. The scan input of the register is connected either to the scan output of another register or to one of the outputs of a PG. Similarly, the scan output of the register is connected either to the scan input of another register or to an input of an RA. This mode is used in some BIST methodologies to apply tests generated by a PG and to shift-out captured responses for compression by an RA. In

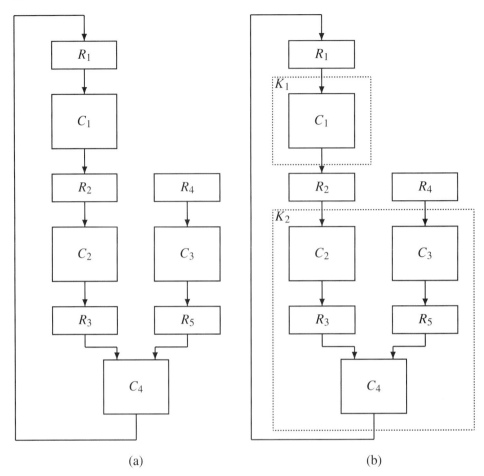

Figure 12.32   Example of kernel identification using partial scan techniques: (a) original circuit, and (b) balanced kernels obtained

other BIST methodologies, this mode is used to initialize PGs and RAs with suitable initial values. Note that, in the scan mode, the outputs of the flip-flops continue to drive the inputs $y_1, y_2, \ldots, y_l$ of the circuit. This is also true for all the modes described next.

Another commonly used test mode is the **pattern generator** (PG) mode shown in Figure 12.33(c). In this mode, the flip-flops are configured as a parallel pattern generator whose outputs drive the inputs $y_1, y_2, \ldots, y_l$ of the circuit. Even though Figure 12.33(c) shows an LFSR PG, CA or any other type of PG may be used. Note that in this mode the state outputs of the circuit, $Y_1, Y_2, \ldots, Y_l$, have no influence on the PG operation.

To capture and compress the response generated at the outputs $Y_1, Y_2, \ldots, Y_l$ of the circuit, register R can be configured in the **response analyzer** (RA) mode shown in Figure 12.33(d). Again, even though an MISR compressor is shown, any other multi-input RA, such as a CA compressor, can be used.

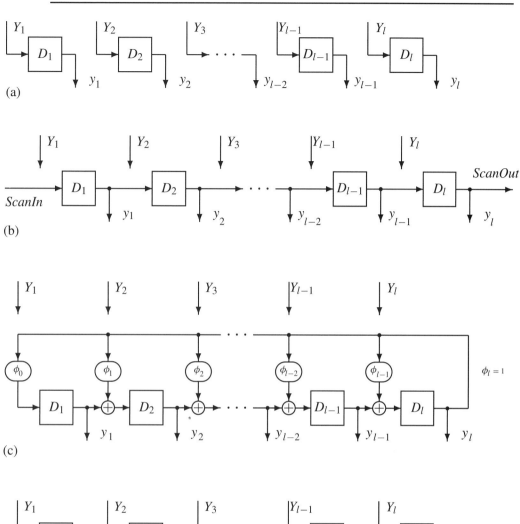

(a)

(b)

(c)

(d)

Figure 12.33 Test mode configurations: (a) normal mode, (b) scan mode, (c) pattern generator mode, and (d) response analyzer mode

In the response analyzer mode, the flip-flops may be used to concurrently capture responses at circuit outputs $Y_1, Y_2, \ldots, Y_l$ and to apply tests to the circuit inputs $y_1, y_2, \ldots, y_l$. The following two test modes are specially used in this manner.

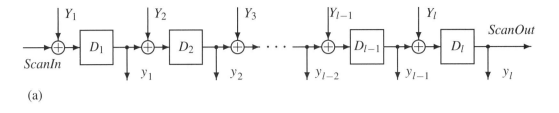

(a)

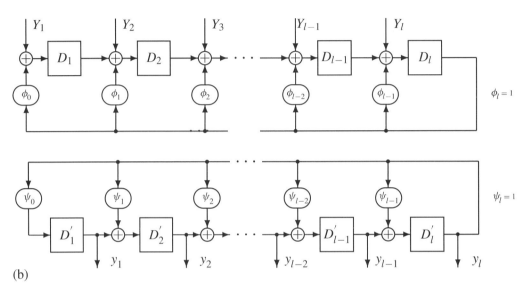

(b)

Figure 12.34 Additional test mode configurations: (a) scan compressor mode, and (b) independent PG and RA mode

The first of these modes is the **scan compressor** (SC) mode shown in Figure 12.34(a). In this mode, the flip-flops are configured as a shift register and XOR gates are inserted between successive shift register stages. One input of the XOR gate inserted between D_{i-1} and D_i is connected to the output Y_i of the circuit. Note that, in this mode, the input to the first XOR gate and the output of the last flip-flop are, respectively, the scan input and output of the register. This structure is identical to the compressor shown in Figure 12.33(d) with its feedback connections removed.

The second test mode that allows concurrent PG and RA functionality is the **independent PG and RA** mode shown in Figure 12.34(b). This can be viewed as a combination of an LFSR PG and an MISR RA, where the RA and PG operate *independently*. This is in contrast with the scan compression mode and the case in which the outputs of an RA are used as test vectors. In both those cases, the vectors applied to the circuit inputs, $y_1, y_2, \ldots, y_l$, depend upon the response observed at the circuit outputs, $Y_1, Y_2, \ldots, Y_l$. However, as will be seen in the following discussion, independent PG and RA operation is achieved at a high area overhead.

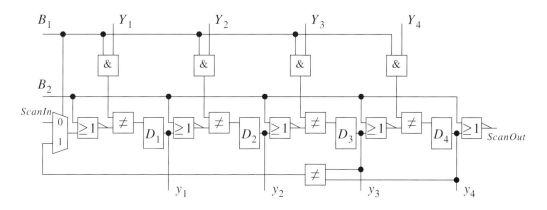

Figure 12.35 Built-in logic-block observer (Konemann *et al.*, 1979) (© 1979 IEEE)

12.8.3.1 Reconfigurable circuitry

In Chapter 11 (Section 11.2), we have seen how a scan function can be added to the circuit by replacing normal D flip-flops by reconfigurable scan flip-flops. Such a use of reconfigurable circuitry reduces the cost of DFT by facilitating sharing of the resource, namely the D flip-flops, between the normal and test modes. We next present some commonly used reconfigurable circuits.

If a BIST methodology only uses the scan mode, then only scan flip-flops are required. Other BIST methodologies may configure a register not only as a scan chain, but also as a PG and an RA. The **built-in logic-block observer (BILBO)** (Konemann *et al.*, 1979), shown in Figure 12.35, is a reconfigurable register that can be configured in normal, scan, reset, and PG/RA modes. In the PG/RA mode, the BILBO register operates as an LFSR PG if the all-zero vector is applied to the inputs $Y_1, Y_2, \ldots, Y_4$; otherwise, it operates as an MISR RA. One can verify that the application of each of the four possible combinations of values to the mode control signals, B_1 and B_2, configures the register in one of the above modes. Since a BILBO-based self-testable design replaces a normal register by a BILBO register, the area overhead caused by the use of BILBO is due to the AND, NOR, and XOR gates and the multiplexers. Also, in the normal mode, each state output of the circuit, Y_i, passes through two gates, an AND gate and an XOR gate, before reaching the input of the D flip-flop. Hence, the use of a BILBO register increases the delay of the block of logic feeding its inputs and may necessitate a reduction in the clock rate at which the circuit is run during its normal operation.

If a BIST methodology uses only the scan compressor mode, then a flip-flop obtained by using a combination of a scan flip-flop and an XOR gate can be used. We refer to this flip-flop as a **CSTP flip-flop**. (We use this name based on the fact that this flip-flop is used in the *circular self-test path* BIST methodology discussed later.)

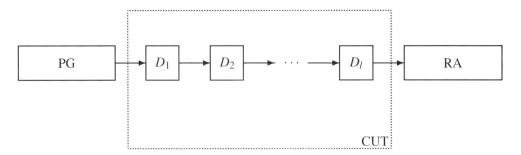

Figure 12.36   A generic scan-based BIST methodology

The independent PG and RA mode of operation (Figure 12.34(b)) requires two flip-flops for each stage of the register, one of which is configured as part of the RA and the other as part of the PG. Hence, additional flip-flops are required, in addition to the gates and multiplexers added to enable reconfiguration. The area overhead required to implement a reconfigurable register that can support this mode is high. An efficient design of such a reconfigurable register, called **concurrent BILBO (CBILBO)**, can be found in Wang and McCluskey (1986).

12.8.4   Classification of BIST methodologies

BIST methodologies can be broadly divided into two main categories, based on the manner in which flip-flops are used for self-test. The first category is **in-situ BIST**, where flip-flops at the inputs and outputs of a kernel are configured as *parallel* PGs and *multi-input* RAs, respectively, to test the kernel. The use of parallel PGs enables the application of one test vector per clock cycle. Hence, such methodologies are also called **test-per-clock BIST**. In contrast, in **scan BIST** methodologies, the flip-flops at the inputs and outputs of kernels are configured as one or more scan chains. As shown in Figure 12.36, a pattern generator is added to the chip and each vector generated by the PG is applied to each kernel via the scan chain. Multiple clock cycles are required to scan each test vector into the flip-flops at the kernel inputs. After a new vector is completely scanned in, the flip-flops in the scan chain (at least those driven by the kernel outputs) are configured in the normal mode and the response to the vector is captured. The flip-flops are again configured in the scan mode and the test response is scanned out of the scan chain and compressed into the RA, which is also added to the chip. Such methodologies are also called **test-per-scan BIST**. The two methodologies differ in their hardware overheads and test application.

Hardware overhead: In-situ BIST requires the replacement of flip-flops at the kernel inputs and outputs by registers such as BILBO. Scan BIST methodologies require these flip-flops to be replaced by scan flip-flops, which require lower area overhead than BILBO registers. However, scan BIST requires the addition of a PG and an RA to the chip. Hence, the area overhead of scan BIST is lower only if the circuit has a large number of kernels that share the added PG and RA.

The use of a BILBO register adds a delay of two additional gates at the input of each flip-flop, while the use of a scan flip-flop adds the delay of a single gate. Hence, the use of scan BIST can be more attractive for high-performance circuits. However, the additional delay overhead of in-situ BIST can be eliminated through the use of alternative designs for the BILBO register.

Test application: The main advantage of in-situ BIST methodologies is that they support testing at the normal clock rate of the circuit, i.e., **at-speed** testing. At-speed testing enables the detection of faults that cause errors only when the circuit is operating at its normal speed. For example, errors caused due to transients in the power/ground lines caused by the switching of circuit lines due to the application of a vector may only be detected by the application of the next vector within the normal clock period of the circuit.

In scan BIST methodologies, application of a single test requires $SCL + 1$ clock cycles, where SCL is the length of the longest scan chain. This precludes the possibility of at-speed testing. Hence scan BIST methodologies cannot be used to detect the types of faults mentioned above. Furthermore, since one test is applied per clock in in-situ BIST, a larger number of tests can be applied in a given test time. This helps attain a higher coverage of random pattern resistant faults using simple PRPGs, such as LFSRs. In contrast, since fewer tests can be applied in scan BIST, it may sometimes become necessary to use more complicated PGs to ensure high fault coverage. This can diminish any area advantage of scan BIST.

One of the key advantages of scan BIST methodologies is that they can be implemented at the chip level even when the chip design uses modules that do not have any BIST circuitry, provided that they have been made testable using scan.

In the following sections, we will describe both types of methodologies, including case studies.

12.9   In-situ BIST methodologies

In this section, we describe several variations of in-situ BIST methodology in detail.

12.9.1   Type of testing

The three sets of flip-flops D^I, D^O, and $D^{S,m}$ shown in Figure 12.30 differ in how they must be used during self-test. The flip-flops in D^I are used solely to generate test vectors and flip-flops in D^O are used solely to capture and compress test responses. In contrast, the flip-flops in $D^{S,m}$ must be used to generate test vectors as well as to

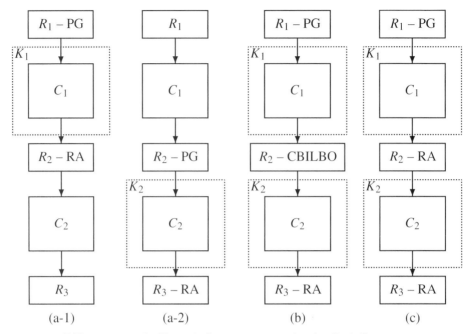

Figure 12.37 Different types of self-test: (a-1) non-concurrent (session 1), (a-2) non-concurrent (session 2), (b) concurrent and independent, and (c) concurrent and dependent

capture test responses. As will become apparent in the following discussion, this is an important issue for in-situ BIST methodologies.

Example 12.33 Consider the circuit shown in Figure 12.37, where sets of flip-flops are collectively shown as registers. In this circuit, register R_1 needs to be used only as a PG. Such a PG generates pseudo-random sequences. In contrast, register R_2 must be configured as an RA to self-test the combinational kernel C_1 and as a PG to self-test C_2. This dual test functionality of R_2 can be dealt with in three ways.

1 R_2 can be replaced by a BILBO register. This register can be first configured as an RA to self-test K_1 using vectors generated by the PG obtained by reconfiguring R_1 (Figure 12.37(a-1)). After K_1 is tested, R_2 can be configured as a PG to test K_2 (Figure 12.37(a-2)). In this case, R_2 generates pseudo-random sequences when it is used as a PG.

2 R_2 can be replaced by a CBILBO register. This would enable its reconfiguration in such a way that it can be used concurrently as an RA for K_1 and a PG for K_2. Since a CBILBO register allows *independent* operation of the PG and the RA, a pseudo-random test sequence is applied to K_2 (Figure 12.37(b)).

3 R_2 can be replaced by a BILBO register, which can be configured as an RA to capture the response generated during the self-test of K_1 and, at each clock cycle, the contents of the RA used as a test vector to concurrently self-test K_2 (Figure 12.37(c)). In this case, since R_2 does not operate as an autonomous FSM

when it is used as a PG to test K_2, the test sequence used to test K_2 may not have the desirable properties possessed by a pseudo-random sequence. Hence, the properties of sequences generated at the outputs of an RA need to be characterized.   □

In configurations where a register must be used as an RA and a PG for *distinct kernels*, allowing the use of RA contents as test vectors helps achieve high test concurrency at low area overhead. In the above example, it helps achieve, using a BILBO register, test concurrency which would otherwise be achieved only by the use of a CBILBO register. In such cases, the use of the contents of an RA as test vectors can be avoided at the expense of decreased test concurrency, while still avoiding the use of a high area overhead CBILBO register. In contrast, in circuit configurations where a register must be used as an RA as well as a PG for the same kernel, such as register R_3 in Figure 12.39(a), the register must either be (1) replaced by a CBILBO register, or (2) if replaced by a BILBO register, the contents of an RA must be used as test vectors. Hence, in such cases, the use of contents of an RA as test vectors can be avoided only by incurring a high area overhead.

The properties of sequences generated by a register when it is being used as an RA are now described. Two cases are discussed corresponding to scenarios where a register is being used as an RA for a kernel that is tested by vectors that are (1) independent of the outputs of the RA, and (2) dependent on the outputs of the RA.

12.9.1.1 Case 1: RA inputs independent of outputs

The first case is exemplified by the third alternative discussed in the above example and depicted in Figure 12.37(c). Note that in such cases, the RA works as a normal compressor and the quality of compression is not altered in any fashion due to the use of its contents as test vectors to concurrently test another kernel. However, since the RA is not an autonomous LFSR, it may generate sequences that are different from LFSR sequences.

Let a register R be configured as an MISR RA with a primitive feedback polynomial to compress the test response at the outputs of kernel K_{in} and its contents simultaneously used to test kernel K_{out}. If the output of kernel K_{in}, which is input to the MISR, is $00 \ldots 0$ for each vector, R would essentially operate like an autonomous LFSR and generate an MLS. R would also generate an MLS if the output of kernel K_{in} is any other vector that remains constant for all the tests. In this case, the only way in which the operation of R will differ from that of an autonomous ML-LFSR with the same feedback polynomial will be that instead of the all-zero vector being excluded from the MLS, a different vector will be excluded.

Example 12.34   Figure 12.38(a) shows a two-stage MISR with feedback polynomial $\phi(x) = x^2 + x + 1$ and Figure 12.38(b) shows the STG of the MISR, where the state is $y_1 y_2$ and the inputs are shown as $Y_1 Y_2$. Note that if the input to the MISR is always

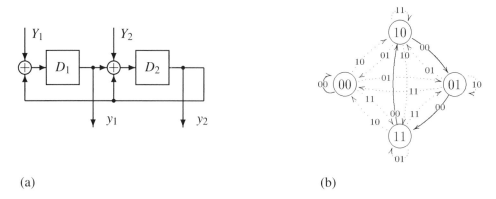

(a)   (b)

Figure 12.38   (a) A two-stage MISR, and (b) its STG (Kim *et al.*, 1988) (© 1988 IEEE)

10, then it will generate an MLS of length three if started at any state other than 01, which is also the vector that is excluded from the MLS.   □

If at least two different input vectors are applied to an n-stage MISR during the self-test of K_{in}, then all possible 2^n vectors can potentially appear at the outputs of the MISR. However, in such a scenario, some vectors may appear multiple times even in a short sequence.

At least two distinct vectors must appear at the output of a kernel during its test, otherwise some SAFs in the kernel will surely not be tested. Hence, the input vector to an MISR is unlikely to remain constant. Consequently, a sequence of μ vectors, where $\mu \leq 2^n - 1$, may contain fewer than μ distinct vectors.

Example 12.35 If the MISR shown in Figure 12.38(a) starts in the initial state 10 and the sequence $(10, 00)$ is applied at its inputs, then the sequence $(10, 11, 10)$ will appear at its outputs. The vector 10 appears twice in this three-vector sequence.   □

Assuming that the inputs of an n-stage MISR PG are independent of its outputs, it has been shown that the number of distinct vectors in a sequence of μ vectors at its outputs is (Kim *et al.*, 1988)

$$2^n \left(1 - \left(1 - \frac{1}{2^n} \right)^{\mu} \right). \tag{12.48}$$

Under the same assumption, it has also been empirically observed that the probability of any n-bit vector being generated rapidly converges to $1/2^n$ after a short initial subsequence. This probability converges to $1/2^n$ more rapidly as (a) the number of stages, n, increases, and (b) the number of distinct vectors applied at the inputs of the MISR increases.

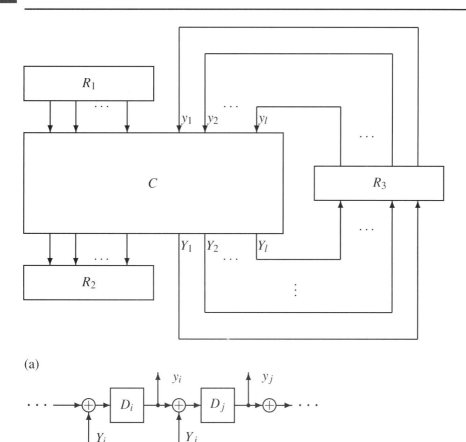

(a)

(b)

Figure 12.39 RA whose inputs depend on its outputs: (a) a block with feedback, and (b) adjacent flip-flops in R_3 when configured as an RA

12.9.1.2 Case 2: RA inputs dependent on outputs

The second scenario in which an RA is used as a PG is where the vectors at the inputs of the RA are dependent on the vectors at its outputs. The circuit in Figure 12.39(a) is an example of this case, where the input to the RA R_3 is the response of logic block C to the tests applied by the RA. In this case, the quality of compression as well as the quality of vectors generated may be compromised due to the simultaneous use of a register as an RA as well as a PG.

Example 12.36 Consider two consecutive flip-flops in the RA formed by configuring register R_3 in Figure 12.39(a), which are shown in Figure 12.39(b). Assume for simplicity that the RA feedback structure is such that no feedback occurs between D_i and D_j. Since the output of D_i drives the previous-state input y_i, which feeds back via the combinational logic block, C, the value of state output Y_j in the next clock cycle is, in general, a function of the output of D_i at this clock cycle. In other words, the response captured at the jth input of the RA in the next cycle can be written as $Y_j = f(y_i)$, where f is a function of y_i and other state and primary inputs. Hence, the state of flip-flop D_j in the next cycle is given by $y_i \oplus f(y_i)$. In an extreme case, when $f(y_i) = y_i$, the next state of D_j becomes $y_i \oplus y_i = 0$. In this case, the content of D_j is always zero.   □

The example scenario described above points to two important facts.

1 The quality of vectors generated at the outputs of the RA is seriously compromised, since a constant value appears at at least one of its outputs.
2 If the error caused by a fault in C can propagate only to state output Y_i, then its presence can never cause any stage of the RA other than D_i to have an erroneous value. Even at D_i, the erroneous value will exist for only one clock cycle. Hence, unless the output of D_i is monitored at the specific clock cycle when the error is captured, the RA will lose the effect of the error. Such **error cancellation** can severely degrade the quality of compression and cause significant aliasing.

Clearly, the above example, where $Y_j = f(y_i) = y_i$, is an extreme case. However, to a lesser extent, the degradation of the quality of compression due to error cancellation as well as that of test pattern generation due to non-occurrence of certain vectors also occurs for other forms of function $f(y_i)$.

Example 12.37   One can verify that when $Y_j = f(y_i) = y_i \cdot \rho$, where ρ is some function of the outputs of flip-flops other than D_i and D_j, the value shifted into D_j can be written as $y_i \oplus f = y_i \oplus (y_i \cdot \rho) = y_i \cdot \overline{\rho}$. This indicates that whenever $\rho = 1$, any erroneous response captured at Y_i will be lost (Avra and McCluskey, 1993).   □

Even though the above examples discuss cases of adjacent flip-flops of an RA, the dependence between the output of flip-flop D_i and the input of flip-flop D_j can degrade the quality of compression and test pattern generation (Carletta and Papachristou, 1994), provided:

1 α flip-flops separate D_i and D_j in the RA, and
2 there exists another path from the output of D_i, y_i, to the input of D_j, Y_j, via the logic blocks that pass through α registers.

These conditions will be referred to as α-**level dependency** between the output of D_i and the input of D_j. Such a dependency can be eliminated in two main ways. The first alternative is to reorder the flip-flops in the RA so that D_i and D_j are not separated by α flip-flops. The second alternative is to replace additional circuit registers by test registers, which can eliminate paths of length α through the circuit.

While α-level dependency can compromise the quality of compression as well as vectors generated by an RA, typically, its detrimental effects decrease with an increasing value of α. Hence, while ideally all such dependencies should be eliminated, in practice, it may be sufficient to eliminate all the dependencies with small values of α.

Next, we discuss two example in-situ BIST methodologies as case studies.

12.9.2 BILBO-based BIST

BILBO-based BIST covers a wide range of methodologies that achieve in-situ self-test by replacing registers in a given circuit by PG, RA, BILBO, and CBILBO registers. Each individual methodology can be characterized by (a) the type of kernels, and (b) the type of testing it allows. To a large extent, choices made on these issues determine the certainty with which high fault coverage can be attained.

12.9.2.1 Types of kernel

Some methodologies only allow *combinational* kernels, others allow *pipeline* kernels (whose depths may be upper bounded by a given constant), still others may also allow special types of *balanced kernels* discussed in Section 12.8.2, and so on.

High fault coverage can be expected for combinational kernels since they are relatively small and an SAF within can be tested by a one-vector test. While an SAF in a pipeline as well as a balanced kernel can also be tested with a one-vector test, the larger cumulative size of combinational blocks in the kernel can decrease the detectabilities of faults. The use of larger kernels may decrease the fault coverage obtained by a pseudo-random sequence of a given length. Also, detection of an SAF in a general acyclic kernel may require a ω-vector test, where $\omega \geq 2$. Hence, special PGs that are capable of generating ω-vector tests may be required.

Clearly, the number of test registers required is maximum for methodologies that allow only combinational kernels and lower for those that allow larger and more general types of kernels. Hence, the choice of the type of kernels can be made by examining the following tradeoffs.

- *Moderate fault coverage target:* In this case, schemes that allow more complicated kernels can be used to obtain BIST solutions with lower area overhead while trading off high fault coverage, or at least by trading off the certainty of obtaining a high fault coverage.

- *High fault coverage target:* Here, allowing more complicated kernels (which include the simpler types of kernels) allows the BIST methodology to consider a larger number of solutions. For example, a methodology that allows the use of general acyclic kernels can examine a wide range of solutions including the following extremes. (1) A solution that uses only combinational kernels. In such a solution, many registers are replaced by test registers but each register has a low area overhead, since high fault coverage may be obtained using low area-overhead PRPGs described before. (2) A solution that uses large acyclic kernels. In such a case, fewer registers are replaced by test registers. However, the area overhead associated with each register may be higher, since special PGs that can generate a rich set of ω-vector tests may be required.

12.9.2.2 Type of testing

The key issue here is whether to allow the contents of an RA to be used as test vectors. Some BILBO methodologies do allow this type of testing, provided that during testing the values at the inputs of the RA are independent of the values at its outputs. Note that the above condition implies that a register cannot be used as an RA as well as a PG for the same kernel. Furthermore, an RA of a kernel K_1 cannot be used as a PG for any other kernel that is tested concurrently and one of whose RAs is being used as a PG for kernel K_1. Under these conditions, the use of an RA as a PG allows the achievement of higher test concurrency at low area overhead.

Once the types of kernels allowed and testing are determined, the BILBO circuitry can be designed in four steps.

1 *Identification of kernels:* One easy way to identify kernels is the use of an appropriate technique to identify flip-flops for partial scan in the manner mentioned in Section 12.8.2.2 and described in Chapter 11 (Section 11.3).

2 *Embedding:* Once the kernels are identified, registers must be identified for re-placement by an appropriate reconfigurable register, such as PG, RA, BILBO, or CBILBO. This process will be referred to as **embedding**.

An embedding must satisfy two main constraints.

(a) At each input of each kernel, there must exist one register that can be configured as a PG.

(b) At each output of each kernel, there must exist one register that can be configured as an RA.

An embedding can be obtained by first identifying registers to be used as PGs and RAs to test each kernel and labeling them $\mathcal{G}$ and $\mathcal{R}$, respectively. If any register is identified as a PG as well as an RA for the same kernel, it is labeled $\mathcal{C}$. The labels can then be used in the following manner.

(a) Registers that are assigned label $\mathcal{C}$ must be replaced by CBILBO registers.

(b) Registers that are assigned labels $\mathcal{G}$ as well as $\mathcal{R}$, but not $\mathcal{C}$, must be replaced by BILBO registers.

(c) Registers that are assigned label $\mathcal{G}$ only must be replaced by PGs.

(d) Registers that are assigned label $\mathcal{R}$ only must be replaced by RAs.

Registers that are not assigned any of these labels are not replaced by any reconfig-urable test register.

If each kernel input is driven by a single register and each kernel output drives a single register, then a simple labeling mechanism provides minimum area-overhead BIST circuitry *for the given set of kernels*. However, in circuits such as the one

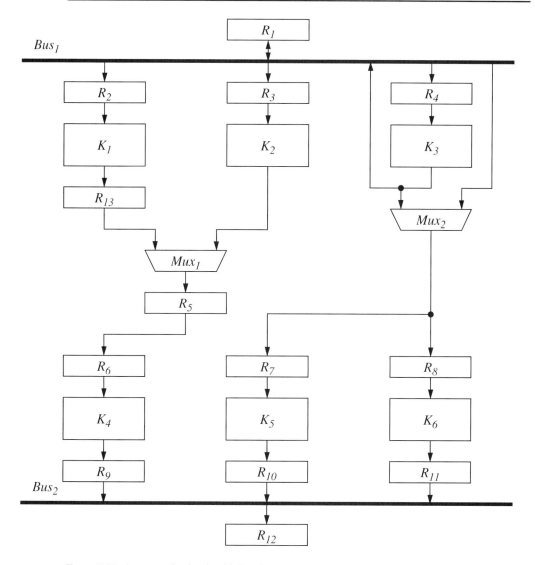

Figure 12.40   An example circuit with I-paths

shown in Figure 12.40, a number of registers may be connected to an input or output of a kernel via multiplexers and busses. Since multiplexers and busses can be used to transport data without alteration, they provide **identity paths (I-paths)** (Abadir and Breuer, 1985) through which test vectors and/or responses can be transferred. If I-paths are exploited, several PG/RA choices may exist at a kernel input/output. This is illustrated next.

Example 12.38   Assume that the example circuit shown in Figure 12.40 is to be tested by a BILBO-based methodology that only allows combinational kernels but exploits I-paths. One possible embedding would be to replace R_1 by a PG, R_3 by an RA, R_5 by a BILBO register, and R_{12} by an RA. Kernel K_1 can be tested by (a) configuring R_1 as a PG, (b) applying each vector via an I-path comprised of Bus_1 and register R_2, (c) configuring R_5 as an RA, and (d) capturing the response observed at the kernel outputs in this RA using the I-path comprised of R_{13} and Mux_1. One can verify that all the other kernels can also be tested by this embedding. A very large number of embeddings exist for this circuit.   □

If a large number of alternative embeddings exist for kernels in a circuit, a labeling algorithm that considers optimization must be used to obtain an embedding with minimum area overhead.

3 *Test scheduling:* Once the test registers are identified, a **test schedule**, which determines when each kernel is tested, can be determined. To minimize the test application time, we would like to test kernels concurrently. However, due to sharing of test resources, i.e., PGs, RAs, and I-paths, certain kernels may not be tested concurrently. Such kernels are said to be **test-concurrency incompatible**.

Example 12.39 Consider the circuit shown in Figure 12.40, where each combinational logic block is to be treated as a separate kernel. Also, consider an embedding where R_1 is replaced by a PG, R_8 and R_{12} by RAs, and R_5 by a BILBO register.

Kernel K_1 is tested by using the resources $(R_1, Bus_1, R_2, R_{13}, Mux_1, R_5)$ where R_1 and R_5 are used as the PG and the RA, respectively, and other elements constitute parts of an I-path. Similarly, K_2 is tested using resources $(R_1, Bus_1, R_3, Mux_1, R_5)$. Both K_1 and K_2 can concurrently use R_1 as a PG. However, they cannot be tested concurrently because both must use R_5 as an RA. (Also, they must use Mux_1 with different settings to use R_5 as an RA.) (K_1, K_2) are hence test-concurrency incompatible. However, K_3 can be tested concurrently with either K_1 or K_2. Hence, (K_1, K_3) as well as (K_2, K_3) are compatible.   □

The above relations, along with all the other pairwise relations, are represented by the **test-concurrency incompatibility graph** shown in Figure 12.41. An edge exists between nodes corresponding to kernel K_i and K_j if they are test-concurrency incompatible.

Any pair or kernels that are test-concurrency incompatible cannot be tested concurrently. Equivalently, any two nodes in the test-concurrency incompatibility graph that have an edge between them must be tested in different **test sessions**. Hence, such kernels should be assigned to different test sessions; equivalently, they should be assigned different *colors*. If equal time is required to test each kernel, a minimum-time test schedule can be obtained by finding the minimum number of colors required to color the test-concurrency incompatibility graph such that no two adjacent nodes have the same color.

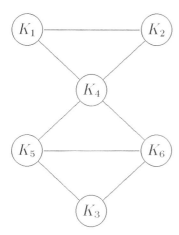

Figure 12.41 Example test-concurrency incompatibility graph

Example 12.40 All kernels in the above example can be tested in three sessions. For example, Kernels K_3 and K_4 can be tested in the first session; K_1 and K_5 in the second, and K_2 and K_6 in the third.   □

The problem of finding a minimum test schedule becomes more complicated if the kernels require different test application times (Craig *et al.*, 1988; Lin *et al.*, 1993b).

4 *Test control:* Finally, a test controller must be designed. In each session, the test controller must configure all the test registers, appropriately initialize the PGs and RAs, apply the requisite number of tests, and check the compressed response(s). The test controller also provides the interface between the BIST circuitry and the RUNBIST command of the boundary scan controller (see Chapter 11 for a description of boundary scan).

12.9.3   Circular self-test path

Circular self-test path (CSTP), also called **circular BIST**, is an in-situ BIST methodology (Krasniewski and Pilarski, 1989). In this methodology, a set of registers is selected and replaced by scan-compression (SC) registers. These registers are configured as a single circular shift-register by connecting the scan-out of one SC register to the scan-in of the next register. This is called the **CSTP register** and is illustrated in Figure 12.42 for an example circuit. In this figure, thick lines are used to indicate the connections between the registers that constitute the CSTP. The details of the configuration of the CSTP in the test mode are shown in Figure 12.43.

The CSTP is used to generate test vectors to test the kernels in the circuit; it is also used to capture and compress test responses at the kernel outputs. Also, as will be seen in the following, it may be used to initialize the CUT prior to self-test.

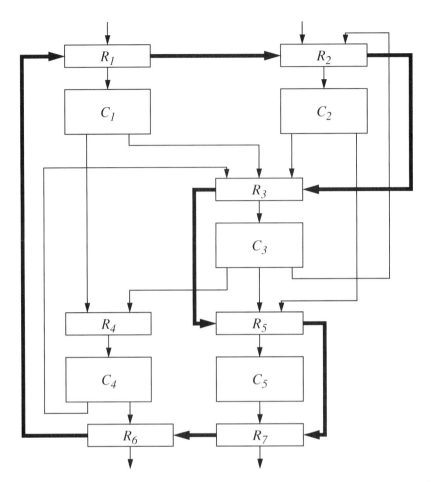

Figure 12.42 An implementation of CSTP for an example circuit (Pilarski *et al.*, 1992) (© 1992 IEEE)

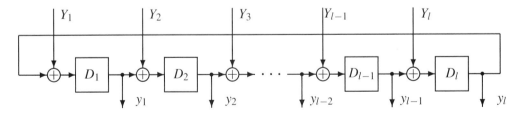

Figure 12.43 Detailed test-mode configuration of CSTP (Pilarski *et al.*, 1992) (© 1992 IEEE)

In this methodology, the circuit is tested by (a) reconfiguring the selected registers as CSTP, (b) initializing all flip-flops in the circuit, and (c) applying the desired number of test vectors to the kernels. The final signature can be observed by simply monitoring the values appearing at the output of any flip-flop in the CSTP for a desired number of clock cycles before the end of test application.

The area overhead incurred due to the use of an SC register is lower than that for a BILBO register, or even a PG or RA register. Also, the entire circuit is tested in a single session, hence the selected registers need to be reconfigured only once. Furthermore, the final circuit signature can be observed without the use of any additional test mode. Hence, the area overhead associated with the test control circuitry is also small. Consequently, the area overhead for this methodology is significantly lower than that for BILBO-based methodologies.

The key issue in this methodology is the quality of testing. In the following, we will first detail these issues and then describe a methodology to design CSTP circuitry.

Issues in testing by CSTP: Self-test performed using CSTP can be viewed in two ways. First, as shown in Figure 12.43, the CSTP can be viewed as an l-stage MISR compressor with feedback polynomial $\phi(x) = x^l + 1$, where l is the *total* number of flip-flops in registers chosen for replacement by SC registers. In each clock cycle, the CSTP captures the response observed at the outputs of the kernels in the circuit, adds the captured response (in a linear fashion) to the current state of the CSTP, and captures the result into the CSTP. The resulting state of the CSTP is applied to the inputs of kernels as a test vector in the subsequent clock cycle. Since all the kernels are tested using a single MISR, the values at the MISR inputs are *dependent* on the values at its outputs. Hence, as discussed earlier in this section, this can lead to α-level dependency (where $\alpha \geq 1$) and can degrade the quality of compression due to error cancellation. It can also degrade the quality of vectors applied to the circuit. As mentioned above, α-level dependency must be eliminated/minimized to avoid these problems.

Alternatively, the CSTP, along with the kernels of the circuit, can be viewed as an FSM. Since during testing, no external inputs are applied to the CSTP or the kernels, this FSM operates autonomously. In this view, the CSTP is a non-linear shift register, with (a) an explicitly added circular feedback connection given by $\phi(x) = x^l + 1$, and (b) the implicit feedback via the kernels of the circuit, which is typically non-linear. The state transitions of the CSTP register are hence dependent on the logic implemented by the kernels.

The STG formed through the combination of the CSTP and circuit kernels is comprised of one or more directed *subgraphs*. Each subgraph is either a simple *cycle* or has a cycle and a set of directed *trees*, where each tree begins at one or more states called *leaves* and ends at one of the states of the cycle. Figure 12.44 shows an example STG. Note that this STG is very different from the STG of an ML-LFSR or an ML-CA PRPG, which only contains a single simple cycle with all the non-zero states and another cycle with only the all-zero state.

The main concern is that the STG of a CSTP implementation for a given circuit may comprise a large number of cycles of short lengths. Since the number of distinct vectors applied to the circuit kernels is limited to the number of states in any subgraph of the STG, this can lead to a low fault coverage. Also, as long as the STG contains

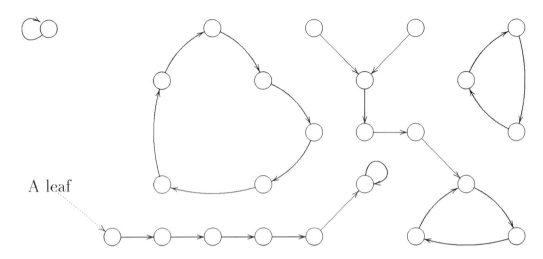

A leaf

Figure 12.44 STG of a CSTP (Carletta and Papachristou, 1994) (© 1994 IEEE)

one or more subgraphs with a small number of states, an inappropriate initial state of the circuit flip-flops may result in poor test quality. Several such cases, sometimes called **system state cycling**, have been observed in real circuits. Hence, one of the key objectives of the design of CSTP circuitry is to ensure that a large number of vectors are applied to the kernels. Furthermore, to ensure a high fault coverage, the kernels should be selected to be one-vector testable.

The quality of compression is also crucial, since significant error cancellation or high aliasing probability can decrease the effective fault coverage.

Design of CSTP: Designing a CSTP involves three types of decision.
1 Selection of circuit registers to be replaced by SC registers.
2 Ordering of flip-flops within registers and ordering of registers in the CSTP. This step may involve insertion of additional flip-flops into the CSTP which are not used during the normal operation of the circuit.
3 Selection of an appropriate seed for testing.

At a minimum, all the registers at the primary inputs and outputs of the circuit must be selected for inclusion in the CSTP; otherwise, some parts of the circuit will not be tested. Typically, additional registers must be selected to ensure that the resulting kernels are acyclic. This becomes essential if the remaining circuit registers do not have any reset mechanism. In such a case, the use of acyclic kernels guarantees that all circuit flip-flops can be initialized to known values through application of a bounded number of vectors to the flip-flops in the CSTP. Since the CSTP is not likely to generate a rich set of ω-vector test sequences for $\omega > 1$, it is advisable to ensure that all the kernels are one-vector testable. Finally, additional registers may be replaced by SC registers to eliminate α-level dependencies and to guarantee low aliasing probability.

The order of flip-flops within each register, as well as the order of registers in the CSTP, must be determined to eliminate or minimize α-level dependencies. If necessary, additional flip-flops, which are not used during the normal operation of the circuit, may be added to eliminate such dependencies. Note that re-ordering of flip-flops and registers changes the distance between flip-flops in the CSTP. Alternatively, by selecting additional registers for replacement, paths between the outputs of some flip-flops and some RA inputs may be eliminated. Either way, some α-level dependencies may be eliminated.

The initial state of the CSTP register must be selected to ensure that a large number of vectors are applied to the kernel inputs before the CSTP state starts repeating. Hence, the initial state must be selected such that during testing the FSM traverses states in one of the larger subgraphs of the STG. Furthermore, the initial state should be such that the sum of the length of the path along the tree leading to the cycle and the length of the cycle is maximized.

Example 12.41 Consider the circuit depicted in Figure 12.42 (first consider the version of the circuit without the CSTP). Assume that it is desired to obtain kernels that are one-vector testable.

Clearly, input registers, R_1 and R_2, as well as output registers, R_6 and R_7, should be a part of the CSTP. The addition of R_3 and R_5 to the CSTP divides the circuit into three kernels, namely (i) C_2 by itself, (ii) C_5 by itself, and (iii) a balanced kernel comprised of C_1, C_3, and C_4. Each of these kernels is one-vector testable.

Assume that the registers selected are connected in the order shown in the figure to form the CSTP. Since a path with zero flip-flops exists between the outputs of flip-flops in R_3 and inputs of flip-flops in R_5, a 0-level dependency exists between the last flip-flop in R_3 and the first flip-flop in R_5. The circuit must be analyzed to ensure that this dependency is not very harmful for the quality of response compression and pattern generation. Other cases of α-level dependency must also be identified and analyzed. Assume that all the identified dependencies are determined to be relatively harmless. In that case, the CSTP organization can be left as shown in the figure. (Otherwise, the registers may be reordered to eliminate the dependencies that were found to be significantly harmful.)

Logic simulation must then be performed to ensure that the design does not suffer from the state cycling problem and that a significant number of patterns are indeed applied to each kernel. □

12.10 Scan-based BIST methodologies

In scan-based BIST methodologies, the subset of state flip-flops that belong to $D^{S,m}$ (see Figure 12.30) are configured into one or more scan chains. In some

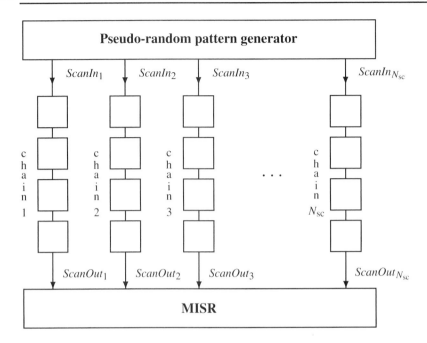

Figure 12.45 STUMPS (Bardell and McAnney, 1982) (© 1982 IEEE)

methodologies, such as **STUMPS (self-test using MISR and parallel SRGS**; where SRGS is a PRPG as shown in Figure 12.45) (Bardell and McAnney, 1982) and **LOCST (LSSD on-chip self-test)** (LeBlanc, 1984), the input and output flip-flops (D^{I} and D^{O}, respectively) are also configured as parts of scan chains, and external PG and RA are added. On the other hand, in other scan-based BIST methodologies, such as **RTS (random test socket)** (Bardell and McAnney, 1982) and **PS-BIST (partial scan BIST)** (Lin *et al.*, 1993a), either all or a subset of state flip-flops are configured as scan chains, while the primary input and output flip-flops are configured as parallel PGs and RAs, respectively. In either case, additional PG and RA are added to the chip. This PG generates test vectors which are serially shifted into the single or multiple scan chain(s) in a manner similar to scan testing described in Chapter 11. If *SCL* is the length of the longest scan chain, then a minimum of *SCL* scan clock cycles are required to scan in the values corresponding to a single vector into the scan chains. The flip-flops in the scan chain are then configured in the normal mode, any parallel PGs used at the primary inputs are used to apply a new vector to the primary inputs, and the response of the circuit kernels is captured in the flip-flops belonging to the scan chains as well as any parallel RAs used at the primary outputs. The flip-flops in the scan chain are then configured in the scan mode and the response is scanned out and shifted into the external RA; concurrently, the next vector is scanned in.

Note that the flip-flops that belong to $D^{\mathrm{S},m}$ are used to apply test vectors as well as to capture responses. However, in this case, a simple scan flip-flop can accomplish both these tasks while keeping each test vector *independent* of the response captured for the previous vector. Hence, for circuits with self-adjacent registers, such as the example shown in Figure 12.39, scan BIST can achieve the type of testing which, in in-situ BIST, can only be achieved through the use of high area-overhead CBILBO registers. Note that scan BIST achieves this reduction in area overhead by time-multiplexing the use of scan flip-flops.

If the number of self-adjacent registers is large, and if it is possible for a large number of kernels to share a single external PG and RA, then the area overhead of scan BIST can be lower than that of BILBO-based BIST methodologies. However, this occurs at the cost of an increase in test time, which, if unacceptable, may be decreased only by an increase in the area overhead. Finally, recall that at-speed testing is not possible in scan BIST.

In the following, we discuss the main issues that must be considered during the design of scan BIST methodologies.

12.10.1 Selection of kernels

The types of kernels used may again range from combinational to arbitrary sequential kernels. However, due to the inability of the external PG to generate a rich set of ω-vector tests for $\omega > 1$, a high fault coverage can be guaranteed only if the kernels are one-vector testable. The kernels can be identified by using an appropriate partial scan design procedure as mentioned in Section 12.8.2.2 and described in Chapter 11 (Section 11.3). In fact, even the scan chains may be configured using the procedures described in Chapter 11 (Section 11.4). As mentioned earlier, the ability to work with existing scan designs is one of the attractive features of these methodologies.

12.10.2 Clocking of the external PG

If an n-stage PG that generates an MLS of length $2^n - 1$ is used, then, ideally, $2^n - 1$ distinct test vectors can be applied to the scan chains. However, the number of vectors that can actually be applied may decrease due the fact that the external PG is clocked *SCL* times during the application of each vector. This manner of use of a PRPG as a serial pattern generator is sometimes called **disjoint sampling**. (See Section 12.2.1.3 and Problem 12.5 for details of this and the *overlapped sampling* modes.)

The reduction in the number of vectors that can be applied, called **decimation**, occurs because the application of N_v vectors via the scan chain causes the external PG to be clocked $SCL \times N_v$ times (Bardell, 1990b). Hence, the external PG returns to its initial state soon after the application of $\lfloor \frac{2^n-1}{SCL} \rfloor$ vectors. Consequently, in an extreme case, if $2^n - 1$ is divisible by *SCL*, the vectors scanned into the chains will start repeating after the application of $(2^n - 1)/SCL$ distinct vectors.

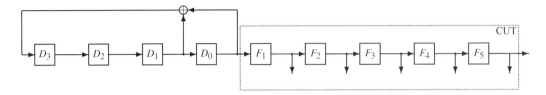

Figure 12.46 A four-stage LFSR driving a five-bit scan chain

Example 12.42 Consider the four-stage external-XOR LFSR PG with the primitive polynomial shown in Figure 12.46 which can generate a sequence of 15 distinct vectors. Assume that this PG drives a five-bit scan chain and hence is clocked five times for the application of each test vector. One can verify that in such a case, only three distinct five-bit vectors are applied to the scan chain before the vectors repeat. □

A decimation that decreases the number of vectors that can be applied to the scan chains is said to be **improper**. Improper decimation can be avoided by ensuring that the number of times the PG is clocked for the application of each test vector, say SCL_c, is co-prime to the length of the sequence generated by the PG, $2^n - 1$. Note that $SCL_c \geq SCL$ to ensure that the test responses are completely shifted out of the scan chain before the application of the next vector, i.e., to ensure that the test vectors are completely independent of the responses. Hence, the problem of decimation can be eliminated by selecting the smallest SCL_c that is (a) greater than or equal to SCL, and (b) is co-prime to the length of the sequence generated by the external PG.

Example 12.43 Again, consider the four-stage external-XOR LFSR PG with the primitive polynomial shown in Figure 12.46. Assume that this PG is clocked seven times for the application of each test vector. One can verify that in such a case, 15 distinct five-bit vectors are applied to the scan chain before the vectors repeat. This is due to the fact that $SCL_c = 7$ and $2^n - 1 = 15$ are co-prime. □

12.10.3 Selection of the PG and chain reconfiguration

In most scan BIST implementations, the sum of the lengths of the scan chains far exceeds the number of stages, n, in the external PG. For example, even in CUT with thousands of scan flip-flops, a PG with around 32 stages may be used. Furthermore, in most typical cases, even the length of the longest scan chain, SCL, is greater than n. These facts make it impossible for all possible vectors to be applied to any given set of $n + 1$ or more scan flip-flops. However, extensive use of scan flip-flops typically divides the circuit into disjoint kernels with a small number of inputs. Furthermore, only the values applied at the inputs of a kernel affect the detection of a fault within the kernel. Hence, high fault coverage can still be obtained using an external PG with few stages, provided that all the

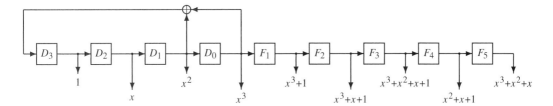

Figure 12.47 Residues and linear dependencies

possible vectors can potentially be applied to the inputs of each kernel. However, all the possible vectors may not be applied even at the inputs of a kernel that is driven by much fewer than n flip-flops. The following example illustrates this fact.

Example 12.44 Consider the scenario shown in Figure 12.47 where a **four**-stage external-XOR LFSR with feedback polynomial $\phi(x) = x^4 + x + 1$ feeds a single scan chain with five stages, $F_1, F_2, \ldots, F_5$. Consider a small example kernel whose inputs are driven by scan flip-flops F_1, F_4 and F_5. One can verify that for any initial LFSR state, some combination of values do not appear at the **three** inputs F_1, F_4, and F_5 of this kernel. In fact, it can be verified that once the values that were initially present in the scan flip-flops are shifted out, the value appearing at the output of F_5 is always the XOR of the values appearing at the outputs of F_1 and F_4. In such cases, the outputs F_1, F_4 and F_5 are said to be **linearly dependent**.   □

Linear dependency between the inputs of a kernel prevents the application of some test vectors at the kernel inputs. This can prevent detection of some faults in the kernel. Hence, ideally, BIST circuitry should be designed to ensure that the inputs of each kernel are linearly independent. A methodology to design BIST circuitry that ensures this is described next.

12.10.3.1 Residues and linear dependence

We use the concept of *residues* to determine whether any given set of kernel inputs are linearly dependent or independent. Associated with the output of each flip-flop in the PG as well as the scan chain is a **residue** which is a polynomial in x. The labeling process can begin by assignment of residue 1 to any PG output. The residues at the other PG outputs, the outputs of scan flip-flops, and the outputs of XOR gates in the circuit, can be computed by using the following rules.

1 The output of a flip-flop is assigned a residue $x \times residue_{in}$, where $residue_{in}$ is the residue assigned to the input of the flip-flop.

2 The output of an XOR gate is assigned a residue that is the linear sum of the residues assigned to its inputs.

In all the cases, the residues can be computed modulo-$\phi^*(x)$, where $\phi(x)$ is the feedback polynomial of the n-stage LFSR PG and $\phi^*(x) = x^n \phi(1/x)$.

Example 12.45 The above procedure can be illustrated by assigning residues to the outputs of all the flip-flops in Figure 12.47. First, residue 1 can be associated with the output of the first flip-flop of the LFSR PG, D_3. Using this, and the above rules for computing residues at the outputs of flip-flops, the outputs of D_2, D_1, D_0 and $F_1, \ldots, F_5$ can be assigned residues x, x^2, x^3, $x^3 + 1$, $x^3 + x + 1$, $x^3 + x^2 + x + 1$, $x^2 + x + 1$, and $x^3 + x^2 + x$, respectively. Note that the residue assigned to the output of F_1, $x^3 + 1$, is simply $x^4 \bmod x^4 \phi(1/x)$. Also, note that the residue assigned to the output of D_3, which is driven by an XOR gate, can also be derived in the following manner. First, the output of the XOR gate can be assigned residue $residue_{xor} = x^3 + x^2$, i.e., the sum of residues assigned to the outputs of D_0 and D_1, which are input to the XOR gate. Then the residue assigned to the output of D_3 can be computed as $x \times residue_{xor} \bmod x^4 \phi(1/x) = x^4 + x^3 \bmod x^4 + x^3 + 1 = 1$. This illustrates that the above rules for computing residues are consistent and the order in which residues are computed does not affect their values *relative to each other*.

An examination of the residues assigned to the outputs of F_1, F_4, and F_5 will now reveal the cause of linear dependency between their outputs. (Recall that $residue_1, residue_2, \ldots, residue_b$ are said to be linearly dependent if there exists a set of coefficients $c_1, c_2, \ldots, c_b$, where $c_i \in \{0, 1\}$ and at least one of the c_is is equal to 1, such that $c_1 residue_1 + c_2 residue_2 + \cdots + c_b residue_b = 0$.) These residues are $x^3 + 1$, $x^2 + x + 1$, and $x^3 + x^2 + x$, respectively, and the linear dependency is indicated by the fact that the sum of the three residues is zero. ☐

12.10.3.2 Design of BIST circuitry

The above analysis can be used to identify whether the inputs to any circuit kernel are linearly dependent. If the inputs to any kernel are found to be dependent, then the BIST circuit design can be modified in the following ways.
1 Change the order of flip-flops within each scan chain.
2 Change the composition of the scan chains.
3 Change how the scan chains are connected to the PG stages.
4 Change the feedback polynomial of the LFSR PG.
5 Replace the LFSR PG by another with a larger number of stages.
The BIST circuit design obtained after each modification must be analyzed, until the inputs of each kernel are found to be linearly dependent. Note that some of the above actions will not be useful under some circumstances. For example, if the inputs of any kernel are connected to the outputs of a single scan chain, then changing how the scan chains are connected to the PG will not help eliminate any linear dependencies between these inputs.

Example 12.46 Again, consider the scenario shown in Figure 12.47. As described above, this configuration fails to apply all possible vectors to an example three-input kernel whose inputs are driven by scan flip-flops F_1, F_4 and F_5. One way in which this situation can be remedied is by re-ordering the flip-flops in the scan chain. One order for which all the non-zero vectors are applied to the above three-input kernel is when the scan flip-flops are ordered as F_1, F_2, F_4, F_5, and F_3. (Since the CUT may have several kernels, it is necessary to ensure that such an order of flip-flops ensures application of all the possible vectors to each kernel.) □

12.11 BIST for delay fault testing

BIST techniques described above focus exclusively on combinational CUTs with target faults that are one-vector testable. In contrast, a delay fault in a static combinational CUT is two-pattern testable in the sense that two vectors must be applied consecutively for its detection (see Chapter 8). In this section, we discuss BIST techniques for delay fault testing of combinational kernels. The key challenge that we focus on is the design of suitable PGs.

12.11.1 PGs for exhaustive two-pattern testing

Consider an n-input combinational CUT. A test for a delay fault in such a CUT is comprised of a sequence of two n-bit vectors, say P_i and P_{i+1}, applied consecutively. Note that in a static CMOS circuit, a delay fault is detected only when $P_i \neq P_{i+1}$. Hence, the total number of useful two-pattern tests is $2^n(2^n - 1)$.

First, consider an n-stage autonomous ML-LFSR. Recall that any such LFSR generates a sequence with $2^n - 1$ unique n-bit PTVs. The above fact has two important consequences. First, a sequence of $2^n - 1$ distinct n-bit vectors includes only $2^n - 2$ two-pattern tests. Clearly, this is a small fraction of all possible two-pattern tests of the type described above. Second, though each non-zero vector is generated, only one two-pattern test is generated that contains any given completely-specified non-zero vector as its second vector.

Example 12.47 The sequences generated by two different five-bit ML-LFSRs are shown in Table 12.2. The reader can verify that the above two facts hold for both these sequences. □

Next, we describe PGs that can generate all two-pattern tests of the above type.

12.11.1.1 Double-length ML-LFSR

We desire generation of $2^n(2^n - 1)$ two-pattern tests. Any sequence of vectors that contains all such two-pattern tests as subsequences must have a length greater than or equal to $2^n(2^n - 1)$. The smallest number of stages required in an LFSR to generate such a sequence is $2n$. We call such an LFSR a **double-length LFSR**.

The first question is whether it is possible to use such an LFSR to generate exhaustive two-pattern tests? Since a $2n$-stage ML-LFSR has $2n$ flip-flops, each with one output, the second question is which n of the LFSR outputs should be connected to the n inputs of the CUT to ensure that all two-pattern tests are applied to the CUT? The second question, referred to as the *tap selection problem*, has been studied in Furuya and McCluskey (1991) and Chen and Gupta (1996). In this section, we describe two simple tap selections that help achieve the above objectives. More general discussion of this problem is deferred to Section 12.11.2.

Let the outputs $s_0, s_2, \ldots, s_{2n-2}$ be collectively called **even taps** and outputs $s_1, s_3, \ldots, s_{2n-1}$ **odd taps** of a $2n$-stage LFSR.

It has been shown in Furuya and McCluskey (1991) that a $2n$-stage ML-LFSR generates exhaustive two-pattern tests for an n-input CUT if its even taps are connected to the CUT inputs $x_1, x_2, \ldots, x_n$. The same is also the case if the CUT inputs are connected to the odd taps of such an LFSR. Finally, similar statements have been proven for a $2n$-stage ML-CA (Chen and Gupta, 1996).

Example 12.48 Table 12.4 shows the sequence generated by a four-stage ML-LFSR as well as that generated by a four-stage ML-CA. One can verify that all the two-pattern tests of the above type are generated at the odd taps of the LFSR. The same is true for the even taps of the LFSR, the odd taps of the CA, and the even taps of the CA. □

In general, in each of the above cases, all the two-pattern tests, with the exception of two-pattern tests of type (P_i, P_{i+1}), where either P_i or P_{i+1} is equal to $(0, 0, \ldots, 0)$, are generated.

The overhead due to the use of additional n flip-flops is high. Next, we describe a PG architecture that has a lower area overhead.

12.11.1.2 Non-autonomous LFSRs in a CUT with multiple blocks

When the inputs of the circuit block under test are driven by the outputs of a different block of logic (see Figure 12.48), which satisfies some given criteria (Vuksic and Fuchs, 1994), then non-autonomous LFSRs can be used. Let us investigate exhaustive two-pattern testing of block CLB_i. Consider a scenario where register R_i at the input of CLB_i is configured as an MISR with a primitive feedback polynomial. Figure 12.38

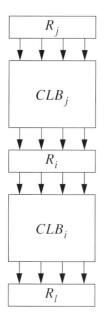

Figure 12.48 An example circuit to illustrate non-autonomous PG for delay fault testing

shows the details of such a design for the case where R_i has two stages. Note that the inputs to the above MISR are driven by the outputs of the logic block CLB_j. Assume that register R_j at the input of CLB_j is configured as an ML-LFSR.

The following steps can now be used to apply two-pattern tests to CLB_i. First, R_i and R_j are initialized to appropriate values. Next, the content of R_j is held constant while the MISR obtained by reconfiguring R_i is clocked 2^n times, where n is the number of inputs to CLB_i. Since R_j is being held constant during this period, a fixed combination of values appears at the outputs of CLB_j and hence at the inputs of the MISR obtained by reconfiguring R_i. Hence, when this MISR is clocked 2^n times, a sequence of $2^n - 1$ distinct vectors are applied at the inputs of CLB_i and, in the last clock cycle, the initial state of R_i is restored.

Example 12.49 Consider the case where R_i is a two-stage register that is configured as a two-stage MISR with a primitive feedback polynomial, as shown in Figure 12.38(a). The STG for the MISR shown in Figure 12.38(b) shows that if the inputs to the MISR are held constant at, say $(1, 0)$, and it has an initial state, say $(1, 0)$, the sequence of vectors $(1, 0)$, $(1, 1)$, $(0, 0)$, and $(1, 0)$ will be applied to the inputs of CLB_i over the next $2^n = 2^2$ clock cycles.

One can verify that for any initial state of the MISR except one $((0, 1)$ in the above case), a sequence of three distinct vectors is applied followed by the application of the first vector and restoration of the initial state of the MISR, as long as the inputs to the MISR are held constant and the MISR clocked four times. ☐

Next, the ML-LFSR obtained by reconfiguring R_j is clocked once to obtain a different vector at the inputs of CLB_j and hence, in general, a different combination of values at the inputs of the MISR obtained by reconfiguration of R_i. Subsequently, the state of R_j is held constant while the MISR is clocked 2^n times. The above process of clocking LFSR R_j once followed by clocking MISR R_i 2^n times can be repeated until the initial state of LFSR R_j is again reached.

Example 12.50 Let us continue with the previous example. After the application of vectors $(1, 0)$, $(1, 1)$, $(0, 0)$, $(1, 0)$ to the inputs of CLB_i, assume that LFSR R_j is clocked once to obtain a different combination of values, say $(0, 0)$, at the inputs of MISR R_i. Now, when R_i is clocked four times while the state of R_j is held constant, the sequence of vectors $(1, 0)$, $(0, 1)$, $(1, 1)$, and $(1, 0)$ is applied to the inputs of CLB_i. Next, if R_j is clocked once to obtain the combination $(0, 1)$ at the inputs of MISR R_i, then the sequence of vectors $(1, 0)$, $(0, 0)$, $(0, 1)$, and $(1, 0)$ is applied to the inputs of CLB_i when R_i is clocked four times while holding the state of R_j constant.

One can verify that the above approach has applied nine out of the 12 desired two-pattern tests.   □

In general, the above approach leads to the application of **nearly** all the desired two-pattern tests if the following condition is satisfied. Each of the 2^n combinations of n-bit values appears at the outputs of CLB_j in response to the values generated by the ML-LFSR obtained by reconfiguring R_j. In other words, the MISR obtained by reconfiguring R_i should be clocked 2^n times, for each of the distinct combination of values at its inputs. Under the above conditions, $2^n(2^n - 3)$ or more out of the $2^n(2^n - 1)$ desired two-pattern tests can be applied to the CUT. Additional modification can be made to obtain all the desired two-pattern tests.

The area overhead of this approach is lower than using a $2n$-stage autonomous LFSR. However, this approach is non-universal, in the sense that it is useful only when the blocks of CUT satisfy certain conditions.

12.11.2 Efficient two-pattern testing

The time required for exhaustive two-pattern testing is prohibitive even for CUTs with a moderate number of inputs. In this section, we describe techniques to reduce the test application time as well as area overhead for testing of a given set of target delay faults. We restrict ourselves to the case where an autonomous LFSR/CA is used as a PG.

One way in which the test application time can be reduced is via pseudo-exhaustive two-pattern testing. In this approach, each circuit cone can be viewed as a separate CUT and all the possible two-pattern tests applied to each cone. One of the key issues in the design of PGs that reduce test application time for pseudo-exhaustive testing is tap assignment.

12.11.2.1 Tap selection

Consider an autonomous linear PG. Flip-flop D_j is said to be a **predecessor** of flip-flop D_i in the PG if the output of D_j drives the input of D_i, either directly or via one or more XOR gates. Let $Pred(D_i) = \{D_{j_1}, D_{j_2}, \ldots\}$ be the set of all the predecessors of D_i other than D_i itself.

Example 12.51 For the four-stage internal-XOR LFSR shown in Figure 12.2(b),

$Pred(D_0) = \{D_3\}$,

$Pred(D_1) = \{D_0, D_3\}$,

$Pred(D_2) = \{D_1\}$, and

$Pred(D_3) = \{D_2\}$.

For the four-stage CA shown in Figure 12.6,

$Pred(D_0) = \{D_1\}$,

$Pred(D_1) = \{D_0, D_2\}$,

$Pred(D_2) = \{D_1, D_3\}$, and

$Pred(D_3) = \{D_2\}$.

□

In general, for an internal-XOR LFSR shown in Figure 12.2(a), for $i \in \{1, 2, \ldots, n - 1\}$, $Pred(D_i)$ contains D_{i-1}. For all $i \in \{1, 2, \ldots, n - 2\}$, $Pred(D_i)$ also contains D_{n-1}, if a feedback connection exists at the input of D_i. Furthermore, $Pred(D_0) = \{D_{n-1}\}$. The sets $Pred(D_i)$ can also be written in a general form for the external-XOR LFSR shown in Figure 12.3(a).

For a CA shown in Figure 12.7, for $i \in \{1, 2, \ldots, n-2\}$, $Pred(D_i) = \{D_{i-1}, D_{i+1}\}$. Furthermore, $Pred(D_0) = \{D_1\}$ and $Pred(D_{n-1}) = \{D_{n-2}\}$.

Without loss of generality, let $\{x_1, x_2, \ldots, x_l\}$ be the set of all the inputs of a circuit cone. Assume that these inputs are connected to the outputs of distinct flip-flops D_{i_1}, $D_{i_2}, \ldots, D_{i_l}$, respectively, of the PG. This connection is what is referred to as the **tap assignment**, since it determines which tap of the PG is connected to which CUT input. Note that a tap assignment is more specific than tap selection; in the latter a subset of PG outputs is selected, while the former also defines how the selected taps are connected to CUT inputs.

The following condition can be used to determine if all the desired two-pattern tests are applied to the cone for the above tap assignment. First, identify the set $Pred()$ for each of the above flip-flops. Next, remove $D_{i_1}, D_{i_2}, \ldots, D_{i_l}$ from each of these sets to obtain $Pred^*()$. The above tap assignment applies all the desired two-pattern tests to the cone if there exist distinct flip-flops $D_{j_1}, D_{j_2}, \ldots, D_{j_l}$, such that $D_{j_1} \in Pred^*(D_{i_1})$, $D_{j_2} \in Pred^*(D_{i_2}), \ldots$, and $D_{j_l} \in Pred^*(D_{i_l})$.

Example 12.52 Consider a circuit cone with two inputs and a four-stage internal-XOR LFSR shown in Figure 12.2(b). In this case, $Pred(D_0) = \{D_3\}$, $Pred(D_1) = \{D_0, D_3\}$, $Pred(D_2) = \{D_1\}$, and $Pred(D_3) = \{D_2\}$. Now, if the even taps are selected, then the inputs of the cone are connected to the outputs of D_0 and D_2. First, we can eliminate each of the above flip-flops from the above sets to obtain $Pred^*(D_0) = \{D_3\}$, and $Pred^*(D_2) = \{D_1\}$. Since distinct flip-flops, namely D_3 and D_1, can be found that satisfy the condition stated above, this LFSR with the given tap assignment applies all the desired two-pattern tests to the two-input cone.   □

The above condition indicates that if a cone has l inputs, then the PG should have $2l$ or more stages. Furthermore, if an LFSR has $2l$ stages then the inputs of an l-input cone can be assigned to the LFSR taps in only two ways, namely to (i) even taps, or (ii) odd taps. For a rule 90/150 CA with null boundary conditions, 2^l distinct tap selections exist for the inputs of an l-input cone (Chen and Gupta, 1996). Note that the number of distinct tap assignments is much larger, since each tap selection leads to several tap assignments.

12.11.2.2 Design of PGs for pseudo-exhaustive testing

One simplistic approach for designing an LFSR PG for efficient pseudo-exhaustive delay fault testing of a circuit can now be formulated. First, a PG with $2l_{max}$ stages, where l_{max} is the number of inputs in the largest cone, is selected. All tap assignments for the n inputs of the circuit can be enumerated (preferably implicitly, to reduce the time complexity of enumeration) to identify a tap assignment that satisfies the above condition for each cone. If no such assignment can be found, then either another PG with the same number of stages, or a PG with one additional stage can be tried until a suitable PG and tap assignment are found.

Alternatively, as described in Section 12.5, the notion of PE-compatibility can be used to combine n CUT inputs into a smaller number of test signals. If α test signals are obtained, then a 2α-stage PG with any of the tap selections described above can be used.

The notions of D-compatibility and I-compatibility have been extended to consider a target set of delay faults to obtain more compact PG designs (Chen and Gupta, 1998b). In addition, two additional notions, namely adjacency-compatibility and inverse-adjacency-compatibility have been defined to accomplish testing of target delay faults using smaller PGs and hence at lower test lengths.

12.12 BIST techniques to reduce switching activity

Pseudo-random sequences containing vectors that have provably low correlation are used in BIST. Due to such low correlation, the number of transitions between

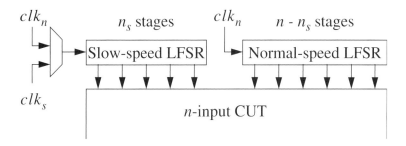

Figure 12.49 Dual-speed LFSR (Wang and Gupta, 1997) (© 1997 IEEE)

consecutive vectors can be high. In many cases, the number of transitions within the CUT caused by these vectors exceeds that caused during normal circuit operation. Coupled with the fact that these tests are applied at-speed, for some circuits, this can cause (i) excessive heat dissipation and damage during test, or (ii) excessive power/ground noise resulting in fault-free circuits failing the test.

In this section, we discuss BIST methodologies that reduce switching activity in the CUT. We first discuss an approach for in-situ BIST; subsequently, we discuss two approaches for scan BIST.

12.12.1 In-situ BIST

The key observation behind this approach is that it is possible to design PGs that generate sequences that possess many of the desirable properties of the vector sequences generated by ML-LFSRs, while reducing the number of transitions between consecutive vectors. Such PGs help reduce the average switching activity within the CUT, since reduction of switching activity between consecutive vectors typically reduces switching activity within the CUT as well. Since the vectors possess the desirable properties, this reduction is attained at little or no reduction in fault coverage.

12.12.1.1 Dual-speed LFSR

Figure 12.49 shows the design of a PG called **dual-speed LFSR (DS-LFSR)** (Wang and Gupta, 1997) for an n-input CUT. The PG is partitioned into two LFSRs, namely (i) an $(n - n_s)$-stage complete LFSR that is clocked using the normal clock clk_n and is called the **normal-speed LFSR**, and (ii) an n_s-stage complete LFSR that is clocked using a slow clock clk_s and is called the **slow-speed LFSR**. For simplicity, we assume that the ratio of the rates of the normal and slow clocks, *clkratio*, is a power of two.

The operation of a DS-LFSR is illustrated next.

Example 12.53 Consider a DS-LFSR for a four-input circuit (i.e., $n = 4$) comprised of a two-stage slow-speed LFSR ($n_s = 2$) and a two-stage normal-speed LFSR ($n - n_s = 2$). Assume that *clkratio* = 4 and that each of these LFSRs is a complete LFSR

Table 12.6. *Sequence of 16 vectors generated by the DS-LFSR in Example 12.53*

	Vector
P_1	$(1, 0, 1, 0)$
P_2	$(1, 0, 0, 1)$
P_3	$(1, 0, 0, 0)$
P_4	$(1, 0, 1, 1)$
P_5	$(0, 1, 1, 0)$
P_6	$(0, 1, 0, 1)$
P_7	$(0, 1, 0, 0)$
P_8	$(0, 1, 1, 1)$
P_9	$(0, 0, 1, 0)$
P_{10}	$(0, 0, 0, 1)$
P_{11}	$(0, 0, 0, 0)$
P_{12}	$(0, 0, 1, 1)$
P_{13}	$(1, 1, 1, 0)$
P_{14}	$(1, 1, 0, 1)$
P_{15}	$(1, 1, 0, 0)$
P_{16}	$(1, 1, 1, 1)$

obtained by modifying an internal-XOR LFSR with feedback polynomial $\phi(x) = x^2 + x + 1$.

Consider a scenario in which the DS-LFSR is used to generate a sequence of 16 vectors ($N_v = 16$). The vectors generated by the LFSR are shown in Table 12.6. One can verify that the average number of transitions between any two consecutive vectors is lower than that for any four-stage complete LFSR. Clearly, this reduction is mainly at the inputs driven by the outputs of the slow-speed LFSR.

Also, note that the sequence generated by the DS-LFSR contains unique vectors. $\square$

12.12.1.2 Properties of DS-LFSR generated sequences

Some of the properties of ML-LFSR generated sequences that are desired for BIST include *uniqueness* of vectors and *signal probability of 0.5* at each circuit input. In addition, if fewer than 2^n vectors are applied to an n-input circuit, then the vectors in the generated sequence should be *uniformly distributed* in the space of all the possible vectors.

Consider a scenario in which $N_v \leq 2^n$ vectors is to be applied to the CUT. For simplicity, assume that N_v is a power of two. In Wang and Gupta (1997), it has been shown that a DS-LFSR generates sequences that satisfy the above properties when

$$n_s = \log_2(N_v/clkratio), \tag{12.49}$$

or, equivalently,

$$clkratio = N_v/2^{n_s}. \tag{12.50}$$

Table 12.7. *Sequence of eight vectors generated by the DS-LFSR in Example 12.54*

	Vector
P_1	$(1, 0, 1, 0)$
P_2	$(1, 0, 0, 1)$
P_3	$(0, 1, 0, 0)$
P_4	$(0, 1, 1, 1)$
P_5	$(0, 0, 1, 0)$
P_6	$(0, 0, 0, 1)$
P_7	$(1, 1, 0, 0)$
P_8	$(1, 1, 1, 1)$

Example 12.54 The DS-LFSR in Example 12.53 satisfies the above conditions for *clkratio* $= 4$ and $N_v = 16$.

If $N_v = 8$ and $n_s = 2$ are desired, then *clkratio* $= 2$ needs to be used. The vectors generated for a four-stage DS-LFSR with these parameters are shown in Table 12.7. It is easy to verify that these vectors are unique and the probability of applying 1 at any input is approximately 0.5. χ^2 tests can be used to determine that these eight vectors are fairly uniformly distributed in the space of all possible 16 vectors.   □

12.12.1.3 DS-LFSR design

In typical cases, the number of circuit inputs n is known and practical constraints on test application time suggests a desired value for N_v. Given these values, different combinations of *clkratio* and n_s values that satisfy the above relations can be computed.

For any given n_s value, n_s-out-of-n circuit inputs must be selected to be connected to the outputs of the slow-speed LFSR. Since a DS-LFSR generates vectors in which the number of transitions is low at the outputs of slow-speed LFSR, circuit inputs at which transitions are likely to propagate to a large number of lines within the circuit should be identified and connected to these inputs. (A heuristic approach has been presented in Wang and Gupta (1997).)

The DS-LFSR design for each combination of n_s and *clkratio* values can be evaluated in terms of the fault coverage and switching activity within the circuit and the most suitable DS-LFSR design selected.

Example 12.55 Consider a circuit with $n = 5$ inputs which must be tested with $N_v = 16$ vectors. The following combination of *clkratio* and n_s values satisfy the above conditions, (i) *clkratio* $= 8$ and $n_s = 1$, (ii) *clkratio* $= 4$ and $n_s = 2$, and (iii) *clkratio* $= 2$ and $n_s = 3$. A DS-LFSR design can be obtained for each of these combination of values, and the most suitable one selected.   □

12.12.2 Scan BIST

We next consider two approaches for reducing switching activity during scan BIST. The first one is based on modification of scan flip-flops and the second one on modification of the PG.

12.12.2.1 Modification of scan cells

As described in Chapter 11, a large fraction of the switching activity occurs during the scan shifting cycles. Switching activity during scan BIST can be reduced if enhanced scan cells, described in Chapter 11, are used in the manner described in Section 11.6.1.3. In this approach, the reduction in switching activity is accomplished at the cost of area and delay overheads due to the use of enhanced scan cells.

12.12.2.2 Low transition PG for scan BIST

In this approach, reduction in the switching activity is accomplished via modification of the PG design.

As described in Chapter 11 (Section 11.6.2), switching activity during scan can be decreased significantly if the number of transitions at the scan input, *ScanIn*, of the scan chain is decreased. One key observation behind the following approach is that it is possible to achieve such a reduction without significantly compromising the desirable characteristics of the serial vectors generated by ML-LFSRs.

Low transition RPGs: The overall architecture of the PGs and CUT is shown in Figure 12.50. The following discussion will focus on the design and characteristics of the **low-transition random pattern generator (LT-RPG)** shown in the figure. An LT-RPG uses an n_{pg}-stage ML-LFSR as its **base PG** with flip-flops $D_0, D_1, \ldots, D_{n_{pg}-1}$. Q or $\overline{Q}$ outputs of n_a of these flip-flops are connected to the inputs of an n_a-input AND gate. The output of this gate drives the *toggle* input of a T flipflop. (The value at the output of a T flip-flop changes in a clock cycle if and only if logic 1 is applied at its *toggle* input.) The output of the T flip-flop drives the scan input, *ScanIn*, of the scan chain. The other PG in this figure is simply an n-stage ML-LFSR used as a parallel PG for the primary inputs of the CUT.

Note that if the output of the AND gate in the LT-RPG is 0 for l consecutive clock cycles, then identical values are applied at *ScanIn* for l clock cycles. Furthermore, the larger the number of inputs to this AND gate, the lower the number of transitions in the sequence of values applied by the LT-RPG at *ScanIn*. This is how LT-RPG reduces the switching activity during scan BIST.

A single vector is applied to the CUT by clocking the base LFSR in the LT-RPG $k - 1$ times while the flip-flops in the scan chain are configured in the scan mode. During this period, the contents of the LFSR at the primary inputs are held constant.

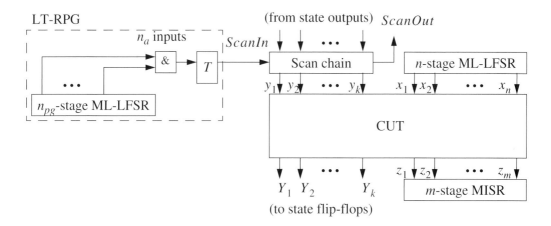

Figure 12.50 Low-transition RPG (Wang and Gupta, 1999) (© 1999 IEEE)

Subsequently, both the LT-RPG and other LFSR are clocked once with the scan chain configured in the scan mode to complete the application of a vector. (Note that during the above k clock cycles, the values captured at the scan flip-flops in response to the previous vector are scanned out.) Next, the flip-flops in the scan chain are configured in the normal mode and the response at the state outputs is captured in the state flip-flops. The entire process is repeated to apply each subsequent vector.

LT-RPG design: Consider a CUT with k state inputs. Let the **span** γ of the CUT be defined as the maximum of the distances between every pair of state flip-flops that drive any one of the cones in the CUT. In such a scenario, LT-RPG should be designed to (a) reduce the number of transitions at *ScanIn*, and (b) be capable of guaranteeing the application of all the possible combinations of values at any consecutive γ flip-flops in the scan chain, if the test application time were not a concern.

    The following procedure can be used to design such an LT-RPG.

1 A $2(\gamma - 1)$-stage ML-LFSR can be used as a base PG, i.e., $n_{pg} = 2(\gamma - 1)$. (In practice, an ML-LFSR with γ or more stages is found to be adequate (Wang and Gupta, 1999).)

2 A value of n_a is chosen. Note that higher the value of n_a, greater the reduction in the number of transitions at *ScanIn*. More precisely, it has been shown in Wang and Gupta (1999) that $n_a = 2$ causes half as many transitions at *ScanIn* as $n_a = 1$, $n_a = 3$ causes half as many as $n_a = 2$, and so on. However, it has also been empirically shown that the vectors generated using $n_a = 2$ provide as good a fault coverage as $n_a = 1$ while the fault coverage by vectors generated using $n_a = 3$ is slightly lower. Hence, in practice, $n_a = 2$ should be preferred. If greater reduction in switching activity is desired, then $n_a = 3$ may be used. If this reduces the fault coverage, then the test length may be increased to obtain the desired coverage.

3 The outputs of the base PG are then chosen and connected to the inputs of the n_a-input AND gate.

If $n_a = 2$ is chosen, then the AND gate inputs are connected to the Q outputs of flip-flops D_i and D_j of the base PG, where $j \geq i + \gamma - 1$. Alternatively, the inputs of the AND gate can be connected to the $\overline{Q}$ outputs of the above flip-flops.

If $n_a = 3$ is chosen, then the AND gate inputs are connected to the Q outputs of flip-flops D_i, D_j, and D_l of the base PG, where (i) $l > j > i$, (ii) $j - i \neq l - j$, and (iii) $l \geq i + \gamma - 1$. Alternatively, the inputs of the AND gate can be connected to the $\overline{Q}$ outputs of the above flip-flops.

Example 12.56   Consider a scenario where a CUT has five state inputs, $y_1, y_2, \ldots, y_5$. The state flip-flops are ordered in the scan chain in such a way that y_1 and y_5 are the outputs of flip-flops closest and farthest from *ScanIn*, respectively. State inputs y_1, y_2, and y_3 drive one cone in the combinational logic and y_2, y_4, and y_5 drive its other cone. For this CUT and chain configuration, the span is $\gamma = 4$.

An LT-RPG can be obtained by using an ML-LFSR with $2(\gamma - 1) = 6$ stages with flip-flops $D_0, D_1, \ldots, D_5$. If $n_a = 2$ is chosen, then the Q outputs of, say, D_0 and D_3 can be connected to the two inputs of the AND gate to obtain a PG.

The above LT-RPG generates a sequence that causes half the number of transitions at the inputs of the scan chain compared to the case where a classical ML-LFSR is used as a serial PG.   □

If allowed to generate $2^{2(\gamma-1)}$ vectors, the above PG can apply all the possible combinations of values at the state inputs of each cone in the combinational logic. Finally, for shorter test sequence lengths, the above PG applies vectors that satisfy many desirable properties, such as signal probabilities of 0.5 and uniform distribution.

Summary

- BIST reduces dependence on ATE for testing, supports at-speed testing even for high-speed circuits, and enables testing even after the chip is deployed in the field. These features help reduce the cost of testing, enable more comprehensive testing, and improve system reliability and availability.

- Area overhead, performance penalty, and reduced support for debugging a new design are the key costs associated with BIST.

- PGs, such as LFSRs and CA, are autonomous FSMs. The sequence of states traversed by a PG can be used as test vectors in a variety of ways, including in parallel, serial, and parallel–serial modes.

- The structure of an n-stage LFSR can be characterized by a degree-n feedback polynomial. The structure of the LFSR can also be captured using a state-transition matrix.

- The feedback polynomial of an LFSR determines its basic properties. In particular, it provides the generating function which describes the sequence of values generated at any output of the LFSR. The feedback polynomial is also equal to its characteristic polynomial, derived from its state transition matrix.

- Starting with a non-zero seed, an n-stage LFSR with a primitive feedback polynomial generates an MLS via a traversal of $2^n - 1$ distinct states before the initial state repeats. Such LFSRs are called ML-LFSRs. When used in the parallel mode or in the serial mode using overlapped sampling, an ML-LFSR generates a near-complete sequence of $2^n - 1$ unique n-bit vectors. The generated sequences satisfy many criteria for pseudo-randomness, including equal distribution of 1s and 0s, distributions on run lengths, and low auto-correlation.

- Additional non-linear feedback can be added to an ML-LFSR to enable generation of a complete sequence of vectors. Also, an LFSR with feedback polynomial $\phi(x)$ generates sequences that are reverse of that generated by one with feedback polynomial $x^n \phi(1/x)$.

- CA with primitive characteristic polynomials also generate an MLS. The MLS generated by such a CA is identical to that generated by an LFSR with the same characteristic polynomial. Hence, ML-CA generate sequences that are near-complete and satisfy all the above criteria for pseudo-randomness. In addition, the phase shift between the sequences generated at the outputs of adjacent stages of an ML-CA is typically higher than that for an ML-LFSR.

- The detectability profile of a circuit provides a distribution of detectabilities of target faults in the circuit, where detectability of a fault is the number of vectors that detect the fault. Faults with detectabilities equal or close to the minimum detectability are called hard-to-detect faults.

- The detectability of a fault can be estimated via circuit traversal or using a combination of statistical simulation – logic simulation performed for randomly generated vectors – and circuit traversal.

- Given an estimate of the detectability profile of a circuit, the expected fault coverage can be estimated as a function of the test length, for random as well as pseudo-random testing. Conversely, a test length that provides a desired level of test confidence can be estimated.

- When standard PRPGs fail to provide the desired coverage at the desired test length for a given CUT, one can either make the CUT more testable by adding test points or can customize the PG to generate vectors that are more suitable for the given CUT.

- Control points and observation points can be added to a circuit to make it more testable. A control point can change the controllabilities of lines in the transitive fanout of its site. It can also change the observability of lines in the transitive fanin of lines in the transitive fanout of its site. A control point may favorably or adversely affect the above-mentioned controllabilities and observabilities. In contrast, an observation point can only enhance the observabilities of lines in its site's transitive fanins.

- The test length required for achieving the desired fault coverage can be decreased by selecting a suitable single seed for the PRPG, selecting the feedback polynomial of the PRPG, or selecting a (small) number of seeds.

- Mapping logic may be used to transform some of the vectors that do not detect target faults in a given CUT into vectors that do.

- Weight sets can be used to capture the characteristics of vectors that are suitable for detection of target faults in a given CUT.

- A weight set can be computed for hard-to-detect faults in a given circuit by analyzing the test vectors for these faults. The quality of a weight set can be estimated using the expected uncoverage metric. A WRPG can be implemented for a given weight set using a PRPG and a set of gates to implement the weights.

- Inputs of a given CUT can be combined into fewer test signals to reduce the size of the space of vectors that can guarantee the coverage of all the target faults in the CUT. In general, a test signal drives multiple inputs, either directly or via logic elements, including inverters, decoders, or l-input logic gates. Such test signals can be identified using notions of PE-, D-, I-, dec-, and C(l)-compatibilities.

- RAs compress the response of a CUT into a signature. The CUT signature is compared with that expected of a fault-free circuit version. Aliasing is said to occur when the signature of a CUT matches that of the fault-free circuit version, even though the responses of the two circuit versions are different. Aliasing causes a faulty CUT to be mistakenly classified as fault-free.

- Count-based compressors count the number of 1s or the number of transitions in the sequence of response bits at the CUT output. The aliasing volume for count-based compressors is low if the fault-free signature is either a very small count or a very large one. If the fault-free count has an intermediate value, then the aliasing volume is high.

- With minor modifications, an LFSR PG can be modified into a linear compressor. Compression using an LFSR can be described using polynomial division where the signature is the remainder of the division of the polynomial representing the CUT response by the feedback polynomial of the LFSR.

- The aliasing volume for LFSR compression is independent of the nature of the fault-free circuit response.

- Aliasing probability for an LFSR/MISR compressor can be computed using a variety of different error models based on (a) Markov models, or (b) coding theory.

- A large circuit can be viewed as an interconnection of combinational logic blocks and registers. During BIST, the entire circuit may be viewed as one kernel or may be divided (typically, partitioned) into a number of kernels. The types of kernels include combinational, linear pipelines, balanced kernels, acyclic kernels, and so on. Techniques described in Chapter 11 (Section 11.3) for the selection of partial scan flip-flops may be used to identify kernels in a large circuit. This process implicitly identifies the registers that may be used for test application and response analysis.

- Since most BIST PGs do not generate a rich set of ω-vector tests when $\omega > 1$, whenever high fault coverage is a requirement, kernels that are one-vector testable are preferred. Furthermore, it is preferred to have one-vector testable kernels where each bit of the vector can be applied to the input of the kernel at the same time.

- Registers that may be used for test application and response analysis may be implemented using reconfigurable cells, such as scan flip-flops, BILBO cells, and so on. They may be configured in one or more of several configurations including normal, scan, PG, RA, scan compressor, and so on.

- In in-situ test methodologies, registers at the inputs and outputs of each kernel are modified into PG and RA, respectively. Such methodologies are also called test-per-clock, since a new test vector can be applied in each clock cycle. Furthermore, at-speed testing is possible. In contrast, in scan BIST methodologies, these registers are configured as scan chains. The scan chains are driven by PGs and drive RAs added to the chip. Several clock cycles are required to scan in each vector, apply it and capture the response. Hence, such methodologies are called test-per-scan methodologies. Clearly, in a given amount of test application time, fewer vectors can be applied in this approach. Also, this methodology typically does not support at-speed testing.

- One of the key questions in in-situ BIST is whether to use a register concurrently as a compressor as well as a PG. Some of the key reasons that compromise the quality of compression and vectors generated include α-level dependency, non-uniqueness of vectors generated, and system state cycling.

- Self-testable versions of a circuit can be obtained under BILBO-based and CSTP approaches. Some of the key issues faced are identification of kernels, enumeration of possible embeddings, exploitation of I-paths, selection of desirable embeddings, and test scheduling, while ensuring high test quality.

- The key issues faced in the design of scan BIST include selection of kernels and design of scan chains to minimize the test application time while ensuring high fault coverage by avoiding decimation and harmful linear dependence.

- Delay fault testing of static combinational circuit requires application of two-pattern tests.

- An exhaustive set of two-pattern tests can be generated by a double-length ML-LFSR, provided that appropriately selected LFSR outputs are connected to the inputs of the CUT. The task of selecting PG outputs to be connected to CUT inputs is referred to as the tap selection problem.

- The manner in which the outputs of a PG are connected to the inputs of a CUT is referred to as a tap assignment. Iterative application of tap assignment algorithm can be used to find an efficient PG for pseudo-exhaustive delay fault testing. The notions of D-, I-, adjacency-, and inverse-adjacency-compatibility can be used to further reduce the test application time for delay fault testing.

- Excessive heat dissipation as well as power-ground noise typically occur during BIST. The former can damage the circuit while the latter can cause a fault-free circuit to fail the test.

- A dual-speed LFSR can be used to reduce heat dissipation during in-situ BIST without reducing fault coverage.

- Enhanced scan cells may be used in a circuit to reduce heat dissipation during scan BIST. Alternatively, a low-transition RPG may be used.

Additional reading

A two-part review of BIST techniques can be found in Agrawal *et al.* (1993a,b).

The properties of LFSR-generated sequences are discussed in detail in Golomb (1967). An interesting discussion of randomness criteria can be found in Knuth (1981).

Techniques that either modify the feedback structure of the PG or select one or more LFSR feedback polynomials and seeds to obtain a high fault coverage in a short test length are given in Akers and Jansz (1989), Dufaza and Cambon (1991), Konemann (1991), Lempel *et al.* (1994), and Mukund *et al.* (1995). Techniques that achieve similar goals by using LFSRs with reconfigurable feedback polynomials and stored seeds for scan BIST can be found in Hellebrand *et al.* (1995) and Venkataraman *et al.* (1993). Techniques to determine, for a given circuit, one or more sets of weights that can help achieve high fault coverage at low test lengths can be found in AlShaibi and Kime (1994), Bershteyn (1993), Hartmann and Kemnitz (1993), Kapur *et al.* (1994), Muradali *et al.* (1990), Pateras and Rajski (1991), Pomeranz and Reddy (1993), Waicukauski *et al.* (1989), and Wunderlich (1990).

The problem of finding a minimum test schedule for cases where kernels in a circuit require different test application times is discussed in Craig *et al.* (1988) and Lin *et al.* (1993b).

Some other specific scan-based BIST methodologies include **LOCST (LSSD on-chip self-test)** (LeBlanc, 1984), **RTS (random test socket)** (Bardell and McAnney, 1982) and **PS-BIST (partial scan BIST)** (Lin *et al.*, 1993a).

Exercises

12.1 Draw an internal-XOR LFSR with feedback polynomial $\phi(x) = x^5 + x^2 + 1$. Simulate the LFSR and obtain the sequence generated at the output of the last stage of the LFSR, s_4. Use the generating function of the LFSR (Equation (12.4)) to also derive the sequence generated at s_4. Are the two sequences identical?

12.2 Derive the generating function of an n-stage external-XOR LFSR shown in Figure 12.3(a). (Hint: first derive the recurrence relation for any LFSR stage. Then substitute this relation in the definition of generating function given in Equation (12.1) and simplify, especially assuming an appropriate initial state.)

12.3 Write the state transition matrix, A, for a three-stage internal-XOR LFSR with feedback polynomial $\phi(x) = x^3 + x + 1$. Derive the characteristic polynomial $M(\lambda)$ by computing $\det(A - \lambda I)$. Is $M(\lambda) = \phi(\lambda)$?

12.4 Prove that, once initialized with any non-zero seed, any non-trivial n-stage LFSR cannot reach the all-zero state.

12.5 An n-stage LFSR can be used to serially generate a sequence containing n-bit patterns using two distinct types of sampling illustrated in the following table.

Pattern	Overlapped sampling	Disjoint sampling
1	$s_0(0)s_0(1)\ldots s_0(n-1)$	$s_0(0)s_0(1)\ldots s_0(n-1)$
2	$s_0(1)s_0(2)\ldots s_0(n)$	$s_0(n)s_0(n+1)\ldots s_0(2n-1)$
3	$s_0(2)s_0(3)\ldots s_0(n+1)$	$s_0(2n)s_0(2n+1)\ldots s_0(3n-1)$
$\vdots$	$\vdots$	$\vdots$

Simulate a six-stage ML-LFSR and determine the number of unique six-bit serial patterns generated under each of the sampling techniques. What additional hardware would be required for overlapped sampling?

12.6 Characterize the proportions of 1s and 0s, numbers of various types of runs, and the numbers of distinct n-bit serial patterns that appear in the sequence generated at each output of an n-stage binary counter. Compare the characteristics of these sequences with those listed in Section 12.2.1.3 for pseudo-random sequences generated at any output of an ML-LFSR.

12.7 Consider a scenario where an n-stage PG must be designed to apply a near-complete set of test vectors to an n-input circuit. Compare an n-stage ML-LFSR and an n-stage binary counter in terms of hardware requirements and ability to generate a near-complete set of vectors for two modes of PG operation: (a) parallel, and (b) serial.

12.8 Modify the design of an n-stage external-XOR ML-LFSR to enable it to generate a complete set of n-bit PTVs.

12.9 Write a program to simulate LFSR and CA PRPGs. The main inputs to your program should include:
(a) *Type of PG:* Internal/external-XOR LFSR or CA.
(b) *Number of stages: n.*
(c) *Configuration:* The feedback/rule polynomial.

(d) *Seed: n*-bit initial state.

(e) *Length of sequence to be generated: N_v*.

Your program should output the sequences generated at each output of the specified *n*-stage PG and the phase-shift between the sequences generated at the outputs of the adjacent PG stages.

12.10 Use the above simulator to:

(a) Verify that the CA rules shown in Table 12.3 generate an MLS and to verify that the sequences generated at any output are identical to those generated by LFSRs with identical (primitive) characteristic polynomials.

(b) Compute and tabulate the phase shifts between sequences generated at each pair of adjacent CA stages. Repeat for corresponding LFSRs.

Compare the CA and LFSR with respect to phase-shifts and hardware requirements.

12.11 Derive the recurrence relation that describes the operation of the four-stage ML-CA with $\rho(x) = 1 + x^2$ shown in Figure 12.6.

12.12 The characteristic polynomial of a three-stage CA has been derived in terms of the coefficients of its rule polynomials, ρ_0, ρ_1, and ρ_2, in Section 12.2.2.1. Use this characteristic polynomial to derive the rule polynomial of a CA with characteristic polynomial $\lambda^3 + \lambda + 1$.

12.13 Estimate probabilistic controllabilities and observability for each line in the circuit shown in Figure 4.24(a) using the circuit traversal approach. Use the above values to obtain an estimate of the detectability profile for this circuit.

12.14 Use a combination of statistical simulation and Equations (12.15) and (12.16) to identify gates in the above circuit at whose inputs the logic values are correlated.

12.15 The following table gives a part of the detectability profile of ALU 181, a 14-input combinational circuit with $n_f = 400$ faults, listing only the detectabilities of the hardest-to-detect faults in the circuit (McCluskey *et al.*, 1988).

κ	h_κ
96	1
128	1
176	1
192	7
216	1
256	2
264	1

(a) For each value of fault detectability, κ, shown in the above table, compute the escape probability of a fault with that detectability assuming that 10 random patterns are applied. Repeat assuming (i) 100, and (ii) 1000, random patterns are applied.

(b) Repeat part (a) assuming the patterns are pseudo-random.

(c) Compute the lower and upper bounds on the expected values of fault coverage for random as well as pseudo-random sequences of the above-mentioned lengths. (Hint: you must make optimistic and pessimistic assumptions above the detectabilities of the remainder of the $n_f = 400$ faults to compute the lower and upper bounds.)

(d) Compute the lower bound on the length of a random test sequence that will guarantee the detection of each fault in the circuit with a probability exceeding 99%.

(e) Compute the minimum and maximum estimates of the minimum value of the length of a pseudo-random test sequence which will guarantee that each fault will be detected with a probability exceeding 99%. (See Section 12.3.4.3.)

12.16 Consider the circuit shown in Figure 4.9 and a target fault list containing all single SAFs in the circuit obtained after fault collapsing.

(a) Apply four random vectors to the circuit generated by a 10-stage LFSR. Perform fault simulation and drop the faults detected by these vectors from the target fault set.

(b) Use a combinational ATPG to generate a vector for each fault that remains in the target fault set.

(c) Compute a weight set using these vectors.

(d) Estimate the quality of this weight set by computing the estimated uncoverage for a test sequence length of 100.

12.17 Design a WRPG that implements (a suitably quantized version of) the weight set identified in the above problem.

12.18 For the circuit in Figure 4.9, generate a PG of minimum size using the notion of PE-compatibility. Repeat using the notion of D-compatibility. Comment on your PG designs.

12.19 Repeat the above problem for the circuit in Figure 4.24.

12.20 Design a test pattern generator for the circuit in Figure 4.9 using the notions of D-compatibility, I-compatibility as well as C(l)-compatibility.

12.21 Consider the case where the D flip-flop in the transition count compressor shown in Figure 12.22 is initialized to r^* before testing commences. Write an expression for the value of TC similar to the one shown in Section 12.6.1.2, which assumes that the flip-flop is initialized to r. Compute the aliasing volume for such a compression assuming that a sequence of N_v test patterns is applied and the good circuit transition count is $TC(Res^*)$.

12.22 Modify the compressor design shown in Figure 12.22 to enable the counting of only the $0 \rightarrow 1$ transitions in the response sequence.

12.23 Consider two partitions of a given test set T for a circuit, T_0 and T_1, which contain all test vectors for which the fault-free circuit response is 0 and 1, respectively. Let T' be a test sequence that contains all the tests in T_0, followed by all the tests in T_1. Do there exist scenarios where a fault in the circuit may cause aliasing when the circuit's response to sequence T' is compressed using a transition count compressor?

12.24 An LFSR compressor with feedback polynomial $\phi(x) = x^4 + x^3 + 1$ is used to compress the CUT response $Res = r_7, r_6, r_5, r_4, r_3, r_2, r_1 = 1, 1, 1, 0, 1, 0, 1$, i.e., $Res(x) = x^6 + x^4 + x^2 + x + 1$. Use polynomial division to compute the value of the signature obtained by using the above compression. Verify by simulating the LFSR compression.

Identify two response sequences different from Res that give the same signature as computed above.

12.25 An m-output circuit is tested using a sequence of N_v vectors and its response compressed using an m-stage MISR compressor. Compute the number of possible error sequences that map to the all-zero equivalent LFSR error sequence.

12.26 An m-output circuit is tested using a sequence of N_v vectors and its response compressed using an m-stage MISR compressor. Compute the number of possible error sequences that map to a given $(N_v + m - 1)$-bit equivalent LFSR error sequence. Use this number, along with the expression for aliasing volume for LFSR compression, to compute the aliasing volume for the above-mentioned MISR compression.

12.27 Equation (12.40) gives the Markov model describing the computation of the state probability vector. Assuming that the response of a single-output circuit is compressed using an m-stage LFSR compressor, and the probability of any response bit being in error is q, *independent* of errors at the other response bits, prove that the entries in each row of M add to 1.

12.28 Design a reconfigurable register that can be configured in the normal and test pattern generator modes. Compare the overhead of your design with that of the BILBO register.

12.29 Assume that the example circuit shown in Figure 12.40 is to be tested by a BILBO-based BIST methodology that only allows combinational kernels but exploits I-paths. Consider an embedding in which register R_1 is replaced by a PG, R_3 by an RA, R_5 by a BILBO register, and R_{12} by an RA. Determine the test incompatibility graph assuming that the test methodology allows use of the contents of an RA as test vectors. Determine the minimum-time test schedule for this embedding, assuming equal test time for all kernels.

12.30 Assume that the area overhead of replacement of a register by a PG or an RA register is 2/3 of the area overhead of its replacement by a BILBO register. Assuming that only combinational kernels are allowed and that the contents of an RA cannot be used as test vectors, compute the minimum area overhead test embedding for the circuit shown in Figure 12.40. Compute the minimum-time test schedule for this embedding, assuming equal test time for all kernels.

12.31 Consider the four-stage, internal-XOR LFSR with feedback polynomial $\phi(x) = x^4 + x + 1$ shown below. Assign residues to the outputs of the four LFSR flip-flops and the five shift register flip-flops.

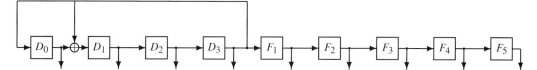

Consider the cones driven by (a) the outputs of D_1, D_2, D_3 and F_1, (b) the outputs of F_1, F_3, F_4 and F_5, and (c) the outputs of D_1, D_3, F_4 and F_5. Which of the above cones cannot be tested exhaustively using a single seed? Modify the LFSR/shift register by reordering the flip-flops and/or changing the LFSR feedback polynomial to ensure that the inputs to all the above cones are linearly independent.

12.32 Consider the scenario described in Problem 12.5 where a six-stage ML-LFSR is to be used to serially generate vectors for a six-input circuit via disjoint sampling. Does the disjoint sampling scheme presented in Problem 12.5 cause improper decimation? If yes, how would you generalize the notion of disjoint sampling to enable the application of all possible vectors to the inputs of the circuit?

12.33 Enumerate all tap selections for a four-stage internal-XOR ML-LFSR that can apply exhaustive two-pattern tests to a two-input CUT. Repeat for a four-stage ML-CA. Compare the two PGs.

12.34 Repeat the above problem for a five-stage ML-LFSR and a five-stage ML-CA.

12.35 Design an ML-LFSR PG of minimum size that applies pseudo-exhaustive delay fault tests to a five-input CUT with inputs $x_1, x_2, \ldots, x_5$ and two outputs z_1 and z_2, where output z_1 depends upon inputs x_1, x_2, x_3, x_4 while z_2 depends on inputs x_2, x_3, x_4, x_5.

12.36 Design a DS-LFSR that can be used when 32 vectors are to be applied to a seven-input CUT.

12.37 Consider a CUT with seven state inputs driven by flip-flops configured as a scan chain in such a manner that the circuit has a span of three. Design a suitable LT-RPG. Are all possible two-pattern tests applied to any three consecutive outputs of the scan chain?

References

Abadir, M. and Breuer, M. (1985). A knowledge based system for designing testable VLSI chips. *IEEE Design and Test of Computers*, **3** (4), pp. 56–68.

Agrawal, V. D., Kime, C. R., and Saluja, K. K. (1993a). A tutorial on built-in self-test. *IEEE Design and Test of Computers*, **10** (1), pp. 73–82.

Agrawal, V. D., Kime, C. R., and Saluja, K. K. (1993b). A tutorial on built-in self-test, part 2: applications. *IEEE Design and Test of Computers*, **10** (2), pp. 69–77.

Akers, S. B. (1985). On the use of linear sums in exhaustive testing. In *Proc. Int. Symposium on Fault-Tolerant Computing*, pp. 148–153.

Akers, S. B. and Jansz, W. (1989). Test set embedding in a built-in self-test environment. In *Proc. Int. Test Conference*, pp. 257–263.

AlShaibi, M. F. and Kime, C. R. (1994). Fixed-biased pseudorandom built-in self-test for random pattern resistant circuits. In *Proc. Int. Test Conference*, pp. 929–938.

Avra, L. J. and McCluskey, E. J. (1993). Synthesizing for scan dependance in built-in self-testable designs. In *Proc. Int. Test Conference*, pp. 734–743.

Bardell, P. H. (1990a). Analysis of cellular automata as pseudorandom pattern generators. In *Proc. Int. Test Conference*, pp. 762–767.

Bardell, P. H. (1990b). Design considerations for parallel pseudorandom pattern generators. *J. of Electronic Testing: Theory and Applications*, **1** (1), pp. 73–87.

Bardell, P. H. and McAnney, W. H. (1982). Self-testing of multichip logic modules. In *Proc. Int. Test Conference*, pp. 200–204.

Bardell, P. H., McAnney, W. H., and Savir, J. (1987). *Built-In Test for VLSI: Pseudorandom Techniques*. John Wiley & Sons.

Bershteyn, M. (1993). Calculation of multiple sets of weights for weighted random testing. In *Proc. Int. Test Conference*, pp. 1031–1041.

Briers, A. J. and Totton, K. (1986). Random pattern testability by fast fault simulation. In *Proc. Int. Test Conference*, pp. 274–281.

Carletta, J. and Papachristou, C. (1994). Structural constraints for circular self-test paths. In *Proc. VLSI Test Symposium*, pp. 87–92.

Chakrabarty, K., Murray, B. T., Liu, J., and Zhu, M. (1997). Test width compression for built-in self-testing. In *Proc. Int. Test Conference*, pp. 328–337.

Chatterjee, M. and Pradhan, D. K. (1995). A novel pattern generator for near-perfect fault-coverage. In *Proc. VLSI Test Symposium*, pp. 417–425.

Chen, C.-A. and Gupta, S. K. (1996). BIST test pattern generators for two-pattern testing – theory and design algorithms. *IEEE Trans. on Computers*, **C-45** (3), pp. 257–269.

Chen, C.-A. and Gupta, S. (1998a). Efficient BIST TPG designs and test set compaction via input reduction. *IEEE Trans. on Computer-Aided Design*, **17** (8), pp. 692–705.

Chen, C.-A. and Gupta, S. K. (1998b). Efficient BIST TPG design and test set compaction for delay testing via input reduction. In *Proc. Int. Conference on Computer Design*, pp. 32–39.

Craig, G., Kime, C., and Saluja, K. (1988). Test scheduling and control for VLSI built-in self-test. *IEEE Trans. on Computers*, **C-37** (9), pp. 1099–1109.

Damiani, M., Olivo, P., Ercolani, S., and Ricco, B. (1989). An analytical model for the aliasing probability in signature analysis testing. *IEEE Trans. on Computer-Aided Design*, **8** (11), pp. 1133–1144.

Dufaza, C. and Cambon, G. (1991). LFSR based deterministic and pseudo-random test pattern generator Structures. In *Proc. European Design and Test Conference*, pp. 27–34.

Frohwerk, R. A. (1977). Signature analysis: a new digital field service method. *Hewlett-Packard Journal*, **28**, pp. 2–8.

Furuya, K. and McCluskey, E. J. (1991). Two-pattern test capabilities of autonomous TPG circuits. In *Proc. Int. Test Conference*, pp. 704–711.

Golomb, S. W. (1967). *Shift Register Sequences*. Holden-Day.

Hamzaoglu, I. and Patel, J. H. (2000). Reducing test application time for built-in-self-test test pattern generators. In *Proc. VLSI Test Symposium*, pp. 369–375.

Hartmann, J. and Kemnitz, G. (1993). How to do weighted random testing for BIST? In *Proc. Int. Conference on Computer-Aided Design*, pp. 568–571.

Hayes, J. P. (1976). Transition count testing of combinational logic circuits. *IEEE Trans. on Computers*, **C-25** (6), pp. 613–620.

Hellebrand, S., Rajski, J., Tarnick, S., Venkataraman, S., and Courtois, B. (1995). Built-in test for circuits with scan based on reseeding of multiple polynomial linear feedback shift registers. *IEEE Trans. on Computers*, **C-44** (2), pp. 223–233.

Hortensius, P. D., McLeod, R. D., Pries, W., Miller, D. M., and Card, H. C. (1989). Cellular automata-based pseudorandom number generators for built-in self-test. *IEEE Trans. on Computer-Aided Design*, **8** (8), pp. 842–859.

Jain, S. and Agrawal, V. (1985). Statistical fault analysis. *IEEE Design & Test of Computers*, **2** (1), pp. 38–44.

Kapur, R., Patil, S., Snethen, T. J., and Williams, T. W. (1994). Design of an efficient weighted random pattern generation system. In *Proc. Int. Test Conference*, pp. 491–500.

Kim, K., Ha, D., and Tront, J. (1988). On using signature registers as pseudorandom pattern generators in built-in self-testing. *IEEE Trans. on Computer-Aided Design*, **7** (8), pp. 919–928.

Knuth, D. E. (1981). *The Art of Computer Programming: Volume 2*. Addison-Wesley Publishing Company.

Konemann, B. (1991). LFSR-coded test patterns for scan designs. In *Proc. European Design and Test Conference*, pp. 237–242.

Konemann, B., Mucha, J., and Zwiehoff, G. (1979). Built-in logic block observation technique. In *Proc. Int. Test Conference*, pp. 37–41.

Krasniewski, A. and Pilarski, S. (1989). Circular self-test path: a low-cost BIST technique for VLSI circuits. *IEEE Trans. on Computer-Aided Design*, **8** (1), pp. 46–55.

LeBlanc, J. J. (1984). LOCST: a built-in self-test technique. *IEEE Design & Test of Computers*, **1** (4), pp. 45–52.

Lempel, M., Gupta, S. K., and Breuer, M. A. (1994). Test embedding with discrete logarithms. In *Proc. VLSI Test Symposium*, pp. 74–80.

Lempel, M., Gupta, S. K., and Breuer, M. A. (1995). Test embedding with discrete logarithms. *IEEE Trans. on Computer-Aided Design*, **14** (5), pp. 554–566.

Lin, C., Zorian, Y., and Bhawmik, S. (1993a). PSBIST: a partial scan based built-in self-test scheme. In *Proc. Int. Test Conference*, pp. 507–516.

Lin, S. and Costello, D. J. (1983). *Error Control Coding: Fundamentals and Applications*. Prentice Hall, Englewood Cliffs, N.J.

Lin, S., Njinda, C., and Breuer, M. (1993b). Generating a family of testable design using the BILBO methodology. *J. of Electronic Testing: Theory and Applications*, pp. 71–89.

Malaiya, Y. K. and Yang, S. (1984). A coverage problem for random testing. In *Proc. Int. Test Conference*, pp. 237–245.

McCluskey, E. (1984). Verification testing – a pseudoexhaustive test technique. *IEEE Trans. on Computers*, **C-33** (6), pp. 541–546.

McCluskey, E. J. and Bozorgui-Nesbat, S. (1981). Design for autonomous test. *IEEE Trans. on Computers*, **C-30** (11), pp. 866–875.

McCluskey, E. J., Makar, S., Mourad, S., and Wagner, K. D. (1988). Probability models for pseudorandom test sequences. *IEEE Trans. on Comptuer-Aided Design*, **7** (1), pp. 68–74.

Mukund, S. K., McCluskey, E. J., and Rao, T. R. N. (1995). An apparatus for pseudo-deterministic testing. In *Proc. VLSI Test Symposium*, pp. 125–131.

Muradali, F., Agarwal, V. K., and Nadeau-Dostie, B. (1990). A new procedure for weighted random built-in self-test. In *Proc. Int. Test Conference*, pp. 660–669.

Parker, K. and McCluseky, E. (1975). Probabilistic treatment of general combinational networks. *IEEE Trans. on Computers*, **C-24** (6), pp. 668–670.

Pateras, S. and Rajski, J. (1991). Generation of correlated random patterns for the complete testing of synthesized multi-level circuits. In *Proc. Design Automation Conference*, pp. 347–352.

Peterson, W. W. and Weldon, E. J. (1972). *Error-Correcting Codes*. MIT Press.

Pilarski, S., Krasniewski, A., and Kameda, T. (1992). Estimating testing effectiveness of the circular self-test path technique. *IEEE Trans. on Computer-Aided Design*, **11** (10), pp. 1301–1316.

Pomeranz, I. and Reddy, S. M. (1993). 3-weight pseudo-random test generation based on a deterministic test set for combinational and sequential circuits. *IEEE Trans. on Computer-Aided Design*, **12** (7), pp. 1050–1058.

Pradhan, D. K. and Gupta, S. K. (1991). A new framework for designing and analyzing BIST techniques and zero aliasing compression. *IEEE Trans. on Computers*, **C-40** (6), pp. 743–763.

Reddy, S. M. (1977). A note on testing logic circuits by transition counting. *IEEE Trans. on Computers*, **C-26** (3), pp. 313–314.

Savaria, Y. and Kaminska, B. (1988). Force-observe: a new design for testability approach. In *Proc. Int. Symposium on Circuits and Systems*, pp. 193–197.

Serra, M., Slater, T., Muzio, J. C., and Miller, D. M. (1990). The analysis of one dimensional linear cellular automata and their aliasing properties. *IEEE Trans. on Computer-Aided Design*, **7** (9), pp. 767–778.

Slater, T. and Serra, M. (1990). Tables of linear hybrid 90/150 cellular automata. Technical Report DCS-105-IR, Department of Computer Science, University of Victoria, Victoria BC, Canada.

Tamarapalli, N. and Rajski, J. (1996). Constructive multi-phase test point insertion for scan-based BIST. In *Proc. Int. Test Conference*, pp. 649–658.

Touba, N. A. and McCluskey, E. J. (1995). Transformed pseudo-random patterns for BIST. In *Proc. VLSI Test Symposium*, pp. 410–416.

Venkataraman, S., Rajski, J., Hellebrand, S., and Tarnick, S. (1993). An efficient BIST scheme based on reseeding of multiple polynomial linear feedback shift registers. In *Proc. Int. Conference on Computer-Aided Design*, pp. 572–577.

Vuksic, A. and Fuchs, K. (1994). A new BIST approach for delay fault testing. In *Proc. European Design Automation Conference*, pp. 284–288.

Waicukauski, J., Lindbloom, E., Eichelberger, E., and Forlenza, O. (1989). A method for generating weighted random patterns. *IBM Jour. of Research and Development*, pp. 149–161.

Wang, L.-T. and McCluskey, E. J. (1986). Concurrent built-in logic block observer (CBILBO). In *Proc. Int. Symposium on Circuits and Systems*, pp. 1054–1057.

Wang, S. and Gupta, S. K. (1997). DS-LFSR: a new BIST TPG for low heat dissipation. In *Proc. Int. Test Conference*, pp. 848–857.

Wang, S. and Gupta, S. K. (1999). LT-RTPG: a new test-per-scan BIST TPG for low heat dissipation. In *Proc. Int. Test Conference*, pp. 85–94.

Williams, T. W., Daehn, W., Gruetzner, M., and Starke, C. W. (1988). Bounds and analysis of aliasing errors in linear feedback shift registers. *IEEE Trans. on Computer-Aided Design*, **6** (1), pp. 75–83.

Wolfram, S. (1984). Universality and complexity of cellular automata. *Physica*, **10D**, pp. 1–35.

Wunderlich, H.-J. (1990). Multiple distribution for biased random test patterns. *IEEE Trans. on Computer-Aided Design*, **9** (6), pp. 584–593.

13 Synthesis for testability

Synthesis for testability refers to an area in which testability considerations are incorporated during the synthesis process itself. There are two major sub-areas: synthesis for full testability and synthesis for easy testability. In the former, one tries to remove all redundancies from the circuit so that it becomes completely testable. In the latter, one tries to synthesize the circuit in order to achieve one or more of the following: less test generation time, less test application time, and high fault coverage. Of course, one would ideally like to achieve both full and easy testability. Synthesis for easy testability also has the potential for realizing circuits with less hardware and delay overhead than design for testability techniques. However, in practice, this potential is not always easy to achieve.

In this chapter, we look at synthesis for testability techniques applied at the logic level. We discuss synthesis for easy testability as well as synthesis for full testability.

We consider both the stuck-at and delay fault models, and consider both combinational and sequential circuits. Under the stuck-at fault (SAF) model, we look at single as well as multiple faults. Under the delay fault model, we consider both gate delay faults (GDFs) and path delay faults (PDFs).

13.1 Combinational logic synthesis for stuck-at fault testability

In this section, we limit ourselves to the SAF model. We discuss testability of two-level and prime tree circuits, logic transformations for preserving single/multiple SAF testability and test sets, synthesis to reduce test set size, synthesis for random pattern testability, and redundancy identification and removal.

13.1.1 Two-level circuits

Two-level circuits are frequently the starting point for further logic optimization. Hence, it is important to consider the testability of such circuits. Consider a Boolean function $z = R_1 + R_2 + \cdots + R_s$, where R_i, $1 \le i \le s$, is a product term. Suppose it is implemented as an AND–OR two-level circuit. This circuit is said to be **prime** if all product terms are prime implicants of z. It is said to be **irredundant** if no product term can be deleted without changing the function. A prime and irredundant single-output

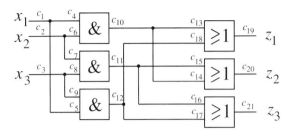

Figure 13.1   A three-output two-level circuit

two-level circuit is fully testable for all single SAFs. In fact, a single SAF test set also detects all multiple SAFs in the circuit (Kohavi and Kohavi, 1972; Schertz and Metze, 1972).

Care must be taken, however, in extending the above results to multi-output two-level circuits. For example, even if each output of such a two-level circuit is prime and irredundant (these type of circuits are sometimes called **single-output minimized multi-output two-level circuits**), a single SAF test set need not detect all multiple SAFs in the circuit, as the following example shows (Jacob and Agrawal, 1992).

Example 13.1   Consider the three-output two-level circuit shown in Figure 13.1. The three outputs implement $z_1 = x_1 x_2 + x_1 x_3$, $z_2 = x_1 x_2 + x_2 x_3$, $z_3 = x_1 x_3 + x_2 x_3$, respectively. Therefore, they are each individually prime and irredundant. Consider the test set T given by $(x_1, x_2, x_3) = \{(0, 1, 1), (1, 0, 1), (1, 1, 0)\}$. T detects all single SAFs in the two-level circuit. However, it does not detect the following multiple SAF: $\{c_6$ stuck-at 1 (SA1), c_{14} stuck-at 0 (SA0), c_8 SA1, c_{16} SA0, c_5 SA1, c_{18} SA0$\}$. The reason is the presence of circular masking as follows: c_6 SA1 is masked by c_{14} SA0, c_{14} SA0 by c_8 SA1, and so on, and finally, c_{18} SA0 is masked by c_6 SA1.   □

In order to obtain multiple SAF testability of multi-output two-level circuits, one can use the following theorem (Jacob and Agrawal, 1992).

Theorem 13.1   In a single SAF testable two-level circuit, a test set for all single SAFs will also detect all multiple SAFs, provided an ordering $z_1, z_2, \ldots, z_m$ can be found among the m outputs such that all single SAFs in the subcircuit feeding output z_j are detected through one or more outputs $z_1, z_2, \ldots, z_i$, $1 \le i \le j \le m$.

If each output is prime and irredundant, then any ordering of the outputs is a valid ordering. However, it is possible that all single SAFs in the circuit are testable even if all outputs are not prime and irredundant. This can happen if multi-output prime implicants have been used in the synthesis process (such circuits are sometimes called **multi-output minimized two-level circuits**). For such circuits, testability of multiple SAFs using this technique depends on being able to find a valid ordering of outputs.

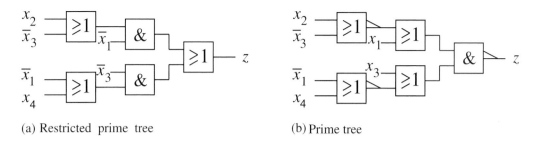

(a) Restricted prime tree   (b) Prime tree

Figure 13.2   Examples of prime trees

Revisiting Example 13.1, since each output is prime and irredundant, (z_1, z_2, z_3) is a valid ordering of outputs. According to the theorem, we need to detect all single SAFs in the subcircuit (logic cone) of output z_1 through z_1. The single SAFs in the cone of output z_2 must be detected through z_2 unless they have already been detected through z_1, and so on. The test set $T = \{(0, 1, 1), (1, 0, 1), (1, 1, 0)\}$ detects c_6 SA1 at z_2 only, even though c_6 belongs to the cones of both z_1 and z_2. Similarly, T detects c_9 SA1 at z_3 only, even though c_9 belongs to the cones of both z_1 and z_3. Since the condition in Theorem 13.1 is violated for these two cases, we need to add another vector $(1, 0, 0)$ to T. This vector detects both c_6 SA1 and c_9 SA1 at z_1. Thus, the modified test set detects all multiple SAFs in the two-level circuit.

The above results are trivially extensible to other types of two-level circuits, such as NAND–NAND circuits.

13.1.2   Prime tree circuits

The concept of prime trees was introduced in Dandapani and Reddy (1974). A **restricted prime tree** is a logic circuit comprised of AND, OR, and NOT gates with the following restrictions: (a) fanouts occur only at the primary inputs, (b) NOT gates are fed only from primary inputs, and (c) if its corresponding sum-of-products obtained by expanding the multi-level switching expression is $z = R_1 + R_2 + \cdots + R_s$ then each R_i is a prime implicant of z (the only simplification allowed in the sum-of-products is based on the following Switching Algebra laws: $x.x = x$ and $x + x = x$). Hence, if terms like $x\bar{x}$, $x + xy$ or $x + \bar{x}y$ occur in the corresponding sum-of-products, then the circuit is not a restricted prime tree. A **prime tree** is a circuit comprised of AND, OR, NOT, NAND and NOR gates which is either a restricted prime tree or can be converted to a restricted prime tree using De Morgan's theorem.

Example 13.2   Consider the circuit in Figure 13.2(a). It obviously meets restrictions (a) and (b) mentioned above. Its corresponding sum-of-products can be obtained by expanding its multi-level switching expression as follows: $z = \bar{x}_1(x_2 + \bar{x}_3) + \bar{x}_3(\bar{x}_1 + x_4) = \bar{x}_1 x_2 + \bar{x}_1 \bar{x}_3 + \bar{x}_1 \bar{x}_3 + \bar{x}_3 x_4$, which reduces to $\bar{x}_1 x_2 + \bar{x}_1 \bar{x}_3 + \bar{x}_3 x_4$ using the $x + x =$

x law. In this equivalent sum-of-products, each product is a prime implicant. Hence, this circuit is a restricted prime tree. Next, let us look at the circuit in Figure 13.2(b). Using De Morgan's theorem, all the inversions can be propagated to the primary inputs, ultimately resulting in the circuit in Figure 13.2(a). Thus, this circuit is a prime tree. □

In general, even when a logic circuit is testable for all single SAFs, it may contain many multiple SAF redundancies. In other words, many multiple SAF faults may be undetectable with any test vector. For prime tree circuits, this cannot happen, as shown by the following theorem.

Theorem 13.2 If a prime tree is testable for all single SAFs then it cannot contain any multiple SAF redundancies.

In order to actually derive a test set for a single SAF-testable prime tree which also detects all multiple SAFs, one needs to obtain the equivalent two-level circuit based on the corresponding sum-of-products. Then the method presented in the previous section can be used to easily derive a multiple SAF test set for the two-level circuit. This test set is also guaranteed to detect all multiple SAFs in the prime tree.

Even if a prime tree is not completely testable for all single SAFs (i.e., it has one or more single SAF redundancies), it is guaranteed that no multiple SAF redundancy can be present in it. Since after removing any single SAF redundancy by placing a constant 0 or 1 on the line, as the case may be, and simplifying the logic (formal logic simplification rules for redundant logic are considered in Section 13.1.7), we get another prime tree, the above result continues to hold. We can continue this process by choosing a single SAF redundancy for removal at each step in the simplification process until no more redundancies are found. The final prime tree would not only be single SAF testable but also multiple SAF testable.

The above properties of a prime tree make it an attractive candidate for being an initial circuit to which transformations preserving multi-fault testability can be applied.

13.1.3   Transformations to preserve single SAF testability

Given an initial circuit which implements the desired functions, one can apply different **transformations** to it to obtain another circuit which meets some desired area, delay, testability and power constraints. In this section, we will look at transformations which can be applied to initial circuits that are testable for all single SAFs and produce a final circuit which are also completely single SAF testable.

A single SAF testability preserving method based on a constrained form of *algebraic factorization* was presented in Rajski and Vasudevamurthy (1992). We have come across this term in Chapter 9 (Section 9.1.1.3) earlier. However, we again define it here for the sake of ease of reading and completeness.

A **cube** is a product of a set C of literals such that if literal $x \in C$ then $\bar{x} \notin C$. Suppose a function z is expressed as $qg + p$. If q and g have no inputs in common, then both q and g are said to be **algebraic divisors** of z, where p is the remainder (Brayton *et al.*, 1987). If an algebraic divisor has exactly one cube in it, it is called a **single-cube divisor**. If it has more than one cube, it is called a **multiple-cube divisor**. For example, if $z = x_1 x_2 x_3 + x_1 x_2 x_4 + x_5$ then $g_1 = x_1 x_2$ is a single-cube divisor of z, whereas $g_2 = x_2 x_3 + x_2 x_4$ is a multiple-cube divisor of z. If we express z as $x_1 g_2 + x_5$ in this example, g_2 is said to be **algebraically resubstituted** in z. By identifying algebraic divisors common to two or more expressions and resubstituting them, one can convert a two-level circuit into a multi-level circuit. This process is referred to as **algebraic factorization**. If the complement of the algebraic divisor is not used in this factorization, it is said to be algebraic factorization without the use of complement. A Boolean expression z is said to be **cube-free** if the only cube dividing z evenly is 1. A cube-free expression must have more than one cube. For example, $x_1 x_2 + x_3$ is cube-free, but $x_1 x_2 + x_1 x_3$ and $x_1 x_2 x_3$ are not. A **double-cube divisor** of a Boolean expression is a cube-free, multiple-cube divisor having exactly two cubes. For example, if $z = x_1 x_4 + x_2 x_4 + x_3 x_4$ then the double-cube divisors of z are $\{x_1 + x_2, x_1 + x_3, x_2 + x_3\}$.

In Rajski and Vasudevamurthy (1992), a method for obtaining multi-level circuits is given which only uses single-cube divisors, double-cube divisors and their complements. These divisors are extracted from functions which are prime and irredundant with respect to every output. The complements are obtained by using only De Morgan's theorem. Boolean reductions such as $a + a = a$, $a + \bar{a} = 1$, $a \cdot a = a$, and $a \cdot \bar{a} = 0$ are not used. Furthermore, for simplicity, only two-literal single-cube divisors are used, and the double-cube divisors are assumed to have at most two literals in each of the two cubes and at most three variables as inputs.

In multi-level circuits, the first level of gates processes primary inputs and produces intermediate nodes. Then successive levels of logic use both primary inputs and intermediate nodes to produce new high-level intermediate nodes and primary outputs. **Single-cube extraction** is the process of extracting cubes which are common to two or more cubes. The common part is then created as an intermediate node. The transformation is as follows: from the expression $z = x_1 x_2 A_1 + x_1 x_2 A_2 + \cdots + x_1 x_2 A_n$, the cube $C = x_1 x_2$ is extracted and substituted to obtain $C A_1 + C A_2 + \cdots + C A_n$. The **double-cube extraction** transformation consists of extracting a double-cube from a single-output sum-of-products expression $AC + BC$ to obtain $C(A + B)$. **Dual expression extraction** transforms a sum-of-product subexpression z in the following ways:

1 $z = x_1 A_1 + x_2 A_1 + \bar{x}_1 \bar{x}_2 A_2$ to $M = x_1 + x_2$ and $z = M A_1 + \bar{M} A_2$.

2 $z = x_1 \bar{x}_2 A_1 + \bar{x}_1 x_2 A_1 + \bar{x}_1 \bar{x}_2 A_2 + x_1 x_2 A_2$ to $M = x_1 \bar{x}_2 + \bar{x}_1 x_2$ and $z = M A_1 + \bar{M} A_2$.

3 $z = x_1 x_2 A_1 + \bar{x}_2 x_3 A_1 + \bar{x}_1 x_2 A_2 + \bar{x}_2 \bar{x}_3 A_2$ to $M = x_1 x_2 + \bar{x}_2 x_3$ and $z = M A_1 + \bar{M} A_2$.

At each step of the synthesis process, the method greedily selects and extracts a double-cube divisor jointly with its dual expression or a single-cube divisor that results in the greatest cost reduction in terms of the total literal-count. If the above transformations are applied to a single-output sum-of-products, then single SAF testability is preserved.

In order to apply this method to a multi-output two-level circuit, one should make each of the outputs prime and irredundant. Also, care must be taken during resubstitution. In a multi-output circuit, many nodes $c_1, c_2, \ldots, c_k$ may be represented by the same expression. Resubstitution is a transformation that replaces each copy of $c_1, c_2, \ldots, c_k$ with a single node. Resubstitution of common subexpressions in a multi-output function preserves single SAF testability if no two subexpressions control the same output.

Because of the simple divisors used in this method, the synthesis process is very fast. In addition it frequently results in circuits with less area compared to those generated by more general synthesis methods, such as MIS (Brayton *et al.*, 1987). For the benchmarks, the speed-up in synthesis time was about nine-fold and reduction in area about 20% compared to MIS.

Suppose that some single SAF testable circuit C_1 is transformed to another circuit C_2 using the above method, then not only is C_2 guaranteed to be single fault testable, but the single SAF test set of C_1 is also guaranteed to detect all single SAFs in C_2. Such transformations are called **test set preserving**. Other such transformations are discussed later on in this chapter.

13.1.4 Transformations to preserve multiple SAF testability

We have seen earlier that if a prime tree is single SAF testable then it cannot contain any multiple SAF redundancies. Also, the single SAF test set for the equivalent sum-of-products also detects all multiple SAFs in the prime tree. Prime trees do not have any internal fanout. A generalization of this method, which allows internal fanout, was given in Hachtel *et al.* (1992). There it was shown that if algebraic factorization without complement is applied to a single-output minimized multi-output two-level circuit then the resultant multi-level circuit is testable for all multiple SAFs using the single SAF test set of the two-level circuit. The latter test set for the two-level circuit has to be derived in such a fashion that it also detects all multiple SAFs in it, as explained in Section 13.1.1. Unlike the method in the previous section, the algebraic divisors in this case need not be limited to single-cube and double-cube divisors.

The proof that algebraic factorization without complement preserves multiple SAF testability and test sets is intuitively quite simple. If we collapse the algebraically factored multi-level circuit to a two-level circuit, we arrive at the original sum-of-products expressions that we began the synthesis process from. Therefore, for every multiple SAF in the multi-level circuit, we can obtain a corresponding multiple SAF in the two-level circuit. Since the test set for the two-level circuit detects all multiple

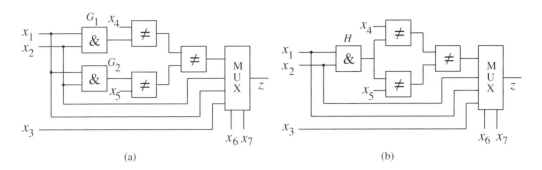

Figure 13.3 Activation of a latent multiple SAF redundancy

SAFs in it, it also detects all multiple SAFs in the multi-level circuit. However, for the benchmarks considered in Hachtel *et al.* (1992), the size of the test set for the two-level circuit was about two to 10 times larger than the size of the single SAF test set for the multi-level circuit. Therefore, increase in the test set size is the price paid for multiple SAF testability. However, by using more sophisticated techniques, the size of the test set for multiple SAF faults can be reduced (Devadas *et al.*, 1993). In addition, it has been shown that algebraic resubstitution with a constrained use of complement also preserves multiple SAF testability (Bryan *et al.*, 1990).

Surprisingly, even though general algebraic factorization without complement preserves multiple SAF testability, it does not preserve single SAF testability. This is owing to the fact that in a single SAF testable circuit, a multiple SAF redundancy may be present, which after algebraic factorization can become a single SAF redundancy.

Example 13.3 Consider the circuit in Figure 13.3(a), which can be verified to be completely single SAF testable. If we replace gates G_1 and G_2 with a single gate, corresponding to factoring out a single cube, we get the circuit in Figure 13.3(b). In this circuit, an SA0 or SA1 fault at the output of gate H is not testable. These single SAF redundancies were a result of the double SA0 (or SA1) fault redundancies at the outputs of gates G_1 and G_2 in the circuit in Figure 13.3(a).   □

One can also derive a multiple SAF/stuck-open fault (SOpF) testable and PDF testable multi-level circuit using Shannon's decomposition (Kundu *et al.*, 1991). This is discussed later.

13.1.5 Transformations to preserve test sets

In the previous two sections, we discussed methods which preserve single or multiple SAF testability as well as test sets. Some additional test set preserving transformations were presented in Batek and Hayes (1992).

Consider a **fanout-free transformation** which transforms a fanout-free circuit into another fanout-free circuit which is functionally equivalent to the first one. Such a

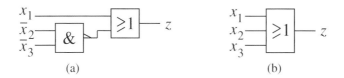

Figure 13.4 Fanout-free transformation

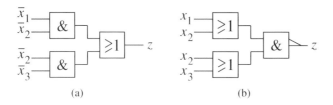

Figure 13.5 De Morgan transformation

transformation preserves the test set. These transformations are primarily useful for grouping items to enable later transformations.

Example 13.4 Consider the fanout-free circuit in Figure 13.4(a). The following test set T at (x_1, x_2, x_3) detects all single SAFs in it: $\{(0, 0, 0), (0, 0, 1), (0, 1, 0), (1, 0, 0)\}$. Another fanout-free circuit which realizes the same function is shown in Figure 13.4(b). T also detects all single SAFs in this circuit. □

Another test set preserving transformation is the **De Morgan transformation**. This transformation converts an AND/OR circuit into a dual OR/AND circuit using De Morgan's theorem, or vice versa. The test set for one circuit remains valid for the other.

Example 13.5 Consider the AND/OR circuit in Figure 13.5(a). The following test set T at (x_1, x_2, x_3) detects all single SAFs in it: $\{(0, 0, 1), (0, 1, 0), (1, 0, 0), (1, 0, 1)\}$. T also detects all single SAFs in the corresponding OR/AND circuit shown in Figure 13.5(b) obtained using De Morgan's theorem. □

Earlier in Section 13.1.3, we mentioned that resubstitution of common subexpressions in a multi-output function preserves single SAF testability if no two subexpressions control the same output. This result was generalized in Batek and Hayes (1992) where conditions were given under which test sets can be preserved even if two or more common subexpressions feed the same output.

13.1.6 Synthesis for reduced test set size

Minimizing the size of the test set is also an important concern. This leads to a reduction in the test application time. When a large number of chips of the same type are to be tested, this could result in significant reduction in the testing cost. Test application time can be targeted in both the deterministic testing and built-in self-test (BIST) environments. Since BIST often employs pseudorandom test sequences, circuits should be synthesized to be easily testable using short test sequences. This also reduces the complexity of the BIST hardware. We discuss test set size reduction in the deterministic environment first.

13.1.6.1 Synthesis for reduced deterministic test set size

Deterministic test set size has been targeted during logic synthesis in (De and Banerjee, 1991). This method uses algebraic factorization.

The decision on which factors to extract at any given step in the synthesis process is based on the concept of *test counts* (Krishnamurthy, 1987). This is done to avoid calculating the actual test set size at each step which is computationally very expensive. We give a simplified version of this method next. In this method, given the test set T for an irredundant circuit, the following two values are associated with each node a of the circuit: a_0^+ and a_1^+. a_0^+ (a_1^+) represents the number of test vectors in T for which node a assumes the value 0 (1) and which detect fault a SA1 (a SA0). a_0^+ and a_1^+ are called **sensitive test counts**. Given a set of single SAFs, the test counting procedure given next gives a method for computing the lower bound for the sensitive test counts of each line in the circuit without knowing T and without performing fault simulation. Assume that we want to detect single SAFs on all the lines in the circuit. First, we initialize each primary input a to have the set of values $(a_0^+, a_1^+) = (1, 1)$, since we need to detect the SA1 and SA0 faults on these inputs. Starting from primary inputs and working towards the primary outputs, we calculate these values for each gate. The method of calculation for a two-input AND gate, with a and b as inputs and c as output, is given below.

$$c_0^+ = a_0^+ + b_0^+$$

$$c_1^+ = \text{MAX}(a_1^+, b_1^+)$$

If a and b are primary inputs, then c_0^+ would be 2 and c_1^+ would be 1. This can be seen to be true from the following test set of the AND gate: $\{(0, 1), (1, 0), (1, 1)\}$. The method can be easily extended to an AND gate with more than two inputs. Similarly, the corresponding equations for a two-input OR gate are as follows:

$$c_0^+ = \text{MAX}(a_0^+, b_0^+)$$

$$c_1^+ = a_1^+ + b_1^+$$

A fanout node requires special treatment in the forward propagation of the sensitive test count values. The difficulty with such nodes is that the counts (i.e., the fault effects) entering the fanout stem can exit through either branch or both. Since we are interested in deriving the lower bound on the sensitive test counts, we will be optimistic and assume that each branch thinks that the other branch will propagate the sensitive values. Thus, we need to reset the sensitive values on the branches to $(1, 1)$. After the sensitive values have been propagated to all the circuit outputs, we derive a **test count** for the circuit C as follows:

$$Test\ Count(C) = \text{MAX}_{v \in C}(v_0^+ + v_1^+)$$

where v is a node in C.

The test count is taken to be an estimate of the test set length required for the circuit. For a fanout-free circuit, the above method yields an optimal test count, i.e., it gives the size of the minimum single SAF test set.

The test counts are used in the synthesis process as follows. Initially, at each step in the process, the best N algebraic divisors are chosen on the basis of literal count. Then among them, the one which gives the best trade-off between area and test set length is chosen.

Example 13.6 Consider the following set of equations:

$$z_1 = x_1 x_3 x_5 + x_2 x_3 x_5 + x_4 x_5 + x_6$$

$$z_2 = x_1 x_3 x_7 + x_2 x_3 x_7 + x_4 x_7 + x_8$$

$$z_3 = x_1 x_9 + x_2 x_9 + x_{10}$$

$$z_4 = x_1 x_{11} + x_2 x_{11} + x_{12}$$

The best two choices for algebraic divisors are as follows: $x_1 + x_2$ and $x_1 x_3 + x_2 x_3 + x_4$. If we choose the first divisor, the set of equations becomes

$$z_1 = g x_3 x_5 + x_4 x_5 + x_6$$

$$z_2 = g x_3 x_7 + x_4 x_7 + x_8$$

$$z_3 = g x_9 + x_{10}$$

$$z_4 = g x_{11} + x_{12}$$

$$g = x_1 + x_2$$

The total literal count is 20 and the test count can be computed to be six.
On the other hand, if we choose the second divisor, the set of equations becomes

$$z_1 = g x_5 + x_6$$

$$z_2 = gx_7 + x_8$$

$$z_3 = x_1x_9 + x_2x_9 + x_{10}$$

$$z_4 = x_1x_{11} + x_2x_{11} + x_{12}$$

$$g = x_1x_3 + x_2x_3 + x_4$$

Now the total literal count is 21 and the test count can be computed to be five. Thus, while the first circuit is better from area of point of view (assuming literal count is a good indicator of area), the second one is better from the test set size point of view (assuming that test count is a good relative indicator of the actual test set size that would be obtained from an actual combinational test generator). $\square$

In order to take into account both the literal count and test count, a cost function $Cost(f)$ to evaluate the quality of a divisor f can be given as follows:

$$Cost(f) = \gamma \Delta T/T + (1 - \gamma)\Delta L/L$$

where T, ΔT, L and ΔL, respectively, denote the test count, decrease in test count, literal count and decrease in literal count, and γ is a relative weighting factor which reflects the importance of test count reduction and literal savings. It has been empirically observed that the best results are generally obtained for $N = 4$ and $\gamma = 0.3$.

Further reduction in test set size can be obtained by making some internal lines directly observable by adding observation points.

13.1.6.2 Synthesis for random pattern testability

Synthesis of multi-level logic circuits for random pattern testability has been considered in Touba and McCluskey (1994) and Chiang and Gupta (1994). Even when all the faults in a circuit are detectable, some faults may have a very few tests. Circuits with such hard-to-detect faults require long pseudorandom test sequences to achieve high fault coverage. Such faults are also, therefore, sometimes referred to as **random pattern resistant**. The aim of work in this area is to reduce or eliminate such faults. Both methods take advantage of algebraic factorization to accomplish this aim.

The **detection probability** of a fault is equal to the number of vectors that detect it, called its **detecting set**, divided by the total number of input vectors, 2^n, where n is the number of primary inputs. Random pattern resistant faults have low detection probabilities. The method in Touba and McCluskey (1994) starts with a two-level representation of the circuit and a specified constraint on minimum fault detection probability, and finds algebraic divisors to eliminate faults whose detection probabilities fall below this specified threshold. Once such faults are eliminated, it uses *random pattern testability preserving transformations* to further optimize the circuit.

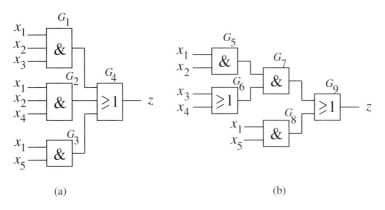

Figure 13.6 Division by single cube x_1x_2

The fault detection probabilities are first calculated for all the faults in the two-level circuit. Then the exact or a lower bound on the detection probability of a fault in the multi-level circuit is computed by exploiting the following result: each SAF in an algebraically factored circuit is equivalent to a set of SAFs in the original two-level circuit. If the multi-level circuit fault is at a primary input or output, then its detecting set is the same as the detecting set for the same fault in the two-level circuit. The detecting set of the remaining faults in the multi-level circuit is assumed to be the union of the detecting sets for the corresponding faults in the two-level circuit. This method requires the computation of the detection probabilities just once, for the two-level circuit, and not for each intermediate multi-level circuit obtained in the synthesis process. For a few cases, it is possible that the detecting set so obtained is a subset of the actual detecting set of the fault in the multi-level circuit. In these cases, the computed detection probability will be a lower bound on the actual detection probability. If this lower bound is above the specified threshold, then we need not bother about the fault. If not, the exact detecting set and detection probability can be obtained by a slightly more time-consuming step, which adds the missing vectors to the detecting set.

Example 13.7   Consider the two-level circuit that implements $z = x_1x_2x_3 + x_1x_2x_4 + x_1x_5$, as shown in Figure 13.6(a). Its algebraically factored three-level circuit obtained by division by cube x_1x_2 is shown in Figure 13.6(b). Denote the detecting set for an SA0/SA1 fault on a literal x_i that feeds gate G_j as $DS[x_i(G_j), \text{SA0/SA1}]$. Then the following detecting sets can be easily derived.

$DS[x_1(G_1), \text{SA1}] = \{(0, 1, 1, x, x)\}$
$DS[x_1(G_2), \text{SA1}] = \{(0, 1, x, 1, x)\}$
$DS[x_1(G_5), \text{SA1}] = \{(0, 1, 1, x, x), (0, 1, x, 1, x)\}$
$DS[x_1(G_1), \text{SA0}] = \{(1, 1, 1, 0, 0)\}$
$DS[x_1(G_2), \text{SA0}] = \{(1, 1, 0, 1, 0)\}$
$DS[x_1(G_5), \text{SA0}] = \{(1, 1, 1, 0, 0), (1, 1, 0, 1, 0), (1, 1, 1, 1, 0)\}$

Consider the SA1 (SA0) fault in $x_1(G_5)$, for which the corresponding fault in the two-level circuit is a double SA1 (SA0) fault on $x_1(G_1)$ and $x_1(G_2)$. We see from the detecting sets that $DS[x_1(G_5), \text{SA1}] = DS[x_1(G_1), \text{SA1}] \bigcup DS[x_1(G_2), \text{SA1}]$. However, $DS[x_1(G_1), \text{SA0}] \bigcup DS[x_1(G_2), \text{SA0}] \subset DS[x_1(G_5), \text{SA0}]$. Thus, in the first case, the detection probability computation is exact, and in the second case, it is a lower bound on the exact probability. □

Define a logic transformation to be **random pattern testability preserving** if the minimum fault detection probability in the transformed circuit is at least as high as the minimum fault detection probability in the original circuit. Then the following theorem can be derived.

Theorem 13.3 If a transformation is test-set preserving for fault class F, then it is also random pattern testability preserving for fault class F.

This theorem implies that the transformations discussed in Sections 13.1.3 and 13.1.5 are also random pattern testability preserving for the single SAF fault class.

Note that in Example 13.7, the detection probability of both the single SAFs on x_1 involved in the single cube division went up after the factorization. The same is true for x_2 as well. Thus, random pattern pattern testability was preserved (in fact, enhanced).

It is possible that algebraic factorization alone may not be able to eliminate all random pattern resistant faults. In such a case, one can use extra **test points**, which are either extra control points or observation points to provide direct controllability or observability of internal lines of the circuit, during logic synthesis to eliminate the remaining such faults. This is usually more efficient in minimizing the number of test points than considering the addition of test points as a post-synthesis step (Krishnamurthy, 1987).

Another interesting result in this area is that redundancy is not necessarily incompatible with testability of a circuit. Clearly, the presence of redundancies can increase the test generation time and decrease the fault coverage in deterministic testing. However, some kinds of redundancies can actually enhance random pattern testability of a circuit (Krasniewski, 1991a). Thus, the conventional wisdom does not hold in this case.

13.1.7   Redundancy identification and removal

Owing to sub-optimal logic synthesis, unintentional redundancies can be introduced into a circuit, which can lead to a larger chip area and increase in its propagation delay. However, identification of redundant faults is computationally expensive since typically the test generation algorithms declare a fault to be redundant only if they have failed to generate a test vector for it after implicit exhaustive enumeration of all the vectors. Furthermore, the presence of a redundant fault may invalidate the test for another fault, or make a detectable fault redundant, or make a redundant fault detectable. Therefore, removal of such redundant faults from a circuit can, in general, help reduce area and delay, while at the same time improve its deterministic testability.

One can categorize the redundancy identification and removal methods as either indirect or direct. If redundancy identification is a byproduct of test generation it is called **indirect**. A **direct** method can identify redundancies without the search process involved in test generation. Such a method can be further sub-divided into three categories: *static*, *dynamic* and *don't care based*. **Static** methods analyze the circuit structure and perform logic value implications to identify and remove redundancies, and usually work as a preprocessing step to an indirect method. **Dynamic** methods work in concert with an indirect method. However, they do not require exhaustive search. **Don't care based** methods involve functional extraction, logic minimization and modification of logic.

13.1.7.1 Indirect methods

If a complete test generation algorithm (i.e., one which can guarantee detection of a fault, given enough time) fails to generate a test for fault l SA0 (l SA1), then l can be connected to logic value 0 (1) without changing the function of the circuit. The circuit can then be reduced by simplifying gates connected to constant values, replacing a single-input AND or OR (NAND or NOR) gate obtained as a result of simplification with a direct connection (inverter) and deleting all gates which do not fan out to any circuit output. The simplification rules are as follows:

1 If the input SA0 fault of an AND (NAND) gate is redundant, remove the gate and replace it with 0 (1).
2 If the input SA1 fault of an OR (NOR) gate is redundant, remove the gate and replace it with 1 (0).
3 If the input SA1 fault of an AND (NAND) gate is redundant, remove the input.
4 If the input SA0 fault of an OR (NOR) gate is redundant, remove the input.

Some efficient algorithms for redundancy identification have been presented in Schulz and Auth (1988), Jacoby *et al.* (1989), Chakradhar *et al.* (1993) and Kunz and Pradhan (1994). Other complete algorithms presented in Chapter 4 (Sections 4.6 and 4.7) can also be used. Since the removal of a redundancy can make detectable faults undetectable or undetectable faults detectable (Dandapani and Reddy, 1974), it is not possible to remove all redundancies in a single pass using these methods, as illustrated by the following example.

Example 13.8 Consider the circuit given in Figure 13.7(a). The following faults in it are redundant: x_1 SA0, x_1 SA1, x_3 SA0, x_3 SA1, c_1 SA0, c_1 SA1, c_2 SA1, and c_3 SA1. If none of these faults is present, we can detect c_4 SA1 by vector $(1, 0, 1, 1)$. However, if the redundant fault x_1 SA0 is present, it makes the above fault redundant too.

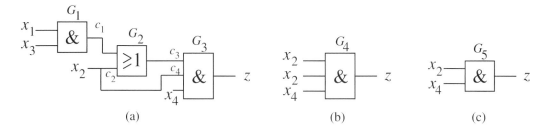

Figure 13.7 Redundancy identification and removal using the indirect method

Suppose we target x_1 SA0 for removal. Using the above simplification rules, we obtain the circuit in Figure 13.7(b) in the first pass. We do test generation for all the faults in this circuit again, and find that both x_2 SA1 faults are redundant. However, if either one of these redundant faults is present, it makes the other one detectable. Targeting either fault for redundancy removal, we get the irredundant circuit in Figure 13.7(c) in the second pass. □

The need to target each fault in each test generation pass makes this method computationally very expensive.

An interesting indirect method is presented in Entrena and Cheng (1995) where some redundancies are deliberately added to an irredundant circuit in order to create yet more redundancies which, upon removal, yield a better optimized circuit. This is done based on the concept of **mandatory assignments**. These are logic value assignments to some lines in the circuit that must be satisfied by any test vector for the given fault. These consist of control and observation assignments which make it possible to control and observe the fault, respectively. If these assignments cannot be simultaneously justified, then the fault is redundant. Using this approach, we can add extra redundant connections (with or without inversions) to the circuit in such a way that the number of connections that become redundant elsewhere in the circuit is maximized. Then after redundancy removal targeted first towards these additional redundancies, we obtain a better irredundant circuit realizing the same functions. This method has also been extended to sequential circuits in Entrena and Cheng (1995).

Example 13.9 Consider the fault c_1 SA0 in the irredundant circuit shown in Figure 13.8(a) (ignore the dashed connection for the time being). Denote the assignment of logic 0/1 at the output of gate G_i as $G_i = 0/1$. The mandatory control assignment for detecting this fault is $G_1 = 1$ and the mandatory observation assignment is $G_2 = 0$. These assignments imply $\bar{x}_1 = 1$, $x_3 = 1$ and $x_2 = 0$, which in turn imply $G_4 = 1$, $G_5 = 0$ and $G_6 = 1$. Since $G_4 = 1$, if we were to add the dashed connection, the effect of c_1 SA0 will no longer be visible at z_1. Thus, this fault will become redundant. However, we still have to verify that adding the dashed connection does not change the input/output behavior of the circuit. In order to test for an SA0 fault

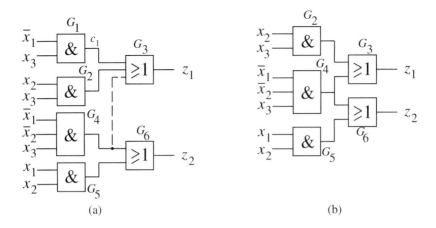

Figure 13.8 Redundancy addition and removal

on this connection, the mandatory control assignment is $G_4 = 1$, and the mandatory observation assignments are $G_1 = 0$ and $G_2 = 0$. However, these three assignments are not simultaneously satisfiable at the primary inputs. Thus, the dashed connection is indeed redundant. After adding this connection, we can use the simplification rules to remove logic based on the fact that c_1 SA0 is now redundant. Finally, we obtain the circuit in Figure 13.8(b), which implements the same input/output behavior as the circuit in Figure 13.8(a), yet requires less area. In the modified circuit, of course, the dashed connection is no longer redundant. □

13.1.7.2 Static methods

Static methods for redundancy identification are very fast since they do not need an exhaustive search (Menon *et al.*, 1994; Iyer and Abramovici, 1996). However, they are usually not able to identify all redundancies. Hence, they can be used as a preprocessing step to an indirect method.

The method in Iyer and Abramovici (1996) uses an illegal combination of logic values to identify redundancies. Suppose logic values v_1, v_2 and v_3 cannot simultaneously occur, respectively, on lines c_1, c_2 and c_3 in a circuit, i.e., this combination is illegal. Then faults for which this combination of values is mandatory are redundant. The problem of finding such faults is decomposed into first finding faults for which each condition is individually mandatory. If $S_{c_i}^{v_j}$ denotes the set of faults that must have value v_j on line c_i for detection, then the faults that require the above combination for detection are in the set $S_{c_1}^{v_1} \cap S_{c_2}^{v_2} \cap S_{c_3}^{v_3}$. To find these faults, the concept of uncontrollability and unobservability analysis is used.

Define a **controlling value** of a gate to be the logic value which can determine its output irrespective of its other input values. Thus, logic 0 (1) is the controlling value for AND and NAND (OR and NOR) gates. Let 0_u (1_u) denote the uncontrollability

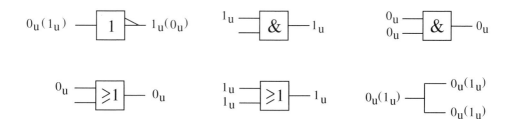

Figure 13.9 Uncontrollability propagation rules

status of a line that cannot be controlled to logic 0 (1). The propagation rules of the uncontrollability status indicators are given in Figure 13.9. Similar rules can be obtained for gates with more inputs. The uncontrollability indicators are propagated forward and backward. Forward propagation of uncontrollability may make some lines unobservable. In general, if a gate input cannot be set to the non-controlling value of the gate then all its other inputs become unobservable. The unobservability status can be propagated backward from a gate output to all its inputs. When all fanout branches of a stem s are marked unobservable, the stem is also marked unobservable if for each fanout branch f of s, there exists at least one set of lines $\{l_f\}$ such that the following conditions are met:

1 f is unobservable because of uncontrollability indicators on every line in $\{l_f\}$; and
2 every line in $\{l_f\}$ is unreachable from s.

These conditions make sure that stem faults that can be detected by multiple-path sensitization are not marked as unobservable. The redundant faults are identified as those which cannot be activated (SA0 fault on lines with 1_u and SA1 fault on lines with 0_u) and those which cannot be propagated (both faults on unobservable lines). The process of propagating uncontrollability and unobservability indicators is called **implication**.

A simple extension of this method based on arbitrary illegal combination of values is as follows. We first form a list L of all stems and reconvergent inputs of reconvergent gates in the circuit. For each line $c \in L$, we imply $c = 0_u$ to determine all uncontrollable and unobservable lines. Let F_0 be the set of corresponding faults. Similarly, we imply $c = 1_u$ to get set F_1. The redundant faults are in set $F_0 \cap F_1$. The reason is that such faults simultaneously require c to be 0 and 1, which is not possible.

Example 13.10 Consider the circuit in Figure 13.10(a). For this circuit, $L = \{x_1, x_2, c_6, c_7\}$. Suppose we target c_6. $c_6 = 0_u$ does not imply uncontrollability or unobservability of any other specific line. Hence, $F_0 = \{c_6\ SA1\}$. $c_6 = 1_u$ implies $x_3 = c_5 = c_1 = c_3 = x_1 = x_2 = c_2 = c_4 = c_7 = z = 1_u$. Hence, from uncontrollability and unobservability analysis, $F_1 = \{c_6\ SA0, x_3\ SA0, c_5\ SA0, c_1\ SA0, c_3\ SA0, x_1\ SA0, x_2\ SA0, c_2\ SA0, c_4\ SA0, c_7\ SA0, z\ SA0, c_6\ SA1, x_3\ SA1, c_5\ SA1, c_1\ SA1, c_3\ SA1, c_7$

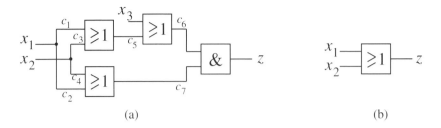

Figure 13.10 An example circuit to illustrate the static method

SA1, c_2 SA1, c_4 SA1}. Since, $F_0 \cap F_1 = \{c_6 \text{ SA1}\}$, c_6 SA1 is redundant. Removing this fault using the simplification rules yields the circuit in Figure 13.10(b).   □

13.1.7.3 Dynamic methods

Dynamic methods also do not require exhaustive search. However, they use a test generator to first identify a redundant fault. Thereafter, they identify additional redundant faults. Such a method has been given in Abramovici and Iyer (1992). This method can remove identified redundancies in just one pass of test generation and fault simulation (however, it cannot guarantee a single SAF testable circuit at the end), as opposed to multiple passes required for indirect methods. It also takes advantage of the uncontrollability and unobservability analysis method introduced in the last subsection.

Define the **region** of a redundant fault to be the subcircuit that can be removed because of it, using the simplification rules mentioned earlier. Also, define the level of a gate in the circuit to be one more than the maximum level of any fanin of the gate, assuming all primary inputs are at level 0. When the region of a redundant fault f_1 is contained within the region of another redundant fault f_2, then it makes sense to target f_2 first. In general, with only a few exceptions, this can be accomplished by targeting the faults at higher levels first for test generation. Once a redundant fault has been removed, we need to identify the newly created redundancies (these are faults that would have been detectable had the removal not occurred). This can be done based on the following theorem.

Theorem 13.4 Let A be an output of a redundant region R and let G be the gate fed by A. Let c be the controlling value and i the inversion of G ($i = 0$ for a non-inverting gate and $i = 1$ for an inverting one). Assume that the combination consisting of $\bar{c}$ values on the remaining inputs of G and $c \oplus i$ value on its output was feasible (legal) in the old circuit. Then this combination becomes illegal as a result of removal.

Once an illegal combination of values is identified, uncontrollability and unobservability analysis can identify the newly created redundancies. In doing so, we need to keep in mind that uncontrollability indicators can be propagated forward and backward everywhere except through gate G. This allows us to identify newly created redundancies, as opposed to redundancies that would be present independently of whether the redundancy removal on the input of G occurred. Of all the newly created redundancies, only the highest level fault is removed and the above process repeated until no more newly created redundancies are found.

Example 13.11 Consider the circuit in Figure 13.7(a) once again. Suppose the test generator has identified x_1 SA0 as redundant. The region R for this fault consists of just gate G_1. This region feeds gate G_2 whose controlling value is 1 and inversion 0. The combination $(c_2 = 0, c_3 = 1)$ was legal in the old circuit. However, once region R is removed, according to the above theorem, this combination becomes illegal. This illegal combination can be translated in terms of uncontrollability indicators as $c_2 = 0_u$ and $c_3 = 1_u$. 0_u on c_2 can be propagated backward. Using the notation and analysis introduced in the previous subsection, we obtain $S_{c_2}^0 = \{c_2 \text{ SA1}, x_2 \text{ SA1}, c_4 \text{ SA1}\}$. Similarly, by propagating 1_u on c_3 forward and recognizing that the side inputs of G_3 become unobservable, we obtain $S_{c_3}^1 = \{c_3 \text{ SA0}, z \text{ SA0}, c_4 \text{ SA0}, c_4 \text{ SA1}, x_4 \text{ SA0}, x_4 \text{ SA1}\}$. Since $S_{c_2}^0 \cap S_{c_3}^1 = \{c_4 \text{ SA1}\}$, c_4 SA1 is the newly redundant fault. After removing this fault as well, we directly obtain the circuit in Figure 13.7(c) in just one pass of test generation. Note that earlier the indirect method required two passes to obtain this final circuit.   □

13.1.7.4 Don't care based methods

A multi-level circuit consists of an interconnection of various logic blocks. Even if each of these logic blocks is individually irredundant, the multilevel circuit can still contain redundancies. These redundancies may stem from the fact that it may not be possible to feed certain input vectors to some of the embedded blocks in the circuit. These vectors constitute the *satisfiability don't care set* (also called *intermediate variable or fanin don't care set*). Also, for certain input vectors, the output of the block may not be observable at a circuit output. These vectors constitute the *observability don't care set* (also called *transitive fanout don't care set*). These don't cares can be exploited to resynthesize the logic blocks so that the multi-level circuit has fewer redundancies. Even if the original multi-level circuit is irredundant, this approach can frequently yield another irredundant circuit implementing the same functions with less area and delay. This method was first presented in Bartlett *et al.* (1988).

Let the Boolean variable corresponding to node j, for $j = 1, 2, \ldots, r$, of the multi-level circuit be y_j and the logic representation of y_j be F_j (here node refers to the output of the logic blocks). The **satisfiability don't care set**, $DSAT$, is common to all nodes, and is defined as

$$DSAT = \sum_{j=1}^{r} DSAT_j$$
$$DSAT_j = y_j \oplus F_j$$

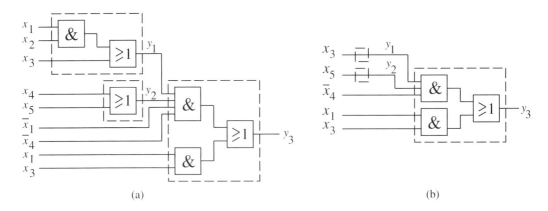

Figure 13.11 Don't care based redundancy removal

$DSAT_j$ can be interpreted to mean that since $y_j = F_j$, the condition $y_j \neq F_j$ is a don't care.

Next, define the **cofactor** of a function f with respect to a literal l, denoted by f_l, as the function f with $l = 1$. Let the set of primary outputs be PO. Then the **observability don't care set**, $DOBS_j$, for each node j is defined as,

$$DOBS_j = \prod_{i \in PO} DOBS_{ij}$$

$$DOBS_{ij} = \overline{(F_i)_{y_j} \oplus (F_i)_{\bar{y}_j}}$$

$DOBS_j$ corresponds to a set of vectors at the primary inputs under which all the primary outputs are insensitive to value y_j that node j takes on.

Since computing the complete satisfiability and observability don't care sets of a general multi-level circuit can be quite time-consuming, efficient heuristics to identify the most useful subsets of these don't care sets have been given in Saldanha *et al.* (1989).

Example 13.12   Consider the circuit in Figure 13.11(a). Suppose this circuit is partitioned into three logic blocks, as shown by the dashed boxes. Even though each of the logic blocks is individually irredundant, the circuit can be easily checked to be redundant. Since $y_3 = y_1 y_2 \bar{x}_1 \bar{x}_4 + x_1 x_3$, $DOBS_1 = \overline{(y_2 \bar{x}_1 \bar{x}_4 + x_1 x_3) \oplus (x_1 x_3)} = \bar{y}_2 + x_1 + x_4$. Therefore, $y_1 = x_1 x_2 + x_3$ can be simplified to just x_3 since x_1 is in $DOBS_1$ which includes $x_1 x_2$ in y_1. Here, the interpretation is that the $x_1 x_2$ term in y_1 is not observable at circuit output y_3. One can think of the don't care minterms in $DOBS_1$ as having been superimposed on y_1, resulting in a new incompletely specified function which needs to be synthesized. Similarly, one can show that $DOBS_2 = \bar{y}_1 + x_1 + x_4$. Hence, $y_2 = x_4 + x_5$ can be simplified to just x_5 since x_4 is present in $DOBS_2$. Using the simplified equations $y_1 = x_3$ and $y_2 = x_5$ we can conclude that $DSAT_1 =$

$y_1\bar{x}_3 + \bar{y}_1 x_3$ and $DSAT_2 = y_2\bar{x}_5 + \bar{y}_2 x_5$. Therefore, y_3 can be simplified with respect to the don't cares in $DSAT_1 + DSAT_2$. In other words, the don't care minterms in $DSAT_1 + DSAT_2$ can be superimposed on y_3, which gives us the simplified expression $y_3 = y_1 y_2 \bar{x}_4 + x_1 x_3$, since the consensus of $x_1 x_3$ and $y_1 \bar{x}_3$ is $x_1 y_1$ which simplifies the term $y_1 y_2 \bar{x}_1 \bar{x}_4$ to $y_1 y_2 \bar{x}_4$. The resultant irredundant circuit is shown in Figure 13.11(b).

13.2 Combinational logic synthesis for delay fault testability

In this section, we concentrate on the GDF and PDF models, with primary emphasis on the latter. We discuss testability of two-level circuits, transformations to preserve or enhance delay fault testability, hierarchical composition rules, and synthesis for delay verifiability.

13.2.1 Two-level circuits

Necessary and sufficient conditions for both general and hazard-free robust PDF and GDF testability of combinational circuits have been presented in Lin and Reddy (1987) and Devadas and Keutzer (1992a). Readers may recall from Chapter 2 (Section 2.4) that if a circuit is fully testable under the hazard-free model, it is also fully testable under the general model (the reverse is not true).

A simple way to check if a single-output two-level AND–OR circuit is robustly PDF testable is to use tautology checking. Suppose we want to test a path starting with literal l going through AND gate G and the OR gate. This path will be hazard-free robust path delay fault testable (HFRPDFT) if and only if after making the side inputs of G equal to 1, the outputs of the remaining AND gates can be made 0 using some input combination without using l. Thus, we can first make the side inputs of G equal to 1, delete l and $\bar{l}$ from the remaining products, and then delete G from the corresponding sum-of-products. If the remaining switching expression becomes a tautology, then the path is not HFRPDFT, else it is. The same conditions are also applicable to general robust path delay fault testable (GRPDFT) two-level circuits.

Example 13.13 Consider the two-level circuit for which the corresponding sum-of-products is $z = x_1 x_2 + x_1 \bar{x}_3 + \bar{x}_1 x_3$. Denote the three AND gates as G_1, G_2 and G_3, respectively. Suppose we want to test the path through literal x_1 in G_1. In order to do this, we first need to enable the side input of G_1 by making $x_2 = 1$. Thereafter, we delete G_1, literal x_1 from G_2 and $\bar{x}_1$ from G_3, obtaining a reduced expression, $z_{\mathrm{red}} = \bar{x}_3 + x_3$. Since z_{red} reduces to 1 (i.e., a tautology), the literal, and hence the path, in question is not HFRPDFT. The reason is that in the transition from the initialization vector to the test vector in the two-pattern test for this path, the outputs of G_2 and G_3 could have a static 0 hazard, thus invalidating the two-pattern test.

Next, consider the path through literal x_2 in G_1. To test this path, after making $x_1 = 1$, we obtain $z_{red} = \bar{x}_3$, which can be made 0 by making $x_3 = 1$. Therefore, a hazard-free robust test (HFRT) for a rising transition on this path is $\{(1, 0, 1), (1, 1, 1)\}$. Reversing the two vectors, we get an HFRT for the falling transition. Readers can check that all the other paths in this circuit are also HFRPDFT. □

An interesting point to note here is that primeness and irredundancy are necessary conditions for HFRPDFT, but not sufficient ones. This stems from the fact that if an untestable SAF is present in a two-level circuit, then the path going through it will not be HFRPDFT. However, the above example showed that even when the circuit is prime and irredundant, the HFRPDFT property is not guaranteed.

To check if a multi-output two-level circuit is HFRPDFT, one can simply check if the above conditions are satisfied for paths starting from each literal to each primary output it feeds.

In order to verify that a two-level circuit is hazard-free or general robust gate delay fault testable (respectively denoted as HFRGDFT and GRGDFT), one just needs to verify that at least one path going through each gate in it is, respectively, HFRPDFT or GRPDFT. This assumes gross GDFs (see Chapter 2; Section 2.2.6). Primeness and irredundancy are neither necessary nor sufficient for obtaining an HFRGDFT or GRGDFT two-level circuit, as the following example shows.

Example 13.14 Consider the two-level circuit based on the expression $z_1 = x_1 + x_2 + \bar{x}_1 \bar{x}_2 x_3$. This circuit is non-prime, yet each gate in it is robustly testable. On the other hand, consider the expression $z_2 = x_1 x_3 + x_1 x_2 + \bar{x}_1 \bar{x}_2 + x_3 x_4 + \bar{x}_3 \bar{x}_4$ which is prime and irredundant. However, the first AND gate with inputs x_1 and x_3 is not robustly testable since neither of the two paths starting from these two literals is robustly testable. □

Finally, let us consider validatable non-robust (VNR) path delay fault testable (VN-RPDFT) two-level circuits (see Section 2.2.6 again for a definition of this property). It has been shown in Reddy *et al.* (1987) that if a single-output two-level circuit is prime and irredundant then it is also VNRPDFT. For a multi-output two-level circuit to be VNRPDFT, a sufficient condition would be to make each output individually prime and irredundant.

13.2.2   Multi-level circuits

Various methods have been presented to obtain nearly 100% or fully robustly testable multi-level circuits. We consider some of them here.

13.2.2.1 Shannon's decomposition

The first method, based on Shannon's decomposition, for guaranteeing completely HFRPDFT multi-level circuits was presented in Kundu and Reddy (1988). It was

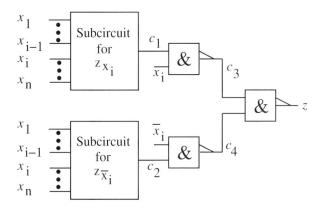

Figure 13.12 Circuit based on Shannon's decomposition

later extended in Kundu *et al.* (1991) to show that the same procedure also results in a circuit, which is testable for all multiple SAFs, multiple SOpFs faults and a combination of these faults, using a particular PDF test set.

Shannon's decomposition theorem states that

$$z(x_1, x_2, \ldots, x_n) = x_i z_{x_i} + \bar{x}_i z_{\bar{x}_i}$$

where z_{x_i} and $z_{\bar{x}_i}$ are cofactors of function z with respect to variable x_i. The corresponding decomposed circuit is shown in Figure 13.12. The importance of this theorem in the present context is that one can show that if z is binate in x_i then the decomposed circuit for z is HFRPDFT if the subcircuits for the two cofactors are HFRPDFT. The reason is that for such a decomposition one can always find at least one vector which, when applied to $x_1, x_2, \ldots, x_{i-1}, x_{i+1}, \ldots, x_n$, results in $z_{x_i} = 1$ and $z_{\bar{x}_i} = 0$ ($z_{x_i} = 0$ and $z_{\bar{x}_i} = 1$), which allows path $x_i c_3 z$ ($\bar{x}_i c_4 z$) to be robustly tested. In addition, making $x_i = 1$ ($x_i = 0$) allows us to fully test the subcircuit for z_{x_i} ($z_{\bar{x}_i}$) by feeding the HFRTs to the corresponding subcircuits. It is possible that the subcircuits are not HFRPDFT after one decomposition. Then the method can be applied recursively to the cofactors until an HFRPDFT circuit is obtained. This method is guaranteed to end up in an HFRPDFT circuit since after at most $n - 2$ Shannon's decompositions, we will get a two-variable cofactor which is guaranteed to be HFRPDFT. In fact, one can stop further decomposition if the cofactor is unate in all its variables since HFRTs can be found for each path in such a subcircuit in which the initialization and test vectors differ in just the literal being tested. Of course, even if the cofactor is binate in some of its variables, further decomposition can be stopped if the corresponding subcircuit is already HFRPDFT. Furthermore, sharing of logic among the cofactor subcircuits does not compromise the HFRPDFT property.

As a useful heuristic to determine which binate variable to target first for decomposition, one can simply choose the variable that appears the most number of times in complemented or uncomplemented form in the given sum-of-products. Another heuristic is to choose a variable which leads to robust untestability in a maximum number of gates.

Example 13.15   Consider the two-level circuit based on the expression $z_2 = x_1 x_3 + x_1 x_2 + \bar{x}_1 \bar{x}_2 + x_3 x_4 + \bar{x}_3 \bar{x}_4$ which we considered earlier in Example 13.14. The only robustly untestable literals are x_1 and x_3 in the first AND gate. Using either of the above two heuristics, we will choose either x_1 or x_3. If we choose x_1, we obtain the decomposition

$$z_2 = x_1(x_2 + x_3 + \bar{x}_4) + \bar{x}_1(\bar{x}_2 + x_3 x_4 + \bar{x}_3 \bar{x}_4).$$

Since the two cofactors are HFRPDFT, so is the decomposed circuit for z_2.

For the benchmark circuits, it has been found that usually applying Shannon's decomposition once results in a multi-level HFRPDFT circuit. To further optimize this decomposed circuit, one can combine Shannon's decomposition with algebraic factorization, as will be explained in the subsections ahead.

13.2.2.2 Algebraic factorization

Readers may not be surprised, having come across various examples of testability preservation based on algebraic factorization for other fault models, that it plays an important role in delay fault testability as well. Many variations of this factorization technique have been shown to be useful, as discussed next.

Simple algebraic factorization: The important result here is that given a completely HFRPDFT circuit, algebraic factorization with a constrained use of the complement produces a completely HFRPDFT circuit, where the complement of the algebraic divisor can be used if the divisor is unate in all its inputs (Devadas and Keutzer, 1992b). A more general result in El-Maleh and Rajski (1995) on algebraic factorization with complement states that if E_1 and E_2 represent two complementary functions then E_2 can be substituted with $\bar{E}_1$ if every robust PDF test set for E_2 is also a complete test set for E_1. Furthermore, the robust test set is also preserved after factorization in both the above methods. The only problem that limits the usefulness of these techniques is that frequently prime and irredundant two-level circuits, that often form the starting point for multi-level logic synthesis, are not completely HFRPDFT. In fact, simple functions exist for which none of the prime and irredundant implementations are HFRPDFT. We have already seen an example of such a function in Example 13.13 earlier: $z = x_1 x_2 + x_1 \bar{x}_3 + \bar{x}_1 x_3$. The only other prime and irredundant implementation for this function is $z = x_2 x_3 + x_1 \bar{x}_3 + \bar{x}_1 x_3$, which is also not HFRPDFT. Function z_2 in Example 13.14 is another example of a function for which no prime and irredundant implementation is HFRPDFT.

It also may happen that one prime and irredundant implementation of a given function is HFRPDFT, but another one is not. For example, $z = x_1 \bar{x}_2 + \bar{x}_1 x_2 + x_1 \bar{x}_3 +$

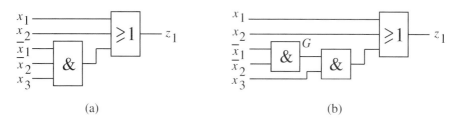

Figure 13.13 Algebraic factorization and gate delay fault testability

$\bar{x}_1 x_3$ is not robustly testable in literal x_1 in $x_1 \bar{x}_2$. However, another implementation of this function, $z = x_1 \bar{x}_2 + \bar{x}_1 x_3 + x_2 \bar{x}_3$, is HFRPDFT. One can use a heuristic to bias the two-level logic synthesizer towards HFRPDFT implementations, whenever possible to do so (Devadas and Keutzer, 1992b). Define a **relatively essential vertex** of a prime implicant in a sum-of-products to be a minterm that is not contained in any other prime implicant of the sum-of-products. Also, define the **ON-set (OFF-set)** of a function to be the set of vertices for which the function is 1 (0). Then this heuristic tries to maximize the number of relatively essential vertices in the prime implicants that are just one bit different from some vertex in the OFF-set of the function. This increases the probability of meeting the necessary and sufficient conditions for robust testability presented in the last section. After that, algebraic factorization can be used to obtain a highly HFRPDFT multi-level circuit. When the above method is applied to single-output minimized multi-output two-level circuits obtained for the benchmarks, it has been shown that on an average about 95% of the paths in the multi-level circuit have HFRTs.

Surprisingly, algebraic factorization does not preserve the HFRGDFT property. Consider the function $z_1 = x_1 + x_2 + \bar{x}_1 \bar{x}_2 x_3$ from Example 13.14. Its HFRGDFT implementation is shown in Figure 13.13(a). However, after one possible algebraic factorization, we obtain the circuit in Figure 13.13(b) which is not HFRGDFT since no path through gate G is robustly testable. To preserve HFRGDFT, we need to use a constrained form of algebraic factorization, where each cube in each factor should have at least one path through it which is robustly testable. Thus, in the above example, if we had used $\bar{x}_1 x_3$ or $\bar{x}_2 x_3$ as a factor instead of $\bar{x}_1 \bar{x}_2$, the HFRGDFT property would have been maintained.

Another important result is that algebraic factorization with a constrained use of the complement preserves VNRPDFT property of a circuit as well as the VNR test set (Devadas and Keutzer, 1992c; El-Maleh and Rajski, 1995). This means that if algebraic factorization is applied to a single-output minimized multi-output two-level circuit then the resultant multi-level circuit will also be VNRPDFT, and the VNR test set for the two-level circuit will be a VNR test set for the multi-level circuit as well.

Extended algebraic factorization: A multi-level circuit obtained by simple algebraic factorization from a set of sum-of-products expressions, when collapsed to a two-level circuit, results in the same set of sum-of-products expressions. However, when a multi-level circuit obtained by **extended algebraic factorization** is collapsed to a two-level circuit, we may get sum-of-products expressions in which some of the cubes are repeated. If one can ensure that all the literals in the repeated cubes are also included in some other cubes which are not repeated, then extended algebraic factorization can be shown to preserve the HFRPDFT property (Pramanick and Reddy, 1990).

Example 13.16 Consider the HFRPDFT sum-of-products $z = x_1x_2 + x_1x_3 + x_2x_5 + x_3x_5 + x_3x_4 + x_1x_6 + x_4x_6$ and its factored expression $z = (x_1 + x_5)(x_2 + x_3) + (x_3 + x_6)(x_1 + x_4)$. This is an extended algebraic factorization since if we collapse the factored expression to a sum-of-products, the cube x_1x_3 will be repeated in it. However, since x_1 appears in another cube which is not repeated, e.g., x_1x_2, and x_3 also appears in another cube which is not repeated, e.g., x_3x_5, the multi-level circuit corresponding to the factored expression is HFRPDFT. □

Targeted algebraic factorization: Since many prime and irredundant sum-of-products expressions are not HFRPDFT, simple or extended algebraic factorization cannot guarantee the HFRPDFT property. In such cases, another method based on **targeted algebraic factorization** can be used, which results in an HFRPDFT multi-level circuit in the vast majority of cases where the original two-level circuit is not HFRPDFT (Jha *et al.*, 1992).

The main idea here is to first convert the two-level circuit, which is not HFRPDFT, into an intermediate circuit (generally with three or four levels) which is HFRPDFT, using targeted use of the distributive law from Switching Algebra. Then algebraic factorization with a constrained use of the complement can be used, as before, to obtain a circuit with more levels which is HFRPDFT. Consider the prime and irredundant expression $z = \sum_{j=1}^{q_1} x_1x_2 \cdots x_n M_j + \sum_{j=1}^{q_2} N_j$, where M_j and N_j are products of literals. Suppose that (a) in each product term in z, all literals in M_j are robustly testable, (b) each literal in the set $\{x_1, x_2, \ldots, x_n\}$ is robustly testable in at least one product term in z, and (c) the other literals in z are not necessarily robustly testable. Then literals $x_1, x_2, \ldots, x_n$ are robustly testable when factored out from the first set of product terms, resulting in the following modified expression: $z = x_1x_2 \cdots x_n(\sum_{j=1}^{q_1} M_j) + \sum_{j=1}^{q_2} N_j$. Furthermore, all literals in $\sum_{j=1}^{q_1} M_j$ remain robustly testable, and the robust testability (or lack thereof) of literals in $\sum_{j=1}^{q_2} N_j$ is not affected. This synthesis rule may have to be applied more than once sometimes to obtain a robustly testable circuit.

Example 13.17 Consider the prime and irredundant expression $z_1 = x_1x_2x_3^* + x_1^*x_3x_4 + \bar{x}_1\bar{x}_2x_4 + \bar{x}_3\bar{x}_4$, where the starred literals are not robustly testable. Using the above synthesis rule, we obtain the modified expression: $z_1 = x_1x_3(x_2 + x_4) + \bar{x}_1\bar{x}_2x_4 + \bar{x}_3\bar{x}_4$, which is completely HFRPDFT.

Consider another expression $z_2 = x_1 x_2^* M_1 + x_1^* x_2 M_2 + x_1^* x_2^* M_3 + x_1^* M_4 + N_1$. Applying the synthesis rule once we get, $z_2 = x_1 x_2 (M_1 + M_2 + M_3) + x_1^* M_4 + N_1$. After applying it once again, we get $z_2 = x_1 \{x_2 (M_1 + M_2 + M_3) + M_4\} + N_1$, where both x_1 and x_2 are now robustly testable.   □

When the synthesis rule is not successful with the sum-of-products z, it frequently becomes successful with the sum-of-products for $\bar{z}$, which can then be followed by an inverter to get z. For 20 out of the 24 multi-output benchmarks considered in Jha *et al.* (1992), the synthesis rule was directly successful for all outputs. For three of the remaining four benchmarks, complementing one of the outputs allowed the application of the synthesis rule. For the last benchmark, targeted algebraic factorization alone was unsuccessful for one of the outputs and its complement. It had to be combined with Shannon's decomposition to obtain a completely HFRPDFT circuit, as explained in the next subsection.

Many other heuristic synthesis rules for enhancing robust testability have also been presented in Jha *et al.* (1992).

It is worth recalling that simple and targeted algebraic factorization also result in a completely multiple SAF testable multi-level circuit based on the results described earlier. The conjecture is that the same is also true for extended algebraic factorization.

Shannon's decomposition and targeted algebraic factorization: Repeated Shannon's decomposition, although it guarantees the HFRPDFT property, is not very area-efficient. On the other hand, area-efficient methods based on the different variations of algebraic factorization cannot guarantee the HFRPDFT property. Hence, a possible compromise to obtain both area efficiency and guaranteed robust testability is to marry these two techniques. For the sake of area efficiency, we should generally do this only if targeted algebraic factorization has failed with the switching expression and its complement. After applying Shannon's decomposition once to such an expression, we first determine if the cofactors are already HFRPDFT or else amenable to targeted algebraic factorization. Only when not, do we consider applying Shannon's decomposition once again. As mentioned before, cases where Shannon's decomposition needs to be applied more than once are extremely rare. Since multiple SAF testability is guaranteed by both Shannon's decomposition and targeted algebraic factorization, it is also guaranteed by their combination.

13.2.3   Hierarchical composition rules for preserving HFRPDFT property

We know from earlier discussions that a composition of two testable circuits, where one feeds the other, may not be highly testable. To avoid this problem, one needs to develop composition rules which preserve testability. The need for such rules is evident from the fact that not all circuits are easily collapsible to two-level circuits,

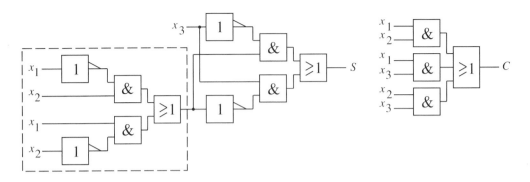

Figure 13.14 Full adder

which is a prerequisite for the synthesis methods presented so far. For example, the sum-of-products for the parity function is of exponential size in the number of inputs.

A simple composition rule is given in the theorem below (Devadas and Keutzer, 1992b).

Theorem 13.5 If a circuit C is created by composing two HFRPDFT circuits C_1 and C_2 such that one output of C_1 is connected to one input of C_2 and the input variables of C_1 and C_2 are disjoint, then C is also HFRPDFT.

Although the rule presented in this theorem is not very general, it can be used to show that some regular structures like the parity tree and ripple-carry adder can be designed to be HFRPDFT. Consider the full adder shown in Figure 13.14 which consists of a sum circuit made up of a cascade of two-input XOR gates and a carry circuit. The two-input XOR gate shown enclosed in the dashed box can be easily checked to be HFRPDFT. Since the cascade of the two XOR gates satisfies the condition in Theorem 13.5, the sum circuit is also HFRPDFT. Since the carry circuit is HFRPDFT, this implies that the complete full adder is HFRPDFT. The above composition can then be applied in a hierarchical fashion to show that a ripple-carry adder is HFRPDFT. Similarly, a binary parity tree consisting of two-input XOR gates can also be seen to be HFRPDFT, based on this theorem and the fact that the two-input XOR gate is HFRPDFT.

More general composition rules which preserve testability of regular structures like carry select adder, carry lookahead adder, carry bypass adder, ripple and parallel comparators, are given in Bryan *et al.* (1991). They are also applicable to adder-incrementers, adder-subtracters, and parallel multipliers.

13.2.4   Synthesis for delay verifiability

A circuit is said to be **delay verifiable** if its correct timing behavior can be guaranteed at the tested speed as well as any slower speed (Ke and Menon, 1995a). In such circuits,

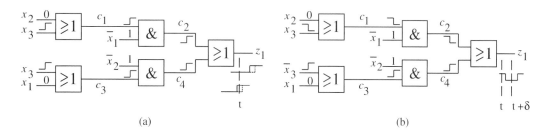

Figure 13.15 Delay verifiability

untestable delay faults are allowed to be present as long as they can be shown to not impact the circuit delay. Verifying the timing of the circuit through testing may require tests which simultaneously detect the presence of more than one PDF, as seen in Chapter 8 (Section 8.2.4) and also illustrated by the example below. From now on, we denote tests which are either general robust or VNR as **RV-tests**. A circuit is said to be **RV-testable** if it has RV-tests for every PDF.

Example 13.18   Consider the circuit in Figure 13.15(a). Denote a PDF by the direction of transition at the primary output, $\uparrow$ (rising) or $\downarrow$ (falling), preceded by the path. In this circuit, all the PDFs except $x_3 c_1 c_2 z_1 \uparrow$ and $x_3 c_3 c_4 z_1 \uparrow$ have general robust tests. Suppose that $x_3 c_1 c_2 z_1 \uparrow$ is present, but $x_3 c_3 c_4 z_1 \uparrow$ is not. The signal values for the two-pattern test $(x_1, x_2, x_3) = \{(0, 0, 0), (0, 0, 1)\}$ are as shown. Fault $x_3 c_1 c_2 z_1 \uparrow$ can produce an error at the output only if the rising transition at c_4 is also late. However, this cannot happen since $x_3 c_3 c_4 z_1 \uparrow$ is not present. Similarly, when only $x_3 c_3 c_4 z_1 \uparrow$ is present, but not $x_3 c_1 c_2 z_1 \uparrow$, no error is produced. When both the untestable faults are present, at sampling time t we get the incorrect value 0. The above two-pattern test detects this incorrect temporal behavior independent of the delays in the rest of the circuit. Hence, it is a general robust test for the double fault. If we include this test in the test set then the circuit becomes delay verifiable.

Next, consider the circuit in Figure 13.15(b). Faults $x_3 c_1 c_2 z_2 \downarrow$ and $\bar{x}_3 c_3 c_4 z_2 \uparrow$ are not RV-testable. When $\{(0, 0, 1), (0, 0, 0)\}$ is applied, a 1-0-1 pulse may appear at the output. At time t, we may sample a 1 which is correct, but with a slower clock at $t + \delta$, we may sample a 0 which is incorrect. Hence, the circuit is not delay verifiable. $\square$

To formalize the concept of delay verifiability, we need the following definitions. A **multiple path** is a set of single paths that share the same circuit output. A line is a **side input** of a multiple path if it is an input to a gate on the multiple path but is not on the multiple path. A **multiple path delay fault** (MPDF) $\Pi \uparrow (\Pi \downarrow)$, where Π is a multiple path, represents the case where every single path $\pi \in \Pi$ has a fault $\pi \uparrow (\pi \downarrow)$. An MPDF $\Pi \uparrow (\Pi \downarrow)$ is said to be **sensitizable** if and only if there exists a two-pattern test which launches $0 \to 1 (1 \to 0)$ transitions at all its path inputs and the final value at each side input of Π is a non-controlling value. A fault is said to be **primitive** if

(a) it is sensitizable, and (b) none of its subsets is sensitizable. One can show that a circuit is delay verifiable if and only if RV-tests exist for each primitive fault in it. In the circuit in Figure 13.15(a), $\{x_3c_1c_2z_1 \uparrow, x_3c_3c_4z_1 \uparrow\}$ is a primitive fault and so are all the other single PDFs. Since RV-tests exist for all these faults, the circuit is delay verifiable.

To synthesize a delay verifiable circuit, one can, as before, start with a delay verifiable two-level circuit and apply **verifiability-preserving transformations** to it to obtain a delay verifiable multi-level circuit. Such circuits are generally more area-efficient than HFRPDFT or VNR testable circuits as there is generally no need to start with a two-level circuit which is prime and irredundant in each output. However, to make the method practical, we need to have an easy way to identify all primitive faults, as discussed next.

Consider a sum-of-products z and its corresponding two-level AND–OR circuit. Let cube $q \in z$. The set of literals $L = \{l_1, l_2, \ldots, l_k\} \subseteq q$ is said to be a **prime literal set** in q if $q - L$ contains a 0-vertex. If this condition is not satisfied, the literal set is called non-prime. A prime literal set L is *minimal* if each proper subset of L is non-prime. For example, consider $z = x_1x_2 + \bar{x}_1\bar{x}_2 + x_1\bar{x}_2x_3$. Both literals x_1 and $\bar{x}_2$ are non-prime in the last cube, but $\{x_1, \bar{x}_2\}$ is a minimal prime literal set in this cube. Similarly, the set of cubes $Q \subset z$ is said to be a **redundant cube set** if $z - Q$ contains the same 1-vertices as z, otherwise it is an irredundant cube set. An irredundant cube set Q is *minimal* if every proper subset of Q is redundant. For example, let $z = x_1\bar{x}_2 + \bar{x}_1x_2 + x_2x_3 + x_1x_3$. Here, cubes x_2x_3 and x_1x_3 are redundant, but $\{x_2x_3, x_1x_3\}$ is a minimal irredundant cube set.

The following theorem gives necessary and sufficient conditions for identifying a primitive fault in a two-level circuit.

Theorem 13.6 In a two-level AND–OR single-output circuit, fault $\Pi \downarrow$ is primitive if and only if (a) Π goes through only one AND gate, and (b) it starts with a minimal prime literal set L of a cube q. If Π goes through a set of cubes Q, then fault $\Pi \uparrow$ is primitive if and only if (a) no two single paths in Π go through the same cube, and (b) Q is a minimal irredundant cube set.

Example 13.19 Consider the following expression once again: $z = x_1\bar{x}_2 + \bar{x}_1x_2 + x_2x_3 + x_1x_3$. Since the circuit is prime, every $\downarrow$ single PDF is primitive. However, since $\{x_2x_3, x_1x_3\}$ is a minimal irredundant set, the four $\uparrow$ faults, consisting of one path from one of these cubes and the other path from the other cube, are primitive. All these four faults can be detected by the RV-test $\{(0, 0, 0), (1, 1, 1)\}$. All other primitive faults also have RV-tests. Hence, the circuit is delay verifiable. $\square$

For the multi-output two-level circuit case, one can make the circuit single SAF fault testable (this does not necessarily mean that each output is prime and irredundant).

Then, modifying the circuit, if necessary, based on Theorem 13.6, one can make it delay verifiable. In order to obtain a multi-level delay verifiable circuit we can use algebraic factorization without complement, as this is known to preserve delay verifiability as well as the delay verifiable test set. Conditions under which delay verifiability is maintained even when the complement of an algebraic divisor is used are given in El-Maleh and Rajski (1995). This synthesis approach typically results in a smaller area for delay verifiable multi-level circuits compared to RV-testable multi-level circuits.

13.3 Sequential logic synthesis for stuck-at fault testability

In this section, we will look at redundancy identification and removal as well as synthesis for easy testability methods for sequential circuits.

13.3.1 Redundancy identification and removal

As we know from Chapter 5 (Section 5.1.3), in sequential circuits an untestable fault is not necessarily a redundant fault. This makes redundancy identification in sequential circuits a very difficult task. Most such methods make unrealistically simplifying assumptions, or are applicable to only small circuits, or identify only special classes of faults (Iyer *et al.*, 1996b). Also, this area contains many pitfalls (Iyer *et al.*, 1996a).

We next describe a method for identifying sequential redundancies without search (Iyer *et al.*, 1996b). This method does not make the simplifying assumption of a global reset mechanism or availability of state transition information. It is applicable to large sequential circuits, but does not guarantee that it can find all sequential redundancies.

Recall from Chapter 5 (Section 5.1.3) that a fault is said to be **partially detectable** if there exists an initial state S_f of the faulty machine and an input sequence I such that for every fault-free initial state S, the response of the fault-free machine to I starting from S, denoted as $z(I, S)$, is different from the response of the faulty machine starting from S_f, denoted as $z_f(I, S_f)$. A fault is redundant if it is not partially testable. We next generalize this conventional definition of redundancy to the concept of c-cycle redundancy. Consider the set of states, S_c, which is reachable after powering up the faulty machine and applying any arbitrary sequence of c input vectors (note that S_c shrinks as c increases). Then a fault is **c-cycle redundant** if it is not partially detectable under the assumption that the initial states of the faulty machine are restricted to S_c. Note that 0-cycle redundancy corresponds to the conventional concept of redundancy. Also, a c-cycle redundant fault is also a $\hat{c}$-cycle redundant fault for any $\hat{c} > c$. A circuit with a c-cycle redundant fault can be simplified by removing the redundant region associated with that fault, provided that c arbitrary input vectors are applied before the application of the existing initialization sequence. The concept of c-cycle

Figure 13.16 Sequential propagation of uncontrollability and unobservability (Iyer *et al.*, 1996b) (© 1996 IEEE)

redundancy also has the **compositional** property, which means that if a fault is c-cycle redundant in a subcircuit of a larger circuit then it is also c-cycle redundant in the larger circuit.

Uncontrollability and unobservability analysis, as described in Section 13.1.7.2 earlier, is also used to identify c-cycle redundant faults. It can be extended to flip-flops as shown in Figure 13.16 (an unobservable line is marked "*"). Uncontrollability propagates both forward and backward and unobservability only propagates backward, as before. However, while in a combinational circuit, unobservability propagates unconditionally from the output of a gate to all its inputs, this is not always the case in sequential circuits. Before marking a gate input as unobservable, we need to ensure that multiple fault effects from that input in different time frames cannot combine to reach a primary output. Hence, conditions for the propagation of unobservability are generalized as follows.

Unobservability propagates onto l^i (the copy of line l at time i) if

1 The fanouts of l^i are marked as unobservable at time i.
2 For every fanout f^i of l^i, there exists at least one set of lines $\{p^j\}$, such that
   (a) f^i is unobservable because of uncontrollability indicators on every line in $\{p^j\}$; and
   (b) there is no sequential path from l^k, $i \leq k \leq j$, to any line in $\{p^j\}$.

The iterative array model of a sequential circuit consists of repeated time frames of its combinational logic. The analysis described above is valid for the fault-free machine only. In order to make sure they are also valid for the faulty circuit, a validation step is needed, as described next. Suppose that the uncontrollability status indicator starts propagating from some fanout stem s in time frame 0. Number the time frames forward of time frame 0 as $1, 2, \ldots, f$ and backward of time frame 0 as $-1, -2, \ldots, -b$. The propagation is allowed for a maximum of T_M time frames, where $f + b + 1 \leq T_M$, in order to prevent an infinite propagation due to feedback loops. A value v_u, $v \in \{0, 1\}$, on line l^i, $-b \leq i \leq f$, is said to be valid in the presence of an SA0 (SA1) fault on line m if and only if v_u propagates onto l^i in the presence of SA0 (SA1) faults on line m in all time frames $j < i$. For example, if the fault in question was SA0 (SA1) and it was necessary to have 0_u (1_u) on line m in a previous time frame to get v_u on l^i, then $l^i = v_u$ is invalid in the presence of the SAF. If this validation step is not carried out, then the fault can only be declared untestable, not redundant.

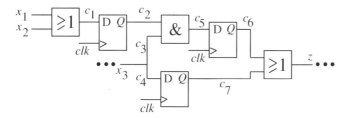

Figure 13.17 Example circuit for sequential redundancy identification (Iyer *et al.*, 1996b) (© 1996 IEEE)

The method processes every fanout stem s in the circuit for 0_u and 1_u values by sequentially propagating uncontrollability and unobservability status indicators for a maximum of T_M time frames (where T_M is predetermined). Then for the assignment $s = v_u$, $v \in \{0, 1\}$, and for each time frame i, $-b \leq i \leq f$, we form a set of faults S_v^i as follows. If a line m has a valid 0_u (1_u) value at time i, we include the uncontrollable fault m SA1 (SA0) in S_v^i. Similarly, suppose m has an unobservability indicator at time i caused by uncontrollability indicators on a set of lines $\{p^j\}$ ($j \geq i$). Then we include the m SA0 (SA1) fault in S_v^i if and only if the uncontrollability indicator on every line in $\{p^j\}$ is valid in the presence of the m SA0 (SA1) fault. The above process is referred to as **sequential implication**. Finally, the set of faults for which a conflict on s is necessary for detection are determined as $S_s = \bigcup_i \{S_0^i \cap S_1^i\}$.

Consider a fault α identified as above in time frame i and let l be the leftmost time frame to which uncontrollability must propagate in order to identify α as redundant. Then fault α is c_α-cycle redundant, where $c_\alpha = 0$ if $l \geq i$ and $c_\alpha = i - l$ if $l < i$.

Example 13.20 Consider the circuit in Figure 13.17. Starting with $x_3 = 0_u$ at time 0 implies $c_3 = c_4 = 0_u$ at time 0, which in turn implies $c_7 = z = 0_u$ at time 1. Since c_6 is unobservable at time 1, lines c_5, c_2 and c_3 at time 0 and c_1, x_1 and x_2 at time -1 become unobservable too. The complete set of sequential implications is given in Table 13.1. By the rules of computing c_α, we see that all the listed faults except c_6 SA0 are 0-cycle redundant whereas c_6 SA0 is 1-cycle redundant. Note that only one of these faults can be targeted for removal at one time. □

13.3.2 Synthesis of sequential circuits for easy testability

Many methods have been presented in the literature for making finite-state machines, which are describable by a state table, easily testable. Many such methods add extra state transitions to the given state table such that it becomes easy to synchronize the modified machine to any state as well as identify every state. However, most such methods are not applicable to large sequential circuits. For tackling large circuits, an alternative approach is to start with the combinational logic of the already-synthesized

Table 13.1. *Sequential implications for Example 13.20 (Iyer et al., 1996b) (© 1996 IEEE)*

Process	Time	Uncontroll.	Unobserv.	Collapsed faults
	0	$x_3 = c_3$ $= c_4 = 0_u$	c_5, c_2, c_3	$S_0^0 = \{x_3 \text{ SA1}, c_3 \text{ SA1}, c_4 \text{ SA1},$ $c_5 \text{ SA0}, c_5 \text{ SA1}, c_2 \text{ SA1}\}$
$x_3 = 0_u$	1	$c_7 = z$ $= 0_u$	c_6	$S_0^1 = \{z \text{ SA1}, c_6 \text{ SA0}\}$
	-1	$-$	c_1, x_1, x_2	$S_0^{-1} = \{c_1 \text{ SA0}, c_1 \text{ SA1},$ $x_1 \text{ SA0}, x_2 \text{ SA0}\}$
	0	$x_3 = c_3$ $= c_4 = c_5$ $= 1_u$	c_2	$S_1^0 = \{x_3 \text{ SA0}, c_4 \text{ SA0},$ $c_5 \text{ SA0}, c_2 \text{ SA1}\}$
$x_3 = 1_u$	1	$c_6 = c_7$ $= z = 1_u$	$-$	$S_1^1 = \{c_6 \text{ SA0}, c_7 \text{ SA0}, z \text{ SA0}\}$
	-1	$-$	c_1, x_1, x_2	$S_1^{-1} = \{c_1 \text{ SA0}, c_1 \text{ SA1},$ $x_1 \text{ SA0}, x_2 \text{ SA0}\}$
Redun.	0			$S_0^0 \cap S_1^0 = \{c_5 \text{ SA0}, c_2 \text{ SA1}\}$
faults	1			$S_0^1 \cap S_1^1 = \{c_6 \text{ SA0}\}$
	-1			$S_0^{-1} \cap S_1^{-1} = \{c_1 \text{ SA0}, c_1 \text{ SA1}, x_1 \text{ SA0}, x_2 \text{ SA0}\}$

sequential circuit, superimpose some testability logic on it and then resynthesize the composite logic. Another approach exploits retiming and resynthesis to impart easy testability to the circuit. We discuss such methods next.

13.3.2.1 Synthesis for parallel scan

In Chapter 11, we saw how scan design can be used to configure the state register as a shift register in the test mode to make it easy to control and observe the state. This reduces the sequential test generation problem to combinational test generation. This is done by incorporating a multiplexer in each flip-flop. However, one could instead add the multiplexer logic (also called the testability logic) directly to the combinational logic of the sequential circuit (Reddy and Dandapani, 1987). This subtle change can lead to many advantages in area/delay overheads and test application time, while still requiring only combinational test generation. The reduction in the overheads comes from heuristics which try to maximally merge the combinational and testability logic (Vinnakota and Jha, 1992; Bhatia and Jha, 1996), reduction in routing area over scan design and reduction in pin count. If the delay overhead is still not satisfactory after the merging, one can apply timing optimization techniques to resynthesize the merged logic to improve its delay (which traditional scan design precludes). The reduction in test application time comes from the fact that primary inputs (outputs) can be used to control (observe) the state register in parallel.

Let the number of primary inputs, primary outputs and flip-flops be n, m and p, respectively. Then $q = \min(n, m, p)$ parallel scan paths can be created through the combinational logic. If $q = p$ then all flip-flops can be controlled and observed in just one cycle, and the synthesis scheme in this case is called **synthesis for fully parallel scan**. Else if $q < p$ the scheme is called **synthesis for maximally parallel scan**.

Denote the input (output) of the ith flip-flop by Y_i (y_i). Consider the logic expression for a flip-flop $Y_i = x_1 x_2 + x_2 \bar{x}_3 + \bar{x}_1 x_4$. If x_2 is chosen to provide the scan input for this flip-flop then we can modify the above expression to $Y_i = T x_2 + \bar{T}(x_1 x_2 + x_2 \bar{x}_3 + \bar{x}_1 x_4) = T x_2 + x_1 x_2 + x_2 \bar{x}_3 + \bar{T} \bar{x}_1 x_4$, where T is the test mode signal. This shows how the testability logic can be merged with the combinational logic. Let the larger of the literal count in the complemented versus uncomplemented form of variable x_j in some logic expression ξ be denoted by $lc_{x_j}(\xi)$. Then clearly, for maximal merging, we should, whenever possible, use that variable x_k to provide the scan input for which $lc_{x_k}(\xi)$ is the highest of all the variables in that expression. Also, if $lc_{x_k}(\xi)$ corresponds to the count of literal $\bar{x}_k$ (x_k), then $\bar{x}_k$ (x_k) should be used as the scan input. A greedy heuristic for logic merging for a more general example is illustrated next.

Example 13.21 Consider a sequential circuit with three primary inputs, two primary outputs and three flip-flops. Thus, the maximum number of scan paths, q, is equal to two. Assume that the switching expressions are as follows:

$$Y_1 = \bar{x}_1 x_2 + \bar{x}_1 x_3$$
$$Y_2 = x_2 y_3 + x_3$$
$$Y_3 = x_1 x_3 + x_2 x_3$$
$$z_1 = y_2 + x_1 x_3$$
$$z_2 = y_2 + y_3$$

Since in the flip-flop expressions, $lc_{x_1}(Y_1)$ and $lc_{x_3}(Y_3)$ are the two largest literal-counts, and these literal-counts correspond to literals $\bar{x}_1$ and x_3 respectively, we choose $\bar{x}_1$ and x_3 to provide the scan inputs of flip-flops Y_1 and Y_3. The output of one of these two flip-flops needs to be selected next to provide the input to flip-flop Y_2 in the test mode. Since Y_2 depends on y_3, but not y_1, we connect y_3 to Y_2 in the test mode when $T = 1$. Now the outputs of flip-flops Y_1 and Y_2 need to be scanned out through the outputs. Since $lc_{y_2}(z_1) = lc_{y_2}(z_2) = 1$ and $lc_{y_1}(z_1) = lc_{y_1}(z_2) = 0$, we can choose either output to scan out y_2, and choose the other output to scan out y_1. Assuming we chose z_2 for y_2 and z_1 for y_1, the two scan paths are formed as shown in Figure 13.18. The modified expressions, after using the simplifications $a + \bar{a}b = a + b$ and $ab + \bar{a}b = b$, are given below.

$$Y_1 = T \bar{x}_1 + \bar{x}_1 x_2 + \bar{x}_1 x_3$$
$$Y_2 = T y_3 + x_2 y_3 + \bar{T} x_3$$
$$Y_3 = T x_3 + x_1 x_3 + x_2 x_3$$
$$z_1 = T y_1 + \bar{T}(y_2 + x_1 x_3)$$
$$z_2 = y_2 + \bar{T} y_3$$

□

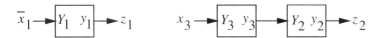

Figure 13.18 Scan paths

After the merged two-level logic is obtained, the standard multi-level logic optimization methods can be applied. In fact, the testability logic can also be directly added to an existing multi-level logic implementation and further optimized if necessary. The testability logic can be tested first by scanning in some particular logic values through the scan paths in the test mode and observing them at the scan outputs. Then, as in traditional scan design, only combinational test generation is necessary. Because of parallel scan paths, test application is sped up over serial scan by roughly a factor of q. The method can, of course, be applied to partial scan as well, resulting in **synthesis for parallel partial scan**, where only the flip-flops selected for partial scan are put in parallel scan paths. Another possibility is to add the testability logic for only parallel scan in (out) for flip-flops which are not easy to control (observe).

13.3.2.2 Retiming and resynthesis for partial scan

An effective way to select flip-flops for partial scan is to choose flip-flops in the minimum feedback vertex set (MFVS) of the S-graph which break all loops, except self-loops. The S-graph has a vertex for every flip-flop and there is an arc from vertex v_1 to vertex v_2 if there is a combinational path from flip-flop v_1 to flip-flop v_2. The MFVS corresponds to a minimum number of vertices whose removal makes the resultant graph acyclic. However, the MFVS of the circuit, which corresponds to the minimum number of gates whose removal makes the circuit acyclic, is a lower bound on and sometimes significantly smaller than the MFVS of the S-graph. Since only flip-flops can be scanned, one can think of a method of repositioning the flip-flops such that in the modified circuit, every circuit MFVS drives at least one flip-flop that can be scanned. It is shown in Chakradhar and Dey (1994) that this can be always accomplished through resynthesis and retiming.

Retiming is a method for repositioning the flip-flops in the circuit without changing its input/output behavior. Figure 13.19(a) shows an example of retiming across a single-output combinational gate G in the forward and backward directions. Figure 13.19(b) shows an example of retiming across a fanout stem in the forward and backward directions.

More formally, consider a circuit with gates, primary inputs and outputs labeled as $u_1, u_2, \ldots, u_n$. Let us model the circuit as a **circuit graph** which has a vertex u_i for every gate, primary input and output u_i. There is an arc e from vertex u_i to vertex u_j if u_i is an input to u_j. A weight $w_e \geq 0$ is associated with every arc, where w_e is the number of flip-flops between u_i and u_j. Retiming of a circuit can be specified as a function $r()$ that assigns an integer $r(u_i)$ to every gate u_i. This integer represents the number of flip-flops to be moved from every fanout of gate u_i to its fanins. The

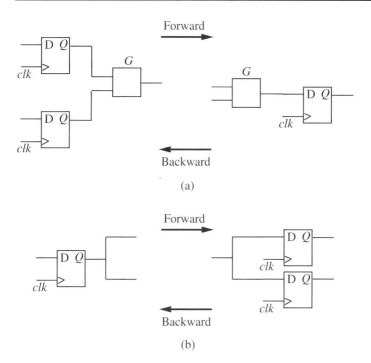

Figure 13.19 Forward/backward retiming across a gate and fanout stem

quantity $r(u_j) - r(u_i) + w_e \geq 0$ is called the **retimed arc weight** and refers to the number of flip-flops on arc e after retiming. A retiming is **legal** if all arcs in the retimed circuit graph have non-negative weights.

The following example illustrates how retiming and resynthesis (which involves duplication of some logic) can reduce the requirements for partial scan.

Example 13.22 Consider the circuit shown in Figure 13.20(a) and its S-graph shown in Figure 13.20(c). In order to eliminate all loops in the S-graph except the self-loops, we need to scan at least two flip-flops. The MFVS of the circuit graph for this circuit consists of gates G_2 and G_3. However, since G_2 is only needed to break a self-loop around flip-flop FF_3, we can just concentrate on G_3. If we could retime the circuit and place a flip-flop at G_3's output c_1 and scan that flip-flop, we would be able to break all the non-self-loops with the scanning of just one flip-flop. However, there does not exist any such retimed configuration since FF_1 cannot be pushed backward to c_1 because of the fanout from gate G_1 to output z, and flip-flops FF_2 and FF_3 cannot be pushed forward through gate G_3 because of the presence of the primary input x_1. Instead, suppose we resynthesize the circuit by duplicating gate G_1 so that there is no path to z blocking the backward movement of the three flip-flops to the desired location c_1. The resynthesized circuit can now be retimed to obtain the circuit shown in Figure 13.20(b). The S-graph of the retimed circuit is shown in Figure 13.20(d). Our aim initially had been to scan the repositioned flip-flop at c_1, i.e., flip-flop FF_6. However,

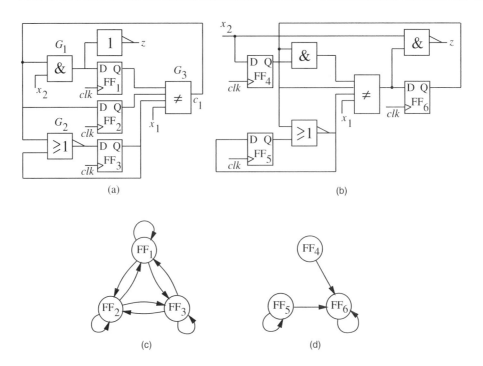

Figure 13.20 Resynthesis and retiming example

since the loops reduce to a self-loop around FF_6, even that is no longer necessary. Thus, the resynthesized/retimed circuit becomes easily testable without any scan. ☐

In Chakradhar and Dey (1994), necessary and sufficient conditions are given which need to be satisfied by the circuit so that legal retiming can position a flip-flop at the output of each gate in the MFVS of the circuit graph. Then resynthesis techniques are used, whenever necessary, to ensure that the modified circuit meets these necessary and sufficient conditions. Experimental results on the benchmarks indicate that the increase in the area of the combinational logic due to resynthesis is usually very small ($< 3\%$ for large circuits) whereas the decrease in the number of flip-flops that need to be scanned is usually two-fold or more.

13.4 Sequential logic synthesis for delay fault testability

In order to synthesize a sequential circuit for PDF or GDF testability, the first step, of course, is to synthesize its combinational logic for full robust testability under the chosen fault model. Then if enhanced scan is used (see Chapter 11; Section 11.6.1.3), the sequential circuit also becomes fully robustly testable. In enhanced scan, an extra latch is used for each state variable so that two bits can be stored for each such variable. This makes it possible to apply all the two-pattern tests derived for the combinational logic. However, enhanced scan requires substantial overheads. Hence, we briefly describe next some of the results which try to avoid the need for complete enhanced scan.

- In Bhatia and Jha (1996), a scheme is given where testability logic is added to the already-synthesized HFRPDFT combinational logic based on the concept of *synthesis for parallel scan* that we discussed earlier. The merged logic is also made HFRPDFT without the need for any scan if the number of primary inputs of the sequential circuit is greater than or equal to the number of flip-flops. If this condition is not met, one has to revert back to enhanced scan.

- In Cheng *et al.* (1993), standard scan methodology is used. For a two-pattern test, all the primary inputs and present state lines are fully controllable in the first time frame. However, only the primary inputs are fully controllable in the second time frame. The required state of the second vector can be produced in two ways: *functional justification* and *scan shifting*, as also explained in Chapter 11 (Section 11.6.1). In **functional justification** the primary inputs and present state lines of the first time frame produce the required state of the second time frame after the clock is activated. In **scan shifting**, assuming the order of the flip-flops in the scan chain is given, the required state of the second time frame is produced using the test mode by shifting one bit into the scan chain through the scan input. If state assignment of the finite-state machine is done in a constrained way, typically very high fault coverage can be obtained for GDFs.

- Another method in Cheng *et al.* (1993) assumes enhanced scan for some state variables and standard scan for others, resulting in a **partial enhanced scan** approach. Then through methods for efficient ordering of scan flip-flops, proper selection of flip-flops to be made enhanced scan, and test vector compaction, relatively small test sets which detect most of the detectable PDFs are obtained.

Summary

- A prime and irredundant two-level circuit is completely testable for all single and multiple SAFs. So is a single-output minimized multi-output two-level circuit, in which each output is prime and irredundant.
- If a prime tree is testable for all single SAFs then it cannot contain any multiple SAF redundancies.
- Single-cube, double-cube and dual expression extraction preserve single SAF testability and test set.
- Algebraic factorization with a constrained use of complement preserves multiple SAF testability and test set.
- Fanout-free and De Morgan's transformations preserve the test set.

- Logic synthesis can be done based on test counts to reduce the size of the single SAF test set.
- Algebraic factorization can also aid random pattern testability by increasing the detection probability of faults in it.
- Redundancy identification and removal methods can be categorized as indirect or direct. Direct methods can be further divided into three categories: static, dynamic and don't care based.
- Tautology checking can be used to check if a two-level circuit is HFRPDFT. Primeness and irredundancy are necessary for a two-level circuit to be HFRPDFT, but not sufficient. Primeness and irredundancy are neither necessary nor sufficient for a two-level circuit to be HFRGDFT or GRGDFT. Primeness and irredundancy are sufficient for a two-level circuit to be VNRPDFT.
- Shannon's decomposition guarantees an HFRPDFT multi-level circuit.
- Simple, extended and targeted algebraic factorization can be used to enhance the HFRPDFT property of a multi-level circuit.
- Some regular structures can be made HFRPDFT by following hierarchical composition rules.
- Delay verification of a circuit may involve simultaneous detection of the presence of more than one PDF.
- In sequential circuits, untestability of a fault does not necessarily imply its redundancy. Uncontrollability and unobservability analysis can be done to identify redundancies in sequential circuit without search.
- Synthesis for parallel scan can be used to make a sequential circuit easily testable. Merging the testability logic with the combinational logic of the sequential circuit allows one to do further delay optimization, if necessary, to minimize the delay overhead.
- Resynthesis and retiming can significantly reduce the amount of partial scan required to make a sequential circuit easily testable.
- A sequential circuit can be made HFRPDFT using enhanced scan by making its combinational logic HFRPDFT. Highly HFRPDFT and HFRGDFT circuits can be obtained using standard scan or partial enhanced scan.

Additional reading

A method to optimize a logic circuit for random pattern testability using recursive learning is given in Chatterjee *et al.* (1995).

In Krasniewski (1991b), pseudoexhaustive testability is targeted during logic synthesis.

A method is given in Ke and Menon (1995b) to make a sequential circuit PDF testable without using scan.

In Cheng (1993), a method is presented to identify and remove sequential redundancies without assuming a reset state.

In Cho *et al.* (1993), a sequential redundancy identification and removal method is given which assumes a global reset state.

In Moondanos and Abraham (1992), logic verification techniques are used to identify sequential redundancies.

In Pomeranz and Reddy (1994), sequential circuits are made fully testable with the help of synchronizing sequences.

Exercises

13.1 Show that a prime and irredundant single-output two-level circuit is fully testable for all single SAFs.

13.2 Give an example of a multi-output two-level circuit which is single SAF testable, but is not multiple SAF testable.

13.3 Give an example of a multi-output two-level circuit in which no output is prime and irredundant, yet the circuit is single SAF testable.

13.4 Show that if each output of a multi-output two-level circuit is prime and irredundant, then any output ordering in Theorem 13.1 is a valid ordering.

13.5 The following switching expressions correspond to multi-level circuits consisting of AND and OR gates, which can be obtained by using the following precedence: ( ), $\cdot$, $+$. Determine if these circuits are prime trees.
(a) $(x_1 + x_2)(x_3 + x_4) + x_1 x_2 (x_1 + x_2) + x_5$
(b) $x_1 (\bar{x}_1 + x_2) + x_3$
(c) $(x_1 + x_2)(x_1 + x_3) + x_4$

13.6 For the restricted prime tree given in Figure 13.2(a), find a test set that will detect all multiple SAFs in it.

13.7 Obtain a multi-level circuit with as few literals as possible using single-cube, double-cube and dual expression extraction starting with a prime and irredundant two-level single-output circuit represented by the expression $z = x_1 x_2 x_4 + x_1 x_2 x_5 + x_3 x_4 + x_3 x_5 + \bar{x}_1 \bar{x}_3 x_6 + \bar{x}_2 \bar{x}_3 x_6$. Obtain a single SAF test set for the two-level circuit and show that it also detects all single SAFs in the multi-level circuit.

13.8 Explain how a logic transformation can preserve multiple SAF testability, but not single SAF testability.

13.9 Prove that the fanout-free and De Morgan logic transformations preserve the test set.

13.10 In the method for deriving the test count of a circuit given in Section 13.1.6, suppose that instead of resetting the sensitive test count values on fanout branches to $(1, 1)$, we propagate these values from the fanout stem to all its

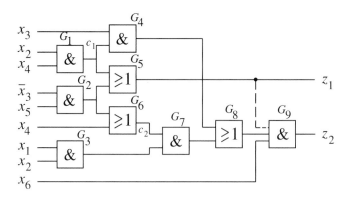

Figure 13.21 Circuit C_1

fanout branches, then would the test count that we obtain be an upper bound on the size of the minimum test set? Prove or give a counter-example.

13.11 For the fanout-free AND–OR circuit that implements $z = x_1x_2 + x_3x_4 + x_5$, compute the test count. Derive a test set of size equal to the test count which detects all single SAFs in it.

13.12 Prove that the test count gives an optimal lower bound on the size of the minimum test set for all single SAFs in a fanout-free circuit.

13.13 In the two-level circuit in Figure 13.6 what is the minimum fault detection probability? What does the minimum fault detection probability become after transformation to the three-level circuit?

13.14 Give an example to show how making an internal line of a circuit observable through an observation point can improve the random pattern testability of the circuit.

13.15 Consider the irredundant circuit C_1 in Figure 13.21 to which a dashed connection is added as shown. Show that this dashed connection is redundant. What are the other faults in the circuit which now become redundant because of the presence of this dashed connection? Simplify the circuit by removing these redundancies and obtain another irredundant circuit which implements the same input/output behavior.

13.16 Obtain a redundant circuit in which the region of a redundant fault at level i also contains the region of a redundant fault at a level greater than i.

13.17 Simplify the AND–OR circuit based on the redundant sum-of-products expression $z = x_1x_2 + x_1x_2x_3 + x_1\bar{x}_2$ using observability and satisfiability don't cares.

13.18 Give an example, other than the one given in Example 13.14, of a prime and irredundant two-level circuit which is not robustly testable for all GDFs.

13.19 Derive an HFRPDFT circuit using Shannon's decomposition for the function $z = x_1x_2 + \bar{x}_1\bar{x}_2 + \bar{x}_3x_4 + x_3\bar{x}_4 + x_1\bar{x}_3$. Obtain a robust test set for this multi-level circuit.

13.20 Find a function of five variables none of whose prime and irredundant implementations is HFRPDFT.

13.21 Show that the constrained use of algebraic factorization, as given in Section
13.2.2, preserves the HFRGDFT property.

13.22 Find the literals in the following expression, paths through which do not have
HFRTs for either rising or falling transitions: $z = x_1x_2x_3x_5 + x_1\bar{x}_3x_4x_5 +$
$x_1x_3\bar{x}_4x_5x_6 + x_1\bar{x}_2x_4\bar{x}_5 + \bar{x}_1x_2x_4\bar{x}_5 + \bar{x}_1x_2\bar{x}_3\bar{x}_4 + x_2\bar{x}_3x_5\bar{x}_6 + x_2x_3x_4x_5 +$
$x_2\bar{x}_3\bar{x}_4\bar{x}_5$. Use targeted algebraic factorization to obtain a three-level HFRPDFT
circuit.

13.23 Assume that each two-input XOR gate in a binary parity tree is implemented as
a two-level OR-AND circuit based on the product-of-sums $(x_1 + x_2)(\bar{x}_1 + \bar{x}_2)$.
Show that the tree is HFRPDFT.

13.24 Show that circuit C_2 given below is delay verifiable, and derive the delay
verification test set for it.

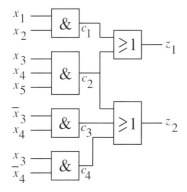

Figure 13.22 Circuit C_2

13.25 In circuit C_3 given below, show that fault c_1 SA1 is 1-cycle redundant, but not
0-cycle redundant.

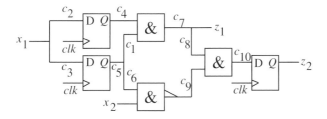

Figure 13.23 Circuit C_3

13.26 Implement a synthesis for parallel scan scheme for a sequential circuit with two inputs, two outputs and three flip-flops, for which the logical equations are as given below.

$$Y_1 = x_1\bar{x}_2 + \bar{x}_1 x_2$$

$$Y_2 = x_1 y_1 + x_1 y_3$$

$$Y_3 = x_1 y_2 + \bar{x}_1 \bar{y}_2$$

$$z_1 = y_1 + y_2$$

$$z_2 = x_2 y_2 + y_1 y_2$$

References

Abramovici, M. and Iyer, M.A. (1992). One-pass redundancy identification and removal. In *Proc. Int. Test Conference*, pp. 807–815.

Bartlett, K.A., Brayton, R.K., Hachtel, G.D., Morrison, C.R., Rudell, R.L., Sangiovanni-Vincentelli, A.L. and Wang, A.R. (1988). Multilevel logic minimization using implicit don't cares. *IEEE Trans. on Computer-Aided Design*, **7** (6), pp. 723–740.

Batek, M.J. and Hayes, J.P. (1992). Test set preserving logic transformations. In *Proc. Design Automation Conference*, pp. 454–459.

Bhatia, S. and Jha, N.K. (1996). Synthesis for parallel scan: applications to partial scan and robust path delay fault testability. *IEEE Trans. on Computer-Aided Design*, **15** (2), pp. 228–243.

Brayton, B., Rudell, R., Sangiovanni-Vincentelli, A.L. and Wang, A.R. (1987). MIS: a multiple-level logic optimization system. *IEEE Trans. on Computer-Aided Design*, **CAD-6** (11), pp. 1062–1081.

Bryan, M.J., Devadas, S. and Keutzer, K. (1990). Testability-preserving circuit transformation. In *Proc. Int. Conference on Computer-Aided Design*, pp. 456–459.

Bryan, M.J., Devadas, S. and Keutzer, K. (1991). Analysis and design of regular structures for robust dynamic fault testability. In *Proc. Int. Symposium on Circuits & Systems*, pp. 1968–1971.

Chakradhar, S.T., Agrawal, V.D. and Rothweiler, S.G. (1993). A transitive closure algorithm for test generation. *IEEE Trans. on Computer-Aided Design*, **12** (7), pp. 1015–1028.

Chakradhar, S.T. and Dey, S. (1994). Resynthesis and retiming for optimum partial scan. In *Proc. Design Automation Conference*, pp. 87–93.

Chatterjee, M., Pradhan, D.K. and Kunz, W. (1995). LOT: Logic optimization with testability – new transformations using recursive learning. In *Proc. Int. Conference on Computer-Aided Design*, pp. 318–325.

Cheng, K.-T., Devadas, S. and Keutzer, K. (1993). Delay-fault test generation and synthesis for testability under a standard scan design methodology. *IEEE Trans. on Computer-Aided Design*, **12** (8), pp. 1217–1231.

Cheng, K.-T. (1993). Redundancy removal for sequential circuits without reset states. *IEEE Trans. on Computer-Aided Design*, **12** (1), pp. 13–24.

Chiang, C.H. and Gupta, S.K. (1994). Random pattern testable logic synthesis. In *Proc. Int. Conference on Computer-Aided Design*, pp. 125–128.

Cho, H., Hachtel, G.D. and Somenzi, F. (1993). Redundancy identification/removal and test generation for sequential circuits using implicit state enumeration. *IEEE Trans. on Computer-Aided Design*, **12** (7), pp. 935–945.

Dandapani, R. and Reddy, S.M. (1974). On the design of logic networks with redundancy and testability considerations. *IEEE Trans. on Computers*, **C-23** (11), pp. 1139–1149.

De, K. and Banerjee, P. (1991). Can test length be reduced during synthesis process. In *Proc. Int. Conference on VLSI Design*, pp. 57–62.

Devadas, S. and Keutzer, K. (1992a). Synthesis of robust delay fault testable circuits: theory. *IEEE Trans. on Computer-Aided Design*, **11** (1), pp. 87–101.

Devadas, S. and Keutzer, K. (1992b). Synthesis of robust delay fault testable circuits: practice. *IEEE Trans. on Computer-Aided Design*, **11** (3), pp. 277–300.

Devadas, S. and Keutzer, K. (1992c). Validatable non-robust delay fault testable circuits via logic synthesis. *IEEE Trans. on Computer-Aided Design*, **11** (12), pp. 1559–1573.

Devadas, S., Keutzer, K. and Malik, S. (1993). A synthesis-based test generation and compaction algorithm for multifaults. *J. of Electronic Testing: Theory and Applications*, **4**, pp. 91–104.

El-Maleh, A. and Rajski, J. (1995). Delay fault testability preservation of the concurrent decomposition and factorization transformations. *IEEE Trans. on Computer-Aided Design*, **14** (5), pp. 582–590.

Entrena, L. and Cheng, K.-T. (1995). Combinational and sequential logic optimization by redundancy addition and removal. *IEEE Trans. on Computer-Aided Design*, **14** (7), pp. 909–916.

Hachtel, G.D., Jacoby, R.M., Keutzer, K. and Morrison, C.R. (1992). On properties of algebraic transformations and the synthesis of multifault-irredundant circuits. *IEEE Trans. on Computer-Aided Design*, **11** (3), pp. 313–321.

Iyer, M.A. and Abramovici, M. (1996). FIRE: a fault-independent combinational redundancy identification algorithm. *IEEE Trans. on Very Large Scale Integration*, **4** (2), pp. 295–301.

Iyer, M.A., Long, D.E. and Abramovici, M. (1996a). Surprises in sequential redundancy identification. In *Proc. European Design & Test Conference*, pp. 88–94.

Iyer, M.A., Long, D.E. and Abramovici, M. (1996b). Identifying sequential redundancies without search. In *Proc. Design Automation Conference*, pp. 457–462.

Jha, N.K., Pomeranz, I., Reddy, S.M. and Miller, R.J. (1992). Synthesis of multi-level combinational circuits for complete robust path delay fault testability. In *Proc. Int. Symposium on Fault-Tolerant Computing*, pp. 280–287.

Jacob, J. and Agrawal, V.D. (1992). Multiple fault detection in two-level multi-output circuits. *J. of Electronic Testing: Theory & Applications*, **3**, pp. 171–173.

Jacoby, R., Moceyunas, P., Cho, H. and Hachtel, G. (1989). New ATPG techniques for logic optimization. In *Proc. Int. Conference on Computer-Aided Design*, pp. 548–551.

Ke, W. and Menon, P.R. (1995a). Synthesis of delay-verifiable combinational circuits. *IEEE Trans. on Computers*, **44** (2), pp. 213–222.

Ke, W. and Menon, P.R. (1995b). Path-delay-fault testable non-scan sequential circuits. *IEEE Trans. on Computer-Aided Design*, **14** (5), pp. 576–582.

Kohavi, I. and Kohavi, Z. (1972). Detection of multiple faults in combinational networks. *IEEE Trans. on Computers*, **C-21** (6), pp. 556–568.

Krasniewski, A. (1991a). Can redundancy enhance testability? In *Proc. Int. Test Conference*, pp. 483–491.

Krasniewski, A. (1991b). Logic synthesis for efficient pseudoexhaustive testability. In *Proc. Design Automation Conference*, pp. 66–72.

Krishnamurthy, B. (1987). A dynamic programming approach to the test point insertion problem. In *Proc. Design Automation Conference*, pp. 695–705.

Kundu, S. and Reddy, S.M. (1988). On the design of robust testable combinational logic circuits. In *Proc. Int. Symposium on Fault-Tolerant Computing*, pp. 220–225.

Kundu, S., Reddy, S.M. and Jha, N.K. (1991). Design of robustly testable combinational logic circuits. *IEEE Trans. on Computer-Aided Design*, **10** (8), pp. 1036–1048.

Kunz, W. and Pradhan, D.K. (1994). Recursive learning: a new implication technique for efficient solutions to CAD problems – test, verification and optimization. *IEEE Trans. on Computer-Aided Design*, **13** (9), pp. 1143–1158.

Lin, C.J. and Reddy, S.M. (1987). On delay fault testing in logic circuits. *IEEE Trans. on Computer-Aided Design*, **CAD-6** (5), pp. 694–703.

Menon, P.R., Ahuja, H. and Harihara, M. (1994). Redundancy identification and removal in combinational circuits. *IEEE Trans. on Computer-Aided Design*, **13** (5), pp. 646–651.

Moondanos, J. and Abraham, J.A. (1992). Sequential redundancy identification using verification techniques. In *Proc. Int. Test Conference*, pp. 197–205.

Pomeranz, I. and Reddy, S.M. (1994). On achieving complete testability of synchronous sequential circuits with synchronizing sequences. In *Proc. Int. Test Conference*, pp. 1007–1016.

Pramanick, A.K. and Reddy, S.M. (1990). On the design of path delay fault testable combinational circuits. In *Proc. Int. Symposium on Fault-Tolerant Computing*, pp. 374–381.

Rajski, J. and Vasudevamurthy, J. (1992). The testability-preserving concurrent decomposition and factorization of Boolean expressions. *IEEE Trans. on Computer-Aided Design*, **11** (6), pp. 778–793.

Reddy, S.M., Lin, C.J. and Patil, S. (1987). An automatic test pattern generator for the detection of path delay faults. In *Proc. Int. Conference on Computer-Aided Design*, pp. 284–287.

Reddy, S.M. and Dandapani, R. (1987). Scan design using standard flip-flops. *IEEE Design & Test*, **4** (1), pp. 52–54.

Saldanha, A., Wang, A.R., Brayton, R.K. and Sangiovanni-Vincentelli, A.L. (1989). Multilevel logic simplification using don't cares and filters. In *Proc. Design Automation Conference*, pp. 277–282.

Schertz, D. and Metze, G. (1972). A new representation for faults in combinational digital circuits. *IEEE Trans. on Computers*, **C-21** (8), pp. 858–866.

Schulz, M.H. and Auth, E. (1988). Advanced automatic test pattern generation and redundancy identification techniques. In *Proc. Int. Symposium on Fault-Tolerant Computing*, pp. 30–35.

Touba, N.A. and McCluskey, E.J. (1994). Automated logic synthesis of random pattern testable circuits. In *Proc. Int. Test Conference*, pp. 174–183.

Vinnakota, B. and Jha, N.K. (1992). Synthesis of sequential circuits for parallel scan. In *Proc. The European Conf. on Design Automation*, pp. 366–370.

14 Memory testing

by Ad van de Goor

A modern workstation may have as much as 256 Mbytes of DRAM memory. In terms of equivalent transistors (assuming one transistor per bit) this amounts to 2×10^9 transistors which is about two orders of magnitude more than the number of transistors used in the rest of the system. Given the importance of system test, memory testing is, therefore, very important.

We start this chapter with a motivation for efficient memory tests, based on the allowable test cost as a function of the number of bits per chip. Thereafter, we give a model of a memory chip, consisting of a functional model and an electrical model.

Because of the nature of memories, which is very different from combinational logic, we define a new set of functional faults (of which the stuck-at faults are a subset) for the different blocks of the functional model.

We describe a set of four traditional tests, which have been used extensively in the past, together with their fault coverage. We next describe march tests, which are more efficient than the traditional tests, together with proofs for completeness and irredundancy.

Finally, we introduce the concept of pseudo-random memory tests, which are well suited for built-in self-test (BIST), together with a computation of the test length as a function of the escape probability.

14.1 Motivation for testing memories

Ongoing developments in semiconductor memories result in a continually increasing density of memory chips (Inoue *et al.*, 1993). Historically, the number of bits per chip has quadrupled roughly every 3.1 (or π) years. One-megabit dynamic random-access memory (DRAM) chips have been available since 1985, 64 Mbit DRAM chips since 1994, and 1 Gbit DRAM chips became available in 2000. For static random-access memory (SRAM) chips, the development follows the same exponential curve albeit with a lag of about one generation (which is about π years) due to the larger number of devices required for SRAM cells.

The exponential increase in the number of bits per chip has caused the area per memory cell and the price to decrease exponentially as well. The consequence of the dramatic decrease in cell area, by a factor of 90 when comparing the technologies of 4 Kbit and 1 Mbit chips, is that the charge stored in the capacitor of a DRAM cell is decreasing rapidly, causing the cell to be more susceptible to disturbances in the manufacturing process (differences in the capacitance or leakage current) and to disturbances due to use (noise, crosstalk, etc.). Currently, the 64 Mbit DRAM chips use a 20 femtoFarad (fF), i.e., 20×10^{-15} Farad, capacitor. In addition, the memory cells are being placed more closely together, which makes them more sensitive to influences of neighboring cells and increases the likelihood of disturbances due to noise on the address and data lines.

For economic reasons, the test cost per chip (which is directly related to the test time) cannot be allowed to increase significantly. At the same time, the number of bits per chip is increasing exponentially and the sensitivity to faults is increasing while the faults become more complex. For example, tests for detecting faults which are dependent on the neighboring cells are more complex and require a longer execution time than those which do not.

The memory tests used in the early days, typically before 1980, will be called *traditional tests*. They require test times of $O(n \cdot \log_2(n))$ or even $O(n^2)$, where $n = 2^N \cdot B$ is the number of bits in the chip; N represents the number of address bits and B represents the number of bits per word. Newer generations of gigabit chips would require test times which are not economically feasible. Table 14.1 lists the required test time in seconds, hours or years, assuming a memory cycle time of 100 ns, as a function of the algorithm complexity and the memory size.

It is because of the above problems that memory testing has been the subject of a large research effort. The result is a new generation of tests which are based on fault models such that for each test it can be proven that it finds all faults of a particular model. By carefully selecting the fault models, the execution times of many of the new tests become $O(n)$.

14.2 Modeling memory chips

This section first presents a functional model of a memory chip. Thereafter, the electrical model for each of the blocks of the functional model is given. Finally, a simplified (reduced) version of the functional model, as needed for testing purposes, is described.

14.2.1 Functional RAM chip model

The memory device of primary concern to memory testing is the RAM chip. Therefore, a model for a RAM chip, and not a memory chip in general, will be developed (van de Goor and Verruijt, 1990). This RAM chip model can easily be transformed into a read-only memory (ROM), erasable programmable read-only memory (EPROM)

Table 14.1. *Test time as a function of memory size*

n	n	Algorithm complexity		
		$n \cdot \log_2 n$	$n^{3/2}$	n^2
1K	0.0001 s	0.001 s	0.0033 s	0.105 s
4K	0.0004 s	0.0049 s	0.026 s	1.7 s
16K	0.0016 s	0.023 s	0.21 s	27 s
64K	0.0066 s	0.1 s	1.7 s	410 s
256K	0.026 s	0.47 s	13 s	1.8 h
1M	0.105 s	2.1 s	107 s	30.5 h
4M	0.42 s	9.2 s	859 s	488 h
16M	1.68 s	40.3 s	1.9 h	0.89 years
64M	6.71 s	174 s	15.3 h	14.3 years
256M	26.8 s	752 s	122 h	228 years
1G	107 s	3221 s	977 h	3655 years

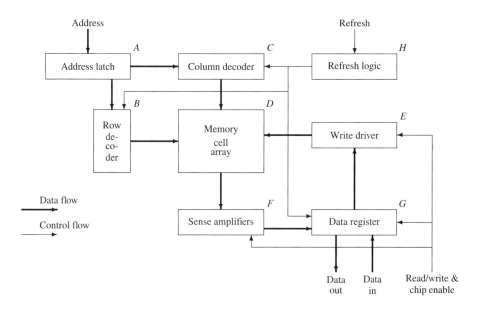

Figure 14.1 Functional model (data path and control) of a DRAM chip

or electrically erasable programmable read-only memory (EEPROM) chip model by
leaving out some of the blocks.

Figure 14.1 shows a general model of a DRAM chip; for an SRAM chip the refresh
logic would be omitted. Before discussing Figure 14.1 in detail, one should be aware
of the difference in the external and internal organization of the memory cell array D.
A 1 Mbit chip may *logically* (as seen from the outside) be organized as 1M addresses
of words which are one-bit wide; i.e., $n =$ 1M and $B = 1$. *Physically* (inside the chip),
the memory cells (each of which contains one bit of data) are organized as a matrix

or a number of matrices. For example, the physical organization could be a matrix of 1K·1K bits (1K rows and 1K bits per row), or four matrices of 512·512 bits, or eight matrices of 128·1K bits, etc. For every read and write operation, a 1 Kbit (or smaller) *row* is read or written internally, while only one bit is made visible to the outside world.

Block A, the address latch, contains the address. The high-order bits of the address are connected to the *row decoder*, B, which selects a row in the memory cell array, D. The low-order address bits go to the *column decoder*, C, which selects the required columns. The number of columns selected depends on B: the number of bits accessed during a read or write operation.

When the read/write line indicates a read operation, the contents of the selected cells in the memory cell array are amplified by the sense amplifiers, F, loaded into the data register, G, and presented on the data-out line(s). During a write operation, the data on the data-in line(s) are loaded into the data register and written into the memory cell array through the write driver, E. Usually, the data-in and data-out lines are combined to form bidirectional data lines, thus reducing the number of pins of the chip.

The chip-enable line enables the data register and, together with the read/write line, the write driver. When the refresh line is activated, the column decoder selects all the columns; the row decoder selects the row that is indicated by the address latch; all bits in the selected row are read and rewritten (*refreshed*) simultaneously. During refresh, the refresh logic, H, disables the data register.

14.2.2   Electrical RAM chip model

The most important blocks of the functional model of Figure 14.1 will be opened next, such that the electrical properties of those blocks become visible. This is important for a later explanation of the reduced functional faults, which have their origin at the electrical or geometrical level of the system. This section covers the following subjects: memory cells (these are used to construct the memory cell array, D, of Figure 14.1), decoders (blocks B and C of Figure 14.1), and the read/write circuitry (blocks E and F of Figure 14.1).

14.2.2.1 Memory cells

A fully static RAM cell (**SRAM cell**) is a bistable circuit, capable of being driven into one of two states. After removing the driving stimulus, the circuit retains its state. Next, the six-device (six-transistor) SRAM cell, and the SRAM cell with polysilicon load devices is discussed. Thereafter, the single-device DRAM cell is described.

A six-device SRAM cell is shown in Figure 14.2(a). It consists of the enhancement mode nMOS transistors Q_1, Q_2, Q_5 and Q_6; and the depletion mode nMOS transistors Q_3 and Q_4. Transistor Q_1 forms an inverter together with depletion load device Q_3, this inverter is cross-coupled with the inverter formed by Q_2 and Q_4, thus forming a

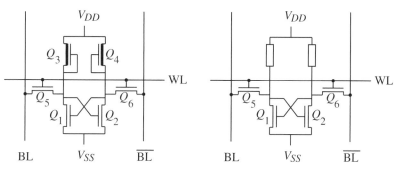

(a) Six-device SRAM cell
   using depletion loads

(b) SRAM cell using
   polysilicon load devices

Figure 14.2 SRAM cell structures

latch. This latch can be accessed, for read and write operations, via the *pass transistors* Q_5 and Q_6.

The addressing of the cell is done using a two-dimensional addressing scheme consisting of a row and a column address. The row decoder, B, of Figure 14.1 allows only one row of cells to be selected at a time by activating the *word line* (WL) of that particular row (note that within the chip a memory word is synonymous with a row). WL is connected to all gates of the pass transistors of all cells in that row, and only one WL should be active at a time. The selection of a particular cell in a row is done under the control of the column decoder, C, of Figure 14.1; it activates the set of complementary *bit lines* (BLs) of that particular cell.

Data can be written into the memory by driving WL high and driving lines BL and $\overline{BL}$ with data with complementary values. Because the bit lines are driven with more force than the force with which the cell retains its information (the transistors driving the lines BL and $\overline{BL}$ are more powerful, i.e., they are larger than transistors Q_1 and Q_2), the cell will be forced to the state presented on BL and $\overline{BL}$. In the case of a read operation, a particular row is selected by activating the corresponding WL. The contents of the cells in a row, accessed by the activated WL, are passed to the corresponding sense amplifiers via the BL and $\overline{BL}$ lines. The data register, D, of Figure 14.1 is loaded by selecting the outputs of the desired sense amplifiers under the control of the column decoder, C.

The SRAM cell with polysilicon load devices is shown in Figure 14.2(b). The depletion loads Q_3 and Q_4 of Figure 14.2(a) have been replaced with resistors in order to decrease the silicon area and the power dissipation of the cell. The resistors are made from polysilicon with a high resistivity (they may have a value of 100 GΩ), causing an asymmetry in the logic 1 and 0 drive power of the latch.

In order to be able to read a cell, the relatively long bit lines with their large capacitance have to be driven. Because of the high ohmic value of the load devices, charging this large capacitance would take a long time, causing a read operation to

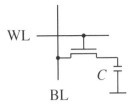

Figure 14.3 Single-device DRAM cell

be very slow. The way to solve this problem is by **precharging** the bit lines BL and $\overline{BL}$ to logic 1 such that a read operation means discharging the charge stored in the capacitance of one of the bit lines via the low ohmic path of a conducting transistor.

The DRAM cell: The most common RAM cell is the **DRAM cell** because of its small size and low power dissipation. Rather than in the state of a latch, the logic value to be stored is represented in terms of the charge stored in a capacitor which is located across the source and gate of a transistor. Because of the reverse-biased junction leakage current of this transistor, the amount of stored charge, which is of the order of 10^6 electrons, will decrease with time (it will leak away). In order to guarantee that a read operation will still produce the logic value originally stored in the cell, the lost charge (due to leakage) has to be replenished periodically, this is called a **refresh operation**. During this operation, which has to be performed typically every 4 ms, the data in an entire row of cells are read, passed to the sense amplifiers to determine the logic value stored in each cell, and thereafter rewritten such that the original charge is restored to all the cells of that row.

The single-device DRAM cell is the most commonly used DRAM cell design (see Figure 14.3). It provides for a high-density, low-cost-per-bit memory cell with a reasonable performance (Rideout, 1979). The single-device cell consists of an enhancement mode transistor and a separate capacitor, C, which may or may not be charged, depending upon the logic value to be stored. A read operation is performed by precharging the bit line, BL, to the threshold level of the sense amplifier; this is a level between logic 0 and 1. Thereafter the word line, WL, is driven high such that the charge from capacitor C is transferred to BL. This causes a voltage swing on BL, whose magnitude is determined by the ratio of the capacitance of C and the capacitance of BL. This ratio is rather small such that very sensitive sense amplifiers are required. The transfer of charge causes the read operation to be *destructive*, requiring a write back in order to restore the original charge. A write operation consists of forcing BL to the desired level and driving WL high.

14.2.2.2 Decoders

Decoders are used to access a particular cell or a group of cells in the memory cell array. A 1 Mbit chip, externally organized as 1M addresses of 1-bit words, would

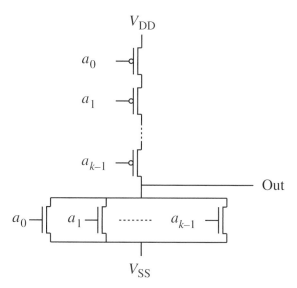

Figure 14.4 Static row decoder circuit

require 1M word lines. The silicon area required for the decoder and for the word lines would be prohibitive. Therefore, two-dimensional addressing schemes are used within the chip, requiring a row decoder with WLs and a column decoder with BLs; the size of each decoder and the area required for the lines is proportional to $\sqrt{n}$, where n is the total number of bits in the chip. For a 1 Mbit chip, this would reduce the required decoder area and the area required for the lines from n to $2\sqrt{n}$, which is a reduction in area and capacitance by a factor of $0.5\sqrt{n} = 0.5 \cdot 1024 = 512$.

An implementation of a simple, *static row decoder* consists of a NOR gate, as shown in Figure 14.4. The inputs to the decoder consist of the address bits a_0 through a_{k-1} or their complements; output (Out) is the WL. All inputs a_0 through a_{k-1} have to be low in order for the output to be high. When a particular address has to be selected, e.g., address 50, which is 110010 in binary, the inputs to gates a_0 through a_5 should be as follows: $\bar{a}_0\bar{a}_1a_2a_3\bar{a}_4a_5$. For a large number of address lines, this decoder becomes too slow because of the long series of pMOS transistors in the V_{DD} path. In this case, dynamic decoders (Mazumder, 1988) are used.

14.2.2.3 Read/write circuitry

The write circuitry of a RAM cell is rather simple (see Figure 14.5). The data to be written on 'Data in' are presented on the BL and $\overline{\text{BL}}$ lines under the control of the 'Write' clock.

The read circuitry may be very simple for ROM-type memories and small SRAMs. Figure 14.6(a) shows a simple read circuit, consisting of an inverter. Figure 14.6(b) shows a differential amplifier which can sense small differences between lines BL

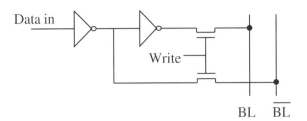

Figure 14.5 RAM write circuit

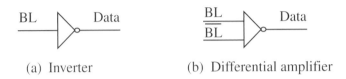

(a) Inverter   (b) Differential amplifier

Figure 14.6 Simple read circuits

and $\overline{BL}$ and allows for fast switching; read circuits are therefore also called **sense amplifiers**. The amplifier is enabled by a column select line, CL.

14.2.3 The reduced functional model

During chip testing, one is not interested in *locating* a fault because a chip cannot be repaired.[1] One is only interested in *detecting* a fault. For this reason, the model of Figure 14.1 can be simplified without loss of information. For functional testing, a model is used that contains only three blocks: the address decoder, the memory cell array and the read/write logic (Thatte and Abraham, 1977; Nair *et al.*, 1978) (see Figure 14.7). The address latch A, row decoder B and the column decoder C of Figure 14.1 are combined as the address decoder because they all concern addressing the right cell or word. The write driver E, sense amplifier F and data register G are combined into the read/write logic because they all concern the transport of data from and to the memory cell array. The refresh part H is left out, because faults in H belong to the class of dynamic faults which are time-dependent (van de Goor, 1998).

14.3 Reduced functional faults

This section describes the faults which can occur in the reduced functional model described in Section 14.2.3. It describes the faults in the memory cell array, followed

---

[1] A chip manufacturer could, however, be interested in locating faults, because of the availability of spare logic with which the chip can be repaired before it is sealed in its package, or for collecting statistics to determine the weak spots within the chip. Some chips can even be repaired within their package. This is done with electrical fusing, which requires special equipment.

Address
⇓
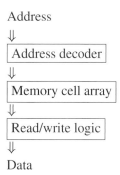
Address decoder
⇓
Memory cell array
⇓
Read/write logic
⇓
Data

Figure 14.7 Reduced functional model

by a subsection which illustrates the validity of the faults. Thereafter, a notation for *march tests* is introduced which is subsequently used to describe the possible effect of combinations of memory cell array faults. Next, faults which may occur in the address decoder are described, followed by a way of mapping read/write logic and address decoder faults to the memory cell array.

14.3.1 Memory cell array faults

This section gives a formal definition for each of the faults which can occur in the memory cell array of Figure 14.7, together with test requirements for detecting and locating the fault. The faults can be classified as: faults in which a single cell is involved, faults in which two cells are involved, and faults involving k cells. To simplify the discussion, the following notation will be used for describing the faults:

$\langle \cdots \rangle$ denotes a particular fault; '...' describes the fault.

$\langle S/F \rangle$ denotes a fault in a single cell.
   S describes the condition for sensitizing the fault; $S \in \{0, 1, \uparrow, \downarrow\}$.
   The values 0 and 1 represent sensitizing states, $\uparrow$ represents an up-transition ($0 \to 1$) write operation, and $\downarrow$ represents a down-transition ($1 \to 0$) write operation. A *transition write operation* can be denoted as a w$\bar{x}$ operation to a cell containing the value x; $x \in \{0, 1\}$. S_T denotes the fact that the effect of sensitizing the fault appears after time T.
   F describes the value of the faulty cell; $F \in \{0, 1, \uparrow, \downarrow, \updownarrow\}$.
   The values 0 and 1 represent faulty values of the cell, $\downarrow$ represents a faulty down-transition (i.e., a $1 \to 0$ change in value), $\uparrow$ represents a faulty up-transition, and $\updownarrow$ represents a $0 \to 1$ or $1 \to 0$ transition.

$\langle S_1, \ldots, S_{m-1}; F \rangle$ denotes a fault involving m cells.

$S_1, \ldots, S_{m-1}$ describe the condition for sensitizing the fault; $S_i \in \{0, 1, \uparrow, \downarrow\}$ for $1 \le i \le m - 1$.

14.3.1.1 Single-cell faults

The memory cell array faults involving only a single cell are: the stuck-at fault, stuck-open fault, transition fault and data retention fault.

Stuck-at fault: The **stuck-at fault (SAF)** can be defined as follows: The logic value of a stuck-at cell or line is always 0 (an SA0 fault) or 1 (an SA1 fault), i.e., it is always in state 0 or in state 1 and cannot be changed to the opposite state. The notation for an SA0 fault is $\langle \forall/0 \rangle$ (Note: $\forall$ denotes any operation; $\forall \in \{0, 1, \uparrow, \downarrow\}$); and for an SA1 fault $\langle \forall/1 \rangle$. A test that has to *detect* all SAFs must satisfy the following requirement: *from each cell, a 0 and a 1 must be read.*

A state diagram for a good memory cell is shown in Figure 14.8(a). The cell contains the logic value '0' in state 0 (denoted by the node labeled S_0), and a '1' value in S_1. When a write '1' (denoted by the arc with label 'w1') operation takes place in S_0, a transition is made to S_1; when a 'w0' operation takes place in S_0, the cell remains in S_0. Figure 14.8(b) shows the state diagram for a memory cell with an SA0 fault, while Figure 14.8(c) shows the state diagram for an SA1 fault; regardless of the type of write operation (a 'w0' or a 'w1' operation) the cell remains in the same state.

Stuck-open fault: A **stuck-open fault (SOpF)** means that a cell cannot be accessed (Dekker *et al.*, 1990), e.g., due to an open WL (see Figure 14.2). When a read operation is performed on a cell, the differential sense amplifier has to sense a voltage difference between the bit lines (BL and $\overline{BL}$) of that cell. In the case of an SOpF, both bit lines will have the same voltage level. Consequently, the output value produced by the sense amplifier (SA) depends on the way it is implemented:

- The operation of the SA is *transparent* to SOpFs.
  When the SA has only a single input (Figure 14.6(a)), an SOpF will always produce a fixed output value. The SOpF will appear as an SAF.
- The operation of the SA is *non-transparent* to SOpFs.
  The implementation of most differential sense amplifiers (see Figure 14.6(b)) is based on a latch. An SOpF may have the effect that the latch is not updated because of the negligible difference between the voltage levels of BL and $\overline{BL}$. A read operation will then produce the value of the last, i.e., previous, read operation. The notation for this fault is $\langle \forall/L \rangle$, where L denotes the value produced by the last read operation.

Transition fault: A special case of the SAF is the **transition fault (TF)**. It is defined as follows: a cell or line which fails to undergo a $0 \rightarrow 1$ transition when it is

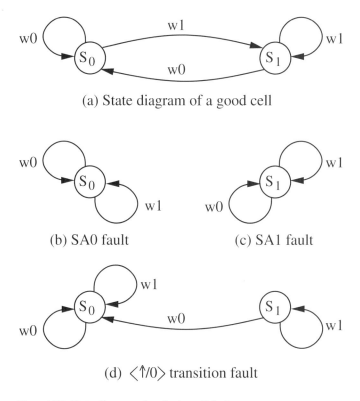

(a) State diagram of a good cell

(b) SA0 fault   (c) SA1 fault

(d) $\langle\uparrow/0\rangle$ transition fault

Figure 14.8 State diagrams for single-cell faults

written is said to contain an up-transition fault; similarly, a down-transition fault is the impossibility of making a $1 \rightarrow 0$ transition. The notation for the up-TF is $\langle\uparrow/0\rangle$, and for the down-TF $\langle\downarrow/1\rangle$.

If a cell has a $\langle\uparrow/0\rangle$ TF and is in state 0 upon power-on, it is effectively an SAx cell (where $x = 0$); when it is in state 1 upon power-on, it can undergo one $1 \rightarrow 0$ transition. The same argument can be used for $\langle\downarrow/1\rangle$ TFs. A test that has to *detect* all TFs, should satisfy the following requirement: *each cell must undergo a $\uparrow$ transition (state of the cell goes from 0 to 1), and a $\downarrow$ transition, and be read after each transition before undergoing any further transitions.*

The state diagram for a memory with a $\langle\uparrow/0\rangle$ TF is shown in Figure 14.8(d). The cell may be in S_1, e.g., after power-on; once the cell has entered S_0 it cannot leave that state any more. The fault is sensitized by a 'w1' operation while in S_0; a successive read operation will detect the fault because a '0' will be read while the expected value is a '1'.

Data retention fault: A **data retention fault (DRF)** occurs when a cell fails to retain its logic value after some period of time (Dekker *et al.*, 1990). A DRF may be caused by a broken (open) pull-up device within an SRAM cell (see Figure 14.2).

Leakage currents then will cause the node with the broken pull-up device to lose its charge, causing a loss of information if a logic value was stored in the cell requiring a high voltage at the open node. Two different DRFs can be recognized (they may be present simultaneously): $\langle 1_T / \downarrow \rangle$ and $\langle 0_T / \uparrow \rangle$. When both are present in one cell, the cell behaves as if it contains an SOpF.

14.3.1.2 Faults involving two cells

The memory cell array faults involving two cells are called **2-coupling faults**, usually simply referred to as coupling faults **(CFs)**. The state of, or an operation applied to, one cell (called the **aggressor cell (a-cell)**), sensitizes a fault in another cell, called the **victim cell (v-cell)**. The following types of CFs can be identified: inversion coupling faults (CFins), idempotent coupling faults (CFids), and state coupling faults (CFsts).

CFins and CFids are assumed to be sensitized by a **transition write operation** applied to the a-cell, i.e., a $w\bar{x}$ operation to an a-cell containing the value x, $x \in \{0, 1\}$. A **non-transition write operation** is a wx operation to an a-cell containing the value x; it will *not* sensitize a CFin or CFid. This assumption has been made for DRAMs in order to allow for read operations to be fault-free; in DRAMs a read operation is destructive and will, therefore, be followed by a non-transition write operation.

The CF caused by a transition write operation, used in this book and in Nair *et al.* (1978), Suk and Reddy (1981) and Marinescu (1982), involves two cells. Note that all definitions talk about 'single-way' faults, i.e., the presence of a CF whereby cell C_i is coupled to cell C_j does not imply one or more other CFs whereby cell C_j is coupled to cell C_i. However, such faults are allowed because an arbitrary number of CFs is allowed.

Inversion coupling fault: An **inversion coupling fault (CFin)** is defined as follows: an $\uparrow$ (or $\downarrow$) transition in the a-cell inverts the contents of the v-cell. The notation, C_v is $\langle \uparrow; \updownarrow \rangle$ coupled to C_a, means that an $\uparrow$ transition in the a-cell causes an $\uparrow$ or a $\downarrow$ transition (i.e., a x to $\bar{x}$ transition) in the v-cell. The two possible CFins are: $\langle \uparrow; \updownarrow \rangle$ and $\langle \downarrow; \updownarrow \rangle$. A test that has to *detect* all CFins should satisfy the following requirement: *for all v-cells, each cell should be read after a series of possible CFins may have occurred (by writing into the a-cells), with the condition that the number of transitions in the v-cell is odd (i.e., the CFins may not mask each other).*

Idempotent coupling fault: An **idempotent coupling fault (CFid)** is defined as follows: an $\uparrow$ (or $\downarrow$) transition in the a-cell forces the contents of the v-cell to a certain value, 0 or 1. The notation, C_v is $\langle \uparrow; \downarrow \rangle$ coupled to C_a, means that an $\uparrow$ transition in the a-cell will cause a $\downarrow$ transition in the v-cell. The four possible CFids are: $\langle \uparrow; \downarrow \rangle$, $\langle \uparrow; \uparrow \rangle$, $\langle \downarrow; \downarrow \rangle$, and $\langle \downarrow; \uparrow \rangle$. A test that has to *detect* all CFids should satisfy the following requirement: *for all v-cells, each cell should be read after a series of possible CFids* *may have occurred (by writing into the a-cells), in such a way that the sensitized CFids do not mask each other.*

State coupling faults: A **state coupling fault (CFst)** (Dekker *et al.*, 1988) is defined as follows: a v-cell is forced to a certain value, x, only if the a-cell is in a given state, y. Four different CFsts can be distinguished: $\langle 0; 0 \rangle$, $\langle 0; 1 \rangle$, $\langle 1; 0 \rangle$, and $\langle 1; 1 \rangle$. CFsts may involve any number of cells or lines and are caused by a logic level rather than a transition write operation.

14.3.1.3 Faults involving k cells

A **k-coupling fault** (Nair *et al.*, 1978) uses the same two cells as the 2-coupling fault and, in addition, only allows the fault to occur when the other $k - 2$ cells are in a certain state, i.e., contain a certain data pattern. This fault model is, therefore, also called a **pattern sensitive fault (PSF)**. Tests for k-coupling faults are very complicated if no restriction is placed on the location of the k cells (Nair *et al.*, 1978; Papachristou and Saghal, 1985; van de Goor, 1998).

One constrained k-coupling fault is the **neighborhood pattern sensitive fault (NPSF)** where the v-cell has $k - 1$ physical neighbors, one of which is the a-cell. The NPSF fault is important for DRAMs due to leakage currents between physically adjacent cells (van de Goor, 1998).

14.3.2 Validity of the fault models

In order to get some appreciation for the functional fault models described in Section 14.3.1, the following study has been included; it shows the relative frequencies of occurrence of these faults for a particular SRAM chip. Because most SRAM chips are based on a similar technology; the results may be generalized to SRAM chips in general.

Dekker *et al.* (1990) has investigated the effect of spot defects, of different sizes, on 16 Kbit SRAM chips manufactured using a 1.5 μm technology. A technique called **inductive fault analysis (IFA)** (Maly, 1985; Shen *et al.*, 1985) has been used for this purpose. As mentioned in Chapter 2 (Section 2.3), IFA is a systematic procedure to predict the faults in an integrated circuit by injecting spot defects in the simulated geometric representation of the circuit. The spots represent inappropriate, extra, or missing material. The faults caused by these spot defects are then translated into faults at the electrical level, which in turn are translated into functional faults.

When the IFA technique was applied to the memory cell array of a 16 Kbit SRAM chip, the fault distributions shown in Table 14.2 resulted. SAFs form about 50% of all faults, CFins did not occur, while TFs and CFids only occur with large spot defects. For additional reading, see also (Naik *et al.*, 1993).

Table 14.2. *Functional faults caused by spot defects*

Fault	Spot size	
class	< 2 μm	< 9 μm
SAF	51.3%	49.8%
SOpF	21.0%	11.9%
TF	0.0%	7.0%
DRF	17.8%	14.8%
CFid	0.0%	3.3%
CFst	9.9%	13.2%
Total	100.0%	100.0%

14.3.3 Notation for march tests

A **march test** consists of a finite sequence of march elements (Suk and Reddy, 1981). A **march element** is a finite sequence of operations applied to every cell in the memory, before proceeding to the next cell. This can be done in either one of two address orders: an increasing ($\Uparrow$) address order (e.g., from address 0 to address $n-1$), or a decreasing ($\Downarrow$) address order which is the opposite of the $\Uparrow$ address order. When the address order is irrelevant the symbol $\Updownarrow$ will be used. An **operation** can consist of: writing a 0 into a cell (w0), writing a 1 into a cell (w1), reading a cell with expected value 0 (r0), and reading a cell with expected value 1 (r1).

The algorithm of the march test $\{\Updownarrow(w0);\ \Uparrow(r0,w1);\ \Downarrow(r1,w0)\}$, consisting of the march elements M_0: $\Updownarrow(w0)$, M_1: $\Uparrow(r0,w1)$ and M_2: $\Downarrow(r1,w0)$, is shown in Figure 14.9, in order to show the notation and concepts. Note that *all* operations of a march element are performed at a certain address, before proceeding to the next address.

Although examples will be used in which the $\Uparrow$ address order goes from address 0 to $n-1$, this is not strictly necessary. The only requirement is that the address orders $\Uparrow$ and $\Downarrow$ be each other's inverse. For example, when the $\Uparrow$ address order is chosen for some reason to be: 0, 1, 7, 6, 2, 5, 3, 4; the $\Downarrow$ address order has to be: 4, 3, 5, 2, 6, 7, 1, 0.

14.3.4 Combinations of memory cell array faults

When multiple faults occur, they can be unlinked or linked. A fault is **linked** when it may influence the behavior of *other* faults (Papachristou and Saghal, 1985); a fault is **unlinked** when that fault does not influence the behavior of other faults. The detection of linked faults makes memory tests more complex because of the possibility of *masking*; see Example 14.1, where two CFids mask each other, and Example 14.2, where a TF is masked by a CFid.

Example 14.1 Consider the two CFids shown in Figure 14.10. The first fault is that

M_0:   **for** $i := 0$ **to** $n - 1$ **do**

       begin

            $A[i] := 0;$

       end;

M_1:   **for** $i := 0$ **to** $n - 1$ **do**

       begin

            read $A[i]$; { Check that 0 is read. }

            $A[i] := 1;$

       end;

M_2:   **for** $i := n - 1$ **downto** 0 **do**

       begin

            read $A[i]$; { Check that 1 is read. }

            $A[i] := 0;$

       end;

Figure 14.9 March test $\{\updownarrow(w0); \Uparrow(r0,w1); \Downarrow(r1,w0)\}$

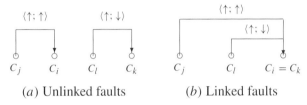

(*a*) Unlinked faults   (*b*) Linked faults

Figure 14.10 Masking of coupling faults

cell C_i is $\langle\uparrow; \uparrow\rangle$ coupled to cell C_j, the second fault is that cell k is $\langle\uparrow; \downarrow\rangle$ coupled to cell C_l. The march test $\{\updownarrow(w0); \Uparrow(r0,w1); \updownarrow(w0,w1); \updownarrow(r1)\}$ will detect both faults if $i \neq k$ (see Figure 14.10(a)). The CFid $\langle\uparrow; \uparrow\rangle$ will be detected by the 'r0' operation of march element $\Uparrow(r0,w1)$, when operating on cell C_i. The CFid $\langle\uparrow; \downarrow\rangle$ will be detected by the 'r1' operation of the last march element, when it operates on cell C_k. However, this test will not detect the combination of faults which occurs when $i = k$ (see Figure 14.10(b)). The 'link' between the faults (caused by the fact that the coupled cells are the same) causes the above test not to find *either* fault; this effect is called **masking**. □

Example 14.2 Figure 14.11 shows a CFid where cell C_i is $\langle\uparrow; \uparrow\rangle$ coupled to cell C_j, and a $\langle\uparrow /0\rangle$ TF in cell C_l. The march test $\{\Uparrow(w0); \Uparrow(w1, r1)\}$ will detect the TF when it is not linked with the CF (see Figure 14.11(a), $i \neq l$).

When $i = l$ (see Figure 14.11(b)), the faults are linked and the march test will not detect the TF. When the march element $\Uparrow(w1,r1)$ operates on cell j, the 'w1' operation will sensitize the CFid $\langle\uparrow;\uparrow\rangle$ in cell l, such that cell l will be forced to a 1. When the march element operates on cell l later on, the TF will not be detected because the 'r1'

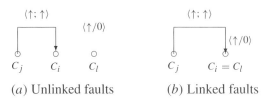

(a) Unlinked faults   (b) Linked faults

Figure 14.11 A TF masked by a CFid

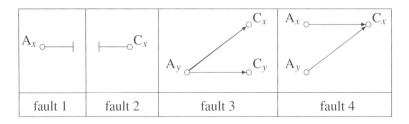

| fault 1 | fault 2 | fault 3 | fault 4 |

Figure 14.12 Address decoder faults

operation will read a 1, which is the expected value. This value is caused by the CFid $\langle\uparrow;\uparrow\rangle$ which masks the TF. $\square$

14.3.5 Address decoder faults

Address decoder faults (AFs) concern faults in the address decoder. It is assumed as in Thatte and Abraham (1977) and Nair *et al.* (1978) that the address decoder is not changed into sequential logic (due to faults within it) and that faults in the decoder logic are the same during read and write operations. These assumptions are essential for the validity of the proof and seem quite reasonable. Only bit-oriented memories, where only one bit of information can be accessed via one address (i.e., $B = 1$), is considered; the extension to word-oriented memories, where more than one bit of information can be accessed via one address, is given in van de Goor (1998).

Functional faults within the address decoder result in the four AFs shown in Figure 14.12:

Fault 1: With a certain address, *no cell* is accessed.

Fault 2: There is *no* address with which this cell can be accessed. A certain cell is *never accessed*.

Fault 3: With a certain address, *multiple cells* are accessed simultaneously.

Fault 4: A certain cell can be accessed with *multiple addresses*.

Because there are as many cells as addresses, none of the above faults can stand alone. When fault 1 occurs, either fault 2 or 3 must also occur. With fault 2, at least fault 1 or 4 must occur; with fault 3, at least fault 1 or 4; with fault 4, at least fault 2 or 3. These four fault combinations are shown in Figure 14.13. In the rest of this chapter, faults A to D will refer to these different combinations of AFs.

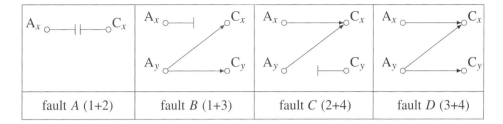

Figure 14.13 Combinations of address decoder faults

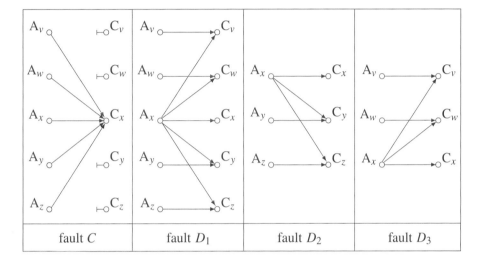

Figure 14.14 Extended faults C and D

Faults A and B are inherently unlinked: it is not possible to mask the fault when one reads A_x. Fault C as well as fault D may be linked: it is possible that writing in A_x masks the fault that occurred when C_x was erroneously written through A_y. Therefore, faults C and D are extended to the general case, with more than two addresses and cells (see Figure 14.14). Any number of cells and addresses with the same fault can be inserted between v and w, and between y and z. The order of the addresses is implied by the sequence of the letters in the alphabet, so that x denotes a lower address than y, etc.

If multiple cells are accessed simultaneously with a single address, as in fault D, a read operation performed on this address can yield one of the following two results:
1 All addressed cells have the same content h; it is assumed that the memory will always return that value.
2 Not all addressed cells have the same content. Two cases can be distinguished:
  (a) The memory returns a value which is a deterministic function of the contents of all addressed cells; for example, the AND or OR function.
  (b) The memory returns a random or pseudo-random result:

- The result can be random as a consequence of noise in the memory. In this case, the memory returns an unpredictable result every time it is brought into exactly the same state, i.e., the error is not repeatable.
- The result can be pseudo-random when it is a complicated deterministic function of the current state of the memory, while this function is not known by the test designer.

In general, it is safest for tests for AFs to expect the memory to return a *random* result, except for the case where all cells contain the same value, h. The tests should therefore be designed such that they are not influenced by (pseudo-) random results.

14.3.6   Mapping read/write logic and address decoder faults

So far a reduced functional model has been defined consisting of three blocks: the address decoder, the memory cell array and the read/write logic (see Figure 14.7). When one wants to test a memory for SAFs, three tests would be necessary: one test which detects SAFs in the address decoder, one test which detects SAFs in the memory cell array, and one test which detects SAFs in the read/write logic.

Faults occurring in the read/write logic and the address decoder can be mapped onto faults in the memory cell array; i.e., to tests for the memory cell array these faults will appear as faults in this array. This further reduces the reduced functional model of Figure 14.7 to only one block: the memory cell array, at the expense of not being able to localize the fault. A proof of correctness of this simplification of the reduced functional model is given below.

14.3.6.1 Mapping read/write logic faults onto memory cell array faults

An algorithm that detects SAFs in the memory cell array also detects SAFs in the read/write logic because an SAF in the read/write logic appears as a large group of cells with SAFs (in the case of a one data bit wide memory, this group contains all cells). The same arguments are valid for SOpFs, TFs, DRFs, and CFs (Nair *et al.*, 1978), (Thatte and Abraham, 1977).

14.3.6.2 Mapping address decoder faults onto memory cell array faults

Faults in the address decoder are detected by tests for the memory cell array provided that the test satisfies certain conditions. This only applies to march tests and not to tests for NPSFs which require a separate test to check for AFs (van de Goor, 1998). The conditions which a march test has to satisfy in order to detect the AFs A to D (see Figure 14.13), are given below; they are shown to be sufficient, as well as minimal.

Conditions for detecting AFs: A march test detects AFs if and only if it fulfills both conditions (i.e., contains the march elements) of Table 14.3. These march elements

Table 14.3. *Conditions for detecting AFs*

Condition	March element
1	$\Uparrow(r.x,\ldots,w\overline{x})$
2	$\Downarrow(r\overline{x},\ldots,w.x)$

need the following clarification: the '...' in a march element indicates the presence of any number of read or write operations. Any number of read operations may be performed between the march elements or concatenated to a march element because read operations do not change the contents of the memory; the value of x can be chosen freely, as long as it is the same in the entire test.

Sufficiency of the conditions for detecting AFs: Below it is proven that the AFs A, B, C and D of Figure 14.13 are detected given a march test which satisfies the conditions of Table 14.3.

- Faults A and B: AFs A and B will be detected by every test which detects SAFs in the memory cell array. When address A_x is written into and then read from, depending on which technology is used, cell C_x will appear to be SA0 or SA1. Thus, either Condition 1 or Condition 2 (depending on whether x is 0 or 1 in the test) will detect the fault.

- Fault C: AF C (see Figure 14.14) is detected by first initializing the entire memory array to a certain value (the *expected* value h, which can be x or $\overline{x}$). Thereafter, any march element which reads the expected value h, and ends with writing the cells with $\overline{h}$, will detect fault C. Thus, any one of the Conditions 1 (where $x = h$) or 2 (where $\overline{x} = h$) will detect fault C.

- Fault D: If AF D is to be detected when the memory may return a random result, case D_1 is not the superset of cases D_2 and D_3, because cells $C_v \ldots C_w$ may not contain the same value as $C_y \ldots C_z$ at all times (see Figure 14.14). When these values differ, the memory may return a random result. Thus, the fault cannot always be detected when A_x is read: the march element may have written $\overline{h}$ into either C_v $\ldots C_w$ or otherwise into $C_y \ldots C_z$, while the other cells still contain h. Generally, the result of reading A_x will then be unpredictable. The fault must be generated when A_x is written, and detected when either A_w or A_y is read. This can be done by a march element marching $\Uparrow$ or $\Downarrow$:
  - Condition 1: $\Uparrow(r.x,\ldots,w\overline{x})$ will detect cases D_1 and D_2. When A_x is written with $\overline{x}$, cells $C_y \ldots C_z$ are also written with value $\overline{x}$. This will be detected when C_y is read, reading $\overline{x}$ while x is expected.
  - Condition 2: $\Downarrow(r\overline{x},\ldots,w.x)$ will detect cases D_1 and D_3. When A_x is written with x, cells $C_w \ldots C_v$ are also written with value x. This will be detected when C_w is read, reading x while $\overline{x}$ is expected.

Minimality of the conditions for detecting AFs: The proof consists of removing an operation from the two conditions of Table 14.3 after which the remaining conditions are shown to be insufficient. The proof is left as an exercise to the reader.

14.4 Traditional tests

Traditional tests have been used extensively in the past; they have been included for historical reasons and because of their capability of being able to detect certain non-functional faults such as parametric and dynamic faults (van de Goor, 1998). The following traditional tests are described below: zero-one test, Checkerboard test, and GALPAT and Walking 1/0 tests.

14.4.1 Zero-one test

This minimal test consists of writing 0s and 1s to the memory. The algorithm is shown in Figure 14.15; it is also known under the name MSCAN (Memory Scan) (Abadir and Reghbati, 1983). The test length is $4 \cdot n$ operations, it is thus an $O(n)$ test (Breuer and Friedman, 1976). The algorithm is very easy to implement, but has very little test strength as explained below.

- Not all AFs are detected because the algorithm does not satisfy the conditions of Table 14.3; this can be seen from the march notation for zero-one: $\{\Uparrow(w0); \Uparrow(r0); \Uparrow(w1); \Uparrow(r1)\}$. It can only be guaranteed that *one* cell is accessed.

- SAFs are detected when it can be guaranteed that the address decoder is fault-free, otherwise one can only guarantee that a single cell in the memory cell array is free of SAFs.

- Not all TFs are detected. Not all $\langle\downarrow/1\rangle$ TFs are detected because not all $\downarrow$ transitions are generated; the memory may contain 0s before Step 1 such that Step 1 does not generate all $\downarrow$ transitions.

- Not all CFs are detected. Since it cannot be guaranteed that all $\downarrow$ transitions are generated in Step 1, it cannot be guaranteed that CFs requiring a $\downarrow$ transition in the coupling cell are detected.

  In Step 3, $\uparrow$ transitions are made. However, $\langle\uparrow;\uparrow\rangle$ CFids are not detected because the expected value of the read operation of Step 4 is the same as the value induced by the CFs. $\langle\uparrow;\downarrow\rangle$ and $\langle\uparrow;\updownarrow\rangle$ CFs are detected for those faults where the a-cell has a lower address than the v-cell (for the other faults, the 'w1' operation of Step 3 masks the fault).

Step 1: **write** 0 in all *cells*;
Step 2: **read** all *cells*;
Step 3: **write** 1 in all *cells*;
Step 4: **read** all *cells*;

Figure 14.15   The zero-one algorithm

1	2	1	2
2	1	2	1
1	2	1	2
2	1	2	1

1	0	1	0
0	1	0	1
1	0	1	0
0	1	0	1

0	1	0	1
1	0	1	0
0	1	0	1
1	0	1	0

Step 1   Step 3

(a) Cell numbering   (b) Test patterns

Figure 14.16   Cell numbering and patterns for the checkerboard test

Step 1: **write** 1 in all *cells-1* and 0 in all *cells-2*;
Step 2: **read** all *cells* (*words*);
Step 3: **write** 0 in all *cells-1* and 1 in all *cells-2*;
Step 4: **read** all *cells* (*words*);

Figure 14.17   The checkerboard algorithm

14.4.2   Checkerboard test

This is another short and simple test. The cells of the memory cell array are divided into two groups, *cells-1* and *cells-2*, forming a checkerboard pattern (see Figure 14.16(a)). The patterns generated by this algorithm are depicted in Figure 14.16(b) and the algorithm is shown in Figure 14.17. The test takes $4 \cdot n$ operations. It is thus an O(n) test (Breuer and Friedman, 1976).

Checkerboard has the following fault coverage:

- Not all AFs are detected because the algorithm does not satisfy the conditions of Table 14.3. It only can be guaranteed that *two* cells are accessed.

- SAFs are detected when it can be guaranteed that the address decoder functions correctly; otherwise, only two cells can be guaranteed to be free of SAFs.

- Not all TFs and CFs are detected for reasons similar to the zero-one test.

The checkerboard test was originally designed, and is still used, to test the maximum refresh period of DRAMs. Each cell containing a 1 is surrounded by cells containing a 0 and the other way around, which maximizes the leakage current between the cells.

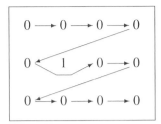

Walking 1/0

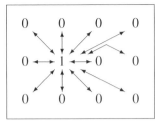

GALPAT

Figure 14.18 Read actions for Walking 1/0 and for GALPAT

14.4.3 GALPAT and Walking 1/0 tests

The algorithms for GALPAT (GALloping PATtern) and Walking 1/0 are similar. The memory is filled with 0s (or 1s) except for the *base-cell*, which contains a 1 (respectively, a 0). During the test, the *base-cell* walks through the memory. The difference between GALPAT and Walking 1/0 is in reading the *base-cell* (see Figure 14.18): with Walking 1/0, after each step all cells are read with the *base-cell* last; with GALPAT, all other cells are read, but after each other cell the *base-cell* is also read. The algorithms for GALPAT and Walking 1/0 are shown in Figure 14.19.

The fault coverage for both tests is as follows:

- All AFs are *detected* and *located*. AFs may cause the *base-cell* not to be written in Step 2 of the algorithm, this fault will be located in Step 5 or Step 7; alternatively, or additionally, other cells may be written in Step 2, these cells will be located in Step 4 or Step 6.

- All SAFs will be located because the *base-cell* is written (Step 2) and read (Step 5 or Step 7) with values 0 and 1.

- All TFs are located because the *base-cell* will make an $\uparrow$ and a $\downarrow$ transition (Step 2) after which it is read (Step 5 or Step 7).

- CFids are located. In Step 2, $\langle\uparrow;\uparrow\rangle$ and $\langle\downarrow;\downarrow\rangle$ CFs may be sensitized (depending on whether the value of d in Step 1 is 0 or 1, respectively) and located in Step 4 or Step 6. In Step 3, $\langle\downarrow;\uparrow\rangle$ and $\langle\uparrow;\downarrow\rangle$ CFids may be sensitized, and located in Step 4 or Step 6.[2]

Note that AFs and CFids can be *located* accurately, because the coupled cell is read immediately after writing the coupling cell, before any further write operations take place. Thus, it is clear which cell is coupled with which other cell(s). The same is true for AFs.

Slow sense amplifier recovery faults (van de Goor, 1998), caused by reading a long sequence of x values (such that the sense amplifier gets saturated) followed by an $\overline{x}$

---

[2] $\langle\uparrow;\downarrow\rangle$ and $\langle\downarrow;\uparrow\rangle$ CFids between any two cells i and i+1 will not be detected. This problem can be solved by inserting the operation 'if ($A[base\text{-}cell\,] = d$) then output ('Error at cell', *base-cell*)' before Step 2.

Step 1: **for** $d := 0$ **to** 1 **do**
        begin
                for $i := 0$ **to** $n - 1$ **do**
                    A$[i] := d$;
                for $base\text{-}cell := 0$ **to** $n - 1$ **do**
                begin
Step 2:   A$[base\text{-}cell] := \overline{d}$;
                    perform READ ACTION;
Step 3:   A$[base\text{-}cell] := d$;
                end;
        end;

READ ACTION for GALPAT:
begin
        for $cell := 0$ **to** $n - 1$ ($base\text{-}cell$ excluded) **do**
        begin
Step 4:   **if** (A$[cell] \neq d$) **then output**('Error at cell',*cell*);
Step 5:   **if** (A$[base\text{-}cell] \neq \overline{d}$) **then output**('Error at cell',*cell*);
        end;
    end;

READ ACTION for Walking 1/0:
begin
        for $cell := 0$ **to** $n - 1$ ($base\text{-}cell$ excluded) **do**
        begin
Step 6:   **if** (A$[cell] \neq d$) **then output**('Error at cell',*cell*);
        end;
Step 7:   **if** (A$[base\text{-}cell] \neq \overline{d}$) **then output**('Error at cell',*cell*);
    end;

Figure 14.19 GALPAT and Walking 1/0 algorithms

value, are detected by the Walking 1/0 test, because the *base-cell* is read immediately after reading all other cells which contain complementary data. GALPAT is also useful for testing for *write recovery faults* (these are AFs that arise due to critical timing of the address decoder circuitry such that the transition from a particular address A_x to another address A_y takes too long), because all address pairs occur within the READ ACTION (van de Goor, 1998).

The test is performed twice: with a background pattern of all 0s ($d = 0$ in Step 1) and all 1s. Writing the background pattern takes 2^N operations. For each background pattern, n operations are performed consisting of two write operations to the *base-cell* and a READ ACTION. The GALPAT READ ACTION takes $2 \cdot (n - 1) = 2n - 2$ operations, while the Walking 1/0 READ ACTION takes n operations. As a result, GALPAT takes $2 \cdot (2^N + n \cdot (2 + 2 \cdot n - 2)) = 2 \cdot (2^N + 2 \cdot n^2)$ operations, Walking 1/0 takes $2 \cdot (2^N + n \cdot (2 + n)) = 2 \cdot (2^N + 2 \cdot n + n^2)$ operations. Both are thus $O(n^2)$ tests (Breuer and Friedman, 1976).

As the above test length is completely unacceptable for any serious testing purposes, the READ ACTION is often only performed on either the rows or the columns within the memory cell array. This reduces the GALPAT READ ACTION to $2 \cdot (\sqrt{n} - 1)$ operations, assuming a square organization of the memory cell array. The total test time is reduced to: $2 \cdot (2^N + n \cdot (2 + 2\sqrt{n - 2})) = 2 \cdot (2^N + 2 \cdot n \cdot \sqrt{n})$ which is $O(n \cdot \sqrt{n})$. These tests are known as GALCOL and GALROW.

14.5 March tests

The simplest tests which detect SAFs, TFs and CFs are part of a family of tests called 'marches' (see Section 14.3.3). Many march tests have been designed (van de Goor, 1993, 1998); the most important march tests (or requirements for march tests) are included in this section:

1 MATS+: test for detecting SAFs (which also covers AFs).
2 March C−: test for detecting unlinked CFids (it also covers AFs, SAFs, CFsts; and TFs and CFins not linked with CFids).
3 March A: test for detecting linked CFids (it also covers AFs, SAFs, TFs not linked with CFids, and certain CFins linked with CFids).
4 March B: test for detecting linked TFs and CFids (which also covers TFs linked with CFids or CFins; certain CFins linked with CFids; AFs and SAFs).
5 Test requirements for detecting SOpFs.
6 Test for detecting DRFs.

14.5.1   MATS+: test for detecting SAFs

The MATS+ algorithm, which stands for modified algorithmic test sequence, is the shortest march test for SAFs and AFs. The algorithmic test sequence (ATS) algorithm was designed by Knaizuk and Hartmann (1977); it has been improved by Nair (1979) and subsequently named MATS. When reading multiple cells may produce a (pseudo-)random result, the MATS algorithm has to be extended to the MATS+ algorithm (Abadir and Reghbati, 1983). First, the MATS+ algorithm is described, thereafter, a verification of the given fault coverage is given.

The MATS+ scheme is shown in Figure 14.20. It consists of the march elements M_0, M_1 and M_2. Figure 14.9 shows the corresponding algorithm which requires $5 \cdot$

$$\{ \updownarrow(w0); \Uparrow(r0,w1); \Downarrow(r1,w0) \}$$
$$M_0 \qquad M_1 \qquad\quad M_2$$

Figure 14.20 The MATS+ scheme

$$\{ \updownarrow(w0); \Uparrow(r0,w1); \Uparrow(r1,w0); \updownarrow(r0); \Downarrow(r0,w1); \Downarrow(r1,w0); \updownarrow(r0) \}$$
$$M_0 \qquad M_1 \qquad\quad M_2 \qquad\quad M_3 \quad\; M_4 \qquad\quad M_5 \qquad\quad M_6$$

Figure 14.21 The March C scheme

$$\{ \updownarrow(w0); \Uparrow(r0,w1); \Uparrow(r1,w0); \Downarrow(r0,w1); \Downarrow(r1,w0); \updownarrow(r0) \}$$
$$M_0 \qquad M_1 \qquad\quad M_2 \qquad\quad M_3 \qquad\quad M_4 \qquad\quad M_5$$

Figure 14.22 The March C− scheme

n operations. MATS+ differs from the zero-One test in march elements M_1 and M_2 which perform two operations on a cell before proceeding to the next cell.

Below it is shown that the MATS+ algorithm detects all SAFs in the memory cell array and read/write logic and all faults in the address decoder:

1 Memory cell array: SAFs in the memory cell array imply that various cells contain SA1 and SA0 faults. Since a 0 and a 1 is written into and read from every memory cell, the MATS+ algorithm will detect SAFs in the memory cell array.

2 Read/write logic: SAFs in the read/write logic cause bit positions to be permanently stuck at some logic level. Since a 0 and a 1 are both written and read, the MATS+ algorithm will detect SAFs in the read/write logic.

3 Address decoder: The MATS+ algorithm satisfies the conditions of Table 14.3 and therefore detects all AFs.

Looking back at the zero-one and the Checkerboard algorithms of Section 14.4 one sees that MATS+ takes about the same number of operations, but has a much better fault coverage.

14.5.2   March C−: test for detecting unlinked CFids

March C is a test for detecting unlinked CFids; it also covers AFs, SAFs, unlinked TFs, CFsts, and unlinked CFins; it is depicted in Figure 14.21 (Marinescu, 1982). It is a redundant test because march element M_3 can be removed without affecting the fault coverage. The resulting, optimal test will be called March C− and is shown in Figure 14.22; it has a test length of $10 \cdot n$.

March C− satisfies Conditions 1 and 2 for AFs in Table 14.3: when $x = 0$ by means of march elements M_1 and M_4, when $x = 1$ by means of march elements M_2 and M_3. March C− detects SAFs and unlinked TFs because all cells are read in states 0, 1, 0, ... Thus, both ↑ and ↓ transitions, and read operations after them, have taken place. March C− also detects CFids, CFins and CFsts with the restriction that these CFs

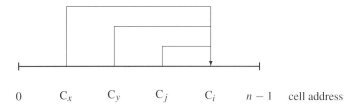

0   C_x C_y C_j C_i $n-1$ cell address

C_i is coupled to C_j, C_x and C_y. C_j is the cell with the highest address C_i is coupled to.

Figure 14.23 Addresses of coupling and coupled cells

are unlinked (it will be shown that not all linked combinations are covered) as proved below.

14.5.2.1 Idempotent coupling faults

The proof that March C− detects unlinked CFids is split into two cases: 1. faults with the addresses of the coupling cells lower than the coupled cell, and 2. faults with the addresses of the coupling cells higher than the coupled cell. The coupled cell will be denoted by C_i, and (one of) the coupling cells with C_j. As a reminder of the notation: C_i is $\langle\uparrow;\downarrow\rangle$ coupled to C_j means that an $\uparrow$ transition in C_j causes a $\downarrow$ transition in C_i.

1 Let C_i be coupled to any number of cells with addresses lower than i, and let C_j be the highest of those cells ($j < i$) (see Figure 14.23). Four cases, corresponding to the four different types of CFids, can be distinguished: a. $\langle\uparrow;\downarrow\rangle$, b. $\langle\uparrow;\uparrow\rangle$, c. $\langle\downarrow;\downarrow\rangle$, and d. $\langle\downarrow;\uparrow\rangle$. These cases are considered below.

(a) If C_i is $\langle\uparrow;\downarrow\rangle$ coupled to C_j, then the fault will be detected in march element M_3 followed by M_4.

M_3 oper. on C_j	C_j	C_i	M_3: $\Downarrow$(r0,w1)
r0	0	1	
w1	1	0	$\langle\uparrow;\downarrow\rangle$ CFid; fault sensitized
M_4 oper. on C_i	C_j	C_i	M_4: $\Downarrow$(r1,w0)
r1	1	0	fault detected

- In M_3, a 1 is written in C_j and due to the $\langle\uparrow;\downarrow\rangle$ coupling fault, C_i will contain a 0.
- In M_4, a read operation is performed on C_i and a 0 instead of a 1 is read.
- Linked CFids are not detected. For example, C_i must not be $\langle\uparrow;\uparrow\rangle$ coupled to cells with a lower address than j, because M_3 operates on them after C_j. In that case, a 1 will be read in M_4, which is the expected value. Thus, the fault will not be detected.

(b) If C_i is $\langle\uparrow;\uparrow\rangle$ coupled to C_j, then march element M_1 will detect the fault.

M_1 oper. on C_j	C_j	C_i	M_1: $\Uparrow(r0,w1)$
r0	0	0	
w1	1	1	$\langle\uparrow;\uparrow\rangle$ CFid; fault sensitized
M_1 oper. on C_i	C_j	C_i	M_1: $\Uparrow(r0,w1)$
r0	1	1	fault detected

(c) If C_i is $\langle\downarrow;\downarrow\rangle$ coupled to C_j, then march element M_2 will detect the fault. The proof is similar to the one above.

(d) If C_i is $\langle\downarrow;\uparrow\rangle$ coupled to C_j, then march elements M_4 followed by M_5 will detect the fault. C_i must not be $\langle\downarrow;\downarrow\rangle$ coupled to cells with addresses lower than j. The proof is similar to the one above.

2 Let C_i be coupled to any number of cells with addresses higher than i and let C_j be the lowest addressed cell among them ($j > i$). The proof is similar to Case 1, whereby M_1 should be replaced by M_3, M_2 by M_4, M_3 by M_1, M_4 by M_2 and M_5 by M_3.

14.5.2.2 Inversion coupling faults

The proof that March C− detects unlinked CFins is similar to the proof for CFids: two cases for the relative positions of the coupling and the coupled cell are considered, and for each case two types of CFins exist.

1 Let C_i be coupled to any number of cells with addresses lower than i and let C_j be the highest of those cells ($j < i$).

(a) C_i is $\langle\uparrow;\updownarrow\rangle$ coupled to C_j; then M_1 will detect the fault, as well as M_3 followed by M_4.

(b) C_i is $\langle\downarrow;\updownarrow\rangle$ coupled to C_j; then M_2, as well as M_4 followed by M_5, will detect the fault.

2 The proof for $j > i$ is similar to above.

14.5.2.3 State coupling faults

All CFsts are detected if the four states of any two cells i and j are reached. Table 14.4 (Dekker *et al.*, 1988) shows the operations performed on two cells, C_i and C_j, by march elements M_0 through M_4 of Figure 14.22, assuming that the address of C_i < address of C_j. The table contains a column which identifies the state S_{ij} before the operation, and a column which identifies the state after the operation; only write operations can change the state. The states of Table 14.4 are represented in the form of a state diagram in Figure 14.24 (Dekker *et al.*, 1988). The arcs of Figure 14.24 are labeled with the step numbers of Table 14.4; in the case of a read operation the step number is followed by the label (i or j) of the cell being read. Figure 14.24 shows that all four states are generated, and verified because in each state the values of cell C_i

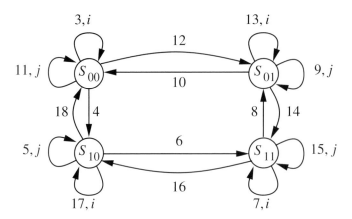

Figure 14.24 State diagram for detecting CFsts

Table 14.4. *State table for detecting CFsts*

Step	March element	State	Operation	State
1	M_0	–	w0 into i	–
2		–	w0 into j	S_{00}
3	M_1	S_{00}	r0 from i	S_{00}
4		S_{00}	w1 into i	S_{10}
5		S_{10}	r0 from j	S_{10}
6		S_{10}	w1 into j	S_{11}
7	M_2	S_{11}	r1 from i	S_{11}
8		S_{11}	w0 into i	S_{01}
9		S_{01}	r1 from j	S_{01}
10		S_{01}	w0 into j	S_{00}
11	M_3	S_{00}	r0 from j	S_{00}
12		S_{00}	w1 into j	S_{01}
13		S_{01}	r0 from i	S_{01}
14		S_{01}	w1 into i	S_{11}
15	M_4	S_{11}	r1 from j	S_{11}
16		S_{11}	w0 into j	S_{10}
17		S_{10}	r1 from i	S_{10}
18		S_{10}	w0 into i	S_{00}

© 1988 Proc. IEEE Int. Test Conference

and C_j are read; for example, in state S_{00}, cell C_i is read in Step 3 and cell C_j in Step 11.

$$\{ \updownarrow(w0); \Uparrow(r0,w1,w0,w1); \Uparrow(r1,w0,w1); \Downarrow(r1,w0,w1,w0); \Downarrow(r0,w1,w0) \}$$
$$M_0 \qquad M_1 \qquad\qquad M_2 \qquad\qquad M_3 \qquad\qquad M_4$$

Figure 14.25 The March A scheme

14.5.3 March A: test for detecting linked CFids

March A (Suk and Reddy, 1981) is the shortest test for detecting linked CFids; it also covers AFs, SAFs, linked CFids, TFs not linked with CFids, and certain CFins linked with CFids. The scheme of March A is shown in Figure 14.25; it has a test length of $15 \cdot n$. Below, the completeness and irredundancy of March A is proven.

14.5.3.1 March A is complete

The proof that March A is **complete**, i.e., all faults that should be covered are detected by the test, is as follows: it can be verified easily that AFs, SAFs and TFs are detected; the proof that CFins and CFids are detected is given below, while the reader can verify easily that not all CFsts will be detected.

CFins are detected: The proof is split into two cases determined by the position of the coupled cell relative to the coupling cell (see Figure 14.23).
1 Let C_i be coupled to an odd number of cells with addresses lower than i and let C_j be the highest of those cells ($j < i$). Then three cases can be distinguished: C_i is $\langle\uparrow; \updownarrow\rangle$ coupled to C_j, C_i is $\langle\downarrow; \updownarrow\rangle$ coupled to C_j, and C_i is $\langle\uparrow; \updownarrow\rangle$ *and* $\langle\downarrow; \updownarrow\rangle$ coupled to C_j.
    (a) If C_i is $\langle\uparrow; \updownarrow\rangle$ coupled to C_j, then M_1 will detect the fault.
    (b) If C_i is $\langle\downarrow; \updownarrow\rangle$ coupled to C_j, then M_2 as well as M_5 will detect the fault.
    (c) If C_i is $\langle\uparrow; \updownarrow\rangle$ and $\langle\downarrow; \updownarrow\rangle$ coupled to C_j, then M_1 will detect the fault.
2 The proof for $j > i$ is similar to the one above. Case (a) will be detected by M_3, Case (b) will be detected by M_4, and Case (c) by M_3.

CFids are detected: It can be proven that March A is a complete march test for the detection of CFids under the condition that *no* SAFs, TFs and AFs are present in the memory. Linked CFids are said to be detected when at least one of the CFids is detected.

The proof that March A will detect linked CFids and certain CFins linked with CFids (see Section 14.3.4) is split into two cases determined by the position of the coupled cell relative to the coupling cell (see Figure 14.23).
1 Let C_i be coupled to any number of cells with addresses lower than i and let C_j be the highest of those cells ($j < i$). Then, four cases, corresponding to the four different types of CFids, can be distinguished: a. $\langle\uparrow; \downarrow\rangle$, b. $\langle\uparrow; \uparrow\rangle$, c. $\langle\downarrow; \downarrow\rangle$, and d. $\langle\downarrow; \uparrow\rangle$. These cases are considered below.

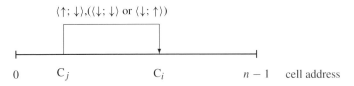

Figure 14.26 Assumed coupling faults for Case (a)

(a) If C_i is $\langle\uparrow;\downarrow\rangle$ coupled to C_j (and possibly *also* $\langle\downarrow;\downarrow\rangle$ or $\langle\downarrow;\uparrow\rangle$ coupled) (see Figure 14.26), then the $\langle\uparrow;\downarrow\rangle$ CFid will be detected by march element M_2.

M_2 oper. on C_j	C_j	C_i	M_2: $\Uparrow(r1,w0,w1)$
r1	1	1	Initial state
w0	0	–	
w1	1	0	$\langle\uparrow;\downarrow\rangle$ CFid

- First, a 0 is written into C_j (w0). The content of C_i is 1, when this transition does affect C_i due to a $\langle\downarrow;\uparrow\rangle$ CFid or in the absence of a CFid; C_i is 0, when there is a $\langle\downarrow;\downarrow\rangle$ CFid.
- Later on, a 1 is written into C_j; C_j thus makes an $\uparrow$ transition. The content of C_i is forced to 0 due to the $\langle\uparrow;\downarrow\rangle$ CFid.

When M_2 operates on C_i the fault is detected because a 0 is read instead of the expected 1.

(b) If C_i is $\langle\uparrow;\uparrow\rangle$ coupled to C_j (and possibly *also* $\langle\downarrow;\downarrow\rangle$ or $\langle\downarrow;\uparrow\rangle$ coupled), then the $\langle\uparrow;\uparrow\rangle$ CFid will be detected by march element M_1.

(c) If C_i is $\langle\downarrow;\downarrow\rangle$ coupled to C_j (and *not* ($\langle\uparrow;\downarrow\rangle$ or $\langle\uparrow;\uparrow\rangle$) coupled to C_j), then the $\langle\downarrow;\downarrow\rangle$ CFid will be detected by march element M_2.

(d) If C_i is $\langle\downarrow;\uparrow\rangle$ coupled to C_j (and *not* ($\langle\uparrow;\uparrow\rangle$ or $\langle\uparrow;\downarrow\rangle$) coupled) then the $\langle\downarrow;\uparrow\rangle$ CFid will be detected by march element M_1.

2 Let C_i be coupled to some cells with addresses higher than i, and let C_j be the lowest addressed cell of them ($j > i$). The proof is similar to Case 1, using M_3 and M_4 instead of M_1 and M_2.

14.5.3.2 March A is irredundant

In an **irredundant test**, no operation of the test can be deleted without making the test incomplete. Figure 14.27 shows six (rather complicated) combinations of CFids[3]. For each combination, the march element required to detect that fault is given. For example, the first combination shows that C_3 is coupled to three other cells (C_0, C_1 and C_2) (see Figure 14.28). A proof will be given that the elements of March A are necessary to detect the six combinations of CFids given in Figure 14.27.

A march element in an irredundant march test can either be of the form

---

[3] Maybe more combinations of coupling faults exist, but it is proven that March A is a complete test; so March A will detect these combinations as well.

Fault	March elements to detect the fault
1 C_3 is $\langle\uparrow;\uparrow\rangle$ and $\langle\downarrow;\downarrow\rangle$ coupled to C_0 C_3 is $\langle\uparrow;\downarrow\rangle$ coupled to C_1 C_3 is $\langle\downarrow;\uparrow\rangle$ coupled to C_2	$\Uparrow(r0,w1,w0,m\cdot wc)$
2 C_3 is $\langle\uparrow;\uparrow\rangle$ and $\langle\downarrow;\downarrow\rangle$ coupled to C_0 C_3 is $\langle\downarrow;\uparrow\rangle$ coupled to C_1 C_3 is $\langle\uparrow;\downarrow\rangle$ coupled to C_2	$\Uparrow(r1,w0,w1,m\cdot wc)$
3 C_2 is $\langle\uparrow;\uparrow\rangle$ and $\langle\downarrow;\downarrow\rangle$ coupled to C_0 and C_1	$\Uparrow(r0,w1,2m\cdot wc)$ or $\Uparrow(r1,w0,2m\cdot wc)$
4 C_0 is $\langle\uparrow;\uparrow\rangle$ and $\langle\downarrow;\downarrow\rangle$ coupled to C_3 C_0 is $\langle\uparrow;\downarrow\rangle$ coupled to C_2 C_0 is $\langle\downarrow;\uparrow\rangle$ coupled to C_1	$\Downarrow(r0,w1,w0,m\cdot wc)$
5 C_0 is $\langle\uparrow;\uparrow\rangle$ and $\langle\downarrow;\downarrow\rangle$ coupled to C_3 C_0 is $\langle\downarrow;\uparrow\rangle$ coupled to C_2 C_0 is $\langle\uparrow;\downarrow\rangle$ coupled to C_1	$\Downarrow(r1,w0,w1,m\cdot wc)$
6 C_1 is $\langle\uparrow;\uparrow\rangle$ and $\langle\downarrow;\downarrow\rangle$ coupled to C_3 and C_2	$\Downarrow(r0,w1,2m\cdot wc)$ or $\Downarrow(r1,w0,2m\cdot wc)$

Note:
- The cell addresses obey the following rule: $C_0 < C_1 < C_2 < C_3$.
- wc stands for write complement (w0, if 1 was last written in this cell; w1, if 0 was last written).
- $m\cdot wc$ stands for m write complement operations, ($2m\cdot wc$ stands for $2m$ write complement operations) ($m \geq 0$).
- When the first write operation is w0 (w1) the cells are assumed to contain all 1s (0s) at the start of the march element.
- No SAFs, TFs or AFs are present.

Figure 14.27 Some coupling faults with march elements

(r0,w1,$m\cdot wc$), (r1,w0,$m\cdot wc$), (w1,$m\cdot wc$) or (w0,$m\cdot wc$). Assume that the CFid of Figure 14.28 is present and no march element of the form $\Uparrow$(r0,w1,w0,$m\cdot wc$) is included in the march test.

1 No march element of the form (w1,$m\cdot wc$) or (w0,$m\cdot wc$) can detect such a fault because of the absence of a read operation.
2 Any march element of the form $\Downarrow(\cdots)$ cannot possibly detect this fault since C_3 would be read prior to any operations on C_0, C_1 or C_2.
3 A march element of the form $\Uparrow$(r1,w0,$m\cdot wc$) cannot detect the fault since the expected state of C_3 before the 'r1' is 1, so the expected state does not differ from the actual state.

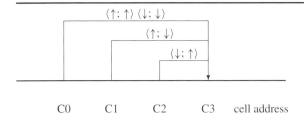

Figure 14.28 Fault 1 of Figure 14.27

$\{ \Updownarrow(w0); \Uparrow(r0,w1,r1,w0,r0,w1); \Uparrow(r1,w0,w1); \Downarrow(r1,w0,w1,w0); \Downarrow(r0,w1,w0) \}$
$\quad M_0 \quad\quad M_1 \quad\quad\quad\quad\quad\quad\quad M_2 \quad\quad\quad\quad M_3 \quad\quad\quad\quad M_4$

Figure 14.29 The March B scheme

4 A march element of the form $\Uparrow(r0,w1)$ cannot detect the fault since the $\langle\uparrow; \downarrow\rangle$ coupling of C_3 to C_1 would force C_3 to 0 (prior to reading C_3).

Note: Proofs for faults 2 to 6 are similar.

The march elements in Figure 14.27 are irredundant in case of $m = 0$. This means that when (with $m = 0$) one operation is eliminated, the combination of faults will not be detected (a combination is detected when at least one of the faults is detected). The six faults given in Figure 14.27 are covered by March A; when one operation is deleted from March A, one of the faults of Figure 14.27 will no longer be covered; hence, March A is irredundant.

14.5.4 March B: test for detecting linked TFs and CFids

March B (Suk and Reddy, 1981) is an extension of March A which detects, in addition to the faults of March A, TFs linked with CFins or CFids. In order to allow for this, march element M_1 of March A is modified to include two extra read operations (see Figure 14.29) such that TFs cannot be masked by CFs because no write operations to other cells, which may be potential coupling cells, take place. The necessity of the other operations can be proven in the same way as for March A (see Section 14.5.3). March B requires $17 \cdot n$ operations.

14.5.5 Test requirements for detecting SOpFs

An SOpF is caused by an open word line which makes the cell inaccessible. To detect $\langle \forall/L \rangle$ SOpFs (that is, SOpFs with a non-transparent sense amplifier), a march test has to verify that a 0 and a 1 can be read from every cell. This will be the case when a march test satisfies the condition of Table 14.5: there must be a march element in which the value x and the value $\bar{x}$ are read from a cell. For example, march element M_1, $\Uparrow(r0,w1,r1,w0,r0,w1)$, of Figure 14.29 satisfies the requirement of Table 14.5. The

Table 14.5. *Condition for detecting SOpFs*

March element
$\ldots, \mathrm{r}x, \ldots, \mathrm{r}\bar{x}, \ldots$

(a)   {Existing march test;Del;⇕(r0,w1);Del;⇕(r1)}

(b)   {Existing march test;Del;⇕(r0,w1,r1);Del;⇕(r1)}

(c)   {⇕(w0);⇑(r0,w1);⇑(r1,w0);⇓(r0,w1);Del;⇓(r1,w0);Del;⇕(r0)}

Figure 14.30 March tests to detect DRFs

MATS+ test of Figure 14.20 and the March C− test of Figure 14.22 can be modified to detects SOpFs by extending M_1 with an 'r1' operation, it will then have the form (r0,w1,r1).

14.5.6 Test for detecting DRFs

Any march test can be *extended* to detect DRFs as well. The detection of a DRF requires that a memory cell can be brought into one of its logic states. A certain time must pass while the DRF develops (the leakage currents have to discharge the open node of the SRAM cell). Thereafter, the contents of the cell are verified. This test must be repeated with the inverse logic value stored into the cell to test for a DRF due to an open connection in the other node of the cell. The amount of waiting time depends on the amount of charge stored in the capacitor of the node and the magnitude of the leakage current (which is difficult to determine). Empirical results (Dekker *et al.*, 1990; Aadsen *et al.*, 1990) show that a wait time (called *delay time 'Del'*) of 100 ms is adequate for the SRAM cells studied.

Figure 14.30(a) shows how an existing march test can be extended to detect DRFs, assuming that the test ends with all cells in state 0. The 'Del' elements represent the delay time before applying the next march element. When one suspects that both pull-up devices may be open, the DRF behaves as an SOpF. When the sense amplifier is non-transparent to SOpFs, the existing march test must be extended according to Figure 14.30(b).

Figure 14.30(c) shows a version of the March C− test that is capable of detecting DRFs. The test is not extended; instead, Del elements are inserted into the existing March C− test of Figure 14.22 to shorten the test (reduce the test time). This has the disadvantage that fault masking may occur; for example, a CF may not be detected when the coupled cell also has a DRF because the DRF may mask the CF.

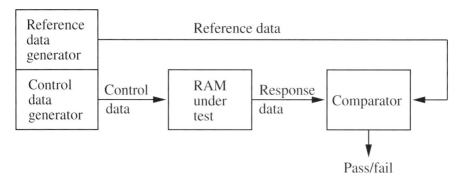

Figure 14.31   Deterministic RAM test

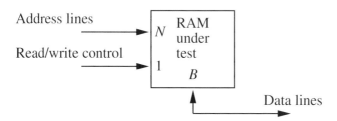

Figure 14.32   RAM chip test model

14.6   Pseudo-random memory tests

This section covers pseudo-random memory tests; a test method very suitable for BIST. It starts with an introduction of the concepts of pseudo-random testing, followed by two subsections covering pseudo-random tests for SAFs and k-coupling faults. The section ends with a summary of pseudo-random tests.

14.6.1   Concepts of pseudo-random memory testing

Sections 14.4 and 14.5 described **deterministic tests** for memories; the control data for the RAM under test and the reference data had predetermined values (see Figure 14.31). The response data of the RAM under test are compared with the reference data to make a pass/fail decision. The control data are applied to the inputs of the RAM under test (see Figure 14.32) to control the N address lines, the read/write control line, and the B bidirectional data lines (during a write operation). The response data are obtained from the B bidirectional data lines (during a read operation).

In case of **pseudo-random testing**, the control data on some or all of the inputs are determined pseudo-randomly; for example, the address and data lines may be controlled pseudo-randomly while the read/write control line may be controlled

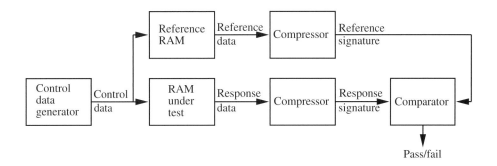

Figure 14.33   Pseudo-random RAM test

deterministically or pseudo-randomly (see Figure 14.32). The reference data, which are the data expected from a read operation, can be obtained from a reference RAM (which is supposed to be fault-free). Rather then using a reference RAM, which may not be available during BIST, the reference data of the complete test may be compressed by a compressor such that only a limited number of bits remain (typically 16 or 32). This compressed result is called a **signature**; it is obtained by applying the control data to a reference RAM (or a simulation model of that RAM) and compressing the reference data (see Figure 14.33). When the test is repeated later on with the same control data, the same signature should be produced such that it can be used as the reference signature of the test; naturally the response data have to be compressed as well and the test has to be repeatable. This requires the application of the same control data as used for generating the reference signature. Therefore, purely random control data cannot be used (such data can only be used if a reference RAM is available); pseudo-random control data are used because it appears to be random but has the property that it is repeatable. Bardell *et al.* (1987) show how pseudo-random control data can be generated.

The fault coverage with pseudo-random testing cannot be guaranteed exactly. For example, with MATS+ it can be guaranteed that 100% of the SAFs will be detected. With pseudo-random testing, an **escape probability**, e, for a particular fault is given (for example, the probability of not detecting SAFs should be less than $e = 0.001$, which means a fault coverage better than 99.9%), and from this the resulting test length is computed.

Faults are detected when reading the RAM under test. They can be reported immediately when the response data are compared with the reference data (when a reference RAM is used, which does not require compression) or alternatively, the faults are reported at the end of the test when the response signature is compared with the reference signature.

In the description of pseudo-random tests, the following notation will be used: p denotes the probability that a particular line has the value 1; p_a applies to an address line, p_d applies to a data line; and p_w is the probability that the control line specifies a write operation. In the remainder of this chapter, it is assumed that the p_a of each address line is independent of the p_a of any other address line.

Section 14.5 described several deterministic (march) tests which differed in the number of march elements together with the addressing order, and the number and sequence of the operations (in terms of a 'r', a 'w0' or a 'w1' operation) performed by the march elements. Similarly, many different pseudo-random memory tests can be designed where the determination of the next address (A), the selection of a read or write operation (W), and the data to be written (D) may be determined deterministically (D) or pseudo-randomly (R). A march test can be described as DADWDD: deterministic address, deterministic write (and therefore also read), and deterministic data values. A pseudo-random test in which all control data are generated pseudo-randomly can be described as RARWRD. This notation will be used to give a global idea of a test.

From the above, it may be clear that many pseudo-random tests can be constructed. In order not to be encyclopedic, only the following two important classes of pseudo-random tests will be described: for SAFs and for k-coupling faults. These two classes have been included because they best illustrate the concepts and the pros and cons of pseudo-random testing.

14.6.2   Pseudo-random test for SAFs (RARWRD)

It has been shown that the most difficult SAF to detect with pseudo-random testing of memories is an SAF in a single storage cell. AFs are as difficult or easier to detect than an SAF in a single storage cell because an AF usually affects more than one cell (Thévenod-Fosse and David, 1978). Hence, a pseudo-random test which is long enough to detect the most difficult fault with a particular escape probability will also detect easier faults with an equal or lower escape probability.

With an RARWRD test, the RAM is tested for SAFs by applying a sequence of pseudo-random vectors of $N + 2$ bits, controlling the N address lines, read/write control line and data line. Faults are detected by comparing the result of reading the RAM under test with that of a reference RAM, or alternatively, by comparing the response signature with the reference signature at the end of the test. Before the pseudo-random test starts, the RAM under test as well as the reference RAM have to be initialized to reproducible states, for example by a $\updownarrow$(w0) or $\updownarrow$(w1) march, or a power-on reset.

This section describes a pseudo-random test for SAFs in the following parts:

- Computation of the test length as a function of the escape probability.

- Analysis of the influence of the values of e, n, p_a, p_w, and the initialization on the test length.

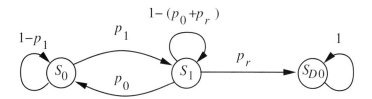

Figure 14.34 Markov chain for detecting an SA0 fault

14.6.2.1 Computation of the test length as a function of the escape probability

The memory has a single read/write control line. If p_w is the probability of a write operation, then $1 - p_w$ is the probability of a read operation. An arbitrary memory address contains z 0s and $N - z$ 1s. If p_a is the probability of each address line being a 1, then the probability of selecting a particular address A, denoted by p_A, is (Bardell *et al.*, 1987)

$$p_A = (1 - p_a)^z \cdot p_a^{(N-z)}. \tag{14.1}$$

The probability of writing a 1 to address A is

$$p_1 = p_d \cdot p_w \cdot p_A. \tag{14.2}$$

The probability of writing a 0 to address A is

$$p_0 = (1 - p_d) \cdot p_w \cdot p_A. \tag{14.3}$$

The probability of reading address A is

$$p_r = (1 - p_w) \cdot p_A. \tag{14.4}$$

Given that a particular address A has been selected, the operation will either be a 'w1', 'w0' or 'r'. Hence,

$$p_A = p_1 + p_0 + p_r. \tag{14.5}$$

Let an arbitrary word in memory have a single cell which is SA0 and assume that there are no other faults in the memory. Then the **test length** $T(e)$, which is a function of the escape probability e, can be computed as follows.

Figure 14.34 describes a Markov chain for detecting an SA0 fault (Bardell *et al.*, 1987). (For texts on Markov chains, the reader is referred to (Isaacson and Madsen, 1976) and (Romanovsky, 1970).) It contains three states: S_0, representing a 0 stored in the cell; S_1, representing a 1 stored in the cell; and S_{D0}, representing the state in which the SA0 fault has been detected. S_{D0} is an **absorbing state** because upon the first detection of the SA0 fault, the Markov chain will stay in that state. S_1 is the state in which the fault is sensitized; this means that this state should have been entered in the absence of an SA0 fault while in fact the memory cell contains the logic '0' value; a read operation in this state will detect the fault. The transition probabilities between the states are labeled along the corresponding edges of the Markov chain. For example, given that the chain is in state S_1, there will be a p_0 probability of entering S_0, a $1 - (p_0 + p_r)$ probability of staying in state S_1, and a probability p_r of entering the S_{D0} state. The equations describing the transitions between the states, as a function of the operation performed at time t, are

$$p_{S_0}(t) = (1 - p_1) \cdot p_{S_0}(t - 1) + p_0 \cdot p_{S_1}(t - 1), \tag{14.6}$$

$$p_{S_1}(t) = p_1 \cdot p_{S_0}(t - 1) + (1 - p_0 - p_r) \cdot p_{S_1}(t - 1), \tag{14.7}$$

$$p_{S_{D0}}(t) = p_r \cdot p_{S_1}(t - 1) + p_{S_{D0}}(t - 1). \tag{14.8}$$

Let the probability that a memory cell is initialized to a '1' be denoted by p_{I1}. Then the initial state probabilities are:

$$p_{S_0}(0) = 1 - p_{I1}, \quad p_{S_1}(0) = p_{I1}, \quad \text{and} \quad p_{S_{D0}}(0) = 0. \tag{14.9}$$

The **cumulative detection probability (CDP)** for an SA0 fault, of a test with length t, is the state probability of S_{D0} after a control data sequence of t test vectors has been applied; it has the value of $p_{S_{D0}}(t)$. Its value can be derived from Equations (14.8) and (14.9) (van de Goor, 1998) to be

$$p_{S_{D0}}(t) = 1 - \frac{C_1 \cdot (A - B)^t + C_2 \cdot (A + B)^t}{2 \cdot B}, \tag{14.10}$$

where $\quad A = 1 - p_A/2 \qquad\qquad B = \frac{1}{2}\sqrt{p_A^2 - 4 \cdot p_1 \cdot p_r}$
$\qquad\qquad C_1 = -1 + A + B + p_r \cdot p_{I1} \quad C_2 = 1 - A + B - p_r \cdot p_{I1}.$

Substituting the value of A, B, C_1 and C_2 in terms of p_{I1}, p_d, p_w, p_A results in van de Goor (1998):

$$p_{S_{D0}}(t) = 1 - \frac{1 - 2 \cdot (1 - p_w) \cdot p_{I1} + \alpha}{2 \cdot \alpha} \cdot \left(1 - \frac{p_A}{2} + \alpha \cdot \frac{p_A}{2}\right)^t$$
$$+ \frac{1 - 2 \cdot (1 - p_w) \cdot p_{I1} - \alpha}{2 \cdot \alpha} \cdot \left(1 - \frac{p_A}{2} - \alpha \cdot \frac{p_A}{2}\right)^t \tag{14.11}$$

where $\alpha = \sqrt{1 - 4 \cdot p_d \cdot p_w \cdot (1 - p_w)}$. $\tag{14.12}$

For tests with a large length, which is typical for pseudo-random testing, one can (assuming $\alpha \neq 0$) conclude that the term with $(1 - p_A/2 + \alpha \cdot p_A/2)^t$ will be dominant, such that the equation can be simplified to

$$p_{S_{D0}}(t) \approx 1 - \frac{1 - 2 \cdot (1 - p_w) \cdot p_{I1} + \alpha}{2 \cdot \alpha} \cdot \left(1 - \frac{p_A}{2} + \alpha \cdot \frac{p_A}{2}\right)^t. \tag{14.13}$$

Given an escape probability, e, the test length T_0 for detecting an SA0 fault is determined by that value of t which satisfies the following equation:

$$ps_{D0}(t) = 1 - e. \qquad (14.14)$$

The equation for the test length to detect an SA0 fault becomes

$$T_0(e) = \left\lceil \frac{\ln\left(\frac{2 \cdot \alpha \cdot e}{1 + \alpha - 2 \cdot (1 - p_w) \cdot p_{11}}\right)}{\ln\left(1 - \frac{(1-\alpha) \cdot p_A}{2}\right)} \right\rceil. \qquad (14.15)$$

Similarly, $T_1(e)$, which is the test length for detecting an SA1 fault with an escape probability e, can be determined to be

$$T_1(e) = \left\lceil \frac{\ln\left(\frac{2 \cdot \beta \cdot e}{1 + \beta - 2 \cdot (1 - p_w) \cdot (1 - p_{11})}\right)}{\ln\left(1 - \frac{(1-\beta) \cdot p_A}{2}\right)} \right\rceil, \qquad (14.16)$$

where $\beta = \sqrt{1 - 4 \cdot (1 - p_d) \cdot p_w \cdot (1 - p_w)}$. $\qquad (14.17)$

The test length which detects *any* SAF, with an escape probability less than or equal to e, is:

$$T(e) = \max(T_0, T_1). \qquad (14.18)$$

14.6.2.2 Analysis of the influence of the values of e, n, p_a, p_w, and the initialization on the test length[4]

This analysis has been included to give the reader a better understanding of the concepts of pseudo-random testing and a 'feel' for the influence of the different parameters on the test length.

The test length as a function of the escape probability and memory size can be seen from inspecting the equations for T_0 (14.15) and T_1 (14.16). One can conclude that the test length is proportional to the logarithm of e and linear with n. Table 14.6 shows the test length as a function of e and n, using $p_a = p_d = p_w = \frac{1}{2}$ and excluding the operations required for initialization. The table entries have the format X/Y. The X numbers show $T(e)$ in terms of the total number of operations required. The Y numbers show the test length in terms of the number of operations per cell; this is called the **test length coefficient**, which is independent of the memory size. The MATS+ test requires five operations per cell to detect all SAFs, while this pseudo-random test requires $48 + 1$ (for initialization) $= 49$ operations per cell to detect SAFs with an escape probability of $e = 0.001$. The pseudo-random test is, therefore, much less efficient in terms of test time. This does not have to be a problem for small embedded memories as found frequently in many VLSI chips and for BIST applications.

[4] The material in this section in based on McAnney *et al.* (1984); © 1984 *Proc. IEEE Int. Test Conference.*

Table 14.6. *Test length T(e) and test length coefficient*

e	Memory size			
	n = 32	*n* = 1K	*n* = 32K	*n* = 1024K
0.1	544/17	17 415/17	557 328/17	17 834 520/17
0.01	1046/33	33 514/33	1 072 539/33	34 321 320/33
0.001	1547/48	49 614/48	1 587 750/48	50 808 110/48
0.0001	2049/64	65 713/64	2 102 961/64	67 294 910/64
0.000 01	2551/80	81 812/80	2 618 173/80	83 781 700/80

The influence of p_a on the test length is such that when $p_a < \frac{1}{2}$, the most difficult address to select is address '111...1'. Conversely, if $p_a > \frac{1}{2}$, the most difficult address to select is address '000...0'. Only when $p_a = \frac{1}{2}$ are all addresses equally likely, resulting in the shortest test length. Similarly, it can be shown that the optimal values for p_w and p_d for detecting both SA0 and SA1 faults is $\frac{1}{2}$, while the influence of p_{11} on the test length coefficient decreases rapidly to the extent that its influence is negligible after $T(e) \geq 8$ (McAnney *et al.*, 1984).

14.6.3 Pseudo-random tests for k-coupling faults

This section describes pseudo-random tests for k-coupling faults of the idempotent type, where the k cells may be located anywhere in memory. For $k = 2$, this will be a test for CFids; while for $k > 2$, it will be a test for PSFs. The set of k cells of the PSF consist of an a-cell and a v-cell and a set of $k - 2$ cells which will be called the *deleted neighborhood cells*. The memory may be sensitive to a particular k-coupling fault only when the deleted neighborhood cells take on a certain value, denoted by G. For example, for a 5-coupling fault, the deleted neighborhood consists of $5 - 2 = 3$ cells. For the case that a deleted neighborhood value of $G = 110 = g_1 g_2 \bar{g}_3$ is required to make the memory sensitive to a particular 5-coupling fault, the probability of this deleted neighborhood pattern is: $p_G = p_d^2 \cdot (1 - p_d)$.

In Savir *et al.* (1989), five pseudo-random tests for k-coupling faults are described; the two most interesting versions will be given below. The first test, called 'explicit memory test with word operations (ETWO)', can be classified as DADWRD; the second test, called 'random memory test', can be classified as RARWRD.

14.6.3.1 Explicit memory test with word operations (DADWRD)

This is a variation on the deterministic march test where the data to be written, denoted by a '?', are determined pseudo-randomly; a 1 is written with probability p_d and a 0 with probability $1 - p_d$. The address sequence is determined deterministically. The DADWRD algorithm is shown in Figure 14.35. The equation for the test length

Step 1: ↕ (w?) {Initialize memory.}
Step 2: **repeat** *t* times
⇑(r,w?) {Perform a 'r' followed by a 'w?' operation on each cell.}
Step 3: ⇑(r) {Perform a final read operation.}

Figure 14.35 The DADWRD algorithm

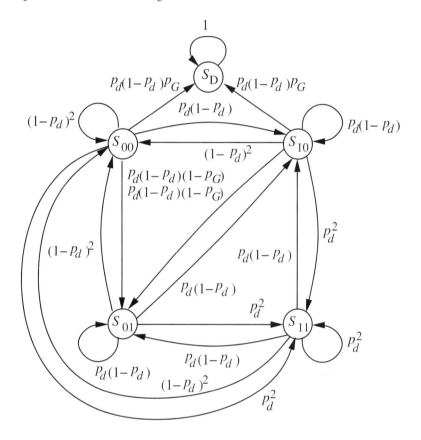

Figure 14.36 Markov chain for the DADWRD algorithm for CFids

required to detect k-coupling faults given a particular escape probability e is given below (Savir *et al.*, 1989).

Figure 14.36 (Savir *et al.*, 1989) shows the Markov chain for the pseudo-random test, using deterministic address sequences; assuming cell i is k-coupled to cell j, which means that a transition in C_j may sensitize a fault in C_i given a deleted neighborhood pattern G. State S_{ij} represents the state with the corresponding values for i and j, S_D represents the absorbing state in which the CFid $\langle\uparrow;\uparrow\rangle$ is detected (different absorbing states are required for the other three CFids), and p_G represents the probability that the deleted neighborhood has a value required for a particular fault to be sensitized.

The state transitions of Figure 14.36 assume that C_i is written before C_j for a fault to be detected, because if C_i is written after C_j the fault effect will be overwritten. (This assumption does not have to be made, as will be shown later on.) Step 2 of Figure 14.35 performs a read operation on each cell followed by a write operation which writes a logic 1 with probability p_d, and a logic 0 with probability $(1 - p_d)$. Each arc in Figure 14.36 shows the probability that cell C_i and C_j are written with the values of the destination node of the arc. State S_D is entered when the requirements for detecting the CFid $\langle\uparrow; \uparrow\rangle$ are met. This will happen during a read operation when the deleted neighborhood contains the required pattern G, and cell C_j makes an $\uparrow$ transition while cell C_i contains the value 0. Next, the transitions from state S_{00} are explained.

S_{00} to S_{01}: a 0 is written into cell C_i, after which a 1 is written into cell C_j, while the deleted neighborhood does not contain a value required for the fault to be sensitized. The probability of this transition is $(1 - p_d) \cdot p_d \cdot (1 - p_G)$.

S_{00} to S_{10}: a 1 is written into cell C_i and a 0 into cell C_j. The probability of this transition is $p_d \cdot (1 - p_d)$.

S_{00} to S_{11}: a 1 is written into cells C_i and C_j. The probability of this transition is p_d^2.

S_{00} to S_{00}: a 0 is written into cells C_i and C_j. The probability of this transition is $(1 - p_d)^2$.

S_{00} to S_D: a 0 is written into cell C_i and a 1 into cell C_j, while the deleted neighborhood contains the value required to allow the fault to be sensitized. The probability for this transition is $p_d \cdot (1 - p_d) \cdot p_G$.

A transition to the absorbing state S_D can only be made from those states in which the logic value of C_j is 0, because otherwise the $\langle\uparrow; \uparrow\rangle$ CFid cannot be sensitized; this transition is, therefore, only possible from S_{00} and S_{10}.

Assuming that the memory is initialized to all 1s (i.e., the initial state is S_{11}), the probability that an $\langle\uparrow; \uparrow\rangle$ k-coupling fault will be detected after t test vectors can be computed to be

$$S_D(t) = 1 + \frac{1-\alpha}{2\cdot\alpha}\cdot\left(\frac{1-\alpha}{2}\right)^t - \frac{1+\alpha}{2\cdot\alpha}\cdot\left(\frac{1+\alpha}{2}\right)^t, \tag{14.19}$$

where $\alpha = \sqrt{1 - 4\cdot p_G\cdot p_d\cdot(1-p_d)^2}$. \hfill (14.20)

For a large number of test patterns, the escape probability $e = 1 - S_D(t)$ can be approximated by

$$e \approx \frac{1+\alpha}{2\cdot\alpha}\left(\frac{1+\alpha}{2}\right)^t. \tag{14.21}$$

Step 1: ↕(w1) {Initialize memory.}

      repeat t times

Step 2: Generate address A pseudo-randomly and perform an 'r' operation

      with probability $1 - p_w$ or a 'w?' operation with probability p_w

Figure 14.37 The RARWRD algorithm

The value of t, which satisfies Equation (14.21), is the required test length T:

$$T(e) = \left\lceil \ln\left(\frac{2 \cdot \alpha \cdot e}{1 + \alpha}\right) / \ln\left(\frac{1 + \alpha}{2}\right) \right\rceil. \tag{14.22}$$

The above test is capable of detecting all k-coupling faults, i.e., those for which the address of C_i < address of C_j, and those for which the address of C_i > address of C_j. For the case 'address C_i < address C_j', the march element '⇑(r1,w?)' of Step 2 of the algorithm of Figure 14.35 detects the $\langle\uparrow;\uparrow\rangle$ k-coupling fault in the following way: a '0' is written into C_i, after which an ↑ transition write operation takes place when the march element operates on C_j; upon the next pass through memory, C_i is read and the fault is detected. For the case 'address C_i > address C_j', the $\langle\uparrow;\uparrow\rangle$ k-coupling fault is detected as follows: an ↑ transition write operation takes place into C_j while the contents of C_i is '0'; when the march element operates on C_i, C_i is read and the fault is detected.

14.6.3.2 Random memory test (RARWRD)

In this test, the memory is initialized, after which t operations are performed on addresses A which are determined pseudo-randomly. The operation to be performed is an 'r' operation with probability $1 - p_w$ and a 'w?' operation with probability p_w. The RARWRD algorithm is shown in Figure 14.37.

Table 14.7 (Savir *et al.*, 1989) shows the test lengths for both tests of this section, for different values of k for $p_a = p_d = p_w = 0.5$ and $e = 0.001$. It should be noted that an increase of the value of k by 1 doubles the number of operations; furthermore, the result applies to memories of any size. Note that the DADWRD test is more efficient because it is less random; only the value of p_d is determined pseudo-randomly.

14.6.4 Summary of pseudo-random memory tests

Table 14.8, which has been taken from David *et al.* (1989), compares deterministic with pseudo-random memory tests. An **active NPSF (ANPSF)** is a fault in which the base cell is forced to a fixed value due to a transition of one of the deleted neighborhood cells (Suk and Reddy, 1981). An active pattern sensitive fault (APSF) is an unrestricted NPSF; the deleted neighborhood cells may be located anywhere in memory. APSF is another name for the k-coupling fault of Section 14.6.3. The numbers in Table 14.8

Table 14.7. *Comparison of the DADWRD and RARWRD tests*

k	p_G	Test	
		DADWRD	RARWRD
2	1	$90 \cdot n$	$228 \cdot n$
3	p_d	$202 \cdot n$	$449 \cdot n$
4	p_d^2	$424 \cdot n$	$891 \cdot n$
5	p_d^3	$866 \cdot n$	$1775 \cdot n$

ⓒ 1989 Proc. IEEE Int. Test Conference

Table 14.8. *Comparison of deterministic and pseudo-random tests*
ⓒ *1989 IEEE Trans. on Computers*

Fault	Test length			
	Deterministic	Pseudo-random		
		$e = 0.01$	$e = 0.001$	$e = 0.000\,001$
SAF	$5 \cdot n$	$33 \cdot n$	$46 \cdot n$	$93 \cdot n$
CFid	$15 \cdot n$	$145 \cdot n$	$219 \cdot n$	$445 \cdot n$
ANPSF $k = 3$	$28 \cdot n$	$294 \cdot n$	$447 \cdot n$	$905 \cdot n$
APSF $k = 3$	$n + 32 \cdot n \cdot \log_2 n$	$294 \cdot n$	$447 \cdot n$	$905 \cdot n$
ANPSF $k = 5$	$195 \cdot n$	$1200 \cdot n$	$1805 \cdot n$	$3625 \cdot n$
APSF $k = 5$	?	$1200 \cdot n$	$1805 \cdot n$	$3625 \cdot n$

differ by a small (negligible) amount from those given earlier. An explanation for this cannot be given because the tests on which the numbers of Table 14.8 are based are not given in David *et al.* (1989).

From inspecting the table it can be concluded that, as expected, pseudo-random tests are less efficient for detecting simple faults (such as SAFs), and for faults for which deterministic tests are rather easy to construct (such as ANPSFs with $k = 5$).

Pseudo-random tests do have an advantage where deterministic tests require excessive test lengths because of the inherent requirement that they have to detect all faults of a particular fault model. This is the case for the general k-coupling fault with $k = 3$; for higher values of k no deterministic tests exist as of this date, such that pseudo-random tests are the only possibility. Pseudo-random tests are also preferred in those cases where *ease of implementation* is important. For example, a deterministic test for ANPSFs is hard to implement (van de Goor, 1998) while a pseudo-random test for even APSFs is very easy to implement. Therefore, pseudo-random tests may be good candidates for BIST and tests for small memories, especially because in such situations the test time is usually less important.

Summary

- The challenge in testing memories is that the test cost is not allowed to increase while the number of bits per chip increases exponentially in time and the fault models become more complex.

- An n-bit memory may be organized logically as $2^N = n$ addresses with $B = 1$ bit words. Physically, the memory cell array is organized as one or more matrices of cells. A cell is addressed using a row and a column address.

- SRAM cells retain their information using a bistable circuit; DRAM cells retain their information in terms of the charge stored in a capacitor, which, due to leakage, has to be refreshed.

- The reduced functional fault model reduces the functional model to three blocks: the address decoder, the read/write logic and the memory cell array. Faults in the first two blocks can be mapped onto those in the last block.

- The fault models for the memory cell array are: SAF, SOpF, TF, DRF, CFin, CFid and CFst. Address decoders have their own fault models; faults A, B, C and D.

- Linked faults have the property that, unless the test has been especially designed for this class of faults, fault masking may occur.

- March tests are efficient tests for detecting faults in SRAMs; they are of $O(n)$ and when they satisfy certain requirements they are guaranteed to detect AFs.

- The checkerboard test has the advantage that it can be used as a refresh test for DRAMs; the GALPAT and Walking 1/0 tests can be used for locating faults. Additionally, GALPAT can be used for detecting write recovery faults in the address decoder, and Walking 1/0 can be used to detect sense amplifier recovery faults.

- March tests can be guaranteed to detect SOpFs, given a non-transparent sense amplifier, when the test contains a march element of the form: $\ldots, rx, \ldots, r\bar{x}, \ldots$. The detection of DRFs requires a cell to be read in both states after a delay (required for the fault to develop).

- Pseudo-random memory tests can be classified according to the way (random or deterministic) the write/read operation, address and data (0/1) are determined. A deterministic test which detects all SAFs has a test length coefficient of five; the RARWRD pseudo-random test has a $T(0.001) = 49$, which is much longer. The advantage of pseudo-random tests is that they are suitable for BIST and that they can cover a high percentage of k-coupling faults.

- The test length coefficient of a pseudo-random test is independent of the size of the memory (n), proportional to the logarithm of e (the escape probability), and doubles when k increases by 1.

Exercises

14.1 Suppose that a certain faulty memory has the property that when multiple cells are accessed simultaneously (in the case of AF D), the returned value of the read operation is the OR of the contents of the read cells (see Section 14.3.5). Determine the conditions a march test has to satisfy in order to detect this type of AF.

14.2 If all march elements in a march test use the same address order (i.e., either the $\Uparrow$ or the $\Downarrow$ address order):

   (a) Which type(s) of unlinked AFs cannot be detected?

   (b) Which type(s) of unlinked CFids cannot be detected?

   Hint: Derive the conditions a march test has to satisfy in order to detect all types of AFs and all types of CFids. These conditions can be expressed in terms of the march element (or pair of march elements) the march test has to contain in order to detect a particular type of AF or CFid. Check then whether the conditions require both the $\Uparrow$ and $\Downarrow$ address orders.

14.3 Modify MATS+ such that

   (a) it is capable of detecting SOpFs (assuming a non-transparent sense amplifier) in addition to AFs and SAFs, and

   (b) it is capable of detecting DRFs in addition to AFs and SAFs.

14.4 Suppose we are given a new fault model: disturb coupling fault (CFds), defined as follows:

   1 Write operations can sensitize faults (regardless of whether they are transition or non-transition write operations).

   2 Read operations also can sensitize faults.

   The notation for this fault model is the following: $\langle w0; \downarrow \rangle$, $\langle w0; \uparrow \rangle$, $\langle w1; \downarrow \rangle$, $\langle w1; \uparrow \rangle$, $\langle r0; \downarrow \rangle$, $\langle r0; \uparrow \rangle$, $\langle r1; \downarrow \rangle$, $\langle r1; \uparrow \rangle$.

   Prove that March C− detects all unlinked CFdss.

14.5 Assume a memory with B-bit words ($B \geq 2$). Modify the MATS+ algorithm such that it detects all AFs and SAFs in a memory with B-bit words. Give an example of a MATS+ algorithm for a memory with $B = 4$.

14.6 Suppose we are given a first-in first-out (FIFO) memory. FIFOs have separate write and read ports. Each port has its own address register which automatically increments upon completion of a read (write) operation; i.e., the read address (RA) is incremented upon the completion of a read operation, and the write address (WA) upon completion of a write operation. A reset operation resets both RA and WA to 0. What are the restrictions for march tests when applied to FIFOs?

14.7 Someone has designed the following test, called the write-address test (WAT), for a memory consisting of 1024 8-bit words:

For I = 0 to 1023 do A[I] := (I mod 256);

This means that address 0, A[0], gets the value 0 (i.e., A[0]:=0), A[1]:=1, ..., A[255]:=255, A[256]:=0, A[257]:=1, etc.

For I = 0 to 1023 do { Read A[I]; A[I]:= 255−(I mod 256) };

i.e., Read A[I] and Write the complement of (I mod 256).

For I = 0 to 1023 do Read A[I]

Verify whether this test detects all unlinked AFs, SAFs, TFs, and CFsts. *Hint:* In order to get a quick feel for the fault coverage, assume initially that the memory consists of 1-bit words (i.e., $B = 1$).

14.8 Design a minimal march test which detects the following faults:

(a) Linked CFids of the form: $\langle \uparrow; \downarrow \rangle a_1 v \# \langle \uparrow; \uparrow \rangle a_2 v$.
Note: This fault consists of two CFids, for which $a_1 < v$ and $a_2 < v$; $a_1 < v$ means that the address of a-cell a_1 is lower than the address of the v-cell.

(b) Linked CFids of the form $\langle \uparrow; \downarrow \rangle \# \langle \uparrow; \uparrow \rangle$. The fault consists of two CFids; the a-cells may take on any position relative to the v-cell.

14.9 The test length of pseudo-random tests is larger than that of deterministic tests (see Section 14.6).

(a) What would be the reasons for implementing pseudo-random tests?

(b) What would be the reasons for not implementing pseudo-random tests?

References

Aadsen, D.R, Scholz, H.N. and Zorian, Y. (1990). Automated BIST for regular structures embedded in ASIC devices. *AT&T Technical J.*, **69** (3), pp. 97–109.

Abadir, M.S. and Reghbati, J.K. (1983). Functional testing of semiconductor random access memories. *ACM Computing Surveys*, **15** (3), pp. 175–198.

Bardell, P.H., McAnney, W.H. and Savir, J. (1987). *Built-In Test for VLSI: Pseudorandom Techniques*. John Wiley & Sons, New York, NY.

Breuer, M.A. and Friedman, A.D. (1976). *Diagnosis and Reliable Design of Digital Systems*. Computer Science Press, Inc., Woodland Hills, CA.

David, R., Fuentes, A. and Courtois, B. (1989). Random pattern testing versus deterministic testing of RAM's. *IEEE Trans. on Computers*, **C-38** (5), pp. 637–650.

Dekker, R., Beenker, F. and Thijssen, L. (1988). Fault modeling and test algorithm development for static random access memories. In *Proc. Int. Test Conference*, pp. 343–352.

Dekker, R., Beenker, F. and Thijssen, L. (1990). A realistic fault model and test algorithms for static random access memories. *IEEE Trans. on Computer-Aided Design*, **C-9** (6), pp. 567–572.

Inoue, M., Yamada, T and Fujiwara, A. (1993). A new testing acceleration chip for low-cost memory tests. *IEEE Design & Test of Computers*, **10** (1), pp. 15–19.

Isaacson, D.L. and Madsen, R.W. (1976). *Markov Chains Theory and Applications*. John Wiley & Sons, New York, NY.

Knaizuk Jr., J. and Hartmann, C.R.P. (1977). An optimal algorithm for testing stuck-at faults in random access memories. *IEEE Trans. on Computers*, **C-26** (11), pp. 1141–1144.

Maly, W. (1985). Modeling of lithography related yield losses for CAD of VLSI circuits. *IEEE Trans. on Computer-Aided Design*, **CAD-4** (3), pp. 166–177.

Marinescu, M. (1982). Simple and efficient algorithms for functional RAM testing. In *Proc. Int. Test Conference*, pp. 236–239.

Mazumder, P. (1988). Parallel testing of parametric faults in three-dimensional random-access memory. *IEEE J. of Solid-State Circuits*, **SC-23** (4), pp. 933–941.

McAnney, W.H., Bardell, P.H. and Gupta, V.P. (1984). Random testing for stuck-at storage cells in an embedded memory. In *Proc. Int. Test Conference*, pp. 157–166.

Naik, S., Agricola, F. and Maly, W. (1993). Failure analysis of high density CMOS SRAMs. *IEEE Design & Test of Computers*, **10** (2), pp. 13–23.

Nair, R., Thatte, S.M. and Abraham, J.A. (1978). Efficient algorithms for testing semiconductor random-access memories. *IEEE Trans. on Computers*, **C-28** (3), pp. 572–576.

Nair, R. (1979). Comments on 'An optimal algorithm for testing stuck-at faults in random access memories'. *IEEE Trans. on Computers*, **C-28** (3), pp. 258–261.

Papachristou, C.A. and Saghal, N.B. (1985). An improved method for detecting functional faults in random access memories. *IEEE Trans. on Computers*, **C-34** (3), pp. 110–116.

Rideout, V.L. (1979). One-device cells for dynamic random-access memories. *IEEE Trans. on Electron. Devices*, **ED-26** (6), pp. 839–852.

Romanovsky, V.I. (1970). *Discrete Markov Chains*. Wolters-Noordhoff Publishing, Groningen, The Netherlands.

Savir, J., McAnney, W.H. and Vecchio, S.R. (1989). Testing for coupled cells in random-access memories. In *Proc. Int. Test Conference*, pp. 439–451.

Shen, J.P., Maly, W. and Ferguson, F.J. (1985). Inductive fault analysis of CMOS integrated circuits. *IEEE Design & Test of Computers*, **2** (6), pp. 13–26.

Suk, D.S. and Reddy, S.M. (1981). A march test for functional faults in semiconductor random-access memories. *IEEE Trans. on Computers*, **C-30** (12), pp. 982–985.

Thatte, S.M. and Abraham, J.A. (1977). Testing of semiconductor random access memories. In *Proc. Int. Symposium on Fault-Tolerant Computing*, pp. 81–87.

Thévenod-Fosse, R. and David, R. (1978). Test aléatoire des mémoires. *Rev. Francaise d'Automatisme, d'Informatique de Rech. Op.*, **12** (1), pp. 43–61.

van de Goor, A.J. and Verruijt, C.A. (1990). An overview of deterministic functional RAM chip testing. *ACM Computing Surveys*, **22** (1), pp. 5–33.

van de Goor, A.J. (1993). Using march tests to test SRAMs. *IEEE Design & Test of Computers*, **10** (1), pp. 8–14.

van de Goor, A.J. (1998). *Testing Semiconductor Memories, Theory and Practice*. ComTex Publishing; Gouda, The Netherlands; http://ce.et.tudelft.nl/~vdgoor/

15 High-level test synthesis

In this chapter, we concentrate on the register-transfer level (RTL) and behavior level of the design hierarchy.

We first discuss different RTL test generation methods: hierarchical, symbolic, functional, and those dealing with functional fault models. We then discuss a symbolic RTL fault simulation method.

Next, we discuss RTL design for testability (DFT) methods. The first such method is based on extracting and analyzing the control/data flow of the RTL circuit. The second method uses regular expressions for symbolic testability analysis and test insertion. These are followed by high-level and orthogonal scan methods.

Under RTL built-in self-test (BIST), we show that some of the symbolic testability analysis methods used for RTL DFT can also be extended to BIST. Then we discuss a method called arithmetic BIST, and a method to derive native-mode self-test programs for processors.

At the behavior level, we first show how behavioral modifications can be made to improve testability. We also present three types of behavioral synthesis for testability techniques. The first type targets ease of subsequent gate-level sequential test generation. The second type deals with ease of symbolic testability using precomputed test sets of different RTL modules in the circuit. The third type is geared towards BIST.

15.1 Introduction

High-level test synthesis refers to an area in which test generation, fault simulation, DFT, synthesis for testability, and BIST are automatically performed at the higher levels, i.e., register-transfer and behavior levels, of the design hierarchy. In recent years, there has been growing interest in high-level test synthesis. This is because evidence suggests that high-level test synthesis can lead to a substantial decrease in the test generation time, test application time, and area/delay/power overheads, while at the same time increasing the fault coverage. Tackling the testability problems at the higher levels also reduces the number of design iterations, which is important from the point of view of time-to-market.

An RTL circuit usually consists of a datapath and a controller. The datapath consists of a network of registers, functional units, multiplexers and buses. The controller governs the data computation in the datapath by generating appropriate load signals for the registers and select signals for the multiplexers (or buses) and arithmetic-logic units (ALUs). An RTL circuit can be manually designed or else obtained through behavioral synthesis from a behavioral description. The behavioral description expresses the design's functionality algorithmically. It does not contain much information on the circuit structure or timing. It is usually given in some high-level hardware description language, such as VHDL or Verilog, which is compiled into a control/data flow graph (CDFG).

Behavioral synthesis is a two-step process consisting of *scheduling* and *allocation*. **Scheduling** involves assigning a control step to each operation for its execution. Values at data arcs or variables crossing a control step boundary need to be stored in registers. Scheduling usually aims at minimizing the number of control steps or hardware resources. **Allocation** transforms a given scheduled CDFG into an RTL structure. It consists of the following three tasks: binding operations in the CDFG to functional units (**module allocation**), mapping variables to registers (**register allocation**), and defining the interconnection among the functional units and registers (**interconnection allocation**).

Since there are fewer functional primitives that one has to deal with at the higher levels, it is usually possible to devise faster methods to analyze the testability of the circuits and derive tests for them. However, to obtain very high fault coverages, this process needs to be aided through appropriate higher-level DFT, synthesis for testability and BIST techniques. The aim of this chapter is to present such techniques. We will step up the hierarchy from the register-transfer to the behavior level. We begin with RTL test generation methods.

15.2   RTL test generation

In this section, we discuss four methods for RTL test generation. The first one is based on hierarchical test generation, which exploits the RTL and gate-level descriptions of a circuit. The second one is based on symbolic test generation for microprocessors. The third one deals with functional test generation for processors. The fourth one uses functional fault models to speed up test generation.

Although significant progress has been made in the area of RTL test generation, it is still not a mature area yet. However, as more and more circuits begin to be synthesized in a hierarchical fashion, the need for efficient higher-level test generation techniques assumes greater importance.

15.2.1   Test generation with a known initial state

A technique for performing hierarchical test generation is described in Ghosh *et al.* (1993). This technique assumes that test generation always starts with a known initial state, i.e., the reset state. The reset state is assumed to be valid for both fault-free and faulty circuits. Another limitation is that it is efficiently applicable to only datapath type circuits, not controllers. However, up to 100-fold speed-up in test generation time is possible using such techniques compared to traditional gate-level sequential test generation.

As in the case of the sequential gate-level test generation technique described in Chapter 5 (Section 5.4.2.1) that assumes a reset state, test generation is divided up into three phases: derivation of the excitation vector, state justification sequence, and state-pair differentiating sequence. The present state part of the excitation vector constitutes the **excitation state**.

The RTL modules supported by this technique are different arithmetic and logic functional units, multiplexer, demultiplexer, encoder, decoder, comparator, finite-state machine, and arbitrary Boolean functions. The excitation vector is derived from the gate-level description of the circuit. However, state justification and differentiation are done at the RTL. Thus, the test generator has to traverse the design hierarchy between RTL and gate level. Wires at the RTL could be single wires or buses. Hence, they may correspond to multiple gate-level wires. This makes it necessary to maintain a correspondence between RTL and gate-level wires. This correspondence is required for the primary inputs, present state wires, primary outputs, and next state wires only. Maintaining correspondence between intermediate wires at the RTL and gate level is neither necessary nor desired. This is owing to the fact that logic synthesis tools may alter the structure imposed by an RTL description such that some intermediate wires may disappear, while some new ones may appear. Integers are used to represent values on wires at the RTL and Boolean values are used at the gate level.

After gate-level combinational test generation for the targeted fault, the excitation vector is examined to determine if its present state part includes the reset state. If it does, the fault can be excited from the reset state. If it does not, backward justification is used to derive the state justification sequence. Backward justification considers the excitation state as a next state and uses the fault-free RTL description to justify the values on the next state wires over multiple time frames. The resulting justification sequence is fault simulated to determine if the excitation state is justified in the faulty machine. If the justification sequence is not valid, then a valid subsequence can be found, as illustrated in Figure 5.19 in Chapter 5.

Consider the RTL circuit shown in Figure 15.1. Wires IN_1, IN_2, IN_3, and s_1 denote primary inputs, PS_1 and PS_2 denote the present state wires, OUT is a primary output, and NS_1 and NS_2 are next state wires corresponding to PS_1 and PS_2, respectively. C_1 and C_2 are intermediate wires. Except for s_1 and C_2, the remaining wires have multiple bits. The functional units are an adder (ADD), multiplier (MUL) and comparator (COMP). The output of the comparator is 0 if the value at PS_1 is greater than the value at PS_2, otherwise it is 1. Suppose that a certain value has to be justified at NS_2. This can be done by assigning values on wires PS_1 and IN_3 so that their sum is the same as the required value at NS_2. Hence, initially, PS_1 and IN_3 are *free* to take any

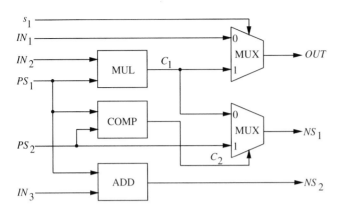

Figure 15.1 Illustration of state justification and differentiation

value. However, once either wire is assigned a value, the other one is forced or *bound* to take a fixed value in order to justify the value at NS_2. A wire in the circuit is said to be **free** if it can take more than one value without causing any *conflict*. A wire that is not free is said to be **bound**. A **conflict** is a situation where a particular wire is required to have two or more different values simultaneously in order to justify the values on the other wires. For example, suppose we need to justify $NS_1 = 20$ and $NS_2 = 30$. Suppose in the first step, C_2 is made 1. This binds PS_2 to the value 20. To justify $NS_2 = 30$, suppose IN_3 is chosen as the free wire and is assigned a value 0. Then PS_1 is bound to 30. However, since the value of PS_1 is greater than the value of PS_2, C_2 is forced to 0. This causes a conflict at C_2 and backtracking is required to consider other choices of values.

Before state justification, the circuit is levelized from the primary outputs backward. Primary output and NS wires are assumed to have level 0. The level of any module is the minimum of the levels of all its fanout wires. The level of any intermediate wire is the minimum of the levels of its fanout modules plus one. Wires at a particular level are assigned values to justify the values on wires at lower levels. The justification procedure is recursively called until the primary inputs are reached. During state justification, only the NS wires, which have level 0, have values to be justified. Justification starts with wires at level 1. Initially, a free wire at a given level is heuristically assigned a value and this decision is stored in a decision tree. All wires which are directly or indirectly bound by the free wire are then assigned values. Then the circuit is simulated to detect any conflicts (this is similar to forward implication performed in the D-algorithm or PODEM). If there is a conflict, a conflict resolution procedure is called to form a system of equations with the wires involved in the conflict. In these equations, all such wires are assigned a variable and the equations represent their interdependence. The system of equations is solved to obtain the correct value assignment for the wires. If no solution can be found, the last free wire is assigned a new value and the process repeated. The backtracking procedure goes through all the values of a free wire before backtracking to the previous node in the decision tree. If all choices in the decision tree are examined without finding a solution, the state is declared to be an invalid (i.e., unreachable) state, and stored along with other invalid states found before.

Example 15.1   Consider the circuit shown in Figure 15.1 once again. Suppose the reset state is the all-0 state. As before, suppose that a justification sequence is needed to justify the value set $NS_1 = 20$ and $NS_2 = 30$. After the conflict mentioned before arises, a system of equations can be formed as follows.

$$PS_1 + IN_3 = 30$$
$$PS_1 \leq PS_2$$
$$PS_2 = 20$$

One solution for this set of equations is $PS_1 = 0$ and $IN_3 = 30$. The values on the present state wires are now 0 and 20. Since this does not include the reset state, this new state has to be justified in the previous time frame, where they become next state values. Once again, assigning 1 to C_2 forces PS_2 to have the value 0. The values at NS_2 can be justified by $PS_1 = 0$ and $IN_3 = 20$. Since the value on the present state wires is now $PS_1 = PS_2 = 0$, the reset state is reached. Therefore, a two-vector justification sequence for the specified state has been found at primary inputs (IN_1, IN_2, IN_3, s_1) as $\{(x, x, 20, x), (x, x, 30, x)\}$, where x denotes a don't care.   □

The excitation vector produces fault-free/faulty next state pairs. The method seeks to derive a differentiating sequence for this state pair in the fault-free machine. If an input vector exists for which a particular output depends on a present state wire to which the fault effect has been propagated, then a single-vector differentiating sequence is obtained. Such a vector does not require the other present state wires to have any particular value. For example, in the circuit shown in Figure 15.1, a differentiating input vector for PS_1 is $IN_2 = 1$, $s_1 = 1$ (other inputs are don't cares). However, there is no differentiating input vector for PS_2. Such vectors are found before test generation begins. This improves the efficiency of test generation.

If a differentiating input vector does not exist, then propagation of the fault effect from the present state wires with the fault effect to a primary output is attempted by taking into account the values on the other present state wires. If successful, the differentiating sequence still has only one vector. If this does not succeed, an attempt is made to propagate the fault effect to a heuristically chosen next state wire. If a particular present state wire has a differentiating input vector, then the corresponding next state wire is chosen. If no present state wire has a differentiating input vector, then the one that requires the least number of side inputs to be set to sensitize a path from that wire to the output is chosen. Once the fault effect is propagated to a next state wire, a single-vector differentiating sequence between the new fault-free/faulty state pairs is obtained, if possible. If not, the above process is repeated.

Example 15.2 Consider the circuit shown in Figure 15.1 once again. Let the fault-free state be $PS_1 = 10$, $PS_2 = 20$, and the faulty state be $PS_1 = 10$, $PS_2 = 5$. Since there is no path from PS_2 to the primary output, the fault effect has to be propagated to the next state wire NS_1. Therefore, C_2 has to be set to 1. For the fault-free state, the value of C_2 is 1, as required. However, for the faulty state, the value of C_2 is 0. Since the fault-free value on NS_1 has been determined to be 20, the procedure has to make sure that the faulty value on NS_1 is different. Therefore, C_1 is set to a value different than 20. Say, the chosen value is 30. The justification procedure is called to obtain this value. This can be done by setting IN_2 to 3. Thus, a two-vector differentiating sequence is obtained at primary inputs (IN_1, IN_2, IN_3, s_1) as $\{(x, 3, x, x), (x, 1, x, 1)\}$.

□

A limitation of this method is that even if a differentiating sequence exists in the fault-free circuit, the sequence may not be valid in the faulty machine, and the fault is not detected. Conversely, a test may exist for a fault, but no fault-free differentiating sequence may exist for it. However, it has been found experimentally that the percentage of times the potential test sequence indeed detects the targeted fault is very close to 100%.

The above procedure is best suited to circuits in which any state can be differentiated from any other within a few time frames. Most datapath-type circuits meet this condition.

15.2.2　Symbolic test generation for microprocessors

Another technique, which works at the RTL and gate level, has been presented in Lee and Patel (1994). This technique is more suitable for microprocessor-like circuits. Its limitation is that in the controller–datapath implementations of such circuits, it cannot handle faults in the controller. Also, cyclic RTL circuits pose a problem for it.

Microprocessors have complex datapaths and embedded control machines to execute instructions. In order to apply a particular vector to the inputs of an embedded module in the datapath, an instruction sequence and internal wire values without conflict have to be derived. This is not easy. In order to make the problem more tractable, the main part of the test generation process can be divided up into two phases: *path analysis* and *value analysis*. In the **path analysis phase**, a sequence of instructions is generated to satisfy the internal test goals. In the **value analysis phase**, exact values are computed for internal wires. The test vector for a fault in the embedded module can be generated through gate-level test generation. However, path and value analyses are performed at the RTL. Also, since only datapath faults are considered, the instructions are assumed to always be valid ones.

The complete test generation process consists of the following phases: preprocessing, test vector generation for the embedded module, path analysis, value analysis. In the preprocessing phase, *symbolic simulation* is used to derive a system of module equations and a *structural data-flow graph* (SDFG) for each instruction. In addition, a *flow-influence model* is used to model the circuit behavior for each instruction. Since the test vector generation phase for the embedded module is performed at the gate level, it is straightforward. We next describe the other three phases in detail.

15.2.2.1 Preprocessing

Different instructions of the microprocessor are implemented by different parts of the datapath, represented as an RTL circuit. **Symbolic simulation** is separately employed for the RTL circuit implementing each instruction. Symbols are first assigned to the different wires in the RTL circuit. Some fixed values, which are either from control words of the instruction or are constant values, are appended to some control lines and wires. All other symbolic values are set to unknown x initially. Symbols are first propagated forward as much as possible using the functionalities of the different modules in the RTL circuit. Then a backward symbol elimination process gets rid of symbols, which are not useful, at the non-selected inputs of some modules such as multiplexers. For each module, none of whose input symbols is the same as its output symbols, a module equation is generated.

The concept of symbolic simulation is illustrated through an example next.

Example 15.3 Consider the RTL circuit shown in Figure 15.2(a) which implements an instruction in one cycle. Before symbolic simulation, each wire in the circuit is assigned a symbol $(v_1, v_2, \ldots, v_{14})$. Select lines s_1, s_2, s_3 have fixed values. All other symbolic values are set to unknown x. The symbols and implied values after symbolic simulation are shown in Figure 15.2(b). Because of the fixed value at s_2, the symbol at wire IN_1 propagates to wire C_8. Similarly, the symbol at wire C_{10} propagates to wire OUT. The symbols at wires IN_4, C_2, C_6, C_9 do not contribute to the output OUT, and are, hence, eliminated from consideration for this instruction. The corresponding wires are indicated by dotted lines. The symbol at wire C_4 is not useful for the multiplexer it feeds. However, since the other output, C_5, of the *Wire Split* module is useful, the symbol at wire C_4 is not eliminated. The module equations are shown below.

$(ALU)(v_2, v_3, v_5) = (v_8)$

$(\textit{Wire Split})(v_8) = (v_9, v_{10})$

$(\textit{Incrementer})(v_1, v_{10}) = (v_{13})$ ☐

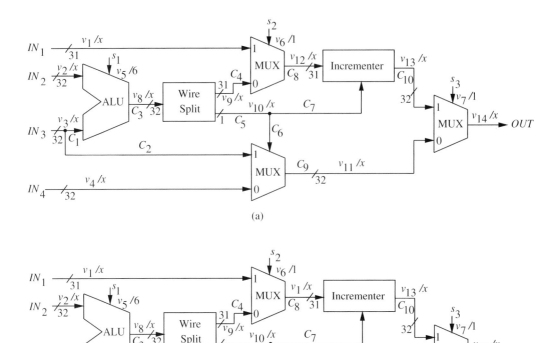

Figure 15.2 Symbolic simulation: (a) before, (b) after (Lee and Patel, 1994) (© 1994 IEEE)

Structural data-flow graph: The SDFG of each instruction is obtained from the system of equations derived in the symbolic simulation step. Each equation contributes to one node in the SDFG. An arc in an SDFG corresponds to a symbol in the equation, and indicates signal flow from one module to another. Extra nodes are added for primary inputs, present state wires, primary outputs and next state wires.

The SDFG of the instruction used in Example 15.3 is derived from the module equations, and is shown in Figure 15.3. The three intermediate nodes correspond to the three module equations. The remaining nodes are input/output nodes.

Flow-influence model: For each type of primitive in the RTL design library, which is a building block of the RTL circuit, a flow-influence model is obtained to help in justification and propagation path selection. Each input of a primitive is classified as *strongly or weakly influencing* as follows.

- An input is **strongly influencing** if it must be justified to justify an output of the module.

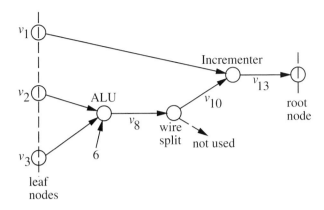

Figure 15.3   An SDFG (Lee and Patel, 1994) (ⓒ 1994 IEEE)

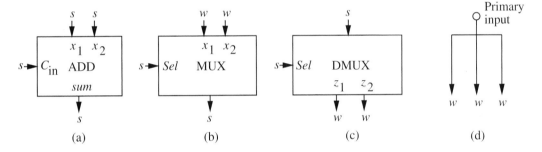

Figure 15.4   Examples illustrating the flow-influence model (Lee and Patel, 1994) (ⓒ 1994 IEEE)

- An input is **weakly influencing** if it need not be justified to justify the output, and exactly one of all weakly influencing inputs is able to affect the output.

These definitions are general enough so that they are applicable to all possible value assignments at the inputs and outputs of the primitive. Some examples shown in Figure 15.4 illustrate the concept. In Figure 15.4(a), all the inputs of an adder are shown to be strongly influencing. This means that all the inputs of an adder have to be justified in order to justify its output. For the multiplexer shown in Figure 15.4(b), the select input and either of the two inputs have to be justified in order to justify the output. Therefore, the select input is strongly influencing and the other inputs are weakly influencing.

Using a similar concept, each output of a primitive is classified as *strongly or weakly influenced* as follows.

- An output is **strongly influenced** if it is always justified by the inputs, i.e., the justifiability does not depend on the input values.
- An output is **weakly influenced** if it is not always justified by the inputs, and only one of all weakly influenced outputs of a primitive is justified.

Consider the demultiplexer shown in Figure 15.4(c). Both its outputs are weakly influenced because only one of them can at a given time be controlled by the inputs.

A weak influence is assigned to the output arcs of a leaf node in an SDFG if the leaf node has a fanout greater than one. This is owing to the fact that the fanout arcs of a primary input or present state wire are not independently controllable. Once one arc assumes a value, the other fanout arcs of that node have to assume the same value. This is depicted in Figure 15.4(d). However, if the leaf node has only one output arc, the arc is assigned a strong influence. Since only one signal can arrive at the root node (primary output or next state wire) of the SDFG at any cycle, each root node has only one input arc. This arc is assigned a strong influence.

15.2.2.2 Path analysis

In the path analysis phase, the functional information of modules is only used to determine the active paths, not the exact values on these paths. The flow-influence model is used to derive a justification cost of each present state wire of the datapath. This cost indicates the minimum number of instructions required to set up a sequence of SDFGs for controlling that state wire only. A propagation cost is similarly obtained to indicate the minimum number of instructions required to observe the particular next state wire. These costs help in assembling the required instruction sequence.

In order to obtain the justification costs, a *minimum dependence cone* of each next state wire is first obtained. This cone contains the minimum number of present state wires that have to be justified before the next state wire in question can be justified. For example, Figure 15.5 shows a minimum dependence cone of next state wire (denoted as NSW) NS_2. This cone consists of one primary input and two present state wires (denoted as PSW) PS_1 and PS_3. For justification purposes, only these two wires are recorded. In addition, only those instructions are useful for justifying NS_i in which a dependence cone does not contain PS_i. Otherwise, a cyclic dependency makes it difficult to justify the state. For example, another dependence cone can be obtained for the above example which contains the present state wires PS_2 and PS_3. However, since we are trying to justify NS_2, we should not use a dependence cone which contains PS_2. The above method can be repeatedly used to justify the targeted next state wire in the SDFG of each instruction in the instruction set. Among all these, that instruction is chosen for this next state wire in which the dependence cone has the least number of present state wires.

The justification costs of all next state wires are initialized to infinity. If a minimum dependence cone of a next state wire consists of only primary inputs, its justification cost is set to zero. If an instruction initializes a next state wire by forcing it to a fixed value, its minimum dependence cone is not defined. The justification costs of all other state wires can be computed iteratively. For example, in the above case, PS_1 and PS_3, which belong to the minimum dependence cone, become next state wires, NS_1 and NS_3, respectively, in the previous time frame. Therefore, the justification cost of NS_2 is computed as the sum of the justification costs of NS_1 and NS_3 plus one. One is added

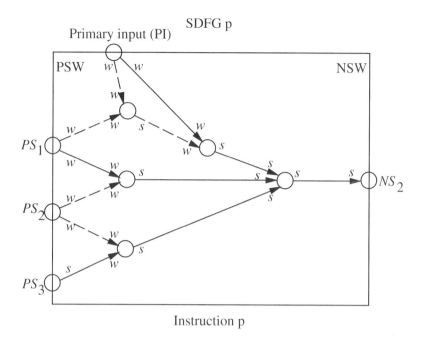

Figure 15.5   Minimum dependence cone of next state wire NS_2

to indicate that it takes one more instruction to justify NS_2 than it takes to justify NS_1 and NS_3.

To justify a test vector at the input of a module under test, an instruction is first obtained that covers all the testing objectives. Then the SDFG of that instruction can be used to obtain the present state wires which are required to justify all the objectives. Thereafter, the instruction sequence assembly method iteratively selects instructions to simultaneously justify all the next state wires. At each step in this method, the SDFG of the instruction is used to check if the instruction can simultaneously justify all objectives left behind by the previous instructions, and the cost of using this instruction based on the accumulated justification costs of present state wires required by the instruction. The justification costs are only used to compute the total cost based on the first cycle of each instruction. All the instructions are retried, and the one requiring least cost is chosen. The instruction assembly process is shown in Figure 15.6.

It may seem odd that justification costs are based on individual controllability of next state wires, even though in reality, most testing objectives require simultaneous justification of next state wires. However, the task of checking the simultaneous justifiability of all the next state wires under the flow-influence model is a very difficult problem. Therefore, the above method serves as a useful heuristic.

Simultaneous justification of the next state wires involves reverse time processing using a branch-and-bound algorithm. All such wires and their input arcs are initially marked. When a node has any of its output arcs marked, the node itself has to be

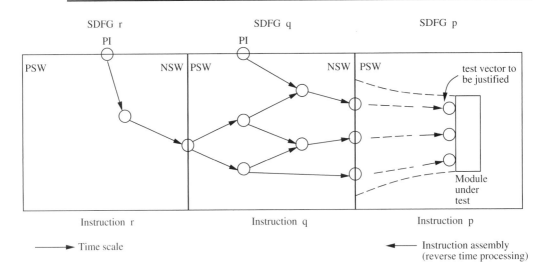

Figure 15.6 Instruction sequence assembly

marked. Whenever a node is marked, all its strongly influencing input arcs are marked. If a node having weakly influencing input arcs is encountered, only one of the arcs is chosen and marked, and this node is put on a decision stack. If more than one weakly influencing input arcs have to be marked under the current status, a conflict happens and backtracking is invoked. This involves trying the next weakly influencing input arc. If all of its weakly influencing input arcs have been tried, the node is popped out and the next node is backtracked to.

The problem of fault effect propagation can be solved in a similar manner by employing forward time processing and using propagation costs instead of justification costs.

15.2.2.3 Value analysis

Given the instruction sequence assembled in the path analysis phase, a complete system of module equations can be obtained. Most of the module equations are non-linear, and cannot be linearized. In addition, the system of equations are *underdetermined*, i.e., the number of variables is larger than the number of equations. Such a system of equations can be solved using discrete relaxation techniques. This is illustrated by the following example.

Example 15.4 Consider the circuit given in Figure 15.2 once again, and the system of three equations presented in Example 15.3. Suppose that the incrementer is the module under test, and a gate-level test generator has derived a test vector for some fault in this module which makes $v_1 = 329\,601$ and $v_{10} = 1$. These two symbols are fixed and their values are implied. Symbol v_{13} is updated to $329\,602$ and also fixed.

An initial guess of 0 is assumed for all unfixed symbols. In the first iteration, the initial guess of 0 at v_8 translates to 0 at v_9 and v_{10}. However, this propagated value of 0 at v_{10} conflicts with the fixed value of 1 at v_{10}. Thus, the value of v_8 is updated to 1. The updated signal strikes back to the first equation. Assume that $v_5 = 6$ corresponds to an addition operation in the ALU. Then to justify $v_8 = 1$, the following values are chosen: $v_2 = 1$ and $v_3 = 0$. The second iteration starts and the method converges. $\square$

15.2.3 Functional test generation for processors

An RTL test generation method for processors based on extraction of functional constraints was presented in Tupuri and Abraham (1997). It targets one embedded module in the RTL circuit at a time. It extracts the functional constraints for the module and synthesizes the constraints at the gate level. The functional constraints consist of justification and propagation constraints. Then a gate-level sequential test generator is used to derive a module-level test sequence. This test sequence is translated to the processor level and fault simulated to verify that the same fault coverage is obtained. The test generation process is illustrated in Figure 15.7.

Processors consist of interconnected modules. A pipeline control logic regulates the data flowing through these modules. Each module can have a local controller and datapath. The **input space** of a module consists of all possible vectors that can be applied to it. Similarly, the **output space** of a module consists of all possible vectors that can be produced by it. The input space of an embedded module is restricted by the output space of the modules feeding its inputs. This provides the justification constraints for the module under test. The output responses of the embedded module are observable at the processor outputs only under certain conditions. These conditions are embodied in the propagation constraints. The constraints are extracted between the module and its surrounding accessible registers (ARs). An AR is an internal register that can be written and read from the primary inputs and outputs of the processor, respectively. A typical processor has a large number of ARs. The justification constraints at the module inputs are a function of the outputs of the ARs. The propagation constraints at the module outputs are a function of the outputs of both the module and ARs. After extracting these functional constraints for a module, they are merged to remove the redundant logic from the gate-level implementation obtained from the merged constraints.

Timing constraints are also extracted in addition to the functional constraints. They represent the number of clock cycles needed to justify module inputs from ARs and propagate module outputs to ARs.

As an example of functional test generation, consider the architectural description of Viper, a 32-bit microprocessor, shown in Figure 15.8. In Viper, the register file and instruction register (not shown) are identified to be ARs. Suppose the embedded module we want to test is the ALU. Since its inputs, *Op1* and *Op2*, are connected to

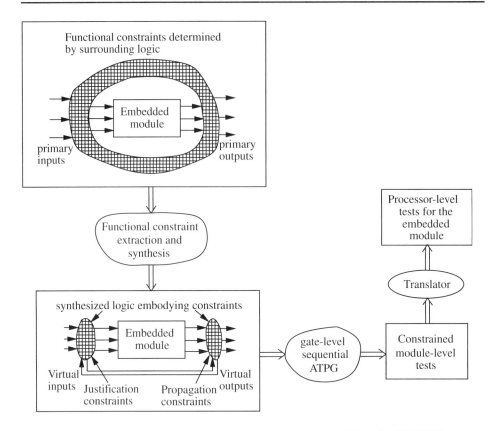

Figure 15.7 Functional test generation process (Tupuri and Abraham, 1997) (© 1997 IEEE)

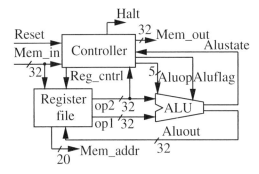

Figure 15.8 Architectural description of Viper

the register file, they are directly controllable. However, since all the ALU operations do not update the ARs, propagation constraints are required at the outputs for *Aluout* and *Alustate*. Similarly, *Aluflag* and *Aluop* need to be justified from the ARs. Thus, justification constraints are needed for these signals. After the constraints are extracted, merged, and synthesized, the transformed module is as shown in Figure 15.9.

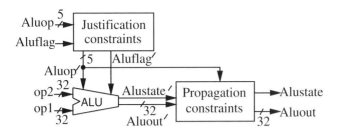

Figure 15.9 Transformed ALU module

The gate-level implementation of the transformed module, generated as above, is used for test generation of the embedded module. A gate-level sequential test generator is used to generate a test sequence for the faults in the embedded module only. Faults in the synthesized constraints are not targeted. Using these test sequences, logic simulation is performed on the transformed module to identify any vectors causing an exception condition. Then a vector translator is used to convert the module-level vectors to processor-level vectors. The other inputs of the vector translator are procedures to load from and store into ARs, and for handling exception conditions and timing constraints. The translated test sequence is fault simulated on the complete gate-level implementation of the processor against the faults in the embedded module to make sure the same fault coverage is obtained. This shows that constraint extraction and translation were performed correctly.

Experimental results show a speed-up of two orders of magnitude, while obtaining a higher fault coverage, compared to purely gate-level sequential test generation of the complete processor.

15.2.4 Test generation with functional fault models

The aim of the test generation method presented in this subsection is to abstract low-level information about a circuit into a functional or high-level form, and generate tests from the functional description (Hansen and Hayes, 1995a). Test generation is limited to one level of the design hierarchy. This is in contrast to hierarchical test generation methods, where various levels of the design hierarchy are traversed by the test generator.

High-level models of circuits contain fewer primitives. If properly exploited, this can result in a significant reduction in test generation time. However, two important issues need to be tackled for the success of such a method: accurate fault modeling and efficient search space management.

Accurate functional fault models, called *physically-induced fault models*, can be obtained to ensure coverage of lower-level faults, such as stuck-at faults (SAFs), which are of interest for both combinational and sequential circuits (Hansen and Hayes,

1995a,b). High-level circuit models can be used to generate tests for SAF-induced faults (SIFs) which can guarantee complete gate-level coverage of SAFs.

The second issue relates to the extremely large search space involved in test generation. Even though it is generally easier to justify and propagate signals functionally, tackling value conflicts still poses a vexing problem. For example, if a logical conflict occurs while propagating an error on an n-bit wide bus, should one try one of the other 2^{n-1} values on the bus or choose another propagation path. Classical test generation algorithms easily get bogged down by this problem. One way to efficiently resolve such conflicts is through *symbolic scheduling*.

15.2.4.1 Functional fault modeling

Consider the relationships among tests, faults and functions, as shown in Figure 15.10. Here, a set of SAFs, Q, for a module produces a set of fault functions, Z_Q. A set of functional faults, F_Q, is defined to establish a one-to-one correspondence between F_Q and Z_Q. Equivalent and dominating faults are removed from F_Q. The test set, T_Q, is then obtained to detect the reduced set of functional faults. In this fashion, SAFs can be detected using only functional information.

Example 15.5 Consider the high-level model of a four-bit carry-lookahead adder shown in Figure 15.11(a). The inputs to module M_2 consist of the four-bit generate (G) and propagate (P) signals from module M_1, as well as the carry-in signal, c_0. The carry-in signal directly propagates through M_2. In addition, M_2 also generates the four carry signals, c_1, c_2, c_3 and c_4. The part of M_2 that implements c_1 is shown in Figure 15.11(b). In this part, denoted as $M_2.c_1$, there are ten SAFs which, upon fault collapsing, reduce to four SAFs. The SIFs derived from these four SAFs are shown in Table 15.1. The entries in row $P_0 G_0 c_0$ indicate the decimal equivalent of the corresponding three-bit binary vector. The bold entries in this table indicate where the faulty functions $c_{1 Fi}$ differ from the fault-free function c_1. An abstract functional fault Fi is identified with each $c_{1 Fi}$ to obtain the induced fault set, F_Q. The Fis have the generic form "Fi permanently enables (disables) function i".

For test generation, the SIF set F_Q can be used with a higher-level "black box" abstraction of this circuit. No low-level details are needed for this type of test generation. Table 15.2 shows the SIF tests, T_Q, derived from the four SIFs. The bold entries represent sensitized inputs from which fault information passes to the output.

Next, consider module M_2. The four SIFs for $M_2.c_1$ can be extended to four composite SIFs for M_2. The composite SIFs are labeled as $CF_1, \ldots, CF_4$. The ten SIF tests derived for these four composite SIFs are shown in Table 15.3. For example, the SIF CF_1 "each P_i cannot propagate with $c_0 = 1$" is detected by the test "propagate each P_i with $c_0 = 1$." As in Table 15.2, the bold entries represent sensitized inputs from which fault information passes to the output. This information is useful when M_2

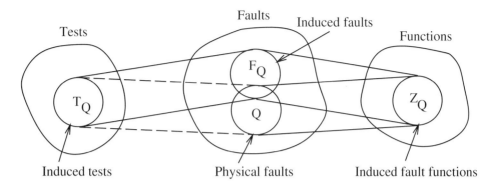

Figure 15.10 Tests, faults and functions (Hansen and Hayes, 1995b) (© 1995 IEEE)

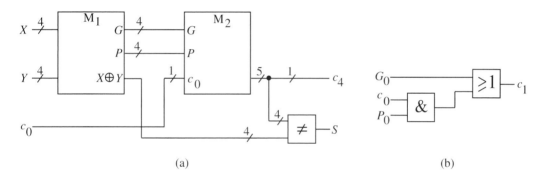

Figure 15.11 (a) High-level model of a carry-lookahead adder, (b) part of module M_2

Table 15.1. *SIFs for circuit $M_2.c_1$*

SAF	$P_0 G_0 c_0$	SIF function								SIF
		0	1	2	3	4	5	6	7	
(Q)	c_1	0	0	1	1	0	1	1	1	(F_Q)
c_0 SA0	$c_{1\,F1}$	0	0	1	1	0	**0**	1	1	no propagate with $c_0 = 1$
c_0 SA1	$c_{1\,F2}$	0	0	1	1	**1**	1	1	1	generate with $c_0 = 0$
P_0 SA1	$c_{1\,F3}$	0	**1**	1	1	0	1	1	1	propagate with $P_0 = 0$
G_0 SA0	$c_{1\,F4}$	0	0	**0**	**0**	0	1	**0**	1	no generate with $G_0 = 1$

is inserted into a larger circuit, as is done in Figure 15.11(a). In general, for an n-bit module M_2, $2n + 2$ tests would be generated.

For the carry-lookahead adder shown in Figure 15.11(a), the testing requirements for module M_2 are simply traced back to the primary inputs of the adder. The testing needs of module M_1 and the XOR word gate are completely covered while testing M_2. Thus, ten tests provide complete SAF coverage for the adder.   ☐

Table 15.2. *Minimal SIF test set for $M_2.c_1$*

SIF	SIF test (T_Q)	c_1	P_0	G_0	c_0
$F1$	Propagate with $c_0 = 1$	1	**1**	0	**1**
$F2$	Stop generate with $c_0 = 0$	0	1	**0**	**0**
$F3$	Stop propagate with $P_0 = 0$	0	**0**	0	**1**
$F4$	Generate with $G_0 = 1$	1	1	**1**	0

Table 15.3. *Minimal SIF test set for the four-bit module M_2*

SIF	SIF test	P_3	G_3	P_2	G_2	P_1	G_1	P_0	G_0	c_0
$CF1$	Propagate with $c_0 = 1$	**1**	0	**1**	0	**1**	0	**1**	0	**1**
$CF2$	Stop generate with $c_0 = 0$	1	**0**	1	**0**	1	**0**	1	**0**	**0**
$CF3$	Stop propagate with $P_0 = 0$	1	**0**	1	**0**	1	**0**	**0**	**0**	1
	Stop propagate with $P_1 = 0$	1	**0**	1	**0**	**0**	**0**	1	1	1
	Stop propagate with $P_2 = 0$	1	**0**	**0**	**0**	1	1	1	1	1
	Stop propagate with $P_3 = 0$	**0**	**0**	1	1	1	1	1	1	1
$CF4$	Generate with $G_0 = 1$	**1**	0	**1**	0	**1**	0	**1**	**1**	0
	Generate with $G_1 = 1$	**1**	0	**1**	0	1	**1**	0	0	0
	Generate with $G_2 = 1$	**1**	0	1	**1**	0	0	0	0	0
	Generate with $G_3 = 1$	1	**1**	0	0	0	0	0	0	0

The above fault modeling and test generation methods frequently result in minimal test sets. They can also be easily extended to sequential modules, and can be used to characterize all members of the RTL design library. The stored composite SIFs can then be used for high-level test generation. We discuss this next.

15.2.4.2 Test generation using symbolic scheduling

The functional test generation method we discuss in this section also divides up test generation into two separate phases of path analysis and value analysis, as in the case of microprocessor test generation discussed in Section 15.2.2. However, path analysis is based on an entirely new approach of symbolic scheduling.

The test generation algorithm contains well-known procedures for state justification and fault effect propagation, with the following new features: symbolic values, multiple-time-step functional operations, and explicit scheduling. Examples of functional operations are load, shift, reset, increment, etc. Functional operations do not receive their absolute order within the tests until the end of processing. This implies that valid parts of a test are allowed to float forward or backward with the sequence of operations until all circuit constraints to propagate the effect of tests are obtained. Thus, low-level processing is deferred until a high-level "road map" is derived.

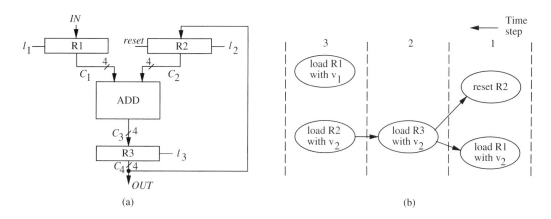

Figure 15.12 (a) Example accumulator, (b) justification operation sequence

The scheduling algorithm is adapted from high-level synthesis applications. It is called with a structured list of operations that are to be performed on the circuit. Its goal is to determine if conflicts occur between two operations in a sequence. If no conflicts occur, the algorithm simply returns this structured list. This completes the path analysis phase. If conflicts occur, the algorithm tries to resolve them. Thereafter, SIFs are used to derive the actual values of signals in the circuit in the value analysis phase. There is no backtracking required during value analysis.

Two well-known scheduling algorithms are as-soon-as-possible (ASAP) and as-late-as-possible (ALAP). ASAP and ALAP determine the earliest and latest time steps, respectively, where a given operation can be scheduled. The difference between these time steps indicates the mobility of the operation. In other words, the operation has the flexibility of being moved forward and backward in this time interval. Using these algorithms, if the scheduler fails to find a valid schedule for the operations, the backtracking mechanism returns to the operations causing the conflict, and tries an alternative sequence. The complexity of the scheduling algorithm is related to the length of the symbolic test sequences, not the size of the overall circuit, which may differ by several orders of magnitude. This is what makes this approach efficient.

Example 15.6 Consider the accumulator circuit shown in Figure 15.12(a). It has three registers and an adder, and the load signals of the registers are also shown. To test the adder, we need symbolic values on wires C_1, C_2 and C_3. Let these values be v_1, v_2 and v_3, respectively. A low-level test generation algorithm immediately runs into problems when applying values on wire C_2. Since the *reset* input cannot provide the arbitrary value needed on this wire, it is not immediately used. Without using *reset*, the values in the $C_2 \rightarrow C_3 \rightarrow C_4$ loop remain undefined. However, the low-level algorithm does not know this.

For functional test generation, the operations that we have to schedule are the load and reset operations. To justify v_1 and v_2, the operations needed are *load R1* and *load R2*, respectively. Since register $R1$ is directly fed by a primary input, no further justification is needed in that direction. To justify v_2, the next operation needed is *load R3*. Further justification requires that register $R1$ be loaded with v_2 and register $R2$ be reset. The **operation sequence graph** derived as above is shown in Figure 15.12(b). In this graph, the operation at the tail of an arc is executed before the operation at its head. Assume each operation takes one cycle. If we apply the ASAP and ALAP algorithms to this graph, we see that the *load R1 with v_1* operation can be scheduled in time step 2 or 3. It cannot be scheduled in time step 1 because register $R1$ cannot be loaded with two different values in the same cycle. There is no conflict in scheduling. Thus, justification can be done in three time steps. Fault effect propagation can be trivially done by simply observing v_3 by loading its value in register $R3$. Thus, four cycles are required to justify and propagate one vector from the test set of the adder. These cycles are shown below.

1 Reset $R2$; load $R1$ with v_2,
2 Load $R3$ with v_2,
3 Load $R2$ with v_2; load $R1$ with v_1,
4 Load $R3$ with v_3.

Note that in the above test sequence, a faulty module is being used to justify its inputs. However, this does not pose a problem. After the first time step, if the output of the adder is not v_2, then in the second time step, after loading $R3$ with this value, the error will be observable at output OUT. Thus, if no error is observed at OUT, it can be assumed that the correct v_2 is available in register $R2$ in time step 3.   □

15.3   RTL fault simulation

An RTL fault simulation technique has been presented in Hsiao and Patel (1995), which is one of only a few such techniques. It accepts circuits described at two levels of the design hierarchy: RTL and gate level. The gate-level description of only the module-under-simulation (MUS) is needed. Multiple symbolic data consisting of groups of faults detected in the MUS are used to exploit the fault list in parallel at the RTL. A gate-level sequential fault simulator is used for the MUS based on the single SAF model. At the RTL, the module's behavioral description is used to propagate the fault effects. The RTL fault simulator is not fully deterministic, yet shows high accuracy when compared to gate-level sequential fault simulation.

Only three data types are used at the RTL: unknown (X), fault-free circuit value (V), and fault effect (FE). X and V are integers, whereas FE is a symbol with a subscript. During fault simulation, the value of an RTL signal is denoted as (fault-free circuit value / fault effect$_i$). The fault-free circuit value is the value when no fault is present,

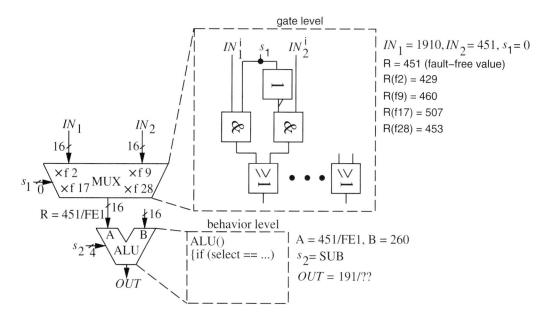

Figure 15.13 Fault effects

whereas the fault effect is a symbolic representation of the faults that were detected by the test sequence being simulated. The actual error caused by the fault is not kept track of.

FEs are used to achieve parallelism among faults at the RTL. Consider the RTL circuit shown in Figure 15.13. Suppose the multiplexer (MUX) is the current MUS. Therefore, it will be simulated at the gate level. Suppose the data inputs of the multiplexer are integers 1910 and 451, respectively. Assume that $s_1 = 0$, and 451 is selected. During gate-level fault simulation, the above vector may detect many single faults in the multiplexer. Suppose the faults detected are $f2$, $f9$, $f17$ and $f28$, with the corresponding erroneous outputs being $R(f2)$, $R(f9)$, $R(f17)$, and $R(f28)$, respectively. Instead of carrying a list of erroneous values, a subscripted symbol FE_1 is used to indicate that a fault effect is present. Different FE_is denote different groups of faults. However, they may cover an overlapping set of faults.

Next, let us consider the ALU. It is represented at the RTL, where simulation is done by invoking its behavioral description. Without knowing the exact erroneous values, it is difficult to predict if FE_1 will propagate through the ALU. In order to make this prediction, when one or more *FEs* arrive at the inputs of an RTL module, several random numbers are generated for these inputs. If the erroneous value indicated by the random number does propagate through the module and produce an error at its output in more than a *threshold* number of trials, the fault effect is considered to have successfully propagated through the module. The threshold is user-selectable. Examples of the threshold are 3-out-of-4 or 4-out-of-5 successes. If more than one *FE* is present at the inputs of a module, each is evaluated independently in the above fashion. The *FE*s are merged after they propagate through the module.

Random numbers are selected for the signals with *FE*s by modifying the fault-free values in one of the following ways: (a) complement one random bit, (b) complement two random bits, (c) complement the bits at all odd positions, (d) complement the bits at all even positions, (e) complement the first half of the bits, (f) complement the second half of the bits, (g) complement all bits.

The *FE* is removed from the RTL circuit when it propagates successfully to its primary output. At that point, the individual faults represented by the *FE* are dropped from the fault list. When *FE*s do not propagate to a primary output, they remain in the circuit in the next time frame.

The above propagation method may lead to both over-detection and under-detection of faults. Over-detections result from the fact that an entire list of faults is represented by a single FE_i and treated as one fault. Therefore, when FE_i is propagated to a primary output, the entire list of faults it represents is declared as detected. This list may contain faults whose effects do not propagate to the circuit output in gate-level simulation. Similarly, one reason for under-detections may be the use of a limited number of data types. For example, while vector $(1, 1, 0, 0)$ will be correctly represented as integer 12, vector $(1, 1, x, 0)$ will be represented as X. Due to this loss of bit-level information, fault detection is not as accurate as in gate-level simulation. However, experimental results show that mispredictions are typically less than 1%. Thus, the fault coverage obtained by RTL fault simulation is very close to that obtained by gate-level fault simulation.

15.4   RTL design for testability

In this section, we discuss four DFT methods for RTL circuits. The first method is based on extraction of a partial or complete behavioral description from the RTL circuit, and then performing symbolic testability analysis of the behavior to determine where the testability bottlenecks are in the RTL circuit and get rid of them. The second method is also based on symbolic testability analysis. However, it works directly at the RTL. It uses regular expressions to perform complete symbolic testability analysis. The third method is based on the concept of RT-scan which tries to modify RTL designs to make them combinationally testable without the high overheads incurred by gate-level scan designs. The fourth method is based on orthogonal scan which also tries to reduce the test overheads by sharing functional and test logic.

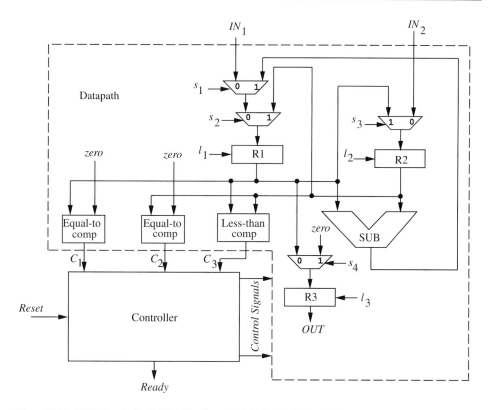

Figure 15.14 RTL circuit for GCD (Ghosh *et al.*, 1998) (© 1998 IEEE)

15.4.1 Testability analysis and optimization based on control/data flow extraction

A technique to extract behavioral information from an RTL circuit in the form of a test control/data flow (TCDF) for DFT purposes has been presented in Ghosh *et al.* (1998). The behavioral information is already available if the RTL circuit is obtained through high-level synthesis. However, the technique is also applicable to RTL circuits which have been manually designed. It is applicable to both data-flow intensive and control-flow intensive circuits. A symbolic testability analysis technique (Bhatia and Jha, 1998) is used to analyze the extracted behavior (this technique is described in greater detail in Section 15.7.2). This analysis is used to feed precomputed test sets to embedded modules in the RTL circuit. When a module is not symbolically testable, test multiplexers are added on off-critical paths to get rid of the testability bottlenecks. Because of its symbolic nature, the complexity of testability analysis and insertion is independent of the datapath bit-width. The test set for the RTL controller–datapath circuit is a byproduct of the analysis and does not require any further search. Hence, the method is very fast. A large part of the test set can be applied at the normal speed of the clock. In addition, the test overheads and test application time are also very small.

    The method is illustrated with the help of a controller–datapath of a control-flow intensive circuit, called *greatest common divisor (GCD)*, as shown in Figure 15.14.

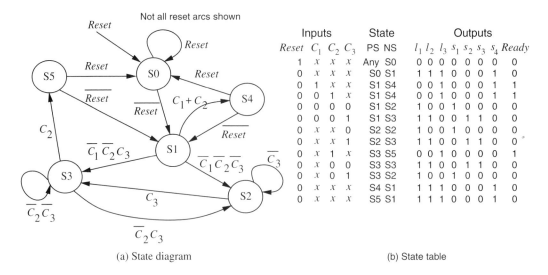

Figure 15.15 State diagram and table for the GCD controller (Ghosh *et al.*, 1998) (© 1998 IEEE)

This circuit computes the GCD of the numbers fed at its two input ports. The controller outputs a *Ready* signal when the GCD is available in output register *R3*. The datapath consists of some registers and multiplexers and four functional units: a subtracter, a less-than comparator, and two equal-to comparators. The outputs of the comparators are the status inputs of the controller.

The state diagram and state table, which can be extracted from the controller circuit, are shown in Figure 15.15. *PS* and *NS* denote the present state and next state, respectively. This is a Moore machine. If the given controller is a Mealy machine, then it is converted to a Moore machine using well-known procedures from Switching Theory (Kohavi, 1978). This is done for analysis purposes only. We first identify the *input states*. These are states where the input registers get loaded from the primary input ports. *S1* is such a state. We next identify the *output states* where an output register is loaded with a datapath value (not a constant). For example, in states *S1* and *S4*, *R3* is loaded with the constant *zero*, whereas it is loaded with a datapath value in state *S5*. Thus, *S5* is the only output state. Finally, we identify the *operation states* where the module under test is exercised. This is determined by identifying the states in which the output of the module gets loaded into a register. For example, the output of subtracter *SUB* is loaded into *R1* in state *S2*. Therefore, *S2* is the operation state of *SUB*.

To obtain a TCDF for a module, we need to derive a sequence of states which goes from an input state to an output state through its operation state. For example, for *SUB* we would need to go from state *S1* to state *S5* through state *S2*. We also need a path from the reset state to the input state, i.e., from state *S0* to state *S1*. The test architecture we present later allows direct controllability of the status signals. Thus,

Inputs

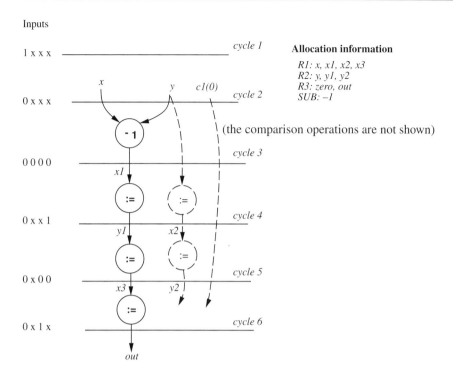

Figure 15.16   TCDF for the subtracter (Ghosh *et al.*, 1998) (© 1998 IEEE)

in the test mode, the desired sequence of states that a controller needs to take can
be dictated to it. The shortest sequence of states from the reset state satisfying the
above condition is $S0 \rightarrow S1 \rightarrow S2 \rightarrow S3 \rightarrow S5$. However, for this sequence of
states, the response of SUB is not propagated to the output port. To obtain a valid state
sequence, we need to backtrack along the path, unroll some loops in the state diagram,
and explore another path which may not be the shortest. The number of loop unrollings
is constrained by a pre-defined limit to reduce search time. For example, consider the
state sequence $S0 \rightarrow S1 \rightarrow S2 \rightarrow S3 \rightarrow S3 \rightarrow S5$. This is obtained by unrolling
the self-loop on $S3$ once. The corresponding TCDF is shown in Figure 15.16. The
six states in the sequence correspond to the six cycles. The dashed lines and nodes are
pruned from the TCDF as they do not affect the final output. During TCDF generation,
the status vectors that are needed to dictate the desired control/data flow are recorded
as shown. The figure also shows how different variables in the TCDF are allocated to
different registers. A TCDF is generated in this fashion for each module in the RTL
circuit.

The TCDF can be used to test SUB by testing operation -1 which is mapped to
it. This can be accomplished by controlling variables x and y from the primary input
ports, and observing variable out when it is loaded into register $R3$. The precomputed
test set of SUB yields the values of x and y. The above process is repeated until all

test vectors are exhausted from this precomputed test set. Then this process is repeated with the TCDFs of other modules. Deriving a precomputed test set for each module in the RTL design library is a one-time cost. This information can be stored in the RTL module library with the other information on area, delay or power of the RTL modules.

It is possible that for some RTL module, a TCDF cannot be obtained no matter which sequence of states is traversed. In order to provide controllability or observability to such a module, an n-bit two-to-one test multiplexer needs to be added to the datapath, where n is the required bit-width. In the test mode, such a multiplexer directly derives one of its inputs from a primary input port or a register fed by such a port to provide extra controllability. If only observability is needed, the output of the test multiplexer is fed to a primary output register in the test mode. A test multiplexer need not be added directly to the module under question, but is usually strategically placed on an off-critical path in the datapath so as to simultaneously solve the controllability/observability problems of as many modules (which do not have a TCDF) as possible. For *GCD*, such a test multiplexer is not needed in the datapath.

Most of the registers and multiplexers in the datapath get tested while testing the functional units. The ones that remain can be separately targeted for symbolic testing in a manner similar to that of a functional unit. The global test application time can be reduced by targeting more than one RTL module at the same time for TCDF-based symbolic testing, whenever this is possible.

The test architecture that is used to test the controller–datapath is shown in Figure 15.17, where the testability hardware is shown shaded. One extra pin is assumed to be available to provide the *Test* mode signal. The controller outputs are multiplexed with a datapath output port to facilitate testing of the controller. The status signals are made directly observable by multiplexing them with an output port as well. They are made directly controllable by feeding the status register input from an input port. A register called the **test reconfiguration register (TCR)** is added to the RTL circuit. Its inputs are fed by the low-order bits of an input port. TCR feeds the select signals of the test multiplexers that are added to the datapath. It also has two bits, B_0 and B_1, to control the loading of the controller state register and select the output multiplexer as well as multiplexer M_s. The load signals of the datapath and controller registers are qualified with the inverted *Test* signal to ensure that they freeze their state while TCR and the status register are being loaded. Writing into TCR results in the circuit being reconfigured to provide controllability and/or observability as required. When we reset TCR, all the test-multiplexer select lines are 0. Hence, the normal datapath configuration exists. The *Test* signal and signals B_0 and B_1 should also be 0 for normal operation.

The datapath and the controller are tested separately. While testing the datapath, the TCDF provides a suitable control flow which is achieved by controlling the status register from a primary input port at appropriate cycles. Signals B_1 and B_0 are set to 1 and 0, respectively, for testing the datapath modules other than the comparators. The

Assuming two test multiplexers in the datapath

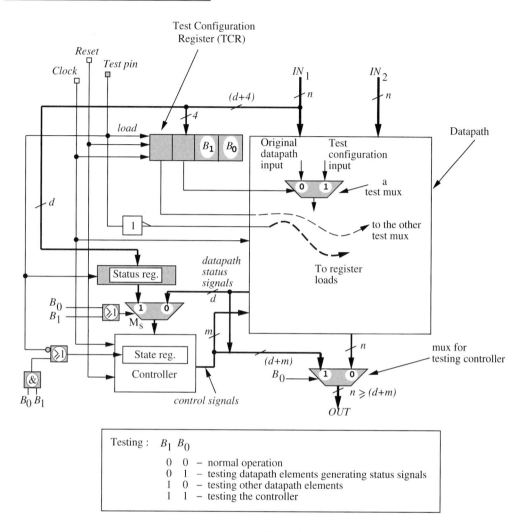

Figure 15.17 Test architecture (Ghosh *et al.*, 1998) (© 1998 IEEE)

Test signal is asserted only at specified cycles as needed. In certain cycles, we may need to load only the status register or TCR. Since we know the value that should be present in each cycle at either of the registers, we can feed the correct default value to the register, which we do not need to load, from the primary input port. To test the datapath comparators, we need to observe the status signals. In this case, we need to keep signals B_1 and B_0 at 0 and 1, respectively.

To test the controller, we make the *Test* signal high to load the status register. Thus, we can directly control the controller inputs using the primary input port. The controller outputs are directly observable at the primary output port. We assert signals

B_1 and B_0 to force the state register to load in all cycles while the controller is being tested.

Experimental results on various benchmarks show that the average area, delay and power overheads for making them testable are only 3.1%, 1.0%, and 4.2%, respectively. Over 99% fault coverage is usually obtained with two-to-four orders of magnitude test generation time advantage over an efficient gate-level sequential test pattern generator and one-to-three orders of magnitude advantage over an efficient gate-level combinational test pattern generator (that assumes full scan). In addition, the test application times obtained for this method are comparable with those of gate-level sequential test pattern generators and up to two orders of magnitude smaller than designs using full scan.

15.4.2   Testability analysis and optimization based on regular expressions

An RTL testability analysis and optimization method called TAO has been presented in Ravi *et al.* (1998a). It exploits the algebra of *regular expressions* to provide a unified framework for handling controller–datapaths of a wide variety of circuits, such as application-specific integrated circuits, application-specific programmable processors, application-specific instruction processors, digital signal processors and microprocessors. It also includes a DFT framework that can provide a low-cost testability solution by examining the trade-offs in choosing from an array of testability modifications like partial scan or test multiplexer insertion in different parts of the circuit. Test generation is symbolic in this method as well, with all the attendant benefits of such a method which were pointed out in the previous section.

We illustrate TAO with the RTL datapath circuit shown in Figure 15.18. Suppose the controller for this circuit behaves like an instruction set processor, i.e., it takes an instruction as its input and outputs the appropriate set of control signals to the datapath. Figure 15.19 shows a hypothetical instruction set for this circuit, consisting of instructions $I0$, $I1$, $I2$ and $I3$. Each instruction consists of one or more micro-instructions, denoted by $a0, \ldots, a5$. Each micro-instruction is a set of micro-operations (MOPs). In Figure 15.18, the MOPs describe the status of registers $R1$, $R2$ and $R3$ in the datapath. The MOPs correspond to load and hold operations in the registers.

The micro-instructions constitute the alphabet of regular expression based testability analysis. A **control path**, CP_{M_i}, is a set, consisting of strings of micro-instructions that can control the output of a module, M_i, to any desired value. A control path, $CP_{M_i \leftarrow c}$, is a set, consisting of strings of micro-instructions that can control the output of module M_i to constant value c. A **path equation** for CP_{M_i} relates CP_{M_i} to the control paths representing the inputs of M_i. The allowed set of operations for manipulating sets of strings consists of *concatenation* ($\cdot$), *star operation* ($*$), *union* ($+$), *intersection* ($\cap$) and *complementation* ($'$). Note that $+$ is also used to denote simple addition. The exact

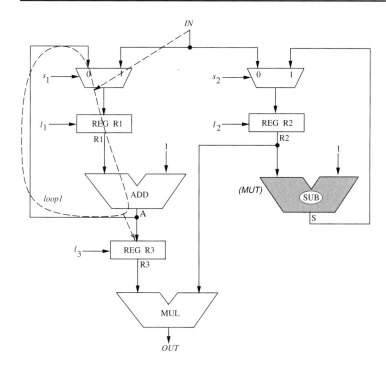

Figure 15.18 An RTL datapath circuit (Ravi *et al.*, 1998a) (© 1998 IEEE)

meaning of this symbol will be clear from the context. Observability of the output of module M_i is achieved if we can derive a set of strings of micro-instructions that can transfer the output of M_i to a primary output port. A **test environment** for a module under test is a set of strings of micro-instructions that describes all possible symbolic paths for testing the module, i.e., control its inputs and observe its subsequent response.

Testing an RTL element (e.g., functional units, multiplexers, registers, etc.) in the datapath involves writing and solving path equations to derive its test environment. Suppose we want to test the decrementer, *SUB*, in the RTL circuit shown in Figure 15.18. To test *SUB*, we will see later that we need to obtain the control paths of registers $R1$, $R2$ and $R3$. Let us consider $R1$ first. The path equation for its control path, CP_{R1}, is given by

$$CP_{R1} = CP_{R1} \cdot HoldR1 + CP_{R1} \cdot LoadR1(A) + CP_{IN} \cdot LoadR1(IN).$$

The above equation implies that there are three ways of controlling $R1$: to control it beforehand and hold the value, through *loop1*, and directly from primary input *IN*. Alphabet Σ for testability analysis consists of the set of micro-instructions $\{a0, a1, a2, a3, a4, a5\}$. From Figure 15.19, we can see that $HoldR1$, $LoadR1(A)$ and $LoadR1(IN)$ are MOPs of micro-instructions $a4$ or $a5$, $a1$ or $a2$, and $a0$ or $a3$, respectively. Let us use ax_y to denote the MOP on register Ry in the datapath, for micro-instruction

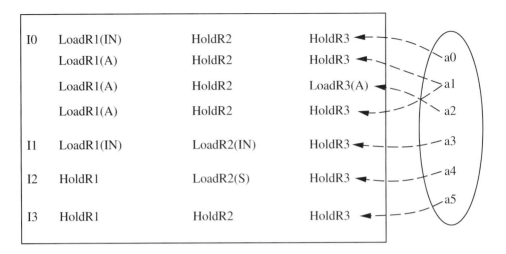

I0	LoadR1(IN)	HoldR2	HoldR3
	LoadR1(A)	HoldR2	HoldR3
	LoadR1(A)	HoldR2	LoadR3(A)
	LoadR1(A)	HoldR2	HoldR3
I1	LoadR1(IN)	LoadR2(IN)	HoldR3
I2	HoldR1	LoadR2(S)	HoldR3
I3	HoldR1	HoldR2	HoldR3

Figure 15.19 Instruction set (Ravi *et al.*, 1998a) (© 1998 IEEE)

ax. For example, $a3_1$ specifies MOP *LoadR1*(IN) for micro-instruction $a3$. Then the above equation can be rewritten as

$$CP_{R1} = CP_{R1} \cdot (a4 + a5)_1 + CP_{R1} \cdot (a1 + a2)_1 + CP_{IN} \cdot (a0 + a3)_1$$
$$= CP_{R1} \cdot (\Sigma - a0 - a3)_1 + CP_{IN} \cdot (a0 + a3)_1.$$

From Arden's rule (Kohavi, 1978), $R = Q.P^*$ is a solution to regular expression $R = Q + RP$ when P does not contain the null string. Thus, the above equation has the following solution:

$$CP_{R1} = CP_{IN} \cdot (a0 + a3)_1 \cdot (\Sigma - a0 - a3)_1^*.$$

CP_{R3} can be obtained from CP_{R1} as follows:

$$CP_{R3} = CP_{R1} \cdot a2_3 \cdot (\Sigma - a2)_3^*.$$

The reason is that once a control path to register $R1$ has been found, to get the desired value in register $R3$, the value at A simply needs to be loaded into $R3$. This is done with $a2_3$. After that, the content of $R3$ can be held for as many cycles as needed. This is done by $(\Sigma - a2)_3^*$.

To test *SUB*, we need to control $R2$ from *IN*, store the response from the output, S, of SUB in $R2$, and propagate this value to *OUT* through multiplier *MUL*. The path equation for controlling $R2$ from *IN* and storing *SUB*'s response in it is given by

$$CP_{R2} = CP_{IN} \cdot a3_2 \cdot (\Sigma - a3 - a4)_2^* \cdot a4_2 \cdot (\Sigma - a3 - a4)_2^*.$$

The term $CP_{IN} \cdot a3_2 \cdot (\Sigma - a3 - a4)_2^*$ loads $R2$ from *IN* and holds the value in it for as many cycles as necessary. The term $a4_2 \cdot (\Sigma - a3 - a4)_2^*$ loads $R2$ from S and holds the value in it.

To observe the content of $R2$ at OUT, we need the following equation:

$$CP_{OUT} = CP_{R3\leftarrow 1} \cap CP_{R2}.$$

Note that the equation for CP_{R3} shows all possible ways to obtain an arbitrary value at $R3$. This includes logic value 1. Thus, the above equation can be expanded simply by plugging into it the expressions for CP_{R3} and CP_{R2}. This gives the symbolic test environment for testing SUB in terms of micro-instructions. However, to obtain a test program, the test environment for testing SUB should be in terms of instructions. Any set of instructions I that is generated from the instruction set Σ_I defined in Figure 15.19 can be written as follows:

$$\Sigma_I = \{I0,\ I1,\ I2,\ I3\},\quad I = \Sigma_I^* - \{\epsilon\}$$
$$\text{or } I = (a0 \cdot a1 \cdot a2 \cdot a1 + a3 + a4 + a5)^* - \{\epsilon\}$$

where ϵ is the null string. Let TP denote the set of test programs for SUB. We can first obtain $TP = I \cap CP_{OUT}$, which is composed of strings defined over the alphabet Σ of micro-instructions. Transforming TP into a sequence defined over Σ_I is simply the task of parsing the sequence of micro-instructions into a set of syntactically correct programs, which can be easily performed by a lexical analyzer. We then obtain

$$TP = \Sigma_I^* \cdot (I0 \cdot (I1 + I2 + I3)^* \cdot I1 \cdot I3^* I2 + I1 \cdot (I3 + I0)^* \cdot I0 \cdot I3^* \cdot I2$$
$$+ I1 \cdot (I3 + I0)^* \cdot I2 \cdot (I3 + I0)^* \cdot I0) \cdot \Sigma_I^*.$$

TP represents many test programs for testing SUB. One possible solution is $I0 \cdot I1 \cdot I2$ which we obtain by parsing CP_{OUT} into the string L given below.

$$L = a0_1 \cdot a1_1 \cdot a2_3 \cdot a1_3 \cdot a3_{2,3} \cdot a4_{2,3} = I0 \cdot I1 \cdot I2$$

Since L is contained in $(a0 \cdot a1 \cdot a2 \cdot a1 \cdot a3 \cdot a4)$, which belongs to I, $I0 \cdot I1 \cdot I2$ is an acceptable solution.

In order to complete the test generation process, we need to determine the values required at the input ports of the datapath in the different clock cycles. Since a control-data flow graph (CDFG) is a convenient model for describing the cycle-by-cycle behavior of a circuit, we map the test environment to a CDFG. This is done by concatenating CDFG descriptions of the micro-instructions that form the test environment. For example, L can be mapped to the CDFG shown in Figure 15.20. From the CDFG, we see that value 1 can be justified at the left input of the multiplication operation $*$ by providing the *all-1's* vector at IN in the first cycle, whereas applying a test vector v to SUB requires v at IN in the fifth cycle. The CDFG can be executed repeatedly by substituting different vectors from the precomputed test set of SUB for v in different iterations of the CDFG. This completely tests SUB. Other elements of the datapath can be similarly tested.

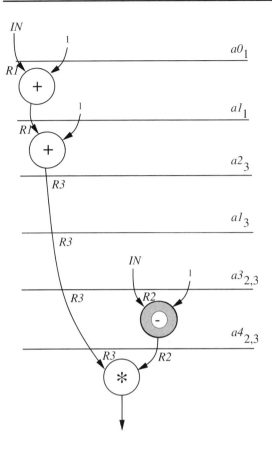

Figure 15.20   CDFG for L (Ravi *et al.*, 1998a) (ⓒ 1998 IEEE)

Sometimes, it may so happen that the test environment for the RTL element under test yields null. This means that this element is not symbolically testable. This problem can be solved by judiciously adding test multiplexers or partial scan to off-critical paths. The controller can be tested by a test architecture similar to the one given in Figure 15.17. The test overheads, fault coverage, test generation times and test application times are similar to those mentioned in the previous section.

15.4.3   High-level scan

Various high-level scan methods, applicable to RTL circuits, have been presented. These include H-Scan (Asaka *et al.*, 1997), RT-scan (Bhattacharya *et al.*, 1997), and orthogonal scan (Norwood and McCluskey, 1996). The basic concept behind these methods is quite similar. They use existing interconnects in the RTL circuit as much as possible to provide parallel scan paths to the registers. The registers themselves do not contain scan flip-flops. This significantly reduces the test application time while making it possible to test the circuit with a combinational test generator. The test overheads are smaller than what full scan would require. We discuss RT-scan and orthogonal scan next.

15.4.3.1 RT-scan

We illustrate this method with the help of the RTL circuit shown in Figure 15.21(a). Its *connectivity graph* is shown in Figure 15.21(b). A **connectivity graph** contains nodes for every primary input, primary output, register, constant and functional unit of the RTL circuit. Its edges connect distinct nodes if there is a path between the corresponding nodes in the RTL circuit that goes through multiplexers only. Since edges connect distinct nodes, self-loops, such as the one around register $R1$, are not represented in this graph. An *RT-scan graph* is derived from the connectivity graph to maximally reuse its connectivity to reduce scan overheads. The **RT-scan graph** has a node for every register node in the connectivity graph. It may also have some or all of the primary input, primary output and functional unit nodes. The edges in the RT-scan graph satisfy the following properties. There is only one edge incident on a node of type register, primary output or a single-input functional unit (e.g., an incrementer). There are two edges incident on a node that represents a two-input functional unit, such as a multiplier or subtracter. For every register node, there is a path from a primary input node and a path to a primary output node. A node can have fanout edges if at most only one of them goes to a node which is not a primary output node. In order to satisfy all these properties, it may be necessary to add an edge in the RT-scan graph that does not exist in the connectivity graph. The RT-scan graph for our example is shown in Figure 15.21(c). The extra edge is shown as a dashed arrow. In general, the RT-scan graph can have loops.

Using the RT-scan graph, the testable RTL implementation can be obtained as shown in Figure 15.22. When test signal T is low, this implementation behaves as the original circuit shown in Figure 15.21(a). When T is high, the paths exercised in this implementation correspond to the RT-scan graph.

To test the circuit, we obtain a combinational test set for it as if it had full scan. Next, let us see how a sequence of primary input vectors can be generated for applying a combinational test vector to the RT-scan implementation. Suppose that the datapath bit-width is four, and the part of the combinational vector that corresponds to the registers is as follows: $R1 = 1110$, $R2 = 1010$, $R3 = 0101$ and $R4 = 0011$, and the part that corresponds to the primary inputs is: $IN_1 = 0001$, $IN_2 = 1111$. The values in the RT-scan graph at time t can be expressed as a function of the values of register nodes and primary input nodes at time $(t - 1)$. For the RT-scan graph shown in Figure 15.21(c), these assignments are as follows:

$$R1(t) \leftarrow IN_1(t - 1)$$
$$R2(t) \leftarrow IN_2(t - 1)$$

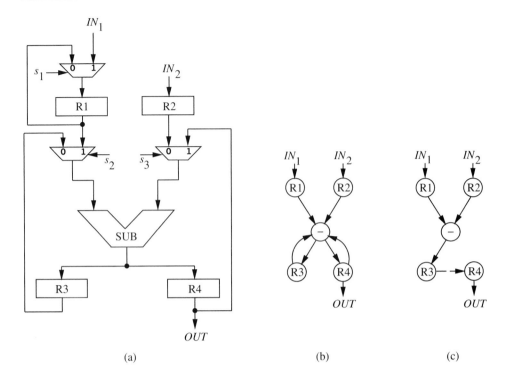

Figure 15.21   (a) An RTL circuit, (b) its connectivity graph, and (c) its RT-scan graph

$$R3(t) \leftarrow R1(t-1) - R2(t-1)$$
$$R4(t) \leftarrow R3(t-1).$$

The vector that needs to be scanned is shown in the fourth row of Table 15.4. To obtain $R4 = 0011$ in cycle 4, the above register assignments tell us that we need $R3 = 0011$ in cycle 3. This value in turn can be obtained with $R1 = 0011$ and $R2 = 0000$ in cycle 2. This implies that we must have $IN_1 = 0011$ and $IN_2 = 0000$ in cycle 1. The other values in the table can be similarly obtained. The blank entries in the table correspond to don't-care values.

The register assignment equations can also be used to propagate the captured fault effect to a primary output. For example, if after applying the above combinational vector, the fault effect is captured in register $R3$, then it can be propagated to primary output register $R4$ in one clock cycle. It is possible that the fault being targeted interferes with the appropriate loading of the registers with the combinational vector or propagation of the error response to a primary output, leading to non-detection of the fault. However, in practice, this situation rarely occurs.

When the RT-scan graph has loops in it, it may be necessary to assume that some or all of the registers are directly resettable.

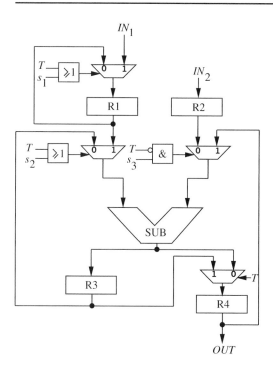

Figure 15.22 The RT-scan implementation

Table 15.4. *Derivation of a test sequence for a combinational vector*

Cycle	R1	R2	R3	R4	IN_1	IN_2
1					0011	0000
2	0011	0000			0101	0000
3	0101	0000	0011		1110	1010
4	1110	1010	0101	0011	0001	1111

15.4.3.2 Orthogonal scan

We illustrate this method with the help of the RTL circuit shown in Figure 15.23(a). Its orthogonal scan implementation is shown in Figure 15.23(b). As in the case of RT-scan, the values are loaded into the registers in a direction orthogonal to the traditional scan path implementation, hence the name. For normal operation, test signal T is kept low. If n is the datapath bit-width, then n AND gates are added at the right input of subtracter *SUB*. The select signals are modified in a manner similar to RT-scan. When $T = 1$, the following path is activated: $IN_1 \rightarrow R1 \rightarrow SUB \rightarrow R2$. Note that when $T = 1$, all the bits at the right input of *SUB* are 0. The above path can be used to load registers $R1$ and $R2$ from primary input IN_1 with values corresponding to a test vector obtained through combinational test generation. After the test vector is applied

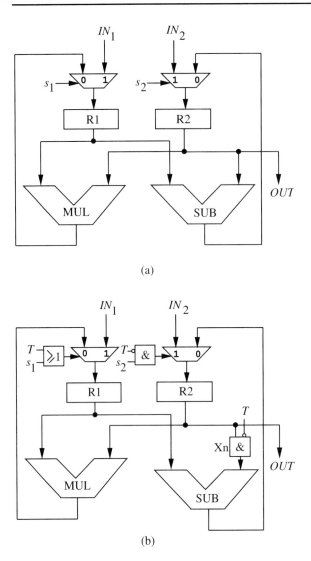

(a)

(b)

Figure 15.23   (a) An RTL circuit, and (b) its orthogonal scan implementation

to the circuit, the fault effect is captured in one of the registers. This effect can be propagated to the primary output *OUT* using the same path. This path can itself be tested beforehand by passing through it a string of 1s and 0s in the test mode.

In general, if a functional unit is used during orthogonal scan of the registers, then logic gates must be added to the bits of the input of the functional unit that is not on the scan path. This additional logic masks the input during scan. For an adder or subtracter, the AND gates, as shown in Figure 15.23(b), make all the bits of that input 0. This enables the value at the other input to pass through unchanged. To pass a value through an input of a multiplier unchanged, the extra logic should make sure that the multiplier's other input has the value $0 \cdots 01$. Whenever possible, the orthogonal scan path should be chosen such that this extra logic is not on the critical path.

Datapaths with many more functional units than registers may require multiple orthogonal scan paths to provide the desired logic values in all the registers. It may also necessitate adding load signals to some registers in order to scan the registers in phases, one orthogonal scan configuration at a time. The load signals allow the values of those registers to be held while the registers are scanned in subsequent orthogonal scan configurations. Multiple orthogonal scan configurations also require more than one test signal.

15.5   RTL built-in self-test

In this section, we discuss four BIST methods which are applicable to RTL circuits. The first method is similar to the method presented earlier in Section 15.4.1, and is based on control/data flow extraction. Previously, its use for deterministic testing of RTL circuits was described. Here, its extension is presented for BIST. Similarly, the second method is a BIST extension of the deterministic testing method presented in Section 15.4.2 which was based on regular expressions. The third method is based on a concept called arithmetic BIST. It is applicable to data-dominated RTL circuits. The last method is based on synthesis of self-test programs for microprocessors.

15.5.1   Built-in self-test based on control/data flow extraction

A method to exploit the TCDF extracted from the RTL circuit for BIST purposes has been presented in Ghosh *et al.* (2000). TCDFs were introduced earlier in Section 15.4.1. In this method, the pseudo-random pattern generators (PRPGs) are only placed at the primary inputs of the RTL circuit, and multiple-input signature registers (MISRs) are only placed at its primary outputs. TCDFs are used to generate test environments for various RTL modules as before, and a few test multiplexers are added to the off-critical paths in the datapath whenever needed. Every module in the RTL design library is made random-pattern testable, whenever possible, using gate-level testability insertion techniques. Finally, a BIST controller is synthesized to provide the necessary control signals to form the different test environments during testing, and a BIST architecture is superimposed on the circuit.

The BIST architecture is shown in Figure 15.24. In this figure, the added test hardware is shaded grey. One extra pin is needed to provide the *Test* mode signal. It is very similar to the test architecture shown in Figure 15.17. Hence, the previous descriptions of the test architecture are applicable here as well. PRPGs and MISRs are added at the primary inputs and outputs, respectively. The function of the TCR shown in Figure 15.17 is taken over by the BIST controller. A path is provided from one

Assuming two test multiplexers in the datapath

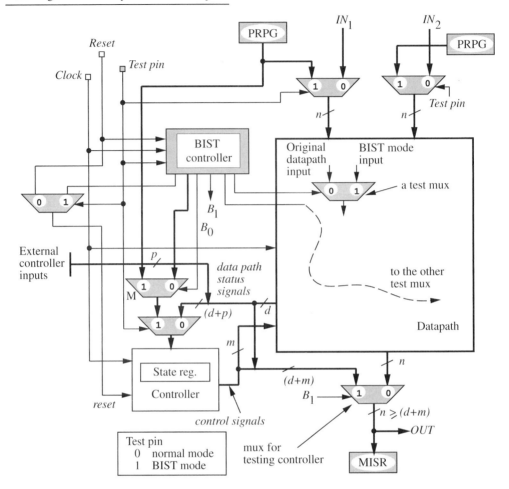

Figure 15.24 An RTL BIST architecture (Ghosh *et al.*, 2000) (© 2000 IEEE)

of the PRPGs to the controller to test it with pseudo-random vectors as well. When the datapath is being tested, the BIST controller provides the status signals to the controller. The BIST controller is activated during the test mode by the *Test* pin. Its *reset* signal is connected to the controller *reset*. The BIST controller feeds: (i) the *select* signals of the test multiplexers that are added to the circuit, (ii) the controller reset, and (iii) two bits B_0 and B_1 that control the select signals of multiplexer M and the output multiplexer, respectively. In the normal mode, the test-multiplexer select lines, *Test* signal, and signals B_0 and B_1 are 0.

The datapath and controller are tested separately. The *Test* pin is set in the test mode. While testing the datapath, the BIST controller provides a suitable control flow by controlling the controller input signals to a desired sequence. This sequence is obtained from the TCDF of the RTL element under test. Since the controller input sequences dictated by TCDFs typically have many don't cares, two or more such sequences can frequently be merged into one. This allows more than one RTL element to be tested simultaneously. It also reduces the area required to implement the BIST controller. Signals B_1 and B_0 are both set to 0 for testing the datapath elements other than the ones that produce the status signals for the controller. At specified clock cycles, we need to reconfigure the datapath using the test multiplexers. The BIST controller also provides the select signals of these test multiplexers. While testing the datapath elements that produce the status signals, such as comparators, we need to observe these status signals. This is accomplished by the BIST controller by making signals $B_1 = 1$ and $B_0 = 0$. The controller is tested by directly controlling the controller inputs using the PRPG at the primary input port and directly analyzing the controller outputs by the MISR at the primary output port. The BIST controller makes $B_1 = B_0 = 1$ to obtain this particular configuration. If the number of controller and status output bits $(d + m)$ exceeds the number of output bits of the datapath (n), the MISR width at the output port is increased to capture these extra bits.

It is frequently possible to convert existing input registers (registers connected to primary input ports) into PRPGs and output registers (registers connected to primary output ports) into MISRs, instead of adding extra PRPGs and MISRs. This reduces the overheads further. This can be done if these registers are not a part of some loop in the RTL circuit.

Experimental results on a number of benchmarks show that high fault coverage ($>99\%$) can be obtained with this scheme in a small number of test cycles at an average area (delay) overhead of only 6.4% (2.5%).

15.5.2 Built-in self-test using regular expression based testability analysis

An extension of TAO, which was described in Section 15.4.2, for BIST purposes has been given in Ravi *et al.* (2000). This extension is called TAO-BIST.

TAO-BIST uses a three-phase approach to address BIST of RTL controller–datapaths. In the first phase, it identifies and adds an initial set of test enhancements (PRPGs, MISRs, and test multiplexers) to the circuit. In the second phase, it uses regular expression based testability analysis to obtain symbolic test paths to test the embedded modules with pseudo-random vectors. Finally, it reduces area overheads under delay constraints by carefully selecting a small subset of registers as test registers (PRPGs and MISRs) and minimizing the number of test multiplexers added, if any.

The BIST architecture is similar to the one given in Figure 15.24, and is given in Figure 15.25. The BIST controller is activated by the *Test* pin which remains at 1 throughout BIST. It feeds: (i) the select signals of the test multiplexers that are added to the circuit, (ii) the controller reset, (iii) two bits B_0 and B_1 that control the

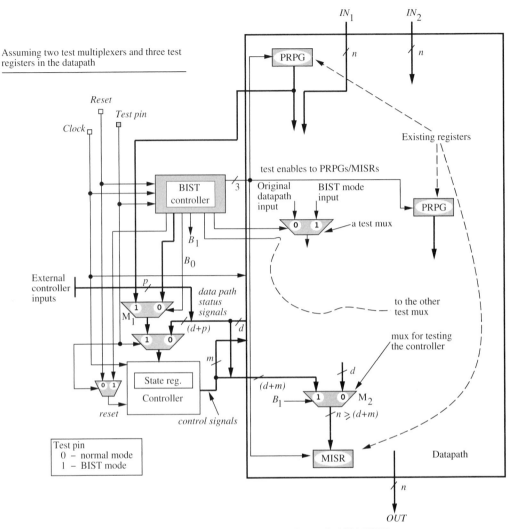

Figure 15.25 An RTL BIST architecture (Ravi *et al.*, 2000) (© 2000 IEEE)

select signals of multiplexers M_1 and M_2, respectively, and (iv) the test enables of the datapath registers which function as PRPGs and MISRs in the test mode. In the normal mode, the *Test* signal and the BIST controller outputs are all 0. The rest of the details is similar to that of the architecture shown in Figure 15.24.

In the first phase, all registers are initially converted into test registers, and test multiplexers are added from primary input registers to internal registers and from internal registers to primary output registers. Using RTL timing analysis, if it is found that some such test enhancements violate the delay constraint then that test enhancement is removed.

In the second phase, symbolic testability analysis is performed on the test-enhanced RTL circuit. The only difference with TAO, which was used for deterministic test generation, is that regular expressions for control paths are now derived as a function of PRPGs/MISRs instead of primary input/output ports. The regular expressions so derived provide many different ways of testing each RTL element.

In the third and final phase, the derived regular expressions help determine a minimal subset of the previously-added test enhancements which allows BIST of all RTL elements.

Experimental results on a number of benchmarks show that high fault coverage (>99%) can be obtained with this scheme as well in a small number of test cycles. The average area (delay) overhead is only 6.0% (1.5%).

15.5.3 Arithmetic built-in self-test

A BIST methodology to test the datapaths of data-dominated architectures, such as those found in the digital signal processing area, has been presented in Mukherjee *et al.* (1995). It uses arithmetic blocks, such as adders, from the datapath to perform test generation and response compaction. Hence, whenever the method is applicable, its overheads are close to zero.

Test generation is accomplished using an n-bit accumulator and a binary adder. It can be characterized by the following equation:

$$AC_j = AC_{j-1} + C \;(\text{mod } 2^n)$$

where AC_j is the content of accumulator AC after j iterations, and C is a constant being accumulated. At each iteration, the new vector from AC is applied to the remaining blocks of the datapath. The initial value AC_0 and the value of C are chosen to make sure that a given number of contiguous bits receive all possible patterns within as few iterations as possible. For example, suppose that $n = 8$, $AC_0 = 120$ and $C = 57$. The first eight vectors produced by the accumulator and adder are {(0, 1, 1, 1, 1, 0, 0, 0), (1, 0, 1, 1, 0, 0, 0, 1), (1, 1, 1, 0, 1, 0, 1, 0), (0, 0, 1, 0, 0, 0, 1, 1), (0, 1, 0, 1, 1, 1, 0, 0), (1, 0, 0, 1, 0, 1, 0, 1), (1, 1, 0, 0, 1, 1, 1, 0), (0, 0, 0, 0, 0, 1, 1, 1)}. The first three bits can be seen to receive all eight patterns. The same is true for the second, third and fourth bits, and so on.

Test responses from the circuit under test are accumulated by rotate-carry addition to generate a signature. In rotate-carry addition, the carry bit generated from the most significant bit position of the result is added to the least significant bit position in the next iteration. This allows erroneous signals to affect all signature bits. Such a response analyzer has an aliasing probability of approximately 2^{-n} for a bit-width of n, which is similar to that required for other response analyzers such as LFSRs, MISRs, etc.

The test generator and response analyzer described above, respectively, feed test vectors to and compact responses from other RTL elements under test. If more than one such test generator and response analyzer exist in the RTL circuit, then multiple RTL elements can be tested simultaneously.

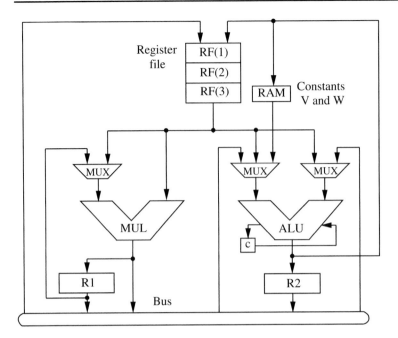

Figure 15.26 A typical RTL datapath

Example 15.7 Consider the RTL datapath shown in Figure 15.26. The following microcode can be used to test it.

RF(1) ← RF(1) + RAM(V);
RF(2) ← RF(2) + RAM(W);
R1 ← RF(1) * RF(2); or R2 ← op[RF(1),RF(2)];
R2 ← R1 + RF(3); or R2 ← R2 + RF(3);
R2 ← R2 + c;
RF(3) ← R2.

The addition operation of the ALU is used to generate the test vectors. These vectors are stored in the register file locations RF(1) and RF(2). They are applied to multiplier *MUL* and the ALU. The same test vector is used in successive iterations to test all the different operations implemented in the ALU. This is represented as op[RF(1),RF(2)] in the microcode. Since two vectors are needed for the two inputs of the functional units, two test generators are needed. These are based on two constants, V and W, stored in the RAM. The test response of each operation is compacted into a signature using rotate-carry addition, and is stored at RF(3). Therefore, the ALU serves both as a test generator and response analyzer. The microcode is run for several iterations based on the number of vectors necessary to obtain a high fault coverage for the circuit. □

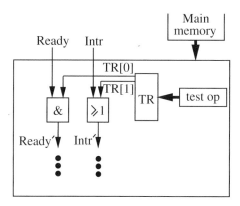

Figure 15.27 Test instruction insertion (Shen and Abraham, 1998) (© 1998 Kluwer)

15.5.4 Self-test programs for microprocessors

A method for generating native mode self-test programs for functional testing of microprocessors has been presented in Shen and Abraham (1998). Such a test is executed at the processor internal speed, i.e., it is an **at-speed test**. The test responses are compressed using the native instructions of the processor. The functional test is produced based on the assembly language instruction set of the processor and the operations performed by the processor in response to each instruction. The test is performed in two sessions. In the first session, each individual native instruction of the processor is targeted and the combination of selected operands and modes selected. In the second session, pseudo-random instructions and random data are used.

Many pins and functional units in a processor, such as control/status inputs, bus interface controllers, and interrupt controllers, cannot be tested by a self-test program since the program cannot directly control them. This method introduces a test instruction to solve this problem. This is done by mapping a new test instruction onto an undefined opcode to enhance the controllability of the control/status inputs. For example, consider the *Ready* and *Intr* inputs of a processor. When *Ready = 0*, the processor cannot access the memory. *Intr = 1* sends an interrupt request to the processor. To control these signals, a new test instruction can be added to load test register *TR*. When its bit 0 is set to 0, the memory becomes unavailable. When its bit 1 is set to 1, an interrupt request is issued. This is shown in Figure 15.27. If an unused opcode is available, implementing a test instruction requires no extra inputs or outputs. Its area overhead is also small – one test register bit and one AND or OR gate for each control/status input.

Modern microprocessors have large register files, memory modules and ALUs implementing many functions, which can be used to generate tests and analyze the responses in a BIST environment. Using the instruction set and architecture of the processor, we can obtain an optimized routine to compute the signature from the state

```
/* General registers: R1, R2, R3, R4 */
/* R4 holds the signature */
load the feedback coefficients into R3
for each register R1 to be compressed{
    save the LSB of R4 in R2
    Rightshift (R4,1)
    R4 = XOR(R4,R1)
    if (R2 = 1){
        R = XOR(R4,R3)
    }
}
```

Figure 15.28   Signature computation (Shen and Abraham, 1998) (© 1998 Kluwer)

of the registers. This routine can be stored in the cache or main memory. One such routine is shown in Figure 15.28. It is the software analog of a MISR. The coefficients stored in register $R3$ correspond to a primitive polynomial to minimize the probability of error masking. The compressed signature can be compared periodically with a golden signature to reduce this probability further. In fact, the tests can be run again with a different set of coefficients corresponding to another primitive polynomial, if necessary, by simply changing the content of register $R3$. This flexibility is obviously not there in a hardware response analyzer. Since compression is done in software, there is no area overhead associated with this approach.

When an instruction in the self-test program involves a program counter, e.g., branches, subroutine calls and returns, or interrupt calls, care must be taken to appropriately manage the memory space of the program. Otherwise, it may branch out of the program or fall into an infinite loop. This can be avoided by reserving a monotonically decreasing counter for each branch and subroutine call in the self-test program. The test is terminated when the content of the counter becomes zero.

The general steps in the development of the self-test program for a processor are *instruction set representation, instruction categorization, test specification* and *instruction sequencing*. These are described next.

A microprocessor's architectural and instruction set description is available in its Programmer's Manual. The instruction set can be represented in an *instruction set description file* for ease of manipulating it with the self-test program generation tool. In this file, the instruction mnemonic, modes and source/destination operands should be included. A manually selected set of operands is also stored in this file, e.g., m-out-of-n codes, and special floating-point numbers (*Infinity, Zero, Indefinite*, etc.).

The general and system registers in the microprocessor store the internal state. Instruction sequences, which can propagate the register contents to the chip pins, are called **observation sequences**. Similarly, instruction sequences, which can assign desired vectors to general registers, are called **control sequences**. To generate control and observation sequences, the instructions are put in different categories depending on their access to primary inputs and outputs.

Under test specification, the user has the option of specifying random operands for the instructions in addition to the manually selected set of operands already available in the library. To make the test more focused, the set of instructions related to the unit under test can also be restricted. For example, for testing a floating-point unit, floating-point instructions can be specified. Otherwise, all instructions are targeted by default.

In order to assemble the instructions into a self-test program, two sets of tests are generated. In the first set, each instruction is tested comprehensively. All combinations of operands in the specified operand set are enumerated. Different source and destination registers are selected. Control and observation sequences are generated for the sources and destinations, respectively. In the second set, random vectors and pseudo-random instruction sequences are used. Of course, the self-test program must contain the instruction sequence to compress the response at periodic intervals. If test instructions have been added to improve testability, they must be included as well.

Experimental results show that reasonable fault coverages (85–90%) can be obtained with this technique for microprocessors. The run-time to generate the self-test program is small.

15.6   Behavioral modification for testability

A method to modify the behavioral description of a circuit in order to improve its testability has been given in Hsu *et al.* (1996). This is achieved by enhancing the controllability of the control flow. After the modification is done, any behavioral synthesis tool can be used to obtain a testable circuit. The test overheads are minimal and the test sequence can be applied at-speed to the circuit.

In this method, testability analysis is performed directly on the behavioral description of the circuit. Consider the behavioral description of the greatest-common-divisor (GCD) circuit shown in Figure 15.29(a), and its flow chart given in Figure 15.29(b). The rectangular nodes in the flow chart contain serial operations that need to be executed and the diamond-shaped nodes are the decision nodes with branching conditions. The true and false branches are denoted by T and F, respectively. The nodes involved in the *while loop* are shown connected in bold lines. A node in the flow chart is said to be the **locus** of execution if it is currently being executed. A decision node is said to be **k-controllable** if the direction of the branch taken can be controlled directly or indirectly by the primary input values k register-transfer statements before the locus of execution reaches the node, where k is the smallest such integer.

```
process
    begin
        X := Port X ;
        Y := Port Y ;
        if (X == 0) or (Y == 0) then
            GCD := 0;
        else
            while (X != Y) loop
                if (X > Y) then
                    X := X − Y ;
                else
                    Y := Y − X ;
                endif ;
            endloop ;
            GCD := X ;
        endif ;
        PortGCD := GCD ;
    end process ;
```

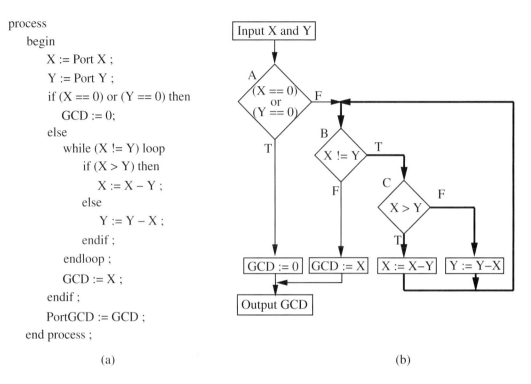

(a)   (b)

Figure 15.29 Behavioral description and flow chart of GCD (Hsu *et al.*, 1996) (© 1996 IEEE)

During normal operation, the GCD behavior first reads the X and Y values from the primary inputs. Depending on what these values are, it either sends the results to the primary output, or spends several iterations in the *while* loop before the results are ready to be sent to the primary output. While the locus stays inside the *while* loop, the behavior becomes uncontrollable and unobservable. By using the controllability measure defined above, the controllability of a *loop-exit* node can be determined. Such a node is a decision node that determines the exit condition of a loop. In the GCD example, decision node A is 1-controllable. However, decision nodes B and C are found to be hard-to-control for the above-mentioned reasons.

Once hard-to-control nodes have been identified, their controllability can be augmented through the use of one or two extra test pins. These pins can control the outcome of the conditional branches by guiding their control flow and thus affecting the controllability of the variables. Four types of control points are used in this work. A control point of type $T1$ is ANDed with the branch condition, which is hard-to-control, to force the condition to *false*. For example, suppose control point $C1$ is an instance of type $T1$. Decision node B in the flow chart shown in Figure 15.29(b) can be made easy to control by modifying it to $(X! = Y)$ AND $C1$. This allows faults activated within the loop to be quickly propagated to the primary output. By making $C1 = 0$, one can easily exit the *while* loop. Similarly, a control point of type $T2$ is ORed with the branch condition, which is hard-to-control, to force the condition to *true*. A control point of type $T3$ uses the XOR function instead with the branch condition. This allows the branch condition to be complemented. If hard-to-control variables remain after using control points of types $T1$, $T2$ or $T3$, a control point of type $T4$ can be added to directly control such a variable from the primary input. Each test pin can be connected to more than one decision node in the behavior as long as only one such node is executed at any given time.

This method can be easily generalized to cascaded loops and nested loops. Control points can be used to easily jump from one loop to another to facilitate the control flow through the circuit. The advantage of controlling the datapath by controlling its control flow is that the test overheads are small. Experimental results show that reasonable fault coverages can be obtained by this method. However, to get closer to 100% fault coverage, additional DFT features may be necessary.

15.7 Behavioral synthesis for testability

In this section, we discuss various ways of incorporating testability considerations into behavioral synthesis. First, we target testability of gate-level designs obtained through behavioral synthesis. Next, we see how hierarchical testability can be promoted during behavioral synthesis. Finally, we look at techniques to incorporate BIST considerations as part of behavioral synthesis.

15.7.1 Behavioral synthesis to improve gate-level testability

As mentioned in Section 15.1, the behavioral description given in some hardware description language is first compiled into a CDFG (see Figure 15.30(a) for an example CDFG) before behavioral synthesis commences.

A method has been given in Lee *et al.* (1993) to evaluate the testability of the circuit based on the *datapath circuit graph* (DPCG), also sometimes called the *register adjacency graph*, of the synthesized RTL architecture. This graph is similar to the S-graph mentioned in Chapter 13 (Section 13.3.2.2). It has a node for each register, and a directed arc connects two nodes if one register feeds the other through combinational logic. An arc from (to) a primary input (output) is also added to the register node in the DPCG which stores that input (output); such a register is called an **input (output) register**. The **sequential depth** between two registers is the shortest number of arcs on a path from one register to the other in the DPCG. A register which stores both a primary input and output is called an **IO register**. To improve controllability and observability and hence testability during allocation, three synthesis rules are targeted: (i) whenever possible, a register is allocated to at least one primary input or output variable, (ii) an attempt is made to reduce the sequential depth from an input

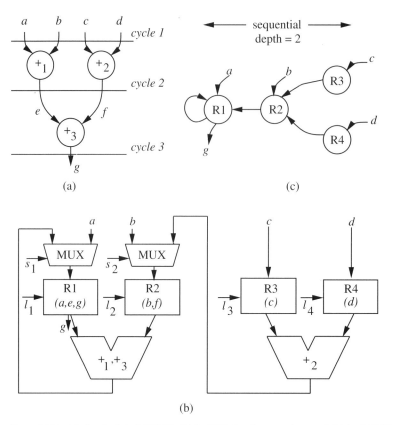

Figure 15.30 (a) A scheduled CDFG, (b) its RTL implementation, and (c) its DPCG (Lee *et al.*, 1993) (© 1993 IEEE)

register to an output register, and (iii) loops in the DPCG are reduced or eliminated by proper resource sharing in acyclic CDFGs and introducing *IO* or scan registers in cyclic CDFGs. Since self-loops on a node generally do not pose problems for subsequent gate-level sequential test generation, they are not targeted. The synthesis rules are mainly embedded in the register allocation phase, and module allocation is done to preserve the testability advantages gained during register allocation. Finally, scheduling is done in such a way as to maximally facilitate the application of the above three synthesis rules during allocation.

Example 15.8 Consider the scheduled CDFG shown in Figure 15.30(a) and its RTL implementation in Figure 15.30(b). The register and module allocation are also indicated in the RTL circuit. Modules and registers are interconnected through multiplexers. Select signals s_1, s_2 and load signals l_1, l_2, l_3, l_4 come from the controller, which is not shown. The DPCG of the RTL circuit, as shown in Figure 15.30(c), indicates that the minimum sequential depth is 2. The self-loop can be ignored for testability purposes, more so because it is on register $R1$ which is an *IO* register.

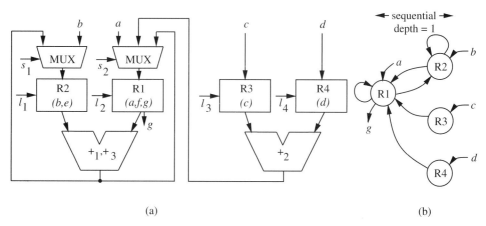

Figure 15.31 (a) Another RTL implementation with testability consideration, and (b) its DPCG (Lee *et al.*, 1993) (© 1993 IEEE)

Another RTL implementation of the same CDFG is shown in Figure 15.31(a) along with its DPCG in Figure 15.31(b). The loop $R1 \rightarrow R2 \rightarrow R1$ in the DPCG is broken because $R1$ is an *IO* register. Ignoring the self-loops and since the sequential depth is now reduced to 1, this is the more testable architecture. This is borne out by subsequent gate-level sequential test generation. The price paid for testability is a slight increase in area. Area overheads, in general, are usually small.   □

The three synthesis rules mentioned above can be aided in other ways also. For example, a method is given to add *deflection operations* to the behavior before synthesizing it in Potkonjak and Dey (1994). This exploits the fact that many operations in a CDFG have an identity element associated with them. For instance, an addition operation has an identity element zero, whereas a multiplication operation has an identity element one. If one of the inputs of such an operation is w and the other is an identity element, the output of the operation remains w. Adding such an operation op between two operations op_1 and op_2 has the effect of deflecting the result of op_1 to op, hence the name **deflection operation**. A deflection operation can be added anywhere in the CDFG without changing its functionality. In order to aid testability, such an operation can be judiciously added so that a scan register can be reused for breaking as many loops in the datapath as possible. Many other behavioral transformations to aid gate-level test generation have been provided in Potkonjak *et al.* (1995).

It has been observed that even if the controller and datapath are separately easily testable at the gate level, they may not be easily testable when they are combined as one circuit. This problem has been addressed in Ravi *et al.* (1998b) through the exploitation of don't cares in the controller specification to provide better controllability and observability to datapath elements. When this is not possible, extra control vectors can be added for this purpose at a minor expense in area.

15.7.2 Behavioral synthesis to improve hierarchical testability

While behavioral synthesis methods, which target gate-level sequential test generation, have shown impressive gains in testability, some shortcomings of the present methods are (i) sequential test generation time usually increases drastically with an increase in the bit-width of the datapath even for testable designs (although, of course, the increase is much more drastic for untestable designs), (ii) direct controllability and observability of the controller–datapath interface is frequently assumed, usually through scan, and (iii) at-speed testing is usually not possible for the controller–datapath.

To overcome most of these shortcomings, one can think of doing behavioral synthesis for hierarchical testability (Bhatia and Jha, 1998). This method assumes that precomputed test sets have been generated for each RTL element (such as adders, multipliers, registers, etc.) in the RTL design library. The synthesis system, which is called *Genesis*, then guarantees and generates a system-level test set which can provide the required test set to all embedded functional units and registers and propagate the error effects to observable outputs. The system-level test set also contains the test set for the controller. This test set is a byproduct of the testability analysis process that drives behavioral synthesis. Some of the salient features of this method are as follows. Since test generation is symbolic, it is independent of the bit-width of the datapath. For several benchmark controller–datapath circuits, this results in two-to-four orders of magnitude reduction in test generation time over efficient gate-level sequential test generators. Even compared to gate-level combinational test generation which assumes full scan, the reduction in test generation time is one-to-three orders of magnitude. The area and delay overheads are close to zero. The fault coverage for the combined controller–datapath is near 100%. At-speed testability is possible for data-dominated circuits. Finally, the test application time is one-to-two orders of magnitude smaller than that required for full scan and comparable to that required for gate-level sequential test generation.

The part of the design hierarchy, which *Genesis* exploits, consists of the behavior level and RTL. It uses the concept of a test environment, which has been alluded to in earlier sections. The test environment for a variable (operation) in a CDFG is a symbolic path (or a set of paths) that consists of a symbolic justification path from the system primary inputs to the variable (inputs of the operation), and a symbolic propagation path from the variable (output of the operation) to a system primary output. A test environment for a variable (operation) is also a test environment for the register (functional unit) that the variable (operation) gets mapped to after allocation. However, if the test environment for the register (functional unit) under test is derived through another variable (operation) mapped to the same register (functional unit), one needs to make sure that the correct test vector is obtained by providing the means to verify its logic values at a primary output (this process is called **verifiability**). This will be clearer through the example given later.

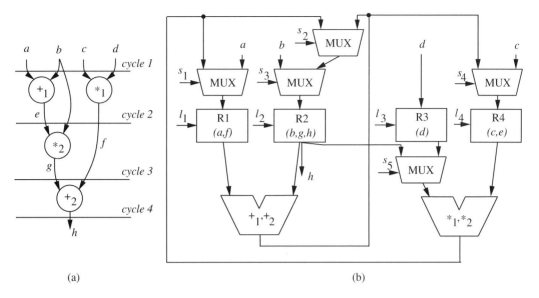

Figure 15.32   (a) A CDFG, and (b) its corresponding datapath (Bhatia and Jha, 1998) (© 1998 IEEE)

In general, an operation may have more than one test environment. For testing purposes, it is sufficient to identify just one. A test environment for a variable or an operation allows one to justify any desired vector at its inputs through system inputs, and propagate any response at its output to a system output. Since each test environment follows the original control flow dictated by the CDFG, there is no need to scan the control signals for datapath testing. Following the control flow also implies that the registers which store primary inputs are directly controllable in only the first control step; however, the registers storing the primary outputs are observable in every control step. This concept is illustrated in the example given next.

Example 15.9   Consider the scheduled CDFG shown in Figure 15.32(a). To test operation $*_2$, we need to control its inputs e and b, and observe its output g. b can be controlled to any desired value, since it is a primary input. Given a value for b, e can be independently controlled to any desired value by controlling primary input a appropriately. g can be observed indirectly by observing the primary output h provided the value of f is known. If c is controlled to 0, then f becomes 0 independent of d. The test environment for operation $*_2$ in that case would constitute the following set of arcs: $\{a, b, c, e, f, g, h\}$. Assume that we need to justify the test vector consisting of w_1 and w_2 at the inputs e and b, respectively, of $*_2$, and w_3 is the test response generated by $*_2$ at g. Then using the test environment for $*_2$ identified above, we simply set $b = w_2$ and $a = (w_1 - w_2) \bmod 2^n$, where n is the bit-width of the datapath. g is indirectly observable at primary output g by setting $c = 0$. This makes $f = 0$ and $h = w_3$.

The same test environment can be used to test the datapath. We simply need a test environment for any one operation allocated to an embedded functional unit in order to test that unit. For example, consider the RTL implementation shown in Figure 15.32(b) of the corresponding CDFG. Suppose we want to test the multiplier by using the test environment for the $*_2$ operation mapped to it. We store the values derived above for a and b in registers $R1$ and $R2$, respectively, in the first control step. The output e of the adder is stored in register $R4$. Now, b from register $R2$ and e from register $R4$ are fed to the multiplier in the second control step. This constitutes one of the desired precomputed test vectors of the multiplier. The output of the multiplier is stored in register $R2$ which is directly observable since primary output h is also mapped to it. The controller is then reset to bring it to the first control step again and this process is repeated for each test vector of the multiplier from its precomputed test set.

Next, suppose that we want to test the adder using the test environment of operation $+_2$ mapped to it. If we want to feed values w_1 and w_2 to inputs g and f, respectively, of $+_2$, one possibility is to make $a = (w_1 - 1) \bmod 2^n$, $b = 1$, $c = w_2$ and $d = 1$. However, since e, and hence g, is derived through operation $+_1$, which is allocated to the adder under test, a fault in the adder could have interfered with the justification of w_1 at g. But this value, if erroneous, is verifiable since the output of $R2$ is observable. If the fault does not interfere with the above justification step then the adder gets the desired test vector and stores its response in register $R2$ where it again becomes observable. Therefore, it takes three control steps to do justification/propagation for each test vector from the adder's precomputed test set.

The registers can be similarly tested using their test environments. A multiplexer usually needs a very small test set (it consists of only four vectors for detecting all single SAFs in its two-level AND–OR logic implementation). The test set derived for the functional units and registers typically also detects nearly all the faults in the multiplexers.   □

Derivation of test environments in the presence of loops in a CDFG is handled in *Genesis* by unrolling the CDFG a few times and using the concept of verifiability. *Genesis* also targets conditionals, multicycling (where an operation is scheduled over multiple control steps), chaining (where multiple operations are chained in just one control step) and pipelined functional units. Datapath allocation is done by simultaneously performing register and module allocation while minimizing the interconnection cost. The synthesis method takes advantage of the test environment based hierarchical testability analysis concept at each step to make sure it is proceeding in the right direction. It backtracks to consider other choices when one choice of test environment does not work. In the extremely rare cases, when all choices fail for some functional unit or register, it adds an extra multiplexer from (to) a primary input (output) to provide direct controllability (observability) to it. This allows a test environment to be found.

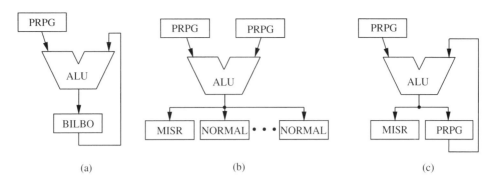

Figure 15.33 BIST configurations (Harmanani and Papachristou, 1993) (© 1993 IEEE)

15.7.3 Behavioral synthesis to facilitate BIST

The area of behavioral synthesis for testability started with work on behavioral synthesis targeting BIST (Avra, 1991; Harmanani and Papachristou, 1993). The advantage of targeting BIST at higher levels of synthesis is that the overheads can be substantially reduced, and various tradeoffs, such as those between area overhead and fault coverage or area overhead and test application time, are easier to evaluate.

From Chapter 12, we know that for pseudo-random BIST, test vectors are generated through a PRPG and the test responses are compressed using a MISR. A built-in logic block observer (BILBO) can be configured in four ways: normal register, serial shift register, PRPG and MISR. One of the problems we face in implementing BIST is when a register is self-adjacent, i.e., an output of the register feeds back to itself through some combinational logic. This is shown in Figure 15.33(a). The problem arises because it is not possible to use a BILBO as a PRPG and MISR at the same time. In this figure, if the BILBO is configured as a PRPG then the output of the arithmetic/logic unit (ALU) cannot be observed. If it is configured as an MISR then the response can be observed, but the ALU may not receive a sufficient number of different input vectors, thus affecting fault coverage. This can be rectified by using a concurrent BILBO (CBILBO) which can simultaneously act as a PRPG and MISR. Its disadvantage is that its area cost is about 1.75 times that of a BILBO.

In Harmanani and Papachristou (1993), the notion of *testable functional block* (TFB) is used. Such a block is shown in Figure 15.33(b) where the two registers at the input ports of the ALU are configured as PRPGs and the output port is connected to a set of registers, one of them configured as an MISR. The number of registers at the output port is optimized during synthesis. The two PRPGs and the MISR can be shared by other TFBs in the datapath. This means that some of these registers may have to be configured as a BILBO, and some of the normal non-BIST registers at the output port may have to configured as a PRPG. The testable datapath allocation begins with a scheduled CDFG. First, a TFB is allocated to each operation in the CDFG. Then the method iteratively reduces the hardware by merging TFBs together, whenever possible to do so, based on a cost function which includes area of ALUs, normal and PRPG/MISR/BILBO registers and multiplexers. Two TFBs can be merged only if their corresponding operations are assigned to different time steps in the schedule and the merging results in a module which exists in the ALU library. Also, self-adjacent registers are allowed only if the resulting structure is self-testable. For example, while the structure shown in Figure 15.33(a) is not acceptable, the one in Figure 15.33(c) is.

At the end of this process, further reduction in BIST hardware is possible. Let p be the number of ALUs in the datapath and $q = 2p$ be the number of distinct input ports of the ALUs. Let r be the number of registers connected to these q input ports. Then solving it as a covering problem, we can identify a minimum subset of these r registers which covers all the input ports. Only the registers in this subset need be made PRPG. One restriction is that it is not allowed to cover both input ports of an ALU with the same register, as this has adverse effects on the fault coverage. Still more reduction of BIST hardware can be obtained based on the notions of *randomness* and *transparency*. Suppose that the output register of some functional unit or ALU A is $R1$, which in turn feeds some functional unit or ALU B whose output register is $R2$. If the vectors produced by A are sufficiently random, then $R1$ need not be made a PRPG. Similarly, if most faulty responses of A can propagate through B and reach $R2$, then $R1$ need not be made an MISR. Here, B is acting like a transparent functional unit for the faulty response. Examples of functional units which have both very good randomness and transparency properties are adders and subtracters. The randomness of the output vectors from a functional unit and the transparency of a functional unit can generally be improved by increasing the testing time.

Another behavioral synthesis method has been given in Kim *et al.* (1998), which trades off BIST area overhead for test application time. If there are m functional units in the resultant RTL circuit, then they can all be tested in parallel (in one test session) or one-by-one (in m test sessions). The required BIST resources, in terms of the number of test registers, are obviously much larger for the former than the latter. Various intermediate solutions are also possible. This method starts with a scheduled CDFG for which module allocation has already been done. Then register allocation is done appropriately to make sure that the RTL circuit thus generated is testable in k test sessions, where k is user-specified, with as small an area overhead as possible.

All the existing behavioral synthesis methods for BIST implicitly assume that the controller–datapath interface is directly controllable.

Summary

- Most hierarchical test generation methods use the register-transfer and gate levels of the design hierarchy, and can significantly speed up test generation.

- Test generation for microprocessors can usually be made more efficient by dividing it into two phases: path analysis and value analysis.

- Test generation for processors can also be sped up by extracting and using functional constraints.

- The low-level information on a circuit can be abstracted into a functional fault model. Tests generated from functional fault models are compact and efficient.

- Symbolic RTL fault simulation is much faster than gate-level sequential fault simulation at a slight expense in accuracy.

- It is possible to extract behavioral information from an RTL circuit in the form of a test control-data flow graph and use it for low-overhead RTL DFT.

- An algebra of regular expressions can provide a unified framework for testing and DFT of a wide range of RTL controller–datapaths.

- RT-scan and orthogonal scan are two high-level scan methods that can significantly improve test application time.

- Test control/data flow graphs and regular expressions can also be used to superimpose a low-overhead, high fault coverage, BIST architecture over an RTL controller–datapath.

- In the datapaths of data-dominated architectures, such as those found in the digital signal processing area, one can use arithmetic blocks, such as adders, from the datapath to perform test generation and response compaction.

- Native mode self-test programs can be generated for at-speed functional testing of microprocessors. The test responses are compressed using the native instructions of the processor. The functional test is produced based on the assembly language instruction set of the processor and the operations performed by the processor in response to each instruction.

- The behavioral description of a circuit can be modified to enhance the controllability of its control flow so that any behavioral synthesis tool can be used to obtain a testable circuit.

- Behavioral synthesis can aid in making a circuit more easily testable through gate-level sequential test generation by increasing the number of primary input-output registers, decreasing the sequential depth and reducing or eliminating loops by proper resource sharing and use of *IO* and scan registers.

- Behavioral synthesis can target hierarchical testability by justifying precomputed test sets at the system level using the concept of test environment. This enables at-speed testability and allows the test generation time to become independent of the bit-width of the datapath.

- If BIST is targeted during behavioral synthesis, the BIST overheads can be substantially reduced by using the concept of testable functional blocks.

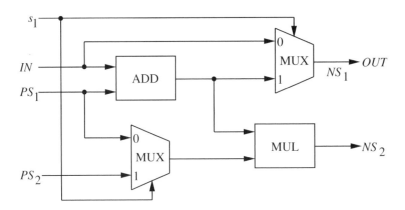

Figure 15.34

Additional reading

A bi-level test generation, as opposed to hierarchical test generation, method has been given in Ravi and Jha (2001). It first performs symbolic test generation at the RTL and passes on the remaining undetected faults to a gate-level sequential test generator. Up to 100X speed-up is obtained.

An RTL symbolic test compaction method, based on test parallelism and test pipelining, has been given in Ravi *et al.* (2002).

An efficient fault simulation technique that targets circuits produced by high-level synthesis has been presented in Kassab *et al.* (1995). It combines hierarchical and functional fault simulation of RTL functional units with regular structures.

A symbolic testability analysis based RTL DFT method for application-specific programmable processors and application-specific instruction processors has been given in Ghosh *et al.* (1999).

A behavioral test generation method has been presented in Ferrandi *et al.* (1998).

Exercises

15.1 Consider the RTL circuit shown in Figure 15.34, where the next state wire NS_1 also serves as an output OUT.

    (a) Find a fault-free state justification sequence for $NS_1 = 20, NS_2 = 125$.

    (b) Find a state-pair differentiating sequence for the fault-free state $PS_1 = 10$, $PS_2 = 30$ and the faulty state $PS_1 = 15, PS_2 = 30$ in the fault-free machine.

    (c) Show that a state-pair differentiating sequence does not exist for the fault-free state $PS_1 = 10, PS_2 = 30$ and the faulty state $PS_1 = 10, PS_2 = 25$ in the fault-free machine.

15.2 In hierarchical test generation for microprocessors, define set S to be the set of multiple next state wires that need to be justified in instruction i. Denote the corresponding SDFG as $SDFG_i$. Prove that S is simultaneously justifiable if all nodes in $SDFG_i$ have strongly influenced arcs at their outputs.

15.3 In Figure 15.2, for some instruction assume that (i) symbols v_5, v_6 and v_7 are fixed to 8, 0 and 1, respectively, (ii) $v_5 = 8$ implies that a multiplication operation will be executed by the ALU, and (iii) C_5 is the least significant bit at the output of the wire split module.

    (a) Perform symbolic simulation on this circuit and derive the system of module equations.

    (b) Derive an SDFG for this instruction.

    (c) For each input of a primitive in the SDFG, indicate if it is strongly or weakly influencing. For each output of a primitive in the SDFG, indicate if it is strongly or weakly influenced.

    (d) Suppose a test vector has been obtained for the incrementer as follows: $v_9 = 294$ and $v_{10} = 1$ (note that after symbolic simulation, the side input of the incrementer has symbol v_9 attached to it). Solve the system of module equations to justify this test vector. Obtain all possible solutions.

15.4 Suppose that the adder shown in Figure 15.12(a) is a four-bit carry-lookahead adder. In Example 15.6, it was shown that a test sequence consisting of four cycles is required to justify and propagate just one vector from the test set of the adder. However, two such consecutive test sequences can be partially overlapped.

    (a) From the minimal SIF test set for the carry-lookahead module M_2 given in Table 15.3, derive the minimal test set for the four-bit adder shown in Figure 15.11(a).

    (b) From the above minimal test set of the adder, derive a minimal test sequence to completely test the circuit shown in Figure 15.12(a) (Hint: the length of the test sequence need be no more than 31).

15.5 Derive the TCDF for the equal-to and less-than comparators in the GCD circuit shown in Figure 15.14. If the less-than comparator has v vectors in its precomputed test set, how many clock cycles will it take to test it when it is embedded within the GCD circuit.

15.6 Derive the TCDF for the GCD circuit shown in Figure 15.14 corresponding to the state sequence $S0 \rightarrow S1 \rightarrow S2 \rightarrow S3 \rightarrow S5$. Show that although this TCDF allows the subtracter to be fed its desired test vectors, it does not allow the propagation of the subtracter's response to the primary output.

15.7 Derive the symbolic test environment for multiplier MUL in the RTL circuit shown in Figure 15.18 using regular expressions.

15.8 Consider the RTL circuit shown in Figure 15.23(a).

    (a) Obtain the RT-scan implementation of this circuit.

    (b) How many clock cycles does it take to load the desired values in the two registers in the RT-scan implementation corresponding to a test vector obtained through combinational test generation? Show how the following combinational test vector corresponding to a four-bit datapath can be applied to the circuit: $R1 = 1001$, $R2 = 0111$, $IN_1 = 0010$, $IN_2 = 1100$.

(c) How many clock cycles does it take to load the desired values in the two registers in the orthogonal scan implementation shown in Figure 15.23(b).

Show how you would apply the above combinational vector to this implementation.

15.9   Consider a test generator based on an accumulator and adder, whose bit-width is four, for arithmetic BIST. Find the initial value of the accumulator, AC_0, and the value of constant C to make sure that each contiguous block of two bits receives all four two-bit patterns with only the first four vectors generated by the test generator.

15.10   For the scheduled CDFG shown in Figure 15.30(a) derive an RTL architecture which does not have any multiplexers (i.e., the registers and functional units are directly connected to each other). What is the sequential depth of the corresponding DPCG? Would you consider this to be a very testable architecture if sequential test generation is used to evaluate its testability? What are the other advantages and disadvantages of this architecture?

15.11   Show how one can test the multiplier shown in Figure 15.32(b) by using the test environment of operation $*_1$ which is allocated to it. If the precomputed test set of the multiplier has v vectors in it, how many clock cycles (i.e., control steps) do we need to accomplish the testing of the multiplier using the above test environment?

15.12   Modify the registers in the datapath given in Figure 15.32(b) so that it becomes testable using the concept of testable functional blocks, and the BIST area overhead is minimized. Assume that the areas of a normal register, PRPG, MISR, BILBO and CBILBO are R, $1.4R$, $1.6R$, $2R$, and $3.5R$, respectively.

References

Asaka, T., Bhattacharya, S., Dey, S. and Yoshida, M. (1997). HSCAN+: a practical low-overhead RTL design-for-testability technique for industrial designs. In *Proc. Int. Test Conference*, pp. 265–274.

Avra, L. (1991). Allocation and assignment in high-level synthesis for self-testable data paths. In *Proc. Int. Test Conference*, pp. 463–472.

Bhatia, S. and Jha, N.K. (1998). Integration of hierarchical test generation with behavioral synthesis of controller and data path circuits. *IEEE Trans. on VLSI Systems*, **6** (4), pp. 608–619.

Bhattacharya, S., Dey, S. and Sengupta, B. (1997). An RTL methodology to enable low overhead combinational testing. In *Proc. European Design & Test Conference*, pp. 146–152.

Ferrandi, F., Fummi, F. and Sciuto, D. (1998). Implicit test generation for behavioral VHDL models. In *Proc. Int. Test Conference*, pp. 587–596.

Ghosh, A., Devadas, S. and Newton, A.R. (1993). Sequential test generation and synthesis for testability at the register-transfer and logic levels. *IEEE Trans. on Computer-Aided Design*, **12** (5), pp. 579–598.

Ghosh, I., Raghunathan, A. and Jha, N.K. (1998). A design-for-testability technique for register-transfer level circuits using control/data flow extraction. *IEEE Trans. on Computer-Aided Design*, **17** (8), pp. 706–723.

Ghosh, I., Raghunathan, A. and Jha, N.K. (1999). Hierarchical test generation and design for testability methods for ASPPs and ASIPs. *IEEE Trans. on Computer-Aided Design*, **18** (3), pp. 357–370.

Ghosh, I., Jha, N.K. and Bhawmik, S. (2000). A BIST scheme for RTL circuits based on symbolic testability analysis. *IEEE Trans. on Computer-Aided Design*, **19** (1), pp. 111–128.

Hansen, M.C. and Hayes, J.P. (1995a). High-level test generation using symbolic scheduling. In *Proc. Int. Test Conference*, pp. 586–595.

Hansen, M.C. and Hayes, J.P. (1995b). High-level test generation using physically induced faults. In *Proc. VLSI Test Symposium*, pp. 20–28.

Harmanani, H. and Papachristou, C.A. (1993). An improved method for RTL synthesis with testability tradeoffs. In *Proc. Int. Conference on Computer-Aided Design*, pp. 30–35.

Hsiao, M.H. and Patel, J.H. (1995). A new architectural-level fault simulation using propagation prediction of grouped fault-effects. In *Proc. Int. Conference on Computer Design*, pp. 628–635.

Hsu, F.F., Rudnick, E.M. and Patel, J.H. (1996). Enhancing high-level control-flow for improved testability. In *Proc. Int. Conference on Computer-Aided Design*, pp. 322–328.

Kassab, M., Rajski, J. and Tyszer, J. (1995). Hierarchical functional fault simulation for high-level synthesis. In *Proc. Int. Test Conference*, pp. 596–605.

Kim, H.B., Takahashi, T. and Ha, D.S. (1998). Test session oriented built-in self-testable data path synthesis. In *Proc. Int. Test Conference*, pp. 154–163.

Kohavi, Z. (1978). *Switching and Finite Automata Theory*, McGraw-Hill, NY.

Lee, T.-C., Jha, N.K. and Wolf, W.H. (1993). Behavioral synthesis of highly testable data paths under nonscan and partial scan environments. In *Proc. Design Automation Conference*, pp. 292–297.

Lee, J. and Patel, J.H. (1994). Architectural level test generation for microprocessors. *IEEE Trans. on Computer-Aided Design*, **13** (10), pp. 1288–1300.

Mukherjee, N., Kassab, M., Rajski, J. and Tyszer, J. (1995). Arithmetic built-in self-test for high-level synthesis. In *Proc. VLSI Test Symposium*, pp. 132–139.

Norwood, R.B. and McCluskey, E.J. (1996). Orthogonal scan: low overhead scan for data paths. In *Proc. Int. Test Conference*, pp. 659–668.

Potkonjak, M. and Dey, S. (1994). Optimizing resource utilization and testability using hot potato techniques. In *Proc. Design Automation Conf.*, pp. 201–205.

Potkonjak, M., Dey, S. and Roy, R.K. (1995). Considering testability at behavioral level: use of transformations for partial scan cost minimization under timing and area constraints. *IEEE Trans. on Computer-Aided Design*, **14** (5), pp. 531–546.

Ravi, S., Lakshminarayana, G. and Jha, N.K. (1998a). TAO: regular expression based high-level testability analysis and optimization. In *Proc. Int. Test Conference*, pp. 331–340.

Ravi, S., Ghosh, I., Roy, R.K. and Dey, S. (1998b). Controller resynthesis for testability enhancement of RTL controller/data path circuits. *J. of Electronic Testing: Theory & Applications*, **13** (2), pp. 201–212.

Ravi, S., Lakshminarayana, G. and Jha, N.K. (2000). TAO-BIST: a framework for testability analysis and optimization for built-in self-test of RTL circuits. *IEEE Trans. on Computer-Aided Design*, **19** (8), pp. 894–906.

Ravi, S. and Jha, N.K. (2001). Fast test generation for circuits with RTL and gate-level views. In *Proc. Int. Test Conference*, pp. 1068–1077.

Ravi, S., Lakshminarayana, G. and Jha, N.K. (2002). High-level test compaction techniques. *IEEE Trans. on Computer-Aided Design*, **21** (7), pp. 827–841.

Shen, J. and Abraham, J.A. (1998). Synthesis of native mode self-test programs. *J. of Electronic Testing: Theory & Applications*, **13** (2), pp. 137–148.

Tupuri, R.S. and Abraham, J.A. (1997). A novel functional test generation method for processors using commercial ATPG. In *Proc. Int. Test Conference*, pp. 743–752.

16 System-on-a-chip test synthesis

In this chapter, we discuss test generation and design for testability methods for a system-on-a-chip. There are three main issues that need to be discussed: generation of precomputed test sets for the cores, providing access to cores embedded in a system-on-a-chip, and providing an interface between the cores and the chip through a test wrapper.

We first briefly discuss how cores can be tested. This is just a summary of the many techniques discussed in the previous chapters which are applicable in this context.

We then present various core test access methods: macro test, core transparency, direct parallel access, test bus, boundary scan, partial isolation ring, modification of user-defined logic, low power parallel scan, testshell and testrail, and the advanced microcontroller bus architecture.

We finally wrap this chapter up with a brief discussion of core test wrappers.

16.1 Introduction

Spurred by an ever-increasing density of chips, and demand for reduced time-to-market and system costs, system-level integration is emerging as a new paradigm in system design. This allows an entire system to be implemented on a single chip, leading to a system-on-a-chip (SOC). The key constituents of SOCs are functional blocks called *cores* (also called intellectual property). **Cores** can be either *soft*, *firm* or *hard*. A **soft core** is a synthesizable high-level or behavioral description that lacks full implementation details. A **firm core** is also synthesizable, but is structurally and topologically optimized for performance and size through floorplanning (it does not include routing). A **hard core** is a fully implemented circuit complete with layout.

A whole range of cores, such as processors, microcontrollers, interfaces, multimedia, and communication/networking cores, are being used in SOCs. They also cover different types of technology, such as logic, memory and analog. Some cores are hierarchical, consisting of one or more simpler cores. A core-based SOC is typically composed of a number of large cores as well as some user-defined logic (UDL) modules connected together by glue logic.

Typically, hard cores come with precomputed test sets from the core provider. For firm and soft cores, such test sets can be obtained by the system integrator. However,

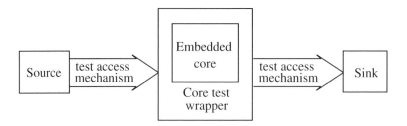

Figure 16.1 Testing of embedded cores (Zorian *et al.*, 1998) (© 1998 IEEE)

once these cores are connected together in a system by the system integrator, it becomes very difficult to justify these precomputed test sets at the inputs of the cores from system primary inputs and propagate their test responses to system primary outputs. A way of getting around this problem is to use extra design for testability (DFT) hardware to facilitate testing. In this chapter, we describe how one can attack the SOC testing problem (Zorian, 1997; Zorian *et al.*, 1998).

The main problems in SOC testing involve the design and development of *core-level test*, *core test access* and *core test wrapper*. **Core-level test** refers to the techniques used in testing cores individually. **Core test access** involves methods for accessing the embedded cores in the SOC from the test vector source and propagating its response to a test vector sink. The **core test wrapper** forms the interface between the embedded core and its SOC environment. Conceptually, these three elements of test can be viewed as shown in Figure 16.1. These elements can be implemented in various ways, as discussed next.

16.2   Core-level test

Core-level testing involves deriving the test set for the core, utilizing the DFT features that have been added to it. This precomputed test set is obtained by the core provider for hard cores and given to the system integrator. This is because the system integrator may have very limited knowledge about the internals of the core. For firm and soft cores, the test set is obtained by the system integrator. The test set consists of both *data vectors* and *protocol vectors*. **Data vectors** consist of the stimulus and response logic values. **Protocol vectors** specify the **test flow**, which contains information on how to apply and capture the data vectors.

Another hurdle facing the core provider is that it is not known a priori which SOC environment the core will be placed in, which testing methods will be adopted at that level, which fault types will be considered, and what fault coverage will be desired. For example, the testing method could consist of one or more of built-in self-test (BIST), scan, sequential test generation, I_{DDQ} test or functional test. The type of faults could be static, dynamic or parametric. Also, different manufacturing processes have different defect type distributions and levels of defect. Hence, if the quality of test set is not high, it puts the quality of the SOC at risk. If the quality level is too high, the test cost consisting of test application time and the test overheads may become high. In addition, the precomputed test set should be adequately described and ported so that it is ready for plug-and-play in any SOC. Thus, the description of the test set should be in a standard core test description language.

A core is said to be **mergeable** if it is soft or firm and can be merged with its surrounding UDL without the need for any intervening circuitry, else it is called non-mergeable. For mergeable cores, the core-level test is not defined at the time of core creation. This becomes the responsibility of the system integrator. For non-mergeable cores, the core provider derives the test set based on the testability technique used and the target technology. For example, for memory cores, the most likely test scenario is to use BIST. Thus, BIST wrappers and control signals would typically be included with such cores.

The test vector source and sink can be either implemented off-chip by an automatic test equipment (ATE) or on-chip by BIST or both. The type of source or sink is governed by the type of circuitry, the type of precomputed test sets, and quality and cost considerations. The type of cores can be broadly classified under logic, memory and analog (or mixed-signal). When random logic cores are prepared for BIST, they may not contain a test generator and response analyzer. They may, however, contain scan and test points (extra control and observation points). The system integrator can choose to share a test generator and response analyzer for multiple cores which are BIST-ready or use multiple test generators and response analyzers. The core test resources can be described indirectly by the core provider, e.g., by specifying the primitive polynomial which characterizes the linear feedback shift register which generates the pseudo-random patterns to attain the specified fault coverage.

If cores come with deterministic or functional tests, the test sets are usually not very regular. To test such cores on-chip, these test sets can be embedded in a large test set generated by an on-chip test vector source. However, such methods are not mature yet. Therefore, in these cases, off-chip testing is usually resorted to. However, because of the increasing cost of ATE and its projected inability to keep up with the high frequencies necessary to detect delay faults adequately, the trend is towards using as much on-chip testing of cores as possible. On-chip testing does result in greater test overheads and hence a possible reduction in yield, which has prevented its wider applicability.

16.3 Core test access

This section deals with mechanisms for transporting test vectors to and from cores embedded in an SOC. Such test access mechanisms are, by definition, implemented

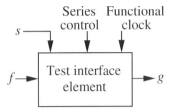

Figure 16.2   The test interface element (Beenker *et al.*, 1986) (© 1986 IEEE)

on-chip. The spectrum of techniques available for this purpose represent a trade-off between the transport capacity (*bandwidth*) of the mechanism and its attendant cost. For example, a wider test access mechanism requires more wiring area, but requires a smaller test application time.

Various options are available while choosing a test access mechanism. For example, the mechanism can either reuse existing functionality or consist of dedicated hardware. It can go through other cores in the SOC or pass around them. It can be either shared across multiple cores or not. It may just directly transport the signals or may contain some test control functions. We next describe various test access mechanisms which cover the above spectrum.

16.3.1 Macro test

An early method, called macro testing, for coupling chip and board testing was presented in Beenker *et al.* (1986). It is also applicable to the testing of core-based SOCs. This method is based on structurally partitioning the design into testable macros (or modules). Access paths are provided to each macro in order that its test set can be directly provided from system primary inputs and its responses directly observed at system primary outputs. In the SOC context, if each core is separately testable, this method can be used to test the core even when it is embedded in the SOC.

Test access is provided by inserting test interface elements on all signals traveling between macros. These elements are transparent in the normal mode. They support both parallel and serial access in the test mode. Such an element is functionally illustrated in Figure 16.2. It has three modes of operation: *functionally transparent*, *functional load* and *serial load*. The **functionally transparent** mode corresponds to normal operation where input f is directly connected to output g. In the **functional load** mode, input f is stored and is made available one clock cycle later at output g. This is helpful during test to collect the output signal of a macro that feeds f. In the **serial load** mode, input s is stored and is made available one clock cycle later at output g. This helps load and unload the macro interfaces using test interface elements as shift registers. These modes can be extended to allow parallel transparent and parallel load modes.

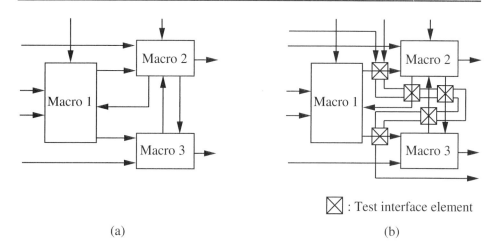

Figure 16.3   A macro test example

As an example, consider the circuit in Figure 16.3(a). In order to facilitate the testing of each macro separately and sequentially, test interface elements can be added to the signals traveling between macros, as shown in Figure 16.3(b). In this figure, the control and clock signals have been combined.

16.3.2   Core transparency

Test access methods, which are based on transparency of cores, have been presented in Ghosh *et al.* (1999, 2000) and Ravi *et al.* (2001). They require the presence of a transparency mode in each core. In this mode, test data can be propagated from the inputs of the core to its outputs without information loss. To test an embedded core C, the transparency modes of the cores from the system inputs to C's inputs and C's outputs to system outputs are exploited to feed the desired precomputed test sequence to C and observe its response. An SOC consisting of an interconnection of testable and transparent cores can be made testable with the addition of a small amount of system-level DFT hardware. In order to ensure that the internal state of a core does not change between the arrival of two consecutive test vectors from its test set, the state of the core is frozen during the intermediate cycles with an external clock enable signal (such a signal may already be available for low power operation).

Providing transparency to cores at the register-transfer level (RTL) usually requires only a slight overhead compared to simply making them testable. For many cores, such transparency paths through various core inputs to core outputs already exist. The RTL DFT methods given in Section 15.4 can be easily extended for this purpose. In fact, the three methods mentioned above extend the RTL DFT methods given in Sections 15.4.1, 15.4.2 and 15.4.3, respectively. We concentrate on the method presented in Ghosh *et al.* (1999) next.

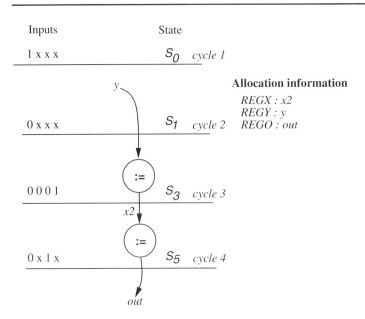

Figure 16.4 A transparency TCDF (Ghosh *et al.*, 1999) (© 1999 IEEE)

The SOC testing solution consists of two parts. The first part consists of core-level DFT and test generation to make each core testable and transparent and generate a precomputed test set for the core. The second part consists of system-level DFT and test generation. For this part, all we need is the precomputed test set of each core and the number of clock cycles required for transparency through various input/output port pairs of the core.

One way to provide transparency to a core is based on symbolic analysis of test control/data flows (TCDFs), which were previously used for RTL DFT in Section 15.4.1. In order to illustrate this method, consider the GCD circuit shown in Figure 15.14, and its controller shown in Figure 15.15. GCD has two primary inputs. Each needs to be provided a transparency path to the primary output. This can be done based on the TCDF extracted from the circuit, as shown in Figure 16.4. This shows the transparency path from y (a variable controllable from input IN_2) to *out* (a variable observable from output OUT). It consists of four cycles. Similarly, there is another path from input IN_1 to OUT that consists of five cycles (the obvious solution of three cycles is not possible because the control signals from the controller do not allow such a data flow). Therefore, for GCD, no test multiplexer is needed within the datapath for either testability or transparency purposes. In general, however, when a transparency path does not exist between an input/output pair, test multiplexers can be inserted at strategic off-critical paths in the circuit to overcome this problem.

In order to activate the transparency paths, the control flow of the controller needs to be dictated by controlling its inputs from an external source. Refer to the test architecture in Figure 15.17 that was used for DFT purposes earlier. This needs to be modified as follows. The inputs of multiplexer M_s that come from the status register can be made external to the core. Signal B_1 also needs to be an external input to act as the select signal of M_s. Hence, TCR of GCD now has only one bit. The status register in the architecture is removed as it is no longer required.

Once the cores are made testable and transparent, the next step is to make the complete SOC testable. This can be done by a system-level extension of the RTL symbolic testability analysis and insertion technique presented in Section 15.4.1. The system-level test set is a byproduct of this analysis. In the system-level test architecture, all required test and transparency inputs of each core are included in a global system test configuration register (STCR). By suitably loading STCR during the testing phase, individual cores can be symbolically tested while other cores are made transparent to propagate the test data. When symbolic testing of an embedded core is not possible, system-level test multiplexers are added to off-critical paths to facilitate symbolic testing. This is done in a fashion analogous to how an RTL circuit is made testable with the addition of test multiplexers using symbolic analysis.

An SOC testing framework has been given in Ravi *et al.* (2001). This exploits regular-expression based symbolic testability analysis, analogous to the RTL testability technique presented in Section 15.4.2. It models the transparency and test requirements of individual cores using finite-state automata. It then uses these models for system-level symbolic test generation and test insertion.

The method in Ghosh *et al.* (2000) provides transparency paths based on an extension of the high-level scan method given in Section 15.4.3. It provides an SOC test overhead versus test application time trade-off based on different versions of cores with different area overheads and transparency latencies.

16.3.3 Direct parallel access

A straightforward way to make embedded cores directly testable from the SOC pins is to provide direct and parallel access to the core from these pins (Immaneni and Raman, 1990). This requires the addition of extra wires connected to the core terminals and multiplexing them with SOC pins, as shown in Figure 16.5. This allows each core to be tested separately. However, this method is not very scalable to an SOC with a large number of cores because of the high area overhead. Design modifications are needed for cores to make them accessible in this fashion. We next describe how these modifications can be made to cores with a small number of inputs and outputs.

The modifications at the input side entail multiplexing a companion test input with each standard input of the core. This is shown in Figure 16.6(a), where $UI1$ and $TI1$ are the user input and corresponding test input, respectively. The test control logic controls the interface to the core. It has two inputs: T_{mode} and T_{sel}. T_{mode} is a global test mode control signal which is routed to each core. T_{sel} is a test select signal. It

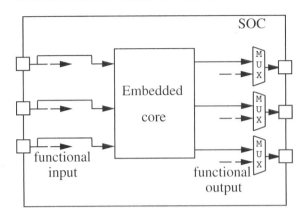

Figure 16.5 Direct parallel access

determines the test mode for a particular core. These two control inputs define three operation modes. $T_{mode} = 0$ corresponds to the **user mode** where the signal at the user input propagates to the core input. $T_{mode} = 1$ and $T_{sel} = 0$ correspond to the **inactive test mode** where the active high core inputs are driven low. Finally, $T_{mode} = 1$ and $T_{sel} = 1$ correspond to the **active test mode** where the signal at the test input propagates to the core input.

The modifications at the output side are shown in Figure 16.6(b). The above three modes are also applicable to the outputs. In the user mode, the core output goes to the user output. In the inactive test mode, the outputs are driven low. This makes it easy to map outputs from different cores onto SOC pins using OR gates. This also isolates the unselected core from the core under test. In the active test mode, the response at the core output can be observed at the user output. Only one core can be in active test mode at any given time. Bidirectional signals are similarly handled.

Design modifications required for cores with a large number of inputs and outputs are more sophisticated. Test registers are used to place the different logic blocks inside the core in various test modes. Thus, different core signals are multiplexed to the SOC pins in different test modes. Each test mode reconfigures the internal logic to test the appropriate logic block using the applied test vectors.

16.3.4   Test bus

A test bus can also be used to provide direct test access to cores and UDLs from SOC pins (Varma and Bhatia, 1998). However, it can be shared among multiple cores and UDLs to reduce the overheads. The test bus width can vary from 1 to n where n is determined by the system integrator. This allows test application time to be traded off for area overhead. An SOC can have multiple test buses of various widths. All cores connected to the same test bus are tri-stated except the one under test.

We next describe the test bus based SOC testing methodology in detail. The hardware components in such a methodology include one or more test data buses,

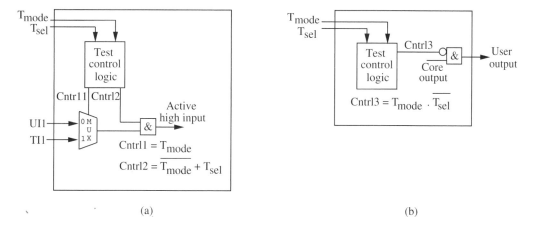

Figure 16.6 Input/output modifications (Immaneni and Raman, 1990) (© 1990 IEEE)

a test control bus, test collar cells, multiple-input signature registers (MISRs), and a test controller.

The test data buses can be input, output or bidirectional. They provide access to the test collar cells at the input/output ports of the cores and UDLs. The use of MISRs at the output ports of the component-under-test is optional. When they are used to compress the test responses, they reduce the test bus requirements and facilitate at-speed testing when the test bus width is restricted. A relatively narrow test control bus enables particular paths through the SOC to be tested. Both test data and control buses can be multiplexed with the SOC pins.

Just like the test data bus, the test control bus can also be of variable width. The control signals can be either static or dynamic. Static control signals, e.g., core test enable signals, change only at the beginning and end of test sessions. Dynamic control signals, e.g., scan enable or select signals, change many times during a test session. The number of static control signals is usually small, whereas the number of dynamic control signals can be quite large. The static control signals can be controlled from a control scan path which may be a simple scan register inserted in the SOC. Control signals can also be multiplexed with the functional pins of the SOC.

This methodology is compatible with boundary scan. It can make use of the boundary scan controller when such a controller is available. An additional test register is then needed for core test control. If the design is not compliant with boundary scan, then a core test controller consisting of a simple control scan path register is added which requires one extra pin. Note that although the methodology supports the use of boundary scan, no SOC or core level boundary scan chain is required.

Test collars are added to cores to facilitate test access and test isolation. They also provide test access and isolation to UDLs and interconnects, as shown in Figure 16.7. The paths labeled $p1$ provide test access to the core ports from the test data bus,

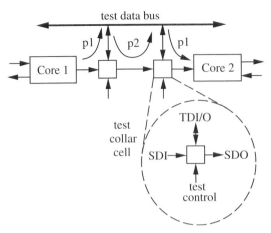

Figure 16.7 Test collar (Varma and Bhatia, 1998) (© 1998 IEEE)

whereas the path labeled $p2$ allows the interconnect to be tested. Providing test access to UDLs in this fashion is cost-effective for large UDLs only. Smaller UDLs can be tested during interconnect test. The delay impact of test collar cells is typically one to two gate delays. However, this delay impact can be avoided with alternative test collar designs when the output enable ports of cores can be tri-stated.

Test collar cells are provided at the input, output and bidirectional ports of the cores that must have test access. The methodology provides different types of test collar cells. A general-purpose universal input/output test collar cell has one or more test data inputs (TDI), one test data output (TDO), one test select (TSEL) and/or test clock (TCLK), a system data input (SDI) and a system data output (SDO), which may be the same as the TDO. If bidirectional test buses are required, the TDI and TDO ports may be combined into one bidirectional port or may be connected to the same bidirectional bus. Such a test collar cell provides a path from each TDI port to the SDO port, from each TDI port to the TDO port, and from the SDI port to the TDO port. One or more of these paths may be latched or registered. The values of the test select signals determine which path is selected. An input test collar cell would generally include a latch in the test datapath, but not the system datapath, in order to minimize the delay overhead. Paths from the SOC pins to the cell TDI/TDO ports may be serial, parallel or a combination of both.

16.3.5 Boundary scan

IEEE standard 1149.1 describes a test architecture for integrated circuits which allows access to internal scan, boundary scan, BIST, or emulation features incorporated into the circuits. The test architecture includes a test access port (TAP) controller, an instruction register, and data registers (see Chapter 11; Section 11.5). The TAP controller is interfaced to test mode select (TMS) and test clock (TCK) pins. The

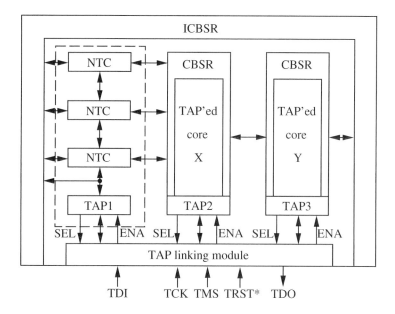

Figure 16.8   Test access architecture (Whetsel, 1997) (© 1997 IEEE)

instruction and data registers are interfaced to test data input (TDI) and test data output (TDO) pins.

The IEEE standard was developed to support only one TAP per integrated circuit. However, many integrated circuits from yesterday are becoming cores today with this test architecture already built in. At the same time, cores such as memories, I/O peripherals and decoders may not subscribe to this standard. Thus, there is a need to provide test access to non-TAPed cores (NTCs) as well as to exploit the test architecture already present in TAPed cores in a hierarchical way to make it easy to test an SOC. Such a method is discussed in Whetsel (1997).

Consider the test access architecture shown in Figure 16.8. It has three TAPs. TAP1 is the traditional standard 1149.1 TAP which provides access to the integrated circuit boundary scan register (ICBSR) and to the NTCs, as shown by the dashed box. NTCs are designed to interface to TAP1 for either scan testing or BIST. NTCs, which are not testable with TAP1, need to be tested through other test access mechanisms, such as direct parallel access.

TAP2 and TAP3 are integral parts of TAPed cores X and Y, respectively. These cores are hard and have a core boundary scan register (CBSR) for internal core-to-core interconnect testing. The TAP linking module (TLM) is an interconnect layer which allows one or more of the three TAPs to be connected to the test pins of the SOC in response to standard 1149.1 scan operations. The connection between the TAPs and the test pins is established through the TLM with the help of select (SEL) and enable (ENA) signals. TRST* provides the test reset signal.

After power up or test reset, TLM initially connects TAP1 to the test pins. This makes it appear that the SOC has a single TAP, as required by standard 1149.1. Later on, other TAPs can also be connected to the test pins through TLM. TLM also allows background BIST operations in the cores independent of the foreground test access operations.

All TAPs are directly connected to the TCK and TMS pins. The SEL signal of a TAP is output in response to a special instruction scanned into its instruction register. This causes TLM to be selected as the data register scan path between the TDI and TDO pins. Thus, TLM acts like a community test data register, accessible by all TAPs. Following the TLM data scan operation, TLM updates the ENA signals to the TAPs, thus changing the SOC's TAP link configuration. The core connected to the enabled TAP has direct test access to the SOC test pins. A currently enabled TAP can select and scan TLM to select and enable another TAP. The previously enabled TAP can either be disabled or remain enabled to operate in conjunction with the newly enabled TAP.

The test architecture can be easily generalized to provide hierarchical test access to future cores. Just as today's TAPed integrated circuits are evolving into TAPed cores, the future TLMed integrated circuits may evolve into TLMed cores. To provide test access to the TLMed cores, the only difference in the test architecture would be that TAP2 and TAP3 would be TLMs instead. Thus, test access will be provided in a two-level hierarchy: from the integrated circuit TLM to the core TLM, and then from the core TLM to the core TAPs.

16.3.6 Partial isolation ring

An isolation ring is akin to boundary scan which is placed around a core to provide full controllability and observability to core inputs and outputs, respectively, as shown in Figure 16.9. It also provides full observability to the UDL driving the core and full controllability to the UDL being driven by the core. However, the area/delay overhead of an isolation ring can be quite high. To reduce the overheads, a partial isolation ring approach has been presented in Touba and Pouya (1997). In such a ring, not all the core inputs and outputs are connected to the isolation ring. It provides the same fault coverage as the full isolation ring, but avoids multiplexers on critical paths.

For the core inputs which are not connected to the isolation ring, the corresponding bits of the precomputed test vectors of the core are justified through the UDL that drives the core. To observe the logic value at such an output of the UDL, space compaction is performed by XORing it with another output of the UDL that does feed the isolation ring. A partial isolation ring, where the third and fifth core inputs are directly connected to the driving UDL, is shown in Figure 16.10(a).

To determine which core inputs should be connected to the isolation ring and which should not, all the test vectors that the UDL can justify at its outputs, from the core precomputed test set, are identified first. If all the test vectors can be justified, there is

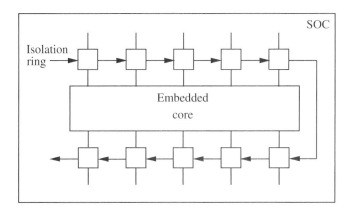

Figure 16.9   Isolation ring (Touba and Pouya, 1997) (© 1997 IEEE)

no need for an isolation ring at the core inputs. If not, the ones that cannot be justified are targeted. If there are n core inputs, there are 2^n possible partial isolation rings since each core input can either be included or not included in the ring. A specific ring enables a test vector to be justified if the subset of bits in the test vector corresponding to excluded core inputs can be fed directly from the UDL. A ring is a solution if it allows all the test vectors in the precomputed test set to be applied to the core. Various search algorithms, ranging from $O(n)$ hill-climbing to $O(2^n)$ branch-and-bound, can be used to obtain the best possible partial isolation ring.

We next need to focus on the isolation ring elements at the core outputs. If the core can supply all the vectors needed to test the UDL that it drives, then isolation ring elements at the core outputs just need to act like a shift register. This enables the core outputs to be observed, but avoids the extra logic required in the isolation ring. In general, the core will not be able to supply all the required vectors to the driven UDL. Then a partial isolation ring needs to be designed at the core outputs in order to feed the required test set to the UDL. This involves replacing some isolation ring elements with shift register elements (or multiplexing ring elements directly with the SOC outputs). This enables justification of a subset of the bits of each test vector through the core and use of the partial isolation ring to shift in the remaining bits. Figure 16.10(b) shows the partial isolation ring at the core outputs where a test vector is generated by shifting the first, third and fifth bits into a partial isolation ring and justifying the second and fourth bits through the core. Since no shift register elements are included here for the second and fourth bits, we have assumed that these core outputs are multiplexed with SOC outputs for direct observability.

Example 16.1   Consider the circuit in Figure 16.10(b). Suppose we want to test the driven UDL and some of the vectors produced by the core are as follows: (0, 1, 0, 0, 1), (1, 0, 0, 1, 0) and (1, 0, 1, 0, 1). Take the first vector. It can be used to apply any test vector to the driven UDL of the form $(x, 1, x, 0, x)$ since the second and fourth bits

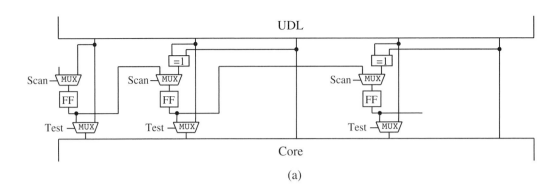

(a)

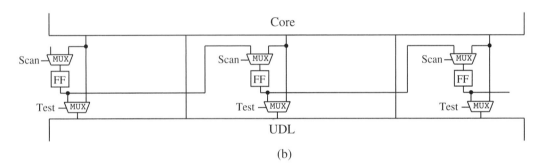

(b)

Figure 16.10 Partial isolation ring

can be applied directly through the core and the remaining three bits can be applied
to the UDL by shifting them into the partial isolation ring. The remaining two core
output vectors can similarly be used to apply test vectors of the form $(x, 0, x, 1, x)$
and $(x, 0, x, 0, x)$, respectively. However, if a test vector of the form $(x, 1, x, 1, x)$
is required to test the UDL, and no vector that can be produced at the core output
contains 1 in both the second and fourth bits, then the fault coverage of the UDL will
suffer.   □

In order to derive an output partial isolation ring which provides full fault coverage
to the driven UDL, all the faults in the UDL detectable by the vectors that the core
can produce are first dropped from the fault list. Redundant faults in the UDL are also
dropped. To detect the remaining faults, different partial isolation rings are tried, where
the core output vectors are transformed in the fashion shown in Example 16.1. Among
the rings that provide full fault coverage to the UDL, the one which has the maximum
number of direct connections to the UDL from the core is the preferred one. A search
algorithm, similar to the one used to identify the partial isolation ring elements on the
input side of the core, can be used for this purpose.

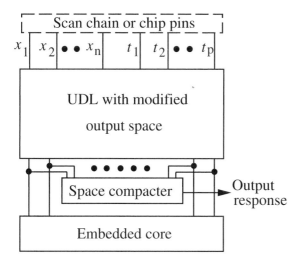

Figure 16.11 Modification of the UDL to provide test access (Pouya and Touba, 1997) (© 1997 IEEE)

16.3.7 Modifying the UDL for test access

Since the UDL driving a core may restrict the set of vectors that can be applied to the core, the complete precomputed test set of the core may not be applicable to it. Another way to solve this problem is by inserting control points in the UDL to modify its output space so that it contains all the vectors in the precomputed test set of the core (Pouya and Touba, 1997), as shown in Figure 16.11. In this figure, $x_1, x_2, \ldots, x_n$ are the normal UDL inputs, and $t_1, t_2, \ldots, t_p$ are its control points. This avoids the need for an isolation ring around the core. Thus, instead of scanning core test vectors into an isolation ring, the core test vectors are justified by controlling the UDL inputs. By placing the control points on off-critical paths, whenever possible, delay overheads can be avoided.

If the UDL contains an internal scan chain, then the contents of the scan chain are also controlled in addition to the UDL inputs. If core X drives core Y through a UDL, then this approach can avoid the need for an isolation ring at the input of core Y. The precomputed test set of core Y can be justified through the UDL by controlling the isolation ring at the output of core X (which also happens to provide the inputs of the UDL). In order to test the UDL itself, a space compactor can be added to provide observability to its outputs.

Example 16.2 Consider the UDL shown in Figure 16.12(a). The following vectors are not justifiable at its outputs: $(1, 1, x, x)$, $(x, 1, 0, x)$ and $(x, 1, x, 0)$. These vectors together represent seven binary vectors. One of these vectors is $(1, 1, 0, 0)$. To try to justify these values at the UDL outputs, the values are driven backwards. However, this ends in a conflict at inputs x_2 and x_3 whose fanout branches need opposite logic

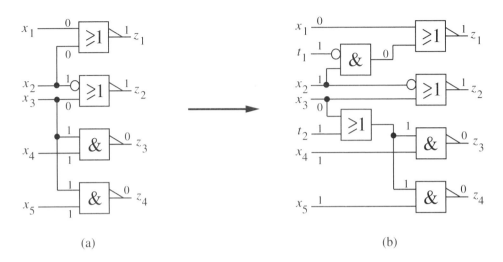

Figure 16.12 Output space modification through control points

values. If some or all of these vectors are part of the precomputed test set of the core that this UDL feeds, then the core will not be completely testable.

The above problem can be taken care of by inserting two control points, t_1 and t_2, and two extra gates, as shown in Figure 16.12(b). Vector $(1, 1, 0, 0)$ can now be shown to be justifiable by making $t_1 = t_2 = 1$, as shown in this figure. The other six output vectors, which were previously not justifiable, can also now be justified by appropriately setting the values of t_1 and t_2. For normal operation, the control points are set to 0. In this case, the UDL incurs a delay overhead due to the modifications. However, in general, it may be possible to add the extra gates and control points on off-critical paths. □

In backtracing through a UDL in order to justify its output vector, two types of gates are encountered: *imply gates* and *decision gates*. If justifying the value at the output of a gate requires all its inputs to be non-controlling, e.g., 0 at the output of an OR gate, then it is said to be an **imply gate**. Else, if justifying the value at the output of a gate requires just one input to be controlling, e.g., 0 at the output of an AND gate, then the gate is said to be a **decision gate**. Each conflict encountered during backtracing is either an *imply conflict* or a *decision conflict*. An **imply conflict** is one that occurs due to backtracing through imply gates only. For example, both conflicts encountered in justifying vector $(1, 1, 0, 0)$ at the outputs of the circuit in Figure 16.12(a) were imply conflicts. A decision conflict is one that occurs due to backtracing through one or more decision gates and any number of imply gates. A decision conflict can be avoided by choosing an alternative path for backtracing through some decision gate, whereas an imply conflict cannot be avoided. Thus, all imply conflicts should be first removed with control points for all output vectors that are needed at the UDL outputs, but not normally produced by the UDL. In the process, some decision conflicts also get

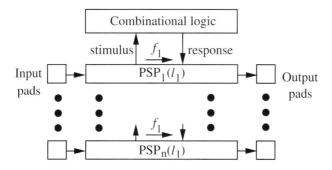

Figure 16.13   A conventional PSP test architecture (Whetsel, 1998) (© 1998 IEEE)

resolved. The remaining decision conflicts need to be targeted one-by-one with some more control points.

16.3.8   Low power parallel scan test access

A test architecture for testing embedded cores, which provides low power parallel scan testing and hierarchical reuse, has been presented in Whetsel (1998). Scan test application time is reduced if flip-flops are arranged in parallel scan paths (PSPs), each with a serial data input and a serial data output.

A conventional implementation consisting of n PSPs is shown in Figure 16.13. Each PSP is designed to have equal or near equal length l_1, and is connected to a pair of integrated circuit pads to input and output test data. Each PSP is also connected to the combinational logic so that the test data that are shifted into the PSPs can be applied to the combinational logic and the response of the combinational logic can be captured into the PSPs and shifted out through the output pads. The number of cycles required to shift in the test data and capture the response of the combinational logic is $l_1 + 1$. While the test data for a new vector is shifted in, the captured response of the previous test vector is shifted out simultaneously. Thus, ignoring the time taken to shift out the response for the last test vector, $l_1 + 1$ is the test application time per test vector, and is referred to as one **scan cycle**. The **test data volume** scanned per scan cycle is nl_1. Thus, the **test data bandwidth** per scan cycle is $nl_1/(l_1 + 1)$. The dynamic power consumed in the combinational logic per scan cycle is $\frac{1}{2}CV_{\text{DD}}^2 f_1$, where C is the effective switched capacitance, V_{DD} is the supply voltage and f_1 is the combinational logic input transition frequency.

Figure 16.14 shows a modification of the PSP test architecture. It includes parallel scan distributors (PSDs) and parallel scan collectors (PSCs). A PSD is a serial-input/parallel-output shift register, whereas a PSC is a parallel-input/serial-output shift register. Between a pair of input/output pads, an m-bit PSD and an m-bit PSC are introduced. Each PSP is divided up into m smaller PSPs each roughly of length l_2,

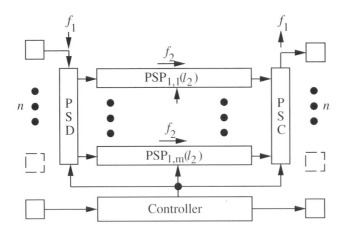

Figure 16.14 A PSD/PSC based test architecture (Whetsel, 1998) (© 1998 IEEE)

where $l_2 = l_1/m$. A controller is also added to control the PSD, PSC and PSPs. The other PSPs in Figure 16.13 are similarly modified. The PSPs maintain their stimulus and response connections to the combinational logic as before. The PSD shifts in data from its input pad and outputs data to the PSPs, while the PSC loads data from the PSPs and shifts out data through its output pad.

The controller is connected to the integrated circuit pads to allow it to be externally controlled. Initially, it is in an inactive mode. While in this mode, it outputs control signals to PSD, PSC and PSPs to enable normal operation. In response to a start test signal, the controller outputs control signals to PSD, PSC and PSPs to execute the steps shown in Figure 16.15. When the test is complete, the controller configures the circuit for normal mode and waits for the next test signal. During test, a tester supplies stimulus data to the PSPs through PSDs, and receives response data from PSPs through the PSCs. It compares the response data to expected data to determine if the circuit passes or fails the test.

For the circuit in Figure 16.14, one scan cycle equals $(m + 1)l_2 + 1$ clock cycles because an m-cycle shift of PSD/PSC is followed by a one-cycle shift of PSP and this is repeated l_2 times. The last cycle is for response capture. The test data volume per scan cycle is nml_2. Thus, the test data bandwidth per scan cycle is $nml_2/[(m + 1)l_2 + 1]$. The dynamic power consumed in the combinational logic is $\frac{1}{2}CV_{DD}^2 f_2$, making the simplifying assumption that the switched capacitance remains the same as before. f_2 is the new combinational logic input transition frequency.

Consider some typical numbers such as $m = 10$, $l_2 = l_1/10$ and $f_2 = f_1/10$. The frequency also becomes a tenth of the earlier frequency since the PSPs are shifted once for each m-cycle shift of PSD/PSC. Then the scan cycle time of the architecture in Figure 16.14 can be seen to be $(10 + 1)l_1/10 + 1 = 1.1l_1 + 1$. The test data volume per scan cycle is $10nl_1/10 = nl_1$. Thus, the test data bandwidth per scan cycle is $nl_1/(1.1l_1 + 1)$. The dynamic power consumed per cycle is $\frac{1}{2}CV_{DD}^2 \frac{f_1}{10}$. Therefore, the

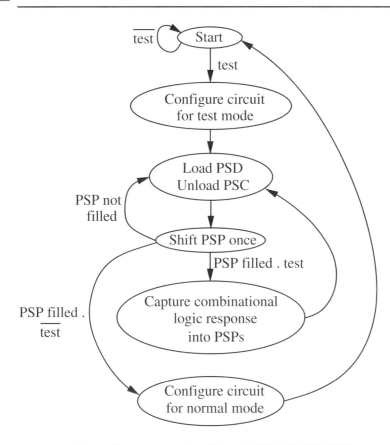

Figure 16.15 Test architecture execution (Whetsel, 1998) (© 1998 IEEE)

scan cycle time, test data volume per scan cycle, and test data bandwidth per scan cycle of the two test architectures are roughly equal. However, the power consumed during test by the second architecture is one-tenth of that consumed by the first. This is likely to be a critical issue in the future.

Next, let us see how the PSD/PSC based test architecture can be used to provide test access to embedded cores. Figure 16.16 shows an SOC with an embedded core enclosed in a dotted box. In this case, both the SOC and embedded core are assumed to have a PSD/PSC based test architecture. For the SOC, PSD/PSC are associated with the chip pads, whereas for the core, PSD/PSC are associated with the core I/O terminals. A multiplexer is present at each core terminal which has a PSD, and a demultiplexer is present at each core terminal which has a PSC. The multiplexer allows either the functional input or the test input to be fed to the core. Similarly, the demultiplexer allows either the functional output or the test output to be selected. The test input of the multiplexer comes from the serial output of the SOC PSD, and the test output of the demultiplexer goes to the serial input of the SOC PSC. The functional inputs and outputs are connected to the remaining parts of the circuit.

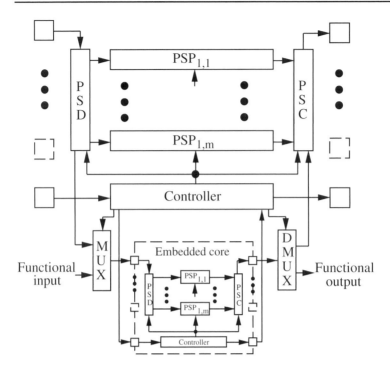

Figure 16.16 Testing embedded cores (Whetsel, 1998) (© 1998 IEEE)

As shown in Figure 16.16, the SOC controller has authority over the core controller so that the former can enable, disable or modify the operation modes of the latter. However, in the test mode, the controllers work in tandem to synchronize the operation of the SOC and core PSD/PSC circuits.

In the set-up shown in Figure 16.16, testing of the non-core and core circuitry can either be done separately or simultaneously. When only the non-core circuitry is being tested, the core controller is disabled by the SOC controller to prevent it from interfering with the testing. When only the core is being tested, the SOC controller enables the core controller, and sets up the core terminal multiplexer to receive test data from the SOC PSD and the core terminal demultiplexer to provide responses to the SOC PSC. When both non-core and core circuitry are being tested, the SOC and core PSD and PSC circuits are serially connected to provide and receive test data. Both the controllers are synchronized to the external control input from the tester for this purpose.

The PSD/PSC test architecture shown in Figure 16.16 can be hierarchically extended. For example, if the core shown in the figure were a part of another embedded core, the PSD/PSC test architecture would be extended to include that core, and so on. Thus, this hierarchical approach can enable massively parallel testing of the SOC. It provides a plug-and-play interface that operates in the same fashion at all levels of the hierarchy. It has also been shown to be compatible with the IEEE 1149.1 boundary scan standard.

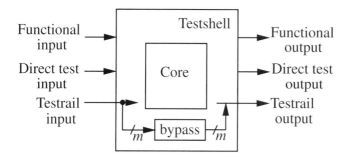

Figure 16.17   SOC–testshell interface (Marinissen *et al.*, 1998) (© 1998 IEEE)

16.3.9   Testshell and testrail based access

A method for wrapping each core in a *testshell* and providing access to them through *testrails* has been given in Marinissen *et al.* (1998). The testshell provides a small interface layer between the core and the SOC.

The inputs/outputs of the testshell are shown in Figure 16.17. The functional inputs/outputs are the normal inputs/outputs of the core. The testrail inputs/outputs transport the test data and can handle synchronous digital tests which include functional tests, scan tests, memory tests and BIST. The testrail has a variable width and a bypass mode. Test signals that cannot be provided by the testrail, e.g., asynchronous signals, clocks, and analog signals, are transported through direct test inputs/outputs.

The testshell has four compulsory modes: functional, core test, interconnect test and bypass. In the functional mode, the testshell is transparent and is meant for normal operation. In the core test mode, the testshell enables the testing of the core by feeding it tests from the SOC inputs and observing its response at the SOC outputs. In the interconnect test mode, the testshell enables the interconnect leading to and from the core to be tested. Finally, in the bypass mode, the test vectors and/or responses for other cores are transported through the testshell.

A conceptual view of the testshell is given in Figure 16.18. The bypass block is not shown for simplicity. In the functional mode, the multiplexer select signals, m_1 and m_2, are set to $(0, 0)$. This allows normal operation. However, the multiplexers may cause some delay degradation. In the core test mode, the select signals are set to $(1, 0)$. In this case, the test vectors for the core come from terminal $s1$ which represents a connection to either the testrail or the direct test input. The test responses from the core are fed to terminal $r1$ which represents a connection to either the testrail or the direct test output. In the interconnect test mode, the select signals are set to $(0, 1)$. Terminal $s2$ feeds the interconnect test vector to the testshell output. It is connected to the testrail. Similarly, terminal $r2$ captures the interconnect test response arriving at the testshell input. It is also connected to the testrail. This figure provides a conceptual description of the testshell only. There is room for area and delay optimization in the actual implementation of the testshell.

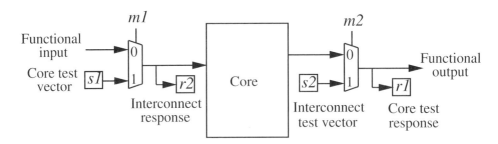

Figure 16.18 Conceptual view of the testshell (Marinissen *et al.*, 1998) (© 1998 IEEE)

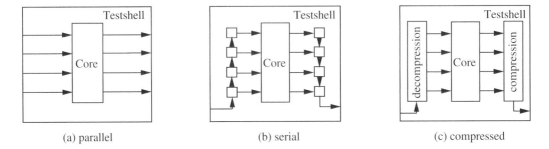

(a) parallel   (b) serial   (c) compressed

Figure 16.19 Core-level testrail connections (Marinissen *et al.*, 1998) (© 1998 IEEE)

Every testshell is connected to a testrail. However, the testrail of various cores can be connected in many different ways. Varying the width of the testrail allows the system integrator the flexibility to trade off test application time for area overhead. The width is also dependent on the number and type of SOC pins available for testing purposes, e.g., input, output or bidirectional pins. On one end of the spectrum, each core can have its private testrail where all testrails are multiplexed onto the available SOC pins. This corresponds to the parallel direct test access scheme discussed earlier. The other end of the spectrum consists of all testrails of cores concatenated into one SOC-level testrail.

The testrail can be connected to the testshell of a core in three different ways: parallel, serial or compressed. This is shown in Figure 16.19. A parallel connection implies a one-to-one connection between the testrail wires and the core input/output terminals. A serial connection indicates that one testrail wire is connected to multiple core terminals via a shift register. A compressed connection implies the presence of decompression circuitry at the core inputs and compression circuitry at the core outputs. Decompression at the core inputs is, in practice, possible for only a regular sequence of test vectors, e.g., for memory tests. Compression at the core outputs can be done through MISRs or XOR trees. Combinations of such connections are also possible, e.g., inputs partly serial, partly parallel, while the outputs are compressed.

The testshell also contains a standardized test control mechanism. This helps control the operation of the testshell by providing it test control signals.

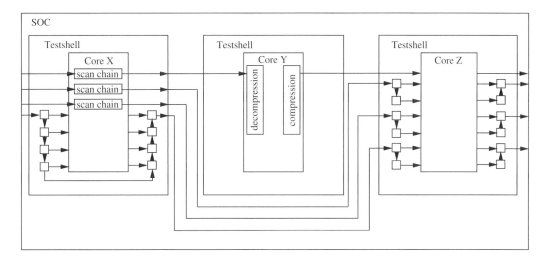

Figure 16.20 Testrail connections for the example

Example 16.3 Consider an SOC with three cores: X, Y and Z. Suppose these cores are to be tested in the following way. X has full scan as the embedded DFT methodology, and has seven inputs, seven outputs and three internal scan chains. Y has BIST as the embedded test methodology, which needs to be initialized, then run autonomously, followed by signature capture. Z has a functional test which involves all its inputs and outputs. Suppose it has seven inputs and seven outputs. In addition, the global interconnect needs to be tested as well.

One possible way to test the SOC is to package all cores in a testshell and testrail of width four. The testrail of all three cores can be concatenated. This is shown in Figure 16.20. Cores X and Z are connected to all the four wires of the testrail. Since the three scan inputs/outputs of core X transport much more test data than the other four inputs/outputs, they are provided with a parallel connection to the testrail, while the other inputs/outputs are provided with a serial connection. Core Y is connected to only one of the testrail wires, the other three bypass it outside its testshell. This is because it needs very little test data for BIST. Since core Z is tested functionally, all its seven inputs/outputs transport an equal amount of test data. However, since the testrail has only four wires, the remaining three inputs/outputs require a serial testrail connection.

To test a core, its testshell is put under the core test mode while the other two cores are put under the bypass mode. For the global interconnect test, the testshells of all cores are put under the interconnect test mode. □

16.3.10 Advanced Microcontroller Bus Architecture (AMBA)

A test access method, called AMBA, has been presented by Advanced RISC Machines Ltd. (ARM). It uses bus transfers for test access (Flynn, 1997; Feige *et al.*, 1998).

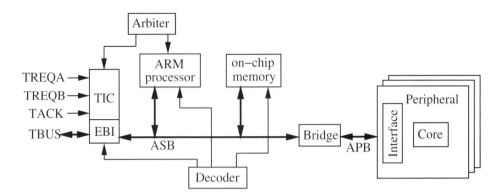

Figure 16.21 The AMBA system (Feige *et al.*, 1999) (© 1999 Kluwer)

It is geared towards SOCs containing 32-bit high-performance embedded RISC microcontrollers. Cores supplied by ARM make use of AMBA and a functional test approach.

A typical AMBA system consists of an advanced system bus (ASB) for high-performance system modules and an advanced peripheral bus (APB) for low-power peripherals, as shown in Figure 16.21. Typical connections to ASB include an ARM processor, an on-chip memory, the test interface controller (TIC), the external bus interface (EBI), and the bridge. An arbiter controls bus arbitration between possible ASB masters such as the processor or TIC. Only one ASB master is active at any given time. It can initiate bus transfers by providing address and control data to the bus. The decoder activates the ASB slave(s) involved in the transfer. The bridge connects the ASB and APB. Hence, its role is that of an ASB slave and an APB master. The APB signals do not change when not in operation in order to save power. APB connects different peripherals to the bridge. An APB peripheral contains an AMBA interface and a core that implements its functionality. The interface has several registers and the corresponding address and decoding logic.

The AMBA test strategy is to test all ASB slaves and APB peripherals independently. This requires an insertion of a test harness in the slaves and peripherals. Figure 16.22 shows the connections in a peripheral. The test harness provides controllability and observability to dedicated peripheral inputs and outputs.

In the test mode, TIC becomes the ASB master. It is controlled by three handshake signals which are connected to SOC-level pins: *TREQA*, *TREQB*, *TACK*. The parallel 32-bit *TBUS* helps transfer the test vectors through the EBI. Depending on the state of the TIC handshake signals, the vectors that can be transferred are of three types: address, write and read. This allows every register within the AMBA system to be accessed by providing the proper sequence of vectors from the SOC-level pins. These vectors usually consist of address/write or address/read combinations.

The AMBA test approach is not intended for structural test techniques. The AMBA

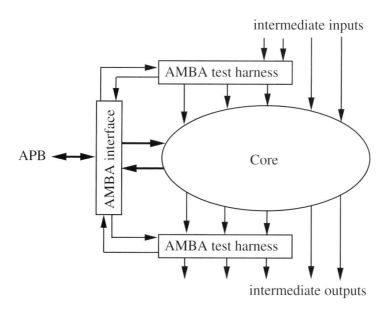

intermediate inputs

intermediate outputs

Figure 16.22   AMBA peripheral with test harness (Feige *et al.*, 1999) (© 1999 Kluwer)

modules supplied by ARM are tested using functional tests. Extensions of this approach to structural test techniques, such as scan, are discussed in Feige (1999).

16.4   Core test wrapper

The core test wrapper connects the core terminals to the rest of the SOC and the test access mechanism (Zorian *et al.*, 1998). It should contain three mandatory modes of operation: *normal*, *core test*, and *interconnect test*. In the **normal mode**, the wrapper is transparent and allows normal SOC operation. In the **core test mode**, the test access mechanism is connected to the core and provides controllability and observability to it. In the **interconnect test mode**, the test access mechanism is connected to the interconnect wires and logic. This facilitates controllability of a core's outputs, so that the interconnect at these outputs can be controlled, and observability of the next core's inputs at which the interconnect terminates.

Apart from the above mandatory modes, a core test wrapper may have several optional modes. For example, in the **detach mode**, the core can be disconnected from its SOC environment and the test access mechanism. Another example is the **bypass mode** of the testrail access mechanism discussed earlier in Section 16.3.9.

It is possible that, depending on how the test access mechanism is implemented, some of the above modes may coincide. For example, if test access is based on the normal functionality of the core, then the normal and core test modes may coincide.

The width of the test access mechanism does not necessarily correspond to the number of core input/output terminals. The width is determined by the system integrator based on a trade-off between area overhead and test application time. If the number of core terminals is larger than the test access width, then *width adaptation* is performed in the core test wrapper by providing serial-to-parallel conversion at the core inputs and parallel-to-serial conversion at the core outputs.

Pre-designed cores come with their internal clock distribution mechanism. Different cores have different clock propagation delays. This may result in a clock skew in communication between cores. The clock skew may corrupt data transfers taking place over the test access mechanism, especially when this mechanism is shared over multiple cores. The core test wrapper can provide means for preventing clock skews in the test access paths. The test collar and testshell, which were discussed in Section 16.3.4 and 16.3.9, respectively, are examples of core test wrappers. Standardization of such a wrapper can make it easy to incorporate interoperable cores of various types in an SOC.

Summary

- The main problems in SOC testing involve the development of core-level test, core test access and core test wrapper.
- Core-level testing involves the derivation of the test set for the core, utilizing its DFT features. This precomputed test set is made available by the core provider for hard cores and given to the system integrator.
- Many methods have been presented for core test access. Examples are macro test, core transparency, direct parallel access, test bus, boundary scan, isolation ring, user-defined logic modification, low power parallel scan, testshell and testrail, and AMBA.
- Macro test is based on the concept of inserting test interface elements on all signals traveling between macros. These elements are transparent in the normal mode of operation and support both serial and parallel access in the test mode.
- Core transparency based test access methods require the presence of a transparency mode in each core in which test data can be propagated through it without information loss.
- The direct parallel test access method, as the name implies, provides direct and parallel access to the cores from the SOC pins. This method is not very scalable because of the high overheads.
- A test bus can also provide direct access to the embedded cores. However, it can be shared among multiple cores and UDLs and its bit-width altered in order to reduce the overheads.
- Boundary scan was initially developed as a concept for providing testability to

embedded chips on a board. However, it can also be extended to provide testability to embedded cores in an SOC.

- An isolation ring is akin to boundary scan. To reduce overheads, a partial isolation ring can be used which provides full controllability and observability to only a subset of the core pins.

- Core test access can also be provided through extra control points by modifying the UDLs that feed it and that are fed by it.

- The concept of parallel scan can be exploited for low power testing and hierarchical reuse of cores.

- Testshells can be used to wrap a core, and access can be provided to them through testrails.

- AMBA is a test access method which uses bus transfers for this purpose.

- The core test wrapper connects the core terminals to the rest of the SOC and the test access mechanism. It consists of three compulsory modes: normal, core test and interconnect test. In addition, it can also have some optional modes such as detach and bypass.

Additional reading

A modular logic BIST architecture for core-based SOCs has been discussed in Rajski and Tyszer (1998).

A genetic algorithm based tool to synthesize an SOC for testability has been given in Ravi and Jha (2001).

Exercises

16.1 Consider the GCD circuit shown in Figure 15.14 and its controller in Figure 15.15. Suppose the GCD circuit is going to be used as a core in an SOC, and we need to provide a transparency path from IN_1 to OUT. Obtain such a path with a minimum number of cycles.

16.2 Suppose a core with five outputs can produce all possible five-bit output vectors except $(0, 0, 1, 1, 1)$, $(1, 0, 1, 1, 1)$, $(0, 0, 0, 1, 1)$ and $(1, 0, 1, 1, 0)$. Also, suppose that each of these vectors is required to provide complete fault coverage for a UDL driven by this core. Design an optimal partial isolation ring (i.e., with a minimum number of ring elements) to be placed between the core and UDL, which enables each of these vectors to be applicable to the UDL. Is your answer unique?

16.3 While backtracing for each of the seven output vectors that the circuit in Figure 16.12(a) cannot produce, indicate which of the encountered conflicts are imply conflicts and which are design conflicts. If the circuit is modified as shown in Figure 16.12(b), give the primary input values and values at t_1 and t_2 to enable the justification of each of these output vectors.

16.4 Suppose an SOC has three cores: A, B, and C. Core A has five inputs and six outputs and is to be tested functionally. Core B is to be tested using BIST. Core C has six inputs and four outputs and is to be tested using two internal scan chains. Suppose the testrails of the three cores need to be concatenated into an SOC-level testrail of width four. Show the testrail connections.

References

Beenker, F.P.M., van Eerdewijk, K.J.E., Gerritsen, R.B.W., Peacock, F.N. and van der Star, M. (1986). Macro testing: unifying IC and board test. *IEEE Design & Test of Computers*, **3** (4), pp. 26–32.

Feige, C., Pierick, J.T., Wouters, C., Tangelder, R. and Kerkhoff, H.G. (1999). Integration of the scan-test method into an architecture specific core-test approach. *J. of Electronic Testing: Theory & Applications*, **14** (1/2), pp. 125–131.

Flynn, D. (1997). AMBA: enabling reusable on-chip designs. *IEEE Micro*, **17** (4), pp. 20–27.

Ghosh, I., Jha, N.K. and Dey, S. (1999). A low overhead design for testability and test generation technique for core-based systems-on-a-chip. *IEEE Trans. on Computer-Aided Design*, **18** (11), pp. 1661–1676.

Ghosh, I., Dey, S. and Jha, N.K. (2000). A fast and low cost testing technique for core-based system chips. *IEEE Trans. on Computer-Aided Design*, **19** (8), pp. 863–877.

Immaneni, V. and Raman, S. (1990). Direct access test scheme – design of block and core cells for embedded ASICs. In *Proc. Int. Test Conference*, pp. 488–492.

Marinissen, E.J., Arendsen, R., Bos, G., Dingemanse, H., Lousberg, M. and Wouters, C. (1998). A structured and scalable mechanism for test access to embedded reusable cores. In *Proc. Int. Test Conference*, pp. 284–293.

Pouya, B. and Touba, N.A. (1997). Modifying user-defined logic for test access to embedded cores. In *Proc. Int. Test Conference*, pp. 60–68.

Rajski, J. and Tyszer, J. (1998). Modular logic built-in self-test for IP cores. In *Proc. Int. Test Conference*, pp. 313–321.

Ravi, S., Lakshminarayana, G. and Jha, N.K. (2001). Testing of core-based systems-on-a-chip. *IEEE Trans. on Computer-Aided Design*, **20** (3), pp. 426–439.

Ravi, S. and Jha, N.K. (2001). Test synthesis of systems-on-a-chip. *IEEE Trans. on Computer-Aided Design*, **21** (10), pp. 1211–1217.

Touba, N. and Pouya, B. (1997). Using partial isolation rings to test core-based designs. *IEEE Design & Test of Computers*, **14** (4), pp. 52–59.

Varma, P. and Bhatia, S. (1998). A structured test re-use methodology for core-based system chips. In *Proc. Int. Test Conference*, pp. 294–302.

Whetsel, L. (1997). An IEEE 1149.1 based test access architecture for ICs with embedded cores. In *Proc. Int. Test Conference*, pp. 69–78.

Whetsel, L. (1998). Core test connectivity, communication, and control. In *Proc. Int. Test Conference*, pp. 303–312.

Zorian, Y. (1997). Test requirements for embedded core-based systems and IEEE P1500. In *Proc. Int. Test Conference*, pp. 191–199.

Zorian, Y., Marinissen, E.J. and Dey, S. (1998). Testing embedded-core based system chips. In *Proc. Int. Test Conference*, pp. 130–143.

Index